Edited by
Ru-Shi Liu, Lei Zhang, Xueliang Sun,
Hansan Liu, and Jiujun Zhang

Electrochemical Technologies for Energy Storage and Conversion

Related Titles

Daniel, C., Besenhard, J. O. (eds.)

Handbook of Battery Materials

2011
ISBN: 978-3-527-32695-2

Aifantis, K. E., Hackney, S. A., Kumar, R. V. (eds.)

High Energy Density Lithium Batteries

Materials, Engineering, Applications

2010
Hardcover
ISBN: 978-3-527-32407-1

Liu, H., Zhang, J. (eds.)

Electrocatalysis of Direct Methanol Fuel Cells

From Fundamentals to Applications

2009
Hardcover
ISBN: 978-3-527-32377-7

Ozawa, K. (ed.)

Lithium Ion Rechargeable Batteries

Materials, Technology, and New Applications

2009
Hardcover
ISBN: 978-3-527-31983-1

Eftekhari, A. (ed.)

Nanostructured Materials in Electrochemistry

2008
Hardcover
ISBN: 978-3-527-31876-6

Nazri, G.-A., Balaya, P., Manthiram, A., Yamada, A., Yang, Y. (eds.)

Advanced Lithium-Ion Batteries

Recent Trends and Perspectives

Series: New Materials for Sustainable Energy and Development
2012
Hardcover
ISBN: 978-3-527-32889-5

Edited by Ru-Shi Liu, Lei Zhang, Xueliang Sun, Hansan Liu, and Jiujun Zhang

Electrochemical Technologies for Energy Storage and Conversion

Volume 1

WILEY-VCH Verlag GmbH & Co. KGaA

The Editors

Prof. Dr. Ru-Shi Liu
Department of Chemistry
National Taiwan University
No. 1, Sec. 4, Roosevelt Road
Taipei 10617
Taiwan

Lei Zhang
Institute for Fuel Cell Innovation
National Research Council Canada
4250 Wesbrook Mall
Vancouver, B.C. V6T 1W5
Canada

Prof. Xueliang Sun
Deparment of Mechanical & Materials
University of Western Ontario
London
Ontario N6A 5B9
Canada

Dr. Hansan Liu
Chemical Sciences Division
Oak Ridge National Laboratory
Oak Ridge, TN 37831
USA

Dr. Jiujun Zhang
Institute for Fuel Cell Innovation
National Research Council Canada
4250 Wesbrook Mall
Vancouver, B.C. V6T 1W5
Canada

All books published by **Wiley-VCH** are carefully produced. Nevertheless, authors, editors, and publisher do not warrant the information contained in these books, including this book, to be free of errors. Readers are advised to keep in mind that statements, data, illustrations, procedural details or other items may inadvertently be inaccurate.

Library of Congress Card No.: applied for

British Library Cataloguing-in-Publication Data
A catalogue record for this book is available from the British Library.

Bibliographic information published by the Deutsche Nationalbibliothek
The Deutsche Nationalbibliothek lists this publication in the Deutsche Nationalbibliografie; detailed bibliographic data are available on the Internet at <http://dnb.d-nb.de>.

© 2012 Wiley-VCH Verlag & Co. KGaA, Boschstr. 12, 69469 Weinheim, Germany

All rights reserved (including those of translation into other languages). No part of this book may be reproduced in any form – by photoprinting, microfilm, or any other means – nor transmitted or translated into a machine language without written permission from the publishers. Registered names, trademarks, etc. used in this book, even when not specifically marked as such, are not to be considered unprotected by law.

Typesetting   Laserwords Private Limited, Chennai, India
Printing and Binding   Fabulous Printers Pte Ltd, Singapore
Cover Design   Formgeber, Eppelheim

Printed in Singapore
Printed on acid-free paper

Print ISBN: 978-3-527-32869-7
ePDF ISBN: 978-3-527-64008-9
oBook ISBN: 978-3-527-63949-6
ePub ISBN: 978-3-527-64007-2
mobi ISBN: 978-3-527-64009-6

Contents to Volume 1

Contents to Volume 2 *XVI*

Preface *XVII*
About the Editors *XIX*
List of Contributors *XXI*

1	**Electrochemical Technologies for Energy Storage and Conversion** *1*	
	Neelu Chouhan and Ru-Shi Liu	
1.1	Introduction *1*	
1.2	Global Energy Status: Demands, Challenges, and Future Perspectives *1*	
1.3	Driving Forces behind Clean and Sustainable Energy Sources *5*	
1.3.1	Local Governmental Policies as a Potential Thrust *6*	
1.3.2	Greenhouse Gases Emission and the Associated Climate Changes *7*	
1.3.3	Public Awareness about Environmental Protection Rose around the World *7*	
1.3.4	Population Growth and Industrialization *8*	
1.3.5	Security and Safety Concerns Arising from Scarcity of Resources *9*	
1.3.6	Platforms Advocating in Favor of Sustainable and Renewable Resources *9*	
1.3.7	Economic Risk Generated from Price Pressure of Natural Resources *10*	
1.3.8	Regulatory Risk from Governmental Action and Legislation *10*	
1.3.9	Fear of Reputational Risk to Strengthen Corporate Social Responsibility *11*	
1.3.10	Operational and Supply Chain Risks from Inefficiencies and Environmental Changes *11*	
1.4	Green and Sustainable Energy Sources and Their Conversion: Hydro, Biomass, Wind, Solar, Geothermal, and Biofuel *11*	
1.4.1	Solar PV Plants *13*	
1.4.2	Wind Power *14*	
1.4.3	Geothermal Power *14*	
1.4.4	Concentrating Solar Thermal Power (CSP) Plants *14*	
1.4.5	Biomass *15*	

1.4.6	Biofuel 15
1.5	Electrochemistry: a Technological Overview 15
1.6	Electrochemical Rechargeable Batteries and Supercapacitors (Li Ion Batteries, Lead-Acid Batteries, NiMH Batteries, Zinc–Air Batteries, Liquid Redox Batteries) 17
1.6.1	Lead-Acid Batteries 19
1.6.2	NiMH Batteries 20
1.6.3	Li-Ion Batteries 21
1.6.4	Zinc–Air Batteries 22
1.6.5	Liquid Redox Batteries 24
1.7	Light Fuel Generation and Storage: Water Electrolysis, Chloro-Alkaline Electrolysis, Photoelectrochemical and Photocatalytic H_2 Generation, and Electroreduction of CO_2 25
1.7.1	Water Electrolysis 26
1.7.2	Electrochemistry of Water Splitting 27
1.7.3	Chlor-Alkaline Electrolysis 27
1.7.4	Photoelectrochemical and Photocatalytic H_2 Generation 28
1.7.5	Carbon Dioxide Reduction 30
1.8	Fuel Cells: Fundamentals to Systems (Phosphoric Acid Fuel Cells, PEM Fuel Cells, Direct Methanol Fuel Cells, Molten Carbon Fuel Cells, and Solid Oxide Fuel Cells) 32
1.8.1	Alkaline Fuel Cells 33
1.8.2	Direct Methanol Fuel Cells 33
1.8.3	Phosphoric Acid Fuel Cells (PAFCs) 34
1.8.4	Proton Exchange Membrane Fuel Cells 35
1.8.5	High-Temperature Molten Carbonate Fuel Cells 36
1.8.6	Solid Oxide Fuel Cells 37
1.9	Summary 38
	Acknowledgments 39
	References 39
	Further Reading 43
2	**Electrochemical Engineering Fundamentals** 45
	Zhongwei Chen, Fathy M. Hassan, and Aiping Yu
2.1	Electrical Current/Voltage, Faraday's Laws, Electric Efficiency, and Mass Balance 45
2.1.1	Current Efficiency 46
2.1.2	Mass Balance 47
2.2	Electrode Potentials and Electrode–Electrolyte Interfaces 48
2.2.1	Potential Difference 49
2.2.2	Electrode–Electrolyte Interfaces 51
2.3	Electrode Kinetics (Charger Transfer (Butler–Volmer Equation) and Mass Transfer (Diffusion Laws)) 53
2.3.1	Limitations of Butler–Volmer Equation 55
2.4	Porous Electrode Theory (Kinetic and Diffusion) 55

2.4.1	Theories of Porous Electrode	56
2.4.1.1	The Single-Pore Model	56
2.4.1.2	The Macrohomogeneous Model	57
2.5	Structure, Design, and Fabrication of Electrochemical Devices	58
2.5.1	Major Types of Hydrogen–Oxygen Fuel Cells	59
2.5.2	Designing Electrochemical Devices	60
2.5.3	Some Structural Aspects of Electrochemical Devices	60
2.5.4	Fabrication Principles of Fuel Cells	61
2.5.5	Fabrication Principles of Batteries	62
2.5.6	Fabrication Principles of Supercapacitors	63
2.6	Nanomaterials in Electrochemical Applications	64
2.6.1	Carbon Nanostructures	65
2.6.2	Inorganic Nanostructures	65
2.6.3	Carbon–Inorganic Nanocomposites	66
2.6.4	Conducting Polymer Nanostructures	66
2.6.5	Application of Nanomaterials Lead to Emerging Devices	66
	References	67
3	**Lithium Ion Rechargeable Batteries**	**69**
	Dingguo Xia	
3.1	Introduction	69
3.2	Main Types and Structures of Li Ion Rechargeable Batteries	70
3.2.1	Structure of Wound Li Ion Cells	70
3.2.2	Structure of Flat-Plate Prismatic Li-Ion Batteries	70
3.3	Electrochemical Processes in Li Ion Rechargeable Batteries	72
3.4	Battery Components (Anode, Cathode, Separator, Endplates, and Current Collector)	73
3.4.1	Anode	74
3.4.1.1	Carbon Anode Materials	74
3.4.1.2	Noncarbonaceous Anode Materials	74
3.4.2	Cathode	76
3.4.2.1	$LiCoO_2$	76
3.4.2.2	$LiNiO_2$	77
3.4.2.3	$LiMnO_2$	78
3.4.2.4	$LiFePO_4$	79
3.4.3	Separator	80
3.4.4	Current Collector	82
3.5	Assembly, Stacking, and Manufacturing of Li Ion Rechargeable Batteries	84
3.5.1	Electrode Manufacturing	85
3.5.1.1	Cathode Materials	85
3.5.1.2	Anode Materials	85
3.5.1.3	Mixing	85
3.5.1.4	Coating	85
3.5.1.5	Drying	86

3.5.2	Assembling Process 86	
3.5.2.1	Core Construction 86	
3.5.2.2	Welding 86	
3.5.2.3	Injecting 87	
3.5.2.4	Sealing 87	
3.5.3	Formation and Aging 87	
3.6	Li Ion Battery Performance, Testing, and Diagnosis 88	
3.6.1	Characteristics of Energy 88	
3.6.1.1	Specific Energy of Cathode Materials 88	
3.6.1.2	Specific Energy of Anode Materials 88	
3.6.2	Working Characteristics 89	
3.6.2.1	Discharge Rate Capability 89	
3.6.2.2	Cycle Life 91	
3.6.2.3	Storage Performance 92	
3.6.2.4	Temperature Effects on Performance 94	
3.6.3	Test and Evaluation of Li Battery Materials 94	
3.7	Degradation Mechanisms and Mitigation Strategies 96	
3.7.1	Degradation Mechanisms 96	
3.7.1.1	The Effect of Anode Materials 96	
3.7.1.2	The Effect of Cathode Materials 97	
3.7.1.3	The Effect of Electrolyte 98	
3.7.1.4	The Effect of Charging and Discharging States 98	
3.7.1.5	The Effect of Current Collector 98	
3.7.2	Mitigation Strategies 99	
3.7.2.1	The Surface Treatment 99	
3.7.2.2	Replacement of $LiPF_6$ by Salts and the Use of Additives 101	
3.7.2.3	Developing a New System of Electrolyte 101	
3.8	Current and Potential Applications of Secondary Li Ion Batteries 101	
3.8.1	Portable Electronic Devices 101	
3.8.2	Applications of Lithium Ion Batteries in Electric Vehicle (EV) Industry 102	
3.8.3	Application Prospect of Lithium Ion Battery in the Aerospace Industry 103	
3.8.4	As Energy Reserves 106	
3.8.5	Application of Military Equipment 106	
	References 107	
4	**Lead-Acid Battery** *111*	
	Joey Jung	
4.1	General Characteristics and Chemical/Electrochemical Processes in a Lead-Acid Battery 111	
4.1.1	General Characteristics 111	
4.1.2	Major Milestones in the Development of the Lead-Acid Battery 111	
4.1.2.1	Chemistry/Electrochemistry 111	

4.2	Battery Components (Anode, Cathode, Separator, Endplates (Current Collector), and Sealing)	115
4.2.1	Battery Grid (Current Collector)	115
4.2.2	Active Material	116
4.2.2.1	Lead Oxide	116
4.2.2.2	Positive Active Material (Cathode Paste)	117
4.2.2.3	Negative Active Material (Anode Paste)	118
4.2.3	Electrolyte	119
4.2.4	Paste Production	119
4.2.5	Pasting	122
4.2.6	Curing	123
4.2.7	Formation	125
4.2.7.1	Tank Formation	125
4.2.7.2	Case Formation	125
4.2.8	Separator	126
4.2.9	Assembly	126
4.2.10	Case to Cover Seal	127
4.3	Main Types and Structures of Lead-Acid Batteries	128
4.3.1	SLI Batteries	128
4.3.2	Deep Cycle and Traction Batteries	133
4.3.3	Stationary Battery	139
4.3.4	VRLA Battery	143
4.4	Charging Lead-Acid Battery	146
4.5	Maintenance and Failure Mode of a Lead-Acid Battery	151
4.5.1	Maintenance	151
4.5.2	Safety	153
4.5.3	Failure Mode of Lead-Acid Battery	154
4.6	Advanced Lead-Acid Battery Technology	154
4.6.1	Negative Current Collector Improvement	154
4.6.1.1	Ultrabattery	154
4.6.1.2	PbC Capacitor Battery	156
4.6.1.3	Firefly Oasis Battery	157
4.6.2	Current Collector Improvement	159
4.6.2.1	Lead-Alloy-Coated, Reticulated Carbon Current Collectors	160
4.6.2.2	Lead-Alloy-Coated Polymer Current Collectors	163
4.6.3	Battery Construction	165
4.6.3.1	Horizon Battery	165
4.6.3.2	Bipolar Battery	166
4.6.4	Electrolyte Improvement	168
4.6.4.1	Gel Silicon Electrolyte	168
4.6.4.2	Liquid Low Sodium Silicate Electrolyte	169
4.7	Lead-Acid Battery Market	169
4.7.1	Automotive (Transportation and Recreation, Original Equipment Manufacturer (OEM) and Replacement)	170
4.7.1.1	SLI Battery (Starting, Lighting, and Ignition)	170

4.7.1.2	Deep Cycle Battery (E-Bike/Marine/Recreational Vehicle/Motor Home) *170*	
4.7.1.3	Microhybrid *170*	
4.7.2	Industrial *171*	
4.7.2.1	Motive (Industrial Trucks, Fork Lifts, Golf Carts, etc.) *171*	
4.7.2.2	Stationary (Utility/Switchgear, Telecommunication, Uninterruptible Power Systems (UPS), Emergency Lighting, Security Systems, Renewable Energy Systems, Cable Television/Broadcasting, Oil and Gas Exploration, Railway Backup, etc.) *172*	
4.7.2.3	Emerging Grid Applications *172*	
4.7.2.4	Distributed Renewable *173*	
	References *173*	
	Further Reading *174*	
5	**Nickel-Metal Hydride (Ni-MH) Rechargeable Batteries** *175*	
	Hua Ma, Fangyi Cheng, and Jun Chen	
5.1	Introduction to NiMH Rechargeable Batteries *175*	
5.2	Electrochemical Processes in Rechargeable Ni-MH Batteries *177*	
5.3	Battery Components *180*	
5.3.1	Anode *180*	
5.3.1.1	Properties of Hydrogen Storage Alloys *180*	
5.3.1.2	The Classification of Hydrogen Storage Alloys *183*	
5.3.1.3	Alloy Preparation *191*	
5.3.2	Cathode *194*	
5.3.2.1	Preparation of $Ni(OH)_2$ *198*	
5.3.3	Separator *199*	
5.3.3.1	Preparation of Separators *200*	
5.3.4	Electrolyte *202*	
5.3.5	Endplates *204*	
5.3.5.1	Conductive Agent *204*	
5.3.5.2	Adhesives *205*	
5.3.6	Cost Distributions for the Components of Ni-MH Batteries *206*	
5.4	Assembly, Stacking, Configuration, and Manufacturing of Rechargeable Ni-MH Batteries *206*	
5.4.1	Electrode Preparation *206*	
5.4.2	Assembly of Ni-MH Batteries *207*	
5.4.3	Battery Pack *207*	
5.4.4	Configurations *207*	
5.4.4.1	Cylindrical Configuration *207*	
5.4.4.2	Prismatic Configuration *208*	
5.4.4.3	Button Configuration *214*	
5.4.4.4	Cell and Pack Configurations *214*	
5.4.4.5	Bipolar Ni-MH Battery *218*	
5.5	Ni-MH Battery Performance, Testing, and Diagnosis *219*	
5.5.1	Capacity *219*	

5.5.2	Voltage	219
5.6	Degradation Mechanisms and Mitigation Strategies	221
5.6.1	Surface Corrosion of Negative Electrode	222
5.6.2	Decrepitation of Alloy Particles	222
5.6.3	Loss of Water in the Electrolyte	222
5.6.4	Crystalline Formation	223
5.6.5	Cell Reversal	223
5.6.6	High Self-Discharge	223
5.6.7	Shorted Cells	223
5.7	Applications (Portable, Backup Power, and Transportation)	224
5.7.1	Electric Tools	224
5.7.2	Electric Bicycles	225
5.7.3	Electric Vehicles (EVs) and Hybrid Electric Vehicles (HEVs)	226
5.7.3.1	The Toyota Prius	226
5.7.3.2	The Honda Insight	227
5.7.3.3	The Ford Escape Hybrid	228
5.7.4	UPS and Energy Storage Battery	229
5.7.5	Recycling of Ni-MH Batteries	230
5.8	Challenges and Perspectives of Ni-MH Rechargeable Batteries	231
	References	232
6	**Metal–Air Technology**	**239**
	Bruce W. Downing	
6.1	Metal–Air Technology	239
6.1.1	Metal–Air Technologies	241
6.2	Introduction to Aluminum–Air Technology	242
6.2.1	Aluminum–Air Cell Structure	243
6.2.2	Electrochemical Processes	244
6.2.2.1	Reaction Products	245
6.2.3	Degradation Mechanisms and Mitigation Strategies	246
6.2.3.1	Anode	246
6.2.3.2	Cathodes	246
6.2.4	Applications	246
6.3	Introduction to Lithium–Air Technology	246
6.3.1	Aqueous Lithium–Air Cell Structure	247
6.3.2	Electrochemical Processes	247
6.3.3	Nonaqueous Lithium–Air Cell Structure	248
6.3.4	Degradation Mechanisms and Mitigation Strategies	248
6.3.4.1	Anodes	248
6.3.4.2	Electrolyte	248
6.3.4.3	Cathodes	249
6.3.5	Manufacturing	249
6.3.6	Applications	249
6.4	Introduction to Zinc–Air Technology	249
6.4.1	Zinc–Air Cell Structure	249

6.4.2	Electrochemical Processes	249
6.4.2.1	Reaction Products	250
6.4.3	Degradation Mechanisms and Mitigation Strategies	250
6.4.3.1	Anode	250
6.4.3.2	Hydrogen Evolution	251
6.4.3.3	Electrolyte	251
6.4.3.4	Cathode	251
6.4.3.5	Regeneration	251
6.4.4	Applications	252
6.5	Introduction to Magnesium–Air Technology	252
6.5.1	Magnesium	252
6.5.1.1	Sources	253
6.5.1.2	Production	254
6.5.1.3	Uses	254
6.5.2	Associations	255
6.6	Structure of Magnesium–Air Cell	255
6.7	Electrochemical Processes	255
6.8	Components	258
6.8.1	Electrolyte	258
6.8.2	Cathode	258
6.8.3	Anode	261
6.8.4	Reaction Products	263
6.9	Manufacturing	263
6.10	Magnesium–Air Battery Performance	267
6.11	Degradation Mechanisms and Mitigation Strategies	269
6.11.1	Hydrogen Problem	269
6.11.2	Magnesium Utilization Efficiency (MUE)	271
6.11.3	Electrolyte	271
6.11.4	Cathode	273
6.11.5	Precipitate	273
6.12	Applications	273
6.13	Challenges and Perspectives of Magnesium–Air Cells	274
	References	275

7 Liquid Redox Rechargeable Batteries 279
Huamin Zhang

7.1	Introduction	279
7.1.1	Summary	279
7.1.2	Background	279
7.1.3	Development	280
7.2	Electrochemical Processes in a Redox Flow Battery	284
7.2.1	Basic Electrochemical Principles	284
7.2.1.1	Apparent Current Density	284
7.2.1.2	Electromotive Force and Open Circuit Potential	284
7.2.1.3	Polarization of Redox Flow Battery	285

7.2.2	Introduction to Electrochemical Processes [7]	286
7.2.2.1	Vanadium Redox Battery (VRB)	286
7.2.2.2	Zinc/Bromine Flow Battery	286
7.2.2.3	Sodium Polysulfide/Bromine Flow Battery	286
7.2.2.4	Vanadium/Bromine Flow Battery	287
7.2.2.5	Fe/Cr Flow Battery	288
7.2.2.6	Zinc/Cerium Flow Battery	288
7.2.2.7	Soluble Lead-Acid Flow Battery	288
7.2.2.8	Zinc/Nickel Flow Battery	288
7.3	Materials and Properties of Redox Flow Battery	288
7.3.1	Introduction	288
7.3.2	Electrodes	289
7.3.3	Bipolar Plates	290
7.3.4	Ion Exchange Membranes	292
7.3.5	Electrolyte Solutions	294
7.4	Redox Flow Battery System	295
7.4.1	Introduction	295
7.4.2	Single Cell	295
7.4.3	Battery Module	296
7.4.4	The Composition of the Battery System	297
7.4.5	Battery System Scale-Up	298
7.5	Performance Evaluation of Redox Flow Battery	298
7.5.1	Introduction	298
7.5.2	Performance Indicators	299
7.5.2.1	Rated Output Power	299
7.5.2.2	Energy Storage Capacity	299
7.5.2.3	Efficiencies of Charge–Discharge Cycles	300
7.5.2.4	Voltage Uniformity of Single Battery	300
7.5.2.5	Self-Discharge	300
7.5.2.6	Cycle Life	301
7.5.3	Evaluation of Battery System Performance	301
7.5.3.1	Evaluation Method	301
7.5.3.2	Evaluation Instruments	301
7.5.4	Factors Influencing Battery Performance	301
7.5.4.1	Battery Materials	301
7.5.4.2	Battery Structures	302
7.5.4.3	Concentration and Composition of Electrolyte	302
7.5.4.4	Charge–Discharge Conditions	302
7.5.4.5	Temperature	303
7.5.4.6	Flow Rate of Electrolyte	303
7.5.5	Control and Management of the Battery System	303
7.5.5.1	Operational Parameter Control	303
7.5.5.2	Safety Management of RFB	303
7.5.5.3	Power Management	304
7.5.5.4	Heat Management	304

7.5.5.5	System Failure Diagnosis	*304*
7.6	Degradation Mechanisms and Mitigation Strategies	*305*
7.6.1	Introduction	*305*
7.6.2	Degradation Mechanisms	*305*
7.6.2.1	Analysis of the Degradation of Materials	*305*
7.6.2.2	Degradation of the System Capacity	*306*
7.6.3	Mitigation Strategies	*307*
7.6.3.1	Material Modification	*308*
7.6.3.2	Controlling Strategies	*308*
7.6.3.3	Optimization of the Battery System	*309*
7.7	Applications of Redox Flow Batteries	*309*
7.7.1	Introduction	*309*
7.7.2	The Application of RFBs in Renewable Energy Systems	*309*
7.7.3	The Application of Peak Shaving of the Power Grid	*310*
7.7.4	The Application of RFB as Emergency and Backup Power	*311*
7.7.5	The Application as Electrical Vehicle Charging Station	*312*
7.7.6	Application as a Communication Base Station	*312*
7.8	Perspectives and Challenges of RFB	*313*
7.8.1	The Breakthrough for Key Materials and Technologies	*313*
7.8.2	System Assembly and Scale-Up	*313*
7.8.3	Cost	*314*
	References	*314*
8	**Electrochemical Supercapacitors**	*317*
	Aiping Yu, Aaron Davies, and Zhongwei Chen	
8.1	Introduction to Supercapacitors (Current Technology State and Literature Review)	*317*
8.1.1	Historical Overview	*319*
8.1.2	Current Research and Industry Development	*319*
8.2	Main Types and Structures of Supercapacitors	*322*
8.2.1	Electric Double-Layer Capacitors (EDLCs)	*322*
8.2.2	Pseudocapacitor	*324*
8.2.3	Asymmetric Hybrid Capacitors	*324*
8.3	Physical/Electrochemical Processes in Supercapacitors	*325*
8.3.1	Physical/Electrochemical Processes for EDLC	*325*
8.3.1.1	Analysis of the Diffuse Layer	*326*
8.3.2	Physical/Electrochemical Processes for the Pseudocapacitor	*328*
8.3.2.1	Thermodynamic Approach	*329*
8.3.2.2	Kinetic Approach	*330*
8.3.2.3	Redox Pseudocapacitance	*331*
8.3.3	Physical/Electrochemical Processes for AHCs	*331*
8.3.4	Electrolyte Processes: Conductance and Dissociation	*333*
8.3.5	Modeling of Electrochemical Behavior	*334*
8.3.5.1	Effects of Pore Size and Pore-Size Distribution	*335*
8.4	Supercapacitor Components	*338*

8.4.1	Electrode	338
8.4.1.1	Carbon as an EDLC Material	339
8.4.1.2	Pseudocapacitive Electrode Materials	346
8.4.2	Electrolytes for Use in Electrochemical Supercapacitors	352
8.4.2.1	Aqueous Electrolytes	353
8.4.2.2	Organic Electrolytes	353
8.4.2.3	Ionic Liquid Electrolytes	353
8.4.3	Separators	354
8.4.4	Current Collector and Sealant Components	357
8.5	Assembly and Manufacturing of Supercapacitors	357
8.5.1	Electrode Fabrication from Carbon Materials	358
8.5.2	Concerns with Cell Assembly	359
8.6	Supercapacitors Stacking and Systems	359
8.6.1	Bipolar Electrodes	361
8.7	Supercapacitor Performance, Testing, and Diagnosis	362
8.7.1	Cyclic Voltammetry	362
8.7.2	Charge or Discharge Chronopotentiometry	363
8.7.3	Impedance Measurement	364
8.7.4	Energy and Power Density	367
8.7.4.1	Energy Density	367
8.7.4.2	Power Density	367
8.7.5	Leakage Current and Self-Discharge Behavior	368
8.7.6	Durability Test and Additional Procedures	369
8.8	Supercapacitor Configurations	369
8.9	Applications	371
8.9.1	Use in Memory Backup	371
8.9.2	Improved Battery Systems	372
8.9.3	Hybrid Electric Vehicles and Transport	372
8.9.4	Military Applications	373
8.9.5	Current Marketed Applications	374
8.9.6	Future Applications	374
8.10	Challenges and Perspectives of Electrochemical Supercapacitors	375
	References	376

Contents to Volume 2

9	Water Electrolysis for Hydrogen Generation *383* Pierre Millet	
10	Hydrogen Compression, Purification, and Storage *425* Pierre Millet	
11	Solar Cell as an Energy Harvesting Device *463* Aung Ko Ko Kyaw, Ming Fei Yang, and Xiao Wei Sun	
12	Photoelectrochemical Cells for Hydrogen Generation *541* Neelu Chouhan, ChihKai Chen, Wen-Sheng Chang, Kong-Wei Cheng, and Ru-Shi Liu	
13	Polymer Electrolyte Membrane Fuel Cells *601* Stefania Specchia, Carlotta Francia, and Paolo Spinelli	
14	Solid Oxide Fuel Cells *671* Jeffrey W. Fergus	
15	Direct Methanol Fuel Cells *701* Kan-Lin Hsueh, Li-Duan Tsai, Chiou-Chu Lai, and Yu-Min Peng	
16	Molten Carbonate Fuel Cells *729* Xin-Jian Zhu and Bo Huang	
	Index *777*	

Preface

In today's world, clean energy technologies, which include energy storage and conversion, play the most important role in the sustainable development of human society, and are becoming the most critical elements in overcoming fossil fuel exhaustion and global pollution. Among clean energy technologies, electrochemical technologies are considered the most feasible, environmentally friendly and sustainable. Electrochemical energy technologies such as secondary (or rechargeable) batteries and fuel cells have been invented and used, or will be used in several important application areas such as transportation, stationary, and portable/micro power. With increasing demand in both energy and power densities of these electrochemical energy devices in various new application areas, further research and development are essential to overcome challenges such as cost and durability, which are considered major obstacles hindering their applications and commercialization. In order to facilitate this new exploration, we believe that a book covering all important areas of electrochemical energy technologies for clean energy storage and conversion, giving an overall picture about these technologies, should be highly desired.

The proposed book will give a comprehensive description of electrochemical energy conversion and storage methods and the latest development, including batteries, fuel cells, supercapacitors, hydrogen generation and storage, as well as solar energy conversion. It addresses a variety of topics such as electrochemical processes, materials, components, assembly and manufacturing, degradation mechanisms, as well as challenges and strategies. Note that for battery technologies, we have tried our best to focus on rechargeable batteries by excluding primary batteries. With chapter contributions from scientists and engineers with excellent academic records as well as strong industrial expertise, who are at the top of their fields on the cutting edge of technology, the book includes in-depth discussions ranging from comprehensive understanding, to engineering of components and applied devices. We wish that a broader view of various electrochemical energy conversion and storage devices will make this book unique and an essential read for university students including undergraduates and graduates, scientists, and engineers working in related fields. In order to help readers to understand the science and technology of the subject, some important and representative figures, tables, photos, and comprehensive lists of reference papers, will also be presented in this book. Through reading this book, the readers can easily locate the latest information on electrochemical technology, fundamentals, and applications.

In this book, each chapter is relatively independent of the others, a structure which we hope will help readers quickly find topics of interest without necessarily having to read through the whole book. Unavoidably, however, there is some overlap, reflecting the interconnectedness of the research and development in this dynamic field.

We would like to acknowledge with deep appreciation all of our family members for their understanding, strong support, and encouragement.

If any technical errors exist in this book, all editors and chapter authors would deeply appreciate the readers' constructive comments for correction and further improvement.

Ru-Shi Liu, Lei Zhang, Xueliang Sun, Hansan Liu, and Jiujun Zhang

About the Editors

Ru-Shi Liu received his bachelor's degree in chemistry from Shoochow University, Taiwan, in 1981, and his master's in nuclear science from the National Tsing Hua University, two years later. He gained one Ph.D. in chemistry from National Tsing Hua University in 1990, and one from the University of Cambridge in 1992. From 1983 to 1995 he worked as a researcher at the Industrial Technology Research Institute, before joining the Department of Chemistry at the National Taiwan University in 1995 where he became a professor in 1999. He is a recipient of the Excellent Young Person Prize, Excellent Inventor Award (Argentine Medal) and Excellent Young Chemist Award. Professor Liu has over 350 publications in scientific international journals as well as more than 80 patents to his name.

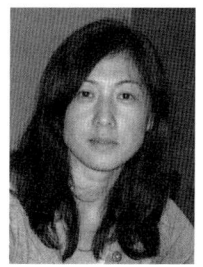

Lei Zhang is a Research Council Officer at the National Research Council of Canada Institute for Fuel Cell Innovation. She received her first M.Sc. in inorganic chemistry from Wuhan University in 1993, and her second in materials chemistry from Simon Fraser University, Canada in 2000. She is an adjunct professor at the Federal University of Maranhao, Brazil and at the Zhengzhou University, China, in addition to being an international advisory member of 7th IUPAC International Conference on Novel Materials and their Synthesis and an active member of the Electrochemical Society and the International Society of Electrochemistry. Ms. Zhang has co-authored over 90 publications and holds five US patent applications. Her main research interests include PEM fuel cell electrocatalysis, catalyst layer/electrode structure, metal-air batteries/fuel cells and supercapacitors.

Xueliang (Andy) Sun holds a Canada Research Chair in the development of nanomaterials for clean energy, and is Associate Professor at the University of Western Ontario, Canada. He received his Ph.D. in materials chemistry in 1999 from the University of Manchester, UK, after which he worked as a postdoctoral fellow at the University of British Columbia, and as a research associate at l'Institut national de la recherche scientifique, Canada. He is the recipient of a number of awards, including the Early Researcher award, Canada Research Chair award and University Faculty Scholar award, and has authored or co-authored over 100 papers, 3 book chapters and 8 patents. Over the past decade, Dr. Sun has established a remarkable track record in nanoscience and nanotechnology for clean energy, mainly in the synthesis and structure control of one-dimensional nanomaterials, as well as their applications for fuel cells and Li ion batteries.

Hansan Liu is a researcher at the Oak Ridge National Laboratory, US Department of Energy. He obtained his Ph.D. in electrochemistry from Xiamen University where he studied cathode materials for lithium ion batteries. After graduation, he worked at the Hong Kong Polytechnic University and the National Research Council Canada on electrophotocatalysis and fuel cell electrocatalysis, respectively. He is currently working on next generation high-energy density batteries at ORNL. Dr. Liu has 14 years of research experience in the field of electrochemical energy storage and conversion. His research interests mainly include battery and supercapacitor materials, fuel cell electrocatalysts, and synthesis and applications of high surface area materials. He has authored and co-authored over 70 publications, including 3 books, 4 book chapters and 3 patent applications relating to batteries and fuel cells. Dr. Liu is an active member of the Electrochemical Society and the International Society of Electrochemistry.

Currently a Senior Research Officer and PEM Catalysis Core Competency Leader at the National Research Council of Canada Institute for Fuel Cell Innovation, **Jiujun Zhang** received his B.Sc. and M.Sc. in electrochemistry from Beijing University, China, in 1982 and 1985, respectively, and his Ph.D. in electrochemistry from Wuhan University in 1988. After this, he took up a position as an associate professor at the Huazhong Normal University, and in 1990 carried out three terms of postdoctoral research at the California Institute of Technology, York University, and the University of British Columbia. Dr. Zhang holds several adjunct professorships, including one at the University of Waterloo and one at the University of British Columbia, and is an active member of The Electrochemical Society, the International Society of Electrochemistry, and the American Chemical Society. He has 240 publications and around 20 patents or patent publications to his name. Dr. Zhang has over 28 years of R & D experience in theoretical and applied electrochemistry, including over 14 years of R & D in fuel cell, and three years of experience in electrochemical sensor.

List of Contributors

Wen-Sheng Chang
Industrial Technology Research Institute
Department of Nano-Tech Energy Conversion
195, Sec. 4, Chung Hsing Road
Chutung
Hsinchu 31040
Taiwan

ChihKai Chen
National Taiwan University
Department of Chemistry
Sec. 4, Roosevelt Road
Taipei 10617
Taiwan

Jun Chen
Nankai University
Key Laboratory of Advanced Energy Materials Chemistry (Ministry of Education)
Chemistry College
Tianjin 300071
China

Zhongwei Chen
University of Waterloo
Department of Chemical Engineering
Waterloo Institute for Nanotechnology
Waterloo Institute for Sustainable Energy
Waterloo
Ontario N2L 3G1
Canada

Fangyi Cheng
Nankai University
Key Laboratory of Advanced Energy Materials Chemistry (Ministry of Education)
Chemistry College
Tianjin 300071
China

Kong-Wei Cheng
Chang Gung University
Department of Chemical and Materials Engineering
259 Wen-Hwa 1st Rd.
Kwei-Shan
Tao-Yuan 33302
Taiwan

Neelu Chouhan
National Taiwan University
Department of Chemistry
Sec. 4, Roosevelt Road
Taipei 10617
Taiwan

and

Government P. G. College
Department of Chemistry
Devpura, Kota Road
Bundi 323001
India

Aaron Davies
University of Waterloo
Department of Chemical
Engineering
Waterloo Institute for
Nanotechnology
Waterloo Institute for
Sustainable Energy
Waterloo, N2L 3G1
Ontario
Canada

Bruce W. Downing
MagPower Systems Inc.
20 – 1480 Foster Street
White Rock, BC V4B 3X7
Canada

Jeffrey W. Fergus
Auburn University
Materials Research and Education
Center
275 Wilmore Laboratories
AL 36849
USA

Carlotta Francia
Politecnico di Torino
Department of Materials Science
and Chemical Engineering
Corso Duca degli Abruzzi 24
Torino 10129
Italy

Fathy M. Hassan
University of Waterloo
Department of Chemical
Engineering
Waterloo Institute for
Nanotechnology
Waterloo Institute for
Sustainable Energy
Waterloo, Ontario N2L3G1
Canada

Kan-Lin Hsueh
National United University
Department of Energy
Engineering, No.2, Lianda Rd.
Miaoli 36003
Taiwan

Bo Huang
Shanghai Jiao Tong University
Institute of Fuel Cells
800 Dongchuan Road
Shanghai 200240
China

Joey Jung
EVT Power Inc.
6685 Berkeley Street
Vancouver, V5S 2J5
Canada

Aung Ko Ko Kyaw
Nanyang Technological
University
School of Electrical and
Electronic Engineering
Nanyang Avenue
Singapore 639798
Singapore

Chiou-Chu Lai
Industrial Technology Research
Institute
Material and Chemical Research
Laboratories, No.195, Sec. 4,
Zhongxing Rd.
Zhudong Township, Hsinchu
County 31040
Taiwan

Ru-Shi Liu
National Taiwan University
Department of Chemistry
Sec. 4, Roosevelt Road
Taipei 10617
Taiwan

Hua Ma
Nankai University
Key Laboratory of
Advanced Energy
Materials Chemistry
(Ministry of Education)
Chemistry College
Tianjin 300071
China

Pierre Millet
Université de Paris-Sud 11
Institut de Chimie Moléculaire et
des Matériaux d'Orsay
UMR 8182 CNRS
15 rue Georges Clémenceau
Bâtiment 410,
91405 Orsay Cedex
France

Yu-Min Peng
Industrial Technology Research
Institute
Material and Chemical Research
Laboratories, No.195, Sec. 4,
Zhongxing Rd.
Zhudong Township, Hsinchu
County 31040
Taiwan

Stefania Specchia
Politecnico di Torino
Department of Materials Science
and Chemical Engineering
Corso Duca degli Abruzzi 24
Torino 10129
Italy

Paolo Spinelli
Politecnico di Torino
Department of Materials Science
and Chemical Engineering
Corso Duca degli Abruzzi 24
10129 Torino
Italy

Xiao Wei Sun
Nanyang Technological
University
School of Electrical and
Electronic Engineering
Nanyang Avenue
Singapore 639798
Singapore

and

Tianjin University
Tianjin Key Laboratory of
Low-Dimensional Functional
Material
Physics and Fabrication
Technology
Weijin Road
Tianjin 300072
China

Li-Duan Tsai
Industrial Technology Research Institute
Material and Chemical Research Laboratories, No.195, Sec. 4, Zhongxing Rd.
Zhudong Township, Hsinchu County 31040
Taiwan

Dingguo Xia
Beijing University of Technology
Department of Environmental and Energy Engineering
Ping le yuan 100
Chaoyang district
Beijing, 100124
China

Ming Fei Yang
Nanyang Technological University
School of Electrical and Electronic Engineering
Nanyang Avenue
Singapore 639798
Singapore

Aiping Yu
University of Waterloo
Department of Chemical Engineering
Waterloo Institute for Nanotechnology
Waterloo Institute for Sustainable Energy
Waterloo, Ontario N2L3G1
Canada

Huamin Zhang
Dalian Institute of Chemical Physics
Chinese Academy of Science, No.457 Zhongshan Road Dilian
Dilian 116023
China

Xin-Jian Zhu
Shanghai Jiao Tong University
Institute of Fuel Cells
800 Dongchuan Road
Shanghai 200240
China

1
Electrochemical Technologies for Energy Storage and Conversion
Neelu Chouhan and Ru-Shi Liu

1.1
Introduction

In this chapter, authors review the contemporary demand, challenges and future prospective of energy resources and discuss the relevant socioeconomical and environmental issues with their impact on global energy status. A sincere effort has been made to explore the better energy options of clean and sustainable energy sources such as hydro, biomass, wind, solar, geothermal, and biofuel as an alternative to the conventional energy sources. Electrolysis, photoelectrochemical, and photocatalytic water-splitting techniques were adopted for green and light fuel generation. Advancement in electrochemical technology for energy storage and conversion devices such as rechargeable batteries, supercapacitors, and fuel cells are also briefed.

1.2
Global Energy Status: Demands, Challenges, and Future Perspectives

World's economy revolves around the axis of energy prices, which are primarily governed by the political consequences, environmental impact, social acceptance, availability, and demand. Nation-wise world's energy consumption plot (1980–2050) is depicted in Figure 1.1, which rated the United States, China, Russia, South Korea, and India as potential energy consumers. Energy consumption rate of our planet in 2007 was 16%, which would be accelerated to an alarming rate of 34% by 2050 (Figure 1.2) [1]. Our severe dependency on oil and electricity makes energy a vital component of our daily life [2]. Soaring prices of oil (starting from $42 per barrel in 2008 to $79 per barrel in 2010, to $108 per barrel in 2020 and $133 per barrel in 2035) as projected in Figure 1.3 [3] and other associated necessary commodities along various burning environmental issues resulted from industrial revolution compel us to give a careful thought on this serious issue. Figure 1.4 assesses the geographical region-wise oil reserve that projects the oil assets and capacities of the different regions [4]. The current global energy scenario

Electrochemical Technologies for Energy Storage and Conversion, First Edition. Edited by Ru-Shi Liu, Lei Zhang, Xueliang Sun, Hansan Liu, and Jiujun Zhang.
© 2012 Wiley-VCH Verlag GmbH & Co. KGaA. Published 2012 by Wiley-VCH Verlag GmbH & Co. KGaA.

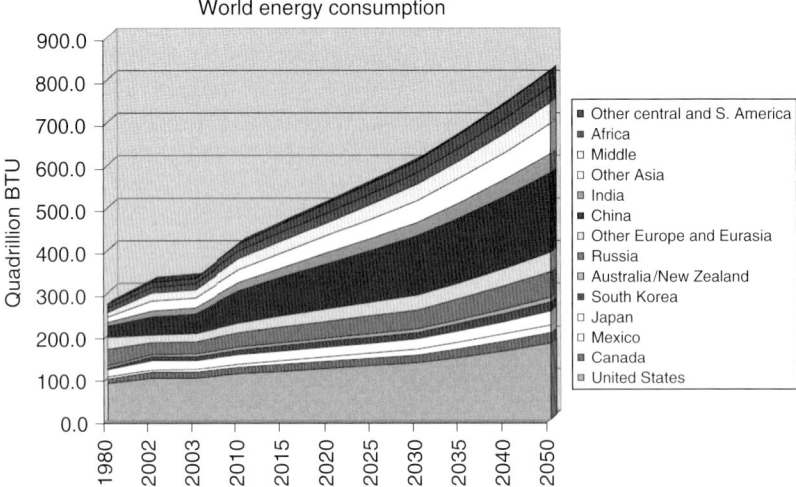

Figure 1.1 Nation-wise world energy consumption in the time interval of 1980–2050. (Energy Information Administration Annual Energy Review, 2007.)

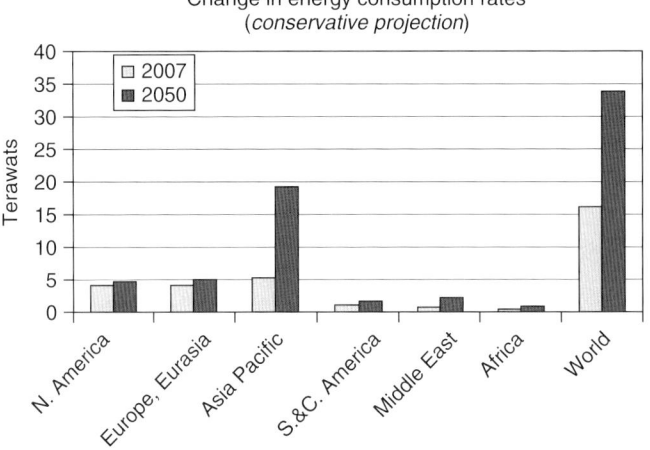

Figure 1.2 Comparative change in energy consumption rates of different zones against the world (actual reported for 2007 vs projected for 2050). (Renewable in global energy supply: an IEA fact sheet, January 2007.)

is full of uncertainty and faces three major energy challenges in the form of energy demand/energy supply ratio and security and their impact on the environment. The present worldwide population of 6.9 billion needs 14 TW annual energy [5] to sustain the current standard of life. Of the total energy production, 45% is required for industries, 30% for transport, 20% for residential and commercial buildings,

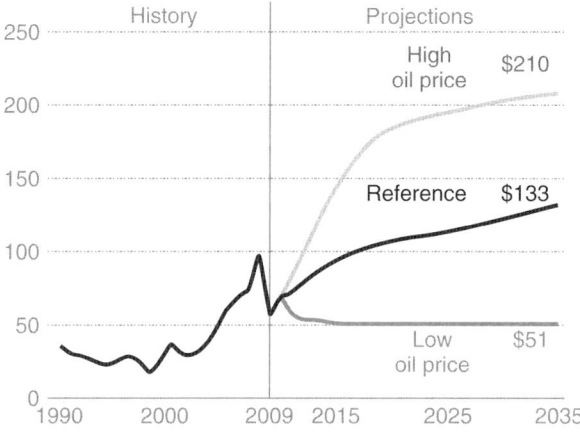

Figure 1.3 World oil prices in three oil price cases on the timescale of 1990–2035 ($2007 per barrel). (1980–2035: EIA, Annual Energy Outlook 2010, DOE/EIA-0383(2010) (Washington, DC, April 2010), web site: www.eia.gov/oiaf/aeo.)

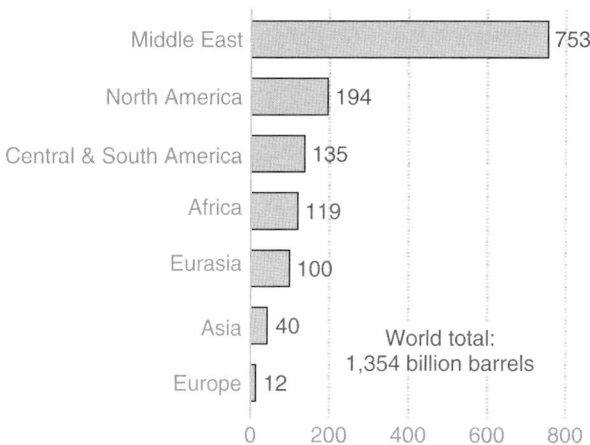

Figure 1.4 World's proved oil reserves by geographic region as of 1 January 2010 [4].

and the rest for services such as education, health, finance, government, and social services. Electricity is the world's fastest growing form of end-user energy consumption. Coal provides the largest share in the world's electricity generation, accounting for 42% in 2007, and its share will be largely unchanged through 2035. Rest share of the world's electricity generation is contributed by water, natural gas, nuclear power, hydropower, wind, and solar power. Economic trends and population growth drive the commercial sector activities and the resulting energy use.

The need for services (health, educational, financial, and governmental) increases as population increases. Slower expansion of gross domestic product (GDP) and declining population growth rate in many organization for economic cooperation and development (OECD membership) nations contribute to slower anticipated rates of increase in commercial energy demand. In addition, continued efficiency improvements moderate the growth of energy demand over time, as energy-using equipment is replaced with newer and more efficient stock. World's projected population would be quadrupled by 2050, the energy use doubled and electricity consumption tripled to our present energy demand. According to Hubbert's bell-shaped curve [6] of the worldwide oil production projection, we have already attained the peak and now observe a downfall and finally, the oil will last for 200 years (Figure 1.5) [7]. Lord Ron Oxburgh, former chairman of Shell, gave the statement on oil production possibilities and price, "It is pretty clear that there is not much chance of finding any significant quantity of new cheap oil. Any new or unconventional oil is going to be expensive." Despite the greenhouse gas concentrations approaching twice as those in the preindustrial period, coal and gasoline are still the major energy sources (34.3% oil, 25.9% coal, 20.9% gas, 13.1% renewables (10.4% combustion renewables and waste, 2.2% hydro, and 1.5% other renewables). Furthermore, alternative sustainable energy sources are still in the experimental stage; for example, some recent studies suggest that biofuels may not be as effective in reducing greenhouse gas emissions as previously thought. As a result, many countries have relaxed or postponed renewal of their mandates [8]. For example, Germany reduced its biofuel quota for 2009 from 6.25 to 5.25%. Therefore, governments, industrialists, and researchers have put their heads together on this leading energy issue with their concerns about the environmental challenges and renewed

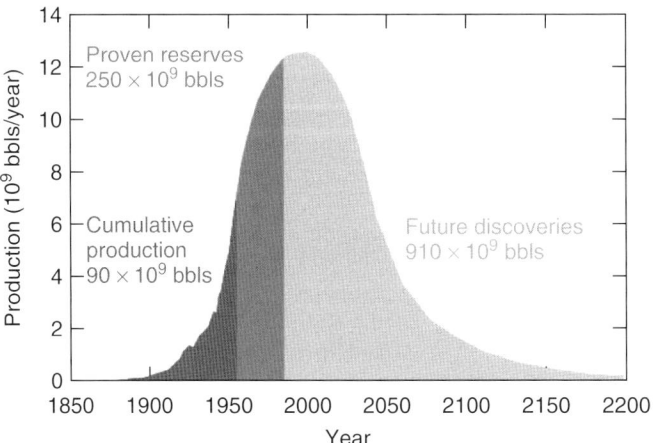

Figure 1.5 Hubbert's bell-shaped curve for time versus production of any exhaustible resources projection plot for the time interval 1850–2200 AC [7].

the interest in development of alternatives to fossil fuels, specifically, nuclear power, and renewable energy sources (wind, solar, biofuel, geothermal, tidal, hydro) using breakthrough concepts (catalysis by design, multielectron transfer) and accelerated application of cutting-edge scientific, engineering, and analytical tools. There are three major options of getting clean energy including carbon neutral energy (fossil fuel in conjunction with carbon sequestration), nuclear, and renewable energy. To satisfy the 10 TW no-carbon energy demands [9], a 38% conservation of energy for the next 50 years via combustion of fossil fuel is required, but the challenge of disposing 25 billion metric ton of CO_2 annually needs to be conquered. The need for nuclear-powered energy required the establishment of 365 GW electric nuclear fission plants per year for 50 years. The amount of annual renewable trappable energy from resources is as follows: the most viable and abundantly sourced solar energy with a capacity of 12 000 TW; integrated overall geothermal energy, 12 TW; globally extractable wind power, 2–4 TW; tidal/ocean current, 2 TW; and hydroelectric energy, 0.5 TW. Among all sources, obviously solar energy stands out as a promising choice of renewable energy, and currently, we are exploiting it only for the satisfaction of 0.1% of the demand. Therefore, by reducing energy demand and emissions accompanied with the use of the diversifying energy sources, we should be able to meet our energy target.

1.3
Driving Forces behind Clean and Sustainable Energy Sources

Our atmosphere is in a constant state of turmoil, and it is never being static. Relatively, internal and external changes in the earth's atmosphere, made by either Nature or man, bring changes in weather and climate. Scientific evidences pointed out the role of man in environmental degradation by insanely exploiting Mother Nature, which causes a disturbance in the delicate balance of Nature by accelerating global warming and associated climate changes, increasing ocean temperature, and bringing out changes in terrestrial geography, rain fall ratio, temperature, and type of soil. These changes fuel the growing consensus about the eminent need for a more pervasive action for environmental protection. Technological advancement attained during the past two decades has provided us a comfortable lifestyle full of facilities on a very high cost of resources consumption and degradation of our environment. The effect of the world's economic development on the environment was defined in the words of Elsa Reichmanis, the former president of the American Chemical Society, "We are past the days when we can trade environmental contamination for economical prosperity that is only a temporary bargain and the cost of pollution both economically and on human health is too high" [10]. However, it has disturbed ecological balance and damaged the environment, which has been proven disastrous for global life and has resulted in tremendous critical issues such as extinction of rare species of flora and fauna from earth, various incurable or semicurable diseases, global warming, acid rain, ozone layer depletion, excessive pollution, nuclear winter, and photochemical smog, especially in and around the urban areas. But still we do not wish to quit the comfortable lifestyle, and we simply cannot afford to continue along this path. Therefore, it is high time we shake hands with Nature to satisfy our energy demands in an eco-friendly manner by utilizing decentralized renewable energy sources such as solar power, wind, geothermal energy, biofuel and biomass, tidal power, wave, and hydropower. Furthermore, these sources are more efficient, abundant, and affordable (available free of cost) and are an environmentally benign solution for getting clean and green energy but on the condition that people master the technology. Both industrialized and developing countries should adopt the above-mentioned sustainable resources to build their energy capacity and improve their regulatory for clean, safe, and renewable energy. Therefore, large-scale transformation in energy policies should be executed with strong willpower along the necessary course of action toward clean and sustainable energy. Volatile energy prices of fossil fuels and increasingly scarce natural resources impend the government legislation that a growing trend of higher corporate social responsibility (CSR) and consumer sentiment favor environmentally friendly products and services. They induce the businesses, industries, and governments to respond in the innovative ways that might have been unimaginable just a few years ago. Executing relatively new wireless technology, networked sensors, management dashboard reporting, and automated alarm management is one way for businesses to reduce waste and optimize their position as environmental stewards in multiple domains (company, government, and geography). Reducing waste, managing scarce natural resources, saving energy, and following efficient operating conditions have always been good business tenets. The vital driving force behind the search for the most powerful, clean, and renewable energy sources is energy security for the future in energy policies of the governments all around the world, assuming renewable energy sources as a *guaranteed growth sector* [11]. Modern renewable energy industry has been hailed by many analysts because of the global trends and drivers underlying its expansion during the past decade. Without widespread improvements in environmental stewardship, impacts from the fundamental drivers will lead to adverse consequences around the world. Among these consequences, strict government legislations, climate changes, water stress, natural resource and raw material scarcity, public pressure, market risk and national security, and safety concerns are highlighted below.

1.3.1
Local Governmental Policies as a Potential Thrust

In 1960s, the slogan *Think Globally and act Locally* was coined by David Ross Brower, a prominent environmentalist and the founder of Sierra Club Foundation, John Muir Institute for Environmental Studies, and Friends of the Earth (1969), and many others that work actively for environment and initiate the worldwide consensus and awareness about environmental issues. In many countries, great progress has been made through awareness programs, proactive guiding principles and policies, new legislation, and government incentives in the form of tax relaxation and subsidies. Environmental risks force the people to go green in a significant way and promote a joint effort at governmental, enterprise, and even individual level to contribute in the awareness programs, proactive guiding principles, and policy making, all together. Government policies for renewable energy are a diverse and growing segment. Hundreds of local governments are setting future targets and adopting a broad array of proactive planning and promotion policies, including new legislation, government incentives for local feed-in tariffs and renewable electricity generation and heating, and mandates for buildings and businesses. Government regulations can play an efficient role in achieving effective changes in favor of environment, but it's only one out of several forces that will drive the needful changes into the future.

1.3.2
Greenhouse Gases Emission and the Associated Climate Changes

Emission of greenhouse gases significantly banks on industries, refrigeration, and transportation because fossil fuels, which are still a primary energy source and responsible for CO_2 emission, are accounted as one of the greenhouse gas. Major driving forces and science behind the *green* movement are focusing primarily on suppressing greenhouse gas emissions and their unwanted effects on global warming and associated climate changes, natural resource scarcity (oil, gasoline, and minerals in particular), and eventual ozone layer depletion, consequences of unabated human-driven pollution. New legislation, community pressure, customer safety concerns encourage proactive actions and strengthen the corporate environmental activities. Environmental improvement is one area of new business activity whose driving forces are so strong, responsibly compelling, and widely appreciated that raise the call to action and can appeal to every industry, enterprise, and organization, from the senior-most executives to the newest entry-level employees.

1.3.3
Public Awareness about Environmental Protection Rose around the World

Baton of the public awareness lights the world to see the real picture of mesmerization of environment done by humans, which push the government and industries to enact on the holistic green strategies to make this world more livable. Layman protesting by chaining themselves to trees or lying in front of bulldozers against deforestation and by organizing boycotts are the most visible environmental stewardships in 1980–1990s against the activities of *big business*, which enforce driving forces behind environmental sustainability. But nowadays the clashes between the corporate world and the environmentalists have become a thing of past. Businesses are adapting and applying suitable methodology to sustain the benefits to business operations without going against Nature. Moreover, they started working in favor of Nature by cross-company, cross-industry, and cross-geography cooperation, which will enable a clear and sustainable focus on improving environment by applying technology in new ways to achieve granular understanding of their operation and their impact on the planet. Public who wants to ride on green wings becomes a major positive driving force for environmental protection. However, public pressure could shift from a positive driving force to a more negative one. Public pressure in the form of unrest is certainly more likely when food production declines while population rises, basic natural resources become scarce, water supplies become more stressed, and there is occurrence of more frequent and more severe natural disasters. Public sympathy is also with endangered species of flora and fauna, an important part of the food chain that might perish because of global warming and associated climate changes. One survey on average American attitude toward the environment found out that more than three-quarter US workers want to have an employer, who is well informed about green movement on a daily basis and contributes accordingly [12]. Furthermore, developments in renewable energy have enough potential to create new industries and generate millions of new jobs around the world. The European Commission published statistics on Europeans' attitudes toward climate change and found that 73% of the Brits feel they were well informed about the causes of climate change and methods to tackle them, placing them in the same group as the citizens of Sweden, the Netherlands, and Finland. Despite all this, only 46% Brits and 50% Europeans but 82% Swedish ranked climate change as the most alarming global problem.

1.3.4
Population Growth and Industrialization

World population continues to grow, and statistical analysis predicts that by the year 2050, it will be increased by 30% of today's population, which will accelerate scarcity of natural resources such as oil, fossil fuel, water, minerals, agricultural land, and clean air. Industrialization further demands additional resources that can create environmental risk and jeopardizes the economy. The Intergovernmental Panel on Climate Change (IPCC) reported in 2007 [13] that the future climate change attributable to global warming is expected to put 50 million extra people at risk of hunger by 2020, which might be increased to 132 and 266 million by 2050 and 2080, respectively, because rising air temperatures could decrease rain-fed rice yields by 5–12% only in China and net cereal production in South Asian countries could decline by 4–10% by the end of this century, making it unable to meet the food demands of that time. Pressure of high consumption of resources because of population explosion, requiring more amount of drinkable water without drought or climate change. Furthermore, the threats such as oil and chemical spills, unmitigated waste, resources exploitation, residential and commercial real estate development, unsafe living conditions, and drainage of polluted industrial waste into river system pose a clear risk to public health and degrade the environment, enhancing the ultimate depletion or extreme scarcity, which would generate high risk to business and Nature.

1.3.5
Security and Safety Concerns Arising from Scarcity of Resources

National security became more relevant when the impact of climate or weather change crosses the country borders and countries rich in natural resources start dominating other countries. Public unrest develops as resource supplies become scarce and global conditions become adverse enough to result in riots, corruption, and military action and is a sufficient cause for disturbing national or international security. Safety for all living organisms and properties is another issue that prominently arises after any natural disaster, which is a result of global warming and associated climate changes, such as severe weather patterns, floods, fires, and hurricanes. Generous cooperation in the form of human help and financial support among countries with proactive measurements can make the affected country or region to be able to overcome the situation effectively. The above-mentioned risks are also a driving force behind environmental protection to mitigate before any of the direst predictions unfold and the current course of global cooperation changes.

1.3.6
Platforms Advocating in Favor of Sustainable and Renewable Resources

Environmental degradation, soaring prices and high consumption of conventional energy sources, perpetual resource wars, catastrophic effect of greenhouse gases on climate change, inextricable link between nuclear weapons and nuclear power, high cost of nuclear plant establishment and nuclear fuel, and problematic disposal of nuclear waste have fostered the international agencies to establish a platform such as United Nations Environmental Program (UNEP) that advocates renewable energy sources to take the place of fossil fuels without resorting to nuclear-powered energy. In 2000, a global network for the elimination of nuclear weapons lobbied nations (accounted 142 in 2009) around the world to institute the International Renewable Energy Agency (IRENA). IRENA opened its headquarters in Abu Dhabi and branch offices in Bonn and Vienna, and it is committed to becoming a principal driving force in promoting a rapid transition toward the sustainable use of renewable energy on a global scale. It included all forms of renewable energy produced in a sustainable manner, including solar power, wind, geothermal energy, hydropower, ocean, and appropriate bioenergy. Another example of such agencies is the Forest Stewardship Council (FSC) certifies some wood as sustainable when it meets the established criteria [14]. The U.S. Green Building Council (USGBC) has created the Leadership in Energy and Environmental Design (LEED, founded in 1993 by Robert K. Watson), a third-party certification program for the design, construction, and operation of green buildings. All agencies working together under the same theme to benefit the environment are a positive driving force.

1.3.7
Economic Risk Generated from Price Pressure of Natural Resources

World economy is a grand driving force to regulate the energy and raw material prices. Speculation on long-term demand, current availability, and cost of resources suggest some substitute raw materials, alternative or renewable energy sources, and conservation of traditional fossil fuels. As long as energy and commodity prices for scarce raw materials remain unpredictable and long-term global demand sustains the heightened levels, the drive for initiatives that reduce energy consumption and raw materials waste or shift to a less risky source with a more stable long-term cost has become stronger. For example, in 2004, crude oil was trading at $40 a barrel and the estimated *fair value* was to be only $27 a barrel [15]. Risk premium has varied up to 30% and potentially higher over time. On the other hand, the price for water in many areas of the world is still much lower than its actual economic value, which is a notable exception for water management. Price pressure alone is a powerful driving force, and as result of this, the world would be destined to experience an endless ebb and flow of cyclical activity without new taxation or *cap-and-trade* mechanisms to sustain the current green movement. That's why some enterprises install straightforward technology devices, such as motion-sensitive light switches, implement new technology to monitor and optimize energy consumption, train employees on energy-saving practices, or appoint new roles in the organization with accountability for achieving business and environmental benefits.

1.3.8
Regulatory Risk from Governmental Action and Legislation

Regulatory risk from new governmental legislations and global agreements is a powerful driving force that can accelerate the trend of environmental sustainability if applied effectively. The United Nations held a conference in 1997 at Kyoto, Japan, on climate change that resulted in an international agreement. The Kyoto Protocol took effect from 2005 to reduce greenhouse gas emissions worldwide, but the protocol's obligations are limited to monitoring and reporting, without actual provisions for enforcement and penalties if reductions are not achieved. Another prominent meeting of leading industrial nations (G8) held at Tokyo, Japan, 8 July 2008, endorsed halving world emissions of greenhouse gases by 2050 but set no near-term targets. The UK Climate Change Act (2008) aims to move the United Kingdom to a low-carbon economy and society, with an 80% cut in emissions by 2050 from a 1990 baseline [16]. The California Energy Commission recognizes the energy efficiency standards for real estate business and made future legislation that would require zero net energy homes and commercial buildings by 2020 and 2050, respectively [17]. In this series, the U.S. Army Energy Strategy set five tenets in 2005 to the strategy: eliminate waste, increase efficiency, reduce dependence on fossil fuels, conserve water resources, and increase energy security. Regulatory proposals being actively developed at the industrial, state, and local levels are more effective than the legislations at global and country levels. Therefore, it is high time for nations to follow international regulations strictly.

1.3.9
Fear of Reputational Risk to Strengthen Corporate Social Responsibility

Reputational risk at CSR is a vital driving force for companies to improve their status as environmental friendly, and it is viewed as an investment that brings financial returns and an opportunity or platform for growth that would increase the visibility in action. To reduce environmental impact, businesses adopted ethical standards to win customers' loyalty and market share and lowered their business risk that led to higher profitability through increased sales or decreased costs, which were often maintained during the adverse environmental events. A large number of companies are enrolled under the CSR network and few of them are (i) Catalyst Paper Corporation (Canadian) uses its own by-products (biomass) to power its operations; (ii) Tesco is a retail chain holder in grocery and industries (>2800 stores in central Europe, Asia, and North America), runs 75% of its delivery fleet on biodiesel fuel, had labeled 70 000 of its products with carbon counts (carbon labeling articulates the total carbon emissions from bringing a product to the store shelf) by 2008, and runs many straw-powered stores with heat and solar photovoltaic (PV) power plant for carbon-neutral electricity; and (iii) Wal-Mart (Latin America) has installed the largest sun-operated PV installation to satisfy 20% of the energy needs of its store. In California, new cars being sold are required to include labeling with *global warming scores.*

1.3.10
Operational and Supply Chain Risks from Inefficiencies and Environmental Changes

Operational and supply chain risks are another driving force generated from extreme adverse weather patterns, environmental hazards, and inefficiencies that will push the businesses to invest in innovative and sustainable energy sources and high-impact renewables, to favor upstream suppliers, and to set examples for other business partners. Although opportunities have been available since the past decades, only recently have all the driving forces aligned in the right direction to prompt the worldwide call to action.

1.4
Green and Sustainable Energy Sources and Their Conversion: Hydro, Biomass, Wind, Solar, Geothermal, and Biofuel

Environmental degradation and soaring prices and high consumption of conventional energy sources (34.3% oil, 25.9% coal, 20.9% gas, 13.1% renewables (10.4% combustion renewable and waste, 2.2% hydro, and 0.5% other renewables),

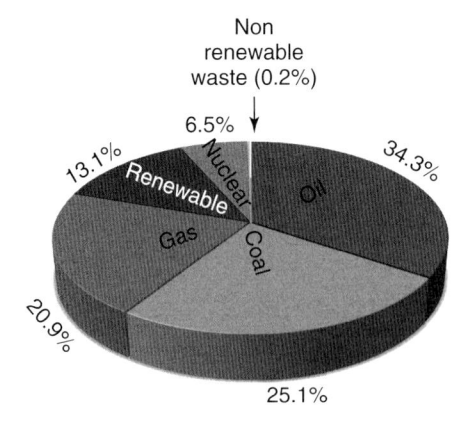

Figure 1.6 World total energy (conventional and renewable) consumption plotted against their percentage contribution (total primary energy, 410 EJ/year) (http://www.iea.org/papers/2006/renewable_factsheet.pdf).

6.5% nuclear, and 0.2% nonrenewables) are illustrated in the world energy consumption chart (Figure 1.6). The perpetual resource wars, catastrophic effect of greenhouse gases on climate change, inextricable link between nuclear weapons and nuclear power, high cost of nuclear plant establishment and nuclear fuel, and problematic disposal of nuclear waste all foster the international agencies to develop the sustainable energy sources to take the place of conventional energy sources without resorting to nuclear power. All time free availability and huge amount of decentralized renewable energy are the principal driving force against the promotion and rapid transition toward the sustainable renewable energy on the global scale. Renewable energy includes a wide spectrum of sustainable and powerful sources of natural energy such as solar, wind, geothermal, hydropower, ocean, and appropriate bioenergy. One can estimate the power of these sources as a single day of sunlight can supply enough energy to satisfy the world's electricity demand for 8 years, whereas wind can meet the world's electricity needs 40 times over and is capable of fulfilling all the global energy demands five times over, and the geothermal energy stored in the top 6 miles of the earth's crust contains 50 000 times the world's energy storage in oil and gas resources. Tidal, wave, and small hydropower can also provide vast stores of energy, available everywhere on earth. Both industrialized and developing countries should start adopting the above-mentioned sustainable resources to build their energy capacity and improve their regulatory for clean, safe, and renewable energy. By the year 2008, the top six countries rated by their total amount of renewable power capacity in use were China (76 GW), US (40 GW), Germany (34 GW), Spain (22 GW), India (13 GW), and Japan (8 GW) (9 September 2009, by Eric Martinot and Janet Sawin, London, UK, Renewables Global Status Report 2009 and update of the Renewable Energy Policy Network for 21st Century (REN21) annual report). Renewable energy capacity in developing countries grew to 119 GW in 2009, or a 43% share (out of which, 18% is

Table 1.1 Renewable energy added and existing capabilities, 2008 (estimated).

	Added during 2008	Existing at the end of 2008
Power generation (GW) electric		
Large hydro power	25–30	860
Wind power	27	121
Small hydro power	6–8	85
Biomass power	2	52
Solar PV grid connected	5.4	13
Geothermal	0.4	10
Concentrating solar power	0.06	0.5
Ocean (tidal) power	0	0.3
Hot water or heating (GW) (thermal)		
Biomass heating	n/a	250
Solar for hot water/space heating	19	145
Geothermal heating	n/a	50
Transport fuels (billion liters/year)		
Ethanol production	17	67
Biodiesel production	3	12

Global Energy Report-2009 REN21.

global electricity supply) of the total global energy capacity. Even though affected by the global economic downturn, the years 2008 and 2009 were remarkably the best for renewable, which is clear from Table 1.1, exhibiting the existing sources, and by 2008, renewable energy was added. The Global Status Report of 2010 on renewables conducted by REN21 shows that all the forms of grid-tied solar PV plants grew annually by 60% from the past decade. On average, the past 5 years' annual growth of wind power was 27%, solar hot water was 19%, and the ethanol and biodiesel production expanded by 34% [18]. Heat and power from biomass and geothermal sources continued to grow, and small hydropower increased by 8%. Globally, the approximate technology share of $120 billion (€85 billion) as renewable capital investment was divided into wind power (42%), solar PV (32%), biofuels (13%), biomass and geothermal power (6%), solar hot water (6%), and small hydropower (5%). Renewable capacity is discussed individually in the following sections.

1.4.1
Solar PV Plants

PV power generation employs solar panels comprising a number of cells containing a PV material, which include monocrystalline silicon, polycrystalline silicon, amorphous silicon, cadmium telluride, and copper indium selenide/sulfide. The cost of PV's has declined steadily since the first solar cells were manufactured, because of the advancement in technology and large-scale manufacturing units

[19]. More than 1800 solar PV plants of 16 GW existed worldwide by the end of 2008; Spain was leading with 2.6 GW of new capacity added, followed by the former PV leader Germany (added 1.5 GW), the United States (310 MW added), South Korea (200–270 MW), Japan (240 MW), and Italy (200–300 MW). Solar PV markets in Australia, Canada, China, France, and India have also continued to grow. Their additions reached a record high of 7 GW in 2009, when Germany topped the market with 3.8 GW added capacity and captured more than half of the global market. Other large markets were Italy, Japan, the United States, Czech Republic, and Belgium.

1.4.2
Wind Power

Wind power is the conversion of wind energy into a useful form of energy, such as using wind turbines to make electricity, wind mills for mechanical power, wind pumps for pumping water or drainage, and sails to propel ships. Around the world, >80 countries had installed commercial wind power by the year 2008 for their energy demand. The global wind power leader since mid-1990s, Germany (24 GW), has handed over its top position to the United States (25 GW), followed by Spain (18 GW), China (12 GW), and India (8 GW). Wind power additions reached a record high of 38 GW, that is, around 60% of the total global energy capacity (80 GW) on renewables utility scale investment in 2009 (excluding small projects). China tops the market with 13.8 GW, the United States was second, with 10 GW added. The share of wind power generation in several countries reached record highs, including 6.5% in Germany and 4% in Spain.

1.4.3
Geothermal Power

Geothermal wells release greenhouse gases trapped deep within the earth, but these emissions yield much lower energy per unit than those of fossil fuels. As a result, geothermal power has the potential to help mitigate global warming if widely deployed in place of fossil fuels. The United States remained the world leader in geothermal power development, with more than 120 projects underdevelopment, representing at least 5 GW. Geothermal projects were underway in over 40 countries, with another 3 GW in the pipeline. Globally, geothermal power capacity reached over 10 GW in 2008 and is increasing yearly [20].

1.4.4
Concentrating Solar Thermal Power (CSP) Plants

Concentrating solar thermal power (CSP) plants employ sunlight concentrated onto PV surfaces for the purpose of electrical power production. The United States and Spain are the leading figures in this field. New projects are also underdevelopment in Abu Dhabi, Algeria, Egypt, Israel, Italy, Portugal, Spain, and Morocco. One of the key trends is that a growing number of these CSP plants will include thermal storage in daytime, allowing power generation in the evening hours. The recently completed Andasol 1 plant in Spain has more than 7 h of full-load thermal storage capability. Overall, it is clear that parabolic trough plants are the most economic, most mature, and efficient thermal storage plants and a promise to solar thermal technology available today, although there are still significant areas for improvement and cost cutting in the near future.

1.4.5
Biomass

Biomass production contributes the energy equivalent of 5% of world gasoline output. Many countries evident the record use of biomass, notably, Sweden, where biomass accounted for a larger share of energy supply than oil for the first time in 2009, which was followed by Brazil.

1.4.6
Biofuel

The United States scored top in biofuels with 31 new ethanol refineries of 40 billion l per year production strength along with an additional 8 billion l per year capacity under-construction plants established by the year 2009. In transport fuels, ethanol production in Brazil ramped up dramatically in 2008 to 27 billion l in 400 ethanol mills and 60 biodiesel mills, after being maintained constant for a number of years, for the first time ever; more than half of Brazilian nondiesel vehicle fuel consumption came from ethanol. Notwithstanding Brazil's achievement, the United States remained the leading ethanol producer, with 34 billion l produced in 2008. Other ethanol fuel–producing countries include Australia, Canada, China, Colombia, Costa Rica, Cuba, the Dominican Republic, France, Germany, India, Jamaica, Malawi, Poland, South Africa, Spain, Sweden, Thailand, and Zambia. The European Union (EU) alone is responsible for about two-thirds of world biodiesel production, with Germany, France, Italy, and Spain with a biodiesel production capacity of 16 billion l per year in more than 200 biodiesel production units, and an additional ethanol production plant with a capacity of over 3 billion l per year is under construction.

1.5
Electrochemistry: a Technological Overview

Electrochemistry serves to illustrate the fundamentals related to the existence and movement of electrons present in bulk, as well as the interfaces between ionics, electronics, semiconductors, photonics, and dielectric materials and their consequences on various fields of science, that is, chemistry, engineering, biology, materials, and environmental [21–23]. It also accomplishes the reverse of above, that is, withdrawal of electricity from energetic chemicals by electrolysis. Electrons are inexpensive redox reagents, as the cost of a mole of electrons is <$0.01 compared to $0.03–$3.00 for common redox reagents [24]. John O'M Bockris described electrochemistry as a subject that deals with the making of substances by means of electricity or making of electricity by consuming substances. In traditional electrolytic techniques, the electric current directly passes between the electrodes (anode and cathode) in contact with the electrolytic phase that contains ions. Since 1972 [25], when the use of semiconductor materials as electrodes came into much closer focus, it widely extended the realm of subjects that can be treated under electrochemistry. Electrodes (anode and cathode) bring about specific chemical changes (oxidation and reduction, respectively), usually in conditions close to ambient temperature and pressure, without the use of any toxic reagents. Electrolysis can be a selective, an easily computer controllable, a convenient, and a cost-effective technology for synthesis, separation, characterization, and pollution control. Sophisticated electrochemical cells and cell components are readily available in market to assist us with high technical expertise. Suitable electrolytic cells are available off the shelf and are capable of being combined with other necessary processor units to construct fully integrated and compact production systems, which may be in a batch or continuous process. Electrochemical phenomenon plays a fundamental role in providing essential materials and devices that contribute significantly to the area of importance in national security and well-being of the mankind. Moreover, humans themselves are bioelectrochemical machines, converting solar energy stored in food via electrochemical reaction into muscle power. On the basis of widely spread occurrence of the electrolytic phenomenon in technology and devices, the arena of electrochemistry is categorized as follows:

- **Materials** of interest includes concrete, ceramic, catalytic materials, composites, colloids, semiconductors, surfactants, inhibitors, biomaterials such as proteins and enzymes, emulsion and foams, metal and alloys, ionic solids, dielectric, polymers, membrane and coating, and aqueous and nonaqueous solvent solutions.
- **Phenomena** that arise in the materials include conduction process, mass transfer by convection, ion exchange, potential field effect, adsorption, electron and ion disorders, colloidal and interfacial activity, wetting, membrane transport, sintering, dendrite formation, electrokinetics, electrocatalysis, passivity, bubbles evolution, and gaseous discharge (plasma) effect.
- **Processes** that critically depend on phenomenon include energy conversion and storage, chlor-alkali industry, pulp and paper, corrosion and corrosion control, membrane separation, surface reactions, desalination, deposition and etching by electrolytic and plasma processes, mining and metallurgy, environmental protection and control, water and wastewater treatment, processing and fabrication, electrochemical synthesis of inorganic and organic chemicals, and pollution detoxification and recovery.
- **Products** resulting from these processes include batteries and fuel cells, microelectronic devices, devices in information technology, ceramics, sensors, membranes, metals, gases, coatings and films, chemicals, pharmaceuticals, and microelectronics

This multidisciplinary field identifies new technological opportunities in widely diverse applications and underpins many technologies. In addition, cutting-edge applications in new areas, including *in situ* characterization, interfacial structures, surface reactions, and plasma, also hold great promise for advancement in the field. Aluminum for building and aircraft and titanium for supersonic aircraft and tanks are made of electrochemical processes. Highly sensitive microsensors implanted in human body can precisely report about the biochemical changes in the body. Electrochemical knowledge has been made feasible to accelerate the healing of tissue and to simulate the action of nerves that have been damaged. Electrochemical life-lasting batteries for pacemakers are also available in market. Coatings for car that would not change in appearance after years of service along with propulsion system for electric vehicles and methods to remove toxic materials selectively from streams of waters have also been made available. The electronics industry underwent a rapid evolution from thick to thin films during the last decade and often played an important and decisive role in the plating through mask technology, plating for thin film heads, plating for high density magnetic thin film, selective etching technology, and so on. New electrochemical approaches have also been playing the prominent roles in the electronics industry, and their activities touch almost all industrial sectors.

While all these technologies are based on the same fundamental principles but their practical manifestations may be quite different with, for example, cell configurations, electrode materials and sizes, electrolytes and separators, each designed to meet the particular demands of the application. Electrolysis should be selected as the preferred method over other competitive chemical routes because redox species are always recycled; hence, only small amounts of redox reagents are used without any stoichiometric by-products. High solubility of reactants required for viable current density is a major limitation of electrochemical technology.

1.6
Electrochemical Rechargeable Batteries and Supercapacitors (Li Ion Batteries, Lead-Acid Batteries, NiMH Batteries, Zinc–Air Batteries, Liquid Redox Batteries)

Electrochemical rechargeable batteries or secondary batteries [26] are energy storage devices, receive electricity that is produced elsewhere, and utilize electricity to derive electrochemical reactions (uphill, a positive ΔG) at both electrodes and become ready to release this energy downhill in a spontaneous manner. A handsome variety of rechargeable batteries is available in commercial market (including lead-acid, nickel cadmium (Ni-Cd), nickel metal hydride (Ni-MH), zinc–air, liquid redox, lithium ion (Li ion), and lithium ion polymer batteries) (Figure 1.7); they come in different shapes and sizes with different energy to weight and energy to volume ratios connected to stabilize an electrical distribution network and can be used several times. Batteries are good at providing high power levels, but the amount of energy they can store per unit weight (50–1000 W kg^{-1}) is not greater

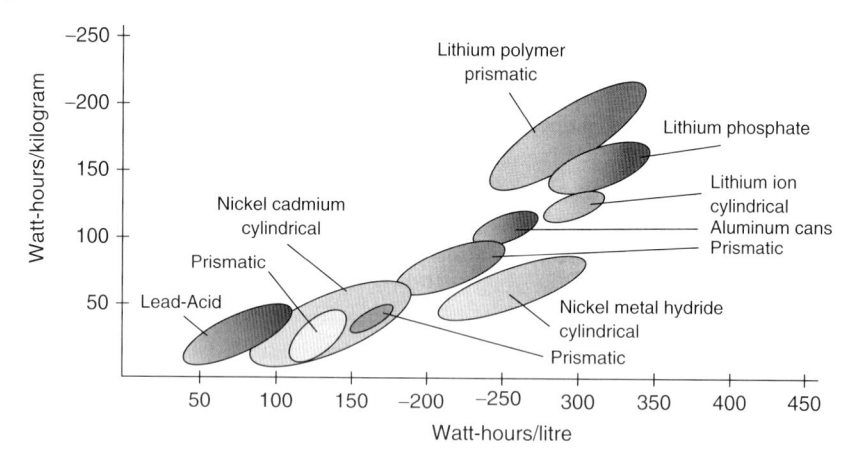

Figure 1.7 Different kinds of batteries rated on the scale of their energy to weight and energy to volume ratios. (http://en.wikipedia.org/wiki/Rechargeable_battery).

than that of fuel cells because they can use up, at best, all the material on their plates, whereas fuel cells simply convert all the available chemical fuel into energy. Most of the present day rechargeable batteries have to be cycled about 100 times when the depth of discharge (DOD) is 90%; moreover, the batteries beyond 1000 recharges with a high DOD are also available. Furthermore, the systems with 50 000 recharges are possible at a 40% DOD. The calendar life of batteries depends on their mode of use. They have made a revolutionary impact on our lives; for example, they can be used in applications such as automobile starters, portable consumer devices, light vehicles (such as motorized wheelchairs, golf carts, electric bicycles, and electric forklifts), tools, providing long-term power to internal artificial organs (heart pace makers, hearing aids, etc.), and providing uninterruptible power supplies. Grid energy storage plants also use industrial rechargeable batteries for load leveling, where they are used to store electric energy for the peak loading periods. Emerging applications in hybrid electric vehicles/electric vehicles are driving the technology to increase the lifetime of vehicles by reducing the cost and weight. Rechargeable batteries have higher initial cost, but the total cost of use and environmental impact are lower than disposable primary batteries, although they are accounted as pseudopollution contributors because the electricity consumed by them for charging is produced from the combustion of fossils or oil. Normally, new rechargeable batteries have to be charged before their use, but the newer low self-discharge batteries hold their charge for many months and are able to supply charges to up to 70% of their rated capacity. The energy used to charge rechargeable batteries usually comes from a battery charger that uses AC mains electricity, and it will take a few minutes (rapid chargers) to several hours to charge a battery. Since the end of twentieth century, in the United States, Japan, and Europe, the market of rechargeable batteries became vibrant and extremely attractive, as the demand for rechargeable batteries is growing twice faster than that for nonrechargeable ones. Still there is no *best battery* in the market, but there are a handsome variety of batteries in the market for different situations, likewise, batteries for torpedoes must be stable during storage and must give high power for a short time, whereas batteries for submarine need giant rechargeable ones when submerged.

Unlike batteries, which store energy chemically, capacitors store energy as an electrostatic field. A typical battery is known for storing a lot of energy and little power, whereas a capacitor can provide large amounts of power, but low amounts of energy. A capacitor is made of two conducting plates and an insulator called the *dielectric*, which conducts ionically but not electrically. Few important batteries are described below.

1.6.1
Lead-Acid Batteries

Lead-acid batteries revolutionize the portable power and fall into the classical category invented by French physicist Gaston Planté in 1959 [27]. They consist of six cells of 2 V nominal voltage, and each cell is composed of a lead dioxide cathode, a sponge metallic lead anode, and about 37% w/w sulfuric acid solution as electrolyte. Its main discharge reaction at anode is

$$Pb + HSO_4^- \rightarrow PbSO_4 + H^+ + 2e^- \tag{1.1}$$

And the corresponding discharge reaction at cathode is

$$PbO_2 + 3H^+ + HSO_4^- + 2e^- \rightarrow PbSO_4 + 2H_2O \tag{1.2}$$

The thermodynamic reversible potential for the overall cell reaction is 1.93 V, meaning that less number of cells is used to attain a given potential. The optimum operating temperature for the lead-acid battery is 25 °C. For higher power applications, lead-acid batteries with intermittent loads are generally too big and heavy, suffer from a shorter cycle life, and typical usable power down to only 50% DOD. These batteries become the technology of choice for automotive starting, lighting, and ignition (SLI) applications because they are robust, tolerant to abuse, and of low cost. Lead-acid batteries have a huge market as the starter battery for internal combustion engines. Although they have one of the worst energy to weight ratios (35–40 Wh kg^{-1}) but quite good power to weight and energy to volume ratios, their life seldom exceeds 4 years and can be recharged for 300–400 cycles. In a valve regulated lead-acid (VRLA) batteries, electrolytes avoid spilling out, and the hydrogen and oxygen produced in the cells largely recombine into water. Since the 1950s, chemical additives such as EDTA and Epsom salts [28] have been used to reduce lead sulfate buildup on plates and improve battery condition when added to the electrolyte of a vented lead-acid battery. EDTA can be used to dissolve the sulfate deposits of heavily discharged plates. Residual EDTA in the lead-acid cell forms organic acids that will accelerate corrosion of the lead plates and internal connectors. Epsom salts reduce the internal resistance in a weak or damaged battery and may allow a small amount of extended life. Heavy metal elements used in their fabrication makes them toxic, and their improper disposal can be hazardous to the environment.

1.6.2
NiMH Batteries

An NiMH battery is similar to the nickel-cadmium cell and was invented in 1967 [29]. The Ni-MH battery uses a hydrogen-absorbing alloy (sintered $Ti_2Ni + TiNi + x$ or the presently used AB_5, where A is a rare earth mixture of lanthanum, cerium, neodymium, praseodymium and B is nickel, cobalt, manganese, and/or aluminum, and AB_2 compounds, where A is titanium and/or vanadium and B is zirconium or nickel, modified with chromium, cobalt, iron, and/or manganese) as the negative electrode (negative electrode of Ti–Ni alloy hydride phases, US patent US 3,669,745 (13 June 1972), inventor: K. D. Beccu of Battelle, Geneva R&D Center); nickel oxyhydroxide (NiOOH) as the positive electrode developed by Dr Masahiko Oshitani; and usually, 28% potassium hydroxide as the alkaline electrolyte. For separation, hydrophilic polyolefin nonwovens are used. Respective cathodic and anodic reactions of the Ni-MH batteries can be written as follows:

Cathode:

$$NiM_{Alloy}H \rightarrow NiM_{Alloy} + H^+ + e^- \quad (1.3)$$

Anode:

$$NiOOH + H_2O + e^- \rightarrow Ni(OH)_2 + OH^- \quad (1.4)$$

Ni-MH batteries can possess 2–3 factors higher capacity (1100–3100 mAh at 1.2 V), the same as an equivalent size nickel-cadmium battery. Its volumetric energy density (140–300 Wh l^{-1}) is similar to that of the lithium ion cell (250–360 Wh l^{-1}), significantly better than that of nickel-cadmium battery at 50–150 Wh l^{-1}, but its self-discharge is higher (30% per month). It can retain specific energy of approximately 30–80 Wh kg^{-1} and a specific power of around 250–1000 W kg^{-1} with a reasonable deep life cycle of 500–1000 cycles (DOD = 100%), and this has led to the new environmentally friendly high-energy NiMH cells [30]. Low internal resistance allows Ni-MH cells to deliver a near-constant voltage until they are almost completely discharged. Modern Ni-MH cells contain catalysts to immediately deal with gases developed as a result of overcharging, without being harmed ($2 H_2 + O_2$ + catalyst → $2 H_2O$ + catalyst). However, this works only with overcharging currents of up to 0.1 C and is used to detect the safe end-of-discharge voltage of the series cells and autoshutdown. Ni-MH cells are used to power the devices such as digital cameras, GPS receivers and personal digital assistants (PDAs), flashlights, and some toys or video games. Improper disposal of Ni-MH batteries poses less environmental hazard than that of Ni-Cd cells because of the absence of toxic cadmium. Although lithium ion batteries (LIBs) have a higher specific energy than NiMH batteries, they also have a much lower shelf life and are significantly more expensive to produce. Currently, more than 2 million hybrid cars worldwide are running with Ni-MH batteries [31], for example, Prius, Lexus (Toyota), Civic, Insight (Honda), and Fusion (Ford). Many of these batteries are manufactured by Panasonic (PEVE) and Sanyo.

1.6.3
Li-Ion Batteries

A LIB was first proposed by M. S. Whittingham of Binghamton University, Exxon, in the 1970s using titanium(II) sulfide as the cathode and lithium metal as the anode, and it belongs to the rechargeable type of batteries. The primary functional components of LIBs are anode, cathode, and electrolyte [32]. Lithium ion cell uses an intercalated lithium compound (Li in porous carbon or graphite) as the anodic material instead of metallic lithium. A layered oxide (such as lithium cobalt oxide) or a polyanion (such as lithium iron phosphate), or a spinel (such as lithium manganese oxide) material is used as cathode. Pure lithium is niche industrial material of very reactive nature. It reacts vigorously with water to form lithium hydroxide and liberate hydrogen gas. Thus, nonaqueous electrolytes are required for LIBs. A typical mixture of nonaqueous organic carbonates such as ethylene carbonate or diethyl carbonate containing the complex of lithium ions (lithium hexafluorophosphate ($LiPF_6$), lithium hexafluoroarsenate monohydrate ($LiAsF_6$), lithium perchlorate ($LiClO_4$), lithium tetrafluoroborate ($LiBF_4$), or lithium triflate ($LiCF_3SO_3$)) is used as the electrolyte. Lithium ions that move from the negative electrode to the positive electrode during the discharge process through the nonaqueous electrolyte and separator diaphragm move back when charging.

The cathodic half reaction of LIB is

$$LiCoO_2 \leftrightarrow Li_{1-x}CoO_2 + xLi^+ + xe^- \tag{1.5}$$

And another anodic half reaction is

$$xLi^+ + xe^- + 6C \leftrightarrow Li_xC_6 \tag{1.6}$$

Recently, novel architectures fabricated using nanotechnology have been employed to improve the performance of LIBs. Chemistry, performance, cost, and safety are the characteristics that are basic to the electrode material used. LIBs became technological of today's choice because of the qualities of the best energy to weight ratios (specific energy density, 150–250 Wh kg^{-1}; volumetric energy density, 250–530 Wh l^{-1}; specific power density, 300–1500 W kg^{-1}), no memory effect, and a slow loss of charge when not in use. Beyond their popularity for portable electronics, LIBs are also the growing interest in military, electric vehicles, and aerospace applications because of their high energy density. There are few disadvantages of LIBs, that is, its internal resistance is higher than other rechargeables such as Ni-MH and Ni-Cd batteries and increases with both cycling and age thus reducing the cell's ability to deliver current. Charging forms electrolyte deposits that inhibit ion transport; high charging levels with elevated temperatures (whether from charging or ambient air) also hasten their capacity. LIBs with a lithium iron phosphate cathode and graphite anode have a nominal open-circuit voltage of 3.2 V and a typical charging voltage of 3.6 V. LIBs could not be charged fast and needed at least 45 min to 2 h to fully charge. In 2007, Cui and colleagues at Stanford University's Department of Materials Science and Engineering discovered that using Si/Ge nanowires as the anode of an LIB increases the volumetric charge density of the anode up to a factor of 10 [33, 34].

Further advancement in LIB is made by replacing lithium salt electrolyte in an organic solvent with a solid polymer composite such as polyethylene oxide or polyacrylonitrile [35]. The advantages of Li ion polymer over the LIBs include potentially lower manufacturing cost, adaptability to a wide variety of packaging (in various shapes), increased life cycle, slow degradation rate, and ruggedness. Lithium ion polymer batteries started appearing in consumer electronics around 1996 [36]. However, in recent years, manufacturers have been declaring upward of 500–10 000 charge–discharge cycles before the capacity drops to 80%. Li poly batteries are gaining favor of the world in radio-controlled aircraft and radio-controlled cars, airsoft gun, PDAs, and laptop computers, such as Apple's MacBook family, Amazon's Kindle, Lenovo's Thinkpad X300 and Ultrabay Batteries, the OQO series of palmtops, the HP Mini, and Dell products featuring D-Bay batteries, where the advantages of both lightweightness and greatly increased run times can be sufficient justification for their price. They can also be found in small digital music devices such as iPods and Zune and other MP3 players, the Apple iPhone, gaming equipment such as Sony's Playstation 3, and wireless controllers.

1.6.4
Zinc–Air Batteries

Zinc–air batteries [37] have some properties resembling the fuel cells, such as zinc electrode consumed as fuel, and the reaction rate can be controlled by varying the air flow. Commercial production began in 1932, when George W. Heise and Erwin A. Schumacher of the National Carbon Company (US patent 1899615(A) dated: 28 January 1933) built these cells. Zinc–air battery consists of a porous *air* electrode (cathode) made of carbon that draws in atmospheric oxygen, which gets absorbed into the electrolyte (aqueous KOH) through a gas-permeable and liquid-tight membrane (also acts as catalyst) to the zinc electrode (anode), where oxidation of zinc releases electrons to generate a current with ZnO production.

Cathode (porous air-carbon electrode):

$$O_2 + 2H_2O + 4e^- \rightarrow 4OH^- \quad (E_0 = 0.4\,V) \tag{1.7}$$

Anode (Zn electrode):

$$Zn + 4OH^- \rightarrow Zn(OH)_4^{2-} + 2e^- \quad (E_0 = 1.25\,V) \tag{1.8}$$

Fluid reaction:

$$Zn(OH)_4^{2-} \rightarrow ZnO + H_2O + 2OH^- \tag{1.9}$$

Overall reaction:

$$2Zn + O_2 \rightarrow 2ZnO \quad (E_0 = 1.65\,V) \tag{1.10}$$

1.6 Electrochemical Rechargeable Batteries and Supercapacitors

The cell voltage for the above theoretical chemistry is 1.65 V; however, almost all designs are optimized for <1.4 or 1.3 V in order to achieve longer lifetimes. Recharging process is the reversal of the above reactions: zinc oxide is converted back to zinc at negative anode and oxygen is released at the positive air electrode. During discharge, Zn is converted to ZnO, which is replaced through mechanical charging in which discharged zinc cartridges are replaced with fresh zinc cartridges. Repeated charge–discharge cycles deactivate (slowing or stopping the oxygen reactions) the air electrode and gradually tend to fill its pores with the liquid electrolyte. The battery can also fail if it dries out or if zinc builds up unevenly, forming branchlike structures that create a short circuit between the electrodes. Like all other battery technology involving Ni-MH, lead-acid, and zebra batteries, zinc–air batteries are safer than LIBs (requires much more extensive electronic monitoring) because they are short-circuit proof, and they do not contain volatile materials, hence, they do not catch fire or explode during a car crash. Zinc–air batteries require filters to remove carbon dioxide from the air for their use; the removal is via passing air through the inexpensive hydroxide scrubbers where it is fixed to carbonate as a support to purify the air. Zinc–air batteries can beat petrol-dependent road transportation for electric propulsion with their unique capability of anthropogenic CO_2 removal. A new breed of rechargeable zinc–air batteries (Figure 1.8, ReVolt, Staefa, Switzerland) will soon to be available in market and may replace LIBs in cell phones, laptops, and other consumer items. LIBs store only one-third of the energy and cost around twice as much as the new breed of rechargeable zinc–air batteries [37]. Ni-MH and LIBs are very expensive ($500 k Wh^{-1}), whereas the relative cost of zinc–air batteries is only $60 k Wh^{-1} but still are a favorite technology for battery industry. The zinc–air systems are quite popular among their category because of the use of nontoxic, inexpensive, and abundantly available raw materials; the high energy density (1480–9780 Wh l^{-1}); the high energy per unit of weight (470 (practical), 1370 (theoretical) Wh kg^{-1}); and an excellent shelf life with a self-discharge rate of only 2% per year when sealed, but they have a low specific power (100 W kg^{-1}). Their few disadvantages are sensitivity to extreme temperature and humid conditions, chemicals tend to dry out, high self-discharge, high internal resistance, and low recycle lifes. In the light of the low cost, material availability, safety, environmental aspects, and financial constraints faced by the world today, governments should prioritize zinc–air batteries to power

Battery cover, which lets in air
Porous air electrode
Interface between electrodes
Zinc electrode
Casing

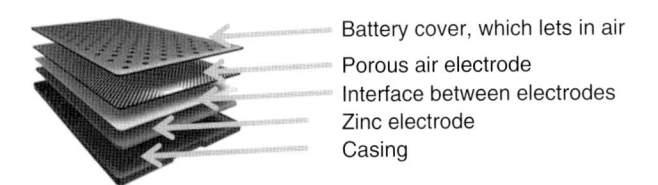

Figure 1.8 Illustration of the multilayered structure of an unpacked ReVolt rechargeable zinc–air battery. (ReVolt, Staefa, Switzerland.)

automobiles [38]. Zinc–air button cells are commonly used for watches and hearing aids. Larger types are employed as prismatic or cylindrical cells for telecoms and railway remote signaling and for safety lamps at road and rail construction sites, as power sources for electric fences, and in film cameras that previously used mercury batteries. Very large batteries are used for electric vehicle propulsion.

1.6.5
Liquid Redox Batteries

Liquid-metal batteries are fabricated by molten anode and cathode materials separated by solid electrolyte (a beta-aluminum oxide ceramic ion conductor). They are of low manufacturing cost, of extremely high power density, and are suitable for grid-scale storage. A classic example of this category is sodium–sulfur batteries (Figure 1.9) [39], discovered in 1980 and consisting of molten sodium anode and molten sulfur cathode, separated by conductive ceramic electrolyte sodium beta-alumina (approximately 11 parts of aluminum oxide Al_2O_3 and 1 part of sodium oxide NaO, with a melting point of 2100 °C). Sodium melts at 98 °C and sulfur at 113 °C, but the operating temperature is around 350 °C. During discharge, electrons are striped off the sodium atoms and flow through the external load to the sulfur cathode. The positively charged sodium ions move through the electrolyte where they react with the sulfur and the electrons to produce sodium polysulfide. During recharge, the applied voltage strips electrons from the sodium polysulfide returning it back to sodium ions. The sodium ions now cross the electrolyte into the sodium where they are reunited with their missing electrons to form sodium atoms. The sodium–sulfur battery can have very high energy (50–200 Wh kg^{-1}) and power densities (100–200 W kg^{-1}) and a limited shelf life, typically of the order of 2–5 years. One of the key advantages of batteries with liquid electroactive materials is that there are no morphological changes on cycling that can offer infinite life cycle to the active materials, but the solid electrolyte is susceptible to mechanical degradation by the phase change that occurs as the sodium ions are reduced to the metallic state. In the new self-assembling liquid-metal batteries,

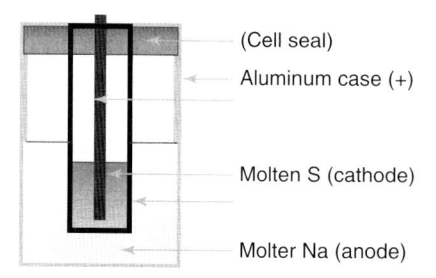

(Cell seal)
Aluminum case (+)
Molten S (cathode)
Molter Na (anode)

Typical sodium–sulfur battery

Figure 1.9 Schematic representation of the typical sodium–sulfur battery, as liquid redox battery.

molten salt serves as both the separator and electrolyte of molten electrodes. Density differences allow two of the three liquid layers to float on the heaviest layer. The very high ionic conductivity of the electrolyte allows for extreme power density, well suited for grid-scale power storage. The safety issues implicit in molten materials and high operating temperature made them not practically attractive for use in electric vehicles, but this problem may be resolved in time. In the proof-of-concept cell by Donald Sadoway's research group at MIT, the three layers, from bottom to top, antimony, sodium sulfide, and magnesium, are expected to be able to handle 10 times the current of typical batteries used in power plants. During discharge, the top and bottom layers are consumed to form magnesium antimonide, which dissolves in the electrolyte. Upon recharge, the metal layers are reformed. New patent pending liquid batteries are designed by Sadoway and his team, including graduate student David Bradwell, using low-cost and abundant materials. The three materials are chosen so that they have different densities that allow them to separate naturally into three distinct layers, with the salt (Mg_3Sb_2) in the middle separating the two metal layers (Mg (top) and Sb (bottom)) and are operated at about 700 °C; if corrosion issues for the electrodes and container can be resolved, the cycle life of these batteries might be nearly infinite. Because of its high energy density, the Na–S battery has been proposed for power grid [40], transport and heavy machinery, and space applications [41].

1.7
Light Fuel Generation and Storage: Water Electrolysis, Chloro-Alkaline Electrolysis, Photoelectrochemical and Photocatalytic H_2 Generation, and Electroreduction of CO_2

Light fuel hydrogen has attracted great attention from environmentalists, scientists, and industrialists as a benign fuel of future because of its capability to produce pollution-free energy and because it has one of the highest energy density, that is, values per mass of 140 MJ kg^{-1}. The majority of hydrogen used in industries is derived from fossil fuels or cleavage of water. Currently, a majority of industrial hydrogen need is satisfied from conventional sources (coal, oil, and natural gas), which contains about 10% CO_2 with hydrogen gas, and only 4% of H_2 comes from electrolysis, which is the cheapest method to generate hydrogen ($3.51 per kg). Central to the success of hydrogen technology is the efficient generation of hydrogen from renewable sources (such as solar/wind) powered by water cleavage. Water splitting by electrolysis, uses the most expensive proton exchange membrane (PEM) and liquid electrolyte (KOH) or thermochemical process (need a temperature of 700–1000 °C). Wind energy, the most cost-effective renewable energy source, is also used to electrolyze water (costs about $6.64 per kg per H_2 if grid back up used). Biomass with solar energy costs $7.05 per kg H_2 production. Solar energy either by electrolysis (PV) or using photoelectrochemical cell now costs $2.82 per kg, but solar energy is available only 20% of the time. Following are the advance techniques to produce and use light fuels (H_2, CO_2).

1.7.1
Water Electrolysis

In 1789, Jan Rudolph Deiman and Adriaan Paets van Troostwijk produced electricity from water using an electrostatic machine by discharge on gold electrodes in a Leyden jar [42]. In 1800, the *voltaic pile* was invented for the electrolysis of water by Alessandro Volta. Finally, the electrolysis of water became a cheap method of hydrogen generation using the Gramme machine invented by Zénobe Gramme in 1869. A method of industrial synthesis of hydrogen and oxygen through electrolysis of water was developed by Dmitry Lachinov in 1888 [43]. Decomposition of pure water into hydrogen and oxygen at standard temperature and pressure is not favorable in thermodynamic terms. Water electrolysis does not convert 100% of the electrical energy into the chemical energy of hydrogen. For a well-designed cell (Figure 1.10), the largest overpotential required for the four-electron oxidation of water to oxygen at the anode. An effective and cheap electrocatalyst to facilitate this reaction has not yet been developed. Platinum alloys are the default state of the art for this oxidation. The simpler two-electron reaction to produce hydrogen at the cathode can be electrocatalyzed with almost no reaction overpotential by platinum or hydrogenase enzyme. In 2008, Kanan *et al.* [44] announced a potentially efficient electrocatalyst (anode) for artificial photosynthesis, composed of a Co^{3+}/Co^{2+} metal and phosphate electrolyte. Other researchers are pursuing carbon-based catalysts for the same. Efficiency of converting electrical energy into hydrogen's chemical energy of water electrolysis varies optimally between 50 and 80% [45].

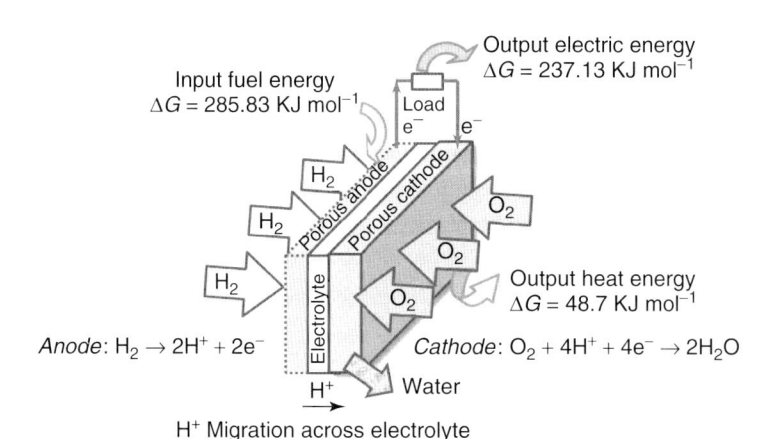

Input fuel energy
$\Delta G = 285.83$ KJ mol^{-1}

Output electric energy
$\Delta G = 237.13$ KJ mol^{-1}

Output heat energy
$\Delta G = 48.7$ KJ mol^{-1}

Anode: $H_2 \rightarrow 2H^+ + 2e^-$

Cathode: $O_2 + 4H^+ + 4e^- \rightarrow 2H_2O$

H^+ Migration across electrolyte

Hydrogen–oxygen fuel cell operation through electrolyte

Figure 1.10 Water electrolysis illustrated in hydrogen–oxygen fuel cell with chemical and energetic electrode operations using electrolyte.

1.7.2
Electrochemistry of Water Splitting

Pure water is considerably a bad conductor of current, 0.055 µS·cm^{-1} (one millionth of that of seawater) because it has a low autoionization coefficient, $K_w = 10 \times 10^{-14}$ at room temperature that leads electrolysis of pure water very slowly or not at all. Thus electrolysis of pure water requires excess energy in the form of overpotential to overcome various activation barriers, which enhance the rate of water autoionization. In a properly designed water-electrolysis cell, two electrodes (anode and cathode) are placed in the water and connected by external electrical power source. Hydrogen appeared at the cathode (the negatively charged electrode, where electrons reduce the water), and oxygen appeared at the anode (the positively charged electrode, where oxidation takes place) in 2 : 1 ratio. The efficacy of electrolysis is increased by the addition of an electrolyte (such as solid polymer membrane, Nafion; strong acids, H_2SO_4; or a strong base, NaOH, KOH) and electrocatalysts. A solid polymer electrolyte can also be used, such as Nafion, and when applied with a special catalyst on each side of the membrane can efficiently split the water molecule with as little as 1.8 V power supply.

Anode reaction:

$$2H_2O(l) \rightarrow 4H^+(aq) + 4e^- + O_2(g) \uparrow \quad (E_A = -1.23 \text{ eV}) \tag{1.11}$$

Cathode reaction:

$$2H^+(aq) + 2e^- \rightarrow H_2(g) \uparrow \quad (E_c = 0.00 \text{ V}) \tag{1.12}$$

Thus, the standard potential of the water-electrolysis cell is -1.23 V at 25 °C at pH 0 ($H^+ = 1.0$ M). It retains -1.23 V at 25 °C even for pH 7 ($H^+ = 1.0 \times 10^{-7}$ M) based on the Nernst equation. This endothermic reaction involves a change in Gibbs free energy $\Delta G = +2.46$ eV or 237.2 kJ mol^{-1}. The negative voltage indicates that the Gibbs free energy for electrolysis of water is greater than zero. The system must be provided with sufficient energy for the dissociation of water plus the energy to expand the produced gases.

1.7.3
Chlor-Alkaline Electrolysis

Chlorine, one of the most important bulk chemicals in the world, is produced by the electrolysis of brine (highly conducting sea water) [46]. Chlorine is used in the day-to-day life as water purifier and is also an essential part in the chemical building block, resulting in a myriad of reactions and products in the major plastic, pharmaceutical, inorganic, and fine chemical and specialty industries [45]. In chlor-alkali electrolysis (Figure 1.11), an air–oxygen gas-diffusion electrode or traditional hydrogen-evolving electrode is used as a cathode, a nickel wire mesh as a counter electrode in 8 M NaOH, and Hg/HgO as reference electrode. Coating of polytetraflouroethylene (PTFE) suspension was performed on gas-diffusion electrode, shielded by an active layer made up of catalysts [46, 47] containing the

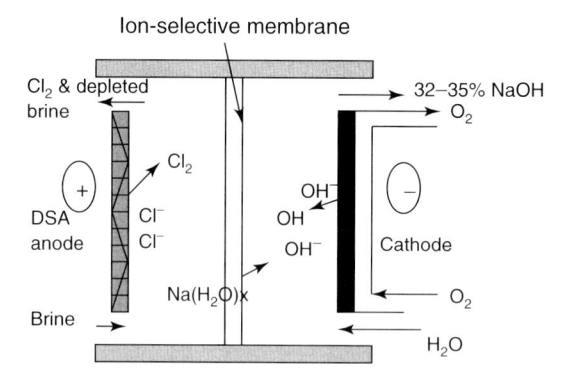

Figure 1.11 Graphical illustration of chlor-alkaline cell.

mixture of perovskite ($La_{0.1}Ca_{0.9}MnO_3$) and the pyrolysis product of cobalt tetramethoxyphenylporphyrin (CoTMPP) with less proofed carbon. In the electrolytic process, during the following overall reaction for every ton of chlorine produced, about 1.1 tons of caustic is generated and 28 kg hydrogen is evolved as a by-product. Four percent of total worldwide hydrogen production is created by electrolysis. The majority of this hydrogen is produced by chlor-alkali electrolysis, where, the chloride ions are oxidized to chlorine and water is reduced to hydrogen as shown in the following reaction.

Overall reaction:

$$2NaCl + 2H_2O \rightarrow Cl_2 + H_2 + 2NaOH \qquad (1.13)$$

It is an energy intensive process, with an electrical power consumption between 2100 and 3300 kWh per unit time. The amount of Cl_2 generation depends on the operating parameters and the type of the process. There are three major processes based on mercury, diaphragm, and membrane cells in use for the electrolysis. In the year 2001, the total world production of chlorine was about 43.3 million tons, and 18% of this production was met by the mercury cell technology, whereas the remaining 82% included diaphragm (49%) and membrane cell (28%) technologies, and others (5%). Hofmann voltametric, high-pressure (120–200 bar), high-temperature electrolyses used for chlor-alkaline electrolysis are also in practice [48, 49].

1.7.4
Photoelectrochemical and Photocatalytic H_2 Generation

Global energy needs for sustainable development demands 50% increase by 2030. The alternative energy source *the sun* is able to obtain the target by using solar energy to split water into hydrogen and oxygen as a cost-effective storage of solar energy in chemical energy for its large-scale utilization. Nature provides us a blueprint of water-splitting reaction in the form of photosynthesis for storing sunlight in form of chemical energy. Laboratory-designed artificial photosynthesis required photocatalyst, water, and sunlight to split the water into hydrogen and oxygen in 2:1 ratio. A range of metal oxides, sulfide, nitrides, (oxy)sulfides, and (oxy)nitrides containing either transition metal cations with d^0 electronic configuration (e.g., Ti^{4+}, Nb^{5+}, and Ta^{5+}) or typical cations with d^{10} electronic configuration (e.g., Ga^{4+}, Ge^{4+}, and Sn^{4+}) as principal cation components have been reported as active photocatalysts for overall water splitting. Photocatalytic reaction involves three basic steps: (i) the photocatalyst absorbs more photon energy (sunlight) than the band gap energy of the material and generates photoexcited electron–hole pairs in the material bulk; (ii) the photoexcited charges separate and migrate to the different sites of the photocatalyst's surface without recombination; and (iii) water is reduced and oxidized by the photogenerated electrons and holes to produce H_2 and O_2, respectively. The first two steps are strongly dependent on the structural and electronic properties of the photocatalyst, whereas the third step is promoted by the presence of a solid cocatalyst (Figure 1.12). The cocatalyst is typically a noble metal or metal oxide or a combination of both, loaded onto the surface of photocatalyst as a dispersion of nanoparticles to produce active sites and reduce the activation energy for gas evolution. To date, various transition metal oxides, including NiO_x, RuO_2, RhO_x, IrO_2, and $RhCr_2O_3$, have been applied as a cocatalyst for photocatalytic overall water splitting. Despite a plentiful library of photocatalytic materials mostly active in UV light (5% of sunlight spectrum), visible light–driven photocatalyst with high efficiency is still one of the *holy grails* of material chemistry. Few photocatalysts reported with remarkable efficiency are $(AgIn)_xZn_{2(1x)}S_2$ [50], $ZnS-CuInS_2-AgInS_2$ [51], La-doped $NaTaO_3$ [52], Ni-doped $InTaO_4$ modified with RuO_2 or NiO as the cocatalyst [53], $(Ga_{1-x}Zn_x)(N_{1-x}O_x)$ [54], $SrTiO_3$, BaO-doped

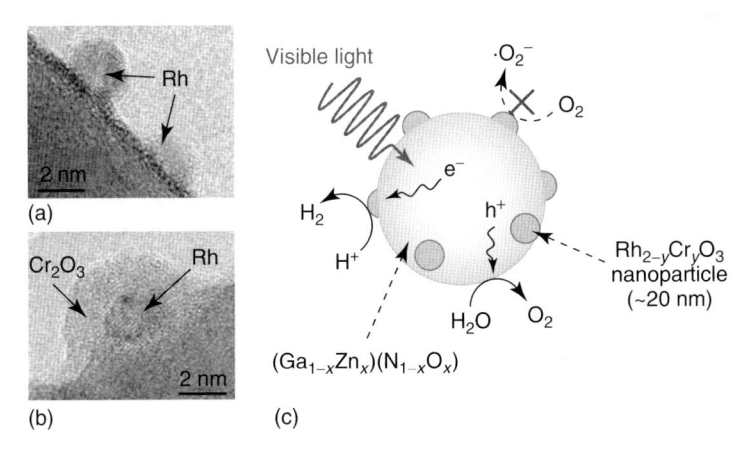

Figure 1.12 Loading of cocatalyst on the surface of the basic photocatalyst and its role in controlling recombination reaction of electrons and protons. HR-TEM images of $(Ga_{1-x}Zn_xN_{1-x}O_x)$ ($x = 0.12$) with photodeposited Rh (a) before and (b) after further photodeposition of a Cr_2O_3 shell. (c) Schematic view of photocatalytic oxidation and reduction sites [54]. (Copyright American Chemical Society Publications.)

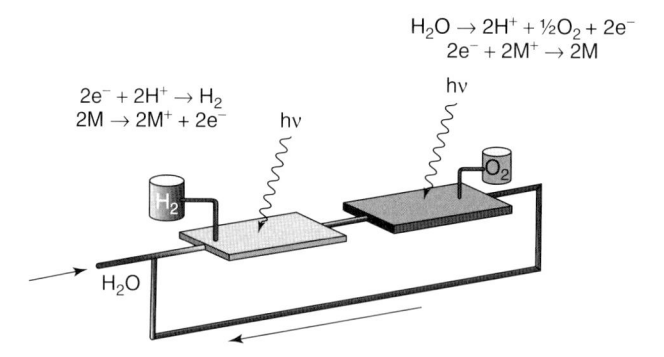

Figure 1.13 Schematic of a dual-bed photocatalytic water-splitting system [60].

La$_2$Ti$_2$O$_7$ [55], and K$_3$Ta$_3$B$_2$O$_{12}$ [56]. Photoelectrochemical water-splitting technology is economically superior to electrolysis of water for electricity production because of the single-plant massive photoproduction of clean hydrogen fuel. Titania (TiO$_2$) was the first material described as a photoelectrochemical water-splitting catalyst, but because of its wide band gap, 3.2 eV, it is able to work in UV light. Fujishima and Honda, in 1972, described an electrochemical cell consisting of an n-type TiO$_2$ (rutile) anode and a Pt black cathode for photoelectrochemical water splitting. Much attention has been given to photoelectrocatalytic splitting of water using metal oxide or nitride semiconductors, for example, ZnO, GaN, TiO$_2$, WO$_3$, and Fe$_2$O$_3$, to supply clean and recyclable hydrogen energy. Advancements such as introduction of nanostructures; elemental doping with N, C, P, S, and dyes (Ru complexes); and quantum dot (CdS, CdSe, InP, CdTe, Bi$_2$S$_3$) sensitization of based material are also in practice. Bockris and Reddy [57] made their effort to produce faceted TiO$_2$ (about 1000-fold increase in surface area) and decorated it with dye with a suitable photoreceptor (bipyridyl complexes of ruthenium), then they became capable of harvesting visible light. Key point is the use of traces of electrocatalyst added to the surface of both photoanode and photocathode [58, 59] to an appropriate extent. Use of a perylene diimide derivative as the O$_2$-evolving photocatalyst and copper phthalocyanine as the H$_2$-evolving photocatalyst in the IO$_3^-$/I$^-$ redox electrolyte demonstrates the feasibility of continuous closed-cycle dual-bed photocatalytic water splitting (Figure 1.13) [60]. Recently, artificial inorganic leafs have been developed as biotemplates [61] by using catalysis modules (Pt/N-TiO$_2$) for enhanced light-harvesting, and photocatalytic water-splitting activities stem from the reproduction of the leafs complex structures and self-doping of nitrogen during synthesis (Figure 1.14).

1.7.5
Carbon Dioxide Reduction

Before the industrial revolution, CO$_2$ concentration in the atmosphere was about 290 ppm. By 1995, it had reached 360 ppm. Today, it is about 410 ppm, and

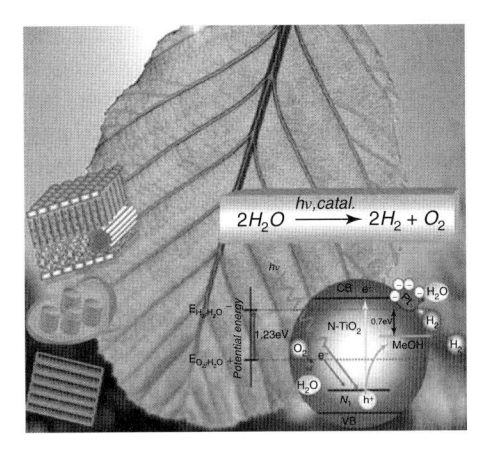

Figure 1.14 Artificial inorganic leafs for efficient photochemical hydrogen production inspired by natural photosynthesis [61].

extrapolation of the present data raised it up to 560 ppm in 2060, accompanied with rise in temperature to about 1.5 K at the equator and a corresponding estimated increase in the sea level by 50 cm and 3.2 K temperature at poles will reduce the reflectivity of the region (water is dark and ice is bright) eventually increasing the earth's average temperature [62]. The situation is becoming more and more critical day by day. One way to deal practically with this problem is to fix this undesirable CO_2 into hydrogen carrier liquid, that is, methanol, to be used in fuel cells [63]. Fisher and Prziza [64] were first to observe the 100% electrochemical reduction of CO_2 to HCOOH. Photoelectrochemical conversion of CO_2 to oxalic acid was performed using InP-electrode (cathode) in two-photon cell having $LaCrO_3$-TiO_2 as photoanode [65]. An 18-crown-6-ethers-decorated-CdTe electrode in the presence of tetraalkylammonium ions reduces CO_2 mainly to CO with a small amount of CH_3OH [66]. A major breakthrough would be the discovery of a suitable effective catalyst or photocatalyst in CO_2 fixation for the reduction of the major greenhouse gas CO_2 into a useful chemical, that is, methanol (MeOH), using H_2 gas supply.

$$CO_2 + 3H_2 + \text{catalyst} \rightarrow CH_3OH + H_2O + \text{catalyst} \qquad (1.14)$$

Inspired by photosynthetic reactions in plants, now a reasonable goal of electrochemical science is to fix planetary CO_2 by photochemical/photoelectrochemical reaction to form MeOH, and the further goal would be based on the production of synthetic food and textile from water, solar light, atmospheric CO_2, and nitrogen (bacteria). However, it is also possible to use CO_2 in electrochemical synthesis of organic compounds that exemplify in the formation of phenyl acetic acid by benzyl chloride [67].

1.8
Fuel Cells: Fundamentals to Systems (Phosphoric Acid Fuel Cells, PEM Fuel Cells, Direct Methanol Fuel Cells, Molten Carbon Fuel Cells, and Solid Oxide Fuel Cells)

Fuel cells are compact electrical energy producer devices that consume fuels (chemical energy) via fuel (e.g., hydrogen, methanol, and hydrocarbon) oxidation at anode and reduce oxygen (usually comes from air and triggered into solution) as oxidant at counter cathode on immerging into electrolyte. The electrolyte is specifically designed so that ions can pass through it, but the electrons cannot. The free electrons travel through a wire, creating the electrical current. The revolutionary advantage of fuel cells over other energy-producing devices is that the free highly dense energy directly comes out as pollution-free electrical energy without planetary warming. As no combustion reactions are involved in fuel cells, they do not produce any of the undesirable products (CO_2, SO_2, oxides of nitrogen, or particulate matters), which are normally associated with the oxidation of fossil fuels in conventional energy conversion systems. Thus, fuel cells are environmentally friendly. However, fuel cells offer the additional advantages of lighter weight, noiselessness, nonpolluting, and vibration-free operation than batteries that require recharging after use. For these reasons, they are beloved candidates to be used as battery's supplements. Formal discovery of fuel cell principle is attributed to Sir William Grove (February 1839) [68] as he passed current through the connecting wire between anode that was bubbled with hydrogen and its counter electrode was blown with oxygen from air, the device was named as *gas voltaic battery* [2] (Figure 1.15). Francis Thomas Bacon made pioneer contribution in the development of fuel cells in 1959. NASA uses Bacon's cells for auxiliary power in space vehicles, as they are three times as effective as any other method of providing energy on board. Fuel cells have the potential to replace the internal combustion engine in vehicles and to provide power in stationary and portable power applications because they are energy efficient, clean, and fuel flexible. Direct fuel cells can feed hydrocarbons directly to the fuel cell stack, without requiring an external *reformer* to generate hydrogen [69]. Fuel cells are thermodynamically open systems that work spontaneously and go down the free energy gradient by

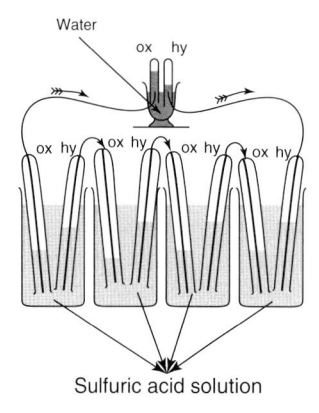

Figure 1.15 Sketch of the first hydrogen–oxygen Grove's fuel cell invented in 1839. The dark lines in the central of the tubes are platinized-foil electrodes [68b].

consuming reactant from an external source, which must be replenished. There are two governing factors in fuel cells. First, is the exchange current density of cathode that causes low overpotential (higher limiting current density $i_0 \sim 10^{-3}$ A cm^{-2}), resulting in high efficiency of chemical energy conversion. Second, is the limiting current density of cell that depends on the type of electrode. Flat-type electrodes show maximum power density, <1 mW cm^{-2}; on the other hand, porous electrodes make the limiting current 100–1000 times greater, with power densities of up to 1 W cm^{-2}. Fuel cells are categorized on the basis of the electrolyte employed and their operating temperature range, that is, low-temperature fuel cells (alkaline, proton exchange membrane fuel cell (PEMFC), direct methanol fuel cell (DMFC)) and high-temperature fuel cells (phosphoric acid fuel cell (PAFC), molten salt, and solid oxide fuel cells). Low-temperature fuel cells use a platinum catalyst. Therefore, these fuel cells are prone to catalyst poisoning by impurities. Although, there are sufficient platinum resources for the future, soaring prices of precious platinum has led recent research to replace platinum with a material that may be less susceptible to poisoning, economic by cost, safe from an environmental point of view with improved lifetime. Considerably, gold and palladium coated [70] iron and sulfur instead of pure platinum through an intermediate conversion by bacteria would lower the cost of a fuel cell substantially from $1500 to $1.50.

1.8.1
Alkaline Fuel Cells

Alkaline fuel cells (AFCs) are economically qualified for providing a power of 10–100 kW, can work optimally at 80 °C with a cell efficiency of 60–70%, and use basic solutions as electrolytes, hence, a wide range of electrode catalysts (relatively inexpensive materials) are available for them, whereas the cells using acid solution or high temperature can use only a noble metal as electrode material. At the time of being switched on, they can produce almost one-fourth of the power that is produced at optimal temperature, whereas intermediate- and high-temperature cells need auxiliary power source to start them and warm them up. However, AFCs have a disadvantage: air contains CO_2 with O_2 and forms carbonates in alkaline solution and blocks the pores of porous electrodes, thus CO_2-free air is used in cells. Bacon cells used in Apollo moon project are AFCs run on pure H_2 and O_2.

1.8.2
Direct Methanol Fuel Cells

DMFCs are qualified for providing a power of 100 mW to 1 kW and 20–30% efficiency and can produce a small amount of power over a long period. Low-volume, lightweight packaging, zero-emission power system, low operating temperature (50–120 °C), and no requirement for a fuel reformer unit made the DMFCs an excellent candidate for very small to mid-sized applications [71]. DMFC systems are used to power portable applications (smaller vehicles such as forklifts and tuggers, small portable power supply units for household use, cell phones, digital camera, laptop, music systems, soldier-carried tactical equipment, battery chargers, and autonomous power for test and training instrumentation) and also as auxiliary power units (APUs) for some niche transport sectors such as marine and submarine vehicles, scooters, and motorbikes [72, 73]. Surya Prakash and Nobel laureate George A. Olah invented the present breed of fuel cells that would directly convert methanol to electricity using a Nafion membrane and platinum electrodes for both half-reactions, in 1992 [74]. DMFC is similar to the PEMFCs in that the electrolyte is a polymer and the charge carrier is the hydrogen ion (proton). However, the liquid methanol (CH_3OH) is oxidized in the presence of water at the anode, generating CO_2, hydrogen ions, and electrons that travel through the external circuit as the electric output of the fuel cell. The hydrogen ions travel through the electrolyte and react with oxygen from the air and the electrons from the external circuit to form water at the cathode, thus completing the circuit.

Anode reaction:
$$CH_3OH + H_2O \rightarrow CO_2 + 6H^+ + 6e^- \tag{1.15}$$

Cathode reaction:
$$\frac{3}{2}O_2 + 6H^+ + 6e^- \rightarrow 3H_2O \tag{1.16}$$

Overall cell reaction:
$$CH_3OH + \frac{3}{2}O_2 \leftrightarrow CO_2 + 2H_2O \tag{1.17}$$

Low concentration of methanol (1–3 M) is preferred in DMFCs because methanol has the tendency to diffuse in high concentrations through the membrane to the cathode, without reacting with anode (methanol cross over), where its concentration becomes zero because of the rapid consumption by oxygen. It limits the maximum attainable current and loses almost half of the methanol. Oxidization of methanol at cathode contributes to the loss of cell voltage potential and produces carbon monoxide that strongly adsorbs onto the platinum catalyst and reduces the active surface area, thus reducing the performance of the cell. Addition of ruthenium or gold to the catalyst tends to ameliorate this problem because according to the most well-established theory in the field, these catalysts oxidize water to yield $OH^\bullet$ radicals: $H_2O \rightarrow OH^\bullet + H^+ + e^-$. These $OH^\bullet$ species oxidizes CO to CO_2: $CO + OH^\bullet \rightarrow CO_2 + H^+ + e^-$. Carbon supported Pt-Cr electroanodes also applied to enhance the activity of DMFCs [75]. Low operating temperatures requiring a high noble metal loading to enhance the kinetics of the methanol electro-oxidation reaction and low efficiency and toxic nature of methanol are the prominent factors that often go against its progress.

1.8.3
Phosphoric Acid Fuel Cells (PAFCs)

Molten phosphoric acid (H_3PO_4) fuel cells are able to produce <100 MW power at a working temperature of 150–220 °C with a cell efficiency of 80% and cost at $4–4.5 per W. Pyrophosphoric acid ($H_4P_2O_7$), a polymer of phosphoric acid usually synthesized at <150 °C, is used as the electrolyte in PAFCs and forms ionic solution of considerably high conductivity than the parent acid. The charge carrier in this type of fuel cell is the hydrogen ion (H^+, proton). This is similar to the PEMFCs where the hydrogen introduced at the anode is split into protons and electrons. The protons migrate through the electrolyte to cathode and combine with the oxygen (which usually comes from air) to produce water. And electrons are routed through an external circuit to cathode where they can perform useful work. This set of reactions in the fuel cell produces electricity and the by-product heat as written below.

Anode reaction:

$$2H_2 \rightarrow 4H^+ + 4e^- \quad (1.18)$$

Cathode reaction:

$$O_2(g) + 4H^+ + 4e^- \rightarrow 2H_2O \quad (1.19)$$

Overall cell reaction:

$$2H_2 + O_2 \rightarrow 2H_2O \quad (1.20)$$

In addition, CO_2 in air does not affect the electrolyte or cell performance. Therefore, it can easily be operated with fossil fuel reformer. Simple construction, low electrolyte volatility, and long-term stability are additional advantages associated with PAFCs. But high concentration of H_3PO_4 creates freezing problem when the cell is turned off, requiring auxiliary heating to maintain a temperature of >40 °C. Alloys of Pt with Ti, Cr, V, and so on, of small particle size [76] and non-noble metal (hence cheap) electrocatalyst were preferred in PAFCs to achieve high performance in hot acidic solution [77]. Graphite from porphyrine ash applied to keep the central atom apart by preventing the aggregation of particles gives better catalysis [78]. PAFCs are increasingly used in big buildings for light and heat supply that utilize 80% of overall energy instead of 40% of others. Some international brands, which bring PAFCs into limelight, are Plug Power (the United States), International Fuel Cells Corporation (the United States), and Japan's Fuji Electric Corporation, Toshiba Corporation, and Mitsubishi Electric Corporation.

1.8.4
Proton Exchange Membrane Fuel Cells

In the archetypal PEMFC design, a proton-conducting, thin, permeable polymer membrane (the electrolyte) separates the anode and cathode sides and was called a solid polymer electrolyte fuel cell (SPEFC) in the early 1970s till the proton exchange mechanism was well understood. In a typical membrane electrode assembly (MEA), the bipolar electrode plates are usually made of noble metals, nickel, or carbon nanotubes and coated with a catalyst (such as platinum, nano iron powders, or palladium) for higher efficiency [79]. Carbon papers are used to separate them from the electrolyte. PEMFCs possess a cell efficiency of about

40–70%, are qualified for providing 100–500 kW power with suitable operating temperature of 50–120 °C (Nafion) and 120–220 °C (polybenzimidazole (PBI)), and are cost worthy for $30–35 per W. Ethanol, butanol, hydrogen, and methanol presently made from either natural gas or biomass are utilized as fuel. PEM units are still considered to be the most prevalent alternative because of its operating on stored hydrogen gas. The first major breakthrough for PEM systems came as part of NASA's space program, Project Gemini. Compact design, light-weighted unit with rapid start up, and the use of solid electrolyte rather than a liquid makes the sealing of the anode and cathode gases far easier and cheaper unit to manufacture and can lead to a longer cell and stack life as it is less prone to corrosion than some other electrolyte materials. However, major disadvantages associated with PEM are low operation temperatures, that is, 80 °C, which are not high enough to perform useful cogeneration. Furthermore, in order to achieve the most effective operation of the electrolyte, it must be saturated with water, otherwise the membrane would be cracked causing the breakdown of the cell. Therefore, control of the moisture at the anode and cathode streams becomes an important consideration. The future of PEMFC systems is certainly developing as a promising technology to be utilized in several key market sectors such as cars, buildings, and smaller stationary applications, which span growth opportunities over the short-, medium-, and long-term time frame. Commercially viable PEMFC vehicles are the need to lower the platinum loading of the MEA without loss of performance and to improve the stability of the catalyst with respect to platinum dissolution and carbon-support corrosion [80].

1.8.5
High-Temperature Molten Carbonate Fuel Cells

Molten carbonate fuel cells (MCFCs) are high-temperature fuel cells, which operate at the temperatures of 600 °C or above and are composed of a molten carbonate salt (potassium carbonate, K_2CO_3) mixture as electrolyte, suspended in a porous and chemically inert ceramic matrix of beta-alumina solid electrolyte (BASE) that transports carbonate ions rather than protons (Figure 1.16). The operating temperature of 600–650 °C is lower than that of solid oxide fuel cells (SOFCs) and could be worked up to 50–60% efficiency with an operating power of 100 MW [81]. On utilization of waste heat, overall fuel efficiencies can become as high as 85%. Improved efficiency of MCFCs over PAFCs is a considerable cause for significant reduction in cost. Unlike alkaline, phosphoric acid, and polymer electrolyte membrane fuel cells, MCFCs do not require an external reformer to convert more energy-dense fuels to hydrogen because MCFCs operate at high temperatures that convert these fuels into hydrogen within the fuel cell itself by a process called *internal reforming*, which also reduces cost. MCFCs are not prone to carbon monoxide or carbon dioxide poisoning because the oxidation reactions that occur at anode at high temperature produce steam that was utilized by CO/CO_2 in reforming hydrocarbon fuel inside the anode [82]. Furthermore, they can even use carbon oxides as fuel, making them more attractive for fueling with gases

Molten carbonate fuel cell

Anode reaction:

$$H_2 + CO_3^{2-} \xrightarrow{\Delta} H_2O + CO_2 + 2e^-$$

Anode internal reformer reactions:

$$CO + H_2O \xrightarrow{\Delta} H_2 + CO_2$$
$$CH_4 + H_2O \xrightarrow{\Delta} 3H_2 + CO$$
$$CO_2 + 2H_2O \xrightarrow{\Delta} CH_4 + 2O_2$$

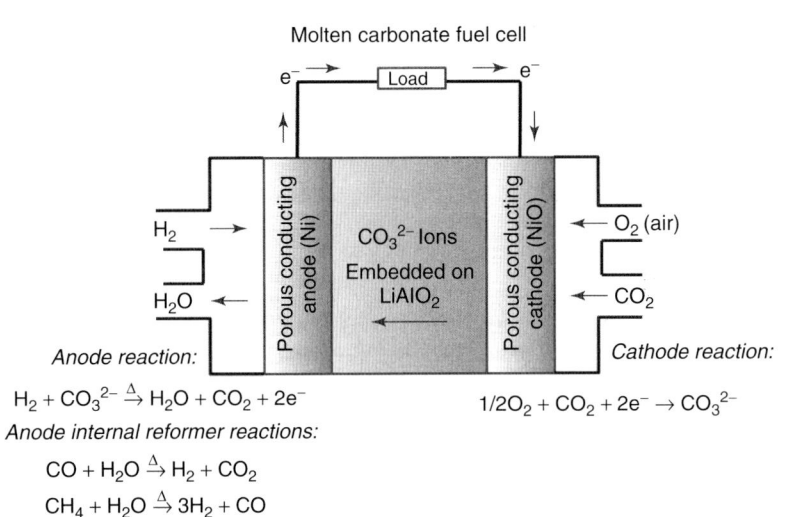

Cathode reaction:

$$1/2 O_2 + CO_2 + 2e^- \to CO_3^{2-}$$

Figure 1.16 Conceptual diagrammatic presentation of the molten carbonate fuel cell, which includes anodic and cathodic chemical reactions with anode internal reformer reactions.

made from coal, and they are highly resistive toward impurities than other fuel cells. Primary disadvantage of current MCFC technology is low durability because the high operating temperatures and corrosive nature of electrolyte accelerate component breakdown that decreases cell life. Scientists are currently exploring corrosion-resistant materials for components as well as fuel cell designs that increase cell life without decreasing performance.

1.8.6
Solid Oxide Fuel Cells

SOFCs were first systematically described by Grubb in 1957. SOFCs work at high temperatures, and ionic conductor or aqueous liquid electrolytes are replaced with well-humidified thin membranes (0.1 mm thick) of perfluorosulfonic acid polymer (Nafion), yttria (Y_2O_3)-stabilized zirconia (ZrO_2), and other such materials between anode and cathode, which can transport protons and oxygen ions, respectively. Their electrode reactions are as follows:

Anode reaction:

$$CH_3OH + H_2O \to CO_2 + 6H^+ + 6e^- \tag{1.21}$$

Internal reformer reaction:

$$CO_2 + 2H_2O + heat \to CH_4 + 2O_2 \tag{1.22}$$

Cathode reaction:

$$\frac{3}{2}O_2 + 6H^+ + 6e^- \rightarrow 3H_2O \tag{1.23}$$

Overall reaction:

$$CH_3OH + \frac{3}{2}O_2 \rightarrow CO_2 + 2H_2O + \text{electrical energy} \tag{1.24}$$

Standard operating temperature of SOFCs is >250 °C. SOFCs are extremely advantageous because of the possibility of using a wide variety of fuels [83] such as hydrogen, butane, methanol, and other petroleum products. Nickel and nickel oxide are used as catalyst instead of the costlier platinum, and they also do not experience catalyst poisoning by carbon monoxide, so they do not require high-purity hydrogen fuel. High temperature decreases i_0 in $\eta = RT/F \ln i_0/i$ reduces the overpotential [62] required for given current densities. High-performance cathodes obtained high power density of SOFCs. For larger stationary applications at the level of 10–100 MW, SOFCs offer the additional advantage over conventional power plants of cogeneration of both electric power and low-grade heating. A major disadvantage of the SOFCs is that as a result of the high operating temperature, a considerable constraint on the choice of the electrode materials and unwanted reactions may occur inside the fuel cell [84]. It is very common for SOFCs to build up carbon dust or graphite on the anode, preventing the fuel from reaching the catalyst. SOFCs with nonfluorine membrane (polystyrenesulfonic acid) was used in the Gemini spacecraft. High-temperature cells are envisaged largely for stationary power plants and electrochemical engines in cars.

1.9
Summary

Global energy-need-projections (based on sound scientific facts) for sustainable development suggest a 50% increase in world energy requirement by 2030. Energy demand accompanied by downfall in conventional energy sources, which is associated with a myriad of environmental issues and their very impact on life, grew a worldwide consensus on clean energy that became the highly witted driving force behind the current trends of transition from conventional to renewable energy resources, that is, hydro, biomass, wind, solar, geothermal, and biofuels. In the foreseeable future, the prospect of using renewables made the strong commitments to expand them as alternate energy sources because of their abundance and vast potential that decrease the cost and hold a strong promise to revolutionize the present energy technology to meet future energy demands. After getting the crystal clear picture of past, present, and future energy scenario, we focus on electrochemical technologies for energy storage and conversion devices. This essay reviews the current energy status and applications of advanced technologies for energy generation and concludes with a discussion of the prospects of a future global-scale energy storage and conversion systems based on electrolysis, PV and photocatalytic hydrogen generation, rechargeable batteries, supercapacitors, and fuel cells. Fundamental issues of electrochemical technologies were scaled-up here with their related safety problems, which have been well understood, and their relevant solutions were short listed, which attracted the world's focus toward sustainable green energy sources.

Acknowledgments

The authors would like to thank the Institute of Atomic & Molecular Sciences, Academia Sinica (Contract No. **AS-98-TP-A05**) and National Science Council of Taiwan (Contract Nos. **NSC 97-2113-M-002-012-MY3** and **NSC 99-2120-M-002-012**) for financial support.

References

1. United Nations (2010) World Economic Situation and Prospects 2010, New York, NY, pp. 125–126, http://www.un.org/esa/policy/wess/wesp2010 files/wesp2010.pdf (accessed 2010).
2. IHS Global Insight (2010) World Overview: First Quarter 2010, Lexington, MA, March 2010, pp. 102–112.
3. Radler, M. (2009) Special Report: Oil, gas reserves rise as oil output declines, Worldwide look at reserves and production. *Oil Gas J.*, **107** (47), 20–21, web site http://www.ogj.com (subscription site) (accessed 2010).
4. Olson, N.P.E., (2007) Worldwide look at reserves and production, Paper: Nuclear energy and the fossil fuels. *Oil Gas J.*, **105** (48), 22–23.
5. Weisz, P.B. (2004) Basic choice and constraints on long term energy supplies. *Phys. Today*, **57** (7), 47–52. http://www.gatsby.ucl.ac.uk/~pel/environment/energy_pt04.html.
6. Hubbert, M.K. (1984) World-energy supply resources, technologies, perspectives. *Am. Sci.*, **72**, 293.
7. Hubbert, M.K. (1956) Paper presented before the Spring Meeting of the Southern District Division of Production, American Petroleum Institute Plaza Hotel. *San Antonio Texas*, **95**, 22–27.
8. Hertel, T.W., Tyner, W.E., and Birur, D.K. (2010) The global impacts of biofuel mandates. *Energy J.*, **31**, 75–100. http://www.iaee.org/en/publications/journal.aspx (subscription site) (accessed 2010).
9. Hoffert, M.I., Caldeira, K., Jain, A.K., Haites, E.F., Harveyk, L.D.D., Potter, S.D., Schlessingee, M.E., Schneider, S.H., Watts, I.R.G., Wigley, T.L., and Wuebbles, D. (1998) Energy implication of future stabilization of atmospheric CO_2 content. *Nature*, **395**, 881–884.
10. Ritter, S.K. (2003) As quoted in "Green Solutions to Global Problems". *Chem. Eng. News*, **81**, 31–33.
11. REN21 Renewables Global Status Report on Energy Transformation Continues Despite Economic Slowdown, Paris, 13 May 2009.
12. Pfeffer, J. (2010) Building sustainable organizations: the human factor. *Acad. Manage. Perspect.*, **24** (1), 34–45.
13. Parry, M.L., Canziani, O.F., Palutikof, J.P., van der Linden, P.J. and Hanson, C.E. (2007) (eds) *Climate Change 2007: Impacts, Adaptation and Vulnerability, Contribution of Working Group II to the Fourth Assessment Report of the Intergovernmental Panel on Climate Change*, Cambridge University Press, Cambridge, pp. 469–506.
14. Putz, F.E. and Pinard, M.A. (1993) Reduced-impact logging as a

carbon-offset method. *Conserv. Biol.*, **7** (4), 755–757.
15. Olson, E.G. (2010) *Better Green Business: Handbook for Environmentally Responsible and Profitable Business Practices*, Chapter 1, Wharton School Publications, p. 15.
16. Tempest, M. (2007) *Climate Change Bill is Revolutionary, Says Blair*, Guardian Unlimited, London. http://politics.guardian.co.uk/green/story/0,2032710,00.html (accessed 2010).
17. Elkind, E., Farber, D., Frank, R., Hanemann, M., Kammen, D., Kantenbacher, A., and Weissman, S. (2010) California at the Crossroads, Proposition 23 AB 32 and Climate Change, White Paper is a Publication of the U.C. Berkeley School of Law's Center for Law, Energy & the Environment (CLEE), p. 5, September, 2010.
18. Freris, L. and Infield, D. (2008) *Renewable Energy in Power Systems*, John Wiley & Sons, Ltd, West Sussex.
19. Swanson, R.M. (2009) Photovoltaics power up. *Science*, **324**, 891–892.
20. Bertani, R. (2009) Geothermal energy: an overview on resources and potential, Proceedings of the International Conference on National Development of Geothermal Energy Use, Slovakia.
21. Pletcher, D. (1991) *A First Course in Electrode Processes*, The Electrochemical Consultancy, Romsey.
22. Walsh, F.C. (1993) *A First Course in Electrochemical Engineering*, The Electrochemical Consultancy, Romsey.
23. Davis, T.A., Genders, J.D., and Pletcher, D. (1997) *A First Course in Ion Permeable Membranes*, The Electrochemical Consultancy, Romsey.
24. Pletcher, D. (1999) *Guide to Electrochemical Technology for Synthesis, Separation, and Pollution Control*, Electrosynthesis Company, Inc., Lancaster, NY, p. 6.
25. Fujishima, A. and Honda, K. (1972) Electrochemical photolysis of water at a semiconductor electrode. *Nature*, **238**, 37–38. (doi: 10.1038/238037a0).
26. Linden, D. and Reddy, T.B. (eds) (2002) *Handbook of Batteries*, 3rd edn, McGraw-Hill, New York, p. 8. ISBN: 0-07-135978.
27. Planté, G. (1959) Recherches sur l'Électricité, Paris.
28. Howard, P.O. and Vinal, G.W. (1951) National Bureau of Standards Circular 504, GPO. - QC100.U555, Battery Additives, Washington, DC.
29. Buchaman, I. (2001) *Batteries in a Portable World: A Handbook on Rechargeable Batteries for Non-Engineers*, 2nd edn, Chapter 10, Cadex Electronics Inc., p. 6. ISBN: 0968211828.
30. Ovshinsky, S.R., Fetcenko, M.A., and Ross, J.A. (1993) A nickel metal hydride battery for electric vehicles. *Science*, **260**, 176–181.
31. Fetcenko, M.A., ECD (2008) In search of the perfect battery, Avicenne Conference, Nice, France.
32. Stanley Whittingham, M. (1976) Electrical energy storage and intercalation chemistry. *Science*, **192** (4244), 1126–1127.
33. Chan, C.K., Peng, H., Liu, G., McIlwrath, K., Zhang, X.F., Huggins, R.A., and Cui, Y. (2008) High performance lithium battery anodes using silicon nanowires. *Nat. Nanotechnol.*, **3**, 31–35.
34. Chan, C.K., Zhang, X.F., and Cui, Y. (2008) High capacity Li-ion battery anodes using Ge nanowires. *Nano Lett.*, **8**, 307–309.
35. Schmutz, C., Tarascon, J.M., Gozdz, A.S., Warren, P.C., and Shokoohi, F.K. (1994) *Rechargeable Lithium and Lithium-Ion Batteries*, Megahed, S., Barnett, B. M. and Xie, L. (eds) vol. 94, The Electrochemical Society Proceedings Series, Pennington, N. J., pp. 330–333.
36. Brodd, R.J. (ed.) (1995) *Lithium Ion Battery Technology, Progress in Batteries and Battery Materials*, Vol. 14, ITEJEC Press, Japan, p. 1.
37. Culter, T. (1996) *A design guide for rechargeable zinc-air battery technology*, Conference paper at Orlando, FL, USA, AER Energy Resources Inc., Smyma, GA, pp. 616–621, ISBN: 0-7803-3268-7.
38. Noring, J. *et al.* (1993) *Proceedings of the Symposium on Batteries and Fuel Cells for Stationary and Electric Vehicle Applications Volumes 93–98 of Proceedings*

(*Electrochemical Society*), The Electrochemical Society, pp. 235–236, ISBN: 1566770556.
39. Oshima, T., Kajita, M., and Okuno, A. (2004) Development of sodium-sulfur batteries. *Int. J. Appl. Ceramic Technol.*, **1**, 269–276. doi: 10.1111/j.1744-7402.2004.tb00179.x
40. Walawalkar, R., Apt, J., and Mancini, R. (2007) Economics of electric energy storage for energy arbitrage and regulation in New York. *Energy Policy*, **35** (4), 2558–2568. doi: 10.1016/j.enpol.2006.09.005
41. Auxer, W. (1986) *32nd International Power Sources Symposium, Cherry Hill, NJ, June 9–12, 1986, Proceedings Volume A88-16601 04-44*, Electrochemical Society Inc., Pennington, NJ, pp. 49–54.
42. de Levie, R. (1999) The electrolysis of water. *J. Electroanal. Chem.*, **476** (1), 92–93.
43. Pauling, L. (1970) General Chemistry, Section 15-2. San Francisco.
44. Kanan, M.W. and Nocera Daniel, G. (2008) In situ formation of an oxygen-evolving catalyst in neutral water containing phosphate and Co^{2+}. *Science*, **321**, 1072–1076.
45. Zittel, W. and Wurster, R. (1996) *Production of hydrogen Part 4: Production from Electricity by Means of Electrolysis*, Ludwig-Bölkow-Systemtechnik GmbH, pp. 40–53.
46. Kruse, B., Grinna, S., and Buch, C. (2002) *Hydrogen – Status and Possibilities*, Bellona Foundation, http://www.bellona.org/filearchive/fil_Hydrogen_6-2002.pdf (accessed 2010).
47. Schmittinger, P. (ed.) (2000) *Chlorine, Principles and Industrial Practice*, Wiley-VCH Verlag GmbH, Weinheim, p. 117.
48. Kiros, Y., Pirjamali, M., and Bursell, M. (2006) Oxygen reduction electrodes for electrolysis in chlor-alkali cells. *Electrochim. Acta*, **51**, 3346–3350.
49. Kiros, Y. and Bursell, M. (2008) Low energy consumption in chlor-alkali cells using oxygen reduction electrodes. *Int. J. Electrochem. Sci.*, **3**, 444–451.
50. Tsuji, I., Kato, H., Kobayashi, H., and Kudo, A. (2004) Photocatalytic H_2 evolution reaction from aqueous solutions over band structure-controlled $(AgIn)_x Zn_{2(1-x)} S_2$ solid solution photocatalysts with visible-light response and their surface nanostructures. *J. Am. Chem. Soc.*, **126**, 13406–13413.
51. Tsuji, I., Kato, H., and Kudo, A. (2005) Visible-light-induced H_2 evolution from an aqueous solution containing sulfide and sulfite over a $ZnS-CuInS_2-AgInS_2$ solid-solution photocatalyst. *Angew. Chem. Int. Ed.*, **44**, 3565–3568.
52. Kato, H., Asakura, K., and Kudo, A. (2003) Highly efficient water splitting into H_2 and O_2 over lanthanum-doped $NaTaO_3$ photocatalysts with high crystallinity and surface nanostructure. *J. Am. Chem. Soc.*, **125** (10), 3082–3089.
53. Zou, Z.G., Ye, J.H., Sayama, K., and Arakawa, H. (2001) Direct splitting of water under visible light irradiation with an oxide semiconductor photocatalyst. *Nature*, **414** (6864), 625–627.
54. Maeda, K., Teramura, K., Takata, T., Hara, M., Saito, N., Toda, K., Inoue, Y., Kobayashi, H., and Domen, K. (2005) Overall water splitting on $(Ga_{1-x}Zn_x)(N_{1-x}O_x)$ solid solution photocatalyst: relationship between physical properties and photocatalytic activity. *J. Phys. Chem. B*, **109** (43), 20504–20510.
55. Song, H., Cai, P., Huabing, Y., and Yan, C. (2007) Hydrothermal synthesis of flaky crystallized $La_2Ti_2O_7$ for producing hydrogen from photocatalytic water splitting. *Catal. Lett.*, **113** (1–2), 54–58.
56. Thaminimulla, C.T.K., Takata, T., Hara, M., Kondo, J.N., and Domen, K. (2000) Effect of chromium addition for photocatalytic overall water splitting on Ni–$K_2La_2Ti_3O_{10}$. *J. Catal.*, **196**, 362.
57. Bockris, J.O'M. and Reddy, A. K. N. (eds) (2006) Morden Electrochemistry: Electrodics in chemistry, Engineering, Biology and Enviornmental Science, Vol. 2B, 2nd edn, (Chapter 10: *Photoelectrochemistry*), Springer, Kluwer Acadic/Plemum, Newyork, pp. 1577–1585.
58. Kainthla, R.C., Zelenay, B., and Bockris, J.O'M. (1987) Significant efficiency

58. increase in self-driven photoelectrochemical cell for water photoelectrolysis. *J. Electrochem. Soc.*, **134**, 841–844.
59. Khaselev, O. and Turner, J. A. (1998) A Monolithic Photovoltaic-Photoelectrochemical Device for Hydrogen Production via Water Splitting. *Science*, **280** (5362), 425–427.
60. Linkous, D.K., Slattery, C.A., Ouelette, A.J.A., McKaige, G.T., and Austin, B.C.N. (1996) Solar photocatalytic H^2 from water using a dual bed photosystem. Hydrogen Energy Progress XI: Proceedings of the 11th World Hydrogen Energy Conference, Vol. 3, pp. 2545–2550.
61. Zhou, H., Li, X., Fan, T., Osterloh, F.E., Ding, J., Sabio, E.M., Zhang, D., and Guo, Q. (2010) Artificial inorganic leafs for efficient photochemical hydrogen production inspired by natural photosynthesis. *Adv. Mater.*, **22** (9), 951.
62. Ohta, K., Kawamura, T., Kuroda, S., and Mizuno, T. (1994) Symp. of Env. Aspects of Electrochemistry and Photoelectrochemistry. *Proc. Electrochem. Soc.*, **93** (18), 85–89.
63. Ohta, K., Ohguchi, Y., Kaneco, S., Mizuno, T., and Hallemann, M. (1992) Proceedings of the international symposium on chemical and electrochemical fixing of carbon dioxide. *Chem. Soc. Jap.*, **A19**, 1378–1381.
64. Fischer, F. and Prziza, O. (1914) Über die elektrolytische Reduktion von unter Druck gelöstem Kohlendioxyd und Kohlenoxyd. *Ber. Dtsch. Chem. Ges.*, **47** (1), 256–260.
65. Guruswamy, V. and Bockris, J.O'M. (1979) Hydrogen and Electricity from Water and Light – Lanthanum Chromite – Titanium Dioxide Anode. *Solar Energy Mater. A.*, **1** (5–6), 441–449.
66. Bockris, J.O'M. and Wass, J. (1989) The Photoelectrocatalytic Reduction of Carbon Dioxide. *J. Electro Chem. Soc.*, **136**, 2521–2528.
67. Uosaki, K. and Nakabayashi, S. (1992) *Chem. Lett.*, **40**, 1447.
68. (a) Grove, W.R. (1839) On voltaic series and the combination of gases by platinum, *Philos. Mag. J. Sci.*, **14**, 127–130; (b) Grove, W.R. (1842) On a gaseous voltaic battery. *Philos. Mag. J. Sci.*, **21**, 417–420.
69. Williams, M.C. (2001) Status and promise of fuel cell technology. *Fuel Cells*, **1**, 87–91.
70. Colin Johnson, R. (2007) Gold is Key to Ending Platinum Dissolution in Fuel Cells, EETimes.com. http://www.eetimes.com/news/latest/showArticle.jhtml?articleID=196901214 (accessed 2010).
71. Dohle, H., Mergel, J., and Stolten, D. (2002) Heat and power management of a direct-methanol-fuel-cell (DMFC) system. *J. Power Sources*, **111**, 268–282.
72. Dillon, R., Srinivasan, S., Aricò, A.S., and Antonucci, V. (2004) International activities in DMFC R&D: status of technologies and potential applications. *J. Power Sources*, **127**, 112–126.
73. Ren, X., Zelenay, P., Thomas, S., Davey, J., and Gottesfeld, S. (2000) Recent advances in direct methanol fuel cells at Los Alamos National Laboratory. *J. Power Sources*, **86**, 111.
74. Olah, G.A., Surya Prakash, G.K., Ellis, R.W., and Olah, J.A. (1986) Remarks on the mechanism of ethylene formation from methyl alcohol. *J. Chem. Soc., Chem. Commun.*, 9. doi: 10.1039/C39860000009.
75. Antolini, E., Salgado, J.R.C., Santos, L.G.R.A., Garcia, G., Ticianelli, E.A., Pastor, E., and Gonzales, E.R. (2006) Carbon supported Pt–Cr alloys as oxygen-reduction catalysts for direct methanol fuel cells. *J. Appl. Electrochem.*, **36**, 355.
76. Cutlip, M.B., Yang, S.C. and Stonehart, P. (1991) Simulation and optimization of porous gas-diffusion electrodes used in hydrogen oxygen phosphoric acid fuel cells-II development of a detailed anode model. *Electrochimica Acta*, **36** (3–4), 547–553.
77. Shim, J.C., Rim, H.R., Yoo, D.-Y., Park, J.-I., Kim, J.-W., and Lee, J.-S. (2000) Development of electrode structure via loading Pt catalyst on PTFE/C powder by colloidal method in PAFC. *J. Ind. And Eng. Chem.*, **16**, 79–84.
78. Steele, B.C.H. and Heinzel, A. (2001) Materials for fuel-cell technologies. *Nature*, **414**, 345–352.

79. Yu, P.T., Gu, W., Zhang, J., Makharia, R., Wagner, F.T., and Gasteiger, H.A. (2009) Carbon Support requirements for highly durable fuel cell operation in *Polymer Electrolyte Fuel Cell Durability* (eds M.I.F.N. Büchi and T.J. Schmidt), Springer, New York, p. 29.
80. Vielstich, W., Gasteiger, H.A., and Yokokawa H. (eds) (2009) *Handbook of Fuel Cells – Fundamentals, Technology and Applications*, vol. **5**, John Wiley & Sons, Ltd, Chichester, p. 18.
81. Sundmacher, K., Kienle, A., Pesch, H.J., Berndt, J.F., and Huppmann, G. (eds) (2007) *Molten Carbonate Fuel Cells*, Wiley-VCH Verlag GmbH & Co. KGaA, Weinheim, Copyright 8, ISBN: 978-3-527-31474-4.
82. Milewski, J., Lewandowski, J., and Miller, A. (2009) Reducing CO_2 emission from a coal fired power plant by using a molten carbonate fuel cell. *Chem. Process Eng.*, **30**, 341–350.
83. Hayashi, K., Yamamoto, O., and Minoura, H. (2000) Portable solid oxide fuel cells using butane gas as fuel. *Solid State Ionics*, **302**, 343–345.
84. Sahibzada, M., Steel, B., Hellgardt, K., Barth, D., Effendi, A., Mantzavinos, D., and Metcalfe, I. (2000) Intermediate temperature solid oxide fuel cells operated with methanol fuels. *Chem. Eng. Sci.*, **55**, 3077–3083.

Further Reading

Holtappels, P. and Steinberger-Wilckens, R. (2009) Realising reliable, durable, energy efficient and cost effective SOFC systems (Real-SOFC). *Fuel Cells*, **9**, 783–784.

Ichinose, O., Kawaguchi, M., and Furuya, N. (2004) Effect of silver catalyst on the activity and mechanism of a gas diffusion type oxygen cathode for chlor-alkali electrolysis. *J. Appl. Electrochem.*, **34**, 55–59.

Mauricio, A.M., Pacheco, A.M.G., Brito, P.S.D., Castro, B., Figueiredo, C., and Aires-Barros, L. (2005) An ionic conductivity-based methodology for monitoring salt systems in monument stones. *J. Cult. Herit.*, **6**, 287–293.

2
Electrochemical Engineering Fundamentals
Zhongwei Chen, Fathy M. Hassan, and Aiping Yu

2.1
Electrical Current/Voltage, Faraday's Laws, Electric Efficiency, and Mass Balance

An electric circuit is generated when a path of free electrons is created and is continuously moving. This continuous movement of free electrons through a conductor in a circuit is called *current*, and it is often referred to in terms of *flow*, just like the flow of a liquid through a hollow pipe.

The force motivating electrons to flow in a circuit is called *voltage*. Voltage is a specific measure of potential energy that is always relative between two points. When we speak of a certain amount of voltage being present in a circuit, we are referring to the measurement of how much *potential* energy exists to move electrons from one particular point in that circuit to another. Without reference to *two* particular points, the term voltage has no meaning.

In case of a simple electric circuit, resistance (R), voltage (V), and current (I) are related by Ohm's law.

$$V = IR \qquad (2.1)$$

In case of electrochemical systems where electrochemical changes occur at the surface of electrodes, Faraday's laws are applied. Faraday's laws of electrolysis are quantitative relationships based on the electrochemical researches published by Michael Faraday in 1834.

- **Faraday's first law of electrolysis**: The mass of a substance altered at an electrode during electrolysis is directly proportional to the quantity of electricity transferred at that electrode. Quantity of electricity refers to the quantity of electrical charge, typically measured in coulomb.
- **Faraday's second law of electrolysis**: For a given quantity of electricity (electric charge), the mass of an elemental material altered at an electrode is directly proportional to the element's equivalent weight. The equivalent weight of a substance is its molar mass divided by an integer that depends on the reaction undergone by the material.

Mathematically, Faraday's laws [1, 2] can be summarized as

$$m = \left(\frac{Q}{F}\right)\left(\frac{M}{z}\right) \quad (2.2)$$

where m is the mass of the substance in grams liberated at an electrode; Q, the total electric charge passed through the substance; F, the Faraday constant (96 485 C mol^{-1}); M, the molar mass of the substance; and z, the electrons transferred per ion (the valency number of ions of the substance).

In simple cases of constant current electrolysis, $Q = I \times t$.

Consequently,

$$m = \left(\frac{I \times t}{F}\right)\left(\frac{M}{z}\right) \quad (2.3)$$

Provided that the number of moles (n) is equal to the mass divided by the molar mass, then,

$$n = \left(\frac{I \times t}{F}\right)\left(\frac{1}{z}\right) \quad (2.4)$$

where t is the total time during which the constant current I is applied.

In the more complicated case of a variable electrical current, the total charge Q is the electric current $I(\tau)$ integrated over time τ:

$$Q = \int_0^t I d\tau \quad (2.5)$$

Note that here the current is a function of time τ, whereas t is the total electrolysis time.

It is important to note that deviation from Faraday's laws can be observed in the case of transient currents. This applies when charge, in addition to electrode reactions, accumulates in near interfaces. Such transient currents are also known as *nonfaradaic*.

2.1.1
Current Efficiency

If, simultaneously, several reactions are taking place at the electrode, a partial electrode *current density* (c.d.) j_k can be assigned to each reaction. It is given by the stoichiometry of the reaction and by the amount of substance reacting (per unit time and per unit electrode area) in the reaction considered. The current efficiency of reaction k, ε_k, is defined as the ratio of j_k to the total current density

$$\varepsilon_k = \frac{j_k}{\sum_m j_m} \quad (2.6)$$

Note that ε_k may be >1 if cathodic and anodic reactions take place simultaneously at the same electrode. However, ε_k still gives correctly the product yield, which is the quantity of industrial interest.

The efficiency of the whole industrial electrochemical process takes into consideration the total energy input and output.

One important example is the electrolysis of water to produce hydrogen.

$$2H_2O\,(l) \rightarrow 2H_2\,(g) + O_2\,(g)\,;\, E_0 = +1.229\text{ V} \tag{2.7}$$

Hydrogen can be used as fuel for hydrogen fuel cells. This has been suggested as one approach to shift economies of the world from the current state of almost complete dependence on hydrocarbons for energy.

Energy efficiency is a measure of what fraction of electrical energy used is actually contained within the hydrogen. Some of the electrical energy is converted to heat, an almost useless by-product. Some reports quote efficiencies between 50 and 70%. This efficiency is based on the lower heating value of hydrogen. The lower heating value of hydrogen is the total thermal energy released when hydrogen is combusted minus the latent heat of vaporization of the water. This does not represent the total amount of energy within the hydrogen; hence, the efficiency is lower than a more strict definition. Other reports quote the theoretical maximum efficiency of electrolysis as being between 80 and 94%. The theoretical maximum considers the total amount of energy absorbed by both the hydrogen and oxygen. These values refer only to the efficiency of converting electrical energy into hydrogen's chemical energy. The energy lost in generating the electricity is not included.

2.1.2
Mass Balance

When current flows in an electrolytic cell, the balance of the charge and the mass of the reactants should be established. Hence, the materials should be brought up to (or carried away from) the electrode surface at the rates at which they are consumed (or produced) by the reaction.

For a redox reaction,

$$O_x + ne^- \rightleftharpoons \text{Red} \tag{2.8}$$

When writing an equation for a reaction at an electrode, we must observe the balance of the reactants, products, and the electronic charges.

$$\sum_{O_x} v_j z_j - \sum_{\text{red}} v_j z_j = n \tag{2.9}$$

where v_j and z_j are the stoichiometric coefficients of, and the charges on, reactant j, respectively. n is the number of electrons involved in the reaction.

The parameter v_j/n mol for each component of the electrode reaction is called the *chemical equivalent* of the reaction named, and the value of $(v_j/n)\,M_j$ is called the *equivalent mass*.

Recalling that the amount of charge corresponds to the conversion of one chemical equivalent of substance, an amount of charge nF/v_j is required to convert 1 mol of substance j.

We can restate Faraday's laws by saying when an amount of charge Q has been consumed at the electrode, the number of moles Δn_j of a substance that has formed or reacted is given by

$$\Delta n_j = \frac{v_j Q}{nF} \qquad (2.10)$$

Since the total amount of substance reacted is proportional to the amount of charge, the rate of a specific reaction u_j is proportional to the current (current density) and is given by

$$u_j = v_j \frac{i}{nF} \qquad (2.11)$$

This rate is a measure of the amount of substance j converted in unit time per unit surface area of the electrode.

Alternatively, we can write the condition for mass balance of the reacting component j as

$$J_j = \frac{v_j}{nF} i \qquad (2.12)$$

where J_j is defined as the flux density of the substance j in the electrolyte stoichiometrically required when the electrode reaction proceeds under steady-state conditions. And the migration flux density $J_{m,j}$ is given by

$$J_{m,j} = \frac{t_j}{z_j F} i \qquad (2.13)$$

where t_j is the transport number of species j.

Uncharged reaction components are transported by diffusion and convection, even though their migration fluxes are zero. The total flux density J_j of species j is the algebraic (vector) sum of densities of all flux types, and the overall equation for mass balance must be written as

$$\frac{v_j i}{nF} = J_j = J_{m,j} + J_{d,j} + J_{kv,j} \qquad (2.14)$$

In the steady state, the diffusion and convection fluxes are always set up in such way as to secure mass balance.

2.2
Electrode Potentials and Electrode–Electrolyte Interfaces

There are three types of absolute potentials that can be defined for any phase.

- Inner (Galvani) potential ϕ_i, which is the electrical potential in the interior of the phase. It is defined as the energy required to transfer a unit of charge from vacuum at infinity to the interior of the phase.
- Outer (Volta) potential φ_i, which is the energy required to transfer unit charge from infinity to the surface of the phase i.
- Surface potential ψ, which is the potential difference generated at the surface as a result of the adsorption of species, for example, Cl^- and SCN^-.

2.2 Electrode Potentials and Electrode–Electrolyte Interfaces

These potentials are related by

$$\phi_i = \varphi_i + \psi \tag{2.15}$$

Only under infinitely high vacuum, where nothing is adsorbed on the surface, does $\psi = 0$, and hence $\phi_i = \varphi_i$.

Chemical systems at constant temperature and pressure proceed spontaneously in the direction that results in a decrease in the Gibbs free energy G or the chemical potential μ.

$$\mu_i^\alpha = \left(\frac{\partial G}{\partial n_i}\right)_{T,P,n_{j\neq i}} \tag{2.16}$$

where μ_i^α is the chemical potential of species i in phase α. If the system includes charged species, an electrochemical potential $\bar{\mu}_i$ can be defined as

$$\bar{\mu}_i = \mu_i + z_i F \emptyset_i = \mu_i^o + RT \ln a_i + z_i F \emptyset_i \tag{2.17}$$

where z_i is the charge of species i (in electron charge units); F, the Faraday constant; $\emptyset_i$, the inner electrical potential of the phase i (the Galvani potential); μ_i^o, a constant; and a_i, the activity of species i.

In electrochemistry, we are only concerned with potential differences. The absolute potential difference between two phases is called the *absolute potential of the electrode* when the two phases are an electrode and an electrolyte.

2.2.1
Potential Difference

Consider a metal–solution interface as shown in Figure 2.1. The absolute electrostatic potentials are ϕ_m and ϕ_s in the metal and solution phases, respectively, that is, the inner Galvani potentials of the two phases. The electrochemical equilibrium across the phase boundary is represented by

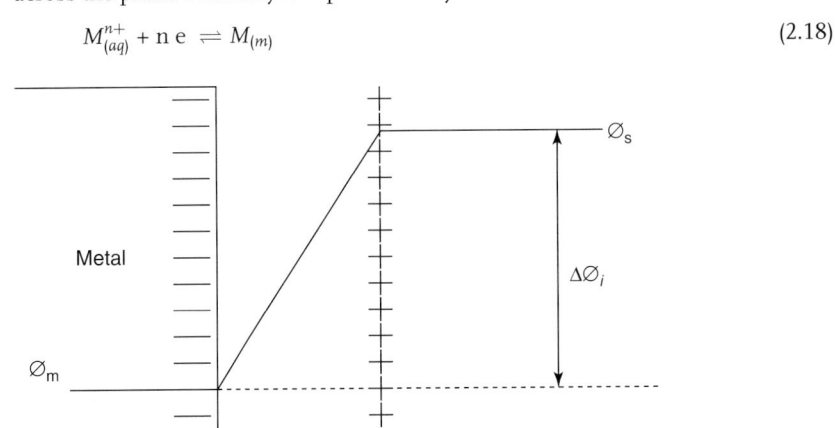

$$M_{(aq)}^{n+} + n\,e \rightleftharpoons M_{(m)} \tag{2.18}$$

Figure 2.1 Inner (Galvani) potential difference across the electrical double layer at electrode–solution phase boundary.

This reaction involves the transfer of electrons between the two phases. As this electron transfer shifts toward equilibrium, a net charge separation must develop between the electrode and the solution. This charge separation creates a potential difference $\Delta\phi_{m/s}$ at the electrode/solution interface.

$$\Delta\phi_{m/s} = \phi_m - \phi_s \tag{2.19}$$

The direct experimental measurement of $\Delta\phi_{m/s}$ at a single interface is not feasible. Measurement of a potential difference requires connection with another electrode to constitute a complete electric circuit. This latter electrode, when inserted in the solution, will also possesses a potential drop ($\Delta\phi_{m/s}$).

To address this problem the introduction of a reference electrode was necessary. The reference electrode is a device that maintains a constant potential drop $\Delta\phi_{m/s}$ across its interface with the solution. In this way, when measurements of potential difference, E, are made between the two electrodes, the observed voltage is given by

$$E = (\phi_m - \phi_s) + \text{constant} \tag{2.20}$$

where $\phi_m - \phi_s$ refers to the electrode of interest, and the constant describes the role of the reference electrode. By convention, the standard hydrogen electrode (SHE) is defined as possessing an absolute potential of zero. This enables us to conveniently report potentials of other half-cells (for example, the Fe^{2+}/Fe^{3+} system) relative to this reference electrode.

For example, the electrode potential at equilibrium for the following reaction

$$Fe^{3+}_{(aq)} + e^-_{(m)} \rightarrow Fe^{2+}_{(aq)} \tag{2.21}$$

is given by

$$E_e = \Delta\phi_{m/s}(Fe^{3+}/Fe^{2+}) - \Delta\phi_{m/s}(SHE) = \Delta\phi_{m/s}\left(Fe^{3+}/Fe^{2+}\right) - 0 \tag{2.22}$$

where E_e is the electrode potential at equilibrium.

An electrode process often involves the transfer of charge across the interface between a metallic electrode (m) and a solution phase (aq) species.

Corresponding to the spontaneous direction of the reaction, a positive ΔE, is defined by

$$\Delta G = -nF\Delta E \tag{2.23}$$

When the cell is under standard conditions and all components are at unit activity, one can write

$$\Delta G^\circ = -nF\Delta E^\circ \tag{2.24}$$

and

$$-\Delta G^\circ = RT \ln K_{eq} \tag{2.25}$$

where K_{eq} is the equilibrium constant of the overall cell reaction.

For the general charge-transfer reaction,

$$O_{(aq)} + ne^-_{(m)} \rightarrow R_{(aq)} \tag{2.26}$$

in which n electrons are transferred, Nernst showed that the potential established at the electrode under equilibrium conditions is given by (approximating the activity equal the concentration)

$$E_e = E^o + \frac{RT}{nF} \ln \frac{[O]}{[R]} \qquad (2.27)$$

where E^o, is the standard electrode potential, [O] and [R] are the equilibrium concentration of the O and R species, respectively.

2.2.2
Electrode–Electrolyte Interfaces

Under equilibrium conditions, the forces acting on a species in the bulk of the electrolyte are isotropic and homogeneous, and no net electric field is developed. At a phase boundary (interface), the symmetry breaks. At the interface, the species experience anisotropic forces that vary with distance from the electrode. This gives rise to net orientation of solvent dipoles and a net excess ionic charge near the phase boundary, on the solution side. At the electrode surface, an induced charge of equal quantity and opposite sign is created. This charge separation results in a potential gradient across the electrode–solution interface; in other words, the interface behaves as a capacitor.

Assuming that the electrode is ideally polarizable, that is, it behaves purely capacitively in the absence of any faradaic reaction, the following models described the double layer created at the interface.

- **Helmholtz model**: Helmholtz considered the double layer as a basic parallel-plate capacitor. He assumes that while all the excess charge in the electrode is located at the surface, a rigidly held layer of oppositely charged ions is created at the solution side. The latter is located in a plane parallel to the electrode surface, which is called the outer Helmholtz plane (OHP). The Helmholtz double-layer capacitance per unit area C_H is given by

$$C_H = \frac{\varepsilon \varepsilon^o}{d} \qquad (2.28)$$

- where ε is the dielectric constant; ε^o, the permittivity of free space (8.85×10^{-12} F m^{-1}); and d, the distance of the closest approach of hydrated ions (in range of 3 A^o).
- **Gouy and Chapman model**: This model assumes a different excess charge distribution on the solution side. Gouy and Chapman coupled Boltzman distribution of ions to account for electrostatic interaction and the Poisson equation for the relationship between potential, distance, and charge density. They obtained the following equation:

$$\frac{d^2\phi_x}{dx^2} = -\frac{4\pi F}{\varepsilon} \sum_i z_i c_i^o \exp\left(\frac{-z_i F \phi_x}{RT}\right) \tag{2.29}$$

- They solved this equation and obtained an expression for the charge on the electrode (which is equal to the total net ionic charge on the solution side), Q

$$Q = \pm \left(\frac{RT\varepsilon}{2\pi}\right)^{\frac{1}{2}} \left\{\sum_i c_i \left[\exp\left(\frac{-z_i F \phi_o}{RT}\right) - 1\right]\right\}^{\frac{1}{2}} \tag{2.30}$$

- where ϕ_o is the potential at $x = 0$ (i.e., at the electrode surface). For a symmetrical electrolyte, where $c_+ = c_- = c$ and $|z_+| = |z_-| = z$, Q is given by

$$Q = \left(\frac{2RT\varepsilon}{\pi}\right)^{\frac{1}{2}} c^{\frac{1}{2}} \sinh\frac{zF\phi_o}{2RT} \tag{2.31}$$

and the differential diffuse double-layer capacitance $C_G = (dQ/d\phi_o)$ is given by

$$C_G = \left(\frac{z^2 F^2 \varepsilon}{2\pi RT}\right)^{\frac{1}{2}} c^{\frac{1}{2}} \cosh\frac{zF\phi_o}{2RT} \tag{2.32}$$

- At $\phi = 0$, that is, at the potential of zero charge, the minimal capacitance is given by

$$C_G(min) = zF \left(\frac{\varepsilon}{2\pi RT}\right)^{\frac{1}{2}} c^{\frac{1}{2}} \tag{2.33}$$

- Gouy–Chapman model is attractive in that it predicts a logical dependence of the double-layer capacitance on the applied potential and on the ionic concentration. However, only in very dilute electrolyte solutions is the measured capacitance in good agreement with the values calculated. At higher concentrations, the measured capacitance is much lower than predicted values.
- **Stern model** combined the Helmholtz and Gouy–Chapman models. He considered the finite distance of closest approach of the ions from the electrode surface, identified as the OHP. The ions located at the OHP are assumed to form a compact double layer, and the potential at this plane is denoted as ϕ_2. The OHP is then taken as $x = 0$ for the Gouy–Chapman model, and thus ϕ decays with x from ϕ_2 toward ϕ_s (the bulk potential), which is defined arbitrarily as $\phi_s = 0$. The total potential drop at the interface is the sum:

$$\phi_M - \phi_S = (\phi_M - \phi_2) + (\phi_2 - \phi_S) \tag{2.34}$$

- where ϕ_M is the potential at the electrode surface. The measured double-layer capacitance, C_d, is given by

$$\frac{1}{C_d} = \frac{1}{C_H} + \frac{1}{C_G} \tag{2.35}$$

- which behaves as the sum of two capacitors connected in series, where C_H is the compact double-layer capacitance and is given by the Helmholtz expression, and C_G is the diffuse double-layer capacitance and is given by the Gouy–Chapman expression (Figure 2.2).

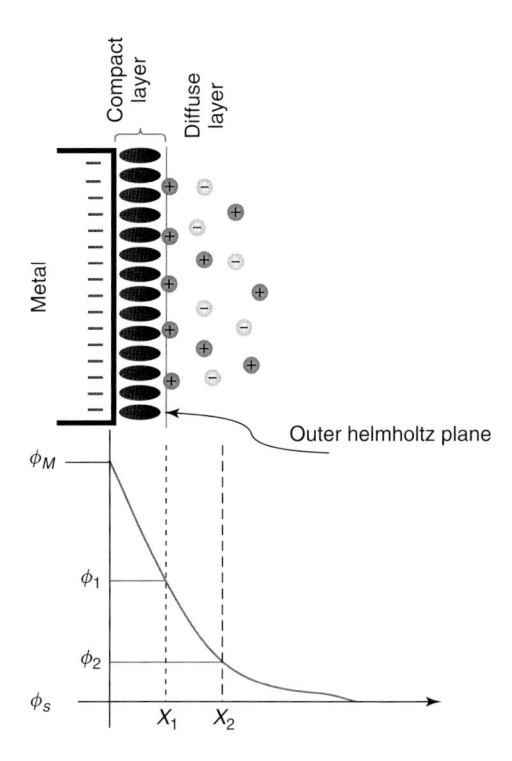

Figure 2.2 Schematic representation of the electric double layer.

2.3
Electrode Kinetics (Charger Transfer (Butler–Volmer Equation) and Mass Transfer (Diffusion Laws))

The following reaction represents two chemical species that are interconnected by a single electron-transfer reaction at the electrode.

$$O_{(aq)} + ne^-_{(m)} \underset{k_{ox}}{\overset{k_{red}}{\rightleftarrows}} R(aq) \tag{2.36}$$

Oxidant (O) and reductant (R) are present in solution, and the first-order heterogeneous rate constants for the forward (reductive) and backward (oxidative) electron-transfer reactions are k_{red} and k_{ox}, respectively. The cathodic current (reductive), i_c, and the anodic current (oxidative), i_a, can be expressed as follows:

$$i_c = -nFAk_{red}[O] \tag{2.37}$$

$$i_a = nFAk_{ox}[R] \tag{2.38}$$

where n is the number of electrons transferred; F, the Faraday constant; A, the surface area; and $k_{red}[O]$ and $k_{ox}[R]$ are the respective fluxes of material to the electrode surface. The net current is given by

$$i = i_a + i_c \tag{2.39}$$

and consequently,

$$i = nFAk_{ox}[R] - nFAk_{red}[O] \tag{2.40}$$

Using transition state theory from chemical kinetics, it is possible to relate the free energies of activation to the rate constants k_{ox} or k_{red}. These are predicted by

$$k = Ae^{\frac{-\Delta G^{\ominus}}{RT}} \tag{2.41}$$

The free energies of activation for the electrode reaction are related to both the chemical properties of the reactants/transition state and the response of both to potential. It is possible to show that the rate constants show the following potential-dependent behavior.

$$k_a = k_o e^{\left[\frac{\alpha_a nF(E - E^{\ominus})}{RT}\right]} \tag{2.42}$$

$$k_c = k_o e^{\left[\frac{-\alpha_c nF(E - E^{\ominus})}{RT}\right]} \tag{2.43}$$

where k_o is the standard reaction rate constant and α_a and α_c are the transfer coefficients for the anodic and cathodic reactions, respectively. $E^{\ominus}$ is the formal potential of the system. At equilibrium, $E - E^{\ominus}$ should equal zero; however, the overpotential (h) is given by $h = E - E^{\ominus}$. For metals, α is approximated to 0.5, which means that the transition state is nearly halfway between reactants and products. For semiconductors, this value can vary significantly.

By using the above approach it is possible to derive the Butler–Volmer equation, which is the fundamental relationship between the current flowing and the applied voltage.

$$i = i_o \left\{ \frac{[R]_o}{[R]_{bulk}} e^{\left(\frac{(1-\alpha)Fh}{RT}\right)} - \frac{[O]_o}{[O]_{bulk}} e^{\left(\frac{-\alpha Fh}{RT}\right)} \right\} \tag{2.44}$$

This is the important Butler–Volmer equation, which is fundamental to electrode kinetics. It predicts how the observed current varies as a function of the overpotential and transfer coefficient, a. If the solution under investigation is well stirred, the surface concentrations of the reactants will be equal to their bulk values, that is, $([R]_\downarrow o = [([R])]_\downarrow bulk$ and $([O]_\downarrow o = [([O])]_\downarrow bulk$. Under these conditions, the Butler–Volmer equation simplifies to

$$i = i_o \left\{ e^{\left(\frac{(1-\alpha)Fh}{RT}\right)} - e^{\left(\frac{-\alpha Fh}{RT}\right)} \right\} \tag{2.45}$$

Without concentration and, therefore, mass transport effects to complicate the electrolysis, it is possible to establish the effects of voltage on the current flowing. In this situation, the quantity $E - E_e$ reflects the activation energy required to force current i to flow.

The equation is valid when the electrode reaction is controlled by electrical charge transfer at the electrode (and not by the mass transfer to or from the electrode surface from or to the bulk electrolyte).

2.3.1
Limitations of Butler–Volmer Equation

There are two limiting cases of the Butler–Volmer equation.

The low overpotential region (called *polarization resistance*), where the Butler–Volmer equation simplifies to

$$i = i_0 \frac{nF}{RT} (E - E_{eq}) \tag{2.46}$$

The high overpotential region, where the Butler–Volmer equation simplifies to the Tafel equation

$$E - E_{eq} = \alpha - b \log(i) \tag{2.47}$$

for a cathodic reaction, or

$$E - E_{eq} = \alpha + b \log(i) \tag{2.48}$$

for an anodic reaction, where a and b are constants (for a given reaction and temperature) and are called the *Tafel equation constants*. It should be noted that the theoretical values of a and b are different for the cathodic and anodic processes.

There is also a region of the limiting current when the electrode process is mass-transfer controlled:

$$i_{limiting} = \frac{nFD}{\delta} C_b \tag{2.49}$$

where D is the diffusion coefficient; δ, the diffusion layer thickness; and C_b is the concentration of the electroactive (limiting) species in the bulk of the electrolyte.

2.4
Porous Electrode Theory (Kinetic and Diffusion)

Research on porous materials has increased considerably because of their wide-ranging applications, for example, catalysis, energy storage, sensing, and gas storage. The term *porous materials* applies to a wide variety of substances, such as porous metals, carbons, metal oxides, and even thin film membranes. These emerging porous materials play an important role in the development of devices such as fuel cells, batteries, and supercapacitors.

In theory, the ideal electrode has a smooth surface of accurately determined surface area, has a known crystal orientation, and is strain free. Practically, electrodes are usually rough and heterogeneous. The need for presenting a large surface area at the electrode–electrolyte interface has resulted in the development of quite porous electrodes.

Basically, the understanding of the behavior at the porous electrode is possible by two approaches. The first considers the porous electrodes as extensions of planar electrodes of known kinetics behavior. This is called the *discrete-pore-model approach*. The second is more successful and is called the *macrohomogeneous model*.

2.4.1
Theories of Porous Electrode

2.4.1.1 The Single-Pore Model

There are two approximations that are basic for the pore model. The first considers the porous rough electrodes as smooth electrodes of enhanced surface area [3]. The second states that the porous electrode is simplified as a parallel array of pores of uniform diameter. This means that the pore is considered one-dimensional, and the resistance of the electrolyte is uniformly distributed along the length. Also, the model assumes that the pore is of uniform cross section and is completely filled with solution. Levie considered the potential drop ($E_0 - E_x$) over the distance x from the mouth of the pore ($x = 0$) within a pore length l. He introduced the following expression:

$$E_x = E_0 \text{Cosh}(\rho x - \rho l)/\text{Cosh}\rho l \tag{2.50}$$

The important constant ρ is defined as $\rho = (R_\Omega/R_D)^{1/2}$, where R_Ω is the ohmic resistance of the solution inside for unit pore length, and R_D is the charge-transfer resistance for unit length. The reciprocal of $\rho(1/\rho)$ has the dimensions of length and is called the *penetration depth*.

On the other hand, the current at the mouth of the pore is given by

$$i_{x=0} = -(R_\Omega)^{-1}(dE/dx)\, x = 0 \tag{2.51}$$

$$= (\rho E_0 \tanh \rho l)/R_\Omega \tag{2.52}$$

If the pore is considered cylindrical with radius a cm, and the solution conductance is $k\ \Omega^{-1}\ \text{cm}^{-1}$, then the value of R_Ω inside the pore is given by

$$1/R_\Omega = \pi a^2 k \tag{2.53}$$

And R_D, expressed in terms of the exchange current density i_0, is given by

$$R_D = RT/2zF\pi a i_0 \tag{2.54}$$

and the penetration depth becomes

$$1/\rho = (akRT/2zFi_0)^{1/2} \tag{2.55}$$

Considering Fick's first law, the current flowing into the porous electrode can be expressed as

$$i_{x=0} = (C_b - C_{x=0})\pi a^2 zFD/\delta \tag{2.56}$$

where C_b is the bulk concentration; $C_{x=0}$, the concentration at the mouth of the pore; δ, the diffusion layer thickness; and D, the diffusion coefficient.

If the electrode front is considered flat, then,

$$i_{x=0} = \rho C_b D' \tanh \rho l / (1 + \rho \delta \tan \rho l) \tag{2.57}$$

where D' is the effective coefficient for diffusion ($D' = \pi a^2 z F D$).

2.4.1.2 The Macrohomogeneous Model

This model was developed by Newman and Tobias. It considers the porous electrode as a superposition of two continua, the porous matrix and the solution filling in all void fraction of the electrode [4]. Both phases are assumed to be homogeneous, isotropic, and spatially complementary. The actual structural detail of the pores was ignored (e.g., distribution and particle size).

The current density in the electrolyte solution is a result of a flux N_i of species i

$$i_z = F \sum_i z_i N_i \tag{2.58}$$

for the material balance

$$\partial \varepsilon C_i / \partial t = a j_{in} - \nabla \cdot N_i \tag{2.59}$$

where ε is the porosity; C_i, the concentration of species i; and j_{in}, the pore wall flux of the species i, averaged over the interfacial area a. For electric neutrality,

$$\sum_i z_i C_i = 0 \tag{2.60}$$

and

$$\nabla \cdot i_1 + \nabla \cdot i_2 = 0 \tag{2.61}$$

where i_1 and i_2 are current densities in the matrix and in the electrolyte in the pores, respectively. Thus, the electrochemistry is expressed by the general relationship

$$\nabla \cdot i_2 = a i_0 \left[\exp\left(\frac{\alpha_a \eta_s F}{RT}\right) - \exp\left(\frac{-\alpha_c \eta_s F}{RT}\right) \right] \tag{2.62}$$

where α_a and α_c are transfer coefficients for the anodic and the cathodic directions, respectively, and η_s is the surface overpotential. The transport processes for the matrix phase are considered using the following equation:

$$i_1 = -\sigma \nabla \emptyset_1 \tag{2.63}$$

where σ and $\emptyset_1$ are the conductivity and electric potential of the matrix, respectively.

For a dilute solution within the pores, the driving forces to move the solutes are diffusion, dispersion, migration, and convection.

$$N_{i/\varepsilon} = -(D_i + D_a) \nabla C_i - z_i u_i F C_i \nabla \emptyset_2 + (v C_{i/\varepsilon}) \tag{2.64}$$

where D_a is the dispersion coefficient; u_i, the mobility of species i; $\emptyset_2$, the electric potential in solution; and v, the fluid velocity.

2.5
Structure, Design, and Fabrication of Electrochemical Devices

An electrochemical cell is a device that produces an electric current from energy released by a spontaneous redox reaction. This device is, basically, composed of two conductive electrodes (the anode and the cathode). The *anode* is defined as the electrode where oxidation occurs, and the *cathode* is the electrode where reduction takes place. Electrodes can be made from any sufficiently conductive material, such as metals, semiconductors, graphite, and even conductive polymers. In between these electrodes is the electrolyte, which contains ions that can freely move. The following are few examples.

- **Battery**: A battery is an electrochemical device where a number of cells are combined, and it is used to supply electrical energy that is stored chemically. In the battery, cells are usually wired in series to increase the supply voltage but are sometimes wired in parallel to allow more current to be supplied. Batteries are optimized to produce a constant electric current for as long as possible. Various battery technologies have been commercialized, including dry cell, lead–acid battery, and lithium rechargeable battery.
- **Lithium rechargeable battery**: It is a solid-state battery that operates using a solid electrolyte. Lithium polymer batteries are an example of this; a graphite bar acts as the anode, a bar of lithium cobaltate acts as the cathode, and a polymer, swollen with a lithium salt, allows the passage of ions and serves as the electrolyte. In this device, the carbon (anode) can reversibly form a lithium–carbon alloy. Upon discharging, lithium ions spontaneously leave the lithium cobaltate cathode and travel through the polymer and into the carbon anode forming the alloy. By charging the device, the lithium dealloys and travels back into the cathode. The advantage of this kind of battery is that lithium possesses the highest negative value of standard reduction potential. It is also a light metal and therefore less mass is required to generate 1 mol of electrons. Lithium ion battery technology is widely used in portable electronic devices because they have high energy storage density and are rechargeable.
- **Fuel cells**: A fuel cell is an electrochemical cell that converts chemical energy from a fuel into electrical energy. The reactants flow into the cell, and the reaction products flow out of it continuously. Electricity is generated from the reaction between the fuel and an oxidizing agent. Various fuels and oxidants are possible. A hydrogen fuel cell uses hydrogen as its fuel and oxygen (or air) as its oxidant. Although fuel cells come in many varieties, they all work in a similar way. They are made of three parts that are assembled together: the anode, the electrolyte, and the cathode. At the anode, a catalyst oxidizes the fuel to positive ions and electrons. Electrons make the current, and the positive ions diffuse through the electrolyte to the cathode where they receive the electrons again along with a third chemical (usually oxygen) to create water or carbon dioxide.

2.5.1
Major Types of Hydrogen–Oxygen Fuel Cells

- **Alkaline fuel cells (AFCs).** The electrolyte is 40–70% KOH. The working temperatures are 60–240 °C. This type is usually used in spacecraft.
- **Polymer electrolyte membrane fuel cells (PEMFCs)** (Figure 2.3). The electrolyte is a polymeric ion exchange membrane (e.g., Nafion). The working temperatures are 60–100 °C.
- **Phosphoric acid fuel cells (PAFCs).** The electrolyte is 85–95% phosphoric acid. The working temperatures are 180–200 °C. Such systems were used to build power plants, with an output power of up to about 250 kW. Some reached up to 4 MW.
- **Molten carbonate fuel cells (MCFCs).** The electrolyte is a molten mixture of the alkali metal (Li, Na, and K) carbonates. The working temperature is about 650 °C. Experimental plants of up to 0.5 MW have been designed.
- **Solid oxide fuel cells (SOFCs).** The electrolyte is a solid electrolyte based on zirconium dioxide doped with oxides of yttrium and other metals. The working temperatures are 800–1000 °C. Experimental plants with up to 100 kW have been designed.

Supercapacitors are highly important in supporting the voltage of a system during increased loads in everything, from portable equipment to electric vehicles. Supercapacitors are similar to batteries in design and manufacture (two electrodes, separator, and electrolyte) but are designed for high power and long cycle life (>100 times battery life), but at the expense of energy density. There are two general categories of electrochemical supercapacitors: electric double-layer capacitors (EDLCs) and redox supercapacitors. In contrast to batteries, where the cycle life is limited because of the repeated contraction and expansion of the electrode on

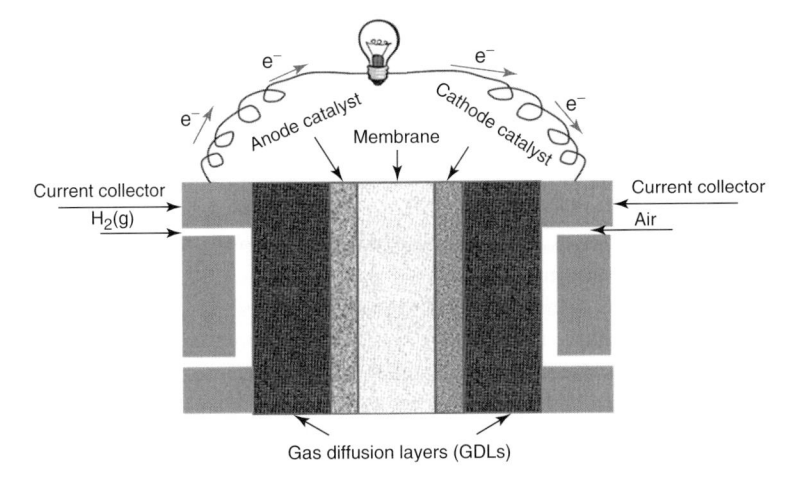

Figure 2.3 Schematic diagram representing polymer electrolyte membrane fuel cell (PEMFC).

cycling, EDLC lifetime is in principle infinite, as it operates solely on electrostatic surface charge accumulation. For redox supercapacitors, some fast faradaic charge transfer takes place as in a battery.

This gives rise to a large pseudocapacitance. Progress in supercapacitor technology can benefit by moving from conventional to nanostructured electrodes. In the case of supercapacitors, the electrode requirements are less demanding than in batteries, at least in terms of electrode compaction, because power prevails over energy density. Thus, the benefits of nanostructures with their high surface area are potentially more important for supercapacitor-based storage sources.

2.5.2
Designing Electrochemical Devices

Electrochemical devices are used for electrochemical generation, or storage, of electrical energy. For designing the device, one must take into account the purpose as well as special features. The most common structure is plane-parallel electrodes in which positive and negative electrodes alternate and all electrodes having the same polarity are connected in parallel. Another method applied for cell or group arrangement is by using bipolar electrodes. One side of the electrode functions as the positive electrode of one cell, and the other side functions as the negative electrode of the next cell. The bipolar electrodes alternate with electrolyte compartments, and both must be sealed carefully along the periphery to prevent outflow of electrolyte and to provide a reliable separation of the electrolyte in neighboring compartments by the electrode plates. The separating plates also function as cell walls and intercell connectors (i.e., the current between neighboring cells merely crosses a thin wall having negligible resistance). This implies considerable savings in the size and mass of the reactor. It is important for the assembly of the device that the blocks of electrodes and separators forming electrolyte compartments are tightened with the aid of end plates and tie bolts and sealed with gaskets.

2.5.3
Some Structural Aspects of Electrochemical Devices

- **The interelectrode gap**: It is the distance between the electrodes, and it determines the relative electrolyte volume. The ohmic losses in the electrolyte increases with the distance between the electrodes. However, when the electrolyte volume is too small, the reactant's concentration changes rapidly.
- **Separators**: They are insulating materials used to separate the electrodes of different polarity. They prevent accidental contact not only between the two electrodes but also between the anolyte and catholyte (the electrolyte volume of the anode and cathode) that is possible without altering the ionic conduction.
- **Scale factor**: When the electrode's area increases and the total current is high, ohmic losses most likely occur within the electrodes. This leads to nonuniform current density distribution. To reduce this effect, special feeders of relatively large cross section are used to reduce the ohmic losses in the electrode. Furthermore,

as the reactor size increases, the difficulty in removing the evolved heat increases. This results in a nonuniform temperature distribution leading to a differentiation of the conditions for the electrochemical reaction in different reactor parts.
- **Selection of corrosion-resistant materials**: The design materials, whether they are metallic or nonmetallic, should have sufficiently high corrosion and chemical resistance. Reference to materials' resistance index should be done to match the suitability of a material to a particular corrosive medium. Corrosion problems are of particular importance, of course, when materials for nonconsumable electrodes (and especially anodes) are selected, which must be stable and at the same time catalytically active.
- **Economic performance**: There are several factors that affect the economics of electrochemical reactors. One factor of particular significance is the current yield. In the field of batteries, it is the ratio between the capacity available on discharge and that theoretically calculated for the amount of active materials present in the battery. The chemical yield is another significant parameter; it is the ratio between the amount of product actually obtained and that which, according to the reaction equation, can theoretically be obtained from a given amount of reactant consumed.

2.5.4
Fabrication Principles of Fuel Cells

The power generators, which are based on fuel cells, consist, generally, of several parts:

- the fuel cell battery or the stack itself;
- vessels for the reactants (hydrogen or methanol, sometimes oxygen);
- special devices controlling the supply of reactants and withdrawal of the reaction products;
- devices for controlling the temperature and humidity in the stack;
- devices for conditioning the power (current and voltage).

In hydrogen-based fuel cells, gas diffusion layer (GDL) is used. It consists of a porous hydrophobic diffusion layer and a catalytically active layer. The GDL is usually made of carbon black and about 35% by mass of polytetrafluoroethylene (PTFE) applied to graphitized cloth. It has a porosity of about 45–60%. In case of the membrane-type fuel cell, the membrane electrode assemblies (MEAs) are used, which consist of a sheet of membrane and the two electrodes (anode and cathode) pressed onto it from either side. The fuel cells are stacked according to a filter-press design. A number of MEAs are combined in series to a block having the desired working voltage. Bipolar plates are arranged between the diffusion layer of the anode in one element and the diffusion layer of the cathode in its neighbor element. They secure electrical contact between the individual elements and separate the gas and electrolyte compartments of neighboring elements. These plates must be impermeable to gases and liquids and must have a good electronic conductivity. The membranes work as the electrolyte for the PEMFCs. It allows the ion diffusion of proton (H^+). The most common membrane is called *Nafion*. It consists of a perfluorinated sulfonic acid polymer having a compact hydrophobic backbone of $-(CF_2)-$ invested with a certain number of hydrophilic groups terminating in $-SO_3H$. Upon proper wetting, the sulfonic acid groups dissociate and secure a relatively high protonic conductivity. A schematic representation of the PEMFC is shown in Figure 2.3.

2.5.5
Fabrication Principles of Batteries

Batteries are electrochemical power sources in which the chemical energy is converted directly to electrical energy. Each battery is composed of one or several galvanic cells. By their principles of functioning, batteries can be classified as follows:

- **Primary (single-discharge) batteries**. This type contains a finite quantity of the reactants participating in the reaction. After complete discharge, this type of battery cannot be used again (i.e., the redox reaction cannot be reversed).
- **Secondary (rechargeable) batteries**. After complete discharge, this type of battery can be recharged again by forcing the electric current through it in the opposite direction. This reverses the redox reaction and regenerates the original reactants. Good rechargeable batteries sustain a large number of such charge–discharge cycles (hundreds or even thousands).

A Li ion battery is a family of rechargeable battery in which lithium ions move from the negative electrode to the positive electrode during discharge and back when charging. Figure 2.4 shows a schematic diagram for a Li ion battery. The three

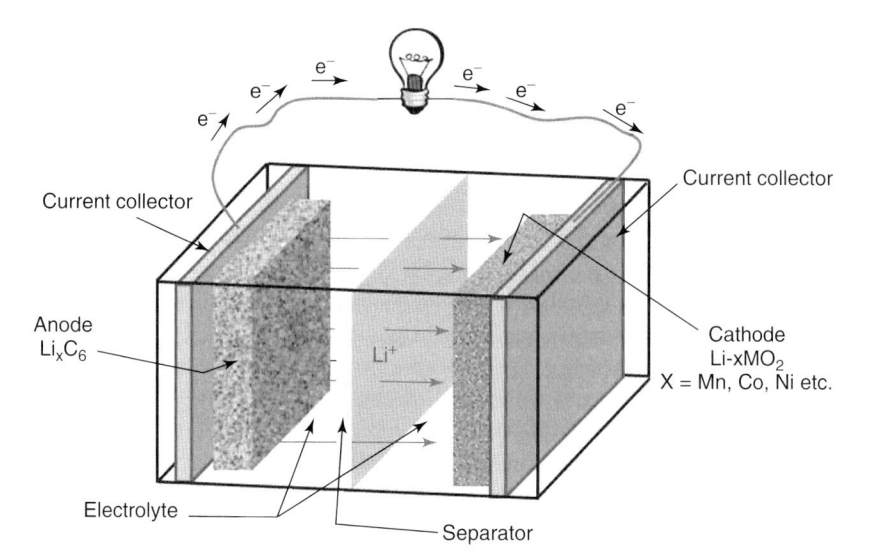

Figure 2.4 Schematic diagram representing Li ion battery.

primary functional components of a lithium ion battery are the anode, cathode, and electrolyte. The anode is generally made of carbon, the cathode is a metal oxide, and the electrolyte is a lithium salt in an organic solvent. The most commercially popular anode material is graphite. The cathode is generally one of the three materials:

- layered oxide, such as lithium cobalt oxide
- polyanion, such as lithium ion phosphate
- spinel, such as lithium manganese oxide.

The electrolyte is typically a mixture of organic carbonates such as ethylene carbonate and diethyl carbonate, containing complexes of lithium ions. These nonaqueous electrolytes generally use noncoordinating anion salts such as lithium hexafluorophosphate ($LiPF_6$), lithium hexafluoroarsenate monohydrate ($LiAsF_6$), lithium perchlorate ($LiClO_4$), lithium tetrafluoroborate ($LiBF_4$), and lithium triflate ($LiCF_3SO_3$). Depending on the choice of material, the voltage, capacity, life cycles, and safety of Li ion batteries can change dramatically. Recently, using nanotechnology has helped to improve the performance.

2.5.6
Fabrication Principles of Supercapacitors

Supercapacitors or what is known as EDLC have a relatively high energy density. The charge is stored physically with no chemical or phase changes taking place, so the process is highly reversible, and the discharge cycle can be repeated virtually without limit. A simple EDLC can be constructed by immersing two electrodes in an electrolyte (for example, two carbon rods in salt water). Figure 2.5 represents a simple schematic illustration of EDLC. Initially, there is no measurable voltage between the rods, but when a current is allowed to flow (by external battery), charge separation at the electrodes' interfaces creates two capacitors that are series connected by the electrolyte. Voltage persists after the circuit is opened, indicating that energy has been stored. The important factor that makes such high capacitance possible is the high porosity of the electrodes. Utilizing breakthrough nanomaterials with excellent properties (e.g., nanoporous carbon, carbon nanotubes (CNTs), graphene, etc.) has enabled fabrication of supercapacitors with excellent capacitance. The operating voltage of an electrochemical capacitor is limited by the breakdown potential of the electrolyte (typically 1–3 V per cell).

Research in EDLCs focuses on improved materials that offer higher *usable* surface areas.

- Graphene has excellent surface area per unit of gravimetric or volumetric densities, is highly conductive, and can now be produced in various laboratories, but it is not available in production quantities. Specific energy densities of 85.6 Wh k^{-1} at room temperature and 136 Wh kg^{-1} at 80 °C (all based on the total electrode weight), measured at a current density of 1 A g^{-1} have been observed.

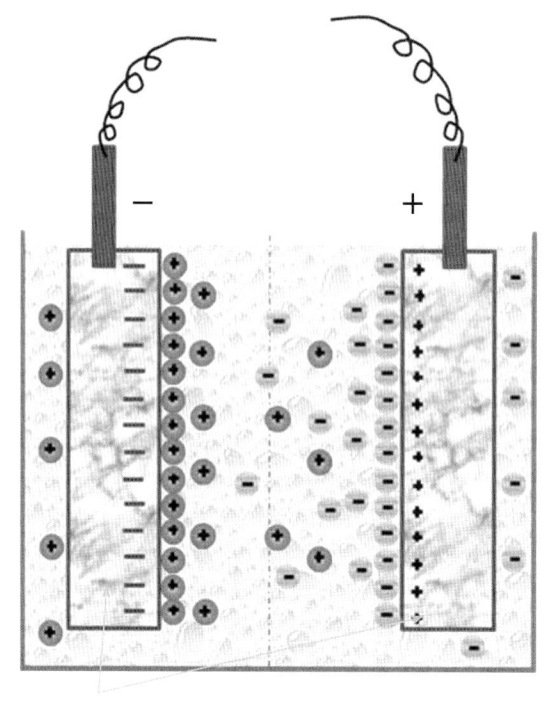

64 | *2 Electrochemical Engineering Fundamentals*

Porous electrodes, for example, porous carbon, graphene, carbon nanotubes, and inorganic material, for example, RuO_2 and $Ni(OH)_2$.

Figure 2.5 Schematic diagram representing double-layer capacitor (supercapacitor).

- CNTs have excellent nanoporosity properties, allowing tiny spaces for the polymer to sit in the tube and act as a dielectric. CNTs can store about the same charge as charcoal (which is almost pure carbon) per unit surface area, but nanotubes can be arranged in a more regular pattern that exposes greater suitable surface area.

2.6
Nanomaterials in Electrochemical Applications

Nanostructured materials have attracted great interest in recent years because of their unusual mechanical, electrical, and optical properties. This is attributed to confining the dimensions of these materials and combining the properties of bulk and surface to the overall behavior. The large number of surface atoms, compared to the bulk, greatly enhances surface activities. The unique surface structure, electronic states, and largely exposed surface area are required for promoting chemical reactions.

2.6.1
Carbon Nanostructures

CNTs, fullerenes, mesoporous carbon structures, and graphene constitute a new class of carbon nanomaterials with properties that differ significantly from other forms of carbon such as graphite and diamond [5–7]. They have unique electrical properties, a wide electrochemical window, and high surface area. Furthermore, the ability to functionalize the nanotubes and graphene or to assemble fullerene (C_{60}) clusters into three-dimensional arrays has opened up new avenues to design high-surface-area catalyst-support materials with high electrochemical activity. Carbon nanostructures possess metallic or semiconductor properties that can induce catalysis by participating directly in the charge-transfer process. This enabled the design of electrode materials for fuel cells, Li ion batteries, and supercapacitors, with high catalytic efficiency. Fullerene clusters, CNTs, and graphene can be deposited as thin films with electrochemical and electrocatalytic activity.

2.6.2
Inorganic Nanostructures

Inorganic nanocrystals [8] comprise inorganic structural units such as nanoparticles, nanorods, nanowires, and nanotubes. Owing to their unique characteristics such as quantum effect, small size effect, and surface effect, inorganic nanocrystals have excellent electronic, optical, magnetic, and catalytic properties, and thus exhibit various applications in electrochemistry.

These nanostructures are proposed to be highly stable and corrosion resistant. Nanotubes, nanoparticles, and nanorods for titanium oxide were prepared. The key issues in considering nanostructure oxide electrode materials are (i) structural and electronic stability of the bulk as well as the surface at the potentials and pH of use and (ii) the bulk and surface electronic conductivity. Examples of oxides that are insoluble at low pH and of high potential are TiO_2, Nb_2O_5, Ta_2O_5, and WO_3. This is based on their Pourbaix diagrams [9]. However, the electronic conductivity of these oxides is still an issue that restricts their application.

Titania and titanate nanostructures in the form of nanoparticles, nanorods, and nanotubes with different morphology and different chemical and physical properties have been prepared. There are three general approaches to the synthesis of TiO_2 and titanate nanotubes, that is, chemical (template) synthesis [10], electrochemical approaches (e.g., anodizing Ti foils or films [11–13]), and the alkaline hydrothermal method [14]. The specific crystal phase of TiO_2 nanotubes and their degree of crystallinity depend on the synthetic routes. The TiO_2 nanotubes produced by templating are usually amorphous or polycrystalline with anatase structure. The crystal structure of electrochemically synthesized polycrystalline TiO_2 nanotubes was reported to correspond to anatase and rutile mixture [11].

2.6.3
Carbon–Inorganic Nanocomposites

CNT-based nanocomposites of inorganic nanocrystals and CNTs can be prepared. For example, TiO_2/CNTs, Co_3O_4/CNTs, Au/CNTs, Au/TiO_2/CNTs, TiO_2/Co_3O_4/CNTs, and Co/CoO/Co_3O_4/CNTs have been prepared [15] by either sonication or simple mechanical stirring via a similar step-by-step self-assembly approach using presynthesized nanoparticles as building units [15]. The nanocrystals are synthesized before assembling with CNTs. Therefore, it is much easier to control both the size and morphology of the nanocrystals because numerous methods have been developed for the preparation of inorganic nanocrystals with controlled size and morphology. However, loading efficiency of nanocrystals onto unmodified CNTs is typically not very high. Functionalization [16] of the CNTs may result in higher loading of nanocrystals on CNTs. The types of functional groups of CNTs play an important role in the interaction with the preformed inorganic nanocrystals. The modification strategy must be carefully designed to avoid excessive interruption to the structure and property of CNTs.

2.6.4
Conducting Polymer Nanostructures

Conducting polymer nanostructures have received increasing attention because of their unique properties. Compared to the bulk, they are expected to display improved performance in energy storage devices [17]. The unique properties arise from high electrical conductivity, large surface area, and improved ion transport. They have been used as electrode materials for lithium ion batteries [18] and supercapacitors [19]. Template synthesis has offered a more efficient route to synthesize well-designed conducting polymer nanostructures with improved electrochemical energy storage [20], for example, preparation of composites of poly(3,4-ethylenedioxythiophene) (PEDOT) and MnO_2 nanowires by a one-step electrochemical codeposition in an AAO (anodic aluminum oxide) template [21]. This nanocomposite could be used as excellent supercapacitor materials with high specific capacitance and the ability to maintain this high capacitance at high current density. Composites of conducting polymers and mesoporous carbon have also been prepared, for example, the preparation of ordered whiskerlike polyaniline on the surface of a mesoporous carbon template and its excellent supercapacitor properties [22].

2.6.5
Application of Nanomaterials Lead to Emerging Devices

In the field of batteries, the introduction of nanomaterials has generated several advantages including [23] (i) better accommodation of the strain of lithium insertion/removal, improving cycle life; (ii) new reactions that are not possible with bulk materials; (iii) higher electrode/electrolyte contact area leading to higher charge–discharge rates; and (iv) short path lengths for electronic transport. For example, for the anode material, introduction of CNTs and graphene provides a specific capacity much higher than that offered by conventional graphite. Silicon-based anodes have, theoretically, the highest capacity for Li. However, the accommodation of so much lithium is accompanied by enormous volume changes. Thus, the mechanical strain generated during the alloying–dealloying processes leads to cracking and crumbling of the electrode and a marked loss of the capacity and durability. The capacity retention was improved by using silicon electrodes prepared with a nanopillar surface morphology [24], because size confinement alters particle deformation and reduces fracturing. For the cathode material, the use of the classical cathode materials, such as $LiCoO_2$ and $LiNiO_2$, in the form of nanomaterials can lead to greater reaction with the electrolyte, and ultimately more safety problems especially at high temperature, than the use of such materials in the micrometer range [23]. Nanostructured materials have also had great impact on fuel cell research. This includes the design of Pt-alloy metal nanocatalysts, the synthesis of carbon and metal oxide nanomaterials as catalyst supports, the development of nonprecious nanocomposite catalysts, and the fabrication of the nanostructured catalyst layer and MEAs. The nanomaterials with enormous surface area, such as nanoporous carbon, CNTs, graphene, and polyaniline–mesoporous carbon, have also resulted in dramatic increase in the performance of the supercapacitors. For example, the specific capacitance of the polyaniline–mesoporous carbon composite was as high as 900 F g^{-1} at a charge–discharge current density of 0.5 A g^{-1} [22]. This was a significant progress in supercapacitor research because the capacitance value was higher than that of amorphous hydrated RuO_2 (840 F g^{-1}), while polyaniline is much cheaper than RuO_2. Moreover, the capacitance retention of this composite was higher than 85% when the charge–discharge current density increased from 0.5 to 5 A g^{-1}. This indicates that this capacitor has high power performance while operating at high charge–discharge rates.

References

1. Ehl, R.G. and Ihde, A.J. (1954) Faraday's electrochemical laws and the determination of equivalent weights. *J. Chem. Educ.*, **31** (5), 226–232.
2. Strong, F.C. (1961) Faraday's laws in one equation. *J. Chem. Educ.*, **38** (2), 98.
3. Levie, R. (1965) The influence of surface roughness of the solid electrodes on electrochemical measurements. *Electrochim. Acta*, **10** (2), 113–130.
4. Newman, J.S. and Tobias, C.W. (1962) Theoretical analysis of current distribution in porous electrodes. *J. Electrochem. Soc.*, **109** (12), 1183–1191.
5. Kamat, P. (2006) Carbon nanomaterials: building blocks in energy conversion devices. *Electrochem. Soc. Interface*, **15** (1), 45–47.
6. Chmiola, J. et al. (2006) Anomalous increase in carbon capacitance at pore sizes less than 1 nanometer. *Science*, **313** (5794), 1760–1763.
7. Gogotsi, Y. et al. (2003) Nanoporous carbide-derived carbon with tunable pore size. *Nat. Mater.*, **2** (9), 591–594.
8. Rao, C.N.R., Müller, A., and Cheetham, A.K. (2004) *The Chemistry of Nanomaterials: Synthesis, Properties and Applications*, John Wiley & Sons, Inc.

9. Pourbaix, M. (1966) *Atlas of Electrochemical Equilibria in Aqueous Solutions*, Pergamon, New York.
10. Liu, S.M. et al. (2002) Synthesis of single-crystalline TiO_2 nanotubes. *Chem. Mater.*, **14** (3), 1391–1397.
11. Balaur, E. et al. (2005) Tailoring the wettability of TiO_2 nanotube layers. *Electrochem. Commun.*, **7** (10), 1066–1070.
12. Hassan, F.M.B. et al. (2010) Functionalization of electrochemically prepared titania nanotubes with Pt for application as catalyst for fuel cells. *J. Power Sources*, **195** (18), 5889–5895.
13. Hassan, F.M.B. et al. (2009) Formation of self-ordered TiO_2 nanotubes by electrochemical anodization of titanium in 2-propanol/NH_4F. *J. Electrochem. Soc.*, **156** (12), K227–K232.
14. Kasuga, T. et al. (1998) Formation of titanium oxide nanotube. *Langmuir*, **14** (12), 3160–3163.
15. Li, J. et al. (2007) Preparation of nanocomposites of metals, metal oxides, and carbon nanotubes via self-assembly. *J. Am. Chem. Soc.*, **129** (30), 9401–9409.
16. Pan, B. et al. (2008) Effects of carbon nanotubes on photoluminescence properties of quantum dots. *J. Phys. Chem. C*, **112** (4), 939–944.
17. Gustafsson, J.C., Liedberg, B., and Inganäs, O. (1994) In situ spectroscopic investigations of electrochromism and ion transport in a poly (3,4-ethylenedioxythiophene) electrode in a solid state electrochemical cell. *Solid State Ionics*, **69** (2), 145–152.
18. Nystrom, G. et al. (2009) Ultrafast all-polymer paper-based batteries. *Nano Lett.*, **9** (10), 3635–3639.
19. Simon, P. and Gogotsi, Y. (2008) Materials for electrochemical capacitors. *Nat. Mater.*, **7** (11), 845–854.
20. Nishizawa, M. et al. (1997) Template synthesis of polypyrrole-coated spinel $LiMn_2O_4$ nanotubules and their properties as cathode active materials for lithium batteries. *J. Electrochem. Soc.*, **144** (6), 1923–1927.
21. Liu, R. and Lee, S.B. (2008) MnO_2/Poly(3,4-ethylenedioxythiophene) coaxial nanowires by one-step co-electrodeposition for electrochemical energy storage. *J. Am. Chem. Soc.*, **130** (10), 2942–2943.
22. Wang, Y.G., Li, H.Q., and Xia, Y.Y. (2006) Ordered whiskerlike polyaniline grown on the surface of mesoporous carbon and its electrochemical capacitance performance. *Adv. Mater.*, **18** (19), 2619–2623.
23. Arico, A.S. et al. (2005) Nanostructured materials for advanced energy conversion and storage devices. *Nat. Mater.*, **4** (5), 366–377.
24. Green, M. et al. (2003) Structured silicon anodes for lithium battery applications. *Electrochem. Solid State Lett.*, **6** (5), A75–A79.

3
Lithium Ion Rechargeable Batteries
Dingguo Xia

3.1
Introduction

Since lithium ion batteries were reported by Sony company, much attention has been attracted because of their potential application from portable electronic devices to vehicles. The first generation of lithium ion batteries using the rocking chair concept is already a commercial success. In its most conventional structure, the carbonaceous material is used in a negative electrode and a lithium-containing compound such as $LiCoO_2$ is applied to a positive electrode. Lithium ions are extracted electrochemically from $LiCoO_2$ during the charge of a cell in aprotic electrolytes and are doped into a carbon negative electrode to form a lithium–carbon alloy. During discharge, lithium ions are extracted from the alloy and doped back into $LiCoO_2$. Electrons are simultaneously extracted from one electrode and injected into another electrode, storing and delivering electrical energy, during which materials are oxidized or reduced in positive and negative electrodes. Lithium ions shuttle between positive and negative electrodes, named lithium ion (shuttlecock, swing, etc.) batteries.

With the development of industry and progress of science, the much higher request is bringing up of lithium ion batteries. Market interest in this cost-effective, high-performance, and safe technology has driven spectacular growth. To meet market demands, an array of designs has been developed, including spiral wound cylindrical, wound prismatic, and flat-plate prismatic designs. Battery performance continues to improve as Li ion batteries are applied to an increasingly diverse range of applications. Li ion batteries offer a low self-discharge rate (2–8% per month), long cycle life (>1000 cycles), and a broad temperature range of operation (charge at 20–60 °C, discharge at 40–65 °C). A wide array of sizes and shapes is now available from a variety of manufacturers. Single cells typically operate in the range of 2.5–4.2 V, approximately three times that of NiCd or NiMH cells, and thus require fewer cells for a battery of a given voltage. The lithium ion batteries currently available in the market range in capacity from 100 mAh to 2.5 Ah for portable applications and up to 200 Ah for motive power and stationary applications. Li ion batteries can offer high rate capability. Discharge at 5C continuous, or 25C pulse, has been demonstrated. The combination of these qualities within a cost-effective, hermetic package has enabled diverse application of the technology. A disadvantage of Li ion batteries is that they degrade when discharged below 2 V and may vent when overcharged, as they do not have a chemical mechanism to manage overcharge, unlike aqueous cell chemistries. Li ion batteries typically employ management circuitry and mechanical disconnect devices to provide protection from overdischarge, overcharge, or high-temperature conditions. Another disadvantage of Li ion products is that they permanently lose capacity at elevated temperatures (65 °C), albeit at a lower rate than most NiCd or NiMH products.

Although a commercial reality, lithium ion batteries are still the object of intense research with the aim of further improving their properties and characteristics. Expected advancements in lithium ion technology include the replacement of carbonaceous materials with alternative low-voltage, Li-accepting anode compounds, with the aim of improving safety characteristics and higher capacity; the replacement of cobalt with nickel or manganese in the cathode structure, with the aim of reducing environmental impact and higher capacity; the replacement of the current electrolyte with a highly stable, wide electrochemical window ion liquid, with the aim of improving the battery's design and reliability.

3.2
Main Types and Structures of Li Ion Rechargeable Batteries

Cylindrical and prismatic Li ion batteries have been developed. Wound designs are typical in small cells (<4 Ah), whereas prismatic configurations with flat-plate construction are more common in large cell designs [1].

3.2.1
Structure of Wound Li Ion Cells

The structure of a cylindrical wound Li ion cell is illustrated in Figure 3.1. The fabrication of wound prismatic cells is similar to cylindrical versions except that a flat mandrel is used instead of a cylindrical mandrel. A schematic diagram of a wound prismatic cell is shown in Figure 3.2. Typically, a single tab at the end of the wind is used to connect the current collectors to their respective terminals. The case, commonly used as the negative terminal, is typically Ni-plated steel. When used as the positive terminal, the case is typically aluminum. Most commercially available cells utilize a header that incorporates disconnect devices, activated by pressure or temperature, such as a PTC device, and a safety vent.

3.2.2
Structure of Flat-Plate Prismatic Li-Ion Batteries

The structure of flat-plate prismatic cells is illustrated in Figure 3.3. As in a wound cell, a microporous polyethylene (PE) or polypropylene (PPE) separator separates

Structure of Lithium–ion Battery

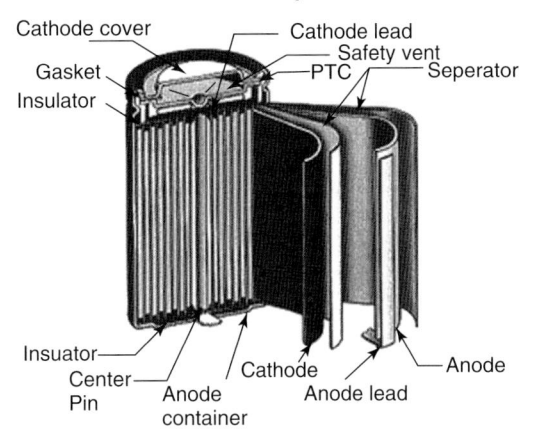

Figure 3.1 Cross-sectional view of a cylindrical Li ion cell.

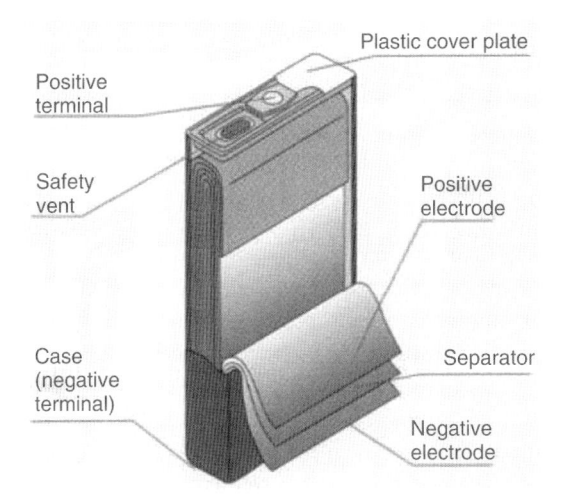

Figure 3.2 Schematic drawing of a wound prismatic cell.

the positive and negative electrodes. Typically, each plate in the cell has a tab, and the tabs are bundled and welded to their respective terminals or to the cell case. Cell cases of either nickel-plated steel or 304L stainless steel have been used. As shown, the cover typically incorporates one or two terminals, a fill port, and a rupture disk. The terminal may be a glass-to-metal seal, for low-cost applications compression type seals have been used, or the terminal may incorporate devices similar to those found in the header of cylindrical products to provide pressure, temperature, and overcurrent interrupt in one component. The case to cover the seal is typically formed either by tungsten inert gas (TIG) or laser welding.

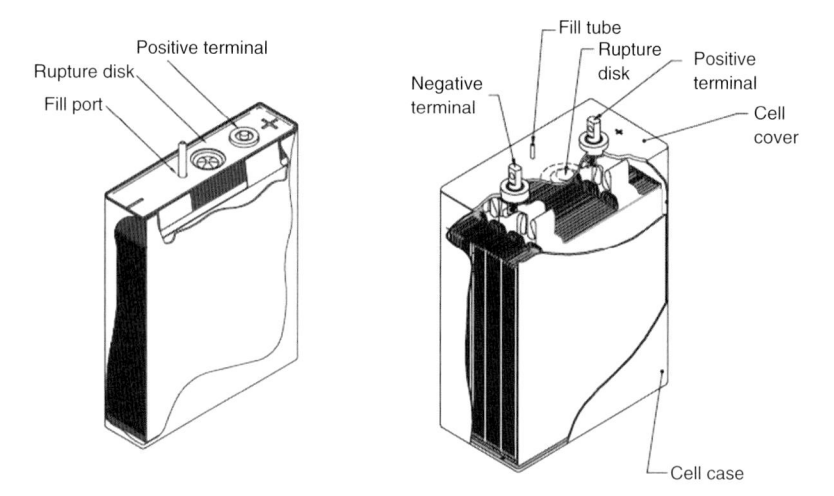

Figure 3.3 Schematic view showing the header and electrodes of 7 Ah (case negative) and 40 Ah (case neutral) flat-plate prismatic Li ion cells.

3.3
Electrochemical Processes in Li Ion Rechargeable Batteries

The three participants in the electrochemical reactions in a lithium ion battery are the anode, cathode, and electrolyte. During discharge, lithium ions, Li^+, carry the current from the negative to the positive electrode, through the nonaqueous electrolyte and separator. During charging, an external electrical power source (the charging circuit) applies a higher voltage (but of the same polarity) than that produced by the battery, forcing the current to pass in the reverse direction. The lithium ions then migrate from the positive to the negative electrode, where they become embedded in the porous electrode material in a process known as *intercalation*. When a Li ion battery is discharged, the positive material is reduced and the negative material is oxidized. The reverse happens on charge. The following equations are in units of moles, making it possible to use the coefficient x. The charge–discharge process in a Li ion battery is further illustrated graphically in Figure 3.4.

The cathode half-reaction is

$$LiMO_2 \underset{\text{discharge}}{\overset{\text{charge}}{\rightleftarrows}} Li_{1-x}MO_2 + xLi^+ + xe^-$$

The anode half-reaction is

$$C + xLi^+ + xe^- \underset{\text{discharge}}{\overset{\text{charge}}{\rightleftarrows}} Li_xC$$

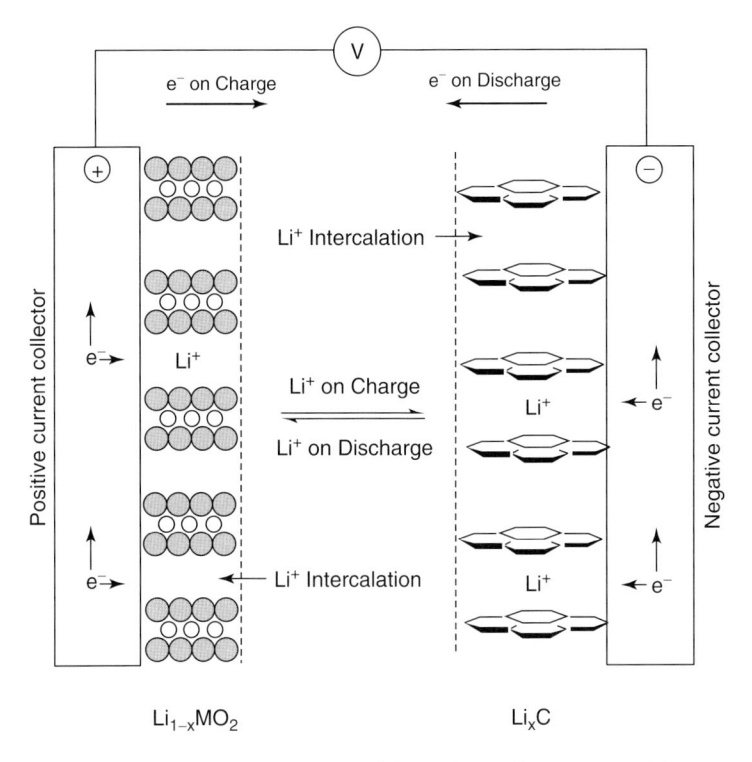

Figure 3.4 Schematic representation of electrochemical process in a lithium ion battery.

The overall reaction is

$$\text{LiMO}_2 + \text{C} \underset{\text{discharge}}{\overset{\text{charge}}{\rightleftarrows}} \text{Li}_x\text{C} + \text{Li}_{1-x}\text{MO}_2$$

3.4
Battery Components (Anode, Cathode, Separator, Endplates, and Current Collector)

The three primary functional components of a lithium ion battery are the anode, cathode, and electrolyte. The most commercially popular anode material is graphite. The cathode is generally one of three materials: a layered oxide (such as lithium cobalt oxide), a polyanion (such as lithium iron phosphate), or a spinel (such as lithium manganese oxide). The electrolyte is typically a mixture of organic carbonates such as ethylene carbonate or diethyl carbonate containing complexes of lithium ions. These nonaqueous electrolytes generally use noncoordinating anion salts such as lithium hexafluorophosphate (LiPF_6), lithium hexafluoroarsenate monohydrate (LiAsF_6), lithium perchlorate (LiClO_4), lithium tetrafluoroborate (LiBF_4), and lithium triflate (LiCF_3SO_3) [2].

3.4.1
Anode

3.4.1.1 Carbon Anode Materials

Researches on Li ion rechargeable batteries were first motivated to conquer the problems that arose from the studies on the development of Li metal batteries.

Li metal is attractive as an anode material for batteries because of its most negative potential and the lowest density. However, it encounters the shortcomings of a Li–liquid electrolyte combination. When Li$^+$ is reduced to metallic Li on the Li metal anode surface during recharge process, it forms a deposit more porous than the original metal (i.e., the dendritic growth), which makes batteries sensitive to thermal, mechanical, and electrical abuse.

Carbonaceous anodes are known to accept Li$^+$ between the graphite layers to form C$_6$Li, which is worth a maximum theoretical capacity of 372 mAh g^{-1}. Especially, carbon anodes offer stable morphology resulting in consistent safety properties over their useful life. By utilizing low–surface area carbons, electrodes with acceptable self-heating rates may be fabricated.

The first Li ion batteries marketed by Sony utilized petroleum coke at the negative electrode. Coke-based materials offer good capacity, 180 mAh g^{-1}, and are stable in the presence of propylene carbonate (PC)-based electrolytes in contrast to graphitic materials. The disorder in coke materials is thought to pin the layers inhibiting reaction or exfoliation in the presence of PC. In the mid-1990s, most Li ion cells utilized electrodes employing graphitic spheres, particularly, a mesocarbon microbead (MCMB) carbon. An MCMB carbon offers higher specific capacity, 300 mAh g^{-1}, and low surface area, thus providing low irreversible capacity and good safety properties. Recently, a wider variety of carbon types has been used in negative electrodes. Some cells utilize natural graphite, available at a very low cost, whereas others utilize hard carbons that offer capacities higher than those possible with graphitic materials.

Many types of carbon materials are industrially available, and the structure of the carbon greatly influences its electrochemical properties, including lithium intercalation capacity and potential. The basic building block for carbon materials is a planar sheet of carbon atoms arranged in a hexagonal array (Figure 3.5). These sheets are stacked in a registered fashion in graphite. In Bernal graphite, the most common type, ABABAB stacking occurs, resulting in hexagonal or 2H graphite. In a less common polymorph, ABCABC stacking occurs, termed *rhombohedral* or *3R graphite*. Performance and physical characteristics of various carbons are included in Table 3.1.

3.4.1.2 Noncarbonaceous Anode Materials

Diverse efforts on the search of novel noncarbonaceous materials are being carried out in hope of realizing innovative enhancement in the anode properties.

Amorphous Sn composite oxide (ATCO) was introduced by Fuji Photo Film Co., Ltd. in 1995. ATCO reacts reversibly with Li at about 0.5 V and has twice the gravimetric capacity advantages over graphite. The high capacity of the ATCO is

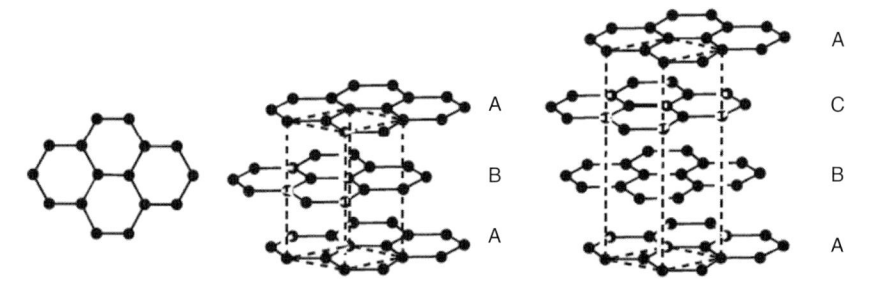

Figure 3.5 The hexagonal structure of a carbon layer and the structures of hexagonal (2H) and rhombohedral (3R) graphite.

Table 3.1 Properties and performance of various carbons (experimental values).

Carbon	Type	Specific capacity (mAh g^{-1})	Irreversible capacity (mAh g^{-1})	Particle size D$_{50}$ (μm)	BET surface area (m^2 g^{-1})
KS6	Synthetic graphite	316	60	6	22
KS15	Synthetic graphite	350	190	15	14
KS44	Synthetic graphite	345	45	44	10
MCMB 25–28	Graphite sphere	305	19	26	0.86
MCMB 10–28	Graphite sphere	290	30	10	2.64
Sterling 2700	Graphitized carbon black	200	152	0.075	30
XP30	Petroleum coke	220	55	45	N/A
Repsol LQNC	Needle coke	234	104	45	6.7
Grasker	Carbon fiber	363	35	23	11
Sugar carbon	Hard carbon	575	215	N/A	40

assigned to the high capacity of Sn that emerged after the formation of Li$_2$O. Sn has a high theoretic specific capacity of 992 mAh g^{-1}. However, this material suffered a huge irreversible capacity loss because of the Li$^+$ consumption in the first cycle to form the Li$_2$O matrix and poor cyclabilty caused by volume changes of electrode. Other metals (M = Al, Si, Ge, Bi, and so on) that can react reversibly with Li$^+$ are faced with the same problem.

Transition metal oxides (MO, M = Co, Ni, Cu, or Fe) as another promising anode material was proposed by Poizot *et al.* in 2000. These materials, containing metal that do not react with Li, can deliver two to three times higher capacity than that of carbon. Such character is realized from the electrochemically driven, *in situ* formation of metal nanoparticles during the initial lithiation process, which enables the substantial formation and decomposition of Li$_2$O on subsequent cycling. However, these materials have often suffered the large initial irreversible capacity loss and poor cyclabilty.

By contrast, Li-Ti-O is attracting attention as a *zero* strain insertion material. The insertion of Li into this cubic spinel electrode is highly reversible, and its lattice constant is almost invariant during charge and discharge. Consequently, this material shows excellent cyclabilty to hundreds of thousands cycles. However, the theoretical capacity is only 175 mAh g^{-1}, which is smaller by half than that of carbonaceous materials.

Other noncarbonaceous materials including nitrides, sulfides, fluorides, and phosphides are also investigated.

3.4.2
Cathode

Positive electrode materials in commercially available Li ion batteries use a lithiated metal oxide as the active material. The first Li ion batteries marketed by Sony used LiCoO$_2$. Goodenough and Mizushima developed this material, as described in a series of patents. Some common cathode materials are included in Table 3.2.

Cathode materials must satisfy a number of requirements. To enable high capacity, materials must incorporate a large amount of lithium. Further, the materials must reversibly exchange lithium with little structural change to permit a long cycle life, high coulombic efficiency, and high energy efficiency. To achieve high cell voltage and high energy density, the lithium-exchange reaction must occur at a high potential relative to lithium. When a cell is charged or discharged, an electron is removed or returned to the positive material. For this process to occur at a high rate, the electronic conductivity and Li$^+$ mobility in the material must be high. Also, the material must be compatible with the other materials in the cell; in particular, it must not be soluble in the electrolyte. Finally, the material must be of acceptable cost. To minimize cost, preparation from inexpensive materials in a low-cost process is preferred.

3.4.2.1 LiCoO$_2$

The most common cathode material for Li ion batteries is the layered structured lithium cobalt oxide, LiCoO$_2$, which in fact is still the most used for commercial

Table 3.2 Some common cathode materials and their properties.

Cathode material	Average voltage (V)	Gravimetric capacity (mAh g^{-1})	Gravimetric energy (kWh kg^{-1})
LiCoO$_2$	3.7	140	0.518
LiMn$_2$O$_4$	4.0	100	0.400
LiNiO$_2$	3.5	180	0.630
LiFePO$_4$	3.3	150	0.495
Li$_2$FePO$_4$F	3.6	115	0.414
LiCo$_{1/3}$Ni$_{1/3}$Mn$_{1/3}$O$_2$	3.6	160	0.576
Li(Li$_a$Ni$_x$Mn$_y$Co$_z$)O$_2$	4.2	220	0.920

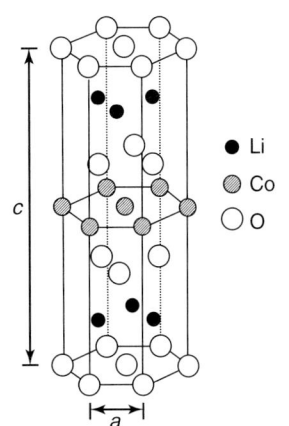

Figure 3.6 Layered LiCoO₂ structure.

production. The structure of $LiCoO_2$ consists of layers of lithium that lie between slabs of octahedra formed by cobalt and oxygen atoms. The crystal structure is denoted Rm in Hermann–Mauguin notation, signifying a rhombuslike unit cell with threefold improper rotational symmetry and a mirror plane. More simply, however, both lithium and cobalt are octahedrally coordinated by oxygen. Each cobalt atom is aligned on a common axis with lithium atoms and separated from each lithium atom by a triangle of oxygen atoms as can be seen in Figure 3.6. The threefold rotational axis is termed improper because the oxygen triangles are antialigned. $LiCoO_2$ is easy to prepare and to cycle with an average capacity of 140 mAh g^{-1}. However, it has various drawbacks, which include cost and environmental risk.

3.4.2.2 LiNiO₂

$LiNiO_2$, another member of the layered lithium metal oxides, potentially offers a higher capacity of 200 mAh g^{-1}. However, it is more difficult to prepare because of its tendency to form Ni-rich, nonstoichiometric phases that affect the cycling capabilities. $LiNiO_2$ also suffers poor thermal stability in its high oxidation state. These drawbacks may be controlled by a partial substitution of Ni by other metals, for example, Mg, Mn, or Al, but despite these progresses, $LiNiO_2$ has not yet reached a large impact in battery manufacture. Interestingly, $LiCoO_2$ and $LiNiO_2$ form complete solid solutions to give rise to a family of rhombohedral layered, structured $LiNi_{(1-y)}Co_yO_2$ compounds, where both Ni and Co are in their oxidation state of III. Thus, the substitution of nickel by cobalt in the $y = 0.2$–1.0 range inhibits the formation of Ni(II) impurities, thus stabilizing the two-dimensional character of the layered structure of $LiNiO_2$. This feature has promoted large attention to the mixed $LiNi_{(1-y)}Co_yO_2$ compounds as improved cathodes in lithium ion batteries. In fact, these compounds are able to offer a stable cyclability on the 180 mAh g^{-1} range. The large replacement of cobalt in the high Ni, $y = 0.2$, range improves the environmental compatibility, and this is an additional bonus for the practical interest of the $LiNi_{0.8}Co_{0.2}O_2$ electrodes.

3.4.2.3 LiMnO$_2$

LiMnO$_2$ is known to exist in several phases. Two phases, whose crystal structures have been well characterized, are a tetragonal spinel-related phase, usually written as Li$_2$Mn$_2$O$_4$, and an orthorhombic phase. The orthorhombic form of LiMnO$_2$ has the advantages of being easy to prepare and air stable, which has a theoretic capacity of 286 mAh g^{-1}. However, the cathodes made with orthorhombic LiMnO$_2$ undergo a phase transition, on cycling, to form the spinel phase Li$_2$Mn$_2$O$_4$, whereas the spinel phase, Li$_{2-x}$Mn$_2$O$_4$, is not air stable at small values of x and is more difficult to make in a fully lithiated form.

Considerable attention has been devoted to the three-dimensional LiMn$_2$O$_4$ spinel. The structure consists of manganese atoms octahedrally coordinated to six oxygen and lithium atoms tetrahedrally coordinated to four oxygen atoms (Figure 3.7). The MnO$_6$ octahedra are colored dark gray, whereas the light gray Lithium ions fill the interstitial sites. There are various reasons for this interest, which include the familiarity of the battery manufacturing industry with manganese compounds, the potential lower cost, and the more friendly impact of Mn on the environment in comparison with Co and Ni. However, LiMn$_2$O$_4$ suffers from some major operational difficulties, such as (i) low specific capacity (i.e., limited to around 120 mAh g^{-1}), which in addition tends to fade on cycling (because of the irreversible structure modifications), and (ii) storage losses especially at high temperature (because of Mn^{2+} dissolution into the electrolyte). Large research efforts have been devoted to improve the electrochemical performance of the LiMn$_2$O$_4$ spinel. The capacity retention has been improved by reinforcing the structure with the modification of the synthesis procedures and with the partial substitution of Mn by several other metals, such as Ni, Cr, and Co.

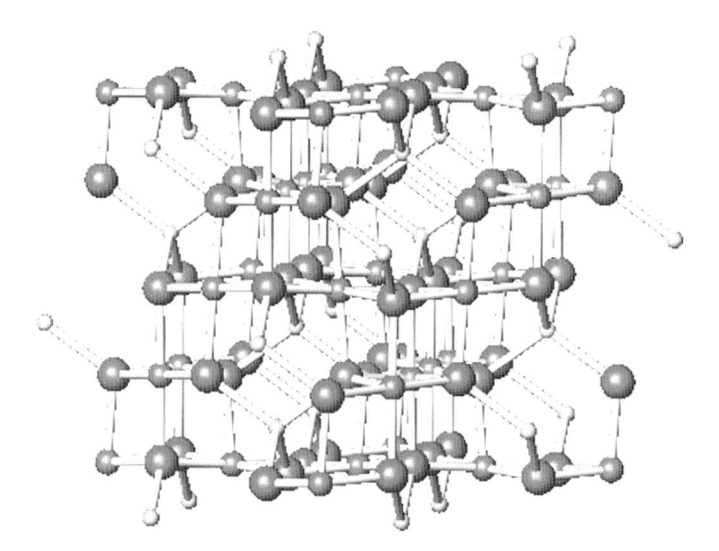

Figure 3.7 Structure of LiMn$_2$O$_4$.

Solid solution materials between layered Li[Li$_{1/3}$Mn$_{2/3}$]O$_2$ (commonly designated as Li$_2$MnO$_3$) and LiMO$_2$ (M = Mn, Ni, and Co etc.) have recently been found to be attractive cathodes for lithium ion batteries as some of them exhibit much higher capacity with lower cost and better safety than the currently used LiCoO$_2$ cathode. For example, layered oxide compositions belonging to the series Li[Li$_{(1/3-2x/3)}$Mn$_{(2/3-x/3)}$Ni$_x$]O$_2$ ($0 \leq x \leq 0.5$), which is a solid solution between Li[Li$_{1/3}$Mn$_{2/3}$]O$_2$ and Li[Mn$_{0.5}$Ni$_{0.5}$]O$_2$, have been found to exhibit capacities as high as 250 mA h/g on cycling them to 4.8 V. The high discharge capacity of these solid solution cathodes is due to the irreversible loss of oxygen from the lattice above 4.5 V during first charge as revealed by *in situ* X-ray diffraction and electrochemical mass spectroscopy, followed by a lowering of the oxidation state of the transition metal ions in the subsequent discharge compared to that in the initial material. During the loss of oxygen (first charge), it has been proposed that the cathode lattice involves diffusion of cations and anions to give a compact layered lattice without any oxygen vacancies, which undergoes lithium insertion/extraction in the subsequent discharge–charge cycles.

3.4.2.4 LiFePO$_4$

Lithium iron phosphate (LiFePO$_4$) was discovered by John Goodenough's research group at the University of Texas in 1997 as a cathode material for rechargeable lithium batteries. The chemical formula of lithium iron phosphate is LiFePO$_4$, in which lithium has +1 valence; iron, +2 valence; and phosphate, −3 valence. The central iron atom together with its surrounding six oxygen atoms forms a corner-shared octahedron (FeO$_6$) with iron in the center. The phosphorus atom of the phosphate forms with the four oxygen atoms an edge-shared tetrahedron (PO$_4$) with phosphorus in the center. A zigzag three-dimensional framework is formed by FeO$_6$ octahedra sharing common O corners with PO$_4$ tetrahedra. Lithium ions reside within the octahedral channels in a zigzag structure. In the lattice, FeO$_6$ octahedra are connected by sharing the corners of the *bc* face. LiO$_6$ groups form a linear chain of edge-shared octahedra parallel to the *b* axis. An FeO$_6$ octahedron shares edges with two LiO$_6$ octahedra and one PO$_4$ tetrahedron. In crystallography, this structure is thought to be the *Pmnb* space group of the orthorhombic crystal system. The lattice constants are $a = 6.008$ A, $b = 10.334$ A, and $c = 4.693$ A. The volume of the unit lattice is 291.4 A^3. The phosphates of the crystal stabilize the whole framework and give lithium iron phosphate good thermal stability and excellent cycling performances (Figure 3.8) [3].

Different from the two traditional cathode materials, LiMnO$_4$ and LiCoO$_2$, lithium ions of LiFePO$_4$ move in the one-dimensional free volume of the lattice. During charge–discharge, the lithium ions are extracted from/inserted into LiFePO$_4$ while the central iron ions are oxidized/reduced. This extraction/insertion process is reversible. LiFePO$_4$ has, in theory, a charge capacity of 170 mAh g^{-1} and a stable open circuit voltage of 3.45 V. The insertion–extraction reaction of the

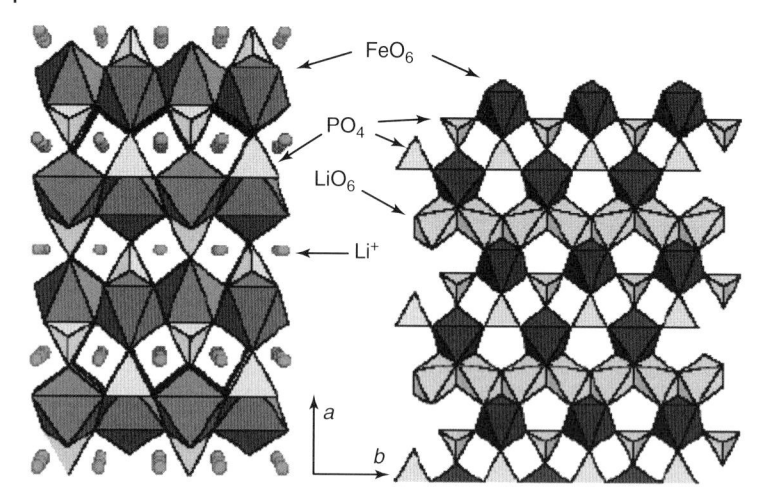

Figure 3.8 Structure of LiFePO$_4$.

lithium ions is expressed as

$$\text{LiFePO}_4 \underset{\text{discharge}}{\overset{\text{charge}}{\rightleftarrows}} \text{FePO}_4 + \text{Li}^+ + e^-$$

Because of its low cost, nontoxicity, high abundance of iron, its excellent thermal stability, safety characteristics, good electrochemical performance, and high specific capacity (170 mAh g^{-1}), it gained some market acceptance. The key barrier to commercialization was its intrinsically low electrical conductivity. This problem, however, was then overcome partly by reducing the particle size and effectively coating the LiFePO$_4$ particles with conductive materials such as carbon and partly by employing the doping approaches.

It is known that LiFePO$_4$ are the very promising cathode materials for large-scale lithium ion batteries, and much attention has been given to improve their basic electrochemical behavior at ambient and high temperatures. Nevertheless, many high-technology applications, such as military and aerospace missions, require Li ion batteries to be capable of operating at low temperatures (e.g. $-20\,°\text{C}$ or even lower) with appropriate energy density and power capability.

3.4.3
Separator

Battery separator is a porous sheet placed between the positive and negative electrodes in a liquid electrolyte, a gel electrolyte, or a molten salt battery. Its function is to prevent physical contact of the positive and negative electrodes while serving as an electrolyte reservoir to enable free ionic transport. According to the structure, the separator can be divided as microporous membrane and nonwoven cloth. The former is a sheet, in which microsize voids are introduced, whereas the latter is a felt or mat, in which fibers are randomly laid down to form numerous voids. The microporous membranes are featured by thinness (about 25 µm or less), small pore size (< 1 µm), and low porosity (~40%). The nonwoven clothes are featured by thickness (80–300 µm), large pore size (10–50 µm), and high porosity (60–80%).

Both natural and synthetic polymers can be used as the separator material. The natural materials mainly are cellulose and its chemically modified derivatives. The synthetic polymers include polyolefins, polyvinylidene fluoride (PVDF), polytetrafluoroethylene, polyamide, polyvinyl alcohol, polyester, polyvinyl chloride, nylon, poly(ethylene terephthalate), and so forth. For specific needs of a battery, the separator or polymer material is often modified, such as by (i) applying a wetting agent to enhance initial wettability of the intrinsically hydrophobic polyolefin separator or (ii) chemically or physically grafting functional groups into polymer chains to increase permanent wettability of the separator or to give the separator special functions. In addition, inorganic ceramics and glasses have been used as the separator material for molten salt batteries, which are normally operated at high temperatures.

The microporous membranes are made either by a dry process or by a wet process. Both processes contain an extrusion step to produce a thin film and employ one or more orientation steps to generate pores. These processes are only applicable to molten or soluble polymers. The dry process generally consists of the following steps: (i) extruding molten polymer to form a film, (ii) annealing the film, and (iii) stretching the film to generate pores. The wet process involves (i) mixing with extractable additives to form a hot polymer solution, (ii) extruding the hot solution to form a gel-like film, and (iii) extracting soluble additives out of the film to form a porous structure. The membranes made by the dry process generally show distinct slit-pore microstructures, whereas those made by the wet process exhibit interconnected spherical or elliptical pores. For the purpose of enhanced safety, two or more layers of membranes with different melting points can be laminated to make a thermal shutdown separator. Owing to the relatively high cost, the microporous membranes are mainly used in high-energy-density primary batteries and secondary batteries.

The nonwoven membranes can be made by the dry-laid, wet-laid, spunbond, or melt-blown process. All these processes consist of three steps: (i) making fabric webs, (ii) bonding webs, and (iii) posttreatment, and in most cases, the web making and bonding are done in one step. Among these processes, the wet-laid process has been used the widest for the manufacture of battery separators. Currently, the nonwoven separators are widely used in primary and secondary alkaline batteries, as well as in polymer gel lithium and lithium ion batteries.

The separator itself does not participate in any electrochemical reaction; however, its properties affect energy density, power density, cycle life, and safety of the battery. Selection of the material and the type of a separator depends on chemistry and application of the batteries. Essentially, the separator must be stable against the battery environments. Compared with primary batteries, secondary batteries have more strict requirements in chemical and electrochemical stability to ensure long cycle life and in mechanical strength to prevent dendrite growth. For long shelf life and operation period, the separator should be able to minimize self-discharge for a primary battery and to prevent penetration of metal dendrites formed during charging for a secondary battery. For high energy and power densities, the separator should be very thin, highly porous, and capable of withstanding high temperature incurred by fast discharge [4–6].

Li ion batteries use thin, microporous films to electrically isolate the positive and negative electrodes. To date, all commercially available liquid electrolyte cells use microporous polyolefin materials owing to their excellent mechanical properties, chemical stability, and acceptable cost. Nonwoven materials have also been developed but have not been widely accepted, in part because of the difficulty in fabricating thin materials with uniform, high strength. Requirements for Li ion separators include the following:

- should have high machine direction strength to permit automated winding;
- should not yield or shrink in width;
- should resist puncture by electrode materials;
- should have an effective pore size of < 1 μm;
- should be easily wetted by electrolyte;
- should be compatible and stable in contact with electrolyte and electrode materials.

Microporous polyolefin materials in current use are made of PE and PPE or laminates of PE and PPE. Also available are surfactant-coated materials designed to offer improved wetting by the electrolyte. These materials are fabricated by either a dry extrusion-type process or a wet solvent-based process.

The low melting point of PE materials enables their use as a thermal fuse. As the temperature approaches the melting point of the polymer, which is 135 °C for polyethylene and 165 °C for PPE, porosity is lost. Tri-layer materials (PPE/PE/PPE) have been developed where a PPE layer is designed to maintain the integrity of the film, while the low melting point of PE layers is intended to shutdown the cell if an overtemperature condition is reached [1].

3.4.4
Current Collector

A current collector is the key component of electron conduction between the electrode materials and the external circuit in lithium ion batteries. As the name implies, the function of current collector is assembling up current from active materials to form a large current output, which should have a low resistance.

A current collector may be considered as an inactive mass and volume in the battery. Therefore, its usage reduces the gravimetric and volumetric energy densities of the system. Hence, thinner and lighter foils would be preferred. However, thinner foils tend to be more expensive depending on the method of preparation and harder to work with because of easy creasing and tearing. Thinner foils also have higher resistance, although placing tabs at regular intervals along a

Figure 3.9 Current collector of Cu and Al.

wound electrode can help offset this problem. Each tab is connected to the battery terminal, thereby reducing the length of electronic pathways.

Current collectors for lithium-based cells are typically in the form of thin foils. Meshes, etched, and coated foils have been used among others to improve adhesion of the electrode material. In addition, microstructured current collectors made from metallized woven polymer fabrics have been investigated to produce flexible electrodes, reduce the amount of additional conductivity additives, and improve handling (Figure 3.9).

Compared with other metals such as Cu, Fe, Ni, or Ti, Al is commonly considered to be the most suitable current collector material for positive electrode in lithium ion batteries on account of its high mechanical strength and excellent ductility, low density, good electrical and thermal conductivity, and low cost. However, it may be subject to corrosion during the charge–discharge cycles, which shorten the lifetime and worsen the safety of lithium ion batteries.

Thermodynamically, Al would be expected to be unstable under highly oxidizing conditions, and the standard redox potential for Al/Al^{3+} is $-1.676\,V$ versus the standard hydrogen electrode in acidic aqueous solution. However, the corrosion stability of Al is well investigated in aqueous solutions, where Al has effective

Table 3.3 Elements that have been examined as positive current collectors for lithium-based cells (shaded boxes).

1	2	3	4	5	6	7	8	9	10	11	12	13	14	15	16	17	18
H																	He
Li	Be											B	C	N	O	F	Ne
Na	Mg											Al	Si	P	S	Cl	Ar
K	Ca	Sc	Ti	V	Cr	Mn	Fe	Co	Ni	Cu	Zn	Ga	Ge	As	Se	Br	Kr
Rb	Sr	Y	Zr	Nb	Mo	Tc	Ru	Rh	Pd	Ag	Cd	In	Sn	Sb	Te	I	Xe
Cs	Ba	Lu	Hf	Ta	W	Re	Os	Ir	Pt	Au	Hg	Tl	Pb	Bi	Po	At	Rn
Fr	Ra	Lr															

corrosion protection from pH 5 to 8 in aqueous environments but is subject to corrosion at high or low pH [7].

Lithium will intercalate into Al at a low potential, so Cu is chosen as the current collector of anode. Two common copper current collectors, electrolytic copper foil and rolled copper foil, are widely used in industry. Copper foam, nanoarchitectured Cu, and nodule-type Cu substrate as current collectors have also been investigated (Table 3.3).

3.5
Assembly, Stacking, and Manufacturing of Li Ion Rechargeable Batteries

Li ion battery is composed of cathode, anode, separator, electrolyte, and shell. The usual process of manufacture of Li ion batteries includes electrodes manufacturing, assembly, and formation. Scheme 3.1 depicts the whole process.

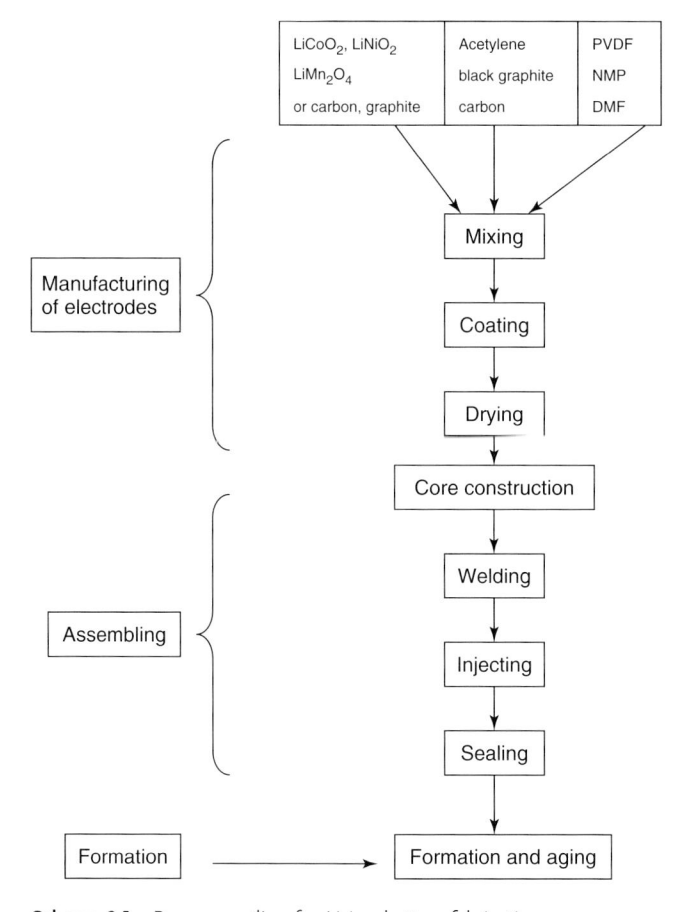

Scheme 3.1 Process outline for Li ion battery fabrication.

3.5.1 Electrode Manufacturing

This process mainly includes preparing electrode materials, mixing (dry blending and wet mixing), coating, and drying.

3.5.1.1 Cathode Materials

Positive active substances usually include $LiCoO_2$, $LiNiO_2$, and $LiMn_2O_4$. Conductive materials are acetylene black or graphite. Binder is PVDF and solvent is NMP. The conductive matrix is aluminum foil of about 25 µm.

3.5.1.2 Anode Materials

Carbon materials are often used as negative active substances, such as MCMB, natural graphite, petroleum coke, asphalt, and carbon fiber. Binder is PVDF, CMC, or SBR latex, and solvent is NMP or water. Conductive matrix is generally a copper foil of 18 µm.

3.5.1.3 Mixing

Before coating operations, it is very important to mix the active materials, polymer, binder, and conductive diluent. The objective of the dry blend is to coat the nonconductive particles of the active material with a thin film of the conductive carbon. Dry mixing often uses ball mill, grinding ball, glass ball, or ceramic balls. Then the polymer is dissolved in the coating solvent in a separate container. The dry mix blend and the solvent solution are then combined to form slurry. Wet mixing uses a planet ball. An intensive mixing procedure is necessary to dry blend fully the nonconducting active material and the carbon before adding the coating solvent and binder. When mixing the cathode, it is essential to coat the nonconductive active material with the conductive diluent, usually carbon black, to ensure that each particle in the porous structure is coated and can connect through the conductive matrix to the current collector. The solvent additions are used to adjust the viscosity of the slurry (or paint) for the coating operation.

3.5.1.4 Coating

The coating process is a critical element ensuring a product of high capacity, high reliability. The slurry from the mixing operation is placed in sealed containers, which serve as the reservoir and transfer medium for the coating operations [8]. Precise amounts of coating slurry are pumped from the storage container with a gear pump, or similar precision pump, to avoid any entrainment of air in the fluid going to the coating head. Common types of coating methods include extrusion, reverse roll, and knife over roll or doctor blade. Both the slot and reverse roll coating can easily handle the viscosity and coating speeds of the anode- and cathode-coating slurries. A common process is the reverse roll coating. In this process, the reverse rolls are raised and coating is interrupted while the foil continues to move. The slot die can perform the same operations with equal precision. The automated manufacturing process requires interrupted coating with precision lengths of the coating area. Both the slot die and reverse roll coating heads can handle the anode- and cathode-coating slurries. Each has sophisticated electromechanical technology for excellent control of the coating thickness across and down the web. Regardless of the choice of coating method, the coated foil passes through an oven to evaporate the solvent and leave a precise amount of active mass on the foil.

It is possible to coat both sides simultaneously, but this is seldom done in practice, as it requires exceptionally good process control. Usually, the electrode stock is passed through the coating operation twice to coat each side of the current collector foil. The coated foils are then calendared to compact the coated layer. Calendaring usually takes two passes to compress the final coating into uniform thickness. Two passes provide better control of pore size and thickness in the final product. The electrode foils are then slit to size for cell assembly.

3.5.1.5 Drying

Regardless of the choice of coating method, the coated foil passes through an oven to evaporate the solvent and leave a precise amount of active mass on the foil.

3.5.2
Assembling Process

3.5.2.1 Core Construction

Both cylindrical and prismatic Li ion batteries use the wound core construction. The strips have the required loading of active material length, width, and thickness to match the cell designs. The coating operation produces interrupted coating to match the length of the coil. The winding machines are designed to operate automatically using strips of anode, cathode, and separator fed from separate reels. The operations start by welding the tabs onto the uncoated section of the foils. The winding machine then cuts the strip to the proper length and winds the combination anode–cathode separator into a tight coil or bobbin like a jelly roll. As the wound core increases in diameter, the winding machine automatically compensates to maintain constant tension as the coil increases in diameter for close tolerance on the diameter. After winding, the coil is checked for internal shorts before being inserted into the can. The bobbin is inserted into the can so that the can provides constant pressure to hold the anode and cathode close together and eliminate any voids between them. All operations are carried out in a dry room or dry box to remove the absorbed water in the active materials before the electrolyte filling process.

3.5.2.2 Welding

Welding is also a very important element in the whole assembly process. After the optimization of welding process, it can decrease and prevent intermetallic formation and growth. Otherwise, these compounds could easily lead to weld embrittlement and some welding materials may crack.

Pulse Nd:YAG laser can be used for termination welding to connect copper foil and aluminum. Because of the different nature of molten material, the size of molten area will change greatly, and only the central parts of a laser beam can permeate the copper. This kind of welding method could produce strong weld strength and durability and has been tested by tensile shear strength and fatigue test.

3.5.2.3 Injecting
Electrolyte is metered into the cell by a precision pump and vacuum filled to ensure that the electrolyte permeates and completely fills the available porosity in the separator and electrode structures.

3.5.2.4 Sealing
After filling the cell with electrolyte, the cell is sealed by controlled compression of a polymer gasket or grommet placed between the cell can and the top plate. Typically, a shoulder or ledge is formed near the top of the cell. This serves as the base for the seal, holds the jelly roll in place, and prevents telescoping or changing the position of the wound bobbin under the influence of vibration and shock. The cell top plate seal contains a vent and PTC and CID safety devices. Both CID and PTC are safety devices designed to activate and prevent dangerous temperatures and pressures from developing internally in the cell. Each lot of devices is checked for proper operation before incorporation into the top assembly. After the seal is applied, the cells can be washed, jacketed, and labeled. They are numbered to trace the day of manufacture and to identify all cell components.

3.5.3 Formation and Aging

After washing and jacketing, it is necessary to measure the voltage and impedance of all cells to sort out any shorted cells. The cells are then charged for the first time. The process is called *formation*. The first charge is important for the following two reasons: first, the solid electrolyte interphase (SEI) layer it forms on the anode and could protect reacting spontaneously with the electrolyte during normal cell operation and second, it ensures a good electrical contact between the active materials and the electrolyte. Cells may continue cycling within the voltage limits for charge and discharge for one or two cycles after formation. After formation or cycling, cell voltage and capacity are measured and stored for later use in the cell selection process. The aging period varies between two weeks and one month depending on the manufacturer.

Here, the cost distribution of components of Li ion battery is also discussed (Figure 3.10). As it is known that Li ion battery is mainly composed of cathode, anode, separator, and electrolyte cathode material is the most significant part in the total cost constitution of Li ion battery, accounting for more than 40%. Anode materials are referred to *graphite* and *graphite fiber materials*. Because production technical barriers are lower, the cost of anode materials in batteries takes up only 5–15% or so. Electrolyte of lithium ion battery is one of the four main components, working as the medium of Li ion transfer between anode and cathode. It has an

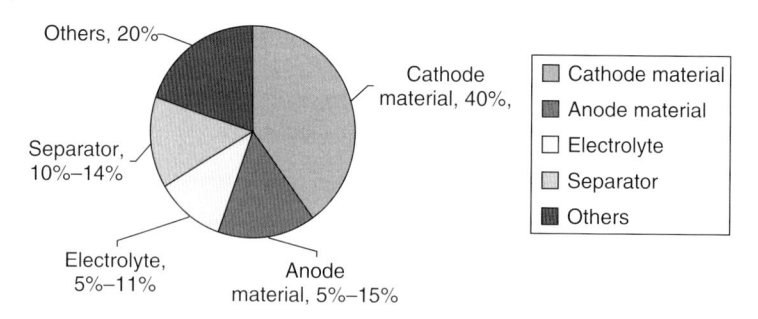

Figure 3.10 Cost distribution of components of Li ion battery.

important influence on Li ion battery capacity, working temperature, circulation efficiency, and security. The cost percentage of electrolyte in lithium battery is 5–11%. The separator, called *third pole* in lithium ion battery, also plays an important role, accounting for more than 10–14% of the total cost.

3.6
Li Ion Battery Performance, Testing, and Diagnosis

Li ion batteries have many advantages, including high voltage, typically in the range of 2.5–4.2 V, high specific energy and energy density (e.g., over 150 Wh kg^{-1} and over 400 Wh l^{-1}, respectively), high rate capability, high power density, and low self-discharge rate, low years of calendar life, no memory effect, and a broad temperature range of operation. Li ion batteries can be charged from 20 to 60 °C and discharged from −40 to 65 °C. All these excellent performances make Li ion batteries active in many fields. The general performance characteristics of Li ion batteries are outlined in Table 3.4.

3.6.1
Characteristics of Energy

The battery energy is related to the specific energy of cathode and anode materials.

3.6.1.1 Specific Energy of Cathode Materials
Currently, the positive electrode materials of lithium ion battery are mainly lithium transition metal oxides. To develop their excellent performance such as high capacity, researchers adopt several measures, such as doping or modifying. Zou et al. [9, 10] prepared LiCoO$_2$ by doping nickel, in which the reversible capacity can reach 156 mAh g^{-1}.

3.6.1.2 Specific Energy of Anode Materials
The first generation of anode materials for lithium ion batteries was petroleum coke. The battery energy and energy density were 80 Wh kg^{-1} and 200 Wh l^{-1},

Table 3.4 General performance characteristics of Li ion batteries [1].

Characteristic	Performance range
Operational cell voltage	4.2–2.5 V
Specific energy	100–158 Wh kg^{-1}
Energy density	245–430 Wh l^{-1}
Continuous rate capability	Typical: 1C
	High rate: 5C
Pulse rate capability	Up to 25C
Cycle life at 100% DOD	Typically 3000
Cycle life at 20–40% DOD	Over 20 000
Calendar life	Over 5 yr
Self-discharge rate	2–10%/month
Operable temperature range	−40 to 65 °C
Memory effect	None
Power density	2000–3000 W l^{-1}
Specific power	700–1300 W kg^{-1}

Abbreviation: DOD, depth of discharge.

respectively. The second generation product adopts hard carbon as anode. The energy and energy density were up to 85 Wh kg^{-1} and 220 Wh l^{-1}, respectively. The third generation of products use graphite as anode, recently. The energy and energy density reached 165 Wh kg^{-1} and 400 Wh l^{-1}, respectively.

Besides carbon materials, polymer arsenic semiconductor (PAS) materials, silicon oxide, tin, and metal nitride are receiving more and more attention because of their high lithium storage capacity. Huang *et al.* [11] prepared Co_3O_4, CoB_xO_z, and $CoB_xAl_yO_z$ anode materials, and the reversible capacity could reach 700–900 mAh g^{-1}.

3.6.2
Working Characteristics

3.6.2.1 Discharge Rate Capability

The rate capability and capacity of Li ion batteries are dependent on the electrode materials and their design and vary considerably. Discharge curves for 18650-type C/LiCoO$_2$ batteries discharged at rates from 0.33 to 3.6 A at 21 °C are shown in Figure 3.11. At low rates (C 2), the battery provided 1.67 Ah, and at 1.65 A, (the 1C rate) 1.60 Ah.

Another example is LiNi$_{0.5}$Mn$_{1.5}$O$_4$ [12]. When it is used as cathode active material in a lithium battery, it delivers a high initial capacity of 134.5 mAh g^{-1} (corresponding to 91.7% of the theoretical capacity) and high rate discharge capability as in Figure 3.12, for example, 133.4, 120.6, 111.4, 103.2, and 99.3 mAh g^{-1} as discharged at 0.5, 1, 5, 10, and 15C rates, respectively. It also shows satisfactory

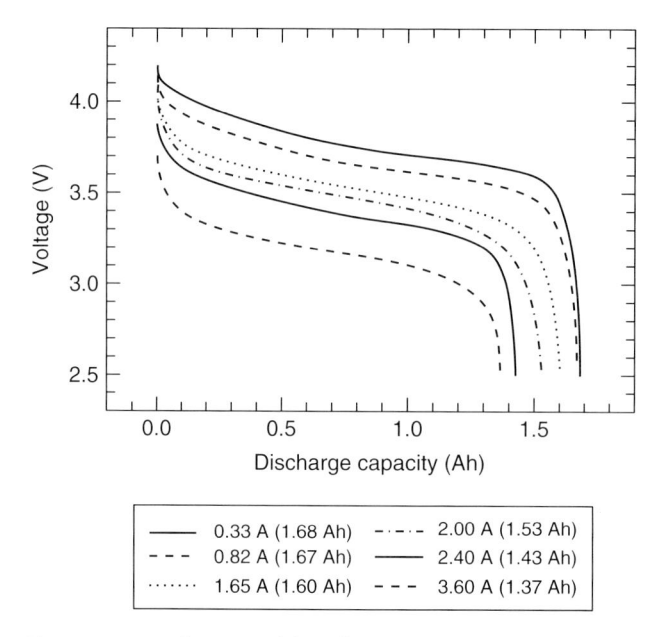

Figure 3.11 Discharge capability of 18 650-type C/LiCoO$_2$ batteries at constant current at 21–25 °C [1]. (Courtesy of the University of South Carolina and NEC Moli Energy.)

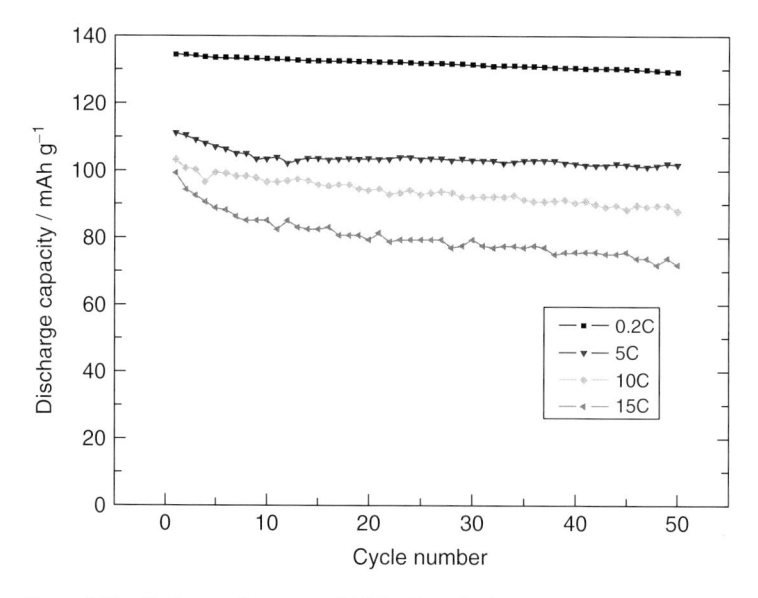

Figure 3.12 Cycling performance of LiNi$_{0.5}$Mn$_{1.5}$O$_4$ heat treated at 850 °C. The cells were charged at the rate of 0.2C and discharged at different rate in voltage range of 3.5–5 V.

capacity retention even at high rate of 5C, which is about 99.83% of the capacity retention per cycle.

3.6.2.2 Cycle Life

A desirable characteristic of Li ion batteries is their long cycle life and ability to cycle continuously at ambient, low, or elevated temperature. Take the commercial product 18650-type $C/LiMn_2O_4$ battery as an example. The capacity of 18650-type $C/LiMn_2O_4$ batteries cycled at 21 or 45 °C at the 1, 0.5, or 0.2C rates is illustrated in Figure 3.13. At 21 °C and 1C rate, after 75 cycles, the fade rate was linear at 0.42% per cycle, and at the 0.5C rate, 0.32% per cycle. After 500 cycles at the 1C rate, the battery delivered 1.01 Ah, or 70% of the initial capacity. For comparison, the cell cycled at the 0.5C rate at 21 °C delivered 1.16 Ah after 500 cycles, or 80% of its initial capacity. Higher fade rates are observed at 45 °C.

For the commercial lithium ion batteries Sony US18650S cells with a nominal capacity of 900 mAh, the discharge capacities charged by the two charging protocols as a function of the number of cycles are shown in Figure 3.14. All the initial discharge capacities of the lithium ion batteries are higher than the nominal capacity of the batteries (900 mAh). Compared with the discharge capacity of the battery charged by DC charging process at the 1C charge–discharge rate, the battery charged by pulse charging shows higher discharge capacity at the same charge–discharge rate. Furthermore, the battery charged by pulse charging process at the 0.5C charge rate followed by 1C rate discharge has the highest discharge capacity. Although exhibiting a higher capacity fade rate during the initial cycles, the capacity fade rate of the batteries charged by pulse charging is more stable than DC charging. The full cycling tests show that 1600 cycles will be completed before the discharge capacity of the battery charged by pulse

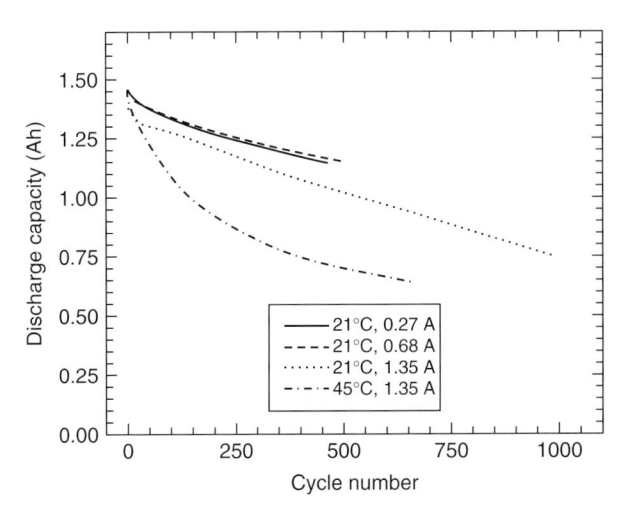

Figure 3.13 Capacity fade of 18 650-type $C/LiMn_2O_4$ batteries when cycled at various rates at either 21 or 45 °C between 4.2 and 2.5 V [1]. (Courtesy of NEC Moli Energy.)

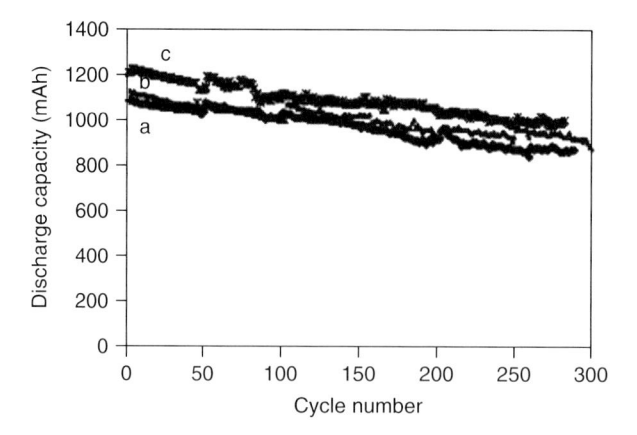

Figure 3.14 Cycling performance of lithium ion batteries using different protocols: (a) DC charging 1C charge–discharge rate, (b) pulse charging 1C charge–discharge rate, and (c) pulse charging 0.5C charge and 1C discharge [13].

charging decreased to 700 mAh, only about 700 cycles brought the DC-charged batteries to this same capacity. It is clear that the higher discharge capacity of the battery charged by pulse charging indicates that the pulse protocol can fully utilize the active materials in the battery without overcharging and sacrificing cycle life.

3.6.2.3 Storage Performance

Because the storage performance of the Li ion batteries has a close relationship with the material production and manufacturing process of the electrodes, it is of vital significance to study the storage properties of the electrode materials and the batteries. Two criteria are often used to characterize the storage performance of batteries: self-discharge rate and capacity recovery after storage. The capacity of a battery may be determined by discharge before storage. After storage in the charged state, discharge enables determination of self-discharge, and subsequent charge and discharge enable determination of capacity recovery, also termed *reversible capacity*. These effects are illustrated by the discharge curves in Figure 3.15. As shown, the voltage is minimally reduced (0.1 V) by extended storage (six months) at 20 °C. While a self-discharge of 10% occurred over six months, after the battery was charged, 97% of the original capacity was delivered.

The cathode material $Li[Ni_xLi_{1/3-2x/3}Mn_{2/3-x/3}]O_2$ was stored in air for one, three, and five months, and the storage performance between 2.0 and 4.8 V at 40 mA g^{-1} current density was tested as illustrated in Figure 3.16 [14]. The newly prepared sample had good electrochemical performance. The reversible capacity reached 200 mAh g^{-1}, and the cycle performance was also excellent. After storing for a month, the first discharge capacity faded sharply, and the discharge capacity was

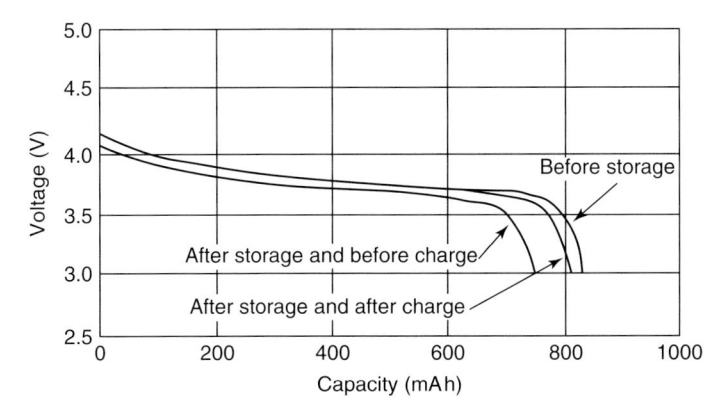

Figure 3.15 Discharge of a 17500-type cylindrical C/LiCoO$_2$ battery before and after six months storage at 20 °C in a fully discharged state. Charge conditions: constant voltage/constant current, 4.2 V, 550 mA (max), 2 h, 20 °C. Discharge conditions: constant current 156 mA to 3.0 V at 20 °C [1]. (Courtesy of Panasonic.)

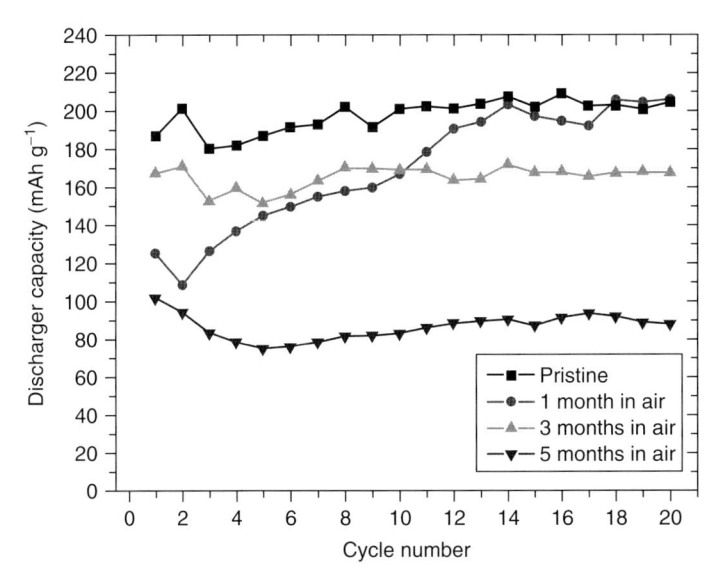

Figure 3.16 Storage performance between 2.0 and 4.8 V at 40 mA g^{-1} current density of Li[Ni$_x$Li$_{1/3-2x/3}$Mn$_{2/3-x/3}$]O$_2$, $x = 1/5$.

only 128 mAh g^{-1}. In the subsequent circle, the capacity increased slowly. After 15 cycles, the capacity reached the level before the storage. Then, as the storage time increased, the capacity became lower. Five months after storage, the discharge capacity was stable at 80 mAh g^{-1}, which was only 40% of the newly prepared sample.

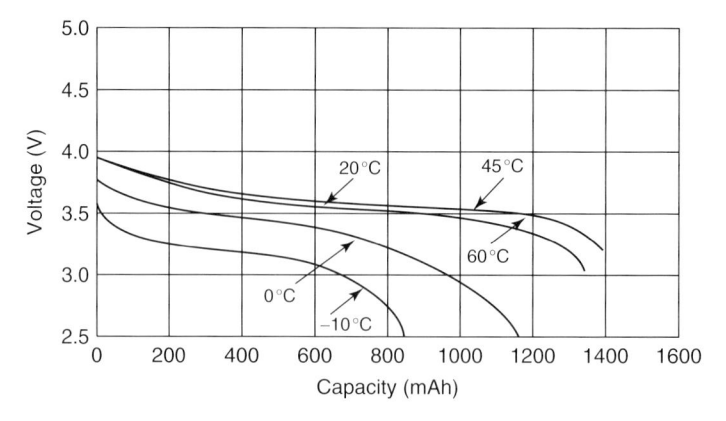

Figure 3.17 High- and low-temperature discharges of a 1.4 Ah CGP345010 C/LiCoO$_2$ wound prismatic battery at 1350 mA. Charged at 945 mA to 4.2 V followed by taper charge for 2 h total at 20 °C [1]. (Courtesy of Panasonic.)

3.6.2.4 Temperature Effects on Performance

It is known that temperature could greatly influence the performance of the batteries. Much progress has been made in improving the performance of Li ion cells at temperature extremes since the last major review in 1994 [15]. The chemistry of liquid electrolytes and their interaction with anode and cathode materials are complex, and methods of improving stabilities over a wide temperature range is a delicate balance of various properties. For example, to ensure high conductivities down to −40 °C, a low viscosity solvent is almost always required. However, most low viscosity solvents, such as esters and ethers, are classified as *aggressive* solvents, as they readily react with anode materials (esters and ethers) or high-voltage cathode materials (ethers). One method proposed by the scientific community regarding the electrolyte solution is to replace with more stable and highly conductive electrolytes. In addition to their interaction with the electrolyte solution, the problems of thermal stability, lithium ion diffusion, irreversible phase changes, and intrinsic electronic resistivity all contribute to limitations in cell performance over the wide temperature range of −40 to ∼71 °C.

Here, we look at the example of the discharge of wound prismatic batteries at high and low temperatures. The discharge capability of wound prismatic batteries is indicated by the discharge curves shown in Figure 3.17 for a 1.4 Ah wound prismatic design. When discharged at 0.945 A (0.7C) and 45 and 60 °C, the battery delivered 1.4 Ah; at 20 °C, 1.34 Ah; at 0 °C, 1.16 Ah; and at −10 °C, 0.85 Ah. The average voltage at 20 °C was 3.56 V; at 0 °CC, 3.43 V; and at −10 °C, 3.2 V.

3.6.3
Test and Evaluation of Li Battery Materials

As already shown, Li batteries are the most advanced secondary electrochemical systems, and therefore, it is necessary to know the test methods and evaluations for

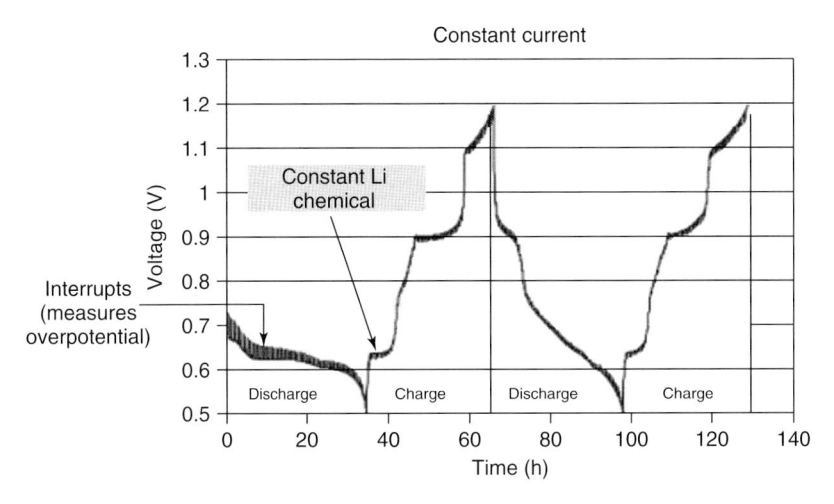

Figure 3.18 Curve obtained from electrochemical cycling [16].

Li batteries. The standard way to test the ability of materials as rechargeable Li ion electrodes is to electrochemically cycle them between the charged and discharged states in order to determine the battery power capability and *lifetime*. The lifetime of a rechargeable battery depends on the stability of the voltage/current/capacity relationship as a function of the number of cycles (Figure 3.18). Such testing can be performed in several ways such as by using flooded and button cells. Button cell testing is more standard and used in most laboratories. The test process is as follows: first, the active electrode material is mixed with a binder (e.g., 10% PVDF) and a small amount (~10%) of carbon to make a laminate. Then the laminate is coated onto a Cu foil; the Cu acts as the current collector. After drying the solvent in the oven, it is punched out into a circular shape and weighted to obtain the active material used in the cell. The cell assembly is then performed in a dry room using a lithium foil as the counter electrode. An example of a liquid electrolyte chemistry is 1 M $LiPF_6$ in 1:1 ratio of EC/DEC. The maximum possible energy in the electrochemical cell is determined by measuring the open circuit voltage (OCV) as a function of the state of discharge. This will act as a benchmark in understanding the amount of internal resistance associated with mass and charge transport. After the OCV measurement, the button cell is put in the cycler (battery test station). The testing procedure starts by discharging (or charging) the cell at a certain current between the maximum and minimum voltages, and once it reaches a minimum voltage, it automatically gets charged, and the process continues. The rate of discharge–charge can be varied by varying the current. Then the software gives a plot between voltage and time as shown in Figure 3.18. Multiple plots of voltage versus time for the charge–discharge cycle can also be obtained. The capacity of the tested cell can then be calculated for each discharge–charge cycle by measuring the area under the curve and dividing by the active material weight to obtain the capacity in milliampere-hour per gram.

3.7
Degradation Mechanisms and Mitigation Strategies

3.7.1
Degradation Mechanisms

With the increasing applications of lithium ion batteries in many fields, a very long cycle and calendar life must be demonstrated. *Degradation of a battery* can be defined as the modification of its properties with time and use. Essential properties are the available energy and power, and cell mechanical integrity (cell dimensions, leakage, etc.). Generally, capacity decrease and power fading do not arise from one single cause but from a number of various processes and their interactions. One main reason leading to energy loss is active materials transformation in inactive phases, and it may reduce the cell capacity at any rate, increase cell impedance, and lower the operating voltage. Degradation mechanisms occurring at anodes and cathodes are different significantly and are therefore discussed separately. The influence of the electrolyte, the separator, the current collector, and the states of charge and discharge will be taken into account.

3.7.1.1 The Effect of Anode Materials

When discussing the aging of anode, graphite is used as an example. In fact, within the bulk of the active material, only minor aging effects can be expected to occur. The volume changes of graphite seem not so drastic (may be 10% or less, depending on the material). Structural changes can cause mechanical stress on defects and on C–C bonds, which might result in cracking or related structural damage, which also has only minor impact on cell aging. However, graphite exfoliation and graphite particle cracking due to several factors, such as solvent cointercalation, electrolyte reduction inside graphite, and gas evolution inside graphite, should never be neglected, and they will certainly lead to a rapid degradation of the electrode [17].

In general, contact loss (e.g., mechanical or electronic) within the composite electrode results in higher cell impedance, thus causing the serious energy-fading problem [18]. One noticeable reason for contact loss is the volume changes of the active anode material. Contact loss may begin between carbon particles, between current collector and carbon, between binder and carbon, or between binder and current collector. Besides, the volume changes of the active material will affect the electrode porosity, which is a key element for good performance, because it avails the electrolyte to penetrate the bulk of the electrode completely. Furthermore, the internal cell pressure will increase [19].

Changes at the electrode–electrolyte interface due to reactions of the anode with the electrolyte are considered to be the major source of aging of the anode by many researchers [20]. It is well known that there will form the protective layers called *solid electrolyte interface* on the electrodes of lithium ion batteries. Usually, SEI formation on the anode is accompanied by the release of gaseous electrolyte decomposition products [21]. The amount of irreversible charge capacity that is consumed during the formation of the SEI was found to be dependent on the specific surface area of the graphite and the layer formation conditions [22]. We know that the unique properties of the SEI layers are that they are permeable to lithium cations but impermeable to other electrolyte components and electrons. Unfortunately, in some cases, other charged (anions, electrons, solvated cations) and neutral (solvents, impurities) species still seem to diffuse through the SEI. Consequently, corrosion of Li_xC_6 (and thus capacity loss) and electrolyte decomposition (and thus electrolyte loss and further SEI formation) are inevitable throughout the entire battery life. Gradually, to make worse, the SEI penetrates the pores of the electrode and in addition may also penetrate the pores of the separator. This may result in a decrease of the accessible active surface area of the electrode.

Many authors have measured the impedance rise of the electrodes, which is directly linked to the power fade of the cell [23]. The increase in electrode impedance is considered to be caused by the growth of the SEI as well as by changes of the SEI in composition and morphology.

As far as the anode side is concerned, aging effects may be mainly attributed to SEI formation and growth, leading to an impedance rise at the anode, graphite exfoliation, and gradual contact loss within the composite anode.

3.7.1.2 The Effect of Cathode Materials

Here, the fading mechanism of $LiMn_2O_4$ as a cathode material is discussed. The capacity fading mechanisms are very complex if $LiMn_2O_4$ is used as a positive material. Cycle life and calendar life are strongly dependent on several factors, such as composition of active material and state of charge. Also, capacity fade in spinel cells is the result of multiple processes, including those related solely to the spinel material and others involving the interaction of spinel with the electrolyte. Most fade mechanisms can be attributed to the dissolution of Mn^{2+} and the Jahn–Teller distortion [24].

- Dissolution of Mn^{2+} into the electrolyte after disproportionation of $Li_xMn_2O_4$.

$$2Mn^{3+}(solid) \rightarrow Mn^{4+}(solid) + Mn^{2+}(solution)$$

- Dissolution of spinel into the electrolyte [25] is promoted by acid-induced delithiation, resulting in disproportionation and formation of $\lambda\text{-}MnO_2$. Li ion electrolytes that use $LiPF_6$ are acidic, as the salt reacts with adventitious water to form HF. This reaction has a secondary effect of reducing electrolyte conductivity:

$$4H_2O + LiPF_6 \rightarrow LiF + 5HF + H_3PO_4$$

- However, studies of the material losses from spinel electrodes, and the resulting Mn^{2+} concentration in the electrolyte, found that dissolution can account for only a fraction of the capacity fade observed in spinel cells [26]. For example, when spinel cells were cycled at 50 °C, an analysis of the electrolyte for manganese showed that spinel dissolution could account for only 34% of the capacity fade observed [27].
- Jahn–Teller distortion in discharged cells ($LiMn_2O_4$)
- Jahn–Teller distortion [28] occurs in $LiMn_2O_4$ at 7 °C (280 K) [29]. This phase transition results in a transformation from the cubic space group $Fd3m$ to the

tetragonal group $I4_1/amd$. The structural distortion results from interaction of the Jahn–Teller active species Mn^{3+} ($t_{2g}^3 - e_g^1$), whereas Mn^{4+} ($t_{2g}^3 - e_g^0$) and Mn^{2+} ($t_{2g}^3 - e_g^2$) are not Jahn–Teller active. Because of the low temperature of this transition, and that modified spinels have, in general, lower transition temperatures, this mechanism may not be as relevant as others for currently used spinel materials. Related mechanisms can cause strain and structural failure [30], resulting in electrically disconnected particles.

3.7.1.3 The Effect of Electrolyte

Electrolyte mainly includes solvents and supporting electrolyte. In general, $LiPF_6$ is used as an electrolyte, with PC and EC, DMC, DEC, and MEC as solvent. The remaining traces of O_2, CO_2, and H_2O in electrolyte could cause disastrous effect for the electrolyte: H_2O reacting with electrolyte will produce HF, LiF, LiOH, and so on; O_2 will cause oxidation of electrolyte; and the Li_2CO_3 impurities result from CO_2. Additionally, the high-oxidizing lithium materials also can make electrolyte undergo oxidation. HF can destroy the stability of SEI, making battery performance deteriorate dramatically.

3.7.1.4 The Effect of Charging and Discharging States

It is well known that the state of charging or discharging of the electrodes has a close relationship with the cycle performance of Li ion batteries. Typically, when the electrodes are overcharged or discharged, much energy loss will occur.

- **Overcharge.** When negative electrodes are overcharged, Li^+ reacts with electrons and generates metal lithium. This will cause lithium plating and make a part of lithium useless. So the lithium for cycle will decrease. Besides, metal lithium could react with trace amounts of HF in the electrolyte and bring about LiF impurities. At last, the generated metal lithium will deposit around the separator and block the pore paths, thus resulting in resistance increase and poor battery performance.
  When the positive electrodes are overcharged, the cathode materials, such as $LiCoO_2$, will release oxygen. The electrolyte will be oxidized by oxygen, and its properties deteriorate, producing some impurities. On the other hand, oxygen can react with metal lithium by oxidation, thus generating Li_2O. At the same time, it may also result in LiOH impurities, as there is trace H_2O in electrolyte.
- **Overdischarge.** When the positive electrodes are overdischarged, they may make active materials generate some electrochemical inactive substances, thereby reducing the content of active substances and causing capacity loss.

3.7.1.5 The Effect of Current Collector

In the cycling process, if electrolyte corrosion, overcharging, or overdischarging of battery happens, the current collector will undergo corrosion or dissolution, thereby affecting the electronic transmission. If the active substances fall off, the contact resistance will increase and so will the total resistance.

To sum up, among all the possible mechanisms for capacity fading on cycling, the following are the most often included:

- Degradation of crystalline structure of positive material (especially for $LiMn_2O_4$).
- Graphite exfoliation and metallic lithium plating.
- Build up of passivation film on both electrodes, especially for the anodes, limiting the active surface area and clogging the electrodes' small pores.
- Mechanical modification of the composite electrode structure due to volume changes during cycling, leading to active particles disconnection from the conductive network.
- Decomposition and oxidation of electrolyte.

3.7.2
Mitigation Strategies

Several mechanisms of capacity fading have been discussed. In the following discussion, some useful mitigation strategies to cope with the problem of aging are presented.

3.7.2.1 The Surface Treatment

Coating and doping are regarded as two useful methods for surface treatment. Coating oxides on $LiMn_2O_4$ could decrease the surface area to retard the side reactions between the electrode and electrolyte and to further diminish the Mn dissolution during the cycling test. As reaction with the electrolyte or dissolution processes occur at the particle surface, materials with lower surface area and specially coated surfaces have been developed, such as $LiCoO_2$- or Li_2CO_3-coated $LiMn_2O_4$.

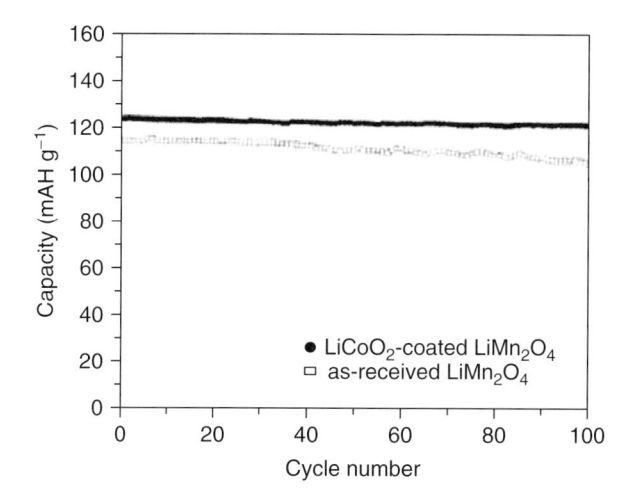

Figure 3.19 Capacity versus cycle number diagram for the cells with active electrodes of as-received $LiMn_2O_4$ and $LiCoO_2$-coated $LiMn_2O_4$ at room temperature.

LiCoO$_2$-coated LiMn$_2$O$_4$ showed a higher discharge capacity (120 mAh g^{-1}) than as-received LiMn$_2$O$_4$ (115 mAh g^{-1}), maintaining the same excellent cycle stability (Figure 3.19) [31]. Especially, the rate capability of LiCoO$_2$-coated LiMn$_2$O$_4$ improved significantly. The rate capability of LiCoO$_2$-coated LiMn$_2$O$_4$ was increased by more than 60% at the 20C rate (2400 mA g^{-1}).

Modified spinels that contain excess lithium, and preferably an admetal (Li$_{1+x}$M$_y$Mn$_{2-x-y}$O$_4$; M = Al^{3+}, Cr^{3+}, Ga^{3+}), offer improved storage stability in the discharged state as manganese disproportionation is inhibited when the Mn^{3+}:Mn^{4+} ratio is reduced. The electrochemical tests demonstrate that Al-doped LiAl$_x$Mn$_{2-x}$O$_4$ can be well cycled at 55 °C without severe capacity degradation (Figure 3.20) [32]. This excellent cycling performance of Al-doped LiMn$_2$O$_4$ samples can be attributed to the higher crystallinity and structural homogeneity.

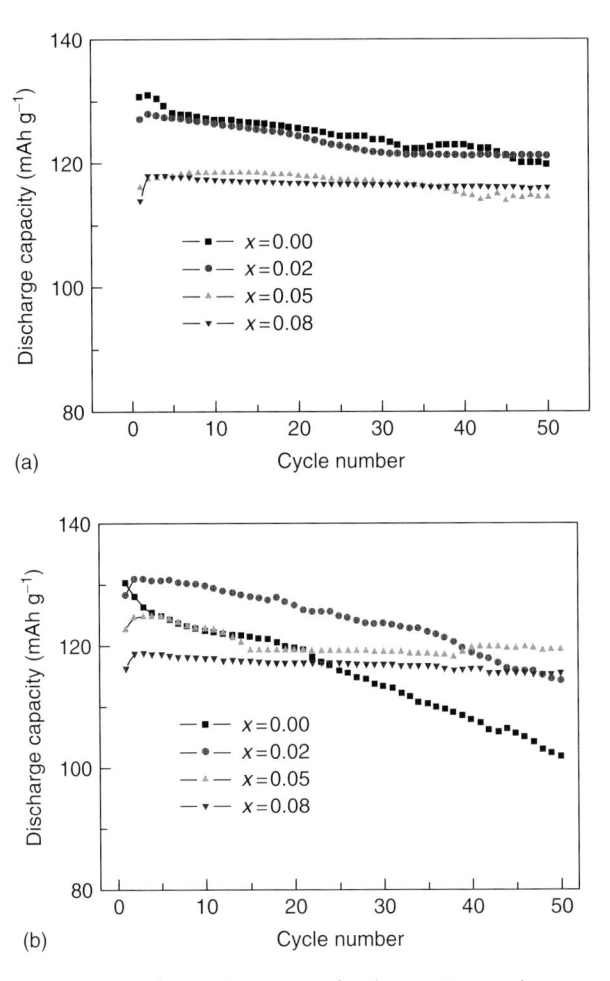

Figure 3.20 Cycling performances of LiAl$_x$Mn$_{2-x}$O$_4$ samples as labeled (a) at room temperature and (b) at 55 °C.

3.7.2.2 Replacement of LiPF$_6$ by Salts and the Use of Additives

It is expected that replacement of LiPF$_6$ with salts, which do not form acidic contaminants, will be beneficial to the mitigation of capacity fading [33]. Or, the use of additives that predominantly react on the electrodes' surface and block reactions of acidic species should improve the performance of Li ion batteries. When polymers forming additives are considered, the polymers have to enable facile Li ion transport through them. VC and Li disalicilato-borate salt were found to be suitable additives.

3.7.2.3 Developing a New System of Electrolyte

Addition of 500–1000 ppm $(CH_3)_3SiNHSi(CH_3)_3$ to the electrolyte solution gives rise to less capacity fading and results in a drastic decrease of Mn dissolution even at such a higher temperature as 80 °C. This compound could work as a suppressant of the reaction mentioned above by dehydration and acid neutralization in the solution [34].

3.8
Current and Potential Applications of Secondary Li Ion Batteries

Various types of batteries currently use lithium ion battery (also known as *lithium ion secondary batteries*), which is a new type of energy storage device growing fast. Since the early 1990s, Sony Energy Devices Corporation and the Canadian Maurice energy company have been developing lithium ion batteries successfully and have been competing research and development and application of the hot spots in countries of the world. The advantages of Li ion batteries over other rechargeable electrochemical cells are they do not exhibit a memory effect, they provide a higher energy density per unit mass, and their self-discharge rate is less than half compared to the next best available solutions such as NiCd and NiMH cells. Furthermore, the lifetime of the current commercial lithium ion batteries is over 1000 cycles, while their shelf life is more than 10 years [35]. The recently developed Li polymer cells are easier to handle with safety [36].

The purpose of this section is to give an overview of the main, current and future, applications that secondary Li batteries have in our society, so as to motivate the reader to focus on the chapters to follow about improving Li battery technology. After the more general application information provided, detailed information will be given concerning the use of secondary lithium batteries as a power source in electric vehicles (EVs), aerospace industry, energy storage, and military equipment.

3.8.1
Portable Electronic Devices

With the continuous development of information society, the development of lithium ion batteries will go deeper, with application in communication, automotive, instrumentation, air and ocean exploration, and other fields.

Since the usage by Armand in 1980, the rocking chair battery (RCB) concept was propound, and Sony and Sanyo in 1985 and 1988, respectively, started the study on applicability of lithium ion batteries, and in June 1991, the world's first lithium ion batteries for mobile phones were listed and inspired the research and development of lithium ion batteries in the world, a time when lithium ion batteries were the most promising chemical power and even called the *limits of the battery* or the *last generation of batteries* [37]. Indeed, the emergence of lithium ion battery and its subsequent continuous improvement and perfection led to the rapid development of related industries, such as mobile phones, video camera, camera, laptop (notebook computer), Pocket PC, mini compact disc, and other small portable electric devices. It is envisaged that the resulting small, light, high-energy lithium ion batteries are the future batteries of choice for EVs requiring high energy, light, and power.

3.8.2
Applications of Lithium Ion Batteries in Electric Vehicle (EV) Industry

With the progress in social civilization, the requirements of environmental awareness have increased. Thus, the ecological environment caused by automobile exhaust and urban air pollution is more and more serious, hence the call for a green battery-powered EVs. It is reported that, until 1 January 1994, the global proven oil reserves is 9999.1 billion barrels (7 barrels equivalent to approximately 1 ton), for an average daily consumption of 6680 barrels from 1993, after 46 years (i.e., by 2043), the world oil will run out. For these reasons, the upgrading of vehicle dynamics is imperative. To this end, the world's advanced countries, such as the United States, Japan [38], Germany [39], Canada [40], and France [41], have developed the EV prototype car research work as early as possible to resolve environmental and noise pollution and energy crisis actively. The form of legislation in California provides that since 1998, California vehicles sold must contain 2% electric car, and by 2003, among the sales of automobiles, electric cars must be accounted for 10% of total percentage of vehicles sold.

In order to promote and support the EV car battery research trial, the United States set up the Advanced Battery Consortium in the early 1990s. The agency supports the development of EV car batteries (mainly lithium ion battery) and has invested US $260 million, of which investment US$11.8 million was to SAFT, to develop lithium ion battery, and the Quebec Company has invested US $85 million to develop lithium ion and lithium polymer batteries. The Japanese government invested US $100 million and developed a plan for EVs and load leveling. With the support of the governments, different varieties of EV cars have appeared. First, in 1995, Japan's Sony Corp. launched lithium-ion-battery-powered EV car. The vehicle weight is 1.7 t, can accomodate four individuals, each charge can travel 200 km (three times as the lead acid batteries), can have a maximum speed of up to 120 km, starting from 0 to 80 km h^{-1}, in only 12 s; the car's power supply is similar to the size of the 1996 Neumann U 67, that is, 410 mm, with $LiCoO_2$ as the cathode of the lithium ion battery. Each battery has a capacity of 100 Ah, the energy of the entire power supply is 110 Wh kg^{-1}, and the energy density is 250 W h l^{-1}. Following Sony, the Japanese Mitsubishi Motors launched the EV car in 1996, the car uses LiMn$_2$O$_4$ as cathode of lithium ion batteries for power, and this battery weight is 15 kg, powering a trip of up to 250 km. Then there are Mitsubishi Heavy Industries, Honda, Nissan EV, and other vehicle manufacturers that have appeared, and in early 1997, they have started official sales of lithium-ion-battery-powered EVs. In addition, Japan has four companies that have developed high-energy lithium ion batteries for EV cars: Hitachi and new Koubo Electric Company developed 200 Wh batteries; Sanyo, 250 Wh batteries; and Japanese Battery, 360 Wh batteries. These 4 are part of 12 companies supported by the Japanese government. After Japan, the United States and some European companies have launched their own lithium ion batteries developed for the power source of the EV car. It can be seen that as the global EV cars industry develops, the high-energy lithium ion battery demand has become very impressive. China's lithium ion battery application development is also very important. Particularly for the development of EV cars, the State Economic and Trade Commission and Department of Mechanical have listed them in 95 major national science and technology industrial projects.

It is not difficult to see that lithium ion batteries for and their impact on the EV car industry are very important. The application and development of lithium ion battery are very promising.

3.8.3
Application Prospect of Lithium Ion Battery in the Aerospace Industry

In 1993, the Lawrence Livermore National Laboratory [42] had a comprehensive assessment of performance and materials testing on 20 500 Sony batteries to prove the possibility of their application in satellites. Analytical work included the battery charge–discharge rate, depth of discharge, and life and characteristics of long-term components. Table 3.5 show that lithium ion batteries can be used for low-orbit satellites, but their capacity, voltage, and structure need to be further improved.

Canadian BlueStar Advanced Technology Corp. is also developing lithium ion batteries to use in EVs and aerospace industry. In 1991, the US Air Force and the

Table 3.5 Test results of 20 500-type lithium ion battery supplied by Sony for satellite.

Depth of discharge (%)	Discharge current (A)	Cutoff conditions	Charge current (A)	Charge time at constant voltage	Cycle life (cycles)
100	0.9	2.75 V	2.4	2 h	650 (650)
44	0.9	0.395 Ah	1.4	2 h	3 800 (2 250)
50	0.36	0.450 Ah	0.36	2 h	14 000 (900)
30	0.36	0.270 Ah	0.9	30 min	40 000 (4 100)

Data in bracket were practically measured before paper publication, and unbracketed data were obtained by prediction based on 2.75 V cutoff voltage.

Table 3.6 The technical characteristics of Canadian BlueStar 20 Ah lithium ion cell.

Case material and dimension (mm)	304SS 26.4 × 128 × 107
Dimension of positive electrode (mm)	6000 × 88 × 0.17
Dimension of negative electrode (mm)	6300 × 91 × 0.11
Separator	Calgary 2300
Cover design	Two GTM terminals, fill tube, rupture disk (1.7 MPa)
Case-cover welding	T IG welding
Specific energy	110 Wh kg^{-1}, 260 Wh l^{-1}

Table 3.7 Canadian BlueStar's objectives on lithium ion battery.

	Near-term target	Long-term target
Specific energy (0.5C, 100% DOD) (Wh·kg^{-1})	135	150
Energy density (0.5C, 100% DOD) (Wh·l^{-1})	325	360
Cycle life (0.5C, 100% DOD) (cycles)	500	1000
Cycle life (0.5C, 60% DOD) (cycles)	2000	5000
Ratio of capacity at 20 °C to initial capacity (0.5C) (%)	50	75
Ratio of capacity at 40 °C to initial capacity (0.5C) (%)	–	50

Abbreviation: DOD, depth of discharge.

Canadian Department of Defense received $300 million to study the 50–100 Ah battery. The 20 Ah battery was completed in 1997 and the 50 Ah battery in 1998. One 20 Ah square lithium ion battery technology indicators are presented in Table 3.6.

The company has also formulated the short-term and long-term development objectives of the battery; the former can be completed in the near future, which can meet space, aviation, and EV requirements in the application of extreme. Table 3.7 lists the two goals.

Tianjin, China Electric Power Research Institute projected in 1992 to study the lithium ion batteries has been engaged in various basic research including positive and negative materials, electrolyte, and shell and has successfully developed AA-, A-, D-type; 18 650; and other small-capacity batteries. The 18 650 battery, especially, has been established since 1997 in the pilot production line. After 1996, with the support of the State Planning Commission, the institute initiated a large-capacity lithium ion batteries exploration for the purpose of EVs, aerospace industry, and military communication. It has been initially developed as listed in the table as 35 and 100 Ah square rectangular cells; Figures 3.21 and 3.22, respectively, show their appearance. Preparations are underway for EVs on six cells of 50 Ah in series and 20 Ah cells for space, with hope for the country's new electric car and satellite services.

Figure 3.21 A 35 Ah lithium ion cell made in TIPS.

Figure 3.22 A 100 Ah lithium ion cell made in TIPS.

In the aerospace industry, the lithium ion battery combined with solar cells provided the power supply. Its characteristics include small self-discharge rate, no memory effect, a large energy, and long cycle life, and its power is much superior to that of the original batteries Cd_2Ni or Zn_2Ag_2O. Especially, in view of a small, lightweight object, it is very important to space devices, because the quality of space device indicators are often calculated in grams and not in kilograms. This application has been reported: lithium ion batteries as the power supply of Telex communications satellite. According to some reports, the US company Alliant Technology Systems has signed an investment of $5.9 million for Naval Underwater Vehicles to use polymer lithium ion batteries in the current Navy underwater vehicles, such as ZnPAg$_2$ used in MK230 Target mine, MK28SDV (closed carriage tools), and ASDS (advanced closed vehicles). The Navy currently uses ZnPAg$_2$O batteries with an alleged cost of no less than $500 million annually. Because ZnPAg$_2$O batteries have limited cycle and wet storage life, they must be replaced in 12–18 months, but the life of lithium ion batteries is 10 times longer.

3.8.4
As Energy Reserves

In the past, peak electricity regulation was a problem; usually, to ensure high peak power, we need to build more power plants; and there is a marked decline for the use of electricity at night. But the generator cannot stop and needs to keep running. Both these have not only caused the burden of increased investment but also resulted in waste of energy.

Morever, wind and solar power generation is growing quickly around the world, mainly to mitigate some of the negative environmental impacts of the electricity sector. However, the variability of these renewable sources of electricity poses technical and economical challenges when integrated on a large scale. Energy storage is being widely regarded as one of the potential solutions to deal with the variations of variable renewable electricity sources and peak electricity regulation.

SAFT France reported that its G3 batteries (anode is LiNi$_{0.75}$Co$_{0.2}$Al$_{0.05}$O$_2$) float for over 1400 days, the capacity changed very little resistance; the test results are shown in Figure 3.23 [43].

3.8.5
Application of Military Equipment

The power supplies for military equipment [44], mainly motor cars start batteries; wireless radio power (in the past mainly dry batteries, Cd$_2$Ni rechargeable silver, and zinc batteries were used); batteries used for special forces, the so-called special

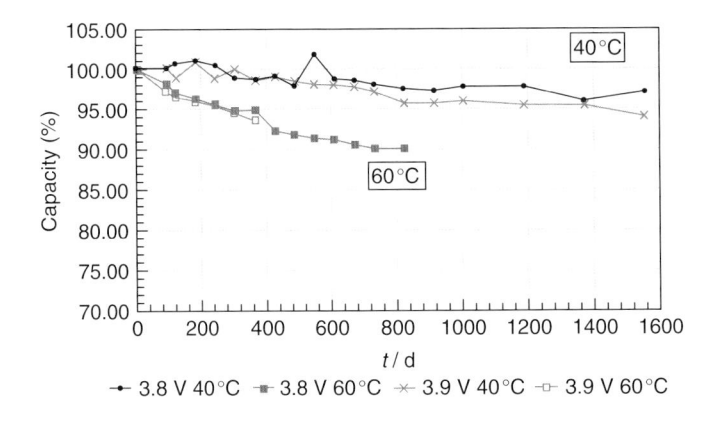

Figure 3.23 Floating test result for G3 lithium ion battery (LIB).

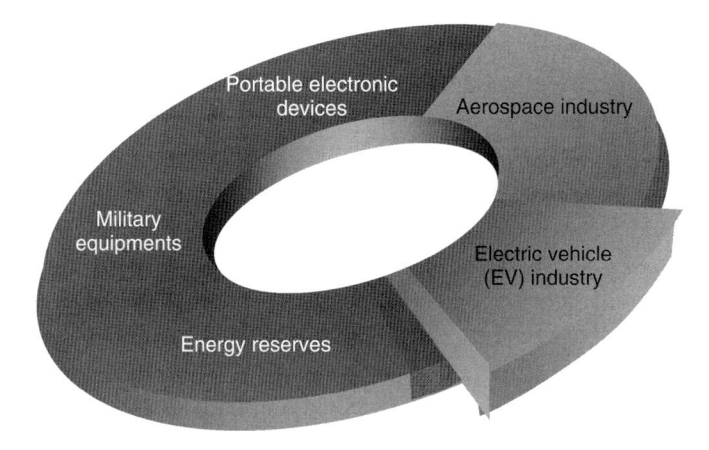

Figure 3.24 Illustration of the application distributions of Li ion rechargeable batteries.

forces power supply such as power supply for weapons in the water (including torpedoes and sonar jammers); micro unmanned reconnaissance aircraft power supply (including video-phase electrical installations); embedded device with a fuse-type power supply; and other types of mines.

Figure 3.24 illustrates the application distributions of Li ion rechargeable batteries. From the figure, we can see that portable electronic devices and energy reserves are widely used. EV industry is the focus of the current study. Aerospace industry and military equipment are the second in-depth research after the electric car.

Lithium ion batteries developed since the early 1990s have become the best overall performance of the battery system. Capacity and cycle performance of lithium ion battery are improving, lithium ion batteries with greater capacity and that are lighter, smaller, and thinner with lower prices continue to stay in the market. Production and sales of lithium ion battery have always maintained a high growth, and it is already widely used in civilian areas. In summary, the preparation of lithium ion batteries for EVs, spacecraft, and storage for power has already made a lot of basic research, especially their applications in EV have been demonstrated. Experiments show that the specific energy of batteries, discharge rate, charge–discharge life, and sealing are formulated to meet interim target of EVs and aerospace basic requirements. In the future, lithium ion batteries of EVs, aerospace industry, storage areas, and military equipment have a very good prospect, and will have a profound impact on people's lives in the future.

References

1. Linden, D. and Reddy, T.B. (2001) *Handbook of Batteries*, 3rd edn, McGraw-Hill Professional.
2. Plichta, E.J. and Behl, W.K. (2000) *J. Power Sources*, **88**, 192.
3. Padhi, A.K., Nanjundaswamy, K.S. and Goodenough, J.B. *et al.* (1997) *J. Electrochem Soc.*, **44** (5), 1609–1613.
4. Arora, P. and Zhang, Z. (2004) *Chem. Rev.*, **104**, 4419–4462.

5. Zhang, S.S. (2007) *J. Power Sources*, **164**, 351–364.
6. Kritzer, P. and Cook, J.A. (2007) *J. Electrochem. Soc.*, **154**, A481–A494.
7. Whitehead, A.H. and Schreiber, M. (2005) *J. Electrochem. Soc.*, **152** (11), A2105–A2113.
8. van Schalkwijk, W.A. (2002) *Advances in Lithium-Ion Batteries*. Springer.
9. Zou, Z.G., Mai, L.Q., Chen, H.Y. et al. (2001) *Battery Bimonthly*, **31** (3), B116–1181.
10. Amine, K., Tukamoto, H., Yasuda, H. et al. (1996) *J. Electrochem. Soc.*, **143** (5), B1607–B1612.
11. Huang, F., Zhou, Y.H., Yuan, Z.Y. et al. (2002) *Battery Bimonthly*, **32** (5), B298–3001.
12. Sun, Y.Y., Yang, Y.F., Zhan, H. et al. (2010) *J. Power Sources*, **195**, 4322–4326.
13. Li, J., Murphy, E., Winnick, J. et al. (2001) *J. Power Sources*, **102**, 302–309.
14. Feng, X.K., Ahu, J.C., and Chen, Y.K. (1999) *Chin. J. Power Sources*, **23**, 186–190.
15. Pistoia, G. (ed.) (1994) *Lithium Batteries: New Materials, Developments and Perspectives*, Elsevier, Amsterdam.
16. Aifantis, K.E., Hackney, S.A., and Vasant Kumar, R. (2010) High Energy Density Lithium Batteries Materials, Engineering, Applications.
17. Anna, M., Andersson, K.E., and Thomas, J.O. (1999) *J. Power Sources*, **81–82**, 8–12.
18. Wang, Y., Guo, X., Greenbaum, S., Liu, J., and Amine, K. (2001) *Electrochem. Solid State Lett.*, **4**, A68–A70.
19. Rubino, R.S., Gan, H., and Takeuchi, E.S. (2001) *J. Electrochem. Soc.*, **148**, A1029–A1033.
20. Aurbach, D., Zinigrad, E., Cohen, Y., and Teller, H. (2002) *Solid State Ionics*, **148**, 405–416.
21. Imhof, R. and Novák, P. (1998) *J. Electrochem. Soc.*, **145**, 1081–1087.
22. Novák, P., Joho, F., Lanz, M., Rykart, B., Panitz, J.C., Alliata, D., Kotz, R., and Haas, O. (2001) *J. Power Sources*, **97–98**, 39–46.
23. Amine, K., Chen, C.H., Liu, J., Hammond, M., Jansen, A., Dees, D., Bloom, I., Vissers, D., and Henriksen, G. (2001) *J. Power Sources*, **97–98**, 684–687.
24. (a) Gummow, R.J., de Kock, A., and Thackeray, M.M. (1994) *Solid State Ionics*, **69**, 59–67; (b) Thackeray, M.M., Horn, Y.S., Kahaian, A.J., Kepler, K.D., Skinner, E., Vaughey, J.T., and Hackney, S.A. (1998) *Electrochem. Solid State Lett.*, **1**, 7–9.
25. (a) Tarascon, J.M., McKinnon, W.R., Coowar, F., Bowmer, T.N., Amatucci, G., and Guyomard, D. (1994) *J. Electrochem. Soc.*, **141**, 1421; (b) Gummow, R.J., de Kock, A., and Thackeray, M.M. (1994) *Solid State Ionics*, **69**, 59; (c) Xia, Y., Zhou, Y., and Yoshio, M. (1977) *J. Electrochem. Soc.*, **144**, 2204; (d) Amatucci, G.G., Schmutz, C.N., Blyr, A., Sigala, C., Gozdz, A.S., Larcher, D., and Tarascon, J.M. (1997) *J. Power Sources*, **69**, 11.
26. Jang, D.H., Shin, Y.J., and Oh, S.M. (1996) *J. Electrochem. Soc.*, **143**, 2204.
27. Xia, Y., Zhou, Y., and Yoshio, M. (1997) *J. Electrochem. Soc.*, **144**, 2593.
28. (a) Thackeray, M.M. (1995) *J. Electrochem. Soc.*, **142**, 2558; (b) Blyr, A., Du Pasquier, A., Amatucci, G., and Tarascon, J.-M. (1977) *Ionics*, **3**, 321; (c) Sun, Y., Jeon, Y., and Lee, H.J. (2000) *Electrochem. Solid State Lett.*, **3**, 7.
29. Yamada, A. and Tanaka, M. (1995) *Mat. Res. Bull.*, **30**, 715–721.
30. (a) Yamada, A., Miura, K., Hinokuma, K., and Tanaka, M. (1995) *J. Electrochem. Soc.*, **142**, 2149; (b) Thackeray, M.M., Shao-Horn, Y., Kahaian, A.J., Kepler, K.D., Skinner, E., Vaughey, J.T., and Hackney, S.A. (1998) *Electrochem. Solid State Lett.*, **1**, 7.
31. Park, S.C., Kim, Y.M., Kang, Y.M. et al. (2001) *J. Power Sources*, **103**, 86–92.
32. Xiao, L.F., Zhao, Y.Q., Yang, Y.Y. et al. (2008) *Electrochim. Acta*, **54**, 545–550.
33. Aurbach, D. (2003) *J. Power Sources*, **119P121**, 497–503.
34. Ritchie, A.G. (2001) *J. Power Sources*, **96**, 1–4.
35. Huneck, S., Schreiber, K., Schulze, C. et al. (1975) Plant growth regulator compositions containing substituted α-methylene-γ-lactones. Ger.(East) Patent 112884.

36. Ma, S. and Gao, W. (2002) *Synlett*, **1**, 65268.
37. Nagaura, T. and Tazawa, K. (1990) *Prog. Batteries Sol Cells*, **9**, 2092210.
38. Sony Energytec Corp (1996) Progress on lithium ion battery for EV application. In Information of First Beijing Electric Vehicle Exhibition.
39. Kochler, U., Kruger, F.J., and Niggemann, E. (1997) Performance of advanced battery system. Proceedings of the 14th International Seminar on Primary and Secondary Batteries, p. 376.
40. Marcoux, L. (1997) BlueStar company, technology and capabilities. Proceedings of the 14th International Seminar on Primary and Secondary Batteries, p. 439.
41. Staniewicz, R.J. (1997) Improvements in SA FT lithiumion cell and batteries. Proceedings of the 13th International Seminar on Primary and Secondary Batteries, p. 321.
42. Feng, X.K., Zhu, J.C., and Chen, Y.K. *J. Power Sources*, **23**, 186–190.
43. Broussely, M., Biensan, Ph., Bonhommeb, F. *et al.* (2005) *J. Power Sources*, **146**, 90296.
44. Owens, B., Osaka, T. *et al.* (1997) *J. Power Sources*, **68**, 173–186.

4
Lead-Acid Battery
Joey Jung

4.1
General Characteristics and Chemical/Electrochemical Processes in a Lead-Acid Battery

4.1.1
General Characteristics

The lead-Acid battery is the first developed secondary (rechargeable) battery. Since French physicist Gaston Planté–invented the lead-Acid battery in 1859, the lead-Acid battery has been developing for over 150 years into numerous applications such as starting batteries, backup power batteries, telecommunication batteries, and traction batteries. Lead-Acid battery sales represent ~70% of the secondary battery market in the world. The major development history of lead acid battery is summarized in Table 4.2

Since 2002, lead-Acid batteries have been widely used to power electric scooters, electric bicycles, and other electric vehicles (EVs). The worldwide lead-Acid battery industry was an approximately US$44.7-billion industry in 2008. Its demand increased to US$60.3 billion in 2010 and will increase to almost US$103.8 billion by 2020 (Figure 4.1).

Despite the recent rapid development of secondary lithium ion/lithium polymer batteries, advanced technology development and design and fabrication processes are still being developed and introduced in the lead-Acid battery industry. Table 4.1 shows major advantages and disadvantages of lead-Acid batteries.

4.1.2
Major Milestones in the Development of the Lead-Acid Battery

4.1.2.1 Chemistry/Electrochemistry
The electrochemistry of lead-Acid batteries in the charged stage consists of a lead dioxide (PbO_2) positive electrode and a lead (Pb) negative electrode in a concentrated (37 wt%, specific gravity of 1.28) sulfuric acid (H_2SO_4) electrolyte. In the discharged stage, the lead dioxide positive electrode and the lead negative electrode will convert

Electrochemical Technologies for Energy Storage and Conversion, First Edition. Edited by Ru-Shi Liu, Lei Zhang, Xueliang Sun, Hansan Liu, and Jiujun Zhang.
© 2012 Wiley-VCH Verlag GmbH & Co. KGaA. Published 2012 by Wiley-VCH Verlag GmbH & Co. KGaA.

4 Lead-Acid Battery

Worldwide secondary battery market (2010)

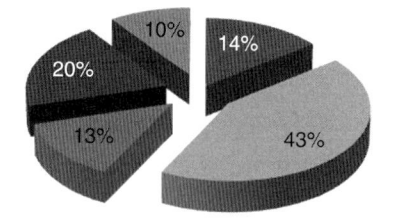

- Lead-acid battery - auto deep cycle (10%)
- Lead-acid battery - auto SLI (43%)
- Lead-acid battery - industrial (14%)
- Lithium ion battery (20%)
- (13%)

Figure 4.1 Worldwide secondary battery market distribution.

Table 4.1 Advantages and disadvantages of lead-Acid batteries [1].

Advantages	Disadvantages
Low-cost secondary battery	Relatively low cycle life (~500 deep cycles)
Recyclable	Low energy density (typically 30–40 Wh kg^{-1})
Available in maintenance-free options	Long-term storage in a discharged condition can lead to sulfation of battery electrode and cause irreversible damage
Easy to manufacture in high-volume production	Grid contains antimony and arsenic, which is a health hazard
Good recharge efficiency (over 70%)	Difficult to make small-sized batteries (<500 mAh, AA size battery)
Available in a variety of capacities, sizes, and designs	Short circuit will cause irreversible battery damage
Good high-rate performance	–
Moderately good operating temperature range (−40 to 60 °C)	–
Good single cell voltage (open-circuit voltage >2 V)	–
Easy state of discharge/charge indication	–

to lead sulfate ($PbSO_4$) and consume the sulfate ions that decrease the specific gravity of the electrolyte to bring it closer to that of water. The positive electrode is the major factor influencing the performance and the cycle life of the lead-Acid battery. The negative electrode determines the cold-temperature performance.

In the charged stage, the lead dioxide on the positive electrode consists of two forms, α-PbO_2 (orthorhombic) and β-PbO_2 (tetragonal). Both forms of PbO_2 have different impacts on the battery performance, as shown in Table 4.3.

Table 4.2 Major milestones in the development of the lead-Acid battery.

Year	Inventor	Development of the lead-Acid battery
1859	Planté	First practical lead-Acid battery; corroded lead foils form active material
1881	Faure	Pasted lead foils with lead oxide mixed with sulfuric acid paste for positive electrode
1881	Sellon	Lead–antimony alloy grid
1881	Volckmar	Perforated lead plates to provide pockets for support of oxide
1882	Brush	Mechanically bonded lead oxide to lead plates
1882	Gladstone and Tribs	Double sulfate theory of reaction in lead-Acid battery; $Pb + PbO_2 + 2H_2SO_4 \leftrightarrow 2PbSO_4 + 2H_2O$
1883	Tudor	Pasted mixture of lead oxide on grid pretreated by the Planté method
1886	Lucas	Formed lead plates in solutions of chlorates and perchlorates
1890	Phillipart	Early tubular construction – individual rings
1890	Woodward	Early tubular construction
1910	Smith	Slotted rubber tube, Exide tubular construction
1920		Research started in materials such as expanders and oxides and manufacture of equipment and technique
1935	Haring and Thomas	Experimental proof of double sulfate theory
1956	Bode and Voss	Clarification of two forms of lead dioxide (α and β)
1960	Ruetschi and Chan	Clarification of two forms of lead dioxide (α and β)
1970	McClelland and Devit	Commercial spiral wound sealed lead-Acid battery; expanded metal grid technology; glass fiber separator; heat sealed plastic case to cover assemblies
1980		Sealed valve-regulated batteries
1990		Electric vehicle battery; bipolar battery
2001	Gyenge and Jung	Lead–tin alloy coated carbon substrate current collectors for lead-Acid battery
2002	Kelly	Bare carbon-foam current collector for negative current collector for lead-Acid battery
2004		Supercapacitor negative electrode for lead-Acid battery
2009	Jung	Multiconductive polymer or carbon-foam substrate current collectors for lead-Acid battery

Lead dioxide is formed through curing and forming a mix of chemicals, including lead, lead oxide, sulfuric acid, and water. The ratios of these chemicals and the curing conditions such as temperature, humidity, and duration have a significant influence on the crystalline morphology and pore structure of the converted chemicals, which consist of lead sulfate, lead oxide, and a small amount of lead (<5%). During formation, the cured materials will electrochemically form lead

Table 4.3 Physical and chemical properties of lead-Acid batteries.

Property	Pb	α-PbO$_2$	β-PbO$_2$
Molecular weight (g mol^{-1})	207.2	239.19	239.19
Composition	–	PbO$_{1.94-2.03}$	PbO$_{1.87-2.03}$
Crystalline form	Face centered cubic	Orthorhombic (larger, more compact)	Tetragonal (ruile)
Lattice parameters (nm)	$a = 0.4949$	$a = 0.4977$	$a = 0.491-0.497$
	–	$b = 0.5948$	$c = 0.337-0.340$
	–	$c = 0.5444$	–
X-ray density (g cm^{-3})	11.34	9.80	~9.80
Practical density @ 20 °C (g cm^{-3})	11.34	9.1–9.4	9.1–9.4
Heat capacity (cal deg mol^{-1})	6.80	14.87	14.87
Specific heat (cal g^{-1})	0.0306	0.062	0.062
Electrical resistivity @ 20 °C ($\mu\Omega$ cm^{-1})	20	~1.709	~1.692
Electrochemical potential (V) @ 4.4 M H$_2$SO$_4$	0.356	~1.709	~1.692
Melting temperature (°C)	327.4	–	–
Electrochemical potential in 4.4 M H$_2$SO$_4$	–	1.709 V	1.692 V
Electrochemical activity	–	Less	Better
Capacity per unit weight	–	Less	Better
Cycle life impact	–	Better	Less

dioxide, which contains a ratio of α-PbO$_2$ and β-PbO$_2$. Both forms of PbO$_2$ are not stoichiometric. The composition, represented by PbO$_x$, has an x value varying between 1.85 and 2.05.

Table 4.3 shows the properties of Pb and PbO$_2$.

The electrochemical reactions of the lead-Acid batteries are as follows:

Positive electrode (cathode):

$$PbO_{2(s)} + 4H^+ + 2e^- \underset{\text{Charge}}{\overset{\text{Discharge}}{\rightleftarrows}} Pb^{2+} + 2H_2O \quad E = +1.7 \text{ V}$$

$$Pb^{2+}_{(aq)} + SO_{4(aq)}^{2-} \underset{\text{Charge}}{\overset{\text{Discharge}}{\rightleftarrows}} PbSO_{4(s)}$$

Negative electrode (anode):

$$Pb_{(s)} \underset{\text{Charge}}{\overset{\text{Discharge}}{\rightleftarrows}} Pb^{2+}_{(aq)} + 2e^- \quad E = -0.3 \text{ V}$$

$$Pb^{2+}_{(aq)} + SO_{4(aq)}^{2-} \underset{\text{Charge}}{\overset{\text{Discharge}}{\rightleftarrows}} PbSO_{4(s)}$$

4.2 Battery Components (Anode, Cathode, Separator, Endplates (Current Collector), and Sealing)

Overall reaction:

$$Pb + PbO_2 + 2H_2SO_4 \underset{\text{Charge}}{\overset{\text{Discharge}}{\rightleftarrows}} 2PbSO_4 + 2H_2O \quad E = 2.0\,V$$

The open cell potential for lead-Acid batteries is 2.10–2.13 V and the nominal voltage of a single practical lead-Acid battery is 2 V.

4.2 Battery Components (Anode, Cathode, Separator, Endplates (Current Collector), and Sealing)

The major components in the lead-Acid battery include a positive grid, a negative grid, a positive active material, a negative active material, an electrolyte, a top lead, a separator, and a plastic container. The weight distribution in a common starting, lighting, and ignition (SLI) battery and traction battery is shown in Figure 4.2.

4.2.1 Battery Grid (Current Collector)

The function of a battery grid is to act as a current collector, which is used to hold the active material and conduct electricity between the active material and the cell terminals.

The battery grid is often made of a lead alloy because pure lead is very soft. One of the most common metals used to harden lead is antimony. The amount

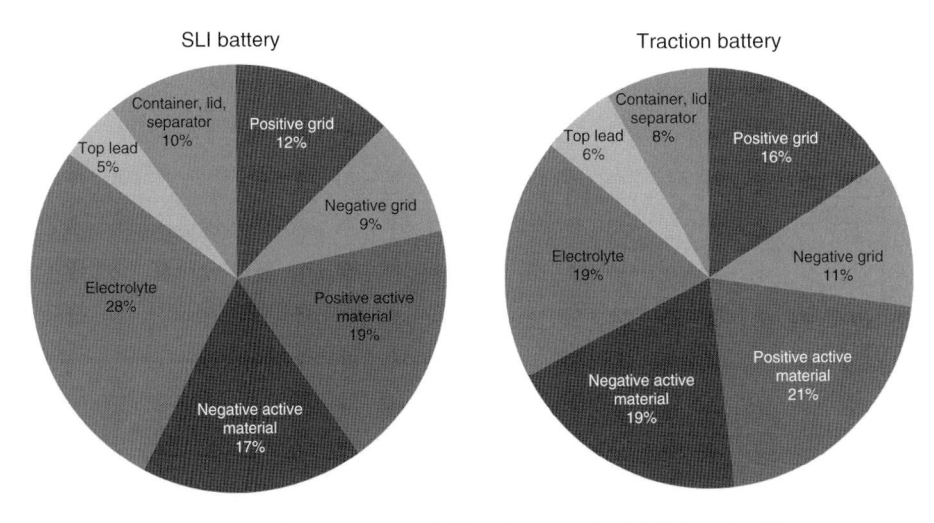

Figure 4.2 Component weight distribution of SLI and traction lead-Acid batteries [1].

of antimony that is used varies between 5 and 12 wt%, depending on the battery applications and the cost of antimony. On the other hand, the modern grid manufacture trend is to lower the antimony content to allow the battery to be maintenance free. This is because the antimony present in the grids dissolves during battery operation and migrates to the negative plates where it redeposits and promotes water splitting to hydrogen and oxygen; therefore, batteries with high antimony content grid require addition of water regularly. As the antimony content goes below 4%, the addition of small amounts of other elements such as cadmium, sulfur, copper, arsenic, selenium, tellurium, and various combinations is necessary to prevent grid fabrication defects and grid brittleness. These added metal elements act as grain refiners and help in decreasing the lead grain size. A widely used lead alloy is lead-antimony-cadmium. Antimony and cadmium react to form an intermetallic compound SbCd. During charge, the positive grid undergoes corrosion and produces antimony in the corrosion layer. The antimony and cadmium give excellent conductivity through the corrosion layer. Although small amounts of cadmium and antimony are leached from grid corrosion and are deposited on the negative grid, research has indicated that cadmium forms a layer over the antimony, raising the potential of the negative plate, which diminishes water loss.

A type of antimony-free lead alloy is developed using calcium or other alkaline earth elements to make lead stiffer. Calcium is added to the lead in the range of 0.03–0.20 wt% depending on different battery manufacturers. The current trend is to use lower calcium levels (0.03–0.05%) for better corrosion resistance. Tin is added in the range of 0.25–2 wt% to the lead–calcium alloy to enhance the mechanical and corrosion-resistant properties. However, owing to the cost of tin, battery manufacturers tend to limit the tin content to the minimum. Some battery manufacturers substitute strontium for calcium. Another metal that is being investigated is barium, but research has shown that it might be detrimental to the performance of the battery.

Another type of lead alloy is the quaternary alloy, which contains lead, calcium, tin, and aluminum. Aluminum is used to stabilize the loss of calcium due to drossing while molten.

4.2.2
Active Material

4.2.2.1 **Lead Oxide**

Lead oxide is the main component of the active material for both the positive and the negative electrode. Lead oxide is made by oxidizing lead by either of the following process: (i) Barton pot process and (ii) ball mill process. The Barton pot process, also called *Barton-like process*, melts lead ingots and then feeds them into a vessel or pot. The molten lead is rapidly stirred and atomized into very small droplets via a rotating paddle in close proximity to the bottom of the vessel. The droplets of molten lead are then oxidized by the oxygen in the air to produce an oxide coating around the droplet. The lead oxidation process is exothermic, in

Table 4.4 Properties of PbO produced from Barton pot and ball mill [2].

Characteristics	Barton pot	Ball mill
Particle size	3- to 4-mm diameter	2- to 3-mm diameter
Stability/reactivity in air	Stable	High reactivity in air
Oxide crystal structures (wt%)	5–30% β-PbO, remaining balance α-PbO	100% α-PbO
Acid adsorption (mg H_2SO_4 g^{-1} oxide)	160–200	240
Surface area (m^2 g^{-1})	0.7	2.0–3.0
Free lead content (wt%)	18–28	25–35
Paste-mixing characteristics	Softer paste	Stiffer paste
Paste curing	Average curing rate	Faster curing rate
Battery performance	Better battery life, low capacity	Good capacity, shorter life
Deep cycle ability	Usually good	Sometimes good
Process control	More difficult	Easier
Production rate (kg h^{-1})	300–900	1000
Operation cost	Low operation and maintenance cost	Higher operation and maintenance cost
Facility requirement	Smaller footprint	Bigger footprint
Energy consumption (kWh ton^{-1})	Up to 100	100–300

which the generated heat is essential for sustaining a continuous reaction as more lead is introduced. The process temperature is critical in determining the degree of oxidation and crystal morphology of lead oxide. The Barton pot process typically produces a product containing lead oxide with 15–30% of free lead, which exists as the core of the spherically shaped particles of lead oxide.

In the ball mill process, lead pieces are fed into a rotating mill and the attrition of the lead pieces produces fine metallic lead flakes. The friction due to the lead flakes tumbling against each other inside the mill chamber creates sufficient heat to oxidize the lead flake surfaces. The degree of lead oxidation is impacted by the airflow through the system. The airflow also moves the lead oxide particles and allows them to be collected in a bag house. The product of the ball mill process also contains 15–30% of free lead in the form of a flattened platelet core surrounded by an oxide coating (Table 4.4).

4.2.2.2 Positive Active Material (Cathode Paste)

The positive active material contains lead oxide and other additives such as red lead (Pb_3O_4), which is more conductive than lead oxide. Red lead is produced by further roasting lead oxide in an airflow chamber until the desired oxidation stage is achieved. Generally, red lead will reduce the free lead content and increase the particle size.

Table 4.5 Conventional positive paste recipe.

Material	Unit	Amount
PbO	kg	25
H_2SO_4	l	1.6
Water	l	4.41
Pb_3O_4	kg	1.25
Graphite powder	kg	0.075
Paste density	$g\ cm^{-3}$	3.7

Table 4.5 shows a conventional positive active material recipe.

4.2.2.3 Negative Active Material (Anode Paste)

Table 4.6 shows the conventional negative active material paste. Organic expanders of lignosulfonates, $BaSO_4$, and active carbon are added to the negative paste. It has been established that, although $BaSO_4$ is isomorphic with $PbSO_4$, it has no effect on the phase composition of the paste. Lignosulfonates fully suppress the formation of $4PbO \cdot PbSO_4$ and orthorhombic-PbO phases. Formation of 4BS is suppressed as a result of the suppressed formation of orthorhombic-PbO, which is a component of the 4BS phase. A family of eight lignosulfonates produced by Borregaard Ligno-Tech (Norway) has been investigated. The beneficial effect of the expander has been found to correlate positively with the low average molecular weight of the lignins and of their derivatives, high COOH content, low OCH_3 group and organic S content, optimum ArOH content, and high lignin purity. Expanders with different chemical compositions should be selected for different types of battery applications. It has also been established that phenol groups of the expander structure react with Sb, Sn, and Ca ions to form organometallic compounds, which influence battery performance.

The function of barium sulfate is to act as a nucleating agent for lead sulfate produced when the plate is discharged. The lead sulfate deposits on barium sulfate particles, assuring homogeneous distribution throughout the active material and preventing coating of lead particles. The term *barium sulfate* represents both blanc fixe and barytes forms of this compound and mixtures thereof in particle sizes from 0.5 to 5 µm. It is desirable that the barium sulfate crystals have a very small particle size, of the order of 1 µm or less, so that a very large number of small seed crystals are implanted in the negative active material. This ensures that the lead sulfate crystals, which are growing on barium sulfate nuclei, are small and of uniform size so that they can be easily converted to lead active material when the plate is charged.

The carbon black increases the electrical conductivity of the active material in the discharged state, which improves its charge acceptance. The carbon is usually in the form of carbon black and/or activated carbon. The amount of carbon in

4.2 Battery Components (Anode, Cathode, Separator, Endplates (Current Collector), and Sealing)

Table 4.6 Conventional negative paste recipe.

Material	Unit	Amount
PbO	kg	25
H_2SO_4	l	1.5
Water	l	3.65
Barium sulfate ($BaSO_4$)	kg	0.2
Carbon black	kg	0.075
Sodium lignosulfonate ($C_{20}H_{24}Na_2O_{10}S_2$)	kg	0.04
Humic acid ($C_{130}H_{140}O_{64}N_9P$)	kg	0.15
Paste density	g cm^{-3}	4

the negative active material of conventional expander formulations is only a small fraction of a percent.

The function of the lignosulfonate is more complex. It is chemically adsorbed on the lead active material, resulting in a significant increase in its surface area. Without lignosulfonate, the surface area is of the order of $\sim$0.2 m^2 g^{-1}, while with 0.50% of lignosulfonate this increases to $\sim$2 m^2 g^{-1}. This high surface area increases the efficiency of the electrochemical process, which improves the performance of the negative plate. The lignosulfonate also stabilizes the physical structure of the negative active material, which retards degradation during the operation of the battery. This property increases the life of the battery in service. The organic material can be any lignosulfonate compound or another suitable organic material that can be adsorbed on the surface of the negative active material and thereby affect its surface area and electrochemical behavior.

4.2.3
Electrolyte

The electrolyte used in lead-Acid battery is sulfuric acid. The properties of sulfuric acid with different specific gravities are given in Table 4.7. The specific gravity of the electrolyte for different type of batteries is shown in Table 4.8, and the specific gravity of the electrolyte for different state of charge is shown in Table 4.9.

4.2.4
Paste Production

Both the positive and negative pastes are prepared into a doughlike material by the following steps using a mechanical mixer:

1) Add lead oxide into the mixer.
2) Add the required additives.
3) Add water.
4) Add sulfuric acid.

4 Lead-Acid Battery

Table 4.7 Properties of sulfuric acid at different specific gravities [3].

Specific gravity		Temperature coefficient α	H_2SO_4			Freezing point (°C)	Electrochemical equivalent per liter (Ah)
@ 15 °C	@ 25 °C		Wt%	Vol%	Mol l^{-1}		
1.000	1.000	–	0	0	0	0	0
1.050	1.049	33	7.3	4.2	0.82	−3.3	22
1.100	1.097	48	14.3	8.5	1.65	−7.7	44
1.150	1.146	60	20.9	13.0	2.51	−15	67
1.200	1.196	68	27.2	17.7	3.39	−27	90
1.250	1.245	72	33.2	22.6	4.31	−52	115
1.300	1.295	75	39.1	27.6	5.26	−70	141
1.350	1.345	77	44.7	32.8	6.23	−49	167
1.400	1.395	79	50.0	38.0	7.21	−36	–
1.450	1.445	82	55.0	43.3	8.20	−29	–
1.500	1.495	85	59.7	48.7	9.20	−29	–

Table 4.8 Specific gravities of various lead-Acid battery electrolytes [1, 3].

Type of battery	Specific gravity	
	Moderate environment	Tropical environment
SLI	1.260–1.290	1.210–1.230
Heavy duty	1.260–1.290	1.210–1.240
Golf cart	1.260–1.290	1.240–1.260
Electric vehicle	1.275–1.325	1.240–1.275
Traction	1.275–1.325	1.240–1.275
Stationary	1.210–1.225	1.200–1.220
Railroad	1.250	1.250
Aircraft	1.260–1.285	1.260–1.285

Table 4.9 Specific gravity of lead-Acid battery electrolytes in various states of charge [1].

State of charge	Specific gravity			
	A	B	C	D
100%	1.330	1.280	1.265	1.225
75%	1.300	1.250	1.225	1.185
50%	1.270	1.220	1.190	1.150
25%	1.240	1.190	1.155	1.115
Discharged	1.210	1.160	1.120	1.000

4.2 Battery Components (Anode, Cathode, Separator, Endplates (Current Collector), and Sealing)

| Pony mixer | Muller | Vertical muller |

Figure 4.3 Conventional lead-Acid battery paste mixer: (a) pony mixer; (b) muller; (c) vertical muller.

Conventional mechanical mixers are batch type. There are three major types of mechanical mixers that are commonly used: the pony mixer, the muller, or a vertical muller. These are shown in Figure 4.3.

The paste viscosity rises at the beginning of paste mixing but gradually decreases. Mixing of the paste releases heat that is generated from the reaction of sulfuric acid and lead oxide. The temperature of the paste needs to be controlled and is often controlled by either cooling the mixer or evaporating the extra volume of water in the paste mixture. The ratio of lead oxide, water, and sulfuric acid varies depending on the type of battery applications. For example, plates for SLI application are generally made at a low $PbO:H_2SO_4$ ratio; and plates for deep cycle application are made at a high $PbO:H_2SO_4$ ratio. On the other hand, the amount of sulfuric acid used affects the plate density as the more the acid used, the lower is the plate density. The paste density is measured by using a cup with a hemispherical cavity and by measuring the paste consistency (viscosity) with a penetrometer.

Teck Inc. has developed a continuous paste mixer to produce lead-Acid battery paste over a broad output and chemistry range. The continuous mixer, as shown in Figure 4.4, has a clamshell with 127-mm-diameter, twin corotating shafts that have paddle and screw mechanisms to mix and drive the material through the chamber. The corotating shafts and the close clearances between the paddles and between the mixer chamber wall and the paddles produce a homogeneous paste by mixing in about 35 s.

The continuous mixer uses a gravimetrical feeder for dry material and a positive displacement pump for liquid material to control the dosage of the ingredients. The continuous paste mixer has the following benefits:

- requires minimum start-up time
- provides a homogeneous paste in 35 s
- fully enclosed and environmentally sealed
- no bag house or cooling ventilation required
- ability to mix high carbon concentrations
- more uniform additive dispersion of fiber and carbon additives
- minimal fiber clumping
- ability to generate tetrabasic crystals of variable sizes and concentration
- ability to regulate paste temperature
- small footprint.

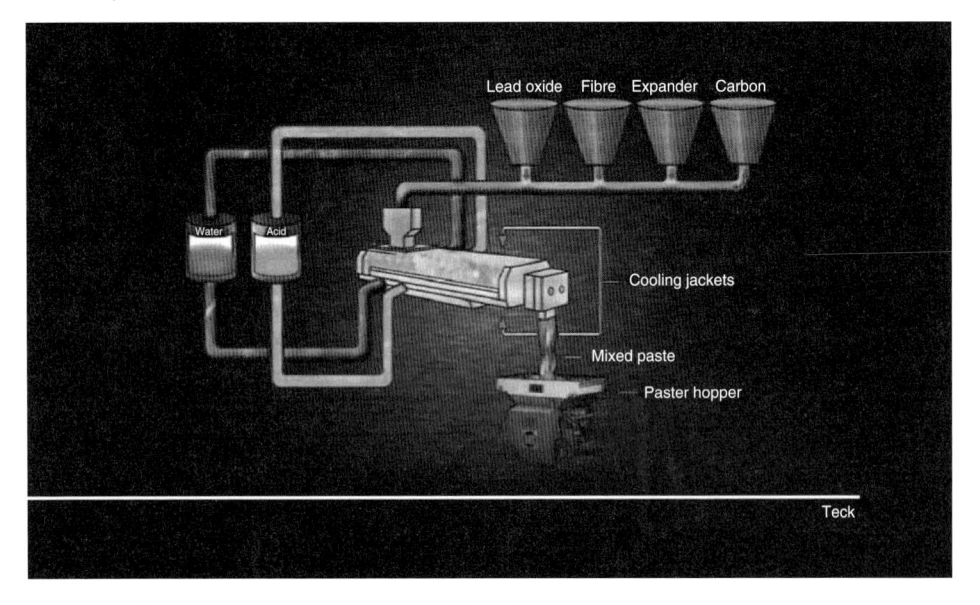

Figure 4.4 Continuous paste mixer by Teck [4].

4.2.5
Pasting

Pasting is a process in which the paste is integrated with the grid to produce a battery plate. This process is carried out through extrusion, and the paste is pressed by a scraper or by a machine into the grid interstices. Two types of pasting machines are used by manufacturers, namely, fixed-orifice paster or belt paster. The orifice paster pushes the paste to both sides of the plate simultaneously, and the belt paster presses the paste to the open side of a grid that is being conveyed past a paste hopper on a porous belt.

The amount of paste that is applied via the pasting machine is regulated by the spacing of the hopper above the grid and the type of troweling such as a roller or a squeegee used at the hopper exit. On the other hand, a trowel roller machine packs the paste thicker and more densely than a squeegee machine.

Grids are automatically or manually placed onto the transfer belt of the pasting machine before being moved under the paste hopper. Most smaller sized plates are made as a double panel that is joined at the bottom or at the top edge, or as multiple panels. Larger industrial stationary or traction battery plates are pasted lengthwise individually.

After pasting, the plates are racked or stacked for curing. Stacked plates contain enough moisture to make them stick together; thus the plates are usually passed through a high-temperature drier or oven to dry the plate surface prior to stacking, the so-called flash-drying process. It has been found that, if the oven is powered by the combustion process, some of the generated carbon dioxide might be absorbed on the plate surface so that the surface becomes harder. The elevated temperature of the flash-drying process also helps start the curing reactions. Small plates are usually stacked horizontally on pallets and thicker, larger plates are usually placed with the long edge upward in racks.

4.2.6
Curing

The curing process is to make the paste bond with the current collector and also to make it into a cohesive, porous material. Generally, the higher the temperature employed during cure, the better is the adhesive bond produced. Curing is an important and time-consuming aspect of the manufacture of the lead-Acid battery as the wet active paste material precursor is cured and converted to a dry porous mass. The battery paste, which has an alkaline pH, reacts with the lead alloy in the grid to partially convert the lead alloy compounds and ultimately to tetrabasic lead sulfate, $4PbOPbSO_4$ (4BS), and tribasic lead sulfate, $3PbOPbSO_4H_2O$ (3BS). The 4BS typically forms at higher temperature (>70 °C) and humidity, whereas the 3BS typically forms at low temperature and humidity. The production of 4BS generally requires a controlled temperature profile and humidity during the curing process. Premature dryout or cooling of the plates inhibits the formation of 4BS. The addition of red lead (Pb_3O_4) can promote the production of 4BS over a larger range of temperature and relative humidity. One of the reasons why the curing process usually takes a long time is that the production of 4BS entails nucleation and growth. Nucleation is carried out at about 70 °C or higher temperature at high humidity. Nucleation during the curing process can have an induction period of about 10 h since 4BS forms as molecules, which slowly coalesce by diffusion into seeds. The seeds react with nearby material to grow into crystals. When curing is adjusted to preclude nucleation and growth of 4BS, a predominance of 3BS is produced. The 3BS has uniform crystal shape and size, that is, 3 μm × 0.5 μm × 0.5 μm. When a plate is pasted with 3BS, the plate has a uniform porosity and high cranking performance.

The discharge capacity of a battery is affected by the porosity and surface area of the porous battery electrode. The 3BS forms as small needlelike crystals, which are about 3-μm-long and less than about 1 μm in width and thickness. The 4BS crystals are larger, and grow in length from several microns to several hundred microns. Longer 4BS crystals have width and thickness in proportion to length. A 300-μm-long 4BS might have a width of 60 μm and a thickness of 50 μm, which will have a surface area of 72 000 μm^2 and a volume of 900 000 μm^3. The 4BS crystal, when packed with 3BS crystals, holds about 10^7 3BS, which corresponds to a total surface area of about 7.2×10^6 μm^2 and represents 1000 times greater surface area. In high-discharge-rate batteries, such as SLI batteries, 3BS is the preferred active material precursor. In deep cycle and long-life stationary batteries, 4BS is the preferred active material precursor. 4BS is also the preferred precursor for use in non-antimony grid batteries as 4BS can help prevent premature capacity loss, which causes shorter battery life.

Lead oxides are yellowish to brownish in color. The color of 3BS is white and that of 4BS is orange. When 3BS is present in cured plates, the 3BS has a pale peach color. The presence of 4BS is revealed by a deep orange color. Sometimes, 4BS in cured pasted plates has a dark greenish color due to the presence of unoxidized free lead. Unoxidized free lead is undesirable because it reacts with sulfuric acid to form lead sulfate during the soak and formation process. Lead sulfate is difficult to convert to lead dioxide on the positive plate, which reduces plate capacity. Unoxidized free lead should not be higher than 2 wt% because this leads to shedding and spalling failure of the positive plates and high self-discharge of the formed lead dioxide plates [5].

The conditions for the curing process vary depending on the paste formulation and the battery applications. The curing process is often carried out using curing ovens with controlled temperature and humidity to ensure that sufficient moisture and heat are available to oxidize free lead in the paste. The curing oven temperature ranges from 65 to 90 °C and the curing oven humidity is maintained above 95%. Curing the paste at higher temperature for extended hours will produce high percentage of tetrabasic lead sulfate and result in a plate with lower energy density. At the end of curing, the free lead content of the paste should be lower than 2%. Table 4.10 shows a typical curing profile.

A conventional curing process for SLI plates is "hydroset" at low temperature and low humidity for 24–72 h. The curing temperature ranges from 25 to 40 °C and the humidity of water from 8 to 20%. The plates are usually covered by canvas, plastic, or paper to retain the moisture. Hydroset is usually carried out in an enclosed room where the temperature and humidity can be controlled by the required climatic conditions. During the hydroset curing process, the plates are under exothermic reaction and reach a peak temperature as the plates cure, and the temperature and humidity decrease. Plates produced by the hydroset curing process will have tribasic lead sulfate, which gives more energy density.

A paste with insufficient curing can be easily broken. The uncured paste is pale in color and soft. If the plate has insufficient curing, it can be rewetted and reheated to force the paste to cure. Another way is to dip the partially cured paste into dilute sulfuric acid, which is called *pickling*. The pickling curing process is also used for curing powder-filled tubular positive plates.

Once the plates are cured, they can be stored and have a long shelf life.

Table 4.10 Typical lead-Acid battery paste curing profile.

No.	Item	Description
1	Loading cycle	Ramp up to 54 °C and 95% RH and hold until all plates are loaded
2	Curing cycle	Ramp up to 80 °C and 95% RH and hold for 72 h
3	Drying cycle	Ramp up to 90 °C and 75% RH and hold for 24 h

4.2.7 Formation

Once the plates are cured, the plates will need to be electrically formed or charged to become functional positive and negative electrodes. During formation, the positive paste is converted to brownish black lead dioxide and the negative paste is converted to soft gray lead. The cured plates can be formed before (tank formation) or after assembly (case formation) into the battery case.

4.2.7.1 Tank Formation

Tank formation is used by some SLI battery manufacturers. The cured plates are formed as doubles, with two to five plates stacked together in a slotted plastic formation tank and facing the counterelectrode in adjacent slots that are spaced 1 inch or less away. In tank formation, all the positive plates are placed on the same side and all the negative plates on the other. All plates with the same polarity are connected by welding a heavy lead bar, and the two bars are connected to a low voltage, constant power supply. The formation tank is filled with electrolyte and current is passed until the plates are formed. Besides forming positive plates together with negatives plates, the plates can also form against dummy pasted plates or bar grids. The common materials used for the formation tank are PVC, polyethylene, or lead. The tanks are arranged so that the acid can be drained and refilled because formation increases the electrolyte acid concentration.

Tank formation conditions depend on battery applications. A variety of formation conditions are used, with variables such as electrolyte density, charging rate, and temperature. The charging electrolyte is typically dilute, in the range of 1.05–1.15 specific gravity. The charging current is fixed by battery manufacturers; some use a sequence of 2–3 charging rates for different periods of time. Tank-formed plates are unstable, as negatives will spontaneously oxidize on contact with air, which require dry charge before use.

4.2.7.2 Case Formation

The more usual method of formation is to completely assemble the battery, fill the electrolyte, and then apply the formation charge.

The two major formation methods used for plates that convert the lead monoxide paste to lead dioxide are as follows: the two-shot formation process (used for stationary and traction batteries) and the one-shot formation process (used for SLI batteries). In the two-shot formation, the electrolyte is dumped to remove the low-density initial formation electrolyte and refilled with a more concentrated electrolyte. The refilled concentrated electrolyte is expected to mix with the dilute initial acid residue, which is absorbed in the plates or trapped in the case.

Table 4.11 Properties of lead-Acid battery separators [1].

Type	Rubber	Cellulose	PVC	PE	Glass fiber	Microglass
Year available	1930	1945	1950	1970	1980	1985
Backweb mils	20+	17–30	12–20	7–30	22–26	10–150
Porosity (%)	55–65	60–70	>40	55–65	80–90	>90
Maximum pore size (µm)	>5	35	25	<1	45	15–30
Average pore size (µm)	3	25	15	<0.1	80	10–15
Electrical resistance in H_2SO_4 (mΩ in^2)	30–50	25–30	15–30	8–40	10	<5
Purity	Good	Fair	Good	Good	Good	Excellent
Corrosion resistance	Good	Fair	Excellent	Excellent	Excellent	Excellent
Flexibility	Brittle	Brittle	Brittle	Excellent	Good	Fair

4.2.8
Separator

Separators are used to electrically insulate between positive and negative plates to prevent short circuit. Separators are porous in order to allow acid transport into or out of the plates. The common separators are microporous polyethylene or absorptive glass mat (AGM). Table 4.11 shows the properties of battery separator materials.

AGM is a nonwoven fabric made with glass microfiber. AGM is widely used for a valve-regulated, lead-Acid battery (VRLA) as it is highly porous (90–95%), can absorb more acid, and has a small pore size to maintain electrolyte levels in the battery.

SLI batteries often use microporous polyethylene in either the leaf or the envelope form, which surround either the positive or the negative plate, or both. The average pore diameters are in the 5- to 30-µm range. Industrial batteries that have deep cycle applications often use AGM separators. Heavy-duty batteries usually have microporous rubber or polyethylene separators, which have finer pores and give longer life. A glass fiber scrim can be used as an additive to some separator material to improve acid retention.

4.2.9
Assembly

A 2-V lead-Acid battery cell consists of at least one positive plate, one negative plate, and a separator between them. Normally, a 2-V cell contains about 3–30 plates with separators in between. Plates and separators are stacked by a stacking machine or manually. The stacked plates are then staged on roller conveyors or carts to the interplate welding operations, where the positive plates and negative plates are welded into individual groups. There are two general methods of welding, the first one is carried out with the lugs of the plates facing upward, and the other is carried out by immersion of the lugs of the plates into a pool of molten lead alloy. The first method is the traditional method for assembling lead-Acid batteries; it is also called *comb mold welding*. In comb mold welding, the plate lugs are fit into slots in a comb mold the shape and the size of the group strap are delineated by the dam and back iron portion of the tooling. A method to increase the efficiency of the comb mold welding is to have premade slotted crowfoot posts to fit over the plate lugs. The second welding process in which the plates are welded with the plate lugs facing downward is called cast-on strap (COS). Most of the SLI Batteries use COS as the standard welding process. In COS, the stacked plates are loaded into slots of the cast-on machine. The cast-on machine places the stacked plates upside down to have the plate lugs facing downward. A mold that has corresponding cutouts to the desired straps and posts is preheated and filled with molten lead alloy. The upside-down stacked plates will immerse into the mold to be welded. External cooling in the cast-on machine then solidifies the strap onto and around each lug. It is important to note that the lead alloy used in the welding process should match with the plate alloy. Lead or lead–calcium alloy should not join with lead–antimony alloy. Comb mold welded plate straps are thicker and smoother than COSs. COSs often show a convex meniscus of metal between adjacent plate lugs on the underside of the strap if the lug is properly cleaned of paste.

Visual examination is important as the first quality check for the weld. A good weld is required between each plate lug and the strap so that it will have minimum voltage drop during high rate discharge performance. Electrical testing for short circuits is the second quality check before further assembly.

Once the weld process is done, the groups are dropped into a battery case for the rest of the assembly process to be carried out. For example, a 12-V, lead-Acid battery consists of six cells. These cells are connected through either a loop over the partition or through the partition wall. The loop over the partition requires long intercell connections, where the cell-to-cell connections travel over the intercell partition wall and are seated in a slot. When connecting the cells through the partition wall, the tabs on the ends of the casted straps are positioned over holes that have been prepunched into the intercell partitions of the battery case and welded together manually using a torch or automatically by a resistance welding machine. The second method also squeezes the tabs and the intercell partition to provide a leak-proof seal.

4.2.10
Case to Cover Seal

Enclosed cells are necessary and important to minimize safety hazards such as internal short circuits from electrolyte mist. There are four different types of processes to seal the case to cover as follows:

1) **Heat (fusion) seal**   Heat seal is carried out by preheating both the case and the cover on a platen, then forcing the two together mechanically or by ultrasonic

welding of the case and cover. Once the battery case and cover is heat sealed, it is almost impossible to repair. If the battery is found damaged, the cover and the case are discarded and replaced with a chance that the grouped plates can be salvaged.

2) **Epoxy cement seal**   Epoxy cement seal is carried out by filling epoxy cement into a groove in the cover. The battery is inserted and positioned so that the case and intercell partition lips fit into the epoxy-filled groove. To activate the epoxy cement, the batteries are often passed through a continuous oven.

3) **Tar (asphalt) seal**   Traditionally, lead-Acid batteries are sealed using tar. Molten tar is dispensed from a heated kettle to fill a groove between the case and the cover. The tar must be hot enough to flow easily but cool and viscous enough to solidify. The tar seal allows easy repair of the battery.

4) **Epoxy glue seal**   The epoxy glue seal employs either solvent cement or thermal seal. Epoxy glue seal is often used for stationary batteries and traction cells so that the coolant can be circulated through the terminals.

4.3 Main Types and Structures of Lead-Acid Batteries

The major types of lead-Acid batteries are SLI batteries, deep cycle batteries, and stationary batteries.

4.3.1 SLI Batteries

SLI batteries are widely used in the automotive industry to start the internal combustion engine. The SLI battery endures a work cycle of discharging briefly at a high current to start the engine; once the engine is running, a generator recharges the battery and then maintains it on float at a full charge or a slightly overcharged status. The parasitic electrical load of lights, motors, and electronics causes a gradual discharge of the battery during engine idle. Because of the working cycle of SLI batteries, they require high-power density and good cycle life while maintaining low cost; the design of the battery depends on the following:

- low internal resistance of the battery
- thickness of the plate and separator
- porosity of the plate and separator
- method of intercell connection
- paste density
- current collector grid material.

High-power density requires the plates and separators to be thin and very porous with a low paste density. On the other hand, high cycle life requires premium separators, high paste density with paste consisting of α-PbO_2 or another bonding agent, and grid containing high antimony (5–7%). In order to maximize the desired performance, optimizations and trade-offs among power density, energy density, cycle life, float service life, and cost are required.

The cranking ability of the SLI battery is directly proportional to the geometric area of the plate surface and the total surface area of the positive plates facing the total surface area of the negative plates. To maximize the cranking ability, usually SLI batteries employ the design of "outside negative," where each 2-V cell consists of $n+1$ negative plates interspersed with $2n$ separators and n positive plates. Typically, the cranking ability is 0.155–0.186 cold crank amperes per square centimeter of positive plate surface at $-17.8\,°C$ ($0\,°F$). The cranking ability is often limited by the positive plate at a higher temperature range ($>18\,°C$) and the negative plate at a lower temperature range ($<5\,°C$). The high cranking capacity of the SLI battery can be obtained by minimizing electrical resistance with a different grid pattern design, employing thin plates to allow more surface area of the positive and negative plates facing each other, and using a higher acid concentration than motive-power batteries or stationary batteries.

Modern SLI batteries are maintenance free; these have several characteristics different from conventional SLI batteries. These characteristics are as follows:

- no addition of water during its life
- higher-capacity retention during storage
- minimal terminal corrosion.

Maintenance-free SLI batteries rely on charge control to prevent electrolysis of water and dry out as compared to the small sealed consumer designs, which rely on oxygen recombination. Maintenance-free SLI batteries have a large acid reservoir, made by using smaller plates and placement of the element directly at the bottom of the container, eliminating the sludge space. The positive plates are usually enveloped in a microporous separator circuit. An important feature of the maintenance-free battery is the use of non-antimony (such as calcium-lead) or low-antimony-lead grids. The use of low or non-antimony-lead grids reduces the overcharge current significantly, reducing water loss during overcharge. The current SLI batteries are also built using different lead alloys for positive and negative plates, where positive plates are built with lead–antimony alloy and negative plates are built with lead-calcium-tin grids.

Generally, SLI batteries have similar designs with small variations depending on the applications. For example, passenger-vehicle SLI batteries use thinner plates, polyethylene separators, and low-density paste. Heavy-duty SLI batteries for trucks, buses, and construction vehicles use thicker plates, glass mat separators, and high-density paste. Figure 4.5 shows a general SLI battery cutaway.

An SLI lead-Acid battery can be discharged at any rate of current without causing damage to the battery as long as the discharge does not go beyond the point where the battery approaches exhaustion or where the voltage falls below a useful value. At a high discharge rate, the capacity is minimized because the electrolyte in the pore structure of the plates becomes depleted and the electrolyte cannot diffuse rapidly enough, which results in the local electrolyte having a lower electrical conductance and higher internal resistance. Cell voltage cannot be maintained because of the

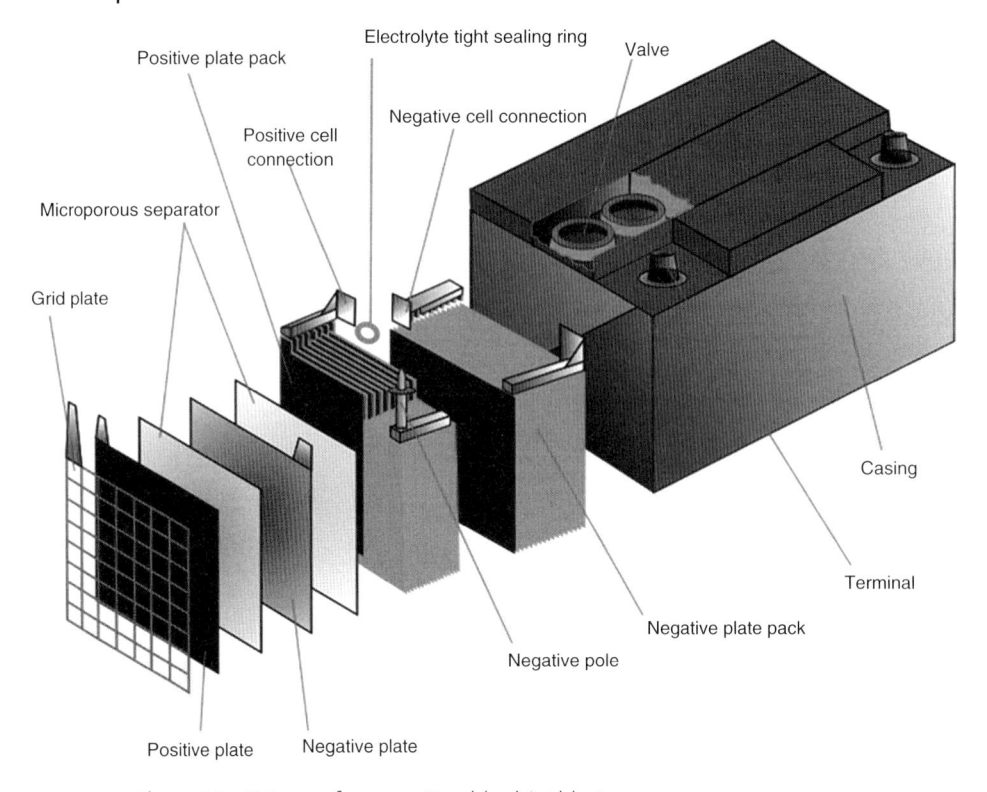

Figure 4.5 Cutaway of a conventional lead-Acid battery.

high internal resistance. Intermittent discharge, which allows time for the local electrolyte to be replenished or recirculated, will improve the performance at a high discharge rate. At a low discharge rate, higher or nominal capacity can be obtained. Figure 4.6 shows a typical SLI discharge curve at various discharge rates.

The performance of the lead-Acid battery is also affected by temperature. Figure 4.7 shows the effect of temperature on the discharge capacity. As shown in Figure 4.7, the capacity of the 20-h discharge rate is greatly affected by lowering the temperature to $-20\,°C$; at $-20\,°C$, it can deliver less than 60% of the battery capacity at $25\,°C$.

SLI batteries deliver high current for engine cranking, which requires low resistance and affects the battery design. As stated before, the plate surface area has a relationship with the cold cranking ampere (CCA), which can be related to the electrochemical double layer of the battery active material. The capacitive reactance component of battery impedance can be determined by the voltage difference at the two levels of the discharge current. The resistance of a lead-Acid battery increases nearly linearly during discharge as the specific gravity of the electrolyte decreases. The resistance difference between a charged battery and a discharged battery is about 40%. The temperature also affects the battery resistance, where the battery resistance increases $\sim$50% between 30 and $-15\,°C$.

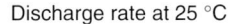

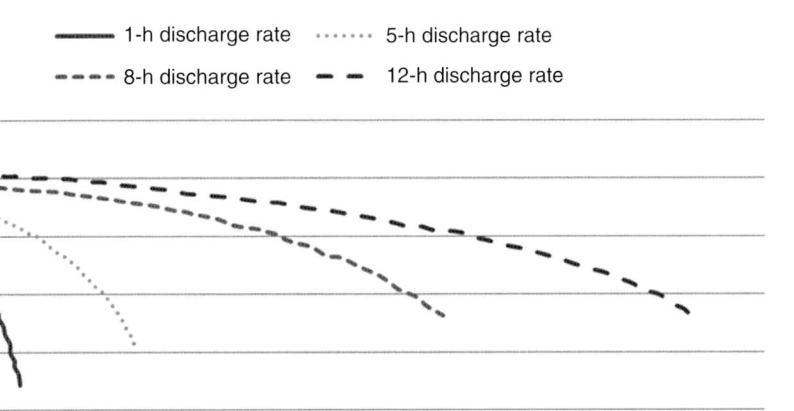

Figure 4.6 Discharge profile of an SLI lead-Acid battery at different discharge rates [1].

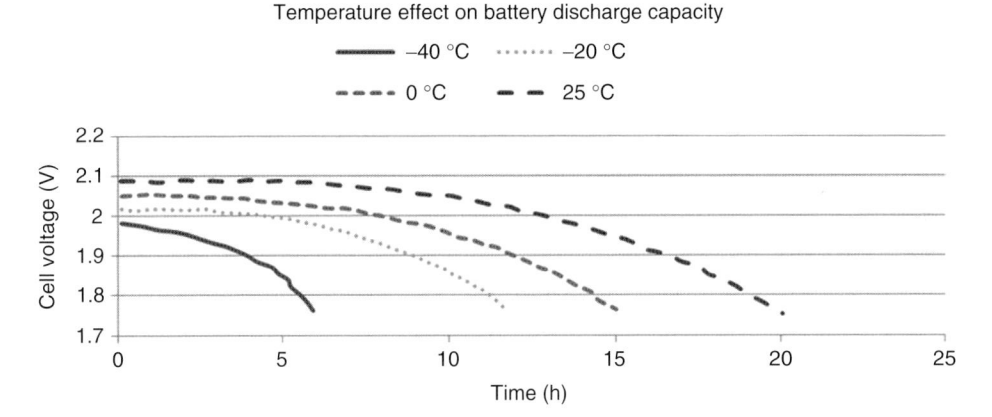

Figure 4.7 Temperature effect on an SLI lead-Acid battery discharge capacity [1].

SLI batteries equipped with lead–antimony grid alloy positive plates endure a severe self-discharge capacity that is lost during an open-circuit stand. Self-discharge can be detected by measuring the voltage drop of the two terminals or the electrolyte specific gravity. Self-discharge occurs as a result of the reaction of sulfuric acid with the active material on the surface of the negative plate to form small particles of lead sulfate; as such, it decreases the sulfate ion concentration in the electrolyte and lowers the electrolyte specific gravity. The high temperature will promote the loss of the electrolyte specific gravity and enhance self-discharge. The phenomenon is more obvious in places where antimony and lead are in contact.

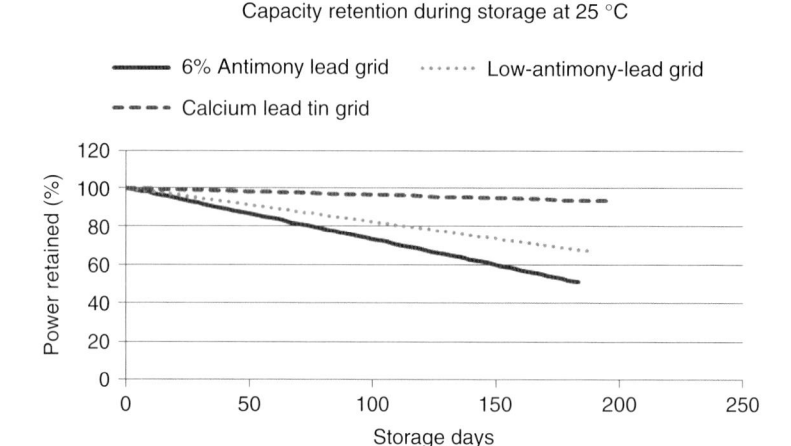

Figure 4.8 Effect of grid alloy on battery capacity [1].

Figure 4.8 shows the self-discharge of three lead-Acid batteries with each one employing different lead grid alloys. Batteries with lead–calcium alloy plates can minimize the self-discharge. For example, batteries using lead calcium (non-antimony) negative and lead antimony positive have less self-discharge than batteries using lead-antimony-lead alloy for both positive and negative plates.

The testing for SLI lead-Acid batteries includes the following tests:

- charge rate acceptance test
- reserve capacity (RC) test
- cranking amps test
- cold cranking amps test
- cycle life test
- vibration test
- capacity test
- charge-retention test
- electrolyte-retention test
- high rate discharge test
- storage test
- overcharge test.

The cranking ampere (CA) test, CCA test, and RC are the most common tests in an SLI battery.

CA is a rating used to describe the discharge load in amperes. The test is conducted with a new, fully charged battery at 32 °F or 0 °C; it can continuously deliver for 30 s and maintain a terminal voltage equal to or greater than 1.2 V per cell. It is sometimes referred to as *marine cranking amps*.

CCA is a rating used in the battery industry to define a battery's ability to start an engine in cold temperatures. The rating is the number of amps a new, fully charged battery can deliver at 0 °F for 30 s, while maintaining a voltage of at least

7.2 V for a 12-V battery. The higher the CCA rating, the greater is the starting power of the battery.

RC is a battery industry rating that defines a battery's ability to power a vehicle with an inoperative alternator or fan belt. The rating is the number of minutes in which a battery at 80 °F can be discharged at 25 A and maintain a voltage of 10.5 V for a 12-V battery. The higher the reserve rating, the longer your vehicle can operate should your alternator or fan belt fail.

Table 4.12 lists the common SLI lead-Acid battery dimensions [3].

4.3.2
Deep Cycle and Traction Batteries

Deep cycle batteries require good cycle life, high energy density, and low cost. The cycle life of the deep cycle batteries is usually longer than that of SLI batteries. The longer cycle life is achieved by employing the following:

- using thick plates with high paste density;
- curing the plates with a high-temperature and high-humidity profile;
- employing a low-specific-gravity electrolyte for formation.

Deep cycle batteries are usually designed to be capacity limited by the amount of electrolyte instead of the active material in the plates in order to protect the plates and maximize the cycle life. Both positive and negative grids will degrade during cycling, but at the end of battery life, the battery capacity is generally limited by the positive plate. The end of deep cycle battery life is when the battery can no longer produce 60–80% of the rated discharge capacity. The end of deep cycle battery life is usually caused by disintegration of the positive active material, PbO_2, and corrosion of the positive grids.

Deep cycle and traction battery cells consist n positive plates and $n + 1$ negative plates. These batteries are often built at high voltage by assembling multiple individual cells. The advantage of this construction is the possibility of replacing one or few cells when the overall battery performance is degraded by a catastrophic failure or a failure of those cells. One indication of the suitability of the traction battery for an application is the change in the specific gravity of the electrolyte during use.

Small traction batteries, such as those used for EVs, are designed to be between full-sized traction batteries and SLI batteries. Small traction batteries employ the traction design concepts that utilize high paste density, and careful control of plate curing and formation to maximize positive utilization. Small traction batteries can also employ SLI design concepts, which include thin-cast radial grids, minimum separator resistance, and through the partition intercell connection. One of the criteria for EVs is high energy density, which results in longer driving range, in which the SLI battery design has prevailed over the traction battery design. In cycling batteries, electrolyte homogeneity occurs by convective flow, where the electrolyte has been found to become stratified during high-discharge-rate applications, such as propulsion in EV application.

Table 4.12 Common SLI lead-Acid battery dimensions [3].

BCI group number	BCI group numbers and dimensional specifications					
	Maximum overall dimensions					
	Millimeters			Inches		
	L	W	H	L	W	H
Passenger car and light commercial batteries 12 V (six cells)						
21	208	173	222	8 3/16	6 13/16	8 3/4
22F	241	175	211	9 1/2	6 7/8	8 5/16
22HF	241	175	229	9 1/2	6 7/8	9
22NF	240	140	227	9 7/16	5 1/2	8 15/16
22R	229	175	211	9	6 7/8	8 5/16
24	260	173	225	10 1/4	6 13/16	8 7/8
24F	273	173	229	10 3/4	6 13/16	9
24H	260	173	238	10 1/4	6 13/16	9 3/8
24R	260	173	229	10 1/4	6 13/16	9
24T	260	173	248	10 1/4	6 13/16	9 3/4
25	230	175	225	9 1/16	6 7/8	8 7/8
26	208	173	197	8 3/16	6 13/16	7 3/4
26R	208	173	197	8 3/16	6 13/16	7 3/4
27	306	173	225	12 1/16	6 13/16	8 7/8
27F	318	173	227	12 1/2	6 13/16	8 15/16
27H	298	173	235	11 3/4	6 13/16	9 1/4
29NF	330	140	227	13	5 1/2	8 15/16
33	338	173	238	13 5/16	6 13/16	9 3/8
34	260	173	200	10 1/4	6 13/16	7 7/8
34R	260	173	200	10 1/4	6 13/16	7 7/8
35	230	175	225	9 1/16	6 7/8	8 7/8
36R	263	183	206	10 3/8	7 1/4	8 1/8
40R	277	175	175	10 15/16	6 7/8	6 7/8
41	293	175	175	11 3/16	6 7/8	6 7/8
42	243	173	173	9 5/16	6 13/16	6 13/16
43	334	175	205	13 1/8	6 7/8	8 1/16
45	240	140	227	9 7/16	5 1/2	8 15/16
46	273	173	229	10 3/4	6 13/16	9
47	246	175	190	9 11/16	6 7/8	7 1/2
48	306	175	192	12 1/16	6 7/8	7 9/16
49	381	175	192	15	6 7/8	7 3/16
50	343	127	254	13 1/2	5	10
51	238	129	223	9 3/8	5 1/16	8 13/16
51R	238	129	223	9 3/8	5 1/16	8 13/16
52	186	147	210	7 5/16	5 13/16	8 1/4
53	330	119	210	13	4 11/16	8 1/4
54	186	154	212	7 5/16	6 1/16	8 3/8
55	218	154	212	8 5/8	6 1/16	8 3/8

Table 4.12 (continued)

BCI group number	BCI group numbers and dimensional specifications					
	Maximum overall dimensions					
	Millimeters			Inches		
	L	W	H	L	W	H

Passenger car and light commercial batteries 12 V (six cells)

BCI group number	L (mm)	W (mm)	H (mm)	L (in)	W (in)	H (in)
56	254	154	212	10	6 1/16	8 3/8
57	205	183	177	8 1/16	7 3/16	6 15/16
58	255	183	177	10 1/16	7 3/16	6 15/16
58R	255	183	177	10 1/16	7 3/16	6 15/16
59	255	193	196	10 1/16	7 5/8	7 3/4
60	332	160	225	13 1/16	6 5/16	8 7/8
61	192	162	225	7 9/16	6 3/8	8 7/8
62	225	162	225	8 7/8	6 3/8	8 7/8
63	258	162	225	10 3/16	6 3/8	8 7/8
64	296	162	225	11 11/16	6 3/8	8 7/8
65	306	190	192	12 1/16	7 1/2	7 9/16
70	208	179	196	8 3/16	7 1/16	7 11/16
71	208	179	216	8 3/16	7 1/16	8 1/2
72	230	179	210	9 1/16	7 1/16	8 1/4
73	230	179	216	9 1/16	7 1/16	8 1/2
74	260	184	222	10 1/4	7 1/4	8 3/4
75	230	179	196	9 1/16	7 1/16	7 11/16
76	334	179	216	13 1/8	7 1/16	8 1/2
78	260	179	196	10 1/4	7 1/16	7 11/16
85	230	173	203	9 1/16	6 13/16	8
86	230	173	203	9 1/16	6 13/16	8
90	246	175	175	9 11/16	6 7/8	6 7/8
91	280	175	175	11	6 7/8	6 7/8
92	317	175	175	12 1/2	6 7/8	6 7/8
93	354	175	175	15	6 7/8	6 7/8
95R	394	175	190	15 9/16	6 7/8	7 1/2
96R	242	173	175	9 9/16	6 13/16	6 7/8
97R	252	175	190	9 15/16	6 7/8	7 1/2
98R	283	175	190	11 3/16	6 7/8	7 1/2

Passenger car and light commercial batteries 6-V (three cells)

BCI group number	L (mm)	W (mm)	H (mm)	L (in)	W (in)	H (in)
1	232	181	238	9 1/8	7 1/8	9 3/8
2	264	181	238	10 3/8	7 1/8	9 3/8
2E	492	105	232	19 7/16	4 1/8	9 1/8
2N	254	141	227	10	5 9/16	8 15/16
17HF	187	175	229	7 3/8	6 7/8	9

(continued overleaf)

Table 4.12 (continued)

BCI group number	BCI group numbers and dimensional specifications					
	Maximum overall dimensions					
	Millimeters			Inches		
	L	W	H	L	W	H
Heavy-duty commercial batteries 12 V (six cells)							
4D	527	222	250	20 3/4	8 3/4	9 7/8
6D	527	254	260	20 3/4	10	10 1/4
8D	527	283	250	20 3/4	11 1/8	9 7/8
28	261	173	240	10 5/16	6 13/16	9 7/16
29H	334	171	232	13 1/8	6 3/4	9 1/8 10
30H	343	173	235	13 1/2	6 13/16	9 1/4 10
31	330	173	240	13	6 13/18	9 7/16
Electric vehicle batteries 6-V (three cells)							
GC2	264	183	270	10 3/8	7 3/16	10 5/8
GC2H	264	183	295	10 3/8	7 3/16	11 5/8

Data excerpted from the BCI (Battery Council International), Battery Replacement Data Book

The relationship of the discharge current to the ampere-hour capacity, up to various end voltages, is shown in Figure 4.9. These data are presented on the basis of the positive plate since cell design and performance data of traction batteries are generally based on the number and size of positive plates that are in the cell. As is typical with most batteries, the capacity decreases with increasing discharge load and increasing end voltage.

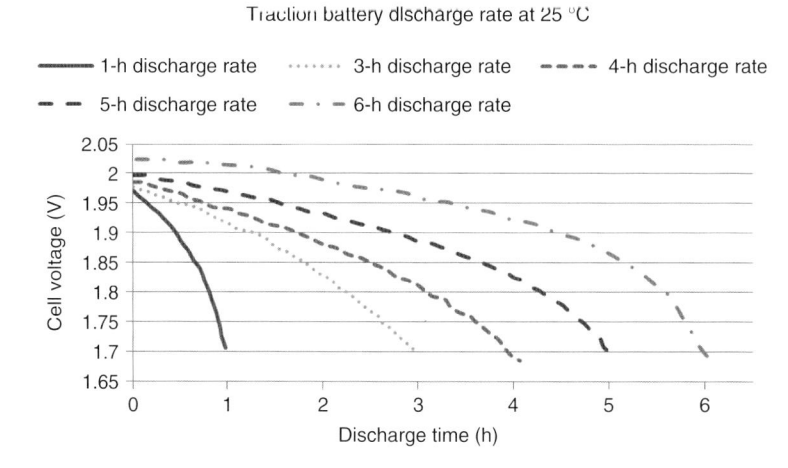

Figure 4.9 Traction lead-Acid battery discharge curve at various discharge rates [1].

4.3 Main Types and Structures of Lead-Acid Batteries | 137

Effect of temperature on traction battery capacity

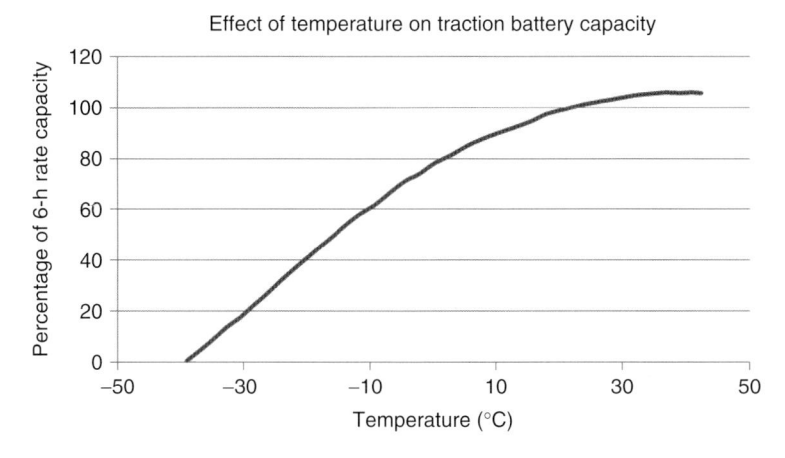

[Graph: Percentage of 6-h rate capacity vs Temperature (°C), ranging from -50°C to 50°C, showing an S-shaped curve rising from near 0% at -50°C to about 105% at 50°C]

Figure 4.10 Effect of temperature on traction lead-Acid battery [1].

The effect of temperature on the discharge performance of traction-type batteries is illustrated in Figure 4.10.

The cycle life characteristics of traction batteries are presented in Figure 4.11. This figure shows the relationship of the cycle life to the depth of discharge at the 6-h discharge rate, *cycle life* being defined as the number of cycles of 80% of the rated capacity. It is quite evident that the deeper the cells are discharged, the shorter is their useful life, and that 80% depth of discharge should not be exceeded if full cycle life expectancy is to be attained. The nominal cycle life of a traction battery is 1500 cycles, which is approximately six years.

Cycle life versus depth of discharge of traction batteries

[Graph: Percentage of nominal cycle life vs Percentage of 6-h rated capacity depth of discharge, showing a decreasing curve from about 300% at 50% DoD to near 0% at 200% DoD]

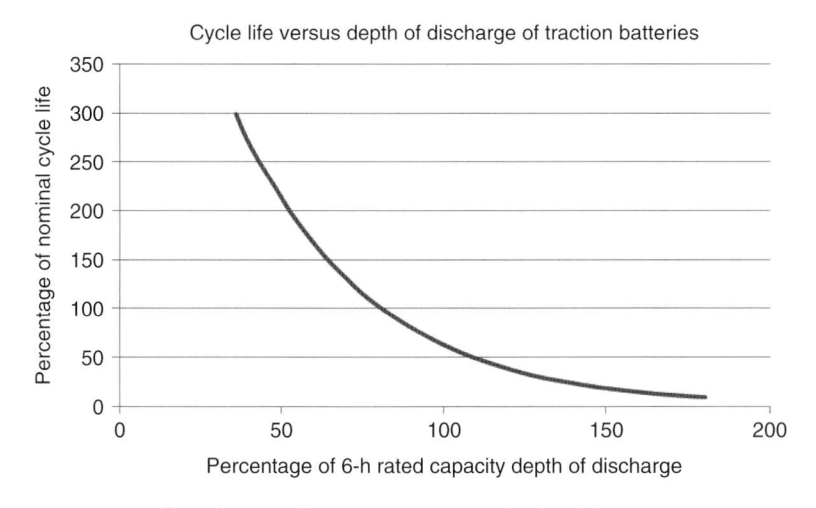

Figure 4.11 Effect of depth of discharge on traction lead-Acid battery [1].

Table 4.13 Traction battery grid dimensions and properties [1].

Positive plate capacity (Ah) at 6-h rate	Plate dimensions (mm)					Positive plate per cell
	Height	Width		Thickness		
		Positive	Negative	Positive	Negative	
45	275	143	138	6.5	4.6	5–16
55	311	143	138	6.5	4.6	5–16
60	330	143	138	6.5	4.6	5–16
75	418	143	138	6.5	4.6	2–16
85	438	146	146	7.4	4.6	3–16
90	489	138	143	6.5	4.6	3–16
110	610	143	143	7.4	4.6	4–12
145	599	200	200	6.5	4.7	4–15
160	610	203	203	7.2	4.7	4–15

Traction or motive-power batteries are made in many different sizes, limited only by the battery compartment size and the required electrical service. The basic rating unit is the positive plate capacity, given in ampere-hours at a 5- or 6-h rate. Table 4.13 lists the typical traction plate sizes using flat-pasted plates. The 2-V cell in the traction batteries is grouped with 2–16 positive plates, or 5–33 positive and negative plates. Traction batteries are often assembled to have battery voltage in 6-V increments (i.e., 6, 12, 18, 24, …, 96 V). The popular traction battery sizes are the 6-cell, 11-plates-per-cell, 75-Ah positive plate (375-Ah cell) and the 6-cell, 13 plates per cell, and 85-Ah positive plate (510-Ah cell) batteries.

Several SLI group sizes have been used for deep cycling applications and EV application. Some of these are listed in Table 4.14.

Table 4.14 Traction lead-Acid battery sizes and properties [1].

BCI group	Volts	Plates per battery	Battery dimensions (mm)			Weight (kg)	Ah C/2	Ah C/3	75 A (min)	Wh kg^{-1} C/3
			Length	Width	Height					
U1	12	54	197	132	186	10	20	22	15	26
24	12	78	260	173	225	22	55	59	39	31
GC2	6	57	264	183	270	26	126	135	100	29
27	12	90	306	173	225	24	62	68	45	32
GC2	6	57	264	183	270	30	150	171	120	33
GC2	6	39	264	183	280	27	158	174	140	37

4.3.3
Stationary Battery

Stationary batteries are designed to have a long service life and are usually under constant float charge. They have the following features in terms of the design and performance characteristics:

- n positive plates and $n - 1$ negative plates (outside-negative design).
  The extra negative is to provide proper support to the positives, which tend to grow or expand during their life. Some manufacturers make the two outermost negative plates thinner than the inside negative plates because the outermost surfaces are not easily recharged.
- Heavy and thick plates with high paste density.
  Unlike SLI and traction batteries, stationary batteries do not focus on high energy and power densities.
- High-humidity plate curing profile to prevent cracks and degradation of the grid–paste interface.
- Excess electrolyte (highly flooded) to minimize maintenance and the watering interval.
  Stationary batteries are designed to endure floating and are moderately overcharged. The overcharge operation of stationary batteries requires a large electrolyte volume and the use of non-antimony-lead grids to minimize gas generation.
- The container is generally overscaled than the positive plate to allow plate growth.
  The overcharge causes some positive grid corrosion, and this is manifested as "growth" or expansion of the grid. Researchers have found that the growth of the positive grids is proportional to their surface area and inversely proportional to their cross-sectional area. The dimensions of the positive plate compared to the inside of the container are scaled higher so that the positives can grow by up to 10% before they touch the container walls. If the growth is greater than 10%, the active material is sufficiently loose on the grid; therefore, the capacity becomes severely positive limited and the battery must be replaced.
- The capacity is generally limited by the amount of positive active material, which increases over time.
  Stationary battery capacity increases over its service life. The capacity increase is due to the lead on the grid surface and grain boundaries being converted to PbO_2. The formation of this PbO_2 on the grid adds to the amount of active paste material of the plate, which increases the plate capacity over time.
- Thermoplastic container
  The containers are usually transparent thermoplastics such as acrylonitrile-butadiene-styrene, styrene-acrylonitrile resin, polycarbonate, PVC, and so on. Other plastic materials such as translucent polyolefin and those used for SLI batteries can also be used for stationary batteries.
- Flame-retardant vent caps
  Batteries for stationary applications may use cells with flat-pasted, tubular, Planté, or Manchester positive plates. The positive plate has a greater influence on the

battery's performance and life. A variety of positives are employed in stationary batteries, depending as much on tradition and custom as on the performance characteristics. The flat-pasted stationary batteries are popular because of their lower costs, lower maintenance, and lower generation of hydrogen. In Europe, Planté and tubular designs are popular because of their longer life. Traditionally, prismatic-shaped stationary batteries have a service life of 5–20 years in telecommunication systems.

The positive plates can be casted with different lead alloys such as lead, lead antimony, lead calcium, lead tin, and lead tin calcium. For most of the standby emergency applications, the grids are cast in a lead–calcium alloy. Some manufacturers use pure lead in place of lead–calcium for the purpose of reducing positive plate growth. The plate is also designed to have a circular and slightly concave shape to counter the effect of growth and ensure good contact between the active material and grid during the life of the battery. At 25 °C, the time to reach 4% growth, an upper limit before the battery's integrity is impaired, is calculated to be 13.8 years for the lead–antimony alloy, 16.8 years for the lead–calcium alloy, and 82 years for pure lead.

The lead–tin alloy (Pb/Sn 72/28 or 65/35) is used to construct the positive terminal to prevent terminal postcorrosion (nodular) and postseal leakage. Batteries constructed with lead–tin alloy show no terminal postcorrosion and no postseal leakage even after being in use for 20 years.

Typical discharge curves for the flat-pasted-type stationary cells at various discharge rates at 25 °C are shown in Figure 4.12. Figure 4.13 shows the effect of the discharge rate on the capacity of the stationary battery. Generally, the discharge rate for a stationary battery is identified as the hourly rate (the current in amperes that the battery will deliver or the rate hours) rather than the C rate used for other types of batteries.

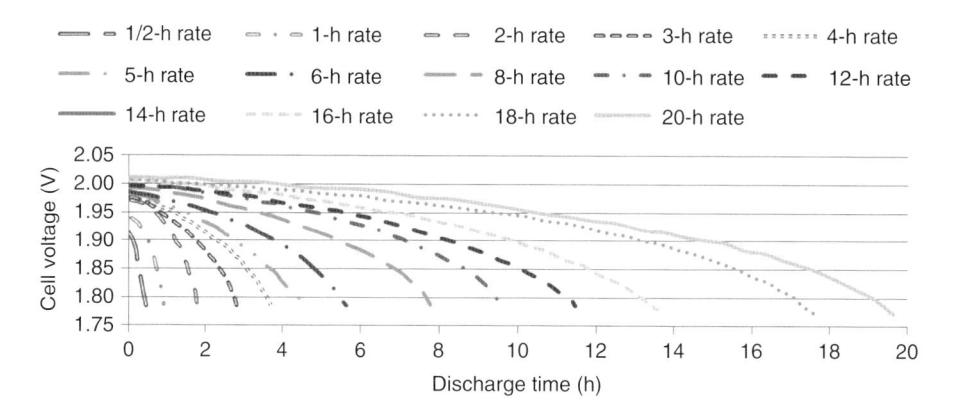

Figure 4.12 Stationary lead-Acid battery discharge curve at various discharge rate [1].

Effect of discharge rate on cell capacity at 25 °C

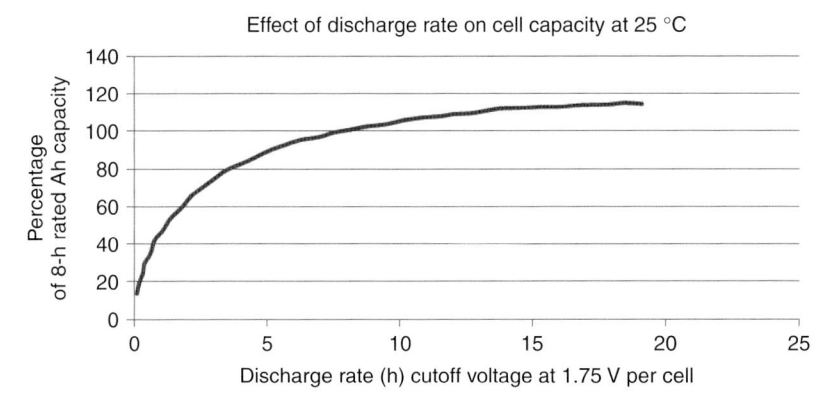

Figure 4.13 Effect of discharge rate on stationary lead-Acid battery [1].

The energy densities of the flat-pasted, positive plate and the tubular positive plate batteries are similar. The energy density is lower for Planté positive plate batteries. The high-rate performance of the flat-pasted positive cells is better because the plates can be made thinner than the tubular or Planté plates.

The battery life is also related to temperature (Arrhenius-type behavior). The optimal temperature for using stationary batteries ranges from 20 to 30 °C, although temperatures from −40 to 50 °C can be tolerated. The percentage of the rated capacity is more available with a higher electrolyte temperature. However, the high-temperature operation promotes self-discharge and reduces battery cycle life.

The rates of self-discharge of various types of stationary batteries are compared in Figure 4.14. Figure 4.14 shows the relative float current at a specified float voltage. The float current under these conditions is a measure of self-discharge. Stationary batteries equipped with lead–calcium alloy pasted positives have the lowest self-discharge and remain low throughout the battery life. The float current is progressively higher for the tubular lead antimony positives, the pasted lead

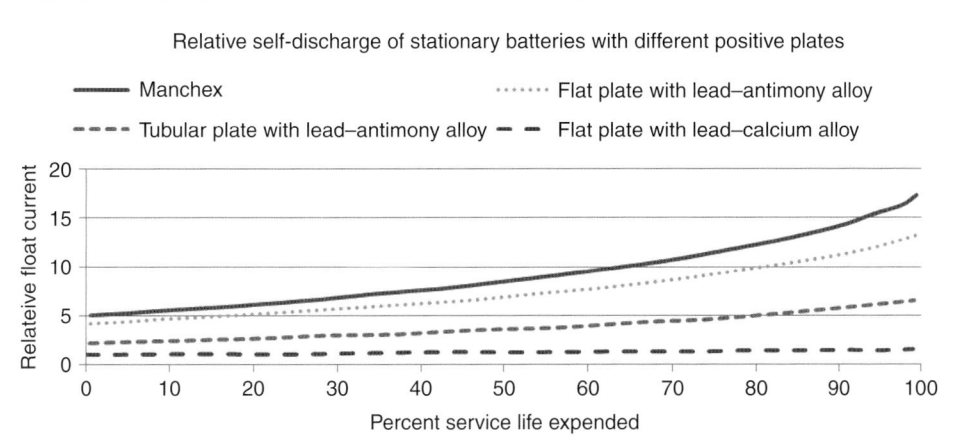

Figure 4.14 Effect of positive plates on stationary lead-Acid battery self-discharge [1].

antimony flat positives, and the Manchester-type positives from the beginning and throughout the battery cycle life.

Researchers have discovered that the float current of the stationary batteries needs to be increased periodically to prevent the lead–antimony alloy plates from becoming progressively self-discharged and sulfated. For fully charged stationary batteries, the self-discharge rate at 25 °C for the lead calcium positive plate cells is about 1% per month, 3% for Planté, and about 7–15% for lead antimony positive cells. At higher temperatures, the self-discharge rate increases significantly, doubling for each 10 °C rise in temperature.

It has been established that a minimum of 50 mV positive and negative overpotential is required to prevent self-discharge. For a stationary battery equipped with lead–calcium alloy grids, a float current of 0.005 A per 100 Ah of battery capacity is required. For a stationary battery equipped with lead–antimony grids, a float current of 0.06 A per 100 Ah of battery capacity is required, but the float current increases to 0.6 A per 100 Ah of battery capacity as the battery is cycled. The higher float current also increases the rate of water consumption and evolution of hydrogen gas.

Generally, stationary batteries equipped with flat, lead–antimony alloy plates have 5–18 years of battery life. Stationary batteries equipped with flat, lead–calcium alloy plates have 15–25 years of battery life. Tubular stationary batteries have 20–25 years of battery life, and the Planté batteries have 25 years of battery life.

Stationary batteries, such as traction batteries, are available in a variety of plate and cell sizes. Tables 4.15 and 4.16 show typical stationary batteries with flat and tubular plates.

The typical stationary battery cell with flat plates has 10 positive 168-Ah plates (a total of 1680 Ah), weighs around 140 kg, and has a dimension of 270 mm × 359 mm × 575 mm (L × W × H).

Table 4.15 Stationary lead-Acid battery plate dimensions and properties [3].

Positive plate capacity (Ah at 8-h rate)	Plate dimensions (mm)				Positive plate per cell
	Height	Width	Thickness		
			Positive	Negative	
5	89	63.5	6.6	4.3	2 or 4
25	149	143	6.6	4.3	1–8
90–95	290	222	7.9	5.3	2–12
150–155	381	304	6.4	4.6	2–17
166	381	304	7.9	5.3	5–16
195	457	338	6.9	5.3	13–18
412	1816	338	7.6	5.5	17–19

Table 4.16 Stationary lead-Acid battery sizes and properties [1].

Positive plate capacity (Ah)		Plate dimensions (mm)				Positive plate per cell
		Height	Width	Thickness		
4-h rate	8-h rate			Positive	Negative	
26	31.25	157	203	8.9	5.6	4
76	96	277P	234P	8.9	–	3–10
		290N	239N	–	6.1	
88	105	277P	234P	8.9	–	3–10
		290N	239N	–	6.1	–
124	152	366	307	8.9	4.8	5–14

The typical stationary battery cell with tubular plates has 11 positive 152-Ah plates (a total of 1672 Ah), weighs around 128 kg, and has a dimension of 272 mm × 368 mm × 577 mm (L × W × H).

4.3.4
VRLA Battery

VRLA battery is the short name for valve-regulated lead-Acid battery; it is also often colloquially called sealed lead-Acid batteries (SLA). The name "valve regulated" does not wholly describe the technology; these are "recombinant" batteries, which implies that the oxygen evolved at the positive plates will largely recombine with the hydrogen ready to evolve on the negative plates, creating water. The water, rather than being released from the cell, is cycled electrochemically to take up the excess overcharge current beyond that used for conversion of active material. Thus, VRLA batteries can be overcharged sufficiently to convert virtually all the active material without loss of water.

VRLA batteries have a safety pressure relief valve. The pressure release valve maintains an internal pressure, and this condition aids recombination by retaining the gases long enough within the cell for diffusion to take place. The oxygen generated on the positive plate reacts with lead at the negative plate in the presence of H_2SO_4 as it can then rapidly diffuse to the lead surface and react with hydrogen to generate water. The valve is a safety feature in case the rate of hydrogen evolution becomes dangerously high, in which case the excess amount of hydrogen is released to the atmosphere through the vent or through the plastic container. The sealable valves are normally closed to prevent the entrance of oxygen from air. The valve allows excess generated hydrogen to vent under a set pressure. The range of venting pressure includes a high of 25–40 psi for a metal sheathed, spirally wound cell to 1–2 psi for a prismatic battery.

In the VRLA battery, the cell is filled with only enough electrolyte to coat the surfaces, thus creating the starved-electrolyte condition. The electrolyte is absorbed either in the separator material or immobilized in a gel. This condition allows for the homogeneous gas transfer between the plates, necessary to promote recombination reactions. The capacity of the VRLA battery is limited by the positive plate or the amount of positive active material. The starved electrolyte and the excess of negative active material facilitate the recombination of oxygen produced during overcharge or float charge with the negative active material.

The electrolyte can be immobilized by two different methods:

1) **Absorption of the electrolyte into the separator** The electrolyte is absorbed into a highly porous AGM that is fabricated from microglass fibers. The AGM acts as a separator between the positive and negative electrodes and also the electrolyte reservoir. The VRLA battery employing an AGM is called an *AGM battery*. AGM batteries were developed by Concorde Aircraft Battery in the late 1980s in San Bernardino, California. The AGM battery technology was pioneered to be a warm-weather, vibration-resistant battery and a chemical alternative to the expensive Ni-Cad batteries in both naval helicopters and fighter aircraft. As production was mechanized, Optima Battery patented a spiral wound method for producing an AGM battery. Owing to the spiral wound cell configuration, they are also sometimes referred to as *spiral wound*.

   AGM batteries are similar to flooded lead-Acid batteries, except that the electrolyte is held in glass mats, as opposed to freely flooding the plates. Very thin glass fibers are woven into a mat to increase the surface area enough to hold sufficient electrolyte on the cells for their lifetime. The fibers that comprise the fine glass fiberglass mat neither absorb nor are affected by the acidic electrolyte they reside in. These mats are wrung out 2–5% after being soaked in acids, prior to manufacture completion and sealing. The AGM battery can accumulate more acid than is available, and never spill a drop.

   The plates in an AGM battery may be of any shape. Some are flat, others are bent or wound. AGM batteries, both deep cycle and starting, are built in a rectangular case to BCI battery code specifications. Optima Battery builds a patented cylindrical AGM series of batteries that are fashioned to fit the same BCI battery size specifications as any other battery.

   All AGM batteries boast of some significant performance enhancement over traditional flooded lead-Acid cells:
   - AGM construction allows purer lead in the plates as each plate does not need to support its own weight based on the sandwich construction with AGM matting. Traditional cells must support their own weight in the bath of acid.
   - AGM batteries are unspillable, keeping lead and acid out of the environment.
   - AGM batteries have high specific power or power density, holding roughly $1.5 \times$ more AH capacity as compared to flooded batteries based on purer lead.
   - AGM batteries have very low internal resistance, allowing them to be charged and discharged quite rapidly without creating heat based on construction and pure lead.
   - AGM batteries are maintenance free, and do not require watering throughout the life of the battery.

- AGM batteries do not corrode their surroundings as the acid is encapsulated in the matting.
- AGM batteries do not freeze and crack, operating well below 0 °F or °C.
- AGM batteries can be UL, DOT, CE, Coast Guard, and Mil-Spec approved to isolate HAZ-MAT.
- AGM batteries are exceptionally vibration resistant based on their sandwich construction.

On the other hand, AGM batteries have the following disadvantages:
- They boast of up to a 10-year lifespan, but must be sized to discharge less deeply than traditional flooded batteries.
- AGM battery depth of discharge for optimal performance is 50%. Flooded battery depth of discharge for optimal performance is 80%.
- AGM batteries do not tolerate overcharging. Overcharging burns its electrolyte, which cannot be replaced, leading to premature failure.

2) **Gelled electrolyte** The electrolyte is immobilized by adding fumed silica. The silica compound hardens the electrolyte into a gel. The battery that uses a gelled electrolyte is called a *gel battery* or a *gel cell*. During formation charge, some water is lost and the gel gets dried up. Upon drying, the gel cracks and fissures are developed between the positive and negative electrodes, which act as the path for the oxygen recombination reaction. Silica reacts with sulfuric acid, and the gelation is a chemical as well as a physical reaction. Unlike a flooded wet-cell, lead-Acid battery, gel batteries do not need to be kept upright. Gel batteries reduce electrolyte evaporation, spillage (and subsequent corrosion issues) common to the wet-cell battery, and boast of greater resistance to extreme temperatures, shock, and vibration. Chemically, they are the same as wet (nonsealed) batteries except that the antimony in the lead plates is replaced by calcium.

The use of VRLA designs is becoming more popular. It accounts for over 75% of telecommunication and uninterruptible power system (UPS) applications. The development of advanced charging techniques has also increased the use of VRLA batteries in cycling applications such as forklift service. New market opportunities in portable electronics, power tools, and hybrid electric vehicles (HEVs) have stimulated the development of new designs of lead-Acid batteries.

On the other hand, VRLA batteries do not have the capability of handling certain types of abuse as well as conventional flooded batteries. The electrolyte, which provides a major internal heat sink in cells, is more limited in VRLA cells. As a result, VRLA designs are more prone to thermal runaway under abusive conditions. This is particularly true when VRLA batteries are subjected to operations at elevated temperatures.

Rapid charging is also not recommended for VRLA batteries. At continuously high overcharge rates (such as $C/3$ and above), gas buildup becomes so rapid that the recombination process is not as highly efficient, and O_2 as well as H_2 gas is released from the cell. In some VRLA batteries, it uses pure lead tin grids, which minimize the evolution of hydrogen on overcharge.

Table 4.17 Various application requirements for VRLA.

No.	Type	Application	Requirement for VRLA
1	Cycle use	Motorcycle	Vibration proof, longer life, higher power
2		Automobile	CCA (cold cranking ampere), higher power (reserve capacity), longer life, vibration proof
3		Electric vehicle, golf cart, forklift, E-bike, invalid wheelchair, and so on	Higher power, longer life, vibration proof
4		Load leveling	Longer life
5	Standby use (float, trickle)	UPS (large, small)	Longer life, higher power, higher capacity

Table 4.17 shows the requirement for VRLA depending on different VRLA applications.

4.4
Charging Lead-Acid Battery

Lead-Acid batteries are charged using DC electric power to reform the active material on the battery electrode back from the discharged state to the charged state. During charging, the lead sulfate on the positive electrode is converted to lead dioxide (PbO_2) and the lead sulfate on the negative electrode is converted to sponge lead (Pb). The electrolyte during charging changes from less concentrated sulfuric acid with a specific gravity of 1.21 to more concentrated sulfuric acid with a specific gravity of 1.30. Charging is temperature sensitive and is affected by different charging methods as the sulfate ions change from the solid to the liquid form during charging. During charging and discharging, the conversion of the sulfate ion state from liquid to solid can result in active material of different morphology, which causes a redistribution of the active material. The redistribution makes the active material get a crystal structure with fewer defects and also helps it become less chemical and electrochemically active. This is a so-called physical degradation. Physical degradation can be minimized by proper charging.

A lead-Acid battery can generally be charged at various charge rates that do not promote excessive gassing, overcharging, or temperature rise. Charging should be controlled so that very little gas is evolved on charge, which minimizes the requirement for watering (less maintenance).

Lead-Acid batteries can take very high charging current during the early stages of the charge. The voltage will rise as the battery is being charged. As shown in Figure 4.15, when a battery is fully discharged, it can take high charging current with the charging voltage remaining low. When the battery becomes charged, the voltage rises if the charging current remains high. The high voltage promotes gas

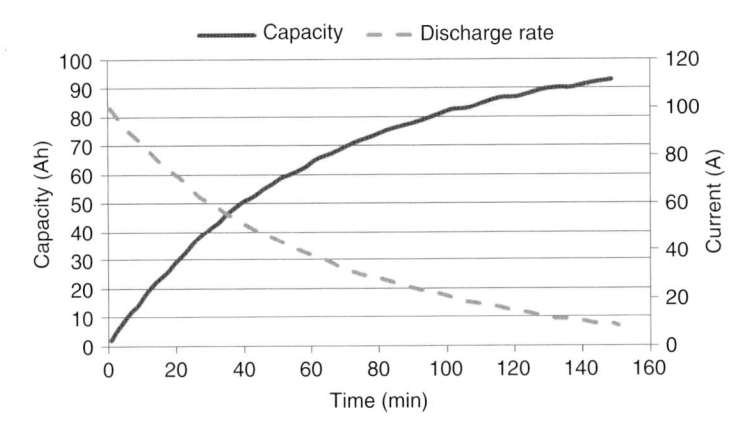

Figure 4.15 Ampere–hour relationship versus charging current [1].

generation and shedding of paste, which results in the battery having capacity loss and a shorter cycle life. It is important to limit the charge current when the battery becomes charged as defined in the following relationship:

$$I = Ae^{-t}$$

where I represents the charging current, A is the number of ampere-hours that had previously discharged from the battery, and t is the time. Figure 4.15 shows the ampere-hour relationship.

The maximum charging voltage is relevant to the current collector, that is, the grid material. When using an antimony–lead alloy as grid material, the maximum voltage is in the range of 14.1–14.6 V for a nominal 12-V lead-Acid battery. When using a lead–calcium alloy grid or other non-antimony grid material, the maximum voltage is in the range of 14.5–15.0 V. The cell voltage at different state of charge under different charge rate is shown in Figure 4.16.

The following are some of the available charging methods:

- **Constant current charge** Constant current charge is used for battery formation and small lead-Acid batteries. Constant current charge is often used in the laboratory because of the convenience of calculating the ampere-hour input and because of using inexpensive equipment. Constant current charging at half the 20-h rate can be carried out in the field to decrease sulfation in batteries that have been overdischarged or undercharged, which means that the batteries that have heavy sulfation can be restored with a long, slow recharge. For example, three to four days at 2–3A for 55 Ah for 12-V SLI batteries. However, a low constant current charge for a normal battery may diminish battery life. A constant current charge with one current charge rate is not widely used because of the need for current adjustment, unless the charging current is kept at a low level throughout the charging process. The disadvantage is that it will prolong the charging time. Multiconstant current charge with multiple decreasing current steps is more common than one current charge. Multiple constant current charges can shorten

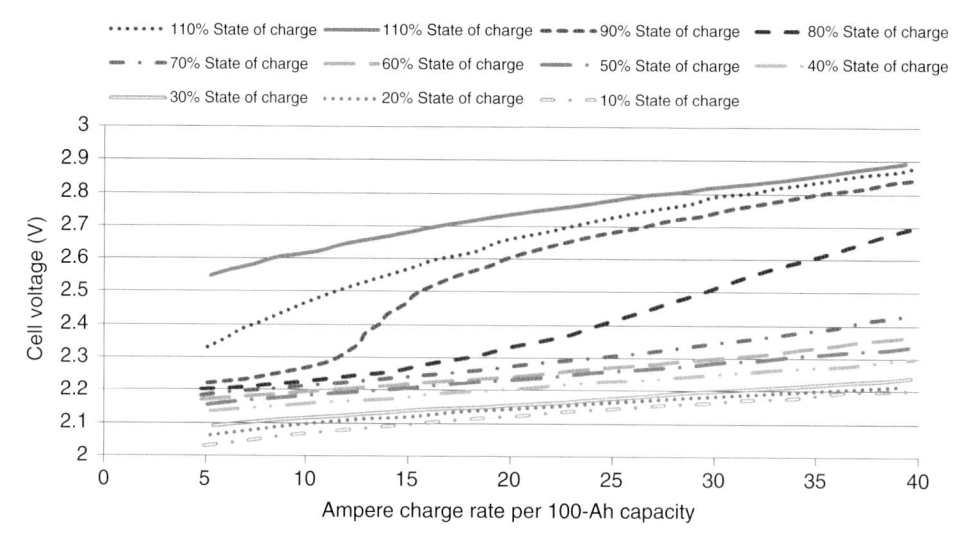

Figure 4.16 Cell voltage at various states of charge [1].

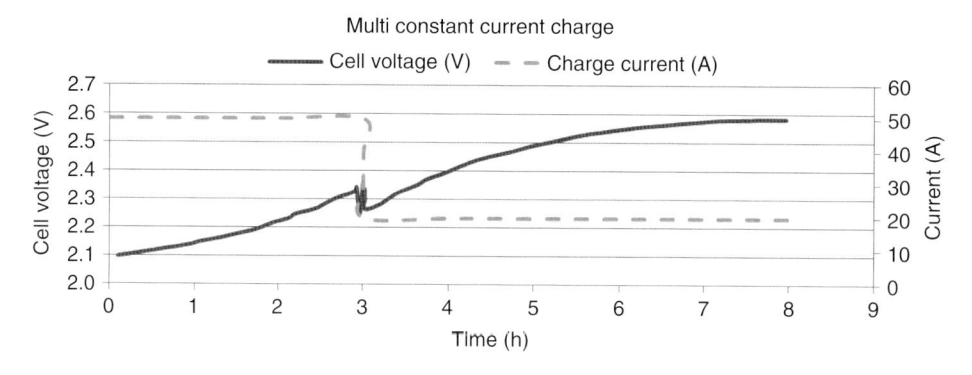

Figure 4.17 Multiconstant current charge profile [1].

the charge time. Figure 4.17 shows the battery cell voltage under two constant current charge stages.
- **Constant voltage charge** Constant voltage (potential) charge is adopted in normal industrial applications to recharge lead-Acid batteries. It is often used for on-the-road vehicles and utility, telephone, and UPS applications where the charging circuit is fixed to the battery. As shown in Figure 4.18, under constant voltage charge, the maximum allowed current is applied until a predetermined voltage is reached. Once the voltage reaches the set level, the current will start to decrease in order to keep the voltage constant until the battery reaches 100% state of charge. During the time interval, the battery is held at a constant voltage and when it reaches 100% state of charge, the battery is charged with a low charge current that can minimize overcharge, grid corrosion from overcharge, water loss by electrolysis of the electrolyte, and maintenance to replenish the water.

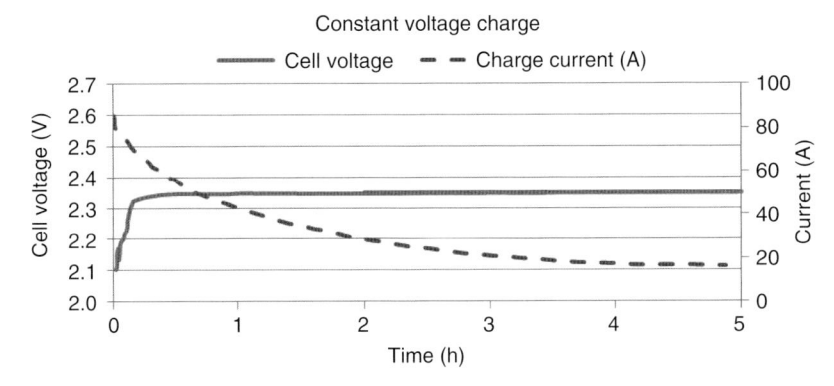

Figure 4.18 Constant voltage charge profile [1].

- **Constant current charge followed by constant voltage charge** The constant current charge followed by constant voltage charge is used for deep cycling batteries that are typically discharged at a 6-h rate to reach a depth of 80% before recharge. The recharge is normally completed in an 8-h period. The charger is set at a constant potential of 2.39 V per cell (the gassing voltage), and the starting current is limited to 16–20 A per 100 Ah of the rate 6 Ah capacity. The initial current is maintained constant until the average cell voltage in the battery reaches 2.39 V per cell. The current decreases at constant voltage to a finishing rate of 4.5–5 A per 100 Ah, which is maintained till the end of the charge. The total charge time is controlled by a timer. The battery, during the total charge time, receives a charge capacity of a predetermined percentage of the ampere-hour output of the previous discharge, normally 110–120%, or 10–20% overcharge.
- **Taper charging** Taper charging is a modified constant voltage charge. Taper charging uses less sophisticated controls to reduce equipment cost. However, the taper charging method reduces the battery life because the charge voltage exceeds 2.39 V per cell at 25 °C before the charge capacity reaches 100% of the prior discharge ampere hours. As the charge voltage exceeds 2.39 V per cell, the battery results in gassing and temperature rising. The gassing voltage decreases with increasing temperature; Table 4.18 shows the voltage-correction factors at various temperatures.

Furthermore, the end of taper charge is controlled by a fixed voltage instead of a fixed current. A new lead-Acid battery has a high counter electromotive force (EMF), which will result in a low final charge rate and the battery will have insufficient charge. When charging an aged lead-Acid battery, which has a low counter EMF, the result is a higher finishing rate and the battery receiving excessive charge [6].
- **Pulse charging** Pulse charging is used for traction application. During charging, the charger is periodically isolated from the battery and the open-circuit voltage of the battery is automatically measured. If the open-circuit voltage is above a present value, depending on the battery temperature, the charger will stop charging the battery. When the open-circuit voltage decays below that limit, the

Table 4.18 Temperature effect on the cell gassing voltage [3].

Electrolyte temperature (°C)	Cell gassing voltage (V)	Correction factor (V)
50	2.300	−0.090
40	2.330	−0.060
30	2.365	−0.025
25	2.390	0
20	2.415	+0.025
10	2.470	+0.080
0	2.540	+0.150
−10	2.650	+0.260
−20	2.970	+0.508

charger delivers a DC pulse for a fixed time period. When the state of charge in the battery is very low, it is connected to the charging current almost 100% of the time because the open-circuit voltage is below the present level or rapidly decays to that level. The duration of the open circuit and the charge pulses are selected so that, when the battery is fully charged, the time for the open-circuit voltage to decay is exactly the same as the pulse duration. When the charger senses this situation, the charger is switched to the finish rate current and short charging pulses are given to maintain the battery at full charge.

If the application has high-voltage batteries, such as forklift batteries, keeping each individual 2-V cell in a balanced condition is difficult, especially when the batteries endure long periods of standby. An equalizing charge can be applied to bring all the battery cells into the same state of charge. The equalizing charge is to completely discharge the battery cells and recharge periodically. In an equalization charge, the normal recharge is extended for 3–6 h at a finishing rate of 5 A per 100 Ah and 5-h rated capacity. This allows the battery voltage to rise in an uncontrolled manner. The equalization charge should be given until cell voltages and specific gravities rise to a constant, adequate value. It is important to check and maintain the electrolyte levels in the cells as some cells might be gassing more than others during charging.

- **Trickle charging** A trickle charge is a continuous constant current charge at a low rate (C/100), which is used to maintain the battery in a fully charged condition by recharging it for losses due to self-discharge as well as for restoring the energy discharged during intermittent use of the battery.
- **Float charging** Float charging is a low rate constant voltage charge also used to maintain the battery in a fully charged condition. The method is used mainly for stationary batteries. One example of float charge is to charge at 2.17–2.25 V per cell when charging a non-antimony grid battery with a 1.21 specific gravity electrolyte.

- **Rapid charging** Rapid charging is desirable in many applications to charge the battery within an hour. It is important to maintain the charge current and battery temperature so that the morphology of the electrode remains the same. If the battery temperature is too high, it will promote grid corrosion, active material conversion to nonconducting oxides, high solubility of active materials, decomposition of active material, gassing, and overcharge.

  Rapid charge can combine with pulse discharge to prevent an excessive rise in the battery temperature during charge. For example, during rapid charge, a brief discharge pulse of a fraction of a second is built-in into the rapid charging protocol.

  In general, charging lead-Acid batteries by any of the charging methods should be able to maintain each individual 2-V cell at a voltage not exceeding the gassing voltage, that is, 2.39 V. When 100% of the previous discharged capacity is reached under this voltage control, the charge rate should decay to the charge finishing rate. The charge finish rate should be at a constant current that is not higher than this rate, normally 5 A per 100 Ah of the rated capacity (or 20 h rate).

4.5 Maintenance and Failure Mode of a Lead-Acid Battery

4.5.1 Maintenance

Proper maintenance of lead-Acid batteries can extend the battery cycle life. The five principles of maintaining lead-Acid batteries are as follows:

1) **Charge the battery with a proper charger that meets the charging requirement** Charging the battery with an inadequate charger is one of the main causes of battery failure. A proper charger will allow the battery to draw only the amount of current that it can accept efficiently, and this current will reduce as the battery approaches full charge. On the other hand, the specific gravity of the electrolyte should also be checked periodically for batteries that have a removable vent and should be adjusted to a correct value.
2) **Do not overdischarge the battery** Overdischarging the battery below the nominal capacity will reduce the electrolyte concentration, which has a deleterious effect on the pore structure of the battery. Battery cycle life has a direct relationship with the depth of discharge as shown in Figure 4.19.
3) **Do not overheat the battery** High temperature (i.e., >55 °C) will promote battery grid corrosion, solubility of metal components, self-discharge, and loss of battery capacity. A high operating temperature during battery cycles will result in the battery requiring higher charge input than room temperature (25 °C) to restore discharge capacity and self-discharge loss. The extra charge input is consumed by the electrolysis of water because of the reduction in gassing voltage at higher temperature. When charging the battery at 35 °C, it will require 10% overcharge to restore the battery capacity compared to

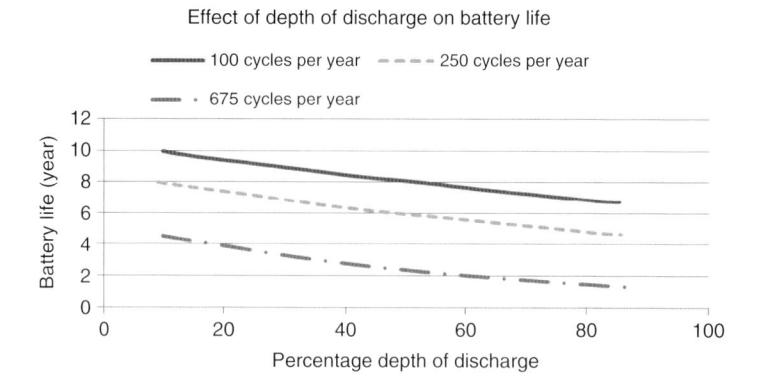

Figure 4.19 Effect of depth of discharge on battery life [1].

charging the battery at 25 °C. When charging the battery at 60–70 °C, it will require 40% overcharge to restore the battery capacity. On floating charging, 11 days' float at 75 °C is equivalent in life to 365 days at 25 °C.

Batteries that are meant for high-temperature applications can start with a lower initial specific gravity electrolyte to allow extra water to be filled into the battery. Research has shown that adding more expander into the negative paste also improves the battery performance at high temperature.

4) **Do not overcharge the battery and maintain the electrolyte level in each individual 2-V cell** The electrolyte level will decrease in each individual 2-V cell from normal operation as a result of water being evaporated or electrolysis into hydrogen and oxygen. Evaporation is considered to be responsible for a small part of water loss, except in hot and dry climates. Electrolysis is considered to be the main factor for water consumption. During overcharge, the water is consumed by electrolysis at a rate of 0.336 ml per Ah of overcharge. A 10% overcharge can consume 0.3% of the water content in each cycle.

The electrolyte inside the battery cells has two functions: the conduction of electricity and heat. If the electrolyte is below the plate level, the area that is not covered by the electrolyte is not electrochemically active. The inactive area causes a concentration of heat in other parts of the cell and promotes grid corrosion. Periodic addition of water to maintain the electrolyte level can also provide indication of the charging efficiency. If the water consumption is high, it indicates poor charging efficiency, and an adjustment of the charging protocol is required.

Water should be replenished after recharge to reach the high acid level line. The replenished water should be distilled water, deionized water, or demineralized water. Gassing during charge will stir the water into the acid uniformly; thus, mixing is not required unless in freezing weather because then the water might freeze before gassing occurs.

5) **Keep the battery clean** Keeping the battery clean will minimize the corrosion of the cell post and connectors. Batteries commonly pick up dry dirt, which needs to be cleaned periodically. If ever there is a spill of electrolyte, the

electrolyte can be neutralized by an alkaline solution such as baking soda with hot water in 1 kg of baking soda to 4 kg (4 l) of water.

4.5.2
Safety

The safety of the lead-Acid battery can be governed by proper precautions such as wearing face shields and rubber gloves when handling. When in contact with lead-Acid batteries, the following are important:

- Wear safety gears to avoid chemical burns from sulfuric acid.
- If eyes, skin, or clothing come in contact with sulfuric acid, flush with plenty of water immediately.
- Obtain medical attention when needed.

Generally, lead-Acid battery safety concerns are as follows:

- **Short circuit and electrolyte spill** An operator error can cause a short circuit; this is a main safety concern in the field. It is important to keep the top of the battery clean to prevent ground short circuits. Instructions should be laid out to educate the operators not to place metallic objects or any electrical conductor on the battery and to insulate all tools used for servicing the batteries. When lifting the batteries, use insulated tools to avoid risks of short circuits between cell terminals and lifting tools.
  Furthermore, when working with lead-Acid batteries that have vent caps, it is important to keep the vent caps tight and level to eliminate acid spill. When there is an electrolyte (acid) spill, an alkaline solution with 1 kg baking soda to 4 kg (4 l) of water can be used to neutralize the acid. The neutralized area should be rinsed with clear water.
- **Potential explosions from the generated hydrogen and oxygen during overcharge**
  One of the common problems that cause batteries to explode is charging the batteries with an unmatched charger, which often causes overcharge and leads to explosion. In spite of the charge currents being low, when lead-Acid batteries are overcharge, the water in the electrolyte is electrolyzed to generate hydrogen and oxygen. The generated hydrogen gas can accumulate and ignite and cause the explosion. One ampere-hour of overcharge can generate 0.42 l of hydrogen and 0.21 l of oxygen at room temperature and pressure. The generated gases are explosive when the content of hydrogen in the air exceeds 4% by volume.
  In telecommunication applications, where a large volume of lead-Acid batteries are assembled in a confined room, exhaust fans should be installed to provide good air ventilation and prevent hydrogen buildup. Hydrogen detectors with the detection limit set at 20% of the lower explosive limit should be installed. As sparks can ignite the hydrogen, electrical sources of arcs, sparks, or flame arrestors should be mounted in explosion-proof metal boxes to eliminate statics.
- **The generation of toxic gases such as arsine and stibine** Antimony and arsenic are used to harden the lead in the grid and to reduce the rate of grid corrosion

Table 4.19 Failure mode of lead-Acid batteries [1].

Battery application	Battery life	Failure mode
SLI	Several years	Grid corrosion
Maintenance-free SLI	Several years	Water loss, grid corrosion, positive material shedding
Golf cart	350–600 cycles	Positive material shedding, grid corrosion, sulfation
Stationary	6–25 yr	Grid corrosion
Traction	Minimum 1500 cycles	Positive material shedding, grid corrosion

during cycling. The batteries containing small quantities of antimony and arsenic form arsine (AsH_3) and stibine (SbH_3) when the hydrogen comes in contact with arsenic or an antimony alloy material. Arsine and stibine are colorless and odorless gases and are extremely dangerous. Exposure to arsine and stibine can cause severe illness and death. In 1978, the United States Department of Labor, Occupational Safety and Health Administration (OSHA) regulated the concentration limits for AsH_3 and SbH_3 to 0.05 and 0.10 ppm, respectively; these can be emitted in any 8-h period [1].

4.5.3
Failure Mode of Lead-Acid Battery

The failure modes of lead-Acid batteries depend on the battery applications and battery construction and design. Lead-Acid batteries usually fail because of positive plate degradation, which is caused by grid corrosion or paste shedding. Table 4.19 shows the common failure mode.

Positive grid corrosion can be caused by grid alloy, grid casting conditions, and active material composition. Shedding of positive active material can be caused by battery construction, active material structure, battery cycles, depth of discharge, and the charging method used. Sulfation can be caused by temperature, the charging method used, and lack of maintenance.

4.6
Advanced Lead-Acid Battery Technology

4.6.1
Negative Current Collector Improvement

4.6.1.1 Ultrabattery
Idling stop, the shutting down of an internal combustion engine instead of idling, is a measure to counteract global warming that has begun to find favor in some

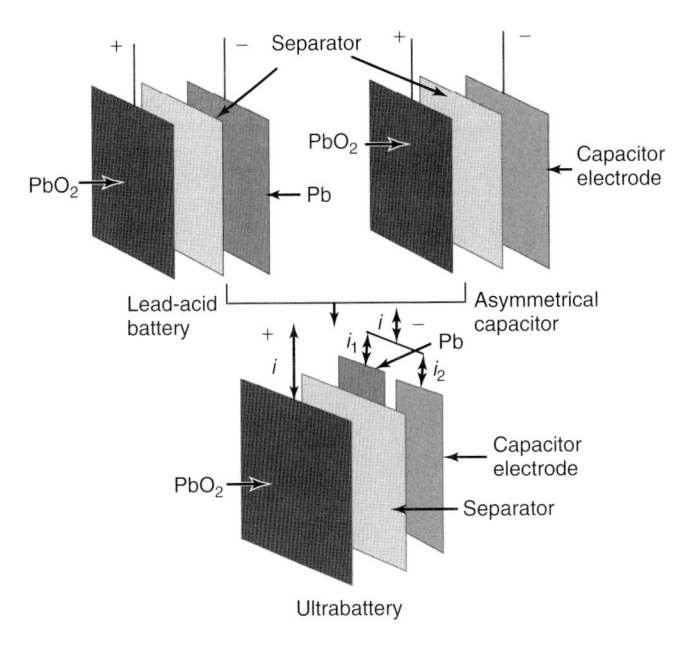

Figure 4.20 Concept of Ultrabattery [7].

taxis and transit buses. It has been reported that in a transit bus idling stop test done by the Tokyo Metropolitan authorities, fuel economy was increased by 14% [7]. In an idling stop cycle life test, reduction in the specific gravity of the electrolyte, sulfation of the negative electrode, and thinning of the lugs are the main factors affecting the service cycle life.

Furukawa Battery in Japan and Commonwealth Scientific and Industrial Research Organisation (CSIRO) in Australia have codeveloped a hybrid lead-Acid battery, called *Ultrabattery* in which the battery and the supercapacitor are integrated at the electrode plate level. Figure 4.20 shows the structure of the Ultrabattery.

The Ultrabattery comprises a lead-Acid battery and an asymmetrical capacitor combined in a single cell. The lead-Acid battery has a positive electrode of lead dioxide and a negative electrode of spongy lead. The asymmetrical capacitor has lead dioxide, the same positive electrode as the battery, and a negative electrode of porous carbon. Since both have a positive electrode of lead dioxide, the lead negative electrode and the capacitor negative electrode can be connected in parallel and housed in the same cell with the positive electrode. The resulting battery has the capacitor electrode bearing a portion of the load of the lead storage electrode.

The capacitor electrodes are made of porous carbon, which would generate hydrogen gas during charge. Hydrogen inhibitors (additives) are important in the Ultrabattery chemistry to suppress the evolution of hydrogen gas during charging. Figure 4.21 shows the amount of hydrogen gas that evolved from the capacitor electrode with hydrogen inhibitors, capacitor electrode without inhibitors, and a conventional lead electrode. As shown in Figure 4.21, the amount of hydrogen

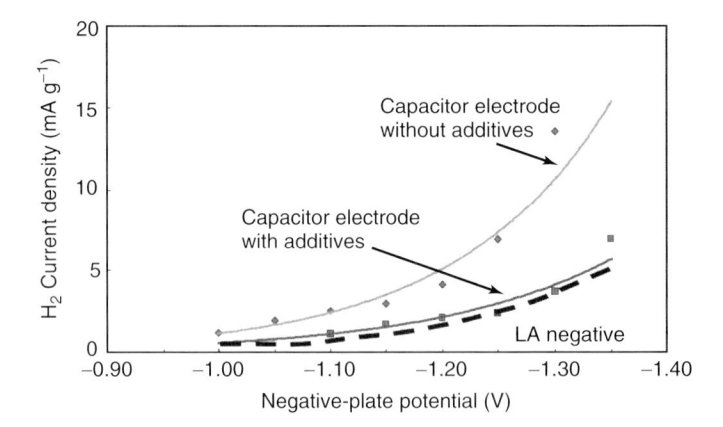

Figure 4.21 Hydrogen inhibitor effect on H_2 gas evolution in capacitor electrode [7].

evolved as potential was varied. In contrast to the capacitor electrode without additives, the amount of hydrogen that evolved from the capacitor electrode with hydrogen inhibitors was the same as that from the lead electrode.

A cycle life comparison between a 2-V VRLA battery and a 2-V valve-regulated Ultrabattery has shown that the life span achieved in an Ultrabattery is about four times longer than that in a lead-Acid battery. The cycle life tests were carried out in a partial state of charge based on a reliable, highly optimized lead-Acid battery (RHOLAB) profile, which was simulated by a combination of high speed and hill climbing running conditions for mild hybrid vehicles.

4.6.1.2 PbC Capacitor Battery

Axion Power International Inc. has developed a PbC capacitor battery as shown in Figure 4.22. A complete technical description of Axion's proprietary *PbC*® technology is a "multicelled asymmetrically supercapacitive lead-Acid-carbon hybrid battery." Similar to a lead-Acid battery, the PbC battery consists of a series of cells. Within the individual cells, however, the construction is more complex. While the negative electrodes in lead-Acid batteries are simple sponge lead plates, negative electrodes in a PbC battery are five-layer assemblies that consist of a carbon electrode, a corrosion barrier, a current collector, a second corrosion barrier, and a second carbon electrode. These electrode assemblies are then sandwiched together with conventional separators and positive electrodes to make the PbC battery, which is filled with an acid electrolyte, sealed, and connected in series to the other cells.

Axion Power International Inc. has revealed test data showing that *PbC* batteries have withstood 1600 cycles before failure. The test protocol calls for a complete charge/discharge cycle every 7 h to a 90% depth of discharge. In comparison, most lead-Acid batteries designed for deep discharge applications can only survive 300–500 cycles under these operating conditions.

Axion has stated that the PbC battery has several key performance advantages over conventional lead-Acid battery including

Activated carbon electrodes

Figure 4.22 Axion PbC battery concept [8].

- significantly faster recharge rates;
- significantly longer cycle lives in deep discharge applications; and
- minimal required maintenance.

4.6.1.3 Firefly Oasis Battery

In 2009, Firefly Energy Inc. (Firefly) revealed its Oasis Battery developed by Kurt Kelly. This battery is equipped with conventional lead grids as positive current collectors and carbon-graphite foams as negative current collectors. Firefly has stated that the carbon-graphite technology is a 3D technology. Owing to the replacement of conventional lead grids with a carbon-graphite foam grid, the battery results in a much improved negative active material utilization (i.e., from 0 to 50% up to the range of 70–90%), as well as enhanced fast recharge capability and greater high-rate-discharge cycle life and low-temperature performance [9].

The negative carbon-foam plate has a structure in which each plate contains hundreds or thousands of spherical microcells, as shown in Figure 4.23. The microcells provide a high surface area, which increases the interface area between

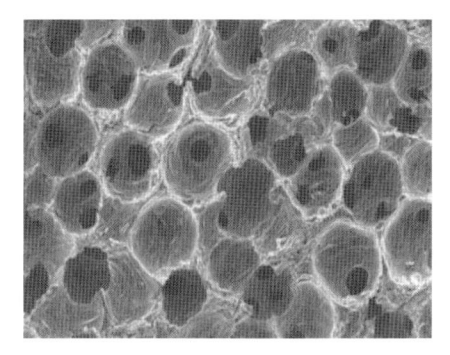

Figure 4.23 Firefly carbon-foam microcell structure [9].

the active chemistry and the current collectors. Each microcell has its full complement of a negative sponge lead paste and a sulfuric acid electrolyte. Liquid diffusion distances in the microcell are reduced from the traditional levels of millimeters over linear paths to the level of micron meters. The diffusion path lengths in the 3D space within the discrete microcells collectively comprise a totally new type of structure. Such a structure results in much higher power and energy delivery and rapid recharge capabilities as compared to conventional lead-Acid batteries. As such, Firefly has stated that the Oasis Battery has driven up both gravimetric and volumetric energy and power values.

Firefly has stated that hydrogen overvoltage levels on the carbon-graphite negative plates are comparable or slightly superior to conventional lead negative grids. This implies that the gassing levels, which correspond to water loss rates, are low. Firefly has published that the self-discharge rates are very low, ~ 0.3 mV/cell-day.

Firefly has stated that its carbon-graphite foam negative current collector has better heat transfer characteristics than conventional lead grids. Thermal images of the heat transfer capability of a Firefly 2-V cell and a comparable commercial VRLA cell were taken. Both cells were subjected to a 3C discharge rate, with thermal images being taken every 15 s.

The thermal images show that even though Firefly's 2-V cell utilizes a standard lead grid positive plate, its carbon-graphite foam negative plate operates at a much cooler level, which also results in a temperature reduction of its lead grid positive. The thermal image also revealed that the carbon-graphite foam negative plate has a cooler, more uniform heat signature throughout the discharge relative to other VRLA cells' positive plates. The carbon-graphite foam negative plate not only dissipates the heat generated on it but also absorbs heat away from the positive plate. The thermal images suggested that the carbon-graphite foam negative plate can transfer heat out of the battery rapidly as it is generated by the electrochemical reactions taking place, thus making thermal runaway less likely, and overall making it a lower temperature operation.

Firefly has stated that its carbon-graphite foam negative plate technology provides a design that accommodates the volume changes of the active material during charge and recharge. The carbon-graphite foam structure allows complete discharge without causing physical stress on the plate structure.

Some applications such as batteries for hybrid vehicles and photovoltaic (PV) energy storage require a partial state of charge. These types of applications often result in the lead-Acid battery having heavy sulfation of the negative plate. The sulfation of the negative plate limits the lead-Acid battery cycle life. Firefly has stated that the lead sulfate deposition on its carbon-graphite foam negative plate is different from conventional lead grids. During deposition, the lead sulfate deposits on the lead grid surface in dense layers of relatively large crystals. The large crystals of lead sulfate are difficult to convert back and cause performance loss. Firefly has claimed that its microcell carbon-graphite foam has a sponge lead active material on the wall of its micro pores in thin layers. When discharged, the lead sulfate deposits comprise small, porous crystal structures that can be easily dissolved on subsequent recharge and will cause less performance loss.

4.6.2
Current Collector Improvement

Dr. Pavlov summarized the relationship between battery-specific energy in watt hours/kilogram and the number of battery discharge/charge cycles for both flooded and valve-regulated type of lead-Acid batteries. For both battery types, the higher the battery-specific energy, the lower is the number of discharge/charge cycles and, hence, the battery cycle life. Typically, a flooded battery with a specific energy of 40 $Wh\,kg^{-1}$ can be used for about 500 discharge/charge cycles, while a battery producing only 30 $Wh\,kg^{-1}$ can be employed for about 850 cycles. Thus, there is clearly a need to improve both the specific energy and cycle life of lead-Acid batteries in order to make them more suitable for electric traction applications.

It is well known that the low utilization efficiency of the active mass, especially on the positive electrode, in conjunction with the heavy weight of the lead current collectors, limits the actual specific energy of the lead-Acid battery. The structure of the current collector plays an important role in determining the utilization efficiency of the positive active mass (PAM). During discharge, on the positive electrode, the structure of the current collector must allow for significant volume increase (e.g., molar ratio of $PbSO_4$ to PbO_2 is 1.88) while maintaining electrical contact with the active material and assuring ionic transport to the electroactive sites. $PbSO_4$ has a significantly larger volume (about 37% more) than the charged active material. The expansion from the volume change stresses the electrode structures and induces mechanical forces that deform the grid. The resulting expansion and deformation of the plates also causes active material to separate from the current collectors.

There are many examples in the literatures describing techniques to increase the specific energy output by improving the porosity and specific surface area of the lead-compound-based paste (active material) applied onto the battery current collector (or grid). For example, Stoilov *et al.* in US Patent No. 5,332,634 state that "there is a need for making lead electrodes with a porous active mass, which has a large active surface area and which strengthens the electrical connection between the active mass and the grid. Such a porous lead electrode would lead to electrochemical cells and accumulators which produce more power per unit of weight and also present very low electrical resistance."

Regarding improvements in the battery current collector structure, Czerwiński and Zelazowska have described the electrochemical behavior of lead deposited on a nonmetallic open pore substrate, namely, reticulated vitreous carbon (RVC) [10]. These researchers prepared small, 1-cm^2-geometric-area collectors by electrodeposition for 10 min of Pb from an alkaline solution containing 20 $g\,l^{-1}$ NaOH to produce the negative electrode and anodic oxidation to form lead dioxide (PbO_2) on the positive electrode using a concentrated lead-nitrate-based solution ($Pb(NO_3)_2$). The amount of the generated active material, Pb and PbO_2, was small at about 19.3 and 22.3 mg respectively. Consequently, if a battery had been assembled with the above-described electrodes, the corresponding capacity would have been extremely low, in the range of 4.5 mAh, which is insufficient for practical use.

Furthermore, the battery structure described by Czerwiński and Zelazowska might not be rechargeable in sulfuric acid, which is the operational electrolyte of lead-Acid batteries, since the recommended active material generation procedure requires alkaline- and nitrate-based electrolyte. Therefore, it creates a onetime-use lead-Acid battery.

Researchers like Das and Mondal [11] have suggested developing lead-Acid current collectors with thin layers of active materials deposited on light-weight, electronically conducting substrates, such as a carbon rod. The rationale was only to reduce the "dead weight" of the lead-Acid system, which would somewhat increase the specific energy.

Snaper, in the United States Patent No. 6,060,198 describes the use of reticulated metal structures for use as current collectors in batteries in which the reticulated structure consists of a plurality of pentagonally faced dodecahedrons. This article does not teach methods for using such a structure to improve the cycle life and performance of a lead-Acid battery and does not envisage the use of nonmetallic, electrically conductive substrates such as reticulated carbon to reduce battery weight. None of the above-mentioned article references regarding reticulated structure suggest any need for combining the reticulated structure with a lead-containing paste to create a rechargeable battery suitable for use in multiple charge/discharge cycles.

4.6.2.1 Lead-Alloy-Coated, Reticulated Carbon Current Collectors

In 2001, Dr. Elod Gyenge and Joey Jung developed a high-surface-area current collector for both positive and negative electrodes. The current collectors were fabricated by depositing a lead–tin alloy on a reticulated carbon plate. Figure 4.24 shows a comparison between the conventional grid and the lead–tin alloy coated current collector.

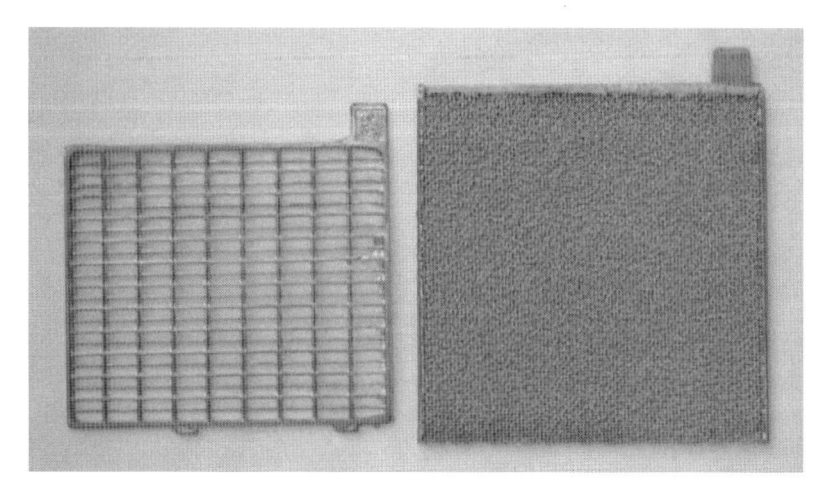

Figure 4.24 Comparison between conventional grid and lead–tin alloy coated current collector.

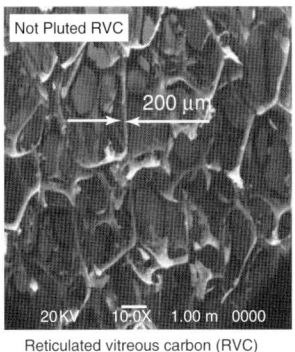

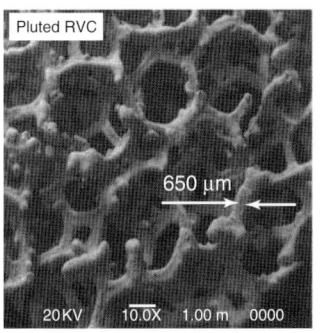

Figure 4.25 Uncoated RVC and lead-tin-coated RVC [12].

The reticulated-carbon-based current collector structure is porous and has a high specific surface area (e.g., $>500\,m^2\,m^{-3}$). The substrate is electronically conductive, 3D RVC characterized by a high specific surface area (i.e., between 5×10^2 and $2 \times 10^4\,m^2\,m^{-3}$) and void fraction (i.e., between 70 and 98%). A number of materials could also serve as the above-mentioned substrate such as reticulated aluminum and reticulated copper.

Furthermore, an integral part of the structure consists of a layer of lead and tin deposited throughout the surface and depth of the 3D reticulated carbon to cover all the ligaments of the substrate matrix as uniformly as possible. Figure 4.25 shows the photograph of an uncoated piece of RVC and a piece of lead–tin alloy coated RVC. The thickness of the deposited lead-alloy layer can range, for example, between 100 and 300 µm. The thickness of the lead tin coating depends on the intended application and battery cycle life. The lead-tin-coated RVC is used as the positive and/or negative current collector in lead-Acid batteries.

Figure 4.26 shows the cyclic voltammogram of the lead tin (1 wt%) electrode-posited RVC recorded at a sweep rate of $5\,mV\,s^{-1}$ for potentials between -1.4 and $1.6\,V$. The voltammogram shows the characteristic electrochemical features of lead–tin alloys. On the anodic scan, the peak at $-0.89\,V$ corresponding to the oxidation of lead (Pb) to lead sulfate ($PbSO_4$) (a1, Figure 4.26) is followed by a small shoulderlike wave at $-0.78\,V$, formed by the oxidation of tin (Sn) (1 wt% in the alloy) to $SnSO_4$ (a2, Figure 4.26). Further, a peak is formed at a potential of $1.6\,V$ corresponding to O_2 evolution and PbO_2 formation, when the current density is about $10\,mA\,cm^{-2}$ (a4, Figure 4.26). For comparison, a reticulated structure without the presence of Sn gave about an order of magnitude higher current density at a potential of $1.6\,V$, as shown in Figure 4.26. Thus, the presence of Sn increases the overpotential for O_2 evolution and PbO_2 formation.

Scanning in the cathodic direction, first an oxidation peak (a3, Figure 4.26) was obtained at $1.2\,V$. The presence of this anodic peak was attributed to either H_2O oxidation by PbO_2 or to cracking associated with an increase in the molar volume when PbO_2 was reduced to $PbSO_4$. Further on the cathodic scan, owing to the presence of Sn in the alloy, the PbO reduction peak potential (c1, Figure 4.26) was

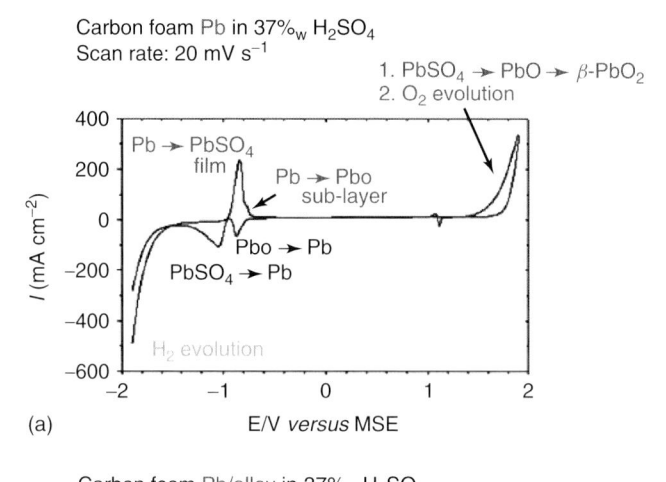

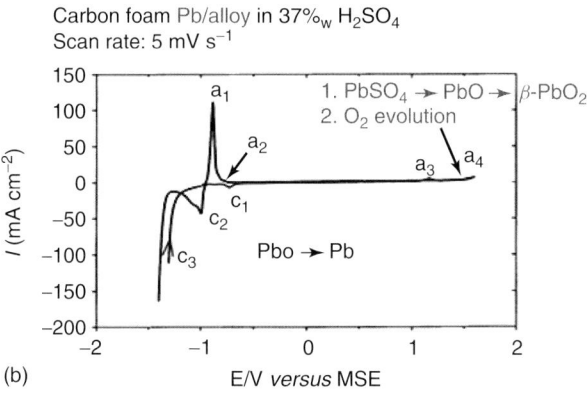

Figure 4.26 Cyclic voltammogram of the lead tin (1 wt %) electrodeposited RVC: (a) Pure lead (Pb) coating on carbon foam; (b) Lead alloy coating on carbon foam [12].

obtained at a more positive potential, that is, −0.74 V (Figure 4.26). Typically, in the absence of Sn, the PbO reduction peak potential is around −0.88 V. The shift of PbO reduction to more positive potentials is part of the well-documented "tin effect" for lead-Acid battery current collectors manufactured from Sn-containing Pb alloys. At more negative potentials, (i.e., −1 V vs MSE), two electrode reactions were observed: reduction of $PbSO_4$ to Pb (c2, Figure 4.26, peak potential −1 V) and H_2 evolution (c3, at potentials < −1.2 V).

Figure 4.27 shows the nominal capacities (Peukert diagram) at discharge rates between 0.5 and 15 h for the positive-limited reticulated battery. At a discharge current of 27 A kg^{-1} PAM, corresponding to a discharge rate of 3 h, the PAM utilization efficiency on the electrodeposited reticulated collector was 47% (specific capacity 107 Ah kg^{-1} PAM) compared to 26% on the conventional lead grid. Discharging the reticulated battery at a 15-h rate (6.5 A kg^{-1} PAM) yielded a PAM

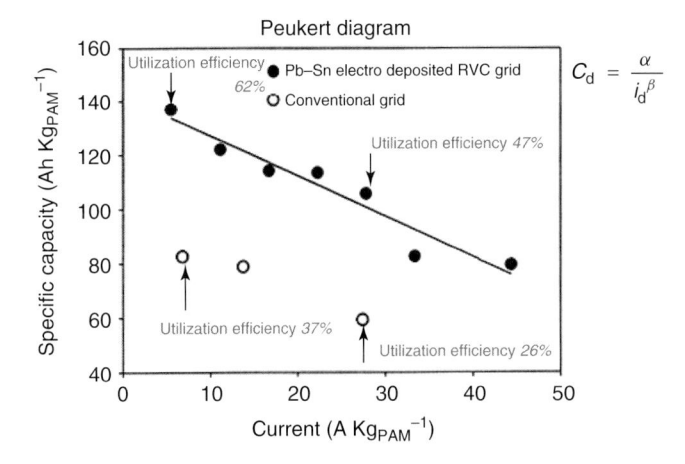

Figure 4.27 Peukert diagram of lead-tin-coated RVC grid comparing with conventional grid [12].

[Figure shows Peukert diagram with Specific capacity (Ah Kg$_{PAM}^{-1}$) vs Current (A Kg$_{PAM}^{-1}$), comparing Pb–Sn electro deposited RVC grid (62% utilization efficiency at low current, 47% at higher current) with Conventional grid (37% and 26% utilization efficiency). Equation shown: $C_d = \dfrac{\alpha}{i_d^{\beta}}$]

utilization of 62% corresponding to a capacity of 135 Ah kg^{-1} PAM, whereas a conventional lead grid yields a 37% PAM utilization.

The reticulated battery ran through 706 cycles, over 2100 h of continuous operation before its capacity dropping below 80% of the nominal capacity. The cycling test was conducted by discharging the reticulated battery at 63 A kg^{-1} PAM and recharging to 105% of the previous discharge capacity.

4.6.2.2 Lead-Alloy-Coated Polymer Current Collectors

Many attempts have been made to find nonmetal matrix substrates that are more suitable for current collectors than lead grids. The goal has been to find a sturdy, light-weight, porous substrate that retains the electroactive paste in hostile environments and operating conditions of a wide range of battery applications. Kelley and Votoupal's US Patent No. 6,979,513 ("Kelley") [13] describes the use of carbon foam to form a battery current collector. Researchers have determined that carbon foam can be used as a negative current collector but not as a positive current collector because of the instability of carbon during battery cycling [14]. Gyenge et al. through US Patent No. 7,060,391 ("Gyenge") [15] teach the use of carbon foam deposited with a layer of the lead–tin alloy in the construction of a current collector for a lead-Acid battery.

These devices can improve the utilization efficiency of active material, and battery energy density. However, current collectors such as those of Kelley and Gyenge employ cast solid lead alloy frames and/or lead grids to enhance electrical current distribution and structural integrity. These lead frames and/or lead grids are relatively heavy, which negates any weight savings achievable from the use of carbon foam, and, consequently, there is basically no or negative gain in power density and little improvement in energy density.

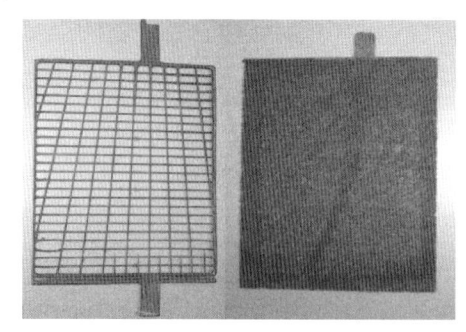

Figure 4.28 Comparison between conventional grid and lead-alloy-coated, multiconductive substrate current collector.

An additional drawback is that carbon foam is fragile and lacks structural integrity, which complicates manufacturing processes such as battery pasting and battery assembling. For instance, while carbon foam is much lighter than metal-based current collectors, carbon-foam current collectors of the Gyenge-type must be thicker than a conventional lead grid because of the need to maintain the structural integrity and strength of the carbon foam. Consequently, the number of such thick current collectors that can be arranged in parallel and series internally is in fact less than that in a battery using conventional lead grids. This means that a lead-Acid battery employing Gyenge-type current collectors have lower power density than the conventional lead-Acid battery.

In 2009, Jung developed a high-surface-area current collector for both positive and negative electrodes with a multiconductive substrate. The current collectors were fabricated by depositing the lead alloy on reticulated, multiconductive substrates [16]. The substrate material can be carbon foam or polymer. Figure 4.28 shows a comparison of the multiconductive-polymer-based current collector and the conventional grid.

The multiconductive current collector consisted of two regions, a frame region and a body region. The frame may also include one or more strips that run through the plate. The frame region is stronger and more rigid than the body region. The frame region has a high specific surface area relative to the body, where the electrical current distribution properties are enhanced and at the same time the frame is denser and stiffer than the body to provide structural support for the current collector. This allows one to dispense with traditional lead frames.

Furthermore, the multiconductive current collector can be manufactured to various thicknesses ranging from 0.8 to 10 mm, which break through the barrier of thick plate requirement in the Gyenge-type current collectors. As such, the total number of positive and negative plates that are assembled in the battery is compatible with the conventional lead-Acid battery. Together, the current collector has approximately a 4.5 times higher surface area per plate, and the power density of the battery employing a multiconductive current collector is improved. Tables 4.20 and 4.21 show the comparison of the battery performances between batteries equipped with conventional grids, carbon-foam-based current collectors, and multiconductive current collectors. Table 4.20 shows the comparison data of 12-Ah, 12-V batteries discharged at a 10-h discharge rate ($C/10$). The battery energy

Table 4.20 Comparison data of 12-Ah, 12-V batteries discharged at a 10-h discharge rate (C/10) [16].

12-Ah battery	Multiconductive	Carbon-foam-based	Conventional grid
Total current collector weight (g)	489	1017	984
Total number of current collectors	78	54	78
Total paste weight (g)	1073	1788	1788
Total battery weight(g)	2710	3300	4400
PAM utilization efficiency (%)	62	56	35
Lead use reduction (%)	38.41	25.00	0
Energy density (Wh kg^{-1})	53	43	33
Power density (W kg^{-1})	239	143	180

Table 4.21 Comparison data of 100-Ah, 12-V batteries discharged at 10-h discharge rate (C/10) [16].

100-Ah battery charge	Multiconductive	Carbon-foam-based	Conventional grid
Total current collector weight (g)	5562	8037	8994
Total number of current collectors	90	54	90
Total paste weight (g)	10 267	17 112	17 112
Total battery weight (g)	24 278	27 943	38 400
PAM utilization efficiency (%)	63	55	25
Lead use reduction (%)	36.78	27.23	0
Energy density (Wh kg^{-1})	50	43	31
Power density (W kg^{-1})	344	215	259

density and power density from multiconductive current collectors are increased by 23 and 67%, respectively, over those from the carbon-foam counterpart, and by 61 and 33%, respectively, over those from the conventional current collector.

Table 4.21 shows the comparison data of 100-Ah, 12-V batteries discharged at a 10-h discharge rate (C/10). The battery energy density and power density from the multiconductive current collector are increased by 16 and 60%, respectively, over those from the carbon-foam counterpart, and by 61 and 33%, respectively, over those from the conventional current collector.

4.6.3
Battery Construction

4.6.3.1 Horizon Battery
Unlike conventional lead-Acid batteries that orient battery electrodes vertically, the Horizon Battery orients the battery electrodes horizontally to provide equal,

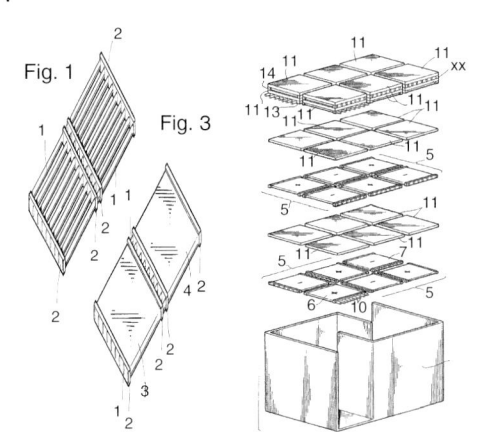

Figure 4.29 Plate and cell structure of Horizon battery [18, 19].

controlled pressure over the whole electrode area; its configuration is shown in Figure 4.29. The current collectors in the Horizon Battery have a high-tensile-strength fiber glass substrate [17]. Lead or lead alloy was extruded onto the fiber glass substrate and the lead-coated fiber glass was woven to have a grid structure using conventional wire-weaving equipment.

The Horizon Battery stacks the plates to share the same electrolyte environment. It also employs biplates design in the construction of lead-Acid battery to eliminate the need for grouping the electrodes and connecting the 2-V cells. The biplate, as shown in Figure 4.29, contains two sections. The upper section is pasted with positive active material to act as the positive electrode, and the lower section is pasted with negative active material to act as the negative electrode. Figure 4.29 also shows the orientation of the biplates inside the Horizon Battery. Owing to the unique orientation, the Horizon Battery does not need to further group the electrodes or connect each 2-V cell to make a 12-V battery [18, 20].

4.6.3.2 Bipolar Battery

The lead-Acid industry has long been aware of the theoretical advantages of bipolar batteries. A bipolar lead-Acid battery eliminates grouping of positive and negative plates and the connections between cells. The bipolar battery construction shortens the current path between the positive and negative terminals of the battery, which reduces the internal resistance of the battery to current flow and improves power delivery. The bipolar battery configuration is particularly suitable for high-voltage batteries as bipolar construction requires less amount of space compared to a conventional lead-Acid battery.

The bipolar design creates a uniform current distribution and load potential over the entire surface of the electrodes, which provides more efficient utilization of paste materials. The bipolar battery is assembled by stacking bipolar electrodes together as shown in Figure 4.30. The bipolar electrode is constructed by pasting positive active material onto one side of the conductive, nonporous substrate; this is followed by pasting the negative active material onto the other side of the same

4.6 Advanced Lead-Acid Battery Technology

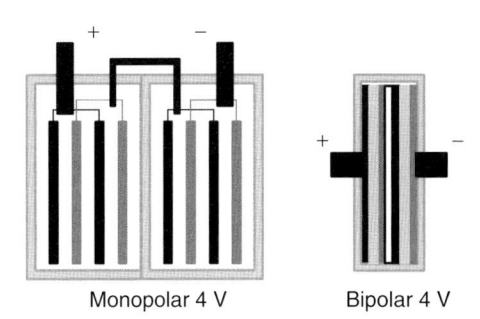

Figure 4.30 Bipolar battery design [21].

Monopolar 4 V   Bipolar 4 V

substrate. Figure 4.30 shows the concept in monopolar (conventional lead-Acid battery construction) and bipolar battery design. The bipolar concept shown in Figure 4.30 shortens the long, complex path of current flow in the monopolar design.

The lead sheet, while an obvious choice for use as a bipolar substrate, is structurally weak and needs to be thick enough for its physical shape to be maintained throughout the battery-manufacture process. The thick and heavy lead plates counter the potential weight savings anticipated in bipolar designs. The lead sheet also suffers from gradual corrosion during normal battery operation and is eventually penetrated by the sulfuric acid electrolyte, causing a short circuit between adjacent cells resulting in catastrophic battery failure. In addition, lead metal presents a notoriously difficult sealing surface. In current SLA batteries, the surface areas of lead that must be sealed are relatively small, but in a bipolar construction the entire perimeter of the substrate must be sealed to prevent the electrolyte from bridging the space between the positive and negative active materials. As a result, the surface area of lead to be sealed is literally orders of magnitude greater than that in a conventional design, creating a significant reliability problem. Thus, the material used as bipolar substrate must have the following properties:

- highly conductive;
- resistant to corrosion by sulfuric acid in a lead battery environment;
- suitable for forming reliable seals to prevent electrolyte leakage;
- orm a variety of shapes;
- allow good adhesion of the active paste materials;
- mechanically robust;
- able to be manufactured in high volumes;
- cost-effective.

In 2009, Atraverda discovered a bipolar element, Ebonex®. To meet the design requirements, he combined Ebonex powder with thermoset resins and lead-alloy foil. Ebonex powder is a titanium suboxide material that has an electrical conductivity comparable to carbon and has a high oxidation resistance. The Ebonex is a bipole element that serves as both the support structure and divider for the positive and negative electrode paste material [21].

Each Ebonex, when assembled with electrode paste material, forms a 2-V cell. This bipolar design brings about a dramatic reduction in the number of components and amount of raw material used in the battery, and hence reduces the weight and size of the battery as compared to the conventional lead-Acid battery.

4.6.4
Electrolyte Improvement

4.6.4.1 Gel Silicon Electrolyte

The gel silicon electrolyte was introduced to eliminate spilling and the need for constant maintenance.

A gel battery (also known as a *"gel cell"*) is a VRLA battery with a gelified electrolyte; sulfuric acid is mixed with silica fume or silica additives, which makes the resulting mass gel-like, stiff, and immobile. Unlike a flooded wet-cell lead-Acid battery, these batteries do not need to be kept upright. Gel batteries reduce electrolyte evaporation, spillage (and subsequent corrosion issues) common to the wet-cell battery, and boast of greater resistance to extreme temperatures, shock, and vibration. Chemically they are the same as wet (nonsealed) batteries, except that the antimony in the lead plates is replaced by calcium.

The recharge voltage on this type of cell is lower than that in other types of lead-Acid batteries. This is probably the most sensitive cell in terms of adverse reactions to overvoltage charging. Gel batteries are best used in a very deep cycle application and may last for a slightly longer time in hot weather applications. If an incorrect battery charger is used on a gel cell, the battery performance is poor and premature failure is certain.

The sealed-gel technology was developed many years ago by the Sonnenschein Company in West Germany. Simply stated, a gel cell is a lead-Acid battery that uses a thick chemotropic gelled electrolyte that is of the consistency of candle wax once it "sets up" and is pressurized and sealed using special valves. It uses the "recombination" technique to replace the oxygen and hydrogen normally lost in a wet-cell battery and is maintenance free and nonspillable.

As noted above, a gel cell is a "recombinant" battery. This implies that the oxygen that is normally produced on the positive plates in all the batteries recombines with the hydrogen given off by the negative plates. The combination of hydrogen and oxygen produces water, which replaces the moisture lost in "wet-cell" batteries. Therefore, the battery is maintenance free, as it does not require any addition of water [22].

The oxygen is trapped in the cell by special pressurized sealing vents. It travels to the negative plates through tiny fissures or cracks in the gelled electrolyte. The sealing vent is critical to the performance of the gel cell. The cell must maintain a pressure of $\sim 1-1/2$ lbs per in^2, otherwise the recombination of the gases will not take place and the cell will not perform. Likewise, the valve must safely release excess pressure that may be produced during charging. Otherwise, the cell could be irreparably damaged. It is important to note that a gel cell must never be opened once it leaves the factory. If opened, the cell loses its pressure and the outside air "poisons" the plates and causes an imbalance that destroys the recombinant chemistry.

4.6.4.2 Liquid Low Sodium Silicate Electrolyte

An electrolyte with low sodium silicate concentration was developed to overcome the disadvantages of conventional lead-Acid batteries and gel lead-Acid batteries. A conventional lead-Acid battery uses a sulfuric acid electrolyte, which pollutes the environment and is a health hazard. In a gel lead-Acid battery, it is difficult to fill the gel electrolyte because of the hardness and crack of the electrolyte, low fluidity, and high internal resistance.

The silicate electrolyte-containing low-concentration sodium is prepared by the magnetization method according to the following steps [23]:

1) Add deionized water into a silica gel containing 40–60 wt% SiO_2.
2) Use a Baume densimeter to measure the concentration and add deionized water until the concentration reaches 0.65–0.85 °Be′.
3) Add inorganic acid into the mixture until the pH value is between 1 and 4. The inorganic acid can be hydrochloric acid (HCl), oxalic acid, or sulfuric acid.
4) Put the solution into a magnetic field of 1000–6000 G for 5–10 min for magnetization.
5) Stir the magnetized solution at 700–1400 rpm for 5–10 min until the viscosity of the solution is decreased to 0.02 P.

Battery manufactures such as Greensaver Corporation that use the low sodium silicate electrolyte in the production of the lead-Acid battery have stated the following advantages over conventional and gel-type lead-Acid batteries.

- The composition of the electrolyte is changed from sulfuric acid.
- The magnetization treatment gives the electrolyte low viscosity and good fluidity, which resolves the difficulty of electrolyte filling.
- The internal resistance of the batteries using the electrolyte containing low concentration of sodium silicate is very low, which contributes to a longer cycle life.
- The batteries can be operated in a wide temperature range from -50 to $+60\,°C$.
- No acidic smog is released during the battery manufacture process.
- The batteries have negligible self-discharge. The storage period is increased from normal 8 to 18 months.

4.7
Lead-Acid Battery Market

The chemistry of the lead-Acid battery has been applied to a variety of capacities, sizes, and designs. Despite other battery chemistry technologies having come out in the market, lead-Acid chemistry remains the technology of choice for automotive and many other applications. Lead-Acid chemistry has been proven to be robust, tolerant to abuse, well understood, and of low cost. It is widely used in industries, portable devices, and transport sectors.

Major market information regarding the lead-Acid battery is given in the following.

4.7.1
Automotive (Transportation and Recreation, Original Equipment Manufacturer (OEM) and Replacement)

4.7.1.1 SLI Battery (Starting, Lighting, and Ignition)

The market share of the lead-Acid battery was annually estimated to be US$38.2 billion in the world market in 2010; the demand will grow moderately as a result of the high degree of maturity in most developed nations and price competition among major suppliers. The estimated world market value in 2020 is US$54.7 billion. The development of hybrid vehicles and EVs could provide significant opportunities to this market in the future.

This sector requires batteries with rapid charge and discharge capability, high-power output for acceleration, long-lasting life, minimal maintenance, safety, and low cost. The major applications include light traction, marine, HEVs, plug-in hybrid electric vehicles (PHEVs), EVs, and neighborhood electric vehicles (NEVs) such as bicycles, motorcycles, buses, and trams.

The key battery performance parameters in this market are CCA and RC. For a 12-V SLI battery, *CCA* is defined as the amperes a lead-Acid battery can deliver at $0\,°F$ in 30 s while maintaining a voltage of at least 7.2 V. The *RC* is defined as the amount of time (in minutes) in which a battery can deliver 25 A at $80\,°F$ while maintaining terminal voltage of at least 10.5 V.

4.7.1.2 Deep Cycle Battery (E-Bike/Marine/Recreational Vehicle/Motor Home)

Lead-Acid batteries were widely used in the electric bicycle (E-Bike) especially in China. The Chinese government estimated $\sim$80 million E-Bikes to be present in China in 2010; 95% of the E-Bikes are powered by lead-Acid battery. The market size of the E-bike battery in 2010 was US$13.6 billion and is expected to grow to US$24.2 billion by 2020.

The key battery performance parameters of E-Bike batteries are cost, cycle life, and weight. E-Bike battery manufacturers currently give a warranty of 8 months for new replacement batteries and from 9 to 15 months for refurbished batteries. The E-Bike batteries weigh between 12 and 26 kg. The high-end batteries in the above range can be problematic in high-rise, thickly populated locations.

4.7.1.3 Microhybrid

The microhybrid technology shuts down the engine at stoplights and then starts it again when the driver releases the brake. The microhybrid technology installs an extra 12-V battery to run the accessories (especially air conditioning) when the engine is shut off.

It is estimated that, by 2015, microhybrid start–stop systems will be used in nearly 20 million cars annually. Urban cars in Europe spend as much as a third of their time on the road not moving. Asian markets, with huge congestion, are likely to gain the most from this technology. On the other hand, US-style highway driving does not benefit much from the above technology; therefore, it has been slow moving in the United States.

The prospects for microhybrids are particularly bright in China, as China could be producing 15 million cars annually by 2020, with half of them using some form of electric drive, and three-quarters of the EV-enabled cars could be microhybrids.

Among the automakers using start–stop in Europe and other world markets (but not yet the United States) are BMW, Mazda (its i-stop in the 2010 Mazda3), Mercedes-Benz, Mini, and PSA/Peugeot-Citroën. PSA stated in 2008 that it intends to make all of its small- and mid-sized cars microhybrids. One Chinese car that already offers the microhybrid technology is the Chery Motors A5 BSG sedan. BSG stands for "belt-starter-generator," which restarts the engine after it shuts off at stops. The "Smart for Two" will also be offered with start–stop technology in its 1-l engine. Indian truck maker Mahindra plans to use the technology in its Scorpio and Bolero SUVs.

The microhybrid technology had a market of US$560 million in 2010. It is estimated that the demand will grow to US$9.02 billion in 2020. The key performance parameters are dynamic charge acceptance and cycle life. HEV applications require good charge acceptance in a dynamic discharge/charge microcycling operation. The main charge has more throughput but only at shallow cycling depths, for example, <1%. A very long stop with high loads might give 2%. A 10 × dynamic charge acceptance is the goal against standard SLI. The cycle life of batteries used in the microhybrid market requires 5000 cycles for a standard SLI battery and over 10 000 cycles at 60% partial state of charge (PSOC) without conditioning for a battery designed for use in start–stop hybrids.

4.7.2
Industrial

4.7.2.1 Motive (Industrial Trucks, Fork Lifts, Golf Carts, etc.)

Increased investment in new manufacturing facilities or upgrading of existing facilities is driving substantial demand for motive lead acid (MLA) batteries. The worldwide annual market was 3.34 billion in 2010 and is expected to grow to 4.54 billion in 2020. The industrial motive batteries form a critical part of heavy-duty equipment, mining, railroad, pallet jacks, fork lift trucks, floor cleaners, automated guided vehicle (AGVs), towing vehicles, as well as wheelchairs and golf carts. Globally, Asia is emerging as a particularly attractive market as a result of the rising demand as well as low costs of production in the region.

The key to good battery performance is the lifetime of the battery. Motive industrial batteries often endure 100% depth of discharge (DOD), thus 100% DOD testing is the industry standard as it is the fastest way to test. The typical benchmark lifetimes are in the 500–600 range.

Designed for applications where the batteries are regularly discharged, it requires rapid charge capabilities, high energy and power output, long cycle life, and minimal maintenance for its applications.

4.7.2.2 Stationary (Utility/Switchgear, Telecommunication, Uninterruptible Power Systems (UPS), Emergency Lighting, Security Systems, Renewable Energy Systems, Cable Television/Broadcasting, Oil and Gas Exploration, Railway Backup, etc.)

Stationary power systems support various critical applications including preventing unscheduled outages, extending sensitive electronic equipment life, reducing maintenance costs, and increasing predictability of equipment failures. Widespread utilization of lead-Acid batteries as standby power sources for several critical applications for over a century has earned the trust of end users. Currently, SLA batteries have become vital components in every industrial application and had a US$3.99 billion market in 2010. The market is projected to grow to US$7.85 billion in 2020. A realization of the need for backup power to sustain the running of critical businesses has resulted in constant demand and revenue growth in the market for these batteries. The technology has been the forerunner in the field of energy storage, thereby increasing the dependence on backup power for both critical and semicritical end-user applications. The key battery parameters for this market are its lifetime and footprint. The expected lifetime of the industrial stationary batteries is 5–20 years with corresponding warranties. UPS and telecommunication batteries often have limited cycling. The footprint is also important as backup power provider would want to have a smaller footprint for the batteries. Some stationary batteries are designed to cycle rather than "float." Applications of cycling stationary batteries include load leveling, utility peak-power shaving, and PV energy-storage systems.

4.7.2.3 Emerging Grid Applications

The calls made by governments in the recent years for alternative/renewable energy power sources such as solar and off/onshore wind power has created a huge demand for stationary batteries. The generated electricity is intermittent by nature, and it is available only when the sun shines or the wind blows. It needs dedicated equipment to stabilize voltages by eliminating irregularities in systems, as well as an efficient and reliable storage source to store the generated electrical power and backup.

Improved energy storage is a key requirement for the deployment of the much touted "Smart Grid." There are many reasons for this. One is that efficiently integrating solar and (especially) wind power into the grid requires buffering, since these renewable energy sources produce energy only at certain times. Another reason is that Smart Grids are intended to be highly resilient; thus, it is critical to build new infrastructure for storing energy for times of peak demand and network outages.

The battery market size for emerging grid applications was around US$22.7 million in 2010. It is anticipated that the market will grow to US$882 million in 2020. The key battery performance requirements are its lifetime and cost. Currently, a traditional lead-Acid battery can deliver 1500–2000 cycles at 90% DOD, which is considered to be insufficient as a lithium ion battery can deliver up to 8000 cycles. The cost of the batteries for emerging grids is important as peak shaving and load-leveling applications are not enormously profitable for utilities. It is believed that the battery cost should be roughly $200 kW h to enable widespread adoption.

4.7.2.4 Distributed Renewable

The integration of distributed renewable generation sources into the electricity grid poses a number of challenges for the industry. Utilities will be faced with issues of enabling high penetration of distributed generation onto both existing and future distribution systems. The battery for this application will primarily serve off-grid PV installations. The worldwide battery market size for this application was US$599 million in 2010 and it is expected to grow to US$2.6 billion in 2020. The key battery performance requirement is a lifetime at deep discharge and cost. The lifetime for the battery is 100-h systems at 20% DOD for five to six years, which is ~1500 cycles.

References

1. Linden, D. and Reddy, T.B. (2002) *Handbook of Batteries*, 3rd edn, McGraw-Hill.
2. Blair, T.L. (1998) Lead oxide technology – past, present, and future. *J. Power Sources*, **73**, 47–55.
3. Mayer, G. (2001) *Battery Reference Book*, 4th edn, Bitrode.
4. www.teck.com/Generic.aspx?PAGE=Teck Site/ProductsPages/BatteryTechnologyPages/ProductInformation&portalName=tc.
5. Mayer, G. (2006) US Patent Application US 2006/0093912 A1.
6. Mayer, G. (2000) Effect of Charging on Life of Float Operated Lead Acid Battery.
7. Nakanno, K., Takeshima, S., and Furukawa, J. (2007) Furukawa Review, Technological Trends in Lead Acid Batteries for Automotive Applications, No. 32, pp. 49–55.
8. www.axionpower.com/Technology.
9. www.fireflyenergy.com/images/stories/pdfs/white%20paper%204.25.08.pdf.
10. Czerwinski, A. and Zelazowska, M. (1997) Electrochemical behavior of lead oxide deposited on reticulated vitreous carbon (RVC). *J. Power Sources*, **64**, 29–34.
11. Das, K. and Mondal, A. (2000) Studies on a lead-Acid cell with electrodeposited lead and lead dioxide electrodes on carbon. *J. Power Sources*, **89**, 112–116.
12. Gyenge, E., Jung, J., and Mahato, B. (2003) Electroplated reticulated vitreous carbon current collectors for lead acid batteries: opportunities and challenges. *J. Power Sources*, **113**, 388–395.
13. Kelley, K. and Votoupal, J. (2005) Battery including carbon foam current collectors. US Patent 6,979,513 B2.
14. Jang, Y., Dudney, N., Tiegs, T., and Klett, J. (2006) Evaluation of the electrochemical stability of graphite foams as current collectors for lead acid batteries. *J. Power Sources*, **161**, 1392–1399.
15. Gyenge, E., Jung, J., and Snaper, A. (2006) Current collector structure and methods to improve the performance of a lead acid battery. US Patent 7,060,391.
16. Jung, J. (2010) Multiply conductive matrix for battery current collectors. PCI Patent Application PCT/CA2010/000006.
17. Blanyer, R. (1985) Method and apparatus for coating a core material with metal. US Patent 4,658,623.
18. Craven, W. (2002) *Horizon Sealed Lead Acid Battery in Electric Vehicle Application*, IEEE, pp. 159–162, ISBN: 0-7803-2994-5-96.

19. Blanyer, R. (1989) Battery grid structure made of composite wire. US Patent 4,865,933.
20. Morris, C. (1990) Lead acid rechargeable storage battery. US Patent 4,964,878.
21. *www.atraverda.co.uk/?q=technology/bipolar-vrla-technology*.
22. Chreitzberg, A. and Chiaccho, F. (1987) Method of making a sealed lead acid battery with a gel electrolyte and sealed lead acid storage battery made according to this method. US Patent 4,687,718.
23. Feng, Y., Han, Y., and Feng, Y. (2008) Liquid low-sodium silicate electrolyte used for a storage battery and manufactured by magnetization process, and the usage thereof. US Patent Application US 2008/0044726 A1.

Further Reading

Varma, P. (1982) Lead-Acid battery with gel electrolyte. US Patent 4,317,872.

5
Nickel-Metal Hydride (Ni-MH) Rechargeable Batteries
Hua Ma, Fangyi Cheng, and Jun Chen

5.1
Introduction to NiMH Rechargeable Batteries

Rechargeable batteries are becoming more and more important in our routine life, especially in the applications of consumer electronic devices such as cellular phones, notebook computers, and compact camcorders [1, 2]. The ideal batteries would feature the attributes including long life, small size, lightweight, high energy density, safety, environmental compatibility, low cost, and worldwide consumer distribution. Aiming at these requirements, nickel-metal hydride (Ni-MH) batteries, the so-called *green* batteries without lead (Pb) or cadmium (Cd) pollution, have been developed and commercialized. The main components of Ni-MH batteries include a positive electrode (cathode) comprised of nickel hydroxide powder, a negative electrode (anode) in which hydrogen storage alloys (and/or their hydrides) are used as the active materials, and an electrolyte with an alkaline solution. The characteristics of Ni-MH batteries depend not only on the performance of the active materials but also on the manufacturing and assembly processes for both the electrodes and the batteries.

As in Ni-Cd cells, the positive electrode of an Ni-MH battery is nickel oxyhydroxide (NiOOH). Dissimilarly, the negative electrode uses a hydrogen storage alloy instead of cadmium. In 1973, people had tried to use $LaNi_5$ as the anode material for secondary batteries, but the rapid capacity decay of $LaNi_5$ alloy limited their application. Until 1984, the Philips Company resolved capacity fading of $LaNi_5$-based alloy by element substitution, resulting in the application of hydrogen storage alloys as the negative electrode material in Ni-MH batteries [3]. The first consumer-grade Ni-MH cells for portable applications appeared in the market in the 1990s [4]. After their commercialization, Ni-MH batteries became important rechargeable power sources for both consumer and industrial applications because of their design flexibility, prominent energy/power density, and environmental acceptability [5]. Early Ni-MH batteries exhibited limited performance beyond ambient operating temperatures, whereas today's Ni-MH batteries can provide excellent power at −30 °C and over 90% capacity at 70 °C as a result of innovation in hydrogen storage alloys and nickel hydroxide materials. An enormous quantity of Ni-MH batteries is

Electrochemical Technologies for Energy Storage and Conversion, First Edition. Edited by Ru-Shi Liu, Lei Zhang, Xueliang Sun, Hansan Liu, and Jiujun Zhang.
© 2012 Wiley-VCH Verlag GmbH & Co. KGaA. Published 2012 by Wiley-VCH Verlag GmbH & Co. KGaA.

manufactured commercially for small portable applications, achieving an annual worldwide production of over 1 billion cells.

Another driving force for the rapid growth of Ni-MH batteries is that they have become the dominant advanced battery technology for electric vehicles (EVs) and hybrid electric vehicles (HEVs) because of their good overall performance in the wide range requirements set by automotive companies [6]. In addition to the essential performance targets of energy, power, cycle life, and operating temperature, Ni-MH batteries have established the technology preeminence for application in HEVs due to the following features [7]: (i) flexible cell sizes from 30 mAh to 250 Ah; (ii) safe operation at high voltage; (iii) high volumetric energy and power density; (iv) flexible vehicle packaging; (v) easy application to series and series/parallel strings; (vi) multiple choice of cylindrical or prismatic cells; (vii) safety in charge and discharge, including tolerance to abusive overcharge and overdischarge; (viii) maintenance free; (ix) high thermal stability; (x) capability to utilize regenerative braking energy; (xi) simple and inexpensive charging and electronic control circuits; and (xii) environmentally acceptable and recyclable materials.

Research and development (R&D) activities in Ni-MH batteries have focused on further improvements in specific power for both transportation and portable applications under both normal and abnormal temperature conditions [7]. For anode materials, early AB_5-type alloys/hydrides could give a moderate capacity of about 310 mAh g^{-1} [8]. In order to improve the capacity, high-rate capability, activation, and cycle life, the AB_2, AB, Mg-based, or composite alloys have been extensively studied as the active materials of the anode [9]. For cathode nickel hydroxide materials, the rapid development has also been made. The positive active materials may contain multiple modified elements to improve the conductivity, utilization efficiency, high-rate capability, and high-temperature performance [10]. Furthermore, a stable α-phase nickel hydroxide with an appropriate amount of metal ion substitution has been used to replace the β Ni(OH)$_2$ with the aim of improving the capacity and stability [11].

The initial cylindrical Ni-MH cell with AA type commercialized in 1991 could merely deliver a specific energy of 54 Wh kg^{-1}, whereas today's similar consumer cell attains a specific energy over 100 Wh kg^{-1}. A large amount of licensed manufacturers have produced different types of Ni-MH batteries ranging from 30 mAh button cells to a wide variety of consumer cylindrical cells with various capacities, prismatic cells of up to 250 Ah for electric buses, and 6 Ah multicell modules for HEVs [12].

In this chapter, we present a brief introduction of the electrochemical process, cell structures and characteristics, as well as the typical application of rechargeable Ni-MH batteries. Progress made in the R&D of the cathode nickel hydroxide materials, the anode hydrogen storage alloys, the electrolytes, the separators, and the battery assembly technologies is summarized. The advantages and limitations of the Ni-MH batteries are discussed, and the outlook of future possible breakthrough is given.

5.2
Electrochemical Processes in Rechargeable Ni-MH Batteries

An Ni-MH battery is composed of a cathode of nickel hydroxide, an anode of hydrogen storage alloy, a separator, an alkaline electrolyte, a metal case, and a sealing plate provided with a self-resealing safety vent [13]. The electrodes are designed with highly porous structures, which have large surface areas to provide low internal resistance for high-rate performance. The positive electrode is a porous, sintered or felt nickel substrate into which the nickel compounds are impregnated/pasted and converted to the active material. Felts and foams have generally been used to replace the sintered plaque electrodes. The negative electrode, similarly, is a highly porous structure using a perforated nickel foil or grid onto which the plastic, bonded active hydrogen storage alloy is coated. The electrodes are separated with a synthetic nonwoven material, which serves as an insulator between the two electrodes and as a medium for absorbing the electrolyte.

The reaction occurring at the positive electrode can be expressed as (Eq. (5.1)) [4, 5]

$$Ni(OH)_2 + OH^- \underset{discharge}{\overset{charge}{\rightleftarrows}} NiOOH + H_2O + e^- \qquad (5.1)$$

The electrode potential for the nickel electrode is about $+0.44$ V versus the Hg/HgO reference electrode. The reversible $Ni(OH)_2$ and $NiOOH$ couple are able to transform from one to the other on charging and discharging. During the charging process, $Ni(OH)_2$ is converted to $NiOOH$, and the electrical energy provided from an external power source is stored as chemical energy in the cell. When discharging, $NiOOH$ is converted back to $Ni(OH)_2$, releasing the stored chemical energy and supplying electricity.

The active material of the negative electrode is an alloy, which can reversibly absorb and desorb hydrogen atoms. The negative electrode has a fairly negative potential of around -0.9 V versus Hg/HgO. The involved reactions during the charge–discharge operations are described as follows:

$$M + H_2O + e^- \underset{discharge}{\overset{charge}{\rightleftarrows}} MH + OH^- \qquad (5.2)$$

where M represents a hydrogen storage alloy and MH is the formed hydride when hydrogen atoms, from the electrolysis of water, are absorbed by the alloy. On discharge, the hydrogen atom is released and converted back to water.

Combining Eqs. (5.1) and (5.2) gives the following overall cell reaction:

$$M + Ni(OH)_2 \underset{discharge}{\overset{charge}{\rightleftarrows}} MH + NiOOH \qquad (5.3)$$

Figure 5.1 schematically illustrates the operation principle of an Ni-MH battery [14]. When the cell is charged, nickel hydroxide $(Ni(OH)_2)$ in the positive electrode combines hydroxide (OH^-) ions from the electrolyte to produce nickel oxyhydroxide

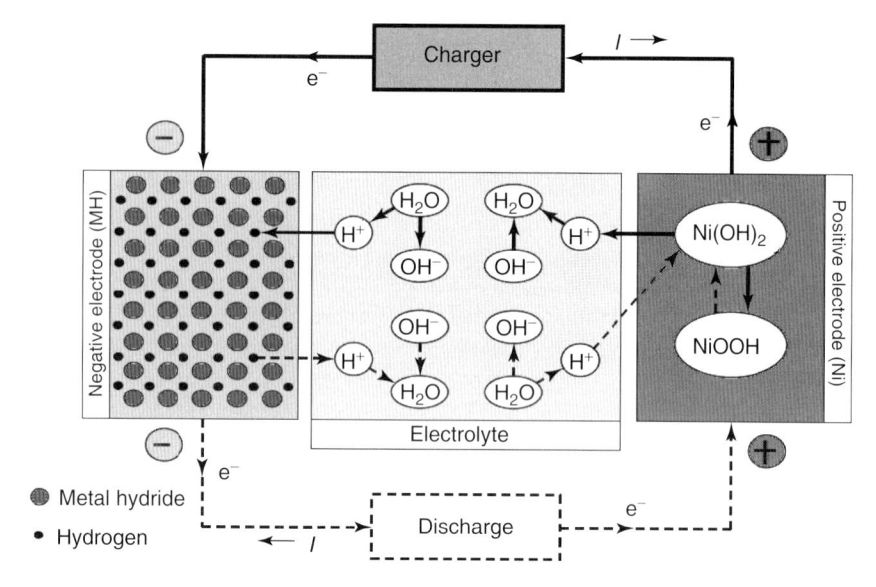

Figure 5.1 Schematic representation of an Ni-MH cell [14].

(NiOOH) within the cathode, water (H_2O) in the electrolyte, and one free electron (e^-). At the anode, the metal alloy (M) in the electrode, water (H_2O) from the electrolyte, and an electron (e^-) react to form metal hydride (MH) and hydroxide ion (OH^-). When the cell is discharged, both electrodes undergo reverse chemical reactions. The overall process presents as a hydrogen transfer between $Ni(OH)_2$ and M, depending on whether the cell is being charged or discharged.

Besides the main charging and discharging processes shown above, overcharge and overdischarge reactions also exist in the Ni-MH battery at abnormal severe operation conditions [15]. As the cell approaches full charge, oxygen gas will start to evolve from the positive electrode through the electrolysis process:

$$4OH^- \rightarrow O_2(g) + 2H_2O + 4e^- \tag{5.4}$$

Overcharged to the point at which the inner cell pressure begins to build up, elevated temperatures are experienced, which may cause the separator to lose electrolyte. The loss of electrolyte within the separator (the so-called *separator dry out*) inhibits the proper transport of hydrogen to and from the electrodes. To protect against the first stage of overcharge, Ni-MH batteries are constructed with the negative electrode having a capacity (or active material) greater than the positive electrode. This design helps to prevent the formation of hydrogen in the negative electrode and to slow the build up of pressure by having more active material available in the negative electrode to effectively recombine the evolved oxygen. The oxygen diffuses through the separator to the negative electrode, where it is consumed in the oxygen recombination reaction via two reaction mechanisms:

$$4MH + O_2 \longrightarrow 4M + 2H_2O \tag{5.5}$$

$$O_2 + 2H_2O + 4e^- \longrightarrow 4OH^- \tag{5.6}$$

The former mechanism represents a direct combination of the O_2 gas with MH, which causes the formation of water. The formed water is consumed by the normal charging reaction (Eq. (5.2)). The latter mechanism is a reverse of the electrolysis reaction that originally generates O_2 at the positive electrode. As a result of the two reactions, the gaseous O_2 is reabsorbed by the negative electrode, thereby preventing unacceptably high internal pressure during the charging reactions.

In the event of deep discharge, depreciation of battery performance may occur. To minimize the possible damage, the excess capacity in the negative electrode also acts as discharge reservation, preventing the negative electrode from being oxidized. There are two situations on overdischarging an Ni-MH cell. The first case involves the full depletion of the active material of the positive electrode and the generation of hydrogen gas (Eq. (5.7)).

$$2H_2O + 2e^- \rightarrow H_2 + 2OH^- \tag{5.7}$$

Since the negative electrode has more active material (namely, metal hydride), it is able to absorb some of the hydrogen gas evolved by the positive electrode. The hydrogen gas is then oxidized at the negative (MH) electrode (Eq. (5.8)).

$$H_2 + 2OH^- \rightarrow 2H_2O + 2e^- \tag{5.8}$$

The net result is no reaction, but heat and some pressure are generated in the cell. The ability of the Ni-MH cells to tolerate overdischarge is very important for large series strings of batteries since capacity mismatch may allow some cells to be overdischarged. The hydrogen not absorbed by the negative electrode starts to build up the inner cell pressure. In this case, the second situation emerges when the entire negative electrode is fully depleted of active material. Once both electrode materials are fully consumed, the negative electrode absorbs oxygen, contributing to the loss of usable capacity. Extreme overdischarge of an Ni-MH cell results in excessive gassing of the electrodes and permanent damage in two aspects. First, the storage capacity of the negative electrode is reduced when oxygen permanently occupies a hydrogen storage site. Second, excess hydrogen is released through the safety vent, decreasing the amount of hydrogen inside the cell. To protect against the detrimental effects of overdischarging, a proper end of discharge terminations should be applied.

For hermetically sealed batteries, if the side reactions are not prevented, the internal pressure may become excessively high. In sealed Ni-MH batteries, the internal pressure is designed to remain at safe levels during operation. The main principle is to ensure that the capacity of the negative electrode exceeds that of the positive electrode. With the appropriate design, the positive electrode is always the capacity-limiting electrode. The excess capacity in the negative electrode is

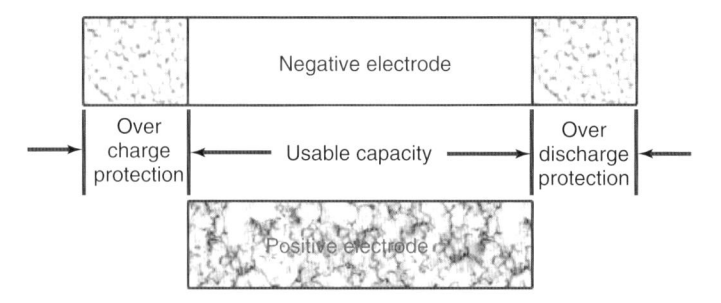

Figure 5.2 Schematic representation of the relationship between the usable capacity, charge reservation, and discharge reservation [14].

referred to as the *charge reservation* and *discharge reservation* of the cell, as shown in Figure 5.2.

5.3
Battery Components

5.3.1
Anode

Ni-MH batteries are similar to sealed nickel-cadmium batteries in cathode materials and differ in that hydrogen rather than cadmium is used as the active element at the hydrogen-absorbing negative electrode (anode). The anode is made from a hydrogen storage alloy that serves as a solid hydrogen source. There are a series of binary A_xB_y alloys that can be employed as potential hydrogen storage materials for MH electrodes [16]. These hydrogen storage alloys combine metal A (whose hydrides generate heat exothermically) with metal B (whose hydride generates heat endothermically) to produce the suitable binding energy so that hydrogen can be absorbed and desorbed at near room temperature and ambient pressure. Generally, the element A controlling the hydrogen storage capacity is the key factor of the hydrogen storage alloys, and the element B controlling the reversibility of the hydrogen adsorption/desorption adjusts the formation of heat and decomposition pressure.

5.3.1.1 Properties of Hydrogen Storage Alloys

The thermodynamic properties of a metal-hydrogen alloy are conveniently illustrated by a pressure–composition–temperature (PCT) curve (Figure 5.3a) [17]. At certain temperatures, increasing hydrogen pressure leads to the increase in the amount of hydrogen dissolving in the metal alloy and finally forms the hydrogen-containing solid solution (α-phase). When the amount of hydrogen reaches the ultimate dissolubility of metals (A point), the α-phase reacts with hydrogen to form a hydride (β-phase). For further reaction, hydrogen is adsorbed

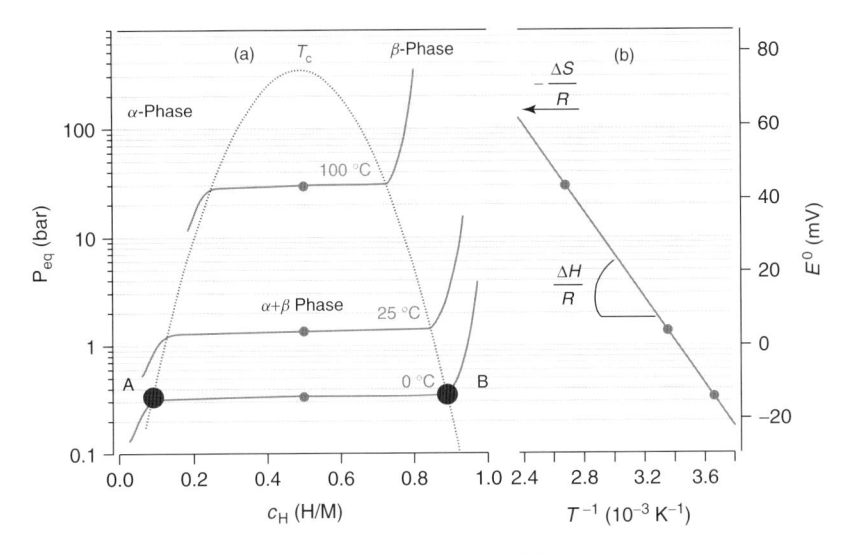

Figure 5.3 (a) Pressure–composition–temperature (PCT) plot and (b) van't Hoff curve for hydrogen absorption in a typical hydrogen storage alloy. The vertical axes indicate the corresponding hydrogen pressure or the equivalent electrochemical potential [17].

by the metal alloy at constant pressure until the α-phase changes completely to the β-phase (B point). When the solid solution and hydride phases coexist (from A to B), there is a plateau in the isotherms, the length of which determines the amount of hydrogen stored. In the pure β-phase, the hydrogen pressure rises steeply with the concentration of hydrogen. When the temperature rises, the plateau moves upward and disappears at a critical point (T_c), above which the transition from the α- to β-phase is continuous. Also, when the temperature drops, the plateau moves downward. Thus, low temperature facilitates adsorbing hydrogen, whereas high temperature is in favor of hydrogen desorbing (releasing).

The equilibrium pressure, P_{eq}, is related to the changes in enthalpy (ΔH) and entropy (ΔS) as a function of temperature according to the van't Hoff equation:

$$\ln\left(\frac{P_{eq}}{P_{eq}^0}\right) = \frac{\Delta H}{R} \times \frac{1}{T} - \frac{\Delta S}{R} \tag{5.9}$$

where P_{eq}^0 is the equilibrium pressure under standard condition, R is the universal gas constant, and T is the absolute temperature. A typical van't Hoff plot is shown in Figure 5.3b. The slope and intercept of the linear line are equal to the enthalpy of formation divided by the gas constant and the entropy of formation divided by the gas constant, respectively.

The PCT curve is important to measure the thermodynamic properties of hydrogen storage alloys. From the curve, one can obtain the ratio of hydrogen to metal alloy and the decomposition pressure. The characteristics of PCT curve, such as the plateau pressure, plateau length, and plateau hysteresis (difference between adsorption and desorption curves), are the key parameters for characterizing hydrogen storage alloys and for developing new hydrogen storage materials. Besides, we can determine the reaction enthalpy and entropy by constructing the reciprocal of temperature with logarithm of plateau pressure, which is not only important for scientific research but also significant for development and application of hydrogen storage alloys.

The hydrogen storage alloy plays an important role in determining the electrochemical performance of an Ni-MH battery [18]. The hydrogen evolution reaction (HER) for hydrogen storage alloys involves the following steps [9]:

$$M + H_2O + e^- \leftrightarrow MH_{ads} + OH^- \quad (5.10)$$

$$MH_{ads} \leftrightarrow MH_{abs} \quad (5.11)$$

$$MH_{abs} \leftrightarrow MH_{hydride} \quad (5.12)$$

$$2MH_{ads} \leftrightarrow M + H_2 \quad (5.13)$$

$$MH_{ads} + H_2O + e^- \leftrightarrow M + H_2 + OH^- \quad (5.14)$$

In the initial step of the HER, the adsorption of hydrogen atoms occurs on the surface of a hydrogen storage alloy by dissociation of water. These atoms can be absorbed to form metal hydride but can also combine together to produce hydrogen gas according to the Tafel reaction (Eq. (5.13)) and the Heyrovsky reaction (Eq. (5.14)).

These steps can also be described in the charge–discharge curve of a hydrogen storage alloy, as shown in Figure 5.4. In stage OA, the potential of the electrode rapidly changes with time or hydrogen content. This stage corresponds to the

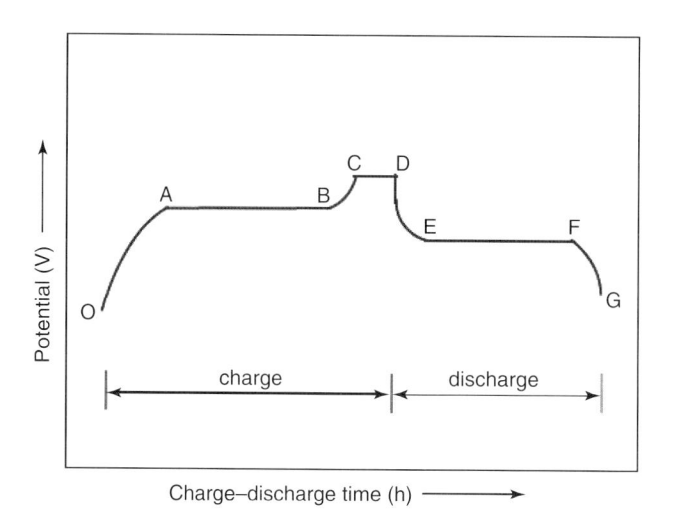

Figure 5.4 Schematic illustration of a charge–discharge curve for a hydrogen storage alloy [9].

formation of a solid solution (i.e., the so-called α-phase). After this stage, the potential remains constant as the hydrogen content increases in the section AB. This constant plateau region indicates a progressive conversion of the α-phase to a metal hydride (i.e., the so-called β-phase). Eventually, the curve begins to slope again, indicating the completion of the conversion of α- to β-phase. Finally, with further increase of the charging time, there is another plateau region corresponding to the process of hydrogen evolution, which indicates that the electrode is in full charge. During discharge, the β-phase decomposes to the α-phase, reversing the charging process. It is obvious that the stage where the two phases coexist determines the performance of the hydrogen storage alloy, including the kinetics of the charge-transfer reaction at the electrode surface, the rate of hydrogen transfer between the absorbed state and the adsorbed state, and the diffusion of absorbed hydrogen between the bulk and the electrode surface.

5.3.1.2 The Classification of Hydrogen Storage Alloys

All hydrogen storage alloys used in Ni-MH batteries should meet the following common requirements [16]:

- high electrochemical hydrogen storage capacity, suitable dehydrogenation equilibrium pressure at operating temperature range (-20–$60\,^{\circ}$C), good catalytic performance for anodic polarization of hydrogen, and the heat of hydride formation (ΔH) of <62 kJ mol^{-1};
- oxidation and corrosion resistance in electrolyte;
- high hydrogen diffusion rate and good reversibility of electrode reaction;
- high stability in alkaline electrolyte solution;
- favorable reversibility of the hydrogenation process;
- good electrical and thermal conductivity;
- low cost.

At present, the widely investigated alloys can be divided in main categories of AB_5-type alloys, AB_2-type alloys, AB/A_2B-type alloys, Mg-based alloys, and V-based solid-solution alloys. Their main characteristics are summarized in Table 5.1.

Among the above five types of hydrogen storage alloys, AB_5-type alloys were first used as the active material, which have been studied most extensively. The other four types of alloys can deliver higher capacities and are currently attracting intensive attention.

AB_5-Type Alloys AB_5-type alloys crystallize in the $CaCu_5$-type structure ($P6/mmm$ space group), and $LaNi_5$ is the prototype alloy. The history of $LaNi_5$ as an electrode material for an Ni-MH battery begins in the 1970s [20]. The most important features of such alloys include easy activation, moderate and flat plateau pressure, slight difference between hydrogen absorption/desorption equilibrium pressure, good dynamic performance, and favorable performance against impurity gas poisoning. At 25 $^{\circ}$C and 0.2 MPa, $LaNi_5$ can absorb hydrogen and form $LaNi_5H_{6.0}$ with a hydrogen storage content of about 1.4 wt% and a decomposition heat of 30 kJ per molar H_2. Thus, the $LaNi_5$ alloy is very suitable for operation at room

Table 5.1 Main characteristics of typical hydrogen storage alloys [9, 16, 19].

Types	Hydrides	Composition	Absorbed hydrogen quantity (%)	Electrochemical capacity (mAh g^{-1})	
				Theoretical value	Experimental value
AB$_5$	LaNi$_5$H$_6$	MmNi$_a$(Mn, Al)$_b$Co$_c$ ($a = 3.5 \sim 4.0$, $b = 0.3 \sim 0.8$, $a+b+c = 5$)	1.3	348	330
AB$_2$	Ti$_{1.2}$Mn$_{1.6}$H$_3$, ZrMn$_2$H$_3$	Zr$_{1-x}$Ti$_x$Ni$_a$(Mn, V)$_b$ (Co, Fe, Cr)$_c$ ($a = 1.0 \sim 1.3$, $b = 0.5 \sim 0.8$, $c = 0.1 \sim 0.2$, $a+b+c = 2$)	1.8	482	420
AB	TiFeH$_2$, TiCoH$_2$	ZrNi$_{1.4}$, TiNi, Ti$_{1-x}$Zr$_x$Ni$_a$ ($a = 0.5 \sim 1.0$)	2.0	536	350
A$_2$B	Mg$_2$NiH$_4$	MgNi	3.6	965	500
Solid solution	V$_{0.8}$Ti$_{0.2}$H$_{0.8}$	V$_{4-x}$(Nb, Ta, Ti, Co)$_x$Ni$_{0.5}$	3.8	1018	500

temperature. However, practical MH electrode has shown large volume expansion after hydrogen absorption, leading to poor cycle stability. The loss of discharge capacity was attributed to the decomposition of LaNi$_5$ into La(OH)$_3$ and Ni. Partial substitution of A or B metal by other transition metals to form pseudobinary alloy has proven to be an efficient strategy to improve the performance of AB$_5$-type alloys [21].

In the composite AB$_5$ hydrogen storage alloys, the A-side element consists of mixed rare earth metals, which are mainly La, Ce, Pr, and Nd as they possess different physical and chemical properties to tune the characteristics of hydrogen absorption. The content and the relative ratio of mixed rare earth are of significant importance in determining the performance of the metal alloy electrode [22]. The current commercial mixed rare earth metals for hydrogen storage alloys can be classified into La-rich (Ml) and Ce-rich (Mm) types. In the AB$_5$ (A = RE metals and B = Ni, Co, Mn, Ti) alloys, NdB$_5$ exhibits the best activity with the highest discharge capacity of 307 mAh g^{-1}. Despite inferior capacity and activation performance, PrB$_5$, LaB$_5$, and CeB$_5$ show better cycling stability.

The presence of La plays a key role in AB$_5$-type alloys. In the La–Ce binary rare earth alloy of La$_{1-x}$Ce$_x$Ni$_{3.55}$Co$_{0.75}$Mn$_{0.4}$Al$_{0.3}$ ($x = 0$–1.0), the cell volume decreases linearly and the hydrogen equilibrium pressure rises with increasing Ce content, which decrease the discharge capacity of the alloy [16]. Also, the cycling stability of the alloy improves significantly with the decrease in Ce content. When x is 0.2, the alloy exhibits good overall performance. Additionally, the surface of the Ce alloy forms a protective film of CeO_2, which improves the corrosion resistance and thus the cycling stability of the alloy. In the La–Pr binary $La_{1-x}Pr_x(Ni, Co, Mn, Ti)_5$ ($x = 0$–1.0) alloys, partial substitution of Pr for La improves the activation performance and cycling stability of the alloy. However, there is no obvious relationship between the discharge capacity of the alloy and the Pr content. For La–Nd binary alloys, partial substitution of Nd for La can also significantly improve the activation performance. In this case, the discharge capacities of the alloys with different amounts of Nd fall in the range of 280–290 mAh g^{-1}, while the cycling ability decreases with the increasing Nd content. Adjusting the ratio of La and Ce in the mixed rare earth elements is the main approach to improve the performance of AB_5-type hydrogen storage alloys.

In the current commercialized AB_5-type alloys, B-side elements are usually Ni, Co, Mn, and Al, considering the overall performance and price of the alloys. Addition of other elements such as Cu, Fe, Sn, Si, and Ti to partially replace Ni is also adapted to enhance the electrode performance. The roles of the B-side elements are outlined below.

- **Co.** Cobalt is the most effective element to improve the cycle life of AB_5-type alloys. Cobalt-based alloys can decrease the hardness, reduce the volume expansion after hydrogenation, increase flexibility, and improve resistance to pulverization. At the same time, during the charge–discharge process, the cobalt alloys can also inhibit the dissolution of Mn, Al, and other elements out of the surface and reduce the corrosion rate, thereby enhancing the cyclability of the alloys. The Co content (atomic number) in commercial AB_5 alloy is generally controlled in the range of 0.5–0.75. On the other hand, developing low-cobalt or cobalt-free alloys has attracted much interest because of the high price of Co [23].
- **Mn.** The partial substitution of Ni by Mn in hydrogen storage alloy can reduce the equilibrium pressure of hydrogen and reduce the hysteresis between hydrogen absorption and desorption processes [24]. In $MmNi_{3.95-x}Mn_xAl_{0.3}Co_{0.75}$ alloy, when the amount of Mn (x) increases from 0.2 to 0.4, the equilibrium pressure of the alloy decreases from 0.24 to 0.083 MPa at 45 °C, and the activation properties and high-rate dischargeability (HRD) of the alloy are improved. However, further increasing the amount of Mn will reduce the cycling stability. During the charge–discharge process, Mn alloy is easy to pulverize. In alkaline solution, Mn on the surface of the alloy is oxidized to form $Mn(OH)_2$ and dissolved, thus accelerating alloy corrosion. Adding an appropriate amount of Co in the alloy can improve the powdering-resistance capacity and inhibit the dissolution of Mn. The amount of manganese (atomic number) in commercial AB_5-type alloy is generally controlled between 0.3 and 0.4.
- **Al and Si.** The partial substitution of Al for Ni in hydrogen storage alloy can reduce the equilibrium pressure, but the hydrogen storage capacity decreases with the amount of Al [25]. Al can mitigate the rate of volume expansion and pulverization during the hydrogenation process. In addition, the surface of Al will form a relatively compact aluminum oxide film, which can prevent the alloy from further corrosion. To trade off the discharge capacity and cycling stability,

the amount (atomic number) of Al in AB$_5$ alloy is generally 0.1–0.3. The effect of Si is similar to that of Al.
- **Cu and Fe.** Adding an appropriate amount of Cu in the alloys can reduce the hardness and volume expansion of alloys, which will help to improve the resistance to pulverization. In this regard, Cu can be used to replace Co. However, Cu alloys require a longer activation cycle, and their surface will form a thick oxide layer during cycling, which causes unsatisfactory high-rate discharge performance. Addition of Fe can lower the equilibrium pressure and also reduce the rate of volume expansion and pulverization. As Fe is cheap and abundant in earth's crust, it is an attractive element to substitute Co. The surface and electrochemical properties of $(La, Mm)Ni_{5-x}(Fe, Al, Mn, Cu)_x$-type alloys with regard to battery applications were investigated. The partial substitution of nickel by iron did not increase the durability of LaNi$_5$. The initial capacity of LaNi$_{4.5}$Fe$_{0.5}$ was 320 mAh g^{-1}, and after 200 charge–discharge cycles, it decreased to 130 mAh g^{-1}. With the combined substitution of iron and aluminum, the durability increased. A more dramatic improvement to the stable alloys was attained for misch-metal-based alloys (e.g., MmNi$_{3.6}$Fe$_{0.7}$Al$_{0.3}$Mn$_{0.4}$) [26].

AB$_2$-Type Alloys Laves-phase alloys belong to the topologically close-packed structure class with the general formula AB$_2$. Since the 1980s, this kind of alloy was studied as anode materials of Ni-MH. The atomic radius ratio (R_A/R_B) in AB$_2$ alloy is close to 1.2, with high symmetry and space-filling density [27]. The AB$_2$-type alloy crystallizes into the structural forms of MgZn$_2$ or MgNi$_2$ (C14, *P*63/*mmc*, hexagonal structure), MgCu$_2$ (C15, *Fd3m*, face-centered cubic), and MgNi$_2$ (C36, *P*63/*mmc*, hexagonal structure). But the AB$_2$-type hydrogen storage alloys only involve C14 and C15 structures. They are characterized by the unique presence of tetrahedral interstices, including the three types AB$_3$ composed of one A atom and three B atoms, A$_2$B$_2$ composed of two A atoms and two B atoms, and B$_4$ composed of four B atoms. In the unit cell of AB$_2$ type, it contains 17 tetrahedral interstices (12 A$_2$B$_2$, 4 AB$_3$, and B$_4$). Because there are lots of tetrahedral interstices (A$_2$B$_2$ and AB$_3$) that supply space for hydrogen storage, the AB$_2$ type of Laves-phase alloys possesses high hydrogen storage density. For example, the hydrogen storage density of ZrMn$_2$ and TiMn$_2$ is 1.8 wt%, and the equivalent theoretical capacity is 482 mAh g^{-1}. The AB$_2$ type of Laves-phase alloys has been considered as the promising hydrogen storage material because of their high hydrogen capacity and long cycle life.

AB$_2$ binary alloys are intermetallic compounds with A = Zr, Ti and B = V, Cr, Mn. Except for TiV$_2$, all the binary compounds adopt either C14 or C15 structure or both. Compared with AB$_5$-type rare-earth-based hydrogen storage alloys, Zr-based AB$_2$ alloys have higher capacity and longer cycle life. Because of their lighter weight, the theoretical discharge capacities are estimated to be two times higher than for AB$_5$ alloys. On the other side, the Zr–Ti based hydrogen storage alloys show low cycle life but high capacity. By studying the binary Zr-based ZrM$_2$ (M = V, Cr, Mn, Fe, Co, Mo, etc.) Laves-phase hydrogen storage alloys, it was found that ZrV$_2$, ZrCr$_2$, and $ZrMn_2$ can absorb a large amount of hydrogen to form $ZrV_2H_{5.3}$, $ZrCr_2H_{4.0}$, and $ZrMn_2H_{3.6}$, respectively, with high hydrogen storage capacity, easy activation, and good dynamic performance [28]. However, the electrochemical properties in alkaline solution are not suitable for electrode material.

Compared with the single-phase structure of AB_5 type, the key character of AB_2 is the multiphase structure. In the AB_2 alloy, C14 and C15 types of Laves phase are both the main phase for hydrogen absorption. The content and proportion of two kinds of Laves phase vary with the component. In addition, the element at the B side also has an important influence on the content of two kinds of Laves phase. Since the phase structure has effect on the electrochemical performance, studying and optimizing the phase structure is an important way for enhancing the electrode performance of AB_2-type alloys.

However, AB_2 alloys are faced with much more difficulties in the synthesis of single-phase compounds, adaptation of the equilibrium pressure to application, and slow activation and high corrosion. Therefore, to satisfy the practical requirements of an Ni-MH battery, the modifications should be processed by element substitution, adjusting the stoichiometric ratio, or surface treatment. For AB_2 alloys, the A-side metal can be substituted mainly by Ti, which has a smaller atomic radius and less affinity for hydrogen, and the B-side metal can be substituted either by other 3D transition metals (e.g., Fe, Co, Ni, and Cu) or even Al [29]. Owing to the reduced affinity for hydrogen and the smaller atomic radii, such substitutions lead to a significant increase in plateau pressure. Besides element substitution, adjusting the stoichiometric ratio (atomic ratio) of sides A and B can also improve the performance of the alloy.

In a highly oxidizing electrolytic medium, AB_2 alloys behave dramatically worse than AB_5. They suffer from slow activation and poor rate capabilities. These drawbacks have been explained in terms of surface passivation because of the presence of a very dense layer formed by Zr and Ti oxidation on the surface, which blocks the electrochemical reaction or increases hindrance to both hydrogen diffusion and electrical resistance [30]. In order to overcome the surface-related problems that plague the electrochemical activity of AB_2 alloys, different surface modification strategies have been investigated, including surface coating, hot charging treatment, and fluorination [31]. Surface coating protects from corrosion by covering the grains of the alloy with a thin nonoxidizable metal layer. Hot charging treatment can dissolve the dense oxide layer covering the surface of the grains in potassium hydroxide and leave the Ni-rich layer on top of the electrode surface to catalyze the charge-transfer reaction. Fluorination techniques help to improve the activation process by dissolving zirconium atoms and allowing precipitation of catalytic nickel.

The capacity of the researched AB_2-type multialloys may reach 380–420 mAh g^{-1}, and these alloys have found application in Ni-MH batteries in the Ovonic Corporation [32]. The Ti-Zr-V-Cr-Ni alloy developed by this company is multistructured with an electrochemical capacity higher than 360 mAh g^{-1} and long cycle life. Compared with the common Zr–Ti based AB_2-type alloys, the main characters of Ovonic alloys are the high content of Ti and V as well as the existence of solid-solution-type non-Laves-phase. Besides C14 Laves phase (main phase), the Ovonic alloy also contains body-centered cubic (bcc) structure of Ti-Cr-Ni solid solution as well as C15 type of Laves phase. By using this kind of alloy as the anode material, Ovonic Corporation has developed several kinds of Ni-MH batteries with cylindrical and square shapes. The energy density can reach 70 Wh kg^{-1} in the produced square Ni-MH battery, which has been used in EVs.

AB-Type Alloys The Ti–Ni alloys are attractive hydrogen storage materials because of their fast activation and low commercial price. Ti–Ni and Ti–Fe alloys were discovered in the early 1970s to form hydrides with reversible hydrogen storage ability [33, 34]. There are two stable intermetallic compounds formed by the Ti–Fe system, namely, Ti–Fe and TiFe$_2$. The application of Ti–Fe alloys in Ni-MH batteries has been limited because of their poor kinetics of hydrogen absorption/desorption. In recent years, more AB-type hydrogen storage alloys consisting of Ti, Zr, or Hf on the A side and Fe, Ni, Al, Co, Mn, or Sn on the B side have been investigated [35–37]. The substitution of Ni for Fe could improve the activation performance and discharge capacity. For example, TiFe$_{0.25}$Ni$_{0.75}$ showed a high discharge capacity of 155 mAh g^{-1} on the third cycle [38]. Martensitic Ti$_{0.64}$Zr$_{0.36}$Ni exhibited a much higher reversible capacity of about 330 mAh g^{-1} [39]. Nevertheless, the electrochemical properties of this system of alloys are unsatisfactory. Ti–Ni based alloys with similar composition but quasicrystalline structure have been described in the past few years [40]. Compared to a normal crystal structure, there are many interstices that could be suitable sites for hosting hydrogen atoms in a quasicrystalline Ti-Zr-Ni alloy. Ti$_{45-x}$Zr$_{35-x}$Ni$_{17+2x}$Cu$_3$ icosahedral quasicrystalline phase (I-phase) had a large discharge capacity of 269 mAh g^{-1} (x was up to 8) without initial activation [41]. Increasing the nickel content improved the electrochemical kinetics and prevented the oxidation of the alloy electrodes. Accordingly, the hydrogen storage alloys with quasicrystalline structure show good hydrogen storage properties. However, direct fabrication of this quasicrystalline phase by mechanical alloying (MA) has not been reported. The MA processes elemental powders mainly by dynamic force or mechanical collision, which leads to chemically inhomogeneous products and makes it difficult to form a single quasicrystalline phase.

Mg-Based Alloys Mg has a high reversible storage capacity of 7.6 wt% hydrogen (corresponding to approximately 2200 mAh g^{-1}), making it a promising candidate as a hydrogen-storage medium in mobile applications. In recent years, more attention has been paid to developing Mg-based alloys for applications in Ni-MH secondary batteries because of their high storage capacity, low cost, and lightweight [42, 43]. However, the Mg-based hydride exhibits strong thermodynamic stability (desorption hydrogen at 250–300 °C) with sluggish kinetics, making it unpractical to satisfy the requirement by Ni-MH batteries.

Previous studies have shown that by making the amorphous Mg–Ni alloy with a high catalytic surface, the thermodynamics and kinetics of Mg-based hydrogen absorption alloy can be significantly improved. MA is a simple and effective method for preparing Mg-based alloys with metastable or nonequilibrium phases, which cannot be found in phase diagrams. The MA method can result in the formation of a homogeneous amorphous structure in the ball grinding process and in the increase of surface areas and defects. A nanocrystalline Mg_2Ni alloy with a grain size of about 4 nm was formed after milling magnesium and nickel powders [44]. Also, amorphous Mg_2Ni alloy can be prepared by MA. The formation and growth of a magnesium oxide and/or hydroxide layer on the surface of Mg_2Ni may increase the electronic resistance at the electrode–electrolyte interface and cause a large discharge overpotential, leading to low discharge capacity and sluggish kinetics of the alloy. In order to improve the electrochemical properties of Mg-based alloys, element substitution of Mg-based alloys has been extensively attempted. Different elements such as Zr, Ti, Co, Al, Ce, Y, Ca, and Fe have been introduced into Mg_2Ni alloy to improve the hydrogen absorption–desorption properties; unfortunately, only limited improvement has been attained [45–47].

The alloy with a composition of $Mg_{0.7}Ti_{0.225}Al_{0.075}Ni$ showed a low decay of discharge capacity because of the formation of $MgTi_2O_4$ instead of $Mg(OH)_2$ [47]. Apart from element substitution, an increase in the Mg–Ni particle size could effectively improve the cycle stability and high-rate dischargeability (HRD) because larger particles had a lower sensitivity to oxidation [48]. Accordingly, a comprehensive work considering both element substitution and particle size demonstrated that an amorphous $Mg_{0.9}Ti_{0.1}NiAl_{0.05}$ alloy with particle size larger than 150 μm displayed comparable electrochemical properties to a commercial $LaNi_5$-based alloy because of the control of the charge input and the cooperative protection effect by Ti and Al [49].

Compared with $LaNi_5$-based alloys, Mg-based alloys are much cheaper and lighter. However, the HRD of Mg-based alloys should be addressed before their practical use.

Composite Hydrogen Storage Alloys Desirable electrochemical properties of hydrogen storage alloys include high storage capacity, easy activation, high resistance to corrosion, favorable kinetic performance, good HRD, and low cost. In practice, it is difficult to simultaneously obtain all these properties in a single alloy system, but using a composite hydrogen storage alloy is an effective way to achieve it. A composite hydrogen storage alloy generally contains two or more hydrogen storage alloys, or a hydrogen storage alloy and another intermetallic compound [9]. Generally, the major component in a composite hydrogen storage alloy is an alloy with good hydrogen storage properties, whereas the minor component is a surface activator to improve the activation properties and the kinetics of hydrogen sorption/desorption.

AB_5- and AB_2-type alloys are the two conventional hydrogen storage alloys. Each one has its own advantages, such as easy activation of AB_5 and high storage capacity of AB_2. However, the hydrogen storage capacity of AB_5 alloy is low (<1.4 wt%), and the catalytic activity of AB_2 alloys is still rather low. Combining the AB_5 and AB_2 units, the AB_3 hydrogen storage alloy is obtained. The structure of AB_3 compounds consists of a long-range stacking arrangement of which one-third is AB_5-like and the other two-thirds are AB_2-like. The relation can be expressed as

$$n(AB_3) = \frac{1}{3}n(AB_5) + \frac{2}{3}n(AB_2) \tag{5.15}$$

The hydrogen storage capacity and stability of AB_3 are better than those of AB_5, and the activation performance outperforms that of AB_2. $LaNi_3$ and $CaNi_3$ are the representative AB_3-type alloys. They can react rapidly with hydrogen under ambient conditions to form $LaNi_3H_5$ and $CaNi_3H_{4.6}$, respectively, with a theoretical hydrogen storage capacity of 425 and 620 mAh g^{-1}, respectively. The A side in AB_3 alloys can contain Mm, Ca, and Mg. Their theoretical hydrogen storage capacity is up to 500 mAh g^{-1}, but generally, it only reaches a practical value of 300–360 mAh g^{-1}. Furthermore, their cycle life needs to be further improved.

A structural characterization of the RMg_2Ni_9 (R = Y, Ca, rare earth) system indicated that substituted AB_3 compounds are promising candidates for reversible hydrogen storage. By partially substituting the A-side element with Ca and Mg, the formed alloy $LaCaMgNi_9$ has good hydrogen storage properties. Its theoretical hydrogen storage capacity is 484 mAh g^{-1} [50, 51]. Partial substitution of Mn or/and Ni in the B side can decrease the volume expansion of the alloy and increase the hydrogen storage capacity [52]. The effect of Al substitution for Ni on the structure and electrochemical properties of $La_2Mg(Ni_{0.8-x}Co_{0.2}Al_x)_9$ ($x = 0$–0.03) alloys was investigated [53]. The increase of Al substitution in the alloys leads to an increase in both the cell volume and the hydride stability but a decrease in unit cell volume expansion rate. As a result, the increase of Al content leads to some decrease in both the discharge capacity and HRD but leads to a significant improvement in cycling stability of the alloys. The alloy with $x = 0.02$ for Al showed a high discharge capacity of 375.9 mAh g^{-1} and the best cycling stability but a poor HRD. A series of Co-free AB_3-type hydrogen storage alloys such as $La_{0.7}Mg_{0.3}Ni_{3-x}Fe_x$ ($x = 0$–0.4) were directly synthesized with vacuum mid-frequency melting method, and the melted alloys were treated by low temperature–short time heat annealing. The content of Fe element affects the phase structure of alloys and leads to different electrochemical properties. Heat treatment has positive effects on the cycle stability of $La_{0.7}Mg_{0.3}Ni_{3-x}Fe_x$ ($x = 0$–0.4) alloys but causes little reduction in the discharge capacity of Fe-substituted alloy. However, the discharge capacity of $La_{0.7}Mg_{0.3}Ni_{2.8}Fe_{0.2}$ alloy increased after heat treatment [54]. In addition, the effect of CuO addition on electrochemical properties of AB_3-type alloy electrodes was investigated. CuO is reduced to Cu during the charging process, and the fine Cu particles are deposited on the surface of the alloy particles. The as-deposited Cu particles form a protective layer to increase electronic and heat conductivity of the electrodes and thus improve maximum discharge capacity, HRD, cycling stability, and high-temperature dischargeability of the electrodes. The maximum discharge capacity increases from 314 mAh g^{-1} (blank electrode) to 341 mAh g^{-1} (3.0 wt% CuO), and the capacity retention rate at the 200th cycle increases from 71.6 to 77.2% (2.5 wt% CuO) [55].

Milling an AB_2-type Zr-Ti-V-Ni alloy with added $LaNi_5$ could successfully coat the AB_2-type alloy with fine nanocrystalline $LaNi_5$, which effectively enhanced the electrochemical activation of the AB$_2$ alloy [9]. The addition of LaNi$_5$ also improved the HRD of the matrix because of a segregated La–Ni phase that provided active sites and pathways for hydrogen diffusion [56]. Another interesting result was that the addition of AB$_5$ alloy could drastically decrease the amount of V being dissolved in the matrix in the KOH electrolyte. On the contrary, an AB$_2$ alloy added to a matrix of LaNi$_5$ alloy improved the discharge capacity of the matrix alloy [57].

Amorphous Mg–Ni alloy can be an excellent surface activator to eliminate the initial activation and substantially improve the kinetic properties of alloys such as $Zr(Ni_{0.6}Mn_{0.15}Cr_{0.1}V_{0.15})_2$ and ZrCrNi, which are usually considered unsuitable for practical application because of their poor kinetics and difficult activation. The amorphous Mg–Ni alloy not only acted as a protective film against corrosion but also formed a uniform layer with high electrocatalytic performance and high elasticity on the surface of the matrix [58]. It is well known that Mg-based alloys have attractively high storage capacity, low cost, and lightweight. Unfortunately, they have poor hydrogen absorption–desorption characteristics, which seriously blocks their practical use in energy storage. The addition of Ti$_2$Ni particles inlaid on the surface of Mg$_2$Ni particles improved both the charge-transfer reaction at the surface of the alloy and hydrogen diffusion in the bulk alloy.

Some intermetallic compounds, such as CoB and TiB, are also effective additives for preparing composite hydrogen storage alloys [59]. In recent years, magnesium has been used as a substitutable element in the system of La-Mg-Ni alloys, which have been extensively studied as negative electrode candidates for Ni-MH secondary batteries because of their high discharge capacities. However, the inferior cycle durability of these alloys limits their practical application. Preparing composite hydrogen storage alloys is also a good method to improve their overall properties.

5.3.1.3 Alloy Preparation

Studies have shown that the structure of hydrogen storage alloys (including alloy solidification, grain size, and grain boundary), which relates to alloy composition, alloy-casting conditions (solidification cooling rate), and the heat-treatment process, has an important effect on the electrochemical performance of alloy electrodes [60]. Alloy solidification and grain size mainly affect the hydrogen absorption alloy pulverization and corrosion rate, which are closely related to the cycle stability of alloy electrodes. The precipitation of different types of alloy elements or the second phase on the grain boundaries may facilitate (or inhibit) the alloy pulverization and corrosion, reducing (or increasing) the cycle stability of alloy electrodes. At the same time, the precipitation of the second phase may also lead to good electrocatalytic activity, and thus, the high-rate discharge performance of alloy electrode is improved. Therefore, besides optimizing the chemical constituents of hydrogen storage alloy, the alloy preparation techniques should also be studied and developed to optimize the structure of the alloy and to further improve the performance of hydrogen storage alloy [61]. A variety of preparation methods for hydrogen storage alloys have been studied, such as the induction melting, arc melting, MA, reduction diffusion method, and combustion synthesis. A brief introduction of the induction melting, arc melting, and MA methods is presented in the following sections.

Induction Melting Currently, the high-frequency electromagnetic induction melting method is the most common method used in industry. The production scale of this method covers the range from a few kilograms to several tons. The disadvantage of this method is high power consumption and difficulty in controlling the alloy structure. The general process of high-frequency induction melting is shown in Figure 5.5.

Vacuum induction-melting furnace is mainly used for melting and casting special steel, high-temperature alloys, and precision alloys in vacuum or a protective atmosphere. This furnace mainly consists of furnace body, bearings, sensors, tilting component, vacuum system, and electrical control system. There are water-cooled mezzanine systems in the furnace body and lid. The furnace is equipped with alloy feeder, observation windows, and thermocouple devices. The electric sensor is made of a spiral coil. Vacuum system is composed of oil diffusion pump, mechanical pumps, traps, control valves, and other components.

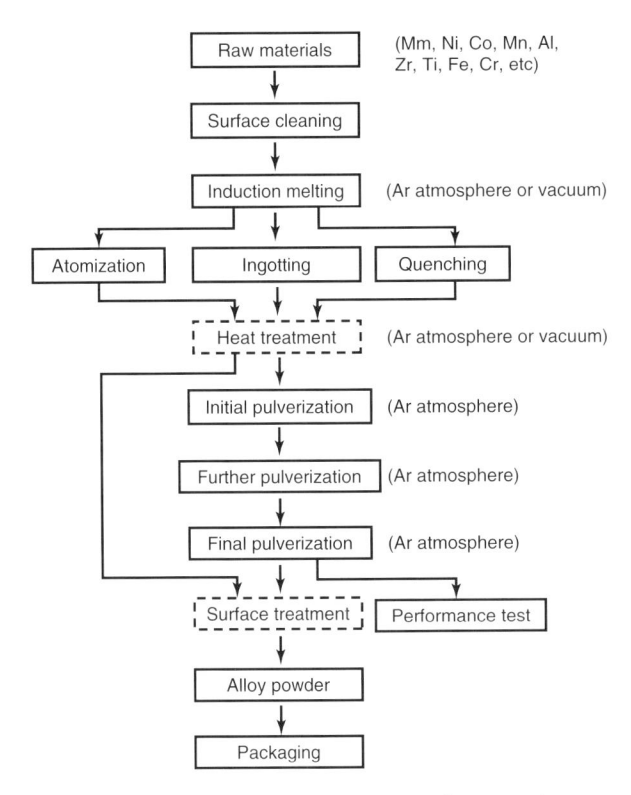

Figure 5.5 Preparation of hydrogen storage alloy (note that the broken line is not necessarily a treatment process) [16].

The basic circuit for the induction furnace involves the start switch, power inverter, capacitor, induction coil, and crucible. The work mechanism of the induction-melting furnace is when the high-frequency current flows through the water-cooled copper coils, induced current is generated from the metal reactant in the furnace because of electromagnetic induction. The flow of induced current in the metal reactant cases the heat, which provides power for heating and melting the metal reactant.

Crucible, which is used for loading, insulation, and energy transformation, is an important component for induction melting. The crucible is divided into (i) basic crucible, including the one made of CaO, MgO, ZrO_2, BeO, and ThO_2; (ii) neutral crucible, including the one made of Al_2O_3, $MgO \cdot Al_2O_3$, $ZrO_2 \cdot SiO_2$, graphite, and other materials; and (iii) acid crucible, which is made of SiO_2. MgO crucible is generally used for melting hydrogen storage materials. The refractoriness of crucible is in the range of 1500–1700 °C. The refractoriness of crucible should be more than 1600 °C for melting Ni-based hydrogen storage materials. Preparation of alloys with the melting method is usually carried out in vacuum or an inert atmosphere.

Arc Melting Compared to the induction melting method, the arc melting method is suitable for experimental and small-scale applications, where it is necessary to melt a great number of alloy compositions [18]. In this method, a pressed pellet is melted by direct contact with an electric arc in argon atmosphere. The arc melting technique provides two main advantages over other melting processes: (i) great versatility in terms of the types of materials and (ii) limited reactivity during melting. The disadvantages of the arc melting method are as follows: (i) low efficiency and hazardous preparation conditions and (ii) for multicomponent alloys, multiple remelting or prolonged annealing at high temperature is necessary in order to obtain a homogenous distribution of components throughout the volume of the pellet. Electrically powered induction furnaces are used for large-scale alloy production. This method is convenient because bulk raw materials can be used instead of powders and step-by-step addition of components can be performed. The molten metal is poured from the furnace by tilting or trapping the furnace, and the metal is then passed to the molds or different vaporizers.

A typical preparation scheme of the AB_5-type alloys includes (i) mixing the chosen materials in the given ratio, (ii) melting in an inert atmosphere (e.g., argon) using arc or induction furnaces, (iii) fast cooling, and (iv) crushing mechanically or using hydrogen gas sorption–desorption cycles.

Production of the AB_2-type alloys involves the following steps: (i) mixing the starting components; (ii) placing into a high-temperature furnace and evacuating it to a pressure of 10^2 Nm^{-2}; (iii) purging with argon and heating to the temperature sufficient to melt the alloys; (iv) mixing the additives with the melt and the removing impurity slag and oxides; (v) prolonged annealing in vacuum or an argon atmosphere at temperatures close to the melting point, where homogenization proceeds; (vi) cooling the solid ingot form before removing from the autoclave; (vii) comminution using hydrogenation in a vacuum pressure vessel followed by cooling in an inert atmosphere (Ar); and (viii) further comminution with ball milling in an inert atmosphere and at low temperature.

Mechanical Alloying The formation of a metal hydride during MA method was first observed by Chen and Williams [62]. They found the formation of a $TiH_{1.9}$ phase in the early stage of Ti milling in ammonia. This suggested that MA could probably be a simple method for the production of metal hydride powders with any composition. The starting materials for ball milling are elemental powders, and the reaction atmosphere used is hydrogen or argon. Especially, hardened steel balls were used, and typical milling times were from 10 to 100 h. The observed decrease in hydrogen pressure during the first hours of milling the pure metals (Zr, Ti, Mg, etc.) and/or their composites showed substantial absorption of hydrogen into the metal particles. Typical dimensions of the alloy particles after prolonged milling (at least 20 h) were under 1 μm. Because as-cast Zr-Ti-Ni based alloys showed slow activation, the composite particles with pure nickel on the surface of a Zr–Ti based Laves-phase compound were prepared by means of mechanical ball milling. X-ray analysis revealed that nickel particles were situated on the surface of the AB_2-alloy grains, and at the same time, some fresh surfaces were simultaneously generated. This increased not only the discharge capacity but also the activation in alkaline electrolyte. Ti–Fe alloys showed an amorphous state after 40 h of ball milling, whereas Zr–Cr alloys revealed a mixture of nanocrystalline $ZrCr_2$ and an amorphous state. Applying the MA method showed that it is possible to make alloys of most metals, including Mg. Electrochemical measurements indicated that only the amorphous phase in Mg_xNi_{100-x} alloys could electrochemically absorb and desorb hydrogen at room temperature, with a capacity over $350\,mAh\,g^{-1}$.

5.3.2
Cathode

The capacity of an Ni-MH battery is limited by the positive electrode. Therefore, it is of great significance to further improve capacities and electrochemical properties of spherical nickel hydroxide positive materials for improving the overall performances of Ni-MH batteries. To improve the discharge capacity and cycle stability, R&D activities in cathode materials have focused on developing materials preparation technology to control the shape, chemical composition, size distribution, structural defects, and surface activity of nickel hydroxides and searching for new additives and molding processes [63].

$Ni(OH)_2$ exists in the α- or β-phase, as shown in Figure 5.6. Correspondingly, there are β- and γ-phases for the charged product (i.e., NiOOH). α-Ni(OH)$_2$ is composed of water-containing layers through hydrogen bonding and is usually unstable in an alkaline electrolyte. In alkaline medium, α-Ni(OH)$_2$ with low crystallinity undergo aging to transform into β-Ni(OH)$_2$ when controlling certain experimental parameters such as temperature, aging time, and pH value [64].

Practically, β-Ni(OH)$_2$ is commonly used in Ni-MH batteries. The well-crystallized β-Ni(OH)$_2$ adapts a layered structure with hexagonal unit cell, each cell

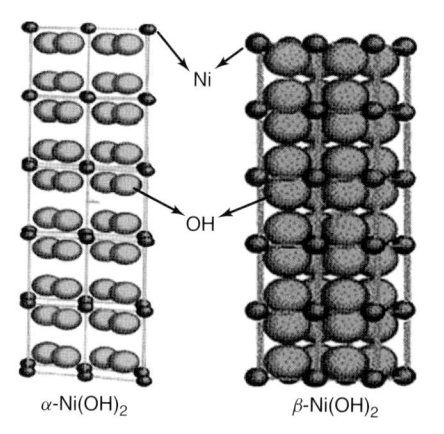

Figure 5.6 Crystal structure of α-Ni(OH)$_2$ and β-Ni(OH)$_2$ [16].

having one nickel atom, two oxygen atoms, and two hydrogen atoms. The distance between two nickel atoms (a_0) is 0.312 nm, and the distance between the two layers of NiO$_2$ (c_0) is 0.4605 nm. The density, Ni oxidation state, and cell parameters of each type of Ni(OH)$_2$ and NiOOH crystalline phases are summarized in Table 5.2.

During the charge–discharge process, the Ni(OH)$_2$/NiOOH couple undergo a complicated change in the crystal phase. Changes in the structural characteristics of the active material in a nickel electrode can be better understood with the help of the Bode diagram [65]. Figure 5.7 presents a useful summary of morphological species involved in nickel electrodes as they are cycled. The β-form of nickel hydroxide (labeled as β(II)) is thermodynamically stable in aqueous KOH solution. Its structure is depicted as a series of evenly spaced platelets. When β-Ni(OH)$_2$ is charged, it is converted to a structurally similar form, in which nickel's valence changes from +2 to +3. If the charging voltage is further increased, the β-form of the charged material (β(III)) is converted to a higher-valence form that has an average valence of 3.67. Some of the ions remain at the +3 valence, whereas others are oxidized to the +4 valence state. This overcharged material is labeled as γ(III). In this crystallographic form, the nickel-to-nickel distance is the same as in the β(III)-form, but the spacing between the platelets is enlarged significantly. Molecules of water and KOH are incorporated into the lattice. The expanded crystal lattice of the γ-phase can result in electrode expansion in weaker plaque structures. Factors that favor γ-phase formation include higher KOH concentrations, lower

Table 5.2 The density, oxidation state, and cell parameter of each type of crystalline phases of Ni(OH)$_2$/NiOOH [64, 65].

Crystalline phase	Ni average oxidation state	Density (g ml^{-1})	a_0 (nm)	c_0 (nm)
α-Ni(OH)$_2$	+2.25	2.82	0.302	0.76–0.85
β-Ni(OH)$_2$	+2.25	3.97	0.312	0.4605
β-NiOOH	+2.90	4.68	0.281	0.486
γ-NiOOH	+3.67	3.79	0.282	0.69

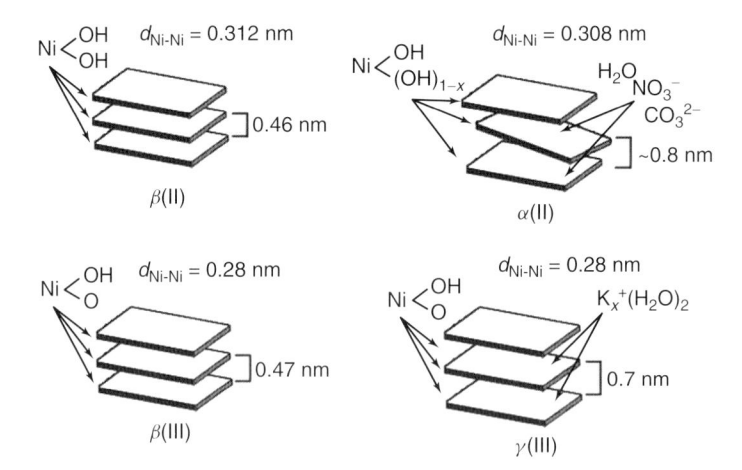

Figure 5.7 The Bode diagram of nickel electrode species [65].

temperatures, and higher cobalt contents. The γ-phase material discharges to the α(II)-phase. This phase structure is referred to as *turbostratic*. As mentioned above, such phase is converted to the more stable β(II)-phase in aqueous KOH solution.

Throughout the discharge process, only concerning the change of valence state of Ni and ignoring the transition among the crystalline phase, the electrode chemical reactions could be described as Eq. (5.1). In other words, during the charging process, Ni(OH)$_2$ transforms to NiOOH and Ni^{2+} is oxidized to Ni^{3+}, whereas during the discharging process, NiOOH transforms reversely into Ni(OH)$_2$ and Ni^{2+} is reduced to Ni^{3+}. According to Faraday's law, the theoretical discharge capacity of Ni(OH)$_2$ is 289 mAh g^{-1} owing to the transition of Ni^{2+}/Ni^{3+}. Generally, owing to the semiconducting properties of nickel oxide electrodes, the charge–discharge reactions cannot proceed completely, resulting in low utilization efficiency of active materials. In order to improve the performance of nickel oxide electrodes, addition of other functional materials is necessary [10]. The effect of additives on the nickel oxide electrode can be summarized in the following four aspects: to improve the utilization efficiency of the active component, to increase the discharge potential, to inhibit nickel electrode expansion and improve the life, and to improve the charge–discharge performance and high current charge–discharge capacity of nickel electrode in a wide temperature range.

Additives can be classified into the following types according to their different load modes: chemical codeposition (such as chemical coprecipitation of zinc, nickel, and cobalt hydroxides), electrochemical codeposition method, surface deposition (also known as *chemical plating*), and mechanical mixing. Additives can be metals, oxides, or hydroxides. Cobalt and lithium compounds are widely used in practical application.

- **Cobalt.** In the cathode, contact resistances of the Ni(OH)$_2$ particles and between the Ni(OH)$_2$ particles and the Ni foam substrate are relatively large. Ni^{2+} cannot be fully oxidized in the charging process, and Ni^{3+} cannot be fully reduced in the discharging process owing to the poor electron transfer. Hence, the active materials cannot be fully utilized, and the capacity of $Ni(OH)_2$ is difficult to improve.

- The addition of Co can enhance the reaction's reversibility between the Ni^{2+}/Ni^{3+} couple in electrochemical process and improve the mass transfer and electrical conductivity, thus increasing the utilization of $Ni(OH)_2$ [66]. By dissolving and precipitating cobalt oxide in an alkaline solution, β-$Co(OH)_2$ is formed and covered uniformly on the surface of $Ni(OH)_2$. After charging, β-$Co(OH)_2$ is oxidized to CoOOH, which has a good conductivity. CoOOH on $Ni(OH)_2$ surface acts as a microcurrent collector, improves the conductivity between $Ni(OH)_2$ and the conductive substrate, and reduces electrode resistance, thus enhancing the utilization of the active materials.

    Notably, $Ni_{1-x}Co_x(OH)_2$ solid solution can be formed by adding cobalt during the synthesis of $Ni(OH)_2$ by the coprecipitation method, where cobalt replaces partial nickel to form cationic impurity defects in the $Ni(OH)_2$ and NiOOH lattices. The defects can facilitate H^+ diffusion in the charge–discharge process and improve the reversibility of the redox reaction between Ni^{2+}/Ni^{3+} couple. Moreover, cobalt doping can enhance the oxygen evolution potential, decrease the internal battery voltage, and improve the utilization efficiency of $Ni(OH)_2$.

- **Zinc.** The main role of doping Zn via the coprecipitation method is to enhance the oxygen evolution potential, elaborate the microcrystalline grain, suppress the production of γ-NiOOH during the overcharge process, and mitigate volume expansion [67]. Usually, the combination of Co and Zn can improve the electrical properties of $Ni(OH)_2$. It is found that the Ni, Co, and Zn coprecipitation into $Ni_{1-x-y}Co_xZn_y(OH)_2$ solid solution reduces crystallinity and results in ideal β-$Ni(OH)_2$ and NiOOH lattice disorders. As a result, the transition between β-$Ni(OH)_2$ and β-NiOOH takes place more facilely in the charge–discharge process, leading to the lengthening of the charge–discharge voltage plateaus of the nickel electrode. Furthermore, the formation of γ-NiOOH is effectively inhibited, which could reduce the volume expansion of the electrode and consequently prevent the battery from short-circuit problem.

- **LiOH.** In long-term cycling of nickel oxide electrode, the $Ni(OH)_2$ grains accumulate gradually and become bigger and thicker, resulting in more difficult charging of the electrode. Also, elevated temperature and alkali concentration can both cause an increase in the size of the $Ni(OH)_2$ particles. Because Li^+ can be adsorbed onto the electrode surface and can effectively prevent the accumulation and growth of grains, adding LiOH to the electrolyte is an effective approach to improve the utilization of active materials [10]. When LiOH coexists with Co, it can reduce the formation of γ-NiOOH and prevent volume variation of the electrode. The addition of lithium can also increase the overpotential of oxygen evolution and improve the charging efficiency of $Ni(OH)_2$.

- **Ba/Ca/Mg.** The purpose of adding $Ba(OH)_2$ to the active substance is to maintain $Ni(OH)_2$ dispersion and promote the oxygen evolution overpotential. In the preparation of $Ni(OH)_2$, adding a certain amount of Ca^{2+}/Mg^{2+} can also alter the potential of oxygen evolution and improve the high-temperature performance of nickel electrode [68]. The existence of Ca/Mg in Ni(OH)$_2$ is mainly in the hydroxide or carbonate form. However, electrochemical experiments show that adding Ca and Mg with content over 0.02 wt% will reduce the activity of Ni(OH)$_2$, block proton transfer in Ni(OH)$_2$, and impede the transition of Ni^{2+}/Ni^{3+}, thus accelerating attenuation of capacity and voltage plateau and exerting negative impact on the cyclability of the electrode.

- **Rare earth oxides.** The addition of rare earth oxides can remarkably improve the high-temperature performances of nickel electrode because of the main reason that these additives alter the difference between the oxidation potential and oxygen evolution potential of the electrode [69].

5.3.2.1 Preparation of Ni(OH)$_2$

The preparation methods for spherical Ni(OH)$_2$ include chemical precipitation, high-pressure catalytic oxidation of nickel powders, electrolytic deposition of nickel, and spraying technique [70]. Spherical Ni(OH)$_2$ prepared via the chemical precipitation method has a relatively good performance and is widely used.

Chemical precipitation is generally carried out in the reaction vessel with a particularly designed structure [71]. The crystal growth parameters such as the amount of nuclei, microcrystalline grain size, grain stacking, crystal growth rate, and crystal defects are adjusted by the reaction temperature, pH value, feedstock quantity, additives, feed rate, and stirring level. All these reaction conditions should be strictly controlled; otherwise, they will affect the structure and composition of Ni(OH)$_2$ and may result in the adsorption of SO_4^{2-} and Na$^+$ in Ni(OH)$_2$. Electrochemical experiments show that changing the Ni(OH)$_2$ crystal structure and increasing the content of sulfate and other impurities in Ni(OH)$_2$ cause the decrease in discharge capacity and increase in electrode polarization.

The average size of spherical Ni(OH)$_2$ particles prepared via chemical precipitation method is generally between 1 and 50 μm. Particle size and size distribution have the major impact on reactivity, specific surface area, and apparent and tap density of Ni(OH)$_2$. Generally, the smaller the size, the larger the specific surface area and the higher the reactivity. Different surface states of Ni(OH)$_2$ will lead to a great difference in surface area, which significantly affects the electrochemical performance. Generally, Ni(OH)$_2$ with a smooth surface and good spherical degree has high tap density and good fluidity but low activity; the product with low spherical degree and rough surface has relatively low tap density and poor fluidity but high activity. At present, the commercial cathode material of Ni-MH batteries mainly focuses on spherical Ni(OH)$_2$, which is doped with additives or coated with active substances. In recent years, nanostructured Ni(OH)$_2$ electrode has increasingly attracted attention. For example, Ni(OH)$_2$ nanotubes exhibit larger reversible capacity, higher charge–discharge efficiency, and better cycling performance than spherical Ni(OH)$_2$ powders.

As nickel hydroxide nanomaterials offer the possibility of further significant improvement in the performance of nickel electrodes, much effort has been dedicated to developing related preparation strategies and technologies.

- **Precipitation transformation.** This method is based on the difference in the solubility (K_{sp}) of the product, by adjusting the preparation conditions such as concentration of precipitating agent and conversion temperature and by resorting to surfactants to control the particle size and growth to obtain nanoparticles with good dispersion [72]. The method is easy to be scaled for industrial production due to its low cost, simple process, and high yield.
- **Coordination precipitation.** In this method, a certain coordination agent is added to the metal salt solution. Then, the nanoparticles are obtained by adding the precipitating agent and controlling the addition of precipitating agent or dropping rate [73].
- **Microemulsion method.** This method is also known as the *reverse micelle method*. Microemulsion is a transparent, low-viscosity, and thermodynamically stable system, which is normally provided by surfactants, cosurfactant, and the composition of the organic solvent and water [74]. In this system, organic solvents act as the dispersion medium, whereas water is the dispersed phase and surfactants work as emulsifiers. Metal salts react with the precipitating agent to form a microemulsion, which controls the nucleation and growth of the initial particles in the water core (known as *microreactor*). The particle size and structure of nanosized $Ni(OH)_2$ prepared by microemulsion method are closely related to the pH value of the solution, water/surfactant ratio, the properties of the surfactant, and reaction temperature. This method has the advantages of easy control of the size of nanoparticles, good monodispersity, easy operation, and so on.
- **Template method.** Template synthesis is based on the use of porous materials with nanometer to micrometer pore size as the template, which is combined with electrochemical deposition, chemical deposition, *in situ* aggregation, sol–gel method, or chemical vapor deposition [75]. The target materials are deposited on the wall of the inner template hole to form the designed structure. The morphology of the nanomaterials produced via this method well reflects the feature of the porous template cavity.

5.3.3
Separator

A separator is a porous membrane placed between the positive and negative electrodes and is permeable to ionic flow but prevents electric contact of the electrodes. Its main function is to keep the positive and negative electrodes apart to prevent electrical short circuits and, at the same time, allow rapid transport of ionic charge carriers that are needed to complete the circuit during the passage of current in an electrochemical cell. As an important part of the cell structure, the separator directly affects the critical properties such as the charge–discharge current and voltage of a battery, specific energy/power, self-discharge, cycle life, and mechanical performance. A number of factors must be considered in selecting the best separator for a particular battery and application. The considerations that are important and influence the selection of the separator include the following [76]:

- electronic insulator;
- minimal electrolyte (ionic) resistance;
- mechanical and dimensional stability;
- sufficient physical strength to allow easy handling;
- chemical resistance against degradation by electrolyte, impurities, and electrode reactants and products;
- effectiveness in preventing migration of particles or colloidal or soluble species between the two electrodes;
- readily wetted by electrolyte;
- uniform in thickness and other properties.

Separators for batteries can be divided into different types, depending on their physical and chemical characteristics. They can be molded, woven, nonwoven, microporous, bonded, papers, or laminates. Separators for sealed Ni-MH batteries should be flexible enough to be wrapped around the electrodes and highly permeable to gas molecules for overcharge protection. At present, the separator membrane for an Ni-MH battery is generally nonwoven [77]. Nonwoven separators have been commonly used as separators in alkaline battery systems since 1960s. Nonwovens are textile products processed directly from fibers. Depending on the method of production, the fibers are used either as raw materials or are produced *in situ*. Fibrous materials used in battery separators are predominantly synthetic polymers such as polyolefins or polyamides, but fibers from natural resources, such as cellulose, are also used. In contrast to woven materials, the fibers in nonwovens are randomly distributed, without an orientated microstructure.

On one hand, the fibrous structure of nonwoven materials offers a high porosity, which is necessary for high electrolyte absorbance and low ionic resistance and results in good charge–discharge acceptance in the battery. On the other hand, the stochastic process of nonwoven separator production with many layers of fibers over the material's thickness avoids pinholes through the separator. Therefore, the risk of electrical short circuits is minimized. This stochastic arrangement of the fibers is the main advantage of a nonwoven material for battery separator applications as compared with woven structures.

5.3.3.1 Preparation of Separators

The nonwoven production process can be separated into three fundamental steps (see Figure 5.8): formation of the web, bonding of the web (fixing of the fibers), and posttreatment. In general, web formation and web bonding are performed in one production step because the nonbonded web is difficult to handle. Posttreatment is normally carried out as a separate step. Additional treatments such as thickness adjustment should be done in-line or in a separate step.

The most challenging new application of industrial batteries is their use in HEVs and as an energy buffer in future fuel cell vehicles. Currently, the Ni-MH technology is the only one that is commercially used in HEVs. In the overall charging–discharging reaction of the Ni-MH system, the electrolyte is not involved. This means that, in principle, the separator need not act as an electrolyte donor or absorber. Consequently, separators used in Ni-MH batteries can be fabricated

```
┌─────────────┐     ┌─────────────┐          ┌──────────────────┐
│    Web      │────▶│    Web      │--------▶ │  Posttreatment / │
│  formation  │     │  bonding    │          │    finishing     │
└─────────────┘     └─────────────┘          └──────────────────┘
```

| One production step; bonding is done in-line | | Posttreatment can be done in-line or separately |

Dry-laid nonwoven	Bonding agents	Wetting agents
Wet-laid nonwoven	Thermo bonding	Corona/plasma
Spunbond nonwoven	Needling (mechanical)	Gas-phase fluorination
Meltblown nonwoven	Hydro entanglement	Acrylic acid grafting
		Wet-phase sulfonation

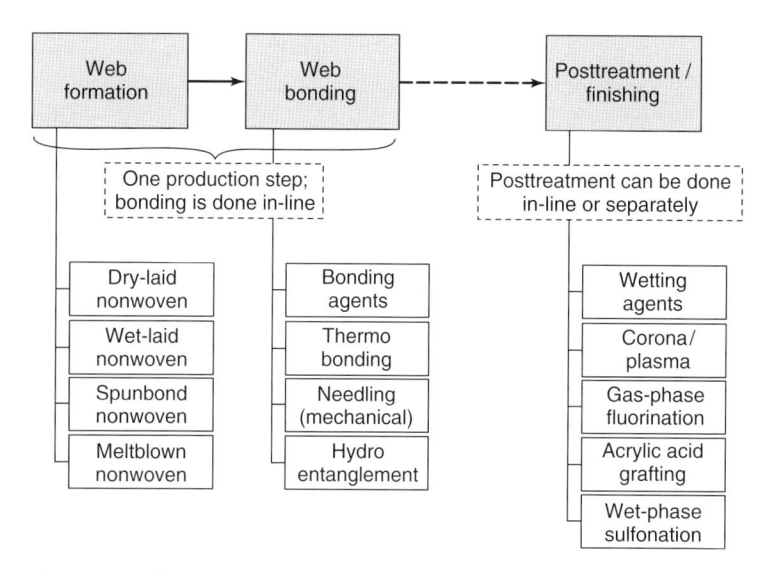

Figure 5.8 Schematic overview of the processes of nonwoven production [77].

thinner than that in Ni-Cd batteries. Current high-capacity consumer batteries use separators with a thickness of or below 120 μm. The decrease in separator thickness is one reason for the higher energy density values (energy per unit volume) of Ni-MH batteries relative to Ni-Cd batteries. However, the separator still needs to act as an electronic insulator between the electrodes, and also, the electrolyte in Ni-MH systems is not completely inert [77].

At a high state of charge (SOC), electrolyte decomposition occurs to liberate oxygen gas. Oxygen is generated at the positive electrode. The released bubbles must pass across the separator to be reduced at the negative electrode. This means that as long as the SOC is low, no electrolyte is decomposed, whereas at high SOC levels, the electrolyte is decomposed and the reservoir effect of the separator becomes necessary. Moreover, the heat generation during high-rate discharge of Ni-MH batteries is less than that in Ni-Cd cells, as the main discharge reaction is endothermic [78]. In an experimental study, only moderate warming to below 50 °C occurs even at high-current discharge of Ni-MH cells. For the separator, this means that it is not exposed to such high temperatures as those that occur during high-rate discharge of Ni-Cd cells. Therefore, polyolefin-based nonwoven separators can be used without any restriction, including in high-rate discharge applications.

One of the largest drawbacks of the Ni-MH system is the high value of self-discharge, especially at elevated temperatures. One of the main reasons for self-discharge is nitrogen-containing impurities, mainly released from nickel hydroxide. In the so-called *nitrite/nitrate-ammonia-shuttle reaction*, nitrate impurities in the nickel electrode are released into the electrolyte and migrate through the separator to the metal hydride electrode. Because of the high activity of the hydrogen atoms in the negative electrode, the anions are rapidly reduced to N(III) in the form of ammonia. Ammonia can then transport back through the separator and reach the nickel electrode where it is oxidized back to nitrate/nitrite. This process can be repeated, and thus both electrodes are slowly discharged [79]. Elevated temperatures worsen the self-discharge of Ni-MH cells as a result of the fact that both the diffusion process and chemical reaction rates increase with temperature. In principle, this self-discharge mechanism is similar to that in Ni-Cd cells. However, because of the lower reactivity of the oxidized species at the Cd electrode than that at the metal hydride electrode, the reduction rate of the oxidized species is much lower, resulting in a lower self-discharge rate for Ni-Cd cells.

Treatments on the surface of the separator, such as acrylic acid grafting, sulfonation, fluorination, and plasma treatment, have been used to obtain a good permanent wettability. Separators possessing these treatments display a reduced self-discharge in Ni-MH batteries [80]. It has been shown that separators able to absorb ammonia strongly can decrease the self-discharge rate. Such so-called *functional separators* are posttreated either by grafting acrylic acid or by chemical sulfonation. The chemical groups introduced by these two processes bind ammonia in the strong alkaline environment. It has also been proven that the introduction of acrylic acid into the polymer before nonwoven processing shows an ammonia-trapping effect [77]. Unlike the Ni-Cd system, short-chain polyamides are not preferentially used as separator materials in Ni-MH systems since they decompose slowly by releasing nitrogen impurities. Separators for Ni-MH batteries can be classified as (i) ammonia-releasing separators, such as polyamides, which lead to an accelerated self-discharge; (ii) ammonia-neutral separators, for example, nonfunctionalized polyolefins, which contribute neither positively nor negatively to the self-discharge; and (iii) ammonia-trapping separators (or ammonia sinks), such as functionalized polyolefins, which can significantly reduce the self-discharge.

Another important issue is the control of self-discharge of a single cell in a multicell battery. In such a system, it is imperative that each individual cell should be on the same SOC level, even after thousands of minicycles. This is a particularly important issue in batteries for HEVs that contain hundreds of single cells. The reduction of self-discharge by functional separators helps to equalize the SOC within the individual cell. The only disadvantage of functional separators is their high cost. Nevertheless, such functionalized separators are standard materials applied in high-end Ni-MH batteries. For the HEVs application, additional requirements regarding quality and the absence of defects must be considered. It must be bore in mind that a single pinhole in the $\sim 10 \text{ m}^2$ of the separator in a HEV battery will lead to a failure of the whole battery system. Therefore, 100% control of the separator integrity is essential. Figure 5.9 shows a fine-fibered melt-blown nonwoven used in industrial Ni-MH batteries.

5.3.4
Electrolyte

The electrolyte used in an Ni-MH battery should have the following properties [81]:
- high ionic conductivity;
- thermal stability in a wide temperature range;

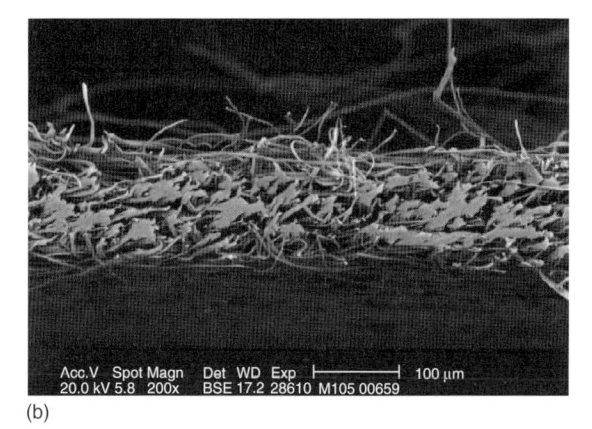

Figure 5.9 SEM images of (a) surface and (b) cross section of the typical polyolefin melt-blown nonwoven separator materials used in industrial Ni-MH batteries (Style FS 2192-11 SG by Freudenberg Vliesstoffe KG, Germany) [77].

- low viscosity;
- suitable concentration.

As an important part of the battery, the electrolyte's composition, concentration, and quantity will have a crucial impact on the battery performance, such as capacity, internal resistance, cycle life, pressure, and other properties.

The electrolyte used in Ni-MH batteries is an aqueous solution of potassium hydroxide (KOH). The electrolyte performs several important functions. The primary function is to form ionically conductive paths for ionic current to flow freely through the separator between the two electrodes [82]. If its volume is too small or its conductivity is too low, poor battery performance will be obtained. One of the known causes of the cell failure is an inadequate amount of electrolyte within the cell components. This is usually referred to as a *separator dryout problem*. Too much electrolyte in the cell cannot be contained within the cell's wettable porous structures and can cause popping damage, leading to premature failure because of the short circuits between the electrodes.

An extensive study of the impact of electrolyte concentration on the cycle life of boilerplate cells has demonstrated the desirability in some situations of using 26% KOH as the electrolyte concentration instead of the more traditional 31% concentration. The use of 26% KOH has resulted in a significantly increased cycle life in flight-weight cells at the expense of a small decrease in energy density, poorer low-temperature performance, and higher freezing temperatures when the cells are fully discharged. A reduction in electrode swelling appears to be the main reason of this increased cycle life at a deep depth of discharge (DOD). The lower electrolyte concentration favors the β–β nickel electrode reaction, whereas the higher concentration promotes the γ–α electrode reaction, which has significantly larger changes in the lattice constants of the active material over the course of a complete cycle. The larger change in the lattice spacing when the active material is charged to the γ-phase is believed to account for the increased degree of volume expansion of the electrode.

In general, a small amount of LiOH and NaOH was added into the KOH aqueous solution as the Ni-MH battery electrolyte [83]. In the electrolyte, by adding LiOH, Li^+ can be adsorbed on the surface of $Ni(OH)_2$ particle, preventing the growth of large grains. Addition of LiOH can also improve the charging efficiency of $Ni(OH)_2$. The addition of NaOH benefits by increasing the utilization of active materials.

Traditional KOH aqueous solution will freeze at low temperatures and will dry up after repeat charge–discharge cycles. Therefore, it was necessary to study the solid electrolyte, the organic gel electrolyte [84]. Compared to an aqueous electrolyte, its cycle life is improved, but because of the lower conductivity of the electrolyte ions, resulting in low current density, it does not meet the actual application.

5.3.5
Endplates

An endplate is the porous substrate that is impregnated with active material, additives, conductive agents, and adhesives. Endplates generally have 74–76% (including skeleton) or 80–85% (without skeleton) porosity with a thickness of 0.5–0.95 mm. A variety of endplates are used in rechargeable Ni-MH batteries, including sintered substrate, nickel foam substrate, and fiber substrate [85].

5.3.5.1 Conductive Agent
Conductive agent is an indispensable part of the electrode. It can improve the electrical contact between the active material and the substrate, provide uniform current distribution in the electrode, and reduce the electrochemical polarization, thereby improving the utilization efficiency of the active material and the electrode charge–discharge efficiency [86].

The main types of conductive agents are acetylene black, graphite powder, and nickel powder. In practice, the mixture of acetylene black and graphite or nickel powder is used as the conductive agent in the electrode. The amount of conductive agent has a strong effect on the electrode capacity and internal resistance. For example, as the amount of nickel powder increases, the discharge capacity of the electrode increases and the resistance decreases. When the amount of conductive agent is <5%, the changes in capacity and resistance are relatively significant, whereas the change is slow when the amount of conductive agent is >15%. Therefore, the amount of conductive agent is controlled within the range of 5–10% for industrial production.

5.3.5.2 Adhesives

Adhesives play a very important role in an Ni-MH battery. First, they help to bind the active materials to the endplates. Second, they prevent deviation of the active materials from the endplates in the charge–discharge process. Finally, the binder is also very important in battery manufacturing process [87].

Reasonable choice of adhesives is important not only for attaining large capacity and reducing resistance and internal pressure, as well as for improving the discharge voltage platform, but also for improving the cycle life and high-rate discharge performance of Ni-MH batteries. The adhesives used in Ni-MH batteries should meet the following requirements:

- better resistance to alkaline electrolyte;
- a certain bond strength and flexibility;
- high electrochemical stability (maintaining stable performance in the battery charge–discharge process);
- permeability;
- being both hydrophilic and hydrophobic;
- low cost.

Adhesives can be classified as hydrophilic and hydrophobic types. Hydrophilic adhesives have high surface activity and can reduce the concentration polarization and internal resistance and increase the conductivity and improve the discharge capacity. On the other hand, the electrode surface quickly absorbs a certain amount of electrolyte to be expanded, resulting in shedding of the active materials and reducing the cycle life. Hydrophobic adhesives link the active materials to the endplate by mechanical bonding, which increases the electrical insulation and the resistance of the electrode reaction, reduces the surface activity, and increases the concentration polarization. However, it exhibits good corrosion resistance and low internal pressure.

The widely used adhesives include polyvinyl alcohol (PVA), methyl cellulose (MC), carboxymethyl cellulose (CMC), polytetrafluoroethylene (PTFE), copolymer of ethylene tetrafluoroethylene and hexachloropropene (FEP), polyvinyl butyral (PVB), and hydroxypropyl methyl cellulose (HPMC). Any single adhesive is difficult to meet the needs of an Ni-MH battery. The combination of hydrophilic and hydrophobic adhesives is an effective way to greatly improve the electrode performance.

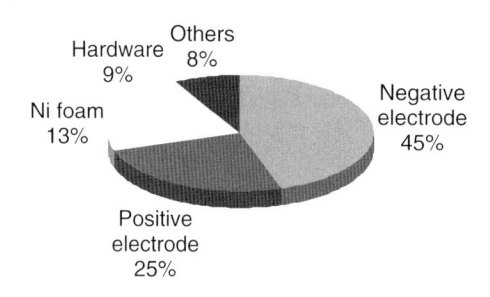

Figure 5.10 Cost distribution of Ni-MH battery components [88].

5.3.6
Cost Distributions for the Components of Ni-MH Batteries

The cost of an Ni-MH battery consists of battery material cost and battery manufacturing cost. The battery material is mainly made up of the negative electrode, positive electrode, Ni foam, and cell hardware. The cost distribution for the components of Ni-MH batteries is shown in Figure 5.10 [88]. The cost of the negative electrode is the highest, which is about 45% of the total material cost. Therefore, R&D of a cheap MH electrode is significant for cost reduction of Ni-MH batteries. Reducing the cobalt content is the most effective way to reduce the cost of the MH electrode. The positive electrode, Ni form, hardware, and the other components occupy about 25, 13, 9, and 8% of the component cost, respectively. On the other hand, increasing production volume and improving manufacturing technology would also reduce the battery manufacturing cost.

5.4
Assembly, Stacking, Configuration, and Manufacturing of Rechargeable Ni-MH Batteries

The fabrication of Ni-MH batteries includes cathode preparation, anode preparation, and cell assembly. The electrode is made by assembling the active materials, additives, conductive agents, adhesives, and conductive substrate through the production process.

5.4.1
Electrode Preparation

The overall performance of an Ni-MH battery depends strongly on the performance of the positive nickel oxide electrode and the metal alloy electrode, which is affected by the electrode preparation process. The cathode manufacturing process is divided into three categories: sintered electrodes, foam nickel electrode, and fiber-type nickel electrode [89]. The anode production process consists basically of two types: slurry anode production process and the sintering production process.

5.4.2 Assembly of Ni-MH Batteries

According to different cell structures, the battery assembly is performed in different processes, but the basic principle is the same. For cylindrical cells, the currently used assembly is the winding process. In this process, the electrode films separated with a separator are coiled into the battery case, the negative plate is connected to the bottom of the case by spot welding, then a groove is made on the top of case to fix the cap, the cathode plate and the battery cap are welded together, a certain amount of electrolyte is infused immediately, and the battery is obtained after sealing. Then, the activation of the batteries is carried out by the charge–discharge process [16].

5.4.3 Battery Pack

Battery pack describes any grouping of individual cells physically connected to one another to achieve the desired dimensional and electrical characteristics. An Ni-MH battery pack can be designed to fit into most compartments, cases, and enclosures, as long as the physical dimensions match the electrical requirements. Before a battery pack is designed and assembled, both electrical requirements and physical parameters need to be determined [14]. The following sections identify the key factors in designing a battery pack that best suit the requirements of a given application and ensure cost-effective reproducibility.

Typically, the electrical requirements needed to run a device or application are already known when the battery pack is to be designed. The primary elements regarding battery pack design are pack voltage, capacity (or runtime), and charging. Both the voltage and capacity of a battery pack are dependent on the conditions in which the battery is discharged and charged. Therefore, the discharge and charge requirements of a device should be identified. The aspects that need to be considered in battery pack design include (i) battery pack configuration, (ii) protective devices, (iii) connectors, (iv) packaging, and (v) labeling.

With the rising demand for portable battery packs, development in the materials and technologies for battery pack assembly has become more and more urgent.

5.4.4 Configurations

Rechargeable Ni-MH batteries are constructed in cylindrical, button, and prismatic configurations.

5.4.4.1 Cylindrical Configuration
The assembly of a cylindrical Ni-MH cell unit is shown in Figure 5.11. The electrodes are spirally wound, and the assembly is inserted into a cylindrical nickel-plated steel can. The electrolyte is filled within the pores of the electrodes and

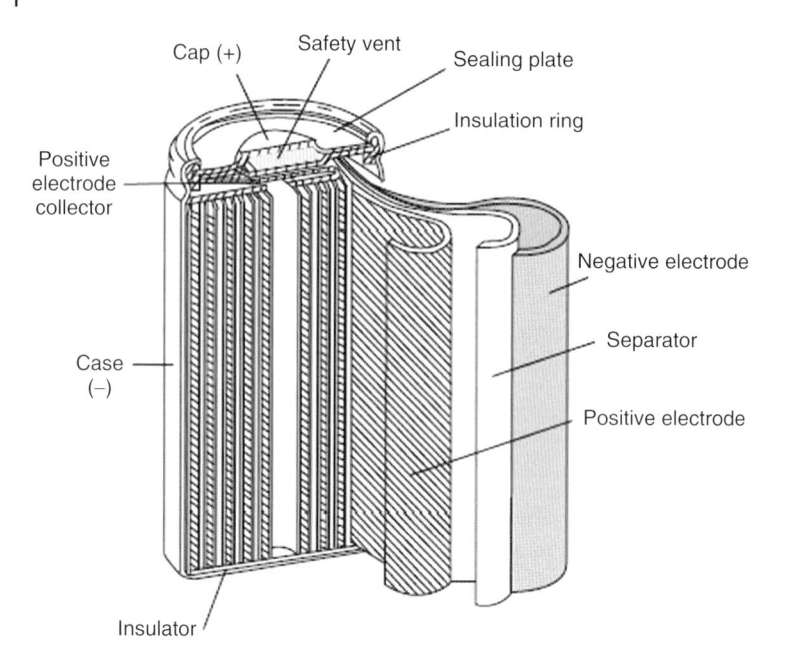

5 Nickel-Metal Hydride (Ni-MH) Rechargeable Batteries

Figure 5.11 Construction of a sealed cylindrical nickel-metal hydride battery [90].

the separator. The cell is sealed by crimping the top assembly to the can. The top assembly consists of a lid, which includes a resealable safety vent, a terminal cap, and a plastic gasket. The can serves as the negative terminal and the lid as the positive terminal, both insulated by each other by the gasket. The vent provides additional safety by releasing any excessive pressure that may build up if the battery is subjected to abuse. Table 5.3 shows the parameters of cylindrical Ni-MH batteries.

5.4.4.2 Prismatic Configuration

The thin prismatic Ni-MH batteries are designed to meet the needs of compact equipment. The rectangular shape permits more efficient battery assembly, eliminating the voids that occur with the assembly of cylindrical cells. The volumetric energy density of the battery can be increased by a factor of about 20%. Figure 5.12 shows the structure of a prismatic battery. The electrodes are manufactured in a manner similar to that for the cylindrical cell, except that the finished electrodes are flat and rectangular in shape. The flat electrodes are assembled, with the positive and negative electrodes being interspaced by separator sheets and welded to the cover plate. The assembly is placed in the nickel-plated steel can, and the electrolyte is added. The cell is sealed by crimping the top assembly to the can. The top assembly is a lid that incorporates a resealable safety vent, a terminal cap, and a plastic gasket, similar to that in the cylindrical cell. The bottom of the metal can serves as the negative terminal and the top lid as the positive terminal. The gasket insulates the terminals by contacting with each other. Table 5.4 lists the parameters of prismatic Ni-MH batteries.

Table 5.3 Parameters of cylindrical Ni-MH batteries [16, 90].

Model	Type	IEC standard	Voltage (V)	Capacity[a] (mAh)	Size		Standard charge		Rapid charge		Weight (g)
					Diameter (mm)	Height (mm)	Current (mA)	Time (h)	Current (mA)	Time (h)	
AAAA	H-4/3AAAAA550	HR09/56	1.2	550	8.6	55.3	55	15	550	1.2	11
	H-5/3AAAAA680	HR09/67	1.2	680	8.6	67.0	68	15	680	1.2	14
AAA	H-2/3AAA280	HR11/29	1.2	280	10.5	29.0	28	15	280	1.2	8
	H-AAA550A	HR11/45	1.2	550	10.5	45.0	55	15	550	1.2	12
	H-AAA550B	HR11/44	1.2	550	10.5	44.0	55	15	550	1.2	12
	H-AAA600A	HR11/45	1.2	600	10.5	45.0	60	15	600	1.2	12
	H-AAA600B	HR11/44	1.2	600	10.5	44.0	60	15	600	1.2	12
	H-AAA650A	HR11/45	1.2	650	10.5	45.0	65	15	650	1.2	13
	H-AAA650B	HR11/44	1.2	650	10.5	44.0	65	15	650	1.2	13
	H-AAA700A	HR11/45	1.2	700	10.5	45.0	70	15	700	1.2	13
	H-AAA700B	HR11/44	1.2	700	10.5	44.0	70	15	700	1.2	13
	H-AAA750A	HR11/45	1.2	750	10.5	45.0	75	15	750	1.2	14
	H-AAA750B	HR11/44	1.2	750	10.5	44.0	75	15	750	1.2	14
	H-AAA800A	HR11/45	1.2	800	10.5	45.0	80	15	800	1.2	15
	H-AAA800B	HR11/44	1.2	800	10.5	44.0	80	15	800	1.2	15
	H-AAA850A	HR11/45	1.2	850	10.5	45.0	85	15	850	1.2	15
	H-AAA850B	HR11/44	1.2	850	10.5	44.0	85	15	850	1.2	15
	H-4/3AAA650	HR11/51	1.2	650	10.5	50.5	65	15	650	1.2	15
	H-4/3AAA700	HR11/51	1.2	700	10.5	50.5	70	15	700	1.2	15
	H-4/3AAA750	HR11/51	1.2	750	10.5	50.5	75	15	750	1.2	16
	H-4/3AAA800	HR11/51	1.2	800	10.5	50.5	80	15	800	1.2	16

(continued overleaf)

Table 5.3 (continued)

Model	Type	IEC standard	Voltage (V)	Capacity[a] (mAh)	Size		Standard charge		Rapid charge		Weight (g)
					Diameter (mm)	Height (mm)	Current (mA)	Time (h)	Current (mA)	Time (h)	
	H-5/3AAA800	HR11/67	1.2	800	10.5	66.5	80	15	800	1.2	18
	H-5/3AAA900	HR11/67	1.2	900	10.5	66.5	90	15	900	1.2	18
AA	H-2/3AA600	HR15/29	1.2	600	14.5	29.0	60	15	600	1.2	14
	H-4/5AA1100	HR15/43	1.2	1100	14.5	43.0	110	15	1 100	1.2	23
	H-4/5AA1200	HR15/43	1.2	1200	14.5	43.0	120	15	1 200	1.2	23
	H-AA800B	HR15/49	1.2	800	14.5	49.0	80	15	800	1.2	24
	H-AA1200A	HR15/51	1.2	1200	14.5	50.5	120	15	1 200	1.2	25
	H-AA1200B	HR15/49	1.2	1200	14.5	49.0	120	15	1 200	1.2	25
	H-AA1200H	HR15/49	1.2	1200	14.5	49.0	120	15	–	–	25
	H-AA1300B	HR15/49	1.2	1300	14.5	49.0	130	15	1 300	1.2	26
	H-AA1400A	HR15/51	1.2	1400	14.5	50.5	140	15	1 400	1.2	28
	H-AA1400B	HR15/49	1.2	1400	14.5	49.0	140	15	1 400	1.2	28
	H-AA1500B	HR15/49	1.2	1500	14.5	49.0	150	15	1 500	1.2	28
	H-AA1500H	HR15/49	1.2	1500	14.5	49.0	150	15	–	–	28
	H-AA1600A	HR15/51	1.2	1600	14.5	50.5	160	15	1 600	1.2	28
	H-AA1600B	HR15/49	1.2	1600	14.5	49.0	160	15	1 600	1.2	28
	H-AA1700A	HR15/51	1.2	1700	14.5	50.5	170	15	1 700	1.2	28
	H-AA1700B	HR15/49	1.2	1700	14.5	49.0	170	15	1 700	1.2	28
	H-AA1800A	HR15/51	1.2	1800	14.5	50.5	180	15	1 800	1.2	28
	H-AA1800B	HR15/49	1.2	1800	14.5	49.0	180	15	1 800	1.2	28
	H-AA1900A	HR15/51	1.2	1900	14.5	50.5	190	15	1 900	1.2	29

5.4 Assembly, Stacking, Configuration, and Manufacturing of Rechargeable Ni-MH Batteries

	Model	IEC	V	mAh	D	H	g		mAh	V	
	H-AA1900B	HR15/50	1.2	1900	14.5	49.5	190	15	1 900	1.2	29
	H-AA2000A	HR15/51	1.2	2000	14.5	50.5	200	15	2 000	1.2	29
	H-AA2000B	HR15/50	1.2	2000	14.5	49.5	200	15	2 000	1.2	29
	H-AA2100A	HR15/51	1.2	2100	14.5	50.5	210	15	2 100	1.2	30
	H-AA2200A	HR15/51	1.2	2200	14.5	50.5	220	15	2 200	1.2	31
	H-AA2300A	HR15/51	1.2	2300	14.5	50.5	230	15	2 300	1.2	31
	H-AA2400A	HR15/51	1.2	2400	14.5	50.5	240	15	2 400	1.2	32
A	H-2/3A900	HR17/28	1.2	900	17.0	28.0	90	15	900	1.2	21
	H-4/5A1600	HR17/43	1.2	1600	17.0	43.0	160	15	1 600	1.2	32
	H-4/5A2100	HR17/43	1.2	2100	17.0	43.0	210	15	2 100	1.2	33
	H-A1800	HR17/50	1.2	1800	17.0	50.0	180	15	1 800	1.2	34
	H-A2000H	HR17/50	1.2	2000	17.0	50.0	180	15	–	–	36
	H-A2100	HR17/50	1.2	2100	17.0	50.0	210	15	2 100	1.2	36
	H-A2700	HR17/50	1.2	2700	17.0	50.0	270	15	2 700	1.2	38
	H-A2800	HR17/50	1.2	2800	17.0	50.0	280	15	2 800	1.2	39
	H-4/3A3000	HR17/67	1.2	3000	17.0	67.0	300	15	3 000	1.2	50
	H-4/3A3000H	HR17/67	1.2	3000	17.0	67.0	300	15	–	–	50
	H-4/3A3300H	HR17/67	1.2	3300	17.0	67.0	300	15	–	–	52
	H-4/3A4000	HR17/67	1.2	4000	17.0	67.0	400	15	4 000	1.2	54
H-1867	H-1867	HR19/67	1.2	4500	18.5	67.0	450	15	4 500	1.2	64
	H-1867H	HR19/67	1.2	3800	18.5	67.0	380	15	–	–	64
H-1872	H-1872	HR19/73	1.2	4500	18.5	72.5	450	15	4 500	1.2	67
SC	H-4/5SC1800P	HR23/34	1.2	1800	23.0	34.0	180	15	1 800	1.2	43
	H-SC2200P	HR23/43	1.2	2200	23.0	43.0	220	15	2 200	1.2	58
	H-SC2500P	HR23/43	1.2	2500	23.0	43.0	250	15	2 500	1.2	58
	H-SC3000P	HR23/34	1.2	3000	23.0	43.0	300	15	3 000	1.2	60

(continued overleaf)

Table 5.3 (continued)

Model	Type	IEC standard	Voltage (V)	Capacity[a] (mAh)	Size		Standard charge		Rapid charge		Weight (g)
					Diameter (mm)	Height (mm)	Current (mA)	Time (h)	Current (mA)	Time (h)	
C	H-C1500A	HR26/51	1.2	1500	26.0	50.5	150	15	1 500	1.2	42
	H-C1800A	HR26/51	1.2	1800	26.0	50.5	180	15	1 800	1.2	42
	H-C2500A	HR26/51	1.2	2500	26.0	50.5	250	15	2 500	1.2	45
	H-C3000A	HR26/51	1.2	3000	26.0	50.5	300	15	3 000	1.2	75
	H-C3000B	HR26/50	1.2	3000	26.0	50.0	300	15	3 000	1.2	75
	H-C3000H	HR26/50	1.2	3000	26.0	50.0	300	15	–	–	75
	H-C3300A	HR26/50	1.2	3300	26.0	50.5	330	15	3 300	1.2	75
	H-C3500H	HR26/50	1.2	3500	26.0	50.0	350	15	–	–	76
	H-C4000B	HR26/51	1.2	4000	26.0	50.5	400	15	4 000	1.2	78
	H-C4000H	HR26/50	1.2	4000	26.0	50.0	400	15	–	–	78
	H-5/3C5500	HR26/71	1.2	5500	26.0	70.5	550	15	5 500	1.2	126
D	H-D1800A	HR33/62	1.2	1800	33.0	61.5	180	15	1 800	1.2	76
	H-D2200A	HR33/62	1.2	2200	33.0	61.5	220	15	2 200	1.2	76
	H-D2500A	HR33/62	1.2	2500	33.0	61.5	250	15	2 500	1.2	76
	H-D3000A	HR33/62	1.2	3000	33.0	61.5	300	15	3 000	1.2	78
	H-D6500A	HR33/62	1.2	6500	33.0	61.5	650	15	4 500	1.7	160
	H-D6500H	HR33/61	1.2	6500	33.0	60.5	650	15	–	–	160
	H-D7000B	HR33/61	1.2	7000	33.0	60.5	700	15	4 500	1.8	160
	H-D7000H	HR33/61	1.2	7000	33.0	60.5	700	15	–	–	160
	H-D9000B	HR33/61	1.2	9000	33.0	60.5	900	15	4 500	2.5	170
	H-3/2D13000	HR33/91	1.2	13 000	33.0	90.5	1 300	15	4 000	4	230

[a] IEC standard: charging at 0.1C for 15 h, then discharging at 0.2C to 1.0 V.

Table 5.4 Parameters of prismatic Ni-MH batteries [16, 90].

Model	Type	IEC standard	Voltage (V)	Capacity[a] (mAh)	Size (mm)	Standard charge		Rapid charge		Weight (g)
						Height (mm)	Current (mA)	Time (h)	Time (h)	
Prismatic	H-PA1200	HF18/07/68	1.2	z11200	17.0 × 6.0 × 67.0	120	15	1200	1.2	26
	H-PA1300	HF18/07/68	1.2	z11300	17.0 × 6.0 × 67.0	130	15	1300	1.2	27
	H-PB750	HF18/07/49	1.2	750	17.0 × 6.0 × 48.2	75	15	750	1.2	18
	H-P8B1000	HF18/07/49	1.2	z11000	17.0 × 8.1 × 48.2	100	15	1000	1.2	23
	H-PC550	HF18/07/36	1.2	550	17.0 × 6.0 × 35.5	55	15	550	1.2	12
	H-P8C850	HF18/09/36	1.2	850	17.0 × 8.1 × 35.5	85	15	850	1.2	20
	H-PD550	HF17/07/35	1.2	550	16.0 × 6.6 × 34.0	55	15	550	1.2	11
	H-PE400	HF18/07/33	1.2	400	17.0 × 6.0 × 32.0	40	15	400	1.2	10
	H-PE500	HF18/07/33	1.2	500	17.0 × 6.0 × 32.0	50	15	500	1.2	10

[a] Charging at 0.1C for 15 h and then discharging at 0.2C to 1.0 V.

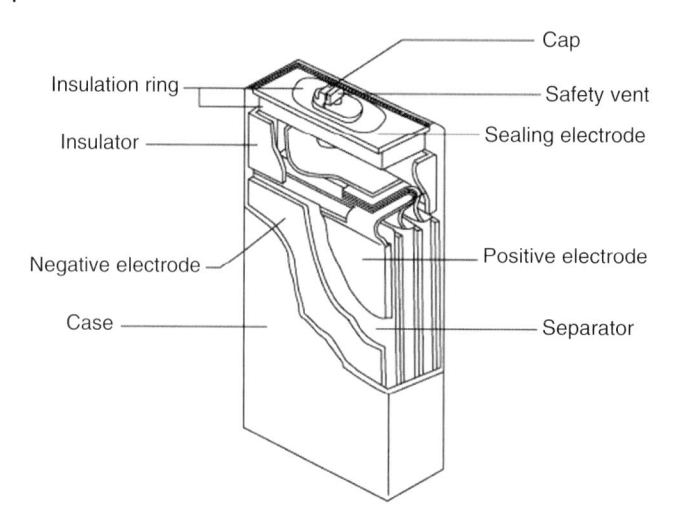

Figure 5.12 Construction of a sealed prismatic Ni-MH battery [90].

5.4.4.3 Button Configuration

The button configuration is illustrated in Figure 5.13. In a button cell, the active materials are pressed into the substrate to form electrodes, while the positive electrode, negative electrode, and separator are alternately stacked to form a layered structure. The button cell is different from the cylindrical and prismatic battery, as it has no safety valve and is particularly suitable to low-current and low-power applications. Table 5.5 shows the parameters of a button Ni-MH cell.

5.4.4.4 Cell and Pack Configurations

The cell configuration refers to the way an individual cell is assembled into a battery pack. The common cell configurations are either a *single* individual cell or an assembled *stick* of two, three, or more cells welded together end to end, as shown in Figure 5.14. The single cell or cell sticks are then assembled into the desired battery pack configuration. It should be noted that the cell stick configurations are sometimes considered as a finished battery pack.

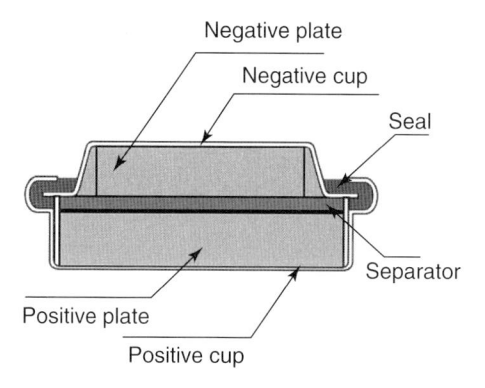

Figure 5.13 Construction of a sealed nickel-metal hydride battery button cell [91].

Table 5.5 The parameters of a button Ni-MH battery [16, 91].

Model	Type	Voltage (V)	Capacity (mAh)	Size		Standard charge		Rapid charge		Weight (g)
				Diameter (mm)	Height (mm)	Current (mA)	Time (h)	Current (mA)	Time (h)	
1131	20 BC	1.2	20	11.6	3.1	2	15	10	3	1.5
1154	40 BC	1.2	40	11.6	5.5	4	15	20	3	2.0
1565	80 BC	1.2	80	15.6	6.1	8	15	40	3	6.5
1575	130 BC	1.2	130	15.6	7.95	13	15	65	3	5.0
2565	200 BC	1.2	200	25.2	6.6	20	15	100	3	11.0
	260 BC	1.2	260	25.2	6.6	26	15	130	3	11.0
	280 BC	1.2	280	25.2	6.6	28	15	140	3	11.5
2575	300 BC	1.2	300	25.2	7.7	30	15	150	3	12.0
2585	320 BC	1.2	320	25.2	8.5	32	15	160	3	13.0
	360 BC	1.2	320	25.2	8.5	36	15	180	3	13.5
	400 BC	1.2	400	25.2	8.5	40	15	200	3	14.0

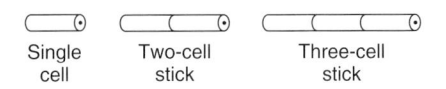

Figure 5.14 Cell configuration [14].

The battery pack configuration is the way the cell configurations are assembled together. Typical pack configurations are shown in Figure 5.15. The *flat* battery pack configuration is the most common one because of its easy assembly. The second common one is the *square*, and the third is the *nested* battery pack configuration. The *staggered* configurations are not common and more difficult to assemble but will sometimes work for battery packs with restrictions in depth. The configuration of a battery pack is almost limitless; however, designing a battery pack that is considered *nontypical* is usually not cost effective. A nontypical battery pack often incorporates a mixture of cell configurations and locates cells at various directions to one another that would inhibit the ease of assembly.

Protective Devices Electrical components of a battery pack refer to the devices incorporated into the battery pack to increase functionality. These components protect a battery pack from damage that may otherwise occur if they are not used. This first group of protective devices includes a variety of ways to sense and protect

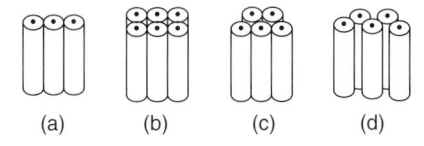

Figure 5.15 Pack configuration: (a) flat, (b) square, (c) nested, and (d) staggered [14].

a battery pack from damage. The following list describes most of the commonly used protective devices [14].

- **Thermistor.** Being used primarily with the more sophisticated methods of charging Ni-MH batteries, this device senses the temperature throughout the charging of the battery, which is then analyzed by the charging circuitry. Thermistors are thermally sensitive resistors that exhibit a large, predictable, and precise change in electrical resistance when subjected to a corresponding change in temperature. The thermistors typically used in battery packs are negative temperature coefficient (NTC) thermistors. This means that they exhibit a decrease in electrical resistance when subjected to an increase in temperature. Thermistors are designed in various types, but the most commonly used and the preferred type is a 10 K 1% glass encapsulated type with axial leads. Other thermistors used are the epoxy-bead-type thermistors, which are less desired because of the increased difficulty to assemble into the battery packs [14].
- **Positive temperature coefficient (PTC) resettable fuse.** The most commonly used component in the assembly of battery packs is the strap-type polymeric PTC device. Strap-type polymeric PTC resettable fuses are easily installed in series inside the battery pack with only a small increase (∼1 mm) in the overall dimensions of the battery pack. PTC resettable fuses limit the flow of dangerously high current during faulty conditions, which include accidental short circuits of battery. Providing over-current protection as well as over-temperature protection, the polymeric makeup of the PTC resettable fuse latches into a high-resistance state when a fault occurs. While only allowing a small amount of current through the battery, this high-resistance state will remain until the fault is repaired. Once the fault and power to the circuit are removed, the PTC resettable fuse will then automatically reset, ready for normal operation [14, 90].
- **Thermostat.** The other commonly used circuit protection device is the bimetallic thermostat or circuit breaker. This type of resettable fuse also protects against short circuit but by using an entirely different technology from that of the PTC resettable fuse. Thermostats use a bimetal disc that senses both heat and current from the battery pack. When the temperature reaches a predetermined level, the disc snaps open the contacts, thus completely breaking the path of the current. When the battery returns to the normal operating temperature, the thermostat resets [14].
- **Thermal fuse.** The one-time use thermal fuse is less common. Thermal fuses open at elevated temperatures caused by runaway current. They are a fail-safe measure since the battery will become inoperative once the fuse has opened. These fuses are recommended, if absolute termination of current is needed for safety concerns only [14].
- **Gas gauge.** Obvious advantage is available with the use of a gauge to inform the user on the available capacity of the battery. Voltage sensing is not accurate for nickel-based batteries. There are several variables to take into consideration in determining the SOC of a pack. These variables include environmental temperature, discharge rate, self-discharge, and time. The electronics that must

be installed in the pack also require the use of the battery, so if the pack is stored on the shelf for an extended period, the gas gauge loses its memory or pulls the pack down below the recommended storage level. Harding does carry standard gas gauge boards and has capabilities to meet most custom requirements. However, the standard gas gauge boards must be configured for each specific pack [14, 91].

Connectors This diverse group of components provides an easy and efficient means of connecting the battery pack to the device. The most commonly used methods are described below.

- **Crimp terminal and crimp terminal housing.** This connector system uses a terminal that is crimped onto the stripped ends of each wire leading from the battery pack. The crimp terminals are then inserted into the crimp terminal housing. These connectors can be used to connect battery lead wires to wires leading to the device or to the printed circuit board of the device. Crimp terminal connectors may have a number of positions, from 2 to over 20, depending on the connector style chosen. Also, these types of connectors have many optional features that aid in the connection of a battery to the device. In addition, a wide range of wire sizes (28–12 AWG) as well as a wide range of connector sizes and styles can be used [14].
- **Quick-connect/fast-on connectors.** This connector style uses a tab-to-tab type connection, with the female tab having rolled edges that receives the flat tab of the male connector. This type of connector is also a crimp-type connection to each wire lead of the battery but is only a single-position connector with an optional protective cover or housing.
- **Contacts.** Some battery pack assemblies have been designed without the use of wires and connectors. As an alternative, they use metal contacts designed within the battery pack to mate directly with the device and charger. The contacts are typically heavy-gauge nickel tabs placed on and/or around the battery pack that line up with the corresponding springs or tabs in the battery compartment of the device and charger. This method of connection can be effective depending on the location of the contacts and overall design of the battery pack. If the contact connection method is preferred, it is recommended to make the contacts on or near the positive and negative ends of the battery pack for ease of assembly [14].

Batteries Matching Matching refers to the grouping of individual Ni-MH cells with similar capacities to be used within a battery pack. Typically, the matching of the cells in a battery pack is within 2%. Matching eliminates the potential of reversing the polarity of one or more cells in a battery pack because of the capacity range of the combined cells being too great. Matching becomes more critical as the number of cells in the battery pack increases. This is due to the potential of one cell having a capacity significantly lower than the average capacity of the other remaining cells. As a result, the cell with the lowest capacity has the potential to reverse polarity, whereas the other cells remain at safe voltage levels before reaching the voltage cutoff. If a battery pack has one or more cells reversed before the voltage cutoff is reached, the performance and cycle life will be reduced [92].

5.4.4.5 Bipolar Ni-MH Battery

Conventional cylindrical coiled electrode and flat-plate prismatic designs have evolved over many years. However, these packaging configurations add considerably to the resultant battery's weight, volume, and cost when compared to the capabilities of the chemicals required for the reaction. In the battery field, it has been well recognized that a bipolar type of construction in which electrodes are stacked in a pile with conductive partitions between cells would be preferred. This design is more compact, exhibits higher power capability, and would be lower in cost than conventional cylindrical and prismatic designs, which is more suitable for its use in EVs and HEVs [93].

Figure 5.16 shows a sketch of the wafer cell design concept. Individual flat wafer cells are constructed with contact faces, one positive electrode, a separator, and one negative electrode. The contact faces serve to contain the cell and make contact with the positive and negative electrodes. The contact faces are sealed around the perimeter of the cell to contain the potassium hydroxide electrolyte. To fabricate multicell batteries, identical cells are stacked one on top of another so that the positive face of one cell makes contact with the negative face of the adjacent cell, resulting in a series connection of the cells. To complete a full battery, current-collecting contact sheets are placed on the end cells to serve as the positive and negative terminals of the battery, and the entire stack is held in compression in an outer battery housing [93].

By fabricating each individual wafer cell as a self-contained unit and sealing the perimeter of each cell, the design overcomes the historic problem of attempting

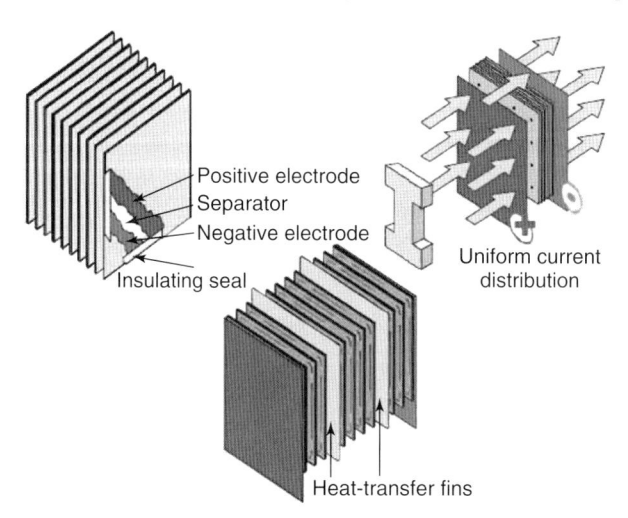

Figure 5.16 Schematic depiction of the wafer bipolar battery design (Electro Energy, Inc (EEI)) [93].

to seal the edge of the multiple cells in a stack. The unit cells can be leak tested, electrically tested, and replaced if necessary to ensure the reliability of a full battery. This design requires an additional conductive layer between cells as opposed to conventional bipolar designs, which utilize one conductive partition between cells. This disadvantage is substantially overcome by the improved reliability of the edge seals of the single wafer cell concept.

A bipolar Ni-MH battery has many advantages, rendering it ideally suitable for powering EVs. However, in its design and manufacturing, there are still some problems, which indicate that the bipolar Ni-MH battery technology is not mature for large-scale production and therefore, further R&D efforts are required.

5.5
Ni-MH Battery Performance, Testing, and Diagnosis

The normal performance of Ni-MH includes the capacity, voltage characteristics, cycle performance, and charge and discharge performance [5, 16].

5.5.1
Capacity

The length of time of a battery supplying the current needed to run a piece of equipment or a device is directly proportional to the capacity of the battery and the discharge rate. Defined as C, *capacity* is the electric current content of a battery expressed in ampere-hour (Ah) or milliampere-hours (mAh). The capacity of a battery is determined by discharging the battery at a constant current until a predetermined end voltage is reached. The time it takes to discharge a battery to the end voltage multiplied by the rate of current at which the battery was discharged is the rated capacity of the battery. Therefore, a battery would be rated at 1500 mAh if it was discharged at a rate of 150 mA for 10 h to an end voltage.

For clarification, the rate of current (charge or discharge) that is applied to a battery is often defined in terms of the rated capacity (i.e., C) of a battery. For example, a battery rated at 1500 mAh that is discharged at a rate of C/2 (or 0.5C) will have 750 mAh discharged from the battery per hour. Thus, a discharge rate of C/2 of a 1500 mAh battery is 750 mA, but this does not mean the battery will last for 2 h. One of the biggest misconceptions regarding Ni-MH cells is that the rated capacity is the capacity that will be received by the user. This would only be true if the user charged and discharged at the same rates of current at which the cell was graded.

5.5.2
Voltage

The discharge voltage profile of an Ni-MH battery is considered flat and varies with the rate of discharge and temperature. As a fully charged battery is discharged, the voltage begins at about 1.5 V followed by a sharp drop to around 1.3 V. The voltage remains between 1.3 and 1.2 V for about 75% of the profile until a second sudden drop in voltage occurs as the useful capacity of the battery begins to deplete. This point is where the discharge current (or load) is terminated at a safe voltage level. At elevated discharge rates, the entire discharge profile is lowered by losses in ohmic polarizations (internal resistance). At high temperatures, the discharge profile is raised by an increase in potential (voltage) between the electrodes. At temperatures below 10 °C, concentration polarization significantly lowers the voltage and usable capacity. This is caused by an increase in energy required to transport molecules within the battery.

The industrial standard for the rated voltage of an Ni-MH cell is 1.2 V. This value is the nominal voltage of a cell that is discharged at a rate of C/10 at a temperature of 25 °C to an end voltage of 1.0 V. This industrial standard is used primarily to call out the rated voltage of battery packs. For example, a battery pack made of three cells in series would be rated as a 3.6 V battery pack. For technical applications and calculations, the nominal voltage of a battery pack provides a useful approximation of the average voltage throughout discharge. The nominal voltage can be simply calculated after the battery has been discharged. To calculate the nominal voltage, the battery's energy (watt-hour (Wh)) is divided by the capacity (ampere-hours (Ah)). This calculation proves beneficial when a battery is discharged at high temperatures since the nominal voltage will increase under the following conditions.

- **State of charge (SOC)**
  The SOC is a measure of the amount of energy remaining in the battery. It is often displayed as a percent of the battery's nominal capacity.
- **Depth of discharge (DOD)**
  The opposite of SOC. The DOD is the amount of energy removed from the battery and is also represented as a percent of nominal capacity.
- **State of health (SOH)**
  The state of health of a battery is a gage of the battery's condition and ability to perform as compared to a new battery. This can also be described as the battery's age.
- **Specific energy**
  The amount of energy contained in a specified amount of mass. It is expressed as the ratio of energy capability to mass, Wh kg^{-1}.
- **Specific power**
  The amount of power contained in a specified amount of mass. It is expressed as the ratio of power capability to mass, W kg^{-1}.
- **Calendar life**
  The length of time the battery is able to provide energy for performance.
- **Cycle life**
  The number of cycles the battery is able to undergo while providing the necessary amount of energy for an application.
- **Internal resistance**
  The internal resistance of Ni-MH cells varies with cell size, construction, and electrode material chemistry. Different materials are used in various Ni-MH cells to achieve the desired performance characteristics. The selection of these

materials also affects the internal resistance of the cell. Since Ni-MH cells of various sizes, construction, and chemistry are different, there is no value of internal resistance that can be defined as a standard.

- **Power**
 The power of the battery is generalized through its ability to provide current. The more powerful the battery, the higher the current it is able to provide.
- **Discharge rate**
 The delivered capacity and nominal voltage of a battery are dependent on the rate of current at which a battery is discharged. For Ni-MH batteries, there is no significant effect on capacity and voltage for the discharge rates below 1C. A reduction in the nominal voltage occurs for discharge rates between 1 and 3C for all sizes of the Ni-MH cells, with the exception of the high-rate series of cells.
- **Charge rate**
 The performance of the battery, as well as the cycle life, depends on effective charging. As the capacity of Ni-MH cells has to be increased, so the demand for faster charging has also increased. This leads to higher charge rates, which requires care to ensure a complete charge while minimizing the potential damage of overcharging. Slow charging is still a reliable method, but not all batteries can be slowly charged without some type of termination. Some active materials of Ni-MH cells have higher capacities and are better suited to the more popular fast charging methods.
- **Charge termination**
 Properly controlling the charging of an Ni-MH battery is critical to achieving optimum performance. Charge control incorporates proper charge termination to prevent overcharging the battery. The overcharging of a battery refers to the state at which the battery can no longer accept (store) the energy entering the battery. As a result, pressure and temperature build up within the cell. If a cell is allowed to remain in the overcharged state, especially at high charge rates, the pressure generated within the cell can be released through the safety vent located within the positive terminal. This may cause damage to the battery by reducing cycle life and capacity. To prevent damage from occurring to the battery, charge termination is one of the most critical elements to be applied to any method of charge control. Charge control may utilize one or more of the following charge termination techniques: time, voltage, and temperature.

5.6
Degradation Mechanisms and Mitigation Strategies

As in all chemical batteries, there are many modes of failure associated with the chemistry inside the battery. In general, these modes cause damage to the battery. Some damages can be rectified, whereas others remain permanent. The main modes of failure for Ni-MH batteries are the following [94]:

1) surface corrosion of negative electrode;
2) decrepitation of alloy particles;

3) loss of water in the electrolyte;
4) crystalline formation;
5) cell reversal;
6) high self-discharge;
7) shorted cells.

5.6.1
Surface Corrosion of Negative Electrode

The primary end-of-life criterion used in many applications is the point at which the battery capacity has faded below 80% of its rated capacity. This occurs because of the corrosion of the negative plate material (metal hydride) throughout the useful life of the battery. The Ni-MH battery does not exhibit a catastrophic *falling off the cliff* type of capacity loss due to this corrosion that some other technologies exhibit. This is because the metal hydride active material is very conductive. Thus, gradual corrosion results in loss of capacity and some loss in power capability because of the consumption of water in the electrolyte during corrosion, but it does not result in failure of the negative plate or loss of conductivity with the active material.

5.6.2
Decrepitation of Alloy Particles

Decrepitation of alloy particles often occurs along with surface corrosion. Normal aging of the battery will cause decrepitation of the negative electrode. This is not technically a failure mode by itself. It only increases the surface area of the electrode, which at first can increase conductivity by increasing the area available for the chemical reactions. However, it also increases the area available for surface corrosion. Some of the literature suggest that alloy decrepitation and surface corrosion of the negative electrode are the main cause of normal battery aging and can be quantified as a function of DOD with electrochemical impedance spectra (EIS).

5.6.3
Loss of Water in the Electrolyte

Another end-of-life criterion is power fade of the battery. Power fade and capacity fade are two different, but somewhat related, phenomena. Power fade is caused by the loss of water in the electrolyte because of a variety of factors. The most severe cause of water loss (sometimes called *separator dry out*) is caused by operating the battery in abusive conditions, causing the safety vent to open and release electrolytic gasses from the cells [82]. A gradual fade of power is also caused by the normal aging and water consumption caused by negative corrosion. Since the Ni-MH battery has considerable power capability, this condition may not affect operation in some applications until the capacity fade deems the battery at end of life. However, power fade may be the limiting factor on life in applications such as HEVs and EVs, UPS applications, and other high-rate applications. Another contribution to water loss in the battery can be water diffusion through the case (in the case of plastic Ni-MH cells); however, this is not a problem with the steel can construction method.

5.6.4
Crystalline Formation

The crystalline formation is the cause of battery memory effect. If the Ni-MH battery is not periodically discharged to 0% SOC, then crystals will form around the electrodes of the battery. A full discharge usually restores the electrode surface. However, if not properly maintained, the crystalline surface will grow and may damage the separator. If the growth reaches through the separator, the battery will simply face short circuit and be rendered useless.

5.6.5
Cell Reversal

Cell reversal only occurs through abusive operation, most notably, overdischarging. This is when the electrodes inside the cell become reversed in charge, rendering the battery unable to store and supply energy.

5.6.6
High Self-Discharge

High self-discharge will not cause catastrophic failure, unless it is at the extreme of an electrical short circuit. All batteries are affected by self-discharge, Ni-MH being one of the highest. Elevated self-discharge occurs along with the growth of crystalline formation. As the crystal growth forms out from the electrode, it begins to mar the separator, thus allowing for an easier path from one electrode to the other. Accordingly, the higher the self-discharge, the lower the capacity of the battery.

5.6.7
Shorted Cells

Shorted cells are generally rare and sometimes unexplainable. It has been suspected that foreign particles may have contaminated the separator during manufacturing and may cause the shorted cells of new batteries. As the manufacturing processes have improved, shorted new batteries have reduced in number. Cell reversal can also cause shorted cells when used in series with other batteries. As mentioned earlier, the growth of crystals through the separator will also cause an electric short.

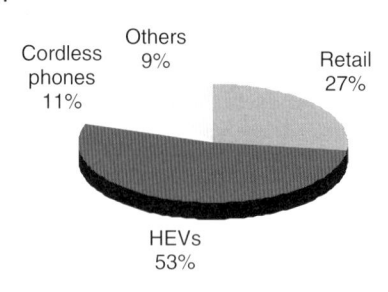

Figure 5.17 Application distributions of Ni-MH batteries in 2008 [95].

- Cordless phones 11%
- Others 9%
- Retail 27%
- HEVs 53%

5.7
Applications (Portable, Backup Power, and Transportation)

The Ni-MH battery has a wealth of applications ranging from portable consumer products, such as digital cameras, laptops, and cell phones, to EV and hybrid vehicle applications, and industrial standby applications, including energy storage for telecommunication, UPS, and distributed electricity-generation applications.

The Ni-MH cell was initially developed to replace sealed Ni-Cd cells for portable electronics. In 1990s, with 30–50% higher energy density, the Ni-MH cell quickly caught on for applications such as cell phones and portable computers. Although the lithium technologies have replaced the Ni-MH cells in the cell phone and laptop markets, the continuous improvement in the specific power of the Ni-MH technology has lead to its adoption in cordless drills and other power tools. The ability to be rapidly charged and deeply discharged at high rates, together with considerable energy density, has resulted in the broad use of Ni-MH batteries in consumer applications and continually leads to new applications that take advantage of their characteristics. The application distributions of Ni-MH batteries in 2008 is shown in Figure 5.17. It can be seen that HEV is the main application filed for Ni-MH batteries, which occupies 53% of the total Ni-MH battery application [95]. Ni-MH HEV batteries accounted for 1.7% of the world rechargeable battery market in 2008, which will grow to hold 4.2% of the market share by 2013 [96]. Currently, Ni-MH batteries dominate about 97% share of the HEV battery market [97].

5.7.1
Electric Tools

The market of the small electric tools has almost been monopolized by Ni-Cd battery for a long time because the Ni-MH battery is not as good as Ni-Cd battery in high power performance. With the development of the manufacturing technology of the Ni-MH battery, its high-power performance has been greatly improved. Currently, by using multipolar current collector, welding edge, and enhancing the catalytic activity and conductivity of the electrode surface, the discharge performance of Ni-MH batteries has been enhanced from 5 to 20C while maintaining a high energy density. The restrictions of Ni-Cd batteries for environment protection have

Table 5.6 Performance comparison at room temperature [16].

Test item	0.2C to 1.0 V	1C to 1.0 V	5C to 0.8 V	10C to 0.7 V
Technical requirement	Discharge time ($\geq$5 h)	Discharge time ($\geq$54 min)	Discharge time ($\geq$9 min)	Discharge time ($\geq$4 min)
Ni-MH	$\geq$5.18	$\geq$58.0	$\geq$10.90	$\geq$5.52
Ni-Cd	$\geq$5.35	$\geq$56.9	$\geq$9.79	$\geq$4.71

also provided Ni-MH batteries with a good opportunity to enter this field. At present time, the high-power Ni-MH battery has entered the market of electric tools and will gradually replace Ni-Cd battery. With the D-SC1300P-type battery as an example, the performances comparison of Ni-MH and Ni-Cd batteries for electric tools is summarized in Table 5.6.

5.7.2 Electric Bicycles

It is generally believed that the electric bicycle is an important travel tool for short distances, and the prospect is cheerful. Electric bicycle can be worked with a speed of up to 20 km h^{-1} and an average mileage of over 30 km, which is able to meet the travel needs of the city. Chinese mainland is the largest market, and the number of electric bikes has reached 120 million in 2010 with an annual growth rate of 30%. The requirements of batteries for electric bicycle are as follows:

- high specific energy and energy density;
- high specific power and power density;
- fast charge, high efficiency, simple equipment;
- long life, maintenance free;
- lower self-discharge rate;
- having a certain resistance to vibration, shock resistance;
- lower production costs;
- no pollutant emission from production;
- safety in the use of process;
- no harm to the environment and good recyclability.

The global battery market for electric bicycles is still dominated by the relatively cheap lead-acid batteries. A typical lead-acid battery module for an electric bicycle weighs about 12 kg, and the whole bicycle weighs about 50 kg. The corresponding driving mileage is about 20 km. Such heavy weight and short mileage are the major drawbacks of the application of lead-acid batteries in electric bicycles. Figure 5.18 shows an electric bicycle equipped with a D-type 24 V/7 Ah Ni-MH battery. The battery box installed in the vehicle basket is about 3 kg, and the total vehicle weight is only 20 kg. The driving distance of the bicycle is up to 50 km after charging,

Figure 5.18 Electric bicycle driven by an Ni-MH battery [98].

indicating the advantages of small size, lightweight, large capacity, and long lifetime characteristics for the D-type Ni-MH battery.

5.7.3
Electric Vehicles (EVs) and Hybrid Electric Vehicles (HEVs)

Ni-MH batteries are one of the most promising candidates for HEV applications. They normally have a specific energy of 60–80 Wh kg^{-1}, a specific power of 200 W kg^{-1}, and an energy efficiency of 65%. The main challenges facing Ni-MH battery development are high cost, low cell efficiency, high self-discharge, the need to control hydrogen loss, and the need for a recycling infrastructure. One of the leaders in Ni-MH technology research is the Ovonic Battery Company. The Ovonic Ni-MH batteries have fast recharge capability, charging up to 60% of full capacity in 15 min. These batteries have drastically increased typical energy density, with ratings up to 80 Wh kg^{-1}. Today's leading application for Ni-MH batteries in the automotive community is the HEV, with examples such as the Toyota Prius, Honda Insight from Japan, Ford Escape Hybrid from the United States, and Dongfeng EQ6110, Chery A5, and Chang'an Jie Xun from China.

5.7.3.1 The Toyota Prius
The Toyota Prius (Figure 5.19) is a five-passenger compact sedan powered by a 52 kW gasoline engine and a 33 kW electric motor. It has a curb weight of 1254 kg. The

Figure 5.19 Picture of the 2010 Toyota Prius [100].

Prius has a complex dual-mode hybrid configuration in which energy to and from the vehicle wheels can travel along several different pathways. Mechanical energy to the wheels passes through a planetary gear set that is coupled to the engine, electric motor, generator, and the final drive. Power to the wheels can be provided solely by a 273.6 V Ni-MH battery pack through the electric motor, directly from the gasoline engine, or from a combination of both. The battery pack can be recharged directly by energy from the wheels powering the motor (regenerative braking) or by excess energy from the gasoline engine, which turns the generator [99].

Starting in 2004 to the current models, the standard Prius battery is a Panasonic Metal Case Prismatic Module. Before 2004 and as early as 2000, the battery used was a Panasonic Plastic Case Prismatic Module. The first generation was sold only in Japan. Furthermore, the battery pack was so large that it took up the entire space between the trunk and the cabin. This led to changes in the Generation II models whose stacks were significantly smaller and far more reliable.

Toyota uses prismatic Ni-MH modules from Panasonic. Each module consists of six 1.2 V cells connected in series. The module has a nominal voltage of 7.2 V and a capacity of 6.5 Ah, weighs 1.04 kg, and has dimensions of 19.6 mm (w) × 106 mm (h) × 275 mm (l). The Prius battery stack consists of 38 prismatic Ni-MH modules connected in series. It delivers a nominal 273.6 V and has a 6.5 Ah capacity. The modules are stacked side by side and then compressed together in a rigid, nonexpandable structure that prevents expansion from internal pressures. The complete battery pack consists of the battery stack, an enclosure for structural support and airflow, battery electronic control unit/monitor, and safety switch. The weight of the complete battery pack is 53.3 kg. The pack is horizontally positioned in the trunk of the vehicle partially under the back seat.

5.7.3.2 The Honda Insight

The Honda Insight is a 65 mile/gallon (27.6 km l^{-1}) parallel HEV. It is a two-seater with a lightweight aluminum body that is powered by Honda's Integrated Motor Assist (IMA) powertrain. The 50 kW, 89.5 N m gasoline engine and a closely coupled 10 kW electric motor comprise the IMA system. Cumulatively, the IMA powertrain produces 54.46 kW and 123.4 N m of torque. At 6.5 Ah rated capacity, the 144 V Ni-MH battery powers the electric motor [101].

The battery module contains 120 individual 1.2 V cells, each with the size of a conventional D-cell battery. The cells are arranged inside the module in groups or sticks. It is series configured with 20 units of 7.2 V modules. Each module consists of 6 D-size cells that are rated at 1.2 V and 6.5 Ah. During start-up and acceleration, an electric motor, located between the engine and the transmission, provides assistance to the engine. During braking and deceleration, the motor acts as a generator, recharging both the high-voltage battery module. Since the battery module is recharged by the electric motor whenever the vehicle decelerates, the battery never needs external charging.

5.7.3.3 The Ford Escape Hybrid

The Ford Escape Hybrid is a gasoline–electric-hybrid-powered version of the Ford Escape SUV developed by the Ford Motor Company, first launched in the United States in 2004, to offset the pollution resulting from nonhybrid SUV use. It was the first hybrid SUV to hit the market [102].

The Escape Hybrid is a full hybrid electric system, meaning the system can switch automatically between pure electric power, pure gasoline engine power, or a combination of both for maximum performance and efficiency at all speeds and loads. When braking or decelerating, the Escape's hybrid system uses regenerative braking, where the electric drive motor becomes a generator, converting the vehicle's momentum back to electricity for storage in the batteries. With 116 kW, the Escape Hybrid has nearly the same acceleration performance as the conventional 150 kW V6 Escape. The high voltage storage battery is located beneath the rear load floor. The high-voltage (330 V) battery pack consists of 250 D-sized Ni-MH cells in a sealed enclosure.

EVs and HEVs in China   After several years of effort, China EVs and related systems has made great strides in development. Under the frame of High-tech (863) Key EV project, hundreds of major companies, universities, and research institutes have developed high-power-type Ni-MH battery and used them on hybrids.

Since the late 1980s, the General Research Institute for Nonferrous Metals (GRINM) has studied the Ni-MH battery and related materials for EVs application. In 1996, GRINM developed a 100 Ah, 120 V Ni-MH battery pack, which is the first battery pack used in EVs in China. After charging once, the EV can run 121 km at the speed of 40 km h^{-1}, the top speed is up to 112 km h^{-1}, and the acceleration time from 0 to 40 km h^{-1} is 6.2 s. GRINM also developed prismatic 80 Ah Ni-MH cells and its 96 V/80 Ah or 384 V/80 Ah Ni-MH high-voltage system [103]. The battery performance could meet such application requirements as pulse charge/discharge from 3 to 10C with high efficiency at −20 to 55 °C and an SOC of 0.2–0.8 V. The specific energy and the maximum specific power of a 12 V module could reach to 61 Wh kg^{-1} and 406 W kg^{-1}, respectively. A city bus driven by proton exchange membrane fuel cell (PEMFC)/Ni-MH hybrid power has been in use and finished a demonstration run in Beijing suburbia.

In addition, the Hunan Shenzhou Science & Technology Co. Ltd. developed high-power 40 Ah Ni-MH battery for hybrid electrical buses (Figure 5.20) [104]. The cell could be charged continuously with 4C rate current and discharged continuously with 8C current. Within the temperature range of −20 to 45 °C, the battery could be operated with continuous high current charge–discharge. The cycle life of the single cell with a 1C rate charge to 80% SOC and 2C rate discharge to 100% DOD was more than 3100 cycles, in which the number of cycles with the capacity higher than 40 Ah could reach 2700. An HEV bus installed with the battery ran for over 10^5 km.

Chery A5 ISG (Integrate Starter Generator) hybrid system combines a 1.3 l gasoline engine and a 10 kW permanent magnet synchronous motor with a control

Figure 5.20 Picture of the Dongfeng EQ6110 hybrid electric bus [105].

Figure 5.21 Chang'an Jie Xun hybrid electric car [107].

system, inverter, and DC–DC converter. The motor is powered by a 144 V Ni-MH battery [106]. This system is based on the complementary of gasoline engine and electric motor, according to the driving conditions. During start-up or at low speed, the vehicle only relies on the electric motor, and gasoline engine is shut down at this time; thus, the vehicle's fuel consumption is zero. When the vehicle speed is higher (typically up to 40 km h^{-1} or above) or the vehicle needs acceleration, the gasoline engine starts to output power. In the vehicle braking, the hybrid system can convert kinetic energy into electrical energy, which is stored in batteries for future driving.

Chang'an Jie Xun is the first mass-product hybrid electric car from China (Figure 5.21). Its power system is composed of a 1.5 l gasoline engine and a 13 kW permanent magnet synchronous brushless motor, which is powered by the 144 V, 6 Ah Ni-MH batteries. The power of this module can be compared to the 2.0 l gasoline engine, but the fuel consumption is reduced by 20%. Jie Xun HEV has successfully served in the Beijing Olympics.

5.7.4
UPS and Energy Storage Battery

Ni-MH batteries have some unique characteristics that make field application of the technology more complicated than simply replacing some other type of battery with Ni-MH and adjusting the charging voltage. The stationary battery landscape has been dominated by lead-acid batteries of various constructions for the past century. It was the obvious choice in many instances because of its comparatively low cost and where space was not a limitation [108]. However, because of the growing requirements for adding capacity to existing installations or the desire of facility management to utilize a new or more *green* technology, other battery technologies are being considered for applications that have been classically the domain of vented lead-acid (VLA) or valve-regulated lead-acid (VRLA) batteries.

To apply Ni-MH batteries to a stationary application, consideration must be given to the duty cycle and specific requirements of the application. Because the stationary market is preexisting, creating a system that will function within the parameters of other battery technologies is a key concern. Since most stationary systems utilizing lead-acid batteries (either VLA or VRLA) are charged with constant potential charging systems, a way of managing this charging method to be compatible with the Ni-MH technology must be provided as a part of the solution.

A Cobasys Ni-MH battery system is designed for a minimum of 15 s DC energy storage for a 750 kV A, 675 kW UPS module [109]. This module occupies $<9\,\text{ft}^2$ of floor space and can be stacked side to side. By comparison, sealed lead-acid batteries require at least four times this floor space, and flooded lead-acid batteries need 10–20 times this space. Similarly, weight is reduced 4- to 10-fold. Ni-MH batteries also cycle almost indefinitely, handle temperature far better than lead acid, and do not need much (if any) maintenance since there are no soft lead retorque issues, watering needs, or other maintenance headaches.

In China, an energy storage station has been installed in the Shanghai 2010 World Expo Park. This station is powered by Chunlan 100 Ah Ni-MH battery with energy density of 84 Wh kg^{-1} and power density of 300 W kg^{-1} [110].

5.7.5
Recycling of Ni-MH Batteries

Ni-MH batteries have been widely used as power sources in the application fields of portable devices, power tools, and HEVs. After long-term utilization, the Ni-MH battery will undergo performance degradation, leading to battery failure. Since consumption of Ni-MH batteries is large, it can be concluded that an appreciable amount of spent batteries is generated every year. Recycling of batteries at the end of their life is necessary both due to the increased requirement for environmental protection and the fact that Ni-MH batteries have a valuable metal content of nickel, cobalt, and rare earth elements. Various recycling processes for recovering valuable metals from Ni-MH rechargeable batteries have been developed [111–113].

There are basically three methods for battery recycling: separation of components through unity operations of mining treatment, pyrometallurgy, and hydrometallurgy [114]. Separation of components through unity operations of mining treatment is especially used for industrial batteries. Batteries are treated in order to separate materials of interest or to concentrate such materials for further recovery through other processes. This type of treatment is often the first phase of the recycling process. Pyrometallurgy essentially consists of recovering materials by using high temperatures. Recycling through hydrometallurgy basically consists of the acid or base leaching of scrap to put the metals in a solution. Once in a solution, metals can be recovered by precipitation, altering the pH of the solution or adding some reaction agent, or by electrolysis. The solution can also be separated by solvent extraction, using an organic solvent, which binds to the metallic ion, separating the metal from the solution. The metal can then be recovered by electrolysis or precipitation. The hydrometallurgical processes can offer reliable, environmental friendly, low-energy routes with low-wastewater generation for the treatment of spent batteries.

5.8
Challenges and Perspectives of Ni-MH Rechargeable Batteries

The Ni-MH battery is an alternative technology to the Ni-Cd battery because of its high specific capacity density, environmental friendliness, and small memory effect, which is welcome by numerous applications. It possesses many advantages such as high capacity, high-rate dischargeability, 50–1000 cycles of recharging, reasonable price, and the compatibility to current facilities that use Ni-Cd batteries. In the early stages of commercialization, Ni-MH batteries are a combination of materials, energy, and information technology. Development of the Ni-MH battery industry is beneficial to improve the urban environment, the sustainable development of the national economy and development of portable electronic products, nonpolluting EVs, and other high-tech industries, as well as driving the development of upstream raw material industry.

Presently, rechargeable Li ion batteries have already replaced Ni-MH batteries in many application domains and occupied the portable electric device market. However, in terms of cost and security considerations, the mainstream of power source for the HEVs is the Ni-MH batteries. With the development of the market, the focus of research and development of Ni-MH batteries has transferred from high capacity to high power and long life. The R&D has mainly been focused on the improvement of existing material and development of new battery assembly technologies.

The mainly technological development trend of the Ni-MH batteries is as follows:

1) Searching for low-cobalt, low-cost, high-performance, and stable hydrogen storage alloys, and new quasi-fast cooling process to produce hydrogen storage alloys.
2) Researching high surface conductive $Ni(OH)_2$ and current collector materials to further improve the operating factor and high-rate discharge performance of the positive materials.
3) Researching new manufacture craft of the positive and negative electrodes, as well as the batteries.
4) Making the shell and dissepiment thinner by technological approaches.

5) Optimal design of the structure of the single cell and the battery pack.
6) Reducing the internal resistance of battery by various kinds of technological approaches.
7) Researching battery test and management system.
8) Following recovery techniques of the battery.

References

1. Winter, M. and Brodd, R.J. (2004) What are batteries, fuel cells, and supercapacitors. *Chem. Rev.*, **104**, 4245–4269.
2. Armand, M. and Tarascon, J.M. (2008) Building better batteries. *Nature*, **451**, 652–657.
3. Willems, J.J.G. (1984) Metal hydride electrodes stability of lanthanum-nickel (LaNi$_5$)-related compounds. *Philips J. Res.*, **39**, 1–94.
4. Ovshinsky, S.R., Fetcenko, M.A., and Ross, J. (1993) A nickel metal hydride battery for electric vehicles. *Science*, **260**, 176–181.
5. Linden, D. and Reddy, T.B. (2002) *Handbook of Batteries*, 3rd edn, McGraw-Hill Inc., New York.
6. Eccles, L. and Morrison, D. (2001) HEV Battery Research Rolls on. http://electronicdesign.com/ (accessed 2010).
7. Fetcenko, M.A., Ovshinsky, S.R., Reichman, B., Toung, K., Fierro, C., Koch, J., Zallen, A., Mays, W., and Ouchi, T. (2007) Recent advances in NiMH battery technology. *J. Power Sources*, **165**, 544–551.
8. Kuriyama, N., Sakai, T., Miyamura, H., Tanaka, H., Ishikawa, H., and Uehara, I. (1996) Hydrogen storage alloys for nickel/metal-hydride battery. *Vacuum*, **47**, 889–892.
9. Zhao, X.Y. and Ma, L.Q. (2009) Recent progress in hydrogen storage alloys for nickel/metal hydride secondary batteries. *Int. J. Hydrogen Energy*, **34**, 4788–4796.
10. Zhu, W.H., Ke, J.J., Yu, H.M., and Zhang, D.J. (1995) A study of the electrochemistry of nickel hydroxide electrodes with various additives. *J. Power Sources*, **56**, 75–79.
11. Hu, W.K. and Noreus, D. (2003) Alpha nickel hydroxides as lightweight nickel electrode materials for alkaline rechargeable cells. *Chem. Mater.*, **15**, 974–978.
12. Fetcenko, M.A. (2005) 22nd International Seminar & Exhibit on Primary and Secondary Batteries, Vol. 16, Ft. Lauderdale, FL.
13. Sakai, T., Matsuoka, M., and Iwakura, C. (1995) *Handbook on the Physics and Chemistry of Rare Earths*, Elsevier, Amsterdam.
14. Harding Energy, INC. (2004) Battery Handbook for Quest Rechargeable Cells and Battery Packs, http://www.hardingenergy.com/index.php/handbook (accessed 2010).
15. Hong, K. (2001) The development of hydrogen storage alloys and the progress of nickel hydride batteries. *J. Alloys Compd.*, **321**, 307–313.
16. Chen, J. and Tao, Z.L. (2006) *Nickel-Metal Hydride Secondary Batteries*, Chemical Industry Press, Beijing.
17. Zuttel, A. (2003) Materials for hydrogen storage. *Materialstoday*, **6**, 24–33.
18. Kleperis, J., Wojcik, G., Czerwinski, A., Skowronski, J., Kopczyk, M., and Beltowska-Brzezinska, M. (2001) Electrochemical behavior of metal hydrides. *J. Solid State Electrochem.*, **5**, 229–249.
19. Lei, Y.Q., Wan, Q., and Shi, Y.K. (2000) *New Energy Materials*, Tianjin University Press, Tianjin.
20. Bronoel, G., Sarradin, J., Bonnemay, M., Percheron, A., Achard, J.C., and Schlapbach, L. (1976) A new hydrogen storage electrode. *Int. J. Hydrogen Energy*, **1**, 251–254.
21. Diaz, H., Percheron-guegan, A., Achard, J.C., Chatillon, C., and

Mathieu, J.C. (1979) Thermodynamic and structural properties of LaNi$_{5-y}$Al$_y$ compounds and their related hydrides. *Int. J. Hydrogen Energy*, **4**, 445–454.

22. Ananth, M.V., Raju, M., Manimaran, K., Balachandran, G., and Nair, L.M. (2007) Influence of rare earth content on Mm-based AB$_5$ metal hydride alloys for Ni-MH batteries-An X-ray fluorescence study. *J. Power Sources*, **167**, 228–233.

23. Tliha, M., Boussami, S., Mathlouthi, H., Lamloumi, J., and Percheron-guegan, A. (2010) Kinetic behaviour of low-Co AB$_5$-type metal hydride electrodes. *Mater. Sci. Eng.*, **175**, 60–64.

24. Mani, N. and Ramaprabhu, S. (2004) Effect of substitutional elements on hydrogen absorption properties in Mm-based AB$_5$ alloys. *J. Alloys Compd.*, **363**, 275–291.

25. Iwakura, C., Oura, T., Inoue, H., and Matsuoka, M. (1996) Effects of substitution with foreign metals on the crystallographic, thermodynamic and electrochemical properties of AB$_5$-type hydrogen storage alloys. *Electrochim. Acta*, **41**, 117–121.

26. Meli, F., Zuttel, A., and Schlapbach, L. (1995) Electrochemical and surface properties of iron-containing AB$_5$-type alloys. *J. Alloys Compd.*, **231**, 693–644.

27. Saldan, I. (2010) Primary estimation of metal hydride electrode performance. *J. Solid State Electrochem*, **14**, 1339–1350.

28. Li, P., Wang, X.L., Wu, J.M., Zhao, W.R., Li, R., and Ma, N. (2000) Present status on research of AB$_2$-type laves phase hydrogen storage electrode materials. *Metal. Funct. Mater.*, **7**, 7–12.

29. Chen, J., Dou, S.X., and Liu, H.K. (1997) Hydrogen desorption and electrode properties of Zr$_{0.8}$Ti$_{0.2}$(V$_{0.3}$Ni$_{0.6}$M$_{0.1}$)$_2$. *J. Alloys Compd.*, **256**, 40–44.

30. Liu, F.J., Ota, H., Okamoto, S., and Suda, S. (1997) Surface properties of the fluorinated La-incorporated Ti/Zr-based AB$_2$ laves phase alloys. *J. Alloys Compd.*, **253–254**, 452–458.

31. Kim, D.M., Jang, K.J., and Lee, J.Y. (1999) A review on the development of AB2-type Zr-based Laves phase hydrogen storage alloys for Ni–MH rechargeable batteries in the Korea Advanced Institute of Science and Technology. *J. Alloys Compd.*, **293–295**, 583–592.

32. Dhar, S.K., Ovshinsky, S.R., Gifford, P.R., Corrigan, D.A., Fetcenko, M.A., and Venkatesan, S. (1997) Nickel/metal hydride technology for consumer and electric vehicle batteries – a review and up-date. *J. Power Sources*, **65**, 1–7.

33. Justi, E.W., Ewe, H.H., Kalberlah, A.W., Saridakis, N.M., and Schaefer, M.H. (1970) Electrocatalysis in the nickel-titanium system. *Energy Convers.*, **10**, 183–187.

34. Reilly, J.J. and Wiswall, R.H. (1974) Formation and properties of iron titanium hydride. *Inorg. Chem.*, **13**, 218–222.

35. Xu, Y.H., Chen, C.P., Wang, X.L., and Wang, Q.D. (2002) The relationship between the high-rate dischargeability and the diffusion coefficient and exchange current for Ti$_{0.5}$Ni$_{0.25}$Al$_{0.25}$ metal hydride alloys. *J. Alloys Compd.*, **335**, 262–265.

36. Cuevas, F., Latroche, M., and Percheron-Guegan, A. (2005) Relationship between polymorphism and hydrogenation properties in Ti$_{0.64}$Zr$_{0.36}$Ni alloy. *J. Alloys Compd.*, **404–406**, 545–549.

37. Drenchev, B. and Spassov, T. (2007) Electrochemical hydriding of amorphous and nanocrystalline TiNi-based alloys. *J. Alloys Compd.*, **441**, 197–201.

38. Jurczyk, M., Jankowska, E., Nowak, M., and Jakubowicz, J. (2002) Nanocrystalline titanium-type metal hydride electrodes prepared by mechanical alloying. *J. Alloys Compd.*, **336**, 265–269.

39. Cuevas, F., Latroche, M., Ochin, P., Dezellus, A., Fernandez, J.F., Sanchez, C., and Percheron-Guegan, A. (2002) Influence of the martensitic transformation on the hydrogenation properties of Ti$_{50-x}$Zr$_x$Ni$_{50}$ alloys. *J. Alloys Compd.*, **330–332**, 250–255.

40. Takasaki, A. and Kelton, K.F. (2006) Hydrogen storage in Ti-based quasicrystal powders produced by mechanical alloying. *Int. J. Hydrogen Energy*, **31**, 183–190.
41. Liu, B.Z., Wu, Y.M., and Wang, L.M. (2006) Crystallographic and electrochemical characteristics of icosahedral quasicrystalline $Ti_{45-x}Zr_{35-x}Ni_{7+2x}Cu_3$ (x = 0–8) powders. *J. Power Sources*, **162**, 713–718.
42. Kalisvaart, W.P., Wondergem, H.J., Bakker, F., and Notten, P.H.L. (2007) Mg-Ti based materials for electrochemical hydrogen storage. *J. Mater. Res.*, **22**, 1640–1649.
43. Zhu, Y.F., Zhang, W.F., Yang, C., and Li, L.Q. (2010) Electrochemical properties of Mg-based hydrogen storage materials modified with carbonaceous materials prepared by hydriding combustion synthesis and subsequent mechanical milling (HGS+MM). *Int. J. Hydrogen Energy*, **35**, 9653–9660.
44. Singh, A.K., Singh, A.K., and Srivastava, O.N. (1995) On the synthesis of the Mg_2Ni alloy by mechanical alloying. *J. Alloys Compd.*, **227**, 63–68.
45. Wang, L.B., Wang, J.B., Yuan, H.T., Wang, Y.J., and Li, Q.D. (2004) An electrochemical investigation of $Mg_{1-x}Al_xNi$ ($0 \leq x \leq 0.6$) hydrogen storage alloys. *J. Alloys Compd.*, **385**, 304–308.
46. Guo, J., Yang, K., Xu, L.Q., Liu, Y.X., and Zhou, K.W. (2007) Hydrogen storage properties of $Mg_{76}Ti_{12}Fe_{12-x}Ni_x$ (x = 0, 4, 8, 12) alloys by mechanical alloying. *Int. J. Hydrogen Energy*, **32**, 2412–2416.
47. Liu, J.W., Yuan, H.T., Cao, J.S., and Wang, Y.J. (2005) Effect of Ti-Al substitution on the electrochemical properties of amorphous MgNi-based secondary hydride electrodes. *J. Alloys Compd.*, **392**, 300–305.
48. Rongeat, C. and Roue, L. (2004) Effect of particle size on the electrode performance of MgNi hydrogen storage alloy. *J. Power Sources*, **132**, 302–308.
49. Rongeat, C., Grosjean, M.H., Ruggeri, S., Dehmas, M., Bourlot, S., Marcotte, S., and Roue, L. (2006) Evaluation of different approaches for improving the cycle life of MgNi-based electrodes for Ni–MH batteries. *J. Power Sources*, **158**, 747–753.
50. Chen, J., Takeshita, H.T., Tanaka, H., Kuriyama, N., Sakai, T., Uehara, I., and Haruta, M. (2000) Hydriding properties of $LaNi_3$ and $CaNi_3$ and their substitutes with $PuNi_3$-type structure. *J. Alloys Compd.*, **302**, 304–313.
51. Chen, J., Kuriyama, N., Takeshita, H.T., Tanaka, H., Sakai, T., and Haruta, M. (2000) Hydrogen storage alloys with $PuNi_3$-type structure as metal hydride electrodes. *Electrochem. Solid State Lett.*, **3**, 249–252.
52. Liu, Y.F., Pan, H.G., Gao, M.X., Zhu, Y.F., Lei, Y.Q., and Wang, Q.D. (2004) The effect of Mn substitution for Ni on the structural and electrochemical properties of $La_{0.7}Mg_{0.3}Ni_{2.55-x}Co_{0.45}Mn_x$ hydrogen storage electrode alloys. *Int. J. Hydrogen Energy*, **29**, 297–305.
53. Liao, B., Lei, Y.Q., Chen, L.X., Lu, G.L., Pan, H.G., and Wang, Q.D. (2006) The structural and electrochemical properties of $La_2Mg(Ni_{0.8-x}Co_{0.2}Al_x)_9$ (x = 0–0.03) hydrogen storage electrode alloys. *J. Alloys Compd.*, **415**, 239–243.
54. Wu, F., Zhang, M.Y., and Mu, D.B. (2010) Characteristics before and after heat treatment of Co-Free AB_3-Type Fe substituted hydrogen storage alloys used for Ni-MH batteries. *Adv. Mater. Res.*, **129–131**, 1049–1054.
55. Li, Y., Han, S.M., Zhu, X.L., and Ding, H.L. (2010) Effect of CuO addition on electrochemical properties of AB3-type alloy electrodes for nickel/metal hydride batteries. *J. Power Sources*, **195**, 380–383.
56. Han, S.M., Zhang, Z., Zhao, M.S., and Zheng, Y.Z. (2006) Electrochemical characteristics and microstructure of $Zr_{0.9}Ti_{0.1}Ni_{1.1}Mn_{0.6}V_{0.3}$–$LaNi_5$ composite hydrogen storage alloys. *Int. J. Hydrogen Energy*, **31**, 563–567.
57. Han, S.M., Zhao, M.S., Zhang, Z., Zheng, Y.Z., and Jing, T.F. (2005) Effect of AB_2 alloy addition on the phase structures and electrochemical characteristics of $LaNi_5$ hydride electrode. *J. Alloys Compd.*, **392**, 268–273.

58. Choi, W.K., Tanaka, T., Miyauchi, R., Morikawa, T., Inoue, H., and Iwakura, C. (2000) Electrochemical and structural characteristics of $TiV_{2.1}Ni_{0.3}$ surface-modified by ball-milling with MgNi. *J. Alloys Compd.*, **299**, 141–147.
59. Feng, Y., Jiao, L.F., Yuan, H.T., and Zhao, M. (2007) Study on the preparation and electrochemical characteristics of MgNi–CoB alloys. *J. Alloys Compd.*, **440**, 304–308.
60. Yang, Y.P., Zhang, P.M., Liu, K.Y., Sang, S.B., Tang, Y.G., and Lu, Z.G. (2002) Effect of preparation technology of AB_5 type hydrogen storage alloys on its electrochemical properties. *Metal. Funct. Mater.*, **9**, 1–4.
61. Wang, J.J., Hou, Y., and Liu, J. (2002) Study on preparing technology of hydrogen storage alloy. *Chin. J. Power Sources*, **26**, 344–345.
62. Chen, Y. and Williams, J.S. (1995) Formation of metal hydrides by mechanical alloying. *J. Alloys Compd.*, **217**, 181–184.
63. Watanabe, K., Kikuoka, T., and Kumagai, N. (1995) Physical and electrochemical characteristics of nickel hydroxide as a positive material for rechargeable alkaline batteries. *J. Appl. Electrochem.*, **25**, 219–226.
64. Cornilsen, B.C., Shan, X.Y., and Loyselle, L.P. (1990) Structural comparison of nickel electrodes and precursor phases. *J. Power Sources*, **29**, 453–466.
65. Singh, D. (1998) Characteristics and effects of gamma-NiOOH on cell performance and a method to quantify it in nickel electrodes. *J. Electrochem. Soc.*, **145**, 116–120.
66. Pralong, V., Delahaye-Vidal, A., Beaudoin, B., Leriche, J.B., and Tarascon, J.M. (2000) Electrochemical behaviour of cobalt hydroxide used as additive in the nickel hydroxide electrode. *J. Electrochem. Soc.*, **147**, 1306–1313.
67. Provazi, K., Giz, M.J., Dall'Antonia, L.H., and Cordoba de Torresi, S.I. (2001) The effect of Cd, Co and Zn as additives on nickel hydroxide opto-electrochemical behaviour. *J. Power Sources*, **102**, 224–232.
68. He, X.M., Ren, J.G., Li, W., Jiang, C.Y., and Wan, C.R. (2006) $Ca_3(PO_4)_2$ coating of spherical $Ni(OH)_2$ cathode materials for Ni-MH batteries at elevated temperature. *Electrochim. Acta*, **51**, 4533–4536.
69. Oshitani, M., Watada, M., Shodai, K., and Kodama, M. (2001) Effect of lanthanide oxide additives on the high-temperature charge acceptance characteristics of pasted nickel electrodes. *J. Electrochem. Soc.*, **148**, A67–A73.
70. Chen, J., Bradhurst, D.H., Dou, S.X., and Liu, H.K. (1999) Nickel hydroxide as an active material for the positive electrode in rechargeable alkaline batteries. *J. Electrochem. Soc.*, **146**, 3606–3612.
71. Ramesh, T.N. and Kamath, P.V. (2006) Synthesis of nickel hydroxide: effect of precipitation conditions on phase selectivity and structural disorder. *J. Power Sources*, **156**, 655–661.
72. Zhang, Y.S., Zhou, Z., and Yan, J. (1998) Electrochemical behaviour of $Ni(OH)_2$ ultrafine powder. *J. Power Sources*, **75**, 283–287.
73. Han, X.J., Xie, X.M., Xu, C.Q., Zhou, D.R., and Ma, Y.L. (2003) Morphology and electrochemical performance of nano-scale nickel hydroxide prepared by supersonic coordination – precipitation method. *Opt. Mater.*, **23**, 465–470.
74. Cao, M.H., He, X.Y., Chen, J., and Hu, C.Y. (2007) Self-assembled nickel hydroxide three-dimensional nanostructures: A nanomaterial for alkaline rechargeable batteries. *Cryst. Growth Des.*, **7**, 170–174.
75. Cai, F.S., Zhang, G.Y., Chen, J., Gou, X.L., Liu, H.K., and Dou, S.X. (2004) $Ni(OH)_2$ tubes with mesoscale dimensions as positive-electrode materials of alkaline rechargeable batteries. *Angew. Chem. Ind. Ed.*, **43**, 4212–4216.
76. Arora, P. and Zhang, Z.M. (2004) Battery separators. *Chem. Rev.*, **104**, 4419–4462.
77. Kritzer, P. and Cook, J.A. (2007) Nonwovens as separators for alkaline batteries. *J. Electrochem. Soc.*, **154**, A481–A494.

78. Araki, T., Nakayama, M., Fukuda, K., and Onda, K. (2005) Thermal behavior of small nickel/metal hydride battery during rapid charge and discharge cycles. *J. Electrochem. Soc.*, **152**, A1128–A1135.
79. Kritzer, P. (2004) Separators for nickel metal hydride and nickel cadmium batteries designed to reduce self-discharge rates. *J. Power Sources*, **137**, 317–321.
80. Bennett, J. and Choi, W.M. (1998) An overview of the effect of separator on the performance of the nickel metal hydride battery. *IEEE*, 153–157.
81. Beck, F. and Restschi, P. (2000) Rechargeable batteries with aqueous electrolytes. *Electrochim. Acta*, **45**, 2467–2482.
82. Geng, M. and Northwood, D.O. (2003) Development of advanced rechargeable Ni/MH and Ni/Zn batteries. *Int. J. Hydrogen Energy*, **28**, 633–636.
83. Bernard, P., Gabrielli, C., Keddam, M., Takenouti, H., Leonardi, J., and Blanchard, P. (1991) Ac quartz crystal microbalance applied to the studies of the nickel hydroxide behaviour in alkaline solutions. *Electrochim. Acta*, **36**, 743–746.
84. Mohamad, A.A. and Arof, A.K. (2008) Ni-MH battery based on plasticized alkaline solid polymer electrolytes. *Ionics*, **14**, 415–420.
85. Metzger, W., Westfall, R., Hermann, A., and Lyman, P. (1998) Nickel foam substrate for nickel metal hydride electrodes and lightweight honeycomb structures. *Int. J. Hydrogen Energy*, **23**, 1025–1029.
86. Jiang, J.J., Gasikb, M., Laineb, J., and Lampinenb, M. (2001) Electrochemical evaluation of sintered metal hydride electrodes for electric vehicle applications. *J. Alloys Compd.*, **322**, 281–285.
87. Fukunaga, H., Kishimi, M., Igarashi, N., Ozaki, T., and Sakai, T. (2005) Non-foam-type nickel electrodes using various binders for Ni-MH batteries. *J. Electrochem. Soc.*, **152**, A42–A46.
88. Ying, T.K., Gao, X.P., Hu, W.K., Wu, F., and Noreus, D. (2006) Studies on rechargeable NiMH batteries. *Int. J. Hydrogen Energy*, **31**, 525–530.
89. Morioka, Y., Narukawa, S., and Itou, T. (2001) State-of-art of alkaline rechargeable batteries. *J. Power Sources*, **100**, 107–116.
90. Panasonic Corporation, Nickel Metal Hydride Batteries, (2005) http://products.panasonic-industrial.com/datasheets/en/Panasonic_NiMH_Technology.pdf (accessed 2010).
91. Power-Sonic Corp. Nickel Cadmium and Nickel-Metal Hydride Rechargeable Batteries and Chargers, (2010) http://www.power-sonic.com/images/powersonic/literature/NiCDNiNH_Batteries_and_Chargers/1277751010_20106027-NiCdCatalog-Lo.pdf (accessed 2010).
92. GPI International Limited Nickel Metal Hydride Technical Hand Book, (2004) http://www.gpbatteries.com.hk/html/front/index.html (accessed 2010).
93. Klein, M.G., Eskra, M., Plivelich, R., and Ralston, P. (2006) Bipolar Nickel Metal Hydride Battery, http://www.haszstudios.com/electro/products/technicalpapers/BipolarNickel.pdf (accessed 2010).
94. Kanda, M., Yamamoto, M., Kanno, K., Satoh, Y., Hayashida, H., and Suzuki, M. (1991) Cyclic behaviour of metal hydride electrodes and the cell characteristics of nickel-metal hydride batteries. *J. Less Common Metals*, **172–174**, 1227–1235.
95. Christophe, P. (2009) Present and Future Market Situation for Batteries. http://v7.kermeet.com/Data/kmclientv7/block/F_4dac49e80345efe84a2836c00cd28c1b4b8f8b5f7d336.pdf (accessed 2010).
96. SBI Reports (2009) Advanced Storage Battery Market: From Hybrid/Electric Vehicles to Cell Phones. http://www.sbireports.com/Advanced-Rechargeable-Battery-1933124/ (accessed 2010).
97. Research and Markets (2010) Future of Global Advanced Batteries Market Outlook to 2020: Opportunity Analysis in Electronics and Transportation. http://www.pr-inside.com/research-and-markets-future-of-global-r1694691.htm (accessed 2010).

98. Chen, J., Tao, Z.L., and Gou, X.L. (2006) *Chemical Power Sources- Principles, Techniques and Applications*, Chemical Industry Press, Beijing.
99. Zolot, M., Pesaran, A.A., and Mihalic, M. (2002) Themal Evaluation of Toyota Prius Battery Park. *http://www.nrel.gov/vehiclesandfuels/ energystorage/pdfs/2a_2002_01_1962.pdf* (accessed 2010).
100. Marc, Carter (2009) Detroit 09': 2010 Toyota Prius Officially Unveiled with 50 MPG Combined Rating. *http://www.thetorquereport.com/2009/01/ detroit_09_2010_toyota_prius_o.html* (accessed 2010).
101. American Honda Motor Co., Inc. Emergency Response Guide for Hybrid Vehicles. *https://techinfo.honda.com/Rjanisis/ logon.asp?Region=US* (accessed 2010).
102. I-Car Advantage Online. Ford escape hybrid. (2005) *http://www.i-car.com/pdf/advantage/ online/2005/051605.pdf* (accessed 2010).
103. Wu, B.R., Tu, H.L., Chen, H., Zhu, L., Jian, X.Y., Liu, M.Y., Jiang, L.J., Du, J., Gong, Z.X., and Zhang, J.Y. (2006) Development of 384 V/80 Ah Ni/MH battery system for hybrid electric vehicles. *Chin. J. Power Sources*, **9**, 701–706.
104. Yang, Y.F., Li, F., Sun, Y.Y., Liu, G.X., Yan, Y., Yang, X.L., and Yu, N.X. (2006) Development of high power Ni/MH battery for HEV bus. *Batt. Bimon.*, **36**, 199–201.
105. DongFeng Electric Vehicle Co., LTD. (2006) *http://www.dfev.com.cn* (accessed 2010).
106. Shaanxi Iseway Group Company Chery A5 HEV performance. (2008) *http://www.bitauto.com/html/news/ 200831/2008318453195314.shtml* (accessed 2010).
107. (2009) Auto.Sina.Com.Cn Chang'an Jie Xun HEV on the market. *http://auto.sina.com.cn/news/2009-06-11/ 1455498989.shtml* (accessed 2010).
108. Kopera, J.J.C. (2005) Considerations for the Utilization of NiMH Battery Technology in Stationary Applications. *http://www.battcon.com/PapersFinal2005/ KoperaPaper2005.pdf* (accessed 2010).
109. DeCoster, D. (2007) NiMH Batteries Projected to Replace Flywheels for Critical Bridge Time Applications, Battery Power, 11, issue of Battery Power Products & Technology.
110. Chunlan corporation (2008) Chunlan high-energy Ni-MH battery is used in Shanghai energy storage power station. *http://www.chunlan.com/* (accessed 2010).
111. Bernardes, A.M., Espinosa, D.C.R., and Tenorio, J.A.S. (2004) Recycling of batteries: a review of current processes and technologies. *J. Power Sources*, **130**, 291–298.
112. Tzanetakis, N. and Scott, K. (2004) Recycling of nickel-metal hydride batteries. I: Dissolution and solvent extraction of metals. *J. Chem. Technol. Biotechnol.*, **79**, 919–926.
113. Muller, T. and Friedrich, B. (2006) Development of a recycling process for nickel-metal hydride batteries. *J. Power Sources*, **158**, 1498–1509.
114. Espinosa, D.C.R., Bernardes, A.M., and Tenorio, J.A.S. (2004) An overview on the current process for the recycling of batteries. *J. Power Sources*, **135**, 311–319.

6
Metal–Air Technology
Bruce W. Downing

6.1
Metal–Air Technology

This chapter is a general introduction to the more common types of metal–air batteries, with emphasis on using the magnesium–air technology as an example in understanding the various components that make up a metal–air battery, and some of the limitations, challenges, manufacturing, and further research. Much of what has been learned can be applied to the other aqueous metal–air technologies. Products that utilize metal–air technology as their power source are or have been developed. Various theoretical calculations of the metal–air types have been compared in many published articles, but they do not reflect the actual operating conditions for commercial products. Similar to any commercial product, first to market is a test of consumer acceptance indicating product and financial returns to the manufacturer. The consumer has accepted battery types such as lead-acid, and zinc alkaline batteries for years and is only being introduced to the concept of metal–air batteries, though the zinc–air button cell for hearing aids has been around for years. Metal–air battery types may have certain niche commercial products, and no one metal–air type reigns supreme.

Metal–air batteries [1–4] are one of the more promising, but less well known, alternatives to conventional and future power sources as primary cells. Metal–air systems are typically very high in energy density but low in power, have an open cell structure, and use oxygen from the air. Great strides have been made and are continuing to be made in the metal–air technology. They have the potential to replace conventional batteries (e.g., zinc alkaline) and high cost hydrogen-based fuel cells because of their high energy density, relatively flat discharge voltage potential, long shelf life, and relatively low manufacturing cost.

Metal–air batteries are an attractive power source for stand-alone power supplies (e.g., for stand-by or emergency power and portable power). They feature the electrochemical coupling of a metal anode to an air-diffusion cathode through a suitable electrolyte. This combination produces a cell with an inexhaustible cathode reactant from the oxygen in atmosphere air. Metal–air batteries also have the versatility of an aqueous or nonaqueous electrolyte. The more common type

Electrochemical Technologies for Energy Storage and Conversion, First Edition. Edited by Ru-Shi Liu, Lei Zhang, Xueliang Sun, Hansan Liu, and Jiujun Zhang.
© 2012 Wiley-VCH Verlag GmbH & Co. KGaA. Published 2012 by Wiley-VCH Verlag GmbH & Co. KGaA.

is the aqueous electrolyte; however, great strides have been made within the last decade to create nonaqueous electrolytes that consist of a gel polymer electrolyte membrane serving as both the separator and ion-transporting medium [5, 6]. This also allows the battery to become rechargeable, albeit from an external source of power. The Li/O_2 and to some extent the Mg/O_2 batteries are being developed for commercialization.

A metal–air battery employs a metal (anode) such as magnesium, aluminum, or zinc as the fuel immersed in an electrolyte coupled with a gas diffusion cathode (GDE) to generate a direct current. The electrolyte is a conducting medium in which the migration of ions creates a flow of electric current.

The GDE consists of (i) a current collector, (ii) reactive carbon layers, one on either side of the current collector with or without a catalyst, and (iii) a microporous hydrophobic film on the airside to allow oxygen-laden air to pass through the electrolyte. The GDE can range from a one- to four-layer structure and serves to provide an interface where O_2 from the air is dissolved in an electrolyte solution and catalytically reduced on the active components of the GDE, which is carbon with or without a catalyst to enhance the rate of O_2 reduction (oxidation reduction reaction (ORR)). The cathode is the positive electrode by which the flow of electrons enters. The structure and chemistry of the air cathode are very important and there are many designs, some of which are commercial and others that are the priority of the battery manufacturer.

The fuel (anode) is a pure metal or metal alloy, which provides the electrons and can be deemed as a sacrificial anode and is replaceable. The metal (anode) is the negative electrode through which the current leaves. During the exothermic reaction, the metal anode undergoes corrosion and the electrolyte is consumed and broken down into separate components that form the reaction products.

$$\text{Metal (fuel)} + \text{Electrolyte} + \text{Oxygen} = \text{Direct current} + \text{Reaction product(s)} + \text{Heat}$$

The metal–air battery is rechargeable in that the metal anode (fuel) needs to be replaced mechanically when it is completely corroded. The electrolyte also needs to be either replenished and/or replaced as it is also consumed and can be open to evaporation. Metal–air battery has also been coined as a semi-fuel cell, which is a hybrid between a fuel cell and a battery; however, this term has not been used in the industry and is notably absent in any codes and standards publications. Though the term *"fuel cell"* has been used for metal–air technology, more so for marketing purposes when fuel cell was the public perception term, the metal–air technology is a closed system as opposed to a fuel cell, which is an open system. The continuous (open) energy carrier in a fuel cell is hydrogen while the metal anode energy carrier in a metal–air is stationary. The simple cell structures of metal–air batteries can be adapted to almost any size for use in many applications, and the cell can be mechanically recharged in seconds. The cathode reaction uses oxygen, a readily available cost-free gas from the atmosphere.

Reaction products will depend on both the metal and electrolyte compositions. An excellent source of determining the various reaction products is from the use of Pourbaix diagrams and X-ray determinations of the products themselves.

The main advantages associated with metal–air batteries are as follows:

- high energy density (on the basis of the metal and electrolyte)
- flat discharge voltage
- long shelf life (on the basis of dry storage)
- relatively low cost (on the basis of metal and air-diffusion cathode used).

Power capacity of aqueous metal–air batteries is a function of several variables:

- air-diffusion cathode area
- catalyst-type used in the air-diffusion cathode (i.e., Mn vs Co)
- air availability and oxygen content
- humidity
- anode-type and anode alloy material (i.e., Al, Mg, Li, and Zn)
- anode mass, morphology, and corrosion mechanisms
- electrolyte-type and concentration
- additives
- operational temperature
- CO_2 content of air
- manufacturing quality and components.

A major challenge for metal–air technology is the development of a continuous anode fuel supply as opposed to the current mechanically recharged anode. The mechanically recharged systems are viable for 12–24 V products but once the voltage (which is a direct result from the number of cells) is greater than 24 V, then a fuel supply needs to be developed nonmechanically.

A fuel rechargeable system has been in the development of a regenerator concept that essentially electrowins the metal from the reaction products oxide/hydroxide compounds. This development has been mainly used in the zinc–air technology. It does require an external source of power for the regeneration process. The regeneration process is conceptually similar to the recharging cycle of a battery.

Metal–air battery technology has been available for several decades and is primarily aimed at portable power and portable power products generally less than 1 kW. Zinc–air has been the most commercialized metal–air battery to date. The most notable application today is in the use of zinc–air "batteries" in hearing aids and pagers.

6.1.1
Metal–Air Technologies

Aluminum–air and zinc–air have had more development in the past three decades with zinc–air being more commercialized. Development of magnesium–air was ongoing in the 1960–1970s but curtailed off owing to various technological barriers but has had a resurgence in the past 10 years. Another type of metal–air

Table 6.1 Magnesium, aluminum, zinc metal–air types.

	Magnesium	Aluminum	Zinc
SG	1.74	2.70	7.13
SHE (V)	−2.363	−1.662	−0.763
Energy (number of electrons)	2	3	2
Open circuit voltage	1.7	1.2	1.3
Anode composition (wt%)	>90	99.999	99.99
Current capacity (Ah/kg)	2200	2500 (alloy)	740
Electrolyte	Saltwater	KOH	KOH
pH	6–8	13–14	13–14
Source(s)	Magnesite, limestone, seawater	Bauxite	Zinc sulfides

Abbreviation: SG, specific gravity

Table 6.2 Properties of metals used in metal–air batteries [7].

Metal anode	Electrochemical equivalent of metal (Ah/g)	Theoretical cell voltage (V)	Theoretical specific energy (of metal) (kWh/kg)
Li	3.86	3.4	13
Ca	1.34	3.4	4.6
Mg	2.2	3.1	6.8
Al	2.98	2.7	8.1
Zn	0.82	1.6	1.3
Fe	0.96	1.3	1.2

is lithium [7], which has the highest theoretical voltage but which has never been available for commercial interests to date. Iron–air technology has not had much development over the past decade, but with advanced research and development, this technology may have some importance as a secondary battery [8]. Published material has been oriented toward Li–air and Zn–air systems [9]. Comparisons of metal–air types are presented in Tables 6.1 and 6.2. Power ratings and costs for various power types are shown in Figure 6.1 [10]. These graphs indicate that the metal–air has certain advantages over other power sources. Metal–air technologies have some limitations and challenges to overcome but will have an impact on alternative energy sources. There is continued research and development for practical metal–air systems because of their potential.

6.2
Introduction to Aluminum–Air Technology

Aluminum–air technology has been known for several decades [11] and several attempts toward commercialization have been made, but no products

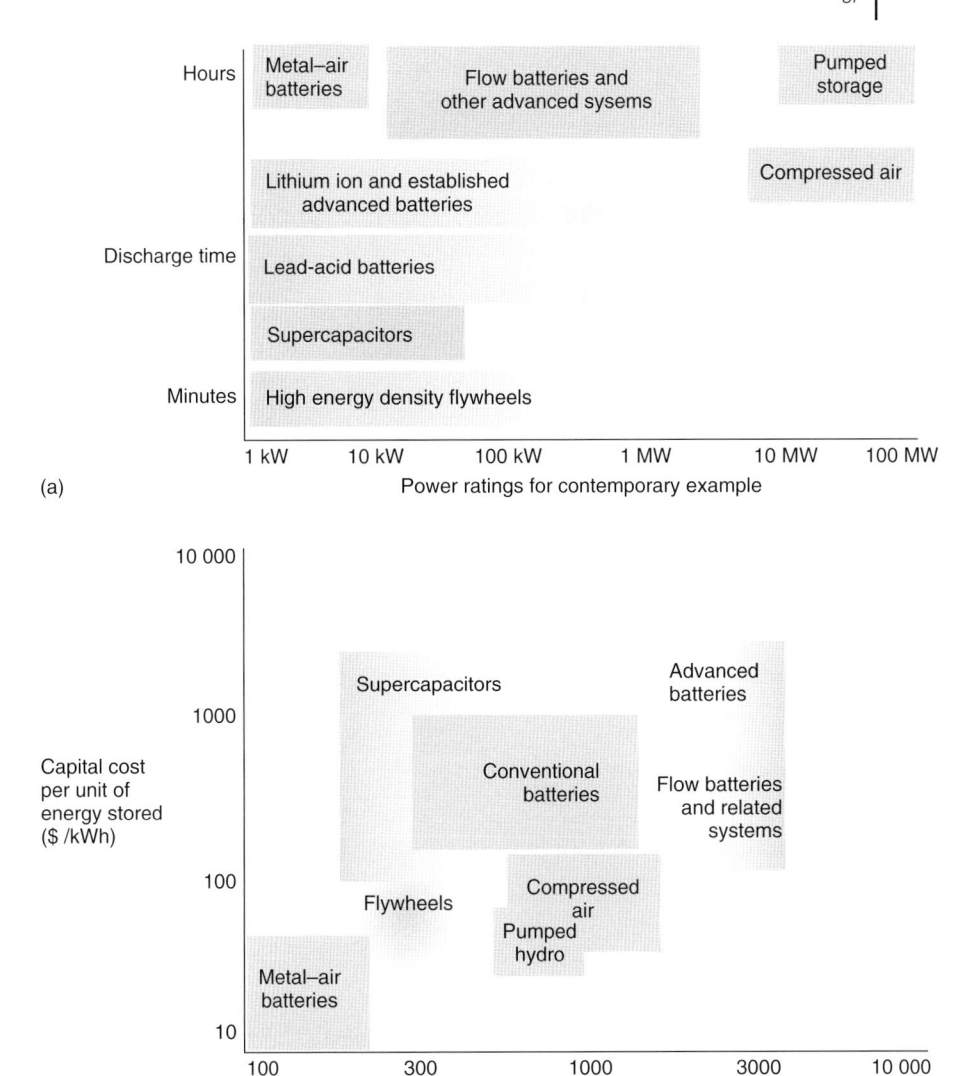

Figure 6.1 (a) Power ratings and (b) costs for various power types [10].

have penetrated the alternative power markets to date. Advancement of the aluminum–air technology was initially pioneered by researchers at Alcan Aluminum Limited [12]. Material used in this chapter is cited from their work.

6.2.1
Aluminum–Air Cell Structure

An aluminum–air cell uses aluminum (metal anode) as the fuel immersed in an electrolyte based on the aqueous solutions of salts and/or bases such as the

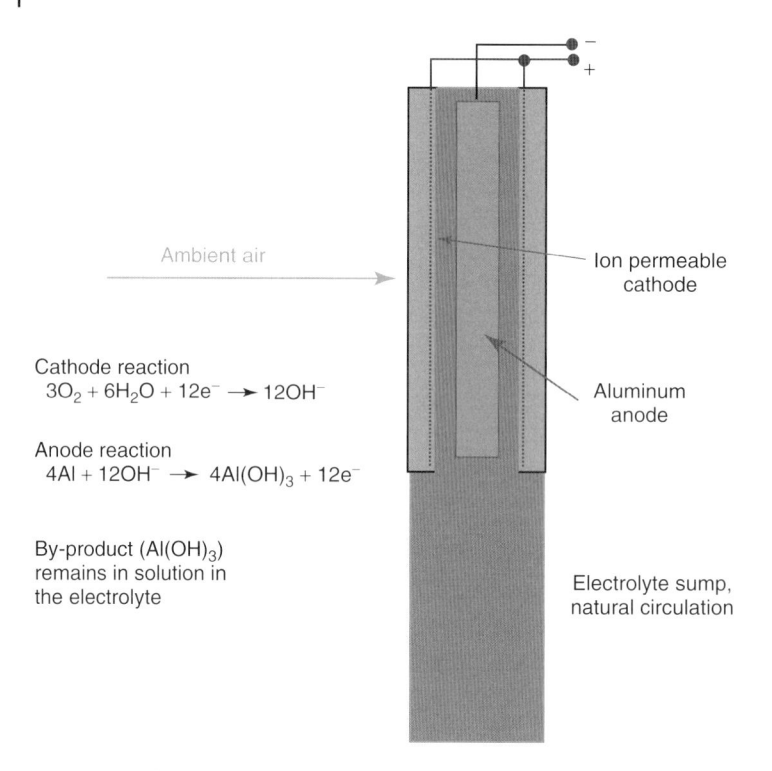

Figure 6.2 Aluminum–air battery [12].

commonly used potassium hydroxide and is coupled with a GDE to generate a direct current (Figure 6.2).

6.2.2
Electrochemical Processes

In the aluminum–air technology, electricity is generated through the combination of aluminum, oxygen, and generally an alkaline (i.e., KOH) electrolyte via the following reactions:

Electrons are formed at the anode:

$$\text{Anode:} \quad Al^{3+} + 3e^- \leftarrow Al$$
$$Al + 4OH^- \rightarrow Al(OH)_4^- + 3e^- \quad E^\circ = -2.35 \text{ V}_{SHE}$$
$$\text{Cathode:} \quad O_2 + 2H_2O + 4e^- \rightarrow 4OH^- \quad E^\circ = +0.40 \text{ V}_{SHE}$$

The net cell electrochemical reaction is

$$\text{Aluminum} + \text{KOH} + \text{Oxygen} \rightarrow \text{Reaction by-products} + \text{DC power} + \text{Heat}$$
$$4Al + 3O_2 + 6H_2O + 2(OH)^- \rightarrow 2Al(OH)_4^- \quad E^\circ = 2.75 \text{ V}_{SHE}$$

Table 6.3 Energy and power capabilities of alkaline and saline aluminum–air systems.

	Energy density (W kg^{-1})	Potential peak power density (W kg^{-1})
Alkaline[a]	400	175
Saline[b]	220	30

[a] Assume KOH but no concentration data.
[b] No data on saline concentration.

As the aluminum dissolves, the aluminate ion concentration increases, which causes the electrolyte solution viscosity and resistance to increase. Eventually, the electrolyte solution becomes supersaturated, which causes aluminum hydroxide to precipitate out.

$$2Al(OH)_4^- \rightarrow Al(OH)_3 + OH^-$$

In concert with the above reactions, aluminum corrosion occurs resulting in the generation of hydrogen. This reaction is also referred to as a *"parasitic"* reaction in that hydrogen is formed with the release of heat energy:

$$2Al + 6H_2O + 2(OH)^- \rightarrow 2Al(OH)_4^- + 3H_2 + heat$$

When saline (e.g., NaCl) electrolytes or electrolytes of pH <9 are used, the products of aluminum hydroxide reaction precipitate out readily and initially cause the electrolyte to gel before forming a more crystalline species. Two hydroxide phases are generated, the initial phase being amorphous pseudoboehmite (AlOOH) and the latter phase being bayerite (Al(OH)$_3$). The overall cell reaction is

$$4Al + 6H_2O + 3O_2 \rightarrow Al(OH)_3 \quad E° = 2.06\ V_{SHE}$$

The following is with a parasitic corrosion reaction

$$2Al + 6H_2O \rightarrow 2Al(OH)_3 + 3H_2 + heat$$

The energy and power capabilities of alkaline and saline aluminum–air systems are shown in Table 6.3.

This table indicates the differences between alkaline- and saline-based water electrolyte systems. Also, hydrogen is generated in both the electrolyte systems.

A deleterious reaction occurs when CO_2 is in sufficient concentration in the air causing the formation of aluminum carbonate reaction products that add resistance to the cell and impede the cell efficiency.

6.2.2.1 Reaction Products

Reaction products (precipitates) occur from the natural reaction of the anode immersed in the electrolyte. These reaction products vary depending on the electrolyte and anode compositions.

6.2.3
Degradation Mechanisms and Mitigation Strategies

6.2.3.1 Anode
Aluminum has an equilibrium potential of $-1.90\,V$. Standard calomel electrode (SCE) and therefore thermodynamically is a very reactive metal. In most natural environments, the formation of a thin oxide layer (Al_2O_3) on aluminum forms a protective coating allowing aluminum to be somewhat stable. This, however, limits the corrosion potential as a sacrificial anode for use in aluminum–air cells. Another limiting factor in aluminum–air technology is the anode itself as it must be of high purity because of its reactions, which lead to hydrogen gassing. This also leads to high anode costs. To help disrupt the protective oxide layer and stabilize the rate of corrosion, aluminum alloys have been developed in order to induce activation, which can also lead to higher costs. Such alloying elements include Hg, In, Sn, Ga, Zn, and Ag.

6.2.3.2 Cathodes
Cathodes are by far the most expensive part of the battery and have limitations (to date) that affect the overall efficiency of the process. There are several types of cathodes being developed, depending on the type of catalyst and cost.

6.2.4
Applications

The aluminum–air technology was directed more to portable alternative energy systems in the 100–1000 W range. There are few commercial products in the market today, specifically in the military applications.

6.3
Introduction to Lithium–Air Technology

The lithium–air technology [13] is a relatively unknown technology in the past but is beginning to make inroads into the metal–air battery. There have been attempts toward commercialization of aqueous lithium batteries (primary) with limited success in the consumer market; however, the lithium nonaqueous (secondary) batteries (e.g., lithium ion) are becoming readily acceptable and are rechargeable.

The lithium–air battery was initially investigated in the 1980s but encountered problems relating to low efficiency owing to the parasitic reaction of lithium with water and major safety concerns owing to excess hydrogen formation from this parasitic reaction [14]. In 1996, various researchers [15, 16] proposed replacing the aqueous electrolyte with a nonaqueous polymer electrolyte, and later liquid aprotic organic solvents for the electrolyte solution. Now, most of the research and commercialization of lithium–air involves nonaqueous electrolytes.

Lithium chemistry provides the best chance for the highest energy batteries and it currently dominates the power supply for portable electronics. Recent developments

Table 6.4 Theoretical specific energy and capacity comparisons for selected systems.

Metal–air and Li ion systems (organic or aqueous electrolyte solution)	OCV (V)	Specific energy (Wh/kg)	Specific capacity (mAh/g)
$2Li + \frac{1}{2}O_2 \rightarrow Li_2O$ (aprotic organic solution)	2.913	11 248[a]	3862
$Li + \frac{1}{2}O_2 \rightarrow \frac{1}{2}Li_2O_2$ (aprotic organic)	2.959	11 425[a]	3862
$2Li + \frac{1}{2}O_2 + H_2SO_4 \leftrightarrow Li_2SO_4 + H_2O$ (aqueous)	4.274	2046[a]	479
$2Li + \frac{1}{2}O_2 + 2HCl \leftrightarrow 2LiCl + H_2O$ (aqueous)	4.274	2640[a]	616
$2Li + \frac{1}{2}O_2 + H_2O \leftrightarrow 2LiOH$ (aqueous)	3.446	5789[a]	1681
$Zn + \frac{1}{2}O_2 \rightarrow ZnO$ (aqueous)	1.65	1353[a]	820
$x6C + LiCoO_2 \leftrightarrow xLiC_6 + Li_{1-x}CoO_2$ (organic)	~4.2	420[b]	139[b]
$Li + H_2O$ (in seawater) $\leftrightarrow LiOH + \frac{1}{2}H_2$	2.512	9701[c]	3862[c]

[a] The molecular mass of O_2 is not included in these calculations because O_2 is freely available from the atmosphere and therefore does not have to be stored in the battery or cell.
[b] Based on $x = 0.5$ in $Li_{1-x}CoO_2$.
[c] The molecular mass of H_2O is not included since it is freely available from seawater (pH 8.2) and does not have to be stored in the battery or cell.

in Li ion battery technology have advanced its application into the area of large format batteries (e.g., automotive markets). This, in turn, has heightened the need for research on higher capacity, safer and less-expensive battery materials [17].

The following is the net cell electrochemical reaction:

Lithium + Oxygen → Reaction byproducts + DC power + Heat

The theoretical specific energy and capacity comparisons for selected systems [16] are shown in Table 6.4. It should be noted that the theoretical calculations are higher than the practical data obtained from actual systems owing to limiting factors such as cathode quality, anode quality, electrolyte quality, resistance, and manufacturing components of the battery.

6.3.1
Aqueous Lithium–Air Cell Structure

A lithium–air cell uses lithium (metal anode) as the fuel immersed in an electrolyte based on the aqueous solutions of salts and/or bases/acids and is coupled with a GDE to generate a direct current. Seawater electrolyte has also been investigated.

6.3.2
Electrochemical Processes

In the lithium–air technology, electricity is generated through the combination of lithium, oxygen, and an aqueous or nonaqueous electrolyte via the following reactions [15]:

Lithium/water:

$$2Li + 2H_2O = 2LiOH + H_2 \quad E° = 2.22 \, V_{SHE}$$

Lithium/seawater (pH = 8.2):

$$2Li + 2H_2O = 2LiOH + H_2 \quad E^\circ = 2.60 \; V_{SHE}$$

Lithium/oxygen in aqueous electrolytes:

$$\text{Basic electrolyte} \quad Li + \frac{1}{2}O_2 + H_2O = 2LiOH \quad E^\circ = 3.45 \; V_{SHE}$$
$$\text{Acidic electrolyte} \quad Li + O_2 + 4H^+ = 2H_2O + Li \quad E^\circ = 4.27 \; V_{SHE}$$

In concert with the above aqueous reactions, lithium corrosion occurs resulting in the generation of hydrogen. This reaction is also referred to as a *"parasitic"* reaction in that hydrogen is formed with the release of heat energy:

$$2Li + 6H_2O + 2(OH)^- \rightarrow 2Li(OH)_4^- + 3H_2 + Heat$$

6.3.3
Nonaqueous Lithium–Air Cell Structure

A lithium–air cell uses lithium (metal anode) as the fuel immersed in an electrolyte based on the nonaqueous and/or organic solutions of salts and/or bases and is coupled with a GDE to generate a direct current.

Lithium/oxygen in nonaqueous electrolytes:

$$Li + O_2 = LiO_2 \text{ (peroxide)} \quad E^\circ = 2.96 \; V_{SHE}$$

6.3.4
Degradation Mechanisms and Mitigation Strategies

6.3.4.1 Anodes

Metallic lithium reacts strongly with water and as such, the anode must be separated from the aqueous electrolyte via an inert membrane (separator) [14].

Lithium anodes in the nonaqueous battery are generally lithium compounds using some common materials such as carbon, titanium, silicon, germanium, tin, or some combination of these.

Anode chemistry, impurities, and morphology play a significant role in the utilization efficiency of the anode itself.

6.3.4.2 Electrolyte

Reaction products, in particular LiO_2 and Li_2O_2, are insoluble in most electrolytes that precipitate and block pores of the cathode, which in turn inhibits the flow of oxygen, thus causing cell degradation. The electrolyte can be contaminated by moisture from the air, which attacks the lithium electrode, thus limiting the battery life. Use of specialized additives/solvents to the electrolyte can inhibit the corrosion and solubility issues [14, 16].

In nonaqueous aprotic electrolytes, the limitations are caused by lithium oxide precipitates and the solubility of oxygen in the electrolyte, which affects oxygen transport within the interior of the air cathode. In aqueous-based electrolytes, the solubility of reaction products is not as severe and there is more oxygen flow into the cell itself.

Depending on the air quality and electrolyte composition, other reaction products are lithium hydroxide, lithium carbonate, and lithium nitrate.

6.3.4.3 Cathodes

Cathodes are by far the most expensive part of the battery and have limitations (to date) that affect the overall efficiency of the process. There are several types of cathodes being developed, depending on the type of electrolyte and cost [18].

6.3.5 Manufacturing

Most manufacturers have their own patented manufacturing process(es) and as such are not generally reported in the literature.

6.3.6 Applications

The nonaqueous lithium–air chemistry has advanced more than the aqueous lithium–air and as such more commercial applications/products have been developed. These include the rechargeable AA to D cell size batteries and larger size batteries for the automotive industry.

6.4 Introduction to Zinc–Air Technology

The zinc–air technology has been known for several decades [19–22] and there have been numerous attempts toward commercialization with success in the consumer market but more specifically in the value-added products in the military market.

6.4.1 Zinc–Air Cell Structure

A zinc–air cell uses zinc (metal anode) as the fuel immersed in an electrolyte based on the aqueous solutions of salts and/or bases such as the commonly used potassium hydroxide and is coupled with a GDE to generate a direct current.

6.4.2 Electrochemical Processes

In the zinc–air technology, electricity is generated through the combination of zinc, oxygen, and generally an alkaline (i.e., KOH) electrolyte via the following reactions:

Electrons are formed at the anode:

Anode: $Zn^{2+} + 2e^- \leftarrow Zn$

$Zn + 4OH^- \rightarrow Zn(OH)_4^{2-} + 2e^- \quad E^\circ = 1.25\ V_{SHE}$

$Zn(OH)_4^{2-} \rightarrow ZnO + 2OH^- + H_2O$

Cathode: $\frac{1}{2}O_2 + H_2O + 2e^- \rightarrow 2OH^- \quad E^\circ = 0.40\ V_{SHE}$

The following is the net cell electrochemical reaction:

Zinc + Oxygen → Reaction byproducts + DC power + Heat

$2Zn + O_2 \rightarrow 2ZnO \quad E^\circ = 1.65\ V_{SHE}$

The practical operating voltage of the zinc–air cell is less than 1.65 V owing to internal loss from resistance.

As the zinc dissolves, the zincate ion concentration increases, which causes the electrolyte solution viscosity and the resistance to increase. Eventually, the electrolyte solution becomes supersaturated and as a result of this, zinc hydroxide precipitates out.

$$2Zn(OH)_4^- \rightarrow Zn(OH)_2 + OH^-$$

In concert with the above reactions, zinc corrosion occurs resulting in the generation of hydrogen. This reaction is also referred to as a *"parasitic"* reaction in that hydrogen is formed with the release of heat energy:

$$2Zn + 6H_2O + 2(OH)^- \rightarrow 2Zn(OH)_4^- + 3H_2 + heat$$

The overall cell reaction is

$$4Zn + 6H_2O + 3O_2 \rightarrow Zn(OH)_3 \quad E^\circ = 2.06\ V_{SHE}$$

With a parasitic corrosion reaction

$$2Zn + 6H_2O \rightarrow 2Zn(OH)_3 + 3H_2 + heat$$

6.4.2.1 Reaction Products

Reaction products (precipitates) occur from the natural reaction of the anode immersed in the electrolyte. These reaction products vary depending upon the electrolyte and anode compositions.

6.4.3
Degradation Mechanisms and Mitigation Strategies

6.4.3.1 Anode

Zhang [21] discusses the relationships between energy density, power density, and zinc anode thickness. The kinetics of the zinc anode is of particular importance to the design of zinc–air batteries. The anode is capable of discharging at a higher current density when a thicker anode is used and hence the battery will have a higher energy density.

6.4.3.2 Hydrogen Evolution

The parasitic reaction of hydrogen evolution decreases the utilization efficiency of the zinc anode (the utilization efficiency is discussed in Section 6.11). Methods to suppress hydrogen evolution revolve around anode chemistry, such as alloying with Hg, Pb, and Cd, and coating the zinc metal with other materials such as lithium boron oxide [9]. Another method is to put additives into the electrolyte.

6.4.3.3 Electrolyte

The dominant alkaline electrolyte used in zinc–air batteries is potassium hydroxide, though sodium hydroxide and lithium hydroxide have been reported [9]. The KOH has a better ionic conductivity compared to the other hydroxides. The major disadvantage of using these electrolytes is their very high pH that is caustic and has health-related issues. High concentrations of KOH can also be detrimental, which leads to increased viscosity in the electrolyte.

A deleterious reaction occurs when CO_2 is in sufficient concentration in the air causing the formation of zinc carbonate reaction products, which clog the cathode pores thereby adding resistance to the cell and reducing the lifetime of the air cathode. Various techniques have been suggested such as CO_2 absorbants but no apparent solution has been effectively developed to date.

6.4.3.4 Cathode

One of the limiting factors in zinc–air is the air cathode. Research for the improvement of the air cathode is ongoing [23] with the purpose of creating a cost-effective manufactured cathode.

6.4.3.5 Regeneration

Various zinc–air companies and researchers [24] have attempted to create a zinc regenerative/recycling system in order to create a closed-loop system. This is important if zinc–air is used for operating on-board a vehicle, which generates electricity when zinc combines with oxygen from the air in the presence of an electrolyte to form zinc oxide, a safe white powder commonly used in skin creams and sunblock. The separate regeneration unit converts the zinc oxide back into fresh zinc, which is then recombined with the electrolyte to be reused as a fresh fuel. A regenerative system incorporates an additional piece of hardware called an *electrolyzer* that uses electricity from an external power source to convert the oxide reaction product back into the fresh fuel. This overall process is ∼30–50% efficient.

A zinc regeneration system that has been developed for the zinc–air system involves the following:

- Reaction of the zinc anode oxidation product (ZnO) with the electrolyte:

$$ZnO + 2KOH + 2H_2O = K_2Zn(OH)_4$$

- Electrowinning of the zincate solution:

$$K_2Zn(OH)_4 = Zn + 2KOH + 2H_2O + \frac{1}{2}O_2$$

- Reassembly of the zinc source containing the electrowon zinc.

6.4.4
Applications

The nonaqueous zinc–air battery technology has been advanced more than the other metal–air aqueous types and as such more commercial applications/products have been developed. These include the button battery used in hearing aids and pagers, and rechargers for cell phones and batteries.

6.5
Introduction to Magnesium–Air Technology

The magnesium–air technology has been known since the 1960s [25–29] but no real advancement has been made toward commercialization owing to technological barriers such as hydrogen generation, precipitate control, anode material, air-diffusion cathode material, and expensive manufacturing processes. Information regarding the magnesium–air technology is largely based on the laboratory and prototype test work conducted by the author [30] and colleagues at MagPower Systems Inc.

The performance of the magnesium–air system depends on the interaction of the following seven major components:

1) cell assembly (configuration and mechanical + electrical integrity of the unit);
2) electrolyte composition;
3) hydrogen suppressant additive to electrolyte (compound and concentration);
4) cathode structure and composition (catalyst type and loading);
5) cell gap (initial + magnesium thickness);
6) anode composition and morphology; and
7) electrolyte pumping and heat exchange.

6.5.1
Magnesium

Magnesium (Mg – atomic number 12) was first isolated by the British chemist Sir Humphry Davy in 1808. The name originates from the Greek word for a district in Thessaly called *Magnesia*. Magnesium is the eighth most abundant element in the earth's crust and although it is not found in its elemental form, it is found in 663 minerals as a component of the mineral (i.e., brucite – $Mg(OH)_2$). It is also the third most plentiful element dissolved in seawater, with an average concentration of $1326\,mg^{-1}$ but can be quite variable in freshwater owing to groundwater interactions with different surfacial material. Seawater varies in trace elements, temperature, salinity, and conductivity depending upon the vertical gradient (depth) and spatial area in the world. Magnesium exceeds calcium in abundance in seawater, but is far less common than calcium in freshwater. Where magnesium content is quite high in freshwater, it occurs naturally in hardwater, which is essentially water that has come in contact with geological material

Table 6.5 Electrochemical series.

Metal	Symbol	
Potassium	K	Most electropositive element
Sodium	Na	
Calcium	Ca	
Magnesium	**Mg**	
Aluminum	Al	
Zinc	Zn	
Iron	Fe	
Lead	Pb	
Hydrogen	*H*	
Copper	Cu	
Mercury	Hg	
Silver	Ag	
Gold	Au	Least electropositive metal

such as dolomite that has a high content of magnesium. In this situation, water softeners that remove calcium and magnesium via ion exchange resin are generally purchased and back flushed into water sewage lines using a salt brine solution. There are no particular standards or guidelines for magnesium in drinking water as compared to Al and Zn.

Magnesium is an important element for plant and animal life and the adult human daily requirement of magnesium is about 0.3 g per day. Magnesium is an essential element, which is involved in energy metabolism, protein and nucleic acid synthesis, formation of bones, and activity of neurotransmitters. When dissolved in water, magnesium carries two positive charges (Mg^{2+}), and because of this relatively high charge has major importance as an electrolyte. Magnesium causes no unwanted side effects for most people and toxicity is rare. Numerous foods contain magnesium in varying amounts ranging from 13 mg (milk) to 490 mg (wheat bran). Magnesium metal is noncarcinogenic and does not have well-characterized toxicity. Magnesium compounds are generally noncarcinogenic and nontoxic.

Magnesium occurs fifth in the electrochemical series (Table 6.5), meaning that any metal higher up in the series will have a greater potential difference (relative to the standard hydrogen electrode (SHE)) than those below (i.e., Mg is more electrochemically reactive in saltwater than either Al or Zn and has a greater electropotential than either of these metals).

6.5.1.1 Sources

Magnesium metal can be recovered from seawater and lake brines, as well as from minerals such as dolomite (calcium–magnesium carbonate), serpentine (hydrous magnesium silicate), olivine (magnesium silicate), magnesite (magnesium carbonate), and brucite (magnesium hydroxide). Recycling has now become an additional source of magnesium metal. Major magnesium producing countries include Peoples Republic of China, Australia, Canada, and Russia.

6.5.1.2 Production

Magnesium can be made commercially by several processes and it can be recovered from seawater (Dow Chemical). There are two major process routes utilized for the production of magnesium metal, namely, electrolytic and thermal routes.

The technologies for the production of magnesium by electrolysis (electrolytic route) are characterized by three essential steps, namely:

- production of magnesium-rich brine (usually $MgCl_2$ or $MgCl_2/KCl$);
- brine drying or crystallization followed by dehydration; and
- electrolytic reduction of $MgCl_2$ or $MgCl_2$-KCl for the production of Mg-metal and chlorine gas.

Nonelectrolytic methods (thermal route) of magnesium production are all based upon thermo-chemical reduction of magnesium oxide that is obtained by processing any one of a number of magnesium resources, for example, magnesite ($MgCO_3$), dolomite ($MgCO_3 CaCO_3$), olivine ($(Mg, Fe)SiO_2$), and so on, although dolomite is utilized most commonly. Newer processes such as Noranda's Magnola [31] are essentially a leaching process of serpentinite material but with high operational costs. The most important factor is the choice of a process, which maximizes energy efficiency in addition to locating a plant where energy sources are not only readily available but also at favorable rates.

6.5.1.3 Uses

Magnesium is a metal that has both major industrial and health uses, and is becoming more important as a result of new technology and markets, especially in the automotive industry.

- Since it is 30% lighter (density of $1.738\,g\,cm^{-3}$ and atomic weight of 24.3050 gmol) than aluminum, it is used in products and alloys such as
  - aircraft and missile construction
  - automotive industry
  - artificial limbs
- It is used as a reducing agent for the production of uranium and other metals from their salts.
- It is used as an anode in hot water tanks.
- It is used in flares and pyrotechnics, including incendiary bombs.
- It is used in computers for radiofrequency shielding.
- It is used in numerous compounds:
  - to make organomagnesium compounds (Grignard reagents), useful in organic synthesis
  - Magnesium oxide
    - since it is refractory it is used as bricks and liners in furnaces
  - Magnesium hydroxide
    - acid neutralizer (i.e., acid rock drainage)

- fire retardant
- cosmetic preparations as an absorbent and pH adjuster, and
- milk of Magnesia, maalox (MgAl hydroxide)
- Magnesium sulfate
  - epsom salts
- Magnesium citrate
  - medicine
  - effervescent beverages
- Magnesium chloride
- vitamins/minerals (a major element in mineral tablets).

6.5.2
Associations

The International Magnesium Association (IMA) is a worldwide organization, headquartered in McLean, Virginia, embracing the interests of those concerned with the metal magnesium. Founded in 1943, the IMA strives to promote the magnesium industry by collecting and disseminating information, encouraging research, and publishing innovative uses of the metal. The magnesium website can be accessed via www.intlmag.org, which discusses in detail the health aspects of magnesium. Another organization dedicated to magnesium can be found at www.magnesium.com.

6.6
Structure of Magnesium–Air Cell

A magnesium–air cell (MAC) is a cell that uses a magnesium (metal anode) as the fuel immersed in a salt electrolyte coupled with a GDE to generate a direct current (Figure 6.3). The important concept in constructing the cell is to minimize resistance where possible. This includes limiting the gap between the anode and cathode, conductivity of the electrolyte, anode material, and precipitate control. The cell can be rectangular or cylindrical in shape and is scalable depending upon the power application.

6.7
Electrochemical Processes

In the magnesium–air technology, electricity is generated by the combination of magnesium, oxygen, and saltwater (i.e., NaCl) electrolyte via the following reactions:

Electrons are formed at the anode:

$$\text{Anode:} \quad Mg^{2+} + 2e^- \leftarrow Mg \quad (6.1)$$

6 Metal–Air Technology

Magnesium–air fuel cell

Figure 6.3 Magnesium–air battery (MAC).

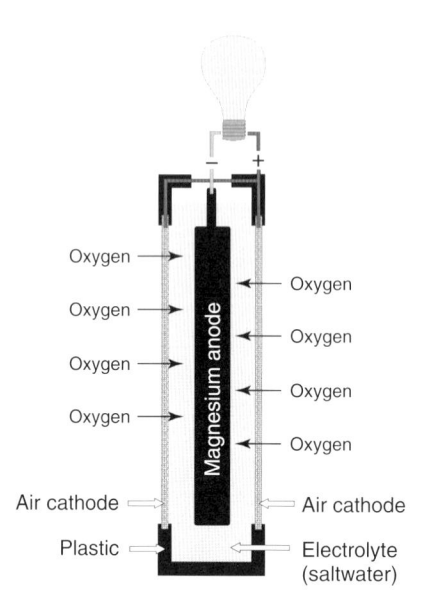

$$Mg + 2H_2O \rightarrow Mg(OH)_2 + H_2 + 353 \text{ kJ} \tag{6.2}$$

Equation (6.2) is also referred to as a *"parasitic"* reaction in that hydrogen is formed with the release of heat energy.

OH^- ions are produced at the cathode and electrons are consumed. An increase in pH occurs due to OH^- generation:

$$\text{Cathode:} \quad \frac{1}{2}O_2 + H_2O + 2e^- \rightarrow 2OH^-$$

$$Mg^{2+} + 2(OH)^- \rightarrow Mg(OH)_2$$

The net cell electrochemical reaction is

$$\text{Magnesium} + \text{Water} + \text{Oxygen} \rightarrow \text{Reaction by-products} + \text{DC power} + \text{Heat}$$

The practical open cell voltage of an MAC is 1.7 V, which is based on the actual measurement. The operating open cell voltage is the product of full cell reaction potential minus the electrode overpotentials and the Ohmic voltage (IR) drop in electrolytes and separators.

The numerical value of the full theoretical reaction potential is 2.356 V (at 25 °C). The formula is as follows:

$$\text{Cathode reaction:} \quad O_2 + 2H_2O + 4e^- \rightarrow 4OH^- \quad E^\circ = 0.40 \text{ V}_{SHE} \quad \text{(I)}$$

$$\text{Anode reaction:} \quad 2Mg^{2+} + 4e^- \rightarrow 2Mg \quad E^\circ = -2.756 \text{ V}_{SHE} \quad \text{(II)}$$

(I)−(II)

$$O_2 + 2H_2O + 2Mg \rightarrow 2Mg(OH) \quad E^\circ = 2.356 \text{ V}_{SHE} \quad \text{(III)}$$

6.7 Electrochemical Processes

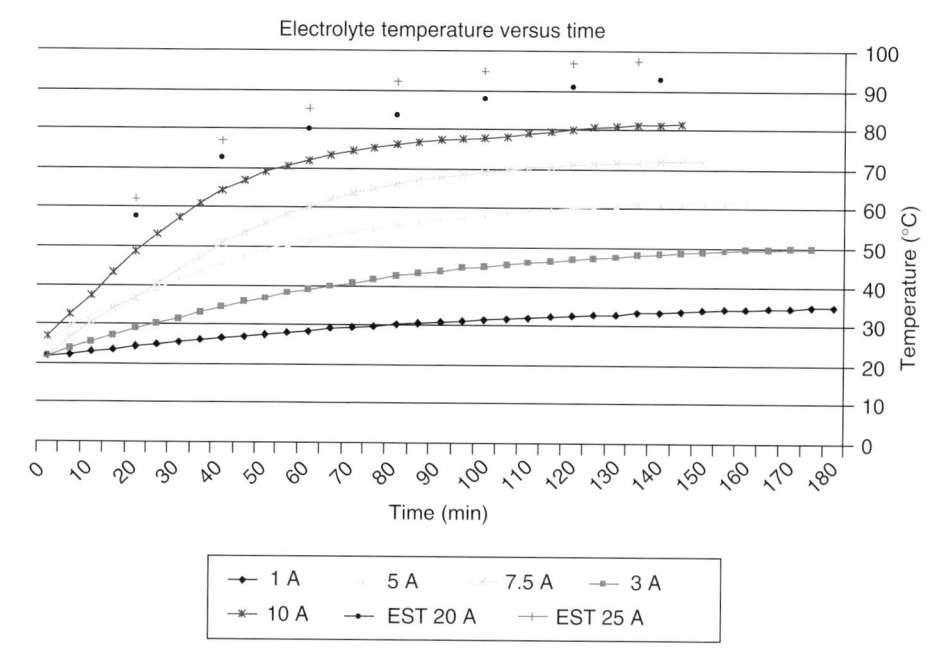

Figure 6.4 Effect of current and temperature using 12 cells (ambient temperature).

Passivation:

The magnesium anode can become passivated by MgO/OH.

This reaction is slow in pure water but is accelerated with high electrical conductive additives such as sodium chloride (NaCl). This reaction is exothermic, the temperature depending upon the load of the power device, i.e., the greater the amperage, the greater the amount of heat generated (Figure 6.4). The magnesium utilization efficiency (MUE) of the cell will in part depend upon the internal cell temperature with 70 °C being the most efficient running temperature.

The electrons produced at the anode pass around the external circuit connected to the fuel cell, and both reactions (anode and cathode) carry on until the magnesium is entirely consumed. When the MAC is under load, it is normal for the water to be consumed and evaporated thus decreasing the volume of the electrolyte solution. As the electrolyte volume decreases, the reaction product(s) mass increases which can lead to electrolyte supersaturation and eventual precipitation of the reaction product(s), and thus clogging of the cell.

Magnesium exhibits an electrochemical corrosion phenomenon known as the negative difference effect (NDE). This effect has been the subject of research dating from the 1950s with varying authors [32, 33] attempting to explain the NDE by means of electrochemical reaction mechanisms. Atrens and Dietzel [34] discusses the two mechanistic approaches to the magnesium NDE which are (i) the unipositive Mg^+ ion NDE mechanism and (ii) the magnesium self-corrosion mechanism. Liu and Schlesinger [35] discusses the NDE behavior based on Tafel kinetics proposing that both Mg^+ and Mg^{2+} result from magnesium dissolution and Mg^{2+} increases faster than Mg^+ with the increase of the overpotential. From experimental work regarding the MAC [30], the decrease of magnesium efficiency with increased power output is probably a consequence of the anomalous NDE on magnesium. The NDE means that the rate of parasitic generation of hydrogen rises when the anode undergoes higher positive polarization as the current density is increased.

6.8
Components

6.8.1
Electrolyte

Research has been mainly based on saline-based electrolytes; however, other aqueous-based electrolytes have been investigated [36, 37]. The electrolyte is a saline solution such as seawater, common salt (NaCl) dissolved in water, or water containing a special salt mixture as prepared by various companies involved in the development of the magnesium–air technology. Other types that should work are naturally occurring brines and alkali pond waters. Urine has also been demonstrated to work effectively in the MAC. The choice of electrolytes is fairly flexible but the power output will differ for each of these electrolytes. The amount of magnesium anode corrosion is dependent on the salt content of the electrolyte (Figure 6.5). Therefore, the current also depends on the conductivity of the electrolyte. Salt is highly corrosive to magnesium, but it (chloride ion) is needed for the conductivity of the water-based electrolyte and activates the magnesium by breaking up the magnesium oxide layer on the anode. Therefore, higher the salt concentration, greater the corrosion rate. Additives may be used in the electrolyte to promote conductivity and help suppress hydrogen gas. However, an overly high salt concentration will lead to unwarranted and potential detrimental corrosion of the cathode.

6.8.2
Cathode

The air-diffusion cathode is a sheet-like member having opposite faces exposed to two different environments, an atmosphere and an aqueous solution or an atmosphere and a solid, respectively. The air-diffusion cathode must form a three-phase (gas–solid–liquid) interface where gas, catalyst/carbon, and electrolyte are in contact to facilitate the reaction of gaseous oxygen. The atmospheric side needs to be permeable to air but substantially hydrophobic in order to avoid electrolyte leakage through the air-diffusion cathode to the atmosphere boundary, and reactant flooding. The current collector embedded in the air-diffusion cathode

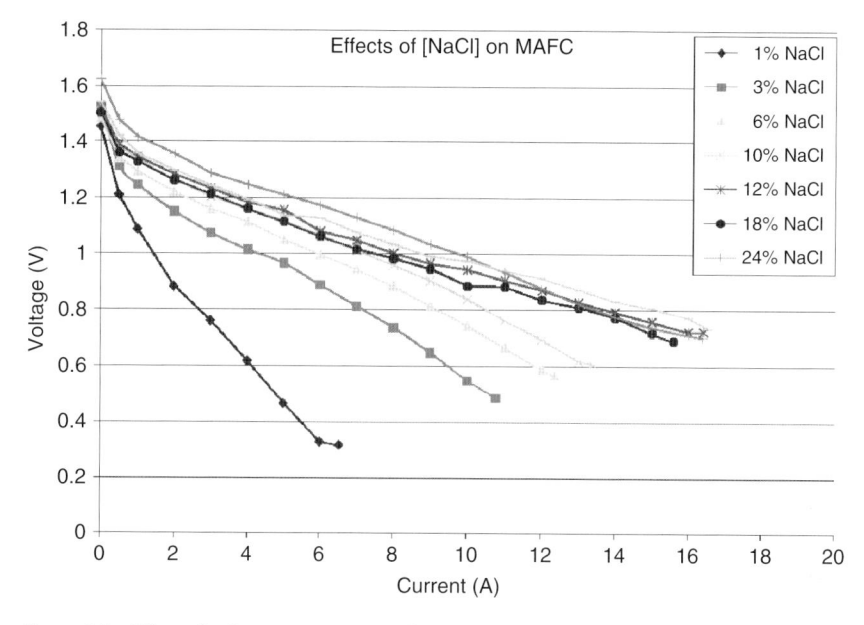

Figure 6.5 Effect of salt concentration (ambient temperature).

is necessary for current flow and a structural support for the air-diffusion cathode. During operation, oxygen passes through the air-permeable and hydrophobic air-diffusion cathode by difference in pressure between the outside and inside of the cell. The catalyst within the carbon cathode facilitates the reduction of oxygen to anion via electrochemical reaction with electrons generated from the oxidation of the magnesium anode. The hydroxyl ions migrate from the air cathode to the magnesium anode to complete the cell reaction.

The discharge reaction mechanism of a metal–air battery is expected as follows if the cathode O_2 reduction is a four-electron process:

$$\text{Anode:} \quad \text{Metal} \rightarrow \text{Metal}^{1+/2+/3+} + e^-/2e^-/3e^-$$
$$\text{Cathode:} \quad \frac{1}{2}O_2 + 2e^- + H_2O \rightarrow 2OH^-$$

where metal is Li, Mg, Zn or Al. When the battery is in operation, air is drawn into and through the air-diffusion cathodes. The oxygen will be used in the oxidation–reduction reaction between the air cathode and magnesium anode. Both oxygen and nitrogen gasses pass through the cathode gas diffusion layer together and react with the catalyst layer. Oxygen will be reduced to H_2O_2 and nitrogen gas will diffuse into the electrolyte and thus form bubbles in the electrolyte and be released into the air. The electrolyte is composed of water that contains air by nature. Air is composed of $\sim$21 vol% oxygen and 78 vol% nitrogen with the other 1% made up of other gasses such as CO_2. When the water is agitated by pumping mechanism, then any air bubbles will disperse from the electrolyte. Hydrogen is generated as a secondary reaction on the anode during the operation

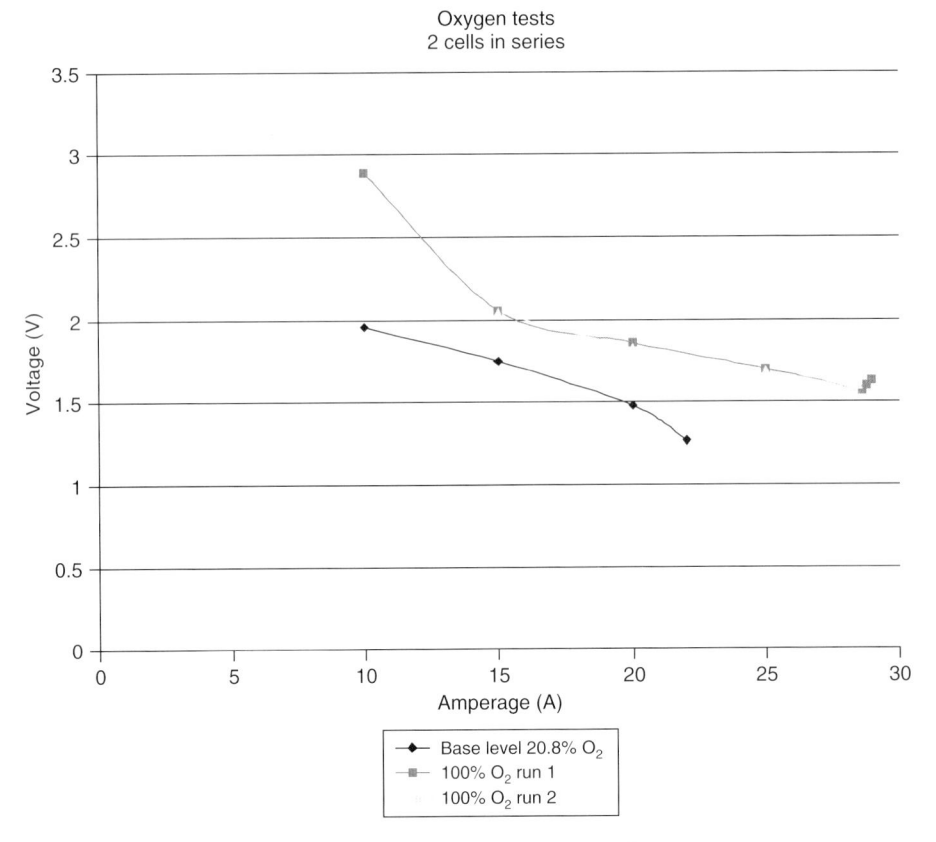

Figure 6.6 Effect of oxygen enrichment using two cells (ambient temperature).

of a magnesium–air cell. The amount of hydrogen generated will depend on the current applied to the cell. The process of hydrogen generation from the electrolysis of the electrolyte is also governed by mass transfer, growth of hydrogen bubbles, and removal of hydrogen from the MAC. Riegel *et al.* [38] discuss the mechanisms for hydrogen removal as (i) hydrogen dissolved in the electrolyte solution is removed by diffusion and convection and (ii) gas bubbles are transported by a two-phase flow. Thus, hydrogen gas will be diluted volumetrically.

The driving force for the gas transfer is the difference in partial pressure of gasses between the bubbles stuck in the air diffuse cathode and the air outside. The bubbles are saturated with water vapors and they are connected with the outside air through the pores in the cathode. The diffusion of nitrogen and oxygen to the bubbles is caused by the difference in partial pressure, resulting in the increase of volume of the gas bubbles. As this new gas volume gets saturated with water vapors, the process will continue (W. Halliop, personal communication).

The effect of oxygen enrichment is shown in Figure 6.6. When the oxygen concentration is high, a higher load can be utilized with minimizing effect on the voltage.

6.8.3
Anode

Magnesium is one of the lightest metals, which together with magnesium alloys have properties that can lead to specific applications. Understanding the electrochemical corrosion behavior and mechanisms of magnesium is important to the commercialization of the MAC. An important and limiting factor to increase power density in the magnesium–air power source is the magnesium anode itself (albeit in conjunction with the electrolyte and air cathode). The anode is the fuel (or fuel carrier) and is consumed and must be replaced. The mass of the anode will influence both the run time and amperage of the MAC. The metal quality of the anode is paramount to the magnesium utilization efficiency. The metal quality [39] may be defined as (i) chemical composition, (ii) inclusions and porosity inside the anode, (iii) the surface appearance, and (iv) consistency. Magnesium anodes are generally specified with respect to strict chemical compositions because of harmful impurities such as iron, copper, and nickel which can impair the corrosion performance of the anode and lead to nonuniform dissolution. These impurities can result in generating local galvanic cells on the anode which has an effect of reducing anode efficiency due to their low solid-solubility limits and their impact as active cathodic sites for reduction of water at the sacrifice of elemental magnesium [40]. Aluminum, manganese [41], and zinc can be added to promote uniform corrosion and reduce the rapid activation of magnesium. Test work to date has indicated that a few commercially available magnesium and magnesium alloy anodes can be used in the MAC but that they have some limitations. They were essentially developed for other commercial applications such as automobiles, cathodic protection for boats, and seawater batteries. The present anodic structure is sheeted and is essentially two-dimensional. The current anode compositions that are commercially available may not be the best for this power source. Therefore, it is possible to further improve the performance of the magnesium–air power source anode in order to achieve useful power for commercial applications.

Understanding corrosion behavior and mechanisms of magnesium [42–44] provides a basis for the design of new alloys and manufacturing techniques. Microgalvanic corrosion reaction between magnesium, alloy elements/impurities, and grain size is important and forms the understanding of passivation and dissolution of magnesium [35, 45]. Another factor to consider in the magnesium alloys is the presence of various phases that can have an impact upon corrosion behavior [44].

Several types of anodes have been tested together with different manufacturing methods in prototype cells such as Mg, MgMn (1 wt% Mn), AZ31, AT61, AM50, and AM60B. The current Mg alloys are designed for their mechanical or physical properties and they may not be the best alloys for use as an anode in the metal–air fuel cell (MAFC). For example, Al and Mn in the Mg alloys are known to reduce the long-term dissolution rate of the metal in various solutions. The current density of Mg anode can be increased by alloying Mg with elements that increase the dissolution rate. On the basis of the current literature, it may be feasible to design

Figure 6.7 DieCast AZ31, Thixomold AM60B, and DieCast AZ31.

the anode chemistry so that microgalvanic cells are formed as current is drawn out of the fuel-cell battery. The anode chemistry should also be assessed as to any impacts it will have on the environment (i.e., eco-material assessment), which is primarily trace element composition. From test work and literature review, the chemistry, especially trace element content, is important in the development of the anode for the MAC. It is imperative that the correct trace element chemistry be initiated and that high purity with very low impurity concentrations is attained. Trace elements and concentrations are ultimately influenced by the raw material at source.

From test work and literature review, the metallurgical factors such as impurities, alloy composition, and grain size influence the manufacturing process which is extremely important in the development of the anode for the MAC (Figures 6.7 and 6.8). This is evident in the grain alignment, porosity, and grain compaction. This in turn impacts the surface quality of the anode (i.e., caking of precipitate on anode) and precipitate quality and color. It is imperative that the correct manufacturing process be initiated as it will influence the structural integrity of the anode material. The magnesium anode commercial manufacturing methods/processes used to date are die casting, thixomolding, extrusion, and cold pressing metal powder. It was found that both thixomold and die cast have a greater amount of porosity than extruded material and that during casting, flow lines and uneven surfaces in the anode have a detrimental effect on the corrosion behavior.

The preferred anode to date is extruded AZ31. The apparent reason that magnesium hydroxide does not cake on the AZ31 versus other anodes is that this material causes a decrease in the magnesium reaction activation as compared to using pure magnesium material that has a high magnesium reaction activation (thus the reason for cleaner anodes and cathodes), which may be due to the presence of zinc.

Figure 6.8 Extruded anode material, Mg, MgMn, and AZ31.

6.8.4
Reaction Products

Reaction products (precipitates) occur from the natural reaction of the anode immersed in the electrolyte. These reaction products vary depending on the electrolyte and anode compositions. The dominant reaction products as identified from X-ray diffraction method are brucite (magnesium hydroxide – $Mg(OH)_2$) and iowaite ($Mg_4Fe(OH)_8OCl\cdot4H_2O$). Other chemical species, such as MgH, MgCl, and MgO, occur as deduced from the overall chemical reactions but have not been identified via X-ray diffraction probably because of very low concentrations. In a seawater electrolyte, the additional reaction product will be aragonite.

6.9
Manufacturing

The design and choice of materials used in the manufacture of the MAC are very important with regard to costs and environmental impact. The cell would encompass an injection moldable plastic and a cathode – bus bar assembled and controlled in an injection mold tool for mass production. Since the cell technology lends itself to being scalable, the use of injection mold tools would be cost effective. Products using the MAC must also be designed for ease of assembly and disassembly that is paramount to recyclability. The smaller cell configurations can be recharged with magnesium by hand, whereas when power requirements involve numerous cells, then an automatic insertion of the anode material is needed. The MAC can be manufactured for stacking (Figure 6.9a,b) or for stand alone

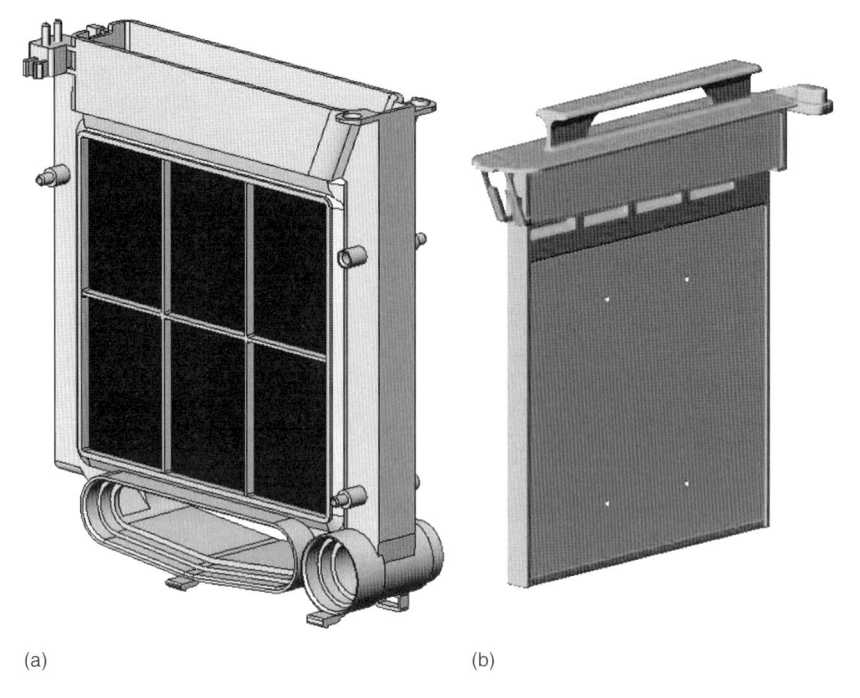

Figure 6.9 (a) Moldable stacking magnesium–air cell. (b) Magnesium anode insert into the above cell.

(Figure 6.10) depending on the power requirements. The cell itself is composed of two air-diffusion cathodes that are connected in parallel configuration and when stacked form a series configuration. The anode is "sandwiched" between the cathodes with particular attention paid to the "gap" between the anode and cathode.

The basic components of a product built using the MAC (or any other aqueous metal–air power sources) in a basic 12-V configuration would include the following:

- basic cell (12 cells in series configuration);
- anode (12 anodes with at least 80 A h capacity);
- electrolyte (water based);
- additives (added to electrolyte);
- cathode (probably with cobalt catalyst);
- tank to hold at least 3 l of electrolyte;
- pump (pump electrolyte through the cells, which helps to maintain constant temperature and prevent precipitate build up on cathodes);
- thermostat (monitor temperature of unit);
- heat sink (absorb heat generated from voltage regulator);
- fan (increase air flow through cathodes and cool unit);
- receptor (12-V receptor for user device connection);

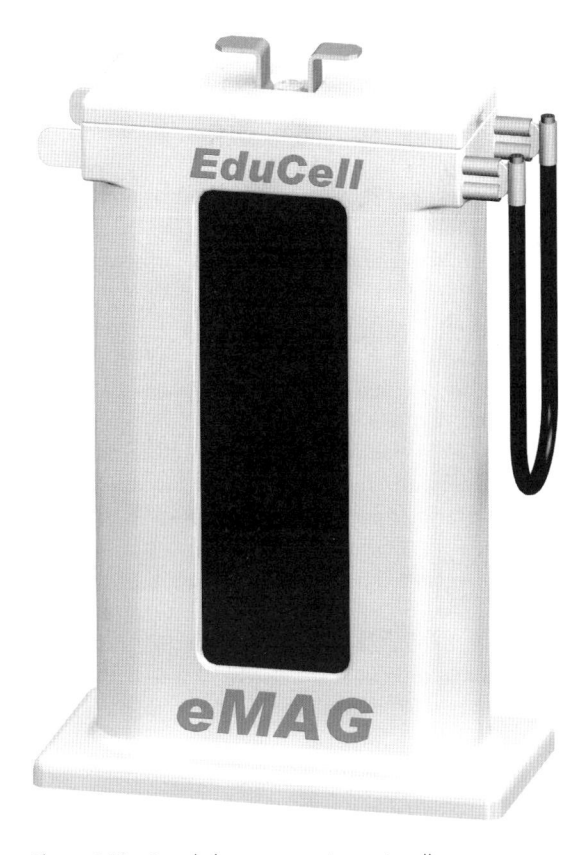

Figure 6.10 Stand-alone magnesium–air cell.

- regulator board that regulates voltage, built-in AC inverter, temperature shut off, internal connectors for on-board power to fan, and 9-V battery recharger;
- casing (internal and external structure strong enough to withstand rugged usage); and
- packaging (labeling with brochures).

The manufacturing model would use worldwide injection mold manufacturers, which will encompass approved manufacturing and assembly facilities. The objective of this is to ensure and maintain quality of product and reduce costs where appropriate.

The goals of all metal–air products should be compact involving the following:

- lightweight
- low cost
- robust
- reliable
- injection moldable.

The manufacturing of the magnesium–air battery and components have some environmental impact concerns that must be addressed for the following reasons:

- Safeguarding of the health and safety of the employees and the public and to protect the environment
  - Potential hazardous materials (material safety data sheets (MSDSs) and workplace hazardous material information system (WHMIS))
    - The MSDS are published information on hazardous material in compliance with the WHMIS.
- Marketing and advertising information
  - Packaging and markings
    - The packaging, labeling, and information materials/brochures/manual(s) should be made using nontoxic materials and recycled materials (where economically possible). This process also will not cause nontoxic materials to be formed and/or dispersed in any form. The whole process will be environmentally friendly from start to finish.
- Manufacturing operations
  - The manufacturers must be ISO registered in their various countries.
- Codes and standards
  - Products using the metal–air technology must undergo certification with respect to the following international codes and standards:
    - IEC – CSA 62282-5-1 (November 2007) that covers construction, markings, and testing for ac- and dc-type portable fuel cells.
    - IEC – CSA 62282-6-100 that covers microfuel cell power systems.
- Quality management plan that will include the following:
  - Quality planning
    - Quality strategy
    - Quality requirements
  - Quality assurance
    - Procedures
    - Methods
  - Quality control
    - Qualify participants
    - Quality control process
    - Inspections/approval
- Recyclability
  - Recycling of the products is imperative to product manufacturing and selling.
- Disposability
  - If recycling is not possible either due to location or other such means, then safe disposal of the products is imperative to product manufacturing and selling.
- Maintenance
  - Generally, a product that utilizes an aqueous metal–air power source will require periodic maintenance that should be conducted by the owner. This maintenance requires neither specialized equipment nor is there any specialized training or service contract required. Maintenance may involve

flushing the cells with water after each usage or when the anodes need replacing.

6.10
Magnesium–Air Battery Performance

Assembly, testing, and diagnosis of the MAC have been conducted in several laboratories using different types of equipment. The following diagrams show typical power performances of the MAC. It should be noted that the cells used in the test work do not represent any new and improved MAC versions that the authors may have access to.

The steep drop in cell potential at low current densities is due in large part to the polarization of the air electrode (cathode); the linear drop at higher current densities is mainly the internal resistance drop of the air electrode and electrolyte. After the initial polarization, the magnesium electrode (anode) shows essentially a constant potential over the current range shown (Figures 6.11–6.14).

The common cause of an unreasonable load is a short circuit of the battery or fuel cell. For example, a lead-acid battery will be damaged, lose its rechargeable capacity, and will experience a shortened lifetime if a short circuit occurs; this is due to generated heat and shedding of paste. Unlike the lead-acid battery,

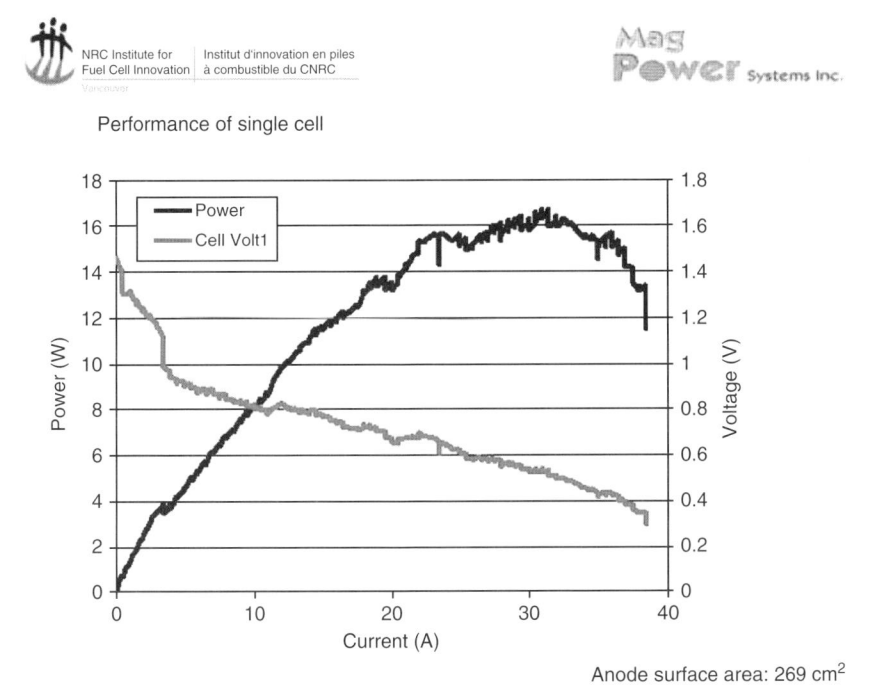

Anode surface area: 269 cm^2

Figure 6.11 Power graph, one cell, NRC Fuel Cell Institute, May 2004.

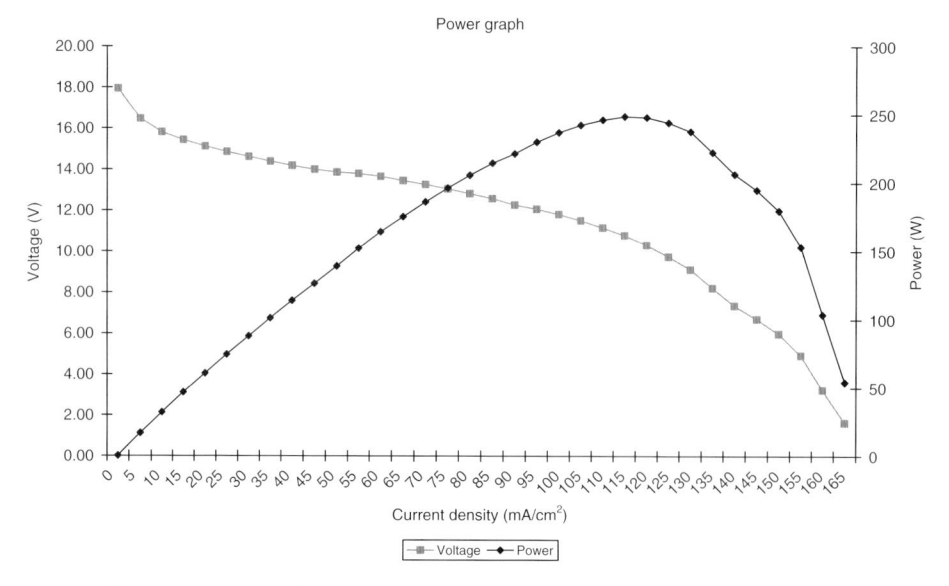

Figure 6.12 Current density data, 12-V MAFC (12-cell stack) Vizon Scitec October 2004. Power graph 12-V MAFC 5 A discharge (12-cell stack) over three days, (Vizon Scitec, November 2004).

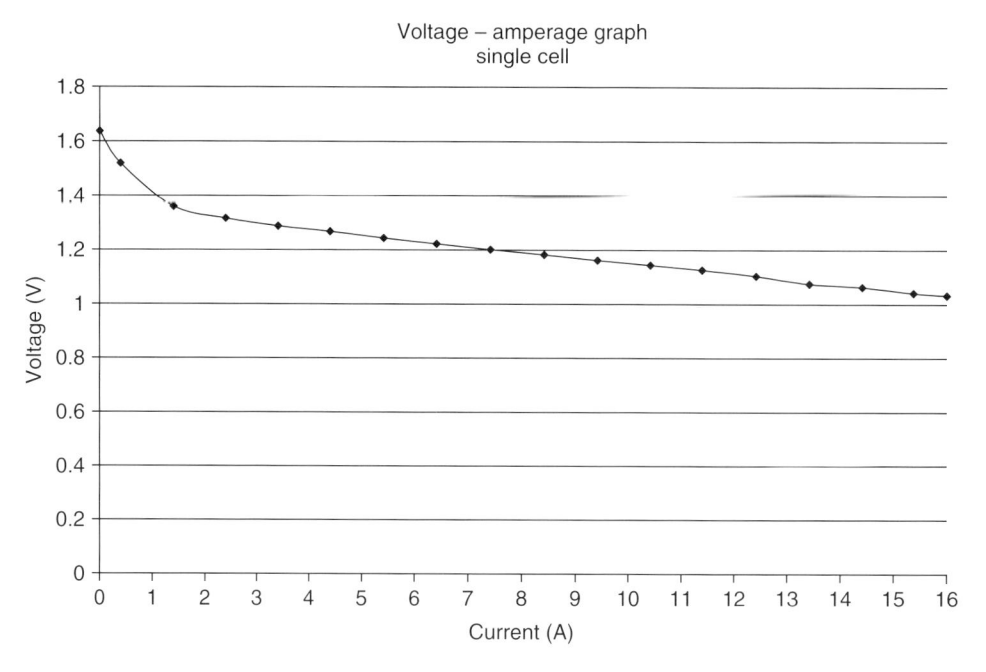

Figure 6.13 Voltage vs current (single cell) (MagPower, January 2005).

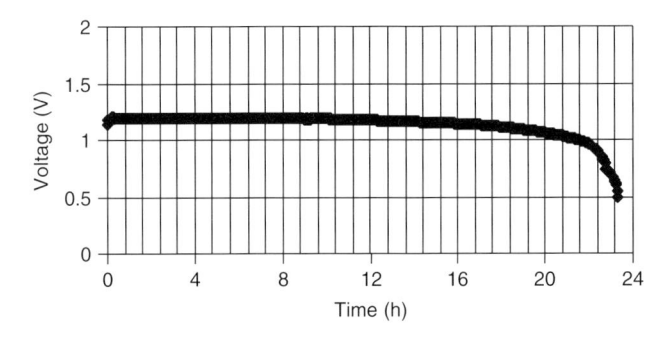

Figure 6.14 Discharge curve – 1 cell at 5 A. Energy = 108 Wh, Temperature = 55 °C, vol = 0.5 l; and wt anode = 140 g (Vizon Scitec, July 2003).

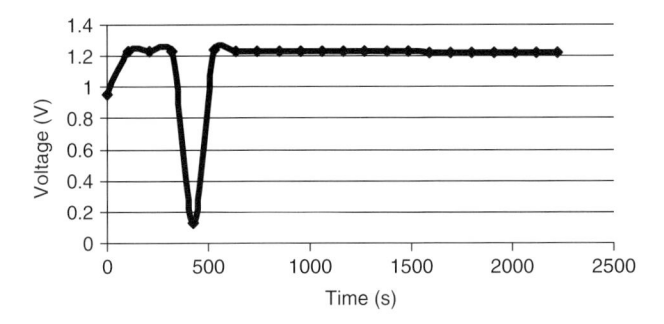

Figure 6.15 Short circuit – 1 cell (5 A discharge) short = 212 s (Vizon Scitec, July 2003).

the MAC will only lose some capacity because of the waste consumption of the magnesium anode during the short circuit. No damage of the MAC components, or loss of future capacity, will be affected. The following graph shows the response of a single MAFC cell under short circuit conditions. The MAC recovers to its operating voltage right after the short circuit is dissolved (Figures 6.15 and 6.16).

6.11
Degradation Mechanisms and Mitigation Strategies

The magnesium–air technology has been limited in the development because of several technological barriers that has a profound impact upon the commercialization of the technology.

6.11.1
Hydrogen Problem

Hydrogen gas is generated by the spontaneous reaction of magnesium with water and magnesium metal which is thermodynamically unstable in water with

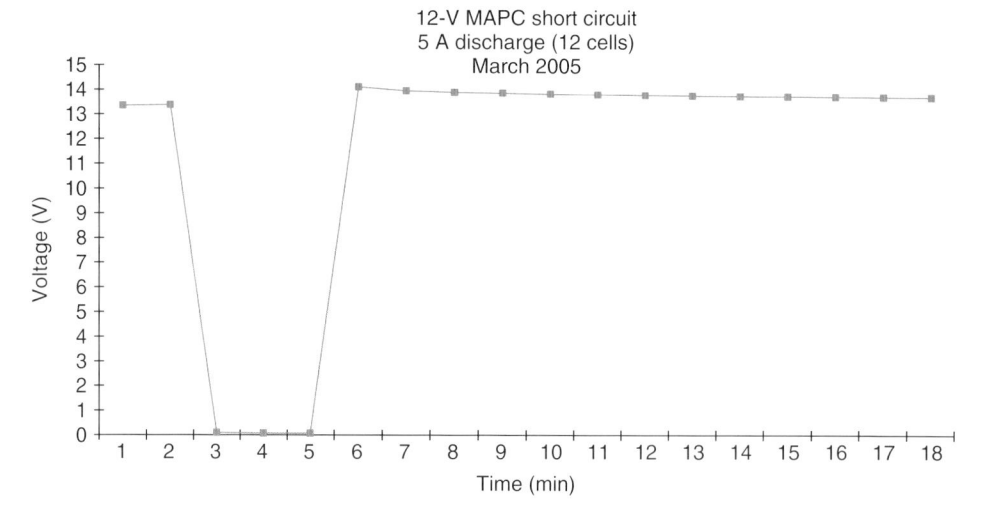

Figure 6.16 Short circuit – 12 cells (5 A discharge) (MagPower, March 2005).

respect to the formation of magnesium hydroxide and hydrogen. This reaction is exothermic which consumes magnesium. Low electrolyte pH may also enhance hydrogen evolution, thus it is desirable to have an electrolyte pH close to neutral.

The sensitivity of hydrogen evolution to pH can be explained by the following reactions:

Anode reaction:

$$Mg - 2e^- + 2H_2O = Mg(OH)_2 + 2H^+ \tag{6.3}$$

Parasitic cathode reaction on anode:

$$Mg + 2e^- + 2H^+ = MgH_2 \tag{6.4}$$

$$MgH_2 = Mg + H_2 \tag{6.5}$$

Overall parasitic reaction:

$$Mg + 2H_2O = Mg(OH)_2 + H_2 \tag{6.6}$$

These reactions suggest that the way to stop parasitic decay of Mg by reaction (6.4) is by removing H^+ species with proton scavengers.

Hydrogen evolution is a secondary reaction that occurs on the anode. The MAC performance improves when hydrogen evolution is diminished resulting in higher magnesium utilization efficiency. The minimization of hydrogen evolution that accompanies the magnesium electro-dissolution is an important issue for the successful development of the MAC. Thus, a hydrogen inhibitor (HI) additive(s) needs to be used in the MAC, which inhibits hydrogen evolution on the anode surface. This can lead to deterioration of the anode surface.

The influence of an HI on the anodic polarization of magnesium is an essential part in the control of hydrogen generation in the magnesium–air–water interactions.

The reduction and management of hydrogen are important for several reasons:

- Reduce internal pressure in the cell and reduce water electrolysis.
- Reduce and control hydrogen from reaction in order to achieve greater energy density. This can be accomplished by reducing hydrogen bubbles and reduction of magnesium hydroxide precipitate, both of which add to the internal resistance of the cell.
- Reduce hydrogen gas bubbles and thereby reduce volume of cell. There is a measurable increase in cell volume in order to accommodate the gas bubbles.
- Hydrogen generation decreases the magnesium utilization efficiency of the MAC.

6.11.2
Magnesium Utilization Efficiency (MUE)

Various papers [46–49] have been published regarding the efficiency of magnesium and magnesium alloy materials.

The magnesium utilization efficiency of the MAC is measured via the amount of electric charge passed (i.e., electricity generated) over the weight loss of the magnesium anode and is referred to as the *magnesium utilization efficiency*.

$$\text{MUE (\%)} = \frac{I \times \text{Ael} \times t \times \text{Mgw}}{2 \times F \times m \times \text{at}} \times 100$$

where

I = current (A)
Ael = anode geometric area (m^2)
t = time (s)
Mgw = Mg atomic weight (24.3)
F = Faraday's number (96485.3)
m = weight loss (g)
at = anode type
that is, anode type:
  AZ31 = 0.96
  AZ61 = 0.93

An HI additive can be used to help suppress hydrogen gas (and increase the magnesium utilization efficiency of the anode) that is naturally formed at the anode. It is added to the electrolyte, and should be nontoxic, water soluble, and nonpolluting to the environment. Figures 6.17 and 6.18 show the effect of an HI added to the saltwater-based electrolyte. With the HI, the magnesium utilization efficiency of anode in the MAC is 70–75%, whereas without the HI the magnesium utilization efficiency of anode is 45–55%.

6.11.3
Electrolyte

The water portion of the electrolyte is consumed and needs to be replaced over a period of time, which depends on the load of the power device. Magnesium

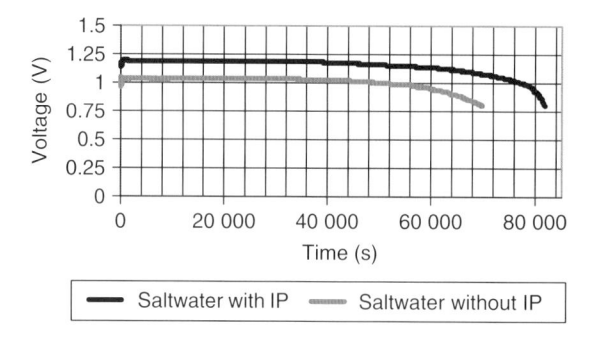

Figure 6.17 Effect of HI additive in electrolyte (one cell at 5-A discharge, anode type – AM60B, active anode face = 99 cm^2, active cathode face = 108 cm^2). Salt concentration 10% (Vizon Scitec, July 2003).

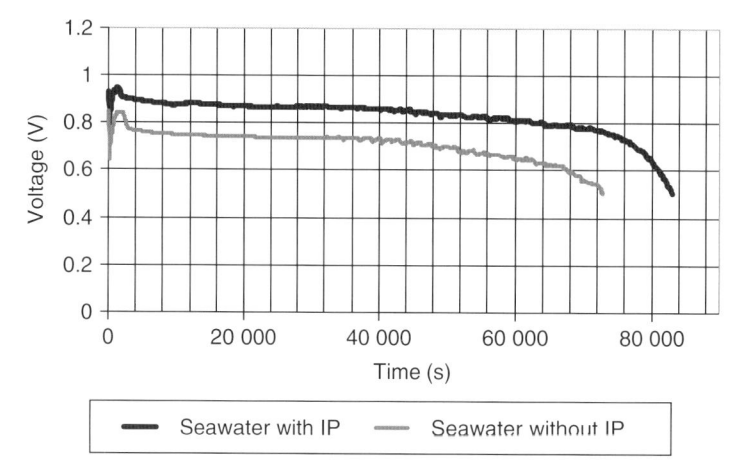

Figure 6.18 Effect of HI additive in seawater 3.2% salt concentration (Vizon Scitec, July 2003) (one cell – 5-A discharge, anode type – AM60B, active anode face = 99 cm^2, active cathode face = 108 cm^2).

discharging into the electrolyte combines with hydroxide and other ions (i.e., sodium, chloride) to form various magnesium reaction products as identified by X-ray diffraction. The electrolyte needs to be replaced before reaching the solubility limits of the magnesium reaction products. There will be an increase in cell resistance with increasing mass of reaction products resulting in an MUE decrease. These reaction products can also adhere to the cathode thereby plugging the pores through which air passes, which will cause a reduction in the amount of oxygen passing through the air-diffusion cathodes and thus a potential decrease in the MUE. Pumping of the electrolyte in certain commercial products will help to alleviate the stagnation problem of stand-alone cells.

6.11.4
Cathode

One of the limiting factors is the difficulty in developing cost-effective, simple, reliable cathode structures, which deliver high performance, optimize cathode catalyst recipe specifications, optimize cathode mass transport architecture structure, and allow economic manufacturing processes. For instance, current commercially developed air-diffusion cathodes typically have problems of high cost, high internal electrical resistance, and corrosion to metal current collector layer in the alkaline or neutral electrolyte environments. Most of the developed air-diffusion cathodes for metal–air batteries are made for the alkaline electrolyte environment, which may not be suitable for neutral or salt (i.e., sodium chloride) electrolyte environments. New one-layer cathodes are being developed, which can be cost effective on large scale manufacturing volume. The cost of the catalyst, whether it is cobalt- or manganese-based, has a direct bearing on the cathode cost. Typically, the cathode cost is approximately one-third the cost of an MAC. The choice of the catalyst should be made on the design and power requirements of the cell. If the current is less than 1 A, then a manganese catalyst should be used.

In the design of larger systems, a fan would provide constant air flow, thereby eliminating any air stagnation problem and improving oxygen reduction reaction.

6.11.5
Precipitate

A significant buildup of precipitate on the anode and cathode surfaces produce poor performance. Precipitate can block air flow through the cathode and create a short circuit between the cathode and anode. Precipitate on the anode can also block electrolyte contact with the magnesium anode, inhibiting conductivity and slowing the anode consumption rate. In this case, the magnesium efficiency could be inaccurate as the precipitate forms a barrier between the electrolyte and anode surface.

Precipitate buildup will reach a critical stage as water is consumed and the electrolyte becomes saturated and reaction product(s) starts to precipitate as a solid. The electrolyte can be refreshed with the addition of water and/or simply changed with new electrolyte. The magic "elixir" would be an additive that binds magnesium ion from combining with the hydroxide ion to forming magnesium hydroxide.

6.12
Applications

The magnesium–air technology as an alternative power source is generally limited up to 5 kW, though no commercial product has been manufactured to date beyond 100 W. The ability to penetrate the power markets with the magnesium–air technology is huge but there are essentially no commercial MAC power products in the market at present. There are numerous theoretical applications but these applications must be transformed into commercially acceptable cost-effective products. This technology will be new to the consumer and as such must be introduced on a "first to market basis," which will include some education and mass marketing. The variable cost to all products is the magnesium, which can vary with world market conditions. There are aqueous batteries [50] in the present market but which have an insignificant market share to date.

The consumer oriented products must be of the following:

- environmentally safe
- nontoxic
- recyclable
- indefinite storage
- easily transportable
- consumer friendly
- cost competitive with other products.

Specific applications that magnesium–air technology will be most applicable to include the following but not limited to:

- value-added power device for commercial applications;
- power device for emergency and disaster situations;
- power device for data loggers and battery chargers;
- camping and recreational vehicle (RV) products;
- education kits and power toys;
- backup power; and
- marine (surface and underwater applications).

Some of the performance claims of the technology are the following:

- indefinite shelf life
- nontoxic electrolyte
- nonhazardous disposal materials
- nontoxic fuel (anode)
- nontoxic cathode
- no green house gas emissions
- fuel flexibility
- electrolyte flexibility
- quiet.

6.13
Challenges and Perspectives of Magnesium–Air Cells

The development progression is to develop a mechanically and automatically rechargeable MAC using

- cost-effective metal anode;
- cost-effective cathode materials and structures as required to achieve reliability and efficiency goals; and
- conductive electrolyte.

This is done to establish the economic competitiveness of the technology, and to develop and integrate systems utilizing the MAC.

Another key challenge is market acceptance. The success of this technology-driven products hinges on whether or not consumers will find the benefits of these technology products to be sufficiently attractive at the price. Because this technology is fundamentally so different from any other power source available, educating potential buyers will be the key promotional challenge.

As with all potential applications, there needs to be a market analysis for the various applications that would include the following:

- existing metal–air market scenario
- existing magnesium–air market scenario
- market metal–air forecasts
  - Market magnesium–air forecasts
- Market drivers
  - market restraints
  - pricing analysis
  - market trends
    - North America
    - Europe
    - Asia
    - Rest of the world
- Identification of metal–air markets (potential applications/products);
- Identification of magnesium–air markets (potential applications/products);
- Identification of industry challenges; and
- Competitive technologies
  - Advantages
  - Disadvantages
- Magnesium market and magnesium producers
  - Forecast/supply/demand.

References

1. (2002) WP ST 9-Metal/Air. Institute for Environment and Sustainability Renewable Energies Unit, Ispra Site, Italy, Storage Technology Report.
2. Linden, T. and Reddy, B. (eds) (2002) *Handbook of Batteries*, 3rd edn, McGraw-Hill, New York.
3. Kaisheva, A. (2005) Metal-air batteries: research, development, application. Proceedings of the International Workshop: Portable and Emergency Energy Sources–from Materials to Systems, Primorsko, Bulgaria, September 2005.

4. Aral, H. and Hayashi, M., (2009) *Encyclopedia of Electrochemical Power Sources*, vol. **4**, Elsevier, pp. 347–365.
5. Abraham, K.M. (2008) A brief history of non-aqueous metal-air batteries. *ECS Trans. (Electrochem. Soc.)*, **3** (42), 67–71.
6. Aurbach, D., Suresh, G.S., Mizrahi, O., Mitelman, A., Levi, E., and Levi, M.D. (2008) New rechargeable magnesium battery systems. 213th ECS Meeting, Abstract #289, The Electrochemical Society.
7. Dobley, A., Joseph, D., and Abraham, K.M. (2004) Non-aqueous lithium-air batteries with an advanced cathode structure. The Proceedings of the 41st Power Sources Conference, Philadelphia, PA.
8. Egashira, M. (2009) *Iron-Air (Secondary and Primary), Encyclopedia of Electrochemical Power Sources*, vol. **4**, Elsevier, pp. 372–375.
9. Lee, J.-S., Kim, S.T., Cao, R., Choi, N.-S., Liu, M., Lee, K.T., and Cho, J. (2011) Metal–air batteries with high energy density: Li–air versus Zn–air. *Adv. Energy Met.*, **1** (1), 34–50.
10. Energy Storage (2002) Refocus, Elsevier Advanced Technology Article, March/April, 2002.
11. Tuck, C.D.S. (ed.) (1991) *Modern Battery Technology*, Ellis Horwood Publishers, pp. 487–502.
12. Fitzpatrick, N. and Scamans, G. (1986) Aluminum is a Fuel for Tomorrow. *New Scientist*, **Jul 17**, 34–39.
13. Visco, S.J., Nimon, E., and De Jonghe, L.C. (2009) *Lithium-Air, Encyclopedia of Electrochemical Power Sources*, vol. **4**, Elsevier, pp. 378–383.
14. Kowalczk, I., Read, J., and Salomon, M. (2007) Li-air batteries: a classic example of limitations owing to solubilities. *Pure Appl. Chem.*, **79** (5), 851–860.
15. Abraham, K.M. and Jiang, Z. (1996) A Polymer electrolyte-based rechargeable Lithium/Oxygen Battery. *J. Electrochem. Soc.*, **143**, 1.
16. Crowther, O., Meyer, B., Morgan, M., and Salomon, M. (2011) Primary li-air cell development. *J. Power Sources*, **196**, 1498–1502.
17. Visco, S.J., Nimon, E., Katz, B., Chu, M., and De Jonghe, L.C. (2009) *Lithium/Air Semi-Fuel Cells: High Energy Density Batteries based on Lithium Metal Electrodes*, PolyPlus Battery Company, Scalable Energy Storage: Beyond Li-Ion Almaden Institute, August 26, 27, 2009.
18. Beattie, S.D., Manolescu, D.M., and Blair, S.L. (2009) High-capacity lithium-air cathodes. *J. Electrochem. Soc.*, **156** (1), A44–A47.
19. Zhang, X.G. (1996) *Corrosion and Electrochemistry of Zinc*, Springer.
20. Neburchilov, V., Wang, H., Martin, J.J., and Qu, W. (2010) A review on air cathodes for zinc–air fuel cells. *J. Power Sources*, **195**, 1271–1291.
21. Zhang, X.G. (2009) *Zinc Electrodes, Section in Encyclopedia of Electrochemical Power Sources*, Elsevier Ltd.
22. Haas, O. and van Mesemael, J. (2009), *Zinc-Air: Electrical Recharge, Encyclopedia of Electrochemical Power Sources*, vol. **4**, Elsevier, pp. 384–392.
23. Li, X., Zhu, A.L., Qu, W., Wang, H., Hui, R., Zhang, L., and Zhang, J. (2010) Magneli phase Ti_4O_7 for electrode oxygen reduction reaction and its implication for zinc-air rechargeable batteries. *Electrochim. Acta*. doi: 10.1016/j.electacta.2010.05.041
24. Smedley, S. and Zhang, X.G. (2009) *Zinc-Air: Hydraulic Recharge, Encyclopedia of Electrochemical Power Sources*, vol. **4**, Elsevier, pp. 393–403.
25. Hamlen, R., Jerabek, E.C., Ruzzo, J.C., and Siwek, E.G. (1969) Anodes for refuelable magnesium-air batteries. *J. Electrochem. Soc.*, **116** (11), 1588–1192.
26. Hasvold, O. et al. (1997) Sea-water battery for subsea control systems. *J. Power Sources*, **65**, 253–261.
27. Sathyanarayana, S. et al. (1981) A new magnesium-air cell for long life application. *J. Appl. Electrochem.*, **11**, 33–39.
28. Wilcock, W. and Kauffman, P.C. (1997) Development of a seawater battery for deepwater applications. *J. Power Sources*, **66**, 71–75.
29. Kent, C.E. and Carson, W.N. (1966) unpublished, inhouse reports *Magnesium-Air Cells*, Technical Information Series, Research and Development Center, General Electric.
30. Downing, B.W. (2003–2008) unpublished MagPower Systems Inc., reports.

31. Ficara, P. et al. (1998) Magnola: a novel commercial process for the primary production of magnesium. *CIM Bull.*, **91** (1019), 75–80.
32. Song, G.L. and Atrens, A. (1999) Corrosion mechanisms of magnesium alloys. *Adv. Eng. Mater.*, **1** (1), 1438–1656.
33. Weber, C.R., Knörnschild, G., and Dick, L.F.P. (2003) The negative-difference effect during the localized corrosion of magnesium and of the AZ91hp alloy. *J. Braz. Chem. Soc.*, **14** (4), 584–593. (Printed in Brazil - ©2003 Sociedade Brasileira de Química).
34. Atrens, A. and Dietzel, W. (2007) The negative difference effect and unipositive Mg+. *Adv. Eng. Mater.*, **9** (4), 292–297.
35. Liu, L.J. and Schlesinger, M. (2009) Corrosion of magnesium and its alloys. *Corros. Sci.*, **51**, 1733–1737.
36. Guadarrama-Munoz, F., Mendoza-Flores, J., Duran-Romero, R., and Genesca, J. (2006) Electrochemical study on magnesium anodes in NaCl and $CaSO_4$–$Mg(OH)_2$ aqueous solutions. *Electrochim. Acta*, **51**, 1820–1830.
37. Vuorilehto, K. (2003) An environmentally friendly water-activated manganese dioxide battery. *J. Appl. Electrochem.*, **33**, 15–21.
38. Riegel, H., Mitrovic, J., and Stephan, K. (1998) Role of mass transfer on hydrogen evolution in aqueous media. *J. Appl. Electrochem.*, **28**, 10–17.
39. Bakke, P., Svalestuen, J.M., and Albright, D. (2002) Metal Quality–The Effects on Die Castors and End Users, Society of Automotive Engineers, paper 2002-01-0078.
40. Shaw, B.A. (2003) Corrosion resistance of magnesium alloys, *ASM Handbook*, Corrosion Fundamentals, Testing and Protection, Vol. **13A**. ISBN: 0871700190.
41. Parthiban, G.T., Palaniswamy, N., and Sivan, V. (2009) Effect of manganese addition on anode characteristics of electrolytic magnesium. *Anti-Corrosion Meth. Mater.*, **56**/2, 79–83.
42. Zanotto, F. (2009) Corrosion behaviour of the AZ31 magnesium alloy and surface treatments for its corrosion protection. PhD thesis. Universit degli Studi di Ferrara, Corrosion Study Centre, Italy.
43. Eliezer, D. and Alves, H. (2002) *Corrosion and Oxidation of Magnesium Alloys, Handbook of Materials Selection*, John Wiley & Sons, Inc., pp. 267–291.
44. Liu, M., Uggowitzer, P.J., Nagasekhar, A.V., Schmutz, P., Easton, M., Guang-Ling, S., and Andrej, A. (2009) Calculated phase diagrams and the corrosion of die-cast Mg-Al alloys. *Corros. Sci.*, **51**, 602–619.
45. Gonzalez-Torreira, M., Fones, A., and Davenport, A.J. (2003) Passivation and dissolution of magnesium. *J. Corros. Sci. Eng.*, **6**, paper C034, pp. 21–26.
46. Gummow, R.A. (2004) *Performance Efficiency of High-Potential Mg Anodes for Cathodically Protecting Iron Water Mains*, CORRENG Consulting Service, Inc., May 2004 Materials Performance.
47. Yunovich, M. (2004) *Performance of High Potential Magnesium Anodes: Factors Affecting Efficiency*, CC Technologies Laboratories, Inc.
48. Campillo, B., Rodriguez, C., Juarez-lslas, J., Genesca, J., and Martinez, L. (1996) An improvement of the anodic efficiency of commercial Mg anodes. Corrosion 96 Conference, paper 201.
49. Di Gabriele, F. and Scantlebury, J.D. (2003) Corrosion Behaviour of Magnesium Sacrificial Anodes in Tap Water, 2003 Corrosion and Protection Centre.
50. Sammoura, F., Bang Lee, K., and Lin, L., (2004) Water-activated disposable and long shelf life microbatteries. *Sens. Actuators A*, **111**, 79–86.

7
Liquid Redox Rechargeable Batteries

Huamin Zhang

7.1
Introduction

7.1.1
Summary

With the development of society, energy plays a crucial role in abating climate change while also facilitating sustainable development in our lifetime. Since most fossil energy resources cannot be recycled and produce serious emissions of polluting gases, new energy sources, especially clean energy, is in great demand. Thus, renewable energy sources, such as solar and wind power, for electricity generation become more and more important.

The random and intermittent nature of renewable energy sources, such as solar, wind, and tide, lead to a lower quality of electricity produced, and even affect the stability of the grid, so energy storage is needed to solve these problems. This is believed to be an important part of the entire power system because of its role in improving electricity utilization efficiency. Most importantly, energy storage is a key technique for increasing the compatibility of renewable energy with the grid and constructing a smart grid.

7.1.2
Background

To date, many kinds of energy storage technologies have been developed. Depending on the type of energy, the storage technologies can be divided into physical energy storage and chemical energy storage, as shown in Table 7.1, which gives a classification of energy storage technologies.

Physical energy storage mainly includes pumping water, compressing air, flywheels, and superconducting magnets. Chemical energy storage options include lead-acid and redox flow battery (RFBs), secondary batteries (lithium ion and nickel metal hydride), and the sodium–sulfur battery. Each type of technology has its own

---

Electrochemical Technologies for Energy Storage and Conversion, First Edition. Edited by Ru-Shi Liu, Lei Zhang, Xueliang Sun, Hansan Liu, and Jiujun Zhang.
© 2012 Wiley-VCH Verlag GmbH & Co. KGaA. Published 2012 by Wiley-VCH Verlag GmbH & Co. KGaA.

Table 7.1 The classification of energy storage technologies.

Classification	Energy transformed forms	Energy storage technologies
Physical energy storage	Potential energy ↔ electricity energy	Pumping water
		Compressing air
	Kinetic energy ↔ electricity energy	Flywheel
	Electromagnet energy ↔ electricity energy	Superconducting magnet
Chemical energy storage	Chemical energy ↔ electricity energy	Sodium–sulfur battery
		Lead-acid battery
		Secondary batteries (Lithium ion battery nickel-metal hydride battery)
		Redox flow battery (RFB)
		Supercapacitors

limitations, which can make it practical or economical for some applications and over a limited range.

Table 7.2 shows the main advantages and disadvantages of these energy storage technologies. Compressed-air and pumped-water storage units are suitable for large-scale systems, but require special sites. In addition, high investment costs and a long construction period both limit their wide application. Lead-acid batteries are a mature, safe, and low-cost technology and are widely used. However, their short lifetime and low deep discharge ability make their use doubtable. The sodium–sulfur battery (NaS) is a new technology that has been given much attention in recent years. Its higher specific power density makes it more competitive. However, its safety is a big concern because this battery is normally operated at temperatures as high as 350 °C with highly combustible sodium as its active ingredient. The RFB represents a new electrochemical energy storage method and has many attractive features, such as independent design of the power and energy capacity, high efficiency, and long cycle life. The RFB is well suited for a variety of applications. This chapter focuses on the RFB, especially the vanadium redox battery (VRB).

7.1.3
Development

Depending on the different active species in the positive and negative half cells, RFBs can be classified into the following main types: the VRB; the sodium polysulfide/bromine, zinc/bromine, and iron/chromium flow batteries; the soluble lead-acid battery; and the zinc/nickel flow battery.

Table 7.2 Advantages and disadvantages of major energy storage technologies.

Energy storage technologies	Advantages	Disadvantages
NaS	High energy density, power density	Thermal management, poor security
Li ion battery	High energy density, power density, and efficiency	High cost, poor security, short cycle life
Lead-acid battery	Mature technology, low cost	Short cycle life, environmental pollution
Supercapacitors	High power density, long cycle life	High cost, low energy density
Pumping water	High capacity, low cost	Special geographical condition
Compressing air	High capacity, low cost	Special geographical condition
Flywheel	High power density, rapid response	Vacuum operation, high cost
Superconducting magnet	High power density, rapid response	Low-temperature operation, high cost
Redox flow battery	Independent design of the power and energy capacity, high capacity, and long cycle life	Low energy density

The development of each of these RFBs is described briefly.

- **Vanadium redox battery (VRB)** Since 1984, the fundamental scientific issues concerning the VRB, such as electrolyte stability, kinetics of the redox couples, and mechanisms of active substrate transmission, have been widely researched by Skyllas-Kazacos's group. They then successfully developed a 1–4 kW VRB system, which provided a good foundation for VRB research and development [1].
- Nowadays, investigation of VRBs is mainly focused on two areas:
   a. Development and production of key materials with higher performance and at lower cost.
   b. Use of the battery for utility peak shaving in combination with wind power generators and solar PV systems.
- **Zinc/bromine flow battery (ZBB)** The zinc/bromine flow battery (ZBB) has also received wide attention because of its good energy density, high voltage, and low cost. However, zinc and bromine are converted to zinc and bromide ions on the respective electrodes during discharge. Two significant problems of the ZBB were overcome by the Exxon and Gould companies between the mid-1970s and early 1980s, namely, the formation of zinc dendrites during electrodeposition and high bromine solubility. The former problem can lead the cell to short circuit, and the latter can depress the battery's efficiency [2]. Since then, Japan, Europe, and the United States have been promoting ZBB research. MITI (Ministry of International Trade and Industry, Japan) initiated *Project Moonlight* in 1980, and the ZBB was listed as an important energy storage technology. Following this, 1, 10, and 60 kW battery modules were assembled step by step to progress the technology. In the mid-1980s, Sandia National Laboratory (the United States) joined Johnson

Controls Inc. (the United States) and the subsequent ZBB Company (the United States) promoted systematic research on key battery materials, system design and optimization, and system validation and demonstration.

- The initial commercial zinc/bromine battery produced by the ZBB Company is available, with an energy density of 70 Wh kg^{-1}. To accelerate commercial development, most researchers have focused on electrode materials with good performance and long lifetime, including high-resistance bromine, a low-electrical-resistance separator, and bromine-complexing agents with high complexing capability.
- **Fe/Cr flow battery** Japan and the United States initiated research into the Fe/Cr flow battery in the 1970s. The concept of the Fe/Cr flow battery was first proposed by L.H. Thaller in the NASA Lewis Research Center. The Fe/Cr flow battery development plans were promoted in 1979 by NASA to verify the feasibility and practical application of this technology. In 1980, Japan began to develop an Fe/Cr flow battery as a part of Project Moonlight. During the period 1985–1990, Sumitomo Electric Industries (SEI) successfully assembled a 10 kW battery module [3].

Although the Fe/Cr flow battery has been under investigation for more than a decade, with the investment of considerable effort and resources, several challenges are still not resolved. Unusually, a follow-up study was carried out on the Fe–Cr flow battery in the 1990s. The main problems are as follows:

   a. First, there is an energy system loss during charge and discharge, derived from the poor reversibility of the Cr^{2+}/Cr^{3+} redox couple. Even though the operating temperature is improved by using an electrocatalyst, it is difficult to obtain good battery performance.
   b. Second, there is no ideal battery separator. Crossover of ions takes place through the membrane separator, reducing the battery's coulombic efficiency, energy storage capacity, and lifetime.

- **Sodium polysulfide/bromine flow battery** The sodium polysulfide/bromine flow battery uses the Br^-/Br_2 redox couple as the positive reactant and the S_x^{2-}/S_{x+1}^{2-} as the negative reactant. The Remick Company in the United States first proposed the concept of a sodium polysulfide/bromine flow battery in 1984. The Innogy Company in the United Kingdom continued research into this technology during the 1990s. This company successfully developed 5, 20, and 100 kW battery modules and registered a trademark.
- The Innogy Company planned to supply 15/120 MWh systems for Little Barford Power Station in the United Kingdom and a 12/120 MWh system for the Tennessee Valley Authority (TVA) in the United States. However, both projects were abandoned, indicating that the technology remains immature. The major problems are as follows:

   a. System performance degradation induced by positive and negative active material crossover.
   b. Deposition of sulfur and sodium sulfate during operation, which reduces system lifetime and increases maintenance costs.

c. Large-scale sodium polysulfide/bromine flow battery systems increase the risk of a bromine vapor leak.
- **Vanadium/bromine flow battery** Skyllas-Kazacos in the University of New South Wales has proposed a vanadium/bromine flow battery. The energy density of this battery can achieve 50 Wh kg^{-1} because of the high electrolyte concentration. Following this, a vanadium/polyhalide flow battery has been proposed. In this battery, polyhalide is used instead of bromine as the positive electrode [4].
- However, with improving energy density of the vanadium/polyhalide flow battery, the cross-contamination by vanadium becomes another problem. The key problem to be solved is how to balance improving the energy density with minimization of cross-contamination. This will determine whether the vanadium/polyhalide flow battery can be of practical use.
- **Zinc/cerium flow battery** The zinc/cerium flow battery proposed by Plurion Systems Inc. in the United States employs the Ce^{3+}/Ce^{4+} redox couple as the positive reactant and the Zn^{2+}/Zn redox couple as the negative reactant. The discharge voltage of the battery is high (>1.8 V) resulting in an ideal energy density, and it can be operated at high current density, further improving the power density [4]. Although the company has designed 65 and 250 kW battery systems, there are still several challenges to be resolved:
  a. How to control the appearance and distribution of zinc.
  b. The development of nonprecious metal electrochemical catalysts.
  c. Inhibition of oxygen and hydrogen evolution during high-voltage operation.
- **Single flow battery** As early as 1946, researchers used perchloric acid as the electrolyte to assemble a soluble lead-acid battery. In the 1970s, the electrolyte was replaced by tetrafluoroborate. In 2004, Plectcher and others proposed the use of methanesulfonic acid as the electrolyte. The energy efficiency of this battery reaches 60% when charged at 20 mA cm^{-2}. This technology shows the advantages of low cost and simple design, since no membrane separator but only an electrolyte tank is needed [4]. However, some complex side reactions, such as phase transitions, usually take place during charge and discharge, which affects the battery reliability and life. Dendrite formation during charge is likely to result in a short circuit, and the cost increases with the use of methanesulfonic acid as electrolyte.
- The zinc/nickel flow battery uses the $Ni(OH)_2/NiOOH$ redox couple as the positive reactant and Zn/Zn^{2+} as the negative reactant, making full use of the high solubility of zinc oxide in an alkaline environment. The advantage of this battery is that it does not need an ion exchange membrane, which reduces the system cost. When the degree of discharge is 80%, the battery can operate for >1000 cycles, and the coulombic and energy efficiencies of the battery can reach 96 and 86%, respectively. Nevertheless, the system capacity is limited by the specific capacity of nickel oxide. When the battery is used as a large-scale energy storage system, the high price of nickel will be a significant issue [5].

7.2
Electrochemical Processes in a Redox Flow Battery

7.2.1
Basic Electrochemical Principles

7.2.1.1 Apparent Current Density

Current density is one of the most important parameters for electrochemical reactions; it directly indicates the rate of reaction. The magnitude of the current density equals the current intensity divided by the electrode area.

RFBs usually use a porous electrode as the catalytic medium. However, the surface area of porous electrodes is difficult to calculate, so the current density is estimated on the basis of geometric area instead, and the current density obtained is termed the *apparent current density*.

The value of the apparent current density is in direct proportion to the reaction rate of the active species. The amount of reacted active species and the charge–discharge quantity of electricity follow Faraday's first law. Taking VRB as an example, when it discharges 1 Ah electricity, 1.9 g V(δ) will be converted to V(χ) at the positive electrode of each battery in the VRB stack and, at the same time, 1.9 g V(χ) is converted to V(χ) at the negative electrode (using the mass of vanadium).

7.2.1.2 Electromotive Force and Open Circuit Potential

Electromotive force is defined as the difference between the equilibrium potentials of the anode and cathode at zero current. For the RFB, the equilibrium potentials of both electrodes change with factors such as the content and chemical composition of the active species in the electrolyte. The change is governed by Nernst's law. Taking VRB as an example,

Anode equilibrium potential:

$$\varphi_+ = \varphi_+^\theta + \frac{RT}{nF} \ln \frac{a_{VO_2^+} a_{H^+}^2}{a_{VO^{2+}}}$$

Cathode equilibrium potential:

$$\varphi_- = \varphi_-^\theta + \frac{RT}{nF} \ln \frac{a_{V^{3+}}}{a_{V^{2+}}}$$

Electromotive potential:

$$\varphi = \varphi_+ - \varphi_- = \varphi^\theta + \frac{RT}{nF} \ln \frac{a_{VO_2^+} a_{V^{2+}} a_{H^+}^2}{a_{VO^{2+}} a_{V^{3+}}} \approx \varphi^\theta + \frac{RT}{nF} \ln \frac{SOC^2 a_{H^+}^2}{(1-SOC)^2}$$

where $a_{VO_2^+}$, $a_{VO^{2+}}$, $a_{V^{3+}}$, $a_{V^{2+}}$, and a_{H^+} are the activities of vanadium ions with different valences and protons; n is the electron transfer number; and SOC is the state of charge.

Open circuit potential (OCP) is the potential difference between the anode and cathode of a battery when it is in the open circuit state. Since the OCP is a measured value, and the anode and cathode may not be in thermodynamic equilibrium, OCP is always smaller than, but still very close to, the electromotive potential of the battery. For VRB, OCP can be used in conjunction with the concentration of hydrogen ions in the positive electrolyte to estimate the battery's SOC.

7.2.1.3 Polarization of Redox Flow Battery

The working voltage of an RFB is always higher than its open circuit voltage (OCV) during charge and always lower than its OCV during discharge. This phenomenon results from the polarization of the battery, which includes ohmic polarization, electrochemical polarization, and concentration polarization. These three types of polarizations are coupled together and lead to the voltage loss of an RFB in practical use [6], which can be summarized as follows:

Battery charging:

$$V = E' + \Delta\varphi_r + \Delta\varphi_t + Ir$$

Battery discharging:

$$V = E' - \Delta\varphi_r - \Delta\varphi_t - Ir$$

where V is the terminal voltage, E' is the open circuit voltage, $\Delta\varphi_r$ is the overpotential caused by electrochemical polarization, $\Delta\varphi_t$ is the overpotential caused by concentration polarization, and Ir is the overpotential caused by ohmic polarization.

Ohmic polarization is caused by the internal resistance of batteries, which is closely related to the dielectric properties of the battery materials (such as ion exchange membrane, current collectors, and electrodes). The ohmic overpotential Ir is numerically equal to the product of the battery internal resistance and current.

Electrochemical polarization is a phenomenon that describes the difference between the electrode potential and the equilibrium potential when a current is applied. It is caused by the low speed of electron transfer between the electrode and the electrolyte.

The overpotential caused by electrochemical polarization can be calculated using Tafel equation:

$$\Delta\varphi_r = -\frac{RT}{\alpha nF} \ln i_0 + \frac{RT}{\alpha nF} \ln i$$

where i_0 is the exchange current density; i, the apparent current density; and α, the symmetry coefficient.

Concentration polarization is caused by the concentration gradients of the reactants near the electrodes, where the concentration changes continuously as a result of the electrochemical process.

For an RFB, the concentration polarization becomes more apparent with the increased consumption of reactants. There are two main kinds of concentration polarization:

1) When the reaction product is insoluble and exists as an independent phase, as in the case of zinc/bromine RFB, where the cathode product precipitates out of the electrolyte and deposits on the electrode surface, the concentration overpotential can be calculated from

$$\Delta \varphi_t \frac{RT}{nF} \ln\left(\frac{i_d}{i_d - i}\right)$$

2) When the product is soluble as in VRB, the overpotential should incorporate both the diffusion of reactants from the electrolyte phase to the electrode surface and the diffusion of products from the electrode to the electrolyte. Thus, the concentration overpotential should be calculated from

$$\Delta \varphi_t = \varphi_{1/2} + \frac{RT}{nF} \ln\left(\frac{i_d - i}{i}\right) - \varphi_e$$

In these two equations, i_d, i, $\varphi_{1/2}$, and φ_e represent the limiting current density, apparent current density, half wave potential, and equilibrium potential, respectively.

7.2.2
Introduction to Electrochemical Processes [7]

7.2.2.1 Vanadium Redox Battery (VRB)

The active species of the VRB consists of four types of vanadium ions dissolved in an aqueous solution of sulfuric acid, with the V(II)/V(III) redox couple employed at the negative electrode and the V(IV)/V(V) couple at the positive electrode.

Figure 7.1a shows the electrochemical processes of the VRB. During charge, V(IV) loses one electron and changes to V(V), while V(III) receives one electron and changes to V(II). At the same time, H^+ is transferred from the anode through the membrane to the cathode. During discharge, the electrochemical process is simply the reverse of the charging process. Both the positive and negative electrolytes are pumped into the batteries and flow back to the storage tank after reaction.

7.2.2.2 Zinc/Bromine Flow Battery

The electrolytes at both electrodes of the ZBB are aqueous solutions of zinc bromide. Figure 7.1b shows the electrochemical processes for the ZBB. During charge, Zn deposits on the cathode surface and Br_2 forms at the anode. During discharge, Zn^{2+} and Br^- are generated at the cathode and anode, respectively. The charge balance is achieved by the migration of Zn^{2+} across a membrane separating the electrolytes.

7.2.2.3 Sodium Polysulfide/Bromine Flow Battery

The active species of sodium polysulfide/bromine flow battery are S_x/S_{x-1} ($x = 2-4$) at the cathode and Br_2/Br_3^- at the anode. Figure 7.1c shows the electrochemical processes for the sodium polysulfide/bromine flow battery. During charge, Br^- is oxidized to Br_2 at the anode and S_x is reduced S_{x-1} at the cathode. During discharge, Br_2 is reduced to Br^- at the anode and S_{x-1} is oxidized to S_x at the cathode. Electrical balance is achieved by the migration of Na^+ across a membrane separating the electrolytes.

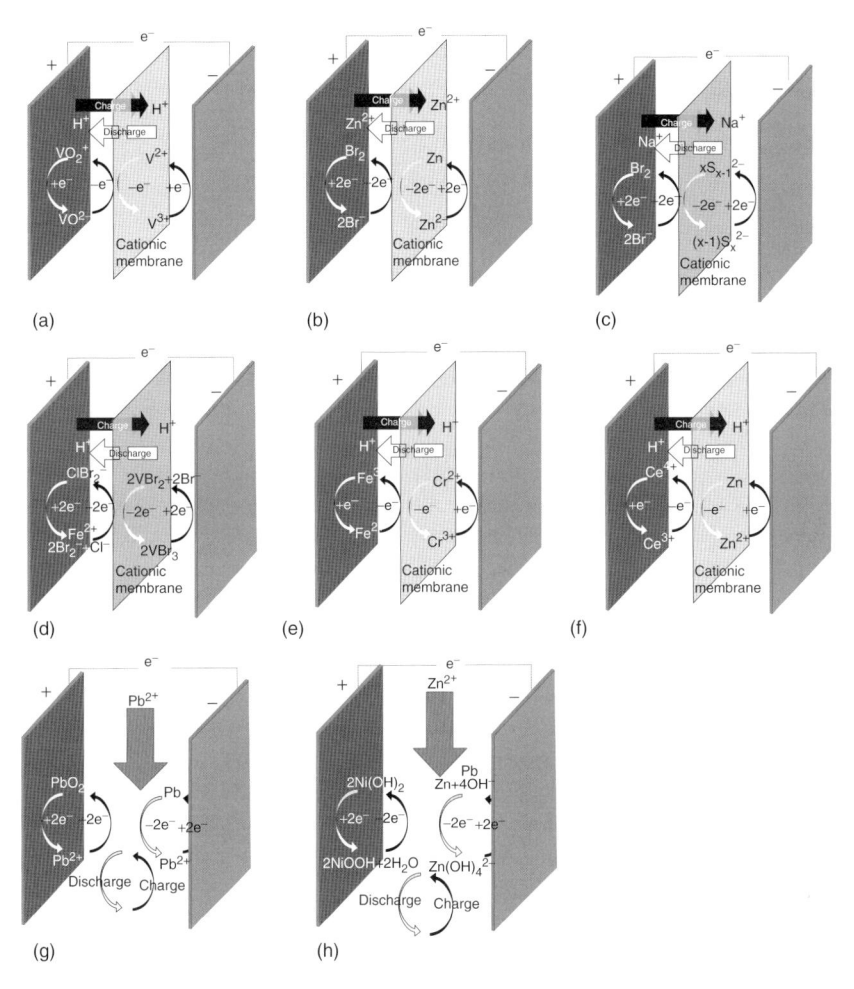

Figure 7.1 The electrochemical processes for different kinds of redox flow batteries. (a) All Vanadium Redox Flow Battery. (b) Zinc/Bromine Redox Flow Battery. (c) Polysulfide/Bromine Flow Battery. (d) Vanadium/Bromine Flow Battery. (e) Fe/Cr Flow Battery. (f) Zinc/Cerium Flow Battery. (g) Lead Acid Flow Battery. (h) Zinc/Nickel Flow Battery

7.2.2.4 Vanadium/Bromine Flow Battery

The electrolytes for the anode and cathode of the vanadium/bromine flow battery are both vanadium bromide, dissolved in aqueous hydrochloric acid or hydrobromic acid solutions. Figure 7.1d shows the electrochemical processes for the vanadium/bromine flow battery. During charge, Br^- is oxidized to Br_2 at the anode and V^{3+} is reduced to V^{2+} at the cathode. The discharge process is simply the reverse of the charging process. Charge balance is achieved by the migration of H^+ across a membrane separating the electrolytes.

7.2.2.5 Fe/Cr Flow Battery

The active species of the Fe/Cr flow battery is a hydrochloric acid solution of iron chloride at the anode and a hydrochloric acid solution of chromium chloride at the cathode.

Figure 7.1e shows the electrochemical processes for the Fe/Cr flow battery. During charge, Fe^{2+} is oxidized to Fe^{3+} at the anode and Cr^{3+} is reduced to Cr^{2+} at the cathode. The charge balance is achieved by the migration of H^+ or Cl^- across the membrane separating the electrolytes.

7.2.2.6 Zinc/Cerium Flow Battery

The active species of the zinc/cerium flow battery is Ce^{3+}/Ce^{4+} and Zn/Zn^{2+} dissolved in methanesulfonic acid, and Figure 7.1f shows the electrochemical processes involved. During charge, Ce^{3+} is oxidized to Ce^{4+} at the anode and Zn^{2+} is reduced to Zn at the cathode. During discharge, the process is reversed. Charge balance is achieved by the migration of H^+ through a membrane separating the electrolytes.

7.2.2.7 Soluble Lead-Acid Flow Battery

There is only one electrolyte storage tank for the soluble lead-acid flow battery, and the active species are Pb, Pb^{2+}, and PbO_2 in methanesulfonic acid solution. What is special is that this kind of battery needs no ion exchange membrane.

Figure 7.1g shows the electrochemical processes for the soluble lead-acid flow battery. During charge, Pb^{2+} is oxidized to PbO_2 at the anode and Pb^{2+} is reduced to Pb at the cathode. During discharge, the process is reversed. Charge balance is achieved by the migration of H^+ inside the battery.

7.2.2.8 Zinc/Nickel Flow Battery

There is also only one electrolyte storage tank for the zinc/nickel flow battery, in which the active species are $Zn/Zn(OH)_4^{2-}$ and $NiOOH/Ni(OH)_2$ in an alkaline solution.

Figure 7.1h shows the electrochemical processes for the zinc/nickel flow battery. During charge, NiOOH is reduced to $Ni(OH)_2$ at the anode and Zn is oxidized to $Zn(OH)_4^{2-}$ at the cathode. During discharge, this process is reversed. The charge balance is achieved by the migration of ions inside the batteries.

7.3 Materials and Properties of Redox Flow Battery

7.3.1 Introduction

Compared to other electrochemical energy storage batteries, the RFB has many advantages. The final battery performance is defined and limited by the battery materials, including the electrodes, electrolyte solutions, membranes, and bipolar plates. Their properties and structural shapes can greatly affect the performances of the RFB. Taking the VRB as an example, this section describes the role and performance requirements of the key materials in the battery, along with the current developments.

7.3.2
Electrodes

VRB electrodes do not take part in the electrochemical reaction directly, but provide the proper electrochemical reaction sites for the active substances. The electrode materials for the VRB require certain characteristics:

- high electrochemical activity and reversibility for the $V(\chi)/V(\delta)$ and $V(\alpha)/V(\beta)$ redox couples;
- good conductivity;
- a stable 3D structure with high active specific area, providing smooth channels for the electrolyte, increasing the mass-transfer rate and decreasing the concentration polarization;
- excellent chemical, mechanical, and electrochemical stability (for the easily oxidized $V(\chi)$) and long life.

VRB electrode materials can be mainly divided into two types, metal and carbon electrodes. Research has been carried out on the use of metals as VRB electrodes, such as Au, Sn, Ti, Pt–Ti, and IrO–Ti. However, it was found that the electrochemical reversibility for the $V(\beta)/V(\beta)$ redox couple was not sufficient on the Au electrode and the Sn and Ti electrodes were easily passivated, with a passivation film covering the surface, which leads to an increase in electric resistance. The problem of the passivation film is successfully avoided while using Pt–Ti and IrO–Ti electrodes, with excellent electrochemical activities for both $V(\chi)/V(\delta)$ and $V(\alpha)/V(\beta)$ redox couples. However, their high price limits them from being widely used. Carbon electrodes consist of many kinds of materials, such as carbon paper, carbon cloth, carbon felt, and carbon nanotubes (CNTs). Of these, carbon felt, prepared from carbon fibers, has become the most promising VRB electrode material owing to its better three-dimensional network structure and larger specific surface area, as well as its higher conductivity and better chemical and electrochemical stability. In addition, with such a rich source of raw materials, the cost is affordable.

Carbon electrodes themselves have some catalytic activity without any supported catalyst. However, modification of the electrode materials is still necessary in order to enhance the reversibility of the electrode reaction by adjusting the material hydrophilicity and pore structure, as well as increasing the concentration of the active group. A variety of modification methods have been studied, which can be classified as physical or chemical methods. The physical methods include air atmosphere heat treatment, plasma treatment, and microwave treatment, while the chemical methods include ion exchange, acid–base treatment, and electrochemical oxidation. Sun and Skyllas-Kazacos dipped graphite felt in a solution containing

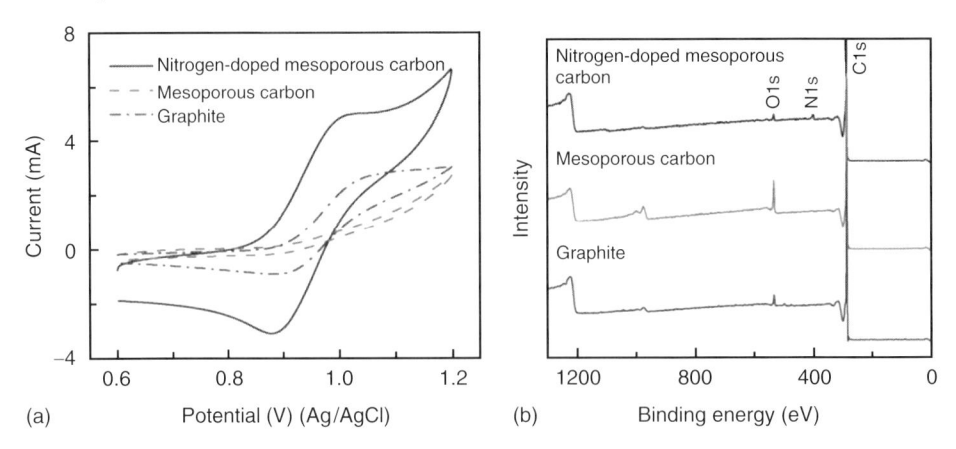

Figure 7.2 (a) CV curves of different electrode materials. (b) X-Ray Photoelectron Spectroscopy (XPS) of different electrode materials 3.0 M H_2SO_4 + 1.0 M $VOSO_4$ (50 mV s^{-1}).

Pt^{4+}, Pd^{2+}, Au^{4+}, and Ir^{3+} metal ions for ion implantation or ion exchange. It was found using cyclic voltammetry measurements that the electrode with the best electrochemical behavior was modified with Ir^{3+} [8]. Skallas-Kazacos at the University of New South Wales, Australia, treated the graphite felt at 400 °C for 30 h in air. He found that the energy efficiency of the battery with this heat-treated graphite felt is 88% at a current density of 25 mA cm^{-2}, about 10% greater than that of an untreated graphite felt electrode [9]. The carbon atoms with lattice defects on a microcrystalline graphite surface have high surface energy, which leads to them being easily oxidized to hydroxyl, carbonyl, carboxyl, and other functional groups in a hot air atmosphere. These functional groups may act as catalytic active sites for the electrode reaction. In 2010, researchers at Pacific Northwest National Laboratory synthesized nitrogen-doped mesoporous carbon materials using phloroglucinol and $EO_{106}PO_{70}EO_{106}$ using a soft template method, followed by heat treatment at 850 °C in an NH_3 atmosphere for 2 h. As shown in Figure 7.2a,b, the reaction activity of the VO_2^+/VO^{2+} redox couple could be increased considerably by introducing nitrogen-containing functional groups [10].

7.3.3
Bipolar Plates

The RFB module is generally assembled according to the assembly procedure for a pressure filter. Bipolar plates in the battery module work as a collection of connected single cells, separated by the electrolyte solution and the current collectors. In addition, bipolar plates are required to provide a pressing force on the carbon felt electrodes in order to maintain their proper shape during assembly. In order to meet these requirements, the bipolar plates must have the following properties:

- good electrical conductivity to reduce the battery internal resistance;
- high mechanical performance, so as not to undergo brittle fracture under the compression force and also to support the electrode materials;
- ability to effectively prevent the penetration of electrolyte, and avoid battery short circuit;
- good chemical resistance; this is a very important property, particularly in the VRB system, in which acid is used as the supporting electrolyte and VO_2^+ ion (a strong oxidizing agent) as the active material, forming a very harsh environment, especially with the addition of the electric field.

Metal plates, graphite plates, and carbon composite panels, which have been widely studied as VRB bipolar plates, are described in more detail below.

- **Bipolar plates** Nonprecious metals can be easily corroded or plated with a passive film having a poor electric conductivity in the VRB system environment. Although the corrosion resistance of precious metals, such as platinum, gold, and titanium, is better, the cost of these materials is much higher. Stainless steel can be exposed to some surface treatment methods, such as electroplating and vapor deposition, in order to enhance its corrosion resistance and durability as bipolar plates.
- **Graphite bipolar plates** Graphite has good electrical conductivity, high chemical stability under acidic conditions, and good corrosion resistance in the potential range applicable for VRB charging and discharging. Hard nonporous graphite plates prepared from pure graphite can effectively prevent the penetration of vanadium electrolyte solution. These characteristics make hard graphite suitable for application as VRB plate materials. However, the preparation of hard graphite plates is very complex and generally includes the mixing of graphite powder, crushed coke, and graphite-based resin or asphalt, heat treatment at 2500–2700 °C following a temperature-programmed procedure in a graphite furnace to obtain nonporous or low-porosity graphite sheet, followed by cutting and grinding [11]. This process is time consuming and costly, limiting the large-scale application of graphite plates.
- **Carbon composite bipolar plates** At present, carbon composite bipolar plates are widely used in engineering. They are commonly prepared by mixing a polymer resin (polyethylene, polyvinyl chloride, etc.) with a conductive filler (graphite, carbon black, carbon fiber, etc.), and then shaping with an injection molding method. Haddadi-Asl *et al.* obtained carbon graphite fiber composite bipolar plates by mixing rubber-modified polypropylene, PVC, nylon, low- and high-density polyethylene with graphite powder and graphite fiber [12]. Carbon composite bipolar plates have the advantages of simpler production process, lower cost, and better corrosion resistance compared with hard graphite bipolar plates. However, as a result of the existence of a certain percentage of the insulating polymer, the electrical conductivity is poorer than that of graphite plates. The major direction of research and development in VRB is the preparation of carbon composite bipolar plates with low resistance, high mechanical strength, and high corrosion resistance by optimizing the preparation method.

7.3.4
Ion Exchange Membranes

In the VRB, ion exchange membranes play the role of separating positive and negative electrolytes to avoid direct contact between them and to supply a pathway for the transfer of protons to form the battery circuit. Ion exchange membranes are an important part of VRB and directly affect the performance of the battery efficiency and cycle life. They require the following characteristics:

- low transmittance of vanadium ions to reduce cross-contamination and self-discharge;
- high ion selectivity and conductivity in order to obtain high coulombic and voltage efficiencies;
- sufficient mechanical strength, chemical resistance, and resistance to electrochemical oxidation to ensure long service life;
- low net water transport rate during the period of battery charging and discharging to maintain the water balance between the positive and negative electrolytes.

So far, no commercial ion exchange membrane dedicated to VRBs has been available. Recent studies have usually focused on the selection of ion exchange membranes from existing materials. Previous investigators have systematically studied many kinds of commercial ion exchange membrane (such as Nafion117, Flemion, Selemion CMV, CMS, AMV, DMV, ASS, and DSV) and found that Nafion and Flemion membranes can be used directly in all VRB systems without further treatment, obtaining good performance. In particular, Nafion membranes showed high energy conversion efficiency as well as high stability in the harsh acidic and strong oxidative environment. Nafion membranes are perfluorinated sulfonic-acid-type proton-exchange membranes produced by the Du Pont Company, the United States, with the structure as shown in Figure 7.3a. It is composed of fluorine-containing

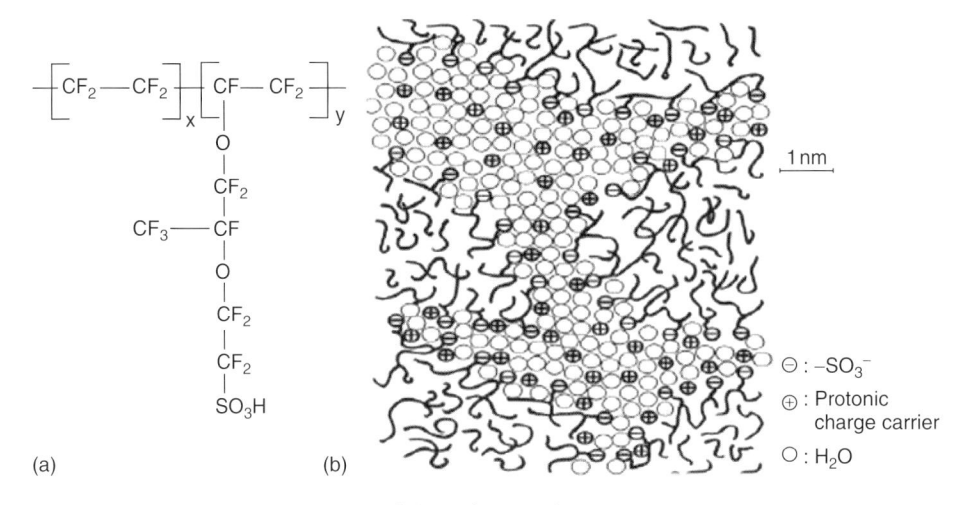

Figure 7.3 (a) The structure of the Nafion membrane. (b) Microstructure of the Nafion membrane.

main chains and sulfonic-containing side chains. The main chain is to ensure the mechanical and chemical stability of the fluoride membrane, while the side chains are capable of transferring protons. Together, these form ion clusters (Figure 7.3b) to ensure good proton conductivity. Although Nafion shows a good performance in the VRB, the high CF bond energy also results in a complex production process, harsh preparation conditions, and a high price (as much as \$500–800 m^{-2}) preventing it from commercialization. Another problem of the Nafion membrane is its low ion selectivity. Cross-contamination easily occurs during the charging and discharging processes between the positive and negative active ions with different valences, resulting in a self-discharge phenomenon and reducing the battery efficiency. In the long run, volume imbalance occurs between the positive and negative electrolyte solutions, resulting in the battery capacity fading. Therefore, the development of high-performance and low-cost ion exchange membranes is particularly important for VRB industrialization.

So far, research on ion exchange membranes have focused on the following aspects.

- **Modification of the perfluorosulfonated ion exchange membranes**. This is mainly conducted using the simple blending method, as well as surface modification methods, to improve ion selectivity and the current efficiency of the battery. For example, Zeng *et al.* decreased the vanadium ion permeability from 2.87×10^{-6} to 5.0×10^{-7} cm^2 min^{-1}, and the water delivery from 0.72 to 0.22 ml/72 h cm^2, by surface modification of Nafion membranes through polymerization of a polypyrrole layer on them. Xi *et al.* [13] prepared Nafion/SiO$_2$ composite membranes by the sol–gel method. The inorganic SiO$_2$ nanoparticles plug the ion cluster holes of the Nafion membrane, thus reducing their vanadium ion permeability. The proton conductivity and ion exchange capacity (IEC) of modified membranes are generally unchanged, whereas the vanadium ion permeability is greatly reduced. Measurement of the battery performance shows that using a Nafion/SiO$_2$ composite membrane, the energy and current efficiencies of VRB are increased, whereas the battery self-discharge rate is decreased. However, high price is the basic problem with this kind of membrane.
- **Modification of some lower-cost fluoride-containing resins**. This method introduces ion exchange groups into polyvinylidene fluoride (PVDF)-, polytetrafluoroethylene (PTFE)-, and fluoride-containing resins to improve membrane ion conductivity. These membranes showed good initial performance, for example, Qiu *et al.* prepared PTFE-g-PS membranes by grafting polystyrene into a PTFE membrane using X-ray radiation grafting method, then copolymerized the resulting membrane with maleic anhydride to synthesize the (PTFE-g-poly(styrene-co-maleic anhydride)PS-CO-PMAn) composite membrane [14, 15], and then chlorosulfonated the two membranes to obtain PTFE-g-PSSA and PTFE-g-poly(styrene sulfonic acid-co-maleic anhydride) (PSSA-CO-PMAc) ion exchange membranes. Compared to the PTFE-g-PSSA membrane, the PTFE-g-PSSA-CO-PMAc composite membrane has higher IEC and lower surface resistance, so the radiation-grafted membranes are expected to be applicable to

VRB. However, the life span of this kind of membranes in VRB has not been reported.
- **Nonfluoride ion exchange membranes**. To reduce the cost of VRB, replacing perfluorosulfonated ion exchange membranes with nonfluorinated ion exchange membranes has become an important research direction. Compared with perfluorosulfonated ion exchange membranes, the greatest advantage of nonfluoride ion exchange membranes lies in their low cost. In addition, nonfluoride ion exchange membranes have other advantages such as reduced manufacturing pollution, adjustable performance, and thermal and mechanical stability; however, its biggest drawback is its susceptibility to degradation during cell operation because of its poor chemical stability. Polyether ether ketone (PEEK) is a wholly aromatic semicrystalline polymer material with good mechanical strength and chemical and electrochemical stability. By sulfonation of PEEK, sulfonated polyether ether ketone (SPEEK) can be obtained. Ion exchange membranes prepared using SPEEK have excellent proton conductivity, so it has great potential for future application to VRB systems. Other polymer materials widely studied include polybenzimidazole (PBI), polysulfone (PS), polyamide to amine (polyimide (PI)), and polyether sulfone (PES). All these compounds can be sulfonated for H^+ transmission by selecting an appropriate sulfonating agent to introduce sulfonic acid groups to the aromatic benzene ring. In recent years, although some new methods, techniques, and materials have been applied in VRB ion exchange membrane preparation and modification, and some progress has been made, the overall performance of the new and the modified membranes are unable to fully meet the requirements of commercial VRB applications. Therefore, the development of ion exchange membrane materials with high conductivity, high selectivity, and high durability and of low cost is still an important area of VRB development.

7.3.5
Electrolyte Solutions

Unlike other types of batteries, the energy storage capacity of the VRB depends on the concentration and volume of electrolyte solutions stored outside. When the battery is running, the electrolyte solutions, driven by the circulating pump, achieve a circular flow in the battery chamber, completing the course of charge and discharge. When not working, the phenomenon of self-discharge does not exist, since the electrolyte solutions are stored in the outside tank. VRB electrolyte solutions include vanadium ions with different valences (positive redox couple $V(\chi)/V(\delta)$ and negative redox couple $V(\alpha)/V(\beta)$) and sulfuric acid solutions, which directly affect the battery performance and operational stability. Activity and stability are the most important indicators for the electrolyte solutions.

As the divalent vanadium ions can be easily oxidized, they can easily react with the oxygen in air and become higher-valence vanadium ions. Sealed storage tanks are normally used to store electrolyte solutions, and they include access to an inert gas such as nitrogen to protect the divalent vanadium ions. In addition, the pentavalent vanadium ions in the electrolyte solution can easily precipitate at high temperature, and the yellow precipitant can plug the circulating pump, obstruct the flow of electrolyte solution, and also become attached to the porous electrode carbon felt, decreasing the effective specific surface area of electrode and causing the battery performance to become attenuated. Wen *et al.* integrated a variety of factors to conclude that the best concentration of electrolyte solutions was 1.5–2.0 mol l^{-1} V^{3+} mol l^{-1} H$_2$SO$_4$ [16]. In addition, the pentavalent vanadium ions may be stabilized by adding a suitable stabilizer.

The stability of electrolyte solutions can be improved by selecting the appropriate concentrations of vanadium ions and sulfuric acid and by adding stabilizers. The stabilizers can play a significant role in enhancing the VRB system stability and cycle life.

7.4 Redox Flow Battery System

7.4.1 Introduction

The RFB is a high energy storage device, with the advantages of long cycle life, high efficiency, design flexibility, and environmental friendliness. They can have output power ranging from a few kilowatts to tens of megawatts. The storage capacity of flow energy storage batteries can be independently designed up to tens of megawatts depending on requirements, which is a good choice for large-scale energy storage. The flow battery system is very complex and consists of the battery module, electrolyte solutions, storage tanks, electrolyte solution circulation pumps, transport pipelines, thermal management systems, electronic control systems, and other components. The battery module is the core of the system and is composed of many single cells, which are the smallest units of the module. This section follows the progression from single cell to battery module to battery system to understand the structure of flow battery energy storage system step by step.

7.4.2 Single Cell

The single cell is the basic unit of the RFB. The performance and durability of the key materials of a flow energy storage battery must first be evaluated from the single cell experimental data. To improve the cell structure and determine the best battery operating conditions, kinetic parameters measured on the single cell are also needed to provide guidance. Understanding the composition of the single cell and its assembly method are the primary requirements of the single cell experiment.

Figure 7.4 shows the structure of a single cell. Its parts are assembled in the following order: end plate – electrode – ion exchange membrane – electrode – bipolar plate – end plate.

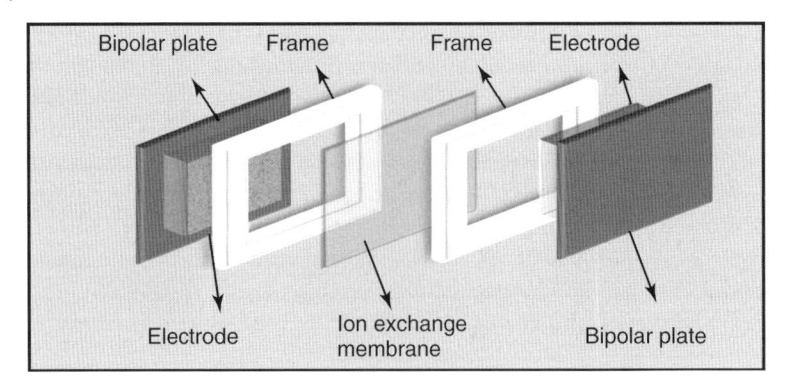

Figure 7.4 The structure of a single cell.

Before a single cell experiment, we should design the structure of the cell with a consideration of the following parameters: working area and thickness of electrode, sealing structure (preventing electrolyte solution leakage) and the flow patterns of electrolyte solution, and so on. VRB requires the sealing material to be corrosion resistant, and fluorine rubber materials are commonly used for this purpose. In the assembly procedure, the first step is to manufacture the components with the specific structure and shape required by the design. The second step is to order the components as shown in Figure 7.4, then finally to compress them by screws or a hydraulic compression machine.

7.4.3
Battery Module

An RFB module is assembled from N single cells. Single cells are connected together by means of the conductive bipolar plate between them. The two ends of the battery module are the current-collecting plates and end plates. The current-collecting plates are the input and output interfaces of the battery module, while the end plates fix the battery module. The electrolyte transport pipeline within the battery module can be of two types, the U-type and the Z-type. In the U-type pipeline, the electrolyte goes into and out of the module at the same side, while in the Z-type, one side is in and the other side out, as shown in Figure 7.5. The selection of the electrolyte flow type in the transport pipeline depends on the system structure, but the U-type transport pipeline is usually more compact. The assembly force or compression needed to seal and contain electrolyte transport in the single cell needs to be considered. This can be designed rationally and calculated by considering factors such as the number of single cells, the mechanical properties of sealing materials, and the electrode pressure drop.

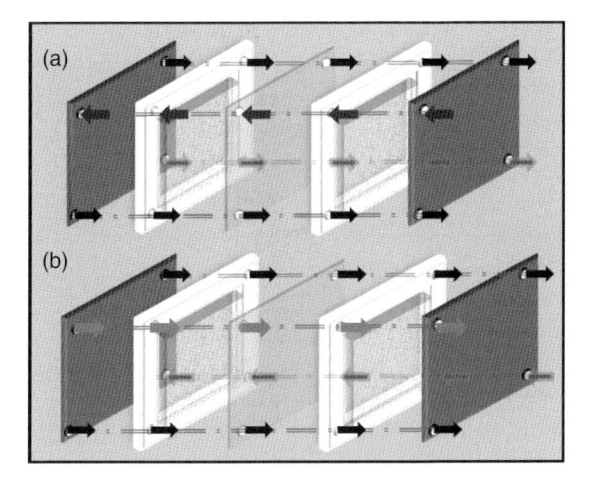

Figure 7.5 (a) U-type electrolyte manifolds for redox flow battery modules. (b) Z-type electrolyte manifolds for redox flow battery modules.

7.4.4
The Composition of the Battery System

The RFB system consists of the electrolyte circulation subsystem, the heat transfer subsystem, the charge and discharge control subsystems, and the battery module subsystem. These are summarized in the RFB system schematic diagram in Figure 7.6. In the RFB system, the electrolyte circulation subsystem, which includes positive and negative electrolyte storage tanks, pumps, and pipeline, plays the role of transporting the electrolyte solution to the module battery. The concentration and volume of electrolyte solution decide the energy storage capacity. The role of the heat-transfer subsystem is to maintain the electrolyte temperature and remove

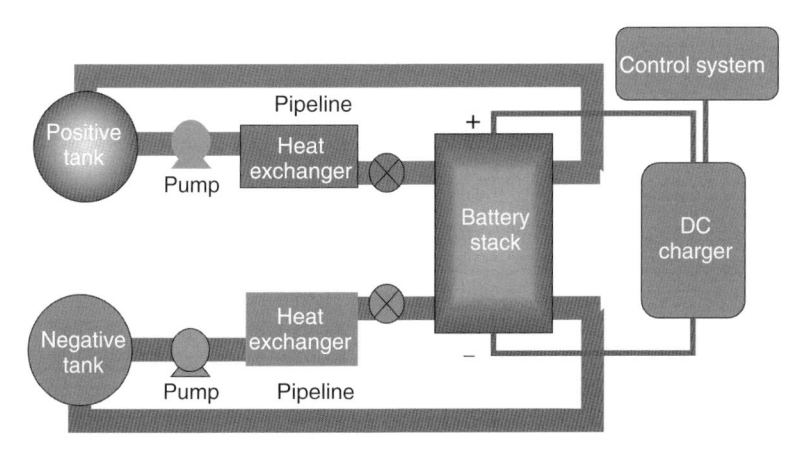

Figure 7.6 Schematic diagram of the RFB system.

waste heat generated in the battery. Heat transfer can be air cooled, water cooled, or a combination of both. The battery module subsystem is the core of the flow battery energy storage system since it is the place where power and chemical energy are transformed. The size and number of battery modules determine the output power. The role of the charge–discharge control subsystem is to complete the conversion between AC and DC and control the input and output of the battery module. This subsystem includes the AC–DC converter, DC charge and discharge devices, relays, and other electrical and electronic equipment.

7.4.5
Battery System Scale-Up

The output power and energy capacity of RFB systems can be scaled up independently by increasing both the number of battery stacks and the concentration and volume of the electrolyte, which renders the task of system designation quite straightforward. According to practical demand, RFB systems are divided into three types. The first type is a high output power, low-energy-capacity battery system. The second is a low output power, high-energy-capacity battery system, and the last is a high output power, high-energy-capacity battery system. However, high output power can be achieved not only by increasing the number of single cells but also by enlarging the electrode area. Similarly, high energy capacity can be achieved by increasing the volume and concentration of the electrolyte. In order to scale-up the battery system, many factors, such as the design of the electrolyte pipeline, the control of electric current leakage, and the bearing capacity, should also be taken into consideration. For one thing, the scaling up of the battery system will worsen the electric current leakage, which not only damages the performance of the battery system but also affects the system safety. For another thing, the scaling up of the battery system proposes a new challenge for distributing electrolyte uniformly in the stack, which is rendered more difficult with an increase in the electrode area and the number of single cells. In addition, the scaling up of the battery system will increase the burden on the charge–discharge equipment, which need to be more durable and of larger capacity.

7.5
Performance Evaluation of Redox Flow Battery

7.5.1
Introduction

Performance indicators for the RFB include output power, energy efficiency, energy capacity, voltage uniformity of single batteries in a stack, and battery life. These indicators can be tested using electrochemical methods and/or by long-term operation of a battery (stack) and are described further in this section.

7.5.2
Performance Indicators

7.5.2.1 Rated Output Power

The power of an RFB is usually referred to as the *rated output power* under specified discharge conditions. The value of the rated output power is closely related to the current density and the number of single batteries assembled in a stack. The instantaneous output power of an RFB is equal to the product of the current and voltage between the electrodes at a specified time.

$$W = I \times V$$

where W is the instantaneous power; I, instantaneous current; and V, instantaneous voltage.

The mean output power can be calculated as

$$\overline{W} = \frac{\int_0^t IV\,dt}{t}$$

where $\overline{W}$ is the mean power; I, instantaneous current; V, instantaneous voltage; and t, time.

The rated output power of a battery stack should be designed for the load for which it provides electricity. If the actual output power is higher than its rated value, the battery is said to be overloaded. However, RFBs have a good compatibility with overload, and their output power directly influences their energy efficiency. The designed energy efficiency of VRBs from the Dalian Institute of Chemical Physics and the Dalian Rongke Power Corp., who are the main manufacturers of VRB in China, is 80%.

7.5.2.2 Energy Storage Capacity

The energy storage capacity of an RFB is the maximum volume of electricity that can be provided by the battery under specified conditions. It is determined by the volume and concentration of the electrolyte used in the battery.

Because the energy capacity is not related to the battery's output power, it can be changed by adjusting the volume of electrolyte, without changing the electrode area or the number of battery stacks.

The electrolyte of an RFB is not always used completely, and this phenomenon, known as the *utilization rate of the electrolyte*, is reflected by the ratio between the practical capacity and theoretical capacity of the battery.

$$\eta = \frac{Q_{practical}}{Q_{theoretical}} \times 100\% = \frac{W_{practical}}{W_{theoretical}} \times 100\%$$

where η is the utilization rate of energy capacity; $Q_{practical}$, the practical energy capacity; $Q_{theoretical}$, the theoretical energy capacity; $W_{practical}$, the practical output power; and $W_{theoretical}$, the theoretical output power.

7.5.2.3 Efficiencies of Charge–Discharge Cycles

The efficiency of a battery charge–discharge cycle is an indication of energy conversion efficiency, including coulomb efficiency, voltage efficiency, and energy efficiency.

Coulomb efficiency is the ratio between the discharged electric quantity and quantity of electricity required for charging under certain charge–discharge conditions, and the higher the coulomb efficiency, the less the loss of battery capacity. The latter is mainly caused by the crossover of electrolyte separated by the membrane, as for the VRB, and any side reactions occurring at the electrode.

$$\eta_C = \frac{Q_{Discharge}}{Q_{Charge}} \times 100\%$$

where η_C is the Coulomb efficiency; $Q_{Discharge}$, the amount of discharged electricity; and Q_{Charge}, the amount of electricity for charging.

The voltage efficiency is the ratio between the mean discharge voltage and the mean charge voltage under certain conditions. In some circumstances, it can be improved by reducing the battery polarization.

$$\eta_V = \frac{\overline{V}_{Discharge}}{\overline{V}_{Charge}} \times 100\%$$

where η_V is the voltage efficiency; $\overline{V}_{Discharge}$, the mean discharge voltage; and $\overline{V}_{Charge}$, the mean charge voltage.

The energy efficiency is the ratio between the output energy and input energy. It is numerically equal to the product of coulomb efficiency and voltage efficiency. A high energy efficiency is important for an RFB, because this is a direct indication of energy conversion capability.

$$\eta_E = \frac{W_{Discharge}}{W_{Charge}} \times 100\% = \eta_C \eta_V$$

where η_E is the energy efficiency; $W_{Discharge}$, output energy; and W_{Charge}, input energy.

7.5.2.4 Voltage Uniformity of Single Battery

A stack of RFBs consists of many single batteries assembled together in a filter press mode, where one liquid pipeline is shared by all single batteries. If the electrolyte is not distributed uniformly among the single batteries, their SOC will lack uniformity, resulting in deterioration of battery performance and life span.

The uniformity of single batteries in an RFB stack can be evaluated by the voltage range, which is the difference between the highest voltage and the lowest voltage. The lower the voltage range, the greater is the uniformity of the single batteries in a stack.

7.5.2.5 Self-Discharge

The capacity of a battery decreases with time when it is left unused, and this phenomenon is called *self-discharge*. The reasons for self-discharge, as in the case of traditional batteries (lead-acid and nickel-cadmium batteries), are the side reactions that occur in the electrodes, such as hydrogen evolution and oxygen evolution. However, self-discharge of RFBs is mainly caused by electrolyte crossover and the leakage of electricity through the electrolyte pipeline. The self-discharge rate can be calculated from the equation below:

$$v = \frac{Q_0 - Q}{Q_0 t} \times 100\%$$

where v is the self-discharge rate; Q_0, the initial electric quantity; Q, the retained electric quantity; and t, time.

7.5.2.6 Cycle Life

An RFB is said to have experienced one cycle when it completes one charge process followed by one discharge process. The *cycle life of an RFB* is defined as the number of cycles that the battery completes before the electrolyte utilization rate drops to a specified value under defined charge–discharge conditions. Cycle life is an important indicator of the performance of RFBs. It can be prolonged by blending the electrolytes of the anode and the cathode.

7.5.3
Evaluation of Battery System Performance

7.5.3.1 Evaluation Method

A battery charge–discharge instrument is usually used to measure the power, capacity, efficiency, and cycle life of RFBs in several ways, including constant current charge–discharge, constant power charge–discharge, constant voltage charge, and constant resistance discharge.

The self-discharge rate of an RFB can be measured using intermittent charge–discharge cycles.

The voltage uniformity of single batteries is evaluated by monitoring the voltage of every battery in the stack.

7.5.3.2 Evaluation Instruments

The instruments usually used for battery performance evaluation are the charge–discharge systems manufactured by companies such as the Arbin Company (the United States), Bitrode Company (the United States), Maccor Company (the United States), and Neware Company (China). Using these systems, different charge–discharge conditions can be set to evaluate battery performance.

7.5.4
Factors Influencing Battery Performance

7.5.4.1 Battery Materials

The performance of a battery is directly influenced by the materials used for its manufacture, including the materials for the bipolar plate, electrode [17, 18], ion exchange membrane [19, 20], and sealants.

The bipolar plate functions to collect current and to connect single batteries in series, so it must have good conductivity, be able to withstand high pressure, and be very flat. These factors are important for reducing the battery resistance and improving the battery performance and life span.

The electrode is one of the key materials within an RFB, because this is where the battery's electrochemical reactions take place. A good electrode should have high conductivity, a high catalytic activity, and a high reaction selectivity, so as to reduce the battery resistance and electrochemical polarization and hence improve the battery efficiency.

The ion exchange membrane is another key material within the RFB. Its main function is to conduct ions and separate the active species of the positive and negative electrodes. As for VRB, the ion exchange membrane needs to have good ion selectivity (allowing proton transport and blocking vanadium ion transport) and a low resistance in order to improve the coulomb efficiency and voltage efficiency. In addition, good mechanical properties and good corrosion resistance (chemical corrosion and electrochemical corrosion) are also desirable to ensure long-term operation of the battery.

7.5.4.2 Battery Structures

Battery structure directly influences the liquid distribution within a battery stack and the internal resistance of the battery. It further influences the voltage uniformity of single batteries in a stack and the overall performance of the stack.

7.5.4.3 Concentration and Composition of Electrolyte

The concentration of electrolyte for RFB is also a very important parameter. A change in concentration may lead to variations in the viscosity, conductivity, and equilibrium potential of the electrolyte solution and affect the performance of the battery. At a fixed volume of electrolyte solution, an increased amount of the active species may increase the capacity of the battery, but it should be held within a reasonable range, instead of being the higher the better. Taking VRB as an example, an unreasonably high concentration of electrolyte solution may lead to instability of the electrolyte. As reported in the literature [21], the optimum concentration of vanadium is $1.6\,\text{mol}\,l^{-1}$ and that of sulfuric acid is $3\,\text{mol}\,l^{-1}$ for the electrolytes of VRB.

7.5.4.4 Charge–Discharge Conditions

The charge–discharge conditions can have a significant influence on the battery's performance. For example, at a fixed charge–discharge cutoff voltage, an increased current density may result in improved coulomb efficiency, but lowered voltage efficiency and battery capacity. When the charge–discharge current density is fixed, a higher cutoff charge voltage will create a larger electricity charge. However, an excessively high voltage may result in side reactions, such as hydrogen evolution and electrode corrosion, after the battery is fully charged. So an optimized cutoff charge voltage should be used.

7.5.4.5 Temperature

An elevated battery temperature is beneficial for improving the kinetics of the electrochemical reactions at the electrodes. It also helps to reduce the viscosity of the electrolyte solution and can thus facilitate mass transfer and lower the internal resistance of battery, resulting in increased voltage efficiency and decreased coulomb efficiency. However, the battery temperature cannot be too high. As for VRB, an excessively high temperature may lead to V(δ) precipitation because of reduced solubility. Meanwhile, the temperature should not be too low because the V(β) tends to precipitate at low temperature. So the operational temperature for VRBs should be chosen carefully, and it typically falls in the range of 5–40 °C [22].

7.5.4.6 Flow Rate of Electrolyte

The electrolyte solution is the medium for energy storage in the RFB. Its flow rate in the battery may influence the working voltage and concentration polarization, and finally the storage capacity and charge–discharge efficiency of the battery.

7.5.5
Control and Management of the Battery System

As an energy conversion device, the RFB has to be well managed and controlled to ensure continuous and steady operation. The management and control of an RFB include control of operational parameters, battery safety management, electrical management, and heat management.

7.5.5.1 Operational Parameter Control

The cutoff voltage for charge is an important factor that determines the utilization of electrolyte for an RFB and can protect the battery from side reactions and overcharge. A higher cutoff voltage for charge means not only better utilization of electrolyte and higher energy storage capacity (for the same amount of electrolyte) but also a higher possibility of side reaction. The cutoff voltage for charge is usually set at 1.5–1.7 V for VRBs developed by the Dalian Institute of Chemical Physics and the Dalian Rongke Power Company.

Another important operational parameter is the cutoff voltage for discharge, which also determines the utilization of electrolyte. The lower the cutoff voltage for discharge, the deeper the battery is discharged. However, the discharge depth should be controlled within a reasonable range because the output power will decrease sharply and damage the battery performance when the discharge depth is close to 100%.

7.5.5.2 Safety Management of RFB

Safe management of RFBs includes voltage protection, leakage protection, and temperature protection.

- **Voltage protection** Voltage protection is accomplished by monitoring the voltage of each single battery in a stack. From the data obtained, one can deduce what

has gone wrong with the battery and take measures to restore the system back to normal operation. For example, when the voltage of one single battery is too high or too low, it probably shows that the electrode is heavily corroded or there is not enough electrolyte in that particular single battery. In this case, the battery should be repaired as soon as possible.
- **Leakage protection** RFBs use pipelines to convey electrolyte solution from a storage tank to the battery stack. If leakage occurs in the pipeline or in the battery stack, the performance of the battery stack will be seriously affected. Therefore, leakage protection must be available within the area of the RFB system. Once a leaking point is found, an alarm will be activated, and the outlet of the electrolyte tank should be turned off in order to avoid large areas of leakage.
- **Temperature protection** The best working temperature for a VRB is 5–40 °C. Once the battery temperature is found to be out of this range, heat exchange should be speeded up, and the battery should even be turned off to avoid damage to the whole battery system.

7.5.5.3 Power Management

The charge–discharge current of an RFB is direct current and the output voltage changes with SOC, so it must be managed before use.

During battery charge, a power management system is needed to ensure that the input power matches the battery power, especially when the input power is from solar energy and wind energy. When the input power is from the grid, an AC–DC inverter is needed to convert the grid current into direct current.

Likewise, when the battery is discharged, the output power should also be managed according to the actual need.

7.5.5.4 Heat Management

Heat is produced continuously during battery operation. For VRBs, the heat produced is mostly below 50 °C, and difficult to use directly, so a heat management system should be used to dissipate the waste heat to protect the battery from high-temperature damage.

Heat management is accomplished through the use of a heat exchanger, and this can be operated using water cooling, forced air cooling, or a combination of the two.

7.5.5.5 System Failure Diagnosis

A good many parameters, including electrolyte temperature, flow rate, single battery voltage, and pump current, can be monitored during battery operation. A variation in these parameters can reflect most system faults or failures. For example, once the electrolyte temperature exceeds the preset range, the control system will give corresponding signals and make an appropriate adjustment, such as increasing cooling water, decreasing the current density, or stopping the service temporarily. Also, when the flow rate is found to have decreased significantly, proper measures must be taken to stop it.

7.6
Degradation Mechanisms and Mitigation Strategies

7.6.1
Introduction

A common problem of rechargeable batteries is the energy capacity loss after long-term operation. The battery's performance, such as the discharge time and the energy efficiency, decreases with the number of charge–discharge cycles. As it is a type of rechargeable battery, the VRB also has the problem of performance degradation. Owing to the special composition and configuration of VRBs, battery module material will be degraded and storage capacity will decrease after the battery has been run for a long time. The degradation of VRB performance is of two main kinds, namely, degradation of materials and loss of storage capacity. And each is discussed separately in this section. Finally, some feasible mitigation strategies are proposed for each of the VRB degradation mechanisms.

7.6.2
Degradation Mechanisms

7.6.2.1 Analysis of the Degradation of Materials
The materials that influence the life of the battery module include the electrode, bipolar plate, ion exchange membrane, and sealing elements. The degradation of these materials will lead to irreversible harm to the VRB. The following list describes the reasons for the degradation of each material in further detail.

- **Corrosion of the electrode and the bipolar plate** Porous carbon materials are usually used as the electrodes for VRBs because of their high durability in strong acid solutions. Carbon plates or composite carbon plates are used as the bipolar plates in VRBs owing to their high electrical conductivity and high durability in strong acid solutions. However, hydrogen evolution will take place during operation, especially under high-current-density conditions. The hydrogen absorbed on the surface of the electrode can infiltrate and diffuse into the electrode, which results in the corrosion of the electrode. Oxygen evolution normally occurs at the positive electrode at the end of the charge cycle. The active intermediate formed during oxygen evolution reactions can oxidize the carbon electrode. Moreover, the specific surface area of the bipolar plate is usually much smaller than that of the porous electrode. Therefore, hydrogen evolution and oxygen evolution can easily occur on the surface of the bipolar plate when the battery is on load and the bipolar plate is in direct contact with the electrolyte.
- **Degradation of the ion exchange membrane** The electrochemical reaction in a VRB is the transference of electric charge between the solid phase and the liquid phase. However, the electrical resistance of the solid phase (porous electrode) is much lower than that of the liquid phase (electrolyte). According to the theory of porous electrodes, the electrochemical reaction occurs mainly in the area close to the side of membrane, that is, the region near the membrane side has the highest

overpotential of the electrode [23]. If there is some oxidant produced during the electrochemical reaction, it will gather at the surface of the membrane. This will lead to degradation of the membrane after long-term operation.

- **Failure of the sealing element** In order to ensure the absolute integrity of the VRB module, high pressure is needed during the assembly process. Usually, line sealing or face sealing is used to seal the VRB module, although sometimes both are used. Because the VRB electrolyte is a strong acid solution and the temperature is constantly changing during operation, the sealing element can easily be corroded by the acid solution and can lose its elasticity. The sealing element can also yield and become deformed after long-term compression and finally lose its elasticity.

7.6.2.2 Degradation of the System Capacity

One of the most important features of VRBs is that the power and energy capacity of the system can be separated. The energy storage capacity is determined by the concentration and volume of the electrolyte [8]. The electroactive species (vanadium ion) with different valences will cross between the positive and negative electrolyte because of the selectivity of the ion exchange membrane. In addition, the diffusion rates of the electroactive species are different, and they can form different hydrated molecules. Therefore, the concentration and volume of the respective electrolytes will change, and the electroactive species in the positive and negative electrolytes will become unbalanced after long-term operation. The degradation of the electrolyte capacity is the primary reason for the loss of the VRB system capacity. Also, side reactions, such as hydrogen evolution, will occur when the battery is operated inappropriately. The side reactions result in the valence imbalance of the electroactive species and lead to system capacity loss.

- **Degradation of the electrolyte capacity** There are two main reasons for degradation of the electrolyte capacity. One reason is self-discharge, that is, cross-contamination of the electroactive species across the ion exchange membrane [24]. The other is the appearance of side reactions, including hydrogen evolution and oxidation of electroactive species during operation [25].
- Self-discharge is a complex process including ion exchange between the membrane and electrolyte and the diffusion transfer and electromigration of the electroactive species. There exist some ion-exchange groups in the membrane, and the mobile ion groups will exchange with the electroactive species until there is balance on the contact surface between the membrane and the electrolyte [26]. However, the selectivity of the membrane between the electroactive species and the ions used to complete the circuit is not usually high enough. Therefore, the electroactive species will be transported to the opposite half cell under the force of the concentration gradient, which will lead to the loss of electrolyte capacity. As the concentration of electroactive species in the electrolyte is different on the two sides of the ion exchange membrane, a concentration gradient exists across the two sides, and the electroactive species will diffuse and transfer across the membrane. If there is a potential difference between the two half cells, the

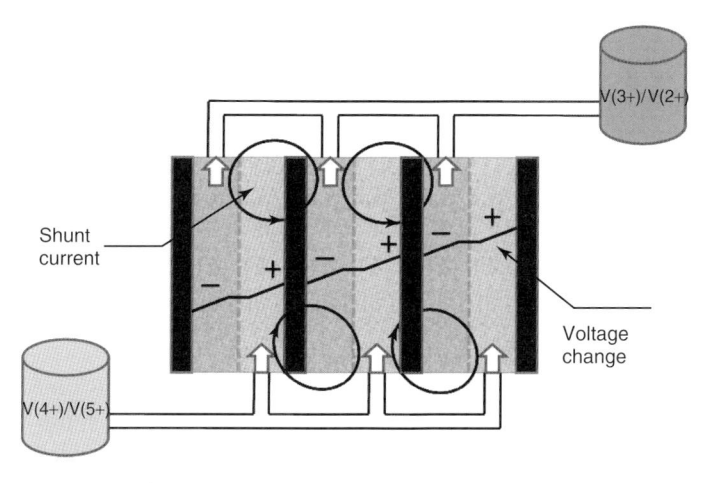

Figure 7.7 Schematic diagram of the generation of shunt current in the charge state of a VRB.

transfer rate of the electroactive species will increase (when the directions of the concentration gradient and potential gradient are the same) or decreases (when the directions of concentration gradient and potential gradient are the opposite).
- The VRB electrolyte contains the V^{2+} ion, which can be easily oxidized. If the seal is not good enough, the V^{2+} ion will be oxidized by oxygen, leading to a valence imbalance of the electroactive species. In addition, the electrolyte of VRB is a strong acid. There are plenty of free hydrogen ions. When the overpotential reaches the potential for the hydrogen evolution reaction, hydrogen evolution will take place and competes with the main reaction. This phenomenon will affect the efficiency of the battery and lead to the valence imbalance, which will decrease the charge–discharge capacity of the battery module [27, 28].
- **Shunt current** It has been proved by many researchers that the presence of a shunt current in the electrochemical reactors was the result of a number of electrolytic cells arranged in a bipolar manner [29, 30]. Figure 7.7 shows the generation of a shunt current for the charge state in a VRB. The electrolytic cells are connected in series with a bipolar plate. At the two sides of the bipolar plate (one side is positive, the other is negative), a circuit loop is formed through the electrolyte distribution pipe and the common shared bipolar plate. Because the voltage of the negative electrode is higher than that of the positive electrode, shunt currents are generated. For the discharge process, the generation of the shunt current is the same as for the charging process.

7.6.3
Mitigation Strategies

This section summarizes the main factors that result in the degradation of VRBs and analyzes the reasons. Some of these factors can be avoided by controlling strategies. Some cannot be avoided, but can be improved by modifying the materials and suitable controlling strategies. This section discusses and analyzes the details of mitigation strategies from three points of view: material modification, controlling strategies, and optimization of the system.

7.6.3.1 Material Modification

Reasonable modification can avoid the degradation of the materials and improve the charge–discharge performance of VRBs, and this section describes ways to modify the key materials, such as the electrode, ion exchange membrane, and the electrolyte.

First, the performance of the porous carbon electrode can be improved by two treatments. One is to improve the hydrophilic property of the porous carbon electrode. The mass transport property at the surface of the electrode determines the polarization of the electrochemical reaction. Therefore, the mass transfer coefficient of the fluid can be increased by improving the hydrophilic property of the electrode. It will reduce the concentration polarization on the electrode surface and increase the charge–discharge performance of batteries [31–33]. The other treatment is to modify the electrocatalyst (metal or metal oxide) on the surface of the electrode [34]. The modified electrocatalyst can improve the reactivity of the electrochemical reaction and reduce the reaction overpotential. At the same time, the hydrogen evolution reaction's overpotential is also reduced, which can reduce the corrosion of the electrode due to hydrogen evolution and slow down the degradation rate of the membrane.

Second, the ion exchange membrane can be modified in three ways. One is to apply surface treatment to improve the antioxidant ability of the membrane [4]. Another is to add an interlayer to the surface of the membrane [35]. In this way, the crossover of the electroactive species can be prevented and the self-discharge rate can be decreased. The third treatment is membrane doping to improve the mechanical intensity and corrosion resistance of the membrane [36–38].

Finally, the properties of the electrolyte can be improved by adding some electrolytic additives. The main function of these additives is to improve the stability of the electrolyte, that is, to avoid the deposition or oxidation of some of the electroactive species. In addition, some electrolytic additives can be added to restrain the occurrence of hydrogen evolution.

7.6.3.2 Controlling Strategies

As well as the modification of key materials, optimization of battery operating conditions can also play a role in reduced degradation of the battery performance. The input–output current of the VRB is direct current. The input–output current density determines the degree of polarization. It is found that the value of the overpotential changes nearly in the same ratio as the current density. Therefore, side reactions, such as hydrogen evolution and oxygen generation, can be avoided by controlling the upper limit of the charge voltage at different current densities [39]. Moreover, the circulation of the electroactive species can be accelerated by increasing the flow rate of the electrolyte, so the concentration polarization of the reaction can be decreased. However, it will increase the power consumption of the fluid delivery device. Therefore, the relationship between the battery performance and the power consumption should be considered carefully.

7.6.3.3 Optimization of the Battery System

When the output power of the battery system is constant, the total shunt current of the battery system will increase as the number of battery modules in series increases; on the other hand, if the number of battery modules increases in parallel, the output current of the battery system also increases. Higher output currents need more advanced electric equipment to transport them; therefore, both shunt current and the endurance of electrical equipment need to be considered when designing the battery system. The usual way of decreasing the shunt current is to increase both the size of the common electrolyte pipe and electric resistance of the distribution manifold and feed port in the battery module. However, this will also increase the flow resistance of the electrolyte and consumes extra power for overcoming this resistance. Therefore, the relationship between shunt current and power consumption should be considered at the same time.

7.7 Applications of Redox Flow Batteries

7.7.1 Introduction

In comparison with other energy storage technologies, RFBs are advantageous because of their independent power and capacity, thus enabling flexible battery design; high energy efficiency (>75%); long lifetime; high stability and reliability; deep discharge with low self-discharge and no damage to the battery; flexible choice of battery location and low operational cost; and high level of security. Therefore, RFBs may have wide applications and vast market prospects as an efficient energy storage technology. The main application areas are

1) in combination with renewable energy sources, to smooth power output, improve efficiency, and then further achieve the popularization of renewable energy;
2) grid peak shaving and load leveling;
3) emergency and standby power sources.

In addition, energy storage systems are considered to be an important part in *smart grid* technology, which has been proposed in recent years. A *Smart Grid* program will further accelerate the commercialization of the RFBs.

7.7.2 The Application of RFBs in Renewable Energy Systems

Renewable energies such as wind energy and solar energy will take larger and larger fractions of the future energy framework. However, they are easily influenced by natural factors such as the season, climate, time, and location, and therefore are unstable and not continuous. In this regard, energy storage constitutes an important part of the overall solution, wherein the frequency and power differences can be adjusted between the renewable energy and the grid, ensuring a continuous and steady power supply. The role of energy storage technology in renewable energy supplies lies in two areas: first, by adjusting the power generation from an independent medium- or small-sized wind, photovoltaic, or ocean energy generator and second, for large-scale wind or solar energy systems assembled with the grid, peak and frequency modulation must be performed before integration to ensure a high standard of grid operation. Therefore, to develop an efficient energy storage technology compatible with renewable energy generation is highly necessary for achieving large-scale utilization of renewable energy [40, 41]. RFB can well satisfy this demand. In recent years, the United States, Japan, and some European countries have, one after another, integrated RFBs with wind or photovoltaic energy supplies for peak shaving of power stations. For example, VRB Power Systems set up a 250 kW/1 MWh all-vanadium RFB system in 2003 for application in the Australian King Island wind power system to provide a stable and reliable power supply [42]. King Island is rich in wind resources. It has four sets of 1500 kW diesel power stacks and three sets of 250 kW and two sets of 850 kW wind power stacks. The VRB system works by stabilizing the transient power transport changes and load fluctuations in wind power generation, controlling the frequency and voltage, enabling *load transfer*, optimizing the performance of diesel–wind hybrid power systems, reducing the workload of the diesel power stack, and thereby reducing fuel costs and waste emission. In 2005, SOLON AG in Australia and Germany purchased the 10 kWh VRB system from the Canadian VRB Power Systems Corporation [43], which is integrated with a photovoltaic system. The 170 kW/1 MWh and 4 MW/6 MWh VRB systems, developed by SEI in 2001 and 2005, respectively, are both used for storage and steady output from wind energy generators [44]. ZBB Energy Corporation of the United States declared in 2008 that they provided Ireland with a 500 kWh zinc/bromine RFB system for energy storage and wind energy technology. From the above examples, one can foresee that RFBs will be an indispensable technology for construction of dynamic and renewable energy storage systems including wind energy and photovoltaics. They will bring significant improvements in the efficiency and profits of power plants, and promote the development of renewable energy industries.

7.7.3
The Application of Peak Shaving of the Power Grid

With the fast development of society and economy, there is an ever increasing demand for power. Meanwhile, demand differs greatly between night and day, and the power plant must be able to satisfy peak demand. In order to satisfy such demand and maintain normal operation, more and more power plants have been built, resulting in larger and larger differences between peak and off-peak loads, and causing a low generator efficiency as well as high waste of fossil fuels during off-peak times. RFBs can store off-peak electricity for use in peak hours, which reduces the power generation requirements at peak hours and thus the scale of the power plant. This is important for improving the efficiency of power generation and maintaining the safety and stability of the grid. Meanwhile, the stored electricity generated in off-peak hours, which is inexpensive, can be used for production and construction, significantly cutting the operational cost of manufacturing. As early as 1996, Kashima-Kita and SEI in Japan successfully developed 200 kW/800 kWh and 450 kW/900 kWh VRB systems, respectively, for demonstration of load leveling and peak shaving of the power grid [43]. Subsequently, SEI developed a series of VRB systems with power/capacity of (200 kW/1.6 MWh)/(500 kW/5 MWh) in 2000 and (1.5 MW/1.5 MWh)/(500 kW/2 MWh) in 2001 and 2003. These are all for the improvement of grid safety and stability [45, 46]. In 2004, Canada's VRB Power Systems Corporation constructed a 250 kW/2 MWh VRB system for Castle Valley, Utah Pacific Corporation, also for load leveling and peak shaving [44]. The British Innogy Company set up the first business-scale sodium polysulfide/bromine RFB, whose energy storage capacity and output power could reach up to 120 MWh and 15 MW, respectively [47]. ZBB Energy Inc. installed a 400 kWh zinc/bromine battery stack in Australia's United Energies Inc., whose main function is load leveling and peak shaving. With the above-mentioned technological developments, it may be anticipated that RFBs will play an important role in load leveling of power grids in the future.

7.7.4
The Application of RFB as Emergency and Backup Power

Sustained economic development sets a high standard for the quality, efficiency, environmental friendliness, safety, and reliability of power supplies, and many industries and sectors require high power and safe, stable, and reliable power supplies for emergency and backup. Among these are the electrolysis, electroplating, and metallurgy industries and transport services including trolleybus, light railway train, and subway, all of which are intensive electricity consumers. These industries and services can use off-peak electricity to charge energy storage batteries, which are then discharged for peak hour use. This can alleviate the workload and lower the operational cost of the power grid. Another important role for large-scale high-efficiency energy storage systems is to provide backup power for government operations, hospitals, transformer stations, and other crucial facilities so that they can operate properly in times of emergency. Compared with lead-acid batteries, VRBs are superior in energy storage capacity, output power, safety, stability, and life span, and thus have a very promising future in the area of energy storage. For example, two VRB systems, one 1.5 MW/1.5 MWh and the other 250 kW/500 kWh, were developed by Canada's VRB Inc. in 2001 and by Japan's SEI in 2003, respectively, and have been used as both emergency and backup power sources [42].

7.7.5
The Application as Electrical Vehicle Charging Station

The world is experiencing a petroleum shortage nowadays because of the excessive consumption of fossil fuels. Therefore, energy-efficient and environmentally friendly vehicles, represented by electrical vehicles, have become the main area of development within the vehicle industry. The construction of charging stations is required to speed up the development of electrical vehicles. However, the impact of the charging station on the electrical network and the high cost of construction and operation of charging stations are significant problems. Currently, storage batteries with large capacity are thought to represent one promising type of economical charging station. They can be charged from the electrical network during off-peak hours, that is, at night, and then the stored electrical energy can be released quickly for use by electric vehicles. The energy powering the electric vehicles is therefore not obtained directly from the electrical network so it will avoid the impact of vehicle charging on the quality and reliability of the network. VRB systems, with the advantages of high efficiency, long life span, independent power and capacity designs, and flexible operation, can meet the requirements of electrical vehicle charging stations. They can provide an efficient means of energy storage, enhance the reliability and efficiency of the electrical network, lower the operating cost, and accelerate the development and popularization of charging stations. The Austrian Cellstrom GmbH Corporation has developed a VRB system in Vienna in 2008. This has a power/capacity of 10 kW/100 kWh and an energy efficiency of 80%, integrated with a photovoltaic system, demonstrating its application as an electrical vehicle charging station [48].

7.7.6
Application as a Communication Base Station

The power supply plays an important role in assuring the stability and reliability of communication base stations, and RFBs can play an important role as a power supply backup. Currently, the inexpensive lead-acid battery is usually used to power communication base stations. However, its low energy efficiency and short life span greatly limit its application in this area. Meanwhile, the lead-acid battery, if scrapped in large numbers, can cause environmental issues. In contrast, a VRB energy storage system can reduce the maintenance cost of the communication station, prolong the life span of diesel engines, and enhance the reliability and safety level of the station. It is known that 5 kW VRB systems developed by Canada's VRB Inc. have been used as the storage energy system for a communication base station. Such usage not only enhances the reliability of the base station but also reduces its operating cost.

7.8
Perspectives and Challenges of RFB

As the development of renewable energy continues, the governments around the world have paid more and more attention to energy storage technologies. There are unprecedented opportunities and also great challenges for the development of RFBs. Although their application has been demonstrated successfully, there is still a long way to go from demonstration to commercialization.

7.8.1
The Breakthrough for Key Materials and Technologies

The extensive applications of RFB rely on breakthroughs for its key materials and technologies. First, the stability of electrolyte has to be improved. RFB electrolytes, including different active ions, sulfate ions, and hydrated ions, are complicated, and factors such as concentration, impurities, temperature, and electric field, can all cause sedimentation or precipitation of the electrolyte. Second, the imbalance of material in the electrolyte needs to be addressed. This problem results from the transfer of active ions and the diffusion of water. Taking VRB as an example, because of the permeation of ions in electrolytes across the membrane, the concentration of vanadium ions and the volume of electrolyte in the positive half cell will increase after the VRB has run for a long time. In this context, membranes with high selectivity should be examined for different RFB systems. Finally, the current density of RFB needs to be increased. The current density of VRBs is normally <100 mA cm^{-2}, which is only 10–20% of that of a proton-exchange membrane fuel cell. Therefore, the volume of the VRB module is large and more battery materials are needed, which causes an increase in battery cost.

7.8.2
System Assembly and Scale-Up

The actual output power of an operating RFB system required is from tens of kilowatts to tens of megawatts, and its capacity should be from several to hundreds of megawatt hours. Consequently, scaling up of the system is not simply an increase in battery size. In fact, several complicated processes, including matter and heat transfer, need to be considered, along with interface reaction kinetics, electrochemical reactions, and so on. The size of a single battery, the number of single batteries in a module, and the number of battery modules should all be increased for scaling up of an RFB system. Meanwhile, the electrolyte should be uniformly distributed at the electrode surfaces within a single battery and among different battery modules to maintain high power density and energy efficiency. The influence of leakage current on the battery performance should be minimized. In addition, problems relating to the sealing of RFBs during prolonged operation should also be solved.

7.8.3
Cost

Cost is an important factor to consider for commercialization of RFBs. Zinc/bromine RFB has the advantage of low cost, but one serious problem is its short cycle life. The VRB, with excellent performance and long life span, exhibits good marketing potential, but its cost is still high and needs to be further reduced. The state-of-the-art membrane currently employed in VRB systems is DuPont's Nafion membrane, whose high cost and low ion selectivity greatly restrict the commercialization of VRB. In addition, the costs of bipolar plates and electrodes are also high. On the basis of above analyses, commercialization of RFBs relies on large-scale applications of the battery. A brisk market is beneficial for larger-scale production of battery components, which will further lead to cost reduction of RFBs.

References

1. An Historical Overview of the Vanadium Redox Flow Battery Development at the University of New South Wales. *http://www.ceic.unsw.edu.au/centers/vrb/overview.htm* (accessed December 18, 2010).
2. Paul, C. B., Eidler, P. A., Grimes, P. G., Klassen, S. E., and Miles, R. C., Zinc/bromine batteries in *Handbook of batteries*, Chapter 39 (eds David Linden and Thomas B. Reddy), McGraw-Hill, pp. 39.1–39.10.
3. Miyake, S. (2001) Vanadium redox-flow battery for a variety of applications. Proceedings of the Power Engineering Society Summer Meeting, Vol. 1–3, pp. 450–451.
4. de Leon, C.P., Frias-Ferrer, A., Gonzalez-Garcia, J., Szanto, D.A., and Walsh, F.C. (2006) Redox flow cells for energy conversion. *J. Power Sources*, **160**, 716–732.
5. Zhang, H.M., Zhang, Y., Liu, Z.H., and Wang, X.L. (2009) Redox flow battery technology. *Prog. Chem.*, **21**, 2333–2340.
6. Linden, D. and Reddy, T.B. (2002) Part I. Principles of Operation in *Handbook of Batteries*, Chapter 39, 3rd edn, (eds Broadhead J. and Kuo, Han C.) Electrochemical Principles and Reactions.
7. Cheng, J., Zhang, L., Yang, Y.S., Wen, Y.-H., Cao, G.P., and Wang, X.D. (2007) Preliminary study of single flow zinc–nickel battery. *Electrochem. Commun.*, **9**, 2639–2642.
8. Sun, B.T. and Skyllas-Kazacos, M. (1991) Chemical modification and electrochemical-behavior of graphite fiber in acidic vanadium solution. *Electrochim. Acta*, **36**, 513–517.
9. Sun, B. and Skyllas-Kazacos, M. (1992) Chemical modification of graphite electrode materials for vanadium redox flow battery application. 2. Acid treatments. *Electrochim. Acta*, **37**, 2459–2465.
10. Shao, Y.Y., Wang, X.Q., Engelhard, M., Wang, C.M., Dai, S., Liu, J., Yang, Z.G., and Lin, Y.H. (2010) Nitrogen-doped mesoporous carbon for energy storage in vanadium redox flow batteries. *J. Power Sources*, **195**, 4375–4379.
11. Yi, B. (2003) *Fuel Cell-principle technology application*, Chemical Industry Press, pp. 45–46.
12. Haddadi-Asl, V., Kazacos, M., and Skyllas-Kazacos, M. (1995) Carbon-polymer composite electrodes for redox cells. *J. Appl. Polym. Sci.*, **57**, 1455–1463.
13. Xi, J., Wu, Z., and Qiu, X. (2007) Nafion/SiO$_2$ hybrid membrane for vanadium redox flow battery. *J. Power Sources*, **166**, 531–536.
14. Qiu, J.Y., Ni, J.F., Zhai, M.L., Peng, J., Zhou, H.H., Li, J.Q., and Wei, G.S.

(2007) Radiation grafting of styrene and maleic anhydride onto PTFE membranes and sequent sulfonation for applications of vanadium redox battery. *Radiat. Phys. Chem.*, **76**, 1703–1707.

15. Qiu, J.Y., Zhao, L., Zhai, M.L., Ni, J.F., Zhou, H.H., Peng, J., Li, J.Q., and Wei, G.S. (2008) Pre-irradiation grafting of styrene and maleic anhydride onto PVDF membrane and subsequent sulfonation for application in vanadium redox batteries. *J. Power Sources*, **177**, 617–623.

16. Wen, Y., Zhang, H., Qian, P., Zhao, P., Zhou, H., and Yi, B. (2006) Investigations on the electrode process of concentrated V(IV)/V(V) species in a vanadium redox flow battery. *Acta Phys. Chim. Sin.*, **22**, 403–408.

17. Li, X.G., Huang, K.L., Liu, S.Q., Tan, N., and Chen, L.Q. (2007) Characteristics of graphite felt electrode electrochemically oxidized for vanadium redox battery application. *Trans. Nonferrous Met. Soc. China*, **17**, 195–199.

18. Haddadi-Asl, V., Kazacos, M., and Skyllas-Kazacos, M. (1995) Conductive carbon polypropylene composite electrodes for vanadium redox battery. *J. Appl. Electrochem.*, **25**, 29–33.

19. Sukkar, T. and Skyllas-Kazacos, M. (2004) Membrane stability studies for vanadium redox cell applications. *J. Appl. Electrochem.*, **34**, 137–145.

20. Oei, D.G. (1985) Permeation of vanadium cations through anionic and cationic membranes. *J. Appl. Electrochem.*, **15**, 231–235.

21. Schreiber, M., Whitehead, A.H., Harrer, M., and Moser, R. (2005) The vanadium redox battery – an energy reservoir for stand-alone ITS applications along motor- and expressways. 2005 IEEE Intelligent Transportation Systems Conference (ITSC), pp. 936–940.

22. Eckroad, S. (2007) Vanadium Redox Flow Batteries – An in-Depth Analysis, pp. 2–6.

23. Sun, Y.-P. and Scott, K. (2004) An analysis of the influence of mass transfer on porous electrode performance. *Chem. Eng. J.*, **102**, 83–91.

24. Sun, C., Chen, J., Zhang, H., Han, X., and Luo, Q. (2010) Investigations on transfer of water and vanadium ions across Nafion membrane in an operating vanadium redox flow battery. *J. Power Sources*, **195**, 890–897.

25. Rahman, F. and Skyllas-Kazacos, M. (2009) Vanadium redox battery: positive half-cell electrolyte studies. *J. Power Sources*, **189**, 1212–1219.

26. Luo, Q., Zhang, H., Chen, J., Qian, P., and Zhai, Y. (2008) Modification of Nafion membrane using interfacial polymerization for vanadium redox flow battery applications. *J. Membr. Sci.*, **311**, 98–103.

27. Hine, F., Yasuda, M., Nakamura, R., and Noda, T. (1975) Hydrodynamic studies of bubble effects on ir-drops in a vertical rectangular cell. *J. Electrochem. Soc.*, **122**, 1185–1190.

28. Boissonneau, P. and Byrne, P. (2000) An experimental investigation of bubble-induced free convection in a small electrochemical cell. *J. Appl. Electrochem.*, **30**, 767–775.

29. Rousar, I. and Cezner, V. (1974) Experimental determination and calculation of parasitic currents in bipolar electrolyzers with application to chlorate electrolyzer. *J. Electrochem. Soc.*, **121**, 648–651.

30. Codina, G. and Aldaz, A. (1992) Scale-up studies of an FE/CR redox flow battery based on shunt current analysis. *J. Appl. Electrochem.*, **22**, 668–674.

31. Kinoshita, S.C. and Leach, S.C. (1982) Mass-transfer study of carbon felt, flow-through electrode. *J. Electrochem. Soc.*, **129**, 1993–1997.

32. Delanghe, B., Tellier, S., and Astruc, M. (1990) Mass transfer to a carbon or graphite felt electrode. *Electrochim. Acta*, **35**, 1369–1376.

33. Vatistas, N., Marconi, P.F., and Bartolozzi, M. (1991) Mass-transfer study of the carbon felt electrode. *Electrochim. Acta*, **36**, 339–343.

34. Marracino, J.M., Coeuret, F., and Langlois, S. (1987) A first investigation of flow-through porous electrodes made of metallic felts or foams. *Electrochim. Acta*, **32**, 1303–1309.

35. Luo, Q., Zhang, H., Chen, J., You, D., Sun, C., and Zhang, Y. (2008) Preparation and characterization of Nafion/SPEEK layered composite

membrane and its application in vanadium redox flow battery. *J. Membr. Sci.*, **325**, 553–558.

36. Xi, J., Wu, Z., Qiu, X., and Chen, L. (2007) Nafion/SiO$_2$ hybrid membrane for vanadium redox flow battery. *J. Power Sources*, **166**, 531–536.

37. Qiu, J., Li, M., Ni, J., Zhai, M., Peng, J., Xu, L., Zhou, H., Li, J., and Wei, G. (2007) Preparation of ETFE-based anion exchange membrane to reduce permeability of vanadium ions in vanadium redox battery. *J. Membr. Sci.*, **297**, 174–180.

38. Jian, X.G., Yan, C., Zhang, H.M., Zhang, S.H., Liu, C., and Zhao, P. (2007) Synthesis and characterization of quaternized poly(phthalazinone ether sulfone ketone) for anion-exchange membrane. *Chin. Chem. Lett.*, **18**, 1269–1272.

39. You, D., Zhang, H., and Chen, J. (2009) A simple model for the vanadium redox battery. *Electrochim. Acta*, **54**, 6827–6836.

40. Fabjan, C., Garche, J., Harrer, B., Jörissen, L., Kolbeck, C., Philippi, F., Tomazic, G., and Wagner, F. (2001) The vanadium redox-battery: an efficient storage unit for photovoltaic systems. *Electrochim. Acta*, **47**, 825–831.

41. Joerissen, L., Garche, J., Fabjan, C., and Tomazic, G. (2004) Possible use of vanadium redox-flow batteries for energy storage in small grids and stand-alone photovoltaic systems. *J. Power Sources*, **127**, 98–104.

42. Shibata, A.S. and Sato, K. (1999) Development of vanadium redox flow battery for electricity storage. *J. Power Eng*, **13**, 130–135.

43. Scalable, affordable clean energy storage is here. http://www.vrbpower.com (accessed December 18, 2010).

44. McDowell, J. (2006) International Renewable Energy Storage Conference, Gelsenkirchen, Germany, IRES 2006, pp. 30–31.

45. Shigematsu, T., Kumamoto, T., Deguchi, H., and Hara, T. (2002) Applications of a vanadium redox-flow battery to maintain power quality. Ieee/Pes Transmission and Distribution Conference and Exhibition 2002: Asia Pacific, Vols 1-3, Conference Proceedings, pp. 1065–1070.

46. Skyllas-Kazacos, M., Kazacos, G., Poon, G., and Verseema, H. (2010) Recent advances with UNSW vanadium-based redox flow batteries. *Int. J. Energy Res.*, **34**, 182–189.

47. Hazza, A., Pletcher, D., and Wills, R. (2005) A novel flow battery – a lead acid battery based on an electrolyte with soluble lead(II) – IV. The influence of additives. *J. Power Sources*, **149**, 103–111.

48. Solar filling station. http://www.cellstrom.at/Solar-filling-station.12.0.html?&L=1 (accessed December 18, 2010).

8
Electrochemical Supercapacitors

Aiping Yu, Aaron Davies, and Zhongwei Chen

8.1
Introduction to Supercapacitors (Current Technology State and Literature Review)

Electrochemical capacitors (ECs) have come to be called "supercapacitors," "ultracapacitors," and "power capacitors" depending on its charge mechanism or active electrode material used for construction. The expression "supercapacitor" was originally coined by Nippon Electric Company (NEC), who first successfully marketed the device as SuperCapacitor™, made from carbon material, while "ultracapacitor" stems from devices designed for use by the U.S. military using ruthenium and tantalum oxides. A clear term has yet to be defined in either research or industry; regardless, the term "electrochemical supercapacitor" will be used in this text to describe all materials that store energy by means of capacitance. Electrochemical supercapacitors use two distinctly different charge mechanisms to develop capacitance: electric double-layer capacitance through an electrostatic charge development between the electrode/electrolyte interfaces, and pseudocapacitance developed from highly reversible Faradaic reactions processes occurring at specific potentials [1–4]. The resulting capacitors are called electric double-layer capacitors (EDLC) and pseudo-capacitors.

In Figure 8.1, a Ragone plot of power with respect to energy densities is shown for the most significant energy storage and conversion systems at present. Each system plays a unique role, with their respective abilities to store and deliver energy, illustrated by their governing area [2].

Energy storage in batteries involves a relatively reversible thermodynamic pathway; however, the electrode reagents undergo phase changes during charge and discharge, which are ultimately irreversible, resulting in a limited cycle life varying among secondary batteries. Advanced Li ion secondary batteries have been produced with energy densities reaching 180 Wh kg^{-1}, yet these high energy density storage devices suffer from a lower power density performance, that is, slow discharge and recharge of the stored energy [5, 6]. The shortcomings of batteries in applications that require high load demands in short pulse intervals

Electrochemical Technologies for Energy Storage and Conversion, First Edition. Edited by Ru-Shi Liu, Lei Zhang, Xueliang Sun, Hansan Liu, and Jiujun Zhang.
© 2012 Wiley-VCH Verlag GmbH & Co. KGaA. Published 2012 by Wiley-VCH Verlag GmbH & Co. KGaA.

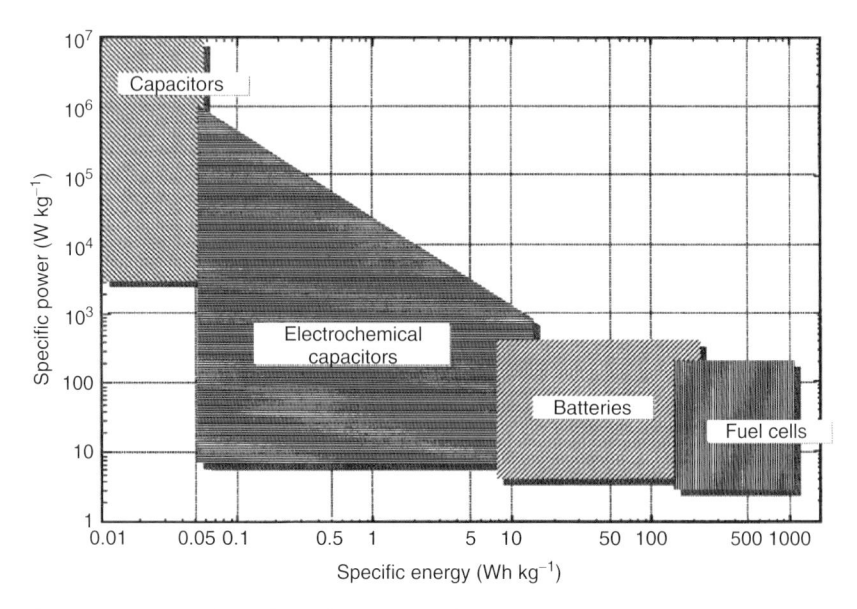

Figure 8.1 Ragone plot with rough guidelines of current energy storage and conversion devices. (Reprinted from Ref. [2], with permission from Elsevier.)

have focused attention to ECs, which contrast the shortcomings of current batteries. As power devices, ECs currently do not possess the high energy densities (commercial products are roughly in the range of 5–10 Wh kg^{-1}) compared to batteries. Nonetheless, they have a number of key attributes, such as the characteristic ability to discharge and recharge stored energy within seconds; near limitless recyclability with a typical cycle life of more than 10^6 cycles; environmental friendliness; maintenance-free operation; and operation over a wide temperature range, which attract research and industry alike, to the extent that the U.S. Department of Energy (DOE) has assigned equal importance to batteries and electrochemical supercapacitors as energy storage devices for future technologies [7].

Initial response from industry was slow to find application for these power devices; however, a rise in interest toward ECs was garnered in the 1990s with the intent of incorporating them into future hybrid electric vehicles (HEVs) [2]. The goals were to utilize ECs in providing the high power necessary for acceleration, as well as providing a means of recovering brake energy through their charging at low load demands. Since then, successful incorporation of ECs in hybrid transport vehicles such as buses [8] has promoted their continued integration with future vehicle designs. Recent advances in nanomaterials and new designs have yielded a distinct improvement in the performance, which has enabled the achievement of a higher energy density together with a significant power density to support extensive utilization for ECs in future applications.

8.1.1
Historical Overview

Storing electrical charge has been observed since antiquity through the rubbing of amber with cloth to generate light or attract small objects. These phenomena remained unexplained over millennia until progressive investigation toward the end of the seventeenth century developed into the study of "electrostatics." Progress in the field materialized with the invention of electrical devices in the mid-eighteenth century such as the Leyden jar or "condenser" in 1745, one of the first real capacitors [9]. Electrostatic devices played an essential part in properly observing and studying electricity and charge storage. Advances made in scientific knowledge and engineering improvements over the following 250 years led to the design of modern capacitors, which consists of two symmetrically sized and oppositely charged metallic plates (typically Si) separated by a dielectric (e.g., vacuum, air).

With the growing understanding of electrochemistry and electrostatic principles, the patents dating to 1957 by Becker for General Electric [10] were the first to use porous carbon flooded with a sulfuric acid electrolyte to develop an electrostatic charge to store electrical energy. The electric double-layer capacitance C_{dl} developed at all carbon surface/electrolyte interfaces along the surface and within the pores was used to charge the device. Pursuing this idea for charge storage, NEC began developing aqueous electrolyte capacitors in 1971 by licensing SOHIO's technology patent [11]. Commercial ECs filled the gap in energy storage needs by introducing devices that had an increased energy density unachievable by either electrolytic or traditional capacitors. In addition to this, they possessed a low intrinsic resistance, thereby enabling it to have a high-power density by releasing the stored energy very quickly.

The first commercial success for EC supercapacitors was achieved by NEC using a high-resistance carbon electrode for memory backup in electronic devices. This encouraged other companies (Panasonic, ELNA) to enter the market with their own commercial ECs. The U.S. military began investing in EC supercapacitor technology with the Pinnacle Research Institute (PRI) in the early 1980s for high-power portable energy storage, furthering the trend to expand the application of ECs in technology [4]. The largest push to promote their development came when the DOE completed a study that concluded the potential for EC supercapacitors in the advancement of HEVs. These successes have since paved the way for research to improve ECs for effective use as energy storage devices in current and future applications.

8.1.2
Current Research and Industry Development

Commercial EC supercapacitors are manufactured by a number of companies around the world. Developers include Ness and Panasonic in Japan, Maxwell and EPCOS in the United States, and ESMA in Russia. Current commercial devices are manufactured using symmetric carbon electrodes, with nonaqueous acetonitrile being the most common electrolyte. A list of various commercial ECs along with their electrochemical performance characteristics is provided in Table 8.1.

320 | 8 Electrochemical Supercapacitors

Table 8.1 Summary of various commercially developed electrochemical devices.

Device	V rated	C (F)	R (mΩ)	RC (s)	Wh/kg[a]	W/kg (95%)[b]	W/kg matched impedance	Weight (kg)	Volume (l)
Maxwell[b]	2.7	2800	0.48	1.4	4.45	900	8000	0.475	0.320
Maxwell	2.7	650	0.80	0.52	2.5	1281	11390	0.20	0.211
Maxwell	2.7	350	3.2	1.1	4.4	1068	9492	0.06	0.05
Ness	2.7	1800	0.55	1.00	3.6	975	8674	0.38	0.277
Ness	2.7	3640	0.30	1.10	4.2	928	8010	0.65	0.514
Ness	2.7	5085	0.24	1.22	4.3	958	8532	0.89	0.712
Asahi Glass (propylene carbonate)	2.7	1375	2.5	3.4	4.9	390	3471	0.210 (estimated)	0.151
Panasonic (propylene carbonate)	2.5	1200	1.0	1.2	2.3	514	4596	0.34	0.245
Panasonic	2.5	1791	0.30	0.54	3.44	1890	16800	0.310	0.24
Panasonic	2.5	2500	0.43	1.1	3.70	1035	9200	0.395	0.328
EPCOS	2.7	3400	0.45	1.5	4.3	760	6750	0.60	0.48
LS Cable	2.8	3200	0.25	0.80	3.7	1400	12400	0.63	0.47
Power Sys. (activated carbon, propylene carbonate)	2.7	1350	1.5	2.0	4.9	650	5785	0.21	0.151
Power Sys. (advanced carbon, propylene carbonate)	3.3	1800	3.0	5.4	8.0	825	4320	0.21	0.15
ESMA-Hybrid (C/NiO/aqueous electrolytes)	1.3	10000	0.275	2.75	1.1	156	1400	1.1	0.547
Fuji heavy industry-hybrid (C/metal oxide)	3.8	1800	1.5	2.6	9.2	1025	10375	0.232	0.143

[a] Energy density at 400 W/kg constant power; V rated. (1/2)V rated.
[b] Power is based on $P = (9/16) \times (1 - EF) \times V^2/R$, EF = efficiency of discharge.
[c] All devices use acetonitrile as the electrolyte, apart from those noted.
Reprinted from Ref. 12, with permission from Elsevier.

To date, most of the portable energy demand addressed by electrochemical storage is met by batteries. This is due to the large amounts of energy batteries can store and the corresponding power densities they can supply relative to their small weight and volume. Thus, the deficiency in energy density of ECs, which can be observed from their placement in the Ragone plot (Figure 8.1), effectively narrows suitable application for such an energy storage device to those requiring short peak power pulses. The endeavors to address the problem of low energy in research and industry alike have focused on the development of electrode and electrolyte materials. The necessity in overcoming this obstacle is motivated by the desire to improve the market response to ECs, where high costs and low-energy storage are the main deterrents.

From early applications of porous activated carbon (AC), carbon continues to demonstrate suitability as an electrode in a variety of distinct material designs. Numerous reviews by Pandolfo and Hollenkamp [4], Frackowiak [13], and Obreja [14], among others, have been written on the developments of carbon supercapacitors, and have since been followed by Pan et al. [15] and Zhang et al. [16] who have respectively reviewed carbon nanotubes (CNTs) and graphene supercapacitor applications, specifically. These relatively new carbon materials have demonstrated excellent suitability for high-power applications due to their unique morphologies and physical/electrical properties; however, the main problem remains in enhancing the energy density when these materials are used independently, which is due to their manner of energy storage (static EDLC) and lower specific surface area (SSA). Mesoporous and hierarchical templated carbon materials have demonstrated their capabilities for their application within ECs by possessing high energy and correspondingly high-power density. Furthermore, accessible submicrometer pores in porous carbon materials have shown great promise in enhancing the capacitance developed, directly increasing the energy storage.

Alternatively, pseudocapacitive materials, exemplified by RuO_2, have demonstrated the appeal for these materials (electrically conducting polymers (ECPs), transition-metal oxides) in attaining higher energy densities – achieved by a distinctly separate mechanism. Unfortunately, these materials also possess a number of undesirable physical characteristics (mainly lower resistivity, instability), and are impractical when applied as separate electrode materials.

Understanding the benefits and detriments of these separate material types have led to designs of composites electrode for ECs or asymmetric hybrid capacitors (AHCs) investigated to optimize energy and power density performances. Lithium ion intercalated battery/capacitor hybrid designs by Wang et al. [17] and Naoi et al. [18] and asymmetric capacitor electrode configurations in aqueous electrolyte investigated by Khomenko et al. [19] illustrate the advantages that asymmetric capacitor hybrids possess in overcoming the difficulty of energy storage faced by current symmetric EC designs. Asymmetric and hybrid ECs being developed by Fuji Heavy Industry, ESMA, and CSIRO [12, 20] are also being pursued to commercialize a high-power-density device with increased energy storage through this design [12, 21].

The importance of developing new electrolytes, particularly ionic liquids (ILs), is supported by the benefits of incorporating a green, solvent-free electrolyte that can increase the operating potential of current commercial devices. Studies performed by Mastragostino and Lazzari have led to an EC using an IL electrolyte that was successful in meeting the target energy and power capabilities prescribed by the DOE for HEV application [22]. Challenges, however, are presented by the low conductivity of these electrolytes, having been discussed by Lewandowski *et al.*, and various combinations of popular IL cations and anions and their mixing within acetonitrile have shown improvements [23].

These distinct strategies have the advantage of being incorporated together to enhance the potential energy storage, and through further developments ECs with higher energy density are expected to be available for commercial application. An improvement in energy density that is expected to approach that of current Li ion batteries and maintain high-power density will have a significant effect on the market strength of EC devices.

8.2
Main Types and Structures of Supercapacitors

Supercapacitors are divided primarily by the charge mechanisms of their respective active material used for electrode construction [3]. Upon their discovery, the initial development of commercial ECs focused on the EDLC that develops in porous carbon materials, which remains the predominant form of industrial EC available. Distinctly separate forms of ECs rely on a pseudocapacitive charge mechanism and a battery-like asymmetric hybrid configuration. The EDLC, psuedocapacitor, and hybrid capacitor have fundamental differences in regard to the electrochemical processes occurring for charge storage and structure design.

8.2.1
Electric Double-Layer Capacitors (EDLCs)

EDLCs are presently the most common ECs sharing a similar principle with a dielectric capacitor. The capacitance of EDLCs stems from the electrostatic charges accumulated from the electrode/electrolyte interface, and the structure is depicted in Figure 8.2 [12]. Carbon is the most commonly used electrode material for EDLC, and AC represents the foremost carbon material used in commercial devices due to its low cost and precursor availability; however, research and development is heavily focused on designing new electrode materials through the employment of CNTs, graphene, and mesoporous or hierarchical templated carbons. These alternative materials are researched in an effort to increase the characteristic energy storage and charge rate capabilities of EDLCs. Commercial ACs currently have a capacitance rating <200 and <100 F g^{-1} in aqueous and organic electrolyte electrolytes, respectively [24]. In addition, their average energy storage of ~5 Wh kg^{-1} is quite low in comparison with that of batteries (35–170 Wh kg^{-1}) [21].

In order to establish any significant capacitance for energy storage, charge accumulation by either the withdrawal or deposition of electrons from one electrode

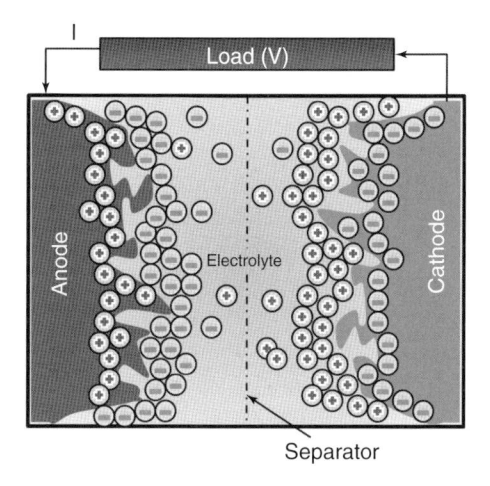

Figure 8.2 Depiction of a charged electric double-layer capacitor.

to another is necessary to induce a charge or potential across the electrode surface. The maximum applied potential is obviously limited by that at which electrolyte decomposition occurs. A number of important physical characteristics of EDLCs affect its overall capacitive performance, which include the following:

- high SSA to ensure a significant active area participates in double-layer charge development;
- high electrical conductivity to minimize power density loss through material resistance;
- distribution of pores of controlled size, ideally chosen to match the size of the electrolyte ions and minimize the thickness of the double layer;
- maintained interconnecting pore structure to ensure high ion mobility throughout the electroactive material, and thus reduce diffusion resistance;
- good wettability of the electrolyte to facilitate pore flooding, which consequently supports the effective development of capacitive charge [4].

Interest has been brought to CNTs and graphene for use as EDLC electrodes, owing to their appealing physical characteristics, particularly their highly accessible surface area and excellent electrical and mechanical properties. In terms of performance, a high specific capacitance of 180 F g^{-1} was obtained for CNTs by An et al. [25], with an energy density between 6.5 and 7 Wh kg^{-1} and a power density of 20 kW kg^{-1}, illustrating their ability to serve as a pure electrode material. Similarly, Wang et al. achieved the highest measured capacitance for graphene at 205 F g^{-1} with a power density of 10 kW kg^{-1} and energy density of 28.5 Wh kg^{-1} [26]. Ordered mesoporous carbons (OMCs) have also been investigated because of their high rate and energy storage capabilities, with a high capacitance of 340 F g^{-1} shown by Ania et al. [27] and a well-maintained capacitance at increasing scan rates [28]. These results justify the efforts for further investigation to improve the performance and expand EDLC applications.

8.2.2
Pseudocapacitor

Pseudocapacitor devices are manufactured from materials that undergo Faradaic oxidation/reduction reaction(s) at specific potentials during charging and discharging due to thermodynamic reasons, unlike a battery which employs Nernstian processes [3]. Depending upon the material design and structure, the charge mechanisms occurring between the electrode and electrolyte can also involve adsorption or intercalation (doping/dedoping) of the electrolyte. Repulsive interactions between these species at the surface or within the material structure itself characterize the expansion of the charge developed over the defined potential, resulting in a larger voltage operating window for useful applications [29]. The significance of pseudocapacitive materials is largely recognized by an enhanced capacitance relative to EDLC (10–100 times the capacitance), owing to the electron transfer reactions that occur during charging, which contrast the electrostatic process defined by a process where no Faradaic reactions take place [3].

The materials mainly utilized for capacitor application as pseudocapacitors are transition-metal oxides (RuO_2, MnO_2, etc.) and ECP (polyaniline (PANI), polypyrrole (PPy), and poly(3,4-ethylenedioxythiophene) and their derivatives). Transition-metal oxides have been known to exhibit high capacitances, with MnO_2 achieving 698 F g^{-1} via a sol–gel process [30], and a range of 600–1000 F g^{-1} attainable with hydrous RuO_2 [31]. However, the use of transition metals as electrodes is often limited by their poor conductivity and region of electroactivity (MnO_2 irreversibility and dissolution as a negative electrode) [32]. Although RuO_2 exhibits excellent performance as a pure electrode material, the high cost and toxicity of this material makes it unsuitable for widespread commercial use [15]. ECPs are used as alternative pseudocapacitive materials that are inexpensive and easily synthesized by *in situ* oxidation or electropolymerization. A range of specific capacitance values between 150 and 400 [33], 115 and 420 [34, 35], and 60 and 250 F g^{-1} [36, 37] have been obtained for PPy, PANI, and poly(3,4-ethylenedioxythiophene), respectively. While these materials possess a relatively high capacitance and good conductivity when doped, they typically suffer poor stability from the mechanical stress of swelling and shrinking during doping/dedoping [38]. As a result of these drawbacks, pseudocapacitors are frequently integrated with carbon materials to overcome their weak mechanical and electronic properties [39].

8.2.3
Asymmetric Hybrid Capacitors

AHCs are constructed with a dissimilar anode and cathode to exploit the characteristic operating voltages of each material. In addition, materials with high hydrogen and oxygen evolution overpotentials are better suited for this design as anode and cathode, respectively, by enabling a larger operating window to be applied in aqueous electrolyte devices [40].

ACs demonstrated their potential in AHCs early on with Pb/PbO_2 designs [41] and have since shown promise as an electrode configurations together with lithium titanate. This design, initially introduced by Amatucci et al. [42], was further developed by Naoi et al. [18], who were successful in achieving a high energy density $\sim$29 Wh kg^{-1} at a power density of 10 kW kg^{-1}. Industry has taken notice of this potential use for ACs, as Fuji Heavy Industries have implemented their own design of a hybrid capacitor utilizing AC predoped with lithium ions as a positive electrode and AC as a negative electrode [41]. OMCs have also been incorporated in hybrid capacitors, as demonstrated by Wang et al. using aligned titania nanotubes (ATNTs) and OMC as the anode and cathode, respectively [17]. The configured hybrid was able to attain high-power density performance of 3000 W kg^{-1} and an excellent energy density of 25 Wh kg^{-1}. An additional advantage of high stability after 1000 cycles indicates the potential these hybrid configurations may have for future devices.

8.3
Physical/Electrochemical Processes in Supercapacitors

Physical and electrochemical processes in electrochemical supercapacitors are described by their contribution to, or direct involvement in, developing and releasing a capacitive charge. Here, considerations are given to theoretical principles involved in three different processes, electrolyte resistance, and modeling. These factors have a direct effect on the key parameters for EC application, specifically the achievable power and energy density, and thus require appropriate consideration. With regard to explaining the development of pore capacitance, the complex electrochemical processes have been considered using equivalent circuit modeling to describe impedance observations.

8.3.1
Physical/Electrochemical Processes for EDLC

A portrayal of the electric double layer was first presented by von Helmholtz in the nineteenth century to describe the development of charge occurring at the interface between a colloidal particle and an ionic solution [43, 44]. The development of an opposite charge $\pm \Delta q$ across an interphase is achieved through the accumulation or deficiency of electrons within electrodes and depends upon an applied potential ΔV. This gives rise to a capacitance $C = dq/dV$. The positive or negative charge developed along the interface of the electrode is balanced by an induced accumulation of oppositely charged solution ions. An insulating separator is included between the two electrodes to prevent short-circuiting and is characteristically porous to allow ion diffusion. The two "layers" of compact, oppositely charged ions resulting at the electrode/electrolyte interphase gave rise to the description of the electric double layer. This model was the first to be applied to electrode interfaces, followed by a number of subsequent models developed by Gouy, Chapman, Stern, and Grahame [1].

Taking into account the effects of thermal fluctuation upon the compact layer of solution ions led to Guoy–Chapman's interpretation of the diffuse layer [45, 46]; however, an overestimation of the developed double-layer capacitance was problematic because of their assumption of ions being point charges with no volume. A model later introduced by Stern [47], and further clarified by Parsons [48], was able to overcome this drawback by recognizing both a compact region of adsorbed ions described by a Langmuir isotherm, and beyond this region a diffusion layer. Each of these layers gave rise to capacitances C_H, named after Helmholtz, and C_{diff}, respectively. These capacitances are conjugate elements of the overall capacitance of the double layer and described by the equation [4, 49]

$$\frac{1}{C_{dl}} = \frac{1}{C_H} + \frac{1}{C_{diff}} \tag{8.1}$$

This theory, named after Stern, was further developed by Grahame's model in 1947 with a distinction made to further divide the compact layer into an inner and an outer Helmholtz layer upon consideration of the variation in ionic radii between cations and anions [50]. Because cations are commonly smaller than their anionic counterparts, they possess a higher retention of their solvation shell. Thus, a difference in the distance of closest approach to the charged surface interphase is present among cations and anions. This gives rise to the inner and outer Helmholtz layers populated by anions or cations, where one or the other dominates depending on the electrode polarization. As anions commonly yield a smaller distance to the electrode, the capacitance at the positive electrode can be twice that developed at corresponding negative electrode depending on the electrolyte ions, solvent, and electrode material. An illustration of these developments in modeling of the electric double layer is shown in Figure 8.3.

8.3.1.1 Analysis of the Diffuse Layer

When analyzing the capacitance of an electrode, the diffuse layer plays less of a role than the compact layers in the typical high-concentration electrolyte solutions employed with capacitors. This has been shown through experimentation and the reciprocal sum relationship alike [1]. Yet, the properties of double-layer capacitance were derived from the analysis of the diffuse layer.

The electrostatic energy applied to cations and anions in solution as a result of the electric field gradient near the electrode surface and their resulting interaction were investigated by Chapman [46]. Poisson's electrostatic equation in combination with Boltzmann distribution law was used by Chapman to explain the behavior of ion distribution within the diffuse layer. Forgoing derivative calculations, for a surface charge density q_s, the difference in potential ψ_a at the distance of closest approach a (i.e., the compact layer) and the bulk solution potential ψ_b is given by the following equation [1, 51, 52]:

$$q_s \left(\frac{kT\varepsilon_1}{2\pi} \sum\nolimits_{+}^{-} c_{\pm} \left\{ \exp\left[\frac{ze(\psi_a - \psi_b)+}{kT} \right] - 1 \right\} \right)^{1/2} \tag{8.2}$$

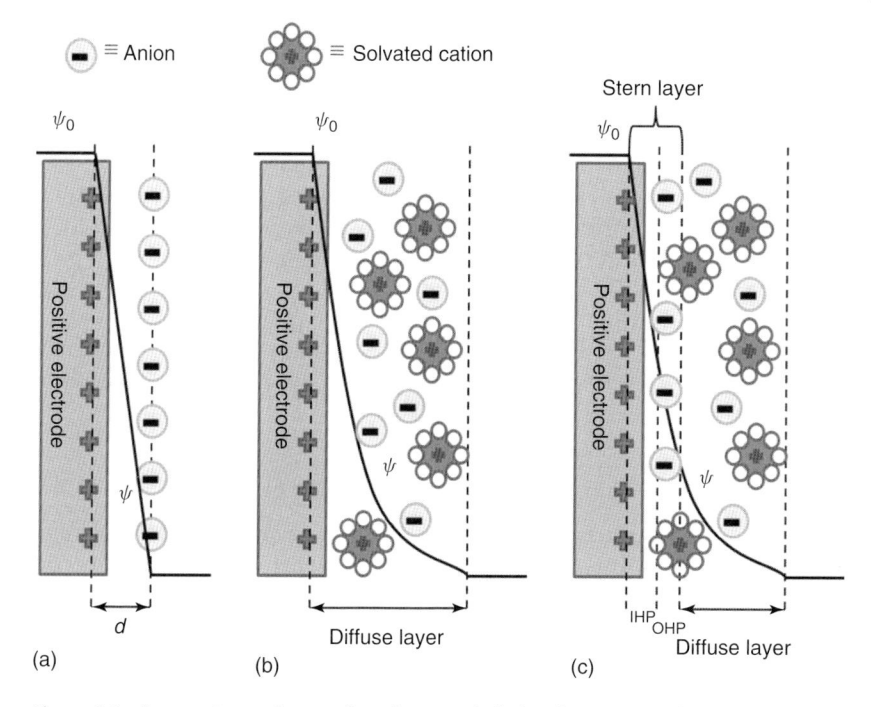

Figure 8.3 Progressive evolution of models describing the electric double layer at a positive electrode: (a) the initial Helmholtz model, (b) the Guoy-Chapman model, and (c) the Stern model, where the division of the inner Helmholtz plane (IHP) and outer Helmholtz plane (OHP) show specifically and nonspecifically absorbed ions. The distance d in the Helmholtz model describes the double-layer distance; ψ_0 and ψ are the surface and electrode/electrolyte interface potentials, respectively.

where k is Boltzmann's constant, T is the temperature, c is the stoichiometric concentration of cations (c_+) and anions (c_-), and ε_l is the local dielectric which may depend upon the polarization of the electrode. With a symmetrical electrolyte, that is, $z = z_+ = |z_-|$, and equal concentration of anions and cations, that is, $c = c_+ = c_-$, Eq. (8.2) reduces [1, 52] to

$$q_s = \left(\frac{2kT\varepsilon_l}{\pi}\right)^{1/2} \sinh\left[\frac{ze(\psi_a - \psi_b)}{kT}\right] \tag{8.3}$$

This expression was used to ascertain the capacitance of the diffuse layer by differentiating q_s with respect to the potential of the diffuse layer ($\psi_a - \psi_b$). The following equation for the diffuse-layer capacitance C_{diff} is given as a result [1]:

$$\frac{\partial q_s}{(\psi_a - \psi_b)} = \left(\frac{z^2 e^2 c \varepsilon_l}{2\pi kT}\right)^{1/2} \cosh\left[\frac{ze(\psi_a - \psi_b)}{kT}\right] \tag{8.4}$$

Using this relationship to determine capacitive behavior at potential differences between the electrode surface and solution ($\psi_s - \psi_b$) is equivalent to any experimental capacitive behavior only when the potential difference approaches the point of zero charge (pzc), which is less than 0.1 V. This is a result of the assumption of point charges that theoretically develop a large charge upto the surface of the electrode. With the recognition by Stern of the necessary existence of a distance of closest approach to distinguish the compact double-layer and diffuse-layer capacitances, he made the necessary separation between these components as discussed earlier. Dividing potential drop across the double layer, $(\psi_s - \psi_b)$ into two parts $(\psi_s - \psi_1)$ and $(\psi_1 - \psi_b)$, where ψ_1 is the mean potential at a distance of closest approach, the dependence of the potential of both layers on a charge q_s along the surface is [1]

$$\frac{\partial(\psi_s - \psi_b)}{\partial q_s} = \frac{\partial(\psi_s - \psi_1)}{\partial q_s} + \frac{\partial(\psi_1 - \psi_b)}{\partial q_s} \tag{8.5}$$

This is analogous to Eq. (8.1). Reorganizing the reciprocals can be done for a better understanding of the contribution made by capacitive components. By replacing C_{diff} in Eq. (8.1) with Eq. (8.4), the double-layer capacitance is given by

$$C_{\text{dl}} = \frac{C_H (z^2 e^2 c \varepsilon_l / 2\pi kT)^{1/2} \cos h[ze(\psi_1 - \psi_b) 2kT]}{C_H + (z^2 e^2 c \varepsilon_l / 2\pi kT)^{1/2} \cos h[ze(\psi_1 - \psi_b) 2kT]} \tag{8.6}$$

The resulting derivation agrees with experimental data. Small deviations in the potential between the surface and the solution will result in a dominant expression of C_{diff}. It is critical to mention that the smaller of the two capacitance components determines the overall capacitance C_{dl}. As well, the mean potential for which ions are at the closest approach to the charged surface is not an equivalent value for both cations and anions. Their differences in polarity and affinity to surface functionalities for adsorption should be taken into account for better accuracy. The potential at the region of *closer* approach can be called ψ_2.

The investigation of the diffuse layer provides a method of investigating the Helmholtz layer capacitance C_H, which is of much more significance in aqueous electrolyte solutions applied in ECs. This will be further explored in the following section.

8.3.2
Physical/Electrochemical Processes for the Pseudocapacitor

Pseudocapacitance has been well described by Conway *et al.* through the study of ruthenium oxide systems. This distinct mechanism of charge arises with the use of metal-oxide films and ECPs through redox reactions and in the electrosorption of ions onto the surfaces of the electrode. Description of pseudocapacitance from a thermodynamic perspective in relation to the potential is given by a Langmuir isotherm by the equation

$$\frac{y}{(1-y)} = K \exp\left(\frac{VF}{RT}\right) \tag{8.7}$$

where the quantity of the y term depends on the

- fractional coverage of the electrode surface θ, through adsorption,
- fractional absorption through intercalation X, and
- conversion of an oxidizing [Ox] or reducing [Re] species in a redox system.

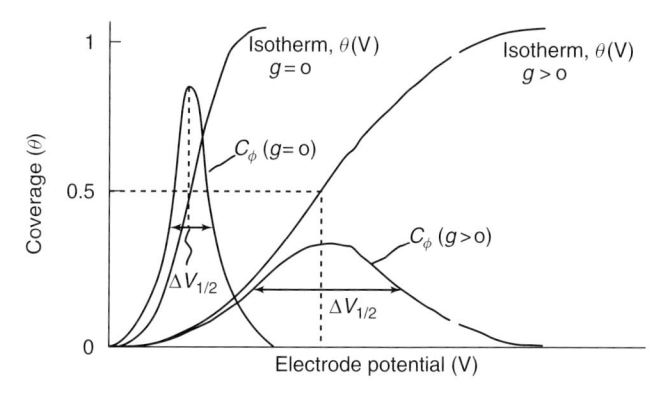

Figure 8.4 The potential dependence of C_Φ relative to the potential dependence of the surface coverage θ of an absorbed electroactive species. (Reprinted from Ref. [3], with permission from Elsevier.)

Each of these mechanisms involves an oxidized and reduced species to which an electron transfer process has taken place to an extent Q. These reactions are a function of potential V such that the extent of their reaction can be equated by dQ/dV. This term is analogous to a form of capacitance derived by the electric double layer, and hence the term pseudo is applied to differentiate the two mechanisms [3, 29, 53]. These processes necessitate both reacting species to undergo a chemical change of state. This detail is of particular importance when considering the stability of the electrode reactant. Taking the differential of quantity y with respect to the potential V gives [3, 53]

$$\frac{dy}{dV} = \frac{F}{RT} \cdot \frac{K \exp\left(\frac{VF}{RT}\right)}{\left[1 + K \exp\left(\frac{VF}{RT}\right)\right]^2} \tag{8.8}$$

Interactions between species adsorbed, electrochemically bonded, or absorbed are explained by incorporating an additional term ($\exp \pm gy$), which helps to explain the more complex behavior that evolves. These complex interactions help to provide a wider potential window as observed in Figure 8.4 for pseudocapacitance to develop, despite lowering the resulting capacitive charge.

When considering pseudocapacitance, two complementary methods have been developed. The first approach is with equilibrium thermodynamics, taking into account the fraction of actively covered (θ) and unoccupied ($1 - \theta$) surface sites, and the second with the kinetic processes occurring beyond the equilibrium state.

8.3.2.1 Thermodynamic Approach

Electrosorption of adsorbed, intercalated, or redox species is proposed to follow an *equivalent* process. Initially, a Langmuir-type isotherm was proposed to describe the electrochemical redox processes such as the adsorption process $H_3O^+ + M + e \rightleftharpoons MH_{ads} + H_2O$. However, Conway and Gileadi considered the repulsion or attraction forces exhibited by the chemisorbed species between each other as well as the charged electrode, and oversimplified these energy effects through a lateral interaction term $g\theta$ [54]. This was described by a Frumkin-type isotherm [3, 53]

$$\frac{\theta}{1-\theta} = K \exp(-g\theta) \cdot C_H + \exp\left(\frac{VF}{RT}\right) \tag{8.9}$$

where c_{H+} is the concentration of hydronium ions, $1-\theta$ and θ are the unoccupied and occupied sites, respectively, and the reaction rate constant K incorporates the repulsion and attraction forces attributed to the adsorbates. When a charge q_1 is required for the formation process of a film or redox reaction to take place, then a pseudocapacitance C_{pc} is derived as

$$C_{pc} = q_1 \left(\frac{d\theta}{dV}\right) \tag{8.10}$$

Therefore, the pseudocapacitance can be described as a function of the quantity θ by [3]

$$C_{pc} = \frac{q_1 F}{RT} \cdot \frac{\theta(1-\theta)}{1+g\theta(1-\theta)} \tag{8.11}$$

By incorporating the interaction term g, repulsion forces ($g > 0$) will result in a capacitance that develops over a wider range of potential that subsequently decreases in value. Attractive interactions occurring between adsorbed particles ($g < 0$) will, in contrast, develop a capacitance that becomes gradually sharper and reaches a singularity at $g = -4$. The electrosorption of a species becomes increasingly difficult with increasing g values, and therefore an increase of potential is required to achieve the desired value of the quantity y of any of the forms of pseudocapacitance that are taking effect. The reverse is true with negative values of g, where a lower potential is required for the electroadsorption of the adsorbed species. A reduction in the working potential range to develop the pseudocapacitance results from this effect.

For the development of a practical maximum pseudocapacitance that is nearly constant over a useful changing potential, the value of quantity y should be in a range close to 0.5 (e.g., $0.3 < y < 0.7$) with an interaction value greater than or equal to 10 ($g \geq 10$) [1].

8.3.2.2 Kinetic Approach

Equilibrium pseudocapacitance exists when a defining Faradaic reaction is in equilibrium at a given potential. Gileadi and Srinivasan [55] made a case to disturb this equilibrium by cyclic voltammetry (CV) and study the relation with the forward and reverse rate constants k_1 and k_{-1} for an electrosorption reaction. They observed that for a value of θ between 0.05 and 0.95, a significantly larger current density i typically develops in contrast to that due to C_{dl}. Only when the system undergoes fast potential shifts, or with a large current pulse, will a nonequilibrium status exist. These irreversible effects can give rise to subsequent kinetic effects, which must be taken into account for pseudocapacitive materials. With this exception, the processes responsible for pseudocapacitance such as the oxidation/reduction [Ox]/[Red] of a species remain kinetically reversible. This behavior is shown for pseudocapacitive electrodes, whose performance is the same as that of a normal capacitor and attain a high pseudocapacitance.

8.3.2.3 Redox Pseudocapacitance

Any species that undergoes oxidation or reduction through the general reaction $Ox + e \rightleftharpoons Red$ can be described by the Nernst equation

$$E = E° + \frac{RT}{F} \ln\left(\frac{[Ox]}{[Red]}\right) \tag{8.12}$$

where E represents the equilibrium potential for the reaction, and square brackets signify concentrations of the species [3]. By incorporating the molar quantity $M = [Ox] + [Red]$ into the equation and taking the exponential of this relation, where $(E - E°) = \Delta E$, the following is obtained:

$$\frac{\left[\frac{Ox}{M}\right]}{\left[1 - \left[\frac{Ox}{M}\right]\right]} = \exp\left(\Delta E \cdot \frac{F}{RT}\right) \tag{8.13}$$

The differential of this equation with respect to ΔE gives a function of capacitance C as

$$\frac{C}{Q} = \frac{\frac{F}{RT} \cdot \exp\left(\Delta E \cdot \frac{F}{RT}\right)}{\left[1 + \exp\left(\Delta E \cdot \frac{F}{RT}\right)\right]^2} \tag{8.14}$$

where the term Q defines the charge related to the material able to undergo oxidation or reduction. As with all quantities of y, the pseudocapacitance is maximized when $[Ox] = [Red] = 0.5$, yielding a rough maximum value for the right-hand side of Eq. (8.14) of approximately 10. Therefore, an approximate maximum capacitive value of $10Q$ is achievable. These redox capacitances can have a limited potential window, and, in addition to reaction kinetics, the reversibility of the reaction may also rely on diffusion, which plays a role in reducing the adsorption or desorption of the adsorbate. When the redox species have a series of couples in a broad potential range, this can provide an increase to the respective operating potential, a behavior well observed in ruthenium dioxide materials. Intercalation processes may also develop pseudocapacitance and can be considered as the fractional occupancy of ions in a three-dimensional lattice through intercalation (X) or a quasi-two-dimensional adsorption (θ) at layer–lattice van der Waal gaps. This can be explained in the same manner as adsorption or reduction–oxidation as previously discussed. Table 8.2 provides an overview of the three types of mechanisms that give rise to pseudocapacitance.

8.3.3
Physical/Electrochemical Processes for AHCs

In general, AHCs utilize both pseudocapacitive and EDLC mechanisms during their charging and discharging due to their incorporation of a carbon electrode

Table 8.2 Systems for materials that give rise to psuedocapacitance in EC supercapacitors.

System type	Essential thermodynamic relations
(a) Redox system $Ox + ze^- \leftrightarrow Red$ and $O^{2-} + H^+ \leftrightarrow OH^-$ in oxide lattice	$E = E_0 + (RT/zF) \ln R/(1 - R)$ $R = [Ox]/([Ox] + [Red])$; $R/(1 - R) \equiv [Ox]/[Red]$
(b) Underpotential deposition $M^{z+} + S + ze^{-1} \leftrightarrow S \times M$ ($S \equiv$ surface lattice sites)	$E = E_0 + (RT/zF) \ln \theta/(1 - \theta)$ $\theta =$ two-dimensional site occupancy fraction
(c) Intercalation system Li^+ into "MA_2"	$E = E_0 + (RT/zF) \ln X/(1 - X)$ $X =$ occupancy fraction of layer-lattice sites (e.g., for Li^+ in TiS_2)

Reprinted from Ref. 3, with permission from Elsevier.

together with a ECP, metal-oxide composite, or battery electrode. While both these mechanisms have been previously discussed (Section 8.3.2.1), the emphasis for these devices is placed on short diffusion pathways for electrolyte diffusion within the pores of carbon materials, and intercalation within the metal oxide or conducting polymer structure [17, 40, 56].

Considerable efforts have gone into the design of AHCs with the employment of lithium ion intercalation through the use of a lithium ion electrolyte as well as by lithium predoping of the electrode to increase the energy storage of EC devices. The latter property has proven particularly useful for high cycling rates which result in slowing ion diffusion and effective intercalation of Li ions necessary to develop a capacitive charge. Wang et al. noted the negative proportionality of tube length to transport ability through comparison of long and short ATNTs, where consideration of both the anode (ATNT) and cathode (OMC) porosity was made

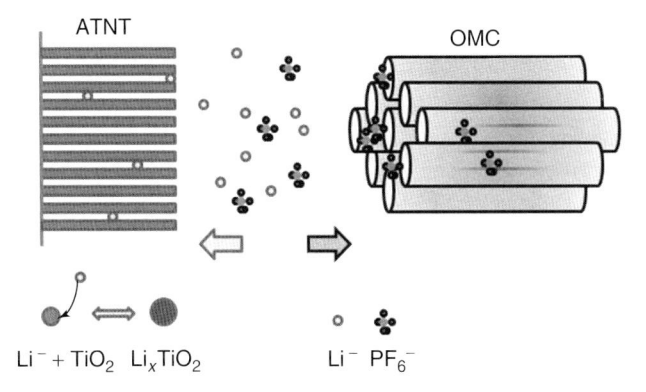

Figure 8.5 AHC: Schematic of ATNT/OMC (anode/cathode) utilizing $LiPF_6$.

to ensure that a high-power density for the AHC was maintained [17]. Figure 8.5 illustrates such an AHC system that utilizes Li ion for charging.

8.3.4
Electrolyte Processes: Conductance and Dissociation

The electrolyte, consisting of a solute dissolved in solvent, influences the electrical behavior of EC supercapacitors in three ways: (i) electrolyte conductance and its equivalent series resistance (ESR); (ii) adsorption of anions (or cations depending upon electrolyte); and (iii) dielectric properties that determine the specific capacitance of the double layer and its dependence on potential.

In any application of EC capacitance, the internal electrolyte resistance R_e within the pores is required to be minimized in order to increase the rates of charge and discharge for reversible processes. This can often be a challenge for highly porous electrode structures where a maximum surface contact between the electrolyte and porous electrode is desirable. In cell assembly, the intrinsic resistance R_c of the electrode material due to interparticle contact has to be taken into account, and when reduced through matrix compression, an adverse reduction of the volume fraction for the electrolyte occurs. This process inevitably increases R_e. Thus, an optimal reduction in these resistances is required to improve the EC power performance. Additional factors including conductance, dissociation, and mobility can have an effect on electrolyte performance, and subsequently influence the overall ESR. The concentration of free charge carriers (i.e., cations, anions) and their ionic mobility will have a direct influence on the conductance of the electrolyte. These in turn are governed by the degree of dissociation α for the dissolved salt to separate into cations and anions, and the solubility of the salt in a given solvent. Effects of salt concentration, temperature, and the dielectric constant of the solution can alter the extent of dissociation of ions and the resulting equilibrium between dissociated and paired ions. In addition, electrostatic interactions among dissociated ions and the temperature dependent viscosity of solution can affect the ionic mobility [57–59]. In order to minimize the electrolyte resistance, maximum dissociation of the ions is desirable to yield the highest achievable concentration of freely mobile ions. The dissociation of ions at equilibrium can be expressed using the following equation:

$$\underset{(1-\alpha)c}{CA} \overset{K_c}{\rightleftharpoons} \underset{\alpha c}{C^+} + \underset{\alpha c}{A^-} \tag{8.15}$$

Therefore, the equilibrium equation in terms of α is then

$$K_c = \frac{\alpha^2 c}{(1-\alpha)} \tag{8.16}$$

with the value of α being in the range of 0–1. Many aqueous electrolytes have a value of $\alpha \rightarrow 1$, while nonaqueous electrolytes used in EC supercapacitors generally have value of $\alpha \ll 1$ owing to weak dissociation and the sharing of solvent shell between the ion pair. This weak ionic behavior is generally what is responsible for an increase in the ESR for nonaqeuous ECs [1, 57].

8.3.5
Modeling of Electrochemical Behavior

Initial modeling behavior of porous electrodes with equivalent capacitive elements was considered by de Levie [60]. The investigation of a single equivalent element of a circuit was done to depict the electrochemical behavior of a pore as shown in Figure 8.6, within a segment dz of pore of length L. The assumptions made were that the pore was cylindrical in shape with constant radius r_0. The solution resistance R existing in series and the double-layer capacitive element C are both defined with respect to pore length z and assumed to be distributed uniformly along the length of the pore, where the opening of the pore pertains to $z = 0$. The intrinsic resistance of the material was not considered by assuming it to be negligible compared to that contributed by the electrolyte. An analogy to the electric power transmission line was considered in terms of the ideal pore circuit. Differential equations of potential V and current I as a function of charging time t and pore length z were evaluated by de Levie at potentiostatic and av (alternating voltage) conditions. Evaluation of the model under av concluded that the charging current of a pore behaves as would the Warburg diffusion impedance element of a flat electrode, giving rise to a 45° phase angle between the applied potential and resulting current at the entrance to the pore [60].

From de Levie, the general conclusions revealed a slow-moving response in either the potential or current relative to the depth of the pore (i.e., the further down the pore, the slower the change) evident in Figures 8.7 and 8.8. The penetration depths to characterize the pore fraction useful in the charging process for both dc (direct current) and ac (alternating current) cases are $z(4t/RC)^{1/2}$ and $z/(1/2\omega RC)^{1/2}$,

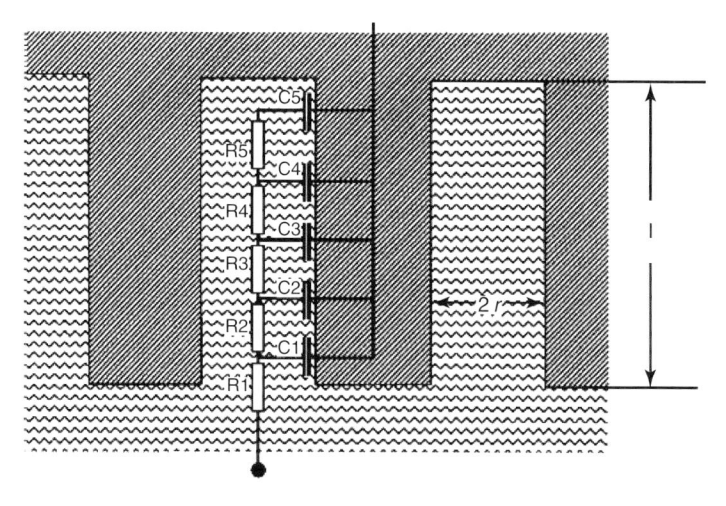

Figure 8.6 Five distributed resistance and capacitance elements representing the equivalent circuit of a constant $2r$ diameter cylindrical pore. (Reprinted from Ref. [2], with permission from Elsevier.)

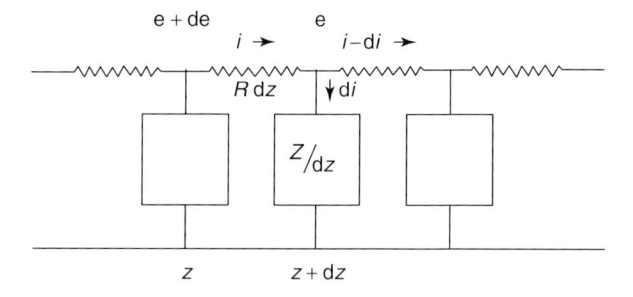

Figure 8.7 Short segment of a transmission line model with resistances and capacitances distributed uniformly. (Reprinted from Ref. [60], with permission from Elsevier.)

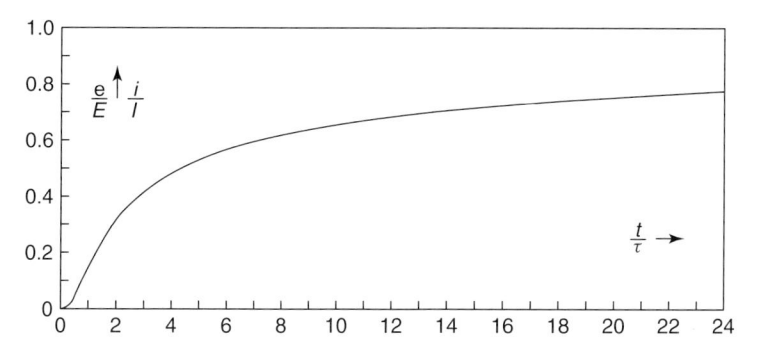

Figure 8.8 Potential (current) response as a function of time when a step function of the current with amplitude $E(I)$ is applied at the mouth of the pore. The characteristic time (τ) is dependent upon the electrolyte resistance (R), and the double-layer capacitance (C) is per unit length of the pore axis coordinate z. (Reprinted from Ref. [60], with permission from Elsevier.)

respectively. From these characteristic lengths, longer periods of charging time or a decrease in frequency can increase the depths of charging for direct or alternating potential/current applications, respectively. The charging of a porous electrode becomes analogous to that of a flat electrode in short periods or high at frequencies.

This model oversimplifies the actual complex electrochemical behavior occurring in porous structures, in addition to making assumptions of a uniform circular cylindrical shaped pore when pores will realistically possess branching and cross-linking. However, it successfully imparts the description of pore charging to distributed RC circuits along their length and generates a means of measuring the active charging depth in ac or potential testing. The model also does well to explain the 45° phase angle existing between an av voltage and resulting ac current, independent of angular frequency ω which shown in Figure 8.9.

8.3.5.1 Effects of Pore Size and Pore-Size Distribution

Recent attention for the modeling of capacitive charge has emphasized the importance of pore size and its distribution through electrode materials, principally

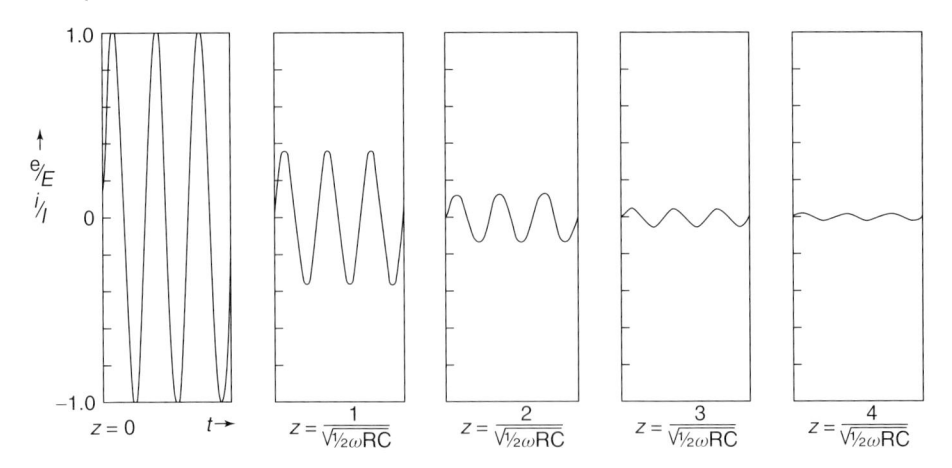

Figure 8.9 Potential (current) response as a function of time when a sinusoidal current with amplitude $E(I)$ and angular frequency ω is applied and maintained at the entrance of the pore ($z = 0$). The electrolyte resistance (R) and double-layer capacitance (C) are per unit length of the pore axis coordinate. (Reprinted from Ref. [60], with permission from Elsevier.)

to evaluate their significance to the energy and power density performance of an electrode or device. Generally, materials incorporating both micropores and mesopores, classified as having <2 and 2–50 nm diameters, respectively, are considered to be important by the particular roles they play in the energy storage and power performance of a resulting electrode [13]. Structures containing a microporous network have been shown to exhibit very high double-layer capacitances in carbon materials, effectively demonstrating how micropores make a significant contribution to the developed capacitance of an electrode material. The importance of micropore inclusion should not, however, dismiss the constraints that arise due to difficulties in wetting the available surface area and cumulative electrolyte resistances. Mesopores are useful in aiding electrolyte transport for micropore distributions, and also in developing a capacitance [61, 62]. For pulse-power demands that are required at high rates or at short time intervals, mesopores provide the necessary reduction in electrolyte resistance contributing to the ESR, and are more accessible to an alternating current or voltage. Thus, an optimization of varying poren sizes with their effective distribution in any application must be considered beforehand to maximize the capacitance [63].

Previous ideas of micropores in charge storage suggested that pores were limited to 0.5 nm diameter for electrolyte accessibility, particularly for larger solvated ions in nonaqueous electrolytes [64]. Work by Salitra *et al.* and further investigation by Raymundo-Pinero *et al.* have shown that this is not the case and an anomalous capacitance arises in materials with pore sizes <1 nm [65, 66]. Partial or complete desolvation of the solvent sheath in these pores yields a considerably shorter distance between the inserted ion and the respective carbon surface. Through experiment, pores sizes of 0.7 and 0.8 nm have been shown to exhibit the largest capacitance for aqueous and nonaqueous electrolytes, respectively. Treatment by Huang et al. to explain the variation in capacitance development in mesopores and micropores describes electric double-cylinder capacitor (EDCC) and electric wire-in-cylinder capacitor (EWCC) models for the respectively classified pore sizes [67, 68]. The capacitance can be calculated by these models as follows for EDCC:

$$\frac{C}{A} = \frac{s_\gamma s_0}{b \ln\left(\frac{b}{b-d}\right)} \tag{8.17}$$

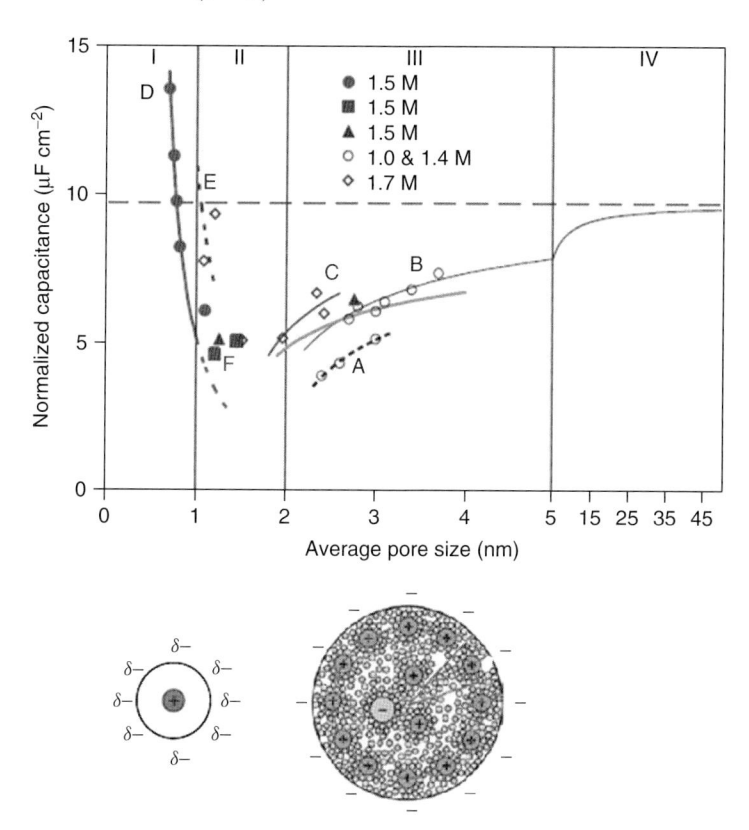

Figure 8.10 Capacitance of various carbons tested with either tetraethylammonium methylsulfronate (TEAMS, 1.7 M) or tetraethylammonium tetrafluoroborate (TEABF$_4$, 1.0, 1.4 and 1.5 M) in acetonitrile and normalized by the specific surface area (SSA) as a function of pore size. The symbols represent experimental data, with lines showing the model fits. Anomalous capacitance is shown in submicropores (area I) where partial or complete desolvation would occur. EWCC modeling applies to areas I and II, area III is modeled using EDCC, and area IV with the traditional EDLC model. Templated mesoporous carbons (A, B), activated mesoporous carbon (C), microporous carbide-derived carbon (D, F), and microporous carbide-derived carbon and microporous activated carbon (E) were used. Model schematics are shown below the respective areas. (From Ref. [67], Copyright Wiley-VCH Verlag GmbH & Co. KGaA. Reproduced with permission.)

and for EWCC

$$\frac{C}{A} = \frac{s_\gamma s_0}{b \ln\left(\dfrac{b}{a_0}\right)} \tag{8.18}$$

where C is the capacitance, A is the active surface area, b is the pore radius, d is approaching distance of the ion to the electrode surface, and a_0 is the effective size of the desolvated ion. These models did well to explain a general trend in experimental data from carbons possessing a range of average pore sizes, as is evident in Figure 8.10.

The results of capacitance development by desolvated ions demonstrates a lack of understanding concerning the interface between an ion and electrode surface at the nanoscale, where the generally accepted model of the double layer containing a diffuse layer does not apply to this situation. Further investigation to explain the behaviors of ions at this scale is expected to aid in the understanding of charge storage for future improvements in material development.

8.4
Supercapacitor Components

The desired performance of an EC supercapacitor ultimately begins with its components. These components can be tailored for specific applications; however, there are requirements for these components that are universal in the development of all EC systems. The aim of minimizing resistances while maximizing the potential to develop capacitive charge requires consideration of a number of factors for each component, including the electrode, electrolyte, separator, and current collector.

8.4.1
Electrode

An electrode for use in EC supercapacitors differs in a number of ways from those used in batteries. Primarily, current commercial supercapacitors are constructed with symmetric electrodes (i.e., both the anode and cathode are of the same material) where the ideal electric double-layer capacitance evolves as pure electrostatic charge with no oxidation–reduction processes taking place. Thus, distinguishing between the anode and cathode electrodes can simply apply which electrode is negatively and positively charged during discharge. This notation can change with the development of asymmetric capacitors; however, the distinction between anode and cathode is typically dependent upon the ideal operating potential window respective to the incorporated materials. A number of properties are required for a material to be of use as an electrode for EC supercapacitors, which are listed below:

- cycling capability ($>10^5$) and resulting highly stable cycle life;

- resistance to irreversible oxidation or reduction reactions along the surface of the electrode, which is important in avoiding loss of stability and performance;
- in some cases, working in opposition to the previous property, the necessity for high SSA to develop the electric double layer, in the range of 1000–2000 $m^2\,g^{-1}$;
- electrochemical stability above the potential limits of electrolyte decomposition;
- controlled pore size distribution to develop high capacitance while minimizing internal resistances;
- surface wettability of the electrolyte;
- mechanical reliability for its integration into a practical EC supercapacitor cell.

Electrodes constructed using material types of either EDLC or pseudocapacitance have the respective benefits and challenges with their application for EC supercapacitors.

8.4.1.1 Carbon as an EDLC Material

Carbon has been incorporated as a base electroactive material for over 40 years because of its unique physical and chemical properties. Benefiting from a relatively low processing cost with a controllable pore structure, its high electrical conductivity and surface area ($\sim$1 to $>$ 2000 $m^2\,g^{-1}$), and high temperature stability and corrosion resistance, it is ideal as an EDLC material. Four existing crystalline allotropes of carbon, namely, diamond (sp^3 bonding), graphite (sp^2), carbine (sp^1), and fullerenes (distorted sp^2), yield a wide variety of structural forms and consequently a range of physical properties [69]. Thus, when describing carbon materials, we refer strictly to its elemental structure.

Activated Carbon Manufactured or engineered carbons, commonly referred to as activated carbon, consist of amorphous carbon, ordered graphite, and a broad range of materials between these two extremes that depend on the level of graphitization (ordering and stacking of disordered parallel graphitic layers). Processing conditions during the carbonization (pyrolysis) of carbon at high temperature ($>$2500 °C) in an inert atmosphere (nitrogen or argon) is one of a list of critical variables ultimately determining the properties of the final product, in conjunction with the precursor material chosen and dominant aggregation state (gas or liquid) [70]. Following the carbonization of precursor materials, further enhancement by way of activation may be required to open existing pores blocked by residual amorphous carbons and, in the process, to create new ones, increasing the overall porosity.

The need for activation of a carbonized material is largely dependent on the chosen precursor, with some (select polymers, coal, and petroleum pitch) undergoing a fluid phase change, called a mesophase. This step permits the alignment of graphite layers, with further carbonization treatment producing a highly ordered graphite material (graphitized carbon) [71]. Non-graphitized carbon is often derived from precursors such as biomass (e.g., wood, nutshell, seaweed, etc.), and thermoset polymers maintain their solid state during carbonization and therefore do not permit the mobility of graphitic layers into an ordered structure. Because of the number of volatile groups lost while maintaining a rigid, complex structure, these

Table 8.3 Evaluation of volumetric capacitance and resistance with variations of carbon A and carbon B with average pore diameters of 1.6 and 1.2 nm, respectively. Using tetraethylammonium tetrafluoroborate $(C_2H_5)_4N^+BF_4^-$ in acetonitrile, a significantly higher resistance is observed when the smaller pore carbon is used as the anode for the larger cation [74].

	Negative electrode	Positive electrode	Volumetric capacity (F cm^{-3})	Internal resistance (mΩ)
Capacitor 1	A	B	26.6	24
Capacitor 2	A	A	20.8	23
Capacitor 3	B	B	27.5	257
Capacitor 4	B	A	18.8	243

nongraphitized carbons can have high porosity and do not require subsequent activation [4].

The porous network of ACs through activation is observed to develop a broad distribution of pore sizes composed of micropores (<2 nm), mesopores (2–50 nm), and macropores (>50 nm). Although high specific surface in theoretically shown to be directly related to high capacitance, efforts to observe this relation in AC materials have established that this is not often the case [72, 73].

Active material pore size and the choice of electrolyte play significant roles in charge accumulation [65]. Aqueous and organic electrolytes are shown to develop capacitances ranging between 150–300 and <150 F g^{-1}, respectively [13]. A difference in capacitance can be thought to intuitively change with respect to the chosen electrolyte. Table 8.3 demonstrates the effect of pore size on the volumetric capacitance through the use of two carbons of dissimilar pore size with tetraethylammonium tetrafluoroborate in acetonitrile. The capacitance increase observed by matching the pore size with the respective ion validates the significance of selecting an appropriate ion size in capacitance charging.

Pore size of ACs, particularly micro- and mesopores, are of significance in developing high-capacitance materials [1, 2]. The EDLC of any material by definition is linearly dependent on the reciprocal of the pore radius, and therefore smaller pores are essential to generate a high capacitance by minimizing the effective distance between charges. Porous carbon with microporous structures (<1.5 nm) [57] has testified to the importance of this relation, having obtained an anomalously high capacitance. However, the importance of mesopores must not be overlooked by this relation. Mesopores are shown to expedite the kinetics of ion diffusion through the electrode, especially at high current densities to improve the power performance. This is particularly important when the energy required to be extracted exceeds a frequency of 1 Hz, often called the frequency response. Accessibility of the constructed micropores is also provided through the inclusion of mesopores, and thus a balance between micro- and mesopores is necessary to enhance both energy and power densities [13]. The proportion of mesoporous content in ACs for optimal EDLC performance has been suggested to be between 20 and 50% [62].

Surface functionalities can be incorporated in ACs to improve the wettability or chemical affinity of the electrolyte toward the carbon material, which results in a higher SSA [75]. Additionally, the inclusion of these functional groups can give rise to the development of pseudocapacitive redox charging. The concentrations of functional groups, however, require optimization. Their presence can also lead to a loss in stability through electrolyte and possible electrode aging in organic electrolytes. Detrimental effects depend on a number of factors, including functional group concentrations, electrode surface area, and operational voltage [76, 77]. Furthermore, an increase in the intrinsic electric resistance between compacted conjugate layers can be expected as a result of the edge sites having a higher reactivity than that of the basal plane. Thus, the bonded heteroatoms will then act as barriers for electron transfer [4, 78].

Commercial AC occupies only a specific market due to their limited energy density and power performance. Further progress to develop production methods with increased control in maintaining an interconnected pore structure with a limited pore size distribution and regulated surface chemistry is needed to expand their current application.

Carbon Nanotubes CNTs possess unique properties that are beneficial in their use as an EDLC electrode material. Superior electrical conductivity, suitable mechanical and thermal stability, and a unique pore structure have stimulated considerable research to develop high-power electrochemical supercapacitors [79, 80]. CNTs can be grouped into single-walled nanotubes (SWNTs) and multiwalled nanotubes (MWNTs), with both types possessing differences in functionalization, pore size distribution, and conductivity depending on their radius, chirality, and, in the case of MWNTs, the number of congruent walls in the structure, offering a wide variety of electrode material designs. In addition, their mechanical resilience and accessible tubular network can provide support for pseudocapacitive materials that suffer degradation due to mechanical stress of swelling and shrinking caused by charging and discharging. A range of specific capacitance from 15 to 200 $F\,g^{-1}$ is obtainable by CNTs, and is reliant on their production (morphology, purity) and treatment methods [81]. While exhibiting high-power capability, its small SSA ($<500\,m^2\,g^{-1}$) yields a lower energy density and difficulties remain in ensuring that the intrinsic properties of individual tubes remain on a macroscopic scale [24]. As with all other nanostructures, CNTs have the tendency to agglomerate, resulting in an entangled arrangement with irregular pore structure. This can alter ion mobility, consequently affecting its power performance. Aligning CNTs through pyrolysis methods, or by the use of liquids' zipping effect, enables bulk CNT materials to retain their intrinsic properties [82, 83]. A desirable arrangement obtained by Futaba *et al.* can provide energy densities of 35 $Wh\,kg^{-1}$ in an organic electrolyte and reduce resistances that occur in entangled networks. Measures to increase the SSA through chemical activation have been taken; however, as previously discussed, any activation can deteriorate the electrical conductivity through the introduction of oxygen functional groups, and therefore balancing these two parameters is essential to improve the overall performance. Another form of carbon, carbon aerogels (CAGs), has been used as an additional material with CNTs to synthesize a CNT–aerogel composite integrating a uniform dispersion of the aerogel within the CNT matrix while maintaining the CNTs original aspect ratio and properties. This composite acquired an SSA of $1059\,m^2\,g^{-1}$ and an exceptionally high specific capacitance for CNTs ($524\,F\,g^{-1}$); however, a lengthy and difficult preparation pathway remains a large restriction for its production [84].

Major deterrents for the extensive use of CNTs in EDLCs, aside from their limited energy storage, include their cost of production and continued difficulties in ensuring high purification. Graphene, a carbon material demonstrating similar physical and chemical properties, is produced through a relatively inexpensive method and may be an essential intermediate concerning the attainment of these desirable characteristics at a lower cost.

Templated Carbons The construction of randomly ordered pore networks containing closed or isolated pores leads to a decrease in the charge storage and rate capability. Templated carbons offer to overcome these drawbacks with controlled pore sizes and a well-defined, interconnected pore network [27, 61]. Production of these materials involves a template (e.g., silica, zeolite Y) infiltrated with a liquid or gas-phase carbon, such as sucrose solution, propylene, or pitch, by flooding of the pore network and into readily accessible pores. Figure 8.11 shows a schematic of the production techniques used for templated materials of various porosities. Carbonization of the treated template at temperatures close to 800 °C is followed by the removal of the template through dissolution by an acid, leaving a highly ordered porous carbon structure. These structures can be engineered with well-controlled micro/meso and/or macropores depending upon the template used, and provide a reliable, highly porous alternative in avoiding the long, tortuous paths of AC pores. A linear correlation between the nanotextural parameters of the templated carbon and its resulting capacitance proves that its effective pore network is successful for the diffusion and subsequent trapping of desolvated ions in micropores to generate high energy storage while maintaining a high rate capability [85].

With the ability to carefully design and control the physical and chemical properties tailored to the chosen electrolyte, templated carbons have proven to perform exceptionally well as a carbon material for use in supercapacitors. Despite their high production cost and difficult preparation methods, they remain an excellent means of evaluating the relationship between pore and electrolyte ion size, and the resulting EDLC.

Graphene By possessing a unique morphology of single-layer two-dimensional (2D) carbon atoms with a high-surface area, graphene distinguishes itself among other competitive carbon materials for EDLCs [86]. Exhibiting high electrical and thermal conductivity, mechanical strength, and good chemical stability, this material has outperformed CNTs as an energy storage device with a high specific

8.4 Supercapacitor Components | 343

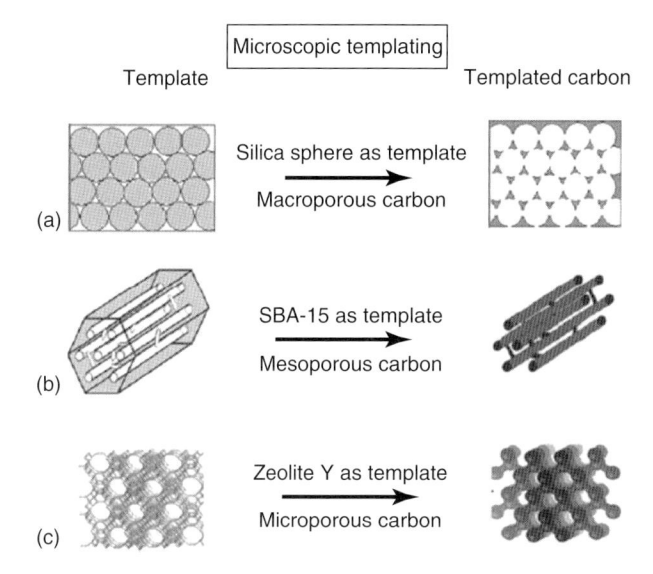

Figure 8.11 Illustration of templating for (a) macroporous carbons, (b) mesoporous, and (c) microporous carbons using respective templates. (From Ref. [24], Reproduced by permission of the Royal Society of Chemistry.)

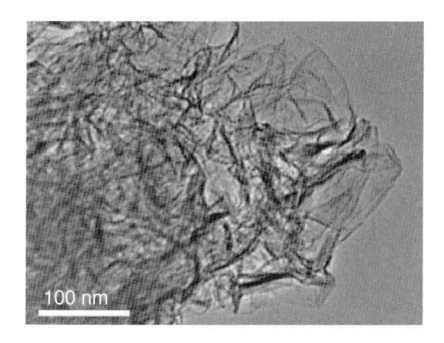

Figure 8.12 Imaging by TEM of individual graphene sheets obtained from chemical modification observed around the edges of the particle. (Reprinted with permission from Ref. [87], Copyright 2008 American Chemical Society.)

capacitance of $\sim$100–200 F g^{-1}, energy density of 28.5 Wh kg^{-1}, and power density of 10 kW kg^{-1} in aqueous electrolytes [26]. Contrasting other EDLC materials that are dependent upon a distributed solid pore structure such as ACs and CNTs, the dispersed graphene layers, shown by the transmission electron microscopy (TEM) image in Figure 8.12, have a greater available surface area (2630 m^2 g^{-1}) than commercialized EDLCs using AC materials [87] through the physical displacement of individual layers in solution. In principle, this would provide a significant total surface area accessible to develop an EDLC. With resilience comparable to that of CNTs, graphene also provides a suitable substrate for the addition of pseudocapacitive materials to enhance the energy charge capability.

Several methods to synthesize graphene are available, including thermal exfoliation of graphite, as well as suspended solution reduction of graphene oxide

(GO) [88]. Doping of graphene by the incorporation of nitrogen groups through initial synthesis procedures, including in-solution reducing agents or chemical vapor deposition (CVD) methods, can further increase the energy storage capacity and increase it solubility in electrolytic solutions; however, distortions in the basal plane or along the edge planes by these functional groups inhibit electron transfer, increasing the material's intrinsic and interparticle electric resistance and decreasing its power capability. Following its discovery, the performance of graphene as an EC supercapacitor material has been investigated with respect to the variety of synthesis procedures available.

A pioneering study of chemically modified graphene (CMG) sheets for use as a EC supercapacitor material was carried out by Stroller *et al.* using graphene synthesized by in-solution reduction of GO by hydrazine hydrate. Good charge propagation exhibited by the material at increasing scan rates demonstrates the excellent electrical conductivity ($\sim 2 \times 10^2$ S m^{-1}) and advantage of a high-surface area material with a short, equal diffusion path length required by electrolyte ions to travel in developing an EDLC. This short path length is made further apparent by the modest Warburg impedance region from ac impedance analysis, commonly associated with the diffusions limitation of the electrolyte within a porous material. Through an alternative chemical synthesis procedure, Wang *et al.* reported similar capacitive behavior of graphene, achieving a high specific capacitance of 205 F g^{-1} [26, 87].

Significance of the electrical and mechanical properties of graphene has raised interest in optical and portable electronic applications requiring flexibility and transparency of high-performance materials. Further CMG performance and transparency as a thin-film material was shown by Yu *et al.* in Figure 8.13, where a $\sim$25 nm thick film obtained a specific capacitance of 135 F g^{-1} at 70% optical transparency (550 nm) [89].

Because of its unique morphology and characteristic properties, graphene is a strong contestant for the development of EDLC deriving from a relatively inexpensive precursor material (graphite). The few current obstacles facing industry application include designing industrial-scale production processes to meet commercial demand [16] and attaining high dispersion while maximizing its electrical

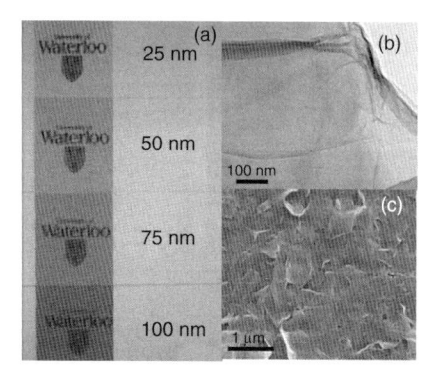

Figure 8.13 Photographs illustrating the transparency of graphene thin films of varying thickness and the corresponding TEM and SEM images of 100 nm graphene films. (Reprinted with permission from Ref. [89], Copyright 2008 American Institute of Physics.)

conductivity. As a nanostructured material, graphene experiences similar agglomeration as that of CNTs, with the restacking of layers in solution remaining a common problem.

Carbon Aerogels and Fibers Additional carbon materials considered for EDLC application include CAGs and activated carbon fibers (ACFs), owing to their high specific surface and good electrical conductivity. CAGs have a highly porous network dominated by mesopores, produced through a sol–gel process that may exclude the often necessary incorporation of a binding agent to produce active electrodes [84]. Because of their mesoporosity, these materials suffer from a poorly developed double-layer capacitance, and a disordered material structure incorporates a high intrinsic resistance. Activation of this material has been less than effective, with the newly developed microporosity remaining largely inaccessible to the electrolyte [75].

Derived from fibrous carbon precursors, ACFs can possess an accessible surface area of $3000 \, m^2 \, g^{-1}$ and the resulting high capacitance of $371 \, F \, g^{-1}$ [90]. As a result of its high surface, functional groups incorporated into the material during activation introduce instability during its lifecycle [1, 5]. Alongside this problem, the higher cost associated with its production with respect to ACs inhibit the future application of this material.

Carbon Composites CNTs as well as graphene are recognized by their unique 2D and 1D morphologies as well as superior mechanical, chemical, and electrical properties [15, 86]. These properties have appeal for their use as a support matrix for active materials in energy storage applications. Initially, the incorporation of CNTs as a backbone for pseudocapacitive materials was done to benefit the composite in a number of ways. By providing high resilience to volumetric change, a notable reduction in composite material degradation can be achieved [38]. As well, being a highly conductive material, CNTs percolation is more efficient than that of carbon black which is traditionally used to enhance material conductivity, in addition to forming a 3D matrix of accessible mesopores structured from an entangled or aligned networks of tubes [39]. These traits serve to reduce the ESR beyond that of commercial ACs owing to a lower intrinsic material resistance and by facilitating ion diffusion. Numerous studies on CNT composite materials synthesized with transition-metal oxides and ECPs in both aqueous and nonaqueous electrolytes have since been conducted [15]. Figure 8.14 shows a ternary $CNT/MnO_2/PPy$ composite with improved performance. Improvements to the energy density of electrochemical supercapacitor electrodes while maintaining a high power rating have been met with some success.

Graphene, exhibiting similar characteristics to CNTs, has been used in composite materials to provide a unique open-pore system, developed as a result of graphene layer interactions forming into aggregated structures. This provides a short diffusion path for electrolyte ions, favoring the power capability of graphene composites. OMC has also been investigated with metals oxides as a carbon support material because of the highly ordered porous structure [92]. By providing a relatively high

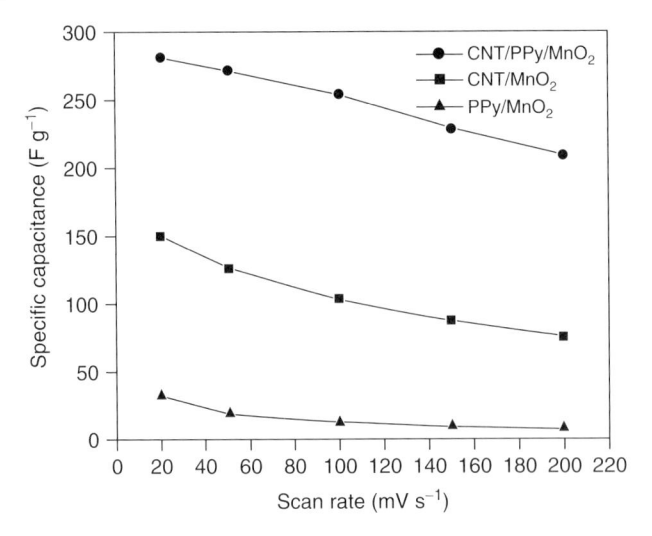

Figure 8.14 Specific capacitance of composites made from polypyrrole (PPy), manganese dioxide (MnO_2), and carbon nanotubes (CNT) in 1 M Na_2SO_4 electrolyte. Evident increases in capacitance and power performance through the incorporation of CNT can be seen. (Reprinted from Ref. [91], with permission from Elsevier.)

accessible surface area interconnected with mesopores to reduce ion diffusion rate, good energy density and high-power densities are achievable.

8.4.1.2 Pseudocapacitive Electrode Materials

In contrast to conventional capacitance of the electric double layer, which evolves principally from electrostatic charge storage, the second method of energy storage employed by ECs is commonly referred to as pseudocapacitance. Faradaic in origin, pseudocapacitance involves the transfer of charge across the established double layer as seen in the charging and discharging of battery electrodes; however, capacitance is derived from a thermodynamic relationship influencing the extent of an accepted charge and the applied change in electrode potential. Pseudocapacitance encompasses a set of processes that can manifest at the electrode/electrolyte interface and depend upon the ionized compounds involved. These processes include the following:

- potential-dependent electrosorption of valence electrons resulting in partial Faradaic transfer of charge of a chemisorbed anion;
- reduction–oxidation reactions of the electrode or oxide film atoms exposed along the interface in either solid or liquid electrolytes where the logged ratio of reductants converted to oxidants (or reversed) directly influences the electrode potential;
- adatom arrays undergoing 2D deposition, often referred to as "underpotential deposition" (UPD) (e.g., H on Pt or Rh) [3, 53].

Pseudocapacitive charge mechanisms are present in all ECs and can represent 1–5% of the measured capacitance in EDLCs, owing to existing oxygen and nitrogen functional groups derived from the precursor atomic makeup or activation procedure. Similarly, capacitance from the electrostatic double layer contributes between 5 and 10% of the measured capacitance in pseudocapacitive materials because of the surface areas present for electrostatic counterion adsorption [1]. Reasoning and discussion of the physical and electrochemical processes of pseudocapacitance can be found in Section 8.3.3. Materials that exhibit pseudocapacitance, including transition-metal oxides and ECPs, have subsequently been used as materials for the construction of electrodes for use in electrochemical supercapacitors with the intention of enhancing the energy storage capability. Owing to the transition between oxidation states that results in coverage or intercalation of the electrode by ions, a greater number of electrons (one to two) are transferable, in principle producing a capacitance 10–100 times larger than that observed in carbon materials that utilize EDLC [3]. Conversely, the life cycle stability has shown to be lacking in a number of pseudocapacitive materials, which is due to the mechanical strain associated with high charging and discharging. Integration of both EDLC and pseudocapacitive materials for composite synthesis in industry applications has been proposed to enhance the specific pseudocapacitance with high active surface areas, in addition to using carbon's good mechanical strength for 3D structural support and added stability [91].

Transition metals have a number of properties making them suitable for use as pseudocapacitors. Metal oxides, including ruthenium dioxide (RuO_2) [93], iron (Fe_3O_4) [94], vanadium pentoxide (V_2O_5) [95], tin dioxide (SnO_2) [96], and manganese dioxide (MnO_2) [92], among others, have been studied extensively for their development of pseudocapacitance [94, 97]. A number of oxidation states defined at specific potentials are what give rise to a highly developed continuous charging of metal oxides that either remains constant or is localized depending upon the number of oxidation/reduction reactions and the interactions among the species of the active coating. It is important to note the potentials where these redox reactions take place, especially when using an aqueous electrolyte, as these electrolytes generally have a potential range of 1.2 V (1.23 V vs. standard hydrogen electrode, simplified as SHE being the potential for water decomposition). Also of particular importance is the conductivity of the metal oxide, where high electrical conductivities enhance the capacitor's ability to propagate charge. These properties not only vary between metals but are also dependent upon the lattice structure (1D, 2D, or 3D structures), with some polymorphs being less conductive than others [98]. Improving the electrical conductance of amorphous, frequently hydrous metal oxides is often achievable through crystallization, yet removal of the water consequentially reduces the mobility of electrolyte ions transported within the inner layers of material, thus reducing the active surface area. In addition, crystallization of metal oxide into forms possessing small, inaccessible pore spacing will further inhibit the effective use of the material.

Ruthenium Dioxide The most widely studied metal oxide RuO_2 [1, 99, 100] is noted as a promising EC material, with highly reversible redox reactions, a wide operating potential, considerably high capacitance, and an excellent cycle life [101, 102]. Both crystalline ruthenium oxide (RuO_2) [103] and amorphous hydrous ruthenium oxide ($RuO_2 \cdot xH_2O$) [104] have been investigated to note a considerable drop in specific capacitance as a result of crystallization, despite the increase made to the intrinsic conductivity. A measurement of the specific capacitance of crystalline RuO_2, which is a d-band metallic conductor, gave 350 F g^{-1} at 100 mV s^{-1} [105]. In comparison, research by Zheng and Jow [101] in 1995 showed for the first time a high specific capacitance greater than 768 F g^{-1} obtained by $RuO_2 \cdot xH_2O$ by annealing the material at a temperature slightly below crystallization, while Hu and Chen in 2004 achieved a maximum capacitance of 1580 F g^{-1} with $RuO_2 \cdot xH_2O$ [106]. Efforts are thus focused on using $RuO_2 \cdot xH_2O$ (analogous to RuO_xH_y) as an electrode material in ECs, with the higher observed capacitance of this form attributed to the better facility provided by the porous structure for proton and electron transport, where the insertion of these into the hydrous oxide surface is described by the following [101]:

$$RuO_x(OH)_y + \delta H^+ + \delta e^- \leftrightarrow RuO_{x-\delta}(OH)_{y+\delta}$$

This equation yields a number of oxidation states that Ru can undergo, from Ru^{4+} to Ru^{3+} and Ru^{2+}. Electron transfer and storage remains in the conduction band of RuO_2, and becomes shared among the various oxidation states of RuO_2 throughout the material. This property allows an overlapping of charge to develop between succeeding potential-dependent redox couples, providing a constant current over a broad operating potential. While their exceptional performance motivates their use, the associated high cost of the rare metal remains a major inhibitor for practical development. To overcome this problem, numerous studies are considering how to improve the utilization of this material through a variety of fabrication techniques, which include sol–gel processes [93, 107], chemical oxidation [108], and cyclic voltammetric deposition [109], among others. Surface treatments that include annealing and electrochemical anodization to further enhance the performance have also been investigated, meeting with intermittent success [110]. More consistent improvements are thought to rest with composite RuO_2/carbon composites by achieving better utilization of RuO_2 films on high-surface, conducting carbon structures. Carbon structures used in these endeavors include OMCs, ACs, and CNTs.

Research has promoted the use of $RuO_2 \cdot xH_2O$ in a number of industries (military, aerospace) for the development of high-performance energy storage devices [31]. Yet, the high costs of ruthenium and the potential safety hazards associated with its use have encouraged investigation into other metal oxides which are more abundantly available, cheaper, and conscientious toward health and environmental concerns.

Manganese Dioxide Among the alternative metals available as substitutes to RuO_2, manganese dioxide (MnO_2) meets the prerequisites of safety concerns and low cost, and has received a great deal of attention by research and industry alike in efforts to improve its capacitive performance. Benefits of MnO_2 use include

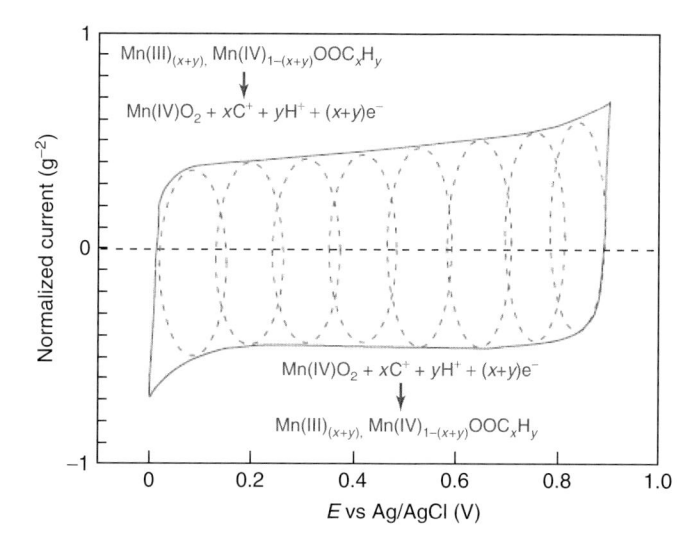

Figure 8.15 Schematic image of MnO$_2$ undergoing cyclic voltammetry analysis in 0.1 M K$_2$SO$_4$ and the consecutive redox reactions that lead to a pseudocapacitive charge. The positive current drives the oxidation reactions of Mn(III) to Mn(IV) (upper part), while the reverse current reduces Mn(IV) to Mn(III) (lower part). (Reprinted by permission from Ref. [5], copyright (2008).)

the many oxidation states of this metal oxide, as shown by a cyclic voltammogram of MnO$_2$ in Figure 8.15 to describe charge behavior in an aqueous electrolyte. Both surface adsorption of cations (alkali solutions) and proton intercalation are hypothesized to contribute to the charging of MnO$_2$, which were described by the following reactions [98]:

$$(MnO_2)_{surface} + M^+ + e^- \leftrightarrow (MnO_2^- M^+)_{surface}$$

where $M^+ = Na^+, K^+$, and so on, and

$$MnO_2 + \delta H^+ + \delta e^- \leftrightarrow H_\delta MnO_2 \ (0 < \delta < 0.5)$$

Syntheses of manganese oxides are largely performed by chemical [111, 112] and electrochemical means [113–115] using a wide array of precursors, while sol–gel methods have made some progress. In terms of a preferred method, electrochemical deposition has been proposed to be superior in avoiding the development of resistive aggregated particles and providing a means of controlling the thickness of the film. A comparative study among multiple methods of electrochemical deposition, including potentiostatic, galvanostatic, and potentiodynamic, has made the claim that potentiodynamic deposition produced MnO$_2$ with optimum capacitive behavior [116]. The conditions and potentials used for deposition are also recognized to have a significant effect on the nanostructure, Brunauer, Emmett and Teller, (BET) surface area, and subsequent capacitive performance. Crystalline MnO$_2$ (α-MnO$_2$, β-MnO$_2$, γ-MnO$_2$, and δ-MnO$_2$), birnessite ξ-MnO$_2$, and amorphous MnO$_2$ have all been investigated, with the last receiving the most attention from the research community. Pore size and structure remain key factors in all these materials, and Brousse et al. [117] have noted that sufficiently large pore sizes are required to provide an accessible high SSA and excellent ion mobility, both of which are principal features for high capacitance.

Loading of MnO_x and the resulting film thickness remain limiting aspects in achieving high capacitance regardless of the preparation method. Materials that obtain high specific capacitance of >1000 F g^{-1} at low loadings of $30\,\mu\,g\,cm^{-2}$ ultimately contribute a negligible surface specific capacitance (~ 0.03 F cm^{-2}); yet, increasing a low loading of MnO_x from 18 to $116\,\mu\,g\,cm^{-2}$ can reduce the capacitance by 55% [118]. These performance limitations are attributed to the longer pathways necessary for proton transport with increasing thickness and the low electrical conductivity. The inaccessibility of protons to the inner layer of thick films has been supported by evaluation of films synthesized from $K_{0.02}MnO_2H_{0.33}0.53H_2O$ of varying thicknesses (5–100 nm), where only the top layers of thick films were observed to change oxidation states similar to the thin films. Thus, a large portion of the material is suggested to remain unused [119]. Improvement of MnO_x use has been observed in carbon composites with ACs and CNTs by providing a high-surface area and excellent electrical conductivity to develop materials with both EDLC and pseudocapacitance. The specific capacitance of MnO_2/CNT composites can range from 110 to 568 F g^{-1} [120], with MWNTs being preferred over SWNTs in providing an open mesoporous network for ion accessibility.

Metal Nitrides Metal nitrides have attracted attention for their use as an EC electrode material, possessing favorable electroactive behavior, high electrical conductivity, and excellent chemical stability. Molybdenum nitrides (MoN and Mo_2N) have been studied by Liu et al. [121] to observe a capacitive behavior similar to that of RuO_2, yet a limiting stable operating potential of 0.7 V versus reversible hydrogen electrode, simplified as RHE effectively inhibits their further use. As an alternative, vanadium nitride (VN) has been studied by Choi and Kumpta as a capacitive material, obtaining specific capacitances of 1340 and 554 F g^{-1} at scan rates of 2 and 100 mV s^{-1} [122, 123], respectively. Methods used for VN preparation include nitridation of oxides using ammonia flow and the more commonly used temperature-programmed reduction of vanadium oxide foams for the production of high-surface area, nanoparticle-sized VN [124].

As mentioned previously, the crystal and pore structures play a crucial role in the diffusion of electrolyte ions required for the redox reactions, and they must be considered in the electrode design. Amorphous structures are believed to provide good ion accessibility and cation diffusion necessary for high performance, with an improved porosity compared with crystalline structures. However, the lack of a defined structure similar to that found in ACs leads to an increased resistance and a capacitance limited by the achievable BET surface area. Observations of the dependence of EDLCs on pore size and structure can be extended to pseudocapacitance, where pore size, structure, and available surface area all have an effect on the accessibility of the electrolyte ions to the metal oxide. Rather than minimizing the effective distance between opposite electrostatic charges with the use of micropores, mesoporosity is observed to enhance pseudocapacitive performance in bulk materials by reducing pore resistance, increasing overall surface area, and allowing better rate capability [98]. Hybridization of metal oxides with carbon materials such as CNTs, graphene, and OMCs has shown promise in overcoming the performance limitations with low loading [125]. The selection of the electrolyte must also be taken into account where the performance and stability of the constructed electrode are of concern, as the partial dissolution of metals at particular oxidation states can occur depending upon the pH of solution and potential of the cell, thereby diminishing the pseudocapacitance and causing cycling instability [32, 126].

Electrically Conducting Polymers Conducting polymers have aroused great interest as a family of synthetic metals due to their rapid and reversible redox reactions and high doping levels [39]. Polymers that have drawn attention in the development as ECs include PPy, PANI, and poly(3,4-ethylenedioxythiophene) (PEDOT) and their derivatives, to name a few, owing to their intrinsic properties and commercial appeal. Benefits of their use include relatively high conductivity, low cost, and ease of synthesis [127]. Through electrochemical deposition or chemical oxidation methods with the option of surfactant use (subsequent recycle processes may then required), nanostructures such as nanotubes, nanowires, and nanospheres can be produced [128]. Similar to metal oxides, a pseudocapacitance mechanism is developed through a redox reaction as the polymer transitions between oxidation states. Through the involvement of the entire volume in accumulating charge, it surpasses other EC materials that are reliant solely on electrostatic surface charging (carbons) and has been able to obtain high specific capacitance values in excess of 400 F g^{-1} [34]. The main advantage of ECPs, however, is responsible for its major drawback as a short-lived material, with poor stability being a result of volumetric changes that occur during doping/dedoping (insertion/extraction) of the counterions, a process illustrated by Figure 8.16.

This process involves the swelling, shrinking, cracking, and/or breaking of the polymer. In addition to this problem, a small working potential can be expected in order to avoid isolating states and/or polymer degradation due to overoxidation of the material [39]. These problems make it necessary to develop composite designs that utilize a carbon support structure with the polymer matrix, or through deposition of the polymer on a carbon surface to reduce these mechanical stresses and improve cyclability. Polymer/carbon composites obtained with high specific capacitance, respective cycle life, and facile synthesis methods illustrate the effectiveness of incorporating carbon to help alleviate this problem and have subsequently been shown to increase porosity of the ECP. Work performed by Yan et al. has shown a high capacitance of 1046 F g^{-1} in 6 M KOH by the incorporation of graphene with PANI. Furthermore, by the addition of ∼1% CNTs, a retention of 6% of the original capacitance after cycling remained; thus improvements in performance through enhanced electrolyte accessibility and diffusion rates are achievable [35].

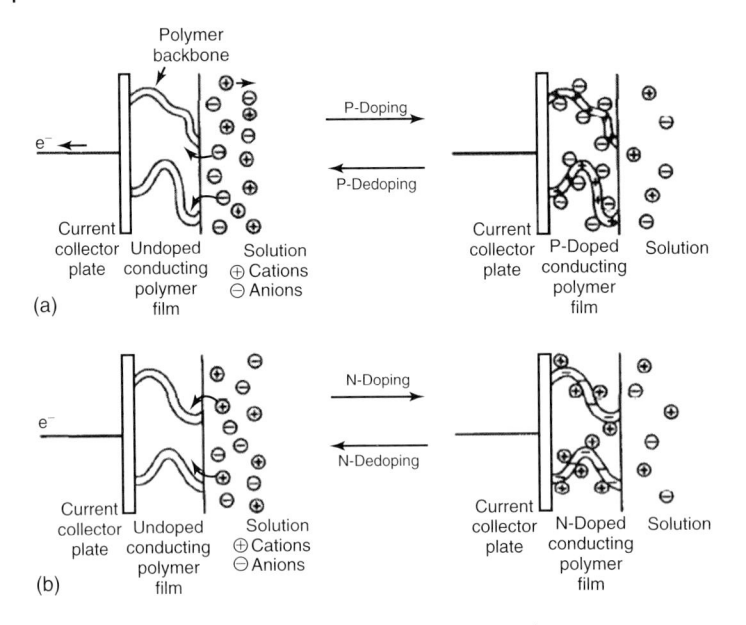

Figure 8.16 Diagram representing the p-doping (a) and n-doping (b) of polymers as they undergo charging and discharging. (Reprinted from Ref. [129], with permission from Elsevier.)

In addition to their high capacitive ability, ECPs have shown good application in asymmetric capacitor configurations by operating within their own optimal potential range to provide their use as either an anodic or cathodic electrode similar to that observed in batteries. The outcome of ECPs for their application in ECs rests on their development as hybrid capacitors using carbon supporting materials for future applications [19].

8.4.2
Electrolytes for Use in Electrochemical Supercapacitors

Electrolytes play a critical role in development of the electric double layer in EDLCs and for the reversible adsorption and intercalation redox processes occurring in pseudocapacitive materials. Hence, their molar conductance and the subsequent dependence on the cell operating temperature are main factors when determining the ESR. Prior mention to the consideration of the cation and anion sizes is necessary in developing a substantial capacitance in porous electrodes, yet microporous systems can cause an increase to the electrolyte resistance [4]. In addition, owing to the quadratic relation of the operating potential to energy density, the electrochemical stability range is significant, with aqueous electrolytes having a smaller 1 V window in comparison to nonaqueous electrolytes' 3.5–4 V range. Each electrolyte inherently has its benefits and disadvantages which ultimately have to be considered when designing a capacitor cell for its desired application.

Corrosion phenomena can be a problem, particularly in aqueous solution, and are dependent upon pH and anion type. Electrodes, including carbon, can undergo corrosion at charged (or overcharged) and open-circuit potentials, which can result in surface oxidation and gas evolution (CO, CO_2). These parallel oxidation reactions are parasitic on the charging current and can occur at potentials below the thermodynamic electrochemical limits of water [1].

8.4.2.1 Aqueous Electrolytes

The benefits of using aqueous electrolytes include its availability and low cost. Tailoring its use in optimizing the ion/pore interaction is easy, and its high conductivity and low health concerns are also notable advantages. Unfortunately, aqueous electrolytes have the disadvantage of being limited by their thermodynamic electrochemical decomposition in a relatively narrow potential range (1.2 V). An operating range of 0.8–1.0 V is employed to avoid the loss of the electrolyte, introduction of oxygen or hydrogen that lead to cell rupturing, and additional losses in current from competing Faradaic reactions. In addition to these problems, consideration must be given to the current collector/electrolyte choice when using electrolytes containing chloride anions. These anions are favored for their increased ion conductivity, yet corrosion can be a problem when using stainless steel (SS) current collectors.

8.4.2.2 Organic Electrolytes

Organic electrolytes are currently used in commercial ECs because of their voltage window, and often operated at potentials above 2 V (2.2–2.7 V), which can be floated as high as 3.5 V for short periods [12]. An increase in operating potential provides a desirable improvement in power and energy performance that industry demands from energy storage devices. Application of ECs in high-potential devices require capacitor stacking to meet these loads; therefore fewer high-operating-potential capacitors would be required, resulting in less material and a lower chance for stack deterioration to occur. Polar aprotic electrolytes used in hightech capacitors include acetonitrile and propylene carbonate; however, the benefits of using propylene carbonate to reduce the safety concerns of acetonitrile are weighed down by an almost threefold increase in electrolyte resistivity [12]. Table 8.4 shows the difference in resistivity and voltage limitations between the common types of electrolytes used in ECs. Organic electrolytes in general suffer from a notable increase in resistivity, which is detrimental to its power performance, and raise a number of safety issues that are associated with incorporating a toxic, flammable material into transportation vehicles. These solvents also require purification processes to remove any residual water content, which can lend to the corrosion of the current collectors at high operating potentials. Efforts to find a replacement nontoxic organic electrolyte yielding good ion conductivity have thus far been unsuccessful.

8.4.2.3 Ionic Liquid Electrolytes

ILs exist as solvent-free molten salts at room temperature and possess high thermal and electrochemical stability in excess of 3 V made possible by the chemical and

Table 8.4 Density, resistivity, and operating voltage window for various electrolytes [12].

Electrolyte	Density (g cm^{-3})	Resistivity (Ω cm)	Cell voltage
KOH	1.29	1.9	1
Sulfuric acid	1.2	1.35	1
Propylene carbonate	1.2	52	2.5–3
Acetonitrile	0.78	18	2.5–3
Ionic liquid	1.3–1.5	125 (25 °C)	4
		28 (100 °C)	3.25

Reprinted from Ref. 12, with permission from Elsevier.

physical properties of the chosen cation and anion [131]. In addition to their wide potential window, these unique solvents are nontoxic and nonflammable, making them attractive alternatives to organic solvents. Synthesized hydrophobic ILs have also been successful in avoiding further purification processes. Being composed entirely of liquid ions at ambient temperatures, this property necessitates large-sized ions. The resulting ionic mobility is therefore reduced and conductivity at ambient temperatures is a few milliSiemens per centimeter. A list representing a number of ILs in Table 8.5 notes their respective potential limits and conductivities at ambient temperature. Increasing the operating temperature to 125 °C can increase their conductivity to be comparable with that of acetonitrile; however, this reduces the electrochemical stability of the ions and reduces the operating voltage window by 0.5–1.0 V. Through the blending of acetonitrile with ILs, a reduction in the flammability is achievable with only a small change in conductivity and operating voltage window [12, 23, 131].

8.4.3
Separators

The choice of a thin, nonconducting separator must be considered when minimizing the electrolyte resistance contribution. Separators made of paper or polymer and ceramic or glass are used with organic and aqueous electrolytes, respectively. Properties of a good separator include good transport of the electrolyte ions and sufficient mechanical resistance to the penetration of charged carbon particles traveling by electrophoresis. The conductance of an electrolyte σ is a function of the porosity of a porous separator p according to the equation

$$\sigma = \sigma_o p^\alpha \tag{8.19}$$

with $1.5 < \alpha < 2$ [2]. Thus, a complex issue arises when optimizing the mechanical durability, thickness, and porosity in combination with the proper electrolyte in reducing the resistances that these two components impart [2, 14].

Table 8.5 Electrochemical stability and conductivity measured at room temperature for a range Ionic liquids.

Ionic liquids	Electrochemical stability					Conductivity
	Cathodic limit (V)	Anodic limit (V)	ΔU (V)	Working electrode	Reference	σ at 25 °C (mS cm^{-1})
Imidazolium						
[EtMeIm]$^+$[BF$_4$]$^-$	−2.1	2.2	4.3	Pt	Ag/Ag$^+$, DMSO	14
[EtMeIm]$^+$[CF$_3$SO$_3$]$^-$	−1.8	2.3	4.1	Pt	I$^-$/I$_3^-$	8.6
[EtMeIm]$^+$[N(CF$_3$SO$_2$)$_2$]$^-$	−2	2.1	4.1	Pt, GC	Ag	8.8
[EtMeIm]$^+$[(CN)$_2$N]$^-$	−1.6	1.4	3.0	Pt	Ag	–
[BuMeIm]$^+$[BF$_4$]$^-$	−1.6	−3	4.6	Pt	Pt	3.5
[BuMeIm]$^+$[PF$_6$]$^-$	−1.9	2.5	4.4	Pt	Ag/Ag$^+$, DMSO	1.8
[BuMeIm]$^+$[N(CF$_3$SO$_2$)$_2$]$^-$	−2	2.6	4.6	Pt	Ag/Ag$^+$, DMSO	3.9
[PrMeMeIm]$^+$[N(CF$_3$SO$_2$)$_2$]$^-$	−1.9	2.3	4.2	GC	Ag	3.0
[PrMeMeIm]$^+$[C(CF$_3$SO$_2$)$_3$]$^-$	–	5.4	5.4	GC	Li/Li$^+$	–
Pyrrolidinium						
[*n*PrMePyrrol]$^+$[N(CF$_3$SO$_2$)$_2$]$^-$	−2.5	2.8	5.3	Pt	Ag	1.4
[*n*BuMePyrrol]$^+$[N(CF$_3$SO$_2$)$_2$]$^-$	−3.0	2.5	5.5	GC	Ag/Ag$^+$	2.2
[*n*BuMePyrrol]$^+$[N(CF$_3$SO$_2$)$_2$]$^-$	−3.0	3.0	6.0	Graphite	Ag/Ag$^+$	–

(continued overleaf)

Table 8.5 (continued)

Ionic liquids	Electrochemical stability				Conductivity
	Cathodic limit (V)	Anodic limit (V)	ΔU (V)	Working electrode	Reference	σ at 25 °C (mS cm^{-1})

Ionic liquids	Cathodic limit (V)	Anodic limit (V)	ΔU (V)	Working electrode	Reference	σ at 25 °C (mS cm^{-1})
Tetraalkylammonium						
[nMe$_3$BuN]$^+$ [N(CF$_3$SO$_2$)$_2$]$^-$	−2.0	2.0	4.0	Carbon	−	1.4
[nPrMe$_3$N]$^+$ [N(CF$_3$SO$_2$)$_2$]$^-$	−3.2	2.5	5.7	GC	Fe/Fe$^+$	3.3
[nOctEt$_3$N]$^+$ [N(CF$_3$SO$_2$)$_2$]$^-$	−	−	5.0	GC	−	0.33
[nOctBu$_3$N]$^+$ [N(CF$_3$SO$_2$)$_2$]$^-$	−	−	5.0	GC	−	0.13
Pyridinium						
[BuPyr]$^+$ [BF$_4$]$^-$	−1	2.4	3.4	Pt	Ag/AgCl	1.9
Piperidinium						
[MePrPip]$^+$ [N(CF$_3$SO$_2$)$_2$]$^-$	−3.3	2.3	5.6	GC	Fe/Fe$^+$	1.5
Sulfonium						
[Et$_3$S]$^+$ [N(CF$_3$SO$_2$)$_2$]$^-$	−	−	4.7	GC	−	7.1
[nBu$_3$S]$^+$ [N(CF$_3$SO$_2$)$_2$]$^-$	−	−	4.8	GC	−	1.4

Reprinted from Ref. 131, with permission from Elsevier.

8.4.4
Current Collector and Sealant Components

Current collectors, typically metals and alloys with high electrical conductivity (e.g., Al, Fe, Cu, steel), are used for collection and efficient transport of current from the electrode to the applied load and, vice versa, recharging of the electrodes after use. Normally, a coating of current collector with the active electrode materials is done to enhance the contact between these two components and to effectively minimize the resistances that arise from the connection of heterogeneous materials. Disruption of this connection is usually due to physical deterioration of the active material upon extensive cycling. This can cause loosening, shrinking, and loss of material to the electrolyte, ultimately leading to degradation in capacitance charge and cycle life. To avoid these undesirable events and ensure optimal contact, methods that involve surface treatments of the collector, such as the addition of a conducting film as well as the incorporation of minute amounts of binding (e.g., polytetrafluoroethylene, simplified as PTFE) and conducting agents (e.g., carbon black) to the active electrode material, are carried out to supplement the electrical conductive and mechanical durability. While the addition of a binding agent is intended to increase the life cycle of the cell and help maintain contact between the electrode and collector, it increases the internal resistance of the electrode and reduces the power capability [14].

The resistances that arise between the current collector and active material may also depend upon whether the EC will be used in a monopolar or bipolar arrangement, and can respectively be contributory or negligible to an ohmic loss. As an alternative to fashioning a cell using two separate collector and electrode components, the growth of the carbon electrode material directly on a metal substrate can be performed to ensure high contact on a molecular level.

Proper sealing of the EC, often with a hermetic seal depending upon the electrolyte, is required to avoid water or gases entering the cell. In a bipolar stacking arrangement, good edge sealing of EC array is particularly important to avoid shunt currents passing from one side of the bipolar electrode to the other, which leads to a loss in charging efficiency and self-discharging through the electrical connection and short-circuiting [2].

8.5
Assembly and Manufacturing of Supercapacitors

The majority of commercial EC electrodes are designed using the same material for both anode and cathode. When preparation of these active electrodes involves a change in volume or mass, or integration with an additional material owing to the fact that materials perform differently when positively or negatively charged, labeling is required for their respective polarity. With respect to the design, assembly, and packaging of EC supercapacitors, these will ultimately depend upon their application. Construction of a capacitor cell is shown in Figure 8.17, with the cell consisting of pair of electrodes separated by a thin, porous, nonconducting

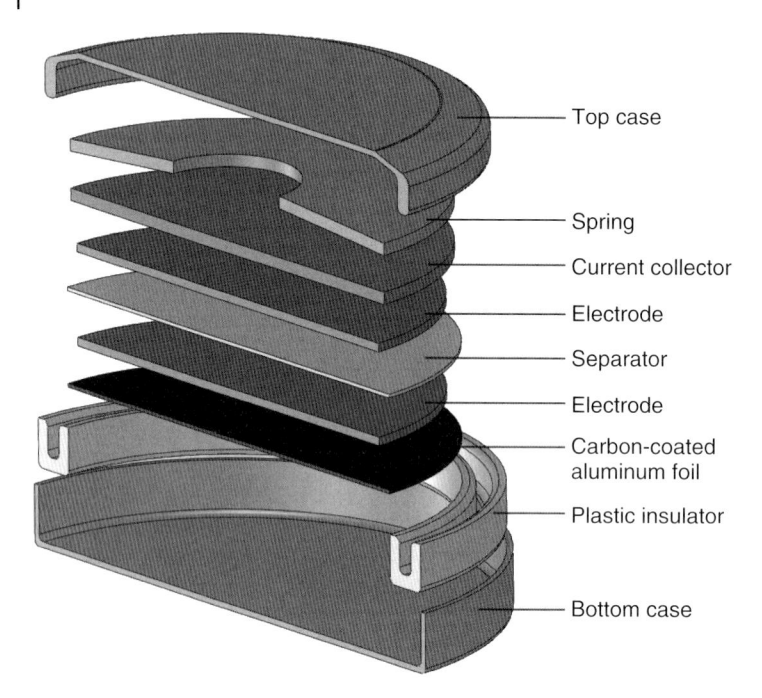

Figure 8.17 Schematic representation of test assembly for EC supercapacitor electrodes.

material and sandwiched between current collectors. Aqueous and nonaqueous electrolytes can be used with this design, provided assembly is performed in a glove box for the latter. Good contact should be maintained between the current collector and active material to minimize the ESR, while moderating compression to avoid pressing out a critical volume of electrolyte. The following presents the methods of electrode fabrication and cell assemblies for EC supercapacitors.

8.5.1
Electrode Fabrication from Carbon Materials

Development of the electrode material in commercial applications is proprietary information and not disclosed in the public domain. Yet, upon synthesizing an electroactive material for use in EC supercapacitors, laboratory and commercial techniques for electrode fabrication and cell preparation can be similar. The electrode material, consisting of conducting particles of carbon powder (carbon black or CNTs) as well as a fluorine-containing polymer (often PTFE) binding agent, can be mixed together by means of a solvent to obtain a paste or slurry material, with a ball mill treatment or ultrasonication often used to ensure proper dispersion and uniformity.

Preparation of the current collector can enhance the conductivity between the active electrode and reduce the loss in power performance during cell cycling. Metal foils of aluminum, copper, and nickel are largely used as a current collector or electrode substrate for their high conductance and relatively low cost, with a foil thickness ranging between 20 and 80 μm. An initial step of etching the foil can be carried out to introduce irregularities along the surface and provide a better contact between the current collector and electrode or intermediate layer when applied. The intermediate layer may be used when adhesion of the electrode material to the collector is a concern by binding both the substrate and material together. Attachment of the electrode can be done through spray deposition, sol–gel method, or by other means; however, in many cases the active electrode is a pellet fashioned using a pellet dye via a dry paste, or through repeated cycles of pressing and drying a slurry mixture, followed by the cutting of a layer electrode into a desired shape. When the pellet or layer is sufficiently dry, the electrodes and separator can then be impregnated with the desired electrolyte for some time to wet all accessible pores prior to assembly. This can take a number of hours, depending on material's hydrophobic nature.

Substitution of the aqueous electrolyte in capacitors utilizing nonaqueous electrolytes (e.g., Li or Na tetrafluoroborate, tetraethylammonium perchlorate dissolved in acetonitrile, or propylene carbonate) can be done; however, preparation of the material must be performed in a dry glove box using dried and distilled electrolyte with no moisture present. Any water within the electrolyte will cause gas to evolve in the high-potential window of 2.5–3.5 V, resulting in disruption in the cell sealant as well as cycling and self-discharging problems.

8.5.2
Concerns with Cell Assembly

Problems pertaining to the current distribution in supercapacitor devices arise in the cell arrangement and assembly. Predominant effects in scaleup are the ohmic drops that occur due to the interparticle resistance within the carbon or oxide matrix, and outer particle surface contacts. When developing large-scale industrial applications, the detrimental effects from these relations can have a substantial effect on their power performance.

The use of bipolar electrodes in devices requiring multiple electrodes for high voltage can avoid these macroscopic resistances caused by the lateral contacts of separate electrode matrices and nonuniform current distributions. Excellent contact between the endplates and respective bipolar electrode surfaces is still required in this arrangement to maintain a minimal ESR during charge and discharge, and can be accomplished through surface treatments of the current collector and application of an optimal compressive force across the cell [4, 132].

8.6
Supercapacitors Stacking and Systems

Stacking of supercapacitors is required for application in commercial devices that operate at voltages beyond the limits of a single cell. By stacking two or more EC supercapacitors, the operating potential of the device can be effectively enhanced. Consideration toward the design of the stack, particularly with proper packaging and a limited tolerance for the engineered electrode materials, is necessary to avoid undesirable effects that lead to a loss in performance or function [1, 2].

There are a sufficient number of applications for supercapacitors that require an operating voltage substantially greater than the 1.2 or 3.5 V windows provided by aqueous and organic electrolytes, respectively, particularly in addressing their application to transportation devices (i.e., HEVs). These high operating voltages are attainable when a bank of individual supercapacitor cells is arranged in series, otherwise known as a "stack."

The overall capacitance C_s of a series of capacitors can be related to the individual capacitances of each cell C_i by the following equation:

$$C_s = \frac{C_i}{n} \tag{8.20}$$

where n is the number of individual capacitors. Therefore, the total capacitance of a stack of capacitors having identical capacitances is simply one-nth of their value. The overall charge q_s accumulated over the stack can be calculated for a given operating potential V using the relation previously defined for capacitance, $C = q/V$, and the known potentials of each cell in series are summed to give the overall potential, $V = \sum V_i = nV_i$, therefore $q_s = nq_i/C_i$.

The energy density of a stack can be obtained by knowing the energy stored in a single cell, $E = (1/2)CV^2$. By incorporating the overall capacitance and voltage of the series, provided each capacitor has identical capacitance, the overall energy stored follows:

$$E = \frac{1}{2} C_n V^2 = \frac{1}{2} C_i n (V_i)^2 \tag{8.21}$$

This is the same as the energy storage of an individual cell multiplied n times and follows the law of conservation of energy, where n charged supercapacitors cannot gain or lose energy stored simply from being in a series combination.

When combining a bank of capacitors with individual capacitors of different capacitances $C_1, C_2, C_3, \ldots, C_n$, and so on, the principles discussed still apply to give an overall capacitance C_s such that

$$\frac{1}{C_s} = \frac{1}{C_1} + \frac{1}{C_2} + \frac{1}{C_3} + \cdots \frac{1}{C_n} = \sum_1^n \frac{1}{C_i} \tag{8.22}$$

While the total stack voltage remains constant, the potential differences across each capacitor are, however, different. As well, the charge components q_i of each unit are distributed with respect to the developed capacitance. Importance is placed on the voltage difference each unit capacitor is subject to, as this value should not surpass the limit potential of the employed electrolyte. This results in an engineering problem to produce capacitor electrodes with equivalent properties such that they perform in an identical manner regardless of what type of material is used (i.e., equivalent capacitance, ESR) [2]. During charging or discharging, overcharging or overdischarging can occur in a unit that differs in its capacitance rating. Hazardous

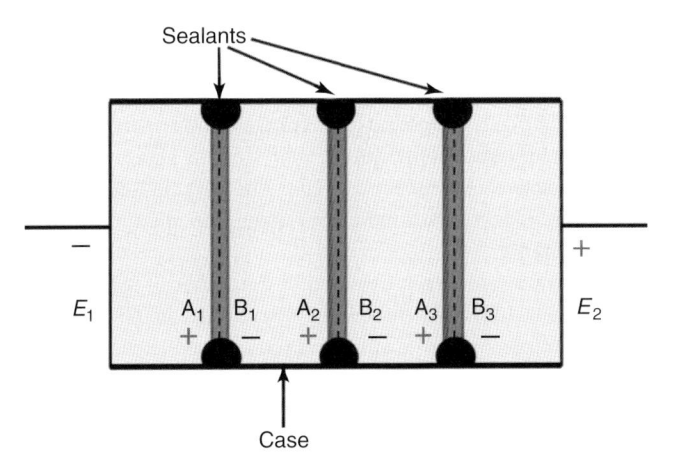

Figure 8.18 Bipolar electrode schematic.

effects can arise when a unit is subjected to a potential difference greater than the limiting potential of electrolyte decomposition. Gassing and cell rupture could possibly occur, resulting in a loss to system functions in addition to health concerns that may arise depending upon the electrolyte. These problems, however, are also of concern to battery systems and can be overcome [1].

8.6.1
Bipolar Electrodes

Each bipolar electrode maintains a double-layer capacitance on either surface of the electrode; however, the polarities of either side are opposite to that in solution. This is done with a progressive drop in potential at each surface/electrolyte interface continuing along the array of cells. A bipolar stack configuration is shown in Figure 8.18, with a series of bipolar electrodes placed between a potential difference V_1 to V_2. The spaces existing between endplate V_1 and A_1, B_1 and A_2, and so on, continuing down the series to endplate V_2 are flooded with electrolyte solution to develop a countercharge layer across each surface. Prevention of any electrolyte crossing the bipolar plate and forming a short-circuit requires that each B_n to A_{n+1} section be electrolytically isolated from its neighboring sections through a seal connecting the ends of each bipolar electrode to the outer casing of the bank. Conducting separators between the electrode pairs AB, AB, and so on, and between the bipolar plates and endplates may be required for the assembly depending upon the stiffness of the material and the mode of construction.

Designing a bipolar electrode stack configuration carries some benefits to the construction and performance of the bank over that of a two-electrode cell configuration. The volume of each cell can be effectively minimized, which circumvents the need to interconnect each cell. In addition to this, it minimizes the overall resistance of the bank and concurrently allows a more uniform current distribution [1].

8.7
Supercapacitor Performance, Testing, and Diagnosis

Measurement of capacitance in the development of electrochemical supercapacitors is necessary to evaluate performance, application, and trouble shooting when designing and redesigning active electrode materials. This evaluation is done through quantitative analysis using experimental variables and their respective relation to a developed capacitance.

Testing methods for capacitors have been explored in a number of informative references and manuals [133, 134]. The following discussion on testing and diagnosis of the results will emphasize some of the main tests used to characterize performance.

8.7.1
Cyclic Voltammetry

CV provides a useful technique in screening potential capacitor materials in a relatively swift procedure. The technique is performed by cycling a linear potential sweep between two defined voltages determined by the electrolyte decomposition limits, and generally aided by a reference electrode in a three-electrode cell. The potential applied to the electrode varies linearly with time, yielding a time-dependent dynamic response current which is recorded by an oscilloscope or computer.

A predetermined constant potential sweep rate $v_s = dV/dt$ is used in the experiment to evaluate the capacitive charging current i passing through the material, described by the following equation:

$$i = C\frac{dV}{dt} = Cv_s \tag{8.23}$$

Ideal capacitance is depicted by a constant and equivalent current $\pm i$ resulting from a constant sweep rate $\pm v_s$, represented by a rectangle within a plot of the current versus the applied potential. ECs exhibiting EDLC approach ideality with constant current through most of the charging process, deviating slightly from this relation due to diffusion resistances during initial charging or discharging of the double layer at the respective voltage limits. Typical CV curves of SWNTs and MWNTs observed approaching near-ideal behavior are exhibited in Figure 8.19. Other materials may not exhibit constant capacitance as a result of material resistances or may display potential specific Faradaic reactions depicted by peaks in the response current due to reversible pseudocapacitive processes typically shown in metal oxides and conducting polymers. Characteristic profiles of these capacitive mechanisms from the CV technique are shown in Figure 8.20. Determining the capacitance from CV experimental results often requires the integration of the current response in a cyclic voltammogram and can be calculated by the following equation [92]:

$$C = \frac{\int i\,dV}{\Delta V v_s} \tag{8.24}$$

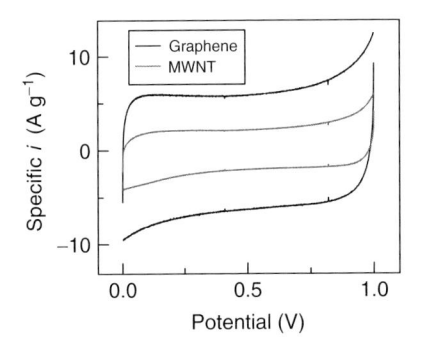

Figure 8.19 Grahene nanoplatelet and MWNT thin-film capacitor electrodes characterized by cyclic voltammetry in 1 M KCl at 200 mV s^{-1}. Approach to ideality is recognized by a rectangular shape.

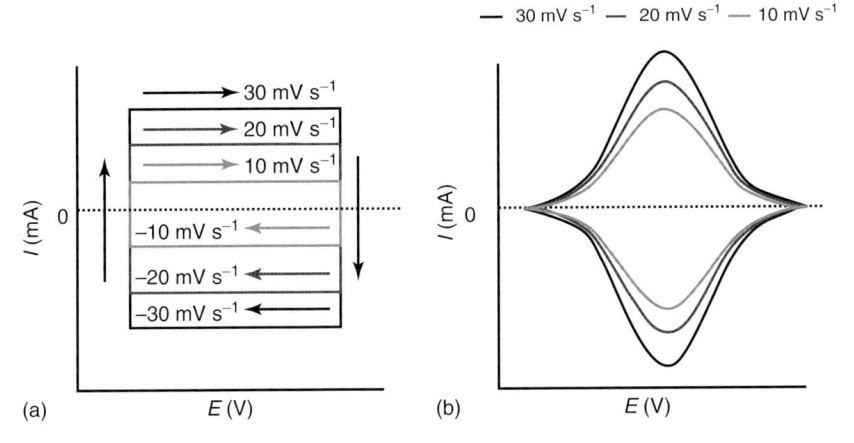

Figure 8.20 Cyclic voltammograms comparing (a) an ideal capacitor and (b) a localized and reversible pseudocapacitive reaction for an electrode thinly coated by a redox material at increasing scan rates.

8.7.2
Charge or Discharge Chronopotentiometry

Interfacial charging of an electroactive material in contact with a solution electrolyte will behave capacitatively when a constant current density i is applied, developing a potential difference linearly with time across the cell. The capacitance, described by $C = \Delta q / \Delta V$ where

$$\Delta q = \int i \, dt \tag{8.25}$$

can be used with charge or discharge constant current testing to determine the capacitance according to the following equation [127]:

$$C = \frac{\int i \, dt}{\Delta V} = i \cdot \frac{\Delta t}{\Delta V} \tag{8.26}$$

where $\Delta t/\Delta V$ is the inverse slope of the response curve obtained from charge or discharge measurements and ΔV is change in potential between the limiting potential difference for electrolyte decomposition.

This is a conventional procedure used in testing electrode materials for use in batteries and supercapacitors alike, involving the charging of the electroactive material at constant current and subsequently discharging at constant current. Discharge across a load of known resistance is often used in the experiment to calculate the input and output charge [q] and energy E provided to the capacitor material. Variations to this testing can involve changing the rate of the charge or discharge current, applying load resistors of different values, and incorporating a hold time at charge as a variable prior to discharge. The charge and energy can be calculated as follows:

$$\Delta q = \int i \, dt = \int \left(\frac{V}{R_L}\right) dt \qquad (8.27)$$

and

$$E = \int iV \, dt = \int \frac{V^2}{R_L} dt \qquad (8.28)$$

where V is the measured voltage across the resistor, and R_L is the applied load resistance. The efficiency of the capacitor to charge and discharge, often called the "round trip efficiency," [134] is obtained using the calculated input energy and resulting output energy determined by experiment. An ideally polarized electrode exhibiting model reversible capacitance is described by a constant linear potential increase and decrease during charge and discharge. During the experiment, however, this may not be the case. Consideration must be given to the Faradaic processes that occur at distinct potentials, including decomposition of the electrolyte near the limiting potentials or diffusion limitations within micropores. When a Faradaic process occurs, the applied current density i will be divided between a double-layer charging current i_{dl} and i_F pertaining to psuedocapacitive chemical processes taking place at the electrode surface. Reversible pseudocapacitive processes can exhibit capacitive charging similar to that of the double layer if there are multiple redox reactions occurring continuously; otherwise, a dynamic change to the slope of the potential will occur with a range dependent upon the interaction term g between the adsorbed species, discussed in Section 8.3.3. The potentials at which pseudocapacitive processes are observed to take place in chronopotentiometry techniques can be supported by CV.

8.7.3
Impedance Measurement

Measured response functions of the potential to changes in time-dependent current charging or vice versa have been shown to be useful in the characterization of capacitance. In addition to the CV and chronopotentiometry techniques, the response function of an ac, generated by application of an av to an electrode interface, is useful in characterizing the pseudocapacitive processes a material exhibits. The imaginary component Z'' of the overall impedance Z response relates to the capacitance of the electrode through the equation $Z'' = 1/j\omega C$ [135]. This relation has placed the ac current impedance technique as a principle method for capacitive behavior analysis. Impedance analysis using ac provides a means of measuring capacitance as a function of the frequency ω. In addition, evaluation of the phase relation between Z and Z'' impedances generated by this procedure can be achieved. Important circuit elements of the electrode structure or constructed device, including the ESR, and potential-dependent Faradaic resistances (e.g., pseudocapacitive, leakage resistance) can be separated and determined using this technique.

The examination of fundamental and applied characteristics of ECs can be performed using impedance spectroscopy over a wide range of frequencies (10 kHz to 10 MHz). ECs yield characteristic observations by plotting the complex plane of impedance (Nyquist), which may include semicircular forms occurring at high frequencies, a 45° region (Warburg region) of transition between high and low frequencies, and nonvertical slopes at low frequencies. The ESR of the electrode or device circuit, comprised of contact, separator, and solution resistances, summarized together as the real impedance, can be obtained from the intercept of real impedance at high frequencies.

The 45° phase angle observed in the complex plane of Z, often termed the "Warburg impedance region," can arise for two separate reasons for an EC electrode undergoing ac impedance testing. Transitioning from the Faradaic resistance parallel to capacitance to diffusion control demonstrates a dependence of both imaginary and real components of Z on the square root of ω^{-1} or $(t)^{\frac{1}{2}}$ from the principles of particle diffusion in a concentration gradient [1, 136].

In a separate case, unrelated to diffusion reason, a phase angle of the complex plane arises, which is independent of ω. This is often observed in high-area, porous electric double-layer materials, where a network of distributed RC elements down the lengths of the pores describes the electrochemical behavior, as discussed in the model presented by de Levie in Section 8.3.3 [60].

A Nyquist plot in Figure 8.21 approaches the ideal electrochemical behavior. At low frequencies, impedance can be reduced to a familiar term $Z_E = 1/j\omega C$. The Warburg impedance region at low frequencies is due to ionic solution resistances present within the pore structures of the electrode and is governed by the electrolyte conductivity and electrode porosity. This has been termed the "equivalent distributed resistance" (EDR). Evidence of this behavior has been shown by Kotz and Carlen [2] (Figure 8.22) using the same electrode material and active electrodes of increasing thickness, thereby increasing the pore number and a subsequent increase in the observed EDR.

Because of nonideal charging processes from macroscopic variations in current path lengths through the electrode (inconsistencies in electrode thickness) or microscopic variations in the electrode structure affecting adsorption processes, a nonvertical slope at low frequencies exists for real ECs. Representing this effect in any model equation requires incorporating a constant phase expression (CPE)

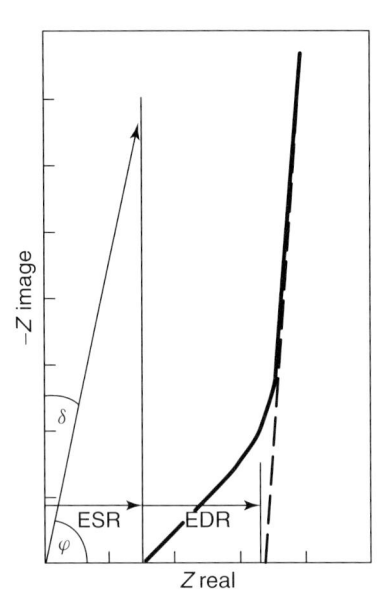

Figure 8.21 Nyquist impedance diagram comparing the ideal vertical impedance of a capacitor (thin line) and that of an EC supercapacitor (thick line). The equivalent series resistance (ESR) is derived from the intercept of the real impedance axis, followed by the equivalent distributed (EDR) of a porous electrode. (Reprinted from Ref. [2], with permission from Elsevier.)

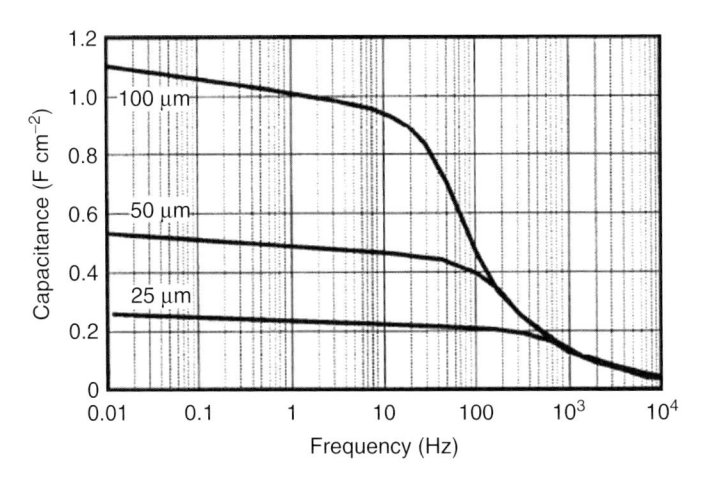

Figure 8.22 Capacitance with respect to frequency. Thicker active layers show the effect of an increasing capacitance at lower frequencies, as well as an earlier dropoff at higher frequencies. (Reprinted from Ref. [2], with permission from Elsevier.)

$(j\omega)^p$ to replace every $j\omega$ expression, where p ranges between 0 and 1, with the ideal capacitor behavior represented with $p = 1$ [2].

At low frequencies, where the imaginary component of impedance dominates the overall impedance, a close approximation can be made to the ideal capacitance of the material achievable under dc. The following equation can then be used to obtain the overall capacitance [137]:

$$C = -\frac{1}{2\pi f z^*} \tag{8.29}$$

8.7.4
Energy and Power Density

The assessment of energy and power density provided by EC supercapacitors is desirable for any application, particularly for those requiring high loads in short time intervals. Specific power density (W kg^{-1}) plotted versus the specific energy density (Wh kg^{-1}), termed the "Ragone plot", is often used to evaluate and compare the respective performance of electrochemical energy storage devices. Through a Ragone plot of ECs and batteries, previously shown in Figure 8.1, a trend in diminishing energy density with increasing power density can be observed for both electrochemical energy storage devices. This can be explained by an increase in current density i leading to polarization losses and a reduction in cell voltage.

8.7.4.1 Energy Density
The theoretical maximum energy storage of an electrode or device can be easily quantified by the Gibbs energy of charging E required to develop an electrostatic charge. Reiterating from Section 8.7.4, the charge of a capacitor Q is directly related to the applied potential ΔV and capacitance C by $Q = CV$ as defined by capacitance. Accumulation of charge along the surface requires additional work to overcome subsequent repulsions between like charges. The resulting energy stored as charge is then

$$E = \frac{1}{2}QV = \frac{1}{2}CV^2 \tag{8.30}$$

Hence, the importance of a large operating potential in the resulting energy stored by the device is apparent by the square voltage term.

It is apparent through the Ragone plots that an increasing energy density lowers the power performance of electrochemical storage devices and vice versa. Thus, the relation between the energy density (E with respect mass or volume) and power density is important to evaluate in order to understand the relative performance between these two properties for practical application.

8.7.4.2 Power Density
The power density defines the rate of power delivered by a device of certain mass at a specified current density. Power densities are influenced by kinetic polarization at overpotentials, ohmic resistances of the electrolyte, and intrinsic material structure.

The latter effect mainly determines the loss in power and is more prevalent in EDLC, where Faradaic reaction processes do not occur (ideally); however, when considering pseudocapacitive materials, kinetic polarization effects come into play, depending on the reversible potential of the cell.

The maximum power deliverable by a device P_{max} has been derived by Miller considering the capacitance of the material in series with an ESR, R_{ESR} [138]. Through this simplified circuit analogy, the following equation can be obtained:

$$P_{max} = \frac{V_i^2}{4R_{ESR}} \tag{8.31}$$

This power is often used to normalize average or normal power deliverable by the device or system with respect to the number of time constants for discharge and energy delivered or dissipated. These factors are important to consider in application, in optimizing the energy and power delivery and minimizing the capacitor energy that dissipates internally.

The relation between the energy density E_d of a system and the respective power density P_d performance is evaluated by the simple ratio

$$\frac{E_d}{P_d} = \Delta t \tag{8.32}$$

where Δt is the rate of discharge of the capacitor. This ratio is independent of any factors detrimental to both the power and energy densities.

8.7.5
Leakage Current and Self-Discharge Behavior

The driving force for self-discharge of an EC, caused by the thermodynamic instability of the charged state, is damaging to the desirable performance characteristics for commercial application. The reliability of energy storage can then become an issue in the design, and requires the self-discharge characteristics of the EC supercapacitor to be evaluated and improved prior to future usage.

The self-discharge of an electrochemical device at open circuit arises for different reasons in batteries and capacitors depending on the mechanism of charge (i.e., pseudocapacitance or EDLC). Typically, charged EDLCs have a greater rate of self-discharge compared to batteries because of the lack of thermodynamic and kinetic mechanisms possessed by battery cell systems [1, 139]. Thus, any surface impurities that can undergo a Faradaic redox coupling reaction at or below the charged potential of the capacitor will cause a loss in electrode polarization [140, 141]. Applied overpotentials beyond the limits of the electrolyte result in a loss to the voltage at thermodynamic stability.

Measuring self-discharge of a system can be performed by charging the device and measuring the open-circuit potential difference across it with respect to time. The time used for evaluation can vary depending on whether the data applies to industrial application or is primarily for research. The measurement periods can be between weeks and months or a few hundred seconds or less, respectively. This test can also be applied to a three-electrode cell configuration, where self-discharge rates can be evaluated for each electrode separately. This can provide valuable information on their respective performances as a designated anode or cathode, which will characteristically yield different results particularly for asymmetric capacitor/battery hybrids [141, 142].

8.7.6
Durability Test and Additional Procedures

Testing of cycle life is an important criterion for performance evaluation and is self-descriptive, where a system or individual capacitor electrode undergoes a long period (upwards of 10 000 cycles in laboratory settings) of charge–discharge cycling followed by a comparison between the subsequent and initial capacitive performance.

A range of other testing procedures can be conducted on EC supercapacitors for performance evaluation, including variations of the ones previously mentioned. Evaluation of charge or discharge performance while maintaining constant power, current, or voltage during discharge through a load is commonly done for practical performance testing. Additionally, these testing procedures can be carried out with changes in temperature to evaluate undesirable effects that may occur as a result. These include electrolyte conductivity, material degradation, and increases in self-discharge rates, among others. This form of testing is given more weight owing to commercial applications that require a minimum device volume, contrasting the common method of heat management through an increase of the surface-to-volume ratio for heat dissipation to the environment.

8.8
Supercapacitor Configurations

Current commercial ECs are fabricated with symmetric carbon materials and an organic electrolyte (polycarbonate or acetonitrile). The use of organic as opposed to aqueous electrolytes provides a larger operating voltage to be applied. However, these advantages come at the price, as organic electrolytes require expensive purification methods, are regarded as environmentally unfriendly, and are considered unsafe in some applications. For these reasons, the employment of aqueous electrolytes would be beneficial for industrial applications, yet the limiting 1 V voltage window is a strong inhibitor and the power and energy densities remain inadequate for commercial use. Thus, asymmetric designs are intended to aid in overcoming this limitation by identifying the most favorable range in potential of each material and employing them as either anode or cathode to improve the operating cell voltage.

High pseudocapacitance has been reported by a number of pseudocapacitive materials; however, deterrents for commercial use owe to material degradation from prolonged cycling (expansion/shrinkage of ECPs), high costs (RuO_2), or poor

Table 8.6 Supercapacitor characteristics for asymmetric capacitor systems configured using polymer (PANI, PPy, PEDOT), active carbon, and MnO_2 [143].

Electrode materials		Supercapacitor characteristics			P_{max} (kW kg^{-1})
Positive	Negative	U (V)	E (Wh kg^{-1})	ESR (Ω cm^2)	
PANI	PANI	0.5	3.13	0.36	10.9
PPy	PPy	0.6	2.38	0.32	19.7
PEDOT	PEDOT	0.6	1.13	0.27	23.8
Carbon Maxsorb	Carbon Maxsorb	0.7	3.74	0.44	22.4
PANI	Carbon Maxsorb	1	11.46	0.39	45.6
PPy	Carbon Maxsorb	1	7.64	0.37	48.3
PEDOT	Carbon Maxsorb	1	3.82	0.33	54.1
MnO_2	MnO_2	0.6	1.88	1.56	3.8
MnO_2	PANI	1.2	5.86	0.57	42.1
MnO_2	PPy	1.4	7.37	0.52	62.8
MnO_2	PEDOT	1.8	13.5	0.48	120.1

properties (electrical conductivity, MnO_2). Thus, the development of composite materials by integrating desirable physical and electrochemical properties of two or three pseudocapacitive and double-layer capacitive materials to improve these effects remains an active research area.

Current hybrid designs favor a negative electrode constructed from AC and positive electrode made of a variety of ECPs, transition-metal oxides, or composites integrated with a carbon. This arrangement can benefit from the reversibility of hydrogen sorption in porous carbon, allowing an increase in voltage beyond the 1.2 V electrochemical decomposition limit. Frackowiak has reviewed a series of asymmetric systems using AC, ECPs (PANI, PPy, and PEDOT), and MnO_2 for both anode and cathode application, listed in Table 8.6, and reported an increase in operating potential and energy density for some configurations while maintaining a high maximum power density [143]. Incorporating CNTs or graphene into composites as a base carbon material (discussed in Section 8.4.1.1) have also been used to improve hybrid asymmetric capacitor electrodes on numerous occasions, particularly their power rating through a reduction in intrinsic resistances and contribution of an entangled or aligned mesoporous structure.

A common problem remains in most cases with the development of hybrid asymmetric capacitors where an increase in energy density through incorporation of some pseudocapacitive or battery electrode materials comes at a cost of cycle life. As a result, these improvements and detriments must be weighed with regard to the intended application. Systems or devices in need of high-power energy storage with a modest life span could be met with hybrid configurations.

8.9
Applications

ECs are still recognized as a relatively new energy storage device that continues to find application in emerging technologies. Prior to their recent developments, they were largely recognized as low-energy storage devices with limited power densities, and used primarily for memory backup and actuator applications [136]. Following in their highly anticipated potential, new EDLC, pseudocapacitive, and hybrid EC supercapacitors are actively researched and have produced substantial improvements to their performance characteristics, bringing attention to their relevance in a wide range of new applications. A selection of commercial EC supercapacitors offered by Maxwell Technologies is shown in Figure 8.23.

For operations that require a large amount of energy for a sufficient length of time, generation systems are more pertinent than storage systems likewise, whereas applications operating in shorter time intervals with more moderate energy demands can be met by the latter. Electrical energy storage by means of batteries (chemical potential), ECs (electrostatic potential), or both are currently being used in a range of applications, chosen appropriately to address their energy requirements.

8.9.1
Use in Memory Backup

A common use for ECs in aiding primary power sources continues to be their ability to provide high pulse power when required within a very short time (milliseconds to seconds). ECs served originally as a power source to protect against disturbance in power supply within the interval between a power failure of the primary power source and initiation of an emergency backup or replacement. Typical power outages last less than a few seconds, but in certain industrial production lines or commercial electronics these disruptions would cause substantial losses [2]. ECs have been incorporated in consumer electronic applications as a short-term power supply since their arrival as a commercial device in 1978. Batteries as a competitor were a more costly alternative and could not provide a competitive

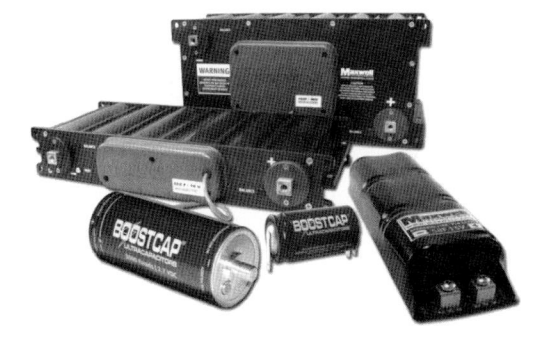

Figure 8.23 Collection of commercial EC supercapacitors with re-print permission from Maxwell Technologies, Inc.

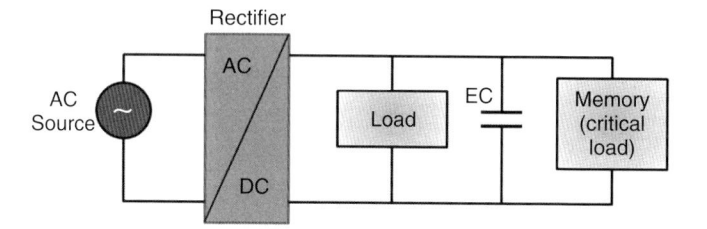

Figure 8.24 Circuit schematic showing the use of ECs in devices powered by an ac source. An EC is incorporated to protect the memory (critical load) from significant voltage drops.

lifetime, advertising the overall cost efficiency of ECs since their arrival. A circuit in Figure 8.24 shows how an EC protects the memory (critical load) against any significant drop in potential in a primary power source system.

8.9.2
Improved Battery Systems

Current and future technology trends are addressing the growing demand for portable or remote applications, which in turn require their own local electrical energy storage or generation. ECs are ideally used in applications that benefit from their high-power delivery and symmetric power uptake, where energy demands are in short pulse intervals ranging from milliseconds to tens of seconds. In order for batteries to meet similar power performance, overdimensioning of the device would be required and further complications follow regarding thermal management as a consequence of such designs. Capacitor devices face similar difficulties in matching the energy density capable of EC supercapacitors. Even the recently commercialized high-power lithium ion batteries operating at high-power release/uptake will suffer severe degradation and loss in cycle life due to the fundamental difference in energy storage and delivery [8].

Using these devices to complement one another allows an optimal design for portable energy storage, where short power pulses can be met by the EC, providing the battery an extended life to address a smaller average load demand. With the addition of an EC, a reduction in battery size and increase in battery life can be achieved. Telecommunications devices, among other mobile devices, stand to benefit from such a configuration, providing them a means to address the increasing demand to reduce their battery size and meet their primary objective of device miniaturization [2, 136].

8.9.3
Hybrid Electric Vehicles and Transport

Interest in EC supercapacitors was reenergized in the early 1990s by the main objective to integrate their high-power delivery rate into transportation vehicles,

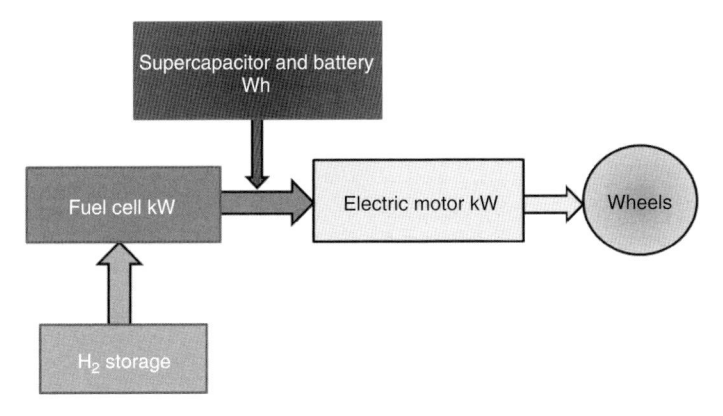

Figure 8.25 Incorporation of EC supercapacitors for break energy storage in a vehicle driveline schematic.

including trains, tramways, and, more significantly, HEVs. Challenges faced by energy storage devices implemented in HEVs are heat management in dissipating the undesirable heat generated by resistances and exothermic reactions during charging/discharging, as well as a required high cycle life. In addition to having high cycle efficiency (>95%), design criteria of ECs address these main issues prior to commercial manufacture, making them highly compatible with such applications. Vehicles that experience significant stop-and-go intervals during operation (buses, garbage disposal trucks, delivery vehicles, and inner-city cars) benefit from the integration of ECs into the energy storage system by allowing recuperation of brake energy to be reused in acceleration phases for additional power boosting, and consequently increase the fuel efficiency of the vehicle, as shown in Figure 8.25.

A recent review on EC supercapacitor applications by Miller and Burke compared high-power batteries to ECs for use in HEVs with respect to the regenerative energy that has been both captured and stored, the latter being available for discharge upon demand. Evaluation with respect to charging time refuted the conventional view of battery superiority vis-à-vis the energy density of the two storage systems at charge times of 10 s or less [8]. EC supercapacitors are shown to have approximately twice the energy density of the high-power battery in these short time intervals and require no heat management system to operate during high-rate cycling.

Train manufacturers recognize the suitable use of ECs in metro and tram transportation, which involves frequent stops across short distances. Recovery of brake energy, load leveling, reduction in operating cost, and reliability as a source for emergency power are a number of key reasons for their integration into these systems.

8.9.4
Military Applications

As the applications for ECs grow, the military market segment has demonstrated an increasing need. Wearable soldier systems that combine sensors, computers, communications, and displays are presently dependent on battery power. ECs provide an opportunity to design power delivery systems with the energy density, form, fit, and function appropriate for applications that require bursts of high power. The suggested areas include rail guns, electromagnetic pulse weapons, radars, and torpedoes.

8.9.5
Current Marketed Applications

Past market use of ECs has largely been in small consumer electronics requiring a relatively small capacitance from a few millifarads to farads for memory backup in computers, cameras, toys, and mobile phones, among other small devices. These applications are able to implement EC supercapacitors in favor of their faster charging and longer cycle life at a cost benefit compared with secondary batteries. Other marketed domestic electric devices taking advantage of the EC storage mechanism are power tools having either a bank of six EC cells (each 100 F 2.7 V) or EC supercapacitors placed in parallel with a rechargeable battery. Although the former stores only approximately 1/20th of the energy capable of lithium ion batteries, the advantages are apparent with a fast recharge time of 90 s and an expected cycle life extending that of the device itself. Performance and dependability of modern EC supercapacitor technology have also been shown by their implementation within the Airbus A380 jumbo jet, with cell banks incorporated into emergency doors for operation in critical events [5, 8].

ECs have appeal to applications that have repetitive time intervals where energy is being squandered as heat. The fast uptake of ECs provides a means to capture this energy for further use. This function is supported by their longevity in making them cost effective (cost recovery) within a relatively short time. Integrating ECs can improve the overall energy efficiency of the application, and, with proper capacitor sizing, peak power demands can be met at short intervals allowing a reduction in size of the primary power generator. This design is already in place with gantry cranes used to load and unload cargo containers in some seaports.

The pulse power required by engine starters can also incorporate ECs for high-power pulses, especially useful at low temperatures, demonstrating EC supercapacitors' large operational temperature range. Actuator applications can be integrated with a battery system of reduced size to increase the battery life.

8.9.6
Future Applications

Integrated systems that incorporate both batteries and capacitors in hybrid cells constituted by a battery electrode (energy storage) and EC electrode (power contributor) demonstrate the effective use of combining both batteries and capacitors by providing a storage system with increased energy density in addition to a higher power rating unattainable by batteries alone. The benefits of integrating a battery and EC electrode together, however, must always be weighed against any decrease in power performance or loss in cycle life, which is typically the case. Despite these considerations, HEVs are one of a range of applications that may stand to benefit from the dual integration of these complementary storage devices.

The growing interests in applying ECs to elevators, cranes or pallet trucks, hand tools or flashlights, pulsed laser, and welding are apparent. Aspiring advances to produce thin, flexible printed electronics that include transistors, photovoltaic cells, and organic LEDs (light-emitting diodes) show a relatively novel technology in need of a complementary energy source. Current developments have been successful in printing a fully integrated EC supercapacitor from SWNTs onto ubiquitous paper as both a substrate and separator. Other works by Yu *et al.* [89] and Moser *et al.* [144] have demonstrated transparent or translucent thin-film electrodes of graphene nanoplatelets (GNPs) and MnO_2, respectively, for EC supercapacitor application. Additional future uses for ECs may include their incorporation with correspondingly low-voltage parallel photovoltaic cells to provide energy storage of solar energy, as well as possible use for electrochromic devices.

8.10
Challenges and Perspectives of Electrochemical Supercapacitors

The importance of electrochemical supercapacitors and their contributions to a growing range of applications are only recently becoming evident, as society and technology continue to explore alternative and ecologically conscious energy storage and generation systems. In addition to the technical benefits ECs can provide, the electrode materials used are recyclable and provide an environmentally conscious alternative to batteries. This comes at a time when considerations toward green energy choices are growing in significance for industry and government regulations and policies are becoming progressively more stringent.

In spite of these benefits, further commercial expansion of this technology is currently limited by the challenges that must be met, specifically pertaining to low-energy density and high costs. At present, the limited number of commercial providers that develop these systems impedes a reduction of their cost. Additional to their high cost, ECs possess substantially lower energy storage compared to high-power batteries, and their appealing cycle life is not enough to persuade a large number of consumers to invest in their purchase.

Two means by which the energy density can be improved upon are enhancement of the specific capacitance of the electrode materials and an expansion to the operating potential window. Increasing the energy storage of future ECs through progressive developments in applicable electrolytes, specifically ILs, can provide a larger range in the operating voltage. The incorporation of ILs can also help to circumvent safety issues that restrict currently marketed nonaqueous (acetonitrile) ECs from being used in some applications due to safety and toxicity hazards. However, a reduction to the resistivity and viscosity of ILs within the desired operating temperature range must be accomplished. Solution mixtures of ILs used in acetonitrile have shown promise in addressing these issues, by achieving a high energy density and acceptable power density. Alternatively, other research and development has been focused on advancing a variety of electrode materials for applicable use in achieving higher capacitance. Promising designs include tailored nanoporous carbons, which correspond to the electrolyte ion size at an atomic scale, pseudocapacitive metal oxides, and ECPs for enhanced energy density. In particular, progress has been shown in asymmetric and hybrid capacitors through the incorporation of transition-metal or ECP composites together with carbon to attain higher energy densities.

These improvements are expected to endow ECs with an energy density comparable to what is presently achieved by marketed Li ion batteries, at no expense to their appreciable power density. Addressing the technical weaknesses of EC capacitors is also helpful in overcoming the costs that stall their marketability. As more applications are found to require high-voltage, series-aligned capacitor banks, an increase to the operating voltage of a cell will lead directly to a reduction in material (electrode, electrolyte, sealant, etc.) cost and subsequent savings to consumers.

The future of ECs is expected to grow in their capacity for energy storage by helping to develop more energy-efficient systems and decreasing hydrocarbon fuel consumption. Progress toward their complementary integration with battery systems are expected to prove useful in large-scale transport vehicles (cars/buses, trains, trams) and mobile devices seeking to minimize volume and weight. Material developments with metal-oxide films and graphene also show a new direction for ECs in future technology applications for printed and optical electronics. From a technical standpoint, ECs will continue to show potential; however, the market will ultimately dictate their commercial success. Efforts to reduce their cost in making them more affordable will go a long way in ensuring that they remain competitive. The ambition to increase the operating potential through improvements in ILs is expected to offset the main obstacles by improving energy storage and simultaneously reducing their cost of production. Meeting these goals will help ECs to remain an attractive energy solution for cases requiring fast power rates, energy efficiency, and a superior lifetime.

References

1. Conway, B.E. (1999) *Electrochemical Supercapacitors*, Plenum Publishing, New York.
2. Kotz, R. and Carlen, M. (2000) Principles and applications of electrochemical capacitors. *Electrochim. Acta*, **45** (15–16), 2483–2498.
3. Conway, B.E., Birss, V., and Wojtowicz, J. (1997) The role and utilization of pseudocapacitance for energy storage by supercapacitors. *J. Power Sources*, **66** (1–2), 1–14.
4. Pandolfo, A.G. and Hollenkamp, A.F. (2006) Carbon properties and their role in supercapacitors. *J. Power Sources*, **157** (1), 11–27.
5. Simon, P. and Gogotsi, Y. (2008) Materials for electrochemical capacitors. *Nat. Mater.*, **7** (11), 845–854.
6. Armand, M. and Tarascon, J.M. (2008) Building better batteries. *Nature*, **451** (7179), 652–657.
7. United States Department of Energy (2007) Basic Research Needs for Electrical Energy Storage.
8. Miller, J.R. and Burke, A.F. (2008) Electrochemical capacitors: challenges and opportunities for real-world

applications. *Electrochem. Soc. Interface*, **17**, 53–57.
9. Robinson, F.N., Kashy, E., and McGrayne, S.B. (2010) *Invention of the Leyden Jar*, Encyclopedia Britannica.
10. Becker, H.E. (1957) in *Low Voltage Electrolytic Capacitor* (ed. U.S.P. Office), General Electric Company, US. Pat. No. 2,800,616 and assignee: General Electric (GE).
11. Boos, D.I. (1970) in *Electrolytic Capacitor Having Carbon Paste Electrodes* (ed. U.S.P. Office), The Standard Oil Company, US. Pat. No. 3,536,963 and assignee: Standard Oil of Ohio (SOHIO).
12. Burke, A. (2007) R&D considerations for the performance and application of electrochemical capacitors. *Electrochim. Acta*, **53** (3), 1083–1091.
13. Frackowiak, E. (2007) Carbon materials for supercapacitor application. *Phys. Chem. Chem. Phys.*, **9** (15), 1774–1785.
14. Obreja, V.V.N. (2008) On the performance of supercapacitors with electrodes based on carbon nanotubes and carbon activated material – A review. *Physica E*, **40** (7), 2596–2605.
15. Pan, H., Li, J.Y., and Feng, Y.P. (2010) Carbon nanotubes for supercapacitor. *Nanoscale Res. Lett.*, **5** (3), 654–668.
16. Zhang, L.L., Zhou, R., and Zhao, X.S. (2010) Graphene-based materials as supercapacitor electrodes. *J. Mater. Chem.*, **20** (29), 5983–5992.
17. Wang, D.W. et al. (2008) Aligned titania nanotubes as an intercalation anode material for hybrid electrochemical energy storage. *Adv. Funct. Mater.*, **18**, 3787–3793.
18. Naoi, K. et al. (2010) High-rate nano-crystalline $Li_4Ti_5O_{12}$ attached on carbon nano-fibers for hybrid supercapacitors. *J. Power Sources*, **195** (18), 6250–6254.
19. Khomenko, V. et al. (2006) High-voltage asymmetric supercapacitors operating in aqueous electrolyte. *Appl. Phys. A*, **82** (4), 567–573.
20. Cooper, A. et al. (2009) The UltraBattery-A new battery design for a new beginning in hybrid electric vehicle energy storage. *J. Power Sources*, **188** (2), 642–649.
21. Burke, A. and Miller, M. (2011) The power capability of ultracapacitors and lithium batteries for electric and hybrid vehicle applications. *J. Power Sources*, **196** (1), 514–522.
22. Lazzari, M., Soavi, F., and Mastragostino, M. (2009) Dynamic pulse power and energy of ionic-liquid-based supercapacitor for HEV application. *J. Electrochem. Soc.*, **156** (8), A661–A666.
23. Lewandowski, A. et al. (2010) Performance of carbon-carbon supercapacitors based on organic, aqueous and ionic liquid electrolytes. *J. Power Sources*, **195** (17), 5814–5819.
24. Zhang, L.L. and Zhao, X.S. (2009) Carbon-based materials as supercapacitor electrodes. *Chem. Soc. Rev.*, **38** (9), 2520–2531.
25. An, K.H. et al. (2001) Supercapacitors using singlewalled carbon nanotube electrodes. Nanonetwork Materials: Fullerenes, Nanotubes and Related Systems, vol. 590, pp. 241–244.
26. Wang, Y. et al. (2009) Supercapacitor devices based on graphene materials. *J. Phys. Chem. C*, **113** (30), 4.
27. Ania, C.O. et al. (2007) The large electrochemical capacitance of microporous doped carbon obtained by using a zeolite template. *Adv. Funct. Mater.*, **17** (11), 1828–1836.
28. Yamada, H. et al. (2007) Electrochemical study of high electrochemical double layer capacitance of ordered porous carbons with both meso/macropores and micropores. *J. Phys. Chem. C*, **111** (1), 227–233.
29. Shukla, A.K., Sampath, S., and Vijayamohanan, K. (2000) Electrochemical supercapacitors: energy storage beyond batteries. *Curr. Sci.*, **79** (12), 1656–1661.
30. Pang, S.C. and Anderson, M.A. (2000) Novel electrode materials for ultracapacitors: structural and electrochemical properties of sol-gel-derived manganese dioxide thin films. Proceedings of New Materials for Batteries and Fuel Cells, vol. 575, pp. 415–421.
31. Sugimoto, W. et al. (2006) Charge storage mechanism of nanostructured anhydrous and hydrous

ruthenium-based oxides. *Electrochim. Acta*, **52** (4), 1742–1748.
32. Messaoudi, B. et al. (2001) Anodic behaviour of manganese in alkaline medium. *Electrochim. Acta*, **46** (16), 2487–2498.
33. Zhang, J. et al. (2010) Synthesis of polypyrrole film by pulse galvanostatic method and its application as supercapacitor electrode materials. *J. Mater. Sci.*, **45** (7), 1947–1954.
34. Zhang, K. et al. (2010) Graphene/polyaniline nanofiber composites as supercapacitor electrodes. *Chem. Mater.*, **22** (4), 1392–1401.
35. Yan, J. et al. (2010) Preparation of graphene nanosheet/carbon nanotube/polyaniline composite as electrode material for supercapacitors. *J. Power Sources*, **195** (9), 3041–3045.
36. Ryu, K.S. et al. (2004) Poly(ethylenedioxythiophene) (PEDOT) as polymer electrode in redox supercapacitor. *Electrochim. Acta*, **50** (2–3), 843–847.
37. Patra, S. and Munichandraiah, N. (2007) Supercapacitor studies of electrochemically deposited PEDOT on stainless steel substrate. *J. Appl. Polym. Sci.*, **106**, 1160–1171.
38. Du, B. et al. (2009) Preparation of PPy/CNT composite applications for supercapacitor electrode material. *Mater. Sci. Forum*, **610–613**, 502–505. (Pt. 1, Materials Research: Eco/Environmental Materials, Energy Materials, Magnesium, Aerospace Materials and Biomaterials for Medical Application).
39. Frackowiak, E. et al. (2006) Supercapacitors based on conducting polymers/nanotubes composites. *J. Power Sources*, **153** (2), 413–418.
40. Malak, A. et al. (2010) Hybrid materials for supercapacitor application. *J. Solid State Electrochem.*, **14** (5), 811–816.
41. Naoi, K. and Simon, P. (2008) New materials and new configurations for advanced electrochemical capacitors. *Electrochem. Soc. Interface*, **17**, 34–37.
42. Amatucci, G.G. et al. (2001) An asymmetric hybrid nonaqueous energy storage cell. *J. Electrochem. Soc.*, **148** (8), A930–A939.
43. Helmholtz, H. (1853) On the laws of the distribution of electrical currents in material conductors with application to experiments in animal electricity. *Ann. Phys. (Leipzig)*, **89**, 211–233.
44. Bockris, J.O.M., Conway, B.E., and Yeager, E.B. (1980) *The Double Layer*, Plenum Press, New York.
45. Gouy, G. (1917) Interferences with large path differences. *Ann. Phys. Paris*, **7**, 88–92.
46. Chapman, D.L. (1913) Theory of electrocapillarity. *Philos. Mag.*, **25'** 475–481.
47. Stern, O. (1924) The theory of the electrolytic double shift. *Z. Elektrochem. Angew. Phys. Chem.*, **30**, 508–516.
48. Parsons, R. (1954) *Modern Aspects of Electrochemistry*, Buttersworth, London.
49. Levine, S. and Outhwaite, C.W. (1978) Comparison of theories of aqueous electric double-layer at a charged plane interface. *J. Chem. Soc., Faraday Trans. 2*, **74**, 1670–1689.
50. Grahame, D.C. (1947) The electrical double layer and the theory of electrocapillarity. *Chem. Rev.*, **41**, 441–501.
51. Levine, S. and Feat, G. (1977) Ion-pair correlation-function in electric double-layer theory. 2. Relevance to discreteness-of-charge effect. *J. Chem. Soc., Faraday Trans. 2*, **73**, 1359–1370.
52. Brockis, J.O.M., Reddy, A.K.N., and Gamboa-Aldeco, M.E. (2000) *Fundamentals of Electrodics 2A*, Kluwer Academic/Plenum Publishers, New York.
53. Conway, B.E. (1990) Transition from "supercapacitor" to "battery" behavior in electrochemical energy storage. Proceedings of the 34th International Power Sources Symposium, pp. 319–327.
54. Conway, B.E. and Gileadi, E. (1962) Kinetic theory of pseudo-capacitance and electrode reactions at appreciable surface coverage. *Trans. Faraday Soc.*, **58**, 2493–2509.
55. Gileadi, E. and Srinivasan, S. (1966) The potential-sweep method: a theoretical analysis. *Electrochim. Acta*, **11** (3), 321–335.
56. Suppes Graeme, M., Deore Bhavana, A., and Freund Michael, S.

(2008) Porous conducting polymer/heteropolyoxometalate hybrid material for electrochemical supercapacitor applications. *Langmuir*, **24** (3), 1064–1069.

57. Arulepp, M. *et al.* (2004) Influence of the solvent properties on the characteristics of a double layer capacitor. *J. Power Sources*, **133** (2), 320–328.

58. Conway, B.E. (1981) *Ionic Hydration in Chemistry and Biophysics*, Elsevier, Amsterdam.

59. Qu, D.Y. and Shi, H. (1998) Studies of activated carbons used in double-layer capacitors. *J. Power Sources*, **74** (1), 99–107.

60. de Levie, R. (1963) On porous electrodes in electrolyte solutions. *Electrochim. Acta*, **8**, 751–780.

61. Fuertes, A.B. *et al.* (2005) Templated mesoporous carbons for supercapacitor application. *Electrochim. Acta*, **50** (14), 2799–2805.

62. Gryglewicz, G. *et al.* (2005) Effect of pore size distribution of coal-based activated carbons on double layer capacitance. *Electrochim. Acta*, **50** (5), 1197–1206.

63. Largeot, C. *et al.* (2008) Relation between the ion size and pore size for an electric double-layer capacitor. *J. Am. Chem. Soc.*, **130** (9), 2730–2731.

64. Qu, D.Y. (2002) Studies of the activated carbons used in double-layer supercapacitors. *J. Power Sources*, **109** (2), 403–411.

65. Salitra, G. *et al.* (2000) Carbon electrodes for double-layer capacitors – I. Relations between ion and pore dimensions. *J. Electrochem. Soc.*, **147** (7), 2486–2493.

66. Raymundo-Pinero, E. *et al.* (2006) Relationship between the nanoporous texture of activated carbons and their capacitance properties in different electrolytes. *Carbon*, **44** (12), 2498–2507.

67. Huang, J.S., Sumpter, B.G., Meunier, V. *et al.* (2008) Theoretical model for nanoporous carbon supercapacitors. *Angew. Chem. Int. Ed.*, **47** (3), 520–524.

68. Huang, J.S., Sumpter, B.G., and Meunier, V. (2008) A universal model for nanoporous carbon supercapacitors applicable to diverse pore regimes, carbon materials, and electrolytes. *Chem. – Eur. J.*, **14** (22), 6614–6626.

69. McEnaney, B. and Burchell, T.D. (1999) *Carbon Materials for Advanced Technologies*, Pergamon.

70. Inagaki, M. and Radovic, L.R. (2002) Nanocarbons. *Carbon*, **40** (12), 2279–2282.

71. Marsh, H. (1989) *Introduction to Carbon Science*, Butterworths.

72. Shi, H. (1996) Activated carbons and double layer capacitance. *Electrochim. Acta*, **41** (10), 1633–1639.

73. Endo, M. *et al.* (2001) Morphological effect on the electrochemical behavior of electric double-layer capacitors. *J. Mater. Res.*, **16** (12), 3402–3410.

74. Okamura, M. (2000) in *Electric Double Layer Capacitor* (ed. U.S.P. Office), JEOL Ltd.

75. Fang, B.Z. and Binder, L. (2006) A novel carbon electrode material for highly improved EDLC performance. *J. Phys. Chem. B*, **110** (15), 7877–7882.

76. Azais, P. *et al.* (2007) Causes of supercapacitors ageing in organic electrolyte. *J. Power Sources*, **171** (2), 1046–1053.

77. Zhu, M. *et al.* (2008) Chemical and electrochemical ageing of carbon materials used in supercapacitor electrodes. *Carbon*, **46** (14), 1829–1840.

78. Leon y Leon, C.A. and Radovic, L.R. (1994) *Chemistry and Physics of Carbon*, Marcel Dekker Inc., New York.

79. Frackowiak, E. *et al.* (1999) Capacitance properties of carbon nanotubes, *Electronic Properties of Novel Materials - Science and Technology of Molecular Nanostructures: XII International Winterschool*, American Institute of Physics Secaucus, College Park, AIP Conference Proceedings Series Vol. 486, pp. 429–432.

80. Jurewicz, K. *et al.* (2001) Supercapacitors from nanotubes/polypyrrole composites. *Chem. Phys. Lett.*, **347** (1–3), 36–40.

81. Liu, C. *et al.* (2010) Advanced materials for energy storage. *Adv. Mater.*, **22** (8), E28–E62.

82. Futaba, D.N. *et al.* (2006) Shape-engineerable and highly densely packed single-walled carbon

83. Zhang, H. et al. (2008) Comparison between electrochemical properties of aligned carbon nanotube array and entangled carbon nanotube electrodes. *J. Electrochem. Soc.*, **155** (2), K19–K22.
84. Bordjiba, T., Mohamedi, M., and Dao, L.H. (2008) New class of carbon-nanotube aerogel electrodes for electrochemical power sources. *Adv. Mater.*, **20** (4), 815–819.
85. Vix-Guterl, C. et al. (2005) Electrochemical energy storage in ordered porous carbon materials. *Carbon*, **43** (6), 1293–1302.
86. Geim, A.K. and Novoselov, K.S. (2007) The rise of graphene. *Nat. Mater.*, **6** (3), 183–191.
87. Stoller, M.D. et al. (2008) Graphene-based ultracapacitors. *Nano Lett.*, **8** (10), 3498–3502.
88. Ruoff, R. (2008) Calling all chemists. *Nat. Nanotechnol.*, **3** (1), 10–11.
89. Yu, A.P. et al. (2010) Ultrathin, transparent, and flexible graphene films for supercapacitor application. *Appl. Phys. Lett.*, **96** (25), 253105–253108.
90. Xu, B. et al. (2008) Highly mesoporous and high surface area carbon: a high capacitance electrode material for EDLCs with various electrolytes. *Electrochem. Commun.*, **10** (5), 795–797.
91. Sivakkumar, S.R. et al. (2007) Performance evaluation of CNT/polypyrrole/MnO_2 composite electrodes for electrochemical capacitors. *Electrochim. Acta*, **52** (25), 7377–7385.
92. Zhang, L.L. et al. (2009) Manganese oxide-carbon composite as supercapacitor electrode materials. *Microporous Mesoporous Mater.*, **123** (1–3), 260–267.
93. Zheng, J.P., Cygan, P.J., and Jow, T.R. (1995) Hydrous ruthenium oxide as an electrode material for electrochemical capacitors. *J. Electrochem. Soc.*, **142** (8), 2699–2703.
94. Wu, N.L. (2002) Nanocrystalline oxide supercapacitors. *Mater. Chem. Phys.*, **75** (1–3), 6–11.
95. Wang, B. et al. (2009) Synthesis of nanosized vanadium pentoxide/carbon composites by spray pyrolysis for electrochemical capacitor application. *Electrochim. Acta*, **54** (5), 1420–1425.
96. Jayalakshmi, M. et al. (2007) Hydrothermal synthesis of SnO_2-V_2O_5 mixed oxide and electrochemical screening of carbon nano-tubes (CNT), V_2O_5, V_2O_5-CNT, and SnO_2-V_2O_5-CNT electrodes for supercapacitor applications. *J. Power Sources*, **166** (2), 578–583.
97. Jayalakshmi, M. and Balasubramanian, K. (2008) Simple capacitors to supercapacitors – An overview. *Int. J. Electrochem. Sci.*, **3** (11), 1196–1217.
98. Zhang, S.W. and Chen, G.Z. (2008) Manganese oxide based materials for supercapacitors. *Energy Mater.*, **3** (3), 186–200.
99. Kim, I.H. and Kim, K.B. (2006) Electrochemical characterization of hydrous ruthenium oxide thin-film electrodes for electrochemical capacitor applications. *J. Electrochem. Soc.*, **153** (2), A383–A389.
100. Lin, Y.S. et al. (2009) Superior capacitive characteristics of RuO_2 nanorods grown on carbon nanotubes. *Appl. Surf. Sci.*, **256** (4), 1042–1045.
101. Zheng, J.P. and Jow, T.R. (1995) A new charge storage mechanism for electrochemical capacitors. *J. Electrochem. Soc.*, **142** (1), L6–L8.
102. Liu, T.C., Pell, W.G., and Conway, B.E. (1997) Self-discharge and potential recovery phenomena at thermally and electrochemically prepared RuO_2 supercapacitor electrodes. *Electrochim. Acta*, **42** (23–24), 3541–3552.
103. Ramani, M. et al. (2001) Studies on activated carbon capacitor materials loaded with different amounts of ruthenium oxide. *J. Power Sources*, **93** (1–2), 209–214.
104. Hu, C., Chen, W.C., and Chang, K.H. (2004) How to achieve maximum utilization of hydrous ruthenium oxide for supercapacitors. *J. Electrochem. Soc.*, **151** (2), A281–A290.
105. Raistrick, I. (1992) *The Electrochemistry of Semiconductors and Electronics-Processes and Devices*, Noyes Publications, Park Ridge, IL.

106. Hu, C.C. and Chen, W.C. (2004) Effects of substrates on the capacitive performance of RuOx center dot nH(2)O and activated carbon-RuOx electrodes for supercapacitors. *Electrochim. Acta*, **49** (21), 3469–3477.
107. Chen, W.C. et al. (2004) Electrochemical characterization of activated carbon-ruthenium oxide nanoparticles composites for supercapacitors. *J. Power Sources*, **125** (2), 292–298.
108. Chang, K.H. and Hu, C.C. (2004) Oxidative synthesis of RuOx center dot nH(2)O with ideal capacitive characteristics for supercapacitors. *J. Electrochem. Soc.*, **151** (7), A958–A964.
109. Hu, C.C. and Chang, K.H. (2002) Cyclic voltammetric deposition of hydrous ruthenium oxide for electrochemical supercapacitors: effects of the chloride precursor transformation. *J. Power Sources*, **112** (2), 401–409.
110. Patake, V.D., Lokhande, C.D., and Joo, O.S. (2009) Electrodeposited ruthenium oxide thin films for supercapacitor: Effect of surface treatments. *Appl. Surf. Sci.*, **255** (7), 4192–4196.
111. Im, D. and Manthiram, A. (2002) Nanostructured lithium manganese oxide cathodes obtained by reducing lithium permanganate with methanol. *J. Electrochem. Soc.*, **149** (8), A1001–A1007.
112. Lee, H.Y., Kim, S.W., and Lee, H.Y. (2001) Expansion of active site area and improvement of kinetic reversibility in electrochemical pseudocapacitor electrode. *Electrochem. Solid State Lett.*, **4** (3), A19–A22.
113. Wu, M.S. et al. (2004) Field emission from manganese oxide nanotubes synthesized by cyclic voltammetric electrodeposition. *J. Phys. Chem. B*, **108** (42), 16331–16333.
114. Broughton, J.N. and Brett, M.J. (2005) Variations in MnO_2 electrodeposition for electrochemical capacitors. *Electrochim. Acta*, **50** (24), 4814–4819.
115. Chen, Y.S., Hu, C.C., and Wu, Y.T. (2004) Capacitive and textural characteristics of manganese oxide prepared by anodic deposition: effects of manganese precursors and oxide thickness. *J. Solid State Electrochem.*, **8** (7), 467–473.
116. Prasad, K.R. and Miura, N. (2004) Electrochemically synthesized MnO_2-based mixed oxides for high performance redox supercapacitors. *Electrochem. Commun.*, **6** (10), 1004–1008.
117. Brousse, T. et al. (2006) Crystalline MnO_2 as possible alternatives to amorphous compounds in electrochemical supercapacitors. *J. Electrochem. Soc.*, **153** (12), A2171–A2180.
118. Nam, K.W. and Kim, K.B. (2006) Manganese oxide film electrodes prepared by electrostatic spray deposition for electrochemical capacitors. *J. Electrochem. Soc.*, **153** (1), A81–A88.
119. Toupin, M., Brousse, T., and Belanger, D. (2004) Charge storage mechanism of MnO_2 electrode used in aqueous electrochemical capacitor. *Chem. Mater.*, **16** (16), 3184–3190.
120. Fan, Z. et al. (2006) Preparation and characterization of manganese oxide/CNT composites as supercapacitive materials. *Diamond Relat. Mater.*, **15** (9), 1478–1483.
121. Liu, T.C. et al. (1998) Behavior of molybdenum nitrides as materials for electrochemical capacitors - comparison with ruthenium oxide. *J. Electrochem. Soc.*, **145** (6), 1882–1888.
122. Choi, D.W. and Kumta, P.N. (2005) Chemically synthesized nanostructured VN for pseudocapacitor application. *Electrochem. Solid State Lett.*, **8** (8), A418–A422.
123. Choi, D., Blomgren, G.E., and Kumta, P.N. (2006) Fast and reversible surface redox reaction in nanocrystalline vanadium nitride supercapacitors. *Adv. Mater.*, **18** (9), 1178–1182.
124. Krawiec, P. et al. (2006) Oxide foams for the synthesis of high-surface-area vanadium nitride catalysts. *Adv. Mater.*, **18** (4), 505–508.
125. Jin, X. et al. (2007) Nanoscale microelectrochemical cells on carbon nanotubes. *Small*, **3** (9), 1513–1517.
126. Hu, C.C. and Tsou, T.W. (2002) Ideal capacitive behavior of hydrous manganese oxide prepared by anodic deposition. *Electrochem. Commun.*, **4** (2), 105–109.

127. Wang, H. et al. (2009) Graphene oxide doped polyaniline for supercapacitors. *Electrochem. Commun.*, **11** (6), 1158–1161.
128. Schulz, B. et al. (2009) Aspects of morphology control during the oxidative synthesis of electrically conducting polymers. *High Perform. Polym.*, **21** (5), 633–652.
129. Rudge, A. et al. (1994) Conducting polymers as active materials in electrochemical capacitors. *J. Power Sources*, **47** (1–2), 89–107.
130. Koch, V.R. (2005) Recent advances in electrolytes for electrochemical double layer capacitors. 1st International Symposium on Large Ultracapacitor Technology and Application, Honolulu, Hawaii.
131. Lewandowski, A. and Galinski, M. (2007) Practical and theoretical limits for electrochemical double-layer capacitors. *J. Power Sources*, **173** (2), 822–828.
132. Zheng, J.P. (2004) Resistance distribution in electrochemical capacitors with a bipolar structure. *J. Power Sources*, **137** (1), 158–162.
133. Mahon, P.J. et al. (2000) Measurement and modelling of the high-power performance of carbon-based supercapacitors. *J. Power Sources*, **91** (1), 68–76.
134. Miller, J.R. and Burke, A.F. (1994) *Electric Vehicle Capacitor Test Procedures Manual*, Idaho National Engineering Laboratory.
135. Hughes, M. et al. (2002) Electrochemical capacitance of nanocomposite films formed by coating aligned arrays of carbon nanotubes with polypyrrole. *Adv. Mater.*, **14** (5), 382–385.
136. Huggins, R.A. (2000) Supercapacitors and electrochemical pulse sources. *Solid State Ionics*, **134** (1–2), 179–195.
137. Wang, J. et al. (2007) Capacitance properties of single wall carbon nanotube/polypyrrole composite films. *Compos. Sci. Technol.*, **67** (14), 2981–2985.
138. Miller, J.E. (1995) 5th International Seminar on Double-Layer Capacitors and Similar Energy Storage Devices, Boca Raton, FL.
139. Niu, J.J., Pell, W.G., and Conway, B.E. (2006) Requirements for performance characterization of C double-layer supercapacitors: applications to a high specific-area C-cloth material. *J. Power Sources*, **156** (2), 725–740.
140. Conway, B.E., Pell, W.G., and Liu, T.C. (1997) Diagnostic analyses for mechanisms of self-discharge of electrochemical capacitors and batteries. *J. Power Sources*, **65** (1–2), 53–59.
141. Andreas, H.A., Lussier, K., and Oickle, A.M. (2009) Effect of Fe-contamination on rate of self-discharge in carbon-based aqueous electrochemical capacitors. *J. Power Sources*, **187** (1), 275–283.
142. Niu, J.J., Conway, B.E., and Pell, W.G. (2004) Comparative studies of self-discharge by potential decay and float-current measurements at C double-layer capacitor and battery electrodes. *J. Power Sources*, **135** (1–2), 332–343.
143. Khomenko, V., Frackowiak, E., and Beguin, F. (2005) Determination of the specific capacitance of conducting polymer/nanotubes composite electrodes using different cell configurations. *Electrochim. Acta*, **50** (12), 2499–2506.
144. Moser, F. et al. (2009) Transparent electrochemical capacitor based on electrodeposited MnO_2 thin film electrodes and gel-type electrolyte. *Electrochem. Commun.*, **11** (6), 1259–1261.

Edited by
Ru-Shi Liu, Lei Zhang, Xueliang Sun,
Hansan Liu, and Jiujun Zhang

Electrochemical Technologies for Energy Storage and Conversion

Related Titles

Daniel, C., Besenhard, J. O. (eds.)

Handbook of Battery Materials

2011
ISBN: 978-3-527-32695-2

Aifantis, K. E., Hackney, S. A., Kumar, R. V. (eds.)

High Energy Density Lithium Batteries

Materials, Engineering, Applications

2010
Hardcover
ISBN: 978-3-527-32407-1

Liu, H., Zhang, J. (eds.)

Electrocatalysis of Direct Methanol Fuel Cells

From Fundamentals to Applications

2009
Hardcover
ISBN: 978-3-527-32377-7

Ozawa, K. (ed.)

Lithium Ion Rechargeable Batteries

Materials, Technology, and New Applications

2009
Hardcover
ISBN: 978-3-527-31983-1

Eftekhari, A. (ed.)

Nanostructured Materials in Electrochemistry

2008
Hardcover
ISBN: 978-3-527-31876-6

Nazri, G.-A., Balaya, P., Manthiram, A., Yamada, A., Yang, Y. (eds.)

Advanced Lithium-Ion Batteries

Recent Trends and Perspectives

Series: New Materials for Sustainable Energy and Development
2012
Hardcover
ISBN: 978-3-527-32889-5

*Edited by Ru-Shi Liu, Lei Zhang, Xueliang Sun, Hansan Liu,
and Jiujun Zhang*

Electrochemical Technologies for Energy Storage and Conversion

Volume 2

WILEY-VCH

WILEY-VCH Verlag GmbH & Co. KGaA

The Editors

Prof. Dr. Ru-Shi Liu
Department of Chemistry
National Taiwan University
No. 1, Sec. 4, Roosevelt Road
Taipei 10617
Taiwan

Lei Zhang
Institute for Fuel Cell Innovation
National Research Council Canada
4250 Wesbrook Mall
Vancouver, B.C. V6T 1W5
Canada

Prof. Xueliang Sun
Deparment of Mechanical & Materials
University of Western Ontario
London
Ontario N6A 5B9
Canada

Dr. Hansan Liu
Chemical Sciences Division
Oak Ridge National Laboratory
Oak Ridge, TN 37831
USA

Dr. Jiujun Zhang
Institute for Fuel Cell Innovation
National Research Council Canada
4250 Wesbrook Mall
Vancouver, B.C. V6T 1W5
Canada

All books published by **Wiley-VCH** are carefully produced. Nevertheless, authors, editors, and publisher do not warrant the information contained in these books, including this book, to be free of errors. Readers are advised to keep in mind that statements, data, illustrations, procedural details or other items may inadvertently be inaccurate.

Library of Congress Card No.: applied for

British Library Cataloguing-in-Publication Data
A catalogue record for this book is available from the British Library.

Bibliographic information published by the Deutsche Nationalbibliothek
The Deutsche Nationalbibliothek lists this publication in the Deutsche Nationalbibliografie; detailed bibliographic data are available on the Internet at <http://dnb.d-nb.de>.

© 2012 Wiley-VCH Verlag & Co. KGaA, Boschstr. 12, 69469 Weinheim, Germany

All rights reserved (including those of translation into other languages). No part of this book may be reproduced in any form – by photoprinting, microfilm, or any other means – nor transmitted or translated into a machine language without written permission from the publishers. Registered names, trademarks, etc. used in this book, even when not specifically marked as such, are not to be considered unprotected by law.

Typesetting   Laserwords Private Limited, Chennai, India
Printing and Binding   Fabulous Printers Pte Ltd, Singapore
Cover Design   Formgeber, Eppelheim

Printed in Singapore
Printed on acid-free paper

Print ISBN: 978-3-527-32869-7
ePDF ISBN: 978-3-527-64008-9
oBook ISBN: 978-3-527-63949-6
ePub ISBN: 978-3-527-64007-2
mobi ISBN: 978-3-527-64009-6

Contents to Volume 2

Contents to Volume 1 *XIII*

Preface *XV*
About the Editors *XVII*
List of Contributors *XIX*

9	**Water Electrolysis for Hydrogen Generation** *383*	
	Pierre Millet	
9.1	Introduction to Water Electrolysis *383*	
9.1.1	Brief Historical Review *383*	
9.1.2	Cell Technologies *383*	
9.2	Thermodynamics *385*	
9.2.1	Gibbs Free Energy of the Water-Splitting Reaction and Cell Voltage *385*	
9.2.2	Role of pH on Cell Voltage *387*	
9.2.3	Role of Operating Temperature on Cell Voltage *388*	
9.2.4	Role of Operating Pressure on Cell Voltage *391*	
9.3	Kinetics *393*	
9.3.1	Current–Voltage Relationships *393*	
9.3.2	Cell Impedance *396*	
9.3.3	Role of Operating Temperature on the Kinetics *398*	
9.3.4	Role of Operating Pressure on the Kinetics *399*	
9.3.5	Cell Efficiencies *399*	
9.3.5.1	Energy Consumption and Cell Efficiency *399*	
9.3.5.2	Faradaic Efficiency *401*	
9.4	Alkaline Water Electrolysis *401*	
9.4.1	Basic Principles *401*	
9.4.2	Cell Components *402*	
9.4.3	Cell Assembly and Manufacturing *403*	
9.4.4	Cell Stacking and Systems *403*	
9.4.5	Cell Performance, Testing, and Diagnosis *404*	
9.4.6	Degradation Mechanisms and Mitigation Strategies *406*	
9.4.7	Challenges and Perspectives *406*	
9.5	PEM Water Electrolysis *406*	

9.5.1	Basic Principles (Electrochemical Processes)	*406*
9.5.2	Cell Components	*407*
9.5.3	Cell Assembly and Manufacturing	*407*
9.5.4	Cell Stacking and Systems	*408*
9.5.5	Cell Performance, Testing, and Diagnosis	*410*
9.5.6	Degradation Mechanisms and Mitigation Strategies	*411*
9.5.7	Challenges and Perspectives	*412*
9.5.7.1	Replacement of Platinum with Non-Noble Electrocatalysts	*412*
9.5.7.2	Replacement of Iridium with Non-Noble Electrocatalysts	*414*
9.5.7.3	Development of Polymeric Proton Conductors for Operation at More Elevated Temperatures	*414*
9.5.7.4	Development of Proton-Conducting Ceramics for Operation at More Elevated Temperatures	*415*
9.6	High Temperature Water Electrolysis	*415*
9.6.1	Basic Principles (Electrochemical Processes)	*415*
9.6.2	Cell Components	*417*
9.6.3	Cell Assembly and Manufacturing	*417*
9.6.4	Cell Stacking and Systems	*417*
9.6.5	Cell Performance, Testing, and Diagnosis	*417*
9.6.6	Degradation Mechanisms and Mitigation Strategies	*419*
9.6.7	Challenges and Perspectives	*420*
9.7	Conclusion	*420*
	List of Symbols and Abbreviations	*421*
	References	*422*
10	**Hydrogen Compression, Purification, and Storage**	***425***
	Pierre Millet	
10.1	Introduction	*425*
10.2	Pressurized Water Electrolysis	*425*
10.2.1	Principles	*425*
10.2.1.1	Thermodynamics	*426*
10.2.1.2	Kinetics	*428*
10.2.2	Technical Developments	*430*
10.2.2.1	HP Stack	*430*
10.2.2.2	Ancillary Equipment	*431*
10.2.3	Electrochemical Performances	*432*
10.2.4	Current Limitations and Perspectives	*434*
10.3	Hydrogen Electrochemical Compression	*438*
10.3.1	Principles	*438*
10.3.1.1	Compression Cell	*438*
10.3.2	Thermodynamics	*439*
10.3.2.1	Kinetics	*440*
10.3.3	Technical Developments	*441*
10.3.4	Electrochemical Performances	*443*
10.3.5	Current Limitations and Perspectives	*446*

10.4	Hydrogen Electrochemical Extraction and Purification	447
10.4.1	Principles	447
10.4.1.1	Concentration Cell	447
10.4.1.2	Thermodynamics	448
10.4.1.3	Kinetics	448
10.4.2	Technical Developments	448
10.4.3	Electrochemical Performances	449
10.4.4	Current Limitations and Perspectives	449
10.5	Hydrogen Storage in Hydride-Forming Materials	450
10.5.1	Principles	450
10.5.1.1	Thermodynamics	450
10.5.1.2	Insertion Mechanisms	451
10.5.1.3	Hydriding Kinetics	454
10.5.2	Electrochemical Storage (see Ni-MH batteries Chapter 5)	457
10.5.3	Gas-Phase Storage	457
10.5.3.1	Selection of Storage Materials	457
10.5.3.2	Hydride Reactor Technology	457
10.5.4	Current Limitations and Perspectives	459
10.6	Conclusion and Perspectives	460
	List of Symbols and Abbreviations	460
	References	461
11	**Solar Cell as an Energy Harvesting Device**	**463**
	Aung Ko Ko Kyaw, Ming Fei Yang, and Xiao Wei Sun	
11.1	Introduction	463
11.2	Solar Radiation and Absorption	463
11.3	Fundamentals of Solar Cells	465
11.3.1	Working Principle of a p-n Junction Solar Cell	465
11.3.2	Equivalent Circuit Diagram (ECD)	466
11.3.3	Current–Voltage Characteristics	468
11.3.4	Quantum Efficiency	470
11.4	Silicon Solar Cell	470
11.4.1	Hydrogenated Amorphous Silicon Thin-Film Solar Cell	471
11.4.1.1	Hydrogenated Amorphous Silicon	472
11.4.1.2	Amorphous Silicon Thin-Film Solar Cell	474
11.4.1.3	Single-Junction Microcrystalline Silicon Thin-Film Solar Cell	475
11.4.2	Polycrystalline Silicon Thin-Film Solar Cell	475
11.4.2.1	Deposition of PolySi Thin Film	476
11.4.2.2	Enhanced Light Absorption and Suppressed Recombination	476
11.4.3	Single-Crystalline Silicon Solar Cell	477
11.4.3.1	Front Texturing	477
11.4.3.2	Diffusion and p-n Junction	478
11.4.3.3	Antireflection Coating (ARC)	479
11.4.3.4	Contact Formation	479

11.5	Other High-Efficiency Solar Cells 479	
11.5.1	Copper Indium Diselenide (CIS or $CuIn_x Ga_{1-x}Se$) Solar Cell 479	
11.5.1.1	Fundamental Characteristics 480	
11.5.1.2	Development History 481	
11.5.1.3	Typical Structure 481	
11.5.1.4	The Active Layer – CIGS Thin Film 483	
11.5.1.5	Future Researches 485	
11.5.2	High-Efficiency III–V Compound Solar Cell 486	
11.5.2.1	Cell Characteristics 486	
11.5.2.2	Growth Method of Single-Crystal GaAs 488	
11.5.2.3	Growth of III–V Thin Film 488	
11.6	Dye-Sensitized Solar Cell 489	
11.6.1	History of Dye-Sensitized Solar Cell 490	
11.6.2	Basic Principle of Dye-Sensitized Solar Cell 492	
11.6.2.1	Charge Transfer and Recombination Kinetics 494	
11.6.2.2	Charge Separation at the Semiconductor–Dye Interface 495	
11.6.2.3	Electron Percolation through the Nanocrystalline Film 497	
11.6.2.4	Photovoltage in Dye-Sensitized Solar Cell 499	
11.6.3	Photoanodes Other Than TiO_2 Nanoparticle Network 502	
11.6.4	Photocathode and p-DSC 509	
11.6.5	Improvement of Spectral Response by Tandem Structure and CoSensitization 513	
11.6.5.1	Tandem-Structure DSC 514	
11.6.5.2	Dye Cocktail or CoSensitization 517	
11.6.6	Flexible Dye-Sensitized Solar Cell 521	
11.7	Routes to Boost the Efficiency of Solar Cells 523	
11.8	Current Ideas for Future Solar Cell 526	
11.9	Summary 528	
	References 529	
12	**Photoelectrochemical Cells for Hydrogen Generation** 541	
	Neelu Chouhan, ChihKai Chen, Wen-Sheng Chang, Kong-Wei Cheng, and Ru-Shi Liu	
12.1	Introduction 541	
12.2	Main Types and Structures of Photoelectrochemical Cells 544	
12.2.1	Types of Photoelectrochemical Devices 544	
12.2.1.1	Photoelectrolysis or Photosynthesis Cell 544	
12.2.1.2	Photo-Assisted Electrolysis Cell 545	
12.2.1.3	Photovoltaic Electrolysis Cell or Regenerative Cell 547	
12.2.2	Photogalvanic Cells 548	
12.2.3	Photoelectrochemical Solar Cells 548	
12.2.4	Photoelectrochemical Cells Based on Inherently Conducting Polymers (ICPs) 549	
12.3	Electrochemical Processes in Photoelectrochemical Cells 550	

12.4	Photoelectrochemical Cell Components	553
12.4.1	Photoreactor	554
12.4.2	Photoelectrode	557
12.4.3	Counter Electrodes	560
12.4.4	Endplate/Current Collector	562
12.4.5	Electrolytes	563
12.5	Assembly of Photoelectrochemical Cells	566
12.6	Photoelectrochemical Cell Performance, Testing, and Diagnosis	572
12.6.1	Design and Construction of the PEC Test Assembly	574
12.6.1.1	Photoelectrodes	574
12.6.1.2	Light Calibration	574
12.6.1.3	Temperature Control	575
12.6.1.4	Solution Preparation	575
12.6.1.5	Gas Collection	575
12.6.1.6	Electrochemical Device Operation	575
12.7	Degradation Mechanisms and Mitigation Strategies	581
12.7.1	Staebler–Wronski Effect (SWE) in Si-Based PEC Devices	581
12.7.2	Degradation of Oxide Semiconductors and Corresponding Mitigation	582
12.7.3	Degradation of Chalcogenide Semiconductors and Corresponding Mitigation	584
12.7.4	Copper Chalcopyrites (CIS or CIGS) Electrode Degradation and Mitigation Strategies	586
12.8	Applications (Portable, Stationary, and Transportation)	586
12.8.1	Portable	588
12.8.2	Stationary	588
12.8.3	Transportation	589
12.9	Conclusions	589
	Acknowledgments	590
	References	590
13	**Polymer Electrolyte Membrane Fuel Cells**	**601**
	Stefania Specchia, Carlotta Francia, and Paolo Spinelli	
13.1	Introduction to PEMFCs	601
13.2	Main Types and Structures of PEMFCs	603
13.3	Electrochemical Processes in PEMFCs	608
13.3.1	Brief Introduction on Electrochemical Kinetics	608
13.3.1.1	Activation Overpotential	609
13.3.2	Concentration Overpotential	611
13.3.3	Open-Circuit Voltage (OCV)	612
13.3.4	Electrocatalysis	612
13.3.5	ORR Mechanism and Electrocatalysis Issues	613
13.3.6	HOR Mechanism and Electrocatalysis Issues	615
13.3.6.1	Carbon Support	616

13.3.7	Other Electrochemical Processes	*617*
13.4	PEMFCs Components	*618*
13.4.1	Electrodes	*618*
13.4.2	Membrane	*620*
13.4.3	End Plates and Sealing	*622*
13.5	Assembly and Manufacture of PEMFCs	*623*
13.6	PEMFC Stacking and System	*627*
13.7	PEM Performance, Testing, and Diagnosis	*629*
13.8	Degradation Mechanisms and Mitigation Strategies	*635*
13.8.1	Membranes	*635*
13.8.2	Electrocatalyst and the Catalyst Layer	*637*
13.8.3	Gas Diffusion Layer	*639*
13.8.4	Bipolar Plates	*640*
13.8.5	Sealing	*641*
13.8.6	Air/Fuel Impurities	*641*
13.8.7	Degradation Effects due to Subzero Conditions	*643*
13.9	Applications	*645*
13.9.1	Transportation	*645*
13.9.2	Auxiliary Power Units	*646*
13.9.3	Backup Power Supplies	*647*
13.9.4	Portable Applications	*647*
13.9.5	Stationary	*648*
13.10	Challenges and Perspectives	*648*
	References	*651*
14	**Solid Oxide Fuel Cells**	***671***
	Jeffrey W. Fergus	
14.1	Introduction	*671*
14.1.1	Principles of Operation	*671*
14.1.2	Types and Structures	*672*
14.2	Fuel Cell Components	*678*
14.2.1	Electrolyte	*679*
14.2.2	Anode	*680*
14.2.3	Cathode	*681*
14.2.4	Interconnect	*682*
14.2.5	Sealant	*683*
14.3	Assembly and Manufacturing	*684*
14.4	Stacking and Balance of the Plant	*685*
14.5	Performance, Testing, and Diagnosis	*688*
14.6	Degradation Mechanisms and Mitigation Strategies	*689*
14.7	Applications	*690*
14.7.1	Large-Scale Power Generation	*690*
14.7.2	Small-Scale Stationary Applications	*691*
14.7.3	Auxiliary Power Units	*692*
14.7.4	Alternative Fuels	*693*

14.8	Challenges and Perspectives 694	
	Acknowledgments 694	
	References 694	

15	**Direct Methanol Fuel Cells** *701*	
	Kan-Lin Hsueh, Li-Duan Tsai, Chiou-Chu Lai, and Yu-Min Peng	
15.1	Introduction to Direct Methanol Fuel Cells 701	
15.2	Main Types and Structures of Direct Methanol Fuel Cells 703	
15.3	Electrochemical Processes in Direct Methanol Fuel Cells 705	
15.4	Fuel Cell Components 709	
15.5	Assembly and Manufacturing of Direct Methanol Fuel Cells 712	
15.6	Direct Methanol Fuel Cell Stacking and Systems 714	
15.7	Direct Methanol Fuel Cells: Performance, Testing, and Diagnosis 718	
15.8	Degradation Mechanisms and Mitigation Strategies 720	
15.9	Applications 721	
15.10	Challenges and Perspectives of Direct Methanol Fuel Cells 724	
	References 725	

16	**Molten Carbonate Fuel Cells** *729*	
	Xin-Jian Zhu and Bo Huang	
16.1	Introduction to Molten Carbonate Fuel Cells 729	
16.2	Current Technologic Status of Molten Carbonate Fuel Cells 730	
16.3	Electrochemical Processes in Molten Carbonate Fuel Cells 733	
16.4	Components of Molten Carbonate Fuel Cells 734	
16.4.1	Anodes and Anode Materials 734	
16.4.2	Cathode and Cathode Materials 735	
16.4.3	The Electrolyte Matrix 738	
16.4.3.1	Electrolyte Matrix Materials 740	
16.4.3.2	Electrolyte Matrix Fabrication 740	
16.4.3.3	Electrolyte Matrix Performance 741	
16.4.4	Fabrication of Electrode–Electrolyte Assembly 743	
16.4.5	Materials and Fabrication of the Bipolar Plate 744	
16.5	Structure and Performance of MCFCs 744	
16.5.1	Structure and Performance of MCFC Single Cell 744	
16.5.2	Structure and Performance of MCFC Stack 747	
16.6	Schematic of MCFC Power Generation Systems 750	
16.7	Fabrication and Operation of MCFCs 752	
16.7.1	Testing System of MCFCs 752	
16.7.2	Performance of MCFCs 753	
16.7.3	Operation of MCFCs 753	
16.8	MCFC Power Plant 754	
16.8.1	MCFC Power Plant for Natural Gas 754	

16.8.2	Hybrid Coal-Gas-Fueled MCFC/GT (Gas Turbine) and ST (Steam Turbine) Systems 756
16.9	Major Factors Affecting the Performance and Lifetime of MCFCs 757
16.9.1	Electrode–Electrolyte Matrix Structure Effect 757
16.9.2	Electrolyte Composition Effect 758
16.9.3	Electrolyte Retention Effect 758
16.9.4	Reactant Crossover Effect 759
16.9.5	Temperature Effect 759
16.9.6	Pressure Effect 760
16.9.7	Effects of Reactant Gas Composition and Utilization 761
16.9.7.1	Oxidant Composition and Utilization 761
16.9.7.2	Fuel Composition and Utilization 762
16.9.8	Effect of Operating Current Density 763
16.9.9	Effect of Gas Impurities 763
16.9.9.1	Sulfides 764
16.9.9.2	Halides 764
16.9.9.3	Nitride 765
16.9.9.4	Particulates 765
16.9.9.5	Other Compounds 765
16.9.10	Consideration of MCFC Design 766
16.10	Challenges and Perspectives of MCFCs 767
16.10.1	High Cost of the MCFC Systems 767
16.10.2	Insufficient Durability of the MCFC Systems 768
16.10.3	Source and Storage of the Fuels for MCFCs 768
16.10.4	Low Volume Power Density of the MCFC Systems 768
16.10.5	Research and Development Directions 768
16.10.5.1	Cathode Material Development for Oxygen Reduction Reaction 768
16.10.5.2	Anode Material Development for Fuel Oxidation Reactions 769
16.10.5.3	Developing New Electrolyte Composition and the Additives 769
16.10.5.4	Improving Corrosion Resistance of MCFCs 769
16.10.5.5	Modeling, Failure Checking, and Automatic Control of MCFC Power Plants 769
	References 770

Index 777

Contents to Volume 1

1. **Electrochemical Technologies for Energy Storage and Conversion** 1
   Neelu Chouhan and Ru-Shi Liu

2. **Electrochemical Engineering Fundamentals** 45
   Zhongwei Chen, Fathy M. Hassan, and Aiping Yu

3. **Lithium Ion Rechargeable Batteries** 69
   Dingguo Xia

4. **Lead-Acid Battery** 111
   Joey Jung

5. **Nickel-Metal Hydride (Ni-MH) Rechargeable Batteries** 175
   Hua Ma, Fangyi Cheng, and Jun Chen

6. **Metal–Air Technology** 239
   Bruce W. Downing

7. **Liquid Redox Rechargeable Batteries** 279
   Huamin Zhang

8. **Electrochemical Supercapacitors** 317
   Aiping Yu, Aaron Davies, and Zhongwei Chen

Preface

In today's world, clean energy technologies, which include energy storage and conversion, play the most important role in the sustainable development of human society, and are becoming the most critical elements in overcoming fossil fuel exhaustion and global pollution. Among clean energy technologies, electrochemical technologies are considered the most feasible, environmentally friendly and sustainable. Electrochemical energy technologies such as secondary (or rechargeable) batteries and fuel cells have been invented and used, or will be used in several important application areas such as transportation, stationary, and portable/micro power. With increasing demand in both energy and power densities of these electrochemical energy devices in various new application areas, further research and development are essential to overcome challenges such as cost and durability, which are considered major obstacles hindering their applications and commercialization. In order to facilitate this new exploration, we believe that a book covering all important areas of electrochemical energy technologies for clean energy storage and conversion, giving an overall picture about these technologies, should be highly desired.

The proposed book will give a comprehensive description of electrochemical energy conversion and storage methods and the latest development, including batteries, fuel cells, supercapacitors, hydrogen generation and storage, as well as solar energy conversion. It addresses a variety of topics such as electrochemical processes, materials, components, assembly and manufacturing, degradation mechanisms, as well as challenges and strategies. Note that for battery technologies, we have tried our best to focus on rechargeable batteries by excluding primary batteries. With chapter contributions from scientists and engineers with excellent academic records as well as strong industrial expertise, who are at the top of their fields on the cutting edge of technology, the book includes in-depth discussions ranging from comprehensive understanding, to engineering of components and applied devices. We wish that a broader view of various electrochemical energy conversion and storage devices will make this book unique and an essential read for university students including undergraduates and graduates, scientists, and engineers working in related fields. In order to help readers to understand the science and technology of the subject, some important and representative figures, tables, photos, and comprehensive lists of reference papers, will also be presented in this book. Through reading this book, the readers can easily locate the latest information on electrochemical technology, fundamentals, and applications.

In this book, each chapter is relatively independent of the others, a structure which we hope will help readers quickly find topics of interest without necessarily having to read through the whole book. Unavoidably, however, there is some overlap, reflecting the interconnectedness of the research and development in this dynamic field.

We would like to acknowledge with deep appreciation all of our family members for their understanding, strong support, and encouragement.

If any technical errors exist in this book, all editors and chapter authors would deeply appreciate the readers' constructive comments for correction and further improvement.

Ru-Shi Liu, Lei Zhang, Xueliang Sun, Hansan Liu, and Jiujun Zhang

About the Editors

Ru-Shi Liu received his bachelor's degree in chemistry from Shoochow University, Taiwan, in 1981, and his master's in nuclear science from the National Tsing Hua University, two years later. He gained one Ph.D. in chemistry from National Tsing Hua University in 1990, and one from the University of Cambridge in 1992. From 1983 to 1995 he worked as a researcher at the Industrial Technology Research Institute, before joining the Department of Chemistry at the National Taiwan University in 1995 where he became a professor in 1999. He is a recipient of the Excellent Young Person Prize, Excellent Inventor Award (Argentine Medal) and Excellent Young Chemist Award. Professor Liu has over 350 publications in scientific international journals as well as more than 80 patents to his name.

Lei Zhang is a Research Council Officer at the National Research Council of Canada Institute for Fuel Cell Innovation. She received her first M.Sc. in inorganic chemistry from Wuhan University in 1993, and her second in materials chemistry from Simon Fraser University, Canada in 2000. She is an adjunct professor at the Federal University of Maranhao, Brazil and at the Zhengzhou University, China, in addition to being an international advisory member of 7th IUPAC International Conference on Novel Materials and their Synthesis and an active member of the Electrochemical Society and the International Society of Electrochemistry. Ms. Zhang has co-authored over 90 publications and holds five US patent applications. Her main research interests include PEM fuel cell electrocatalysis, catalyst layer/electrode structure, metal-air batteries/fuel cells and supercapacitors.

Xueliang (Andy) Sun holds a Canada Research Chair in the development of nanomaterials for clean energy, and is Associate Professor at the University of Western Ontario, Canada. He received his Ph.D. in materials chemistry in 1999 from the University of Manchester, UK, after which he worked as a postdoctoral fellow at the University of British Columbia, and as a research associate at l'Institut national de la recherche scientifique, Canada. He is the recipient of a number of awards, including the Early Researcher award, Canada Research Chair award and University Faculty Scholar award, and has authored or co-authored over 100 papers, 3 book chapters and 8 patents. Over the past decade, Dr. Sun has established a remarkable track record in nanoscience and nanotechnology for clean energy, mainly in the synthesis and structure control of one-dimensional nanomaterials, as well as their applications for fuel cells and Li ion batteries.

Hansan Liu is a researcher at the Oak Ridge National Laboratory, US Department of Energy. He obtained his Ph.D. in electrochemistry from Xiamen University where he studied cathode materials for lithium ion batteries. After graduation, he worked at the Hong Kong Polytechnic University and the National Research Council Canada on electrophotocatalysis and fuel cell electrocatalysis, respectively. He is currently working on next generation high-energy density batteries at ORNL. Dr. Liu has 14 years of research experience in the field of electrochemical energy storage and conversion. His research interests mainly include battery and supercapacitor materials, fuel cell electrocatalysts, and synthesis and applications of high surface area materials. He has authored and co-authored over 70 publications, including 3 books, 4 book chapters and 3 patent applications relating to batteries and fuel cells. Dr. Liu is an active member of the Electrochemical Society and the International Society of Electrochemistry.

Currently a Senior Research Officer and PEM Catalysis Core Competency Leader at the National Research Council of Canada Institute for Fuel Cell Innovation, **Jiujun Zhang** received his B.Sc. and M.Sc. in electrochemistry from Beijing University, China, in 1982 and 1985, respectively, and his Ph.D. in electrochemistry from Wuhan University in 1988. After this, he took up a position as an associate professor at the Huazhong Normal University, and in 1990 carried out three terms of postdoctoral research at the California Institute of Technology, York University, and the University of British Columbia. Dr. Zhang holds several adjunct professorships, including one at the University of Waterloo and one at the University of British Columbia, and is an active member of The Electrochemical Society, the International Society of Electrochemistry, and the American Chemical Society. He has 240 publications and around 20 patents or patent publications to his name. Dr. Zhang has over 28 years of R & D experience in theoretical and applied electrochemistry, including over 14 years of R & D in fuel cell, and three years of experience in electrochemical sensor.

List of Contributors

Wen-Sheng Chang
Industrial Technology Research Institute
Department of Nano-Tech Energy Conversion
195, Sec. 4, Chung Hsing Road
Chutung
Hsinchu 31040
Taiwan

ChihKai Chen
National Taiwan University
Department of Chemistry
Sec. 4, Roosevelt Road
Taipei 10617
Taiwan

Jun Chen
Nankai University
Key Laboratory of Advanced Energy
Materials Chemistry (Ministry of Education)
Chemistry College
Tianjin 300071
China

Zhongwei Chen
University of Waterloo
Department of Chemical Engineering
Waterloo Institute for Nanotechnology
Waterloo Institute for Sustainable Energy
Waterloo
Ontario N2L 3G1
Canada

Fangyi Cheng
Nankai University
Key Laboratory of Advanced Energy Materials Chemistry (Ministry of Education)
Chemistry College
Tianjin 300071
China

Kong-Wei Cheng
Chang Gung University
Department of Chemical and Materials Engineering
259 Wen-Hwa 1st Rd.
Kwei-Shan
Tao-Yuan 33302
Taiwan

Neelu Chouhan
National Taiwan University
Department of Chemistry
Sec. 4, Roosevelt Road
Taipei 10617
Taiwan

and

Government P. G. College
Department of Chemistry
Devpura, Kota Road
Bundi 323001
India

Aaron Davies
University of Waterloo
Department of Chemical
Engineering
Waterloo Institute for
Nanotechnology
Waterloo Institute for
Sustainable Energy
Waterloo, N2L 3G1
Ontario
Canada

Bruce W. Downing
MagPower Systems Inc.
20 – 1480 Foster Street
White Rock, BC V4B 3X7
Canada

Jeffrey W. Fergus
Auburn University
Materials Research and Education
Center
275 Wilmore Laboratories
AL 36849
USA

Carlotta Francia
Politecnico di Torino
Department of Materials Science
and Chemical Engineering
Corso Duca degli Abruzzi 24
Torino 10129
Italy

Fathy M. Hassan
University of Waterloo
Department of Chemical
Engineering
Waterloo Institute for
Nanotechnology
Waterloo Institute for
Sustainable Energy
Waterloo, Ontario N2L3G1
Canada

Kan-Lin Hsueh
National United University
Department of Energy
Engineering, No.2, Lianda Rd.
Miaoli 36003
Taiwan

Bo Huang
Shanghai Jiao Tong University
Institute of Fuel Cells
800 Dongchuan Road
Shanghai 200240
China

Joey Jung
EVT Power Inc.
6685 Berkeley Street
Vancouver, V5S 2J5
Canada

List of Contributors | XXI

Aung Ko Ko Kyaw
Nanyang Technological
University
School of Electrical and
Electronic Engineering
Nanyang Avenue
Singapore 639798
Singapore

Chiou-Chu Lai
Industrial Technology Research
Institute
Material and Chemical Research
Laboratories, No.195, Sec. 4,
Zhongxing Rd.
Zhudong Township, Hsinchu
County 31040
Taiwan

Ru-Shi Liu
National Taiwan University
Department of Chemistry
Sec. 4, Roosevelt Road
Taipei 10617
Taiwan

Hua Ma
Nankai University
Key Laboratory of
Advanced Energy
Materials Chemistry
(Ministry of Education)
Chemistry College
Tianjin 300071
China

Pierre Millet
Université de Paris-Sud 11
Institut de Chimie Moléculaire et
des Matériaux d'Orsay
UMR 8182 CNRS
15 rue Georges Clémenceau
Bâtiment 410,
91405 Orsay Cedex
France

Yu-Min Peng
Industrial Technology Research
Institute
Material and Chemical Research
Laboratories, No.195, Sec. 4,
Zhongxing Rd.
Zhudong Township, Hsinchu
County 31040
Taiwan

Stefania Specchia
Politecnico di Torino
Department of Materials Science
and Chemical Engineering
Corso Duca degli Abruzzi 24
Torino 10129
Italy

Paolo Spinelli
Politecnico di Torino
Department of Materials Science
and Chemical Engineering
Corso Duca degli Abruzzi 24
10129 Torino
Italy

Xiao Wei Sun
Nanyang Technological
University
School of Electrical and
Electronic Engineering
Nanyang Avenue
Singapore 639798
Singapore

and

Tianjin University
Tianjin Key Laboratory of
Low-Dimensional Functional
Material
Physics and Fabrication
Technology
Weijin Road
Tianjin 300072
China

Li-Duan Tsai
Industrial Technology Research Institute
Material and Chemical Research Laboratories, No.195, Sec. 4, Zhongxing Rd.
Zhudong Township, Hsinchu County 31040
Taiwan

Dingguo Xia
Beijing University of Technology
Department of Environmental and Energy Engineering
Ping le yuan 100
Chaoyang district
Beijing, 100124
China

Ming Fei Yang
Nanyang Technological University
School of Electrical and Electronic Engineering
Nanyang Avenue
Singapore 639798
Singapore

Aiping Yu
University of Waterloo
Department of Chemical Engineering
Waterloo Institute for Nanotechnology
Waterloo Institute for Sustainable Energy
Waterloo, Ontario N2L3G1
Canada

Huamin Zhang
Dalian Institute of Chemical Physics
Chinese Academy of Science,
No.457 Zhongshan Road Dilian
Dilian 116023
China

Xin-Jian Zhu
Shanghai Jiao Tong University
Institute of Fuel Cells
800 Dongchuan Road
Shanghai 200240
China

9
Water Electrolysis for Hydrogen Generation
Pierre Millet

9.1
Introduction to Water Electrolysis

9.1.1
Brief Historical Review

The word electrolysis comes from a combination of two Greek words: *εlektron* (amber) and *lýsis* (to separate). Electrolysis is therefore the process in which a direct electric current passing between two electrodes through an ionic substance that is either molten (molten salt) or dissolved in a suitable solvent (electrolyte) is used to perform a nonspontaneous chemical reaction and to separate reaction products. It is generally admitted that the principles of electrolysis were discovered in 1800 by two English scientists, William Nicholson (1753–1815) and Anthony Carlisle (1768–1842), who used a voltaic pile to liberate hydrogen and oxygen from water [1] and thus initiated the science of electrochemistry:

$$H_2O(liq) \rightarrow H_2(g) + \frac{1}{2}O_2(g)$$

The laws of electrolysis determined by Faraday were reported in 1834 [2]. Since the nineteenth century, electrolysis has been used for the synthesis of several chemical elements. Sir Humphry Davy discovered potassium, sodium, barium, calcium, and magnesium in 1807. Henri Moissan discovered fluorine in 1886. Aluminum was obtained in 1886 using the Hall–Héroult process and sodium hydroxide was obtained in 1890 using the Castner–Kellner process. There are various extremely important electrochemical processes in the industry. In terms of economic activities, many key chemicals produced in the world are obtained from electrochemical processes, for example, chlorine, soda, and aluminum [3, 4].

9.1.2
Cell Technologies

The device used to perform electrolysis is called an *electrolyzer*. An electrolyzer is made by the interconnection of several elementary electrolysis cells. The key component of an electrolysis cell is the interface between a metal (an electron conductor or electrode) and an electrolyte (an ionic conductor). Electrons are highly reactive species and cannot exist freely in the electrolyte. Therefore, the charge (electron) transfer process at the interface, driven by an external power supply, leads to redox chemical transformations on the electrolyte side of the interface. A second metal/electrolyte is provided, face-to-face with the first one, to collect/supply electrons and equilibrate charges. The anode is the electrode where an oxidation takes place and the cathode is the electrode where a reduction takes place. Thus, basically, an electrolysis cell is a galvanic (electrochemical) chain made, in the simplest and most common case, by the series connection of three different phases: one electronic conductor (anode), one ionic conductor (electrolyte), and a second electron conductor (cathode). In electrolysis technological developments, planar geometry is preferred over cylindrical geometry (more largely found in electrochemical generators) because it is more appropriate for the design of large systems.

From a thermodynamic viewpoint, an electrolyzer can be seen as a device that can convert electrical energy into chemical energy. Electrolysis is sometimes referred to as an *"electricity-driven chemical reaction."* This means that the chemical transformation is a nonspontaneous process. In most cases, the reverse reaction, however, is spontaneous and it is therefore necessary to prevent the spontaneous recombination of chemical products produced at the anode and the cathode during electrolysis. Liquid electrolytes are not efficient separators because chemicals can be transported rapidly from one electrode to the other by diffusion and/or convection. Solids are more efficient and, in practical cases, a charge-conducting separator (a membrane) is inserted between the anode and the cathode.

Basically, three different cell concepts are used in the industry (Figure 9.1). The first concept is called a *"gap cell"* (Figure 9.1a).

This is the most conventional configuration. Two planar electrodes (in fact, a cheap electron-conducting substrate surface covered by a thin layer of electrocatalyst) are placed face-to-face in a liquid electrolyte and a membrane is inserted into the interpolar gap to prevent recombination of products. The membrane can be

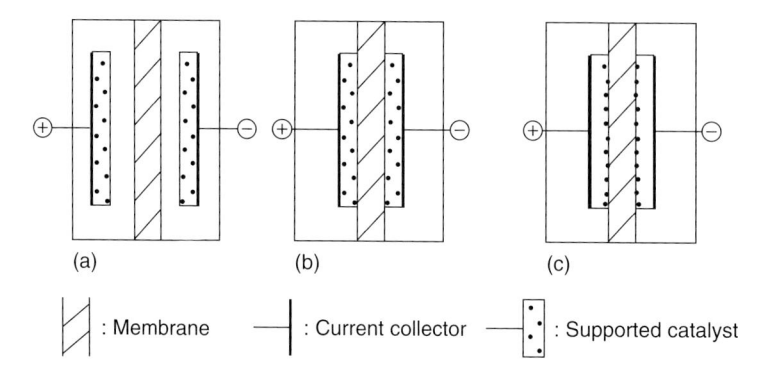

Figure 9.1 Two-dimensional schematic diagrams of (a) a gap cell; (b) a zero-gap cell; and (c) an SPE cell.

an inert and porous material (sometimes called a *diaphragm*), impregnated with the liquid electrolyte. Since transport of electric charges in an electrolyte follows Ohm's law, the larger the distance between the two electrodes, the larger the ohmic losses and the lower the cell efficiency. Therefore, cell design is of uppermost significance. When electrolysis leads to the formation of gaseous products (for example, H_2 and O_2 during the electrolysis of water), the maximum current density is limited to a few hundred milliamperes per square centimeter. This is because nonconducting gaseous films tend to form over the electrode surfaces. Therefore, the concept of "gap cell" pictured in Figure 9.1a is not efficient and appropriate for gas-evolving electrodes. The second concept is called a "*zero-gap cell*" (Figure 9.1b). In this case, porous electrodes are pressed onto the membrane separator to reduce the distance between the anode and the cathode (and corresponding ohmic losses) as much as possible. Products are released through the pores at the backside of the electrodes. This is an interesting concept when gases are produced, at least at one electrode. Significant larger current densities can be achieved without significant extra losses. However, a liquid electrolyte is still needed. The third concept is called a "solid polymer electrolyte (SPE) cell" (Figure 9.1c). SPE, which is a registered trademark of Hamilton Standard Corp., stands for solid polymer electrolyte. In such cells, ions that convey electric charges from one electrode to the other are immobilized inside the membrane, which acts as a solid electrolyte. There is no liquid electrolyte in circulation in the electrolyzer. This concept was first proposed for application in H_2/O_2 fuel cells in 1950, at the dawn of the US space program, to solve the problems associated with operation in a low-gravity environment and those due to the corrosion of metals in acidic media. The idea was used later for the development of water electrolyzers. In the literature, it is referred to as *SPE water electrolysis* or PEM (proton exchange membrane or polymer electrolyte membrane) water electrolysis. In a PEM water electrolysis cell, electrodes are deposited on the surface of the membrane and porous current collectors are pressed on each side of the cell. There is no liquid electrolyte. Only deionized water is circulated in the anodic chamber to feed the electrochemical reaction. The same concept is used at high temperatures, with oxide-ion-conducting membranes.

9.2
Thermodynamics

9.2.1
Gibbs Free Energy of the Water-Splitting Reaction and Cell Voltage

From a thermodynamic viewpoint, the water-splitting reaction is a nonspontaneous transformation. It can be driven externally by providing energy to the system, for example, electricity. The technical device used to convert electricity into chemicals by performing a nonspontaneous transformation is called an *electrolyzer*. An electrolyzer is made of at least one electrolysis cell, consisting of two electrodes (electronic conductor) placed face-to-face and separated by a layer of electrolyte (ionic conductor). Therefore, in a water electrolysis cell, electricity is used to split water molecules into gaseous hydrogen and oxygen. In an acidic PEM cell, water splitting occurs according to the following half-cell reactions:

$$\text{Anode:} \quad H_2O(\text{liq}) \rightarrow \frac{1}{2}O_2(g) + 2H^+ + 2e^- \quad (9.1)$$

$$\text{Cathode:} \quad 2H^+ + 2e^- \rightarrow H_2(g) \quad (9.2)$$

$$\text{Full reaction:} \quad H_2O(\text{liq}) \rightarrow H_2(g) + \frac{1}{2}O_2(g) \quad (9.3)$$

Equation (9.1) is the oxygen evolution reaction (OER) and Eq. (9.2) is the hydrogen evolution reaction (HER). Solvated protons formed at the oxygen-evolving anode of the PEM cell migrate through the SPE membrane to the cathode where they are reduced into molecular hydrogen. Liquid water is consumed at the anode but also released at the cathode (so-called electro-osmotic drag). At equilibrium, the amount of electricity (nFE) required to split 1 mol of water is equal to the change in Gibbs free energy ΔG_d of the water dissociation reaction (Eq. (9.6)):

$$\Delta G_d - nFE = 0 \quad \text{where} \quad \Delta G_d > 0 \quad (9.4)$$

$n = 2$ (number of electrons exchanged during the electrochemical splitting of water)
$F = \approx 96\,485\,\text{C mol}^{-1}$ (Faraday).
$E =$ thermodynamic voltage (V) associated with reaction (Eq. (9.6)).
$\Delta G_d =$ free energy change in J mol^{-1} associated with reaction (Eq. (9.6)).

ΔG_d is a function of both operating temperature and pressure and thus

$$\Delta G_d(T, P) = \Delta H_d(T, P) - T\Delta S_d(T, P) > 0 \quad (9.5)$$

$\Delta H_d(T, P)$ and $\Delta S_d(T, P)$ are, respectively, the enthalpy change (mol^{-1}) and entropy change (J mol^{-1} K^{-1}) associated with reaction (Eq. (9.6)). To split 1 mol of water, ΔG_d (J mol^{-1}) of electricity and $T\Delta S_d$ (J mol^{-1}) of heat are required. The thermodynamic electrolysis voltage E in volt is defined as

$$E(T, P) = \frac{\Delta G_d(T, P)}{nF} \quad (9.6)$$

The thermoneutral voltage V in volt is defined as

$$V(T, P) = \frac{\Delta H(T, P)}{nF} \quad (9.7)$$

Under standard conditions of temperature and pressure ($T = 298$ K, $P = 1$ atm), water is liquid and H_2 and O_2 are gases. Standard free energy, enthalpy, and entropy changes for reaction (Eq. (9.6)) are

$$\Delta G_d^\circ(H_2O) = 237.22\,\text{kJ mol}^{-1} \Rightarrow E^\circ = \frac{\Delta G_d^\circ(H_2O)}{2F} = 1.2293\,\text{V} \approx 1.23\,\text{V}$$

$$\Delta H_d^\circ(H_2O) = 285.840\,\text{kJ mol}^{-1} \Rightarrow V^\circ = \frac{\Delta H_d^\circ(H_2O)}{2F} = 1.4813\,\text{V} \approx 1.48\,\text{V}$$

$$\Delta S_d^\circ(H_2O) = 163.15 \text{ J mol}^{-1} \text{ K}^{-1}$$

$\Delta H_d^\circ(H_2O)$ kJ of energy are required to electrolyze 1 mol of water under standard conditions according to reaction (Eq. (9.6)). This energy is provided as electricity (ΔG_d° kJ) and heat ($T\Delta S_d^\circ$ kJ). In other words, a cell voltage $E = \Delta G_d^\circ/(2F) = 1.23$ V is required and an additional voltage term $T\Delta S_d^\circ/(2F) = 0.25$ V must be added to the thermodynamic voltage E to provide the heat required by reaction (Eq. (9.6)).

9.2.2
Role of pH on Cell Voltage

As indicated before, in acidic media, the following reactions take place

$$\text{Anode}(+): \quad H_2O(\text{liq}) \rightarrow \frac{1}{2}O_2(g) + 2H^+ + 2e^- \quad (9.1)$$

$$\text{Cathode}(-): \quad 2H^+ + 2e^- \rightarrow H_2(g) \quad (9.2)$$

$$\text{Full reaction}: \quad H_2O(\text{liq}) \rightarrow H_2(g) + \frac{1}{2}O_2(g) \quad (9.3)$$

From the Nernst equation, it follows:

$$E^+ = E_{H_2O/O_2}^\circ + \frac{R_{PG}T}{nF} \ln \frac{(a_{H^+}^2)(f_{O_2}^{1/2})}{a_{H_2O}} \quad (9.8)$$

At 298 K, when the pressure of oxygen is 1 bar (ideal gas), $E^+ \approx 1.23 - 0.06$ pH

$$E^- = E_{H_2/H^+}^\circ + \frac{R_{PG}T}{nF} \ln \frac{a_{H^+}^2}{f_{H_2}} \approx -0.06 \text{ pH} \quad (9.9)$$

At 298 K, when the pressure of hydrogen is 1 bar (ideal gas), $E^- \approx -0.06$ pH. Therefore, under the same conditions, the cell voltage $E_{\text{cell}} = E^+ - E^- = 1.23$ V.
In an alkaline electrolyte, the following reactions take place:

$$\text{Anode}(+): \quad 2OH^- \rightarrow H_2O + \frac{1}{2}O_2(g) + 2e^- \quad (9.10)$$

$$\text{Cathode}(-): \quad 2H_2O + 2e^- \rightarrow H_2(g) + 2OH^- \quad (9.11)$$

$$\text{Full reaction}: \quad H_2O(\text{liq}) \rightarrow H_2(g) + \frac{1}{2}O_2(g) \quad (9.12)$$

From the Nernst equation, it follows:

$$E^+ = E_{H_2O/O_2}^0 + \frac{R_{PG}T}{nF} \ln \frac{(a_{H_2O})(f_{O_2}^{1/2})}{a_{HO^-}^2} \quad (9.13)$$

At 298 K, when the pressure of oxygen is 1 bar (ideal gas), $E^+ \approx 1.23 + pK_w - 0.06 \, pH$

$$E^- = E^0_{H_2O/H_2} + \frac{R_{PG}T}{nF} \ln \frac{a^2_{H_2O}}{f_{H_2} a^2_{HO^-}} \quad (9.14)$$

At 298 K, when the pressure of hydrogen is 1 bar (ideal gas), $E^- \approx pK_w - 0.06 \, pH$. Therefore, under the same conditions, the cell voltage $E_{cell} = E^+ - E^- = 1.23$ V.

Therefore, the thermodynamic voltage required to split water into hydrogen and oxygen is the same, whatever the pH. The only difference is that the potential of each electrode is shifted along the potential axis as a function of electrolyte pH. Consequences are mostly on electrode material stability Figure 9.2.

9.2.3
Role of Operating Temperature on Cell Voltage

At atmospheric pressure, liquid water is electrolyzed in the 0–100 °C temperature range and water vapor above 100 °C. $\Delta H(T, 1)$ and $\Delta G(T, 1)$ are obtained from

$$\Delta H(T, 1) = \frac{V(T, 1)}{2F} = H_{H_2}(T, 1) + \frac{1}{2} H_{O_2}(T, 1) - H_{H_2O}(T, 1) \quad (9.15)$$

$$\Delta G(T, 1) = \frac{E(T, 1)}{2F} = G_{H_2}(T, 1) + \frac{1}{2} G_{O_2}(T, 1) - G_{H_2O}(T, 1) \quad (9.16)$$

Enthalpy and entropy changes of $H_2(g)$ and $O_2(g)$ species are obtained from

$$H_i(T, 1) - H_i^0 = a(T - T_0) + \frac{b}{2} 10^{-3} (T^2 - T_0^2) - c 10^5 \left(\frac{1}{T} - \frac{1}{T_0}\right)$$

$$- \frac{e}{2} 10^8 \left(\frac{1}{T^2} - \frac{1}{T_0^2}\right) \quad (9.17)$$

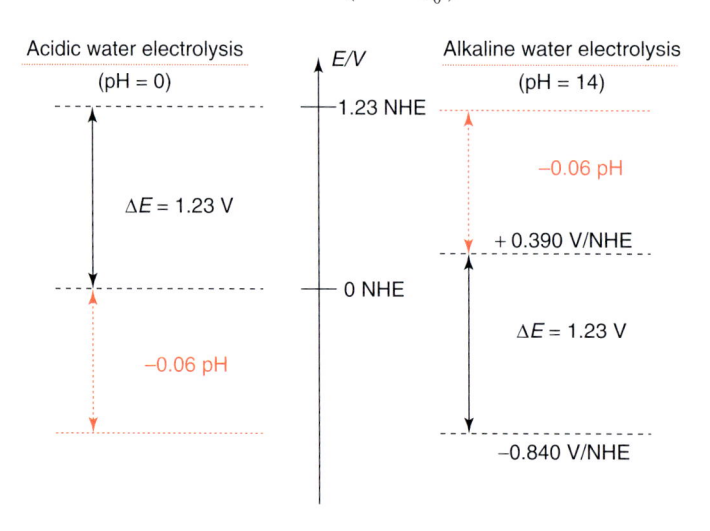

Figure 9.2 Electrode potential versus pH for the water-splitting reaction.

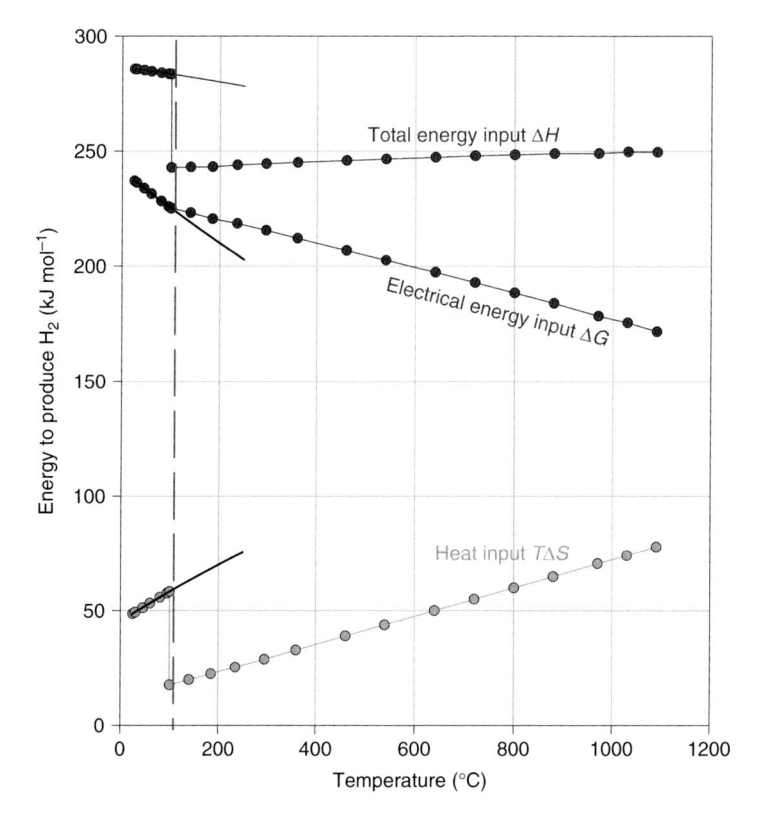

Figure 9.3 $\Delta G(T)$, $\Delta H(T)$, and $T\Delta S(T)$ of the water-splitting reaction at $P = 1$ bar. (—) data for liquid water up to 250 °C.

$$S_i(T, 1) - S_i^0 = a(\ln T - \ln T_0) + b10^{-3}(T - T_0) - \frac{c}{2}10^5 \left(\frac{1}{T^2} - \frac{1}{T_0^2}\right)$$
$$- \frac{e}{3}10^8 \left(\frac{1}{T^3} - \frac{1}{T_0^3}\right) \tag{9.18}$$

Coefficients a, b, c, and e used in Eqs. (9.17) and (9.18) are compiled in Table 9.1. Enthalpy and entropy changes of $H_2O(g)$ are obtained from the steam tables [5]. Data are plotted in Figures 9.3 and 9.4. The discontinuity at 100 °C is due to the vaporization of water. Above 100 °C, the entropy change in the water-splitting reaction is reduced and the slope of $T\Delta S(T, 1)$ is less than that for T < 100 °C. An increase in the operating temperature facilitates the dissociation of water by decreasing the electrolysis voltage. At room temperature, 15% of the energy required for electrolyzing water comes from heat and 85% from electricity. At 1000 °C, one-third comes from heat and two-third from electricity. This is why high temperature water electrolysis is interesting when heat is available at such temperatures: it requires less electricity. Usually, water vapor is electrolyzed at medium (250–500 °C)

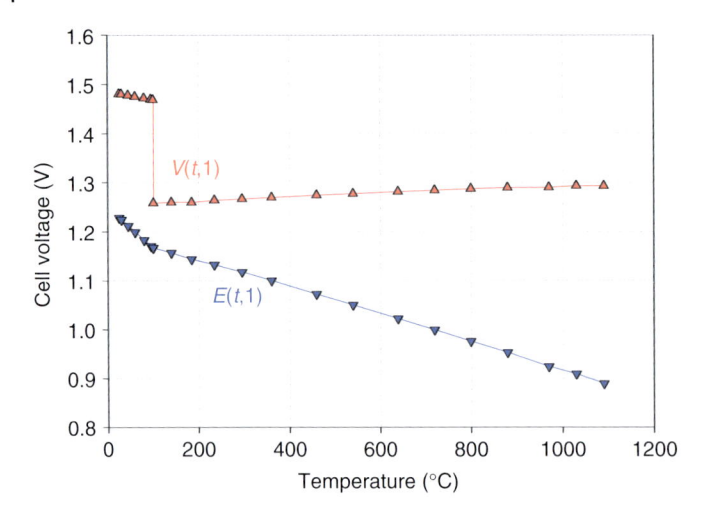

Figure 9.4 Role of the operating temperature on the thermodynamic E (blue ▲) and the enthalpy V (red ▲) cell voltage of electrolysis of water at $P = 1$ bar. (—) data for liquid water up to 250 °C, Eqs. (9.19) and (9.20).

Table 9.1 Coefficients used in Eqs. (9.17) and (9.18).

Element	a	b	c	e
H_2 (gas)	26.57	3.77	1.17	–
O_2 (gas)	34.35	1.92	−18.45	4.06

and high (500–1000 °C) temperatures. In the low-temperature domain (0–250 °C), liquid water is electrolyzed. Operating temperatures in the 100–250 °C range are obtained by increasing the operating pressure. Some empirical expressions of $V(T,P)$ and $E(T,P)$ have also been proposed in the specialized literature [6]. They apply only to the electrolysis of liquid water at a pressure sufficient to avoid water vaporization:

$$V(T, P) = 2F\Delta H(T, P) = 1.485 - 1.49 \times 10^{-4} \times (T - T^0) \\ - 9.84 \times 10^{-8} \times (T - T^0)^2 \tag{9.19}$$

$$E(T, P) = 2F\Delta G(T, P) = 1.5184 - 1.5421 \times 10^{-3} \times T + 9.523 \\ \times 10^{-5} \times T \times \ln(T) + 9.84 \times 10^{-8} T^2 \tag{9.20}$$

As can be seen from Figure 9.4, when liquid water is electrolyzed, both $V(T)$ and $E(T)$ are continuous at 100 °C.

In practical cases, the maximum temperature in conventional PEM and alkaline water electrolysis cells is limited to 100–130 °C. The limitation does come from the ebullition of water: the same situation prevails even when electrolysis is performed under pressure. The reason comes from material stability above 100 °C. This is especially true in PEM technology where hydrated polymers (such as perfluorosulfonic Nafion®) are used as electrolytes and lose all mechanical stability above 100 °C. Even if the domain of the operating temperature can be extended to +150 °C using more stable materials, this is always detrimental to the stability of the membrane–electrode assembly (MEA). Many R&D activities are currently being carried out to find a more stable and efficient solid electrolyte that can operate satisfactorily at temperatures up to 1000 °C.

9.2.4
Role of Operating Pressure on Cell Voltage

In view of the so-called "hydrogen economy," there is a need to store hydrogen and/or oxygen. Pressurized storage is, at the moment, the simplest and less expensive technique. Two approaches can be followed: (i) water electrolysis at atmospheric pressure followed by compression (either mechanical or electrochemical) and (ii) direct pressurized water electrolysis. Current research on pressurized storage investigates the technical possibility of reaching pressures as high as 700 bar. This can be achieved directly using pressurized water electrolyzers. The role of operating pressure on cell voltage can be inferred from Eqs. (9.5) and (9.6) as follows:

$$\left(\frac{\partial E}{\partial P}\right)_T = \frac{1}{nF}\left(\frac{\partial \Delta G}{\partial P}\right)_T = \frac{\Delta \overline{V}}{nF} \tag{9.21}$$

where $\Delta \overline{V}$ is the change in the molar volume associated with the transformation. Considering the fact that the effect of pressure on the volume of condensed phases is small, $\Delta \overline{V}$ can be written by considering only gaseous species:

$$\Delta \overline{V} = \sum_i v_i \frac{R_{PG} T}{p_i} \tag{9.22}$$

where p_i is the partial pressure of species i and v_i is the stoichiometric factor. Equation (9.22) is derived from the law of ideal gas and is applicable on a limited pressure domain:

$$\left(\frac{\partial E}{\partial P}\right)_T = \sum_i v_i \frac{R_{PG} T}{nF p_i} \equiv \sum_i \left(\frac{\partial E}{\partial p_i}\right)_T \tag{9.23}$$

Integration of Eq. (9.23) gives

$$E = E^\circ + \sum_i v_i \left(\frac{R_{PG} T}{nF}\right) \ln\left(\frac{p_i}{p^\circ}\right) \tag{9.24}$$

In the case of water electrolysis, Eq. (9.24) can be rewritten as

$$E = E° + \frac{R_{PG}T}{2F} \ln\left(\frac{p_{H_2}}{p°} \times \sqrt{\frac{p_{O_2}}{p°}}\right) \quad (9.25)$$

Let P be the total pressure of the system. Neglecting the saturating pressure of water vapor in H_2 and O_2 and assuming that $p_{O_2} = P$ at the anode and $p_{H_2} = P$ at the cathode (this is in general the case because PEM water electrolyzers are operated at constant pressure in both anodic and cathodic compartments; management of a differential of pressure across the polymer SPE is nontrivial), the following relationship is obtained:

$$E = E° + \frac{3R_{PG}T}{4F} \ln(P) \quad (9.26)$$

When the electrolysis pressure changes from P_1 to P_2, the cell voltage changes by ΔE:

$$\Delta E = \frac{3R_{PG}T}{4F} \ln\left(\frac{P_2}{P_1}\right) \quad (9.27)$$

The thermodynamic voltage (E) calculated from Eq. (9.27) for different operating pressures at 298 and 373 K is plotted in Figure 9.5. The additional energy cost due to pressure remains significantly lower than interface overvoltages which appear during operation at high current density. More accurate calculations can be made using more elaborated state equations, in particular at pressures above 10 bars, when significant differences appear with the law of ideal gas (details are provided in Chapter 10) [7].

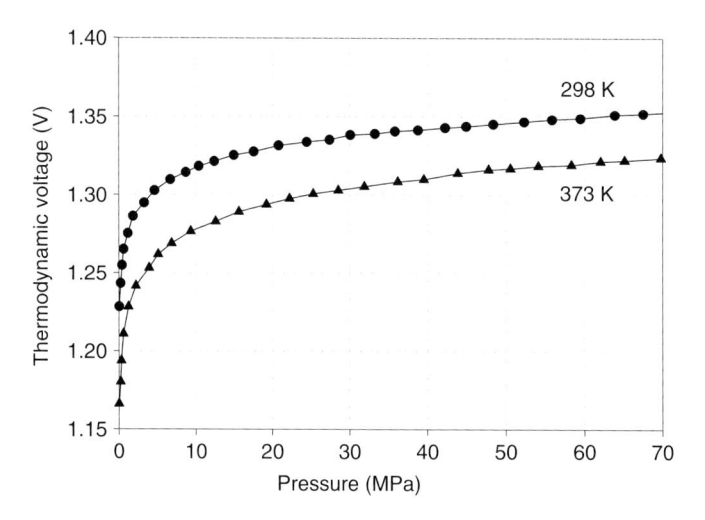

Figure 9.5 Plot of the thermodynamic voltage of water electrolysis as a function of pressure for three different operating temperatures.

9.3
Kinetics

9.3.1
Current–Voltage Relationships

The free energy change ΔG_d of the reaction corresponds to the minimum energy required to transform 1 mol of reactant. There is no practical interest to perform electrolysis close to equilibrium. On the contrary, it is necessary to significantly increase the operating current density in order to reduce investment costs. In conventional electrolyzers, current densities ranging from 100 mA cm^{-2} up to several amperes per square centimeter are commonly achieved. To reach such values, it is necessary to overcome different kinds of current-dependent inefficiencies such as interface overvoltages associated with charge-transfer processes (η in volt) or ohmic drop across the electrolyte due to the resistance (R_e) of the electrolyte. As a result, an increasing portion of the electrical energy is transformed into heat and the efficiency of the electrochemical cell decreases. Let us consider a single water electrolysis cell using a gap cell. Let V (in volt) be the voltage applied to the cell. When a current flows across the cell, V can be decomposed into several voltage contributions:

$$V = \varphi_a - \varphi_c + \eta_A^a + \eta_T^a + \eta_A^c + \eta_T^c + \Sigma RI \qquad (9.28)$$

where

$\varphi_a - \varphi_c$ is the thermodynamic voltage of water electrolysis
η_A^a and η_A^c are the anodic and cathodic activation overvoltages
η_T^a and η_T^c are the anodic and cathodic mass transport overvoltages (for example, due to diffusion of reactants and/or products)
ΣRI is the sum of the different ohmic drops across the cell: in the anodic and cathodic compartments, across the electrolyte (membrane), and in electronic conductors (electrodes and external wiring).

The variation in the electric potential along the interpolar gap (between anode and cathode) in a "gap cell" with a diaphragm is schematically plotted in Figure 9.6:

Depending on the operating conditions, some of the terms in Eq. (9.6) are more or less significant, and in some cases equal to zero. For example, in acidic (PEM) water electrolysis, mass transport overvoltages are inexistent. The electrochemist has to find the most appropriate operating conditions that will reduce the value of V for a given current density i as much as possible. The shape of potential lines can be calculated by solving Laplace's equation:

$$\frac{\partial^2 \phi}{\partial x^2} + \frac{\partial^2 \phi}{\partial y^2} = 0 \qquad (9.29)$$

A typical plot of electric potential, obtained in two dimensions for two linear electrodes placed face-to-face, is provided in Figure 9.7 [8]. Current lines are perpendicular to the potential lines. Most of the current flows directly from the

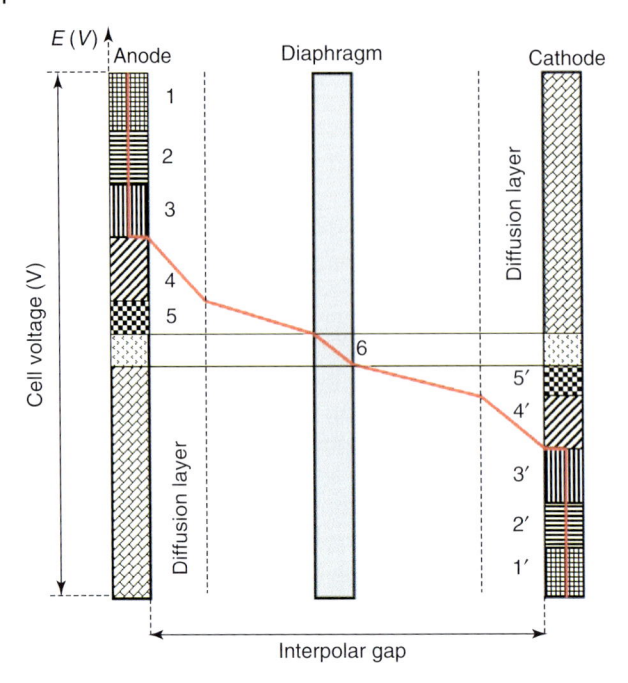

Figure 9.6 Qualitative distribution of electric potential between anode and cathode during electrolysis: (1) anode ohmic drop ($R_a \times I$), (1′) cathode ohmic drop ($R_c \times I$), (2) ($\Delta G/nF$) anodic reaction, (2′) ($\Delta G/nF$) cathodic reaction, (3) anodic charge-transfer overvoltage (η_A^a), (3′) cathodic charge-transfer overvoltage (η_A^c), (4) anodic mass transport overvoltage (η_T^a), (4′) cathodic mass transport overvoltage (η_T^c), (5) anodic electrolyte ohmic drop ($R_{el} \times I$), (5′) cathodic electrolyte ohmic drop ($R_{el} \times I$), and (6) diaphragm ohmic drop ($R_{dia} \times I$).

anode (+2 V) to the cathode (0 V) where the resistance of the electrolyte is minimum, but some current can also flow from the backside of the anode to the backside of the cathode. Interfacial overvoltages, which are located over very small distances (a few nanometers away from the electrodes), play a significant role in the secondary repartition of the electrode potential. This role is even more significant when the kinetics of charge transfer is slow (for example, for a weakly reversible or an irreversible reaction).

A typical polarization curve (a plot of the cell voltage as a function of the operating current density) measured during PEM water electrolysis is given in Figure 9.8. As for the many electrochemical processes, the thermodynamic voltage is the most significant contribution to the energy cost (inefficiencies are less than 50%, i.e., the energy efficiency is at least 50%). Then, charge-transfer overvoltages produce the most important energy loss (electric energy is dissipated as heat at the interfaces). In the case of water electrolysis, the anodic overvoltage is significantly larger than the cathodic one because the kinetics of the OER is much lower than that of the HER. The expression of the ohmic drop across the electrolyte is obtained by taking the product of the electrolyte resistance R_e and the current density i: $R_e \times i$. In

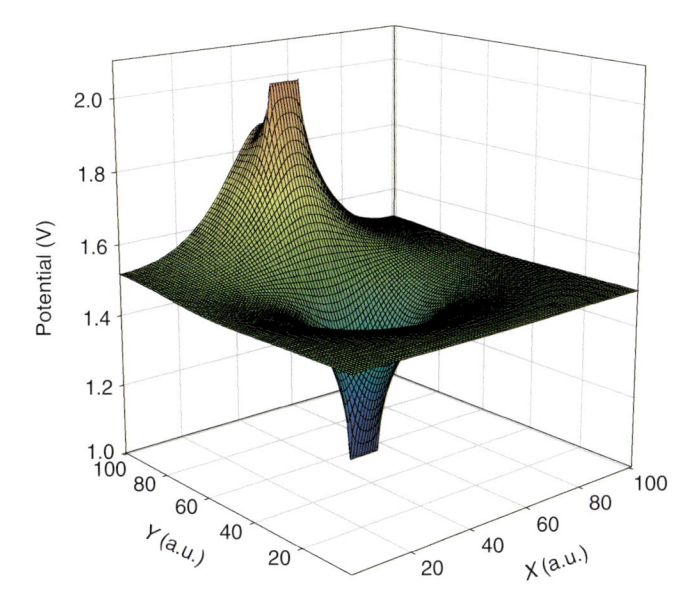

Figure 9.7 Plot of the electric potential between two line electrodes (one dimension) when a voltage of 2 V is applied.

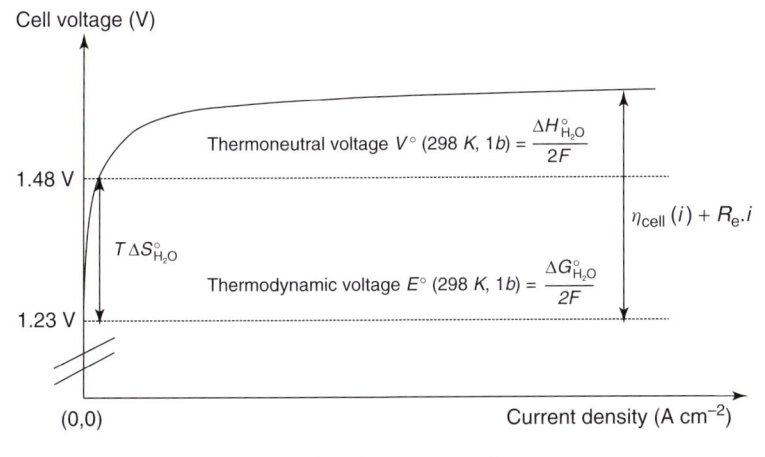

Figure 9.8 Schematic diagram of a voltage–current density polarization curve during PEM water electrolysis.

general, the ohmic drop across the electrolyte is the less significant term, even at high operating current density. This is the situation commonly observed at the laboratory scale. In industrial electrolyzers, additional parasitic ohmic losses can appear due to inadequate electric contacts between cell components (bipolar plates, current collectors, etc.). This can lead to unacceptable cell voltages. The quality of an electrolyzer must be appreciated by its energy efficiency and its ability to maintain good electrochemical performances in the long term (in the upper range of the 10^4–10^5 h time interval).

When the operating current density is such that Eq. (9.30) is satisfied,

$$2F(\eta_{cell} + R_e \times i) = T\Delta S_d \tag{9.30}$$

then there is no need to transfer heat from the surroundings to the electrolysis cell. This is the reason why the operating point $V_d(T, P) = \Delta G_d/2F + T\Delta S_d/2F = \Delta H_d/2F$ is called the *"thermoneutral voltage"* of the cell.

9.3.2
Cell Impedance

Electrochemical impedance spectroscopy (EIS) can be used to measure the global cell impedance during water electrolysis. Such measurements do not require the use of an internal reference electrode, a technique that is not always easy to implement for "zero-gap" or "SPE" cells (especially on large MEAs operating at high pressure) due to the solid polymer electrolyte. The interest in measuring global cell impedances is that they provide *in situ* information about the electrolysis cell operating in real water electrolysis conditions. Cell impedances measured on a 23-cm² PEM cell at 90 °C and different cell voltages are plotted in Figure 9.9 (curves 1–6). There are two different time constants. In a first approximation, each time constant can be attributed to each electrode process: the HER at the cathode and the OER at the anode. Impedances are real at low frequencies and there is no mass transport limitation. Even if the transport of molecular hydrogen and oxygen away from the interfaces is involved in the process, this does impact the overall kinetics.

Experimental impedances of Figure 9.9 can be modeled using an electrical analogy in which two RC circuits are connected in series with the resistance of

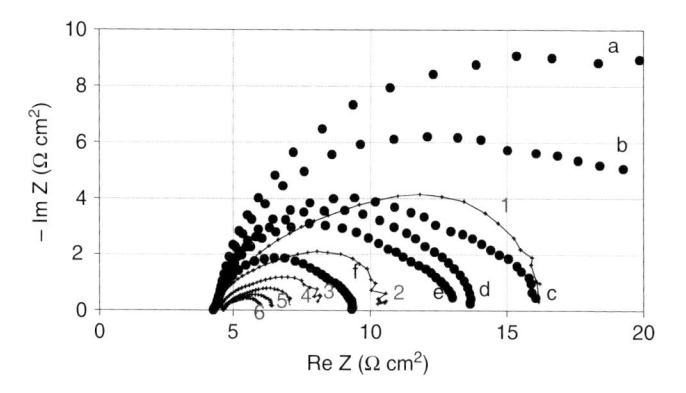

Figure 9.9 Nyquist plots of impedance diagrams measured at 90 °C on (—) a Pt/Nafion117/Ir reference MEA and (•) a Co(dmg)/Nafion117/Ir MEA at increasing cell voltage values: 1 and a: 1400 mV; 2 and b: 1425 mV; 3 and c: 1450 mV; 4 and d: 1475 mV; 5 and e: 1500 mV; 6 and f: 1525 mV.

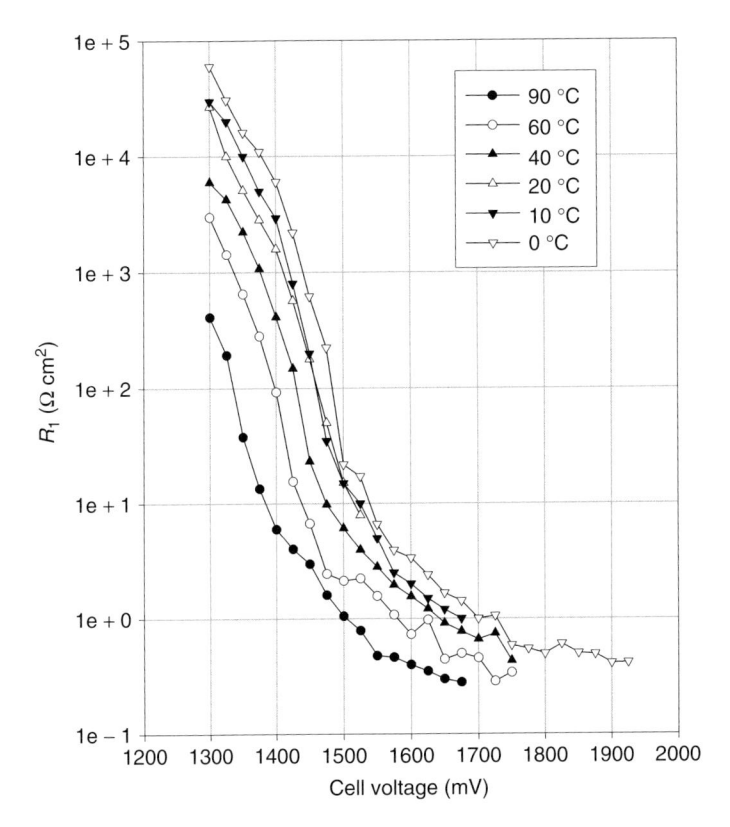

Figure 9.10 Experimental charge-transfer resistance R_1 measured on the 23-cm² reference PEM cell as a function of cell voltage and operating temperature.

the electrolyte. From the best fits, charge-transfer resistances R_1 and R_2 associated with each time constant can be determined as a function of the cell voltage. Results obtained over the 0–90 °C temperature range are plotted in Figures 9.10 and 9.11, respectively. As expected, both R_1 and R_2 decrease with the cell voltage (the exponential decay with cell voltage can be related to the kinetics of charge-transfer processes and Butler–Volmer equation) and operating temperature. R_1 is found to be larger than R_2 by almost 3 orders of magnitude. Cyclic voltammetry experiments indicate that the roughness factors of both anode and cathode are similar. Therefore, the different charge-transfer resistances are due to different charge-transfer kinetics. It can be postulated that the smallest time constant $\tau_1 = R_1 \times C_1$ (high-frequency semicircle) is associated with the OER and that the second one $\tau_2 = R_2 \times C_2$ (low-frequency semicircle) is associated with the HER. These resistance values obtained at the laboratory scale on a 23-cm² Pt/Nafion117/Ir MEA are used as reference values to determine whether MEA impedances in the PEM stack are correct or not. EIS can thus be used as a diagnostic tool.

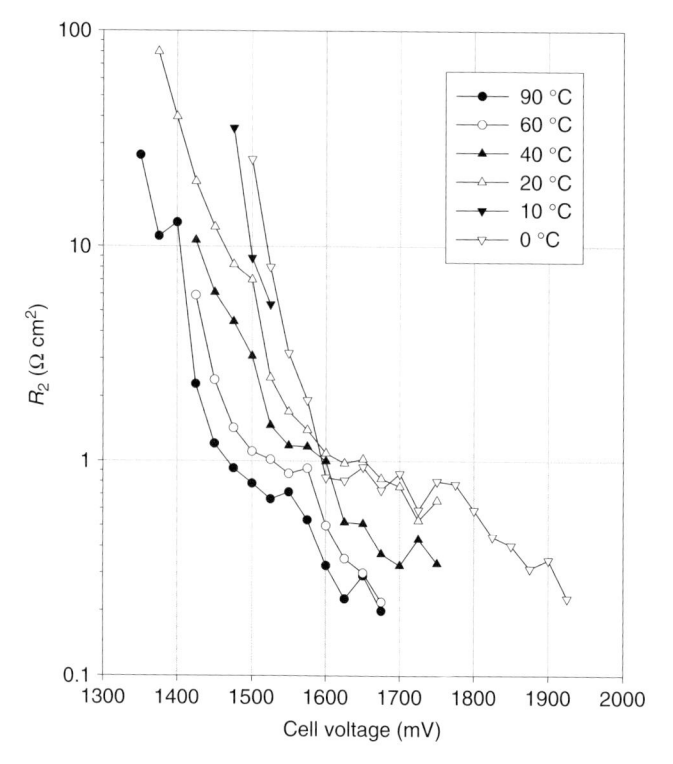

Figure 9.11 Experimental charge-transfer resistance R_2 measured on the 23-cm² reference PEM cell as a function of cell voltage and operating temperature.

9.3.3
Role of Operating Temperature on the Kinetics

As can be seen from Figure 9.12, the water-splitting reaction becomes more and more reversible as the operating temperature increases. At 90 °C (PEM technology), the change in the current sign (transition from water electrolysis to fuel cell operation) takes place over an extended region of potential ($\approx$500 mV) but at 1000 °C (solid oxide water electrolysis) there is a continuous transition between water electrolysis and fuel cell operation. The equilibrium potential is circa 1.23 V at 90 °C ($\Delta G \approx 238$ kJ mol^{-1}) and circa 0.9 V at 1000 °C ($\Delta G \approx 177$ kJ mol^{-1}).

PEM water electrolysis technology can be used over the -10 to $+100$ °C temperature range. Over this limited temperature range, the different terms of the cell voltage (anodic and cathodic overvoltages and ohmic drop across the SPE vary by circa 30% [9] (Figure 9.13)). Tafel parameters for the OER are compiled in Table 9.2. When the current densities are corrected for the roughness of the electrodes, their values are similar to those measured on polished electrodes.

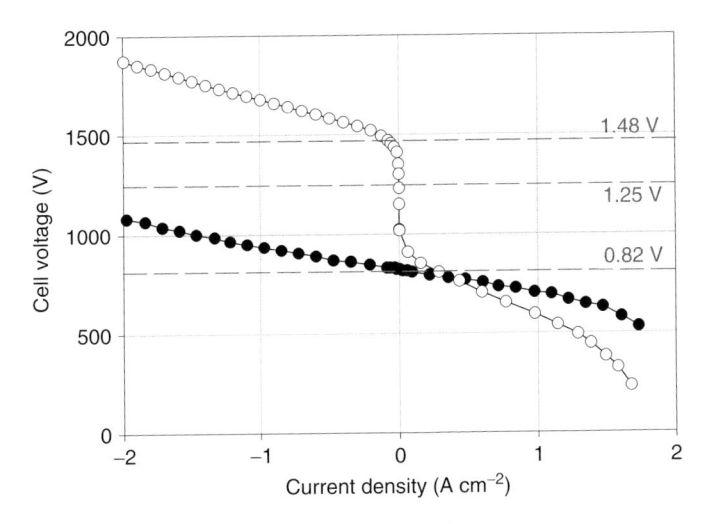

Figure 9.12 Experimental water electrolysis polarization curves measured at (○) 90 °C and (●) 1000 °C.

9.3.4
Role of Operating Pressure on the Kinetics

From the thermodynamics viewpoint, temperature and pressure play an opposite role on the electrolysis voltage: the thermodynamic voltage decreases by circa 0.7 mV °C^{-1} and increases by circa 0.5 mV bar^{-1}. A higher operating pressure should negatively impact the half-cell reactions and reduce the kinetics. Experiments show that this is not the case. This is probably due to the fact that charge-transfer processes at both electrodes are rate determining, not gas transport away from the interfaces. A more detailed discussion on such issues is provided in Chapter 10.

9.3.5
Cell Efficiencies

9.3.5.1 Energy Consumption and Cell Efficiency

The efficiency of a water electrolysis cell relates the theoretical amount of energy W_t required to split 1 mol of water to the real amount of energy W_r. Because different sources of inefficiency appear when current flows across the cell (ohmic drops, electrochemical overvoltages), $W_r > W_t$. The cell efficiency is defined as

$$\varepsilon = \frac{W_t}{W_r} \tag{9.31}$$

where

$W_r = (U_{cell} \times I \times t)$, U_{cell} is the actual cell voltage in volt, I is the current in amperes, and t is the duration in seconds.

W_t can be defined from the thermodynamic voltage E: $W_{t,\Delta G} = (E \times I \times t)$

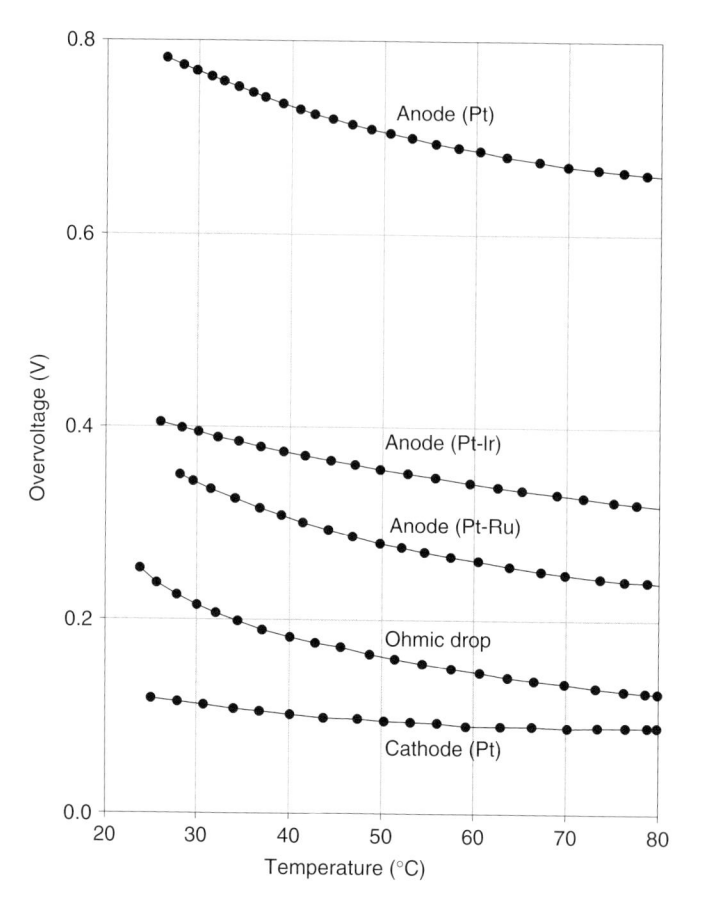

Figure 9.13 Role of operating temperature on the different terms of the cell voltage in a PEM water electrolysis cell.

Table 9.2 Tafel parameters for oxygen-evolution reaction.

Electrode	$d\eta^{anodic}/d(\log i)$ (mV)	α	Measured i_0 (A cm^{-2}) (geometrical area)	Measured (i_0 A cm^{-2}) (real area)	i_0 (A cm^{-2}) [10]
Pt					
25 °C	110	0.54	1×10^{-7}	6×10^{-10}	1×10^{-9}
80 °C	130	0.45	2×10^{-5}	1×10^{-7}	–
Ir					
25 °C	110	0.54	2×10^{-4}	1×10^{-6}	1×10^{-6}
80 °C	130	0.45	6×10^{-3}	4×10^{-5}	–

W_t can also be defined from the thermoneutral voltage V: $W_{t,\Delta H} = (V \times I \times t)$.

In water electrolysis technology, it is possible to electrolyze liquid or vapor water. Therefore, two different definitions can be used to calculate the efficiency of a water electrolysis cell. Since E and V are both functions of the operating temperature T and operating pressure P, and since U_{cell} is also a function of the operating current density j, the two different cell efficiencies can also be expressed as a function of T, P, j:

$$\varepsilon_{\Delta G}(T, P, j) = \frac{E(T, P)}{U_{cell}(T, P, j)} \quad \varepsilon_{\Delta H}(T, P, j) = \frac{V(T, P)}{U_{cell}(T, P, j)} \quad (9.32)$$

At low current densities, cell efficiencies close to 100% are obtained. In conventional PEM water electrolyzers, $\varepsilon_{\Delta H} \approx 70\%$ at 1 A cm^{-2}, $T = 90\,°C$, and $P = 1$ bar. The efficiency of a PEM water electrolysis cell is a critical parameter responsible for the energy cost of the process. Operation at high current density is necessary to reduce investment costs but since efficiency decreases when current density increases, a compromise has to be found between energy and investment costs.

9.3.5.2 Faradaic Efficiency

The faradaic efficiency ε_F relates the current flowing across the cell to the chemical production at electrodes (dm/dt):

$$\varepsilon_F = 2F\frac{\left(\frac{dm}{dt}\right)}{i} \times 100 \quad (9.33)$$

Ideally, ε_F should be as close to unity as possible. In an ideal PEM water electrolysis cell, the membrane would act as an ideal gas separator. In a real situation, the solubility of gases (hydrogen and oxygen) in the solid polymer electrolyte is limited but not zero. As a consequence, hydrogen can diffuse across the SPE from the cathode to the anode and oxygen from the anode to the cathode. Because the hydrogen diffusion coefficient is larger (by a factor of 2) than the oxygen diffusion coefficient, a decrease in faradic efficiency is mostly due to hydrogen crossover. From a global viewpoint, it is as if a parasitic current was decreasing the chemical or faradaic efficiency of the cell. In general, $\varepsilon_F \approx 1$ (no parasitic reaction) but gas cross-permeation tends to reduce it, more and more with pressure.

9.4 Alkaline Water Electrolysis

9.4.1 Basic Principles

In alkaline water electrolysis technology, an aqueous solution of potassium hydroxide is used as a liquid electrolyte. The following half-cell reactions take place:

$$\text{Anode:} \quad 2OH^- \rightarrow H_2O + \frac{1}{2}O_2(g) + 2e^- \quad (9.10)$$

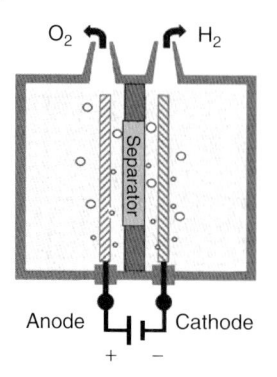

Figure 9.14 Schematic diagram of an alkaline water electrolysis cell (gap cell).

$$\text{Cathode:} \quad 2H_2O + 2e^- \rightarrow H_2(g) + 2OH^- \qquad (9.11)$$

$$\text{Full reaction:} \quad H_2O(\text{liq}) \rightarrow H_2(g) + \frac{1}{2}O_2(g) \qquad (9.12)$$

In a conventional alkaline electrolyzer ("gap-cell" technology), hydrogen is produced at the cathode facing the diaphragm and oxygen at the anode facing the diaphragm (Figure 9.14).

Water molecules are reduced into molecular hydrogen and hydroxyl ions at the cathode of the alkaline cell. Solvated hydroxyl ions migrate through the electrolyte to the anode where they are oxidized into molecular oxygen.

9.4.2
Cell Components

One advantage of the alkaline process is that only cheap materials are required. The strongly alkaline pH of the electrolyte leads to the passivation of the electrodes and their protection against anodic dissolution. Most catalysts used in commercial systems are made of nickel and/or cobalt. Asbestos has been largely used as a separator (diaphragms) in alkaline water electrolysis. Asbestos is a naturally occurring silicate mineral exploited commercially for its physical properties. Asbestos became increasingly popular among manufacturers and builders in the nineteenth century because of its sound absorption, average tensile strength, and its resistance to heat, electrical, and chemical damage. It is used in electrochemical cells because of its properties such as wettability and its chemical resistance to electrolytes. It is formed of characteristic fibrous crystals. The inhalation of asbestos fibers can cause serious illnesses (lung cancer). This is why all use of asbestos and extraction, manufacture, and processing of asbestos products have been banned by the European Union. Less hazardous materials can be used alternatively, although ion-conducting polymers are increasingly used. For example, diaphragms of porous sinter metal oxidized to metal oxide are used for alkaline water electrolysis [11]. Conventional plastics are used as cell sealant.

9.4.3
Cell Assembly and Manufacturing

In modern electrolyzers, the filter-press configuration using bipolar electrodes is preferred. As can be seen from Figure 9.15, the alkaline electrolysis stack is composed of a succession of bipolar electrodes, cell spacers, and diaphragms. The geometry of the unit is a critical issue. Care must be taken to reduce, as much as possible, parasitic ohmic contacts, which could degrade the efficiency and disrupt the distribution of current lines.

Also, a sufficient space is necessary between electrodes and diaphragm for the circulation of the electrolyte and the collection of the liquid–gas biphasic mixtures. Finally, since the electrolyte is liquid, the cell must be designed to avoid parasitic current lines, which could lead to the production of H_2/O_2 hazardous mixtures in the stack. Insulated spacers made of polysulfone are placed between electrodes and diaphragms to satisfy these different constraints.

9.4.4
Cell Stacking and Systems

The photograph of a commercial alkaline water electrolyzer is provided in Figure 9.16. The electrolysis stack on the left side of the picture is made of circa 70 circular cells of $3.5\,m^2$. The ancillary equipment (right side of the picture) is made of circulation pumps, liquid–gas separators, connectors to the power supply, and safety equipment.

A typical process flow sheet is provided in Figure 9.17. After liquid–gas separation, the anodic and cathodic KOH electrolytes are mixed and recycled. Hydrogen and oxygen produced by the electrolyzer are cooled, washed, and compressed for final use.

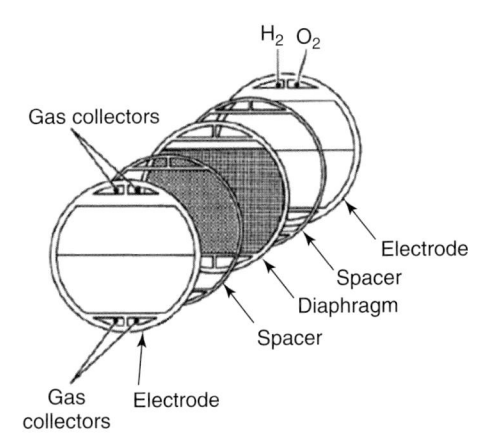

Figure 9.15 Schematic diagram of an alkaline water electrolysis cell.

Figure 9.16 Photograph of an alkaline water electrolyzer operating at 5 bars (Norsk–Hydro. 485 Nm^3 H$_2$ h^{-1}; 4.1 kWh Nm^{-3}).

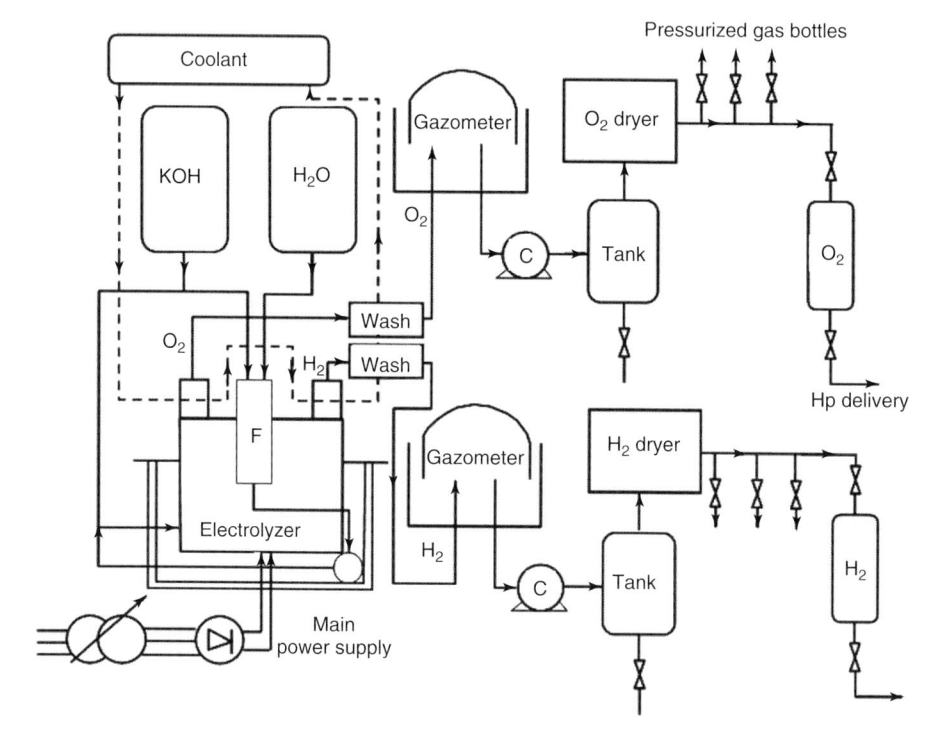

Figure 9.17 Membrane cell process flow sheet. F – filter and C – compressor.

9.4.5
Cell Performance, Testing, and Diagnosis

A typical polarization curve is plotted in Figure 9.18. In industrial processes, current densities of 400–500 mA cm^{-2} are commonly achieved. At such values, ohmic drop across the electrolyte and interface overvoltages are equally the main sources of power loss. More efficient concepts

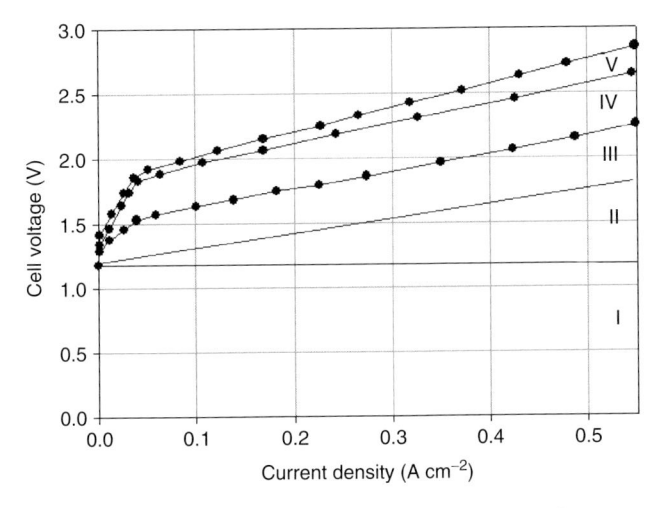

Figure 9.18 Electrochemical performances of a conventional alkaline "membrane cell" water electrolyzer. I: ohmic drop in the main power line; II: cathodic overvoltage associated with the HER; III: anodic overvoltage associated with the OER; IV: ohmic drop in the electrolyte; and V: thermodynamic voltage.

(so-called advanced alkaline water electrolysis) have been developed over the past decades. Prototypes delivering up to 25 Nm^3 H$_2$ h^{-1}, operating at 1.25 A cm^{-2}, 120 °C, 5–40 bars, with an electrical power consumption (at 200 mA cm^{-2}) of 3.81 kWh Nm^{-3} H$_2$ at 90 °C ($\approx$78% efficiency) and 3.65 kWh Nm^{-3} H$_2$ at 120 °C ($\approx$80% efficiency) have been reported [12].

The energy consumption and efficiency of some commercial apparatus are compiled in Table 9.3 for comparison.

Table 9.3 Main characteristics of some commercial alkaline water electrolyzers.

Manufacturer (model)	H$_2$ production (Nm^3 h^{-1})	Production pressure (bar)	Total power (kW)	Energy system (kWh Nm^{-3})	Energy electrolyzer (kWh Nm^{-3})	Efficiency (%)
Stuart (IMET 1000)	60	25	290	4.8	4.2	73
Teledyne (EC 750)	42	4–8	235	5.6	–	63
Proton (Hogen 380)	10	14	63	6.3	–	56
Norsk Hydro (5040)	485	30	2330	4.8	4.3	73
Avalence (hydrofiller 175)	4.6	up to 700	25	5.4	–	64

9.4.6
Degradation Mechanisms and Mitigation Strategies

Corrosion by caustic electrolyte as well as the durability of electrochemical performances is a matter of concern in alkaline water electrolysis technology. Hydroxyl-ion-conducting polymers that could prevent the use of liquid and corrosive electrolytes are increasingly being studied for application in alkaline fuel cell technology (so-called alkaline anion-exchange membranes fuel cells) and water electrolysis.

9.4.7
Challenges and Perspectives

Industrial alkaline water electrolysis is a mature technology. Several mega watt (MW) industrial electrolyzers (e.g., those developed at Hydroquebec in Canada or Norsk-Hydro in Norway) can be used for the large-scale production of hydrogen. Applications are limited by the fact that hydrogen from water ($\Delta H° = +285$ kJ mol^{-1}) requires more energy than hydrogen from methane ($+160$ kJ mol^{-1}). From the process viewpoint, some improvements are still possible. In the open literature, attempts made to identify alternative electrocatalysts such as transition metal macrocycles [13] and to develop alternative diaphragms have been reported. Challenges concern lifetime of systems and maintenance costs.

9.5
PEM Water Electrolysis

9.5.1
Basic Principles (Electrochemical Processes)

In a PEM water electrolysis cell, there is no liquid electrolyte. Sulfonated tetrafluoroethylene-based fluoropolymer-copolymer (W. Grot, E. I. DuPont Co., Nafion products [14]) is used as the solid electrolyte. The following reactions take place:

$$\text{Anode:} \quad H_2O(\text{liq}) \rightarrow \frac{1}{2}O_2(g) + 2H^+ + 2e^- \tag{9.1}$$

$$\text{Cathode:} \quad 2H^+ + 2e^- \rightarrow H_2(g) \tag{9.2}$$

$$\text{Full reaction:} \quad H_2O(\text{liq}) \rightarrow H_2(g) + \frac{1}{2}O_2(g) \tag{9.3}$$

A schematic representation of a PEM cell is provided in Figure 9.19. In such "SPE-cells," electric current is used to split liquid water into gaseous oxygen and protons (anodic reaction, Eq. (9.1)). Solvated protons migrate to the cathode where they are reduced into molecular hydrogen (cathodic reaction, Eq. (9.2)) and where liquid water is released (electro-osmosis drag). Anode and cathode are separated by

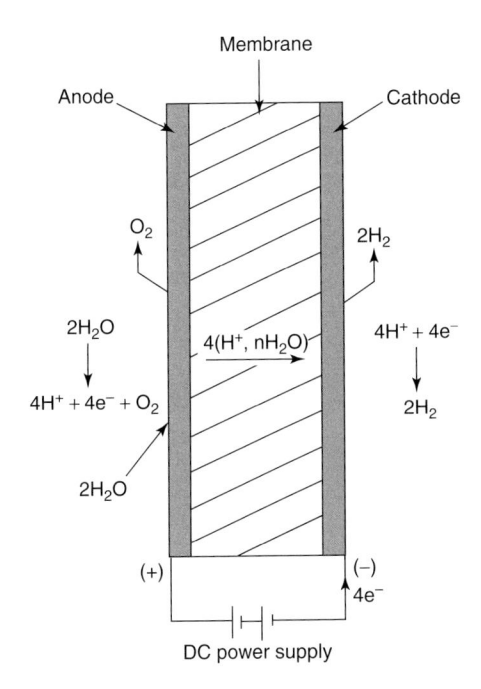

Figure 9.19 Schematic diagram of a PEM water electrolysis cell.

a thin (0.2 mm thick) membrane of proton-conduction polymer electrolyte. Such cells are very compact and the efficiency is high.

9.5.2
Cell Components

The different components of a PEM cell are described in Figure 9.20. The SPE membrane (1) is coated with two porous layers of electrocatalyst: (2) one at the anode (a) and the other at the cathode (c). MEA thus obtained is clamped between two porous current collectors (3). Two hollow bipolar plates (4) are used to convey electric current to the cell and separate two adjacent cells. Channels are used to carry feed water to the anode and to collect liquid–gas mixtures in each cell compartment. Contacts points between current collectors and electrocatalytic layers (5) are critical; the average distance between contact points must be sufficiently small (typically several micrometers) to obtain a good distribution of current lines. Current collectors are made of sintered titanium particles in a hexagonal, closely packed lattice of particles.

9.5.3
Cell Assembly and Manufacturing

A large variety of processes can be used to plate electrocatalysts onto the SPE to obtain the MEAs. For example, platinum and iridium can be deposited onto the membrane using an impregnation–reduction method (Figure 9.21). As a result, a

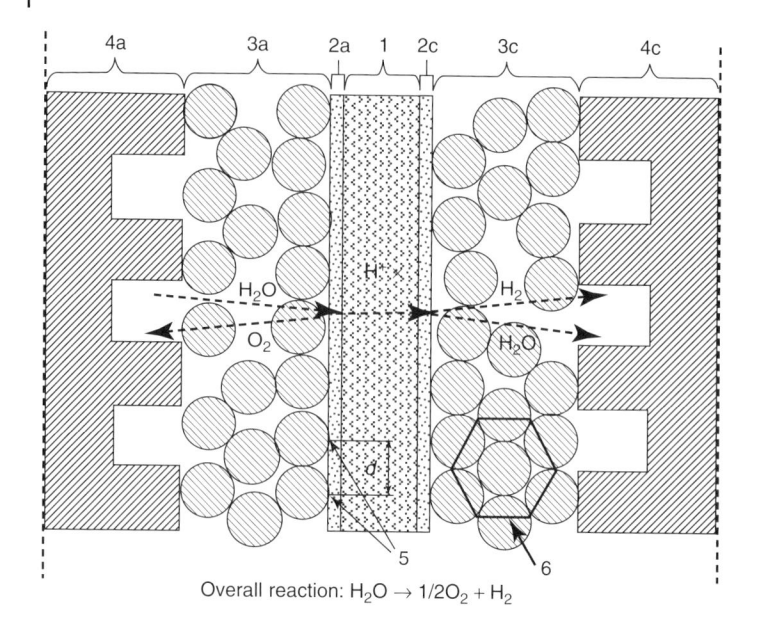

Figure 9.20 Schematic representation (cross-sectional view) of a PEM water electrolysis cell. 1 – PEM; 2 – electrocatalytic layers (a – anodic, c – cathodic); 3 – porous current collectors (a – anodic, c – cathodic); 4 – bipolar plates (a – anodic side, c – cathodic side); 5 – contacts points between current collectors and electrocatalytic layers; d – average distance between contact points between current collector and catalytic layers; and 6 – a fragment of a layer with hexagonal closely packed lattices of particles.

Figure 9.21 Photograph showing the process used to plate platinum directly onto the SPE.

thin and homogeneous platinum layer is deposited over the active section of the SPE. Then, the MEA is clamped between two porous current collectors and two bipolar cells to form a unit cell (Figure 9.22). Several unit cells are then stacked in a filter-press system to form the electrolyzer.

9.5.4
Cell Stacking and Systems

To increase the production rate of the electrolyzer, several cells can be stacked in series. A stack that can produce up to 1 Nm^3 H$_2$ h^{-1} is shown in Figure 9.23.

9.5 PEM Water Electrolysis

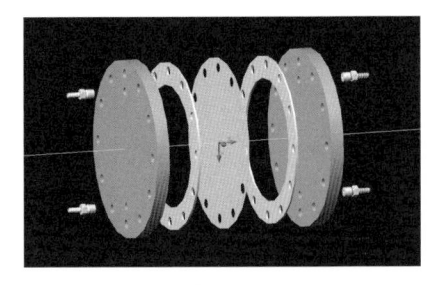

Figure 9.22 Schematic representation (cross-sectional view) of a PEM water electrolysis cell.

Figure 9.23 Photograph of a GenHy1000 PEM water electrolyzer developed in the course of the GenHyPEM project.

For high-pressure experiments, the stack is encapsulated in a high-pressure case and the electrolysis unit is mounted on the production setup (Figure 9.24). A process flow sheet is provided in Figure 9.25. Electrolysis cells are either self-pressurized by gases produced during electrolysis or using nitrogen. Two high-pressure pumps (50 bar, PCB Co.) are used for the circulation of pressurized water in both anodic and cathodic loops. Liquid–gas mixtures

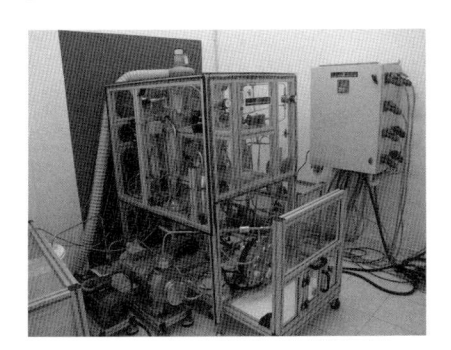

Figure 9.24 Photograph of a high-pressure (50 bars) test station used for the characterization of PEM water electrolyzers.

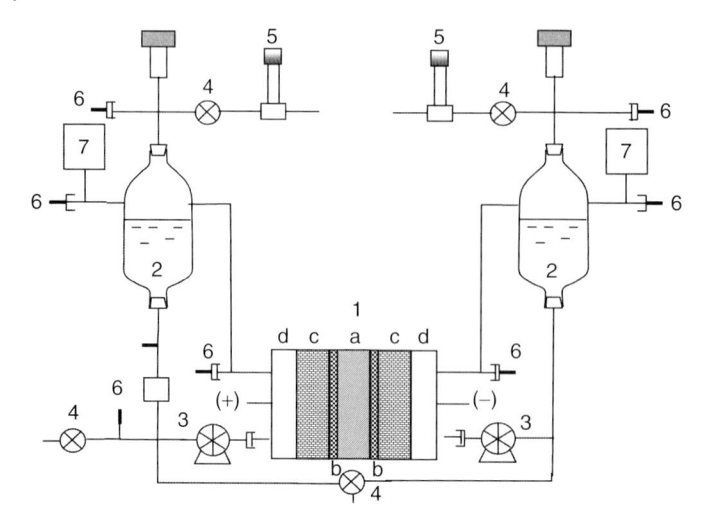

Figure 9.25 Process flow sheet : (1) the PEM electrolysis cell; (1-a) PEM; (1-b) catalytic layers; (1-c) porous titanium current collectors; (1-d) gas collection compartments. Ancillary equipment: (2) liquid–gas separators; (3) pumps; (4) valves; (5) production valves; (6) thermocouples; and (7) pressure transducers.

are collected and separated in 0.5-l stainless steel separators. Water levels are monitored using a magnetic level-measuring device (Kubler Co.). Pressures are measured using 0–50 bar pressure transducers (Keller Co.). Data acquisition and remote control are performed using specific software (GenHy_soft, CETH Co.). Data are displayed on a tactile screen, enabling local control of the process (Horner Co.).

A typical process flow sheet is diagramed in Figure 9.25. Liquid water is circulated on separate anodic and cathodic circuits.

9.5.5
Cell Performance, Testing, and Diagnosis

In conventional PEM water electrolyzers, metallic platinum is used as the electrocatalyst for the HER and iridium (metal or oxide) is used for the OER [15]. Best electrolysis efficiencies ($\varepsilon_{\Delta H} \approx 70\%$ at 1 A cm^{-2}) can be obtained using noble metal loadings of circa 0.5 mg cm^{-2} of Pt for the HER and circa 2 mg cm^{-2} of Ir (or iridium oxide) for the OER. To a certain extent, PEM water electrolysis remains an expensive technology due to the use of noble metals. As for the PEM fuel cell technology, noble metal loadings must be reduced because of cost requirements. Regarding the HER, platinum loadings can be significantly lowered by coating Pt nanoparticles at the surface of electronic carriers of large surface areas. Pt can be coated over carbon nanotubes or over carbon powders. Typical polarization curves

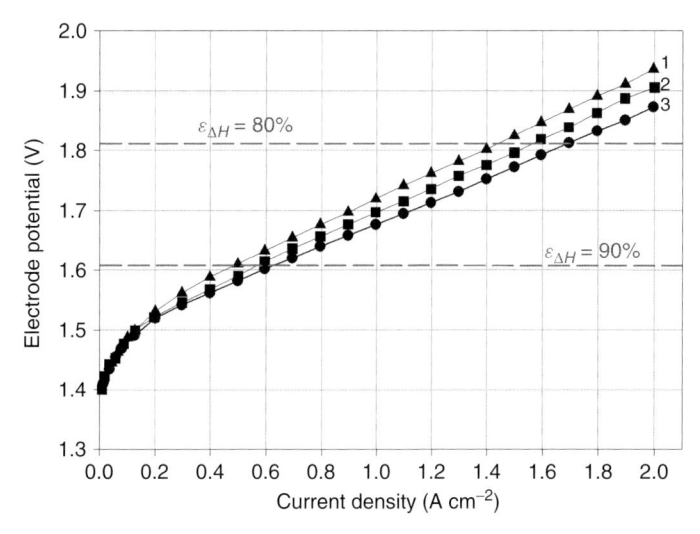

Figure 9.26 Current–voltage performances measured on a single cell (operating area of 7 cm^2) with different noble metal catalysts. Cathode: 1 – Pd20/Vulcan® XC-72 (0.35 mg cm^{-2} of Pd); 2 – Pd40/Vulcan® XC-72 (0.7 mg cm^{-2} Pd); 3 – Pt40/Vulcan® XC-72 (0.7 mg cm^{-2} Pt). Anode: iridium black (2.0 mg cm^{-2}). Nafion®-115 SPE membrane. Operating conditions: $T_{cell} = 90\ °C$, $P = 1$ bar.

obtained using different MEAs at 90 °C and in the 0–2 A cm^{-2} current density range are plotted in Figure 9.26.

Pt40/VulcanXC-72 denotes a catalyst made of Pt nanoparticles (40 wt%) deposited onto Vulcan carbon powder XC72. Electrochemical performances measured using Pt and Pd nanoparticles deposited on the surface of Vulcan XC72 are compared. Slightly more efficient results were obtained using Pt in place of Pd at the cathode. A cell efficiency $\varepsilon_{\Delta H} \approx 90\%$ is obtained at 0.6 A cm^{-2} and $\varepsilon_{\Delta H} \approx 80\%$ is obtained at 1.6 A cm^{-2}. Slightly more efficient results can be obtained by using thinner SPEs, thus reducing the ohmic drop across the membrane. However, gas cross-permeation effects increase more rapidly when the operating pressure is raised, and such membranes should be used only for operation at atmospheric pressure. Noble metal loadings can also be reduced at the anode by using mixed oxides [16] but, in view of the large-scale application of PEM water electrolysis technology, cheap electrocatalysts are required for both HER and OER.

9.5.6
Degradation Mechanisms and Mitigation Strategies

The economic interest of water electrolysis requires technologies with stable electrochemical performances in the upper range of 10^4–10^5 h. Regarding PEM water electrolysis, durability tests have been reported over 50 000 h of intermittent operation without significant performance losses. In PEM technology, there are two main sources of performance degradation. The first one encompasses reversible processes such as a gradual loss of electrochemical performances due to the incorporation, into the membranes, of trace amounts of miscellaneous metallic cations released by the stainless steel tubing of the electrolysis module. Such cations incorporate the SPE by ion exchange with the protons of the membrane and then migrate to the cathode where they can form underpotential deposits onto platinum particles. As a result, the cathodic overvoltage increases with time and the energy consumption of the stack increases. Such problems can be limited by a strict monitoring of the quality of the input water and by occasional maintenance operations. The second source encompasses irreversible processes such as the formation of hot points due to uneven repartition of current lines across the cells. Such problems are very critical because they can lead to the perforation and destruction of the MEAs and the electrolyzers. Irreversible degradations can be avoided by some maintenance operations made at regular time intervals to detect precursor signs and remove defecting MEAs.

9.5.7
Challenges and Perspectives

PEM water electrolysis technology is an efficient process: efficiencies of 85% (calculated from the higher heating value of hydrogen) have been demonstrated at the laboratory scale at 1 A cm^{-2}. The main drawback comes from the fact that this is an expensive technology. The replacement of platinum-family catalysts and expensive SPE remains challenging issues. Some recent advances have been made at the cathodes of PEM cells.

9.5.7.1 Replacement of Platinum with Non-Noble Electrocatalysts

Results Obtained with Cobalt Clathrochelates for the Hydrogen Evolution Reaction
In recent years, different compounds have been identified as possible electroactive materials, which could be used in place of platinum for the HER [17, 18]. Boron-capped tris(glyoximato) cobalt complexes adsorbed at the surface of an electronic carrier of large surface area have been successfully experimented at the cathode of PEM water electrolysis cells. Some typical results obtained during PEM water electrolysis are plotted in Figure 9.27. In all cases, iridium black (2.5 mg cm^{-2}) was used at the anode. First (curve 1), a mixture of bare Vulcan XC72 and Nafion (5 wt%) was used as catalyst for the HER. Then (curve 2), a mixture of Co(dmg) and Nafion (5 wt%). Then (curve 3), a mixture of Co(dmg) adsorbed onto Vulcan XC72 and Nafion (5 wt%). For comparison, a conventional polarization curve measured with Pt for the HER and Ir for the OER is also plotted (curve 4). Comparison of curves 3 and 4 reveals that, although the cobalt clathrochelate remains significantly less efficient than Pt for the HER, interesting electrochemical performances are nevertheless obtained. A cell efficiency $\varepsilon_{\Delta H} \approx 80\%$ is obtained at a current density of 500 mA cm^{-2} and 90 °C.

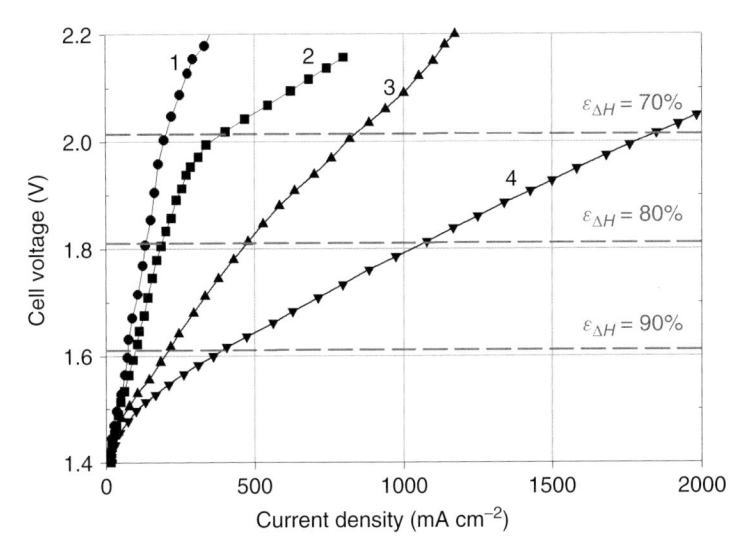

Figure 9.27 Polarization curves measured at 90 °C using different MEAs (23-cm² PEM cell). Anode: 2.5 mg cm^{-2} Ir°. (1) 1 mg cm^{-2} Vulcan®XC72/Nafion115/Ir°; (2) 1 mg.cm^{-2} Co(dmg)/Nafion115/Ir°; and (3) 1 mg cm^{-2} Co(dmg)/Vulcan®XC72 /Nafion115/Ir; (4) 1 mg cm^{-2} Pt/Nafion115/Ir.

Results Obtained with Polyoxometalates for the Hydrogen Evolution Reaction Electroactivity of polyoxometalates (POMs) with regard to the HER in liquid electrolytes had been reported in the literature a few decades ago [19]. These catalysts have been successfully tested at the cathode of PEM water electrolyzers in place of platinum for the HER. Some typical polarization curves, obtained with the simple and commercially available tungstosilicic acid hydrate (α-H$_4$SiW$_{12}$O$_{40}$) adsorbed at the surface of the cathodic titanium current collector, are plotted in Figure 9.28. On the top (black circles) is plotted a reference curve obtained when Pt is used for both the HER and the OER. At the bottom (green circles) is plotted a second reference curve obtained when platinum is used at the cathode for the HER and metal iridium is used at the anode for the OER. Between these two limiting curves are plotted three additional curves (red circles, curves labeled 1, 2, and 3) obtained with increasing loadings of α-H$_4$SiW$_{12}$O$_{40}$ for the HER and Ir metal for the OER. Although the electrochemical performances are not as good as those obtained with Pt at the cathode, those obtained with optimized POMs (curve 3) cathodes are already very efficient: $\varepsilon_{\Delta H} \approx 70\%$ has been obtained at 1 A cm^{-2}, significantly higher than those obtained with the above-mentioned clathrochelates. These results offer new and interesting perspectives to PEM water electrolysis technology. By getting rid of noble metals, substantial cost reductions can be expected, opening the way to new applications in the industry.

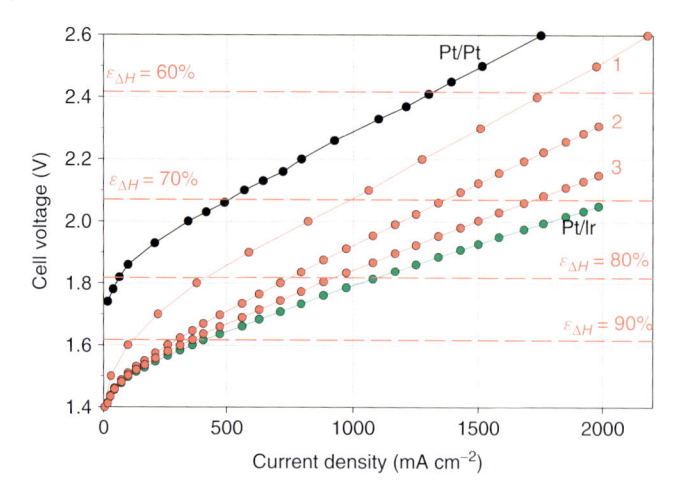

Figure 9.28 Polarization curves measured on the laboratory scale (7 cm^2) PEM cells at 90 °C. (black ●) Pt for both HER and OER. (green ●) Pt for the HER and Ir for the OER. (red ●) α-H$_4$SiW$_{12}$O$_{40}$ for the HER: 1–0.2 mg cm^{-2}; 2–0.5 mg cm^{-2}; and 3–0.8 mg cm^{-2}. Ir° for the OER (2.5 mg cm^{-2}).

9.5.7.2 Replacement of Iridium with Non-Noble Electrocatalysts

In natural photosynthesis, the oxygen reaction center (ORC) is located in the photosystem II [20]. Oxidation of water involves a manganese–calcium cluster [21]. Manganese-based di- and trinuclear complexes such as perchlorate salts of mixed valence species [Mn$_2$O$_2$L$_2$]$^{3+}$ (L = dipyridyl (bpy) and phenanthrolin (phen)) have been tested with regard to the OER in PEM water electrolysis cells [22]. The chemical stability of such compounds was found to be sufficient, whereas their efficiency as a catalyst for promoting the OER in acidic media is still inadequate and requires alternative materials.

9.5.7.3 Development of Polymeric Proton Conductors for Operation at More Elevated Temperatures

Regarding the solid polymer electrolyte, research is driven by potential applications in the H$_2$/O$_2$ fuel cell technology for automotive applications. Since most of the energy delivered by such fuel cells is released as heat, operation at conventional (80–90 °C) temperatures requires a larger heat exchanger compared to that used in internal combustion engine (ICE) driven cars. An operating temperature close to 150 °C would be more appropriate. Many R&D activities are currently underway to find alternative and less expensive materials. For example, different perfluorinated polymers (Hyflon® from Solvay Solexis Co.) are stable at such temperatures. An additional problem in fuel cell technology comes from water management issues. Feed gases must be appropriately hydrated and the water produced at the cathode from the reduction of oxygen must be removed to avoid condensation and loss of performances. Such problems are critical, and a new trend in research is to develop new polymers involving new proton conduction mechanisms and operating in quasi-dry conditions. It is expected that the results of all this research will also benefit PEM water electrolysis.

9.5.7.4 Development of Proton-Conducting Ceramics for Operation at More Elevated Temperatures

As discussed in section 9.2.3, significant interest has been shown in performing water electrolysis at elevated (800–1000 °C) temperatures: from a thermodynamic viewpoint, the amount of electricity required or the process is lower and the kinetics of the reaction is favored. Therefore, the energy consumption is reduced and the efficiency is higher. However, state-of-the-art technology is not satisfactory (Section 9.6). Research in material science is facing very critical issues due to important corrosion effects. Operation in the intermediate temperature range (500 °C) is now receiving more attention. Some proton-conducting electrolytes are available for operation in such conditions. For example, proton-conducting perovskites are oxide materials that transport protons at elevated (400–1000 °C) temperatures. In particular, some perovskites (simples or mixes) present a high ionic conductivity when exposed to water vapor. They can therefore be used as thin (<100 μm) solid electrolytes for manufacturing water electrolysis cells. For example, $Sr[Ce_xZr_{1-x}]_{0.95}Yb_{0.05}O_3$ shows a maximum conductivity for $x = 0.8$ [23]. Another example is given by $La_{0.6}Ba_{0.4}ScO_{2.8}$.

9.6 High Temperature Water Electrolysis

9.6.1 Basic Principles (Electrochemical Processes)

A high temperature solid oxide fuel cell (SOFC) can be operated in a reverse mode to electrolyze steam and generate hydrogen. The concept of high temperature (900–1000 °C) electrolysis of water vapor originates from studies carried out on SOFCs. It is sometimes referred to as *solid oxide water electrolysis*. In a high temperature water electrolysis cell, electricity is used to split water molecules into gaseous hydrogen and oxygen, according to the following half-cell reactions (Figure 9.29):

$$\text{Anode:} \quad O^{2-} \rightarrow \frac{1}{2}O_2(g) + 2e^- \quad (9.34)$$

$$\text{Cathode:} \quad H_2O(\text{vap}) + 2e^- \rightarrow H_2(g) + O^{2-} \quad (9.35)$$

$$\text{Full reaction:} \quad H_2O(\text{vap}) \rightarrow H_2(g) + \frac{1}{2}O_2(g) \quad (9.36)$$

The electrolyte is an oxide-ion-conducting ceramic, typically an yttrium-stabilized ZrO_2 membrane. As for PEM technology, the role of the membrane is double. It first acts as a solid electrolyte that conveys electric charges from one electrode

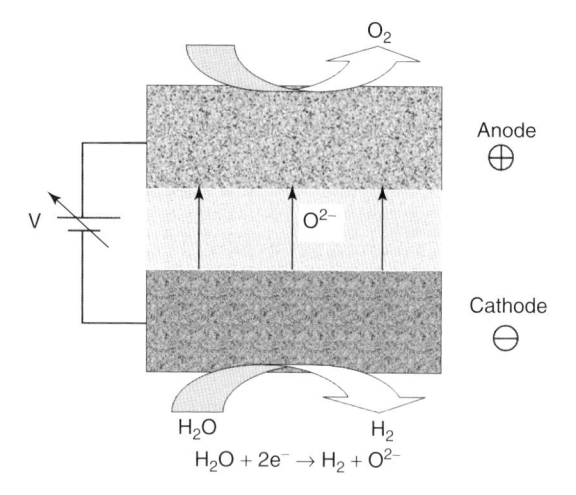

Figure 9.29 Schematic diagram of an HT water electrolysis cell.

$H_2O + 2e^- \rightarrow H_2 + O^{2-}$

to the other and second as a separator to prevent the recombination of reaction and products. Water vapor is introduced into the cathodic compartment. The main advantages of the high temperature process are (i) lower energy consumption ($\approx$3 kWh Nm^{-3} H$_2$ at 900 °C against $\approx$4 kWh Nm^{-3} H$_2$ at 80 °C) due to a reduced ΔG and better kinetics; (ii) reduced costs because no electrocatalyst is required at such temperatures; and (iii) the technique is well suited for the centralized production of large amounts of hydrogen, using the high temperature heat released by nuclear reactors. On the less positive side, main drawbacks come from reduced lifetime of cell components because of the high temperature of operation: material degradation (phase transformation, decreasing ionic conductivity) in relation to temperature and thermal cycling, and gas tightness. The technique is not appropriate for intermittent operation. The same concept of cell can also be used for the oxidation of methane or for the co-electrolysis of water vapor and CO$_2$ for the production of syngas (Figure 9.30), opening the way to the production of synthetic fuels.

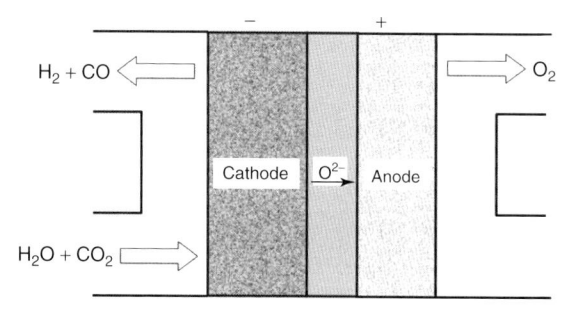

Figure 9.30 Schematic diagram of a cell used for the co-electrolysis of CO$_2$ and water vapor.

9.6.2
Cell Components

Solid oxide water electrolysis (SOWE) cell components are (i) stainless steel bipolar plates and (ii) manganite-coated stabilized zirconia as solid electrolyte. A photograph of main cell components is provided in Figure 9.31.

9.6.3
Cell Assembly and Manufacturing

As for the alkaline and PEM technologies, SOWE cells are stacked in a filter-press configuration to increase power units. A photograph of a prototype SOWE stack is provided in Figure 9.32.

9.6.4
Cell Stacking and Systems

The productivity of the electrolysis unit can be further increased by interconnecting several SOWE stacks. The photograph of a system developed at CERAMATEC Co. is provided in Figure 9.33.

9.6.5
Cell Performance, Testing, and Diagnosis

The concept of SOWE has been known and studied for many years. In the 1980s, a tubular prototype capable of producing $1\, Nm^3\, H_2\, h^{-1}$ at $1000\,°C$ has been developed and tested at Dornier Systems Co. This prototype has been operated for more than 500 h, with an efficiency close to 100% [24, 25]. The

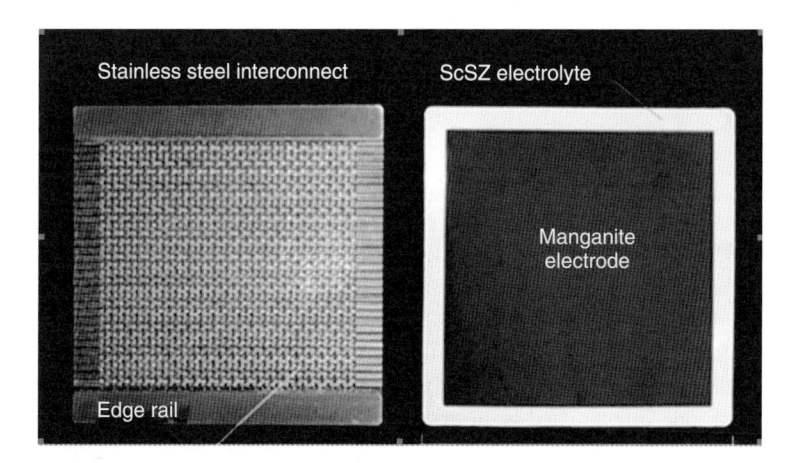

Figure 9.31 Photographs of bipolar plates and MEAs used in SOFC technology. (Source: CERAMATEC Co.)

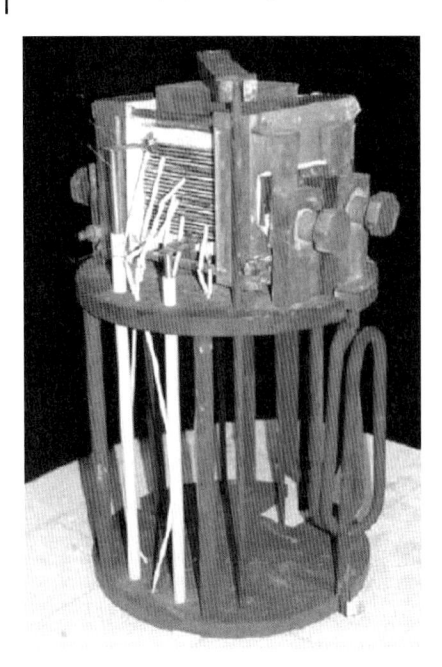

Figure 9.32 Mono-stack (25 cells) and 17-kW unit. (Source: CERAMATEC Co.)

Figure 9.33 Multi-stack solid oxide water electrolysis unit. (Source: CERAMATEC Co.)

experiments have been stopped because of prohibitive costs and lack of short-term commercial applications. A similar approach was followed at the same time by Westinghouse. The situation is now more favorable for such developments, and new systems are being tested [26]. SOWE is now the subject of several R&D programs based on recent advances in SOFC technology with planar geometries [27]. Very high efficiencies close to 100% have been measured at the laboratory scale at high current densities (a current density of 3.6 A cm^{-2} has been reported at 1.48 V at 950 °C [28]). As mentioned above, the co-electrolysis of water vapor and carbon dioxide can be used to improve the efficiency of the process and to produce a syngas ($H_2 + CO$), which in turn can be used for the production of synthetic fuels. Typical polarization curves measured in these systems are plotted in Figure 9.34.

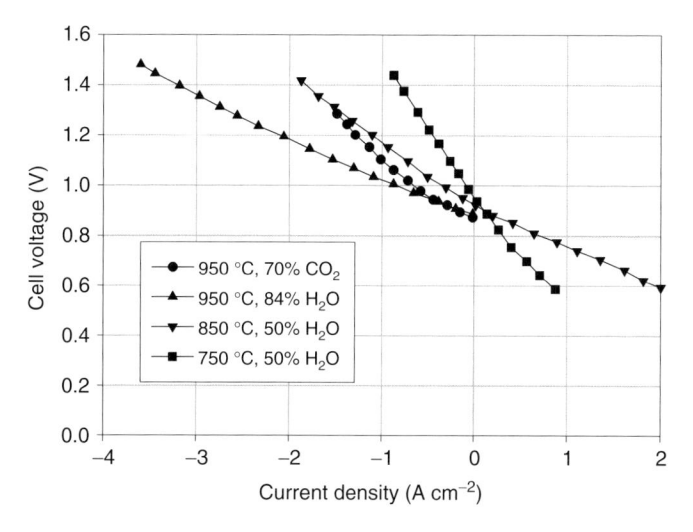

Figure 9.34 Polarization curves measured on a 5 × 5 cm² SOFC cell [25].

9.6.6
Degradation Mechanisms and Mitigation Strategies

At such elevated temperatures, the most significant challenges concern material science (material dilatation, material stability, and corrosion issues). In current technology, material resistance to thermal gradients and thermal cycling remains rather poor, and largely contributes to high investment costs. An option would be to reduce the operating temperature down to the 700–800 °C range, where the ionic conductivity of yttrium-stabilized ZrO_2 is still insufficient. In this temperature range, catalysts are required. Some nickelates (Ruddlesden–Popper compounds) such as $Ln_{1+n}M_nO_{3n+1\pm\delta}$, where $n = 1, 2$, or 3; Ln = La, Nd, or Pr; and M is preferably Ni, are currently investigated [29]. The chemical composition can be adjusted by performing simple or double cationic substitutions for higher anodic efficiency. Such compounds can be easily powdered and used in catalytic ink compositions for the automated production of large-scale and homogeneous catalytic layers.

The electrochemical performances plotted in Figure 9.34 cannot be maintained in the long term. Classical degradation tests performed at constant current density (between 0.25 and 0.5 A cm^{-2}) show a rapid increase in the cell voltage over the first 100 h of operation due a rapid increase in internal resistances, followed by a plateau over several hundred hours of operation. The same situation prevails on SRUs and stacks, and sometimes is amplified. Tests carried out on a five-cell unit over 2000 h at 800 °C and 0.3 A cm^{-2} have revealed a degradation rate of cell voltage close to 15% 1000 h^{-1} [30]. Results obtained at CERAMATEC [27] and Idaho National Laboratory (INL) [31] indicate a degradation rate close to 20% 1000 h^{-1} at 800–900 °C.

9.6.7
Challenges and Perspectives

These high rates of degradation of cell performances are due to a steady increase in cell resistances. In the 800–850 °C temperature range, resistances of ≈ 1 kW stacks are 1.5 Ω cm^2 at the beginning of tests and rapidly reach 2.5 Ω cm^2, whereas results obtained at the laboratory scale [32] are of circa 0.5 Ω cm^2. The nature of such degradations remains unclear and few experimental data are available on the subject. Some EIS measurements have been reported using conventional cells (Ni-YSZ/YSZ/LSM), but data interpretation is somewhat cumbersome because cathode and anode have very similar responses. At the same time, manufacturers of SOFC stacks (between 1 and 10 kW) announce continuous improvements in efficiency and durability [33, 34]. Some SOFC stacks (having 50–75 cells and an output power of circa 1 kW) have been operated with 92% fuel consumption and degradation rates of 0.5% over 1000 h. Such improvements have been obtained by choosing a more appropriate alloy used for the interconnection plates with a better protective oxide layer [35]. These results show that a compromise has to be found between performance and lifetime. Among various causes for performance degradation, the highly corrosive environment in the electrolyzer and mechanical and thermal constraints are responsible for most losses.

9.7
Conclusion

Currently, hydrocarbons (fossil fuels) are the main source of hydrogen, and the dominant technology used for the large-scale production of molecular hydrogen is steam reforming of methane from natural gas. Although water is as cheap and abundant as methane, water is not used as a hydrogen source in the industry, except for specific niche applications. This is mainly because the energy required to extract hydrogen from methane is almost two times less than the energy required for its extraction from water. As long as the massive emission of carbon dioxide in the atmosphere will be permitted at zero cost, this situation will not change. Other chemical sources such as ammonia and urea are seldom used. Among different processes, hydrogen can be extracted from water using the water electrolysis technology. Alkaline water electrolysis technology is the most mature process. PEM technology is receiving a growing attention because of its inherent advantages such as higher safety and efficiency. High temperature water electrolysis is considered for the large-scale production of hydrogen in the longer term. Even if water electrolysis is appropriate to produce hydrogen from green (produced by renewable sources such as solar photovoltaics and wind energy) electricity, energy, and investment costs are such that the process is not competitive, except for special applications. The situation can be balanced to some extent when the oxygen produced by the reaction can be exploited (for example, for oxy-combustion). Oxygen by-product credits are another way by which electrolytic hydrogen can be made more economical. A better understanding of the current and future oxygen market is needed to justify such a credit.

List of Symbols and Abbreviations

A	membrane area (m²)
a_i	activity of species i
C_i	concentration of species i in mol m^{-3}.
D_i	diffusion coefficient of species i (m² s^{-1})
E	thermodynamic electrolysis voltage (V)
F	Faraday constant (96 485 C mol^{-1})
f_i	fugacity of species i
G	Gibbs free energy (J mol^{-1})
H	enthalpy (J mol^{-1})
I	current (A)
i	current density (A cm^{-2})
i_0	exchange current density (A cm^{-2})
m	mole number
n	number of electrons exchanged during the electrochemical reaction
P	pressure (Pa)
p_i	partial pressure of component i in a gas mixture (Pa)
R	resistance (Ω)
r	specific resistance (Ω m²)
R_{PG}	constant of perfect gas (0.082 J K^{-1} mol^{-1})
S	entropy (J mol^{-1} K^{-1})
T	absolute temperature (K)
t	time (s)
U_{cell}	cell voltage (V)
V	thermoneutral electrolysis voltage (V)
$\overline{V}$	molar volume (m³ mol^{-1})
W	electrical work (J)
x,y,z	space coordinates

Greek Symbols

δ	membrane thickness (m)
Δ	difference
ε	cell efficiency (%)
η	overvoltage (volt)
φ	electrode absolute potential (V)
ρ	electrical resistivity (Ω m)
σ	electrical conductivity (S m^{-1})

Subscripts or Superscripts

°	standard conditions (298 K, 1 bar)
cell	electrolysis cell
d	dissociation
e	electrolyte
i	species i
r	real
t	theoretical

Acronyms

EW	equivalent weight of membrane (eq g^{-1})
PEM	proton exchange membrane
SPE	solid polymer electrolyte
SOFC	solid oxide fuel cells
SOWE	solid oxide water electrolysis

References

1. Encyclopedia Britannica, Inc., (2010) *Encyclopedia Britannica*, Chicago, IL.
2. Rosemary Gene, E. and Aaron, I. (1954) Faraday's electrochemical laws and the determination of equivalent weights. *J. Chem. Educ.*, **31**, 226–232.
3. Goodridge, F. and Scott, K. (1995) *Electrochemical Process Engineering: A Guide to the Design of Electrolytic Plant*, Plenum Publishing Corporation.
4. Wendt, H. and Kreysa, G. (1999) *Electrochemical Engineering: Science and Technology in Chemical and Other Industries*, Springer.
5. Meyer, C.A. (1993) *ASME Steam Tables: Thermodynamic and Transport Properties of Steam*, American Society of Mechanical Engineers, New-York.
6. LeRoy, R.L., Bowen, C.T., and Leroy, D.J. (1980) The thermodynamics of aqueous water electrolysis. *J. Electrochem. Soc.*, **127**, 1954.
7. Onda, K., Kyakuno, T., Hattori, K., and Ito, K. (2004) Prediction of production power for high-pressure hydrogen by high-pressure water electrolysis. *J. Power Sources*, **132**, 64–70.
8. Millet, P. (1996) Electric potential distribution in an electro-chemical cell. *J. Chem. Educ.*, **73**, 956.
9. Millet, P., Durand, R., and Pinéri, M. (1990) Preparation of new solid polymer electrolyte composites for water electrolysis. *Int. J. Hydrogen Energy*, **15** (4), 245–253.
10. Damjanovic, A., Dey, A., and Bockris, J.O'M. (1966) Electrode kinetics of oxygen evolution and dissolution on Rh, Ir, and Pt-Rh alloy electrodes. *J. Electrochem. Soc.*, **113**, 739.
11. Divisek, J. and Murgen, J. (1983) Diaphragms for alkaline water electrolysis and method for production of the same as well as utilization thereof. US Patent 4, 394,244, 1983.
12. Vandenborre, H., Leysen, R., Nackaerts, H., Van der Eecken, D., Van Asbroeck, Ph., Smets, W., and Piepers, J. (1985) Advanced alkaline water electrolysis using inorganic membrane electrolyte (I.M.E.) technology. *Int. J. Hydrogen Energy*, **10** (11), 719–726.
13. Pile, D.L. and Doughty, D.H. (2005) Sandia National Laboratory, report on DOE contract DE-AC04-94AL85000.
14. Mauritz, K.A. and Moore, R.B. (2004) State of understanding of Nafion. *Chem. Rev.*, **104**, 4535–4585.
15. Millet, P., Durand, R., and Pineri, M. (1989) New solid polymer electrolyte composites for water electrolysis. *J. Appl. Electrochem.*, **19**, 162–166.
16. Marshall, A., Børresen, B., Hagen, G., Tsypkin, M., and Tunold, R. (2009) Hydrogen production by advanced proton exchange membrane (PEM) water electrolyzers-Reduced energy consumption by improved electrocatalysis. *Int. J. Hydrogen Energy*, **34**, 4974–4982.
17. Pantani, O., Anxolabéhère, E., Aukauloo, A., and Millet, P. (2007) Electroactivity of cobalt and nickel glyoximes with regard to the electro-reduction of protons into molecular hydrogen in acidic media. *Electrochem. Commun.*, **9**, 54–58.
18. Pantani, O., Naskar, S., Guillot, R., Millet, P., Anxolabéhère, E., and Aukauloo, A. (2008) Cobalt clathrochelate complexes as hydrogen-producing catalysts. *Angew. Chem. Int. Ed.*, **120**, 10096–10098.

19. Keita, B. and Nadjo, L. (1987) New aspects of the electrochemistry of heteropolyacids: part II. Coupled electron and proton transfers in the reduction of silicotungstic species. *J. Electroanal. Chem.*, **217**, 287–304.
20. Ferreira, K.N., Iverson, T.M., Maghlaoui, K., Barber, J., and Iwata, S. (2004) Architecture of the photosynthetic oxygen-evolving center. *Science*, **303**, 1831–1838.
21. Mishra, A., Wernsdorfer, W., Abboud, K., and Christou, G. (2005) The first high oxidation state manganese-calcium cluster: relevance to the water oxidizing complex of photosynthesis. *Chem. Commun.*, 54–56.
22. Medhoui, T., Anxolabehere, E., Aukauloo, A., and Millet, P. (2006) Biomimetic approaches for the development of non-noble metal electrocatalysts. Application to PEM water electrolysis. Proceedings of the 16th World Hydrogen Energy Conference, Lyon, June 13–16, 2006.
23. Sata, N., Yugami, H., Akiyama, Y., Sone, H., Kitamura, N., Hattori, T., and Ishigame, M. (1999) Proton conduction in mixed perovskite-type oxides. *Solid State Ionics*, **125**, 383–387.
24. Doenitz, W. and Schmidberger, R. (1982) Concepts and design for scaling up high temperature water vapour electrolysis. *Int. J. Hydrogen Energy*, **7**, 321.
25. Quandt, K.H. and Streicher, R. (1985) Concept and design of a 3.5MW pilot plant for high temperature electrolysis of water vapor. *Int. J. Hydrogen Energy*, **11**, 309–315.
26. Mogensen, M., Jensen, S.H., Hauch, A., Chorkendorff, I., and Jacobsen, T. (2008) Proceedings 7th European SOFC Forum (2006), Lucerne, Switzerland, July 2008.
27. Stoots, C., O'Brien, J.E., Hawkes, G.L., Herring, J.S., and Hartvigsen, J.J. (2006) High temperature steam and carbon dioxide electrolysis experiments at INL. Workshop on High Temperature Electrolysis, Roskilde, Denmark, September 18–19, 2006.
28. Jensen, S.H., Larsen, P.H., and Mogensen, M. (2007) Hydrogen and synthetic fuel production from renewable energy sources. *Int. J. Hydrogen Energy*, **32**, 3253–3257.
29. Stevens, P., Bassat, J.-M., Mauvy, F., Grenier, J.-C., and Lalanne, C. (2006) Matériaux d'anode pour SOEC. French Patent EDF/CNRS WO 2006/008,390.
30. Brisse, A., Schefold, J., and Zahid, M. (2008) High temperature water electrolysis in solid oxide cells. *Int. J. Hydrogen Energy*, **33**, 5375–5382.
31. Herring, J.S., O'Brien, J.E., Stoots, C.M., Hawkes, G.L., Hartvigsen, J.J., and Shahnam, M. (2007) Progress in high temperature electrolysis for hydrogen production using planar SOFC technology. *Int. J. Hydrogen Energy*, **32**, 440–450.
32. O'Brien, J.E., Stoots, C.M., Herring, J.S., Lessing, P.A., Hartvigsen, J.J., and Elangovan, S. (2004) "Proceedings of ICONE12", 12th International Conference on Nuclear Engineering, Virginia, 2004.
33. Vinke, I.C., Erben, R., Song, R.-H., and Kiviaho, J. (2008) Installation and operation of kW-class stacks from Juelich in external laboratories. Proceedings 7th European SOFC Forum; (2006), Lucerne, Switzerland, July 2008.
34. de Haart, L.G.J., Mougin, J., Posdziech, O., Kiviaho, J., and Menzler, N.H. (2008) Stack degradation in dependence of operation parameters; the Real-SOFC sensitivity analysis. 8th European SOFC Forum, Lucerne, Switzerland, July 2008, paper B1002.
35. Christiansen, N., Hansen, J.B., Holm-Larsen, H., Linderoth, S., Larsen, P.H., Hendriksen, P.V., and Mogensen, M. (2008) Proceedings 7th European SOFC Forum (2006), Lucerne, Switzerland, July 2008.

10
Hydrogen Compression, Purification, and Storage
Pierre Millet

10.1
Introduction

In view of the so-called "hydrogen economy," hydrogen is considered as an energy carrier that can be used for the promotion of renewable energy sources. With conversion efficiencies above 70% at current densities of 1 A cm^{-2} (this is a common good practice, at least at the laboratory scale), proton exchange membrane (PEM) water electrolysis has been identified in the European Hydrogen and Fuel Cell program as a key process for transforming zero-carbon electricity sources into the supply of zero-carbon hydrogen and oxygen for miscellaneous end-uses [1], and the European Commission is actively supporting R&D activities in this field [2]. New technologies are required for pressurizing, purifying, and storing hydrogen before it can fulfill its role of new energy carrier. The purpose of this chapter is to provide an overview of some electrochemical technologies that can be used for this purpose. The principles and main characteristics of pressurized water electrolysis are presented and discussed in Section 10.2. Electrochemical devices that can be used for compressing and purifying hydrogen are presented and discussed in Sections 10.3 and 10.4. A comparison of chemical and electrochemical hydrogen storage techniques is provided in Section 10.5.

10.2
Pressurized Water Electrolysis

10.2.1
Principles

The general principles and main characteristics of low temperature (alkaline and PEM) liquid water electrolysis are described in Chapter 9. Basically, there are two different approaches for producing pressurized hydrogen of electrolytic grade: first, by performing water electrolysis at atmospheric pressure and then pressuring the gas in a second step, for example, using a conventional gas compressor; second, by

Electrochemical Technologies for Energy Storage and Conversion, First Edition. Edited by Ru-Shi Liu, Lei Zhang, Xueliang Sun, Hansan Liu, and Jiujun Zhang.
© 2012 Wiley-VCH Verlag GmbH & Co. KGaA. Published 2012 by Wiley-VCH Verlag GmbH & Co. KGaA.

performing water electrolysis under pressure (self-pressurization or pressurization using an inert gas). Since the compression of liquid water requires less power than the compression of hydrogen, it can be expected that the second process will require less power than the first one. Also, in terms of investment cost, when hydrogen is pressurized directly in the electrolyzer, the additional cost due to the high pressure (HP) structure is less than the need for an external hydrogen compressor. To compare more accurately the two processes, it is necessary to evaluate the role of operating pressure on the thermodynamics and kinetics of the water-splitting reaction and to take into account the energy required for the injection of liquid water at atmospheric pressure into the pressurized electrolyzer. Although pressurized alkaline water electrolysis is commonly used in the industry (with operating pressure in the 1–50 bar range), the discussion provided in the following subsections is mainly based on PEM water electrolysis for which prototypes operating at very HP (several hundred bars) have been developed.

10.2.1.1 Thermodynamics

The role of operating pressure on the thermodynamic voltage E of liquid water electrolysis has been approximated in the previous chapter using the following relationship:

$$\Delta E = \frac{3 R_{PG} T}{4F} \ln\left(\frac{P_2}{P_1}\right) \tag{10.1}$$

The reference pressure being $P_1 = 1$ bar, Eq. (10.1) can be rewritten as

$$E(T,P) = E(T,1) + \frac{3 R_{PG} T}{4F} \ln(P) \tag{10.2}$$

Equation (10.2) provides a simple and good approximation for $E(T,P)$. The domain covered by this relationship is (273–1273 K) and (1–1000 bar). However, some discrepancies appear in the upper range of the (T,P) domain. More accurate calculations can be made using more elaborated state equations. Technological processes can be classified as a function of the operating temperature. Liquid water is electrolyzed in the low temperature domain (0–250 °C) and water vapor is electrolyzed at medium (250–500 °C) and high (500–1000 °C) temperatures. Electrolysis of liquid water in the 100–250 °C temperature range is performed by increasing the operating pressure. Some empirical expressions of $V(T)$ and $E(T)$ have been proposed in the specialized literature [3]. They apply only to the electrolysis of liquid water, at a pressure sufficiently high to avoid water vaporization:

$$V(T) = 2F\Delta H(T) = 1.485 - 1.49 \times 10^{-4} \times (T - T^0)$$
$$- 9.84 \times 10^{-8} \times (T - T^0)^2 \tag{10.3}$$

$$E(T) = 2F\Delta G(T) = 1.5184 - 1.5421 \times 10^{-3} \times T + 9.523 \times 10^{-5} \times T$$
$$\times \ln(T) + 9.84 \times 10^{-8} T^2 \tag{10.4}$$

However, Eqs. (10.3) and (10.4) do not provide accurate expressions for the electrolysis of water vapor. Such expressions can be obtained from the values of

$V(298,1)$ and $E(298,1)$ as follows:

$$\Delta H(T,P) = H_{H_2}(T,P) + \frac{1}{2}H_{O_2}(T,P) - H_{H_2O}(T,P) \tag{10.5}$$

$$\Delta S(T,P) = S_{H_2}(T,P) + \frac{1}{2}S_{O_2}(T,P) - S_{H_2O}(T,P) \tag{10.6}$$

$$\Delta G(T,P) = \Delta H(T,P) - T\Delta S(T,P) = G_{H_2}(T,P)$$
$$+ \frac{1}{2}G_{O_2}(T,P) - G_{H_2O}(T,P) \tag{10.7}$$

Values for $H_i(T,P)$, $S_i(T,P)$ and $G_i(T,P)$ are obtained in two steps: first, by calculating the effect of temperature at a constant pressure of 1 bar and then by calculating the pressure effect at a constant temperature:

$$H_i(T,P) = [H_i(T,P) - H_i(T,1)] - [H_i(T,1) - H_i(298,1)] \tag{10.8}$$
$$S_i(T,P) = [S_i(T,P) - S_i(T,1)] - [S_i(T,1) - S_i(298,1)] \tag{10.9}$$
$$G_i(T,P) = [G_i(T,P) - G_i(T,1)] - [G_i(T,1) - G_i(298,1)] \tag{10.10}$$

Enthalpy and entropy changes of $H_2(g)$ and $O_2(g)$ species at 1 bar are obtained from (Table 10.1)

$$H_i(T,1) - H_i^0 = a(T - T_0) + \frac{b}{2}10^{-3}(T^2 - T_0^2)$$
$$- c10^5\left(\frac{1}{T} - \frac{1}{T_0}\right) - \frac{e}{2}10^8\left(\frac{1}{T^2} - \frac{1}{T_0^2}\right) \tag{10.11}$$

$$S_i(T,1) - S_i^0 = a(\ln T - \ln T_0) + b10^{-3}(T - T_0)$$
$$- \frac{c}{2}10^5\left(\frac{1}{T^2} - \frac{1}{T_0^2}\right) - \frac{e}{3}10^8\left(\frac{1}{T^3} - \frac{1}{T_0^3}\right) \tag{10.12}$$

Enthalpy and entropy changes of $H_2O(l)$ at 1 bar are obtained from the steam tables [4]. The role of operating pressure is taken into account as follows. Usually, the classical virial expansion (introduced in 1901 by H. Kamerlingh Onnes as a generalization of the ideal gas law) is used to express the pressure of a many-particle system at equilibrium as a power series in the density. Virial coefficients B_i appear as coefficients in the virial expansion of pressure, thus providing systematic corrections to the ideal gas law. They are characteristic of the interaction potential between the particles and in general depend on the temperature. The second virial coefficient depends only on the pair interaction between the particles, the third depends on two- and nonadditive three-body interactions, and so on. The pressure

Table 10.1 Coefficients used in Eqs. (10.11) and (10.12).

Compounds	a	b	c	e
H_2 (gas)	26.57	3.77	1.17	–
O_2 (gas)	34.35	1.92	−18.45	4.06

Table 10.2 Main characteristics of some commercial alkaline water electrolyzers.

Manufacturer (model)	H_2 production $(Nm^3 h^{-1})$	Production pressure (bar)	Total power (kW)	Energy system (kWh Nm^{-3})	Energy electrolyzer (kWh Nm^{-3})	Efficiency (%)
Stuart (IMET 1000)	60	25	290	4.8	4.2	73
Teledyne (EC 750)	42	4–8	235	5.6	–	63
Proton (Hogen 380)	10	14	63	6.3	–	56
Norsk Hydro (5040)	485	30	2330	4.8	4.3	73
Avalence (hydrofiller 175)	4.6	up to 700	25	5.4	–	64

change of ΔH and ΔG is given by the following state equations:

$$H(T,P) - H(T,1) = \left(B - T\frac{\partial B}{\partial T}\right)P + \frac{C - B^2 - 0.5T\left[\frac{\partial C}{\partial T} - 2B\frac{\partial B}{\partial T}\right]}{RT}P^2 \quad (10.13)$$

$$G(T,P) - G(T,1) = R_{PG}T \ln P + BP + \frac{C - B^2}{2R_{PG}T}P^2 \quad (10.14)$$

The virial constants B and C can be approximated as a function of temperature:

$$B = b_1 + \frac{b_2}{T} \; (\text{cm}^3 \, \text{mol}^{-1}) \quad (10.15)$$

$$C = c_1 + \frac{c_2}{T^{1/2}} \; (\text{cm}^6 \, \text{mol}^{-2}) \quad (10.16)$$

$$\frac{dB}{dT} = -\frac{b_2}{T^2} \; (\text{cm}^3 \, \text{mol}^{-1} \text{K}^{-1}) \quad (10.17)$$

$$\frac{dC}{dT} = -\frac{c_2}{T^{3/2}} \; (\text{cm}^6 \, \text{mol}^{-2} \text{K}^{-2}) \quad (10.18)$$

The coefficients of viral constant for $H_2(g)$ and $O_2(g)$ are (for pressures up to 1000 atm) given in Table 10.2.

Enthalpy and entropy changes of $H_2O(l)$ with pressure are obtained from the steam tables. $E(T,P)$ and $V(T,P)$ obtained from the above equations are plotted in Figures 10.1 and 10.2 [5].

Although the effect of pressure on the thermodynamic voltage E is significant (+200 mV at 25 °C when the pressure is raised from 1 to 700 bar), its effect on the enthalpy voltage V remains limited. The magnitude of the additional energy cost due to higher operating pressure remains significantly lower than the charge transfer overvoltages that appear during operation at high current density.

10.2.1.2 Kinetics

From the thermodynamic viewpoint, temperature and pressure play an opposite role on the electrolysis voltage: the thermodynamic voltage decreases by circa

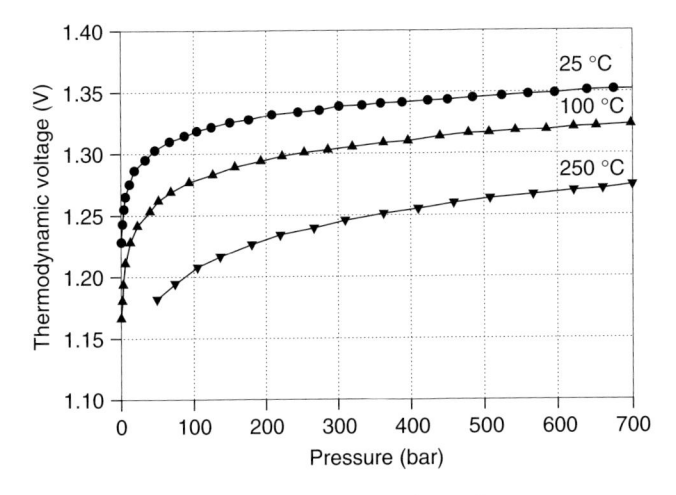

Figure 10.1 Plot of the thermodynamic voltage of water electrolysis as a function of pressure for three different operating temperatures.

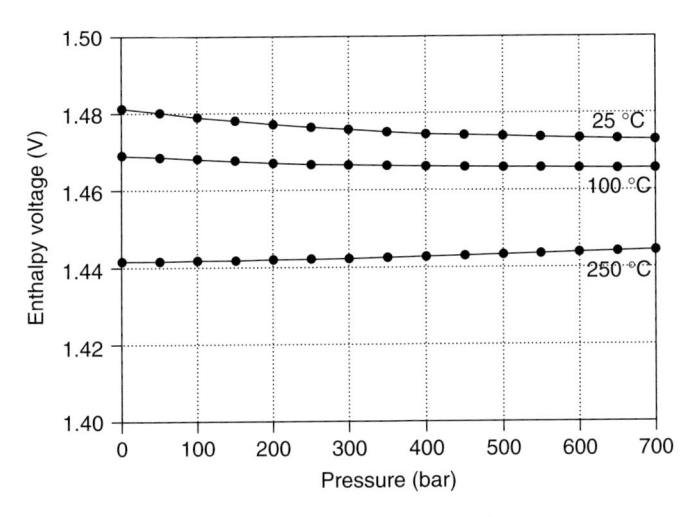

Figure 10.2 Plot of the enthalpy voltage of water electrolysis as a function of pressure for three different operating temperatures.

-0.8 mV C^{-1} and increases by circa $+0.3$ mV bar^{-1}. From the kinetic viewpoint, a higher operating pressure could negatively impact half-cell reactions and reduce the kinetics because gaseous H$_2$ and O$_2$ are formed at the cathode and anode, respectively. Experiments show that this is not the case (Figure 10.3). This is probably due to the fact that charge transfer processes at both electrodes are rate

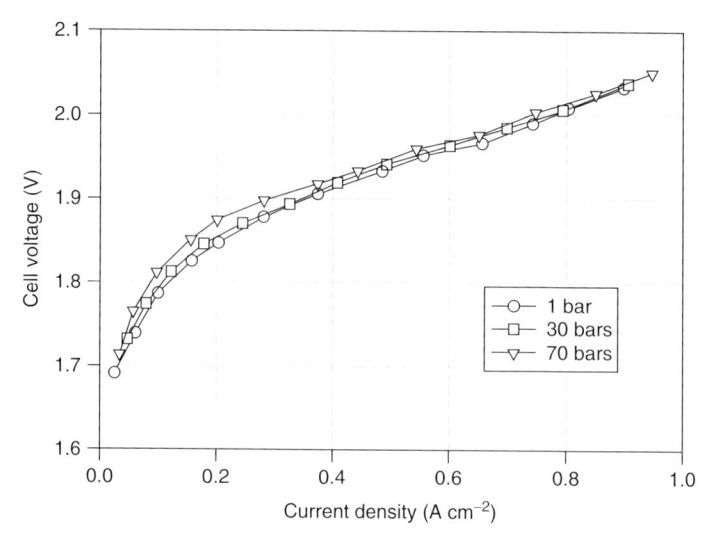

Figure 10.3 Role of operating pressure on the polarization curves of a PEM water electrolysis cell using platinum as cathodic catalyst and iridium as anodic catalyst.

determining, not gas transport away from the interfaces. Therefore, it can be concluded that pressurized water electrolysis does not require additional energy costs compared to atmospheric water electrolysis.

A comparison of the performances of commercial electrolyzers reveals that the situation is similar to alkaline water electrolysis (Table 10.2 [6]), the operating current density being the most significant parameter responsible for system efficiency.

10.2.2
Technical Developments

10.2.2.1 HP Stack

To reach high operating pressures, different stack designs can be used. A first concept is that of self-pressurized PEM cells. A second concept is that

Figure 10.4 Photograph of two individual PEM stacks.

Figure 10.5 Photograph of a high-pressure bi-stack with the pressurization vessel.

of an electrolysis stack introduced in a (stainless steel) pressurization vessel. The picture of a PEM water electrolysis bi-stack (each stack can deliver up to 1.0 Nm^3 H$_2$ h^{-1} at 1.2 A cm^{-2}) is given in Figure 10.4.

Each stack is made of eight membrane–electrode assemblies (MEAs), 250 cm^2 active area each, series connected in a filter-press configuration to reach the targeted hydrogen production capacity. MEAs contain conventional noble metal electrocatalysts: ≈0.4 mg cm^{-2} of metallic platinum for the hydrogen evolution reaction (HER) and ≈2.0 mg cm^{-2} of mixtures of metallic platinum and iridium for the oxygen evolution reaction (OER). Nafion®-reinforced membranes from E. I. DuPont Co. are used as solid polymer electrolyte (SPE). Porous titanium disks (1.2 mm thick, 40% open porosity) are used as current collectors, titanium grids as cell spacers, and titanium foils as bipolar plates. Carbon-based gaskets are used as cell sealants. The two stacks are placed in the stainless steel pressurization vessel shown in Figure 10.5. The dimensions of the unit are adjusted to minimize the open volume inside the pressurization cylinder. During electrolysis experiments, the pressure inside the stacks (due to the gaseous production of hydrogen and oxygen) is counterbalanced by the continuous flow of nitrogen in the pressurization cylinder. The pressure of nitrogen is set at a few hundred millibar above the internal pressure in the stacks in order to reduce the effect of pressure on the MEAs, to reduce gas leaks during the high-pressure measurements and to avoid H$_2$ and/or O$_2$ accumulation in the pressurization vessel.

10.2.2.2 Ancillary Equipment

The photograph of a pressurized PEM water electrolysis unit is given in Figure 10.6. The corresponding process flow sheet is detailed in Figure 10.7. Two high-pressure pumps are used, one for the oxygen circuit and one for the hydrogen circuit. Biphasic water–gas mixtures are separated in two different separation units. Gases are then purified (traces of hydrogen in the oxygen main stream and traces of oxygen in the hydrogen main stream are removed by catalytic recombination into water) and dried. Gas purity on each line is monitored continuously for safety purposes.

Figure 10.6 Photograph of a pressurized PEM water electrolysis unit.

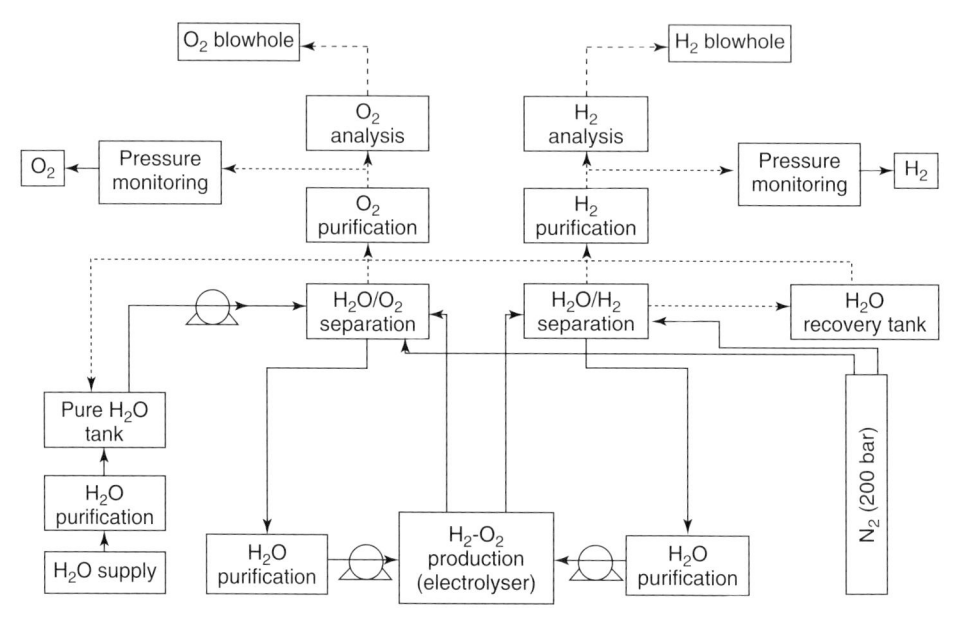

Figure 10.7 Pressurized PEM water electrolysis process flow sheet.

The gas (O_2 and H_2) exhaust pressures are monitored continuously. Compressed feed water is injected into the oxygen–water separator. The electro-osmotic flow of water collected in the cathodic loop is recycled. Pressurized nitrogen is initially used to pressurize the stack and for dilution purpose in case of emergency process shutdown.

10.2.3
Electrochemical Performances

Typical polarization curves measured at 40 °C on the two PEM stacks of Figure 10.4 (in the limited 0–400 mA cm^{-2} current density range) are plotted in Figure 10.8.

For safety reasons, Nafion-reinforced membranes are used as SPE. As a result, the ohmic drop across the SPE increases compared to laboratory-scale results

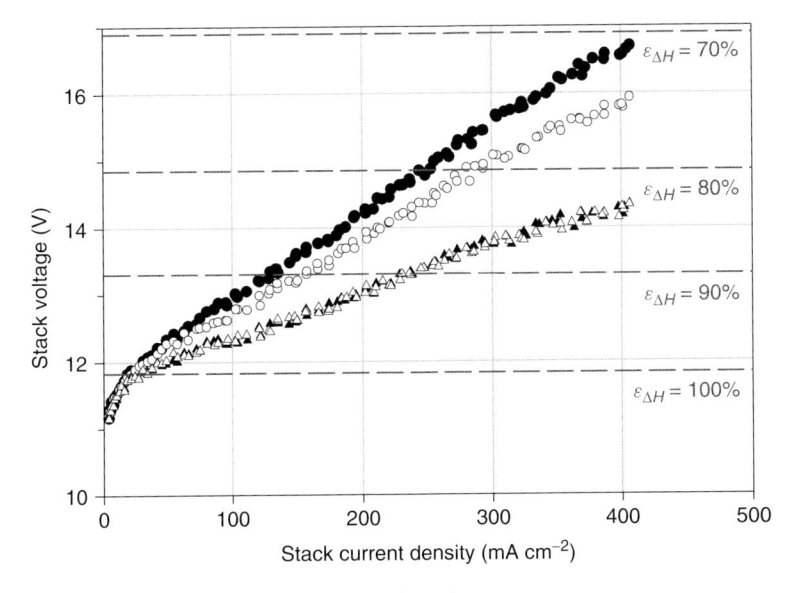

Figure 10.8 Polarization curves measured on the two stacks of Figure 10.4 at (●, ○) 40 °C and (▲, △) 80 °C. $P = 1$ bar.

obtained with conventional, non-reinforced, Nafion 117 membranes. An efficiency of 70% (calculated on the higher heating value of hydrogen) is obtained at 400 mA cm^{-2} and 40 °C. A higher efficiency close to 80% is obtained with the same current density, at an operating temperature of 80 °C. Some typical results obtained in the 1–130 bar pressure range are compiled in Table 10.3. Tests have been performed under stationary conditions, at a constant current density of 500 mA cm^{-2} and a mean operating temperature of 88 °C. The main operating parameter is the operating pressure. As can be seen from Table 10.3, individual cell voltages are homogeneous from one cell to the other and remain stable during operation. Some differences between individual cells are directly related to the distribution of current lines that are not totally homogeneous over the active area. A mean stack efficiency $\varepsilon_{\Delta H} \approx 85\% (\varepsilon_{\Delta G} \approx 72\%)$ is obtained at 500 mA cm^{-2} from the value of the stack voltage.

As can be seen from Figure 10.9, the mean energy consumption (for one stack) increases from 4.06 to 4.14 kWh Nm^{-3} H$_2$ when the operating pressure is raised from 1 to 130 bars.

In addition, the hydrogen production rate decreases from 0.42 to 0.36 Nm^3 h^{-1} at a constant current density and individual cell voltages slightly increase with the operating pressure. These effects can be attributed to the increase of hydrogen crossover across the membrane with pressure. This interpretation is further supported by the fact that the current (faradic) efficiency ε_j drops to 90% at 130 bars (Table 10.3). Therefore, the value of the stack efficiency inferred from the stack voltage is overestimated and must be corrected for current efficiency, the value of which is determined from the measurement of the gas production.

Table 10.3 Main test results obtained with the high-pressure PEM water electrolysis stack under stationary operating conditions.

Parameter	Unit	Measured value at operating pressure (bar)					
		1	25	50	75	100	130
Electric current	A	123.5	126.5	125.5	124.5	125.0	124.0
Operating temperature	°C	86	87	85	89	84	88
H_2 production rate	$Nm^3 h^{-1}$	0.42	0.43	0.42	0.40	0.38	0.36
H_2 purity before recombiner	vol%	99.98	99.76	99.18	98.56	98.01	97.34
H_2 purity after recombiner	vol%	99.999	99.999	99.997	99.995	99.993	99.991
Individual cell voltage	V						
1		1.70	1.70	1.71	1.71	1.72	1.73
2		1.68	1.69	1.70	1.71	1.71	1.72
3		1.71	1.71	1.71	1.72	1.71	1.71
4		1.70	1.71	1.71	1.73	1.74	1.74
5		1.71	1.70	1.71	1.71	1.73	1.74
6		1.69	1.69	1.70	1.71	1.73	1.73
7		1.70	1.69	1.69	1.70	1.72	1.73
8		1.68	1.70	1.70	1.73	1.74	1.74
Stack voltage	V	13.57	13.59	13.63	13.72	13.80	13.84
Power consumption	kW	1.676	1.719	1.711	1.708	1.725	1.716
Specific energy	$kWh\ Nm^{-3}$	3.99	4.00	4.07	4.27	4.54	4.77
Current efficiency	%	99.98	99.68	98.13	96.57	92.91	90.45
Energy efficiency $\varepsilon_{\Delta G}$	%	72.5	72.4	72.2	71.7	71.3	71.0

From these results it can be concluded that, whereas a negative effect of operating pressure has been observed on the energy consumption of the stack, the increase remains significantly lower than the energy required for compressing hydrogen from atmospheric pressure using a mechanical (piston or membrane) compressor [5]. Therefore, the direct production of pressurized hydrogen with a pressurized electrolyzer remains an interesting process from the energy viewpoint.

10.2.4
Current Limitations and Perspectives

The major impact of the high operating pressure is on gas purity. Owing to the increasing gas-crossover phenomena across the membrane with pressure, the hydrogen purity fell from 99.98% at atmospheric pressure to 97.34% at 130 bars (due to oxygen transport from the anode) and the oxygen purity is also strongly altered. Nafion perfluorinated membranes used as SPE in PEM technology can be considered as a homogeneous two-phase medium, a mixture of hydrophobic regions concentrating fluorocarbon backbones and hydrophilic regions containing water, where proton conductivity takes place. According to the models of Gierke et al. [7] and Yeager and Steck [8], their microstructure consists of more or less

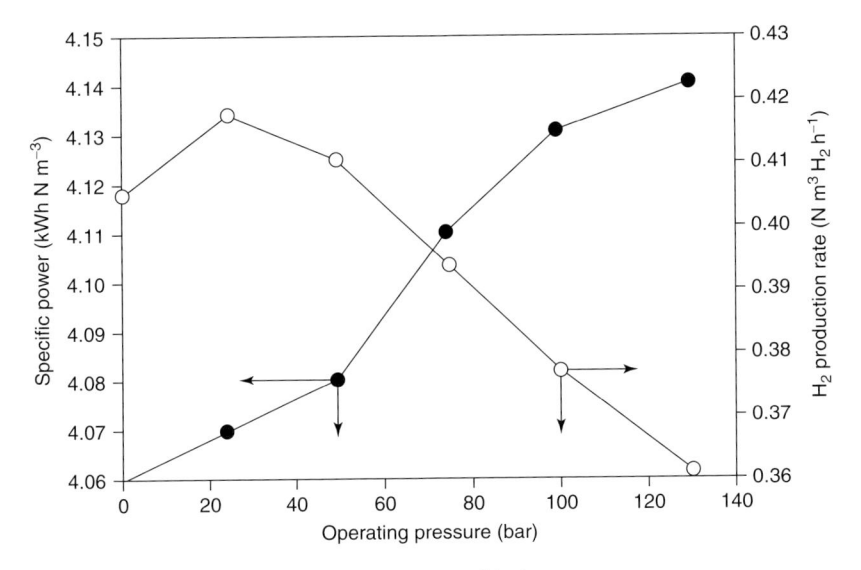

Figure 10.9 Mean specific power consumption and hydrogen production rate versus operation pressure.

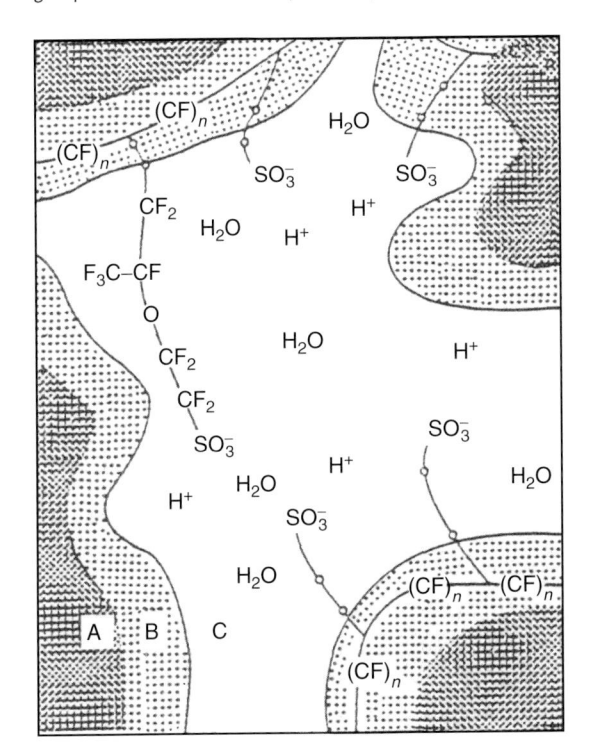

Figure 10.10 Three-phase model proposed by Yeager et al. for the microstructure of Nafion materials.

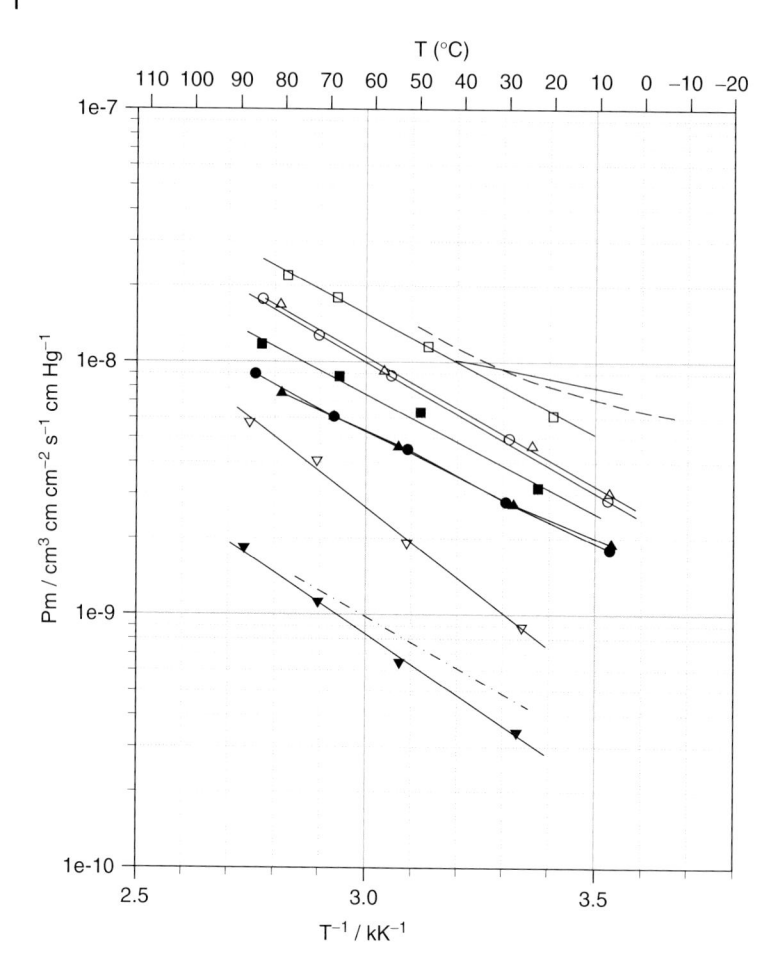

Figure 10.11 Hydrogen and oxygen permeability of Nafion 117 with temperature and water content [9].

distorted ($\approx$40 Å in diameter) clusters of sulfonate-ended perfluoroalkyl ether groups that are organized as inverted micelles and arranged on a lattice. These micelles are connected by pores or channels that are $\approx$10 Å in size (Figure 10.10). These SO_3^--coated channels are invoked to account for intercluster transfer of cations, ionic conductivity, and also gas permeability (Figure 10.11).

H_2 and O_2 are soluble and mobile in both hydrophobic and hydrophilic microdomains. The driving force for the transfer of dissolved gases from one side of the cell to the other during HP water electrolysis is the gradient of chemical potential set across the SPE by the gradient of pressure (concentration). The mass transport mechanism is mostly diffusion controlled (convection has also been reported but found less significant since the size of ionic clusters in the SPE is very small). The most striking point concerning cross-permeation phenomena in

PEM cells is that dissolved gases can diffuse across the porous electrodes placed at the surface of the thin polymer membranes and be released in the gaseous stream without being totally reoxidized (hydrogen at the anode) or reduced (oxygen at the cathode). In fact, a portion of gases is consumed at the electrodes (this is the reason why the faradaic yield at each interface is less than unity). The reason why they are not totally consumed is due to the fact that these gases are dissolved in liquid water inside the SPE, and the kinetics of oxidation (H_2) or reduction (O_2) is much lower than that in the gas phase (as in fuel cells). As a result, a significant amount of these gases can pass through the porous electrodes and be released in the opposite compartment of the PEM cell, thus polluting the gaseous production. A model which takes into account gas solubility in perfluorinated SPEs and diffusion-based mass-transport equations has been recently proposed to account for pressure-induced cross-permeation phenomena [10]. Some typical results obtained at pressures up to 30 bars are plotted in Figure 10.12. Current collectors are platinized to promote the catalytic recombination of H_2 and O_2 into water and reduce pollution levels.

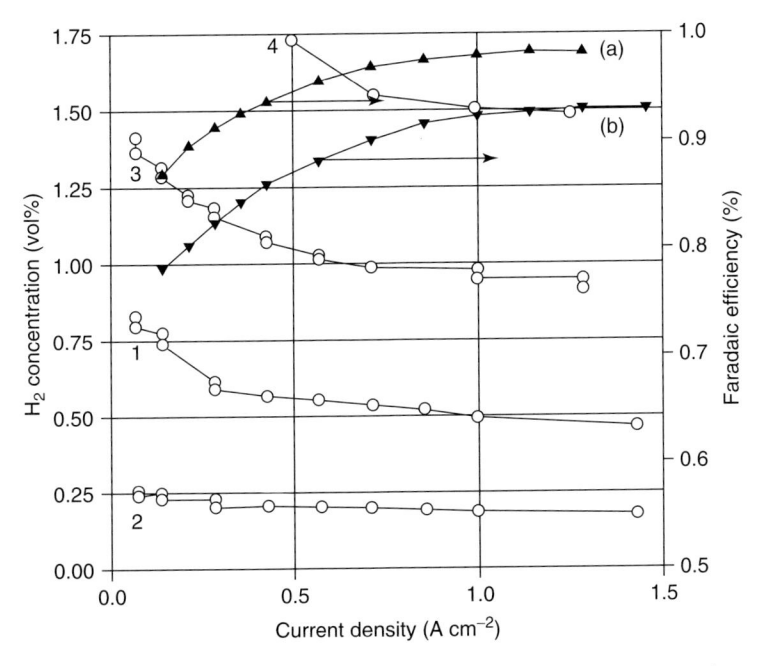

Figure 10.12 (Left) hydrogen content (vol %) in the anodic oxygen-water vapor mixture, measured at different pressures as a function of operating current density. 50 cm² mono-cell. Pt for the HER, Ir for the OER and Nafion 117 as SPE. $T = 85\,°C$. (1) 1 bar (current collector is platinized); (2) 1 bar (current collector not platinized); (3) 6 bar (current collector is platinized); (4) 30 bar (current collector is platinized); Full lines: best fits using Eq. (10.11) (see text). (right) current efficiency at the cathode at (a) 2 bar and (b) 30 bar, as a function of operating current density.

To reduce gas transport by cross-permeation through the SPE, different approaches can be followed. First, thicker SPEs can be used. In particular, Nafion-reinforced membranes (in which a polytetrafluoroethylene (PTFE) cloth is clamped between two conventional perfluorinated membranes) developed for the chlorine-alkali industry are commercially available. Second, it is possible to reduce either the diffusion coefficient or the solubility of the gas. This can be obtained by incorporating inorganic fillers, such as silica or zirconium oxide, into the SPE. For example, results obtained with zirconium dioxide in the course of the GenHyPEM project show that both diffusion coefficient (D_i) and solubility (H_i) can be reduced by a factor varying from 0 to about 10, depending on the filler concentration added to the SPE. Usually, such fillers are incorporated by impregnation of cationic precursor species followed by chemical precipitation using an adequate chemical reactant. Third, it is possible to promote the catalytic recombination of sub-stoichiometric H_2 with O_2 (anodic compartment) and sub-stoichiometric O_2 with H_2 (cathodic compartment) to remove trace amounts of hydrogen and oxygen. This can be done using internal gas recombiners. However, the price to pay for such improvements is always a reduction in the ionic conductivity of the SPE, an additional ohmic term during electrolysis, and a lower cell efficiency. Alternatively, it is possible to maintain the final purity of hydrogen at levels >99.99%, that is, at values well above the thresholds of explosion mixtures, by using hydrogen recombiners placed at the exhaust of the electrolysis stack. The use of internal gas recombiners offers a simple and efficient way to manage explosion hazards.

10.3
Hydrogen Electrochemical Compression

10.3.1
Principles

10.3.1.1 Compression Cell

Electrochemical hydrogen compressors are usually based on PEM technology. A schematic diagram of a unit PEM cell is given in Figure 10.13. It operates like an electrolysis cell. Hydrogen compression is a nonspontaneous process ($\Delta G > 0$). When electric power is supplied to the cell, hydrogen is transferred from the low pressure (LP) anodic compartment (left side) to the HP cathodic compartment (right side). Electrochemical reactions taking place in this process at the electrode–SPE interfaces are those of Eqs. (10.19) and (10.20). Equation (10.21) is the overall cell reaction:

$$\text{Anode}: H_2(LP) \rightarrow 2H^+ + 2e^- \quad (10.19)$$

$$\text{Cathode}: 2H^+ + 2e^- \rightarrow H_2(HP) \quad (10.20)$$

$$\text{Overall reaction}: H_2(LP) \rightarrow H_2(HP) \quad (10.21)$$

This type of compressor has no moving parts and is very compact. MEAs can be stacked to form a multistage compressor. Pressure of several hundred bars can

10.3 Hydrogen Electrochemical Compression

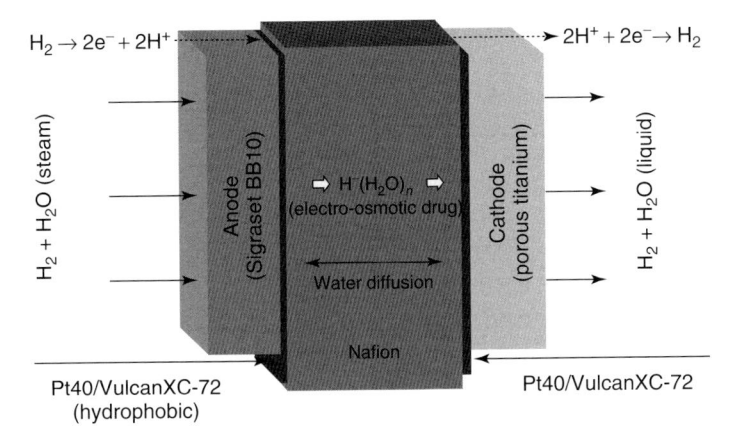

Figure 10.13 Schematic diagram of a PEM hydrogen compression cell showing MEA components.

potentially be reached. Water vapor partial pressure in the feed gas, current density, operating temperature, and hydrogen back diffusion due to the pressure gradient are the main parameters affecting the maximum output pressure.

10.3.2
Thermodynamics

Assuming that hydrogen behaves as an ideal gas, the thermodynamic voltage of the compression process can be approached by the ratio of input and output pressures. A quantitative expression is easily derived from the Nernst equation:

$$E(V) = \frac{R_{PG} T}{2F} \ln \frac{P_{H_2}^{HP}}{P_{H_2}^{LP}} \tag{10.22}$$

When the cell is used only for hydrogen purification (and not for compression), $\Delta P = P_{H_2}^{HP} - P_{H_2}^{LP} = 0$ bar and, according to Eq. (10.22), $E = 0$ V. When the cell is used for pressurizing hydrogen, a thermodynamic cell voltage of 30 mV is required at 298 K for a 10-fold compression factor. Potentially, hydrogen compressors can be used to reach pressures of several hundred bars. More accurate calculations can be made in place of the simplified Eq. (10.22). The compression cell always operates at constant temperature on both sides. Therefore, only the difference in pressure between HP and LP compartments generates a non-zero ΔH, ΔG, and ΔS change. When $\Delta P = (P^{HP} - P^{LP}) = 0$, no energy is required, whatever the temperature. The cell is symmetrical. This is a major difference with the water-splitting reaction. The effect of pressure on the enthalpy and free enthalpy of hydrogen is given by (see Chapter 9)

Table 10.4 Coefficients used in Eqs. 10.15–10.18.

Compound	b_1	b_2	c_1	c_2
H_2 (g)	20.5	−1 857	−351	12 760
O_2 (g)	42.6	−17 400	−2604	61 457

$$H(T,P) - H(T,1) = \left(B - T\frac{\partial B}{\partial T}\right)P + \frac{C - B^2 - 0.5T\left[\frac{\partial C}{\partial T} - 2B\frac{\partial B}{\partial T}\right]}{RT}P^2 \tag{10.23}$$

$$G(T,P) - G(T,1) = R_{PG}T \ln P + BP + \frac{C - B^2}{2R_{PG}T}P^2 \tag{10.24}$$

The virial constants B and C can be approximated as a function of temperature:

$$B = b_1 + \frac{b_2}{T} \; (\text{cm}^3 \, \text{mol}^{-1}) \tag{10.25}$$

$$C = c_1 + \frac{c_2}{T^{1/2}} \; (\text{cm}^6 \, \text{mol}^{-2}) \tag{10.26}$$

$$\frac{dB}{dT} = -\frac{b_2}{T^2} \; (\text{cm}^3 \, \text{mol}^{-1} \text{K}^{-1}) \tag{10.27}$$

$$\frac{dC}{dT} = -\frac{c_2}{T^{3/2}} \; (\text{cm}^6 \, \text{mol}^{-2} \text{K}^{-2}) \tag{10.28}$$

The coefficients of viral constant for H_2(g) for pressures up to 1000 atm are given in Table 10.4. The energy required for compressing hydrogen from atmospheric pressure up to 700 bar is plotted in Figure 10.14 for two operating temperatures. The associated thermodynamic and enthalpy voltages (see definition in Chapter 9) are plotted in Figure 10.15.

From Figure 10.14, it can be seen that the compression reaction (10.21) is increasingly endothermic with pressure. As mentioned above, $\Delta G > 0$ and therefore reaction (10.21) is a non-spontaneous reaction, requiring an external source of power. $\Delta S < 0$ because the compression process reduces distances between gaseous molecules and contributes to increase order. This is a main difference with water electrolysis. When $\Delta P = P^{HP} - P^{LP}$ is less than 200 bars, $\Delta H \approx 0 => \Delta G = -T\Delta S. \Delta H < \Delta G$ over the entire pressure range. Therefore (Figure 10.15), the thermodynamic voltage E is larger than the enthalpy (or thermoneutral) voltage V (V is the voltage at which no heat is released by the compression cell).

10.3.2.1 Kinetics

Would reactions (10.19) and (10.20) take place at low current density (close to the thermodynamic voltage), small amounts of electric power would be needed to pressurize hydrogen. However, high current densities are required in practical situation to reduce investment costs. Hence, increasing amounts of electric power

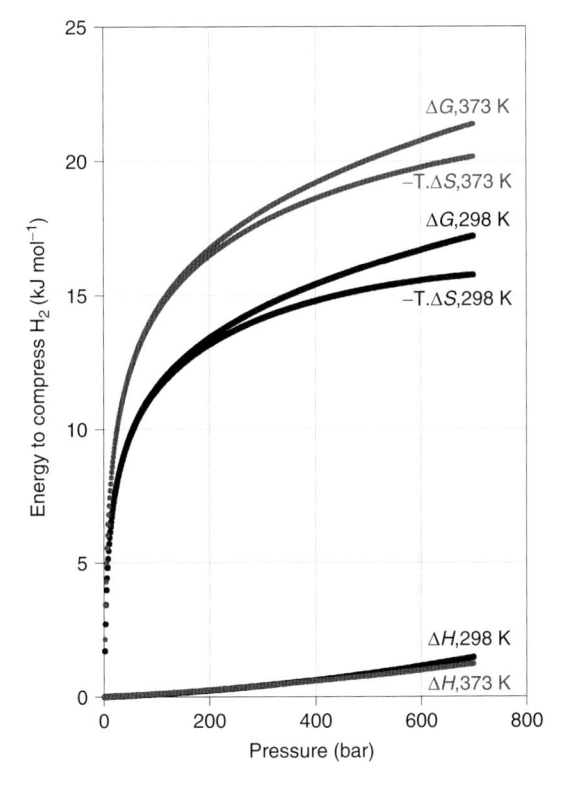

Figure 10.14 ΔH, ΔG, and T.ΔS associated with the electrochemical compression of H_2 at two different temperatures.

are required with increasing current density to overcome electrode overvoltages associated with reactions (10.19) and (10.20) and the resistance of the SPE. Since the kinetics of interfacial charge transfer processes at platinum–SPE interfaces is high, the electric power consumption of the cell is predominantly due to the ohmic drop inside the SPE during operation, as discussed below.

10.3.3
Technical Developments

The photograph of a laboratory-scale hydrogen monocell compressor is given in Figure 10.16. Similar to SPE, the MEA is also made of Nafion-117 (0.183 mm thick; equivalent weight of dry membrane = 1.01–1.03 meq g^{-1}). Carbon-supported platinum catalyst (0.4–0.8 mg cm^{-2} of Pt) is used at both anode and cathode. PTFE (10 wt%) is added to the anodic electrocatalyst for purposes of increasing the hydrophobic property of the electrode and improving the kinetics of hydrogen oxidation. Porous titanium sheets (0.94 mm thick and porosity of 30%) are used as cathode current collectors. Carbon paper of Sigracet 10bb (SGL Carbon Group)

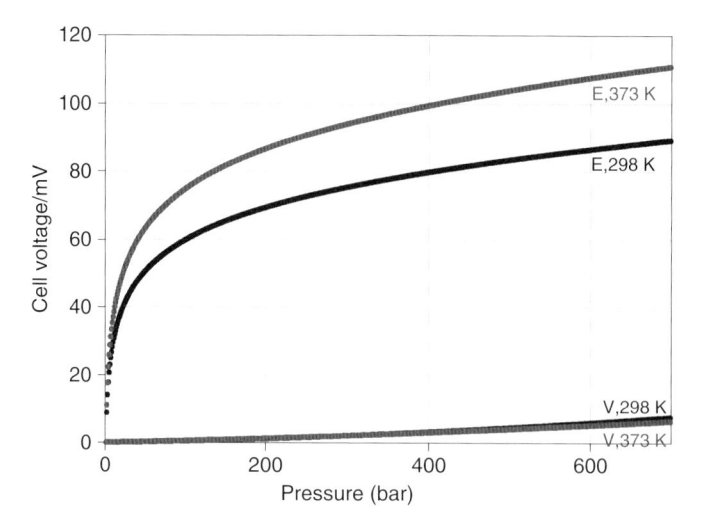

Figure 10.15 V and E associated with the electrochemical compression of H_2 at two different temperatures.

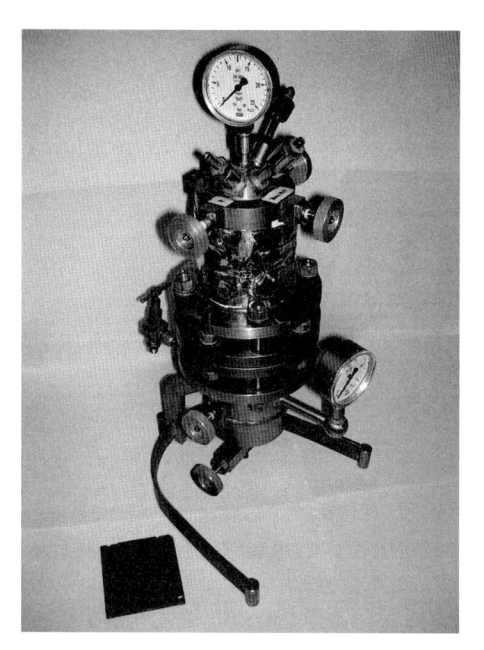

Figure 10.16 Photograph of a labscale hydrogen compressor. (Courtesy of Kurchatov Institute, Moscow, Russian Federation).

with a microporous sublayer (overall porosity of 84% and 320 μm thick) is used as an anode current collector.

The process flow sheet is diagrammed in Figure 10.17. To carry out the tests, hydrogen of electrolytic grade is produced by a PEM water electrolyzer (E) connected to an oxygen–water liquid–gas separator (C1) and a hydrogen–water liquid–gas

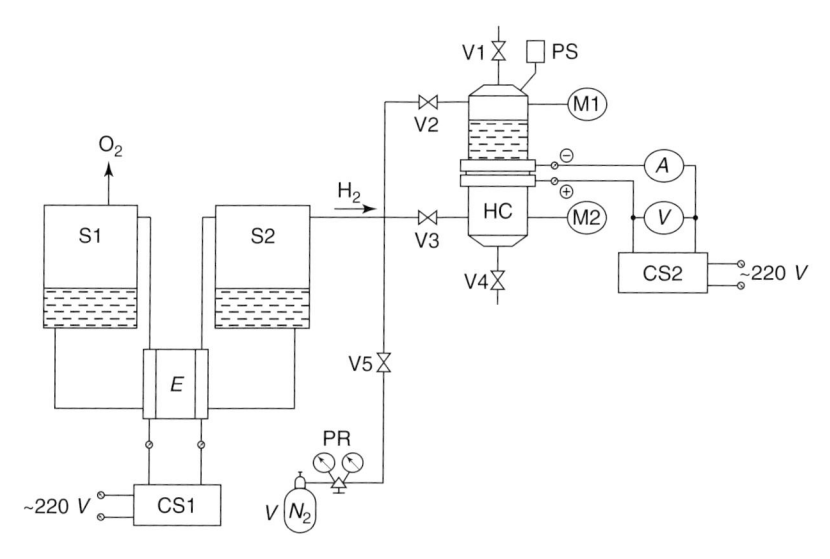

Figure 10.17 Process flow sheet.

separator (C2). An electric power supply (CS1) is used as power source. The electrochemical hydrogen compressor/concentrator (HC) and its holder are connected to a power supply (CS2) equipped for the measurement of current (A) and cell voltage (V). Different valves (V1 ... V5), manometers (M1, M2), pressure reducers (PRs), pressure sensors (PSs), and a pressurized nitrogen reservoir (V) are used as ancillary equipment. A data acquisition unit is used for remote control of the experiments and to make the measurements.

10.3.4
Electrochemical Performances

Current–voltage polarization curves measured for three different values of relative humidity (R_H) of the hydrogen–water feed mixture are plotted in Figure 10.18. R_H values were set by bubbling hydrogen into thermostated water located in the admission chamber. Therefore, $I - V$ measurements were made at different temperatures. Results obtained for $\Delta P = 0$ bar are plotted in Figure 10.18.

$I - V$ plots are mostly linear in shape at low current densities whereas significant mass-transport limitations appear at high current densities. The contribution of the different cell components to the total cell voltage can be estimated as follows (data are compiled in Table 10.5). According to Eq. (10.4), the Nernstian voltage of reaction (3) is 0 V when $\Delta P = 0$ bar. Therefore, the increase in the cell voltage with current density observed in Figure 10.18 is only due to polarization terms (hydrogen oxidation overvoltage at the anode and hydrogen reduction overvoltage at the cathode) and ohmic drop across the SPE. The slopes (dV/di in Ω cm^2) of the curves in Figure 10.18 are a measure of these different contributions (Table 10.5).

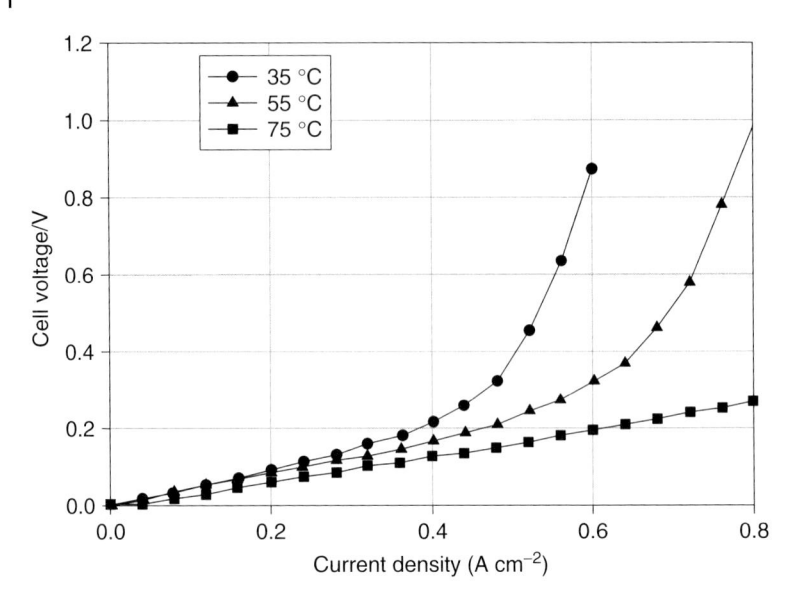

Figure 10.18 I–V characteristics measured on the compression cell for three different temperatures of the thermostat: $\Delta P = 0$ bar.

The ionic conductivity of perfluorinated materials varies with temperature. Therefore, the ohmic drop ρ_M in $\Omega\ cm^2$ across such membranes is given as the product of the membrane resistivity in $\Omega\ cm$ [11] and the membrane thickness in cm ($\approx 200\ \mu m$ for a fully hydrated Nafion 117 membrane). The algebraic difference between dV/di and ρ_M gives the ohmic drop associated with the anodic (η_a) and cathodic (η_c) overvoltages. As can be seen from the data of Table 10.5, the contribution of the SPE resistance to the total cell voltage is about 40% and the contribution of charge transfer processes is about 60%. Mass transport limitations observed at higher current density are probably due to a local dehydration of the SPE, particularly in the anodic region, where water molecules are required to insure

Table 10.5 dV/di = slopes of the experimental curves of Figure 10.18; ρ_M = Nafion 117 resistance; $i(\eta_a + \eta_c)$ = charge transfer resistance.

T (°C)	(a) 35 °C	(b) 55 °C	(c) 75 °C
$dV/di\ (\Omega\ cm^2)$	0.5	0.4	0.35
$\rho_M\ (\Omega\ cm^2)$	0.18	0.16	0.14
$i(\eta_a + \eta_c)\ (\Omega\ cm^2)$	0.32	0.24	0.21
$\rho_M/dV/di\ (\%)$	36	40	40
$i(\eta_a + \eta_c)/dV/di\ (\%)$	64	60	60

the transport of hydrated proton across the membrane up to the cathode (so-called electro-osmosis drag, similar to PEM water electrolysis and PEM fuel cells). This interpretation is consistent with experimental observations: as can be seen from Figure 10.18, the value of the current density at which a sharp increase in the cell voltage is measured depends markedly on the water content of the hydrogen–water feed gas mixture.

For practical applications, it is necessary to find a compromise between investment cost (the investment cost can be reduced by increasing the operating current density) and the energy cost (the energy cost can be reduced by decreasing the operating current density). A current density in the 0.5–1 A cm^{-2} range is commonly used in PEM water electrolysis and fuel cell technologies. This situation is similar to a PEM compressor. To avoid undesirable mass-transport limitations and therefore to extend the range of operating current density, the water content of the hydrogen–water gas mixture must be adjusted by monitoring the temperature of water in the admission chamber, in order to match the amount of water required by the electro-osmotic drag for each operating current density. Some results obtained at a constant current density of 0.2 and 0.5 A cm^{-2} are plotted in Figure 10.19 [12]. Up to ≈48 bars, stable and stationary pressure values have been measured on the exhaust side of the compressor. At larger pressures (up to 150 bars), it becomes difficult to accurately measure output pressure because of the increasing leak rates.

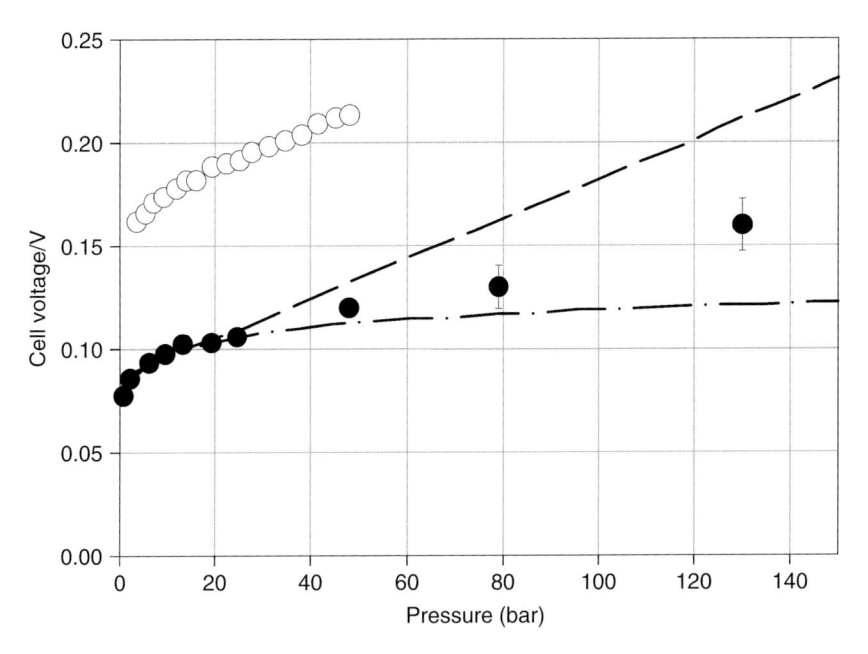

Figure 10.19 Experimental cell voltage vs. pressure at (●)$i = 0.2$ A cm^{-2}; (o)$i = 0.5$ A cm^{-2}.

At $i = 0.2$ A cm^{-2}, the energy consumption for pressurizing hydrogen at 48 bars was determined to be ≈ 0.26 kWh N m^{-3} (i.e., $\approx 9\%$ of the low heating value (LHV) of hydrogen which amounts to 3.0 kWh N m^{-3}). It is expected that, when hydrogen leaks are reduced (e.g., using a pressurization vessel), PEM technology can be used to deliver hydrogen at much higher pressures. Results obtained on water electrolysis at pressure up to 150 bars have been recently reported [13]. At such elevated pressures, cross-permeation effects are still acceptable. Therefore, electrochemical compression can potentially compete with mechanical compression, at least in terms of output pressure. To estimate the energy consumption of the electrochemical compression at 150 bars ($i = 0.2$ A cm^{-2}), experimental data of Figure 10.19 obtained in the limited 1–50 bar pressure range have been extrapolated. Two limiting extrapolation cases are plotted in dotted lines in Figure 10.19: on top, a linear relationship (less favorable case) and at bottom, a logarithmic relationship (most favorable case). It can be assumed that the true value of the voltage–pressure relationship lies somewhere in the interval bracketed by these two extrapolation lines. Data-points obtained at ≈ 80 and ≈ 130 bars are in agreement with this assumption. The upper cell voltage required for compressing H_2 at 150 bars can be estimated from the linear extrapolation. A value of circa 0.24 V is obtained, corresponding to a power consumption of circa 0.57 kWh N m^{-3} (i.e., $\approx 19\%$ of the LHV of hydrogen). The lower cell voltage required for compressing H_2 at 150 bars can be estimated from the logarithmic extrapolation. A value of circa 0.13 V is obtained, corresponding to a power consumption of circa 0.31 kWh N m^{-3} (i.e., $\approx 10\%$ of the LHV of hydrogen). That means that the energy required for the compression of hydrogen up to 150 bars at a current density of 0.2 A cm^{-2} lies somewhere between 10 and 20% of the LHV of hydrogen. However, the investment cost for a device operating at 0.2 A cm^{-2} may be prohibitive. Since the compression efficiency is inversely proportional to the current density, a detailed cost analysis of PEM electrochemical compressors and mechanical compressors is still required to compare the two technologies. In practical cases, to reach pressures higher than 50 bars, a possible alternative is the interconnection of several individual PEM hydrogen compressors [12] or to use different SPEs with low gas permeability.

10.3.5
Current Limitations and Perspectives

The energy required for the compression of hydrogen as a function of pressure, obtained for four different hydrogen compression processes, is plotted in Figure 10.20. Electrochemical hydrogen compression is the most effective process. However, the technological maturity of electrochemical compressors is far from being the same as mechanical compressors. Such systems are seldom found in the industry – only prototypes being tested at the laboratory scale. Demonstration has been made up to 150 bars. Investment costs and durability issues are still limiting factors requiring additional R&D.

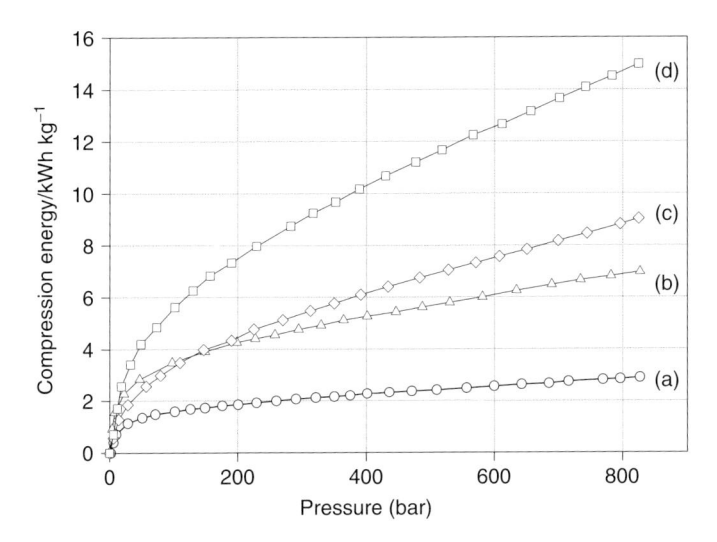

Figure 10.20 Comparison of different hydrogen compression processes. (a) Electrochemical (isothermal); (b) 3 stages (60%); (c) adiabatic; (d) adiabatic (60%).

10.4
Hydrogen Electrochemical Extraction and Purification

10.4.1
Principles

10.4.1.1 Concentration Cell

Electrochemical hydrogen extraction (purification) from a gas mixture is a process very close to hydrogen compression described in the previous section. The two processes can also be combined into a single one and the same device can deliver purified and compressed hydrogen. Electrochemical hydrogen purification is mostly based on PEM technology although high temperature processes are also possible in principle. A schematic diagram of a unit PEM cell is given in Figure 10.21. It also operates like an electrolysis cell because hydrogen purification is a non-spontaneous process ($\Delta G > 0$). When electric power is supplied to the cell, diluted hydrogen is transferred from the left anodic compartment to the HP cathodic compartment (right side). Electrochemical reactions taking place in this process at the electrode–SPE interfaces are those of Eqs. (10.19) and (10.20). Equation (10.21) is the overall cell reaction:

$$\text{Anode:} \quad H_2 \text{ (dilute, LP)} \rightarrow 2H^+ + 2e^- \quad (10.29)$$

$$\text{Cathode:} \quad 2H^+ + 2e^- \rightarrow H_2 \text{ (HP)} \quad (10.30)$$

$$\text{Overall reaction:} \quad H_2 \text{ (dilute, LP)} \rightarrow H_2 \text{ (HP)} \quad (10.31)$$

As for the compressor, the hydrogen concentrator has no moving parts and is very compact. MEAs can be stacked to form a multistage concentrator.

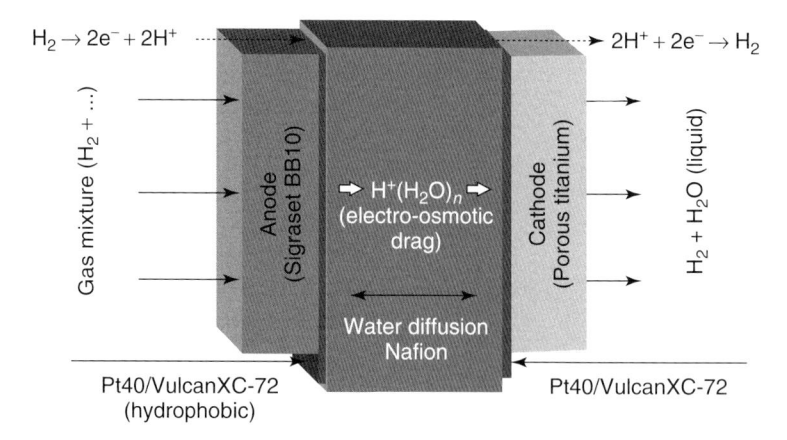

Figure 10.21 Schematic diagram of a PEM hydrogen concentration cell showing MEA components.

10.4.1.2 Thermodynamics

The thermodynamics of hydrogen extraction and purification is the same as that in the hydrogen compressor (Figures 10.14 and 10.15).

10.4.1.3 Kinetics

As discussed in the previous section for the hydrogen compressor, the kinetics of hydrogen oxidation at the anode and the kinetics of proton reduction at the cathode are fast. The main kinetic limitation in the hydrogen concentration cell comes from mass-transport limitation in the feed compartment. An additional problem comes from the possible deleterious effect of poisonous chemicals present in the feed mixture, which can alter the noble metal catalyst used as the anode for the hydrogen oxidation reaction (HOR).

10.4.2
Technical Developments

Such devices are seldom found in the industry. Only prototypes have been tested at the laboratory scale. A potential application of electrochemical hydrogen concentrator has been evaluated in the course of R&D programs supported by the European Commission. In the NaturalHy project (FP6), the possibility of incorporating significant amounts of green hydrogen into the pipes of natural gas extracted from North-African regions and distributed to Euro-Mediterranean countries (rising its concentration up to 20%) has been evaluated. The idea was that hydrogen-enriched gas could still be burned for heating purposes, thus reducing emissions of carbon dioxide, but could also serve as a large-scale hydrogen source for electricity generation with PEM fuel cells. The performances of electrochemical concentrators have been compared to those of classical gas permeators using metallic membranes [14].

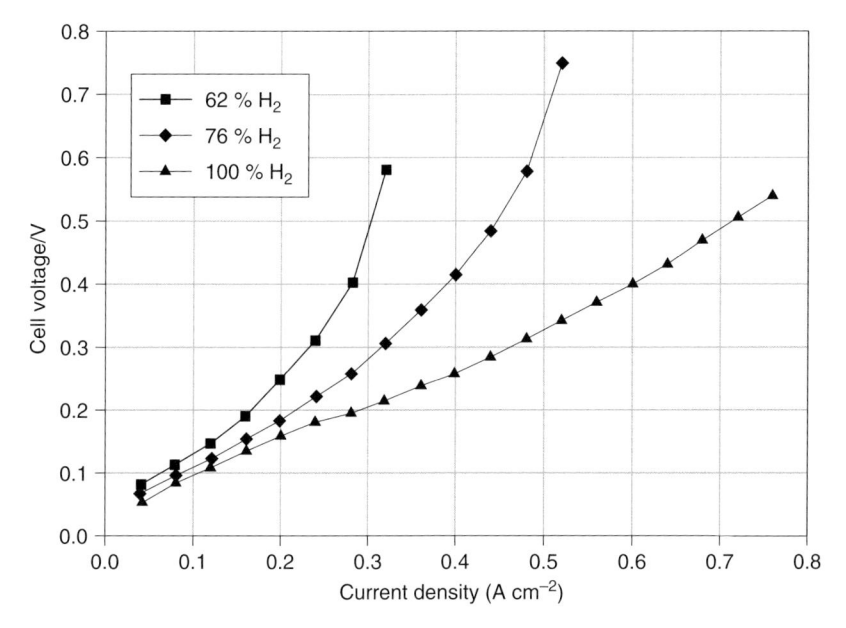

Figure 10.22 Current-voltage performances measured on the concentration cell for different compositions of anodic gas (H_2 concentrations are expressed in vol%).

10.4.3
Electrochemical Performances

Results obtained with different hydrogen–nitrogen gas mixtures are plotted in Figure 10.22. As the hydrogen content of the mixture decreases, increasing mass transport limitations appear and the compression efficiency decreases. Diffusion- or convection-controlled processes such as hydrogen transport in the gas phase through the gas diffusion electrode to the anode take place and introduce strong rate limitations. As a consequence, there is an important increase in energy consumption and as expected, it can be concluded that the system cannot be used to extract hydrogen efficiently from diluted sources. As the hydrogen concentration in the feed gas mixture decreases, the current density of the electrochemical concentrator must be reduced to match mass-transport limitations and avoid significantly low concentration/compression efficiencies.

10.4.4
Current Limitations and Perspectives

Electrochemical extraction of hydrogen from diluted sources is potentially an alternative to hydrogen extraction based on permeation across metallic membranes. In both processes, the kinetic limitations are similar: hydrogen mass transport to the active surface is such that, even when forced convection techniques are used to improve the situation, process efficiencies are low. The techniques remain interesting when hydrogen must be extracted from sufficiently concentrated gas sources.

10.5
Hydrogen Storage in Hydride-Forming Materials

There are different processes for storing hydrogen: (i) pressurization at pressures up to several hundred bars and storage in pressurized containers; (ii) liquefaction at very low temperatures (20 K); (iii) adsorption on various substrate of large surface area, under the combined effect of HP and low temperature; (iv) chemical hydriding; and (v) formation of reversible metal hydrides, using either chemical (gas phase) or electrochemical processes. The discussion in this section is restricted to the formation of reversible metal hydrides.

10.5.1
Principles

There is a large diversity of metallic materials capable of forming reversible metal hydrides with hydrogen and hence contribute to the storage of hydrogen and other miscellaneous applications (detection, purification, thermal cycles, etc.). Among elemental metals, palladium is probably the most-cited and most-studied one, although others such as zirconium, titanium, and so on present similar properties. In practical applications, a large variety of alloys and intermetallic compounds (IMCs) are used. In this section, the simple $Pd-H_2(g)$ system is used as a model system to introduce the basic principles of hydriding reactions.

10.5.1.1 Thermodynamics

Hydrogen insertion into palladium (and most hydride-forming metals) is a reversible process. Hydrogen can be inserted directly from the gas phase (chemical insertion obtained by exposing the metal to an atmosphere of pure or even diluted hydrogen) or from an electrolytic solution, by reducing the solvent or the electrolyte at the surface of the metal (electrochemical insertion). The thermodynamic features of both techniques are very close. Two isotherms (a chemical isotherm on top measured at 25 °C and an electrochemical isotherm at bottom measured at 60 °C) on the same 150 μm thick Pd foil (99.95%, Good Fellow Co.) are plotted in Figure 10.23.

A first characteristic of metal hydrides comes from the shape of the hydrogen insertion isotherms. Isotherms are plots of pressure (electrode potential) as a function of hydrogen content in the hydride-forming host material, often expressed as the dimensionless factor H/M which is the ratio of the mole number of hydrogen to the mole number of metal (or unit metal cell). At low hydrogen content ($0 < H/M < 0.02$), a solid solution (αPdH) of interstitial hydrogen in palladium is formed. According to Gibbs' rule of phase, the system is divariant: as

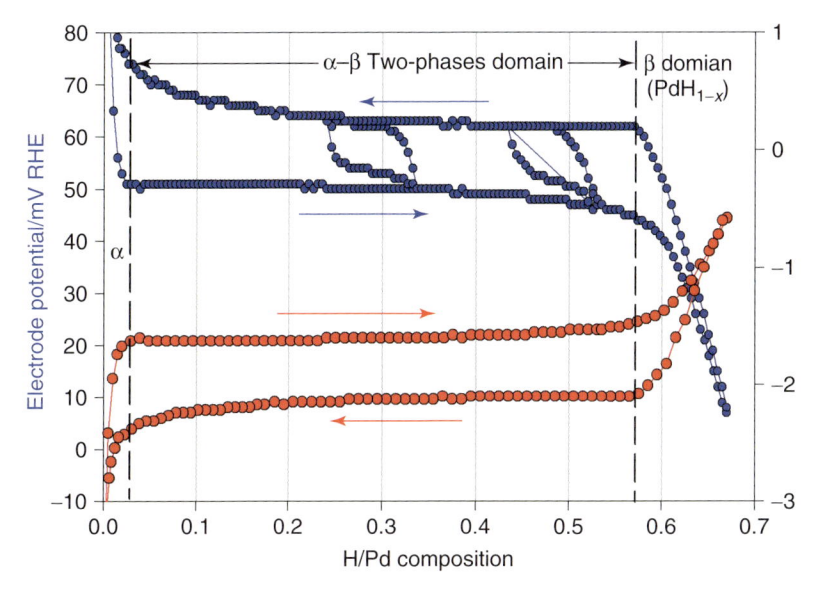

Figure 10.23 Chemical (blue •) and electrochemical (red •) isotherms measured on the same 150 μm-thick Pd foil. $T = 298$ K; electrolyte is H_2SO_4 0.1 M.

the hydrogen concentration increases, the equilibrium pressure (or potential) varies. When the maximum solubility of hydrogen in the metal is reached, an irreversible phase transformation takes place, leading to the precipitation of a hydride phase. In the case of palladium, at 298, this hydride phase is the substoichiometric $PdH_{0.6}$ phase called βPdH. Since a new phase appears, according to Gibbs' rule of phase the system becomes monovariant and the pressure (potential) becomes independent of composition. A flat pressure (potential) plateau develops over an extended composition range. In this intermediate composition range, both saturated αPdH and βPdH phases coexist. The limit of this two-phase domain depends very markedly on the nature of the metallic host. At high hydrogen content $(H/M > 0.6)$, a solid solution of hydrogen in the sub-stoichiometric hydride phase is formed. The system becomes divariant and the pressure (potential) varies again.

A second characteristic of metal-$H_2(g)$ systems is the presence of a significantly large hysteresis observed between the absorption and desorption pressure plateaus. For Pd at 298 K, the difference is circa 10 mbar (15 mV).

10.5.1.2 Insertion Mechanisms

Chemical Insertion Hydrogen insertion in metals is a multistep process involving surface and bulk steps. Chemical and electrochemical insertions only differ by the surface step. The different steps of the gas-phase insertion are as follows:

1) Non-dissociative surface physisorption: the molecule of di-hydrogen adsorbs weakly at the surface of the metal.

$$M + H_2 \leftrightarrow M\text{-}H_2 \tag{10.32}$$

2) Surface dissociative chemisorption leading to the formation of surface hydrogen ad-atoms.

$$2M\text{-}H_2 \leftrightarrow 2M\text{-}H_{ad} \tag{10.33}$$

3) Surface absorption, a process during which surface-adsorbed hydrogen ad-atoms are absorbed by the metal, leading to the formation of subsurface ad-atoms.

$$H_{ad} \leftrightarrow H_{ab} \tag{10.34}$$

4) Transport from subsurface to bulk regions by diffusion:

$$H_{ab}^{sub} \leftrightarrow H_{ab}^{bulk} \tag{10.35}$$

5) Phase transformation leading to the formation of hydride:

$$H_{ab}^{bulk} \rightarrow M_x H_y \tag{10.36}$$

A quantitative expression for the isotherms in the solid solution domain can be derived as follows. When the metal is at equilibrium with the gas phase, the chemical potential of hydrogen is the same in both phases:

$$\mu_H(ab) = \mu_H(H_2) \tag{10.37}$$

The chemical potential of hydrogen in the metal (M = Pd) is

$$\mu_H(ab) = \mu_H^0 + R_{PG} T \ln\left[\frac{H/Pd}{(1 - H/Pd)}\right] + \Delta\mu_H \tag{10.38}$$

where μ_H^0 is the standard chemical potential of hydrogen and H/Pd is the concentration of hydrogen. $\mu_H(ab) = \mu_H^0 + R_{PG} T \ln[(H/Pd)/(1 - (H/Pd))]$ is the chemical potential of hydrogen in the metal host when an ideal solid solution (no interaction between H atoms or with the metallic matrix) is formed. The chemical potential of hydrogen in the gas phase is

$$\mu_H(H_2) = \frac{1}{2}\mu_{H_2} = \frac{1}{2}\left(\mu_{H_2}^0 + R_{PG} T \ln p_{H_2}\right) \tag{10.39}$$

where $\mu_{H_2}^0$ is the standard chemical potential of H_2 and p_{H_2} is the partial pressure of hydrogen at equilibrium. From Eqs. 10.37–10.39, it follows that

$$K = \exp\left[\left(\frac{\mu_H^0 - \frac{1}{2}\mu_{H_2}^0}{R_{PG} T}\right)\right] \tag{10.40}$$

K is the Sieverts' constant used to compare hydrogen solubility in different metals. This is a function of temperature and material. From there, the relationship between the partial pressure of hydrogen and the hydrogen content H/M can be derived for solid solutions:

$$\text{Ln}\left(\sqrt{p_{H_2}}\right) = \ln\left[K\left(\frac{H/Pd}{1 - H/Pd}\right)\right] + \frac{\Delta\mu_H}{R_{PG}T} \quad (10.41)$$

$\Delta\mu_H/R_{PG}T$ is negligible for ideal solid solutions. This condition is satisfied for real solid solutions at very low hydrogen contents. In such cases, Eq. (10.41) simplifies into the so-called Sieverts relationship:

$$\sqrt{p_{H_2}} = K\frac{H/Pd}{(1 - H/Pd)} \quad (10.42)$$

For H/Pd $\ll 1$, Eq. (10.42) simplifies into

$$\sqrt{p_{H_2}} = K \times H/Pd \quad \text{because H/Pd} \ll 1 \quad (10.43)$$

Electrochemical Insertion The mechanism of the electrochemical hydrogen insertion differs from the chemical insertion by the surface step. Two different surface steps are evoked in the literature:

$$\text{Volmer adsorption step: } H^+ + e^- \leftrightarrow H_{ad} \quad (10.44)$$

$$\text{Frumkin absorption step: } H_{ad} \leftrightarrow H_{ab} \quad (10.45)$$

Equation (10.45) is also used to describe the direct insertion of hydrogen in one step, without surface adsorption:

$$H^+ + e^- \leftrightarrow H_{ab} \quad (10.46)$$

Relationship Between Chemical and Electrochemical Isotherms A quantitative relationship between the chemical and the electrochemical isotherms has been proposed by Lewis [15]. In the case of electrochemical insertion (Eq. (10.46)), the equilibrium condition writes

$$\tilde{\mu}_{H^+} + \tilde{\mu}_{e^-} = \tilde{\mu}_H(ab) \quad (10.47)$$

$$\mu_{H^+} + F\varphi_{sol} + \mu_{e^-} - F\varphi_{Pd} = \mu_H(ab) \quad (10.48)$$

where φ_{Pd} and φ_{sol} are the Galvani potential of the metal electrode and F is the Faraday constant (96 485 C mol^{-1}). Since $E = \varphi_{Pd} - \varphi_{sol}$:

$$\mu_{H^+} + \mu_{e^-} - FE = \mu_H(ab) = \frac{1}{2}\left[\mu_{H_2}^0 + RT\ln\left(\frac{p_{H_2}}{p^0}\right)\right] \quad (10.49)$$

The relationship between the chemical and the electrochemical isotherms is derived from Eq. (10.49):

$$E = Cste - \frac{R_{PG}T}{2F}\ln\left(\frac{p_{H_2}}{p^0}\right) \quad (10.50)$$

where $Cste = \frac{1}{2F}\left(2\mu_{H^+} + 2\mu_{e^-} - \mu_{H_2}^0\right)$ is a constant.

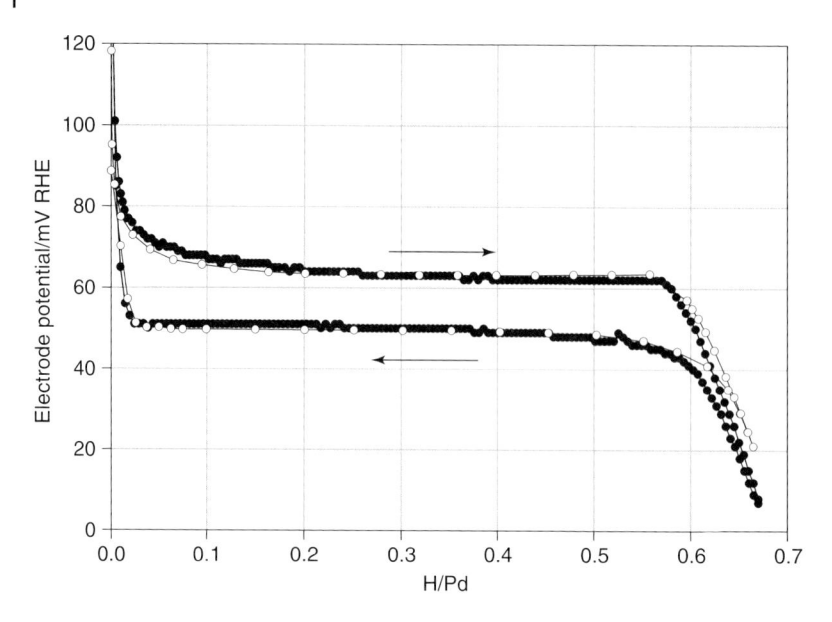

Figure 10.24 (■) Electrochemical isotherm from Figure 10.23; (○) chemical isotherm from Figure 10.23 converted using Eq. (10.51).

When potentials are expressed with regard to the normal hydrogen electrode (NHE), the following relationship is finally obtained:

$$E = -\frac{R_{PG}T}{2F}\ln\left(\frac{p_{H_2}}{P^0}\right) \qquad (10.51)$$

where $P^0 = 1$ bar.

The chemical and electrochemical isotherms of Figure 10.23 both satisfy Eq. (10.51). The chemical isotherm of Figure 10.23 and the electrochemical isotherm of Figure 10.23 converted into a chemical isotherm using Eq. (10.51) are both plotted in Figure 10.24. The electrochemical isotherm of Figure 10.23 and the chemical isotherm of Figure 10.23 converted into an electrochemical isotherm using Eq. (10.51) are both plotted in Figure 10.25. Results support the validity of Eq. (10.51) and demonstrate the fact that electrochemical and chemical isotherms are similar.

10.5.1.3 Hydriding Kinetics

As described in the previous subsection, the hydriding reaction of metals, alloys, and IMC is a multistep process. A proper analysis of the hydriding reaction requires the identification of the different steps involved in the global reaction mechanism and the determination of individual rate parameters. Concerning the electrochemical process, such information can be obtained using electrochemical impedance spectroscopy (EIS). The concept has been extended to solid–gas reactions to measure separately surface and bulk rate contributions [16]. Typical impedance

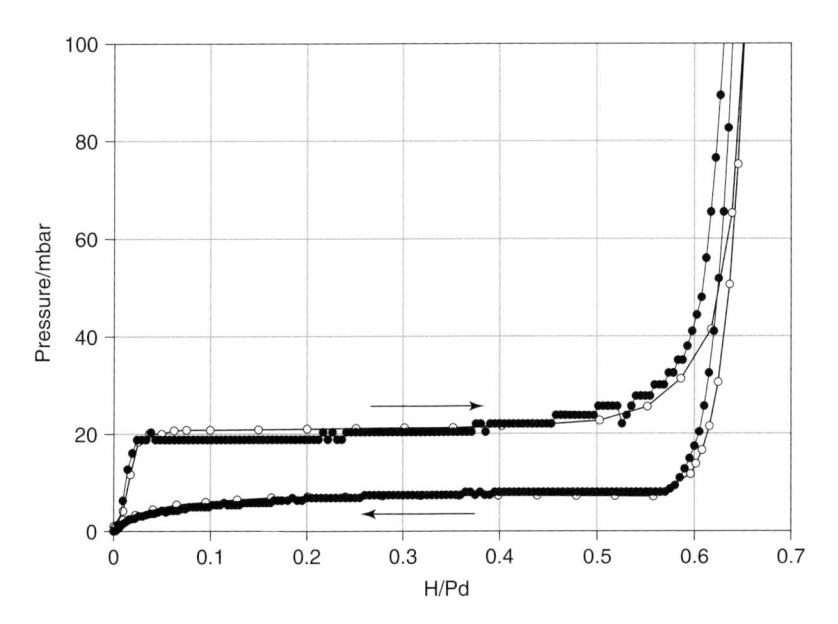

Figure 10.25 (■) Chemical isotherm from Figure 10.23 ; (○) electrochemical isotherm from Figure 10.23 converted using Eq. (10.51).

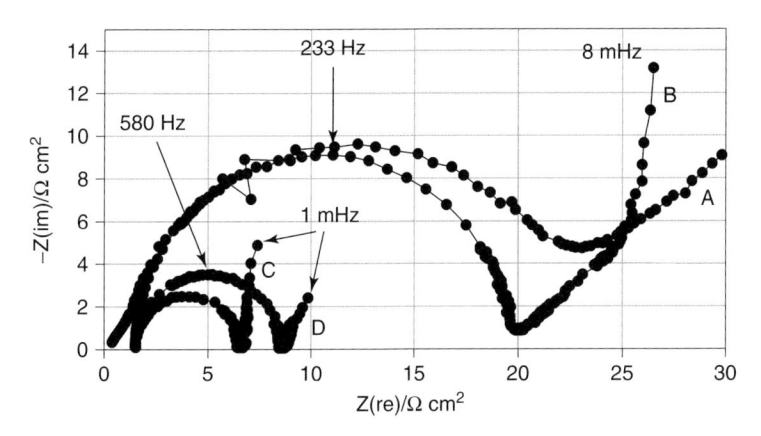

Figure 10.26 Experimental electrochemical impedance diagrams measured in the two-phase domain of the PdH system during absorption at A: $H/Pd = 0.01$; B: $H/Pd = 0.02$, C: $H/Pd = 0.05$; D: $H/Pd = 0.2$.

diagrams measured in the two-phase domain of the PdH system are plotted in Figure 10.26 (chemical insertion) and Figure 10.27 (electrochemical insertion).

In Figure 10.26, spectra C and D are measured in the two-phase domain. Spectra A and B obtained at $H/Pd < 0.05$ (single-phase domain) are also plotted for comparison. The difference between single- and two-phase domains is that in

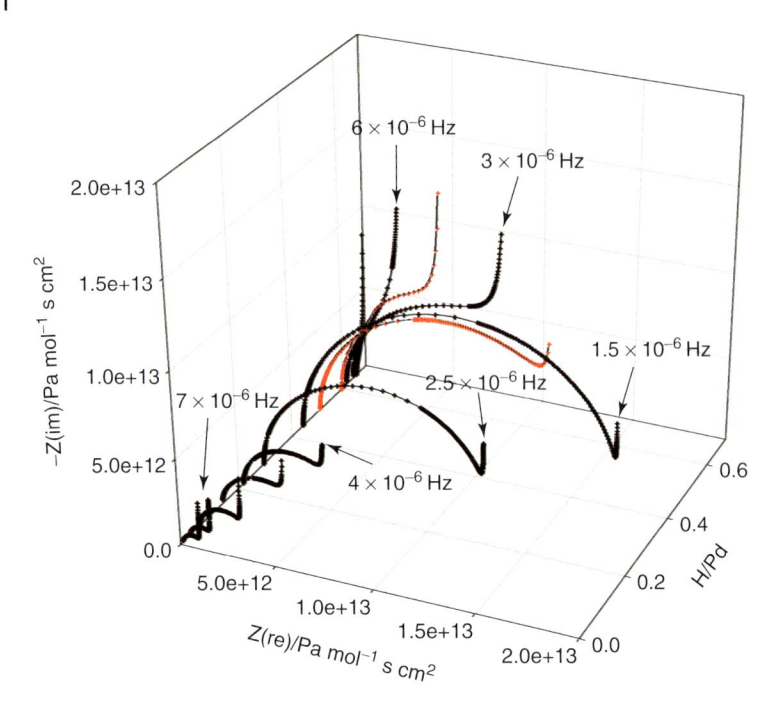

Figure 10.27 Experimental chemical impedance diagrams measured at 298 K during absorption (black) and desorption (red) along the isotherm of the PdH system.

the second case, the resistance associated with the phase transformation process is added to the surface resistance. As can be seen from spectra C and D of Figure 10.26, there is a slight tendency for the high frequency (HF) semicircle to increase with H/Pd. Also, the diffusion behavior observed in the medium frequency (MF) region in the solid solution domain (spectra A and B) disappears. At low frequencies, spectra C and D show that there is an additional tendency to form a second semicircle, which can be attributed to the parasite HER. Because of this parasite reaction of hydrogen evolution, it is difficult to analyze the phase transformation reaction during the electrochemical hydriding of the Pd foil. During gas-phase experiments, there is no parasite HER (Figure 10.27). Compared to the spectra measured in the single-phase domain, HF and low frequency (LF) contributions are well separated. The increasing value of the HF semicircle with composition is associated with the phase transformation step. Another observation that can be made from Figure 10.27 is that the diameter of the HF semicircle increases with composition, suggesting that the rate of phase transformation step decreases with composition. Since we are using a metal foil, this may be related to the precipitation of β-PdH, which induces strain and stresses inside the foil and makes the precipitation more difficult in terms of phase accommodation in the metallic matrix. The same tendency was observed during electrochemical experiments (Figure 10.26) but because of the parasite

H$_2$ evolution reaction placed in parallel with direct insertion, results were not so convincing. However, during desorption, the diameter of the HF semicircle increases up to values similar to those measured during absorption, suggesting that strains inside the foil do not facilitate the decomposition of the hydride phase through step.

10.5.2
Electrochemical Storage (see Ni-MH batteries Chapter 5)

10.5.3
Gas-Phase Storage

10.5.3.1 Selection of Storage Materials

A major characteristic of metal hydride systems is that both absorption and desorption plateau pressures vary with the operating temperature. The value of the plateau pressure is related to operating temperature through the van't Hoff relationship (Eq. (10.52)) which is derived as follows. Considering the hydriding reaction:

$$xM + H_2 \rightarrow xMH_{2/x}$$

At equilibrium, the change in free Gibbs energy is zero:

$$G_{MH} - G_{H_2} = \Delta G = 0 = G^0_{MH} - G^0_{H_2} - R_{PG} T \ln\left(\frac{p}{p_0}\right)$$

It follows that

$$\Delta G^0 = R_{PG} T \ln\left(\frac{p}{p_0}\right) = \Delta H^0 - T\Delta S^0 \Leftrightarrow \ln\left(\frac{p}{p_0}\right) = \frac{\Delta H^0}{R}\frac{1}{T} - \frac{\Delta S^0}{R_{PG}}$$

(10.52)

where P_{H_2} is the plateau pressure, P^0 the standard pressure, R the gas constant, T the temperature, and ΔH and ΔS the enthalpy and entropy changes of desorption for Pd-H. Between two points on the van't Hoff plot, Eq. (10.52) can be written as

$$\ln\left(\frac{P_2}{P_1}\right) = \frac{-\Delta H}{R_{PG}}\left[\frac{1}{T_2} - \frac{1}{T_1}\right]$$

(10.53)

Equation (10.53) can be used to evaluate the equilibrium pressure of the system in two-phase domains as a function of temperature. A schematic diagram showing the relationship between isotherms and van't Hoff plots is given in Figure 10.28 and experimental van't Hoff plots of different metal–hydrogen systems are plotted in Figure 10.29. Ideally, hydrides should form/decompose at atmospheric temperature and pressure. AB$_5$ such as LaNi$_5$ or MmNi$_5$ and FeTi compounds can be used in these temperature and pressure domains.

10.5.3.2 Hydride Reactor Technology

There are a large number of patented hydride reactor designs. A typical example is given in Figure 10.30. A first problem is caused by possible mass-transfer limitation

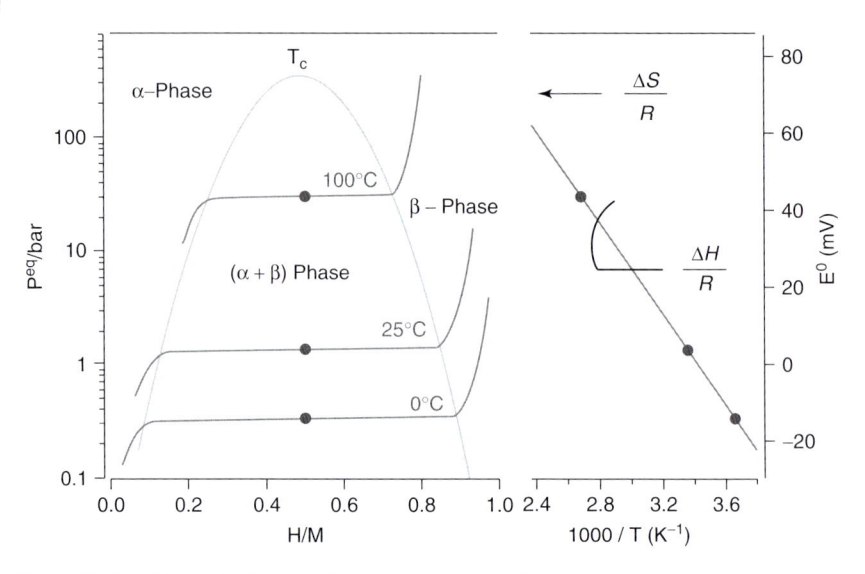

Figure 10.28 Schematic diagram showing the relationship between isotherms and Van't Hoff plots [17].

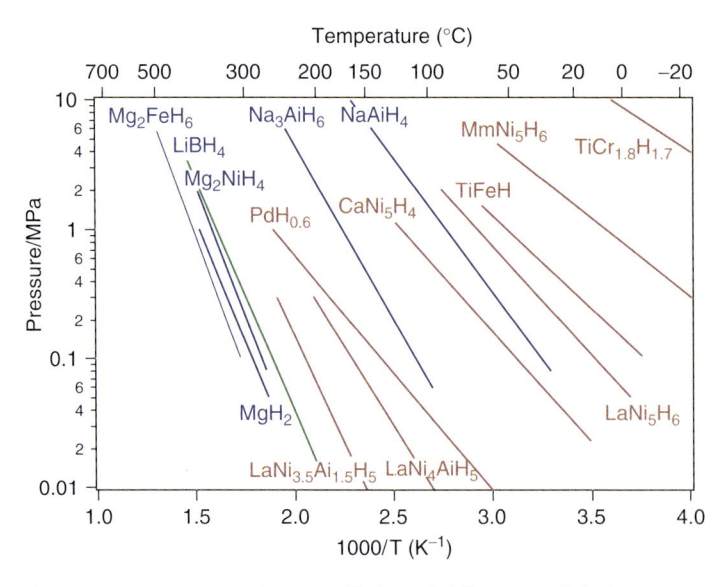

Figure 10.29 Experimental van't Hoff plots of different metal–hydrogen systems [17].

in the powdered hydride bed. This is usually dealt with by disposing the metallic powder in separate compartments and providing a direct access of hydrogen to the different compartment cells. A second problem is caused by the management of the heat of hydride formation. To insure isothermal conditions of operation and avoid heat transfer problems to the surroundings, the hydride reactors are placed

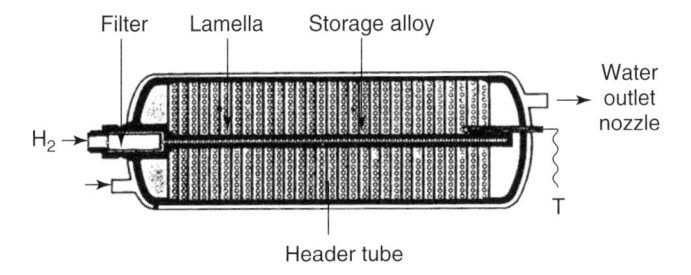

Figure 10.30 Diagram of a hydrogen storage reactor (source Siemens).

Figure 10.31 Photograph of a hydrogen reactor. (source Siemens).

in thermostated baths. Several hydride reactors can be stacked to adjust the storage capacity to miscellaneous stationary end-uses (Figure 10.31).

10.5.4
Current Limitations and Perspectives

Hydrogen storage by the formation of metal hydrides is appropriate for stationary applications but not for mobile applications. On-board storage of hydrogen in the automotive still requires higher energy densities (kJ kg^{-1}). A lot of R&D has been performed over the past decades to find alternative solutions. Most successes have been obtained using chemical hydrides. Some of them offer energy densities higher than 8%, but unfortunately chemical hydriding is not a reversible process. The perspectives of finding light metal hydrides remain limited. Mg-based compounds still offer some interesting prospective and R&D efforts are being made to reduce the operating temperature and manage corrosion issues.

10.6
Conclusion and Perspectives

Among various synthetic fuels, pure hydrogen is considered as a potential energy carrier for the storage of renewable energy sources. Hydrogen transportation, purification, and storage require compressed gas at pressures up to several hundred bars. A large variety of electrochemical processes can be used for pressurizing, purifying, and storing hydrogen. Direct pressurized water electrolysis remains a highly efficient process for producing hydrogen from water sources. Alternatively, hydrogen compression can be realized separately using electrochemical hydrogen compressors. Hydrogen can be purified electrochemically and stored in hydride-forming materials. However, hydrogen storage remains a bottleneck process on the way to the so-called "hydrogen economy," in particular for automotive applications. This is mostly due to the fact that hydrogen remains gaseous with temperature and pressure. The need for liquid synthetic fuels is now opening the way to the chemical or electrochemical reduction of carbon dioxide, which can potentially be used for the synthesis of low-carbon fuels [18] and act as a hydrogen carrier. To turn carbon dioxide into an energy carrier and build a short-term carbon cycle, it will be necessary to develop new technologies among which hydrogen will still play an important role.

List of Symbols and Abbreviations

A	membrane area (m²)
a_i	activity of species i
C_i	concentration of species i in mol m^{-3}
D_i	diffusion coefficient of species i (m²s^{-1})
E	thermodynamic electrolysis voltage (V)
F	Faraday constant (96 485 C mol^{-1})
f_i	fugacity of species i
G	Gibbs free energy (J mol^{-1})
H	enthalpy (J mol^{-1})
I	current (A)
i	current density (A cm^{-2})
i_0	exchange current density (A cm^{-2})
m	mole number
n	number of electrons exchanged during the electrochemical reaction
P	pressure (Pa)
p_i	partial pressure of component i in a gas mixture (Pa)
R	resistance (Ω)
r	specific resistance (Ωm²)
R_{PG}	constant of perfect gas (0.082 J K^{-1} mol^{-1})
S	entropy (J mol^{-1} K^{-1})
T	absolute temperature (K)
t	time (s)
U_{cell}	cell voltage (V)

V	thermoneutral electrolysis voltage (V)
$\overline{V}$	molar volume (m^3 mol^{-1})
W	electrical work (J)
x, y, z	space coordinates

Greek Symbols

δ	membrane thickness (m)
Δ	difference
ε	cell efficiency (%)
η	overvoltage (V)
μ	Chemical potential (J mol^{-1})
φ	electrode absolute potential (V)
ρ	electrical resistivity (Ωm)
σ	electrical conductivity (S m^{-1})

Subscripts or Superscripts

∘	standard conditions (298 K, 1 bar)
ad	adsorbed species
ab	absorbed species (in bulk regions)
cell	electrolysis cell
d	dissociation
e	electrolyte
e$^-$	electron
i	species i
r	real
t	theoretical

Acronyms

EW	Equivalent weight of membrane (eq g^{-1})
HF	High Frequency
H/M	dimensionless ratio of hydrogen mole number to metal mole number
LF	Low Frequency
MF	Medium Frequency
PEM	Proton Exchange membrane
PTFE	polytetrafluoroethylene
SPE	Solid Polymer Electrolyte

References

1. European Commission (2003) Hydrogen Energy and Fuel Cells, Directorate General for Research. Report EUR 20719 EN, European Commission, Brussels.
2. Millet, P., Dragoe, D., Grigoriev, S.A., Fateev, V., and Etievant, C. (2009) GenHyPEM: a research program on PEM water electrolysis supported by the European Commission. *Int. J. Hydrogen Energy*, **34**, 4974–4982.
3. Leroy, R.I., Bowen, C.T., and Leroy, D.J. (1980) The thermodynamics of aqueous water electrolysis. *J. Electrochem. Soc.*, **127**, 1954.
4. Meyer, C.A. (1993) *ASME Steam Tables: Thermodynamic and Transport Properties of Steam*, American Society of Mechanical Engineers, New York.
5. Onda, K., Kyakuno, T., Hattori, K., and Ito, K. (2004) Prediction of production

power for high-pressure hydrogen by high-pressure water electrolysis. *J. Power Sources*, **132**, 64–70.

6. Vandenborre, H., Leysen, R., Nackaerts, H., Van der Eecken, D., Van Asbroeck, P., Smets, W., and Piepers, J. (1985) Advanced alkaline water electrolysis using inorganic membrane electrolyte (I.M.E.) technology. *Int. J. Hydrogen Energy*, **10**, 719–726.

7. Gierke, T.D., Munn, G.E., and Wilson, F.C. (1981) The morphology in Nafion perfluorinated membrane products, as determined by wide- and small-angle x-ray studies. *J. Polym. Sci., Part B: Polym. Phys.*, **19** (11), 1687–1704.

8. Yeager, H.L. and Steck, A. (1981) Cation and water diffusion in Nafion ion-exchange membranes: influence of polymer structure. *J. Electrochem. Soc.*, **128**, 1880–1884.

9. Sakai, T., Takenaka, H., Wakabayashi, N., Kawami, Y., and Torikai, E. (1985) Gas permeation properties of Solid Polymer Electrolyte (SPE) membranes. *J. Electrochem. Soc.*, **132**, 1328–1332.

10. Grigoriev, S.A., Porembskiy, V.I., Korobtsev, S.V., Fateev, V.N., Aurêtre, F., and Millet, P. (2011) High pressure PEM water electrolysis and corresponding safety issues. *Int. J. Hydrogen Energy*. **36**, 2721–2728.

11. Millet, P. (1990) Water electrolysis using EME technology: temperature profile inside a Nafion membrane during electrolysis. *Electrochim. Acta*, **36**, 263–267.

12. Grigoriev, S.A., Shtatniy, I.G., Millet, P., Porembsky, V.I., and Fateev, V.N. (2011) Description and characterization of an electrochemical hydrogen compressor/concentrator based on solid polymer electrolyte technology. *Int. J. Hydrogen Energy*, **36**, 4148–4155.

13. Millet, P., Ngameni, R., Grigoriev, S.A., Mbemba, N., Brisset, F., Ranjbari, A., and Etivant, C. (2010) PEM water electrolyzers: from electrocatalysis to stack development. *Int. J. Hydrogen Energy*, **35**, 5043–5052.

14. "NaturalHy, using the Existing Natural Gas System for Hydrogen", Integrated R&D project n° SES6/CT/2004/502661, European Commission (FP6, 2002–2006), *http://www.naturalhy.net/*.

15. Lewis, F.A. (1967) *The Palladium–Hydrogen System*, Academic Press, London, New York.

16. Millet, P. (2005) Pneumato-chemical impedance spectroscopy: (I) principles. *J. Phys. Chem.*, **109**, 24016.

17. Schalpbach, L. and Zttel, A. (2001) Hydrogen-storage materials for mobile applications. *Nature*, **414**, 353–358.

18. Olah, G.A., Goeppert, A., and Surya Parkash, G.K. (2009) *Beyond Oil and Gas: the Methanol Economy*, Wiley-VCH Verlag GmbH.

11
Solar Cell as an Energy Harvesting Device
Aung Ko Ko Kyaw, Ming Fei Yang, and Xiao Wei Sun

11.1
Introduction

In the history of mankind, every single step of innovation in energy generation technology has been proven to be the most robust driving force for industrial productivity. Exploration, transportation, utilization, and storage of energy have been accepted worldwide as the leading force for social civilization and advancement. However, we are building our civilization at the cost of burning the earth and placing burdens on environmental conditions. In the twenty-first century, it has been for the first time in human history that the whole world has been put into one shoe to search for a solution toward a better future. Slogans like "Change Energy Infrastructure" have been coined to enlighten the public on the future of sustainable development, with the incorporation of new types of energy generation technology.

Solar radiation is enormous, and its potential unlimited. The energy arriving at the earth's atmosphere is tens of thousands of times more than the average energy consumption of the whole world in a year. To accept this incredible "gift" from the sun, the most promising and dominating technology is solar cell technology. In this chapter, the fundamentals, advanced applications, and future perspectives of solar cell technology are addressed.

11.2
Solar Radiation and Absorption

The fact we can observe is the interaction between light and objects, for example, reflection, refraction, deflection, and so on. We may interpret these phenomenona as the energetic interaction between light, which is intrinsically an electromagnetic wave, and the excitons inside the material. Figure 11.1 shows the incident light energy density as a function of photon energy [1]. The solid heavy line shows the condition just outside the earth's atmosphere, while the thin line shows a blackbody at a temperature near that of sun's surface.

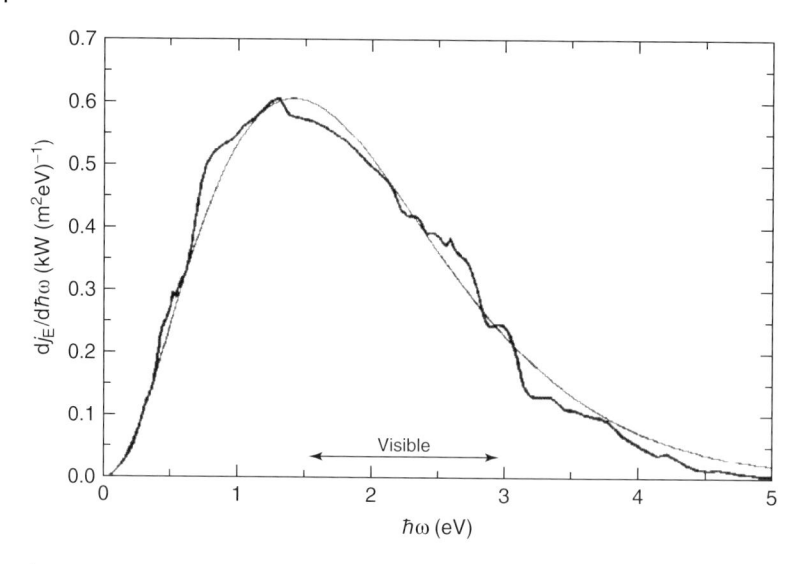

Figure 11.1 Light energy density as a function of photon energy.

When sunlight arrives at the outer atmosphere, it possesses an average power per distance in the free space of around 1350 W m^{-2}. The atmosphere surrounding the earth has a huge influence on the light absorption of an object placed on ground. And this influence is defined as the "air mass" (AM). When the AM is zero (AM0), it represents the condition of absorption outside our atmosphere, which is suitable for application in satellites or space shuttles. When the AM equals 1, it represents vertical direct illumination on the planet's surface after passing through the air, and the power of AM1 is around 925 W m^{-2}. The main difference between AM0 and AM1 is the reduction of light intensity in several categories: ultraviolet reduction due to absorption by ozone, infrared (IR) reduction due to absorption by water vapor, and reduction due to diffraction from suspended particles in air. When the AM is 1.5, it demonstrates a typical bright light radiation from an angle of 48.2°, with a power of around 1 kW m^{-2}. And AM1.5 is typically used for testing solar cell devices [1].

There are several issues we need to consider when we apply photovoltaic (PV) technology to convert sunlight into electricity. First of all, sunlight consists of radiation of different wavelengths, corresponding to different photon energies. So it is difficult for a single device to cover most of the energy range. To combat this issue, bandgap engineering of solar cell material is considered, in order to provide a trade-off between light absorption and the terminal output voltage. Secondly, the radiation from sun differs significantly with change of weather conditions, geographical location, atmospheric composition, and so on. A detailed analysis based on geographical data is not provided in this chapter.

When light falls on the surface of a semiconductor, a part of the light is reflected back, a part is transmitted through the material, and the remaining part is absorbed. Photon absorption is a process dependent on the wavelength of light (photon energy), therefore photons that are absorbable need to be sufficiently energetic: that is, they should possess an energy equal to or larger than the bandgap of the semiconductor. Light intensity suffers a characteristic exponential decay inside the semiconductor material, characterized by a term called the *"skin depth."* The skin depth is denoted $1/\alpha$, where $\alpha(\lambda)$ is the absorption coefficient with respect to different wavelengths. Because of this decay, the photon flux at one skin depth will fall to 37% of the intensity at the surface. Taking silicon as an example, it strongly absorbs in the ultraviolet region while it is almost transparent in the IR. For the visible spectral range, the skin depth of silicon is 0.1 µm for 400 nm wavelength and 5.3 µm for 7 nm wavelength. As a result, the characterization of solar cell performance must be conducted with a universally agreed standard, that is, AM1.5.

11.3
Fundamentals of Solar Cells

The conversion of solar energy into electric power requires the absorption of light, producing electron pairs in a semiconductor, separation of charge carriers, and collection of charges at the electrodes. A p-n junction, either homo or hetero, is used for the separation of electron pairs. In fact, a solar cell may be imagined as a solar-energy-driven "electron pump" [2]. The maximum height the electrons can be "pumped" is equal to the highest voltage the solar cell can develop. The maximum current is analogous to the pumping rate. As all generations of solar cells are based on the concept of p-n junctions, it is important to understand the basic principle of a conventional p-n junction solar cell. The understanding of the p-n junction solar cell provides general guidelines to the further improvement of newer generation solar cells in regard to efficiency, manufacturing cost, and energy consumption for device fabrication.

11.3.1
Working Principle of a p-n Junction Solar Cell

Conventional PV cells are built on the p-n junction of silicon or other inorganic semiconductors. A p-type semiconductor consists of holes (mobile charge) and acceptors (ionized impurity atom), while an n-type semiconductor consists of electrons and donors. When a p- and an n-type semiconductor are joined together, holes diffuse into the n region and electrons to p region due to the concentration gradient between the p- and n-type semiconductors. The electrons (holes) crossing the junction are compensated with holes (electrons) in the p region (n region), leaving only ionized impurity atoms (acceptors and donors). Because of the ionized impurity atoms, a layer without mobile charge carriers called the *"space charge region"* or *"depletion region"* is formed near the junction. This space charge sets up a built-in electric field which opposes the diffusion across the junction since the direction of the electric field is opposite to that of carrier diffusion. At equilibrium,

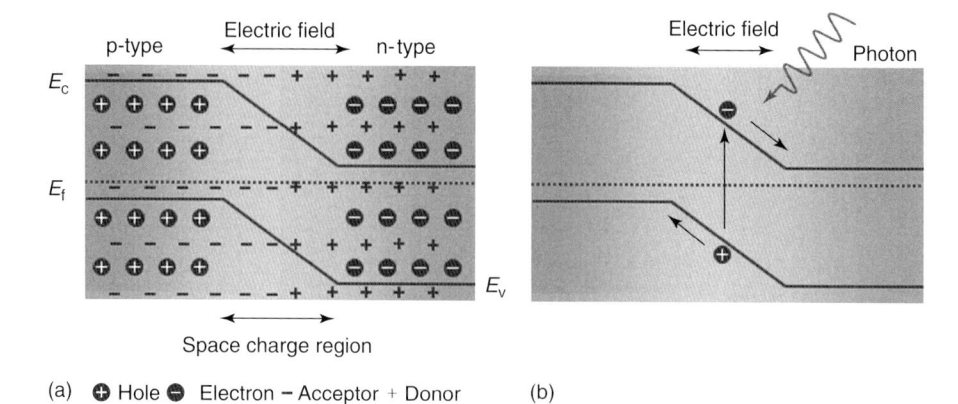

Figure 11.2 (a) Schematic structures of the p-n junction and its energy band diagram in equilibrium. (b) Illustration of charge carrier generation and separation under illumination.

the diffusion current due to concentration gradient is balanced by the drift current due to the electric field and no net current flows in the semiconductor. Figure 11.2a depicts the charge distribution in the p-n junction and the energy band diagram at thermal equilibrium.

When a semiconductor is illuminated, the absorption of photons with energy higher than a threshold, that is, the bandgap of the semiconductor, results in the generation of electron–hole pairs. The number of generated electron–hole pair is proportional to the light intensity. Because of the electric field in the depletion region, the drift of electrons toward the n side and that of holes toward the p side occur in the depletion region, as shown in Figure 11.2b. This separation of charges results in a current flow from the n side to the p side when an external wire is connected. Since the electron–hole pairs will recombine after a certain time via emission of photons or phonons (heat), only those created in or close to the depletion region can be separated by the built-in electric field and contribute to the photocurrent. Since both electrons and holes coexist in the same type of semiconductor, conventional p-n junction cells are called *"minority carrier devices"* and their efficiencies are highly dependent on the ability of the photogenerated minority carriers: for example, electrons in a p-type material to be collected out of the device before recombining with the majority carriers (holes in this case).

11.3.2
Equivalent Circuit Diagram (ECD)

Equivalent circuit diagrams (ECDs) are frequently used to model the electronic behavior of complex semiconductor devices with a network of ideal electrical components such as diodes, current or voltage sources, resistors, and capacitors. Silicon-based p-n junction cells are well studied, and a circuit model governing

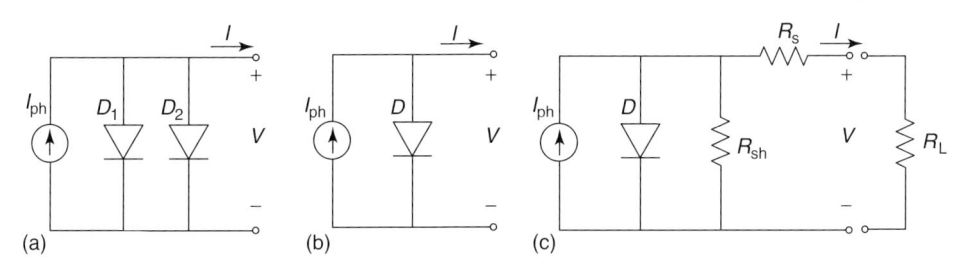

Figure 11.3 Equivalent circuit diagram of a typical solar cell. (a) Ideal solar cell with two-diode model; (b) Ideal solar cell with one-diode model and (c) non-ideal solar cell with an external load.

the current–voltage (I–V) characteristic has been established. Because of the p-n junction nature of silicon solar cells, the general expression for the current produced by the cells can be derived by solving the minority carrier equation across the junction (Eq. (11.1)). According to the equation, the ECD can be modeled by an ideal current source in parallel with two diodes, as shown in Figure 11.3a [3].

$$I = I_{ph} - I_{S1}\left(e^{\frac{qV}{kT}} - 1\right) - I_{S2}\left(e^{\frac{qV}{2kT}} - 1\right)$$
$$= I_{ph} - I_{S1}\left(e^{\frac{V}{V_T}} - 1\right) - I_{S2}\left(e^{\frac{V}{2V_T}} - 1\right) \quad (11.1)$$

The photocurrent I_{ph} resulting from the generation of free electron/hole pairs after absorption of photon is modeled as a current source. It does not take into account any recombination but only considers the charge carrier creation efficiency. The second and third terms in the expression are the recombination current in the quasi-neutral regions and in the depletion regions; they are modeled as currents flowing through diodes D_1 and D_2, respectively. Currents passing through the diodes are governed by their saturation currents (I_{S1}, I_{S2} for diodes 1 and 2, respectively) and ideality factors ($n = 1$, $n = 2$ for diodes 1 and 2, respectively, since the two diodes represent different conducting mechanisms). V_T is often referred to as the *"thermal voltage"* which is equal to kT/q.

For simplicity, often the two diodes are combined into one (D), as shown in Figure 11.3b, and Eq. (11.1) is rewritten as Eq. (11.2). The new term n is called the *"diode ideality factor"*, which typically has a value between 1 and 2. For a diode dominated by recombination in the quasi-neutral regions, n is close to 1. If recombination in the depletion region dominates, n is close to 2. In cells where the recombination in each region is comparable, n is somewhere in between [3].

$$I = I_{ph} - I_S\left(e^{\frac{V}{nV_T}} - 1\right) \quad (11.2)$$

In practice, there is no ideal solar cell, hence a shunt resistance (R_{sh}) and a series resistance (R_s) are added to the model, as shown in Figure 11.3c. R_{sh} represents the loss of carriers via leakage paths; these may include pinholes in the film and the recombination and trapping of the carriers during their transit through the cell. R_{sh} can be derived by taking the inverse slope of the $I-V$ curve around 0 V: $R_{sh} = \frac{dV}{dI}\big|_{V=0}$. This is because, at very small voltages, the diode is not conducting and the current driven by the external voltage (positive or negative) is only determined by $R_{sh} + R_s$ with typically R_{sh} being much larger. R_s is attributed to the finite conductivity of the semiconducting materials, the contact resistance between the semiconductors and the adjacent electrodes, and the resistance associated with the electrodes and interconnections. R_s also increases with a longer distance of travel of the charge, for example, thicker transport layers. R_s can be estimated from the inverse slope at a positive voltage greater than the open-circuit voltage (V_{oc}) where the $I-V$ curves become linear: $R_s = \frac{dV}{dI}\big|_{V \geq V_{oc}}$ (V_{oc} is elaborated in the next section.) This is because, at a high positive external voltage, the diode becomes much more conducting than R_{sh}, and thus R_s dominates the shape of the $I-V$ curve.

Incorporating R_{sh} and R_s into the the ECD, the $I-V$ relationship can be expressed as

$$I = I_{ph} - I_S\left(e^{\frac{V+IR_S}{nV_T}} - 1\right) - \frac{V+IR_S}{R_{sh}} \tag{11.3}$$

In principle, R_s needs to be minimized (ideally zero) so that most of the electrical potential is applied to the load. R_{sh} needs to be maximized (ideally infinity) so that all the photocurrent passes through the load. The external load resistor R_L shows that the solar cell is connected to an external load. Alternatively, a source meter can be connected to the cell when the $I-V$ characteristic of the cell is measured.

11.3.3
Current–Voltage Characteristics

Standard $I-V$ measurement of a solar cell determines the current–voltage response of the device in the dark and under illumination. In the dark, the applied voltage, *forward bias*, on the device generates a current that flows in the direction opposite to that of the photocurrent. It will increase substantially when the forward bias is larger than the built-in potential of the device. This reverse current is usually called the *"dark current"*, which is related to the applied voltage through Eq. (11.4)

$$I_{dark} = I_S\left(e^{\frac{V}{nV_T}} - 1\right) \tag{11.4}$$

In fact, dark current refers to the current passing through the diode in the ECD, which is the second term in Eq. (11.2). Under illumination, the current–voltage characteristic follows Eq. (11.3). A typical $I-V$ curve for a cell in the dark and under illumination is shown in Figure 11.4. The shape of the curve is similar to that of a p-n diode, but crosses the first, third, and fourth quadrants.

The short-circuit current (I_{sc}) is measured at the condition when the applied voltage equals zero. From Eq. (11.2), $I_{sc} \approx I_{ph}$ and I_{sc} increases linearly with the light intensity. The open-circuit voltage (V_{oc}) is measured under the condition when there is no external load connected: that is, the circuit is broken or open. In this condition, there is no current flow between the two terminals of the device,

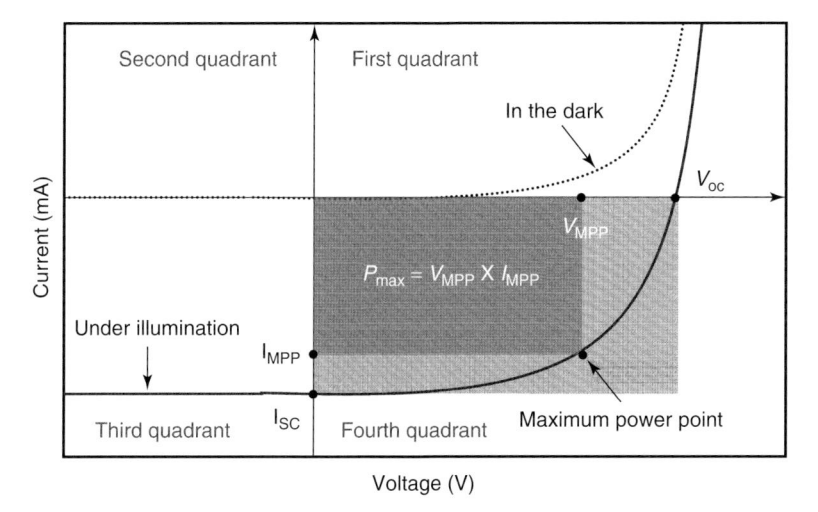

Figure 11.4 Current–voltage characteristics of a typical solar cell.

that is, $I = 0$ and $V = V_{oc}$. From Eq. (11.2)

$$I_{ph} - I_S \left(e^{\frac{V_{oc}}{nV_T}} - 1 \right) = 0$$

$$V_{oc} = nV_T \ln\left(\frac{I_{ph}}{I_S} + 1\right) \cong nV_T \ln\left(\frac{I_{ph}}{I_S}\right) \qquad (11.5)$$

Hence, V_{oc} increases logarithmically with the photocurrent and light intensity.

At a particular current and voltage among the data points in the fourth quadrant, the product of the current and voltage gives the maximum power (P_{max}). This point is called the *"maximum power point"*, while the current and voltage at this point are the current and voltage at the maximum power point (I_{MPP} and V_{MPP}). The larger the area of the rectangle ($I_{MPP} \times V_{MPP}$), the more the I–V curve resembles a rectangle with area $V_{oc} \times I_{sc}$. The ratio of these two areas is called the "fill factor" (FF), which relates P_{max} with V_{oc} and I_{sc}. The FF also represents the quality of the shape of the I–V curve. Mathematically, the FF can be written as

$$FF = \frac{V_{MPP} I_{MPP}}{V_{oc} I_{sc}} \qquad (11.6)$$

The higher the FF, the more the I–V curve resembles a constant current source with maximum voltage, and the higher the electric power that can be extracted.

The most important parameter that determines the performance of a solar cell is the power conversation efficiency (PCE or η), which is defined as the ratio of the maximum output power to the input power (P_{in}).

$$\eta = \frac{P_{max}}{P_{in}} = \frac{V_{MPP} I_{MPP}}{P_{in}} = \frac{FF \times V_{oc} \times I_{sc}}{P_{in}} \qquad (11.7)$$

P_{in} is an important experimental parameter to compare the results of the PCE, and hence standard test conditions should be used. This condition is set as AM1.5 spectrum illumination with an incident power density of $100\,mW\,cm^{-2}$ at $25\,°C$.

11.3.4
Quantum Efficiency

Since the sensitivity of a solar cell varies with the wavelength of the incoming light, another important parameter that determines the performance of a solar cell is the quantum efficiency (QE), which shows the spectral response of the device at an individual wavelength over the solar irradiation. QE basically measures the ratio of the electron count over the photon count. External quantum efficiency (EQE or $\eta_{ext}(\lambda)$) is defined as the ratio of photogenerated electrons (holes) collected at the cathode (anode) to the number of incident photons at certain wavelength. Mathematically, it can be expressed as

$$\eta_{ext}(\lambda) = \frac{N_{electrons}(\lambda)}{N_{photons}(\lambda)} = \frac{\frac{I(\lambda)}{q}}{\frac{P_{in}(\lambda)}{\frac{hc}{\lambda}}} = \frac{I(\lambda)hc}{\lambda P_{in}(\lambda)q} = \frac{1240 \cdot I(\lambda)}{\lambda(nm)P_{in}(\lambda)} \tag{11.8}$$

where q is the charge of the electron, c is the speed of light, h is the Plank constant, $I(\lambda)$ is the current generated by the cell at wavelength λ, and $P_{in}(\lambda)$ is the incoming power at wavelength λ. EQE is also known as the "incident photon-to-current conversion efficiency" (IPCE). For a given spectrum of the incident light $S(\lambda)$, the photocurrent can be calculated as

$$I_{ph} = \int \frac{q\lambda}{hc} \eta_{ext}(\lambda) S(\lambda) d\lambda \tag{11.9}$$

In some cases, it is preferable to express the efficiency of a solar cell without considering its absorption efficiency. Internal quantum efficiency (IQE or $\eta_{int}(\lambda)$) is a more appropriate parameter to represent the efficiency in such cases. IQE is the ratio of the number of collected charges at the electrode to the number of absorbed photons. Thus, IQE is also known as the "absorbed photon-to-current conversion efficiency" (APCE). It is related to EQE through the absorption efficiency η_A:

$$\eta_{ext} = \eta_A \eta_{int} \tag{11.10}$$

11.4
Silicon Solar Cell

Until now, PV industry has been built mainly on silicon-based substrates. The main reasons are its nontoxicity, abundant supply, and performance stability. Silicon has different phases, and a large number of the PV devices are working on crystalline silicon. In this section, different forms of silicon materials related to PV applications are discussed. We will briefly discuss each type of material, from basic characteristics to applications and limitations.

First of all, let us take a look at some facts about silicon:

- Silicon ranks number 2 in the elemental abundance on the earth's shell. It is chemically stable at room temperature and does not dissolve in any single type of acid, while a layer of several nanometers of silicon dioxide is easily formed at the surface isolating and protecting the basic structure. It is physically stable at room temperature and not stretchable. It has intriguing semiconductor properties with high intrinsic carrier concentrations and carrier mobility. Doping technology of silicon with group IV or III materials has matured.

- Single-crystalline silicon, also known as "monocrystalline Si" (mono-cSi) was the first ever used material for solar cell applications, and it still continues to play a crucial role. Mono-cSi can be manufactured with high purity, and its bandgap is 1.12 eV, one of the most ideal energy bandgaps for a single-junction cell. However, it is an indirect bandgap semiconductor and thus its absorption coefficient is low. As a result, bulky silicon materials have to be used for effective light absorption. Moreover, with the high cost of purification and manufacture, the cost of commercial mono-cSi modules is still way beyond that of its competitor regular energy sources such as oil and gas. Without subsidies from the government, bulky mono-cSi modules are hardly able to challenge the other energy generation sources.
- Amorphous silicon (a-Si) is normally fabricated by a deposition process that is simple, low-cost, and large- area-compatible. It is easier to design different structures based on a-Si materials and subsequently integrate them into electronic devices. Therefore, small power consumption devices like watches, calculators, and so on, mainly use a-Si solar cells. Nevertheless, the well-known Staebler–Wronski effect [4], which is a degradation mechanism for a-Si solar cell under prolonged illumination, has been the main obstacle for its applications, and this problem has not been completely resolved yet.
- Another form of silicon is polysilicon (polySi). After its discovery in the 1970s, the market share of polySi in the whole PV industry has rocketed from only 10% in the 1980s to about 50% nowadays. The reason behind it is its relatively low manufacturing cost compared to mono-Si. It is now the most dominant material used for commercial solar cells.

11.4.1
Hydrogenated Amorphous Silicon Thin-Film Solar Cell

Despite being the dominant solar cell materials in commercial market, single-crystalline and polycrystalline silicon in the bulky state have limitations in expanding their competence in the twenty-first century. These technologies are closely related to those of semiconductors and the supply of silicon raw materials, and therefore scaling up of these technologies for the purpose of competing with regular energy source is difficult. From a macroeconomic perspective, solar cell fabrication plants need to continuously produce large-area, efficient, and low-mass (material cost) solar panels. Thin-film cells have been found to be the most promising solution to this challenge. The hydrogenated a-Si thin film (a-Si:H) is a very attractive PV material. It has good physical and mechanical properties, although its stability suffers from the Staebler–Wronski degradation mechanism [4]. Additionally, the newly invented microcrystalline silicon (μc-Si) is receiving tremendous attention, as well. In this section, we address the fundamentals and applications of these two members of the silicon family.

11.4.1.1 Hydrogenated Amorphous Silicon

Hydrogenation can greatly improve the quality of a-Si. a-Si:H thin films can be deposited by plasma chemical vapor deposition (CVD), thermal CVD, or vapor–liquid–solid growth. The advantages and drawbacks of this film are summarized in Table 11.1.

Development History In 1975, Spear and his colleagues successfully prepared doped the a-Si:H film [5]. One year later, the p-i-n structure was introduced which improved the cell efficiency up to 2.4% [6]. In 1980, Carlson further boosted the efficiency to 8% [7]; as a result, more attention from industry was devoted to further its commercialization. Currently, the highest efficiency achieved for a-Si:H is around 9.5% [8]. In the development of the a-Si:H film, the efficiency was boosted mainly by the use of innovative materials and improvements in the deposition technique. In 1977, Staebler observed the so-called "light-induced degradation" of the a-Si:H film; since then, suppressing the negative effect due to this degradation mechanism has been a key challenge [4]. Plasma-enhanced chemical vapor deposition (PECVD) was successfully used to deposit high-quality a-Si:H films with good stability [9]. Moreover, silicon alloys such as a-SiC:H were developed because their higher bandgap could be used to form the p-type window layer in order to increase the open-circuit voltage and short-circuit current [10, 11]. Furthermore, a-SiGe:H with a shorter bandgap was also found to be useful to increase the light absorption range of the cell [12].

Fundamentals Amorphous silicon has long been considered as an obsolete material in solid-state electronic technology because of its poor electrical property.

Table 11.1 Summarized key properties of a-Si:H film.

Positive attributes	Negative attributes
Large absorption coefficient ($>105 \text{ cm}^{-1}$) at visible range, which makes realization of thin film realistic and economic	Suffers from degradation after long-term illumination or electrical bias, although degradation is reversible under certain conditions
Deposition process is at low temperature ($\sim 250\ °C$), improving substrate tolerance Uniformity can be easily achieved during deposition even for large-area cell	Electrical property of a-Si:H film does not match the standard of crystalline Si, but still useful for solar cell application
Thin film significantly reduces material cost for manufacturing the cell Process can be easily integrated to form alloy or other bandgap-engineered film	

This is because, although the bonding structure of a-Si is the same as that of crystalline silicon (both being sp^3), the bond length and angle between Si–Si atoms are different as a result of the lack of strict long-range order in a-Si; therefore, a-Si only exhibits the electrical properties of short-range-ordered silicon. High concentrations of structural defects are present in a-Si film due to high constraints of periodic lattices. These "incorrect" arrangements are created when growth is carried out under unbalanced thermal conditions. These defects create defect levels located largely between the valence and conduction bands (VBs and CBs), which promotes carrier recombination. The optical-electrical properties of a-Si:H are closely related to its deposition parameters. For PV application, normally ~10% hydrogen content makes the optical bandgap of the film up to ~1.8 eV. Dark conductance of the a-Si:H film is smaller than $10^{-10}(\Omega\ cm)^{-1}$, and the light conductance can increase by an order of 5.

Deposition CVD is commonly used for depositing a-Si for PV application. The CVD process involves the decomposition of SiH_4 or Si_2H_6 under certain conditions (of heat and pressure), and the silicon atoms or silicon-containing functional groups are forced to deposit on the substrate. The n-type a-Si:H film is formed by doping silicon with phosphine (PH_3). Diborane (B_2H_6), trimethylborane ($B(CH_3)_3$), or boron trifluoride (BF_3) is used for obtaining p-type a-Si:H film. Figure 11.5 shows a schematic layout of an equipment setup. Electrical power is supplied to silane gas or other Si-containing gaseous compounds, and a plasma is formed using significant electric power. Collision with electrons with energy ranging from several electronvolts to tens of electronvolts can decompose the silicon compounds and further ionize them. After several stages of absorption, deionization, decomposition, and surface diffusion, the a-Si:H film is formed.

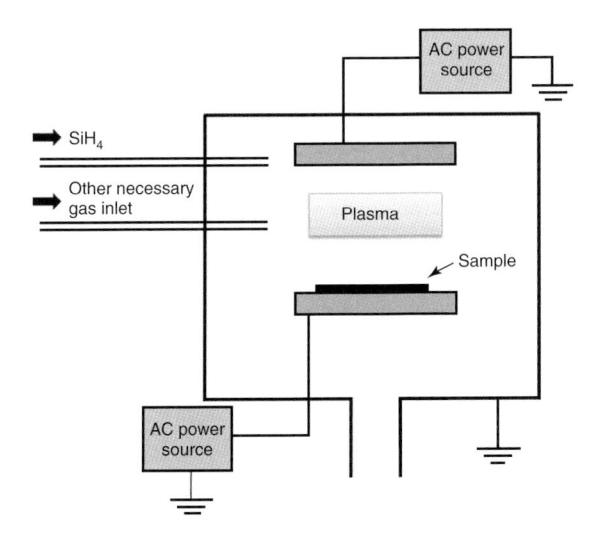

Figure 11.5 Schematic of typical setup for a-Si:H thin-film deposition.

Another common deposition method is thermal decomposition. In the case of thermal decomposition, the substrate is placed in a chamber which is filled with the reactant gas. The gas is heated to a very high temperature (easily higher than 1000 °C) until decomposition takes place. The decomposed components will arrive and settle on the sample surface. Silicon films fabricated by thermal decomposition normally contain a large number of defects and a small amount of hydrogen atoms. To compensate for these drawbacks of thermal decomposition, another type of excitation energy source must be provided to decompose the gaseous compound at a relatively low temperature. Glow discharge method, hot wire CVD, and photoinduced CVD are some of the substitutes. The deposition speed is a target industry needs to pursue, and therefore high-frequency plasma-based CVD is the common choice. Very high frequency (~100 MHz) is also utilized to provide fast enough excitation for the gaseous reactants. Nevertheless, plasma CVD described above is the most preferred solution nowadays.

11.4.1.2 Amorphous Silicon Thin-Film Solar Cell

The carrier diffusion length is very short in a-Si:H film, even after doping. This characteristic will directly lead to high carrier recombination in the device. For compensation, thinner films may be used. However, a smaller thickness will deteriorate light absorption and lead to a smaller photocurrent eventually at the external load. Finally, a solution has emerged to solve this problem – using a p-i-n structure for a-Si:H thin-film solar cell. The p and n layers are doped with boron and phosphorus, respectively, while the i layer is intrinsic a-Si:H. Under illumination, the carriers are mainly generated inside the i layer, and then diffuse into respective p and n layers. V_{oc} is directly related to the bandgap of the intrinsic layer and the Fermi level difference between the n and p layers. Therefore, optimization of the doping of n and p layers is crucial, especially for the p layer. In order to increase V_{oc}, an a-SiC:H alloy is used to replace the common a-Si:H p layer to increase its bandgap. As a result, it not only lets more light to pass through, but also boosts the open-circuit voltage. The so-called p-i-n single-junction structure is a sequential deposition of these three layers on top of a glass superstrate. A transparent conducting oxide (TCO) such as indium tin oxide is normally coated on glass before deposition of the active layers. The addition of a TCO increases the utilization of the incoming light and also serves as a top contact. The bottom contact after the n layer deposition can be directly connected on the n layer, and the common materials are Al and Ag. Silver is used mainly in the laboratory, as it has higher reflection, which boosts the short-circuit current. Aluminum is preferred in by the industry because of its low cost.

Another typical structure of a-Si:H solar cell is the n-i-p layout. Since holes possess a much lower (about 2 times lower than that of the electron) mobility than electrons in silicon thin film, the p-type film needs to be placed on the incoming light side. Therefore, an n-i-p structure uses a substrate that is not transparent. Typical structure can be Ag/n/i/p, deposited in sequential order. On top of the p layer, a conductive oxide is used. However, this layer must permit more light to go through; therefore, its thickness is well controlled, and "finger shaped" metal wires may be added on top for better conductivity.

11.4.1.3 Single-Junction Microcrystalline Silicon Thin-Film Solar Cell

During the thermal growth of the a-Si thin film, it is possible to obtain silicon crystals with size varying from several nanometers to tens of nanometers. And this film is not a pure a-Si film, as it shows some crystalline properties. Nanocrystalline silicon (nc-Si) is the name that has been frequently used worldwide for silicon films with crystal size from 1 to 100 nm. However, nc-Si does not have strict distinction from μc-Si, as they both refer to almost same range of crystal sizes. The atoms inside the μc-Si film have a longer range of order compared to that in a-Si, and it has a smaller bandgap. μc-Si film can achieve higher conductance and better doping efficiency, but a smaller light absorption coefficient. These are the main distinguishing features of the μc-Si film. It can be used as the n layer with a thin a-Si p layer in a cell, but using it as the p layer is not preferred. Power conversion efficiency of the μc-Si solar cell is around 10%. Further boost of efficiency of this type of solar cell is not easy even if focus is put on material optimization. Placing two or three materials in stack with different light absorption ranges is a solution that can effectively increase the power efficiency. Since the deposition techniques of the μc-Si film and the a-Si:H film are similar, series connection of solar cells made on the basis of these two materials can achieve high performance. Incoming light passes through the bigger bandgap a-Si-based solar cell, and remaining absorbable part will be absorbed by the μc-Si-based cell. This design suggests a series connection where two solar cells must match their current generation capabilities. In other words, the top and bottom cells must have the same current flowing during operation. More details about stacked cells are given in a later part of this chapter.

11.4.2
Polycrystalline Silicon Thin-Film Solar Cell

While the total solar power generation of the world is approaching gigawatts per year, the consumption of Si raw material has increased to hundreds of tons. Although it is still arguable whether we have sufficient supply of this semiconductor raw material, it is essential to establish a supply chain of low cost. Therefore, it is essential to establish a cell production process of low cost, leading to inexpensive thin-film-based devices. Polycrystalline silicon (polySi) is an alternative to the c-Si for thin films. Its application has been rapidly expanding because of its low cost of fabrication and large-scale productivity while maintaining high power conversion efficiency (e.g., 13%). In the industry, growth of bulk polySi material is mostly done by casting nowadays. The polySi has to be maintained at high quality, and this can be ensured from several testing techniques that feed back the characteristics of the material. Photoconductivity decay measures the minority carrier lifetime in the material, and laser-beam-induced current measures the grain boundary carrier recombination velocity.

11.4.2.1 Deposition of PolySi Thin Film

Power conversion largely depends on the crystal grain size; therefore, the deposition conditions of polySi are stricter than those of a-Si deposition. However, the utilization of a "thin" film effectively reduces the distance between the photogenerated carriers and the collection contact. As a result, the quality of the film has a larger tolerance region because the diffusion of carriers sees a shorter path and so the recombination is suppressed. The most commonly used deposition setup is PECVD, where the hydrogen content can be easily controlled. PECVD costs less to set up the equipment, while the process parameters can be easily controlled. Repeatability and large-scale application can also be easily achieved. In addition, liquid-phase epitaxy (LPE) and solid-phase crystallization (SPC) are also used widely. However, high-temperature thermal nonequilibrium cannot be used in deposition if a cheap substrate such as glass is included. (Glass has a phase transition temperature of 600–700 °C.) Therefore, use of LPE is limited because the silicon alloy is kept at 800–100 °C. However, the benefit of using LPE is its high-quality control, resulting from the slow growth. If a Si substrate is applied, thin-film solar cells can be grown of very high quality. Doping can be easily incorporated into LPE to form the p-n junction. To obtain a polySi thin film, two different routes are available. The first method is direct deposition by PECVD, LPE, and so on, where the process is similar to that of a-Si fabrication. Another method is an extension of the a-Si thin-film deposition, where a post-recrystallization process is applied to the film.

11.4.2.2 Enhanced Light Absorption and Suppressed Recombination

In thin-film polySi solar cells, the light absorption length is high because of their poor absorption coefficient. Therefore, the first and most important target is to boost light trapping in the film. Top surface texturing is necessary to trap more light into the film. Grain orientation is multiphased for polySi compared to monoc-Si, so a simple chemical etching (e.g., using KOH) is not possible to create the desired dimensions. Use of mechanical methods to engrave multiple V grooves on the film is an idea not suitable for future generation of thin films. A more appropriate way is to use reactive ion etching (RIE) for surfacing the samples. RIE is a typical dry-etching method suitable for large-area processing, which is able to control the aspect ratio of the texture. The main benefit of this technology is its independence from crystal orientations. During processing, dangerous gases such as chloride-related compounds are avoided, although they are efficient in performing the task. Other methods using a SF_x and O_2 mixture are also possible to create textured surfaces. Another way to improve cell performance is by surface recombination reduction. Surface recombination is directly related to the carrier lifetime and efficiency of carrier collection by the external circuit. PECVD is used to grow a very thin layer of SiN_x for surface passivation. This deposition has two key features: (i) defect states at grain boundaries are passivated; and (ii) possibilities of carrier recombinations are reduced by surface passivation, leading to improvement in both short-circuit current and open-circuit voltage.

11.4.3
Single-Crystalline Silicon Solar Cell

Extracted from refined quartz sand, mono-cSi can be grown using mainly two methods: Czochralski growth (CZ) and float zone (FZ) melting. In the former, high-purity polysilicon is melted in a quartz crucible, and then a seed is injected into the melt to lead the formation of the crystal. By slowly withdrawing the seed out of the molten silicon, an ingot of mono-cSi can be obtained. This Czochralski method incorporates two major categories of impurities into the silicon material: carbon and oxygen. However, solar cell applications can tolerate lower grade materials. For FZ silicon, higher purity can be easily obtained since the melting zone is "detached" from the seed. Without direct contact, carbon and oxygen concentration in the FZ process is substantially lower than in the CZ method. For large-scale production, the CZ process is normally preferred because of its cost effectiveness.

Next, a standard fabrication process of a commercial wafer-based solar cell will be introduced. After the growth of the high-purity ingot, wafers can be obtained by a sequential process: cutting (into blocks), shaping (into the required dimension), sawing (into slices of wafers), and chemical cleaning. Figure 11.6 shows a schematic of the process flow from a wafer into a functional cell that is generally adopted in industry. It includes several stages in sequence.

11.4.3.1 Front Texturing
Most commonly, NaOH or KOH is used to etch the surface of mono-cSi. This chemical etching is crucial because it removes most of the sawing damage, and it gives a pyramidal texture to the cell's font surface. The texturing makes the Si wafer capable of "guiding and trapping" light incident on the cell material with minimum reflection. The great difference in light reflection will increase the light path traveled inside the material beneath the textured structure. Typically, for a mono-cSi wafer of (100) surface orientations, KOH etching generates square-based pyramids with (111) oriented side surfaces.

However, texturing is not as simple as it seems. The reaction is normally conducted at 80–90 °C by immersing the wafer in a chemical solution. Repeatability has to be ensured with reasonable tolerance while trying to control the pyramid's size and to avoid the nucleation of the pyramid. Excessive solution turbulence may be present because of the generation of hydrogen gas. Through appropriate

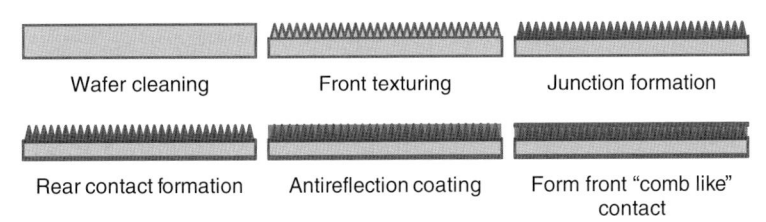

Figure 11.6 Schematic showing the fabrication procedures for Si wafer-based commercial solar cell.

additives, that is, isopropanol, which control the etching rate, the aspect ratio of the pyramid, and the solution turbulence, texturing quality can be improved.

Besides chemical etching, texture at the front surface can also be obtained using mechanical means, for example, laser-based techniques or plasma treatment. All these techniques serve one purpose: creating a surface structure to reduce light reflection, thereby increasing the absorption and utilization of the radiation.

11.4.3.2 Diffusion and p-n Junction

After texturing, wafers are cleaned by rinsing with deionized water. The wafers used for PV application are commonly p-type doped with boron atoms, which can be incorporated during the CZ growth stage, and the n-type junction is made with phosphorus diffusion, using different phosphorus sources. Three types of diffusion are available: gas phase, liquid phase, and solid phase. Most commonly, gaseous and liquid-based diffusion are utilized in commercial fabrication.

Gas-phase n diffusion takes place in a chamber filled with phosphine gas (PH_3). More commonly, liquid $POCl_3$ is adopted as the phosphorus source. $POCl_3$ is a liquid at room temperature, but it is vaporized and forced down into the diffusion tube with a carrier gas, which is normally nitrogen. The p-type wafers are loaded into the furnace chamber in a quartz boat. The boat moves slowly to prevent the wafers from suffering large temperature gradients. The diffusion temperature is normally about 900 °C, with suppressed temperature variations across the boat. N_2 is kept flowing during loading and unloading to prevent the wafer contacting any oxidizing agents.

Liquid-phase diffusion uses a phosphorus source such as H_3PO_4 which is dissolved in an organic solvent such as ethanol. The liquid is first sprayed or spun onto the wafers at room temperature, and then wafers are sent for further annealing in a belt furnace. Edge and backdiffusion or "contamination" is deadly, because it provides a shunt path between the front and rear of the cell. Edge diffusion is prevented by plasma etching of the round edges, while backdiffusion is mostly not of concern, as contact firing can be used to neutralize this effect. Organic compounds are burned off at around 400 °C before annealing at about 900 °C.

In both gas- and liquid-phase diffusion, there is another gas crucial to be introduced during heating in the furnace – oxygen. O_2 is introduced to oxidize the P compound into phosphosilicate glass P_2O_5 on the wafer surface. Phosphorus immediately starts to diffuse into the wafer after its formation. As a result, the dopant profile is shallow, with a high surface concentration. Furthermore, silicon is consumed and turned into a SiO_2 layer. With the diffusion continuing, the SiO_2 layer grows. After turning off the flow of the dopant gas, the doping profile will proceed as a drive-in and the SiO_2 layer will stop to grow when it actually shields the bulk silicon from further diffusion. After heating, surface oxide can be easily removed by dipping into a dilute HF solution for a short period. The drive-in process deepens the junction position and reduces the surface concentration as time increases.

11.4.3.3 Antireflection Coating (ARC)
Generally, a Si wafer reflects about 30% or more of the solar radiation. Light absorption can be enhanced by incorporating specific surface-texturing techniques. To capture more radiation into the cell, a reflection-reducing layer can also be deposited to further boost efficiency by 4–5% (relative increase, not absolute value). For mono-cSi wafers, it is easy to etch the front surface into the required pattern. However, for polySi or multicrystalline Si, it is not straightforward to create the surface texture. Therefore, an antireflection coating (ARC) layer is essential in that case. It is normally done chemically by depositing a layer of TiO_2 or silicon nitride. The principle can be simply understood in terms of the interference of the reflected rays from the front and rear surfaces of the ARC film. With the help of ARC, light reflection can be effectively reduced to less than 10%.

11.4.3.4 Contact Formation
Both the front and rear surfaces need to be connected with the external load, and therefore a good metal contact should be made. Before the invention of printing technology, contacts were normally evaporated in a vacuum, which made the process expensive and increased complexity. Furthermore, the evaporated metal film blocks the incoming sunlight at the front surface. Nowadays, metal contacts are made using screen-printing, where a "comb-shaped" contact is utilized. For contact material, commonly used is high-purity silver and lead. Ag is preferred, as Pb poses environmental concerns. The printing includes dissolving the Ag powder into a frit (lower melting point composite), making a design pattern, drying with heat treatment, and firing at around 700 °C. The firing condition is crucial for a good ohmic contact formation, while the existence of metal silicide greatly improves the quality. For the top contact, less blocking of incoming light is pursued. For the rear contact, a thin aluminum layer is commonly deposited before the contact to form a p+ back surface field (BSF) for better carrier extraction. The steps described above are essential to make a good solar cell. However, for commercial cells to be used outdoors, further steps are needed to connect multiple cells into modules and panels.

11.5 Other High-Efficiency Solar Cells

11.5.1 Copper Indium Diselenide (CIS or $CuIn_xGa_{1-x}Se$) Solar Cell

Solar cells can be made from special chemical compounds. The ternary compound $CuInSe_2$ thin film is a very important available option for PV applications. Copper indium diselenide (CIS) was invented as a single-crystal bulk material initially, and then developed into the thin-film phase. There are three main dominating advantages about $CuIn_xGa_{1-x}Se$ (CIGS) materials: (i) The absorption coefficient can be as high as 10^5 cm^{-1}, which means that a film as thin as 1.5–2.5 μm would be enough to collect most of the incoming illumination; (ii) The ternary compound has a direct bandgap of around 1.04 eV. By adjusting the concentration of Ga in replacement of In, it can be modified and becomes tunable between 1.04 and 1.67 eV; (iii) High conversion efficiency can be achieved with excellent stability. Assuming that we can adopt the techniques from mature Si-technology, it will be beneficial to apply the technology to even residential use, and a short "cost payback" time might also be achieved.

11.5.1.1 Fundamental Characteristics

There are two phases of the CIS crystals, γ and δ. The γ-phase has a phase-change temperature at around 810–980 °C, and it is stable below the phase transition temperature. The δ-phase has a phase-change temperature at about 1050 °C. Therefore, the only used material for outdoors PV applications is the γ-phase. Table 11.2 summarizes the basic physical properties of CIS. Since CIS is a ternary compound material, it is not easy to accurately control the concentration of each element. And it is the "missing" or "interstitial" atoms that make the film into p- or n-type alike. Figure 11.7 summarizes the conditions regarding the formation of n- and p-type CIS materials. CIGS material is made by substituting some In atoms with Ga. CIGS is a direct bandgap material and its bandgap can be tuned by adjusting the concentration of Ga, ranging from 1.02 to 1.68 eV. The tunable bandgap enables bandgap engineering for higher light absorption. The bandgap is

Table 11.2 Physical properties of CuInSe$_2$.

Density (g cm^{-3})	5.77	Dielectric constant	15.2
Heat conductivity (W cm^{-1} K^{-1})	7.1	Lattice constant (Å)	$a = 5.773$
			$c = 11.55$
Melting point (°C)	986	Thermal expansion coefficient (10^{-6} K^{-1})	2.9
Bandgap (eV)	1.04		

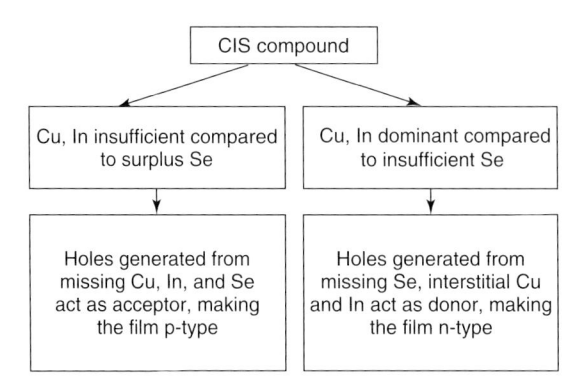

Figure 11.7 Details about the doping of CIS.

calculated on the basis of the x value [13]:

$$E_g(x) = (1 - x)E_g(\text{CIS}) + xE_g(\text{CGS}) - bx(1 - x) \qquad (11.11)$$

11.5.1.2 Development History
In 1974, Wagner from the Bell Labs developed a CIS-based single-crystal solar cell successfully and for the first time reported a power conversion efficiency of up to 5% [14]. And he further boosted the efficiency to 12% in 1975 by making a single-crystalline CIS–CdS p-n junction [15]. This effort demonstrated the great potential of CIS material. Boeing Company used the coevaporation method to replace In with Ga, showing a good control of the concentration of each element and an attractive road to thin-film technology [16]. In the 1980s, ARCO Solar Commercial from the United States invented a co-sputtering method to make the CIS thin-film solar cell, which consisted of depositing a preliminary film followed by a selenization process. The cell showed an efficiency of 14.1% [17]. In 1994, the National Renewable Energy Laboratory (NREL) in United States made a small-area copper indium gallium diselenide (CIGS) cell which attained the world record of 15.9% energy conversion efficiency [18]. The main success was due to the adoption of "three-step coevaporation" technique. Since 1994, NREL has been holding the best efficiency record for small-area lab-based CIGS thin-film solar cell [8].

There are several points worth noticing about the small-area CIGS thin-film solar cell: (i) Tuning of Ga and S (in replacement of Se) concentration could effectively increase bandgap of the absorber, at the same time forming a bandgap gradient to optimize the whole structure of the cell. (ii) A chemical bath deposition (CBD) technique was used to deposit the CdS layer and ZnO window layer, which improved the quality of the p-n junction and enhanced the light response in short wavelength region. (iii) Sodalime glass, which contains Na, was used in replacement of normal Na-free glass. Na diffusion into the CIGS layer, through molybdenum (Mo) contact, improved the open-circuit voltage and FF of the cell.

Although NREL leads the competition in small-area lab-based cells, the Swedish Ångström Solar Center at the Uppsala University is in the forefront of lab-based module fabrication. They demonstrated a 16.6%, 19.59 cm^2 module efficiency in 2003 [19]. In terms of large-area fabrication, different companies employ different approaches. Table 11.3 briefly summarizes the development of large-area CIGS thin-film technology [20].

11.5.1.3 Typical Structure
Sequential layers are deposited on sodalime glass substrates. One crucial benefit of using sodalime glass is the matching of the thermal expansion coefficient of the CIGS thin film and that of the substrate. Furthermore, sodium atoms inside sodalime diffuse into the CIGS layer during the subsequent manufacturing process, and this has a potential but debatable benefit: sodium replaces certain Cu atoms to form the more stable NaInSe$_2$. Theoretically, the bandgap increases by 0.11 eV, and thus open-circuit voltage of the cell is increases. From the rear contact to the

Table 11.3 Short summary of the CIGS solar cell made by different commercial manufacturers.

Company	Absorber	Cell area (cm^2)	Efficiency (%)	Power (W)	Release time (month/year)
Global Solar	CIGS	8390	10.2	88.9	May 2005
Shell Solar	CIGSS	7376	11.7	86.1	October 2005
Wurth Solar	CIGS	6500	13	84.6	June 2004
Shell Solar GmbH	CIGSS	4938	13.1	64.8	February 2004
Shell Solar	CIGSS	3626	12.8	46.5	March 2003
Showa Solar	CIGS	3600	12.8	44.15	May 2003

front contact, materials commonly used are the Mo metal layer, CIGS absorber, CdS-based buffer, ZnO-based window layer, and top-contact Ni–Al alloy.

Bottom Contact Mo Mo is proven to be the best choice for the CIGS thin-film solar cell [21]. Mo can form a 0.3 eV low-barrier surface with the CIGS material, thereby ensuring a good ohmic contact. This metal layer is normally DC-sputtered, while the electrical property of Mo is closely related to the sputtering pressure. High argon pressure leads to a Mo layer under compressive stress and low adhesive capability but with high electrical conductivity, whereas low argon pressure leads to a Mo layer under tensile stress and with high adhesive capability and low electrical conductivity. In practice, it is common to first deposit a thin Mo layer of good adhesion on the substrate, followed by a low-resistance Mo layer. In this way, the film is made suitable for high-performance CIGS solar cell applications.

CdS Buffer It is common practice to introduce a buffer layer between the CIGS active film and the window layer. CdS has been used most often because of its direct -2.4 eV bandgap and n-type property. It forms a smooth transition from CIGS to ZnO by reducing the mismatch between their crystal lattice and bandgap barrier. Furthermore, CdS can protect CIGS from the damage caused from the deposition of ZnO, while the S atoms diffuse into CIGS and passivate the surface to reduce surface recombination, thus increasing the carrier lifetime. A thin layer of CdS can be evaporated or chemical bath deposited, but CBD has attracted the most attention. This is simply due to the high demand for this buffer layer. To make a good buffer layer, it is necessary to make the layer as thin as possible so as to ensure good conductance between the CIGS and the top contact. Additionally, the CdS layer needs to be pore-free so it can shield the CIGS material from ZnO deposition damage. Normal evaporation cannot meet these requirements, but CBD can. Furthermore, ammonia used in the CBD process can clean the natural oxide on top of the CIGS. Besides all these benefits, it is also worth mentioning that CBD costs much less and the procedures are easier to handle.

Nevertheless, it is also important to notice the most significant drawback of this CdS buffer layer – pollution. Although the absolute amount of Cd used for residential-use CIGS solar cell is small, hardly producing any pollutants, the waste liquid after fabrication still contains harmful Cd elements. So it is a common understanding to avoid using Cd in fabrication, by replacing it with, most commonly, zinc. Different Zn compounds such as ZnO, ZnS, or $Zn_{1-x}Mg_xO$ are being researched. Besides replacing Cd with Zn, it is also a target to avoid bringing the sample out of CIGS deposition vacuum chamber for buffer layer deposition, which slows down the whole fabrication process.

Window Layer ZnO The ZnO layer on top of n-type CdS buffer is called the *"window layer."* It includes generally two layers of intrinsic ZnO and Al-doped ZnO. This layer is crucial in transporting minority carriers into the external circuit, so it must be made (i) with high electrical conductance, (ii) with high light transparency, and (iii) with suitable Fermi energy level and high minority carrier lifetime. Therefore, it is common to use a low-resistance layer and a high-resistance layer stack. The high-resistance layer is thin (~50 nm), while low-resistance layer is thicker (~300 nm) to keep the series resistance of the solar cell small. ZnO has a bandgap of ~3.4 eV, which is much larger than that of CIGS, allowing only photons with energy less than 3.4 eV to pass through the "window." Although CdS has a bandgap of 2.4 eV, which seems to be a "second window" on CIGS, it is possible that the thin CdS layer is completely depleted inside the p-n junction, therefore only ZnO serves as a window layer.

Top-Contact Ni–Al Alloy and Antireflection Layer The addition of nickel into the Al contact makes a better ohmic contact between the metal layer and the Al-doped ZnO window. In addition, the Ni–Al alloy also prevents the aluminum atoms from diffusing into ZnO, thus enhancing the long-term stability of the cell. The thickness of the layer is 1–2 μm. Around 10% of the light is reflected back at the surface. To save this portion of energy, a layer of ARC is provided on the Al:ZnO. This layer needs to be transparent in the useful light range. And it should be thermally, mechanically, and chemically stable on its substrate. MgF_2 is the answer mostly agreed upon. It has a refractive index of 1.39, which is close to 1.4, a value ideal for the antireflection layer on ZnO with refractive index of 1.9.

11.5.1.4 The Active Layer – CIGS Thin Film

There are different methodologies in depositing the CIGS layer, and the CIS layer is the basic structure for substitution of indium atoms with gallium atoms. Two techniques are commonly used: evaporation and selenization. First of all, evaporation is conducted with different sources in a vacuum chamber to deposit the $CuInSe_2$ film. This technique is very useful in small-scale laboratory-based fabrication for high-efficiency solar cell. But it is hard to achieve uniformity over a large area while keeping control of the elemental concentrations. Next, selenization involves a predeposition stage of a metal layer (Cu, In, and Ga), and followed by heating in a Se-containing gaseous environment. Post-selenization consists of several simple processes, and each of them is easy to control, with mature technology. Therefore, selenization is generally preferred in industrial applications.

Coevaporation for CIS Three types of evaporation are available for obtaining a CIS thin film:

- **Single source**: High-purity copper, indium, and selenium powders are mixed together to make the source. It might be easier to first obtain mixture of powders of Cu_2Se and In_2Se_3, and then combine them to ensure a better control of the elemental concentrations. Furthermore, due to intrinsic defects inside the crystal, films grown from the mixture generally manifest a lower concentration of Se compared to theoretical values, leading to an n-type CIS film. In order to obtain a p-type film, a slightly larger portion (∼0.02%) of Se is added compared to the theoretical value.
- After the preparation of mixture powder, it is loaded into a quartz boat inside vacuum. Heat treatment at around 1050 °C is used to convert the powder into the γ-phase crystal. The crystal is then used as the source for evaporation in the next stage. Electron-beam evaporation (Ebeam) or thermal evaporation is commonly used to make the final p-type CIS layer based on the γ-phase crystal obtained.
- **Two sources**: The invention of two-source evaporation was prompted by the lack of accurate control of elemental concentration in single-source evaporation. Therefore, in addition to the CIS crystal, Se powder is used as another source in the evaporation. By Ebeam or thermal evaporation, Se powder can be controlled to tune the film electrical properties: that is, carrier concentrations.
- **Three sources**: Different from the two methods mentioned above, three-source evaporation uses three independent powders, namely, Cu, In, and Se. By evaporating the materials in steps, good-quality CIS films can be made at only 350–400 °C thermal budget. Three-source evaporation has been very popular in laboratory experiments because of the good control in crystal quality and film electrical properties it offers.

Coevaporation for CIGS The bandgap of CIS is only 1.02 eV, which is not ideal for a single junction. Therefore, gallium is added to widen the bandgap. Deposition of CIGS is slightly different from the direct deposition of a CIS film, as coevaporation techniques are categorized on the basis of the number of steps involved in process.

- **Single step**. By evaporating the four elements at the same time, and controlling the concentration of each element, a CIGS film can be made with a certain concentration gradient inside.
- **Two steps**. First, the CIGS film is made by evaporation of the four elements. This film is generally rich in Cu, and therefore a second step evaporation of In, Ga, and Se is necessary.
- **Three steps**. The first step includes evaporation of Ga, In, and Se at a low temperature of around 350 °C. The result of first step is the formation of In_2Se_3

and Ga_2Se_3. Next, Cu is evaporated onto the substrate which is kept at around 550 °C. This step injects a great amount of copper into the film and makes it Cu-rich. Finally, evaporation of In and Ga is controlled to form the final CIGS thin film. The last step is crucial because it creates a group-III elemental rich top surface, thus creating a concentration gradient of Ga inside the film. Furthermore, the last evaporation step also reduces surface recombination centers.

Three-step coevaporation is able to achieve high-quality CIGS thin film with good PV performance. Nevertheless, this technique introduces difficulty in controlling uniformity of elemental concentration in large-area applications. And this technique imposes high standards on the equipment setup, and therefore it is not preferred for industrialized fabrication.

Fabrication CIGS Film by Selenization Selenization method is mainly used for large-area fabrication. Cu, In, and Ga with controlled elemental concentration are first deposited using sputtering, evaporation, or chemical deposition. Next, the sample is kept in a H_2Se gas flow while being heated up. As long as good concentrations of Cu, In, and Ga are predeposited, selenization process is able to generate high-quality CIGS films. H_2Se gas is sometimes avoided due to its harmful nature, and a solid Se source is used instead. The solid is evaporated by heat to adjust the selenization environment. Furthermore, to assist forming a Ga concentration gradient profile inside the film, part of the selenization process is replaced by annealing in a sulfur-containing environment instead of selenium. The prepared film is annealed in a H_2S gas or S powder environment, with argon or hydrogen as carrier.

11.5.1.5 Future Researches
As the active material of CIGS solar cell is obtained by deposition, it is an inevitable challenge to achieve uniformity for large-area CIGS cells. When using either selenization or direct deposition, the variation of performance among the modules must be within the tolerance range.

The absolute amount of Cd used for residential application is very small, and the use of a CdS buffer actually enhances the cell performance. However, besides the potential harm to the environment, the use of Cd is also avoided also for operational reasons. The chemical bath normally requires the unloading of samples out of the vacuum chamber. This is surely a drawback for industry-scale fabrication. Zn-based buffer layers are researched intensively. A solution-based coating method of ZnS buffer has been developed [22], but the performance is not catching up with that of CdS. Dry-environment deposition of zinc indium diselenide is another good invention for the complete vacuum-based process. Another challenge comes from the substrate. The new generation of solar cells requires the active materials to be incorporated onto a soft material substrate. Flexibility is an attractive property for solar cells, but the matching of physical and mechanical properties between the active layer and the substrate is important.

Furthermore, multijunction high-efficiency CIGS solar cells are possible to further boost efficiency (Figure 11.8). By connecting two or even more solar cells

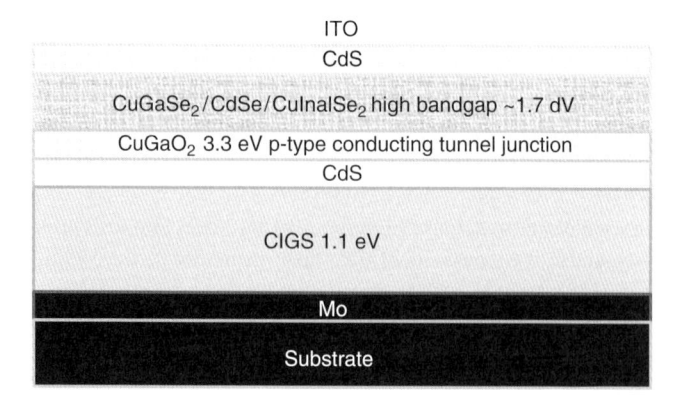

Figure 11.8 Example of multijunction CIGS-based stacked solar cell.

in series, the open-circuit voltage can be improved significantly. But there are several issues: (i) Matching of the short-circuit current of the top and bottom cells is important to attain the proper I–V characteristic. (ii) Tuning of the elemental concentration, exploration of new chemical compounds, and advanced deposition techniques are important to fabricate bandgap-engineered films. (iii) The large-bandgap, p-type conducting film is worth researching to serve its "tunnel" function between the top and bottom cells, since metal contacts are not allowed.

In conclusion, the CIGS solar cell is another commercial cell that has been well developed and marketed. However, boosting the efficiency further and bringing down the cost are the feasible options to bring this material into competition with the market's dominant player – the Si solar cell.

11.5.2
High-Efficiency III–V Compound Solar Cell

Solar cells can be made from semiconductors that consist of group-III (Ga, In) and group-V (P, As) elements in the periodic table. Owing to the availability of different combinations of these elements, III–V solar cells offer many choices in terms of material composition. The most typical compounds, GaAs and InP, have superior radioactive stability and long-term performance stability, but excessive experimental cost. Therefore, III–V technology is not comparable to silicon and they have been widely used in space-quality solar cells and concentrator cells. GaAs and InP are intensively used for single-junction solar cells because of their optimum bandgaps, which are 1.42 and 1.34 eV, respectively.

11.5.2.1 Cell Characteristics
The most intriguing feature of III–V compounds is their tunability while changing the chemical compositions. More than two elements can be combined together in the film; therefore, the physical characteristics of the film are largely tunable. Generally, III–V solar cells have high efficiency and stability, and this is achieved by controlling the elemental concentration in the crystals. Almost all III–V compounds

are direct bandgap materials, and have high absorption coefficients. As a result, they are very useful for thin-film solar cells.

Generally, III–V compound solar cells have the following features: (i) High efficiency. The power conversion efficiency of III–V compounds is much higher than that of other single-junction materials. The reasons are, first, their tunable bandgap for optimum light absorption, and, second, their direct bandgap where phonon generation is suppressed. (ii) Large absorption coefficient. Thin-film technology only utilizes several micrometers of the materials, so III–V compound solar cells not only save material cost but also maintain high performances. (iii) Supreme thermal and radioactive stability. The cells are very suitable for different applications, for example, outer space. (iv) The possibility of making tandem cells based on III–V compounds to enlarge the absorption range of light. By stacking cells of different bandgaps, very high efficiency (>40%) is achievable. For instance, on January 16, 2009, researchers at the Fraunhofer Institute reported a record high 41.1% efficiency from a GaInP/GaInAs/Ge tandem cell. Figure 11.9 shows a typical combination of a tandem cell. The top cell, which is InGaP-based, is able to absorb light of wavelength smaller than 650 nm, the middle GaAs single-junction cell can utilize light with wavelength from 650 to 850 nm, and bottom Ge cell is capable of absorbing wavelength from 850 nm and up to 1800 nm. By using this combination – although the experimental procedures are extremely complicated – light absorption is significantly enhanced, which directly improves the cell performance. Besides the tandem structure, incorporation of quantum wells or quantum dots in a III–V compound cell is a promising choice. They introduce differential energy states inside the bandgap, which enlarges the absorption range of the active region. Thus the photocurrent is increased. Multi-quantum-well structures are mainly based on lattice-matching compounds, for example, AlGaAs/GaAs. Quantum-dot-based III–V solar cells work on almost the same principle as quantum-well-based solar cells. Quantum dots can be

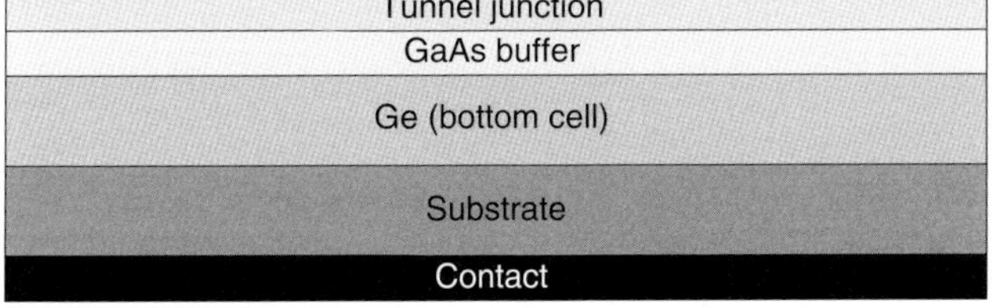

Figure 11.9 Example of multijunction III–V stacked solar cell.

embedded in the active layers, and they have a strong coupling effect that enables the excited carriers to tunnel through energy barriers. Thus the IQE can be boosted.

11.5.2.2 Growth Method of Single-Crystal GaAs

We shall discuss the typical growth methodologies for GaAs materials. GaAs single crystals can be grown by the Bridgman method [23], in which melting takes place locally. There are two main categories of the Bridgman technique: vertical orientation and horizontal orientation. In the following, we shall discuss briefly single-crystal GaAs growth using horizontal orientation.

1) A vacuum chamber made of quartz is separated into two subchambers, where high-purity Ga and As are stored. The two chambers share a quartz window in between.
2) Oxides on the surfaces of each material are eliminated by thermal annealing at a specific pressure and temperature. For Ga, 2 h of annealing at 700 °C and around 10^{-2} Pa is used; for As, 2 h of annealing at about 250 °C is preferred.
3) Both sides of chambers are sealed, and the quartz window is broken.
4) The chamber storing Ga is heated to a temperature around 1200 °C, and the chamber with As is kept around 610 °C. Vaporized As tunnels through the quartz window and reacts with high-temperature Ga to form GaAs.
5) The GaAs is in the multicrystalline phase, and single-crystalline GaAs is obtained by a further stage of growth. Attaching a single-crystal GaAs seed at the head of quartz boat is a method for further single-crystal bulk growth.

One problem of the Bridgman method is that the single-crystal GaAs easily sticks to the quartz boat. The high reaction temperature allows GaAs to invade the quartz boat, which leads to significant amounts of defects after cooling down. This problem has been effectively controlled by pretreatment of the boat in a Ga-containing environment.

Another method is called the "liquid encapsulation crystal" (LEC) method. LEC is similar to CZ growth of single-crystalline Si. High-purity GaAs material is melted in a crucible, and a seed head is used to guide the growth. The seed crystal rod is very slowly pulled upward with constant rotation. By accurately controlling the temperature gradients, pulling speed, and rotation rate, LEC is able to obtain large single-crystal GaAs ingots from the melt.

11.5.2.3 Growth of III–V Thin Film

Although the main target of III–V compound solar cells is high-efficiency application at outer space or under concentrated irradiation, thin-film technology to grow III–V material is still attractive for cost-saving purposes. The most important techniques in the growth of III–V thin films are LPE, metal oxide chemical vapor deposition (MOCVD), and molecular beam epitaxy (MBE).

In LPE, high-purity metallic Ga and single or polycrystalline GaAs materials are mixed under high temperature (~ 780 °C) to form a saturated solution. While cooling down slowly, the solution is brought into contact with a single crystal of GaAs substrate. Lowering the temperature of the saturated solution makes the extra GaAs to precipitate onto the substrate. Because of the thermally balanced condition during precipitation, the single-crystal thin film would be matching with the substrate perfectly in the crystal lattice. In addition, the thin film is of extremely high purity. Doping of the film is easily achieved by incorporating IV or VI group elements such as Si, Sn, and so on, for n-type film. p-Type doping is formed by including group II elements such as Mg. The main benefit of LPE is its comparatively low equipment cost and safer reactants and products. However, it is extremely difficult to grow a III–V thin film on Si or Ge substrates, because of the high solubility of Si or Ge in high-temperature Ga. Furthermore, the surface morphology cannot be precisely controlled because of the difficulty in thickness control. Because of the drawbacks mentioned above, LPE is gradually being replaced by MOCVD and MBE, though some researchers still use it for deposition.

MOCVD is a key deposition technique for III–V thin films. It involves a chamber where carrier gas, H_2 being generally the case, brings in the group-III metal oxide and the group V compound for decomposition onto the substrate. The Ga source is normally trimethyl gallium, and the As source is normally arsine (AsH_3). They are decomposed in the chamber following the equation

$$(CH_3)_3Ga + AsH_3 \rightarrow GaAs + 3CH_4$$

The real chemical reactions are more complicated and are not discussed here. By controlling the gas flow rate, concentration, chamber pressure, and temperature, different film characteristics can be achieved. In practice, deposition is normally done at low temperatures. Compared to LPE, MOCVD gives better GaAs films with neat control of film thickness and composition. The drawbacks of the MOCVD technique are its high cost and the toxicity and flammability of metal oxides used in this system. Therefore, waste gas treatment is needed. High-efficiency III–V solar cells are mostly made by MOCVD nowadays.

Lastly, the other type of important deposition technique is MBE. The principle is nearly the same as vacuum evaporation, but with stricter requirement on the vacuum level (easily $<10^{-9}$ Torr). Sources of Ga and As are heated slowly for evaporation, thus keeping the growth rate extremely slow. As a result, the quality of the film can be monitored for improvement. For example, doping, composition, and thickness are strictly controlled. Furthermore, it is even possible to grow a monoatomic layer with MBE. This technique is also capable of growing GaAs on a foreign substrate, which is a key advantage for III–V application because it enables interaction with Si technologies. In addition, thin films grown by MBE have highly smooth surface profiles. Nevertheless, the obstacles to commercializing MBE into mass-scale production are the high cost equipment and slow growth rate.

11.6
Dye-Sensitized Solar Cell

Despite a being major market player with high PCE, single-crystal-silicon-based solar cells find it hard to compete in the energy market because the cost of solar electricity generated from silicon solar cell is still an order of magnitude higher than electricity generated from fossil fuel (<US$0.05 kWh^{-1}). To reduce the price per kilowatt per hour, either the efficiency of solar cell has to be increased or the production cost must be reduced. However, the current efficiency of crystal-silicon-based solar cell (>24%) is approaching to the theoretical limit of 30% and it is difficult to further improve it. Generally, the production of silicon solar cell suffers from high cost because of the energy-intensive processes at high temperature (400–1400 °C) and high-vacuum conditions with numerous lithography processes. The increased demand for silicon solar cells in recent years, moreover, has inflated the price of raw silicon, causing even harder to reduce the solar electricity cost. These driving forces have led researchers to find novel ways to design PV cells using inexpensive materials with uncomplicated processes. Nanotechnology provides another approach to obtaining cheap, high-efficiency devices, leading to the emergence of new-generation solar cells.

Dye-sensitized solar cells (DSCs) belong to the group of thin-film solar cells and are a promising alternative to conventional silicon PV technology owing to the high photon-to-electricity conversion efficiency, low cost, and easy fabrication. The capability of large-scale manufacture from abundant, environment friendly, and nontoxic materials without expensive and elaborate apparatus makes this technology particularly attractive. The cell concept is also believed to reduce the energy payback time significantly compared to traditional silicon cells because it does not require energy-intensive processing at high temperatures. On top of that, DSCs can be engineered into flexible and mechanically robust cells through roll-to-roll production. In contrast to silicon solar cells, they completely depart from the classical solid-state p-n junction device. In fact, it is a photoelectrochemical system in which an electrolyte in the form of liquid, gel, or solid is filled between a photosensitized semiconductor electrode and a metallic counter electrode. A sensitizer that chemically adsorbs on the surface of a wide-bandgap semiconductor is responsible for light absorption. Charge separation takes place at the interface of the sensitizer and the semiconductor via photoinduced electron injection from the excited state of the sensitizer into the CB of the semiconductor. The regeneration of the sensitizer is possible by electron donation from the redox system of the electrolyte to the oxidized state of the sensitizer. Up to now, DSCs based on nanocrystalline TiO$_2$ electrode, Ru-polypyridyl-complex dye, liquid electrolyte with iodide/triiodide redox couple, and a Pt counter electrode offer the best efficiency. The highest reported efficiency of DSCs (~11%) is even higher than that of the a-Si solar cell (9.5%), which is appealing to the manufacture of inexpensive and highly efficient solar cells. However, further improvements in the efficiency, stability, and scalability are still required to realize them into commercialization.

11.6.1
History of Dye-Sensitized Solar Cell

The discovery of dye-sensitized semiconductors is considered to be forerunner in the emergence of DSCs. The first case of the dye-sensitized semiconductor was revealed as early as in 1873 by Vogel via the dye sensitization of silver halide emulsion applied to photography [24]. In 1887, Moser applied this concept to the photoelectric effect by using the dye erythrosine on silver halide electrodes [25]. Various cyanine and related dyes are adsorbed onto silver halide microcrystals to extend the spectral response to the visible region for photography applications. Some semiconductors studied for spectral sensitization are ZnO, CdS, and so on. Putzeiko and Trenin first reported sensitization of pressed ZnO powder by adsorbed rhodamine B, eosin, erythrosine, and cyanine dyes in 1949 [26, 27]. Nevertheless, the mechanism of electron injection from the photoexcited dye molecule into wide-bandgap semiconductor was not clear until the late 1960s: whether it was through energy transfer from the dye to defect centers in the underlying crystal or through direct electron transfer from the excited states of the dye [28]. The injection of electrons from an organic dye into semiconductor was discovered only in 1964 by Namba and Hishiki [29] and in 1968 by Tributsch and Gerischer [30, 31] with their work on ZnO. This early electrochemical work demonstrated that excited dyes on the surface of ZnO were oxidized to produce photocurrents that could be sustained through the addition of reducing agents to the solution that would reduce the photooxidized dye. In 1971, conversion of photon to electric current by mean of electrochemical reactions of excited molecules was first demonstrated by Tributsch *et al.* employing chlorophyll molecules adsorbed on a single-crystal ZnO electrode in contact with an electrolyte containing suitable electron donors [32, 33]. This kind of reaction is analogous to the primary steps of photosynthesis and chlorophyll-sensitized reactions in photosynthetic membranes, which can be applied to elucidate the concept of electrochemical kinetics. Later, fundamental researches were carried out on ZnO and TiO_2 single-crystal electrodes but the efficiency of the devices was still poor [34, 35]. The main problem was that the dye anchorage on a flat surface was poor and only 1–2% of the incident light could be absorbed by dye. Thicker dye layers increased the resistance of the cell without contributing to additional current generation [31]. In the following years, the chemisorption of the dye on the surface of the semiconductor was developed to enhance the linkage and dye loading [36, 37]. While the investigations on photoelectrode materials and the semiconductor–electrolyte interface were being carried out [38–41], new dyes were also developed. Ruthenium(II)–polypyridine complex was started to be used to sensitize a SnO_2 electrode in 1975 by Gleria and Memming and a TiO_2 electrode in 1977 by Gleria and Memming [42] and Clark and Sutin [43]. Introduction of dispersed particles with a high roughness factor, instead of a planar semiconductor, not only provides a large surface area but also generates multiple reflection at the rough surface, allowing the capture of more incident light by a monolayer of dye with a high molar extinction coefficient [44, 45]. With these improvements in light harvesting and photoelectrode materials, a conversion efficiency of 7.1–7.9% under simulated solar light and 12% under diffuse daylight was achieved in 1991 [46]. This achievement is really exciting and further incites more research activities in photoelectrochemical cell technology. Through continuous improvement, today state-of-art technology has driven the efficiency of solar cells to more than 11% [47]. Another evolution to this classic system has been the implementation of solid-state dye-sensitized cells by replacing the liquid electrolyte with (quasi) solid-state electrolytes such as hole transport materials [28, 32], polymeric matrix electrolytes [33, 34], quasi-solid molten salts (ionic liquids) [25], and nanocomposite gel electrolytes [24, 35, 37] to prevent leakage and evaporation of volatile liquid electrolytes. However, the efficiency of this solid-state DSC is lower than that of the traditional DSC mainly due to difficulties in filling pores and the low conductivity.

11.6.2
Basic Principle of Dye-Sensitized Solar Cell

As DSCs are developed on the basis of the photoelectrochemical system, their working principle is much different from that of conventional p-n junction solar cells. In a p-n junction of silicon or an inorganic semiconductor, the electronic contact between the p and n components form a depletion layer or space charge region near the metallurgical junction where the photogenerated electron–hole pairs are separated by an internal electric filed. Both majority and minority carriers coexist in the same volume of the semiconductor. Because the separated charge carries can recombine in the bulk via defects, high-purity material is crucial in the p-n junction solar cell to minimize the recombination.

On the contrary, in a DSC, an electron-conducting phase (n-type semiconductor) and a hole-conducting phase (redox species or hole conductors) build a junction

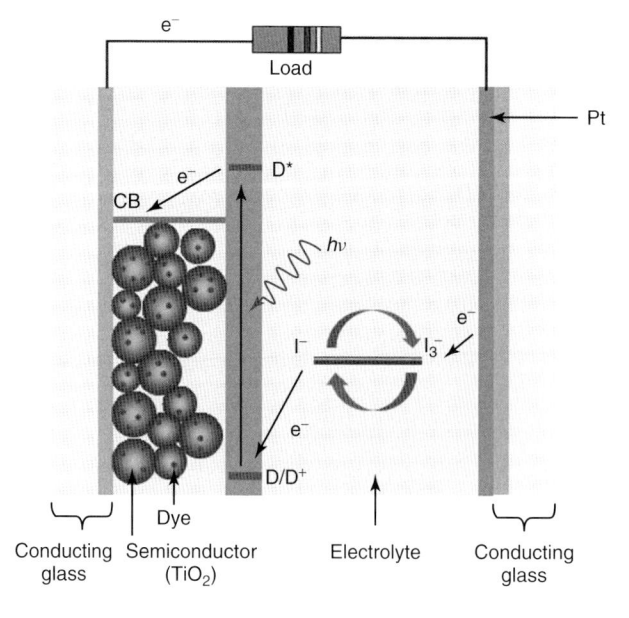

Figure 11.10 Main components of a dye-sensitized solar cell (DSC) and schematic illustration of the working principle of the DSC.

in a bulk heterojunction form. The main components of a typical DSC are a conducting glass (usually glass coated with fluorine-doped SnO_2 (FTO)), a semiconductor network, dye molecules attached to the semiconductor, iodide/triiodide (I^-/I_3^-) redox electrolyte, and platinized conducting glass (shown from left to right in the schematic shown in Figure 11.10). A porous network of interconnected nanometer-sized crystallites of a wide-bandgap semiconductor is used as the electron-conducting phase. Anatase (TiO_2) crystallites have been a material of choice for the electron-conducting phase, although alternative wide-bandgap oxides such as ZnO [43], SnO_2 [48], and Nb_2O_5 [49] have also been scrutinized. A monolayer of a dye chemically adsorbed on the anatase is used as the sensitizer. The whole network of anatase together with dye is immersed in an electrolyte. Because the individual size of the particles is too small compared to the Debye length of the material, the whole part of the particle is depleted upon contact with electrolyte and no significant space charge region is formed [50]. Upon illumination, the dye is sensitized to the excited state (D*) by light absorption and rapidly injects electrons into the CB of semiconductor, while the holes, at least initially, remain on the ground state of the dye (D). In contrast to the p-n junction solar cell, the significant role of the electric field in charge separation is not observed, although an electric field exists at the electrolyte/semiconductor interface, which is not due to the space charge in the semiconductor. Moreover, in a DSC, injection of only one type of carrier (electrons, in the case of most of common DSCs) from the excited dye into the semiconductor is energetically possible [50]. The ground state of the dye (D) is subsequently restored by electron donation from iodide (I^-) to the oxidized state of the dye (D^+). The iodide is regenerated in turn by the reduction of triiodide (I_3^-) at the counter electrode with the aid of a Pt catalyst. Concurrently, the recuperation of the redox system is realized by transporting holes to the counter electrode through either diffusion or a hopping mechanism depending on the transporting mediator. The circuit is finally completed by electron migration via the outer circuit [51]. The comparison between the DSC and the conventional p-n junction solar cell is summarized in Table 11.4.

Table 11.4 Comparison between dye-sensitized solar cell and conventional p-n junction solar cell [50].

p-n Junction solar cell	Dye-sensitized solar cell
Space charge region is formed near the junction and charge separation happens there	No significant space charge region due to small particle size
Internal built-in electric field separates the electron-hole pairs	Existence (and importance) of an electrical field is not observed
Both minority and majority carriers coexist in the semiconductor	Carriers of only one type are present in the semiconductor
Grain boundaries act as recombination centers and hence, must be avoided or minimized	Grain surface maximization is necessary to increase dye adsorption

11.6.2.1 Charge Transfer and Recombination Kinetics

In addition to the charge transfer process briefly discussed in the previous section, there are a series of undesirable back reactions (recombination) opposing the forward reactions. Because the photogenerated current is the result of the dynamic competition between the charge transfer and recombination processes [53, 54], the kinetics of these processes is important in the DSC. In a simple model, the electron extraction efficiency at the short-circuit condition can be defined as

$$\eta_{ext} = \frac{k_{trans}}{k_{trans} + k_{rec}} \tag{11.12}$$

where k_{trans} and k_{rec} are the first-order kinetic constant for the electron transfer and recombination process, respectively, indicating that an efficient cell should facilitate the forward reactions and suppress the back reactions.

The schematic of the kinetics in the charge transfer and recombination process is illustrated in Figure 11.11. The discussion in this section is based on a DSC with a nanocrystalline TiO_2 film, ruthenium–polypyridyl complex dye, and I^-/I_3^- electrolyte, which has yielded the highest energy conversion efficiency to date. The photoexcitation of dye (F1) occurs within a few femtoseconds upon illumination. The timescale of injection of electron from the exited dye into the CB of TiO_2 (F2) is tens of femtoseconds to hundred of picoseconds, depending on the electronic coupling and energetic overlap between the donor (D^*) and acceptor (semiconductor)

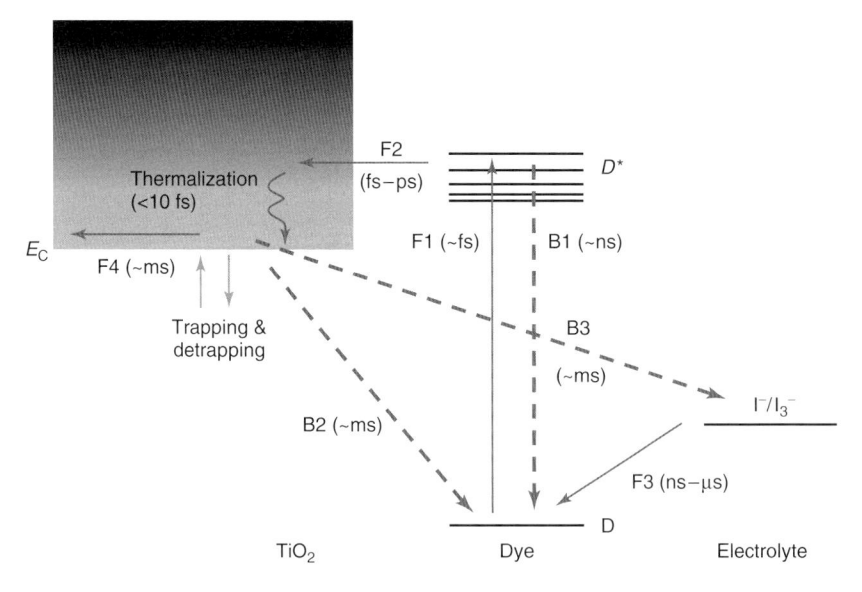

Figure 11.11 Kinetics of the charge transfer and recombination process in a typical DSC containing ruthenium complex-sensitized TiO_2 nanocrystalline film and I^-/I_3^- redox electrolyte. The forward reactions (charge transfer) and back reactions (recombination) are indicated by solid and dash arrow, respectively.

states [26, 27, 42]. The injected electrons in TiO_2 are rapidly thermalized by lattice collisions and phonon emissions within less than 10 fs. Since the exited state of the dye (D*) eventually returns to its ground state (D) radiatively or nonradiatively after some time, the injection process (F1) has to be competed with the relaxation of the dye (B1). This intramolecular relaxation of the excited states of the dye also might complicate the injection process and change the timescale. However, the relaxation of the dye from the excited state to the ground state happens in the nanosecond range, which is 3 orders of magnitude slower compared to the injection [53, 55]. As a result, the injection efficiency is close to unity. The reduction of the oxidized dye, therefore, is facilitated by the electron donation from I^- in the electrolyte (F3), which has to compete with recombination of injected electrons in TiO_2 with the oxidized dye (B2). The half-time for regeneration of most ruthenium complex dyes, in presence of about 0.5 M iodide, is in the range of 100 ns to 10 µs depending on the precise composition of the electrolyte [52, 55–60], whereas the recombination of electrons in TiO_2 with the dye happens in the milliseconds range [51], ensuring that the forward reaction is orders of magnitude faster than the back reaction for the regeneration of the dye. The reaction between the oxidized dye and I^- leads to the formation of diiodide radicals ($I_2^{-\bullet}$) [59, 61, 62]. These radicals were observed in DSCs using nanosecond-laser spectroscopy [59, 61, 62] and pseudo-steady-state photoinduced absorption spectroscopy [63].

Another important process is the electron percolation along the nanocrystalline film toward the electrode after being injected (F4). The injected electrons from the excited dye experience 10^6 times of trapping and detrapping at the grain boundary of TiO_2 nanoparticles and the nanoparticle/electrolyte interface [64–66]. The electron diffusion coefficient in TiO_2 nanocrystalline film, therefore, is considerably low (5×10^{-5} cm^2 s^{-1}) and it takes about 10 ms to reach the conducting electrode [67, 68]. Meantime, the electrons percolating along the nanocrystalline film can be recaptured by the oxidized species (I_3^-) in the electrolyte (B3) within the milliseconds time range [51, 55]. The recombination of electrons in TiO_2 with I_3^- in the electrolyte has been mostly studied using steady-state [69, 70] or dynamic [71–73] measurements of the open-circuit potential. Since the kinetics of this back reaction (B3) is comparable to that of the forward reaction (F4), the recombination of injected electrons with I_3^- in the electrolyte is more serious than the other two back reactions (B1 and B2). Hence, it must be suppressed as much as possible to improve the charge conversion in the DSC. This recombination can happen at the interface between the electrolyte and TiO_2 as well as that part of the conducting electrode which is exposed to the electrolyte. Nevertheless, the latter route is usually less important in the case of I^-/I_3^- and can be further reduced by employing a compact blocking layer of metal oxide between the conducting electrode and a mesoporous film [70]. The reactions for electron-transfer process and competing back reactions are summarized in Table 11.5.

11.6.2.2 Charge Separation at the Semiconductor–Dye Interface

Although the photogenerated charges are separated by an internal electric field at the space charge region in the p-n junction solar cell, there is practically no

Table 11.5 Reactions for electron-transfer process and competing back reactions.

$D + h\nu \rightarrow D^*$	Photoexcitation (F1, femtosecond range)
$D^* \rightarrow D^+ + e^-$ (TiO$_2$)	Electron injection (F2, femto- to picoseconds)
$D^* \rightarrow D +$ raditive (or) nonradiative emission	Relaxation (B1, nanosecond range)
$D^+ + 2I^- \rightarrow D + I_2^{-\bullet}$	Regeneration (F3, nano- to microseconds)
$D^+ + e^-$ (TiO$_2$) $\rightarrow D$	Recombination (B2, millisecond range)
$2e^-$ (TiO$_2$) $+ I_3^- \rightarrow 3I^-$	Recombination (B3, millisecond range)

depletion layer within the semiconductor owing to the small size of the particles and low doping level. Therefore, charge separation upon dye excitation must occur for other reasons. A main mechanism for charge separation is the band offset at the acceptor/donor interface as in other excitonic solar cells. Normally, the lowest unoccupied molecular orbital (LUMO) of the dye is above (less negative referenced to the vacuum level) the CB of the semiconductor. The potential difference between E_{LUMO} and E_{CB} creates an enthalpic driving force for charge separation. Similarly, the highest occupied molecular orbital (HOMO) of the dye is below the redox potential of electrolyte, making a driving force for hole injection into the electrolyte. In contrast to other excitonic solar cells such as organic and hybrid (polymer and inorganic composite) solar cells, the photogenerated charges (excitons) do not need to diffuse to reach the acceptor/donor interface. The excitons are generated right at the interface and they readily dissociate there and electrons are rapidly injected into the CB of the semiconductor.

Another driving force for charge separation is the surface field at the semiconductor and the dye/electrolyte interface. According to Schlichthorl et al. [66] and Zaban et al. [74], a surface field is established spontaneously as a result of a charged dipole layer despite no depletion layer inside the semiconductor. The dyes usually used are acidic because of carboxylic or phosphate groups and release protons upon binding with the semiconductor film [50]. In addition, if the film is immersed in a protic solvent such as the electrolyte, the solvent also acts as a proton donor. In protic media, Li^+ or Mg^{2+} ions are potential determining ions for TiO$_2$ and charge the surface positively [51, 75]. Thus, the protons together with other cations in the electrolyte adsorb on the surface of TiO$_2$ and a dipole is formed across the Helmholtz layer between the negatively charged (iodide and dye) species and the cations. This dipole layer establishes a surface electric field pointing away from the TiO$_2$ surface, facilitating the separation of the charges. The same field also prevents the injected electrons in TiO$_2$ from re-entering the dye or electrolyte after charge separation and makes the recombination kinetics very slow. Under illumination, the field will be reduced compared to the dark condition because the electrons injected into the TiO$_2$ will neutralize some of the positive charge screening the surface.

Charge separation in the DSC is also explained by photoinduced electron injection, which is an ultrafast electron transfer between a discrete and a localized molecular state of the sensitizer and electronic states in the semiconductor.

According to Marcus and Gerischer [49, 58], the electron-transfer rate depends on the energetic overlap of the donor and acceptor states, which are related to the density of states at energy E relative to the CB edge ($\rho(E)$), reorganization energy (λ), and temperature (T). At a given temperature, the overlap increases with a higher $\rho(E)$ and a smaller λ and reaches a maximum when the entire donor distribution function lies above the CB edge. A significant overlap is more favorable to the electron transfer than a small overlap. In general, a functional group, such as carboxylate, hydroxamate, or phosphonate moieties, is attached to the dye to attain a strong chemical bonding between the dye and the semiconductor surface. In addition to chemical adsorption, these functional groups also enhance the electronic coupling of the LUMO of the dye with the CB of the semiconductor via metal-to-ligand charge-transfer (MLCT) transitions by channeling the excitation energy into the ligand [51]. For instance, in the case of $RuL_2(SCN)_2$ dye adsorbed on a TiO_2 nanocrystalline film, the injecting orbital is the π^* wave function of the carboxylated bipyridyl ligand. The direct interaction between the carboxylate groups of dye with the surface Ti (IV) ions results in good electronic coupling of the π^* wave function with the 3d orbital manifold of the CB of TiO_2, leading to an extremely rapid injection of electrons. On the other hand, the back reaction of injected electrons in TiO_2 with the oxidized ruthenium complex involves a d orbital localized on the ruthenium metal whose electronic overlap with the CB of TiO_2 is small and is further reduced by the spatial contraction of the wave function upon oxidation of ruthenium, as illustrated in Figure 11.12. Thus, as discussed in the previous section, the back reaction of electrons with oxidized dye is orders of magnitude slower than the electron injection.

11.6.2.3 Electron Percolation through the Nanocrystalline Film

A perplexing phenomenon related to the DSC is its efficient charge transport although (i) the conductivity of oxide semiconductor film is inherently low, (ii) it lacks the depletion region and built-in electric filed due to small particle size, and (iii) the porous network of interconnected nanocrystallites creates huge grain boundaries and contact area [76]. Thus, a fundamental question of whether the electrons merely diffuse along the porous network or they are driven by an electric field needs to be addressed. Soedergren suggests that diffusion is the operative mechanism for charge transport in nanocrystalline films [77]. When the experimental results are interpreted as a diffusion model, the diffusion coefficient is in the range of 10^{-5} to 10^{-4} cm^2 s^{-1} [78, 79], and found to increase with light intensities [80]. However, these values are much lower than the diffusion coefficient for electrons in the CB of TiO_2 (0.2–0.5 cm^2 s^{-1}) [68, 81]. The discrepancy can be attributed to the trapping and detrapping of electrons. Different depths of traps make the process even more complex, leading to a distribution of trapping and detrapping times. Which type of trap the electron falls into while traveling randomly along the semiconductor film depends on its quasi-Fermi level under light illumination and, in turn, the intensity of light. At low light levels, electrons fall into deep traps and a slow transport with a corresponding low diffusion coefficient is expected. When the light intensity is increased, the deeper trap states are filled

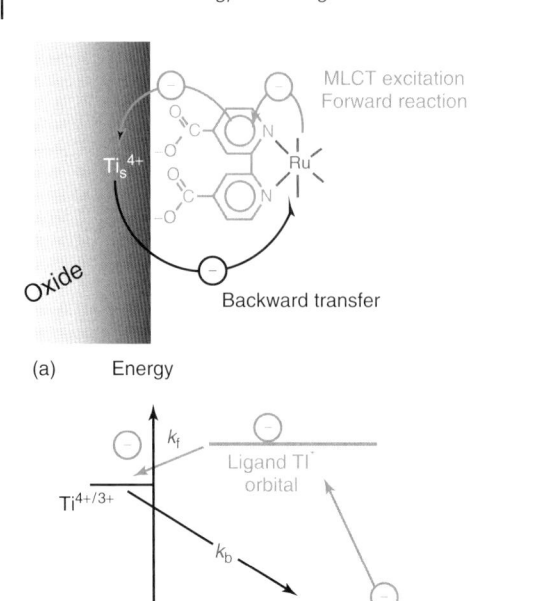

Figure 11.12 (a) Schematic of metal-to-ligand charge-transfer transitions (MLCTs) in ruthenium complex–TiO$_2$ system, where electron is transferred from ruthenium complex to the TiO$_2$ surface via a carboxylated bipyridyl ligand. (b) Orbital diagram for the forward electron injection with rate constant k_f from the π^* orbital of ligand into the 3d orbitals of the CB of TiO$_2$ and the backward electron transfer (rate constant k_b) from the CB of oxide into the Ru(III) d orbital. (From Ref. [51]. Copyright American Chemical Society. Reproduced with permission.)

under steady-state conditions. As the electrons fall only into shallow traps, the transport will be faster and the diffusion coefficient is higher accordingly [76]. Nevertheless, whether the origin and the location of the traps occur at the surface or near the surface remains unclear [82]. The charge transport in nanocrystalline film is also still under debate due to high current density despite a low conductivity of the film. It was also suggested that the charge transport in nanocrystalline TiO$_2$ is assisted by photodoping, which increases the conductivity of the film sharply because of an insulator–metal (Mott) transition in the donor band of anatase [83]. Schlichthorl et al. have calculated the steady-state carrier concentration in full sunlight and found that there is one electron per TiO$_2$ particle [66]. Goossens et al. have shown that, on average, there is one electron per particle over a wide range of illumination intensities. Based on this calculation, they suggested that the transport in a porous network is a "caterpillar"-like transport mechanism, where injection of an additional electron "pushes out" the electron that was occupied in that particle [50]. However, Hagfeldt pointed out that the resistance of the illuminated nanocrystalline film calculated using this result is at least 1000 times higher than the experimentally observed value and the discrepancy was explained as due to the photodoping of the nanocrystalline film upon illumination. Thus, more experimental data and transport models are required to obtain a clear picture on the electron transport along the nanocrystalline network.

11.6.2.4 Photovoltage in Dye-Sensitized Solar Cell

The photovoltage in any photovoltaic device ultimately depends on the ability of the illumination to effect, in the steady state, a difference in work function of the two electrodes through which the photoeffects are measured. [84]

According to the above statement, generally, photovoltage or open-circuit voltage (V_{oc}) is a function of the work function difference between the two electrodes. It also depends logarithmically on the light-generated and diode saturation current as mentioned in Section 11.3.3 (Eq. (11.5)). Sinton et al. suggested that, in a p-n junction silicon solar cell, V_{oc} can also be determined from the excess carrier concentration due to illumination [85].

$$V_{oc} = \frac{kT}{q} \ln\left[\frac{(N_A + \Delta n)\Delta n}{n_i^2}\right] \tag{11.13}$$

where N_A is the doping concentration, Δn is the excess carrier concentration, and n_i is the intrinsic carrier concentration. This means that the more the excess carriers injected into the semiconductor, the larger the photovoltage attained. The concepts mentioned above are also generally applicable to the DSC.

Now, let us take a look at energetics of the DSC (Figure 11.13) to better understand the concept of V_{oc} in the DSC. The energetics is referred to a DSC with an FTO electrode, a TiO_2 nanocrystalline film, the ruthenium–polypyridyl complex dye, I^-/I_3^- electrolyte, and a Pt counter electrode. Figure 11.13a reveals the energy levels of the individual layers in the DSC before contact. After TiO_2 makes contact with the liquid electrolyte in the dark, the Fermi level of semiconductor (E_{FS}) is aligned to the redox potential of electrolyte (E_{redox}) as well as the Fermi level of the two electrodes (E_F) under equilibrium condition (Figure 11.13b). Strictly speaking, a band shift in the CB of TiO_2 (E_C) is also observed upon contact with the electrolyte after dye adsorption because of solvation effects [86], and the Helmholtz layer dipole arises from the cation adsorption on the semiconductor surface due to the carboxylate or phosphate groups of the dye [66]. This band shift and band bending at the interface are neglected in this figure for simplicity. Under strong illumination at open circuit, the equilibrium condition is disturbed and E_{FS} moves up to electron quasi-Fermi level (E_{Fn}^*), which is close to the CB minimum of TiO_2 due to the accumulation of photogenerated electrons in the E_C of TiO_2 (see Figure 11.16c). Electron injection will also minimize any electric potential difference across the TiO_2/FTO interface. The electrochemical potential of electrons in FTO and TiO_2 will be the same under the open-circuit condition, resulting in an upward shift of the CB of FTO to that of TiO_2. Likewise, E_{redox} is shifted to hole quasi-Fermi level (E_{Fp}^*) owing to virtual hole injection into the electrolyte. The difference between the two

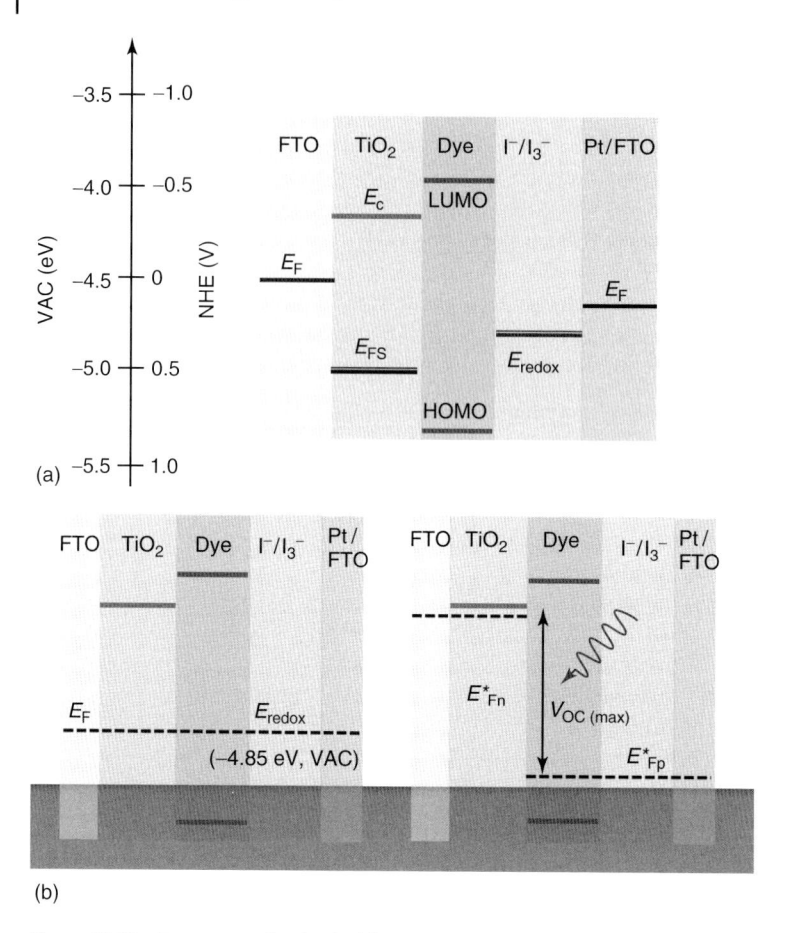

Figure 11.13 Energetics of individual layers in a typical DSC (a) before contact each other, (b) after contact under the equilibrium condition (in the dark), and (c) after contact in nonequilibrium condition (under illumination). The vertical axis on the left is the potentials referred to the vacuum level (VAC) and on the right is the potentials referred to the normal hydrogen electrode (NHE).

quasi-Fermi levels ($E_{Fn}^* - E_{Fp}^*$) determines the maximum attainable photovoltage of the DSC. As E_{Fn}^* cannot rise significantly above E_C of the semiconductor under normal injection conditions, and because E_{Fp}^* is close to E_{redox}, the maximum photovoltage in the DSC is generally expressed as $V_{oc(max)} = E_C - E_{redox}$.

However, as the experimental value of V_{oc} is smaller than expected value ($E_C - E_{redox}$) for all DSCs, there must be other limiting factors for the attainable V_{oc}. Turrion et al. proposed that the maximum attainable photovoltage is conditioned by the quasi-Fermi level of FTO (E_F^*(FTO)) rather than the CB of TiO$_2$ [87]. The injection of photogenerated electrons shifts upward the equilibrium Fermi level of TiO$_2$ and FTO to the quasi (nonequilibrium) Fermi level. When V_{oc} is reached, any additional increase of photon flux will raise the Fermi level at TiO$_2$ but not at FTO.

Thus $V_{oc} = [E_F^*(FTO) - E_{redox}]/e$ does not further increase. In fact, a potential barrier occurs at FTO/TiO$_2$ interface for applied bias close to or larger than V_{oc}. This barrier hinders the transport of electrons from TiO$_2$ to FTO, making $V_{oc(max)}$ smaller than the expected value of $E_C - E_{redox}$.

Because V_{oc} depends on the photogenerated current and recombination current as discussed above, it can also be expressed mathematically using these parameters [69]. As the photocurrent is the result of competition between the forward and back reaction, the observed photocurrent density J_{ph} is given by

$$J_{ph} = J_{inj} - J_r \tag{11.14}$$

where J_{inj} is the electron injection current resulting from dye sensitization and J_r is the recombination current arising from back reactions. J_{inj} is related to the incident photon flux I_0 (cm^{-2} s^{-1}) by the expression

$$J_{inj} = qAI_0 \tag{11.15}$$

where q is the charge of the electron and A is the ratio of the absorbed photon flux to I_0. We consider two main back reactions for recombination current: (i) the back reaction between the injected electrons in TiO$_2$ and the oxidized dye molecule and (ii) the back reaction between the injected electrons and the oxidized species in the electrolyte. The recombination current depends on the kinetic constant of the back reactions, the concentration of I_3^- in the redox couple $[I_3^-]$ and the oxidized dye $[D^+]$, and the difference between the electron concentration in the semiconductor in the dark n_0 and under illumination n. Thus, the recombination current density can be expressed as

$$J_r = qk_1[I_3^-](n - n_0) + qk_2[D^+](n - n_0) \tag{11.16}$$

where k_1 and k_2 are the kinetic constants of the back reaction with I_3^- in redox couple and oxidized dye, respectively. The electron concentration n is exponentially related to the photovoltage V by the equation

$$n = n_0 \exp\left(\frac{qV}{kT}\right) \tag{11.17}$$

where qV is the difference between the Fermi level of TiO$_2$ under illumination and in the dark. At the open-circuit voltage ($V = V_{oc}$), J_{ph} become zero, and from Eqs. (11.15) and (11.16) we obtain

$$qAI_0 = qk_1[I_3^-](n - n_0) + qk_2[D^+](n - n_0) \tag{11.18}$$

Combining Eqs. (11.17) and (11.18) yields the expression of V_{oc} with light intensity, the concentration of the oxidized dye molecule and triiodide in electrolyte, and kinetic constant of back reactions as in following equation:

$$V_{oc} = \frac{kT}{q} \ln\left(\frac{AI_0}{k_1[I_3^-]n_0 + k_2[D^+]n_0} + 1\right) \tag{11.19}$$

Usually, $AI_0 \gg k_1[I_3^-]n_0 + k_2[D^+]n_0$ and simply V_{oc} can be expressed as

$$V_{oc} = \frac{kT}{q} \ln\left(\frac{AI_0}{k_1[I_3^-]n_0 + k_2[D^+]n_0}\right) \tag{11.20}$$

The order of the rate of reaction and electron-transfer coefficient are considered in some expressions but they are taken as unity here. In Eq. (11.16), both localized electrons in the surface states and free electrons in the CB contribute to the recombination current. Moreover, in contrast to conventional solid-state devices, n_0 is not the same as the doping density of the semiconductor. Because the TiO_2 nanocrystalline film is in contact with the electrolyte, n_0 depends on the difference between the CB of TiO_2 (E_c) and redox potential of electrolyte (E_{redox}) and can be expressed as $n_0 = N_c \exp[(E_c - E_{redox})/kT]$ where N_c is the effective density of states in the CB. Thus, a band shift in the dark due to solvation effects upon contact with electrolyte and Helmholtz layer dipole that arises from dye adsorption will alter n_0 and affect V_{oc}.

11.6.3
Photoanodes Other Than TiO$_2$ Nanoparticle Network

A photoanode with a titania anatase nanoparticle film that provides a large surface area for anchoring sufficient sensitizers offers the highest efficiency, up to now. Despite exhibiting efficient charge collection efficiency, the diffusion coefficient of electrons in a randomly oriented nanoparticle TiO_2 network is several orders of magnitude smaller than in a single-crystal film due to millions of trapping and detrapping events along the network, which was discussed in an earlier section. As the diffusion length of electrons in a TiO_2 nanoparticle network is in the range of 4–30 μm, depending on the intensity of light [80, 88–90] an increase in the nanoparticle film thickness significantly above 10 μm (in order to increase light absorption) does not increase the energy conversion efficiency. Owing to this limitation, TiO_2 with other morphologies as well as other wide-bandgap n-type semiconductors with various morphologies are also paid attention to in DSC research.

A vertically aligned titania nanotube array is a promising alternative to a random nanopatricle network for the potential enhancement of the electron percolation pathway. The diffusion length of electrons in a titania nanotube array on a titanium foil is estimated to be 100 μm [91], which is at least 3 times longer than that in a random nanoparticle film. A longer diffusion length allows the use of a higher thickness of film, leading to high light harvesting using either a single dye or multiple dyes covering different regions of the nanotubes without sacrificing the charge transport. In addition to efficient electron percolation and long diffusion length, it has been shown that DSCs fabricated using TiO_2 nanotube arrays grown on a Ti foil have 20% higher light-harvesting efficiency owing to the enhanced light-scattering properties of TiO_2 nanotubes [92]. Various methods such as sol–gel [93], sonochemistry [94], surfactant templating [95], molecular assemblies [96], and anodization [97] have been used to fabricate the nanotube structure but only molecular assemblies and anodization methods have been demonstrated in DSC applications. Adachi et al. have fabricated DSCs with a 4-μm thick photoanode composed of TiO_2 single-crystalline nanotubes (10 nm diameter, 30–300 nm length) using molecular assemblies and obtained an efficiency of 4.88% [96]. Later on, TiO_2 nanotubes fabricated by anodization followed by oxygen annealing, however, became more popular for DSC applications. Presently, anodic TiO_2 nanotube-based DSCs have two configurations: backside-illuminated foil-based, and front-side illuminated, as shown in Figure 11.14 [98]. In the former, TiO_2 nanotubes are directly anodized on the Ti foil and are illuminated from the transparent platinized counter electrode. A PCE of 6.89%, with V_{oc} of 0.817 V, J_{sc} of 12.72 mA cm^{-2}, and FF of 0.663, has been achieved with the optimized nanotube length of 20 µm [99]. The efficiency was improved approximately 62% compared to the DSC with a 6-µm long nanotube array. Although this configuration of DSC enables the fabrication of very long nanotubes, the drawback is a loss of ~25% of the incident light because the platinized counter electrode partially reflects light while iodine in the electrolyte absorbs photons in the near-UV region. In order to take advantage of front-illuminated configuration, freestanding nanotubes are prepared by immersion in 10% H_2O_2 solution for several hours after anodization of the Ti foil to separate the nanotubes and Ti substrate [100], and then transferred to the FTO glass with the aid of binding agents such as an alkoxide solution [101] or a titania nanoparticle paste [100]. The front-illuminated DSC fabricated with 25-µm long freestanding nanotubes gives J_{sc} of 12.4 mA cm^{-2}, V_{oc} of 0.7 V, and FF of 0.633, producing an overall efficiency of 5.5%, which is even lower than that of the backside illuminated cell. The challenges for this method are in preventing macro/micro crack formation in the film and the collapse of nanotube arrays into small debris during the transfer to FTO and annealing treatment. Long-term stability of binding between freestanding nanotubes and FTO glass is also another issue for such kind of devices. Another method employed in the frontside-illuminated cell is the fabrication of nanotubes on Ti-sputtered FTO glass. Mor *et al.* have prepared 500-nm thick titanium films on FTO by RF sputtering processed at 500 °C, followed by anodization and annealing in oxygen, resulting in 360-nm long nanotube arrays with high optical transparency [102]. The front-illuminated DSC fabricated with $TiCl_4$-treated nanotube arrays yields an efficiency of 2.9% with J_{sc} of 7.87 mA cm^{-2}, V_{oc} of 0.75 V, and FF of 0.49. The low current density and FF are attributed to the less photoabsorption due to short nanotube length and high series resistance arising from the thick barrier layer left over during anodization, respectively. However, increasing the length of nanotubes is limited by the difficulties in fabrication of high-quality, uniform Ti film above 500 nm [102], nonuniform optically transparent nanotube array due to different anodization rates at different regions [103], and debris or clumping of the nanotubes [104]. The first problem was overcome by bombarding the growing metal film with energetic ions during sputtering, and the second and third problems by conditioning the organic electrolytes with applied electric field, resulting in up to a 33-µm long and debris-free nanotube array from 20-µm Ti film [105]. (Note that anodization in organic electrolyte can yield thickness up to 3 times the thickness of the starting metal film thickness, partially due to lateral motion of anodized material into the nanotube walls.) The front-illuminated DSC with 17.6-µm long nanotubes, fabricated from 20-µm thick Ti film, yielded J_{sc} of 15.8 mA cm^{-2} (which is higher than that of backside-illuminated DSC), $V_{oc} = 0.73$ V and FF $= 0.59$, leading to a power conversion efficiency of 6.9%. However, a shortcoming of the deposition of

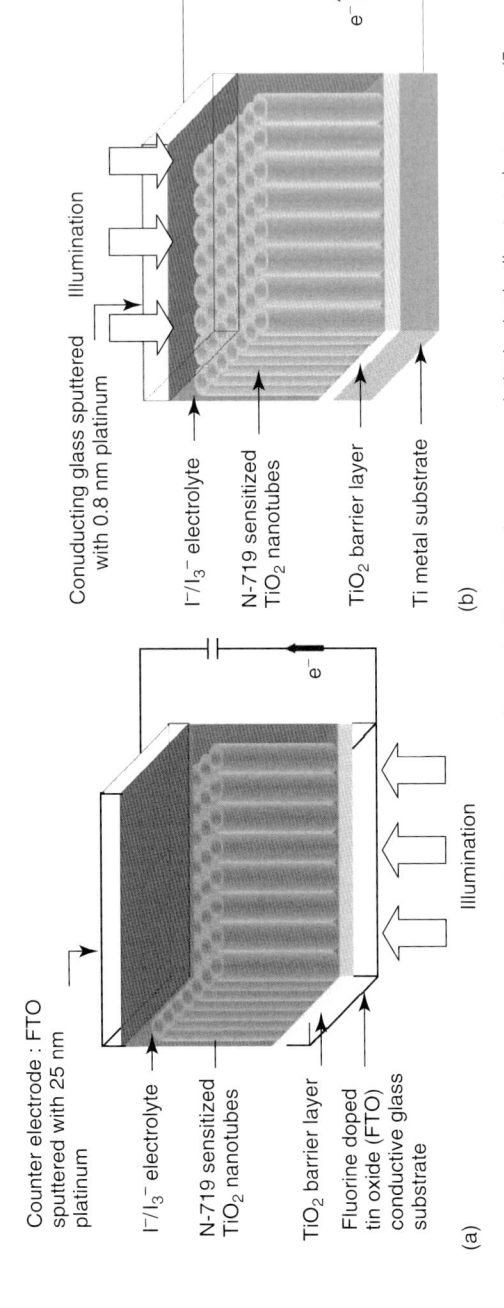

Figure 11.14 Integration of TiO_2 nanotube array into (a) front-side illuminated structure and (b) backside illuminated structure. (From Ref. [98]. Copyright Elsevier. Reproduced with permission.)

a thick Ti film by sputtering is the extremely long processing time. It took about 2.5 days to attain a 20-μm film with a deposition rate 5.4 nm min^{-1} and it is hard to realize in mass production.

ZnO is an alternative n-type semiconductor to TiO$_2$ in applications of photoanode because of its wide bandgap (3.37 eV), favorable CB energy for electron injection, and high electron mobility. Its amenability to large-scale and low-cost fabrication with controlled complexity through template-less and surfactant-free aqueous method is also appealing to low-cost solar cell manufacture. In fact, ZnO as photoanode in DSCs was applied earlier than TiO$_2$ and the first photon-to-electrical current conversion from a dye-sensitized semiconductor was demonstrated with a ZnO single crystal [32, 33]. The lack of technology to fabricate large-area ZnO films at that time, however, made TiO$_2$ ideal as a photoanode. The fabrication of nanostructures with low-cost aqueous solution processes revitalized ZnO in DSC research again. Various nanostructures such as nanoparticles [106], nanowires/nanorods [52], nanotubes [107], nanosheets [108], nanoflowers [109], and nanofiber mats [110] have been investigated for DSC. In 1994, Redmond et al. demonstrated an overall efficiency of 0.4% using 10-nm large and 0.4-μm thick ZnO nanoparticles fabricated by a sol–gel process [111]. A few years later, Rensmo et al. attained efficiency an of 2% by employing 30-μm thick ZnO nanoparticle films produced from ZnO powder [112]. By controlling the dye sensitization to avoid dye aggregation, efficiency up to 5% was achieved with 14-μm thick ZnO nanoparticle photoanode [106]. Even a higher efficiency of 5.4% was reported by employing polydisperse submicrometer-sized ZnO aggregates consisting of nanoparicles [113]. The nanoparticles effectively enhance the surface area of photoelectrode while submicrometer-sized aggregates facilitate the light scattering inside the film. Saito et al. fabricated a photoanode consisting of two layers: a 16.1-μm thick bottom layer composed of 20-nm particles and a 6.7-μm thick top layer composed 600-nm particles for light scattering from a nanoparticle paste followed by sintering at 450 °C and an efficiency of 6.58% was achieved [114]. This is more than half of the highest efficiency of TiO$_2$-based DSC and the highest efficiency reported so far for ZnO-based DSC, to our best knowledge.

While progress in ZnO nanoparticle-based DSC was continuing, nanowire/nanorod-based DSCs were also introduced, since it was expected that a dense network of ZnO nanowires and nanorods should be favorable for efficient electron collection because these morphologies provide direct conduction paths for electrons to be transported from the point of injection to the electrode. Unfortunately, the efficiency of DSCs with ZnO nanowires/nanorods turned out to be lower than that of nanoparticle-based DSCs. Baxter and Aydil attained an efficiency of 0.5% using ZnO nanowires of several micrometers long with secondary nucleation of tiny additional nanowires on the surface grown by MOCVD [115]. The low efficiency was attributed to the low dye loading of 9×10^{-9} mol cm^{-2}, which is an order of magnitude lower than for the nanoparticle film, and in turn low absorption of light: only 9% of incident light at the dye's absorption maximum. Thus, growing longer wires with a smaller diameter was recommended to increase the surface area for higher dye loading. Law et al. pushed the PCE to 1.5% by increasing the aspect ratio (length to diameter ratio) of hydrothermal-grown nanowires to more than 125 [52]. Incorporation of polyethylenimine in a precursor aqueous solution to control the lateral growth of nanowire produced ZnO nanowires with length of 20–25 μm and diameter of ~200 nm. Law also pointed out the enhanced charge collection efficiency of ZnO nanowires, comparing the J_{sc} of the nanowire cells to those of TiO_2 and ZnO nanoparticle cells as a function of the surface area (roughness factor), as shown in Figure 11.15. The rapid saturation and subsequent decline of J_{sc} from the DSCs built with 12-, 30-, and 200-nm ZnO particles was observed, but a linear increase in J_{sc} was found in the ZnO nanowire cell. The J_{sc} of DSCs with a thin TiO_2 nanoparticle film (2.2 μm) is equivalent to that of DSCs with long ZnO nanowires (18–24 μm) at the same dye loading (3×10^{-8} mol cm^{-2}), suggesting that the charge collection in the ZnO nanowires is more efficient than in TiO_2 nanoparticles. Moreover, the nanowire cells generated considerably higher currents than either ZnO particle cells over the accessible range of roughness factors. Another important feature of nanowires is the radial electric field (depletion layer) distributed inside the wire, which is not possible in nanoparticles due to the small size of the particles. The Debye–Hückel screening length of ZnO is about 4 nm for a carrier concentration of 10^{18} cm^{-3}, making the nanowire thick enough to

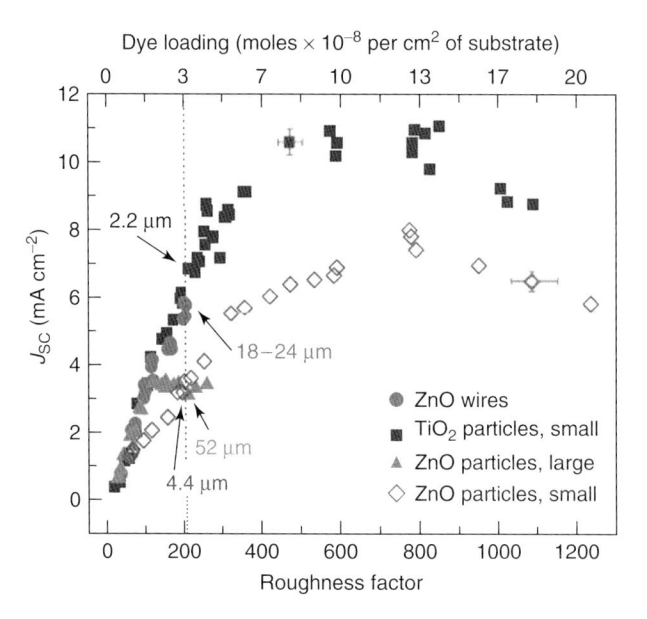

Figure 11.15 Comparison of the performance of the ZnO nanowire cell and ZnO/TiO_2 nanoparticle cell as a function of the roughness factor and dye loading. Error bars are provided on only two points for the clarity of the figure. (From Ref. [52]. Copyright Nature Publishing Group. Reproduced with permission.)

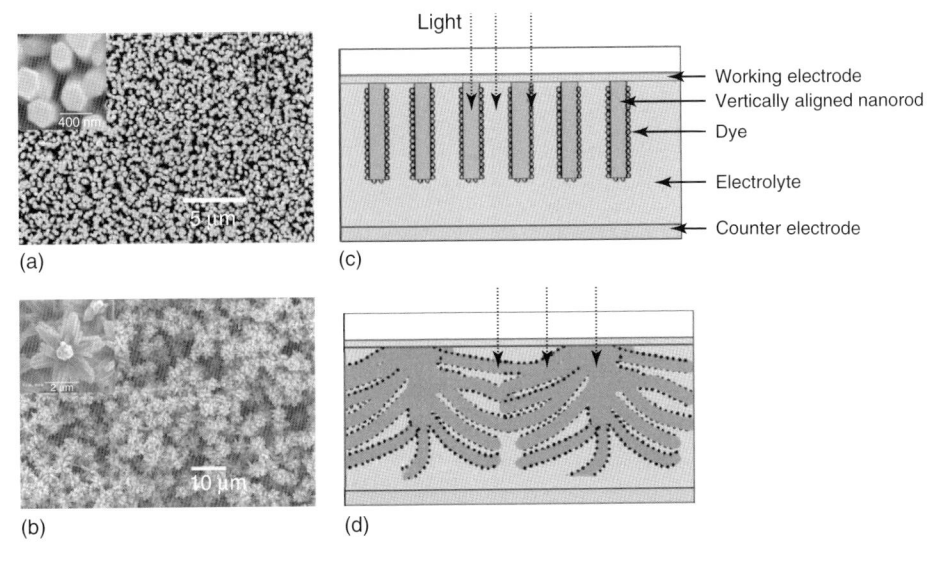

Figure 11.16 SEM images of (a) ZnO-nanorod array and (b) nanoflower film grown on FTO glass by hydrothermal synthesis. Schematic of (c) upstanding nanorod and (d) flowerlike arrays illuminated under light. (From Ref. [116]. Copyright American Institute of Physics. Reproduced with permission.)

support the electric field. The axial electric field along each nanowire not only drives the injected electrons toward the electrode but also encloses the carriers within the wire cores because of the upward band bending at the nanowire surface.

Despite enhanced charge collection efficiency, the main challenge for ZnO nanowire-based DSC is to increase surface area for higher dye loading. Thus, morphologies that can provide a higher surface area have become appealing to ZnO-based DSC research. We have shown that the efficiency of a hydrothermal-grown nanoflower cell (1.9%) is nearly twice that of a nanorod cell (1%) because nanoflower morphology increases both the surface area and light–dye interaction, promoting light harvesting of the cell, as illustrated in Figure 11.16 [116]. Kakiuchi *et al.* also fabricated large-surface-area ZnO films composed of mesoporous nanosheets accumulating with submicrometer-order spacing in between, through an aqueous solution method followed by annealing at 300 °C in air (Figure 11.17) [108]. A 4-µm thick nanosheet film achieved the dye loading of 1.1×10^{-7} mol cm^{-2}, which is comparable to 1.3×10^{-7} mol cm^{-2} of a 10-µm thick TiO$_2$ nanoparticle film. In addition, an effective discharging of Zn^{2+}/dye aggregates through the macropores present in the film also facilitates the increase of the current density, leading to an efficiency of 3.3%.

Regardless of morphology, the reported efficiency of ZnO-based DSCs is still lower than that of TiO$_2$-based DSCs. One problem associated with ZnO-based

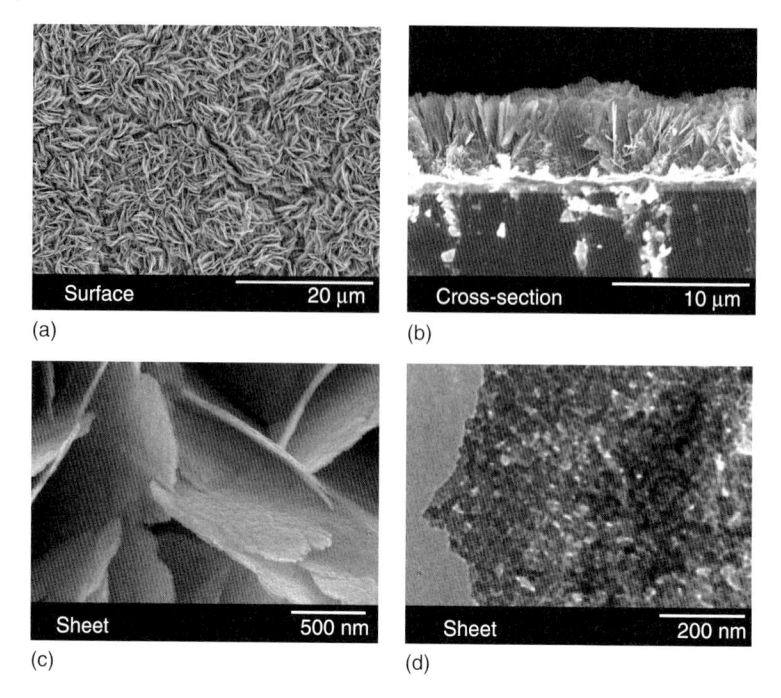

Figure 11.17 (a) Surface and (b) cross-sectional field emission scanning electron microscope (FESEM) images of the ZnO nanosheet film heat-treated at 300 °C, (c) high-magnification FESEM image, and (d) transmission electron microscope image showing the ZnO nanosheets. (From Ref. [108]. Copyright Elsevier. Reproduced with permission.)

DSCs is that the protons derived from the ruthenium complexes make the dye solution relatively acidic and dissolve the surface of ZnO, generating Zn^{2+}/dye aggregates. These aggregates are detrimental to the photoanode because they block the nanoscale pores and lower the electron injection efficiency. Thus, the dye loading time for the ZnO film must be controlled to within 1–2 h, which is much shorter than dye loading time for the TiO_2 film. The controlled sensitization time to avoid Zn^{2+}/dye aggregates may lead to low dye loading and, in turn, to low efficiency. Thus, for an efficient dye loading process of ZnO electrodes, Keis et al. suggested dyes with no acidic protons or adding a base (KOH) to the dye solution [118]. They proved that a ZnO film sensitized in the dye solution containing KOH for 24 h yielded higher efficiency than in the dye solution without KOH for the same period (3.4 vs 1.1%); yet the overall efficiency was still low [106]. Another factor responsible for the low efficiency is probably related to the slow electron injection from ruthenium complexes to ZnO, but the experimental data for the electron injection rate in the ruthenium–ZnO system are ambiguous. Lian et al. observed multiexponential injection of electrons from Ru N3 dye to the ZnO film with a small <100 fs injection component and multiple slower components with time constants ranging to hundreds of picoseconds, which is more than an order of magnitude slower than injection from Ru N3 to TiO_2 [119–121]. Despite the similarities in semiconductor energetics, TiO_2 and ZnO exhibit different electron injection rates probably due to differences in CB structure, electronic coupling strength, and density of semiconductor states. Bauer et al. argued that the electron injection time from Ru N719 to ZnO occurs in the subpicosecond regime (<300 fs), which is comparable to that of the N719–TiO_2 system [122]. However, Lian et al. pointed out that Bauer's observation was based on a fast rise in the visible transient absorption at 770 nm, which can be observed regardless of injection rate, and hence the rise time probed at this wavelength alone cannot reliably indicate electron injection rates [120]. On the other hand, Katoh et al. suggested that, although the electron injection rate in ruthenium–ZnO system may differ from that in ruthenium–TiO_2 system, the injection efficiency from the excited dye to ZnO is still very high because the injection rate is much higher than relaxation rate of the excited state of the dye which is in the nanosecond range [123]. Therefore, the electron injection rate does not limit the photocurrent generation. As the regeneration of dye is also insensitive to the nature of the semiconductor, the possible limiting factor for the low efficiency in ZnO-based DSCs is either the electron transport in semiconductor or the back reaction between CB edge electrons and the molecules in the solution. As the charge collection efficiency in ZnO nanowires is better than in a TiO_2 nanoparticle network, the latter reason is more probable for the low efficiency, yet a recombination kinetics study for the dye–ZnO system is necessary to confirm this argument.

Other n-type semiconductors such as SnO_2 [42, 117], Fe_2O_3 [124], Nb_2O_5 [123, 125, 126], In_2O_3 [123, 125], WO_3 [125], and ZrO_2 [123, 125] were also investigated for their use as photoanodes in the DSC. Except ZrO_2, in which the bottom of the CB is located above the LUMO level of the ruthenium complex, the rest show PV characteristics, but the efficiency of the cells with these materials is far below the current highest efficiency of TiO_2- and ZnO-based cells.

11.6.4
Photocathode and p-DSC

The conventional DSC developed by Grätzel can be considered as an n-DSC because the excited states of the dye inject electrons into the CB of a wide-bandgap n-type semiconductor. On the other hand, p-DSC, in which holes are injected into the VB of a p-type semiconductor after photoexcitation, can be fabricated by replacing the traditional photoanode with a p-type semiconductor photocathode. Such a standalone p-DSC does not offer special benefits in terms of device performance or cost, yet a simple tandem cell is viable by combining a p-DSC and an n-DSC, which was highlighted by He et al. [127]. While the theoretical upper limit of electrical conversion efficiency for a cell with a DSC with only one photoanode is around 30%, the corresponding limit for a tandem cell with two photoactive semiconductors can reach up to 43% (tandem DSCs are discussed in detail in Section 11.6.5.1). With the aim to improve the efficiency of the DSC by replacing the passive electrode with a dye-sensitized photocathode, the development of efficient p-DSCs becomes important in DSC research.

Several p-type semiconductors such as CuI [128], CuSCN [129, 130], and NiO [131] have been tested as dye-sensitized photocathodes. Among these materials, NiO is a typical material for photocathode in p-DSCs as it is a large-bandgap (3.6–4.0 eV) semiconductor with stable p-type semiconducting property. The HOMO level of the dye used in p-DSCs must be low enough below the VB of NiO, while the LUMO level should be high enough above the redox potential of the iodide–triiodide system. After photon absorption and the optical excitation of the dye, two processes occur: (i) electron transfer from excited dye to the oxidized species (I_3^-) in the electrolyte; and (ii) electron transfer form the VB of the semiconductor to the HOMO level of the dye (Figure 11.18) [131]. In early days, it was not clear which process took place first. However, later on, ultrafast spectroscopy measurement revealed that (ii) is followed by (i) [132–134]. The hole in the VB diffuses toward the conducting electrode. I_3^- in the electrolyte is reduced to I^- as a result of the injection of electrons from the excited dye. I^- is regenerated back to I_3^- at the counter electrode by donating the electron there. It is expected that, under illumination, the Fermi level of NiO will be lowered and approach the upper edge position of the VB at open circuit, and the V_{oc} is governed by the difference between the valence band potential of NiO and the redox potential of the electrolyte.

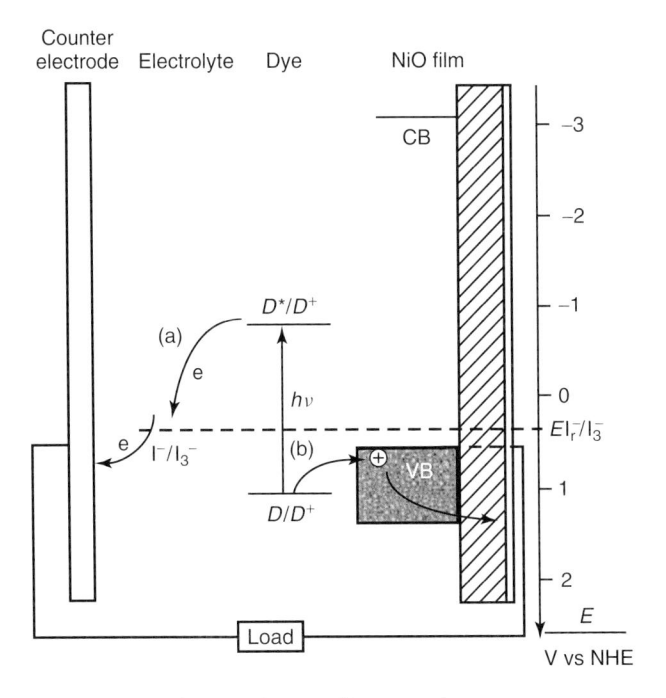

Figure 11.18 Schematic diagram illustrating the energetics of p-DSC based on NiO semiconductor and charge-transfer process in the system (From Ref. [131]. Copyright American Chemical Society. Reproduced with permission.)

Generally, NiO p-type semiconductor is sensitized with erythrosine [131, 136, 137], coumarin [132, 135, 137, 138], and porphyrin [131, 133] dyes whose energetics meets the requirement for the electron-transfer process. However, in contrast to n-DSC, the reported efficiency of p-DSC with these dyes is very low, with V_{oc} around 0.1 V and J_{sc} less than $1\,\text{mA}\,\text{cm}^{-2}$. A low V_{oc} is caused by the inherent property of the material used, that is, the small difference between the VB potential of NiO (0.5 V normal hydrogen electrode (NHE)) and the redox potential of the electrolyte (0.35 V NHE). A low J_{sc} would be due to a low charge separation (injection) efficiency and/or to a short hole diffusion length. The charge separation process at the interface of the NiO semiconductor and the commonly used dyes is rather fast. For instance, the photoinduced electron transfer from the VB of NiO to coumarin dye takes ~200 fs [132], which is comparable to the electron injection time in n-DSCs. Similarly, the light excitation of porphyrin adsorbed on NiO induces a rapid hole injection into the VB of NiO in the 2–20 ps time scale [133]. An ultrafast charge injection suggests that a small J_{sc} is not limited by the charge separation efficiency. Thus, a short diffusion length might result in a low photocurrent. Since $L = \sqrt{D \times \tau}$ where L, D, and τ are the diffusion length, diffusion coefficient, and lifetime, respectively, the short diffusion length could be caused by a low diffusion coefficient and fast recombination. The hole diffusion coefficient in a NiO electrode is estimated to be $4 \times 10^{-8}\,\text{cm}^2\,\text{s}^{-1}$, which is an order of magnitude smaller than that in a TiO_2 photoanode. Moreover, unlike in the n-type semiconductor, in n-DSCs the diffusion coefficient of NiO is almost independent of the light intensity but strongly dependent on the type of cation in the electrolyte. This is probably due to the fact that different cations adsorb to different extents at the NiO surface and affect the hole transport. This suggests that hole conduction occurs mainly at the NiO/electrolyte interface, most likely by hopping of the charges at the interface. Nonetheless, charge transport in the photocathode is less studied, and more conclusive data is required. A more interesting finding in the p-DSC is its fast recombination process. The back reaction between the reduced dye and the injected hole in the VB of the semiconductor can be as fast as ~20 ps, meaning that recombination occurs just after the hole injection. Therefore, even though the diffusion coefficient of holes in NiO is comparable to that of electrons in TiO_2, most of the carriers will be lost through recombination. The rapid recombination also hinders an efficient regeneration of dye to ground state from iodide/triiodide redox mediator. Thus recombination appears to be a major cause for the low photocurrent in a p-DSC.

A possible strategy to tackle the recombination problem is to provide an efficient electron-transfer pathway from the dye to the electrolyte. Morandeira *et al.* designed a sensitizer–acceptor dyad composed of the peryleneimide (PI) sensitizer and a naphthalenediimide (NDI) unit which acts as an electron acceptor [139]. The PI sensitizer was bonded to the semiconductor surface, while the secondary electron acceptor unit was located further away from the interface. With an appended secondary electron acceptor unit, the reduced dye can be rapidly regenerated via triiodide in electrolyte and the distance between the hole and electron is significantly increased, resulting in an efficient, long-lived charge separation.

The electron injection to the dye from the VB of NiO occurs with an average time constant of 0.5 ps in both systems (pristine sensitizer and sensitizer–acceptor dyad), whereas the latter exhibits a substantial retardation of the charge recombination between the hole and the reduced sensitizer compared to the former system, leading to an APCE of approximately 45%, which is a threefold increase over the pristine sensitizer system. Qin et al. also adopted a similar strategy, synthesizing an acceptor–donor-type dye using the triphenylamine moiety as the electron donor, malononitrile moiety as the electron acceptor, and a thiophene unit as the conjugated chain [134]. In contrast to the normal organic dyes in n-DSCs, the attaching group is now on the electron donor part. By introducing two electron acceptor groups and having the anchoring group on the donor moiety, light excitation is associated with electron flow from the donor part of dye to the acceptor part, providing an efficient electron-transfer pathway from the dye to the electrolyte.

Another approach to suppress recombination is increasing the channeling distance between the LUMO of the dye and the VB in NiO. Nattestad et al. synthesized acceptor–donor dye types (**1–3**) comprising perylenemonoimid (PMI) as the acceptor and an oligothiophene coupled to triphenylamine as the donor, while the length of the region regularly alkylated oligo-3-hexylthiophene unit was systematically varied from a bithiophene ($n = 1$; **1**) to a quaterthiophene

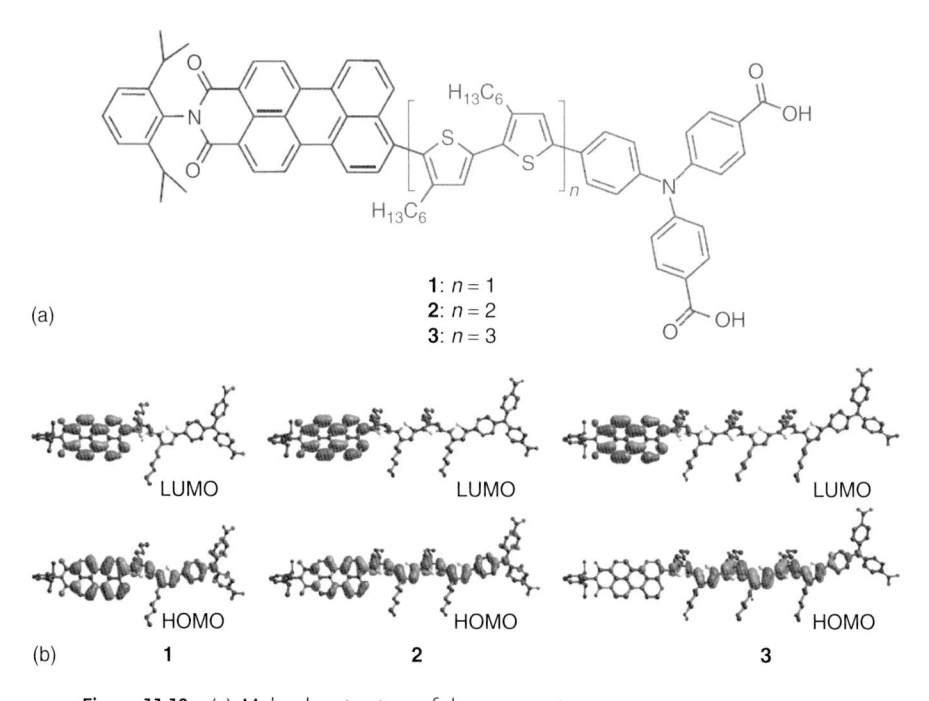

Figure 11.19 (a) Molecular structure of donor–acceptor dyes **1–3**. (b) Electron density distributions for LUMOs and HOMOs of the dyes. (From Ref. [140]. Copyright Nature Publishing Group. Reproduced with permission.)

($n = 2$; **2**) and a sexithiophene ($n = 3$; **3**) (Figure 11.19) [140]. Increasing the length of oligothiophene bridge does not alter the energetic position of the LUMO significantly but increases the channeling distance between the electrons occupying the dye's LUMO and the vacancies located in the NiO electrode, decelerating the recombination kinetics of the photoreduced dyes with vacancies present in the VB of NiO by several orders of magnitude. The increased hydrophobicity of the dyes with increasing length of the oligothiophene linker also helps to shield the NiO surface from electrolyte and hinder the back reaction between I^- in electrolyte and holes in VB of NiO. The APCE of the p-DSC employing acceptor–donor-type dye with long oligothiophene bridge ($n = 3$) yielded up to 96%.

11.6.5
Improvement of Spectral Response by Tandem Structure and CoSensitization

Light absorption in DSC is dictated by the molar extinction coefficient of the dye, the surface coverage of the dye (dye molecules per square nanometer), and the total surface area of the semiconductor film [142]. Dye molecules generally anchor to the semiconductor surface in a close-packed manner with a density of 0.5–1 dye molecule nm^{-2} [142]. Furthermore, replacing the flat-surface semiconductor film with a porous film comprising of nanoparticles increases the surface area effectively. The surface area of the porous film is more than 1000 times greater than that of a flat surface for a mere 10-μm thick film. However, the molar extinction coefficients of conventionally used ruthenium-based dyes are low (5000–20000 M^{-1} cm^{-1}). For example, N719 and N3 dyes can absorb solar radiation up to 750 nm with fairly broad absorption spectra ($\Delta\lambda \approx 350$ nm) but the molar extinction coefficients of these dye are not enough to absorb photons efficiently at longer wavelengths (>600 nm) [143]. Because the solar spectrum has a large photon flux in a wavelength region 500–1000 nm, the absorbance of photons in the longer wavelength region is a key to enhance the photogenerated current. Thus, sensitizing dyes that can extend absorption to the near-IR region of spectrum were developed. Ruthenium complex with terpyridine as a ligand, called the *"black dye"* or N749, extends the absorption up to 1000 nm with very broad absorption spectra ($\Delta\lambda > 500$ nm) [144, 145]. Unfortunately, dyes having near-IR absorption tend to have a lower molar extinction coefficient due to lower excited-state excess free energies to the electron injection. On the other hand, organic dyes have recently been developed with significantly higher molar extinction coefficients (50000–200000 M^{-1} cm^{-1}) but with narrow spectral responses ($\Delta\lambda \approx 250$ nm) [141, 146, 147]. The absorption spectrum can be broadened through dye aggregation, which has been explored with both the blue-shifted H-aggregates and red-shifted J-aggregates of merocyanine dyes [148, 149] but the spectral response is still narrower than that of metal coordination complexes. As a general rule, dyes that absorb light strongly do not have broad absorption spectra. Therefore, one limiting factor to further increasing the efficiency of DSCs is the inability of the commonly used dyesju to efficiently absorb light over a broad range of the solar spectrum. This limitation can be overcome only by the combination of several dyes having different absorption spectra, leading to the emergence of cosensitized and tandem-structure DSCs.

11.6.5.1 Tandem-Structure DSC

A tandem cell formed by physically stacking two individual cells and then connecting externally in parallel is the simplest way to enhance light absorption and extend the absorption range of DSCs. This concept was demonstrated by Kubo et al. in 2004, using the structure shown in Figure 11.20a, where D1 and D2 refer to N719 and black dye, respectively [150]. With a 7.7-μm thick TiO_2 in the front cell and a 9.9-μm thick TiO_2 in the back cell, the fabricated tandem cell revealed improved spectral response (especially at the higher wavelength region), higher photocurrent, and higher conversion efficiency. The photocurrent generated by the tandem cell is even higher than that produced from single cell with a light-scattering layer (TiO_2 particle with diameter ca. 300 nm), which has been proven to be due to the increase in light absorption as a result of increase in the path length for light travel [151–153]. The highest efficiency can be attained by a tandem cell with light scattering incorporated in the back cell. Durr et al. also fabricated a similar structure to realize the tandem concept for DSCs [154]. A TiO_2 film of 10 μm thickness and comprising of 14-nm diameter particles was used in the front cell, whereas mixed small and large particles (80 wt% of 20-nm diameter particles and 20 wt% of 300-nm diameter particles) were employed in the back cell to increase the light scattering. Similar observations in the result of such tandem cell are (i) current density is enhanced as a result of the contribution from the second cell; and (ii) V_{oc} of tandem cell is smaller than that of subcell with higher value owing to the compensation from subcell with lower value. Instead of stacking two subcells back to front, Murayama proposed a tandem structure with two subcells which are stacked face to face using a transparent platinum-mesh sheet as a common counter electrode and electrically connected in parallel as shown in Figure 11.20b [155]. Although the authors claimed that this structure harvests the incident light more efficiently than previous structures, it seems that the back cell is still suffering from loss of light because the incident light still needs to pass through the Pt layer and electrolyte in the back cell before reaching the sensitizers. The only advantage over the previous structure is that it can reduce the weight and probably the cost of the cell from eliminating a conducting glass. For both structures, a tandem cell with a short-wavelength-harvesting dye located in the front cell while long-wavelength-harvesting dye placed in the back cell gives the highest efficiency because the absorption of triiodide in the electrolyte is considerably strong in the short wavelength region (<500 nm) yet almost transparent in long wavelength region. Principally, it is hard to achieve the current density of a tandem DSC equal to the sum of the densities of individual separate cells mainly because the absorption spectra of the different dyes overlap and a significant fraction of light is lost after passing through the electrolyte and platanized counter electrode. Therefore, despite an enhanced efficiency, a tandem cell formed by stacking two complete cells is not beneficial in terms of cost per electricity output, as the material cost is nearly doubled yet the efficiency is not doubled.

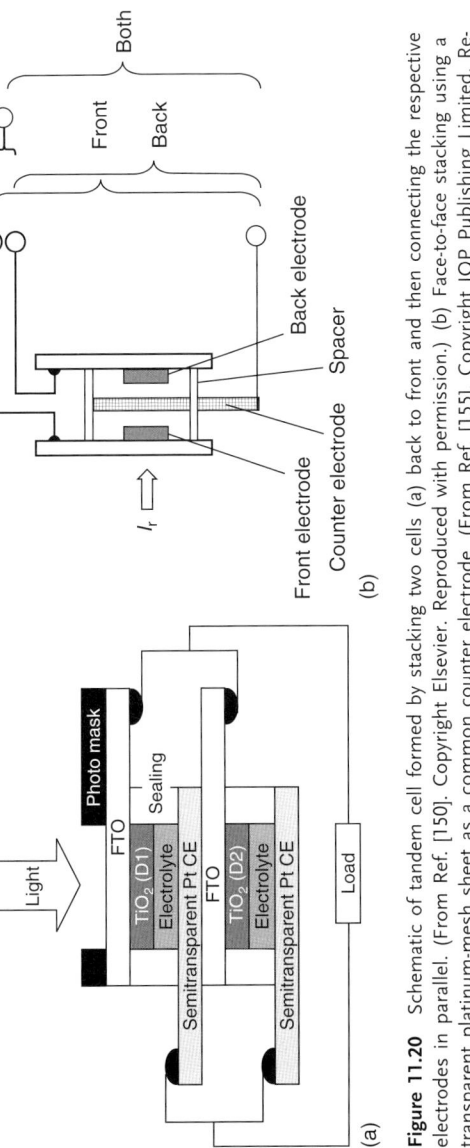

Figure 11.20 Schematic of tandem cell formed by stacking two cells (a) back to front and then connecting the respective electrodes in parallel. (From Ref. [150]. Copyright Elsevier. Reproduced with permission.) (b) Face-to-face stacking using a transparent platinum-mesh sheet as a common counter electrode. (From Ref. [155]. Copyright IOP Publishing Limited. Reproduced with permission.)

An alternative architecture of tandem DSCs other than stacking two complete cells is a p-n DSC consisting of an n-DSC and a p-DSC connected in series by an electrolyte as the intermediate layer. From a commercialization perspective, this concept appears to be more attractive, as it requires only a marginal extra manufacturing cost compared to a single DSC. This kind of architecture is similar to tandem organic PVs in which top and bottom subcells absorbing complementary spectra are connected with an intermediate layer where charge carriers are recombined. Because of the series connection of the photoanode and photocathode in the p-n DSC, the resulting photocurrent is governed by the weakest photoelectrode, whereas the photovoltages add up. The first p-n tandem DSC was demonstrated by He *et al.* in 2000 [127]. Despite the low efficiency (0.39%), the photovoltage of the tandem cell was larger than that of an individual cell. The main challenge in constructing a highly efficient p-n DSC is to improve the performance of the p-DSC, which is currently far lower than that of the n-DSC due to the fast recombination kinetics. Recently, Nattestad *et al.* suppressed the charge recombination in p-DSC by increasing the tunneling distance between the electron occupying the dye's LUMO and the VB of the photocathode, thereby significantly boosting the efficiency of the p-DSC. As a consequence, they successfully demonstrated a highly efficient p-n tandem DSC with a TiO_2 photoanode and a NiO photocathode, as illustrated in Figure 11.21 [140]. N719 dye was used in the n-DSC, and a

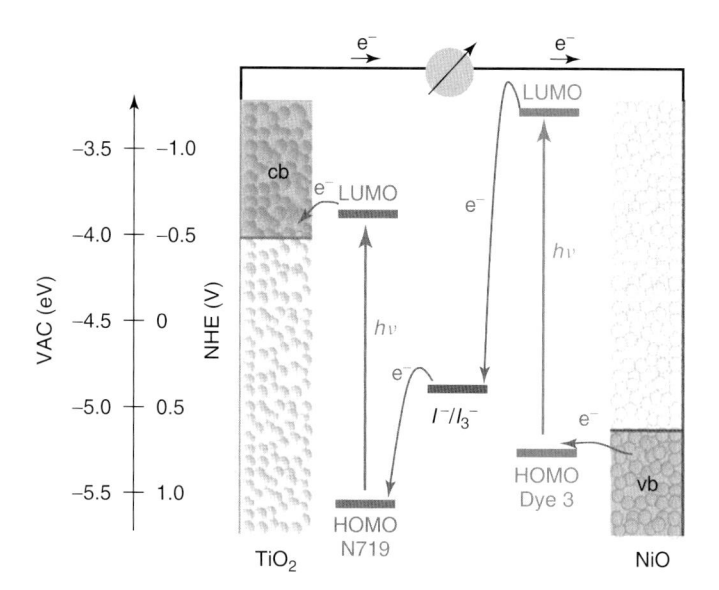

Figure 11.21 Schematic of a p-n tandem cell showing the electron-transfer process occurring in the cell. The vertical axis on the left is the potentials referred to the vacuum level (VAC) and that on the right is the potentials referred to the normal hydrogen electrode (NHE). Dye **3** refers to a donor–acceptor-type dye comprising perylenemonoimid (PMI) as the acceptor and oligothiophene coupled to triphenylamine as the donor. (From Ref. [140]. Copyright Nature Publishing Group. Reproduced with permission.)

donor–acceptor-type dye comprising a PMI as the acceptor and an oligothiophene coupled to triphenylamine as the donor (Figure 11.19) was used in the p-DSC. The reduction of triiodide occurs at the counter electrode in conventional DSC, whereas this process is facilitated by the injection of electrons from the LUMO level of acceptor–donor-type dye after photoexcitation of the dye in the p-n tandem DSC. The injected electrons at the photoanode migrate to the photocathode via the load. The regeneration of the dye in the p-DSC is accomplished by injection of electrons at the photocathode to the HOMO level of the dye through the VB of NiO (Figure 11.21). The elimination of expensive platinized counter electrode in this p-n DSC is also appealing to commercialization. The optimization of the thickness of photoelectrodes is essential to achieve photocurrent-matching in series-connected tandem cells. In this p-n tandem DSC, the optimized performance ($J_{sc} = 2.4$ mA cm^{-2}, $V_{oc} = 1.079$ V, FF = 74%, and $\eta = 1.91$%) was obtained with a 0.8-µm thick TiO$_2$ and 3.38-µm thick NiO for illumination from photoanode side. The V_{oc} of the tandem cell (1.079 V) is almost the sum of the V_{oc}'s of the individual subcells (0.905 V for n-DSC and 0.186 V for p-DSC). However, for illumination from the photocathode side, the optimized thicknesses are 12 and 1.55 µm for TiO$_2$ and NiO, respectively, resulting in an efficiency of 2.42%, which is considerably less than that of a standalone n-DSC with the same thickness of photoanode (5.9%). The low efficiency was attributed to the similar spectral response of the acceptor–donor-type dye and N719 dye and hence competition of the two dyes for the photons in the same spectral range. Further improvement can be realized by sensitizing the acceptor–donor-type dye which has complementary absorption spectrum with commonly used ruthenium complexes and a p-type semiconductor with a more favorable VB edge.

11.6.5.2 Dye Cocktail or CoSensitization

The main idea of cosensitization is the combination of several dyes, each with a narrow spectral width yet with a high extinction coefficient, together in a spectral sensitization package to cover the whole visible region of the solar spectrum. This approach was attempted by both simultaneous [156, 158] and sequential [159, 160] adsorption of multiple dyes on the porous semiconductor film. As early as 1981, the extension of IPCE spectrum was observed by the cosensitization of rhodamine B and Ru complex simultaneously on a sintered ZnO electrode [161]. However, the efficiency and photocurrent of cosensitized DSCs were not always improved at the initial attempts. Zhao et al. reported that a mixture of Ru(dcbpy)$_2$(NCS)$_2$ and squarylium cyanine (SQ) in 1 : 1 (molecular ratio) worsened the photocurrent and efficiency of cell with a single Ru(dcbpy)$_2$(NCS)$_2$ dye, whereas a slight improvement in current (15.0–15.2 mA cm^{-2}) and efficiency (5.87–6.62%) was observed when the ratio of the two dyes was 100 : 1 [156]. Spitler et al. loaded multiple cyanine dyes on a TiO$_2$ electrode and studied its effect. It was observed that the photocurrent was enhanced when special combinations of two different dyes were used, compared to the photocurrent generated by single-dye cell [157]. The cosensitization of three dyes (yellow, blue, and red cyanine dyes) was also employed to harvest the near-IR region of the spectrum. The spectral response of the cell with the cosensitization of three dyes was in accordance with their absorption spectrum on TiO_2 but the QE of current production was only 10%. Later on, Sayama demonstrated that the current density and efficiency were improved with simultaneous cosensitization of three cyanine dyes (yellow, red, and blue) compared to a single-dye cell, and even higher efficiency (J_{sc} = 11.5 mA cm^{-2}, η = 3.1%) was achieved by simultaneous sensitization of yellow and red cyanine dyes with the SQ dye [158].

However, the influence of photocurrent by cosensitization of multiple dyes can be complicated by dye aggregation [157, 158], rate of electron injection from different dyes [158], and the interaction between dyes such as electron transfer and/or energy transfer [158, 162]. The multiple organic dyes adsorbed on the nanocrystalline surface can be of both monomeric and aggregate forms, depending on the structure of the chromophore as well as the nature of the functional group attached to the dye [157]. The alignment of the dye (aggregation) on the semiconductor surface also affects the sensitization efficiency. The disorder of the chromophore arrangement in the aggregate decreases the efficiency owing to the self-quenching between neighboring dyes [163]. Spitler *et al.* suggested that aggregates of dyes can play a dual role in the spectral sensitization: (i) energy transfer in which they serve to absorb energy from photon and channel this energy to the reddest absorbing dyes on the surface; and (ii) electron transfer in which they transfer electrons from each chromophore in the assembly of the adsorbed dye [157]. In the former case, the relative efficiency of the aggregate and monomer would be equal only if the energy transfer from the aggregate to the monomer is 100% efficient. In the latter case, the quantum yield for electron transfer of the monomer and the aggregate form will be equal if the electron transfers for both forms are sufficiently fast. Nonetheless, the efficiency of the monomer form may not be equal to that of aggregate form because of the losses in transfer processes to a certain extent. The shape of IPCE spectrum of the cosensitized cell is also influenced by the electron injection from the dye into semiconductor and by the interactions between aggregate dyes such as electron transfer and charge transfer [158]. If the rate of electron injection from the dye into semiconductor is faster than the interaction between dyes, the IPCE spectrum is controlled by the QE of the individual dyes. On the other hand, if the interaction between dyes occurs rapidly and efficiently compared to the electron injection, the IPCE in the whole wavelength will be controlled by the QE of the reddish dye.

To control the formation of molecular aggregates, a zinc carboxyphthalocyanine (TT1) cosensitized with 3-{5′ [*N,N*-bis-(9,9-dimethylfluorene-2-yl)phenyl]-2,2′-bisthiophene-5-yl-}-2-cyano-acrilic acid (JK2 dye) was introduced and an impressive result was attained [160]. The presence of *tert*-butyl groups in the TT1 dye prevents the formation of molecular aggregates, arranges the excited states to provide the directionality of charge transfer from the LUMO level of the dyes to Ti 3d orbital, and increases the solubility of dye in organic solvents owing to their push-pull character. The photoresponse of the cosensitized cell covers from 400 to 750 nm with an IPCE of 72% at the absorption maximum of the TT1 dye (690 nm), providing a PCE of 7.08%. Neither single cell with JK2 nor TT1 yields this high efficiency. Instead of organic dyes, we also cosensitized a flowerlike ZnO-nanorod structures with CdS quantum dots, which have a higher molar extinction coefficient than ruthenium complexes but narrow absorption spectra, and the commercial N749 (black dye) [164]. The spectral response of the cosensitized DSC extends to more than 800 nm due to incorporation of N749 dye, and IPCE at the absorption peak of CdS (406 nm) is about 64% while the maximum value is only 13% for a single cell with the N749 dye.

In addition to simultaneous and sequential adsorption of dyes, another route toward cosensitization is selective positioning of multiple dyes in a nanocrystalline oxide film employing chromatographic principles to control the proper positioning of each dye on the desired site in the oxide film [165]. Three different dyes, 2-cyano-3-(5-(4-ethoxyphenyl)thiophen-2-yl)acrylic acid (P5), N719, and N749, were attached to a TiO_2 mesoporous film at different locations as shown in Figure 11.22. To achieve this selective positioning, first, the dye was attached to the whole TiO_2 film, and then the pore size of TiO_2 film was reduced by a homogeneous polystyrene coating. The reduced pore size in the film was effective in controlling the subsequent adsorption depths of the dye molecule. P5 dyes adsorbed on the upper layer of TiO_2 were selectively desorbed by the penetration of an aqueous NaOH solution. The desorption could proceed rapidly owing to the very small Na^+ and OH^- ions without the retardation agent. Therefore, to slow down the penetration rate of the aqueous solution and prevent excessive desorption, the aqueous solution was mixed with the polypropylene glycol (PPG) polymer. The desorption depth was precisely controlled by the concentration of PPG and NaOH, as well as the penetration time. The second dye adsorbed again on the selectively desorbed region in the TiO_2 film. For this step, the PPG polymer was not required in the dye solution because the polystyrene-filled TiO_2 film was small enough to retard the flow rate of the dye. The steps were repeated for the second dye desorption and third dye adsorption. Finally, polystyrene was removed by dissolving the coiled polymer chain using an ethyl acetate solution. The thicknesses of TiO_2 film attached

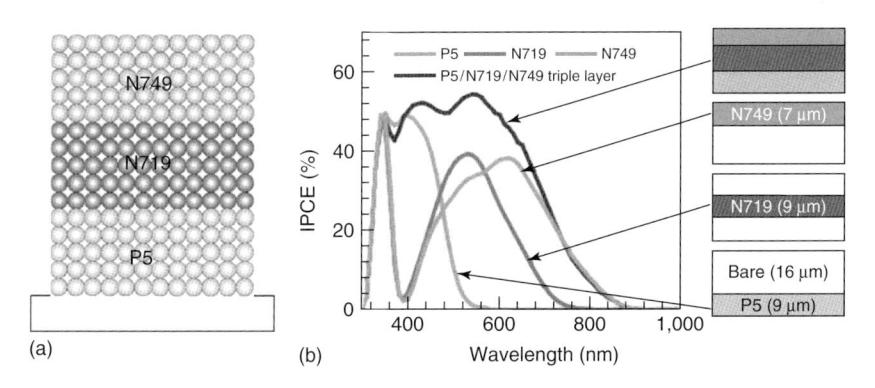

Figure 11.22 (a) Schematic of TiO_2 nanocrystalline film selective-position cosensitized with P5, N719, and N749 dye. (b) The IPCE spectra of cosensitized cell compared with individual single cell. (From Ref. [165]. Copyright Nature Publishing Group. Reproduced with permission.)

with P5, N719, and N749 dye were 9, 9, and 7 µm, respectively. The photocurrent density of the multidye cell was higher than that of any single-dye cell with the same thickness yet not equal to the sum of the current densities of the individual cells. This is probably due to the incomplete desorption of the dye, which blocks the adsorption of subsequent dyes. It is worth mentioning that V_{oc} of the multidye cell is close to the average of those of the single-dye cells and not likely to rely on those of the single-dye cells.

Despite the enhanced light absorption and broad spectral response, dye cosensitization still places a constraint on the light absorption because only a limited number of sites are available on the porous surface and dyes still need to compete with each other to anchor to the surface. To overcome this limitation, one approach is to incorporate a high concentration of an energy-relay dye inside the liquid electrolyte without attaching it to the semiconductor surface. Highly luminescent energy-relay dyes inside the electrolyte absorb high-energy photons and efficiently transfer the energy to the near-IR sensitizing dye anchored to the semiconductor surface by Förster resonant energy transfer (FRET) as shown in Figure 11.23. This approach is analogous to the photosynthesis in purple bacteria in nature, where aggregates of light-harvesting pigments transfer their energy to the reaction center, without in direct contact with the reaction center, leading to charge separation [167]. Siegers *et al.* observed efficient quenching of the donor-chromophore emission when alkyl-functionalized aminonapthalimide was used as the energy-donor chromophore and ruthenium dye complex as energy acceptor and sensitizing chromophore [168]. An energy transfer efficiency as high as 85% was achieved in this particular system. Adopting this concept, 26% increase in PCE of the cell

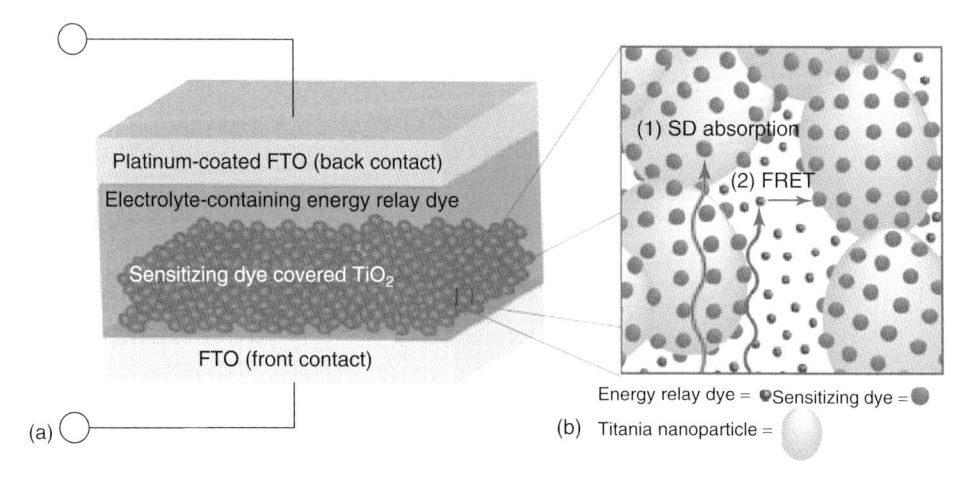

Figure 11.23 (a) Schematic of a DSC containing an energy-relay dye. (b) Illustration of the absorption process in the DSC: light is absorbed by the sensitizing dye (1), transferring an electron into the titania. The higher energy photons, however, are first absorbed by the energy-relay dye, which undergoes Förster energy transfer (2) to the sensitizing dye. (From Ref. [162]. Copyright Nature Publishing Group. Reproduced with permission.)

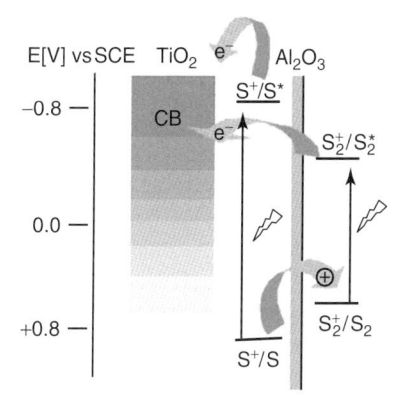

Figure 11.24 Illustration of multistep electron-transfer process in multilayer cosensitized films. The electron injection into TiO$_2$ nanocrystalline film is observed from both dye S and S2, which refer to RuL$_2$(CN)$_2$ and RuPc, respectively. (From Ref. [169]. Copyright American Chemical Society. Reproduced with permission.)

was demonstrated when the energy-relay dye PTCD1, which transfers energy to the anchored sensitizing zinc phthalocyanine dye TT1, was incorporated into the electrolyte [162]. Another approach to overcome the limited number of sites for multiple dyes is employing a multistep electron-transfer process rather than an energy transfer process [169]. The deposition of a secondary metal oxide layer between the adsorption of the first and second sensitizer dyes, as illustrated in Figure 11.24, not only increases the numbers of sites for dye adsorption but also suppresses the charge recombination by increasing the distance between the oxidized dye species and the injected electrons as the holes from the inner dye closer to the TiO$_2$ surface are transferred to the outer dye. These approaches offer a feasible pathway to further increase the efficiency of DSCs.

11.6.6
Flexible Dye-Sensitized Solar Cell

With high demand in mobile electronic devices, light-weight and flexible DSC technology has emerged as an important topic in DSC research. Flexible DSCs also have advantages in transportation and PV power supply system equipment. The replacement of a rigid substrate with a flexible substrate leads to drastic cost reduction through a roll-to-roll assembly process. In a rigid DSC fabrication, a binder-containing TiO$_2$ paste is deposited on FTO glass and sintered at 450–500 °C to improve the interconnection between the particles and charge transport. In the case of flexible DSCs, indium tin oxide (ITO) coated polyethylene terephthalate (PET), which can withstand a temperature of not more than 200 °C, is used as the substrate. Thus, the key technology required for flexible DSCs is the low-temperature fabrication of mesoporous photoanodes. Previously, the TiO$_2$ layer was deposited on the ITO/PET substrate by spin-coating of a TiO_2 colloid [171] and doctor-blade coating [172], followed by low-temperature sintering (100–150 °C). However, simply reducing the sintering temperature to be compatible with the plastic substrate leads to inferior electrical properties of TiO_2. The problems associated with low-temperature sintering or without sintering are an increase in series resistance due to lack of adhesion at the interface between the conducting substrate and TiO_2 layer and/or slow electron diffusion due to the imperfections of interparticle binding. Thus, low-temperature fabrication methods, such as electrophoretic deposition [166, 173, 174] and spray coating [170] followed by surface activation methods, were introduced to improve the efficiency. Various methods such as compression [172, 175], microwave irradiation [166, 170], UV irradiation [173], hydrothermal crystallization [176], and electron-beam annealing [177] have been reported for surface activation. The highest efficiency of flexible DSCs with low-temperature sintering on a plastic substrate was 4.1% fabricated with electrophoretic deposition followed by additional TiO_2 particle impregnation [174]. Nonetheless, the efficiency of DSCs fabricated on plastic substrates was much lower than that of DSCs on an FTO-glass substrate mainly due to the poor connectivity of particles caused by low-temperature sintering.

To take the advantage of high-temperature sintering process, a lift-off and layer transfer technique was adopted by Dürr et al. [178]. The TiO_2 layer on a Au-coated glass substrate was pre-sintered at 450 °C to ensure good interconnection between particles, and then it was lifted off by dissolving the Au underlayer in iodine solution and pressed onto the ITO/PET substrate using high pressure. This high-temperature sintering and transfer method leads to an efficiency of 5.8%. An alternative route to use high-temperature sintering in flexible DSCs is to construct a front-illuminated configuration in which a flexible metal foil is used as substrate and light is allowed to enter from a transparent counter electrode [179, 180]. Kang et al. demonstrated a high-temperature-processed, flexible, dye-sensitized TiO_2 solar cell on a stainless steel (SS) substrate. J_{sc} of only $\sim 7\,mA\,cm^{-2}$ was obtained for the SS substrate alone, whereas it increased to more than $\sim 9\,mA\,cm^{-2}$ when an additional ITO layer was sputtered on the SS substrate. Even a higher J_{sc} of $11.2\,mA\,cm^{-2}$ was achieved by inserting a SiO_x isolation layer between the sputtered ITO layer and the SS substrate to reduce the recombination, resulting in an overall efficiency of 4.2%. Ito et al. also fabricated flexible DSCs on Ti-metal foils and attained an efficiency of up to 7.2%, which is comparable to the efficiency of the rigid solar cell on an FTO-glass substrate. About 20% drop in efficiency compared to DSC on FTO glass is mainly due to the absorption of near-UV light by the electrolyte.

In addition to TiO_2, ZnO photoanodes were also engineered for flexible DSCs. A nanoporous structure consisting of interconnected nanosized ZnO crystals fabricated by cathodic electrochemical deposition and low-temperature treatment (150 °C) was demonstrated for the realization of a ZnO-based flexible DSC [181]. Liu reported room-temperature fabrication of a porous ZnO photoanode for flexible DSCs by doctor-blading a ZnO gel [182]. To achieve good adhesion to substrate and interparticle connection, the gel was prepared by ball-milling the ZnO powder in a dilute acetic acid (HAc) solution (0.1 M). The $ZnAc_2$ formed during etching of ZnO surface by HAc acted as the binder. After deposition of film, $ZnAc_2$ could be easily removed by ammonia surface activation without using high temperature. The flexible DSCs using this ZnO photoanode on ITO-PET substrates obtained an efficiency of 3.8%, which is comparable to that of low-temperature-fabricated TiO_2 DSCs. Despite a high efficiency, one concern for the porous film fabricated from powder is its capability to withstand repeated bending stress. The bending stress in a mesoporous network has to be released through network deformation, leading to cracks or peeling. In contrast, vertically aligned ZnO nanowire arrays can efficiently release the bending stress in the film through the gap between adjacent nanowires. We compared the mechanical stability of ZnO porous network films and nanowire films by a bending test [183]. After bending the nanowire films to an extreme radius of 2 mm for several cycles manually and further 2000 cycles with a radius of 5 mm using a bending machine, no visible cracks or signs of peel-off were observed in the film. However, visible cracks and peel-off were observed in the porous network film after several cycles of bending test with a radius of 10 mm. Moreover, the efficiency of flexible DSCs made from nanowire films and subjected to several cycles of bending was comparable to that made from non-bent nanowire films, showing that the nanowire film is a probable candidate for flexible DSC application due to its excellent mechanical stability. Nevertheless, the flexible DSCs made from nanowire films exhibit a low efficiency of only 0.4% because of the low surface area. When we filled up the gap between the nanowires with a nanopowder, the efficiency increased by more than double to 0.9%. In this flexible DSC, nanowires serve as nanopillars to support the mechanical stability of the film while the nanopowder provides additional anchoring areas for the sensitizer. In future, similar hybrid structures, combining supporting pillars and a porous layer, will be useful for realization of mechanically stable and efficient flexible DSCs.

11.7
Routes to Boost the Efficiency of Solar Cells

In general, the performance of solar cells is influenced by several key factors: (i) matching between material bandgap and incident photon energy, which leads to thermalization loss and loss of photons with energy smaller than semiconductor bandgap; (ii) photon loss from surface reflection, carrier loss from surface recombination, and bulk recombination; and (iii) energy dissipated due to cell resistance. Therefore, solar cell researches are mainly focused on addressing these problems to increase the efficiency of the cells.

The following techniques are adopted to enhance the efficiency of silicon and other thin-film solar cells (CIGS, III–V compounds). First, it is desirable to introduce light into the active absorber as much as possible, which is greatly improved by the application of surface texturing and ARC technology. Surface-texturing traps light inside semiconductor materials and prevents it from escaping into ambient. Multilayered ARC is used in practice, and typical examples are MgF_2^-, $Al_2O_3^-$, and TiO$_2$-based materials. Besides coatings on the top surface, back surface reflection is also used to prevent light transmitting through the active layer. Titanium oxide painted with a metal powder can be used to introduce back surface diffuse reflection, which seals off the "exit" gate of photons. Second, if light cannot be absorbed with a single junction, multijunction solar cells are stacked together to maximize light absorption. A larger bandgap cell is placed closer to side of light incidence, while a smaller bandgap cell is placed closer to the back contact. Third, in order to extract photogenerated carriers, bulk and surface recombination need to be suppressed. Surface recombination can be passivated before the top contact is deposited, while bulk recombination can be effectively reduced by a technology called "back surface field (BSF)." A high p-type doping at the BSF reduces back surface recombination and enhances the extraction of carriers at the back contact. Nonetheless, the efficiency of the silicon solar cell is approaching its theoretical limit of 30% and there is only a small room to further increase its efficiency.

In contrast to silicon and other thin-film solar cells, DSCs are currently operating far below their theoretical limit and we can further increase their efficiency from many aspects. Like in other types of solar cells, J_{sc}, V_{oc}, and FF are the three main parameters to be considered for improvement of efficiency. Because the highest possible FF is 0.86 assuming a minimum diode quality factor of 1 and V_{oc} of 0.8 V, the current FF of ~0.75 does not have much room for improvement. Without considering dye loading and the molar extinction coefficient of the dye, the simple way to increase J_{sc} is to extend the absorption spectrum as much as possible, that is, to absorb the photons in the near-IR region. The optical gap of ruthenium dye that gives the highest efficiency is 1.8 eV, limiting the photoresponse of the cell to ~700 nm. Hagfeldt and Grätzel suggested that a photoresponse close to 920 nm absorption threshold would increase J_{sc} from the current 20.5 mA cm^{-2} to about 28 mA cm^{-2}, leading to an overall efficiency of over 15% [76]. Thus, from a molecular engineering perspective, the optical gap of the sensitizer must be narrowed down by either lowering the energy level of the LUMO or raising the energy level of the HOMO to increase J_{sc}. As the entire donor distribution function must lie above the CB edge of the semiconductor, lowering the LUMO of the dye to extend the light absorption will be limited by the efficient charge injection from the dye to the semiconductor. Fortunately, the ultrafast charge injection rate (femtosecond range) and slow excited-state decay time (nanosecond range) would seem to have enough room for lowering the LUMO of the dye before the injection efficiency considerably deteriorates. The raising the HOMO of the dye, an alternative route to increase J_{sc}, is limited by the minimum potential difference between the redox potential and the HOMO of the dye necessary for efficient dye regeneration. This potential difference is approximately 550 mV in a current typical DSC and there is much room for improvement (Figure 11.25a).

The V_{oc} is another major component that can boost the efficiency of DSCs. It was proposed that even a relatively small increment of 200 mV in V_{oc} would allow the overall solar conversion yield to be raised from current 10–11% to nearly 15% [76]. As discussed in Section 11.6.2.4, V_{oc} is determined by the difference between the redox potential of the electrolyte and the quasi-Fermi level of the semiconductor.

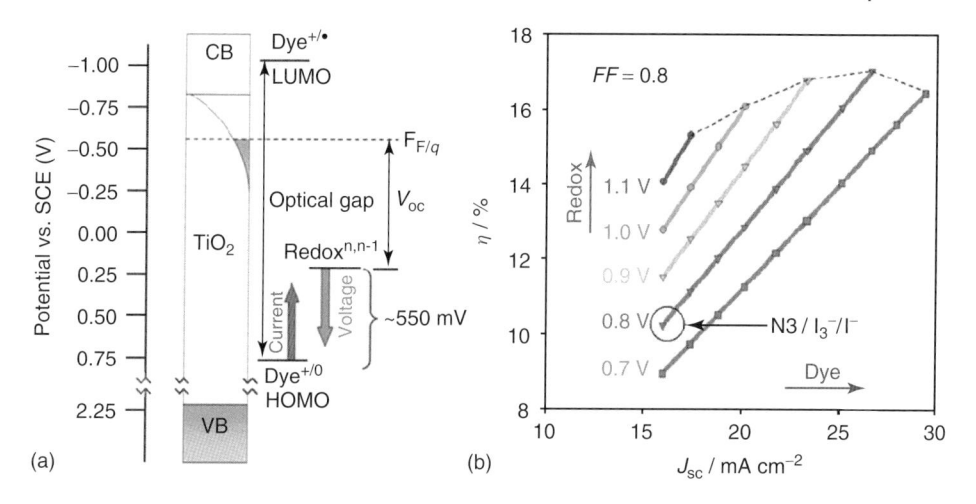

Figure 11.25 (a) Strategies to improve the efficiency of DSC: (i) raising the HOMO of dye to extend the absorption of light, that is, to increase current; (ii) lowering the redox potential to increase the photovoltage. (b) Estimated efficiencies of DSC using these strategies. Each data point on the same line represents a 50-mV shift in the HOMO of the dye. (From Ref. [184]. Copyright The Royal Society of Chemistry. Reproduced with permission.)

Thus, raising the quasi-Fermi level of the semiconductor or lowering the redox potential is a possible route to increase the V_{oc}. Theoretically, approximately 200 mV can be increased if we can lift up the quasi-Fermi level close to the maximum level, that is, the CB of the semiconductor. Unfortunately, this is unlikely because, according to previous studies, an order of magnitude increase in injection rate or reduction in recombination is required for just an increase of 60 mV [69, 70]. Thus, a more feasible way to increase V_{oc} is to lower the redox potential of the electrolyte. Similar to the J_{sc} case, the lowering the redox potential is limited by the minimum potential difference between the redox potential and the HOMO of the dye necessary for efficient dye regeneration. Based on this discussion, we can increase the efficiency of the current DSCs by (i) raising the HOMO level of the dye, (ii) lowering the redox potential, or (iii) the simultaneous action of (i) and (ii). Assuming that the minimum overpotential of 200 mV is necessary between the HOMO of the dye and the redox potential to ensure efficient dye regeneration, Hamann et al. estimated the possible efficiencies of DSCs by shifting the HOMO level of the dye and the redox potential in various combinations [184]. Efficiencies beyond 16% are reasonably attainable with various combinations of dyes and the redox couple as illustrated in Figure 11.25b. This estimation, however, is solely based on the assumption that there are no changes in transport of the redox media and recombination at semiconductor/electrolyte interface due to the shift in redox potential as well as excited-state decay time and molar extinction coefficient due to the shift in the HOMO level of the dye. In reality, these undesirable side effects may be caused by variation in the energy level of the dye or the redox couple. If we can minimize the undesirable side effects upon varying the energetics of the dye and the redox couple, it would be a promising route to achieve a commercial target of 15% efficiency.

Increasing the surface area of the semiconductor (dye loading) and improving the molar extinction coefficient of the dye are the other two ways to increase J_{sc}. Upon intensive investigations of various morphologies of TiO_2 and ZnO photoelectrode, it appears that TiO_2 nanocrystalline film is the optimum choice for high surface area and charge collection. The molar extinction coefficient can be increased with an organic dye, but the spectral response suffers. Thus, an alternative route to boost the efficiency is the tandem cell or the dye cosensitization concept, as discussed in detail in Section 11.6.5. Tandem cell or cosensitization with multiple dyes with high molar extinction allows extending the absorption spectrum as well as increasing light absorption. The development of new materials or modifying the existing materials to fulfill the energetics as outlined above, and constructing tandem cells or cosensitization of multiple dyes with complementary absorption spectra, will lead to DSCs with an efficiency exceeding 15%.

11.8
Current Ideas for Future Solar Cell

As conclusion of this chapter, some ideas proposed by researchers are summarized, while several ideas have been under extensive trails. However, most of them are still far from commercialization. Figure 11.26 is the most famous "generation map" made by Green [185]. In the figure, three generations of solar cells are classified with different cost and efficiency performances. The second-generation solar cell (mainly thin film) has been under development for more than 20 years. Although its

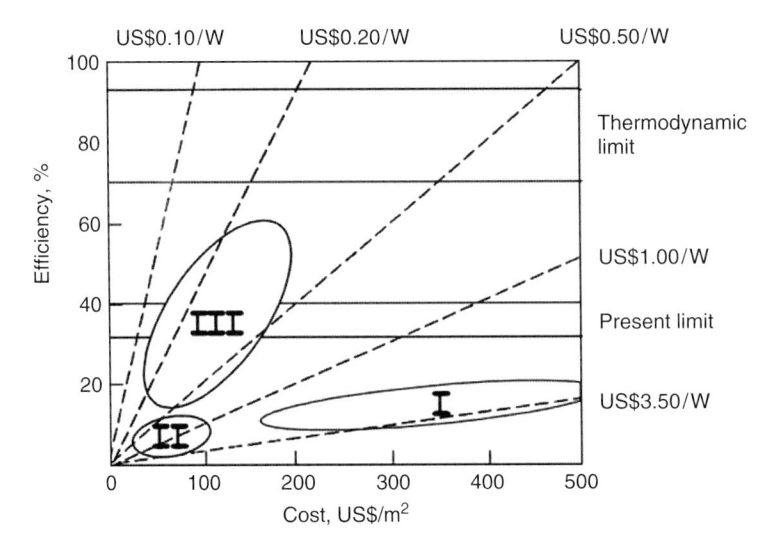

Figure 11.26 Efficiency–cost trade-off for three generations of solar cell technology.

efficiency is still low, the low-temperature processing technology and low demand on material equip this generation with huge potential to lower the cost to 1 USD W^{-1}, which is a benchmark identified for solar cells to be actively competitive with regular energy sources. In 1960, Shockley and Queisser published their results on calculating the extremely optimized performance for single-junction solar cell, with a power conversion efficiency of 31% [186]. Along the development of first- and second-generation solar cells, performances have been enhanced by several key milestone technologies, for example, surface texturing for increased light absorption. Future technologies are expected to further optimize the cell structure, material utilization, and even the operation principle of the cells.

Tandem cells have been described in this chapter, while there are still more topics to be covered in this technology. It is not necessarily true that a stacked structure has to be used for a tandem cell; we can actually split the incoming light radiation into different parts. Special mirrors that can reflect specific light wavelengths can be used for the purpose. Figure 11.27 shows an idea in which different wavelengths are reflected to devices with different absorption bandgaps. However, this technology is difficult to realize because of optical reasons. The separation of wavelengths is not easy in practice.

One new idea for future solar cells is hot-carrier solar cells. Excited carriers with high energy will lose energy through thermalization, while phonons are generated. Hot-carrier solar cells are based on the idea of extracting high-energy carriers before they actually "cool down." By doing so, significantly higher open-circuit voltage can be expected. Furthermore, if ionization takes place during extraction, a QE <1 is also possible, which leads to larger current output.

Another idea for future solar cells is the intermediate-band solar cell. By introducing an intermediate energy band in the bandgap of the cell, low-energy photons

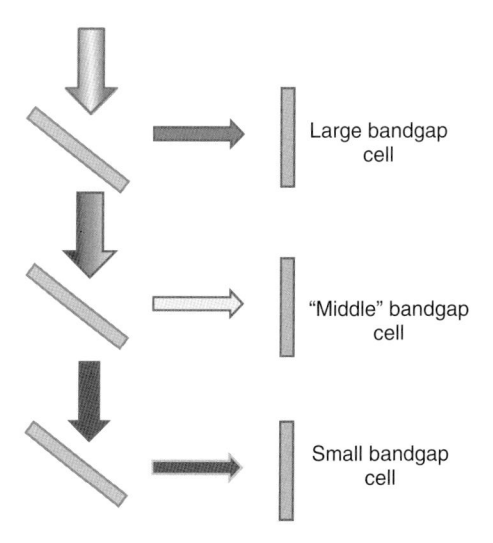

Figure 11.27 Incoming light reflected to three solar cells with corresponding bandgaps.

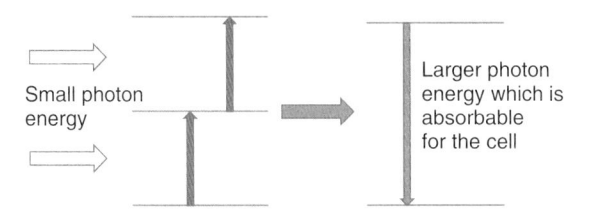

Figure 11.28 The mechanisms of upconversion and downconversion solar cells.

can be absorbed by the carriers. The operation principle of this new technology is similar to that of the tandem cell: enlarging the absorption range of the cells. However, the intermediate-band solar cell is able to achieve a higher efficiency. Stacked tandem cells need to consider the matching of current in all individual cells, and one extractable carrier requires the generation to take place in all individual cells, as they are connected in series. The realization of intermediate-band solar cells faces several obstacles. For instance, the intermediate state has to be at least "half-filled" with electrons and holes to facilitate electron's energy-loss-free transition into the CB. The incorporation of these materials into original active layers will affect its electrical properties significantly.

In addition, upconversion and downconversion solar cells are one of the hot topics in third-generation solar cell research. Figure 11.28 shows the mechanisms. An upconversion device is placed behind the solar cell and the device has a bandgap equal to or larger than that of the cell. There is an intermediate state in the upconversion material, which converts two or more photons with smaller energy into a high-energy exciton. And the exciton will combine radiatively by emitting a photon with high energy. Light is totally reflected back by the reflector at the back of the upconversion device. As a result, the top cell will have another chance to absorb the original input light because low-energy photons have been converted into high-energy ones. The downconversion device works on a similar mechanism, but it is placed in front of the solar cell. A downconversion device absorbs high-energy photons and emits two lower-energy photons. By making sure that the two low-energy photons are still absorbable by the cell, power conversion efficiency of the cell can be increased.

11.9
Summary

The role of solar cells for energy harvesting is becoming more and more important in the twenty-first century due to increasing energy demand, depletion of oil reserves, and climate change issues. In this chapter, we have discussed a wide range of topics related to solar cells: from solar radiation to device operation. The various kinds of solar cells such as the silicon solar cell, CIS/CIGS, and high-efficiency III–V cells were elaborated in terms of material deposition and device fabrication. Despite drawbacks such as high manufacturing cost and high-energy payback time, the silicon-based solar cell is still dominating the market because the electronic manufacturing infrastructure is rooted in silicon technology. Therefore, the a-Si solar cell has become more important than ever before as a result of reducing cost and compatibility with silicon manufacturing technology. CIS/CIGS is also appealing for future PV technology due to their high absorption coefficient, tunable bandgap, and adoptability by mature silicon technology. However, the challenges for CIGS technology are the scarcity of In and toxicity of the CdS buffer layer. DSC, a new generation of solar cells, cannot be excluded from the PV area as they can be manufactured from low-cost material without elaborate equipment. However, the commercialization of DSC is retarded by issues of long-term stability of the cell and complete departure from established silicon manufacturing facility. The various aspects of DSCs from single n-DSC and p-DSC to tandem cells and dye cosensitized cells were also discussed in this chapter. As the last part of the chapter, we also included the directions and ways to improve the efficiency of each of type of solar cell and some new ideas for the future.

References

1. Würfel, P. (2005) *Physics of Solar Cells – From Principles to New Concepts*, Wiley-VCH Verlag GmbH & Co. KGaA, Weinheim.
2. Petritsch, K. (2000) Organic solar cell architectures. PhD Thesis. Technische Universität Graz, Austria.
3. Gray, J.L. (2011) The physics of the solar cell, in *Handbook of Photovoltaic Science and Engineering* (eds A. Luque and S. Hegedus), John Wiley & Sons, Ltd., Chichester, pp. 82–129.
4. Staebler, D.L. and Wronski, C.R. (1977) Reversible conductivity changes in discharge-produced amorphous Si. *Appl. Phys. Lett.*, **31** (4), 292–294.
5. Spear, W.E. and Lecomber, P.G. (1975) Substitutional doping of amorphous silicon. *Solid State Commun.*, **17** (9), 1193–1196.
6. Wronski, C.R., Carlson, D.E., and Daniel, R.E. (1976) Schottky-barrier characteristics of metal-amorphous-silicon diodes. *Appl. Phys. Lett.*, **29** (9), 602–605.
7. Carlson, D.E. (1980) Photovoltaics. 5. Amorphous-silicon cells. *IEEE Spectr.*, **17** (2), 39–41.
8. Green, M.A., Emery, K., Hishikawa, Y., and Warta, W. (2010) Solar cell efficiency tables (version 36). *Prog. Photovoltaics*, **18** (5), 346–352.
9. Guha, S., Narasimhan, K.L., and Pietruszko, S.M. (1981) On light-induced effect in amorphous hydrogenated silicon. *J. Appl. Phys.*, **52** (2), 859–860.
10. Tawada, Y., Okamoto, H., and Hamakawa, Y. (1981) A-SiC:H-a-Si:H heterojunction solar-cell having more than 7.1-percent conversion efficiency. *Appl. Phys. Lett.*, **39** (3), 237–239.
11. Tawada, Y., Tsuge, K., Kondo, M., Okamoto, H., and Hamakawa, Y. (1982) Properties and structure of alpha-SiC:H for high-efficiency alpha-Si solar-cell. *J. Appl. Phys.*, **53** (7), 5273–5281.
12. Ovshinsky, S.R. (1985) Chemical and configuration basis of high efficiency amorphous photovoltaic cells. Proceedings of the 18th IEEE Photovoltaic Specialists Conference, Las Vegas, pp. 1365–1371.
13. Wei, S.H., Zhang, S.B., and Zunger, A. (1998) Effects of Ga addition to CuInSe$_2$ on its electronic, structural, and defect properties. *Appl. Phys. Lett.*, **72** (24), 3199–3201.
14. Wagner, S., Shay, J.L., Migliora., P., and Kasper, H.M. (1974) CuInSe$_2$-CdS heterojunction photovoltaic detectors. *Appl. Phys. Lett.*, **25** (8), 434–435.

15. Shay, J.L., Wagner, S., and Kasper, H.M. (1975) Efficient CuInSe$_2$/CdS solar cells. *Appl. Phys. Lett.*, **27** (2), 89–90.
16. Mickelsen, R.A., and Chen, W.S. (1981) Development of A 9.4% efficient thin film CuInSe$_2$/CdS solar cell. Proceedings of the 15th IEEE Photovotaic Specialist Conference, Orlando, pp. 800–804.
17. Mitchell, K.W., Eberspacher, C., and Ermer, J. (1988) Single and tandem junction CuInSe$_2$ cell and module technology. Proceedings of the IEEE Photovoltaic Specialists Conference, Las Vagas, pp. 1384–1389.
18. Gabor, A.M., Tuttle, J.R., Albin, D.S., Contreras, M.A., Noufi, R., and Hermann, A.M. (1994) High efficiency CuIn$_x$Ga$_{1-x}$Se$_2$ solar-cells made from (In$_x$, Ga$_{1-x}$)$_2$Se$_3$ precursor films. *Appl. Phys. Lett.*, **65** (2), 198–200.
19. Kessler, J., Wennerberg, J., Bodegard, M., and Stolt, L. (2003) Highly efficient Cu(In,Ga)Se-2 mini-modules. *Sol. Energy Mater. Sol. Cells*, **75** (1–2), 35–46.
20. Noufi, R. and Zweibel, K. (2006) High-efficiency CdTe and CIGS thin-film solar cells: highlights and challenges. Proceedings of the IEEE 4th World Conference on Photovoltaic Energy Conversion (WCPEC-4), Waikoloa, Hawaii, pp. 317–320.
21. Wada, T., Kohara, N., Nishiwaki, S., and Negami, T. (2001) Characterization of the Cu(In,Ga)Se$_2$/Mo interface in CIGS solar cells. *Thin Solid Films*, **387** (1–2), 118–122.
22. Hashimoto, Y., Kohara, N., Negami, T., Nishitani, N., and Wada, T. (1998) Chemical bath deposition of Cds buffer layer for GIGS solar cells. *Sol. Energy Mater. Sol. Cells*, **50** (1–4), 71–77.
23. Rudolph, P. and Kiessling, F.M. (1988) The horizontal bridgman method., *Cryst. Res. Technol.* **23** (10), 1207–1224.
24. West, W. (1974) First hundred years of spectral sensitisation. *Photogr. Sci. Eng.*, **18**, 35–48.
25. Moser, J. (1887) Notiz über Verstärkung photoelektrischer Ströme durch optische Sensibilisirung. *Monatsh. Chem.*, **8**, 373.
26. Putseiko, E.K. and Terenin, A.N. (1949) Photosensitization of the internal photoeffect in zinc oxide and other semiconductors by adsorbed dyes. *Zh. Fiz. Khim. - Russ. J. Phys. Chem. A*, **23**, 676.
27. Jing, B., Zhang, M., and Shen, T. (1997) Advanced in dye-sensitized solar cell. *Chin. Sci. Bull.*, **42**, 1937–1948.
28. Spitler, M.T. and Parkinson, B.A. (2009) Dye sensitization of single crystal semiconductor electrodes. *Acc. Chem. Res.*, **42** (12), 2017–2029.
29. Namba, S. and Hishiki, Y. (1965) Color sensitization of zinc oxide with cyanine dyes. *J. Phys. Chem.*, **69** (3), 774–779.
30. Gerischer, H. and Tributsch, H. (1968) Electrochemical studies on the spectral sensitization of zinc oxide single crystals. *Ber. Bunsen-Ges. Phys. Chem.*, **72**, 437–445.
31. Tributsch, H. and Gerischer, H. (1969) Electrochemical studies on the mechanism of sensitization and supersensitization of zinc oxide single crystals. *Ber. Bunsen-Ges. Phys. Chem.*, **73**, 251–260.
32. Tributsch, H. and Calvin, M. (1971) Electrochemistry of excited molecules: photo-electrochemical reactions of chlorophylls. *Photochem. Photobiol.*, **14** (2), 95–112.
33. Tributsch, H. (1972) Reaction of excited chlorophyll molecules at electrodes and in photosynthesis. *Photochem. Photobiol.*, **16**, 261–269.
34. Mark, S. and Melvin, C. (1977) Adsorption and oxidation of rhodamine B at ZnO electrodes. *J. Chem. Phys.*, **67** (11), 5193–5200.
35. Mark, T.S. and Melvin, C. (1977) Electron transfer at sensitized TiO$_2$ electrodes. *J. Chem. Phys.*, **66** (10), 4294–4305.
36. Tsubomura, H., Matsumura, M., Nomura, Y., and Amamiya, T. (1976) Dye sensitised zinc oxide: aqueous electrolyte: platinum photocell. *Nature*, **261** (5559), 402–403.
37. Anderson, S., Constable, E.C., Dare-Edwards, M.P., Goodenough, J.B., Hamnett, A., Seddon, K.R., and Wright, R.D. (1979) Chemical modification of a titanium (IV) oxide electrode

to give stable dye sensitisation without a supersensitiser. *Nature*, **280** (5723), 571–573.
38. Gerischer, H. (1972) Electrochemical techniques for the study of photosensitization. *Photochem. Photobiol.*, **16** (4), 243–260.
39. Memming, R. (1972) Photochemical and electrochemical processes of excited dyes at semiconductor and metal electrodes. *Photochem. Photobiol.*, **16** (4), 325–333.
40. Jayadevaiah, T.S. (1974) Semiconductor-electrolyte interface devices for solar energy conversion. *Appl. Phys. Lett.*, **25** (7), 399–400.
41. Hamnett, A., Dare-Edwards, M.P., Wright, R.D., Seddon, K.R., and Goodenough, J.B. (1979) Photosensitization of titanium(IV) oxide with tris(2,2'-bipyridine)ruthenium(II) chloride. Surface states of titanium(IV) oxide. *J. Phys. Chem.*, **83** (25), 3280–3290.
42. Gleria, M. and Memming, R. (1975) Photoelectrochemical studies of tris [2, 2'-bipyridyl] ruthenium (II) at semiconductor electrodes. *Z. Phys. Chem.*, **98**, 303.
43. Clark, W.D.K. and Sutin, N. (1977) Spectral by polypyridine ruthenium (II) complexes. *J. Am. Chem. Soc.*, **99**, 4676–4682.
44. Duonghond, D., Serpone, N., and Grätzel, M. (1984) Integrated systems for water cleavage by visible light; sensitization of TiO_2 particles by surface derivatization with ruthenium complexes. *Helv. Chim. Acta*, **67** (4), 1012–1018.
45. Desilvestro, J., Gräetzel, M., Kavan, L., Moser, J., and Augustynski, J. (1985) Highly efficient sensitization of titanium dioxide. *J. Am. Chem. Soc.*, **107** (10), 2988–2990.
46. O'Regan, B. and Grätzel, M. (1991) A low-cost, high-efficiency solar cell based on dye-sensitized colloidal TiO_2 films. *Nature*, **353** (6346), 737–740.
47. Nazeeruddin, M.K., De Angelis, F., Fantacci, S., Selloni, A., Viscardi, G., Liska, P., Ito, S., Takeru, B., and Grätzel, M. (2005) Combined experimental and DFT-TDDFT computational study of photoelectrochemical cell ruthenium sensitizers. *J. Am. Chem. Soc.*, **127** (48), 16835–16847.
48. Clifford, J.N., Palomares, E., Nazeeruddin, M.K., Grätzel, M., and Durrant, J.R. (2007) Dye dependent regeneration dynamics in dye sensitized nanocrystalline solar cells: evidence for the formation of a ruthenium bipyridyl cation/iodide intermediate. *J. Phys. Chem. C*, **111** (17), 6561–6567.
49. Gerischer, H. (1969) Charge transfer processes at semiconductor-electrolyte interfaces in connection with problems of catalysis. *Surf. Sci.*, **18** (1), 97–122.
50. Cahen, D., Hodes, G., Grätzel, M., Guillemoles, J.F., and Riess, I. (2000) Nature of photovoltaic action in dye-sensitized solar cells. *J. Phys. Chem. B*, **104** (9), 2053–2059.
51. Grätzel, M. (2005) Solar energy conversion by dye-sensitized photovoltaic cells. *Inorg. Chem.*, **44** (20), 6841–6851.
52. Law, M., Greene, L.E., Johnson, J.C., Saykally, R., and Yang, P. (2005) Nanowire dye-sensitized solar cells. *Nat. Mater.*, **4** (6), 455–459.
53. Grätzel, M. (2003) Dye-sensitized solar cells. *J. Photochem. Photobiol. C: Photochem. Rev.*, **4** (2), 145–153.
54. Haque, S.A., Palomares, E., Cho, B.M., Green, A.N.M., Hirata, N., Klug, D.R., and Durrant, J.R. (2005) Charge separation versus recombination in dye-sensitized nanocrystalline solar cells: the minimization of kinetic redundancy. *J. Am. Chem. Soc.*, **127** (10), 3456–3462.
55. Boschloo, G. and Hagfeldt, A. (2009) Characteristics of the iodide/triiodide redox mediator in dye-sensitized solar cells. *Acc. Chem. Res.*, **42** (11), 1819–1826.
56. Haque, S.A., Tachibana, Y., Klug, D.R., and Durrant, J.R. (1998) Charge recombination kinetics in dye-sensitized nanocrystalline titanium dioxide films under externally applied bias. *J. Phys. Chem. B*, **102** (10), 1745–1749.
57. Kuciauskas, D., Freund, M.S., Gray, H.B., Winkler, J.R., and Lewis, N.S. (2000) Electron transfer dynamics in nanocrystalline titanium dioxide solar

cells sensitized with ruthenium or osmium polypyridyl complexes. *J. Phys. Chem. B*, **105** (2), 392–403.

58. Marcus, R.A. (1956) On the theory of oxidation-reduction reactions involving electron transfer. *J. Chem. Phys.*, **24** (5), 966–978.

59. Montanari, I., Nelson, J., and Durrant, J.R. (2002) Iodide electron transfer kinetics in dye-sensitized nanocrystalline TiO_2 films. *J. Phys. Chem. B*, **106** (47), 12203–12210.

60. Mori, S.N., Kubo, W., Kanzaki, T., Masaki, N., Wada, Y., and Yanagida, S. (2007) Investigation of the effect of alkyl chain length on charge transfer at TiO_2/Dye/electrolyte interface. *J. Phys. Chem. C*, **111** (8), 3522–3527.

61. Bauer, C., Boschloo, G., Mukhtar, E., and Hagfeldt, A. (2002) Interfacial electron-transfer dynamics in $Ru(tcterpy)(NCS)_3$-sensitized TiO_2 nanocrystalline solar cells. *J. Phys. Chem. B*, **106** (49), 12693–12704.

62. Nogueira, A.F., De Paoli, M.-A., Montanari, I., Monkhouse, R., Nelson, J., and Durrant, J.R. (2001) Electron transfer dynamics in dye sensitized nanocrystalline solar cells using a polymer electrolyte. *J. Phys. Chem. B*, **105** (31), 7517–7524.

63. Boschloo, G. and Hagfeldt, A. (2008) Photoinduced absorption spectroscopy as a tool in the study of dye-sensitized solar cells. *Inorg. Chim. Acta*, **361** (3), 729–734.

64. van de Lagemaat, J. and Frank, A.J. (2001) Nonthermalized electron transport in dye-sensitized nanocrystalline TiO2 films: transient photocurrent and random-walk modeling studies. *J. Phys. Chem. B*, **105** (45), 11194–11205.

65. Boschloo, G. and Hagfeldt, A. (2005) Activation energy of electron transport in dye-sensitized TiO_2 solar cells. *J. Phys. Chem. B*, **109** (24), 12093–12098.

66. Schlichthorl, G., Huang, S.Y., Sprague, J., and Frank, A.J. (1997) Band edge movement and recombination kinetics in dye-sensitized nanocrystalline TiO2 solar cells: a study by intensity modulated photovoltage spectroscopy. *J. Phys. Chem. B*, **101** (41), 8141–8155.

67. Dloczik, L., Ileperuma, O., Lauermann, I., Peter, L.M., Ponomarev, E.A., Redmond, G., Shaw, N.J., and Uhlendorf, I. (1997) Dynamic response of dye-sensitized nanocrystalline solar cells: characterization by intensity-modulated photocurrent spectroscopy. *J. Phys. Chem. B*, **101** (49), 10281–10289.

68. Kopidakis, N., Benkstein, K.D., van de Lagemaat, J., Frank, A.J., Yuan, Q., and Schiff, E.A. (2006) Temperature dependence of the electron diffusion coefficient in electrolyte-filled TiO_2 nanoparticle films: evidence against multiple trapping in exponential conduction-band tails. *Phys. Rev. B*, **73** (4), 045326.

69. Huang, S.Y., Schlichthorl, G., Nozik, A.J., Grätzel, M., and Frank, A.J. (1997) Charge recombination in dye-sensitized nanocrystalline TiO_2 solar cells. *J. Phys. Chem. B*, **101** (14), 2576–2582.

70. Cameron, P.J. and Peter, L.M. (2005) How does back-reaction at the conducting glass substrate influence the dynamic photovoltage response of nanocrystalline dye-sensitized solar cells. *J. Phys. Chem. B*, **109** (15), 7392–7398.

71. Zaban, A., Greenshtein, M., and Bisquert, J. (2003) Determination of the electron lifetime in nanocrystalline dye solar cells by open-circuit voltage decay measurements. *ChemPhysChem*, **4** (8), 859–864.

72. Nakade, S., Kanzaki, T., Kubo, W., Kitamura, T., Wada, Y., and Yanagida, S. (2005) Role of electrolytes on charge recombination in dye-sensitized TiO_2 solar cell (1): the case of solar cells using the I^-/I_3^- redox couple. *J. Phys. Chem. B*, **109** (8), 3480–3487.

73. Boschloo, G., Häggman, L., and Hagfeldt, A. (2006) Quantification of the effect of 4-tert-butylpyridine addition to I^-/I_3^- redox electrolytes in dye-sensitized nanostructured TiO_2 solar cells. *J. Phys. Chem. B*, **110** (26), 13144–13150.

74. Zaban, A., Ferrere, S., and Gregg, B.A. (1998) Relative energetics at the semiconductor/sensitizing dye/electrolyte

interface. *J. Phys. Chem. B*, **102** (2), 452–460.

75. Redmond, G., O'Keeffe, A., Burgess, C., MacHale, C., and Fitzmaurice, D. (1993) Spectroscopic determination of the flatband potential of transparent nanocrystalline zinc oxide films. *J. Phys. Chem.*, **97** (42), 11081–11086.

76. Hagfeldt, A. and Grätzel, M. (2000) Molecular photovoltaics. *Acc. Chem. Res.*, **33** (5), 269–277.

77. Soedergren, S., Hagfeldt, A., Olsson, J., and Lindquist, S.-E. (1994) Theoretical models for the action spectrum and the current-voltage characteristics of microporous semiconductor films in photoelectrochemical cells. *J. Phys. Chem.*, **98** (21), 5552–5556.

78. Solbrand, A., Lindstrom, H., Rensmo, H., Hagfeldt, A., Lindquist, S.-E., and Sodergren, S. (1997) Electron transport in the nanostructured TiO_2- electrolyte system studied with time-resolved photocurrents. *J. Phys. Chem. B*, **101** (14), 2514–2518.

79. Tirosh, S., Dittrich, T., Ofir, A., Grinis, L., and Zaban, A. (2006) Influence of ordering in porous TiO_2 layers on electron diffusion. *J. Phys. Chem. B*, **110** (33), 16165–16168.

80. Fisher, A.C., Peter, L.M., Ponomarev, E.A., Walker, A.B., and Wijayantha, K.G.U. (2000) Intensity dependence of the back reaction and transport of electrons in dye-sensitized nanocrystalline TiO_2 solar cells. *J. Phys. Chem. B*, **104** (5), 949–958.

81. Savenije, T.J., de Haas, M.P., and Warman, J.M. (1999) The yield and mobility of charge carriers in smooth and nanoporous TiO_2 films. *Z. Phys. Chem. - Int. J. Res. Phys. Chem. Chem. Phys.*, **212**, 201.

82. Kopidakis, N., Neale, N.R., Zhu, K., van de Lagemaat, J., and Frank, A.J. (2005) Spatial location of transport-limiting traps in TiO_2 nanoparticle films in dye-sensitized solar cells. *Appl. Phys. Lett.*, **87** (20), 202106.

83. Wahl, A. and Augustynski, J. (1998) Charge carrier transport in nanostructured anatase TiO_2 films assisted by the self-doping of nanoparticles. *J. Phys. Chem. B*, **102** (40), 7820–7828.

84. Pankove, J.I. (1971) *Optical Processes in Semiconductors*, Dover, New York.

85. Ronald, A.S. and Andres, C. (1996) Contactless determination of current-voltage characteristics and minority-carrier lifetimes in semiconductors from quasi-steady-state photoconductance data. *Appl. Phys. Lett.*, **69** (17), 2510–2512.

86. Bastide, S., Gal, D., David, C., and Kronik, L. (1999) Surface photovoltage measurements in liquids. *Rev. Sci. Instrum.*, **70** (10), 4032–4036.

87. Turrión, M., Bisquert, J., and Salvador, P. (2003) Flatband potential of $F:SnO_2$ in a TiO_2 dye-sensitized solar cell: an interference reflection study. *J. Phys. Chem. B*, **107** (35), 9397–9403.

88. Oekermann, T., Zhang, D., Yoshida, T., and Minoura, H. (2004) Electron transport and back reaction in nanocrystalline TiO_2 films prepared by hydrothermal crystallization. *J. Phys. Chem. B*, **108** (7), 2227–2235.

89. Krüger, J., Plass, R., Grätzel, M., Cameron, P.J., and Peter, L.M. (2003) Charge transport and back reaction in solid-state dye-sensitized solar cells: a study using intensity-modulated photovoltage and photocurrent spectroscopy. *J. Phys. Chem. B*, **107** (31), 7536–7539.

90. Jennings, J.R. and Peter, L.M. (2007) A reappraisal of the electron diffusion length in solid-state dye-sensitized solar cells. *J. Phys. Chem. C*, **111** (44), 16100–16104.

91. Jennings, J.R., Ghicov, A., Peter, L.M., Schmuki, P., and Walker, A.B. (2008) Dye-sensitized solar cells based on oriented TiO2 nanotube arrays: transport, trapping, and transfer of electrons. *J. Am. Chem. Soc.*, **130** (40), 13364–13372.

92. Zhu, K., Neale, N.R., Miedaner, A., and Frank, A.J. (2006) Enhanced charge-collection efficiencies and light scattering in dye-sensitized solar cells using oriented TiO_2 nanotubes arrays. *Nano Lett.*, **7** (1), 69–74.

93. Zhang, M., Bando, Y., and Wada, K. (2001) Sol-gel template preparation

of TiO$_2$ Nanotubes and nanorods. *J. Mater. Sci. Lett.*, **20** (2), 167–170.

94. Zhu, Y., Li, H., Koltypin, Y., Hacohen, Y.R., and Gedanken, A. (2001) Sonochemical synthesis of titania whiskers and nanotubes. *Chem. Commun.*, (24), 2616–2617.

95. Peng, T., Hasegawa, A., Qiu, J., and Hirao, K. (2003) Fabrication of titania tubules with high surface area and well-developed mesostructural walls by surfactant-mediated templating method. *Chem. Mater.*, **15** (10), 2011–2016.

96. Adachi, M., Murata, Y., Okada, I., and Yoshikawa, S. (2003) Formation of titania nanotubes and applications for dye-sensitized solar cells. *J. Electrochem. Soc.*, **150** (8), G488–G493.

97. Gong, D., Grimes, C.A., Varghese, O.K., Chen, Z., Hu, W., and Dickey, E.C. (2001) Titanium oxide nanotube arrays prepared by anodic oxidation. *J. Mater. Res.*, **16**, 3331–3334.

98. Mor, G.K., Varghese, O.K., Paulose, M., Shankar, K., and Grimes, C.A. (2006) A review on highly ordered, vertically oriented TiO$_2$ nanotube arrays: Fabrication, material properties, and solar energy applications. *Sol. Energy Mater. Sol. Cells*, **90** (14), 2011–2075.

99. Karthik, S., Mor, G.K., Prakasam, H.E., Yoriya, S., Paulose, M., Varghese, O.K., and Grimes, C.A. (2007) Highly-ordered TiO$_2$ nanotube arrays up to 220 μm in length: use in water photoelectrolysis and dye-sensitized solar cells. *Nanotechnology*, **18** (6), 065707.

100. Chen, Q. and Xu, D. (2009) Large-scale, noncurling, and free-standing crystallized TiO$_2$ nanotube arrays for dye-sensitized solar cells. *J. Phys. Chem. C*, **113** (15), 6310–6314.

101. Park, J.H., Lee, T.-W., and Kang, M.G. (2008) Growth, detachment and transfer of highly-ordered TiO$_2$ nanotube arrays: use in dye-sensitized solar cells. *Chem. Commun.*, (25), 2867–2869.

102. Mor, G.K., Shankar, K., Paulose, M., Varghese, O.K., and Grimes, C.A. (2005) Use of highly-ordered TiO$_2$ nanotube arrays in dye-sensitized solar cells. *Nano Lett.*, **6** (2), 215–218.

103. Mor, G., Varghese, O., Paulose, M., and Grimes, C. (2005) Transparent highly ordered TiO$_2$ nanotube arrays via anodization of titanium thin films. *Adv. Funct. Mater.*, **15** (8), 1291–1296.

104. Paulose, M., Shankar, K., Yoriya, S., Prakasam, H.E., Varghese, O.K., Mor, G.K., Latempa, T.A., Fitzgerald, A., and Grimes, C.A. (2006) Anodic growth of highly ordered TiO$_2$ nanotube arrays to 134 μm in length. *J. Phys. Chem. B*, **110** (33), 16179–16184.

105. Varghese, O.K., Paulose, M., and Grimes, C.A. (2009) Long vertically aligned titania nanotubes on transparent conducting oxide for highly efficient solar cells. *Nat. Nanotechnol.*, **4** (9), 592–597.

106. Keis, K., Magnusson, E., Lindström, H., Lindquist, S.-E., and Hagfeldt, A. (2002) A 5% efficient photoelectrochemical solar cell based on nanostructured ZnO electrodes. *Sol. Energy Mater. Sol. Cells*, **73** (1), 51–58.

107. Martinson, A.B.F., Elam, J.W., Hupp, J.T., and Pellin, M.J. (2007) ZnO nanotube based dye-sensitized solar cells. *Nano Lett.*, **7** (8), 2183–2187.

108. Kakiuchi, K., Saito, M., and Fujihara, S. (2008) Fabrication of ZnO films consisting of densely accumulated mesoporous nanosheets and their dye-sensitized solar cell performance. *Thin Solid Films*, **516** (8), 2026–2030.

109. Jiang, C.Y., Sun, X.W., Lo, G.Q., Kwong, D.L., and Wang, J.X. (2007) Improved dye-sensitized solar cells with a ZnO-nanoflower photoanode. *Appl. Phys. Lett.*, **90** (26), 263501–263503.

110. Il-Doo, K., Jae-Min, H., Byong Hong, L., Dong Young, K., Eun-Kyung, J., Duck-Kyun, C., and Dae-Jin, Y. (2007) Dye-sensitized solar cells using network structure of electrospun ZnO nanofiber mats. *Appl. Phys. Lett.*, **91** (16), 163109.

111. Redmond, G., Fitzmaurice, D., and Gräetzel, M. (1994) Visible light sensitization by cis-bis(thiocyanato) bis(2,2′-bipyridyl-4,4′- dicarboxylato)ruthenium(II) of a transparent nanocrystalline ZnO film prepared by sol-gel techniques. *Chem. Mater.*, **6** (5), 686–691.

112. Rensmo, H., Keis, K., Lindstrom, H., Sodergren, S., Solbrand, A., Hagfeldt, A., Lindquist, S.E., Wang, L.N., and Muhammed, M. (1997) High light-to-energy conversion efficiencies for solar cells based on nanostructured ZnO electrodes. *J. Phys. Chem. B*, **101** (14), 2598–2601.

113. Zhang, Q., Chou, T., Russo, B., Jenekhe, S., and Cao, G. (2008) Aggregation of ZnO nanocrystallites for high conversion efficiency in dye-sensitized solar cells. *Angew. Chem. Int. Ed.*, **47** (13), 2402–2406.

114. Saito, M. and Fujihara, S. (2008) Large photocurrent generation in dye-sensitized ZnO solar cells. *Energy Environ. Sci.*, **1** (2), 280–283.

115. Jason, B.B. and Eray, S.A. (2005) Nanowire-based dye-sensitized solar cells. *Appl. Phys. Lett.*, **86** (5), 053114.

116. Jiang, C.Y., Sun, X.W., Lo, G.Q., Kwong, D.L., and Wang, J.X. (2007) Improved dye-sensitized solar cells with a ZnO-nanoflower photoanode. *Appl. Phys. Lett.*, **90** (26), 263501.

117. Bedja, I., Hotchandani, S., and Kamat, P.V. (1994) Preparation and photoelectrochemical characterization of thin SnO_2 nanocrystalline semiconductor films and their sensitization with Bis(2,2'-bipyridine)(2,2'-bipyridine-4,4'-dicarboxylic acid)ruthenium(II) complex. *J. Phys. Chem.*, **98** (15), 4133–4140.

118. Keis, K., Lindgren, J., Lindquist, S.-E., and Hagfeldt, A. (2000) Studies of the adsorption process of Ru complexes in nanoporous ZnO electrodes. *Langmuir*, **16** (10), 4688–4694.

119. Asbury, J.B., Hao, E., Wang, Y., Ghosh, H.N., and Lian, T. (2001) Ultrafast electron transfer dynamics from molecular adsorbates to semiconductor nanocrystalline thin films. *J. Phys. Chem. B*, **105** (20), 4545–4557.

120. Anderson, N.A., Ai, X., and Lian, T. (2003) Electron injection dynamics from Ru polypyridyl complexes to ZnO nanocrystalline thin films. *J. Phys. Chem. B*, **107** (51), 14414–14421.

121. Asbury, J.B., Wang, Y., and Lian, T. (1999) Multiple-exponential electron injection in $Ru(dcbpy)_2(SCN)_2$ sensitized ZnO nanocrystalline thin films. *J. Phys. Chem. B*, **103** (32), 6643–6647.

122. Bauer, C., Boschloo, G., Mukhtar, E., and Hagfeldt, A. (2001) Electron injection and recombination in $Ru(dcbpy)_2(NCS)_2$ sensitized nanostructured ZnO. *J. Phys. Chem. B*, **105** (24), 5585–5588.

123. Katoh, R., Furube, A., Yoshihara, T., Hara, K., Fujihashi, G., Takano, S., Murata, S., Arakawa, H., and Tachiya, M. (2004) Efficiencies of electron injection from excited N3 dye into nanocrystalline semiconductor (ZrO_2, TiO_2, ZnO, Nb_2O_5, SnO_2, In_2O_3) films. *J. Phys. Chem. B*, **108** (15), 4818–4822.

124. Bjoerksten, U., Moser, J., and Grätzel, M. (1994) Photoelectrochemical studies on nanocrystalline hematite films. *Chem. Mater.*, **6** (6), 858–863.

125. Sayama, K., Sugihara, H., and Arakawa, H. (1998) Photoelectrochemical properties of a porous Nb_2O_5 electrode sensitized by a ruthenium dye. *Chem. Mater.*, **10** (12), 3825–3832.

126. Guo, P. and Aegerter, M.A. (1999) RU(II) sensitized Nb_2O_5 solar cell made by the sol-gel process. *Thin Solid Films*, **351** (1–2), 290–294.

127. He, J., Lindström, H., Hagfeldt, A., and Lindquist, S.-E. (2000) Dye-sensitized nanostructured tandem cell-first demonstrated cell with a dye-sensitized photocathode. *Sol. Energy Mater. Sol. Cells*, **62** (3), 265–273.

128. Tennakone, K., Fernando, C.A.N., Dewasurendra, M., and Kariappert, M.S. (1986) Dye-sensitized cuprous iodide photocathode. *Jpn. J. Appl. Phys.*, **26**, 561.

129. Kirthi, T., Mahendra, K., Cyril, K., Parakrama, A., Ranmuthu, H.W., and Padmananda, K. (1984) Dye sensitization of cuprous thiocyanate photocathode in aqueous KCNS. *J. Electrochem. Soc.*, **131** (7), 1574–1577.

130. O'Regan, B. and Schwartz, D.T. (1995) Efficient photo-hole injection from adsorbed cyanine dyes into electrodeposited copper(I) thiocyanate thin films. *Chem. Mater.*, **7** (7), 1349–1354.

131. He, J., Lindstrom, H., Hagfeldt, A., and Lindquist, S.-E. (1999) Dye-sensitized

nanostructured p-type nickel oxide film as a photocathode for a solar cell. *J. Phys. Chem. B*, **103** (42), 8940–8943.
132. Morandeira, A., Boschloo, G., Hagfeldt, A., and Hammarström, L. (2005) Photoinduced ultrafast dynamics of coumarin 343 sensitized p-type-nanostructured NiO films. *J. Phys. Chem. B*, **109** (41), 19403–19410.
133. Borgström, M., Blart, E., Boschloo, G., Mukhtar, E., Hagfeldt, A., Hammarström, L., and Odobel, F. (2005) Sensitized hole injection of phosphorus porphyrin into NiO: toward new photovoltaic devices. *J. Phys. Chem. B*, **109** (48), 22928–22934.
134. Qin, P., Zhu, H., Edvinsson, T., Boschloo, G., Hagfeldt, A., and Sun, L. (2008) Design of an organic chromophore for P-type dye-sensitized solar cells. *J. Am. Chem. Soc.*, **130** (27), 8570–8571.
135. Mori, S., Fukuda, S., Sumikura, S., Takeda, Y., Tamaki, Y., Suzuki, E., and Abe, T. (2008) Charge-transfer processes in dye-sensitized NiO solar cells. *J. Phys. Chem. C*, **112** (41), 16134–16139.
136. Vera, F., Schrebler, R., Muñoz, E., Suarez, C., Cury, P., Gómez, H., Córdova, R., Marotti, R.E., and Dalchiele, E.A. (2005) Preparation and characterization of Eosin B- and Erythrosin J-sensitized nanostructured NiO thin film photocathodes. *Thin Solid Films*, **490** (2), 182–188.
137. Andrew, N. et al. (2008) Dye-sensitized nickel(II)oxide photocathodes for tandem solar cell applications. *Nanotechnology*, **19** (29), 295304.
138. Zhu, H., Hagfeldt, A., and Boschloo, G. (2007) Photoelectrochemistry of mesoporous NiO electrodes in iodide/triiodide electrolytes. *J. Phys. Chem. C*, **111** (47), 17455–17458.
139. Morandeira, A., Fortage, J., Edvinsson, T., Le Pleux, L., Blart, E., Boschloo, G., Hagfeldt, A., Hammarstrom, L., and Odobel, F. (2008) Improved photon-to-current conversion efficiency with a nanoporous p-type NiO electrode by the use of a sensitizer-acceptor dyad. *J. Phys. Chem. C*, **112** (5), 1721–1728.
140. Nattestad, A., Mozer, A.J., Fischer, M.K.R., Cheng, Y.B., Mishra, A., Bauerle, P., and Bach, U. (2010) Highly efficient photocathodes for dye-sensitized tandem solar cells. *Nat. Mater.*, **9** (1), 31–35.
141. Campbell, W.M., Jolley, K.W., Wagner, P., Wagner, K., Walsh, P.J., Gordon, K.C., Schmidt-Mende, L., Nazeeruddin, M.K., Wang, Q., Grätzel, M., and Officer, D.L. (2007) Highly efficient porphyrin sensitizers for dye-sensitized solar cells. *J. Phys. Chem. C*, **111** (32), 11760–11762.
142. Grätzel, M. (2004) Conversion of sunlight to electric power by nanocrystalline dye-sensitized solar cells. *J. Photochem. Photobiol. A: Chem.*, **164** (1–3), 3–14.
143. Wang, Z.-S., Kawauchi, H., Kashima, T., and Arakawa, H. (2004) Significant influence of TiO_2 photoelectrode morphology on the energy conversion efficiency of N719 dye-sensitized solar cell. *Coord. Chem. Rev.*, **248** (13–14), 1381–1389.
144. Nazeeruddin, M.K., Péchy, P., Renouard, T., Zakeeruddin, S.M., Humphry-Baker, R., Cointe, P., Liska, P., Cevey, L., Costa, E., Shklover, V., Spiccia, L., Deacon, G.B., Bignozzi, C.A., and Grätzel, M. (2001) Engineering of efficient panchromatic sensitizers for nanocrystalline TiO_2-based solar cells. *J. Am. Chem. Soc.*, **123** (8), 1613–1624.
145. Islam, A., Sugihara, H., Yanagida, M., Hara, K., Fujihashi, G., Tachibana, Y., Katoh, R., Murata, S., and Arakawa, H. (2002) Efficient panchromatic sensitization of nanocrystalline TiO_2 films by [small beta]-diketonato ruthenium polypyridyl complexes. *New J. Chem.*, **26** (8), 966–968.
146. Yum, J.-H., Walter, P., Huber, S., Rentsch, D., Geiger, T., Nüesch, F., De Angelis, F., Grätzel, M., and Nazeeruddin, M.K. (2007) Efficient far red sensitization of nanocrystalline TiO_2 films by an unsymmetrical squaraine dye. *J. Am. Chem. Soc.*, **129** (34), 10320–10321.
147. He, J., Benkö, G., Korodi, F., Polívka, T., Lomoth, R., Åkermark, B., Sun, L.,

Hagfeldt, A., and Sundström, V. (2002) Modified phthalocyanines for efficient near-IR sensitization of nanostructured TiO$_2$ electrode. *J. Am. Chem. Soc.*, **124** (17), 4922–4932.

148. Nüesch, F., Moser, J.E., Shklover, V., and Grätzel, M. (1996) Merocyanine aggregation in mesoporous networks. *J. Am. Chem. Soc.*, **118** (23), 5420–5431.

149. Liu, D. and Kamat, P.V. (1995) Electrochemically active nanocrystalline SnO$_2$ films: surface modification with thiazine and oxazine dye aggregates. *J. Electrochem. Soc.*, **142** (3), 835–839.

150. Kubo, W., Sakamoto, A., Kitamura, T., Wada, Y., and Yanagida, S. (2004) Dye-sensitized solar cells: improvement of spectral response by tandem structure. *J. Photochem. Photobiol. A: Chem.*, **164** (1–3), 33–39.

151. Kay, A. and Grätzel, M. (1996) Low cost photovoltaic modules based on dye sensitized nanocrystalline titanium dioxide and carbon powder. *Sol. Energy Mater. Sol. Cells*, **44** (1), 99–117.

152. Usami, A. (1997) Theoretical study of application of multiple scattering of light to a dye-sensitized nanocrystalline photoelectrochemical cell. *Chem. Phys. Lett.*, **277** (1–3), 105–108.

153. Ferber, J. and Luther, J. (1998) Computer simulations of light scattering and absorption in dye-sensitized solar cells. *Sol. Energy Mater. Sol. Cells*, **54** (1–4), 265–275.

154. Durr, M., Bamedi, A., Yasuda, A., and Nelles, G. (2004) Tandem dye-sensitized solar cell for improved power conversion efficiencies. *Appl. Phys. Lett.*, **84** (17), 3397–3399.

155. Murayama, M. and Mori, T. (2007) Dye-sensitized solar cell using novel tandem cell structure. *J. Phys. D: Appl. Phys.*, **40** (6), 1664.

156. Zhao, W., Jun Hou, Y., Song Wang, X., Wen Zhang, B., Cao, Y., Yang, R., Bo Wang, W., and Rui Xiao, X. (1999) Study on squarylium cyanine dyes for photoelectric conversion. *Sol. Energy Mater. Sol. Cells*, **58** (2), 173–183.

157. Ehret, A., Stuhl, L., and Spitler, M.T. (2001) Spectral sensitization of TiO2 nanocrystalline electrodes with aggregated cyanine dyes. *J. Phys. Chem. B*, **105** (41), 9960–9965.

158. Sayama, K., Tsukagoshi, S., Mori, T., Hara, K., Ohga, Y., Shinpou, A., Abe, Y., Suga, S., and Arakawa, H. (2003) Efficient sensitization of nanocrystalline TiO$_2$ films with cyanine and merocyanine organic dyes. *Sol. Energy Mater. Sol. Cells*, **80** (1), 47–71.

159. Kuang, D., Walter, P., Nüesch, F., Kim, S., Ko, J., Comte, P., Zakeeruddin, S.M., Nazeeruddin, M.K., and Grätzel, M. (2007) Co-sensitization of organic dyes for efficient ionic liquid electrolyte-based dye-sensitized solar cells. *Langmuir*, **23** (22), 10906–10909.

160. Cid, J.J., Yum, J.H., Jang, S.R., Nazeeruddin, M., Martínez-Ferrero, E., Palomares, E., Ko, J., Grätzel, M., and Torres, T. (2007) Molecular cosensitization for efficient panchromatic dye-sensitized solar cells. *Angew. Chem. Int. Ed.*, **46** (44), 8358–8362.

161. Alonso, N., Beley, M., and Chartier, P. (1981) Dye sensitization of ceramic semiconducting electrodes for photoelectrochemical conversion. *Rev. Phys. Appl., Paris*, **16**, 5.

162. Hardin, B.E., Hoke, E.T., Armstrong, P.B., Yum, J.-H., Comte, P., Torres, T., Frechet, J.M.J., Nazeeruddin, M.K., Grätzel, M., and McGehee, M.D. (2009) Increased light harvesting in dye-sensitized solar cells with energy relay dyes. *Nat. Photonics*, **3** (7), 406–411.

163. Sayama, K., Tsukagoshi, S., Hara, K., Ohga, Y., Shinpou, A., Abe, Y., Suga, S., and Arakawa, H. (2002) Photoelectrochemical properties of J aggregates of benzothiazole merocyanine dyes on a nanostructured TiO$_2$ film. *J. Phys. Chem. B*, **106** (6), 1363–1371.

164. Chen, J., Zhao, D.W., Lei, W., and Sun, X.W. (2010) Cosensitized solar cells based on a flower-like ZnO nanorod structure. *IEEE J. Sel. Top. Quantum Electron.*, **16** (6), 1607–1610.

165. Lee, K., Park, S.W., Ko, M.J., Kim, K., and Park, N.-G. (2009) Selective positioning of organic dyes in a mesoporous inorganic oxide film. *Nat. Mater.*, **8** (8), 665–671.

166. Miyasaka, T., Kijitori, Y., Murakami, T.N., Kimura, M., and Uegusa, S. (2002) Efficient nonsintering type dye-sensitized photocells based on electrophoretically deposited TiO_2 layers. *Chem. Lett.*, **31** (12), 1250–1251.
167. Hu, X. and Schulten, K. (1997) How nature harvests sunlight. *Phys. Today*, **50** (8), 28–34.
168. Siegers, C., Hohl-Ebinger, J., Zimmermann, B., Würfel, U., Mülhaupt, R., Hinsch, A., and Haag, R. (2007) A dyadic sensitizer for dye solar cells with high energy-transfer efficiency in the device. *ChemPhysChem*, **8** (10), 1548–1556.
169. Clifford, J.N., Palomares, E., Nazeeruddin, M.K., Thampi, R., Grätzel, M., and Durrant, J.R. (2004) Multistep electron transfer processes on dye co-sensitized nanocrystalline TiO_2 films. *J. Am. Chem. Soc.*, **126** (18), 5670–5671.
170. Uchida, S., Tomiha, M., Takizawa, H., and Kawaraya, M. (2004) Flexible dye-sensitized solar cells by 28∼GHz microwave irradiation. *J. Photochem. Photobiol. A: Chem.*, **164** (1–3), 93–96.
171. Pichot, Fo., Pitts, J.R., and Gregg, B.A. (2000) Low-temperature sintering of TiO_2 colloids: application to flexible dye-sensitized solar cells. *Langmuir*, **16** (13), 5626–5630.
172. Lindström, H., Holmberg, A., Magnusson, E., Lindquist, S.-E., Malmqvist, L., and Hagfeldt, A. (2001) A new method for manufacturing nanostructured electrodes on plastic substrates. *Nano Lett.*, **1** (2), 97–100.
173. Murakami, T.N., Kijitori, Y., Kawashima, N., and Miyasaka, T. (2003) UV light-assisted chemical vapor deposition of TiO_2 for efficiency development at dye-sensitized mesoporous layers on plastic film electrodes. *Chem. Lett.*, **32** (11), 1076–1077.
174. Tsutomu, M. and Yujiro, K. (2004) Low-temperature fabrication of dye-sensitized plastic electrodes by electrophoretic preparation of mesoporous TiO_2 layers. *J. Electrochem. Soc.*, **151** (11), A1767–A1773.
175. Yum, J.-H., Kim, S.-S., Kim, D.-Y., and Sung, Y.-E. (2005) Electrophoretically deposited TiO_2 photo-electrodes for use in flexible dye-sensitized solar cells. *J. Photochem. Photobiol. A: Chem.*, **173** (1), 1–6.
176. Zhang, D., Yoshida, T., and Minoura, H. (2003) Low-temperature fabrication of efficient porous titania photoelectrodes by hydrothermal crystallization at the solid/gas interface. *Adv. Mater.*, **15** (10), 814–817.
177. Kado, T., Yamaguchi, M., Yamada, Y., and Hayase, S. (2003) Low temperature preparation of nano-porous TiO_2 layers for plastic dye sensitized solar cells. *Chem. Lett.*, **32** (11), 1056–1057.
178. Durr, M., Schmid, A., Obermaier, M., Rosselli, S., Yasuda, A., and Nelles, G. (2005) Low-temperature fabrication of dye-sensitized solar cells by transfer of composite porous layers. *Nat. Mater.*, **4** (8), 607–611.
179. Kang, M.G., Park, N.-G., Ryu, K.S., Chang, S.H., and Kim, K.-J. (2006) A 4.2% efficient flexible dye-sensitized TiO2 solar cells using stainless steel substrate. *Sol. Energy Mater. Sol. Cells*, **90** (5), 574–581.
180. Ito, S., Ha, N.-L.C., Rothenberger, G., Liska, P., Comte, P., Zakeeruddin, S.M., Pechy, P., Nazeeruddin, M.K., and Grätzel, M. (2006) High-efficiency (7.2%) flexible dye-sensitized solar cells with Ti-metal substrate for nanocrystalline-TiO2 photoanode. *Chem. Commun.*, (38), 4004–4006.
181. Yoshida, T., Oekermann, T., Okabe, K., Schlettwein, D., Funabiki, K., and Minoura, H. (2002) Cathodic electrodeposition of ZnO/eosinY hybrid thin films from dye added zinc nitrate bath and their photoelectrochemical characterizations. *Electrochemistry (Tokyo, Jpn.)*, **70**, 470.
182. Liu, X., Luo, Y., Li, H., Fan, Y., Yu, Z., Lin, Y., Chen, L., and Meng, Q. (2007) Room temperature fabrication of porous ZnO photoelectrodes for flexible dye-sensitized solar cells. *Chem. Commun.*, (27), 2847–2849.
183. Jiang, C.Y., Sun, X.W., Tan, K.W., Lo, G.Q., Kyaw, A.K.K., and Kwong, D.L. (2008) High-bendability flexible dye-sensitized solar cell with a nanoparticle-modified ZnO-nanowire

electrode. *Appl. Phys. Lett.*, **92** (14), 143101.

184. Hamann, T.W., Jensen, R.A., Martinson, A.B.F., Van Ryswyk, H., and Hupp, J.T. (2008) Advancing beyond current generation dye-sensitized solar cells. *Energy Environ. Sci.*, **1** (1), 66–78.

185. Green, M.A. (2006) *Third Generation Photovoltaics – Advanced Solar Energy Conversion*, Springer, Berlin.

186. Shockley, W. and Queisser, H.J. (1961) Detailed balance limit of efficiency of p-n junction solar cells. *J. Appl. Phys.*, **32**, 510.

12
Photoelectrochemical Cells for Hydrogen Generation

Neelu Chouhan, ChihKai Chen, Wen-Sheng Chang, Kong-Wei Cheng, and Ru-Shi Liu

12.1
Introduction

Energy, as a major driving force behind world economy, is integral to industry, transport, and everyday life. Fossil fuels are the prime source of energy, but they create a large amount of pollution. Consequently, they produce ill effects on the health of both humans and the ecosystem. Besides the soaring prices of petroleum products, our demand for gasoline is increasing day by day and it has become harder to find new oil sources, which raises several issues such as environmental degradation, need for more sophisticated instrumentation and advanced techniques for marking new oil/gas sources, layman's deep concern for drastic environmental changes, and political instability in key oil-producing regions. Our major reliance on fossil fuels for energy also has negative political and socioecological impacts in the form of economic dependence on oil-producing countries. Furthermore, electricity, which is a main source of energy, is generated either from coal or portable sources such as batteries, supercapacitors, and fuel cells, which do not have sufficient energy/power densities and/or efficiencies owing to poor charge- and mass-transport properties, or are too expensive as a result of the material and manufacturing costs [1, 2]. A new realm of possibilities comes from the new frontiers area of renewable and sustainable energy sources. Sun is the most powerful and richest renewable and sustainable source of energy. Cyrus Smith has stressed the importance of water as a fuel resource by his famous saying "Water is the coal of the future." The efficient combination of water and solar energy has enormous capacity to fulfill the world's present and future demands of energy in an ecofriendly manner by producing hydrogen fuel in photocells. That requires efficient light-harvesting systems, for example, semiconductors known as *"photocatalysts,"* to split water under the action of sunlight. Photocatalysts are smart materials to exploit energy from sunlight, but the poor charge-carrier mobility and narrow absorption range of currently used semiconductors limit the energy conversion efficiency of photocells [3–15].

Electrochemical Technologies for Energy Storage and Conversion, First Edition. Edited by Ru-Shi Liu, Lei Zhang, Xueliang Sun, Hansan Liu, and Jiujun Zhang.
© 2012 Wiley-VCH Verlag GmbH & Co. KGaA. Published 2012 by Wiley-VCH Verlag GmbH & Co. KGaA.

Hydrogen fuel is a leading contender to solve energy problems. High fuel efficiency, no carbon emission, and a useful by-product of hydrogen-fuel combustion, which is only water (but in practice small amounts of nitrogen oxides (NO_x) can be produced), are the advantages associated with hydrogen fuel. Herman Kuipers, Manager, Exploratory Research, Royal Dutch Shell, has commented on hydrogen fuel, which is quite contemporary and relevant in present energy scenario: "We are at the peak of the oil age but the beginning of the hydrogen age. Anything else is an interim solution. The transition will be very messy, and will take many technological paths.... but the future will be hydrogen-fuel cells." Nature abundantly stores hydrogen in the form of water and it has great potential as a future fuel candidate. Visible light (50–60% of the solar radiation) has enough power to split water (H_2O) into hydrogen (H_2) and oxygen (O_2).

Fortunately, water is transparent and does not absorb visible light energy directly, and so it requires the help of photocatalysts. Water splitting can be accomplished via various techniques such as electrolysis, photocatalysis, photoelectrolysis, or photoelectrochemical (PEC) and thermochemical reactions, and biophotolysis. In order to meet the goal of an ideal fuel, a PEC system must be of low cost and should be operated at high solar-to-chemical conversion efficiency (greater than 10%) with commercial viability for a long operating life. A variety of approaches employed on a series of semiconductors have pointed to several technologies but, to date, none of them have successfully satisfied the criteria of performance, design, operation, cost, efficiency, and stability. Therefore, an ideal photocatalyst still remains one of the "holy grails" of the chemistry for renewable hydrogen production. A major breakthrough in this field is the discovery of the photovoltaic (PV) effect by Henri Becquerel [16] in 1839, which laid the foundation of modern photoelectrochemistry but was initially developed to support photography and took its own road through the work of Brattain and Garret [17]. Subsequently, Gerischer undertook the first detailed electrochemical and PEC studies of the semiconductor–electrolyte interface [18]. After the oil crisis of the 1970s, research on PEC cells was stimulated worldwide as a quest for alternative energy sources. Fujishima and Honda's discovery of water photolysis for the production of hydrogen and oxygen fuels by using the most favorite semiconductor titanium dioxide as photoanode established a milestone in this field [19]. Furthermore, Nozik's concept of "photochemical diodes" (termed *antenna catalysts*) stimulated the research on PEC cells, which consisted of two electrodes from a PEC cell directly fused together [20]. This produced either p-n-type (e.g., p-GaP/n-TiO_2) or Schottky-type devices (e.g., n-CdS/Pt or p-GaP/Pt). The efficiency of such dual-component catalysts was expected to improve because a space charge layer at the material interface enhances the electron–hole separation. Unfortunately, the wide bandgap (3.2 eV) of the most studied materials (TiO_2, ZnO) restricts them to be active under the ultraviolet (UV) portion of the solar emission and, therefore, has low conversion efficiencies. The high voltage requirement to dissociate water as well as the corrosiveness of semiconductors in aqueous electrolytes had been major hurdles for the achievement of high efficiency in PEC cells. The maximum practical solar-to-hydrogen conversion efficiency (STH) for water splitting of bulk TiO_2 is below 0.5% by UV irradiation [21]. Numerous efforts have been made to shift the spectral response of TiO_2 into the visible region or to develop alternative oxides (ZnO, Fe_2O_3, SnO_2, Nb_2O_5, Ta_2O_5, ZrO_2, metal titanates, oxynitrides, and oxysulfides) [22–32] to increase the water-splitting efficiency. Doping with anions (N, C, S, P) also increase the efficiency of PEC cells for solar-to-hydrogen production, such as nitrogen doping of ZnO (Yang, 2009), which increases the efficiency from 0.11 to 0.15% [33]. Khan et al. modified TiO_2 photocatalysts by carbon doping, which absorbs light at wavelengths below 535 nm and enhances the STH efficiency of up to 8.35% [34]. Metal-doped (Cd, Cu, Pd, In, etc.) semiconductors also modify the efficiency of PEC cells. Anpo et al. [35] prepared TiO_2 photocatalysts by advanced metal-ion implantation and radio frequency (RF) magnetron sputtering deposition methods, which resulted in significant visible light responsiveness of TiO_2 photocatalysts for water splitting. For example, the incident photon-to-current conversion efficiencies of three $Nb:SrTiO_3$ photoanodes (0, 0.07, and 0.69 mol% of Nb) are reported to be 0.92, 15.67, and 4.32%, respectively, at 298.2 nm wavelength and an applied potential of 1.5 V (vs the saturated calomel electrode) [36]. Furthermore, Licht et al. presented the concept of multiple bandgap tandem cells for PEC water splitting, and demonstrated a PEC STH efficiency of 18.3% using illuminated AlGaAs/Si and RuO_2/Pt photoelectrodes (PEs) [37]. The advantage of the Z-scheme tandem approach comes from two photoelectrode systems (dye-sensitized nanocrystalline TiO_2 and thin film of WO_3) that absorb complementary parts of the solar spectrum with single-junction cells, which can achieve an overall solar light to chemical energy conversion efficiency of 4.5% [38]. Researchers of Toledo University have successfully achieved PV efficiencies of up to 12.7% for their triple-junction cells deposited on a stainless steel (SS) foil. Wang et al. [39] and Khaselev et al. have reported an STH efficiency of over 16% for the $GaAs/GaInP_2$ system [40]. Nanostructuring of the photocatalyst also introduces profound changes in its PEC properties. Moreover, conversion efficiencies of up to 19.6% have been reported for multijunction regenerative cells [41]. Recently, Mark Wanlass of NREL (National Renewable Energy Laboratory) invented the triple-junction metamorphic solar inverted cell of 40.8% efficiency under concentrated light exposure of 326 Suns [42].

In this chapter, we review the contemporary issues related to PEC hydrogen generation with a brief overview of the periodic development of the PEC cells. Specific categorization of PEC cells on the basis of their performance and materials used is discussed along with their primary components. A unique attempt has been made to describe the chemistry behind the electrochemical processes of semiconductor–electrolyte interface and solar cleavage of water in an electrolyte. Relative merits of PEC cells are discussed on the basis of their performance, testing, and diagnosis techniques. Efforts has been made to explain the degradation mechanisms and their mitigation strategies for various PEC cells. And, finally, some light is thrown on commercial applications of PECs in portable, stationary, and transportation devices.

12.2
Main Types and Structures of Photoelectrochemical Cells

PEC cells can be easily used for the electrolysis of water to produce hydrogen, which is the fuel of future green economy. Unlike fossil fuel reactions, no carbon dioxide, carbon monoxide, sulfur dioxide, or suspended particles, which are the major contributors to greenhouse gases, are produced by the hydrogen fuel. Hence, hydrogen economy will result in the reduction of greenhouse gas emissions and improved air quality; in turn, it will greatly improve the environmental health. There are several methods for utilizing solar radiation in splitting water for hydrogen production.

- **Solar to thermal route:** possesses two steps, which are the photon-to-thermal energy conversion step and the thermal-to-chemical energy conversion.
- **Photovoltaic electrolysis:** comprises two steps, namely, photon-to-electric energy conversion and electric energy-to-chemical energy conversion.
- **Concentrate solar thermal (CST)-electrolysis route:** includes three steps, namely, photon-to-thermal energy, thermal energy-to-electric energy, and, finally, electric energy-to-chemical energy conversion. This multiple energy conversion process suffers efficiency loss at each step.
- **Photoelectrochemical water splitting:** single-step, direct path that proceeds by the low-temperature conversion of photon energy to chemical energy. It has the inherent advantage of good performance.
- **Photobiological water splitting:** involves a two-step conversion process, namely, photon energy to biological energy (light reaction) and biological energy to chemical energy (dark reaction) conversion. Photobiological systems utilize microbes that absorb sunlight and give out hydrogen as a metabolic waste product. Some photoheterotrophic bacteria (*Rhodospirillum rubrum*) survive in the dark via the water–gas shift reaction. These bacteria generate ATP (adenosine triphosphate) using carbon monoxide by utilizing electrons produced during the oxidation of CO and reduction of H^+ to H_2. Photosynthetic bacteria, such as *Rhodobacter spaeroides*, can also generate hydrogen through photofermentation of organic wastes. Another promising process is the dark fermentation of carbohydrate-rich substrates by anaerobic bacteria and green algae. For example, glucose and sucrose from sugar refinery waste water can be fermented to produce hydrogen.

12.2.1
Types of Photoelectrochemical Devices

In principle, there are four general types of PEC devices using semiconductor electrodes for the conversion of water into hydrogen.

12.2.1.1 Photoelectrolysis or Photosynthesis Cell
These cells convert radiant energy into net chemical energy so as to produce hydrogen as a useful fuel and are classified as photosynthesis or photoelectrolysis

cells. Radiant energy provides a required Gibbs energy ($\Delta G < 0$) to the system to drive a particular reaction such as $H_2O \rightarrow H_2 + \frac{1}{2}O_2$, with additional electrical or thermal energy, which may be recovered by allowing the reverse and spontaneous reaction to proceed. In this type of PEC cells, both electrodes (working and counter) with the reference electrode are immersed in the same solution of constant pH. n-SrTiO$_3$ photoanode/9.5–10 M NaOH electrolyte/Pt cathode is a typical example of this type [43, 44]. These cells work on the principle of "internal electric field generated at the semiconductor–electrolyte interface used for efficient separation of the photogenerated electronhole pairs (excitons)." Subsequently photogenerated holes and electrons are readily available for oxidation and reduction of water, at the anode and cathode, respectively, for O_2 and H_2 production. Usually, the anode and the cathode are physically separated in a PEC cell, but they can be combined together into a monolithic structure known as a *photochemical diode*.

12.2.1.2 Photo-Assisted Electrolysis Cell

A single-junction photosystem configuration (n-type or p-type) is not very effective owing to the inappropriate bandgap or nonmatching band edges of semiconductors. Thus, additional voltage or bias is required to fit the energy gap when the band edges do not overlap for water splitting. The photo-assisted electrolysis cells operated under light are coupled with an external or internal bias (electrical bias (grid) – fossil based; or chemical bias – pH based; or PV cell bias or dye-sensitized solar cell (DSSC) bias, or internal bias), which serves either to drive electrolytic reactions for which the photon energy is insufficient or to increase the rate of chemical energy conversion by suppressing electron–hole recombination in the bulk semiconductor (Figure 12.1a–d). Different kinds of bias are discussed in the following sections [45].

Electrical Bias (Grid) – Fossil Based Electrical bias is an external bias used for the PEC water splitting and uses electricity obtained from the grid, which is not an attractive option because of the use of the notorious fossil-based energy source for the electricity generation (Figure 12.1a). A very small percentage of the electricity comes from renewable sources such as thermal, hydro, wind, nuclear, or solar sources.

Chemical Bias – pH-Based Chemical bias for PEC water cleavage is basically produced when the anode and cathode electrodes are dipped in different chambers filled with electrolyte of different pH and separated by an ion exchange membrane (Figure 12.1b). A typical example is an n-TiO$_2$ photoanode/4 M KOH/4 M HCl/Pt-cathode setup [47]. Each unit of pH difference between the electrolyte chambers provides a potential of 0.06 V [48]. However, fossil-based energy sources are used to manufacture the acid or alkali, and raw materials required for their production. The bias will decrease as H^+ (acid) and OH^- (alkali) are consumed during the progress of the PEC reactions, and the pH on both sides tends toward equilibrium [47, 49]. Hence, a constant supply of acid and alkali is needed to maintain the desired bias. This method of providing the additional voltage is not favorable because the system requires additional input of chemicals along with sunlight.

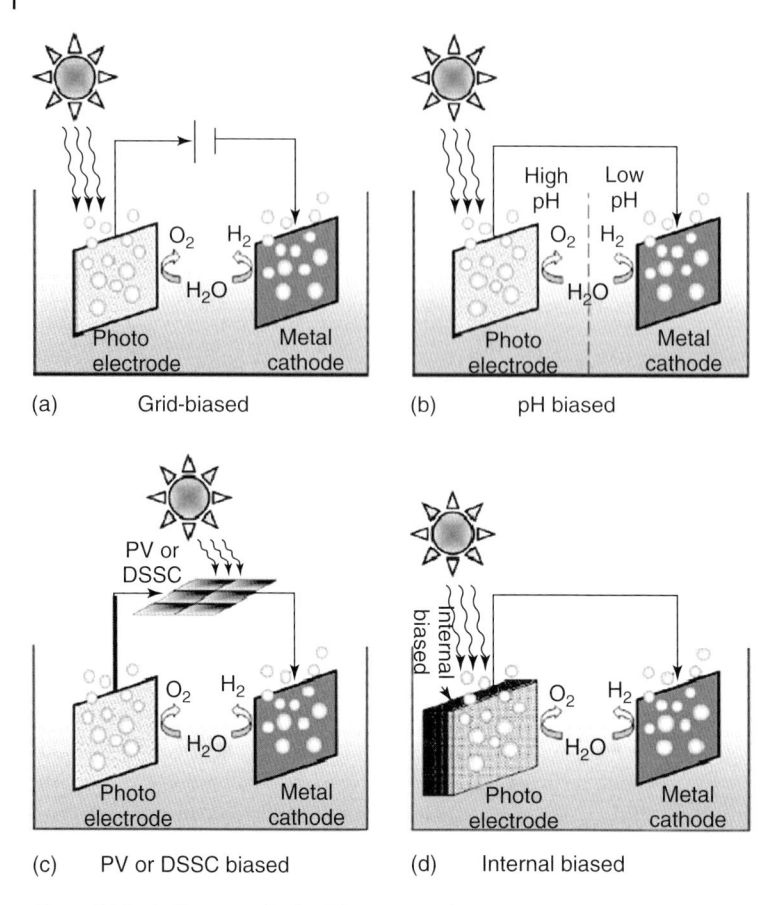

Figure 12.1 Different methods of biasing used in photo-assisted electrolysis cell, for PEC water splitting [46].

PV Cell Bias When a PV cell is illuminated by sunlight, current is generated by the creation, separation, and movement of the electron–hole pairs, which is utilized as the external bias for PEC cells. The solar PV cell is connected directly to the PEC water-splitting device without going to the grid first (Figure 12.1c). In the conventional method, the solar PV panel converts solar energy into electricity, which goes to the PEC through an electrolyzer to perform water electrolysis. However, in PV-cell-biased PEC water splitting the current generated from PV panel directly goes to the PEC cell instead of going via electrolyzer. The PV cell normally consists of multijunction semiconductor layers of various n and p combinations.

Dye-Sensitized Solar Cell (DSSC) Bias DSSC is categorized under third-generation solar cells based on cheap materials (TiO_2, ZnO, etc.) known as *Grätzel cells*. The principle is similar to the PV-cell-biased system (Figure 12.1c). Instead of solar panels of the conventional solar cell, which is used to generate electricity, a dye sensitization is used to capture sunlight and convert it to current, which then flows to the PEC cell for water splitting [50]. The DSSC itself is a PEC system because it also involves a regeneratable electrolyte. The conventional solar PV cell, on the other hand, is a solid-state device. In DSSC, the photoanode consists of the wide-bandgap semiconductor TiO_2 cast on the substrate and sensitized by a dye that acts as the antenna to capture visible light, and a Pt-coated substrate forming the cathode, both immersed in the electrolyte (I_2/KI redox couple). The dye molecule (D) is excited (D^*) under exposure to light and injects electrons into the conduction band of TiO_2. The reduced dye (D) is regenerated by oxidizing the I_2/KI electrolyte. The electron from the conduction band (CB) of TiO_2 then flows out to do work and travels up to the counter electrode where the I_2/KI electrolyte is reduced. The DSSC is capable of continuous current generation under light irradiation [51].

Internal Bias Internal bias refers to the bias potential implemented on the PE itself by its unique structure as layered, stacked, or hybrid (Figure 12.1d), which involves arranging several different semiconductor films on top of each other so that the total bandgap is large enough for water splitting but small enough to absorb the visible portion of the light spectrum which has a high photon concentration. This also allows the arrangement of the resultant band edges to straddle the water-splitting redox potential. The internal-biased PE structure can be of several types such as PV/PEC (PV–a-SiGe, PEC–WO_3) [52] PV/PV (PV1–GaInP, PV2–GaAs) [40] and PEC/PEC [53] (PEC1–DSSC, PEC2–WO_3). There are numerous possible combinations to achieve the appropriate bandgap and band edges for direct water splitting by internal biasing, but the main drawback of this system is its complexity [54]. Improved efficiency for the PV/PEC, PV/PV, or PEC/PEC structures can be achieved by matching the potential and current for each layer. Since these systems are capable of splitting water directly, internally biased structures are essentially zero bias systems.

12.2.1.3 Photovoltaic Electrolysis Cell or Regenerative Cell

PV electrolysis cells [55] employ solid-state PV cells to convert light into electric power without leaving a net chemical change in the cell, which is then passed to a commercial-type water electrolyzer for water splitting, as shown in Figure 12.2a,b. An alternative system involves the semiconductor PV cell configured as a monolithic structure and immersed directly in the aqueous solution. This cell involves a solid-state p-n or Schottky junction to produce the required internal electric field for efficient charge separation and the production of a photovoltage sufficient to decompose water. Most of the work on regenerative cells has been focused on electron-doped (n-type) II/VI or III/V semiconductors using redox-coupled electrolytes such as sulfide/polysulfide, vanadium(II)/vanadium(III), or I_2/I^-. Conversion efficiencies of multijunction regenerative cells have been reported up to 19.6% [41].

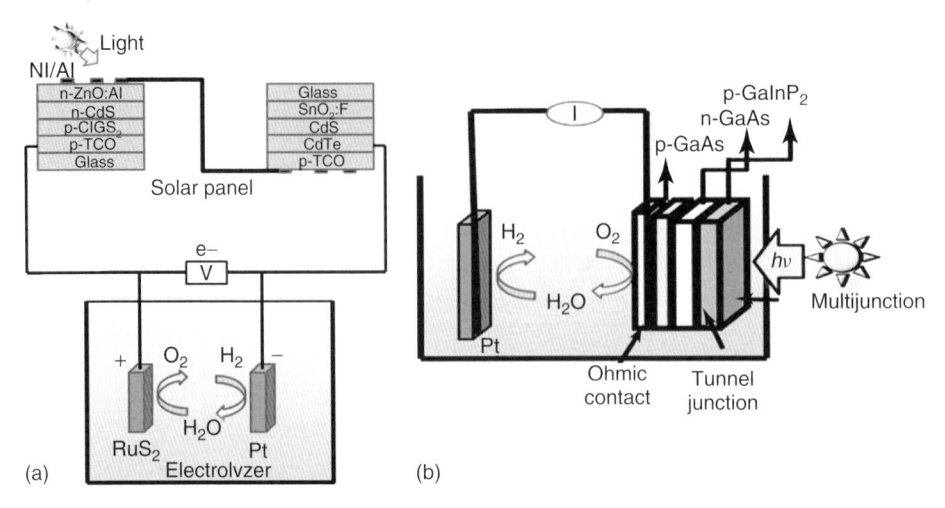

Figure 12.2 Illustrative photovoltaic electrolysis or regenerative cell setup consisting of (a) two thin-film PV cell and RuS$_2$ photoanode for oxygen evolution and Pt cathode for H$_2$ evolution and (b) monolithic photovoltaic photoelectrochemical p-GaAs/n-GaAs/p-Ga$_{0.52}$In$_{0.48}$P device [56].

12.2.2
Photogalvanic Cells

Photogalvanic cells are used for the conversion of the solar energy into the electrical energy. Here, for instance, two inert electrodes are dipped into an electrolyte with a dye solution and the light is absorbed by the electrolyte. In this case, an electron transfer takes place between the excited dye molecules, and electron donor or acceptor molecules are added to the electrolyte. A photovoltage difference develops between the two electrodes if the light is absorbed by the electrolyte near one of the two inert electrodes. Accordingly, the photogalvanic cell is essentially a concentration cell.

12.2.3
Photoelectrochemical Solar Cells

The major difference between PEC solar cells and PEC cells for water splitting is that in the former case free energy (electrical energy) is produced but the net gain is zero, whereas in the latter there is a net gain in free energy from H$_2$ and O$_2$ production. Figure 12.3 demonstrates the configuration of a typical PEC solar cell that uses a Cd(Se, Te)/S$_x$ conversion half cell with a Sn/SnS storage system [57], resulting in a high solar-to-electric conversion efficiency solar cell with continuous output. Under illumination, as shown in Figure 12.3a, the photocurrent drives an external load. Simultaneously, a portion of the photocurrent

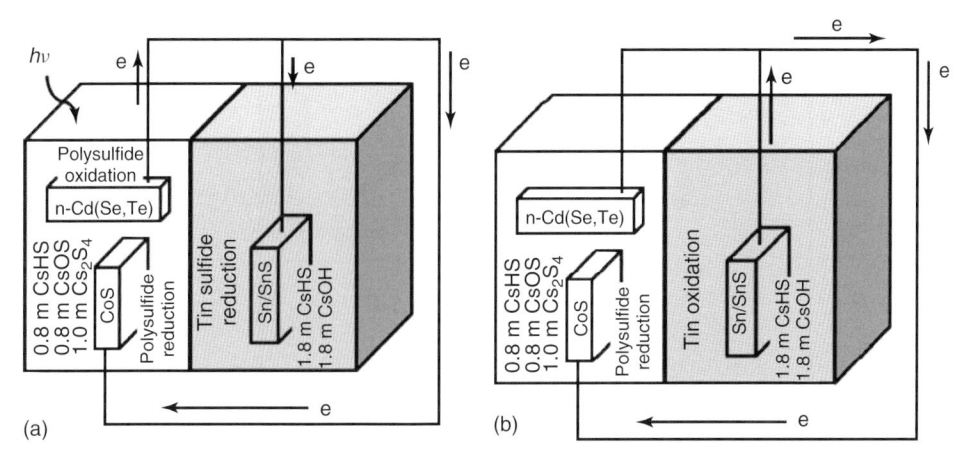

Figure 12.3 Schematic of a photoelectrochemical solar cell combining both solar conversion and storage capabilities: (a) under illumination and (b) in the dark [58].

is used in the direct electrochemical reduction of metal cations ($Sn^{2+} \rightarrow Sn$) in the device storage half cell. In the dark or below a certain level of light, the storage compartment spontaneously delivers power by metal oxidation ($Sn \rightarrow Sn^{2+}$), as seen in Figure 12.3b.

12.2.4
Photoelectrochemical Cells Based on Inherently Conducting Polymers (ICPs)

Conducting polymers are polymers with a framework of alternating single and double bonds, and delocalization of the double bonds over the entire polymer molecule produces a bandgap between the valence and conduction band of 1.5–3 eV, which leads to broadband visible light absorption, making inherently conducting polymers (ICPs) particularly well suited for photoelectrodes in PECs [59]. ICP semiconductors inherit a nice blend of the electronic and optical properties of semiconductors and the mechanical and physical properties of polymers. In contrast to the oxide semiconductors, ICP semiconductors (polyterthiophene) are generally of p-type materials. Thus, an ICP-based PEC works in a direction reverse to that of Grätzel cells when the light-excited semiconductor electrons pass on to the redox electrolyte and the charge is conducted through the exciton mechanism in conducting polymers (Figure 12.4). They possess a wide, stable electrochemical potential window with extremely low vapor pressure, which means that stable electrochemical devices containing conducting polymers can be produced. Traditionally used electrolytes limit the lifetime of the devices but, on the other hand, ICP-based PECs (e.g., polypyrrole) exhibit excellent compatibility with selected ionic liquids (e.g., imidazolium-based ones) with even improved electromechanical properties [60, 61].

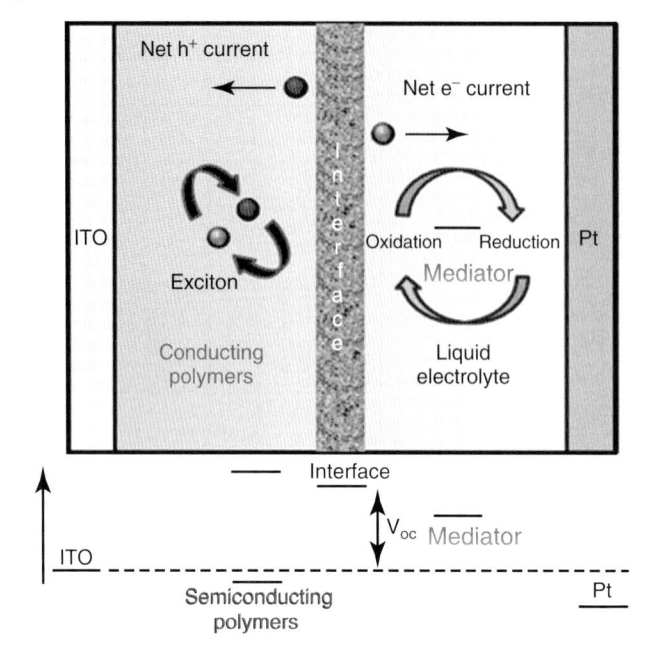

Figure 12.4 Schematic diagram of a typical polymer photoelectrochemical cell structure consisting of a ITO/polymer/electrolyte/Pt sandwich structure. Different electrochemical interfaces (between the conducting polymer–liquid electrolyte and ITO–polymer interface) present in ICP-based PEC induce a photovoltage, which is generated by the difference in standard electrode potential (work function) between interfaces. However, the electrochemical potential of this interface may be equal to the redox potential of the mediator [59].

12.3
Electrochemical Processes in Photoelectrochemical Cells

Metals do not absorb solar radiation because they do not have a bandgap; therefore, they can either reduce or oxidize the substrates depending on the band position. Likewise, insulators are also not suitable for absorbing light because of their very wide bandgap (>5 eV), which demands expensive, high-energy photon sources or requires energy-intensive methods to create the appropriate excitons to excite electron from the valence band (VB) to the conduction band (CB). Hence metals and insulators cannot be employed for the purpose of water-splitting reactions. Therefore, semiconductors with their suitable bandgaps (<3.5 eV) remain the only choice to promote electrons from the VB to the CB upon the solar radiation, which creates holes and electrons for oxidation and reduction reactions, respectively. Therefore, semiconductors alone are the suitable materials to catalyze the complete water-splitting reaction in presence of sunlight. The photoelectrochemistry in the PEC cell is quite complex by nature, as it involves not only the electrochemical interactions between the chemical and ionic species but also interfacial interactions between ionic conductors (such as electrolyte solutions (ESs)) and solid-state

12.3 Electrochemical Processes in Photoelectrochemical Cells

electronic conductors (such as semiconductors). Following are the important steps necessary for overall water cleavage (Eqs. (12.1–12.4)):

1) Absorption of light near the surface of the semiconductor creates electron–hole pairs.

$$\text{Photocatalyst} + h\nu \rightarrow \text{Photocatalyst}\,(e^-) + \text{Photocatalyst}\,(h^+) \quad (12.1)$$

2) Holes (minority carriers) drift to the surface of the semiconductor (the photoanode) where they react with water to produce oxygen.

$$2h^+ + H_2O \rightarrow \tfrac{1}{2}O_2\,(g) + 2H^+ \,(1.23\text{ V vs NHE at pH} = 0 \text{ and}$$
$$+ 0.82\text{ V vs NHE at pH} = 7) \quad (12.2)$$

3) Electrons (majority carriers) are conducted toward a counter metal electrode (typically Pt) where they combine with H^+ ions in the ES to generate H_2.

$$2e^- + 2H^+ \rightarrow H_2\,(g) \quad (0.00\text{ V vs NHE at pH} = 0 \text{ and}$$
$$- 0.41\text{ V vs NHE at pH} = 7) \quad (12.3)$$

Transport of H^+ from the anode to the cathode through the electrolyte completes the electrochemical circuit. Following is the overall reaction:

$$2h\nu + H_2O \rightarrow H_2\,(g) + \tfrac{1}{2}O_2\,(g)\,(1.23\text{ eV}, \quad \Delta G^\circ = +237\text{ kJ/mol}) \quad (12.4)$$

For water splitting, the theoretically required bandgap is $E_g = 1.23$ eV (electrolysis), a value derived from the following relationship: $\Delta G^\circ = -nF\Delta E^\circ$, where n is the number of electrons transferred in above reaction, F is the Faraday constant, and ΔG° and ΔE° are standard Gibbs free energy change and standard electric potential of the reaction, respectively. The optimum required bandgap for an effective photocatalyst is 1.8–2 eV (due to overpotential (0.6–1.2 V) and the losses associated with the cathode, anode, ionic conductivity, and system in PEC cells). Furthermore, after suitable electronic energies and chemical stability, the band edges of semiconductors (the top of the valence band and the bottom of the conduction band) need to straddle the redox potentials of the water, that is, 0.00 V (H^+/H_2) versus normal hydrogen electrode (NHE) at pH = 0 (−0.41 V at pH = 7) and 1.23 V (OH^-/O_2) versus NHE at pH = 0 (+0.82 V at pH = 7), so that the shifting of the band edges in opposite directions can facilitate the reduction and oxidation reactions. The charge transfer from the surface of the semiconductor must be fast enough to prevent photocorrosion and shifting or bending of the band edges, resulting in loss of photon energy. Therefore, suitable electrolyte or doping is adopted to prevent the above phenomena.

The photocatalytic semiconductors are often used with the addition of small amounts of metals, nonmetals (C, N, S, P, etc.) or other hole-trapping agents, so that the lifetime of the excitons created can be increased and the recombination would be effectively reduced. The bond polarity, which is also a deciding factor for photocatalytic water splitting, can be estimated from the following expression:

$$\text{Percentage ionic character (\%)} = \left(1 - e^{-\frac{(x_A - x_B)^2}{4}}\right) \times 100 \quad (12.5)$$

Oxide semiconductors with more ionic M–O bonds (50–60%) are less suitable for the photocatalytic splitting of water because the top of the VB (mainly contributed by the p orbitals of oxide ions) becomes less and less positive (since the binding energy of the p orbitals will be decreased as a result of the negative charge on the oxide ions) and the bottom of the CB will be stabilized to higher binding energy values because of the positive charge on the metal ions, which is not favorable for the hydrogen reduction reaction. A very low value of the ionic character also is not suitable since these semiconductors do not have the necessary bandgap value of 1.23 V. Therefore, it is deduced that systems that have an ionic bond character of about 20–30% with suitable positions of the VB and CB edges may be appropriate for the water-splitting reaction. Much research has been undertaken in the last three decades to engineer a suitable semiconducting photocatalyst material for water splitting in an electrochemical cell. Various concepts such as a flat-band potential, band bending, and Fermi level pinning can also be understood in terms of the bond character and the redox chemical aspects by which the water-splitting reaction has been dealt. Metal is incorporated into typical photocatalytic systems so as to make them responsive to longer wavelength radiations. Doping changes the distribution of electrons within the solid and hence changes the Fermi level. For an n-type semiconductor, the Fermi level lies just below the CB, whereas for a p-type semiconductor it lies just above the VB, as shown in Figure 12.5.

In addition to doping, as with metal electrodes, the Fermi level of a semiconductor electrode varies with the applied potential and electrolyte properties; for example, moving to more negative potentials will raise the Fermi level. If the redox potential of the solution and the Fermi level do not lie at the same energy, movement of charge between the semiconductor and the solution takes place in order to equilibrate the two phases. Excess charge located on the semiconductor does not lie at the surface but extends into the electrode for a significant distance (100–10 000 Å) – referred to as *space charge region* or *depletion region* because the majority charge carriers have left the electrode when they came into the contact with solution. Hence, there are two double layers to consider: the interfacial (electrode/electrolyte) double layer, and the space charge double layer. The width of the depletion layer (W) at the semiconductor/solution interface can also be derived from the Mott–Schottky

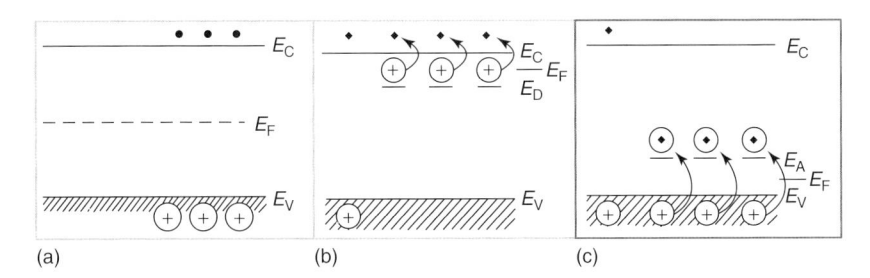

Figure 12.5 Schematic diagram demonstrating the effect of doping on the energy levels of an (a) intrinsic semiconductor, (b) an n-type semiconductor, and (c) a p-type semiconductor.

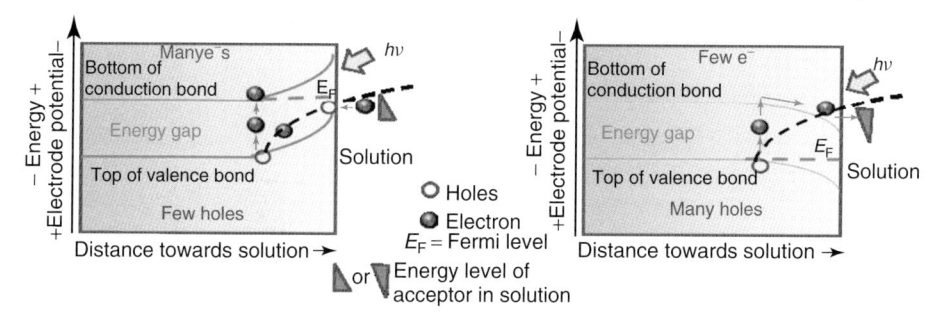

Figure 12.6 Mechanism of the production of photocurrent in an electrolyte solution by (a) an n-type photoanode and (b) a p-type photocathode.

relationship and expressed as follows [60]:

$$W (nm) = \left[\frac{2\varepsilon\varepsilon^0 (V - V_{FB})}{e^0 N_d} \right]^{\frac{1}{2}} \quad (12.6)$$

where e^0 is the electron charge, ε is the dielectric constant of ZnO, ε^0 is the permittivity of vacuum, N_d is the dopant density, V is the electrode applied potential, and V_{FB} is the flat-band potential.

For an n-type semiconductor electrode at open circuit, the Fermi level is higher than the redox potential of the electrolyte; hence electrons will be transferred from the electrode into the solution and electric current flows across the junction until electronic equilibrium is reached, when the Fermi energy of the electrons in the solid (E_F) is equal to the redox potential of the electrolyte (E_{redox}), and leads to the positive charge associated with the space charge region, which is reflected in an upward bending of the band edges as the majority charge carrier is removed from this region. This region is also referred to as a *depletion layer* as illustrated in Figure 12.6a. For a p-type semiconductor, the Fermi layer is lower than the redox potential of the solution; hence electrons must transfer from the solution to the electrode, which generates a negative charge in the space charge region, causing a downward bending in the band edges (Figure 12.6b). Since the holes in the space charge region are removed by this process, this region is again a depletion layer. The effect of illumination when an n-type photocatalyst and electrolyte come into contact is reflected by Figure 12.7a–c.

12.4
Photoelectrochemical Cell Components

The ultimate goal of this chapter is to review all aspects related to the various components of the PEC cell and their fabrication techniques to develop a system of high efficiency (greater than 10%), stability, scalability, and low-cost manufacturing of semiconductor PE for PEC reactors to directly produce hydrogen from water

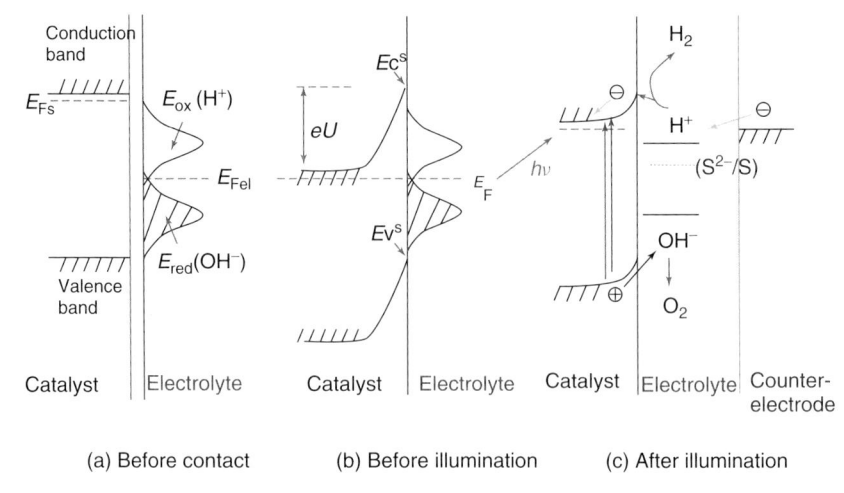

Figure 12.7 Band energetics of production of photocurrent by an n-type photoanode in aqueous electrolyte [62].

using sunlight as the energy source. To reach this goal, parallel investigations have been undertaken in the direction of (i) reactor modeling, (ii) extensive experiments in electrode materials synthesis and testing to identify critical materials issues (such as stability and catalytic activity), and (iii) PE fabrication and testing to identify and address critical PE operations issues. To date, none of the PEC devices have been found to satisfy all aforementioned criteria. The most basic setup of a photocell for PEC water splitting is a transparent photoreactor with a flat surface area as optical window, which allows light to reach the PE, and sufficient space to immerse all the electrodes (working electrode – WE/PE, reference electrode – RE, counter electrode – CE) in the electrolyte, as illustrated in Figures 12.8a,b. We now discuss these components in what follows.

12.4.1
Photoreactor

The photoreactor is a container to house all the PEC components (photoanode, counter electrode, reference electrode, electrolyte, etc.) for PEC water splitting, and takes various shapes and configurations. Flat surfaced reactors are important during measurement to avoid distortion of the impinging light, whereas spherical or hemispherical transparent containers (Figure 12.9a) [64] provide light focusing on the working PE surface. Focusing of light provides higher light concentration in a lab setting, which promotes more photons to reach the PE surface for the required reaction. Reactor vessels having separate ports for the PE, the counter electrode, and the reference electrode (Figure 12.9b) are advantageous to ensure a constant distance between the electrodes for each measurement [65, 66]. Quartz vessels are very expensive; therefore the alternative to this is a horizontal cylindrical vessel of glass or robust material such as Perspex and Teflon, or other plastic material with

12.4 Photoelectrochemical Cell Components

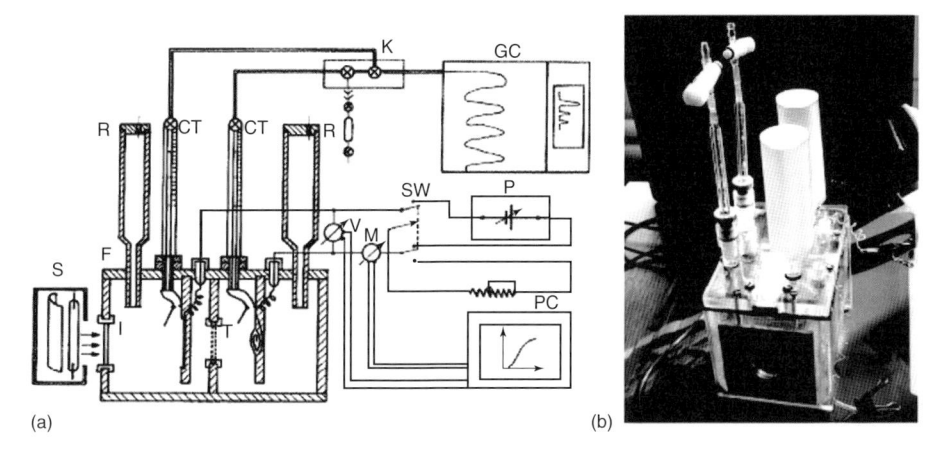

Figure 12.8 (a) Typical experimental arrangement of the photoelectrochemical cell. F, the photoelectrochemical cell; GC, the gas chromatograph; P, the power supply; K, the three-way valve system; S, the light source; I, the quartz window; T, Nafion proton transfer membrane; R, electrolyte reservoirs; CT, calibrated tubes; SW, double electric switcher; P, variable DC stabilized power supply; V, digital voltmeter; M, digital milliampere meter; and PC, computer. (b) Snapshot of a photoelectrochemical cell [63].

optical window [67], where both or one end is open to attach a quartz piece; but glass and quartz do not fused easily. The flat quartz piece can be fitted to the side of the vessel with the opening by sandwiching it with a holder plate that holds the quartz with an O-ring or silicone gasket to form a tight seal (Figure 12.9c) using nuts and bolts. Another design [68, 69] consists of a rectangular plate-type photoreactor that is totally separated from the aqueous electrolyte (as illustrated in Figure 12.9d) with an SS substrate and PV semiconductor material (triple-junction Si) deposited on one side and hydrogen evolving catalyst (CoMo) on the other side. The electrons generated from the PV-based PE move through the SS substrate to reach the electrolyte to reduce H^+ to H_2 and interconnect to the other substrate deposited with an oxygen evolving catalyst ($Fe:NiO_x$), where H_2O is oxidized to O_2 by the holes collected on the surface of the PV. The membrane for the gas separation is placed in the middle on a porous support between the H_2 evolution chamber and the O_2 evolution chamber. Hence, wiring is not required and the monolithic device can be completely enclosed. In the PEC water-splitting system, it is important to separate H_2 and O_2, as a mixture of these two gases is potentially explosive. In a small-scale lab photocell, this does not pose a major concern, but this separation should be essentially made [70] for large-scale production by holding the two separate vessels together by nuts and bolts with a membrane (proton exchange membrane Nafion), glass frit, and diaphragm acting as a separator for the electrolytes (Figure 12.9e,f). The H-type reactor allows gas separation without a membrane separator, but an ion-permeable divider such as glass frit can be placed in middle of the connection tube to allow different electrolytes to be used in the compartments (Figure 12.9g) [71]. Another interesting reactor is the dual-bed system (Figure 12.9h) [72], which basically consists of two photosystems in which

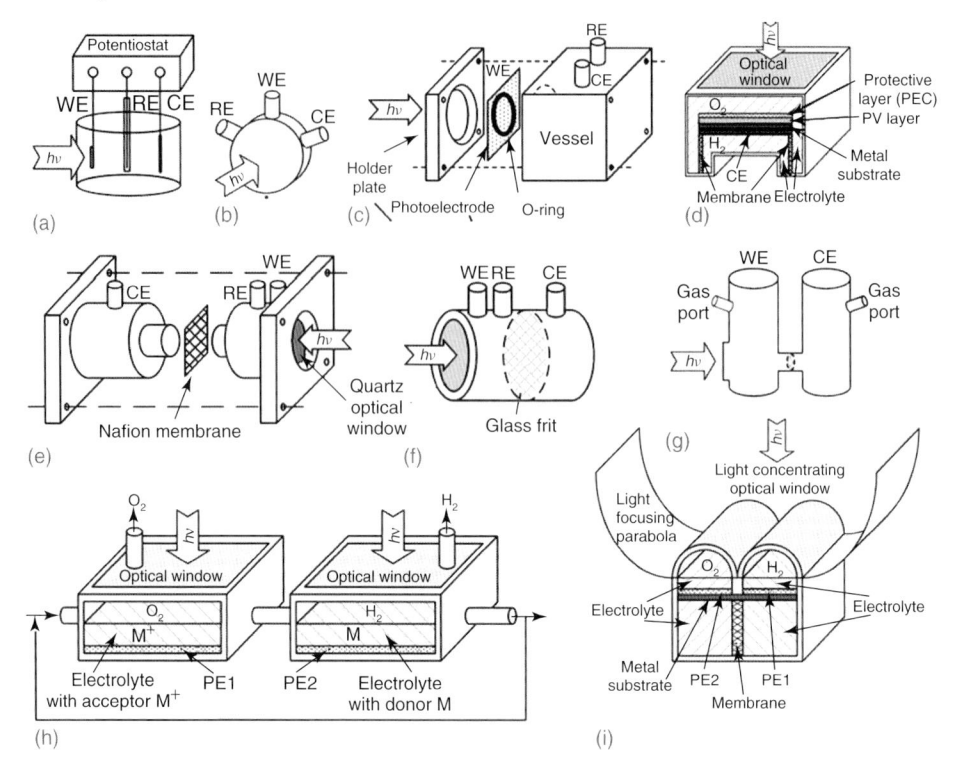

Figure 12.9 Variety of photoreactors.
(a) Schematic of PEC water-splitting setup;
(b) single-chamber circular vessel; (c) sandwich assembly; (d) nonbiased monolithic photoreactors with electrolyte/gas separation; (e) separate vessels; (f) single vessel (vessel with optical window); (g) H-vessel reactor; (h) dual-bed photoreactor; and (i) bi-photoelectrodes reactor [46].

two PEC cells are linked together and each cell has half electrochemical potential for water splitting. The apparatus consist of two shallow and flat sealed containers, where photocatalytic particles are immobilized on the bed at the bottom of the container or grids. Then aqueous solution (alkaline) containing the redox mediator ($M = IO_3^-/I^-$) is pumped between the two chambers. In large-scale application, as an alternative to pumping a large volume of solution in the chamber, the membrane between the two beds is used to allow the ions of the mediator and electrolyte to cross over. Since the two beds are now isolated, particles can be suspended in the solution instead of immobilized on the bed. In this system, both oxidation and reduction reactions occur on each PE. To eliminate wiring, a metal substrate can be used, and both the PE semiconductor materials (n-type and p-type semiconductor) can be deposited on the surface of the metal substrate side by side. For the separation of the gases evolved, an arrangement to separate the compartment is required, and the ion exchange route is provided as shown by Aroutiounian et al. (Figure 12.9i) [73]. In this design, n-type PE (n-doped TiO_2 or n-Fe_2O_3) and p-type PE (p-doped TiO_2 or p-Cu_2O) [74] are placed on the top of the reactor; therefore, they are only submerged slightly in the electrolyte. The underlying backside of the metal substrate has to be covered with an insulating material so the current does not leak into the electrolyte. Both PEs are kept in separate compartments by placing a divider along the length of the middle of the two PEs, thereby separating the evolved gases. Along the divider, a membrane is placed to allow ion movement, and the light concentrator parabola and the focusing optical window are also placed on top of the photoreactor.

12.4.2
Photoelectrode

In the basic instrumentation of the PE (cathode or anode), a thin layer of the semiconductor is cast on the top of the conductive side of the substrate. $SrTiO_3$, $KTaO_4$, GaN, CIGSS (copper indium gallium selenium sulfide), ZrO_2, Nb_2O_5, SnO_2, SiC, polycrystalline Fe_2O_3, ZnO, WO_3, and TiO_2 are the most commonly studied in PEC materials. The substrate is normally a transparent conductive glass (indium-doped tin oxide (ITO) or fluorine-doped tin oxide (FTO)) or a metal foil. The potential of the PE can be judged on the criteria of cost, efficiency, weight, size, durability, and feasibility to be easily scaled up to larger sizes. A small area, usually a strip at the top or side of the conductive substrate, is left uncoated to allow electrical connection. The ohmic contact is normally made by attaching a copper wire to the exposed part with silver conductive glue and covering the connection with epoxy resin to provide insulation and also to strengthen the connection (Figure 12.10a,b). If the substrate is metal, then the ohmic contact is usually made on the backside. The copper wire can be further protected by inserting it into a glass tube which also functions as a handle for the PE. All the four edges and the back of the PE are then covered with epoxy resin for insulation [75]. The transparent epoxy layer (e.g., 20–3004 HV resin and 20–3004 C catalyst, Epoxies, etc., Cranston, NJ, USA) is very thin and leaves the area on the front and back exposed, covering only the sides. The area of the exposed PE produced in this way is nondefined, and is in contact with the electrolyte, and can be estimated using the method given by Kelly and Gibson [65], in which a piece of paper with known area and weight is used as a

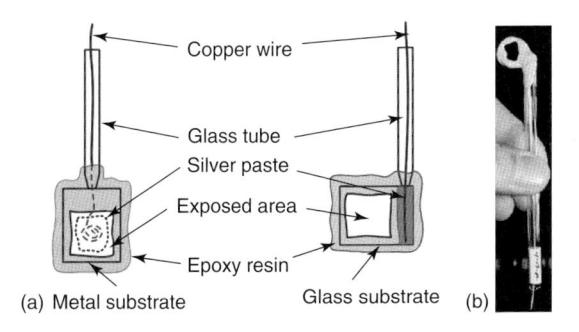

Figure 12.10 (a) Preparation of PE and (b) photograph of photoelectrode [46].

standard. The surface of the PE is photocopied and the exposed area is carefully cut along the outline. The weight of the cutout of the exposed PE area is compared with the weight of a standard piece of paper (1 cm × 1 cm) and, using the weight/area ratio, one can get the required area, using the relation unknown PE area = weight of standard paper × known standard paper area (g cm^{-2})/weight of PE cut out.

It is difficult to achieve high STH conversion efficiencies in single-junction PEC PE systems due to the inherent electrochemical and solid-state losses. Both photopotential and photocurrent are detrimental for single junctions. For example, if the minimum water-splitting potential (based on redox separation with overpotentials) amounts to 1.6 eV, and the quasi-Fermi-level separation in the semiconductor can achieve 50% of the bandgap level, then the minimum bandgap for the onset of photoelectrolysis would be 3.2 eV. For any appreciable level of hydrogen production, overpotential and other system losses increases, requiring even higher bandgaps. For bandgaps greater than 3.2 eV, photocurrent density is limited to ∼1 mA cm^{-2}. This places the upper STH efficiency limit at 1.23%.

Multijunction devices are well known for their improved the solar-to-electricity conversion efficiency. Conceptually, in a multijunction (tandem) system, sunlight is partly absorbed in the higher bandgap top junctions, while the remaining filtered light is absorbed in the lower bandgap bottom junctions. All junctions generate photovoltage and photocurrent. Since they are stacked in a series-connected configuration, the photovoltages V_1, V_2, V_3, and so on, will add, but the net photocurrent will be the minimum of J_1, J_2, J_3, and so on. For optimal performance, it is important to match the currents J_1, J_2, J_3, and so on, by optical tuning of the two junctions. In PEC water-splitting applications, multiple junctions can be stacked to take advantage of the photopotential enhancement, but the reduced photocurrent directly limits the hydrogen production rates. The device design must strike the right balance to maximize conversion efficiency. The NREL manufactured a multijunction hybrid photoelectrode (HPE) that holds the 16% STH efficiency world record for PEC water splitting (Figure 12.2b) [56]. The hybrid tandem device incorporates high-quality III–V crystalline semiconductor materials, with a GaAs (E_g = 1.44 eV) PV junction as the bottom cell and a Ga-InP (1.83 eV) PEC junction as the PEC top cell. The device fabricated uses extremely expensive III–V materials and suffers from limited durability. Therefore, it is still a great challenge to fabricate a durable PEC device with high STH efficiency at low cost.

Morphological features can also affect the energetics and kinetics of the nanocrystalline semiconductor electrodes for use in PEC cells because of two critical reasons: (i) a large surface area and small size of the crystals are required to anchor a large amount of dye molecules on the semiconductor surfaces and (ii) the electrons need to be effectively transferred before they are stopped by bulk or surface defects and the grain boundaries. The diffusion of the electrons in the nanocrystalline materials may be limited by the slow diffusion through different grains and by the trapping states on the grain boundaries. This perspective would provide a useful basis to optimize the desired PEC properties [76]. However, Pal and coworkers introduced a new concept by replacing the nanocrystalline films with oriented, long, high-density ZnO nanowires prepared from a seeded substrate–solution synthesis [77]. The high surface area is favorable for trapping the dye molecules, and the electron transport in oriented nanowires should be orders of magnitude faster than the percolation in polycrystalline films. Still, this approach produced a full sun efficiency of 1.5%. More recently, Yang and coworkers reported that the efficiency could be increased to 2.25% by applying a thin crystalline TiO_2 coating [78]. The increase was attributed to the passivation of the surface trap sites and the energy barrier to repel the electrons from the surfaces. Gradual improvement in PEC electrode materials and fabrication techniques, in term of solid-state junctions, integrate the PE, as shown in Figure 12.11. Figure 12.11a represents a liquid junction formed at the interface of the electrode deposited on a glass substrate and solution composition, resulting in a charge layer formation at the interface. Each material has a characteristic energy which is needed to remove an electron from its surface to infinity. This is called the *work function of the material*. When a metal with a high work function comes into contact with a semiconductor with a lower work function, electrons flow from the semiconductor to the metal. This results in the band configuration shown in Figure 12.11b. The band structure is dictated by conditions that are always met: the Fermi level is always constant in a system if it is in contact, and the work function must be discontinuous, otherwise electron flow will result, removing the discontinuity. Therefore, a typical application of

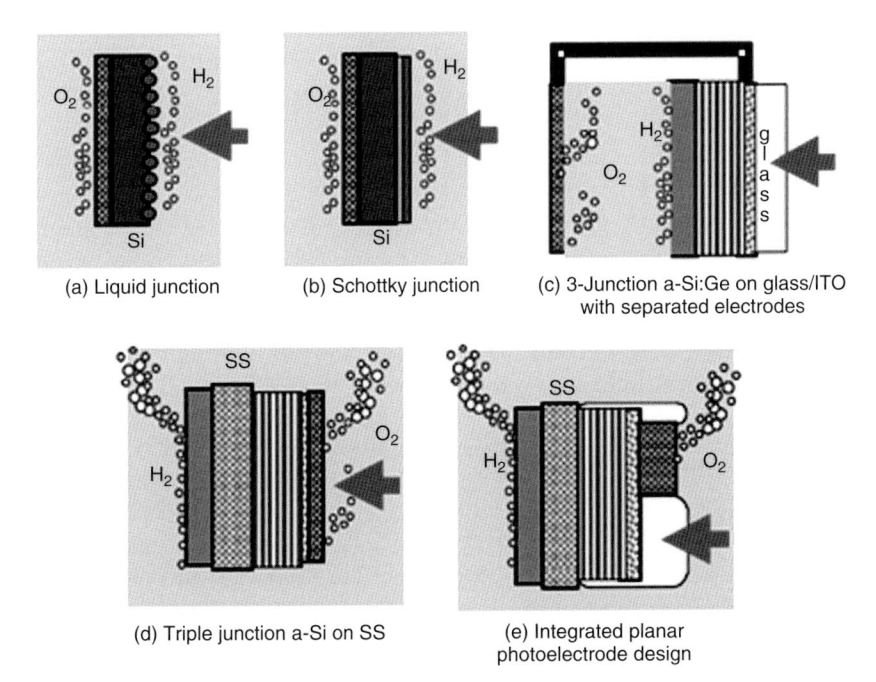

Figure 12.11 Different types of photoelectrode assemblies for photoelectrochemical hydrogen production [79].

power Schottky junction is discharge protection for the solar cells connected. The three-junction amorphous Si–Ge on glass/ITO with a separate anode is displayed in Figure 12.11c, whereas a monolith triple junction (energy conversion devices) on SS substrate is shown in Figure 12.11d. The back surface of the SS substrate was coated with the CoMo hydrogen evolution reaction (HER) catalyst, while the front surface of the electrode was coated with the Fe:NiO$_x$ oxygen evolution reaction (OER) catalyst. Unfortunately, the relatively thick (1 μm) Fe:NiO$_x$ layers introduce high optical losses, while thinner and more transparent film layers suffer from the loss of catalytic activity and reduced corrosion protection. Therefore, an appropriate thickness of the film must be chosen. The PE depicted in Figure 12.11e was further modified by the introduction of an optical window on the electrode surface on the OER side.

A typical monolith PE structure developed onto 1″ × 1″ substrate squares is described using the following procedure:

1) Sputter a TiO$_2$ film (about 1 μm thick) passivation on the cleaned substrate (conducting glass/metal) at the non-PV front side.
2) Sputter-deposit the ITO top contact film (about 1200 Å thick) through a shadow mask over entire active area.
3) Sputter-deposit a NiFe/Fe:NiO$_x$ HER catalyst film (about 1 μm thick) through mask on the non-PV front side.
4) Screen-print the TiO$_2$ encapsulant (about 3 μm thick) over the PV front side.
5) Coat the CoMo OER catalyst over the entire back side. Clean the device using methanol and DI (deionized) water between processing steps.

12.4.3
Counter Electrodes

CEs are made using either Pt or Ni, in the form of both mesh and foil. Platinum is chosen since it has a low overpotential for the HER [37] and Ni because it also has a low overpotential for the HER in basic solutions and offers a lower price alternative to Pt [81]. Both Pt and Ni electrodes resist corrosion and have long lifetimes in basic solutions. Both Pt mesh and foil, with purity exceeding 99.9%, are obtained from several sources used as the CE. Aiming to replace Pt with less expensive materials, based on previous reports of general similarities between Pt and transition-metal carbides, tungsten monocarbide (WC) comes out as a promising CE material for PECs on the basis of its activity and stability [82, 83]. The counter electrode area is purposefully made larger than the PE area so that it would not limit the photocurrent. With the object to develop a model for CEs in nanocrystalline solar cell design, to enhance the electrocatalytic properties of the CE, it is desirable to minimize the energy losses around the maximum power point (a low fill factor due to porosity). As most of the electrode materials, except the costlier Pt cluster catalyst, exhibit slow intrinsic kinetics for triiodide electrolyte reduction, other materials for producing higher performance CEs are sought. For a given CE length a and "intrinsic" exchange current of i_o, higher exchange currents

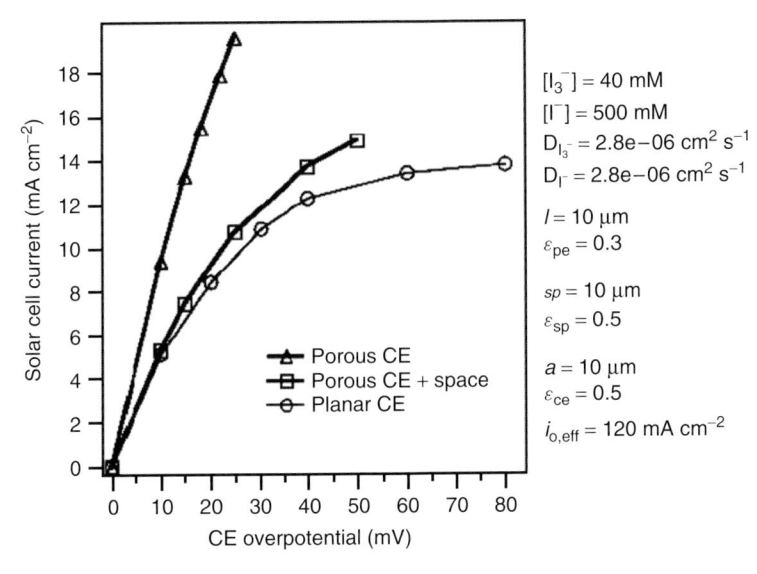

Figure 12.12 Model configuration of cell current density versus total CE overpotential for three different selections of cell (planar CE, porous CE with space, and porous CE) in 40 mM I_3^-/500 mM I^- solution and under the corresponding layer parameter conditions shown. All calculations were done by assuming a colloid diameter of 25 nm and 193 mM concentrated dye of diameter of 1.2 nm. The defining extinction coefficient ε_{540} of the dye at $\lambda = 540$ nm is 12.5 mM^{-1} cm^{-1} [80].

can be achieved through surface area (porosity or specific surface), with an effective exchange current $i_{o,\text{eff}}$:

$$i_{o,\text{eff}} = i_o \left(\frac{\text{surface}}{\text{volume}} \right) \times a \tag{12.7}$$

An excessively high kinetic overpotential on a CE can be compensated by increasing the effective surface area, and therefore the effective exchange current density. Effective exchange current values increase by orders of magnitude with the activity toward triiodide reduction, as their exposed surface area is enhanced by increasing the porosity of the electrode material (Figure 12.12). On the other hand, another aim of the porous CE would be to form a better conducting electrode to replace the conducting glass current collector. For example, an electrode comprised of colloidal particles with diameter $d = 25$ nm and "sintered" into a film of porosity $\varepsilon_{ce} = 0.5$ and thickness $a = 10$ μm would have a specific surface of $A = 6(1 - \varepsilon_{ce})/d = 1.2 \times 10^8$ m^{-1} (randomly packed sphere model) and therefore, an effective exchange current density $i_{o,\text{eff}} = 120$ mA cm^{-2}, given an "intrinsic" exchange current of $i_o = 0.1$ mA cm^{-2} for the smooth material. In other words, the charge-transfer resistance $R_{ct} = 1/(2fi_o) = 128$ Ωcm^2 reduces to $R_{ct,\text{eff}} = 1/(2fi_{o,\text{eff}}) = 0.107$ Ωcm^2, merely by the introduction of a surface effect [84].

12.4.4
Endplate/Current Collector

In a PEC cell, the endplate assembly provides an axial directional support to the electric lead and minimizes the contact resistance between the end fluid separator plate of a cell stack and the current collector (CC), which comprises an electrically conductive flat plate on a transparent substrate assembly. The current collector layer (ITO, FTO, conductive material) is sandwiched between the transparent glass and a solid electrode material. The CC provides a firm contact to the electrolyte separator plate to maintain good contact across the entire area between the two components. Firm contact is made by molding or bonding the separator/endplate and current collector together, or by applying pressure to the separator plate and current collector such that firm contact is maintained between the two components. The CC material should have good electrical conductivity, optical transparency with high light transmission in the visible and near-infrared, high-quality glass substrate, SiO_2 barrier layer, low roughness, high sheet resistance homogeneity, uniform transmission homogeneity, and reflectivity in the infrared region. A conductive glass substrate for use in a photoelectric conversion element should comprise a glass plate coated with a transparent conductive film that is provided so as to be conductive with the transparent conductive layer, and an insulating circuit protective layer that is formed on the conductive circuit layer, wherein a metal CC that can easily form a passive state is provided in pinhole portions, which are present in the circuit protective layer. Other factors to be considered in designing and evaluation of the endplate or current collector for PEC devices are (i) the ordering of the semiconductor layers in the solar cell, that is, whether the outer semiconductor layer is an n-type semiconductor (which would produce a photocathode) or a ''p-type'' semiconductor (which would produce a photoanode); (ii) the electrolyte pH (acidic, basic, or neutral); and (iii) the type of coating on the outer surface of the PE. The coating is of utmost importance since it needs to be transparent, conducting, catalytic, and corrosion resistant in an electrolysis environment.

A common current collector considered for the optimum solar cell configuration is indium-tin oxide (In_2O_3:SnO_2, referred to as *ITO*, which is normally used as the antireflection coating for solar cells). The lifetime of an a-Si-based, ITO-coated photoanode is a few hours due to pitting corrosion of the ITO and silicon layers and the resultant failure of the solar cell. However, all ITO-coated PEC cells have STH efficiencies of only 2–3%, indicating that ITO has poor catalytic activity for the oxygen evolution reaction. Fluorine-doped tin oxide (SnO_2:F) is a very corrosion resistant yet transparent and conducting material. This resulted in a major improvement in the lifetime, that is, more than 31 days at a hydrogen production efficiency of 5–6%, and this device remained operative at the end of the testing period. This design is a significant step forward in realizing a practical PEC device for large-scale hydrogen production [84, 85]. Recently, Nb-doped anatase TiO_2 ($Ti_{0.94}Nb_{0.06}O_2$ (TNO) of resistivity 4.6×10^{-4} Ωcm at room temperature and optical transmittance of 60–80% in the visible region) films with high electrical conductivity and transparency were fabricated on non-alkali glass using pulsed laser deposition and subsequent annealing in a H_2 atmosphere [86]. Results indicate that the TNO films are able to protect the photoanode of the PEC device from corrosion so that the lifetime improvements of over 100 h could be achieved, in addition to being capable of giving a PEC device with good efficiency. Therefore, TNO films have the potential to be practical transparent conducting oxides that could replace ITO and FTO in the near future.

12.4.5
Electrolytes

The electrolyte composition and its pH play a potential role in affecting the PEC response, onset potential, and flat-band potential of the semiconductor electrode in PEC cells. The rate of transfer of carriers at the junction is enhanced by proper band-edge matching of the semiconductor with the redox level of water. However, the positions of the CB and VB of semiconductors are dependent on the pH of the electrolyte. A shift in the CB and VB positions to more cathodic potentials by 59 mV per unit pH at 25 °C (Nernstian behavior) has been observed. Figure 12.13 demonstrates the percentage incident photon-to-current conversion efficiency (%IPCE) of a Fe_2O_3 nanorod electrode, which increases with increase in the electrolyte's pH [87]. Better overlap between the VB and the reduced form of the electrolyte energy levels causes this increase. In PEC studies on the solar splitting of water, usually an aqueous solution of acid, neutral, or alkali substance is employed as the electrolyte.

Typical effect of the electrolyte on the photoresponse of $ZnIn_2S_4$ electrode of different [Zn]/[In] ratios versus Ag/AgCl for the electrolytes 0.5 M K_2SO_4 and 0.6 M K_2SO_3, respectively, is shown in Figure 12.14 [88]. The highest PEC response in

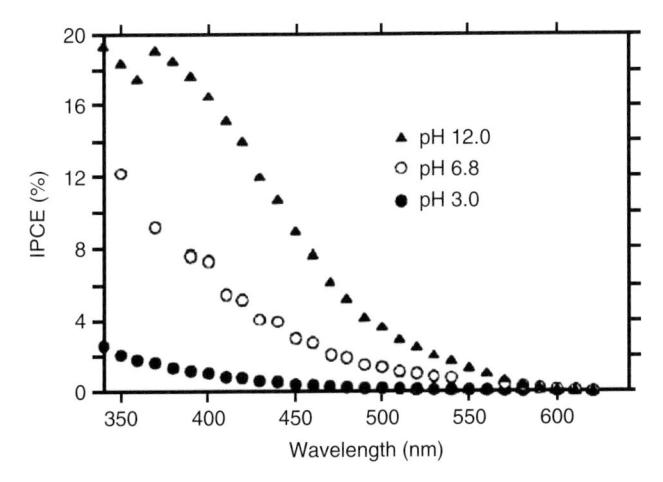

Figure 12.13 Action spectrum of a 0.6 mm thick Fe_2O_3 electrode, consisting of nanorods oriented in a perpendicular direction with the conducting substrate in electrolytes of different pH values under the same illumination [87].

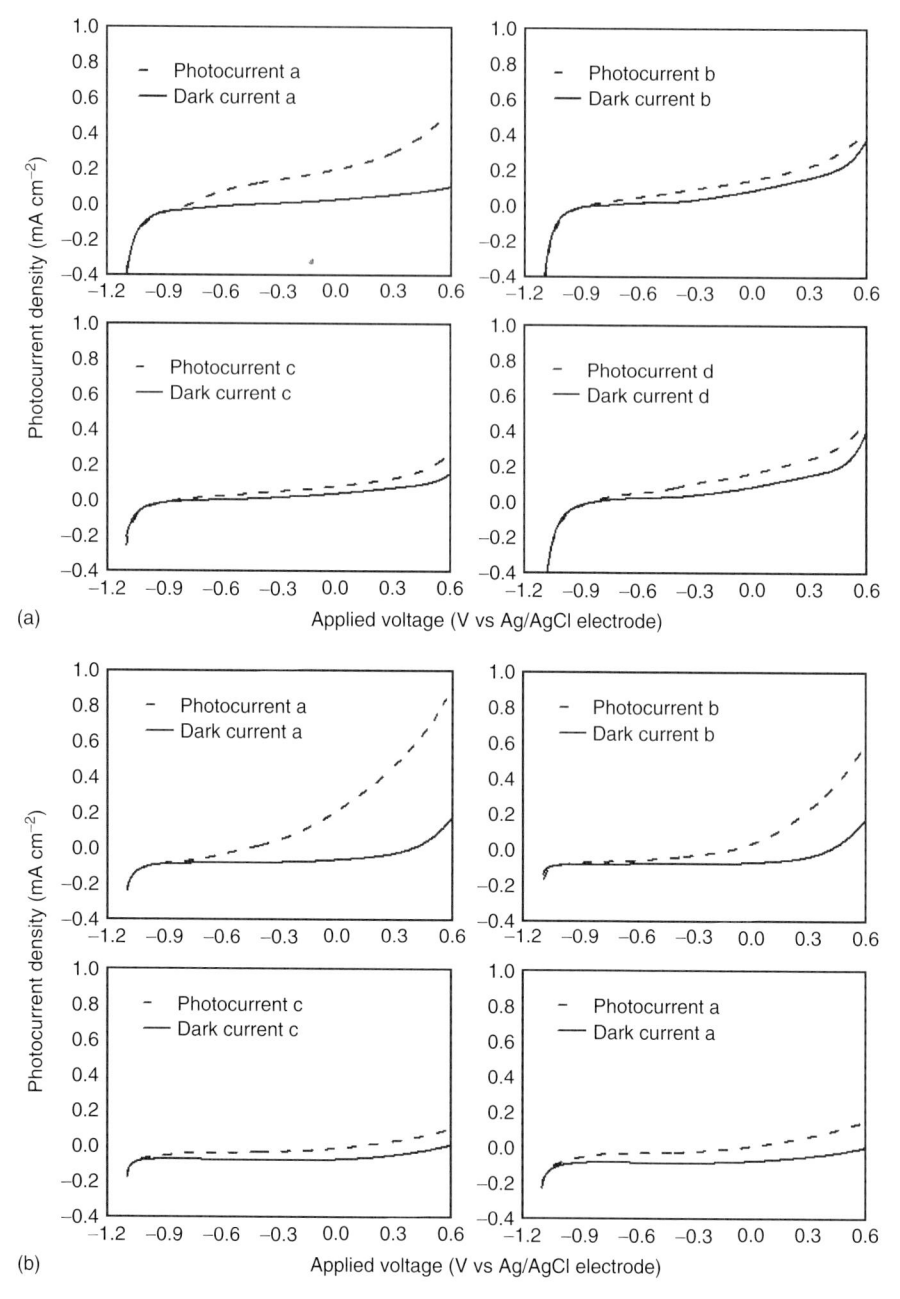

Figure 12.14 Photocurrent density – applied voltage plots in the range of −1.0 to + 0.6 V versus an Ag/AgCl electrode for $ZnIn_2S_4$ in 0.5 M K_2SO_4 and 0.6 M K_2SO_3 aqueous solution, respectively, for (a) [Zn]/[In] = 0.125, (b) [Zn]/[In] = 0.25, (c) [Zn]/[In] = 0.50, and (d) [Zn]/[In] = 0.75 in precursor solution using chemical bath deposition [88].

terms of the photocurrent density at external applied voltage of +0.6 V relative to the Ag/AgCl electrode, was found to be 1.31 mA cm^{-2} in a 0.6 M K$_2$SO$_3$ solution. All samples show anodic enhancement under illumination in K$_2$SO$_4$ and K$_2$SO$_3$ aqueous solutions. The experimental results indicate that aqueous K$_2$SO$_3$ solution can act as a photoabsorber in PEC devices. In aqueous K$_2$SO$_3$ solution, the holes produced in the VB of the samples oxidize SO$_3{}^{2-}$ to form SO$_4{}^{2-}$ under irradiation, when the applied potential is more positive than the flat-band potential of the sample. Water is reduced to hydrogen by the electrons produced in the CB of the samples, accompanied by the oxidation of SO$_3{}^{2-}$ ions, under illumination. With K$_2$SO$_3$ solution as the electrolyte, the following reactions may occur at the photoctalyst surface, photoanode, and cathode, respectively.

$$\text{samples} \xrightarrow{h\nu} e^- \text{ (CB)} + h^+ \text{ (VB)} \tag{12.8}$$

$$SO_3{}^{2-} + 2OH^- + 2h^+ \rightarrow SO_4{}^{2-} + H_2O \tag{12.9}$$

$$2H_2O + 2e^- \rightarrow H_2 + 2OH^- \tag{12.10}$$

Luo et al. [89] studied WO$_3$/Fe$_2$O$_3$ nanoelectrodes in a PEC cell using Na$_2$SO$_4$ aqueous solution (pH = 7.5) as the electrolyte and concluded that the charge separation is not solely provided by the photocatalyst, but works in concert with the electrolyte. Popular redox-couple electrolytes generally used in PEC cells are K$_3$ Fe(CN)$_6$/K$_4$ Fe(CN)$_6$, iodide/triiodide, FeCl$_2$/FeCl$_3$, sulfide salt/sulfur, 0.01 M K$_2$HPO$_4$/KH$_2$PO$_4$, [87] Na$_2$B$_2$O$_4$/Na$_2$B$_4$O$_7$ [90], 0.5 M H$_2$O$_2$ (50%) and 1 M NaOH (50%) [91], and ethylene carbonate/propylene carbonate [87]. A ferricyanide/ferrocyanide redox couple was employed as a model electrolyte, since it possesses a high standard rate constant for an electrode reaction. Figure 12.15 shows the J–V characteristics measured for CdS quatum dots (QD)-sensitized films (five coating cycles) under irradiation of light (wavelength 450 nm) using a series

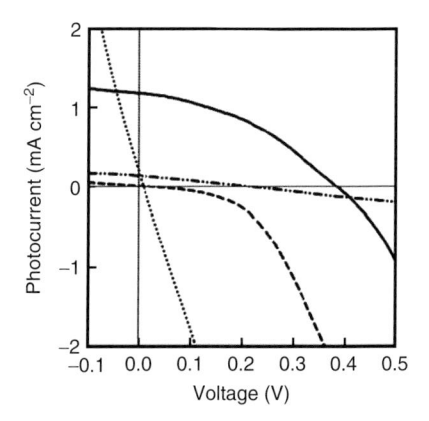

Figure 12.15 J–V characteristics obtained for CdS QD-sensitized solar cells under 450 nm light illumination, using various electrolytes: Na$_2$S$_x$/Na$_2$S (——), I^{3-}/I$^-$ (—••), FeCl$_3$/FeCl$_2$ (– – ––), or Fe(CN)$_6^{3-}$/Fe(CN)$_6^{4-}$ (••••) [96].

of electrolytes and compares the incident photon-to-current conversion efficiency (IPCE) spectra. As seen from Figure 12.15, both the short-circuit photocurrent (J_{sc}) and open-circuit voltage (V_{oc}) are small because of the large redox potential: for the electrolytes I^{3-}/I^- E_{red}: +0.45 V versus NHE and Fe^{3+}/Fe^{2+} electrolytes E_{red}: +0.6 V versus NHE. A fast reversible reaction between $Fe(CN)_6^{3-}/Fe(CN)_6^{4-}$ (E_{red}: −0.15 V vs NHE) was observed at the F–SnO$_2$/electrolyte interface, being identical to the data reported by Gregg et al. for the ferrocene/ferrocenium couple [92]. J_{sc} and V_{oc} were found to be remarkably higher for the Na_2S_x/Na_2S electrolyte due to their low redox potential (E_{red}: −0.45 V vs NHE). These results, therefore, demonstrate that the polysulfide electrolyte can be used as the active redox electrolyte for stabilizing the QDs. Recently, cobalt-complex redox electrolytes have been employed as an alternative redox couple for dye-sensitized solar cells [93], because the standard rate constant is slightly higher than that for the I^{3-}/I^- couple [94] and therefore the electron-blocking layer effectively functions to prevent electron leakage. The ionic liquids were successfully used as solar cell electrolytes, especially those originating from the diethyl and dibutyl–alkylsulfonium iodides [95].

For example, owing to the nanoporous structure of TiO$_2$, there always exists a part of the conducting glass surface that is exposed to the redox electrolyte. If the redox electrolyte possesses a significantly high standard rate constant at a conducting glass electrode, the electrons collected through the TiO$_2$ nanoporous structure may react with the oxidized electrolyte. This electron leakage then becomes an important cause of efficiency loss. Therefore, an electron-blocking layer of the conducting layer in ruthenium dye-sensitized wet solar cells was introduced to minimize electrons leaching into the I^{3-}/I^- electrolyte [97]. Stability tests are very important because, apart from the sensitizer, other components of the device, such as the redox electrolyte and the seal, may fail under long-term illumination. Indeed, a problem emerged with electrolytes based on cyclic carbonates, such as propylene or ethylene carbonate, which were found to undergo thermally activated decarboxylation in the presence of TiO$_2$, rendering these solvents unsuitable for practical use. These were therefore, replaced by highly polar and nonvolatile solvents, such as methoxypropionitrile (MPN), that do not exhibit this undesirable property. Using an MPN-based electrolyte in conjunction with a surfactant ruthenium dye, a dye-sensitized solar cell passed the critical 1000 h stability test at 80 °C recently for the first time [98].

12.5
Assembly of Photoelectrochemical Cells

In the past several decades, photochemists have put their heads together to revolutionize PEC fabrication by adopting various techniques to enhance its stability and efficiency at low cost. Modification of PEC using noble metal loading or ion doping to engineer the required bandgap; dye or QDs sensitization for making them visible light-harvesting; use of nanofilms of electrode materials to exploit quantum confinement and quantized charge effects; nanostructures of core–shell (i.e., Ag-TiO$_2$,

12.5 Assembly of Photoelectrochemical Cells

etc.) or a of variety of shapes for storing photogenerated electrons; composite metal-oxide semiconductors to influence the energetic and catalytic activities; and multicomponent semiconductors for photoinduced charge separation and refining energetics, followed by meticulous designation, fabrication, and then investigation for application to split water are some areas of development. Liquid-phase and gas-phase synthetic routes appear as the most common approaches for the production of modified semiconductor nanoparticles as photocatalysts, some of which will be considered here for illustration of typical cases.

Mainstream STH production from water is an uphill chemical reaction ($\Delta G > 0$ or $+\Delta G$), which necessarily needs external energy input. The source of external energy can be thermal, solar, or electrical. For the water cleavage process, the standard Gibbs free energy of water splitting ($\Delta G°$) is ~237 kJ mol^{-1}, which corresponds to 1.23 eV. Therefore, the minimum external energy input required in a PEC cell is 1.23. Hence, the most abundant and freely available solar energy is the most obvious choice for water-breaking, which requires photons of wavelength shorter than 1100 nm. It is seen from Figure 12.16 that the maximum value of solar energy conversion efficiency is about 32% with the energy gap of samples kept at 1.23 V. The practically due to the various overpotentials of the system and the required high input external energy that might be larger (1.8–2.5 eV) than the theoretical energy limit [97]. From the economic point of view, the conversion efficiency should be lie in the range of 10–15%, so the energy gap of the photocatalyst materials should be in the range of 1.8–2.5 V. Although the material issue is vital for water splitting in a PEC cell, the nature of electrolytes used and fabrication technology are also important concerns for their industrial applications.

Figure 12.16a illustrates the relationship between the photocatalyst energy bandgap (E_g) and solar energy conversion efficiency (η), and Figure 12.16b shows the comparative efficiency versus bandgap for different photosensitive materials, assuming that the quantum efficiency of the materials is 100%.

The important PEC materials that are under current worldwide investigations include following primary classes: tungsten oxide ($E_g \sim 2.6$ eV, $J = 3$ mA cm^{-2}, and STH efficiencies over 3%), iron oxide and related modified compounds ($E_g \sim 2.1$ eV, $J = 2.2$ mA cm^{-2}, STH efficiencies over 3%) [100], amorphous silicon compounds including silicon carbides and nitrides, copper chalcopyrite compounds ($E_g = 1.0$ eV in CuInSe$_2$ to $E_g = 1.6$ eV in CuGaSe$_2$, and up to $E_g = 2.43$ eV in CuGaS$_2$, $J = 13$ mA cm^{-2}, 19% STH efficiency), tungsten and molybdenum sulfide nanostructures ($E_g = <1.2$ eV but can be increased up to 2.5 eV by quantum confinement effect), III–V semiconductor classes (phosphide, nitride, and arsenide of gallium, indium, STH efficiency = 12–16%), oxide semiconductors (ZnO, TiO$_2$, ZrO$_2$, SnO$_2$, Nb$_2$O$_5$, highly stable but low efficiency), and so on. Generally, the PEC performance measurements of the samples are carried out in the electrolytic cell with a quartz window (Figure 12.8). The sample and a Pt wire or plate electrode are the working and CEs, respectively, and are inserted into PEC cell. The reference electrode used in PEC performance measurements can be an Ag/AgCl or a saturated calomel or standard hydrogen electrode. Aqueous H$_2$SO$_4$, HCl, Na$_2$CO$_3$,

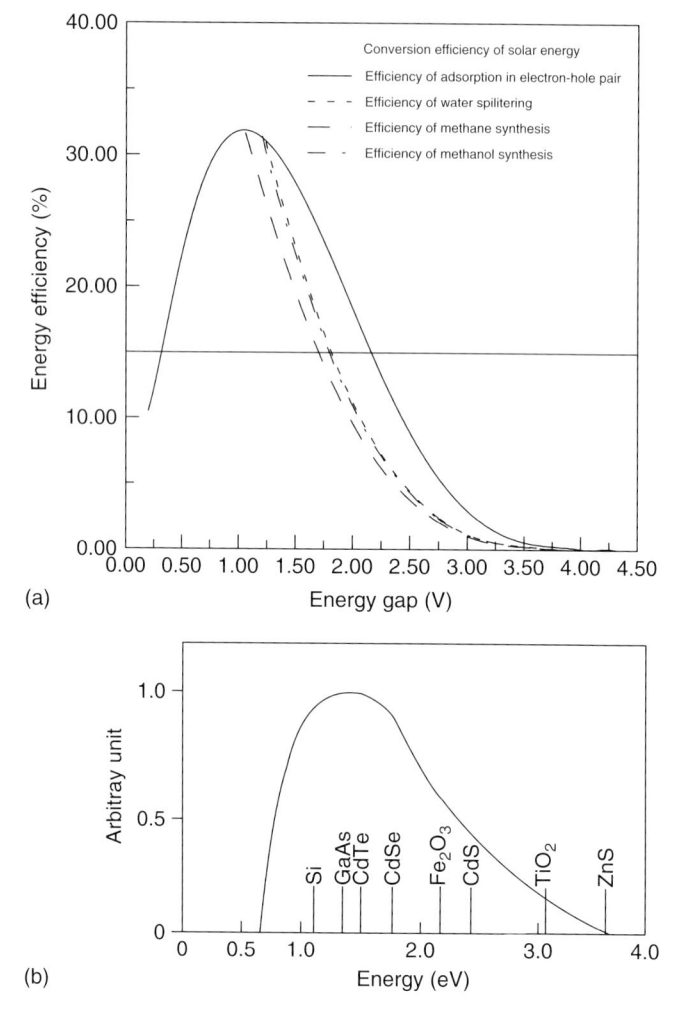

Figure 12.16 (a) Relationship between the solar energy conversion efficiency and energy gap of photocatalysts exhibited for water splitting, methane synthesis, and methanol synthesis. (b) Intensity of solar energy absorption by semiconductors of different bandgap energies. It is seen that the low-bandgap materials absorb more of solar radiation, but are easily photodegradable [99].

NaOH, KOH, K_2SO_3, $Na_2S + K_2SO_3$, Na_2SO_4, $NaClO_4$, or K_2SO_4 solution can be used as the electrolyte. In basic media, the negative ion concentration is enhanced, so it helps to compensate the deficiency of electrons in the VB, thus increasing the photocurrent and ultimately the efficiency of the cell. This is the reason for preferring alkaline media to neutral and acidic media.

The PEC measurements of samples should to be carried out in a nitrogen atmosphere. Current densities, as a function of applied potential for the samples, are monitored using a computer-controlled potentiostat under light illumination.

The light systems used for PEC responses of samples can be a Xe short-arc lamp [101], high-pressure mercury lamp [102], or high-pressure sodium lamp [103]. Figure 12.8a,b shows the schematic diagram of a PEC measurement system [63, 104].

In the process of the development of high-efficiency photocatalyst materials for PEC cells, one must know the relationship between the energy bandgap (E_g) of photocatalysts and solar energy conversion efficiency (η). The solar energy conversion efficiency as a function of energy bandgap of photocatalysts or Gibbs free energy of reaction (ΔG) is given by Eqs. (12.11) and (12.12):

$$\eta_{adsorption}\ (\%) = \frac{E_g \int_0^{\lambda = \frac{hC}{E_g}} np(\lambda)d\lambda}{\int_0^\infty \frac{hC}{\lambda} np(\lambda)d\lambda} \quad (12.11)$$

where, h is the Planck constant, λ is the wavelength of incident light, C is the velocity of light, and np is total numbers of photons having wavelength between λ and $\lambda + d\lambda$. The wavelength of the incident light should be shorter than hC/E_g, which means the incident photon energy must be greater than the energy bandgap of the photocatalyst, whereby the photons adsorbed on photocatalyst's surface generate electron–hole pairs. The solar energy conversion efficiency as a function of the Gibbs free energy of reaction (ΔG) given by Eq. (12.12):

$$\eta_{reaction}\ (\%) = \frac{\frac{\Delta G^{reaction}}{n} \int_0^{\lambda = \frac{hC}{E_g}} np(\lambda)d\lambda}{\int_0^\infty \frac{hc}{\lambda} np(\lambda)d\lambda} \quad (12.12)$$

where, $\frac{\Delta G^{reaction}}{n}$ is equal to 1.23 eV for the water-splitting process, 1.20 eV for methanol synthesis, and 1.05 eV for methane synthesis. The photoconversion efficiency of light into chemical energy in the presence of external applied voltage is calculated using Eq. (12.13):

$$\%\eta = (\text{Total power output} - \text{electrical power output}) \times 100$$
$$= j_p \left[\frac{(E_{rev}^0 - |E_{app}|)/I_0}{I_0} \right] \times 100 \quad (12.13)$$

where $E_{app} = E_{meas} - E_{aoc}$, and j_p is the photocurrent density in mA cm^{-2}. E_{rev}^0 is the standard reversible redox potential of the water and equal to 1.23 versus NHE. E_{meas} is the electrode potential (vs Ag/AgCl or saturated calomel electrode (SCE)) of the working electrode at which photocurrent is measured under illumination of light, and E_{aoc} is the electrode potential (vs Ag/AgCl) of the same working electrode at open-circuit condition under same illumination and in the same ES. I_0 is the intensity of incident light (mW cm^{-2}). Incident photon-to-current conversion efficiencies (IPCE) are calculated using Eq. (12.14):

$$IPCE(\%) = \frac{(1240\ eV\ nm)\ (\text{photocurrent density } \mu A\ cm^2)}{(\lambda\ nm)\ (\text{irradiance } \mu W\ cm^{-2})} \quad (12.14)$$

Dividing the IPCE by the fraction of incident photons absorbed at each wavelength gives the absorbed photon-to-current efficiency (APCE). Various factors including nanotube length and Fe composition go into APCE calculation through the absorbance A, as shown by Eq. (12.15):

$$\text{APCE} = \frac{\text{IPCE}}{(1 - 10^A)} \tag{12.15}$$

The STH (%) of the water-splitting reaction [66] can be determined using Eq. (12.16):

$$\text{STH (\%)} = \frac{j_p (V_{WS} - V_{Bias}) \times 100}{e E_S} \tag{12.16}$$

where j_p is the photocurrent produced per unit irradiated area, $V_{WS} = 1.23$ eV is the water-splitting potential per electron, E_S is the photon flux, V_{Bias} is the bias voltage applied between the working and the counter electrodes, and e is the electronic charge. According to Eq. (12.16), every electron contributing to the current produces a half H_2 molecule. The IPCE provides a way to measure of the efficiency of conversion of the photons to the current flowing between the working and counter electrodes in a PEC cell. Equation (12.17) is used to calculate the IPCE:

$$\text{IPCE (\%)} = \frac{J_p(\lambda)}{e E_S(\lambda)} \tag{12.17}$$

Photons provided by solar energy have enough energy to split water into hydrogen and oxygen but the efficiency of STH conversion depends on how the reaction is carried out. It is possible to split water into hydrogen and oxygen using the thermal energy given by a solar concentrator or a solar furnace by heating water to 1500–2000 K [105]. However, the efficiency of this process is still low, and the cost of the system and material stability are still problematic [101]. Converting solar energy into hydrogen energy in a PEC system is thus an important route to obtain clean energy economically [106]. The performance of hydrogen production from water using a TiO_2 photocatalyst is limited due to its large bandgap of 3.0–3.2 eV; therefore, various modifications in TiO_2 or development of new materials are under progress [107, 108], which focus on the development of new visible-light-active photocatalysts with better and durable efficiency for water splitting under visible irradiation. On the other hand, several new approaches for assembling the PEC cell have been reported in the literature, such as (i) water splitting in a commercial water electrolyzer using electricity provided by a solid PV solar cell (Figure 12.17a), (ii) water splitting using a p/n PV solar cell immersed in an aqueous electrolyte (Figure 12.17b), (iii) water splitting using visible-light-active semiconductor photocatalyst in a PEC cell (Figure 12.17c), and (iv) water splitting using a dye-sensitized metal-oxide semiconductor in a PEC cell (Figure 12.17d) [105].

Solar water splitting using a TiO_2 PE and a platinum counter electrode in a PEC assembly was first reported by Fujishima and Honda [19], as shown in Figure 12.18a. Ichikawa and Doi designed a PEC cell using a perforated TiO_2 photocatalyst, a proton separator, and a platinum electrocatalyst in a single unit [109]. Figure 12.18b shows the schematic diagram of their PEC cell. The maximum photocurrent in their PEC cells was around 6 mA cm^{-2} at an external bias of +2.0 V

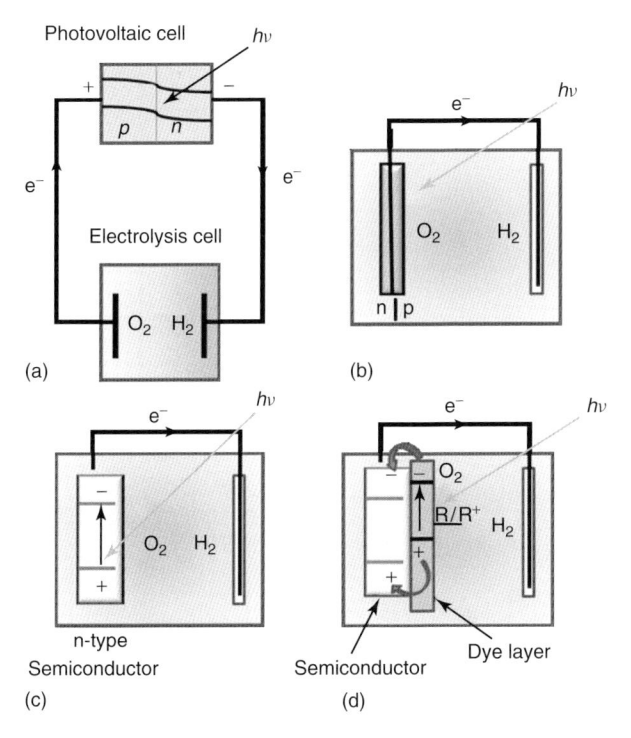

Figure 12.17 Schematic representation of various semiconductor-based system for water splitting. (a) Water splitting using electricity provided a solid photovoltaic solar cell in commercial water electrolyzer; (b) water splitting using a p/n photovoltaic solar cell immersed in an aqueous solution; (c) water splitting using visible-light-active semiconductor photocatalyst in a PEC cell; and (d) water splitting using sensitized metal-oxide semiconductor in a PEC cell [105].

versus SHE (standard hydrogen electrode) using a 500 W mercury lamp with a light intensity of 200 mW cm^{-2}. The hydrogen production in their PEC cell was 26 l h^{-1} m^{-2}. Hydrogen production from seawater using their PEC cell has also been reported in the literature [110]. The photocurrent of their cell using seawater as electrolyte was 94.7 μA cm^{-2} without external bias. The hydrogen production from seawater is around 0.4 l h^{-1} m^{-2} without external bias using a sun light intensity of 36.2 mW cm^{-2}. Kitano et al. reported an H-type glass photoreactor (Figure 12.18c) for hydrogen production from water with different pH electrolytes/chambers in the cell. [111] Visible-light-active TiO$_2$/Ti/Pt thin-film reactor (shown in Figure 12.18c) was used for the hydrogen production from water without making any special arrangement for H$_2$ and O$_2$ separation. Grätzel proposed a tandem cell for hydrogen production [112]. This tandem PEC cell consists of a front cell with a glass window containing the ES and the photocatalytic film deposited on the rear wall of this cell on conducting glass. The external bias for the water-splitting process was provided by a dye-sensitized solar cell. Figure 12.18d shows a schematic of the tandem cell.

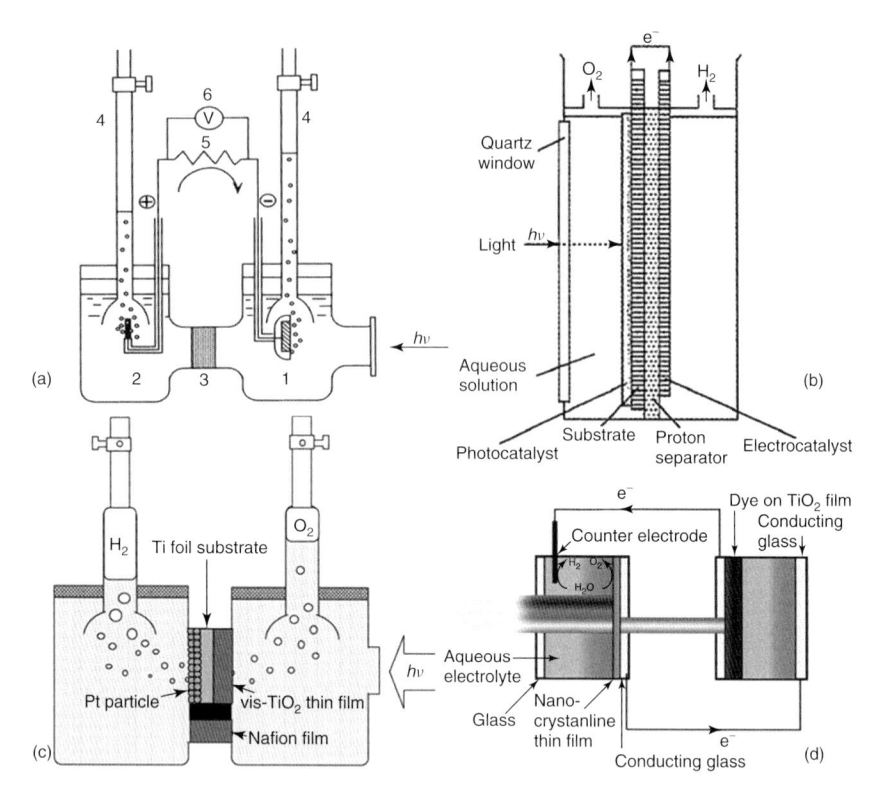

Figure 12.18 (a) Fujishima and Honda's PEC cell for overall water splitting. (b) single unit photoelectrocatalysis (SUPEC) cell, reported by Ichikawa and Doi. Thin-film titania photocatalyst are combined into one unit where protons can pass through from the surface of the photocatalyst to the electrocatalyst. (c) H-type PEC cell reported by Kitano et al. (d) Schematics of tandem PEC cell for water cleavage reported by Grätzel [19, 109, 111, 112].

Khaselev and Turner proposed a novel PEC–PV cell using p-type gallium-indium phosphide (p-GaInP$_2$) as the semiconductor photocatalyst, shown in Figure 12.3 [113]. Because the energetics of the band edge for GaInP$_2$ is not suitable for water splitting, an external bias was provided by a GaAs solar cell. The hydrogen production efficiency of this cell was 12.4% under exposure to light intensity of 1190 mW cm^{-2} using a 150 W tungsten-halogen lamp.

12.6
Photoelectrochemical Cell Performance, Testing, and Diagnosis

One of the Holy Grail's of the materials science is the search for a PE material of high efficiency (at least 10%) for water splitting using solar energy and a PEC cell that will not be consumed or degraded under irradiation for at least 10 years [99]. In traditional solid-state PEC devices, efficiency, technological cost, and the stability of PE materials remain difficult constraints to be overcome altogether. Technological innovations can significantly resolve the above constraints, but we need a balanced portfolio of options. Nobel Laureate Richard Smalley, a distinguished professor of chemistry and physics, Rice University, coined the term "Tetra Watt (TW) challenge" to meet the future energy challenge of 20 TW for 2025. Furthermore, ballooning prices of traditional fuels and their massive and ugly fingerprints in the form of environmental degradation are driving the world in the quest for green and sustainable energy sources to meet its energy needs. Researchers, industrialists, and politicians have seen tremendous possibilities in hydrogen as a future fuel. Although sunlight may be free, the technology to convert water into hydrogen by sunlight is still expensive. Electricity supplied from the power grid is not very costly, and solar-assisted electricity can be favorable for hydrogen generation from water. Therefore, the investment of mind and money for developing high-efficiency PEC cells for hydrogen production from water splitting is quite reasonable. In this view, some well-known research institutes and industries, such as MV System (amorphous silicon solar cells and fabrication process scale-up), University of Toledo (amorphous silicon/germanium solar cell design and fabrication), South West Research Institute (film modification and roll processes), Intematix (combinatorial materials discovery), Duquesne University (pyrolytic oxide research), National Renewable Energy Laboratory (NREL, USA, for materials R&D guidance), and Swiss Federal Institute of Technology, Hawaii Natural Energy Institute (HNRI, solar cells), the Institute for Energy Conversion (IEC) Delaware, UNAM Mexico, IEA annex 14 (University of Geneva), have conducted research on PEC/photocatalyst technology to generate high-efficiency PEC devices for hydrogen production using low-cost and photostable materials (such as TiO_2, ZnO, WO_3, and Fe_2O_3) for electrodes. These improved electrode materials need rigorous testing for their suitability in PEC cells in respect of their performance and durability [114, 115]. Desirable features for a PEC test cell are the following:

- It must have a flat, quartz window for uniform illumination of the cell PEs.
- The construction material (borosilicate glass) of the cell should not degrade in solution (acids/bases) under solar exposure and be easy to clean.
- It should have the capability to measure two-electrode (PE and CE) and three-electrode (reference (RE), auxiliary (AU), and working electrodes (WEs)) systems.
- It should be an enclosed cell with gas purging capabilities.
- It should have a modular design allowing for easy exchange of electrode components.
- It should be capable of working under controlled temperature conditions (double-walled cell with inlet and outlet arrangements) and be able to afford the desired pressure of purge gases and gaseous products.
- It should be capable of measuring and capturing the amount of gaseous products.
- It should be a totally leak (gas and liquid) proof system.

12.6.1
Design and Construction of the PEC Test Assembly

12.6.1.1 Photoelectrodes

PEs assembled in a PEC cell are either positioned as a substrate Figure 12.19a or as a superstrate Figure 12.19b [116]. In the substrate configuration Figure 12.19a, light passes through a transparent substrate before reaching a thin-film PE, and the back of the PE should be squarely placed against the back of the cell window to eliminate absorption by the electrolyte. The thickness of the PE layer should be optimized to ensure good light absorption and charge-carrier collection throughout the PE, which are governed by the specific conductivity, photoconductivity, and optical absorption properties of the PE material being used. In the superstrate configuration of illumination, light passes through a layer of solution before reaching the PE, so the PE should be placed close to the window to minimize light absorption by the solution but far enough away to avoid diffusion-limited reaction conditions. A spacer may be attached to the inner side of the window or on the front edge of the PE, as shown in Figure 12.19b.

PEs should be annealed at the optimum temperature before being used in PEC cell to improve efficiency of the assembly.

12.6.1.2 Light Calibration

Reproducibility of the PEC performance is of utmost importance, which is commonly achieved using a reference solar cell and proper calibration of the light source, and should be conducted according to standard practices [117, 118]. The light source must be calibrated prior to every PEC measurement to a desired intensity of $100\,mW\,cm^{-2}$ using a power/energy meter (e.g., Gentec (ε) Solo2 Model-UP12E105-H5) based on the measured J_{SC} of a Si reference cell. During calibration, care must be taken to ensure that the distance between the light source and the reference cell is maintained at the same as that used between the light source and the PE in PEC measurements.

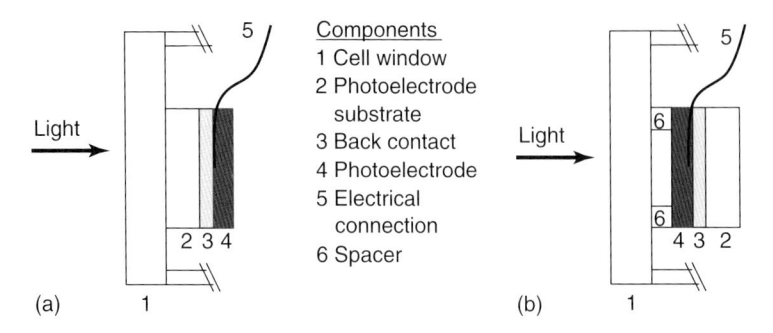

Components
1 Cell window
2 Photoelectrode substrate
3 Back contact
4 Photoelectrode
5 Electrical connection
6 Spacer

Figure 12.19 Schematic side views of (a) substrate and (b) superstrate PE configurations [116].

12.6.1.3 Temperature Control

Electrode kinetics, PE quantum efficiency, and electrode stability are sensitive to temperature variation. Therefore, a constant temperature should be maintained to avoid thermal effects on the PEC reaction. There are several ways to mitigate thermal effects, such as (i) placing a high wavelength IR absorber such as Tec15 glass (Pilkington) between the light source and the PEC; (ii) controlling the solution temperature within the cell, using a heating tape wrapped around the cell body and adjusting the power input to the tape; and (iii) using a reactor with a water jacket connected to an electric water bath. For temperature monitoring during cell operation, a thermometer or thermocouple can be inserted into one of the cell ports.

12.6.1.4 Solution Preparation

ESs should be dearated and preheated to the desired temperature of operation before measurements are made. The electrolyte material should be kept and weighed in a glove box because most electrolytes are air-sensitive. The pH of the ES must be adjusted to the desired value before running the experiment and checked after completion of the experiment. These tasks can be performed inside or outside the test cell. Preheated and dearated solution can be poured or pumped directly into the PEC device until the electrode surfaces are fully submerged in the solution. A gas line may be connected directly to the cell stopcock to allow for a small flow rate of inert gas over the electrolyte during experiments.

12.6.1.5 Gas Collection

For the study of photosynthetic cells, it may be necessary to collect the gases produced by the PEC cell for chemical analysis. This may be achieved by placing a rubber septum over one of the ports 1, 3, or 4 and collecting the gas samples by inserting a gas syringe into the septum and taking it from the region above the electrolyte. Alternatively, the produced gases can be directly channeled from the electrode surface to a gas chromatograph. In the chromatograph, the amount of the collected gas is measured after a set period (1 or 5 or 10 h).

12.6.1.6 Electrochemical Device Operation

PEC analyses are made by compiling the PEC cell in modified two-electrode and three-electrode PEC setups, in which the reference electrode with PE (working electrode) and CE are inserted into the appropriate cell ports and connected to a potentiostat as shown in Figure 12.8a,b. This allows measurement of hydrogen photocurrent for the calculation of hydrogen efficiency (Eqs. (12.13–12.17)), PV properties, and stability. Photoresponses of the WE recorded as current density (J) versus the applied electrochemical potential (V against given reference electrode) measurements are reported by the schematic linear sweep voltammograms (LSVs, Figure 12.20a), and stability of the PE is monitored using amperometric $I-t$ curves (Figure 12.20b) which are recorded at a constant applied voltage (vs reference electrode) for the desired period for repeated light on/off cycles or observed separately for light or dark PEC reactions [119, 120].

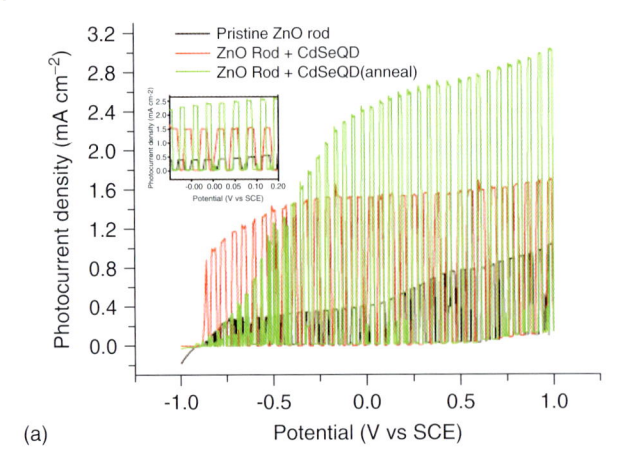

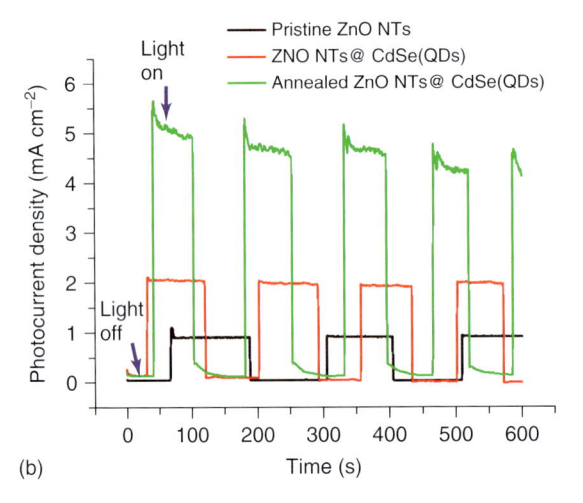

Figure 12.20 (a) Linear sweep voltammograms of CdSe (QD)/ZnO (nanorods (NR)) arrays (exposure area = 1 cm², in 0.35 M Na$_2$S and 0.25 M K$_2$SO$_3$ electrolyte solution, pH 13.3) under AM1.5 G illumination of power density 100 mW cm^{-2} at applied potential range −1 to +1 V. Inset shows enlarged view of J–V (vs SCE) around zero bias current. (b) Amperometric I–t curves of pristine ZnONTs, ZnO NTs@CdSe(QDs), and annealed ZnO NTs@CdSe(QDs) at applied voltage of +0.35 V at 100 mW cm^{-2} for 600 s of repeated light on/off cycles [119, 120].

The standard potential of the desired electrochemical reaction E_o and the flat-band potential V_{FB} of the PE have also been labeled in Figure 12.21. For a device operating at open circuit, the difference between E_o and V_{FB} gives the expected open-circuit voltage for the cell. For a device operating at short circuit, the operating point is given as no voltage between the two electrodes. In order to understand the intrinsic electronic properties of PEs in the ES, electrochemical impedance measurements in the dark are performed to determine the capacitance, dopant density, and width of the space charge layer. Nanowire carrier density and flat-band potential at nanowire/electrolyte interface can be quantified by the Mott–Schottky equation (Eq. (12.18)) [121].

$$1/C^2 = (2/e_0 \varepsilon \varepsilon_0 N_d)[(V - V_{FB}) - kT/e_0] \tag{12.18}$$

where e_0 is the electron charge, ε is the dielectric constant of electrode material, ε_0 is the permittivity of vacuum, N_d is the dopant density, V is the electrode applied potential, V_{FB} is the flat-band potential, and kT/e_0 is a temperature-dependent correction term. The Mott–Schottky plot presented by Figure 12.21 was collected for a 3.7% N-doped ZnO:N nanowire in a PEC cell. In contrast to planar samples, they exhibit a nonlinear behavior due to the cylindrical geometry of the nanowires [123]. The V_{FB} of the nanowires was determined from the extrapolation of the X-axis intercepts in Mott–Schottky plots ($1/C^2$ vs V, Figure 12.21) at various frequencies, which was found to be -0.58 V (vs NHE). The V_{FB} required for a water-splitting reaction to proceed should be more cathodic than the reduction potential of hydrogen (E_h), which varies as a function of the pH value and can be quantified by the equation $E_h = 0 - 0.059$ (pH). In addition, the slopes determined from the analysis of Mott–Schottky plot are used to estimate the carrier density using Eq. (12.19):

$$N_d = (2/e_0 \varepsilon \varepsilon_0) \left[d\left(1/C^2\right)/dV \right]^{-1} \tag{12.19}$$

The positive slope obtained for ZnO:N nanowires indicates that the material is n-type with electron conduction. With an ε value of 10 for ZnO [123], the electron density of ZnO:N nanowires calculated from Eq. (12.18) is found to be in the range of $10^{17} - 10^{18}$ cm^{-3} at a frequency of 10 kHz. The width of the space charge

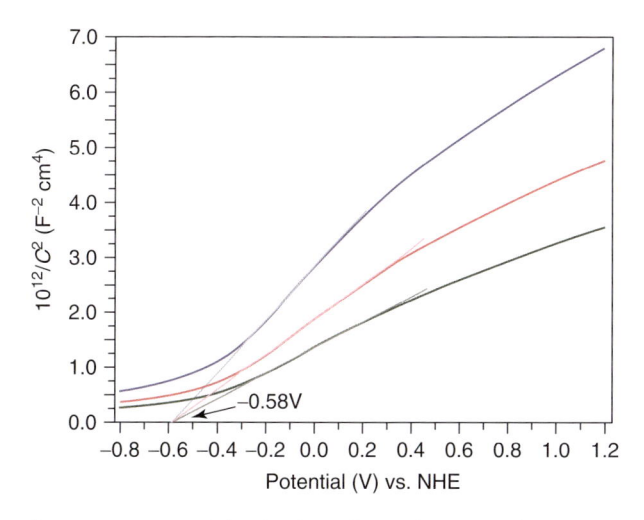

Figure 12.21 Mott–Schottky plots of a 3.7% ZnO:N nanowire in the dark at frequencies of 5 (black line), 7 (red line), and 10 kHz (blue line) and an AC current of 7 mV with a three-electrode system. The potential was measured against an Ag/AgCl reference and converted to NHE potentials by using $E_{NHE} = E_{Ag/AgCl} + 0.197$ V. Dashed lines represent the extrapolated lines from the linear portion of the Mott–Schottky plots [122].

layer (W) at the semiconductor/solution interface can also be derived using the Mott–Schottky plot and the following relationship [124]:

$$W = [2\varepsilon\varepsilon_0 (V - V_{FB})/e_0 N_d]^{1/2} \tag{12.20}$$

At a potential of $+1.0\,V$, W has been calculated to be 22 nm, which is much smaller than the radius of ZnO nanowires, indicating that the nanowire is not fully depleted yet.

Functional HPEs [79], using unoptimized tandem amorphous silicon junctions coated with reactively sputtered WO_3 films, have proved to be an important milestone in the development of the multilayered PEs. The testing assembly of the HPE is shown in Figure 12.22a. The performance of the component devices and the integrated HPE are shown in Figure 12.22b. Figure 12.22c shows (i) 0.7% STH efficiency using unoptimized solid-state and sputtered WO_3 films; (ii) 1.3% STH obtained using the solid-state and sputtered WO_3 films; (iii) >3% STH possible from a system using WO_3 without bandgap modification; and (iv) >5% STH requiring both bandgap modification of the WO_3 and further optimization of the solid-state junction. PEC device operating points (i.e., curve intersections) travel linearly to STH efficiencies with optimization of the device.

NREL achieved the highest 12.4% STH efficiency by developing a remarkable PEC electrode with a multilayer (GaAs/GaInP) structure in 1998, as shown in Figure 12.2b [125]. However, its life was less than 20 h because of the corrosion in aqueous solution. To resolve this problem, they spent more than a decade seeking alternative PE materials with high stability and efficiency via N-doping (such as GaPN, GaAsPN) [126] and a combinatorial approach [127], but the results are far from meeting both requirements. Simultaneously, Rocheleau *et al.* of the Hawaii Natural Energy Institute (HNEI) have developed a multilayer HPE thin-film oriented structure based on the CB and VB positions in a semiconductor for hydrogen generation in a PEC reactor [128]. HNEI demonstrated their first device with a triple-junction solar cell reactor for external bias and electrodes coated with thin films of OER ($NiFe_yO_x$) and HER (CoMo) catalysts on opposite sides for direct photoelectrolysis of water [128]. Furthermore, in order to verify the actual STH efficiency and stability of this integrated system, they also developed a simple outdoor test system in which the STH efficiency of the system was maintained at 7.8% without any degradation for 2 h in the basic KOH electrolyte. But this highly efficient system utilized separate solar cell and PEs and still required an external wiring for connection. The Swiss Federal Institute of Technology fabricated a dye-sensitized tungsten oxide (WO_3, $E_g = 2.6$ eV) PEC system of 4.5% solar conversion efficiency, in which they combined the concept of photocatalyst and solar cells into a PEC device. Because the energy level of the CB of WO_3 was insufficient for reduction of H_2O to H_2, DSSC was introduced as a chemical bias to tailor the energy level for hydrogen generation [50].

In another representative class of PV-biased tandem cell of excellent efficiency of 11.99%, $CIGS_2/CdS$ and CdTe thin-film solar cells connected in series with a RuO_2 anode and a platinum cathode were prepared on glass with a transparent conducting p-type back contact (Figure 12.23a). The I–V characteristics of this

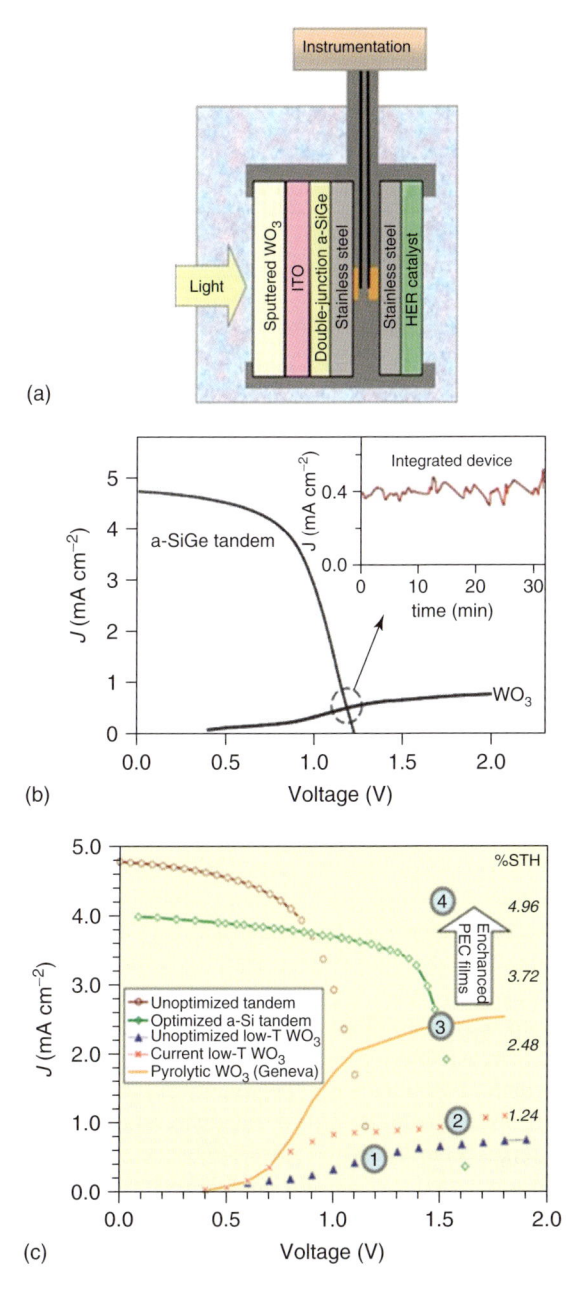

Figure 12.22 Hybrid photoelectrode (HPE). (a) Test structure. (b) Performance of WO_3 component films versus a-SiGe tandem cell. Stability test for of the integrated device shown in the inset. (c) Progressive current–voltage characteristics of a WO_3 based HPE systems after optimization of the solid-state junction layer, recorded under simulated AM 1.5 G illumination. Current levels at the device operating points (i.e., curve intersections) translate linearly to STH efficiencies [79].

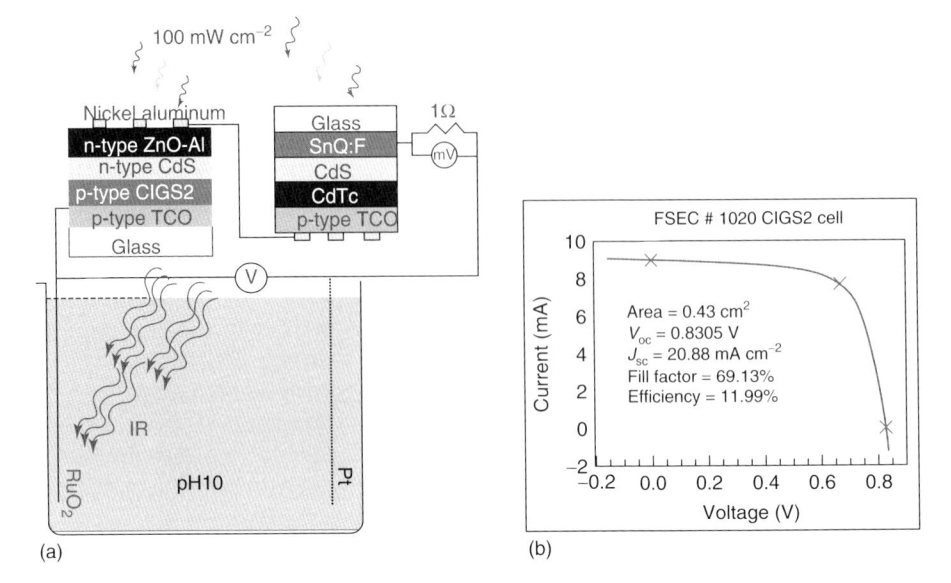

Figure 12.23 (a) Two-cell PEC setup with p-type transparent conductive oxides (TCO) back contact. (b) Low-sweep voltametric I–V characteristic curve for CIGS2/CdS thin films measured at NREL under the illumination of 1.5 AM visible light [130].

$CIGS_2$ cell were measured at NREL (Figure 12.23b) under an illumination power density of $100\,mW\,cm^{-2}$ and 1.5 AM G conditions [129].

Alex et al. also reported a PEC device with a configuration of $SS/ni_2pni_1p/ZnO/WO_3$ as PE [131]. The system exhibited a photocurrent density of $2.5\,mA\,cm^{-2}$ with the STH conversion efficiency limited to >3%. Another interesting example based on a new tandem cell structure of decreasing bandgaps was created in the above-mentioned structure. Feng et al. modified the Alex et al.'s device by substituting WO_3 with an a-SiC:H (p-type) and controlling the intrinsic layer thickness up to 100 nm in 0.33 M H_3PO_4 electrolyte; the system gave 8.61% efficiency with $7\,mA\,cm^{-2}$ photocurrent density [132].

The same group further put their effort to enhance the stability of the device by reducing the acidity of the electrolyte, varying the type of substrate, and changing the Pt coating. When the materials were loaded on glass/SnO_2 and glass/ZnO substrates and the PEC device was tested under acidic conditions for 100 h, the experimental result established the superiority of the glass/SnO_2 sample over glass/ZnO substrate [133]. Takabayashi et al. configured a low-cost composite electrode system which can work with more than 10% conversion efficiency utilizing the combination of doped-TiO_2 and polycrystalline Si thin films [134].

Although the STH efficiency crossed the threshold of 10% by a monolithic PEC/PV device using an integrated or HPE/tandem cell configuration, it also suffered from multiple issues such as fabrication complication, price, and interface integration. Furthermore, the low stability of III–V semiconductors due to the corrosion of PE in the electrolyte under illumination is another prime area that needs to be worked on. In fact, PEC technology is still in the development stage and a little far away from the large-scale commercial application to compete with the traditional energy field. But there is no doubt on the enormous potential of the environmentally friendly PEC technology for hydrogen production, which will soon overtake its competitors. Moreover, world energy will be in a win–win position when the triangular issues of technological cost, stability, and efficiency are resolved altogether.

12.7
Degradation Mechanisms and Mitigation Strategies

Survivability of PEC cells under intense light exposure and an alkaline electrolyte (causes cascading failure of the device) is a major challenge for utilizing them in clean-fuel energy applications. A series of experimental and theoretical studies have been conducted to clarify the degradation mechanisms governing the gradual or abrupt decrease in device performance, which would definitely help develop durable and efficient PEC systems. Various mitigation strategies are being tried, including appropriate material band-engineering, improving the substrate, raising the working temperature, cyclic exposure to light, stacking one or more thinner layers of materials to form a multijunction PEC solar cell, exposure to light intensity of less than $100\,\text{mW cm}^{-2}$, and corrosion-resistant conducting polymer Ni–P–PTFE (polytetrafluoroethylene) coatings [135], all of which would lead to reduced degradation and possible cell damage. Some important degradation processes of PEC devices and their mitigation techniques are discussed below.

12.7.1
Staebler–Wronski Effect (SWE) in Si-Based PEC Devices

In Si-based PEC devices, the attached hydrogen atom enhances the electrical properties of the a-Si. But when these hydrogen atoms are pushed away from silicon, a process energized by the presence of photons, defects occur in the atomic structure and the performance of the a-Si cell gradually reduces. Metastable light-induced degradation of hydrogenated amorphous silicon called the "Staebler–Wronski effect" (SWE) takes place in which the dark conductivity and photoconductivity of the hydrogenated amorphous silicon decrease owing to the prolonged illumination with intense light. As a result of the SWE, the efficiency of hydrogenated a-Si PEC drops during the first six months of operation. This drop may be in the range of 10–30% and the fill factor falls from over 0.7 to about 0.6 depending on the material quality and device design. When the effect reaches equilibrium, it causes little further degradation. Annealing of the amorphous silicon at a few hundred degrees Celsius will reverse this effect. That is why in winter the Staebler–Wronski effect is relatively strong and module efficiency is relatively low as a result; warmer weather results in relatively improved efficiencies [136]. The proposed mechanisms of SWE include the following steps (Eqs. (12.21–12.23)) [137]:

$$\text{Si–Si} + \text{Si–H} \rightarrow \text{Si–db} + \text{Si–H–Si}$$
$$\text{defect} + \text{interstitial (dissociation)}, \quad (12.21)$$

$$\text{Si–Si} + 2\text{Si–H} \rightarrow 2\text{Si–db} + \text{Si–H–H–Si}$$
$$\text{defects} + (\text{Si–H bonds}$$
$$+ \text{clustered H or H}_2^* \text{ molecules})(\text{dissociation}) \quad (12.22)$$

$$\text{Si–H–Si} + \text{Si}^*\text{–Si}^* \rightarrow \text{Si–Si} + \text{Si}^*\text{–H–Si}^*$$
$$(\text{weak Si–Si bond})(\text{mobile hydrogen}) \quad (12.23)$$

In the above, Si–db denotes the dangling bonds when hydrogen has left Si. It is proposed that an electron–hole pair formed by the incident light may recombine near a weak Si–Si bond, releasing sufficient energy to break the bond. A neighboring H atom then forms a new bond with one of the Si atoms, leaving a dangling bond. These dangling bonds can trap electron–hole pairs, thus reducing the current that can pass through. The defect creation kinetics in an a-Si PE has been studied by Stutzmann [138], who showed that the defect density D depends on the illumination intensity G and the time t: $D(t) = \text{constant } G^{2/3} \cdot t^{1/3}$. The total dangling bond concentration (but not the individual ones) is proportional to $t^{1/3}$ during exposure [139]. The following techniques are adopted to reduce SWE degradation.

- Exposure to high intensity illumination causes deeper degradation than 1 sun illumination; therefore illumination levels should be below $100\,\text{mW cm}^{-2}$.
- Cyclic exposure results in stabilization at the higher efficiency than a continuous exposure.
- Elevated operating temperatures (60–90 °C) stabilize at a higher efficiency than those operated at a room temperature or below.
- High impurity concentration (above $10^{18}\,\text{cm}^{-3}$) in the intrinsic layer degrades more than that in high-purity intrinsic layers.
- Thick intrinsic layers degrade more than thin intrinsic layers.
- Microcrystalline Si and Ge films instead of a-Si additionally reduce the SWE.
- The general trend of improving stability of a-Si:H suggests that it is most important to decrease the presence of weak bonds that give rise to deep traps for mobile hydrogen species in the density of H states.
- Stacking one or more thinner layers of amorphous silicon together with other materials to form a multijunction solar cell appears to reduce the SWE through the higher electric field that is applied in the thinner layers.

12.7.2
Degradation of Oxide Semiconductors and Corresponding Mitigation

Oxide semiconductors (TiO_2, ZnO, WO_3, ZrO_2, $SrTiO_3$, Nb_2O_5, etc.) are quite popular (because of their low cost and stability) among PE materials for photo-assisted water splitting. However, they suffer from low photoconversion efficiency due to the losses that are inherent from their wide bandgap and the transportation of the photogenerated electrons or H^+ ions between the anode and the cathode. Schrauzer and Guth used photocatalytic decomposition of water vapor, instead of liquid water, under UV light illumination via a TiO_2-based PE to stimulate the cogeneration of oxygen and hydrogen [140]. Subsequently, modified doped (anion or cation) TiO_2 surfaces were investigated for efficient photocatalytic splitting of water vapor [140–143]. Effectively, rather than using a bulk semiconductor particle electrode in the electrochemical cell, where each particle behaves as its own electrochemical cell, the logical extension of this idea is to use one-dimensional "nanoparticles" of various geometries (e.g., sphere, tube, ribbon, or wire) to enhance the photoresponses of the PEC cell. The concept offers an appealing and, of course, simple and low-cost technical idea that has gained considerable academic and industrial interest. Doping of transition-metal (V, Cr, Fe, Mo, Ru, Os, Re, V, Rh, etc.) and rare-earth metal (La, Ce, Er, Pr, Gd, Nd, Sm, etc.) ions has been extensively used in the attempts to enhance the visible-light-induced photocatalytic activities. Several experimental results confirm that metal-ion doping extends the photoresponse of TiO_2 into the visible region [144]. Doping titanium dioxide with anions such as N, C, P, S, and F [145–147], also increases the efficiency of the PEC system. Khan et al. modified TiO_2 photocatalysts by carbon doping, which absorbs light at the wavelengths below 535 nm and enhances photoconversion efficiency up to 8.35% [34]. Sensitization of oxides with dyes and other anchored molecular species (QDs) has been suggested as an alternative to extend the wavelength region of absorption. These coupled system gives rise to multijunctions, which is another approach being pursued in recent times with some success to increase the efficiency of the PEC system. Efficiency of the photoanodic heterosystems can also be tailored by means of "semiconductor sensitization," the unique example of this category being the WO_3-Pt/CdS PE in an aqueous solution of methyl viologen (MV^{2+}) which serves as an electron relay and has been found the most efficient to date in terms of water splitting [148]. The following reactions are representative (Eqs. (12.24–12.31)) of the electronic transfer mechanism:

$$CdS + h\nu \rightarrow e^- (CdS) + h^+ (CdS) \quad (12.24)$$

$$WO_3 + h\nu \rightarrow e^- (WO_3) + h^+ (WO_3) \quad (12.25)$$

$$h^+ (CdS) + e^- (WO_3) \rightarrow CdS + WO_3 \quad (12.26)$$

$$e^- (CdS) + Pt \rightarrow CdS + e^- (Pt) \quad (12.27)$$

$$e^- (Pt) + MV^{2+} \rightarrow Pt + MV^{\bullet +} \quad (12.28)$$

$$MV^{\bullet +} + H^+ \rightarrow MV^{2+} + \tfrac{1}{2} H_2 \quad (12.29)$$

$$h^+ (WO_3) + H_2O \rightarrow WO_3 + OH^\bullet + H^+ \quad (12.30)$$

$$2OH^\bullet \rightarrow H_2O + \tfrac{1}{2} O_2 \quad (12.31)$$

The conduction band electrons of CdS move to Pt, which in turn reduces MV^{2+}. The reduced methyl viologen, in turn, reduces a hydrogen ion to a hydrogen molecule. The VB holes of WO_3 then oxidize water to O_2 molecules.

12.7.3
Degradation of Chalcogenide Semiconductors and Corresponding Mitigation

Chalcogenide (sulfide, phosphide, nitride, etc.) semiconductors show great efficiency toward hydrogen generation in PEC cells. But, unfortunately, they are not very stable in an aqueous medium and undergo self-anodic dissolution under visible light leading to the formation of sulfur, phosphide, or nitrogen, as shown in the reaction. Such photoanodic dissolution can be quenched by adding reducing agents such as S^{2-}, SO_3^{2-}, and $S_2O_3^{2-}$ to PEC water splitting using CdS without addition of the reducing agent [149–151], as shown in Eqs. (12.32–12.36).

$$CdS + h\nu \rightarrow e^- \text{ (CB of CdS)} + h^+ \text{ (VB of CdS)}$$
$$\text{(photoelectron and photohole generation)} \quad (12.32)$$

$$H_2O \rightarrow H^+ + OH^- \quad E^\circ \text{ (vs NHE at pH 7)} = +1.23\,V$$
$$\text{(water splitting)} \quad (12.33)$$

$$2e^- \text{ (CdS)} + 2H^+ \rightarrow H_2 \quad \text{(cathode reaction)} \quad (12.34)$$

$$2h^+ \text{ (CdS)} + 2OH^- \rightarrow 2H^+ + \frac{1}{2}O_2 \quad \text{(anode reaction)} \quad (12.35)$$

$$2h^+ \text{ (CdS)} + CdS \rightarrow Cd^{2+} + S, \quad E^\circ \text{ (vs NHE at pH 7)} = +0.32\,V$$
$$\text{(photocorrosion)} \quad (12.36)$$

The following reactions (Eqs. (12.37–12.41)) decide the role of the electrolyte (S^{2-}/SO_3^{2-}) in preventing photodegradation of CdS in PEC cell.

$$2S^{2-} + 2h^+ \rightarrow S_2^{2-}, \quad E^\circ \text{ (vs NHE at pH 14)} = -0.52\,V \quad (12.37)$$

$$SO_3^{2-} + 2OH^- + 2h^+ \rightarrow SO_4^{2-} + H_2O \quad E^\circ = -0.92\,V \quad (12.38)$$

$$2SO_3^{2-} + 2h^+ \rightarrow S_2O_6^{2-}, \quad E^\circ = -0.25\,V \quad (12.39)$$

$$S_2^{2-} + SO_3^{2-} \rightarrow S_2O_3^{2-} + S^{2-} \quad E^\circ = -0.493\,V \quad (12.40)$$

$$0.5S_2O_3^{2-} + 1.5H_2O + 2h^+ \rightarrow SO_3^{2-} + 3H^+ \quad E^\circ = -0.2855\,V \quad (12.41)$$

By consuming holes, the poly chalcogenide electrolytes somehow protect (to some extent) the photocorrosion of the chalcogenides. But the photoinduced platinization of CdS, as well as coating with electrically conducting polymers (polypyrrole-Pt), has been found more effective in protecting the material against photodecomposition while in an electrolyte. The stabilization mechanism involves the electroactive redox species (generated by the high conductivity of the polymer) that efficiently trap the photogenerated holes before they can react with the semiconductor [151–153]. CdS surface with a polymer-catalyst coating (polypyrrole-Pt) may engage almost 68% of the photogenerated holes in O_2 production, but the rest of holes still lead to the destruction of semiconductor lattice [151]. However, in the absence of the catalyst or polymer-catalyst films, nearly 99% of all the photogenerated holes go to destroy the CdS lattice. CdSe photoanodes are also prone to photocorrosion in a similar manner [154]. Another most popular chalcogenide exploited for splitting water is n-GaN, which is used as the working electrode in PEC cells with Pt counter electrode and Ag/AgCl/NaCl reference electrode (SSSE), immersed in 1 mol l^{-1} aqueous KOH solution [155]. The CB and VB edge potentials in KOH solution are -1.66 and $+1.73$ V versus the SSSE, respectively, and are appropriate for water splitting. The wide bandgap of n-GaN (the working electrode) requires an external bias of 1.0 V at the Pt counter electrode to generate H$_2$ gas. Evolution of 94% hydrogen and 6% nitrogen at the Pt counter electrode after 5 h indicates the occurrence of photo-assisted electrolysis as well as n-GaN decomposition due to the following reaction (Eq. (12.42)):

$$2GaN + 6h^+ \rightarrow 2Ga^{3+} + N_2 (g) \tag{12.42}$$

Beach et al. studied p-GaN (p-GaN made by Mg doping of GaN using a high-temperature (1000 °C) metal–organic vapor phase) and found that, because the reduction potential of water is located below the CB edge of p-GaN, photogenerated electrons can be spontaneously injected into the ES that reduces H$^+$(aq) to H$_2$ [156]. Similarly, the anodic corrosion of GaAs PEs in aqueous electrolytes results from either preferential dissolution of the electrode materials or the oxidation reaction at its surface (Eqs. (12.43) and (12.44)). The stability of elemental arsenic on a GaAs surface depends on the electrolyte pH as well as the applied potential.

$$GaAs + 3h^+ \rightarrow Ga^{3+} (aq) + As \tag{12.43}$$

$$GaAs + 3H_2O + 6h^+ \rightarrow \tfrac{1}{2}Ga_2O_3 + \tfrac{1}{2}As_2O_3 + 6H^+ \tag{12.44}$$

Arsenic as the degradation product is oxidized to stable As$_2$O$_3$ (Eq. (12.45)) [157].

$$2As + 3H_2O \rightarrow As_2O_3 + 6H^+ + 6e^-$$
$$E° \text{ (vs NHE at pH 2, 3, or 4)} = -0.45 \text{ V} \tag{12.45}$$

The existence of the oxide layer shifts the onset potential of the photocurrent toward more negative values. And, due to oxide layer formation, a decrease in photocurrent amplitude resulted because of the following reasons: (i) newly generated oxide surface states consume the photogenerated electrons; (ii) the surface transmits less light through the oxide layer due to scattering and absorption; and (iii) the electrolyte can consume the anions, which retards the movement of the H$^+$ ion, and hence a higher negative potential would be required to displace the anions [157]. Use of nonaqueous organic and inorganic solvents and concentrated acidic solutions as electrolytes, as well as coating of metals and/or metal oxides, are some of the common mitigation practices that have been adopted to make GaAs PEs enabling their use in PEC cells [158–161]. For example, the increase in photogenerated current density from 4 to 18 mA cm^{-2} during irradiation at $+0.4$ V (vs SCE) applied potential in 2.5 M H$_2$SO$_4$ + 6.0 M LiCl ES confirms the suppression of photocorrosion of GaAs electrode, presumably due to the formation of an arsenic and chlorinated arsenic layer on the electrode surface [158]. Coating of the photoanode with a 5 nm gold layer drastically reduces photodecomposition activity under visible light illumination, with the anodic photocurrent onset shifting toward a cathodic potential with increasing pH [162]. Upon irradiation of n-type GaP in an alkaline solution, electron flow from GaP ($E_g = 2.24$ eV) to the counter electrode leads to photoanodic dissolution by the following reaction (Eq. (12.46)):

$$GaP + 4OH^- \rightarrow GaO_2^- + P + 2H_2O + 3e^- \tag{12.46}$$

To protect GaP from photocorrosion, suitable redox species such as Se^{2-}/Se_n^{2-} and Te^{2-}/Te_n^{2-} can be added to the ES [163] Alternatively, layers of chemically stable GaPN ($E_g \approx 2.0$ eV) can be grown or deposited on GaP, for example, using molecular beam epitaxy [126], resulting in a slight lattice mismatch due to replacement of the P with N. At low nitrogen concentrations, GaPN has been found to be chemically stable in aqueous ESs.

12.7.4
Copper Chalcopyrites (CIS or CIGS) Electrode Degradation and Mitigation Strategies

CIS or CIGS PEs are corrosion-prone in KOH but, unfortunately, significant shortcomings in both efficiency and stability were observed over the course of these tests. Experiments indicate that hydrogen and oxygen gases are evolved at the PE surfaces, but performance is limited by device instabilities originating around the edges, where the films begin to degrade even within 15 min of operation. The plausible explanations behind this degradation is the occurrence of high electric fields at the abrupt edges between the cathode and anode surfaces, resulting in outer structural damage in the protective films or development of the microcracks between the different layers. Therefore, a suitable PE geometry is required to minimize the edge interfaces and to eliminate the high-field edge effects. This can be mitigated by mechanically fabricated electrodes in side-by-side connected series configuration. Efficiency can be further improved for CIS/CIGS films deposited on molybdenum foil, developed by the IEC at the University of Delaware, which also have reasonably good corrosion resistance in KOH electrolyte as depicted in Figure 12.24a,b [164], as demonstrated by the scanning electron microscopic images of the CIS surface taken before and after 48 h of immersion of PE films in 1 N KOH. It shows that, except for few scattered vacancies, the basic surface morphology remains intact.

12.8
Applications (Portable, Stationary, and Transportation)

The use of hydrogen as an energy carrier has enough potential to reduce the dependence on gasoline as well as to prevent pollution and greenhouse gas emissions. Therefore, advancement in hydrogen technologies is fast progressing.

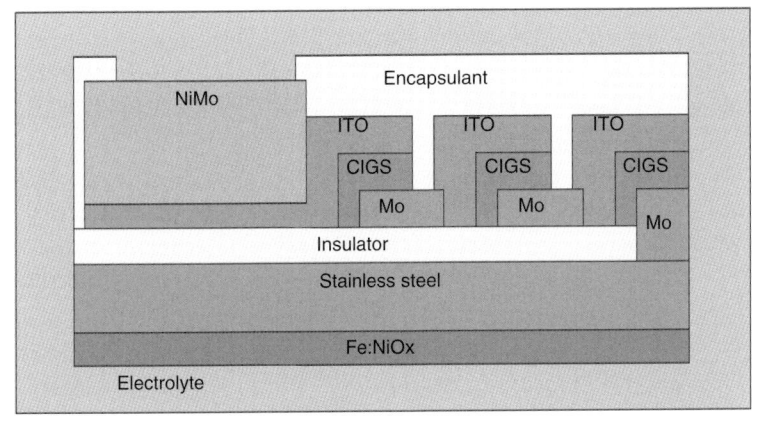

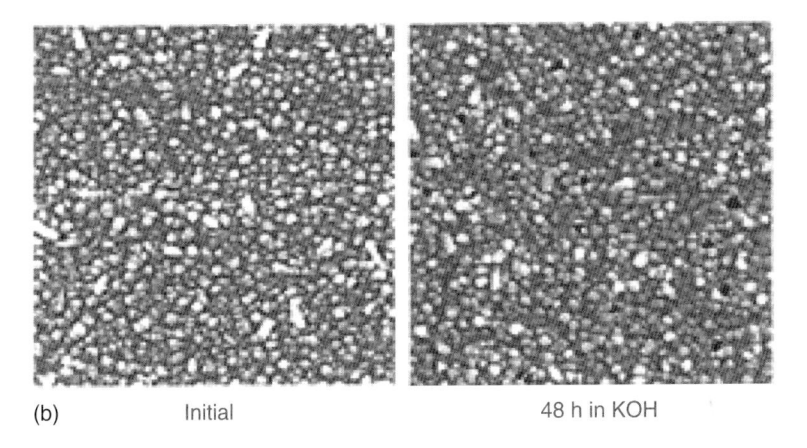

Figure 12.24 (a) PEC cell with CIS/CIGS films deposited on a molybdenum foil. (b) SEM images of CIS films before and after 48 h of KOH exposure [164].

Hydrogen economy can provide renewable energy solutions in many situations, such as emergency backup power, heating and electricity for commercial and residential purpose, and hybrid electric vehicles (in spite of on-board storage of hydrogen remaining a "critical path" barrier, it is one of the primary focus areas). PEC water splitting for hydrogen production is a promising approach for terrestrial (versatile applications in portable, stationary, and transportation) and space applications. To advance the commercialization of PEC-generated hydrogen technologies, market transformation activities aim at promoting their use in stationary, portable, and specialty vehicle applications, such as forklifts, municipal vehicles, and lawn mowers. Space photovoltaic is another field of possibilities for the design and fabrication of PEC devices for hydrogen production.

12.8.1
Portable

PEC technology is unique in terms of portable supply of clean-fuel hydrogen to run electrical and other utility appliances (heater, air-conditioner, fan, etc.). Hydrogen burner is one more interesting example of a portable hydrogen appliance, which can be used indoors safely without risk [165] and the reaction product is water which is actually beneficial for the room climate. Fundamental technical experience will be attained through thoughtful concerns on stand-alone power systems (SAPs), the use of hydrogen as future energy carrier, the durability, storage systems, as well as the use of hydrogen for cooking and heating purposes. A commercial barbecue has been adapted by replacing the propane burner by a series of hydrogen burners connected to a hydrogen source. A panel of evaluators confirmed that hydrogen-roasted meat was undistinguishable in taste from propane-roasted meat. Likewise, an ecofriendly portable hydrogen house is another use of PEC hydrogen, in which hydrogen is used as the energy source for most of the appliances.

12.8.2
Stationary

Technological advancement drives meaningful reductions in the cost, such as in the installation of a UNI-SOLAR solar energy system, and will reduce the cost of generating solar electricity by over 20%. Ultimately, the company expects to offer solar energy systems that are capable of providing electricity at a cost below that of the utility grid. Comparative conventional PV modules, which require an electrolyzer as additional system component, affect the cost factor of electricity plants that directly load feed at day time and accumulator storage of electricity harvest surplus for night times, as well as innovative PEC cells with long-term hydrogen storage and gas-engine-driven generator for the seasonal compensation (Figure 12.25) [166]. Direct solar hydrogen production using the newly available dynamic PEC cells eases and simplifies both plant design and operation. These PEC cells utilize their great potential in exemplary application of autonomous electricity supply for industrial plants. Another most promising application of hydrogen produced by PEC cells is in micro combined heat and power (CHP) supply of residential houses [167], exploiting both gas-engine-generated electric power and waste heat [168]. It seems to be an interesting perspective for which the employment of PEC cells is currently being investigated. Initial results have proved that, for a typical single family house in Germany with up-to-date thermal insulation, a share of $\sim$9% (1 MWh per year) of yearly natural gas consumption could be covered by hydrogen produced by $14\,m^2$ of PEC cells installed on the roof. Rather, under present prototype conditions, the investment cost, which is supposed to be considerably reduced in the future, is still being questioned under a pure economic aspect, as the PEC cells are still at the very beginning of their development phase.

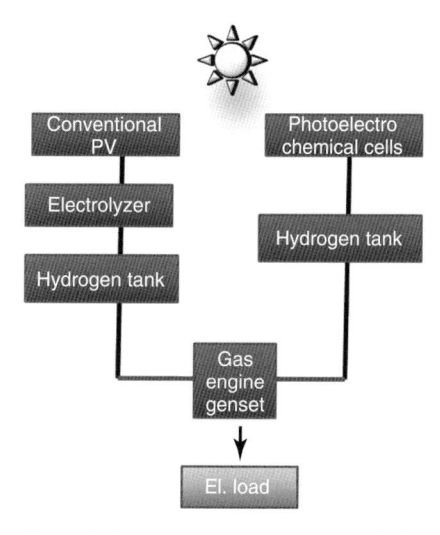

Figure 12.25 Autonomous energy supply based on PV and PEC devices [166].

12.8.3
Transportation

Highly efficient clean hydrogen fuel is the undisputed fuel for spacecraft. In 1990, the world's first solar-powered hydrogen production plant (a research and testing facility) became operational at Solar-Wasserstoff-Bayern, Southern Germany. In 1994, Daimler Benz demonstrated its first NECA I (New Electric CAR) fuel cell vehicle at a press conference in Ulm, Germany. In 1999, Europe's first hydrogen fueling stations were opened in the German cities of Hamburg and Munich. United Solar's technology roadmap will take this ground-breaking technology into commercial production in 2012. Hydrogen-driven city buses, lawn mowers [169], municipal vehicles, and so on, are some commercial means of transportation that utilize PEC hydrogen.

12.9
Conclusions

This chapter reviewed the various aspects of PEC hydrogen generation, using the renewable resources – water and a perpetual source of energy, Sun. Meticulous efforts were made to survey the successive progress of materials-related issues to develop high-efficiency PEC cells. The cell requirements for photoreactors, material energetics of PEs, various assemblies of PEC, and electrolytes were stepwise overviewed, in terms of their semiconducting and electrochemical properties and their impact on the performance of PECs. Different categories of photo-assisted water splitting were discussed, and thoughtful comments were made on the impact of the PEC structure and materials selection on the STH. The important issues related to the processes occurring in PEC cells, such as the solid/solid interfaces (e.g., grain boundaries) and solid/liquid interfaces (e.g., electrode/electrolyte interface), excitation of electron–hole pair in photoelectrodes, charge separation in photoelectrodes, electrode processes and related charge transfer within PECs, and generation of the PEC voltage required for water decomposition. Degradation mechanisms that retard the efficiency and stability of the variety of PEC assemblies with their corresponding mitigation strategies were also briefly explained. Utilization of PEC hydrogen in the application of portable, stationary, and transportation was also reviewed. Notably, this chapter established PEC cell technology as the brightest star in the galaxy of the contemporary energy harvesting techniques.

Acknowledgments

The authors would like to thank the Institute of Atomic and Molecular Sciences, Academia Sinica AS-98-TP-A05), the National Science Council of the Republic of China, Taiwan NSC 97-2113-M-002-012-MY3 and NSC 99-2110-M-002-012), and the Bureau of Energy, Ministry of Economic Affairs (8455DC6100), for financial support.

References

1. Flipsen, S.F.J. (2006) Power sources compared: the ultimate truth? *J. Power Sources*, **162** (2), 927–934.
2. Dyer, C.K. (2004) Fuel cells and portable electronics. Symposium on VLSI Circuits, Digest of Technical Papers, pp. 124–127.
3. Luque, A.H.S. (2003) *Handbook of Photovoltaic Science and Engineering*, John Wiley & Sons, Ltd, New York.
4. Merrill, J. and Senft, D.C. (2007) Directions and materials challenges in high-performance photovoltaics. *Jom*, **59** (12), 26–30.
5. Compaan, A.D. (2007) Materials challenges for terrestrial thin-film photovoltaics. *Jom*, **59** (12), 31–36.
6. Kazmerski, L.L. (2006) Solar photovoltaics R&D at the tipping point: a 2005 technology overview. *J. Electron Spectrosc. Relat. Phenom.*, **150** (2–3), 105–135.
7. Alam, M.M. and Jenekhe, S.A. (2004) Efficient solar cells from layered nanostructures of donor and acceptor conjugated polymers. *Chem. Mater.*, **16** (23), 4647–4656.
8. Gregg, B.A. (2003) Excitonic solar cells. *J. Phys. Chem. B*, **107** (20), 4688–4698.
9. Peumans, P., Uchida, S., and Forrest, S.R. (2003) Efficient bulk heterojunction photovoltaic cells using small-molecular-weight organic thin films. *Nature*, **425** (6954), 158–162.
10. Shaheen, S.E., Radspinner, R., Peyghambarian, N., and Jabbour, G.E. (2001) Fabrication of bulk heterojunction plastic solar cells by screen printing. *Appl. Phys. Lett.*, **79** (18), 2996–2998.
11. Dennler, G. and Sariciftci, N.S. (2005) Flexible conjugated polymer-based plastic solar cells: from basics to applications. *Proc. IEEE*, **93** (8), 1429–1439.
12. Jenekhe, S.A. and Yi, S.J. (2000) Efficient photovoltaic cells from semiconducting polymer heterojunctions. *Appl. Phys. Lett.*, **77** (17), 2635–2637.
13. Alam, M.M. and Jenekhe, S.A. (2001) Nanolayered heterojunctions of donor

and acceptor conjugated polymers of interest in light emitting and photovoltaic devices: photoinduced electron transfer at polythiophene/polyquinoline interfaces. *J. Phys. Chem. B*, **105** (13), 2479–2482.

14. Xue, J.G., Rand, B.P., Uchida, S., and Forrest, S.R. (2005) A hybrid planar-mixed molecular heterojunction photovoltaic cell. *Adv. Mater.*, **17** (1), 66–71.

15. Miles, R.W., Zoppi, G., and Forbes, I. (2007) Inorganic photovoltaic cells. *Mater. Today*, **10** (11), 20–27.

16. Bequerel, E. (1839) Recherches sur les effets de la radiation chimique de la lumière solaire, au moyen des courants électriques. *C. R. Acad. Sci.*, **9**, 145–149.

17. Brattain, W.H. and Garrett, C.G.B. (1955) Experiments on the interface between germanium and an electrolyte. *Bell Syst. Tech. J.*, **34** (1), 129–176.

18. Gerische, H. (1966) Electrochemical behavior of semiconductors under illumination. *J. Electrochem. Soc.*, **113** (11), 1174–1182.

19. Fujishima, A. and Honda, K. (1972) Electrochemical photolysis of water at a semiconductor electrode. *Nature*, **238**, 37–38.

20. Plumb, I. (2005) Photocatalytic materials for hydrogen production. Industrial Physics, CSIRO, http://www.cip.csiro.au/IMP energysustain/hydrogenproduction/ materials.htm (accessed 2010).

21. (a) Nozik, A.J. (1977) Photochemical diodes. *Appl. Phys. Lett.*, **30** (11), 567–569; (b) Nozik, A.J. and Memming, R. (1996) Physical chemistry of semiconductor-liquid interfaces. *J. Phys. Chem.*, **100** (31), 13061–13078.

22. Kudo, A. (2006) Development of photocatalyst materials for water splitting. *Int. J. Hydrogen Energy*, **31** (2), 197–202.

23. Sayama, K. and Arakawa, H. (1994) Effect of Na_2CO_3 addition on photocatalytic decomposition of liquid water over various semiconductor catalysts. *J. Photochem. Photobiol., A*, **77**, 243–247.

24. Arakawa, H. (2002) Water Photolysis by TiO_2 Particles: Significant Effect of Na_2CO_3 Addition on Water Splitting, in *Photocatalysis Science and Technology* (eds M. Kaneko and I. Okura), Springer, New York, pp. 235–248.

25. Inoue, Y. (2002) Water Photolysis by Titanates with Tunnel Structures, in *Photocatalysis Science and Technology* (eds M. Kaneko and I. Okura), Springer, New York, pp. 249–260.

26. Domen, K. (2002) Water Photolysis of Layered Compounds, in *Photocatalysis Science and Technology* (eds M. Kaneko and I. Okura), Springer, New York, pp. 261–278.

27. Yamasita, D., Takata, T., Hara, M., Kondo, J.N., and Domen, K. (2004) Recent progress of visible-light-driven heterogeneous photocatalysts for overall water splitting. *Solid State Ionics*, **172** (1–4), 591–595.

28. Maeda, K., Teramura, K., Saito, N., Inoue, Y., Kobayashi, H., and Domen, K. (2006) Overall water splitting using (oxy)nitride photocatalysts. *Pure Appl. Chem.*, **78** (12), 2267–2276.

29. Maeda, K. and Domen, K. (2007) New non-oxide photocatalysts designed for overall water splitting under visible light. *J. Phys. Chem. C*, **111** (22), 7851–7861.

30. Zou, Z.G. and Arakawa, H. (2003) Direct water splitting into H-2 and O-2 under visible light irradiation with a new series of mixed oxide semiconductor photocatalysts. *J. Photochem. Photobiol., A*, **158** (2–3), 145–162.

31. Kato, H. and Kudo, A. (1998) New tantalate photocatalysts for water decomposition into H-2 and O-2. *Chem. Phys. Lett.*, **295** (5–6), 487–492.

32. Shangguan, W.F. (2007) Hydrogen evolution from water splitting on nanocomposite photocatalysts. *Sci. Technol. Adv. Mater.*, **8** (1–2), 76–81.

33. Yang, X., Wolcott, A., Wang, G., Sobo, A., Carl Fitzmorris, R., Qian, F., Zhang, J.Z., and Li, Y. (2009) Nitrogen-doped ZnO nanowire arrays for photoelectrochemical water splitting. *Nano Lett.*, **9** (6), 2331–2336.

34. Khan, S.U.M., Al-Shahry, M., and Ingler, W.B. Jr. (2002) Efficient photochemical water splitting by a chemically modified n-TiO$_2$. *Science*, 297 (5590), 2243–2245.
35. Anpo, M., Dohshi, S., Kitano, M., Hu, Y., Takeuchi, M., and Matsuoka, M. (2005) The preparation and characterization of highly efficient titanium oxide-based photofunctional materials. *Annu. Rev. Mater. Res.*, 35, 1–27.
36. Yin, J. and Ye, J. (2004) Enhanced photo electrolysis of water with photoanode Nb:SrTiO$_3$. *Apl. Phys. Lett.*, 85 (4), 689–691.
37. Licht, S., Wang, B., Mukerji, S., Soga, T., Umeno, M., and Tributsch, H. (2000) Efficient solar water splitting, exemplified by RuO$_2$-catalyzed AlGaAs/Si photoelectrolysis. *J. Phys. Chem. B*, 104 (38), 8920–8924.
38. Santato, C., Ulmann, M., and Augustynski, J. (2001) Photoelectrochemical properties of nanostructured tungsten trioxide films. *J. Phys. Chem. B*, 105 (5), 936–940.
39. Wang, W., Liao, X.B., Han, S., Povolny, H., Xiang, X.B., Du, W., and Deng, X. (2001) Triple-Junction a-Si-Based Solar Cells with All Absorber Layers Deposited at the Edge of a-Si to µc-Si Transition, http://www.physics.utoledo.edu/~dengx/ref3.pdf (accessed 2010).
40. Khaselev, O., Bansal, A., and Turner, J.A. (2001) High-efficiency integrated multijunction photovoltaic/electrolysis systems for hydrogen production. *Int. J. Hydrogen Energy*, 26 (2), 127–132.
41. Licht, S. (2001) Multiple band gap semiconductor/electrolyte solar energy conversion. *J. Phys. Chem.*, 105 (27), 6281–6294.
42. News Release *NR-2708* dated August 13, 2008, on NREL Solar Cell Sets World Efficiency Record of 40.8 Percent, http://www.nrel.gov/news/press/2008/625.html (accessed 2010).
43. Wrighton, M.S., Ellis, A.B., Wolczanski, P.T., Morse, D.L., Abrahamson, H.B., and Ginley, D.S. (1976) Strontium titanate photoelectrodes. Efficient photoassisted electrolysis of water at zero applied potential. *J. Am. Chem. Soc.*, 98 (10), 2774–2779.
44. Mavroides, J.G., Kafalas, J.A., and Kolesar, D.F. (1976) Photoelectrolysis of water in cells with SrTiO$_3$ anodes. *Apl. Phys. Lett.*, 28 (5), 241–243.
45. Giordano, N., Antonucci, V., Cavallaro, S., Lembo, R., and Bart, J.C.J. (1982) Photoassisted decomposition of water over modified rutile electrodes. *Int. J. Hydrogen Energy*, 7 (11), 867–872.
46. Minggu, L.J., Daud, W.R.W., and Kassim, M.B. (2010) An overview of photocells and photoreactors for photoelectrochemical water splitting. *Int. J. Hydrogen Energy*, 35 (13), 5233–5244.
47. Bak, T., Nowotny, J., Rekas, M., and Sorrell, C.C. (2002) Photoelectrochemical properties of TiO$_2$-Pt system in aqueous solutions. *Int. J. Hydrogen Energy*, 27, 19–26.
48. Bard, A.J. and Faulkner, L.R. (2001) *Electrochemical Methods: Fundamentals and Applications*, John Wiley & Sons, Inc, New York.
49. Selli, E., Quartarone, E., Chiarello, G.L., Mustarelli, P., Rossetti, I., and Forni, L.A. (2007) Photocatalytic water splitting device for separate hydrogen and oxygen evolution. *Chem. Commun.*, 5022–5024.
50. Gratzel, M. (2001) Photoelectrochemical cells. *Nature*, 414 (6861), 338–344.
51. Oregan, B. and Gratzel, M.A. (1991) Low-cost, high-efficiency solar cell based on dye-sensitized colloidal TiO$_2$ films. *Nature*, 353 (6346), 737–740.
52. Miller, E.L., Paluselli, D., Marsen, B., and Rocheleau, R.E. (2005) Development of reactively sputtered metal oxide films for hydrogen producing hybrid multijunction photoelectrodes. *Sol. Energy Mater. Sol. Cells*, 88 (2), 131–144.
53. Gratzel, M., and Augustynski, J. (2005) Tandem cell for water cleavage by visible light. Patent No. US 6936143.
54. Miller, E.L., Rocheleau, R.E., and Deng, X.M. (2003) Design considerations for a hybrid amorphous silicon/photoelectrochemical multijunction cell for hydrogen production. *Int. J. Hydrogen Energy*, 28 (6), 615–623.

55. Siegel, A. and Schott, T. (1988) Optimization of photovoltaic hydrogen production. *Int. J. Hydrogen Energy*, **13**, 659–675.
56. (a) Avachat, U.S. and Dheere, N.G. (2006) Preparation and characterization of transparent conducting ZnTe:Cu back contact interface layer for CdS/CdTe solar cell for photoelectrochemical application. *J. Vac. Sci. Technol., A*, **24** (4), 1664–1667; (b) Khaselev, O. and Turner, J.A. (1998) A monolithic photovoltaic photoelectrochemical device for hydrogen production via water splitting. *Science*, **280** (5362), 425–427.
57. Licht, S., Hodes, G., Tenne, R., and Manassen, J. (1987) A light-variation insensitive high efficiency solar cell. *Nature*, **326** (6166), 863–864.
58. Wei, D., Andrew, P., and Ryhanen, T. (2010) Electrochemical photovoltaic cells – review of recent developments. *Chem. Technol. Biotechnol.*, **85** (12), 1547–1552.
59. Wallace, G.G., Too, C.O., Officer, D.L., and Dastoor, P.C. (2005) Photoelectrochemical cells based on inherently conducting polymers. *MRS Bull.*, **30** (1), 46–49.
60. Ding, J., Zhou, D., Spinks, G., Wallace, G., Forsyth, S., Forsyth, M., and MacFarlane, D. (2003) Use of ionic liquids as electrolytes in electromechanical actuator systems based on inherently conducting polymers. *Chem. Mater.*, **15** (12), 2392–2398.
61. Wang, P., Zakeeruddin, S.M., Exnar, I., and Grätzel, M. (2002) High efficiency dye-sensitized nanocrystalline solar cells based on ionic liquid polymer gel electrolyte. *Chem. Commun.*, **21** (24), 2972–2973.
62. Photo-Chemical H_2 Production Group KRICT (2008) 306th Korean Science and Engineering Foundation Exchange Seminar held on May 5, 2008.
63. Ardelean, P., Indrea, E., Silipas, T.D., Ardelean, C., Mihailescu, G., Suciu, R.C., Dreve, S.V., Moldovan, Z., and Rosu, M.C. (2009) A photoelectrochemical cell for the study of the photosensitive materials used in solar-hydrogen energy. *J. Phys.: Conf. Ser.*, **182** (1), 012046.
64. Wolcott, A., Smith, W.A., Kuykendall, T.R., Zhao, Y., and Zhang, J.Z. (2009) Photoelectrochemical water splitting using dense and aligned TiO_2 nanorod arrays. *Small*, **5** (1), 104–111.
65. Kelly, N.A. and Gibson, T.L. (2006) Design and characterization of a robust photoelectrochemical device to generate hydrogen using solar water splitting. *Int. J. Hydrogen Energy*, **31** (12), 1658–1673.
66. Murphy, A.B., Barnes, P.R.F., Randeniya, L.K., Plumb, I.C., Grey, I.E., Horne, M.D., and Glasscock, J.A. (2006) Efficiency of solar water splitting using semiconductor electrodes. *Int. J. Hydrogen Energy*, **31** (14), 1999–2017.
67. Bae, S., Kang, J., Shim, E., Yoon, J., and Joo, H. (2008) Correlation of electrical and physical properties of photoanode with hydrogen evolution in enzymatic photo-electrochemical cell. *J. Power Sources*, **179** (2), 863–869.
68. Jahagirdar, A.H. and Dhere, N.G. (2007) Photoelectrochemical water splitting using $CuIn_{1-x}Ga_xS_2$/CdS thin-film solar cells for hydrogen generation. *Sol. Energy Mater. Sol. Cells*, **91** (15), 1488–1491.
69. Gratzel, M. (2005) Mesoscopic solar cells for electricity and hydrogenproduction from sunlight. *Chem. Lett.*, **34** (1), 8–13.
70. Rozendal, R.A., Hamelers, H.V.M., Molenkamp, R.J., and Buisman, C.J.N. (2007) Performance of single chamber biocatalyzed electrolysis with different types of ion exchange membranes. *Water Res.*, **41** (9), 1984–1994.
71. Allam, N.K., Shankar, K., and Grimes, C.A. (2008) Photoelectrochemical and water photoelectrolysis properties of ordered TiO_2 nanotubes fabricated by Ti anodization in fluoride-free HCl electrolytes. *J. Mater. Chem.*, **18** (20), 2341–2348.
72. Linkous, C.A. and Slattery, D.K. (2001) Solar photocatalytic H_2 production from water using a dual bed photosystem. *Proceedings of the 2001 DOE Hydrogen Program Review*, NREL/CP-570-30535. U.S.

Department of Energy National Renewable Energy Laboratory, pp. 1–8. ww1.eere.energy.gov/hydrogenandfuelcells/pdfs/30535v.pdf (accessed 2010).

73. Aroutiounian, V.M., Arakelyan, V.M., and Shahnazaryan, G.E. (2005) Metal oxide photoelectrodes for hydrogen generation using solar radiation-driven water splitting. *Sol. Energy*, **78** (5), 581–592.

74. Seok, N.W., Eun, Y.K., and Young, H.G. (2008) Photocatalytic production of oxygen in a dual bed system using a reversible redox mediator on Ir-TiO$_2$ catalyst. *Korean J. Chem. Eng.*, **25**, 1355–1337.

75. Madan, A. (2008) Photoelectrochemical Hydrogen Production: MV Systems Incorporated. US DOE Hydrogen Program, 2008 Annual Progress Report. II.E.9, http://www.hydrogen.energy.gov/pdfs/progress08/ii_e_9_madan.pdf (accessed 2010).

76. Choi, K.-S. (2010) Shape effect and shape control of polycrystalline semiconductor electrodes for use in photoelectrochemical cells. *J. Phys. Chem. Lett.*, **1** (15), 2244–2250.

77. Pradhan, B., Batabyal, S.K., and Pal, A.J. (2007) Vertically aligned ZnO nanowire arrays in Rose Bengal-based dye-sensitized solar cells. *Sol. Energy Mater. Sol. Cells*, **91** (9), 769–773.

78. Greene, L.E., Law, M., Yuhas, B.D., and Yang, P. (2007) ZnO-TiO2 core-shell nanorod/P3HT solar cells. *J. Phys. Chem. C*, **111** (50), 18451–18456.

79. Miller, E.L., Paluselli, D., Marsen, B., and Rocheleau, R. (2004) Final Report of Photoelectrochemical-Hydrogen-Production for Reporting Period: 05/01/2000 – 06/30/2004 DE- FC36-00GO10538 presented at Hawaii Natural Energy Institute, University of Hawaii, Manoa, pp. 1–13, web address: http://www.scribd.com/doc/42288251/Photoelectrochemical-Hydrogen-Production-Final-Report (accessed 2010).

80. Papageorgiou, N. (2004) Counter-electrode function in nanocrystalline photoelectrochemical cell configurations. *Coord. Chem. Rev.*, **248** (13), 1421–1446.

81. Kibria, M.F., Mridha, M.S., and Khan, A.H. (1995) Electrochemical studies of a nickel electrode for the hydrogen evolution reaction. *Int. J. Hydrogen Energy*, **20** (6), 435–440.

82. Weigert, E.C., Stottlemyer, A.L., Zellner, M.B., and Chen, J.G. (2007) Tungsten monocarbide as potential replacement of platinum for methanol electrooxidation. *J. Phys. Chem. C*, **111** (40), 14617–14620.

83. Weigert, E.C., Esposito, D.V., and Chen, J.G. (2009) Cyclic voltammetry and X-ray photoelectron spectroscopy studies of electrochemical stability of clean and Pt-modified tungsten and molybdenum carbide (WC and Mo2C) electrocatalysts. *J. Power Sources*, **193** (2), 501–506.

84. Rottkay, K.V. and Rubin, M. (1996) Optical indices of pyrolytic tin oxide glass. *Mater. Res. Symp. Proc.*, **426**, 449–456.

85. Lewis, B.G. and Paine, D.C. (2000) Applications and processing of transparent conducting oxides. *MRS Bull.*, **25**, 22–27.

86. Hitosugi, T., Ueda, A., Nakao, S., Yamada, N., Furubayashi, Y., Hirose, Y., Shimada, T., and Hasegawa, T. (2007) Fabrication of highly conductive Ti[sub 1 - x]Nb[sub x]O[sub 2] polycrystalline films on glass substrates via crystallization of amorphous phase grown by pulsed laser deposition. *Appl. Phys. Lett.*, **90** (21), 212106.

87. Beermann, N., Vayssieres, L., Lindquist, S.-E., and Hagfeldt, A. (2000) Photoelectrochemical studies of oriented nanorod thin films of hematite. *J. Electrochem. Soc.*, **147** (7), 2456–2461.

88. Tsuji, I., Kato, H., Kobayashi, H., and Kudo, A. (2004) Photocatalytic H2 evolution reaction from aqueous solutions over band structure-controlled (AgIn)xZn2(1-x)S2 solid solution photocatalysts with visible-light response and their surface nanostructures. *J. Am. Chem. Soc.*, **126** (41), 13406–13413.

89. Luo, W., Yu, T., and Wang, Y. (2007) Enhanced photocurrent voltage

characteristics of WO_3/Fe_2O_3 nanoelectrodes. *J. Phys. D: Appl. Phys.*, **40** (4), 1091–1096.
90. Miyake, H. and Kozuka, H. (2005) Photoelectrochemical properties of Fe_2O_3-Nb_2O_5 films prepared by sol-gel method. *J. Phys. Chem. B*, **109** (38), 17951–17956.
91. Prakasam, H.E., Varghese, O.K., and Paulose, M. (2006) Synthesis and photoelectrochemical properties of nanoporous iron (III) oxide by potentiostatic anodization. *Nanotechnology*, **17** (17), 4285–4291.
92. Gregg, B.A., Pichot, F., Ferrere, S., and Fields, C.L. (2001) Interfacial recombination processes in dye-sensitized solar cells and methods to passivate the interfaces. *J. Phys. Chem. B*, **105** (7), 1422–1429.
93. Nakade, S., Makimoto, Y., Kubo, W., Kitamura, T., Wada, Y., and Yanagida, S. (2005) Roles of electrolytes on charge recombination in dye-sensitized TiO_2 Solar Cells (2): The case of solar cells using cobalt complex redox couples. *J. Phys. Chem. B*, **109**, 3488–3493.
94. Cameron, P.J., Peter, L.M., Zakeeruddin, S.M., and Grätzel, M. (2004) Electrochemical studies of the Co(III)/Co(II)(dbbip)2 redox couple as a mediator for dye-sensitized nanocrystalline solar cells. *Coord. Chem. Rev.*, **248** (13), 1447–1453.
95. Gamstedt, H. (2005) Ionic liquids of organic sulphonium and imidazolium iodides. *Ionic Liquid Electrolytes for Photoelectrochemical Solar Cells*, Chapter 6, Doctoral Thesis, Universitetsservice US AB, Stockholm, pp. 69–89. (ISSN 0348-825X, ISBN 91-7178-122-6) http://www.dissertations.se/dissertation/15b4939e8f/.
96. Vayssieres, L. (ed.) (2009) *On Solar Hydrogen and Nanotechnology*, John Wiley & Sons (Asia) Pte Ltd, Singapore, p. 253.
97. Bak, T., Nowotny, J., Rekas, M., and Sorrell, C.C. (2005) Photo-electrochemical hydrogen generation from water using solar energy. Materials-related aspects. *Int. J. Hydrogen Energy*, **27**, 991–1022.
98. Wang, P., Zakeeruddin, S.M., Moser, J.E., Nazeeruddin, M.K., Sekiguchi, T., and Gratzel, M. (2003) A stable quasi-solid-state dye-sensitized solar cell with an amphiphilic ruthenium sensitizer and polymer gel electrolyte. *Nat. Mater.*, **2** (6), 402–407.
99. Cheng, K.W., Huang, J.C., Kuo, J.C., and Yen, P.S. (2004) The Study of Hydrogen Generation with Photoreactor, Hydrogen and Fuel Cells Conference and Trade Show, September, Canada.
100. Sivula, K., Zboril, R., Formal, F.L., Robert, R., Weidenkaff, A., Tucek, J., Frydrych, J., and Gratzel, M. (2010) Photoelectrochemical water splitting with mesoporous hematite prepared by a solution-based colloidal approach. *J. Am. Chem. Soc.*, **132** (21), 7436–7444.
101. Cheng, K.-W., Huang, C.-M., Pan, G.-T., Chen, P.-C., Lee, T.-C., and Yang, T.C.K. (2008) Physical properties of AgIn5S8 polycrystalline films fabricated by solution growth technique. *Mater. Chem. Phys.*, **108** (1), 16–23.
102. Kudo, A., Nakagawa, S., and Kato, H. (1999) Overall water splitting into H_2 and O_2 under UV irradiation on NiO load $ZnNb_2O_6$ photocatalysts consisting of d^{10} and d^0 ions. *Chem. Lett.*, **28** (11), 1197–1197.
103. Li, X.Z., Li, F.B., Yang, C.L., and Ge, W.K. (2001) Photocatalytic activity of WOx-TiO2 under visible light irradiation. *J. Photochem. Photobiol., A: Chem.*, **141** (2), 209–217.
104. Wu, C.-C., Cheng, K.-W., Chang, W.-S., and Lee, T.-C. (2009) Preparation and characterizations of visible light-responsive (Ag-In-Zn)S thin-film electrode by chemical bath deposition. *J. Taiwan Inst. Chem. Eng.*, **40** (2), 180–187.
105. Bard, A.J. and Fox, M.A. (1995) Artificial photosynthesis: solar splitting of water to hydrogen and oxygen. *Acc. Chem. Res.*, **28** (3), 141–145.
106. Huang, C.-M., Chen, L.-C., Pan, G.-T., Yang, T.C.K., Chang, W.-S., and Cheng, K.-W. (2009) Effect of Ni on the growth and photoelectrochemical properties of ZnS thin films. *Mater. Chem. Phys.*, **117** (1), 156–162.

107. Kudo, A. (2003) Photocatalyst materials for water splitting. *Catal. Surv. Asia*, **7** (1), 31–38.
108. Ye, J., Zou, Z., Arakawa, H., Oshikiri, M., Shimoda, M., Matsushita, A., and Shishido, T. (2002) Correlation of crystal and electronic structures with photophysical properties of water splitting photocatalysts $InMO_4$ ($M = V^{5+}, Nb^{5+}, Ta^{5+}$). *J. Photochem. Photobiol., A: Chem.*, **148** (1), 79–83.
109. Ichikawa, S. and Doi, R. (1997) Photoelectrocatalytic hydrogen production from water on transparent thin film titania of different crystal structures and quantum efficiency characteristics. *Thin Solid Films*, **292** (1), 130–134.
110. Ichikawa, S. (1997) Photoelectrocatalytic production of hydrogen from natural seawater under sunlight. *Int. J. Hydrogen Energy*, **22** (7), 675–678.
111. Kitano, M., Tsujimaru, K., and Anpo, M. (2006) Decomposition of water in the separate evolution of hydrogen and oxygen using visible light-responsive TiO_2 thin film photocatalysts: effect of the work function of the substrates on the yield of the reaction. *Appl. Catal., A: Gen.*, **314** (2), 179–183.
112. Grätzel, M. (2003) Tandem cell for water cleavage by visible light. European Patent EP 1 198 621.
113. Khaselev, O. and Turner, J.A. (1998) A monolithic photovoltaic-photoelectrochemical device for hydrogen production via water splitting. *Science*, **280** (5362), 425–427.
114. Esposito, D.V., Dobson, K.D., McCandless, B.E., Birkmire, R.W., and Chen, J.G. (2009) Comparative study of tungsten monocarbide and platinum as counter electrodes in polysulfide-based photoelectrochemical solar cells. *J. Electrochem. Soc.*, **156** (8), B962–B969.
115. Murakami, T.N. and Gratzel, M. (2008) Counter electrodes for DSC: application of functional materials as catalysts. *Inorg. Chim. Acta*, **361** (3), 572–580.
116. Esposito, D.V., Goue, O.Y., Dobson, K.D., McCandless, B.E., Chen, J.G.G., and Birkmire, R.W. (2009) A new photoelectrochemical test cell and its use for a combined two-electrode and three-electrode approach to cell testing. *Rev. Sci. Instrum.*, **80** (12), 125107.
117. Osterwald, C.R., Anevsky, S., Bucher, K., Barua, A.K., Chaudhuri, P., Dubard, J., Emery, K., Hansen, B., King, D., Metzdorf, J., Nagamine, F., Shimokawa, R., Wang, Y.X., Wittchen, T., Zaaiman, W., Zastrow, A., and Zhang, J. (1999) The world photovoltaic scale: an international reference cell calibration program. *Prog. Photovolt.*, **7** (4), 287–297.
118. Mullejans, H., Zaaiman, W., Merli, F., Dunlop, E.D., and Ossenbrink, H.A. (2005) Comparison of traceable calibration methods for primary photovoltaic reference cells. *Prog. Photovolt.*, **13** (8), 661–671.
119. Chouhan, N., Yeh, C.L., Hu, S.F., Huang, J.H., Tsai, C.W., Liu, R.S., Chang, W.S., and Chen, K.H. (2010) Array of CdSe QD-sensitized ZnO nanorods serves as photoanode for water splitting. *J. Electrochem. Soc.*, **157** (10), B1430–B1433.
120. Chouhan, N., Yeh, C.L., Hu, S.F., Liu, R.S., Chang, W.S., and Chen, K.H. (2011) CdSe(QDs)@ ZnO nanotubes as an effective photoanode. *Chem. Commun.*, **47**, 3493–3495.
121. Cardon, F. and Gomes, W.P. (1978) Determination of flat-band potential of a semiconductor in contact with a metal or an electrolyte from mott-schottky plot. *J. Phys. D: Appl. Phys.*, **11** (4), L63–L67.
122. Yang, X.Y., Wolcott, A., Wang, G.M., Sobo, A., Fitzmorris, R.C., Qian, F., Zhang, J.Z., and Li, Y. (2009) Nitrogen-doped ZnO nanowire arrays for photoelectrochemical water splitting. *Nano Lett.*, **9** (6), 2331–2336.
123. Mora-Sero, I., Fabregat-Santiago, F., Denier, B., Bisquert, J., Tena-Zaera, R., Elias, J., and Levy-Clement, C. (2006) Determination of carrier density of ZnO nanowires by electrochemical techniques. *Appl. Phys. Lett.*, **89** (20), 203117.
124. Schottky, W., (1942) Vereinfachte und erweiterte Theorie der Randschichtgleichrichter. *Z. Phys.*, **118**, 539.

125. Oscar, K. and Turner, J.A. (1998) A monolithic photovoltaic- photoelectrochemical cell device for hydrogen production via water splitting. *Science*, **280**, 425.
126. Deutsch, T.G., Koval, C.A., and Turner, J.A. (2006) III-V nitride epilayers for photoelectrochemical water splitting: GaPN and GaAsPN†. *J. Phys. Chem. B*, **110** (50), 25297–25307.
127. Turner, J.A. (2007) Photoelectrochemical Water Splitting, http://www.nrel.gov (accessed 2010).
128. Rocheleau, R.E., Miller, E.L., and Misra, A. (1998) High-efficiency photoelectrochemical hydrogen production using multijunction amorphous silicon photoelectrodes. *Energy Fuels*, **12** (1), 3–10.
129. Gaillard, N., Chang, Y., Kaneshiro, J., Deangelis, A., and Miller, E.L. (2010) Proceedings of the SPIE Conference on Solar Hydrogen and Nanotechnology, p. 7770.
130. (a) Jahagirdar, A.H. and Dhere, N.G. (2007) Photoelectrochemical water splitting using CuIn1-xGaxS2/CdS thin-film solar cells for hydrogen generation. *Sol. Energy Mater. Sol. Cells*, **91** (15–16), 1488–1491; (b) Jahagirdar, A.H., Kadam, A.A., and Dhere, N.G. (2006) Proceedings of the 4th World Photovoltaic Solar Energy Conference, Hawaii, pp. 557–559.
131. Alex, S., Augusto, K., Jian, H., Ari, F., Bjorn, M., Brain, C., Eric, L.M., and Arun, M. (2006) Proceedings of the SPIE Conference on Solar Hydrogen and Nanotechnology, p. 6340.
132. Feng, Z., Jian, H., Augusto, K., Ilvydas, M., Bjorn, M., Brian, C., Eric, M., and Arun, M. (2007) Proceedings of the SPIE Conference on Solar Hydrogen and Nanotechnology, p. 6650.
133. Jian, H., Feng, Z., Ilvydas, M., Augusto, K., Todd, D., Eric, M., and Arun, M. (2008) Proceedings of the 23rd European Photovoltaic Solar Energy Conference.
134. Takabayashi, S., Nakamura, R., and Nakato, Y. (2004) A nano-modified Si/TiO$_2$ composite electrode for efficient solar water splitting. *Photochem. Photobiol., A: Chem.*, **166** (1–3), 107–113.
135. Zhu, G. and Zhao, Q. (2010) *Anti-scratch Material for Surgical Instruments*, Electronic Engineering, Physics and Renewable Energy Division and Mechanical Engineering and Mechatronics Division, University of Dundee, Scotland, UK, pp. 44–62. http://www.dundee.ac.uk/elecengphysics/Cat%20-20pages%20complete%20pdf.pdf (accessed March 31, 2010).
136. Uchida, Y. and Sakai, H. (1986) Light induced effects in a-Si:H films and solar cells. *Mat. Res. Soc. Symp. Proc.*, **70**, 577–586.
137. Kolodziej, A., (2004) Staebler-Wronski effect in amorphous silicon and its alloys. *Opto-Electron. Rev.*, **12**, 21–32.
138. Stutzmann, M. (1992) Metastability in amorphous and microcrystalline semiconductors, in *Amorphous and Microcrystalline Semiconductor Devices: Materials and Device Physics* (ed. J. Kanicki), Artech House, Norwood, MA, pp. 129–187.
139. Carlson, D.E., Chen, L.F., Ganguly, G., Lin, G., Middya, A.R., Crandall, R.S., and Reedy, R. (1999) A comparison of the degradation and annealing kinetics in amorphous silicon and amorphous silicon-germanium solar cells. *Mat. Res. Soc. Symp. Proc.*, **557**, 395–400.
140. Schrauzer, G.N. and Guth, T.D. (1977) Photocatalytic reactions. 1. Photolysis of water and photoreduction of nitrogen on titanium dioxide. *J. Am. Chem. Soc.*, **99** (22), 7189–7193.
141. Kawai, T. and Sakata, T. (1980) Photocatalytic decomposition of gaseous water over TiO$_2$ and TiO$_2$–RuO$_2$ surfaces. *Chem. Phys. Lett.*, **72** (1), 87–89.
142. Vandamme, H. and Hall, W.K. (1979) Photoassisted decomposition of water at the gas-solid interface on TiO$_2$. *J. Am. Chem. Soc.*, **101** (15), 4373–4374.
143. Domen, K., Naito, S., Soma, M., Onishi, T., and Tamaru, K. (1980) Photocatalytic decomposition of

water-vapor on an NiO-SRTIO$_3$ catalyst. *J. Chem. Soc., Chem. Commun.*, (12), 543–544.
144. Choi, W., Termin, A., and Hoffmann, M.R. (1994) The role of metal ion dopants in quantum-sized TiO$_2$: correlation between photoreactivity and charge carrier recombination dynamics. *J. Phys. Chem.*, **98** (51), 13669–13679.
145. Xu, A.W., Gao, Y., and Liu, H.Q. (2002) The preparation, characterization, and their photocatalytic activities of rare-earth-doped TiO$_2$ nanoparticles. *J. Catal.*, **207** (2), 151–157.
146. Yin, J., Ye, J.H., and Zou, Z.G. (2004) Enhanced photoelectrolysis of water with photoanode Nb:SrTiO$_3$. *Appl. Phys. Lett.*, **85** (4), 689–691.
147. Asahi, R., Morikawa, T., Ohwaki, T., Aoki, K., and Taga, Y. (2001) Visible-light photocatalysis in nitrogen-doped titanium oxides. *Science*, **293** (5528), 269–271.
148. Ashokkumar, M. and Maruthamuthu, P. (1991) Photocatalytic hydrogen production with semiconductor particulate systems: an effort to enhance the efficiency. *Int. J. Hydrogen Energy*, **16** (9), 591–595.
149. Hodes, G., Manassen, J., and Cahen, D. (1976) Photoelectrochemical energy-conversion and storage using polycrystalline chalcogenide electrodes. *Nature*, **261** (5559), 403–404.
150. Ellis, A.B., Kaiser, S.W., Bolts, J.M., and Wrighton, M.S. (1977) Study of n-type semiconducting cadmium chalcogenide-based photoelectrochemical cells employing polychalcogenide electrolytes. *J. Am. Chem. Soc.*, **99** (9), 2839–2848.
151. Frank, A.J. and Honda, K. (1982) Visible-light-induced water cleavage and stabilization of n-type cadmium sulfide to photocorrosion with surface-attached polypyrrole-catalyst coating. *J. Phys. Chem.*, **86** (11), 1933–1935.
152. Honda, K. and Frank, A.J. (1984) Polymer-catalyst modified cadmium sulfide photochemical diodes in the photolysis of water. *J. Phys. Chem.*, **88**, 5587–5582.
153. Frank, A.J., Glenis, S., and Nelson, A.J. (1989) Conductive polymer-semiconductor junction: characterization of poly(3-methylthiophene):cadmium sulfide based photoelectrochemical and photovoltaic cells. *J. Phys. Chem.*, **93** (9), 3818–3825.
154. Bhattacharya, C. and Datta, J. (2005) Studies on anodic corrosion of the electroplated CdSe in aqueous and non-aqueous media for photoelectrochemical cells and characterization of the electrode/electrolyte interface. *Mater. Chem. Phys.*, **89** (1), 171–175.
155. Fujii, K., Karasawa, T., and Ohkawa, K. (2005) Hydrogen gas generation by splitting aqueous water using n-type GaN photoelectrode with anodic oxidation. *Jpn. J. Appl. Phys.*, **44**, L543–L545.
156. Beach, J.D., Collins, R.T., and Turner, J.A. (2003) Band-edge potentials of n-type and p-type GaN. *J. Electrochem. Soc.*, **150** (7), A899–A904.
157. Menezes, S. and Miller, B. (1983) Surface and redox reactions at GaAs in various electrolytes. *J. Electrochem. Soc.*, **130** (2), 517–523.
158. Khader, M.M. (1996) Surface arsenic enrichment of n-GaAs photoanodes in concentrated acidic chloride solutions. *Langmuir*, **12** (4), 1056–1060.
159. Ginley, D.S., Baughman, R.J., and Butler, M.A. (1983) BP-stabilized n-Si and n-GaAs photoanodes. *J. Electrochem. Soc.*, **130** (10), 1999–2002.
160. Allongue, P. and Cachet, H. (1988) Charge transfer and stabilization at illuminated n-GaAs/aqueous electrolyte junction. *Electrochim. Acta*, **33**, 79–87.
161. Yener, D.O. and Giesche, H. (2001) Synthesis of pure and manganese-, nickel-, and zinc-doped ferrite particles in water-in-oil microemulsions. *J. Am. Ceram. Soc.*, **84** (9), 1987–1995.
162. Baba, R., Nakabayashi, S., Fujishima, A., and Honda, K. (1985) Investigation of the mechanism of hydrogen evolution during photocatalytic water decomposition on metal-loaded semiconductor powders. *J. Phys. Chem.*, **89** (10), 1902–1905.
163. Ellis, A.B., Bolts, J.M., Kaiser, S.W., and Wrighton, M.S. (1977) Study of

n-type gallium arsenide- and gallium phosphide-based photoelectrochemical cells. Stabilization by kinetic control and conversion of optical energy into electricity. *J. Am. Chem. Soc.*, **99**, 2948–2853.

164. (a) Miller, E.L. and Rocheleau, R.E. (1999) Photoelectrochemical hydrogen production, in *Proceedings of 1999 the U.S. Department of Energy Hydrogen Program Annual Review Meeting* (ed. C. Elam), U.S. Department of Energy, Lakewood, CO pp. 1–17; (b) Rocheleau, R., Miller, E. and Misra, A. (1998) High-efficiency photoelectrochemical hydrogen production using multijunction amorphous silicon photoelectrodes. *Energy and Fuels*, **12**, 3–10.

165. Vogt, U. (2010) *Hydrogen Technology in the Self Sufficient and Substainable Space Unit*. (ed. A. Züttel), Hydrogen report Switzerland 2010/2011 Major Achievements published by Hydropole + (Wasserstoff Hydrogene Idrogeno Hydrogen) May 12, 2010, pp. 30–31. *http://issuu.com/hydropole/docs/hsr11* (accessed July, 2010).

166. Wingens, J., Krost, G., (German Member, IEEE) Ostermann, D., Damm, U., and Hess, J. (2008) Hydrogen Production for Autonomous Solar Based Electricity Supply, DRPT2008, Nanjing China, April 6–9, 2008.

167. Hollmuller, P., Joubert, J.-M., Lachal, B., and Yvon, K. (2000) Evaluation of a 5 kWp photovoltaic hydrogen production and storage installation for a residential home in Switzerland. *Int. J. Hydrogen Energy*, **25** (2), 97–109.

168. Matics, J. and Krost, G. (2007) Prospective and adaptive management of small combined heat and power systems in buildings. Proceedings of the 9th REHVA World Congress CLIMA, Helsinki, Finland, June, 2007.

169. Yvon, K., and Lorenzoni, J.L. (2006) Hydrogen-powered lawn mower: 14 years of operation. *Int. J. Hydrogen Energy*. **31** (12), 1763–1767.

13
Polymer Electrolyte Membrane Fuel Cells

Stefania Specchia, Carlotta Francia, and Paolo Spinelli

13.1
Introduction to PEMFCs

Fuel cells (FCs) are energy conversion devices that directly transform the energy stored in a fuel into electricity and heat, and in which the fuel is not burned in a flame but oxidized electrochemically. There are several types of FCs depending on the operation principles and conditions. Proton electrolyte membrane fuel cells (PEMFCs) are based on a proton-conducting polymer electrolyte membrane (PEM).

The basic operating principle of a PEMFC is illustrated in Figure 13.1. A PEMFC consists of two porous electrodes, with a proton-conducting PEM between them. The hydrogen at the anode gives up electrons to the electrode and enters the electrolyte as a positive ion (H^+), while the oxygen at the cathode takes electrons and reduces to negative ions (O_2^- or OH^-). The respective ions combine to form water at the cathode, while the electrons move through the external circuit to produce an electric current. The involved reactions, both accelerated by platinum-based catalysts, can be written as follows:

$$\text{Anode Reaction}: 2H_2 \longrightarrow 4H^+ + 4e^-$$
$$\text{Cathode Reaction}: O_2 + 4H^+ + 4e^- \longrightarrow 2H_2O$$
$$\text{Overall Cell Reaction}: 2H_2 + O_2 \longrightarrow 2H_2O$$

FCs have been known in science for more than 150 years. Sir William Robert Grove (1811–1896), a Welsh lawyer and scientist, developed an improved wet-cell battery in 1838. The "Grove cell" used a platinum electrode immersed in nitric acid and a zinc electrode immersed in zinc sulfate to generate about 12 A of current at about 1.8 V. Grove discovered that, if two platinum electrodes were arranged so that one end of each was immersed in a container of sulfuric acid and the other ends separately sealed in containers of oxygen and hydrogen, a constant current would flow between the electrodes. The sealed receptacles contained water as well as the gases, and he noted that the water level rose in both tubes as the current flowed. Grove published his first results in the *Philosophical Magazine and Journal*

Electrochemical Technologies for Energy Storage and Conversion, First Edition. Edited by Ru-Shi Liu, Lei Zhang, Xueliang Sun, Hansan Liu, and Jiujun Zhang.
© 2012 Wiley-VCH Verlag GmbH & Co. KGaA. Published 2012 by Wiley-VCH Verlag GmbH & Co. KGaA.

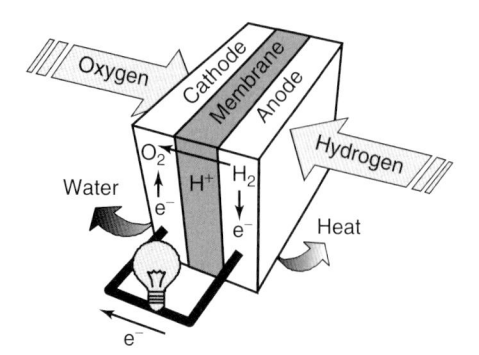

Figure 13.1 Schematic operating principle of a PEM FC.

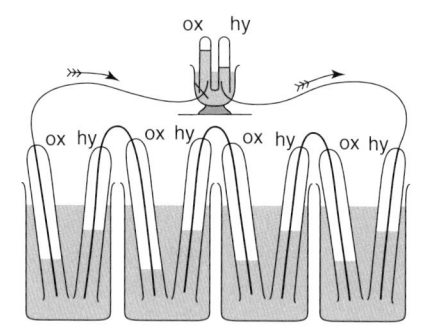

Figure 13.2 First scheme of an FC, published by Sir William Robert Grove in 1842 [3].

of Science in 1839 [1, 2]. Using his research, and the knowledge that electrolysis used electricity to split water into hydrogen and oxygen (first described by the British scientists William Nicholson and Anthony Carlisle in 1801), he concluded that the opposite reaction must be capable of producing electricity. On the basis of this hypothesis and by combining several sets of platinum electrodes in a series circuit, Grove developed a device that could combine hydrogen and oxygen to produce electricity. This device, which he named "gas battery," and which is shown in Figure 13.2, was the first FC, and it was made in 1842 [3, 4].

In 1893, Friedrich Wilhelm Ostwald (1853–1932), one of the founders of physical chemistry, experimentally determined the connected roles of the various components of the FC: electrodes, electrolyte, oxidizing and reducing agents, anions, and cations, thus providing much of the theoretical understanding of how FCs operate. Grove speculated that the action in his gas battery occurred at the point of contact between the electrode, gas, and electrolyte [5, 6].

In the nineteenth century, the gas battery stimulated research and the formulation of theories, but no practical device emerged, although several attempts were made in this field. Recent PEM technology was first ascertained at General Electric in the early 1960s through the work of Thomas Grubb and Leonard Niedrach. General Electric announced an initial success in the mid-1960s when the company developed a small FC within a project for the US Navy's Bureau of Ships and the US Army Signal Corps. The unit was fueled by hydrogen, which was generated by mixing water and lithium hydride. This fuel mixture was contained in disposable canisters that could be easily supplied to the operational staff. The cell was compact and portable, but its platinum catalysts were expensive. The first commercial use of an FC was conceived by General Electric in cooperation with the National Aeronautics and Space Administration (NASA) and McDonnell Aircraft. PEM technology was, in fact, involved in NASA's Gemini Project in the early days of the US piloted space program. Batteries had provided spacecraft power in the earlier Mercury Project missions. Gemini's main objective was to test the equipment and procedures for the Apollo mission, and any missions lasting up to 14 days included operational tests on FCs. General Electric's PEMFCs were selected, but the first models frequently encountered technical difficulties, including internal cell contamination and oxygen leakage through the membrane. The PEMFCs were redesigned and a new model, despite various malfunctions and poor performance, served adequately for the remaining Gemini flights. However, alkali FCs were preferred for the subsequent Apollo missions for both the command and lunar modules [7–11].

Work on the science of FCs continues: current work concerns the development of better materials and more efficient designs in order to reduce costs and increase their durability, and consequently their commercialization.

13.2
Main Types and Structures of PEMFCs

A PEMFC is a galvanic system that is capable of converting the energy associated with a chemical reaction into electrical energy. PEMFCs usually operate in a temperature range of 60–80 °C and the theoretical voltage of a single cell, in standard conditions, is about 1.23 V. The cell consists of an anode and a cathode, which are externally supplied by the gases, the fuel, and the oxidant, respectively, and a solid polymer electrolyte that separates the electrodes and provides ionic conduction. This system is known as the membrane electrode assembly (MEA) and is the "core" of the PEMFC. There are flow fields outside the MEA that direct the flow of gaseous reactants along the surface of the electrodes. The performance of PEMFCs depends on many different parameters, such as temperature, pressure of the gas stream, and relative humidity. The kinetics of the electrochemical reaction, involving the oxidation of hydrogen (the fuel) and the reduction of oxygen (the oxidant) to produce electricity, is promoted by the presence of a catalyst inside the electrodes. Thus, a PEMFC consists of porous gas diffusion electrodes (GDEs) in which the reactants are supplied to the active centers, where a noble metal catalyst must simultaneously be in contact with the ionic and electronic conductors [12].

As described in Chapter 2, the practical efficiency of an FC is always lower than the theoretical one (usually about 50%) because of a number of losses due to irreversible phenomena associated with the rate of the overall reaction, which is proportional to the current intensity. Such losses result in a decrease in the cell voltage V(i) and the energy efficiency can therefore be assumed as the ratio of the cell voltage and the theoretical reversible voltage under the same temperature and pressure conditions. Voltage losses are frequently called *"polarization effects"* and can be analyzed or experimentally determined as activation polarizations (electrochemical kinetics), ohmic losses (electrical resistance), and mass transport polarizations. The total loss is the sum of each of these terms and markedly increases with the current intensity of the cell [12].

Polarization losses are above all due to the slow kinetics of the reduction of oxygen at the cathode; in addition, internal resistances arise as a consequence of the ionic and electronic connections between the electrolyte and the electrodes and within the electrode material and the cell, respectively. Finally, a cell operating at high current densities is subjected to a series of mass transport or diffusion limitations [13, 14]. These limitations can be observed by following the trend of the potential of the FC with the current load.

The core of the PEMFC is the MEA, which is basically an assembly of the electrocatalysts (anodic and cathodic) and the polymeric membrane. The solid polymeric electrolyte is an ion-exchange membrane that is capable of transferring protons. In this particular system, the polymer electrolyte not only plays the role of ionic conductor (and of electronic insulator), as occurs in conventional batteries, but must also possess the characteristics of a separator, which means low permeability to the gas flows. A membrane for PEMFC applications should satisfy many other requirements, ranging from good chemical and thermal stability at quite high temperatures, to low water drag, good mechanical properties such as strength and flexibility, to low cost and easy availability. Nevertheless, PEMFC membranes suffer from attacks from radicals, strong oxidants, and temperature fluctuations [15]. Perfluorinated sulfonic acid (PFSA) polymers are the best type of membrane currently available, and the perfluorosulfonic acid membrane, Nafion®, manufactured by the DuPont Co., stands out as standard target for PEMFC applications [16]. Similar proton-conducting membranes developed by the Dow Chemical Company and the Asahi Chemical Company were also employed in PEMFCs showing higher performances with respect to Nafion; but the latter still remains the best known and most popular membrane for its lower cost and easy availability combined with a durability of 60 000 h at 80 °C [17]. Nafion possesses an aliphatic perfluorinated backbone with ether-linked side chains ending in sulfonate cation exchange sites [18]. The backbone of this structure, which resembles Teflon, imparts a hydrophobic nature to the membrane and gives it its long-term stability toward oxidative and reductive environments. Ion conduction in Nafion (and generally in PFSA) membranes is a direct consequence of its hydrated state, since the polymer does not have the ability to transfer protons along its side chains in the dried state. A well-hydrated Nafion membrane shows an ionic conductivity > 0.1 S cm^{-1}. As the water content falls, the conductivity also falls almost linearly. During PEMFC operation, an almost dry membrane is more exposed to temporary or permanent material degradation, which limits the overall performances and durability. In many PEMFC applications, the direct consequence of the conduction behavior of the PFSA (polyperfluorosulfonic acid) membranes implies humidification of the fuel and oxidant before the cell inlet. Although external humidification of the reactants has become a common practice to ensure the hydration of the electrolyte, it requires a heat supply for the humidifier and a careful control of the heat management, which complicate the overall PEMFC system [19]. Furthermore, possible accumulation of liquid water in the porous gas diffusion layers (GDLs) can lead to electrode flooding and gas starvation phenomena. The use of thin polymeric membrane electrolytes (25–50 μm thick) usually improves the water management, as it prevents anode dehydration at high current densities but, unfortunately, these thin membranes are not recommended for some PEMFC applications, such as in the automotive field, in which the drastic load cycles can increase the probability of mechanical failure [20]. Apart from the high cost of Nafion membranes, which amounts to US$700 m^{-2} [17], the maintenance of a high level of hydration inside the polymer structure has led to considerable research efforts directed toward developing new polymeric materials for PEMFC applications, an aspect which we will discuss later on.

As previously mentioned, PEMFC electrodes are porous electrodes (GDEs) which contain an electrocatalyst. Electrocatalysts usually consist of platinum or platinum alloy nanoparticles dispersed on a high-surface-area carbon support. In this way, the active catalytic surface area per unit mass of platinum increases together with the reaction rate, while low levels of platinum are maintained at the electrodes (loadings of about 0.6–0.8 mg$_{Pt}$ cm$^{-2}_{MEA}$) [21]. The most important requirement of a PEMFC electrode is the accomplishment of a three-phase boundary (TPB) among the gas, the catalyst particles, and the ionic conductor. This ensures both ionic and electronic conductivity and an efficient catalyst utilization. For this reason, the carbon-supported platinum catalyst is generally bonded with solubilized Nafion ionomer to form a catalyst ink which is spread on both sides of the membrane to form the catalytic layer (CL). This ensures good contact between the catalyst and the ionomer which is in ionic contact with the membrane. Since water is produced at the cathode during the electrochemical reaction, electrodes must have a certain degree of hydrophobicity in order to avoid flooding. This is usually achieved by employing porous teflonized carbon backing layers (or gas diffusion layers, GDLs), which are hot pressed onto the membrane, and covered by the catalyst ink, to form the MEA. Gas diffusion occurs through the GDL and the CL, i.e., from the carbon support to the catalyst platinum particles. Single cells are connected in series to obtain the required voltage. In PEMFCs, the connection of individual cells is achieved through the use of bipolar plates (BPs) with flow fields that allow gases to pass along and inside the electrodes. A BP not only has the function of a current collector, but it must also provide a uniform transport of the reactants through its channels and keep the hydrogen and air separate while preventing leakage. Graphite is still the standard material for BP production because of its high electronic conductivity. Flow-field production on graphite plates implies an extremely costly and long machining procedure followed by resin impregnation in order to reduce the permeability of the gases inside the graphite.

Apart from cost and management issues (dependence of electrolyte conductivity on the cell temperature and relativity humidity, possible flooding or delamination of the electrodes, corrosion, aging of the components), another problem limits the widespread use of this technology on the market and this is related to the fuel and oxidant feeding of the stack. Even trace amounts of CO, sulfur, or ammonia in the fuel result in severe contamination (poisoning) of the platinum electrocatalyst. It is known that chemisorbed CO on active Pt sites decreases the kinetics of the hydrogen oxidation reaction (HOR) and increases the electrode overpotential.

At 80 °C, low levels of CO (10–20 ppm) dramatically reduce the platinum electrocatalytic activity [22]. Ammonia levels as low as 13 ppm in the fuel affect the PEMFC performance [23]. It has also been reported that 1 ppm of H_2S at the anode of a PEMFC operating at a constant voltage of 0.5 V caused severe loss after 21 h of exposure, but H_2S poisoning effects were already noticed after only 4 h of operation [23]. Furthermore, the air that feeds the cathode can contain several impurities, such as sulfur dioxide, nitrogen oxides, salts, and particulates in different proportions. As far as hydrogen is concerned, although it is an attractive fuel for PEMFCs, and the most abundant element in the universe, it is unfortunately not present in the free form, but in various compounds, mainly water and hydrocarbons. Therefore, hydrogen is an energy carrier and must be extracted from natural resources [24, 25]. Because of the prohibitive production costs and the difficulty of not being readily available, the use of pure hydrogen and oxygen in FC technology cannot be considered realistic. Oxygen can be taken directly from the air, while hydrogen can be produced only indirectly through the reforming of natural gas, methanol gasoline, diesel, or any other kind of commercially available fuel [26, 27]. One of the consequences is that the overall system becomes more complex and expensive due to the additional gas clean-up units that are necessary for the reformate [28]. An alternative has been sought with the development of MEAs with contaminant-tolerant catalysts capable of sustaining high levels of impurities in the fuel stream. At present, the effectiveness of such electrocatalysts is not entirely satisfactory.

Finally, some considerations should be given to the heat management in PEMFCs working in the 60–80 °C temperature range, which are usually known as low-temperature (LT) PEMFCs. Waste heat is generated when the current flows through the FC and this accounts for about the 40–50% of the total energy produced [29]. Heat is generated as a result of the overvoltage at the electrodes, ohmic drop, and the entropy change associated with the chemical reaction. The entropic heat accounts for 55% of the total heat released, and the remaining 35 and 10% are due to irreversible reaction heating and ohmic heating, respectively [30]. The heat transfer is linked to water transport and phase changes in the PEMFCs. In most cases, an improper thermal management leads either to the dehydration of the electrolyte or to electrode flooding. The excess heat produced by the cell must be removed because temperature fluctuations could lead to an incorrect water balance. A PEMFC stack at 80 °C, with 40–50% of efficiency, produces a large amount of heat which can only be dissipated through the cooling system, while internal combustion engine vehicles dissipate less than 40% of the waste heat through the cooling system and the remaining 60% is removed with the exhaust gases. The cooling of LT PEMFC stacks is therefore a very complicated issue and, in most cases, requires heat exchangers with a large surface area [22]. In spite of the several difficulties that can be encountered when operating an LT PEMFC, this technology has been improved considerably in terms of materials, performance, and durability, especially for transport applications, where high efficiency, high power, fast startup, and zero emissions are important requirements.

Some of the problems mentioned above could be solved, if not entirely at least partially, if PEMFCs could work at temperatures above 100 °C. There is currently great interest in alternative materials that are able to extend the operating temperature range of PEMFCs, which would lead to an appreciable cost reduction of the overall system. These FCs are usually termed *high-temperature* (HT) PEMFCs. Ninety percent of the current research on HT PEMFCs is aimed at exploring alternatives to Nafion membranes that can be used in low relative humidity (25–50%) and HT (120–150 °C) operating conditions [31]. The ideal situation for these applications would be an HT membrane that delivers protons without involving water, while retaining the same chemical and mechanical stability as Nafion at low temperatures. The chemical structure of Nafion is in fact affected at temperatures of around 150 °C (for Nafion the glass transition temperature is at about 80–120 °C); if the temperature is increased above 200 °C, a loss in the number of sulfonate groups begins to occur [15]. One of the first strategies adopted to obtain alternative polymers was to modify Nafion. In most cases, the modified Nafion composite membranes showed enhanced water uptake compared to the unmodified counterpart. The disadvantages were related to the high production costs and to a low ionic conductivity. Thus, self-humidifying membranes were developed in order to increase the proton conductivity of the electrolyte in dry conditions. Other types of materials are non-fluorinated hydrocarbon polymers. Hydrocarbons are cheaper than PFSA membranes and are available on the market. Furthermore, those containing polar groups have shown a high water uptake over a large temperature range. The third kind of membranes developed for HT PEMFC applications are the acid–base complex polymers that are derived from the introduction of an acid component into an alkaline polymer base. These electrolytes showed enhanced proton conduction at high temperatures, along with good mechanical and thermal stability.

HT PEMFCs represent a challenge in the field of research of new materials for FC systems, but some other benefits may also be provided by extending the temperature range of operation. Advantages are obtained since the kinetics at both electrodes increase with temperature. On the other hand, the increase in the kinetics with temperature is also accompanied with a decrease in the reversible cell potential, as a result of the enhancement of the water partial pressure [22]. The tolerance to CO is also expected to increase at higher temperatures. Li *et al.* [32] investigated CO poisoning on carbon-supported platinum catalysts in PEMFCs in the 125–200 °C temperature range. They found a temperature dependence of CO adsorption and, by defining the CO tolerance as a voltage loss of less than 10 mV, also found that 3% CO in hydrogen can be tolerated at current densities of up to 0.8 A cm^{-2} at 200 °C, while 0.1% CO in hydrogen can be tolerated at current densities of less than 0.3 A cm^{-2} at 125 °C.

Improvements can also be achieved from the heat and water management point of view. Under high temperatures, the heat produced by the PEMFC can be recovered as steam, which in turn can be employed for direct heating, steam reforming, or pressurized operation [31]. On the other hand, the water of a PEMFC operating above 100 °C and at atmospheric pressure is steam. Thus, the transport of the reactants and water will be enhanced and there will be no flooding at the cathode side [22].

When the first membranes were employed for NASA's Gemini Space program, it was found that they operated for approximately 500–1000 h inside the FC. The analysis of the water exiting from the FC highlighted a great deal of degradation products [10, 11]. The same membranes had previously been used for electrodialysis, but only minimal degradation had been observed at the same temperature. This was the first indication of the hostile environment of the FC. A temperature increase makes it even more aggressive [15]. Thus, it is generally recognized that the durability issue of the materials and components of the PEMFC will be more stringent for HT PEMFCs.

13.3
Electrochemical Processes in PEMFCs

13.3.1
Brief Introduction on Electrochemical Kinetics

PEMFCs, like all types of FCs, are electrochemical devices that are used to convert the free energy change of a chemical reaction into electrical energy. Thus the basic operating principle is based on electrochemical reactions that occur within the system. The fundamental aspects of the thermodynamics and the kinetics of electrochemical reactions have been outlined in Chapter 2; here only some specific topics of the electrochemical processes will be considered in order to obtain a deeper insight into electrocatalysis, electrical performance and efficiency, and electrode degradation.

Because of the nature of the electrochemical system, which requires at least the presence of two separate contacts between an electronic conductor (the electrode) and an ionic conductor (the electrolyte), the most important feature of electrochemical reactions, compared to chemical reactions, is their heterogeneous character. This implies two important facts: (i) the reaction occurs at the surface of the two (or three) phase boundary because electrons, ions, and uncharged species are needed at the same site; and (ii) the kinetic mechanism, in addition to adsorption and chemical steps, includes one or more charge-transfer steps, whose activation energy depends on the electrode potential.

13.3.1.1 Activation Overpotential

In the case in which a single charge-transfer step is involved in the kinetic mechanism, the well known Butler–Volmer equation can be applied to describe the relationship between the current density i (which is proportional to the reaction rate) and the overpotential η (activation overpotential):

$$i = i_0 \left[\exp\left(\frac{\beta F}{RT}\eta\right) - \exp\left(-\frac{(1-\beta)F}{RT}\eta\right) \right] \qquad (13.1)$$

where i_0 is the exchange current density and β is the so-called symmetry factor, that is, the fraction of the overpotential that affects the anodic reaction. The other symbols have the usual meaning.

A more complex situation occurs when the whole reaction consists of multiple elementary reactions, including charge-transfer steps and chemical steps. The reaction rate (current density) is determined by the slowest step (rate-determining step, RDS). The derivation of the reaction rate for multistep reactions is usually complicated, but a similar equation to the previous one is obtained when the RDS is an electrochemical step:

$$i = i_0 \left[\exp\left(\frac{\alpha_{an} F}{RT}\eta\right) - \exp\left(-\frac{\alpha_{cat} F}{RT}\eta\right) \right] \qquad (13.2)$$

α_{an} and α_{cat} are the so-called charge-transfer coefficients for the anodic and the cathodic terms of the $i(\eta)$ equation. These coefficients include other parameters such as the symmetry factor and the number of electrons being transferred before and after the RDS. For an interesting discussion on this point, see [33].

For a simple charge transfer reaction of the type

$$\text{Ox} + n e^- \longrightarrow \text{Red}$$

where the oxidized species Ox is reduced to Red by n electrons, it is possible to write the rate of the forward (cathodic) reaction as

$$i_- = -nFk_- C_{ox} \exp\left[-\frac{\Delta G_-}{RT} - \frac{\alpha_{cat} F}{RT}\eta\right] \qquad (13.3)$$

Similarly, the rate of the backward (anodic) reaction is

$$i_+ = nFk_+ C_{red} \exp\left[-\frac{\Delta G_+}{RT} + \frac{\alpha_{an} F}{RT}\eta\right] \qquad (13.4)$$

where C_{ox} and C_{red} are the reacting species concentrations at the electrode surface, k_- and k_+ are the rate constants for the forward and backward reactions, ΔG_+ and ΔG_- are the activation energies for the forward and backward reactions when the electrode is at equilibrium ($\eta = 0$), and α_{an} and α_{cat} represent the fraction of the activation energy due to the change in the electrode potential. For a positive overpotential, this fraction decreases the activation energy of the forward (anodic) reaction as well as that of the backward (cathodic) reaction. It should be observed that, for a one-step, single-electron transfer reaction, $\alpha_{an} = \beta$ and $\alpha_{cat} = 1 - \beta$.

By considering Eqs. (13.2–13.4), and taking into account that $i = i_+ + i_-$, the following expression can be obtained for the value of the exchange current density i_0:

$$i_0 = (i_+)_{\eta=0} = (|i_-|)_{\eta=0} = nFk_+ C_{red} \exp\left(-\frac{\Delta G_+}{RT}\right)$$
$$= nFk_- C_{ox} \exp\left(-\frac{\Delta G_-}{RT}\right) \quad (13.5)$$

It can be seen that the exchange current density includes the concentration of reagents in the pre-exponential term and the activation energy in the exponential term. It must be observed that in Eq. (13.5) the concentration of the reagents at the reaction site (electrode surface) must be used. These surface concentrations differ from bulk concentration or pressure of reagents as a consequence of mass transfer processes. In most cases, one (or more) adsorption step occurs in the kinetic mechanism, thus the surface concentration of the reacting species is effectively described by the number (concentration) of active sites of the electrode surface. Since the adsorption of intermediate species, of products, and of possible poisoning substances decreases the number of active sites available for the charge transfer reaction, a coverage factor θ is usually introduced in the form of a function $f(\theta)$ in the pre-exponential term, depending on the applicable adsorption mechanism (Langmuir–Hinshelwood, Temkin, etc.). In simple cases, such a function can be expressed as θ/θ_0 or $(1-\theta)/(1-\theta_0)$, where θ_0 is the fractional surface coverage at equilibrium.

Another practical observation concerns the numerical value of i_0, which is usually computed as the apparent exchange current density (A cm^{-2}), by dividing the current intensity by the geometric surface area of the electrode. The true exchange current density refers to the real electrochemical active surface (EAS), which can be completely different from the geometrical one due to the electrode morphology and its constitution (roughness of the electrode surface, particle size, and distribution of the catalyst on the supporting material, etc.). In many instances, particularly for comparison purposes, the evaluation of EAS is not considered.

When the electrochemical reaction occurs sufficiently far from equilibrium, that is, when the reaction rate is rather high, one of the two exponential terms in Eq. (13.2) can be neglected. The second term is negligible for high positive overpotentials; conversely, the first term is close to zero for high negative overpotentials. A linear semilogarithmic plot η versus $\log(|i|)$ is obtained under such circumstances [Tafel equation: $\eta = a + b\log(|i|)$]. If the cathodic reaction is considered to be far from equilibrium, rearranging Eq. (13.2) and neglecting the first exponential term it is possible to obtain

$$\ln|i| = \ln i_0 - \frac{\alpha_{cat} F}{RT}\eta$$
$$\eta = \frac{RT}{\alpha_{cat} F}\ln i_0 - \frac{RT}{\alpha_{cat} F}\ln|i|$$
$$\eta = a + b\log|i|$$
$$b = -2.03 RT/\alpha_{cat} F, \, a = -b\log(i_0)$$

where the b parameter (Tafel slope) is usually computed by considering the logarithm to base 10 of the current density.

A similar derivation is obtained for the anodic reaction far from equilibrium, which provides

$$\eta = -\frac{RT}{\alpha_{an}F}\ln i_0 + \frac{RT}{\alpha_{an}F}\ln |i|$$

$$\eta = a + b\log|i|$$

$$b = 2.03 RT/\alpha_{an}F, a = -b\log(i_0)$$

The Tafel equation, which was introduced on an experimental basis, is important because when the activation overpotential exceeds all the other dissipation terms (concentration overpotential, ohmic drop), the two kinetic parameters, that is, the exchange current density i_0 and the charge-transfer coefficient α, can be obtained by plotting the experimental polarization curve on a semilogarithmic diagram.

The Tafel relationship is also frequently used in FC modeling because of its simplicity. One disadvantage of this approach is that, because of the logarithmic form of the current density, the activation overpotential cannot be computed for very low values of i. In fact, the Tafel equation only holds for $|\eta|$ values larger than about 150 mV.

13.3.2
Concentration Overpotential

Owing to the fact that the reacting species for the charge-transfer step need to reach the electrode surface from the bulk of the phase in which they are contained, mass transport phenomena may limit the overall reaction rate. These effects are usually considered in terms of the so-called concentration or diffusion overpotential.

The concentration overpotential arises from the concentration gradient that is established when a given current flows within an electrochemical cell because of the consumption of the reagents at the electrode surface and the consequent mass transfer (diffusional) mechanism that occurs to replace them. The simplest way of expressing this type of overpotential is to consider the value of the limiting current density that corresponds to the flux of the reagent supplied to the system. In the case of gaseous reagents, such as oxygen or hydrogen, if the flux (e.g., q_{H_2} mols^{-1}cm^{-2}) is known, the current corresponding to such a flux, $|i_l| = 2Fq_{H_2}$ A cm^{-2}, is the maximum possible value, that is, the limiting current density. If one considers only the concentration overpotential, by neglecting the other dissipation terms (namely the activation overpotential), the relationship between overpotential and current density is easily obtained:

$$|\eta_{conc}| = -\frac{RT}{nF}\ln\left(1 - \frac{i}{i_l}\right) \quad (13.6)$$

Equation (13.6) is derived for simplified systems in which the diffusion process of the reagent species occurs through a diffusion layer (e.g., in a liquid electrolyte) on a planar electrode surface. For three-dimensional (3D) porous electrodes, such as the

CLs of MEAs, or when the mass transport occurs through a multilayer structure, such an approach is oversimplified and Eq. (13.6) does not describe the mass transfer overvoltage correctly, and therefore accurate modeling of such situations have been proposed [34]. Having indicated the absolute value of the concentration overpotential, Eq. (13.6) is valid for both cathodic and anodic reactions, since the term in the logarithmic expression is always negative.

13.3.3
Open-Circuit Voltage (OCV)

It is known that the measured open-circuit voltage (OCV) of single PEMFCs markedly differs from that computed according to the equilibrium (Nernst) equation. OCV is usually lower than the cell reversible voltage by about 200 mV, but it depends on a number of factors, such as temperature, pressure of the gases, type and thickness of the membrane, relative humidity, and so on. Such a difference is attributable to the oxygen electrode, for which equilibrium conditions cannot be attained at open circuit (OC). The reasons for this behavior have been interpreted, from the basic studies of Tarasevich *et al.* [35], in terms of the parasitic reactions that are responsible for a mixed-potential condition. Among the various parasitic reactions that have been assumed to explain the attainment of the mixed potential, those that have received most attention are Pt oxidation, oxidation of organic impurities, and hydrogen peroxide reactions. In PEMFCs, in addition to such parasitic reactions, which occur independently of the hydrogen electrode, the main effect on the open circuit potential is related to the small amount of hydrogen that crosses the membrane, which is rapidly oxidized at the cathode, resulting in the so-called hydrogen crossover current [36]. The strong correlation between the OCV and hydrogen crossover through the membrane in LT PEMFCs has been the subject of numerous papers dealing with MEAs degradation (see, for example, Vilekar and Datta [37] and references therein). It is now accepted that the OCV values in such systems correspond to a mixed potential condition at the oxygen electrode, due to the very low value of the exchange current density for the oxygen reduction reaction (ORR). The aforementioned paper represents an extensive approach to the subject, and it encompasses the theoretical background of this complex situation.

13.3.4
Electrocatalysis

In its simplest definition, electrocatalysis can be considered as the heterogeneous catalysis of electrochemical reactions by electrode materials [38]. Thus, the enhancement of the electrochemical reaction rate is obtained as a consequence of the nature and surface structure of the electrode. A key point in understanding the electrocatalytic mechanism is that the activation energy, which is decreased by the (electro)catalyst, depends on the change in the electrode potential.

Even though there has been some dispute about which authors first introduced the term *"electrocatalysis"* into the scientific community (the Russian school in the

1930s), the term has been recognized and widely utilized starting from the 1960s, particularly in the field of FCs [39, 40].

Electrocatalysis has been investigated and advantageously applied in different engineering fields (industrial electrochemistry), but there is no doubt that one of the most studied reactions, with the aim of enhancing its kinetics, is the ORR, since it is the main cause of voltage losses in FCs and any action to increase FC efficiency is of outmost economic importance.

As mentioned in the introductory section, in order to increase the rate of an electrochemical reaction at a given overpotential, see Eq. (13.2), different actions can be undertaken: the increase in the exchange current density i_0, and the increase in the charge transfer coefficient α (which corresponds to a decrease in the Tafel slope b).

For the sake of clarity, the following should be distinguished (i) the action by which it is possible to increase the reaction rate by increasing the pre-exponential term in Eq. (13.5) and (ii) the action of increasing the exponential terms in both Eqs. (13.5) and (13.2). Case (ii) corresponds to a true electrocatalytic effect, as the enhancement of the reaction rate is obtained through a decrease in the activation energy, while case (i) is not referable to a catalytic mechanism. However, from a practical point of view, both situations are important and have been widely investigated to improve the so-called electrocatalytic performance. To take advantage of case (i) enhancement, the following aspects are considered: the increase in the true active surface of the electrode (e.g., by means of a better dispersion of the catalyst particles and a more effective contact between the catalyst particles and the support, possibly reducing the inactive fraction of the catalyst).

Since the introduction of PEMFCs in the 1960s, Pt-based electrocatalysts have been considered the most effective materials for such applications, because they are effective for both ORR and HOR [41]. It was known well before the 1960s that Pt was an excellent catalyst for many electrochemical reactions, see, for example, review paper [42].

At the present stage of the technology for LT PEMFC, Pt-based catalysts, particularly carbon supported Pt particles, are still considered the most effective and successful catalysts in spite of their well known drawbacks, such as their high costs, hydrogen peroxide formation, and their sensitivity to impurities and contaminants (CO, H_2S, organic compounds). All these aspects have been widely investigated in the literature and various Pt alloys or possible alternatives to Pt have been proposed. The number of papers published in the last few years on this subject is so high that it is not possible to provide a complete list here. We shall here limit ourselves to summarizing the specific issues, and to indicating some relevant reviews or specific papers.

13.3.5
ORR Mechanism and Electrocatalysis Issues

The kinetic mechanism for ORR is very complex compared to HOR, mainly because (i) the strength of the O–O bond is high, which leads to stable Pt–O or

Pt–OH species, (ii) the total process is a four-electron transfer reaction, and (iii) partially oxidized species (H_2O_2) can be formed. Because of this complex situation, ORR is known to have important kinetic limitations; the overall reaction

$$O_2 + 4H^+ + 4e^- \longrightarrow 2H_2O \tag{13.7}$$

could be kinetically subdivided into the following steps:

$$O_2 + H^+ + M + e^- \longrightarrow MHO_2$$
$$MHO_2 + H^+ + e^- \longrightarrow MO + H_2O$$
$$MO + H^+ + e^- \longrightarrow MOH$$
$$MOH + H^+ + e^- \longrightarrow M + H_2O$$

where M is the metal electrocatalyst, and MHO_2, MO, and MOH are intermediate oxidized species which form on the metal surface. The previous mechanism cannot be considered a real description of the reaction paths for two main reasons: (i) the chemical nature of the intermediate species must be defined and (ii) the path leading to hydrogen peroxide (H_2O_2) formation, which is particularly important for Pt, is not indicated. The interest in such a simplified description of the kinetic mechanism for ORR is due to the clear evidence of its complexity. In fact, even after about 60 years of intense research, no definitive conclusions concerning the mechanism for the intermediate and the RDSs for ORR on different electrocatalysts have been reached. For a deeper understanding of all the different aspects involved in ORR kinetics, see, for example [43].

Owing to its high catalytic activity, Pt has been the most extensively investigated electrode for ORR. As explained above, ORR kinetics is a multielectron process that consists of various elementary steps, involving different reaction intermediates. In addition to the four-electron reduction (Eq. (13.7)), a two-electron reduction can occur that leads to the formation of H_2O_2 [44, 45]:

$$O_2 + 2H^+ + 2e^- \longrightarrow H_2O_2$$

The two-electron reaction proceeds via the intermediate $(H_2O_2)_{ad}$, which can be further reduced to H_2O or form H_2O_2. Theoretical investigations on the ORR mechanism on Pt [46], based on the electronic structure of the metal, have shown that the formation of H_2O_2 depends on how O_2 is adsorbed on the metal surface. If a dissociative mechanism takes place, that is, atomic O is adsorbed as a consequence of the breaking of the O–O bond, then the reaction proceeds with the formation of water and no H_2O_2 is produced. If, on the contrary, an associative mechanism is considered, O_2 is adsorbed, the bond between the two oxygen atoms is not broken and H_2O_2 can be formed.

Another difficulty in understanding the kinetics of ORR in acidic electrolytes is related to the value of the charge-transfer coefficient α_{cat} in Eq. (13.2), which, as previously shown, is inversely proportional to the Tafel slope b. A Tafel slope of 60 mV dec^{-1} (at 25 °C) was experimentally observed on Pt at low current densities, indicating a RDS involving two electrons. For higher current densities, usually at

($V < 0.8$ V), the Tafel slope changes to 120 mV dec^{-1}, and a different mechanism should therefore be assumed with a one-electron RDS [43]. This two-slope behavior is observed by performing polarization curves on Pt electrodes in laboratory cells with a three-electrode configuration, while it is not clearly visible on polarization curves on FC, because of the importance of the other dissipative terms, which mask the change in slope. From a practical point of view, lower values of the Tafel slope and high values of the exchange current density are desirable to enhance the kinetics of ORR. It should be noted that the first action is much more difficult to control, while the second one, being related not only to the metal catalyst, but also to the value of the active surface of the electrode, has been the main subject of the numerous investigations on ORR electrocatalytic materials, which have taken into account, for example, the size and dispersion of the catalyst particles, the contact between the catalyst particles and the support, and different types and treatments of the carbon support.

In addition to Pt, the kinetics of ORR was extensively investigated in the past on other metals such as Au, Ir, Rh, and so on [47]. However, Pt has shown higher catalytic activity toward ORR and higher stability.

One promising area concerns the alloying of Pt with transition metals such as Fe, Co, Ni, Cr, Co–Cr, and Co–Ni. Possible explanations for the improvement in activity due to the addition of less noble metals to Pt have been discussed in the literature and they can be summarized as: (i) the lowering of the Pt oxidation state; (ii) the suppression of Pt oxide formation; (iii) the electronic structure; and (iv) a decrease in the Pt–Pt distance, and therefore a more favorable adsorption of O_2.

In order to reduce the catalyst cost, the investigation of non-noble catalysts can be considered an alternative to decrease the noble metal loading. In spite of the interest in this approach, the work done so far on these catalysts has shown that they have not reached the performance of a Pt-based catalyst concerning either catalytic activity or stability [48].

13.3.6
HOR Mechanism and Electrocatalysis Issues

As previously mentioned, the kinetic mechanism for the HOR is simpler than the ORR one, and its rate is markedly higher. In order to illustrate the difference, it can be pointed out that the literature values of the exchange current density on Pt are several orders of magnitude higher for HOR than ORR.

It is generally accepted that the kinetic process in acidic environments involves an H_2 dissociative adsorption step (the so-called Tafel step) followed by the electrochemical oxidation of adsorbed hydrogen (Volmer step):

$$H_2 + 2Pt \longrightarrow 2Pt\text{-}H \tag{13.8}$$

$$2Pt\text{-}H \longrightarrow 2Pt + 2H^+ + 2e^- \tag{13.9}$$

These reactions are very fast and almost independent of the metal catalyst; correspondingly, the exchange current density is very high. A Tafel slope b close to 30 mV dec^{-1} has been measured on various catalysts [49]. This is why a very low catalyst loading on the anode of a PEMFC can be adopted. Unfortunately, in the presence of CO, which is unavoidable when a reformate fuel is fed, a poisoning effect takes place because a large part of the surface is covered with CO, which makes reaction (13.8) the RDS. Under these conditions, only at very low overpotentials (<50 mV) a Tafel slope of 20–30 mV dec^{-1} can still be observed. At higher overpotentials, a kinetic limitation, due to hydrogen adsorption, occurs, and this leads to a limiting current density in the polarization curve. The problem of CO poisoning has been dealt with by adopting different approaches. An increase in CO tolerance can be obtained by performing a slight oxidation or by increasing the temperature. Another possibility is to increase the catalyst loading, but this was immediately recognized to be completely uneconomical. Therefore, for more than two decades, interest has been focused on finding CO-tolerant catalysts. To this end, various Pt alloys have been investigated, such as Pt–Ru, Pt–Rh, Pt–Mo, Pt–Sn, or Pt–Ir [50–54]. Pt with mixed metal oxides (Co, Mo, W) was also tested in the past [55].

In spite of the variety and numerous catalysts investigated, most of the efforts in preparing an efficient CO tolerant catalyst have concerned the Pt–Ru system, whose action has been explained by the so-called bifunctional mechanism. In simple words, Ru is capable of producing some oxy/hydroxy species at a lower potential than Pt. Thus, the oxidation of CO is possible at rather low potentials. Recently, a Pt-supported RhO_2 catalyst has been reported to have very good CO tolerance [56].

Even though CO poisoning is the most investigated problem concerning the use of reformate fuels in PEMFCs, other contaminants are also important. If the reformed hydrocarbon contains sulfur, a dramatic poisoning effect, due to H_2S, occurs on a PEMFC anode, with a high degree of irreversibility [57]. An interesting review paper has analyzed the different effects of various impurities and contaminants, and has indicated the impact on the PEMFC performance and possible mitigation strategies [58].

13.3.6.1 Carbon Support

In order to increase the EAS, PEMFC electrodes usually contain the catalyst in the form of small highly dispersed metal particles (about 2–6 nm) supported on carbon particles with an average diameter of about 15–50 nm. The carbon support must provide high electron conductivity (with a good electrical connection between the particles) to the 3D electrode layer. Moreover, its morphology should allow the reagents and products to have access to the catalyst sites. Other important features of the carbon support are the corrosion resistance and, for an optimal water management, the hydrophilic character. Different types of carbon supports have been used (carbon black, activated carbon, graphite, and graphitized materials). Carbon black is the most commonly used support for PEMFCs because of its excellent characteristics, which are normally improved by means of a graphitization process.

Even though some electrochemical activity has been found for ORR on different types of carbon, particularly on graphite [59], on carbon nanotubes (CNTs) [60], and on pretreated carbons [61], the use of carbon supports in PEMFCs has not been introduced to take advantage of this feature. However, the carbon support plays a major role in the overall electrochemical performance of an MEA. To understand this point, in addition to the important characteristics outlined above (high electronic conductivity, high surface area, water interaction), it is necessary to take a closer look at the electrode structure. The very small catalyst metal particles are deposited on carbon particles (or on a carbon nanostructure) with an optimal size and distribution. It is also very important to achieve a good electric contact between the electronic conductors (carbon support and metal catalyst, i.e., the CL) and the ionic conductor (i.e., the PEM); this is usually obtained by brushing the electrodic material with a soluble form of the ionomer, thereby increasing the number of reacting sites (TPB) at which the electrochemical reaction takes place. It is particularly important that the amount of inactive catalyst is limited. This undesirable but unavoidable situation occurs when the catalyst particles are not in contact with carbon and ionomer and it depends to a great extent on the catalyst/carbon/ionomer ratio in the CL.

In recent years, other types of carbons have been investigated, in particular to increase the specific surface area, to achieve some controllable porosity, and to obtain specific functionalization. Different carbon structures have been investigated: ordered mesoporous carbons [62], carbon aerogels [63], and cryogels [64]. Even though some interesting results have appeared from these investigations, these materials do not seem practically ready to be used as substitutes for the standard carbon supports.

13.3.7
Other Electrochemical Processes

This section has presented various important aspects concerning the electrochemical processes in a PEMFC, and the problems regarding the kinetics of the ORR and HOR fundamental reactions have been emphasized. For the sake of completeness, other electrochemical processes have to be mentioned, namely those involved in the so-called corrosion of the carbon support and of platinum. The topic of MEA degradation and the related issues will be described in another section in this chapter; here some electrochemical aspects are briefly presented.

In thermodynamic terms, the corrosion of carbon [65] can occur at potentials above 0.207 V (vs. NHE, normal hydrogen electrode) through the following electrochemical reaction:

$$C + 2H_2O \longrightarrow CO_2 + 4H^+ + 4e^-$$

There is no full agreement on the kinetic mechanism, but it can reasonably be assumed that the process proceeds via the parallel formation of surface and gaseous carbon oxides by disproportion of oxygen functional groups [66]:

$$C + H_2O \longrightarrow C\text{-}O_{ad} + 2H^+ + 2e^-$$
$$C\text{-}O_{ad} + H_2O \longrightarrow CO_2 + 2H^+ + 2e^-$$

The corrosion of the carbon support at the cathode side is higher than at the anode side because of the more aggressive conditions due to high oxygen concentration, low pH, and a high potential. Particular operational conditions, such as shutdown and restart, can enhance carbon corrosion.

The carbon support attack induces a high instability of the Pt/C electrocatalyst [67], which involves Pt corrosion and which may proceed by (i) Pt particle agglomeration, triggered by corrosion of the carbon support and (ii) Pt dissolution and redeposition. These works clearly indicate the need for a fundamental investigation of the electrochemical processes in order to understand the durability issues and degradation mechanism of MEAs.

13.4
PEMFCs Components

13.4.1
Electrodes

As with any other electrochemical power source, PEMFC performances are closely linked to the electrode design. This issue is even more crucial for PEMFCs since the electrodes must be the result of a delicate balance of transport media: the transport of the electrons, reactants, and reaction products must be optimized to reduce losses as much as possible [68]. Furthermore, the electrochemical reaction that occurs inside a PEMFC is an electrocatalyzed electrochemical reaction in which the cathode reaction kinetics (the ORR) contributes about 80% of the total cell voltage losses [69]. Of all the metals, platinum has the highest catalytic activity toward ORR in an acidic medium. Pt-alloys could also be used as an alternative to the Pt catalyst. Alternative catalysts, based on inexpensive non-noble metals, might provide an answer to some of the problems concerning catalyst costs. Metal elements smaller than Pt (Cr, V, Ti, W, Al, Ag) can enter into the crystal structure by means of a substitution mechanism and cause lattice contractions. This leads to a decrease in the interatomic distance between the Pt atoms and an increase in the electrocatalytic activity of the alloy, since a shorter Pt–Pt bond enhances the dissociative adsorption of oxygen [70]. Furthermore, the d-band vacancies of the Pt 5d orbital are higher for alloys than for Pt on its own and the change in the d-orbital vacancies follows the same trend as the electronegativity variations in the alloying element [71]. Gasteiger et al. [21] analyzed the Pt_xCo_{1-x}/C alloy system in small single H_2–O_2 cells as well as in H_2–air short stacks and found a threefold enhancement of the specific activity and a 2.5-fold enhancement of the mass activity of the alloy compared to the Pt catalyst supported on the same carbon. Duration tests with Pt_xCo_{1-x}/C alloy lasted 1000 h, a lifespan comparable to that of the standard catalyst. Although Pt alloys often show high and specific catalytic activity toward ORR, many of the tested ones have proved to be unstable in the FC environment. Thus, even though Pt alloys or Pt-free catalysts for ORR have been subjected to extensive research at a laboratory scale [72, 73], their use in commercial PEMFCs is still very limited, and Pt/C catalysts remain the standard.

In the first types of PEMFCs, the catalyst loading amounted to 4 mg cm^{-2} of Pt in the electrodes [68]. It is easy to understand why the reduction in platinum in the CLs has always been the driving force behind R&D programs. In 1993, Wilson [74] described a new electrode construction method called the "thin-film method." The design involved the use of a Nafion ionomer solution for the electrode preparation, which was used to bind the catalyst powder and to create conduction paths for the proton transport through the catalyst and the polymeric membrane. The novelty of this method was that the binding material in the CL had the same composition as the electrolyte membrane; before this, polytetrafluoroethylene (PTFE) was the conventional binder of GDE. By suitably adjusting the weight ratios of the Nafion ionomer, catalyst, and solvents, high-performance thin-film electrodes could be obtained with catalyst loadings of about 0.35 mg cm^{-2} [75]. Laboratory-scale processes and other commercial techniques for electrode preparation employ vacuum [76, 77], electrodeposition [78], electrospray [79–82], or impregnation methods [83]: each of these techniques has the purpose of enhancing the platinum utilization and reducing the catalyst content. The subsequent preparation of the MEA could be achieved according to two pathways: one consisted of the so-called catalyst-coated membrane (CCM) in which the CL is deposited onto the polymer and then sandwiched between two GDLs; the second is the catalyst-coated substrate (CCS) where the CL is applied to the substrates, which are laminated on each side of a membrane [21].

The platinum catalyst is generally supported on a high-surface-area carbon substrate (namely, Pt/C catalyst). The size of the Pt particles in the Pt/C is usually in the 2–6 nm range. Carbon supports with high surface area ensure a good dispersion of the metal particles and also offer good electronic conductivity. Carbon supports have several effects on the metal particles: they increase the electronic density in the catalyst and decrease the Fermi level, and therefore the electron transfer is enhanced as is the electrocatalytic activity [84]. The interaction between the metal and its substrate takes place through electron transfer from the platinum particles to the oxygen functional groups on the carbon surface [69]. This electronic effect can modify the properties of the catalyst. There is also a geometric effect: the carbon support can change the shape of the catalyst particles during their growth; thus modifications in the number of active sites can be expected [85]. Oil furnace and acetylene carbon blacks are the standard carbon supports for PEMFC catalysts. The surface areas range from 50 to 1500 m^2 g^{-1} for oil furnace carbon blacks, while it is below 100 m^2 g^{-1} for acetylene carbon blacks. Carbon blacks are often activated before their use as catalyst supports by heat or oxidative treatments. The first treatment is used to remove the impurities on the carbon surface, the second to encourage the formation of surface acidic sites at the expense of the basic ones, since the surface oxygen functional groups on carbon affect the dispersion of the metal particles. It has been reported that carbon corrosion is one of the main causes of catalyst degradation [86]. In the PEMFC environment, the carbon substrate is subjected to corrosive conditions such as high water content, low pH, high temperature, high oxidative potential, and high oxygen concentrations. The platinum particles themselves are also responsible for carbon corrosion. For example, carbon atoms can react with the oxygen atoms that are produced by the catalyst particles and give rise to CO and CO_2 by-products. Not only the substrate but also the metal is subjected to degradation in the FC environment. Pt nanoparticles tend to agglomerate because of their high specific surface energy, and this leads to a decrease in the number of catalytic Pt active sites and consequently to a performance loss of the PEMFC [87].

The last component of the electrode is the substrate on which the CL is deposited, namely the GDL. PEMFC performances are influenced to a great extent by the structure of the GDL, which, in turn, depends on the perfect equilibrium between hydrophobicity and hydrophilicity. The GDL has a carbon-based multilayered structure. A macroporous layer, which is essential for the electrical conductivity, mass transport, and mechanical strength, is connected to a microporous layer. The microporous layer is made of carbon black and a hydrophobic agent. It allows the reactants to be transported from the flow fields toward the CL on one hand, and it pushes away the liquid water from the CL to the flow fields on the other. The microporous layer at the cathode is essential since the removal of liquid water must occur simultaneously to the passage of oxygen molecules toward the catalytic sites through the GDL microchannels. Material optimization and the structural characteristics of the GDL are essential to ensure the proper functioning of the FC and to avoid cathode flooding at high current densities. For this reason, the porosity, permeability, and the amount of compression of the GDLs must be carefully controlled. Changes in the porosity of the GDL imply a nonuniform mass transport of both the gases and water, which in turn affects the overall performances of the cell. Nevertheless, conductivity of the GDL is essential because the electrical contact resistance between the GDLs and the BPs constitutes a significant part of the whole FC electrical resistance [88].

13.4.2
Membrane

The use of solid polymer electrolytes in PEMFCs dates back to the 1690s, when General Electric employed an FC system in the NASA Gemini Space Program [7, 9]. Such a system was the result of the development of a US patent that dealt with the design of an FC that incorporated a solid polymer (ion exchange resin) as the electrolyte. PEMFCs became a commercial reality in the 1970s, when the joint efforts of DuPont and General Electric led to the creation of the more reliable Nafion [89]. Nafion (DuPont) and Dow® (Dow Chemicals) products consist of a perfluoroalkyl side chain and a perfluoroalkyl ether side chain with a sulfonic acid group at its end. The difference between the two types of membranes lies in the fact that the Dow side chain is shorter than that of Nafion [90]. A lifespan of 60 000 h at 80 °C has been achieved with these materials.

Proton conduction along the side chains of the PFSA membranes reaches values of 10^{-2}–10^{-1} S cm^{-1} in fully humidified polymer condition. Proton conduction through the membrane has always been described on the basis of two mechanisms: "proton hopping" and "diffusion mechanism." In the proton hopping mechanism, protons jump from one SO_3^- H_3O^+ site to another, along the membrane. Protons produced at the anode interact with water to form hydronium ions. Once the hydronium ion has been formed, its proton is transferred (hops) toward another water molecule. The diffusion mechanism is related to the transport of the hydronium ions through the aqueous medium under the effect of the electrical field. The electroosmotic drag and concentration gradients can carry the water molecules through the membrane, and protons are transferred with them as $(H^+(H_2O)_x)$ species. This mechanism implies the existence of free volumes (pores) in the polymeric material of the membrane that allow the transport of hydrated protons [17]. The mechanism that prevails over the other depends on the state of hydration of the polymer. The conductivity and stability with temperature of the membranes are essential for applications in the automotive industry. Temperature fluctuations affect the water content within the membrane and, in some cases, lead to irreversible dryness. According to the U.S. Department of Energy (US DOE) guidelines, membranes for automotive applications must be able to withstand cold and hot temperatures and tolerate thousands of cycles between -40 and $+80\,°C$ [91]. Mc Donald et al. [92] checked the stability of MEAs assembled with Nafion 112 membranes subjected to 385 temperature cycles between $+80$ and $-40\,°C$ over a period of three months, simulating the conditions that might occur in electric vehicles: they suggested that some kind of rearrangement occurs at a molecular level in the polymer after repeated freeze/thaw cycles which affects the percentage elongation of the membrane until failure. The level of hydration, temperature resistance, and thickness are of primary importance when selecting the membrane. Nafion 117 and 115 (with an equivalent repeat-unit molecular weight of 1100) have thicknesses of 175 and 125 μm, respectively, in the dry state. The Dow membrane (with an equivalent weight of about 800) has a thickness of 125 μm in the wet state. Asahi Chemical Industries have developed a series of Aciplex-S membranes with equivalent repeat-unit molecular weights of 1000–1200 and dry state thicknesses ranging from 25 to 100 μm [93]. Membranes with less thickness hydrate faster, show lower resistance, and avoid the water drag more efficiently. However, there is a limit below which the membrane thickness cannot be reduced, because of gas crossover and durability issues. Thus, a membrane for PEMFC applications must satisfy many requirements such as high ionic (protonic) conductivity to sustain high current density even at low relative humidity levels, but low electronic conductivity; high mechanical, thermal, and chemical stability; low fuel and oxidant permeability to maximize the coulombic efficiency; and costs that meet the application target and durability [94].

Mechanical degradation is often the cause of early failures, especially for very thin membranes. This is related to fluoride loss, which is a direct measure of the degradation rate of the PFSA membranes [15]. Other mechanical degradation factors can be observed in many different forms, such as cracks, tears, punctures, and pinhole formations. The Gore Company has developed Gore-Select® membranes which are microreinforced composite polymers in which the pores of an expanded-PTFE (e-PTFE) host membrane are filled with perfluorinated ionomers. Such a reinforcement allows the production of membranes as thin as 5 µm with proton conductivities 10 times higher than that of Nafion [95].

Some approaches have been followed to try to overcome some of the limits that affect the Nafion polymer by acting on modifications of the membrane structure, such as the inclusion of inorganic oxides, the preparation of self-humidifying membranes, the modification of Nafion membranes with polymeric acid–base complexes, and so on. Hygroscopic inorganic particles such as SiO_2, TiO_2, and ZrO_2 have been found to improve water retention and decrease gas permeability. It has been reported that [96] the ZrO_2–SiO_2 binary oxide increases the proton conductivity and the water uptake of the polymer matrix at high temperature (120 °C) and low relative humidity (50%). Nafion/MO_2 (where M stands for Si, Ti, Zr) composite membranes have been shown to better tolerate the high temperatures (120 °C) more easily than unmodified Nafion, since the degradation temperature and the glass transition temperature have improved for all the nanocomposites. The enhanced water uptake that was observed was related to the hydrophilic nature of the inorganic additives within the pores of the Nafion and to the increased acidity of the nanoparticles [97]. However, there were also some disadvantages associated with a relatively low proton conductivity compared to unmodified membranes and to the high cost of production. Self-humidifying membranes have been developed for both HT and LT PEMFCs with the aim of improving the proton conductivity of the membranes under dry operation conditions. Many materials such as Pt [98], hetero-poly acid (HPA) [99], Pt/SiO_2 [100, 101], and ZrP [102] were dispersed in Nafion or sulfonated poly ether ether ketone (SPEEK) resins [103]. Modifications in Nafion membranes have also been obtained on the basis of the acid–base polymer complex concept [104]. Nafion–polybenzimidazole (PBI) composite membranes were prepared by casting a blend solution of Nafion-Na and PBI. These membranes were characterized by low gas permeability, good conductivity, and excellent thermochemical stability; in fact, Nafion acted as a cross-linking agent in the polymer composite and improved the mechanical stability [105].

13.4.3
End Plates and Sealing

End plate (EP) materials should satisfy several requirements, such as good electrical conductivity and low permeability to gases. They should also possess shock and vibration endurance, be of low thickness and costs, and should be easy to manufacture. During stack assembly and operation, EPs must ensure a good sealing; therefore the material that they are made of should have a sufficient degree of rigidity and strength. Since EPs are exposed to an aggressive environment, the materials must be resistant to corrosion attacks, because the leaching of the metal ions can pollute the MEA and lead to a decrease in the PEMFC performance [106]. Nonmetallic materials, such as polysulfone and engineering plastics, usually possess low thermal stability and are seldom employed for this reason. Furthermore, stainless steel, gold-plated aluminum, titanium, and some aluminum alloys are also often used in EP manufacturing. Metals usually have the required rigidity and thermal stability. Some problems could be encountered with corrosion and electrical insulation. Stainless steel is widely used in EP manufacturing because of its low cost and easy availability. It offers high stability against corrosion but is very dense, and this affects the gravimetric power density. It tends to form a passive layer which increases the interfacial resistance at the contact with the other components. Hentall *et al.* [107] proposed a new type of material named Grafoil for EP manufacturing: this material is a sort of exfoliated graphite that is usually employed in HT gaskets. This material satisfies most of the requirements for EPs, such as electrical conductivity and easy availability. It is inexpensive and corrosion resistant, but it is not mechanically robust; therefore further protection of the stack is required.

Good sealing of the BPs prevents leakage of the fuel and oxidant, and it also ensures good electrical contact between the MEA and BPs. O-rings, based on different elastomers, are usually employed to seal the stack components. Problems can be encountered when the PEMFC operates at temperatures above 100 °C, since most of the elastomers employed do not possess the required thermal stability. Silicon rubber, tetrafluroethylene–propylene, fluorocarbon, and ethylene propylene rubber should be recommended for HT PEMFC applications [22].

13.5
Assembly and Manufacture of PEMFCs

The electrochemical reaction between oxygen and hydrogen in a single PEMFC produces only about 0.7 V, which is far from sufficient to power any end user (a vehicle, for example). Many separate cells must be combined to form an FC stack to get this voltage up to a reasonable level. The potential power generated by an FC stack depends on the number and size of the individual FCs that make up the stack and the surface area of the MEA. BPs are used to connect one FC to another and they are subjected to oxidizing and reducing conditions and potentials.

BPs are therefore vital components of PEMFCs, as they supply the fuel and oxidant to reactive sites, remove reaction products, collect the produced current, and provide mechanical support for the cells in the stack. The BP acts as a separator between the individual FCs, whose function is to provide a series of electrical connections between different cells in the FC stack and to direct the fuel and oxidant gas streams to individual cells [108]. BPs constitute more than 60% of the weight and 30% of the total cost in an FC stack. For this reason, the weight, volume, and cost of the FC stack should be reduced significantly by improving the layout configuration of the flow field and by using lightweight materials [109, 110].

The overall efficiency of an FC depends on the performance of the PBs/EPs in the FC stack. These plates are conventionally made up of graphite with machined gas flow-field channels. Graphite BPs are currently being used and they offer relatively good performance, thanks to their high corrosion resistance and good surface contact resistance. However, owing to cost restrictions, the low mechanical resistance toward shocks and vibrations, and the porous nature of graphite, its commercial use in FCs still remains a challenge [111, 112]. Moreover, the machining required to make flow-field channels in graphite plates is quite expensive and, consequently, alternative materials and concepts are required to produce these plates. Materials such as coated stainless steel, aluminum alloys, and composites, with or without coatings, are good viable options for BPs/EPs in PEMFCs [113]. The use of metallic components for BPs/EPs offers several advantages over conventional graphite plates: higher mechanical strength, better resistance to shocks and vibration, no permeability, and much superior manufacturability and cost effectiveness compared to carbon-based materials, namely, carbon–carbon and carbon–polymer composites. However, the main handicap of metals is the lack of ability to combat corrosion in the harsh acidic and humid environment inside the PEMFC without forming oxidants, passive layers, and metal ions, which can cause considerable power degradation. Stainless steel plates, for example, have relatively high strength, but they self-passivate on exposure to air: the cathodic potential imposed on the plate during FC operation forms an insulating passive film on the surface. This reduces the electrical conductivity and results in high contact resistance, which leads to a high ohmic overpotential [114, 115]. Composite BPs constitute a small fraction of the cost of machined graphite plates and less than two-thirds the cost of etched stainless steel plates, but offer lower electrical conductivity [116], due to the partial polymer content. They also exhibit good electrochemical stability in the aggressive PEMFC environment, and long-term durability is not a major concern [117]. For example, the use of metal foams with low permeability values has resulted in an increased pressure drop across the flow-field, which enhanced cell performance [108].

An important issue with BPs is stability. Metallic BPs can corrode, and the by-products of corrosion (mainly iron and chromium ions) can decrease the effectiveness of FC membranes and electrodes [118, 119]. Coating the metal BPs with a conductive polymer [120], carbon [121], a novel metal [122], a metal nitride [123, 124], or a conductive oxide [125] is another viable method that is used to increase the corrosion resistance and decrease the contact resistance. Among the various options, transition-metal nitrides, especially TiN, have attracted considerable attention owing to their high chemical stability and conductivity [126]. The (Ti,Cr)N coating serves as a barrier coating that prevents corrosive media from penetrating the substrate; furthermore, it prevents oxide formation at the coating surface.

In LT PEMFCs, nonmetal graphite composites and carbon/thermoplast composites with additional additives can be adopted as BP material [127]. Composites generally show high corrosion resistance in the PEMFC environment; additional additives allow the conductivity to be increased.

The BP material that is suitable for transportation applications should have the following characteristics [112]:

- high corrosion resistance (corrosion current < 16 µA cm^{-2} at 0.1 V and hydrogen purge or at 0.6 V and air purge);
- interfacial contact resistance of 20 mΩ cm^2 at 140 N cm^{-2}
- should not dissolve and produce metal ions;
- steady low Ohmic resistance throughout the operation;
- high surface tension with water contact angle close to 90°, that is, high dehydration;
- light weight; and
- high mechanical strength (<200 N m^{-2}).

As far as the shape of the BPs/EPs is concerned, several models have been developed to optimize the design and channel dimensions [128–131]. The geometry of the flow plate has a significant effect on the nature of the fluid flow as well as the heat and mass transfer, and therefore also influences the performance of the FC. The most "primitive" designs (e.g., parallel channels) were dismissed in the past by most manufacturers and substituted by serpentine or interdigitated flow patterns [10, 132, 133]. In some cases, investigations were conducted on fractal flow fields for both PEMFCs and direct methanol fuel cells (DMFCs) [134]. In the serpentine design (Figure 13.3b), the flow-field structure forces the flow into a serpentine that spans the membrane surface. Instead, in the interdigitated pattern (Figure 13.3a), the feed is forced to pass under the graphite ribs and to permeate the porous electrode structure. This enhances the mass transfer toward the electrode active sites, at the price of considerably high pressure drops. However, this flow structure was found to lead to gas trapping in portions of the matrix with a consequent reduction in the FC power [135]. However, the channel and graphite rib geometries are of critical importance [130, 136, 137]. The rib width in particular should be kept as low as possible to enhance the achievable specific power. Conversely, as the channel width is increased, the cell effectiveness is enhanced and the pressure drop is reduced. The pressure drop per MEA is particularly important in the case where the cells are fed in series. This situation leads to pile up all the MEA's pressure drops. Conversely, a parallel MEA feed scheme generally entails a lower overall pressure drop, even though it may pose problems of uneven distribution. If,

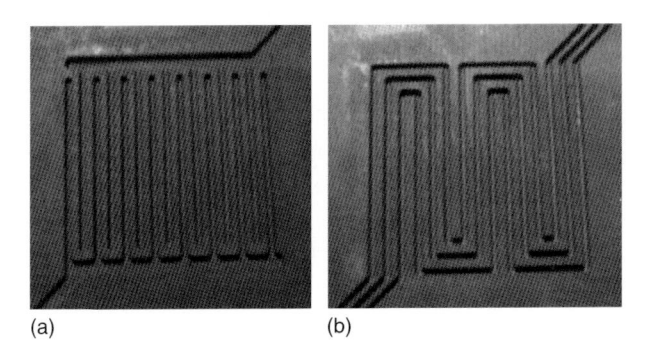

Figure 13.3 Flow-field design: (a) interdigitated and (b) serpentine. (Courtesy of Centro Ricerche FIAT, Italy.)

for some reason, one of the flow fields becomes blocked, the parallel feed scheme would lead to an enhancement of the specific flow rate to the other MEAs, which could in part recover the negative effect of the blocked MEA. The choice of the correct geometry is therefore a question of the compromise that can be made on the basis of modeling calculations and experimental checks. The current BP topologies include straight, serpentine, interdigitated, diagonal, or cascade flow fields, internal manifolding, internal humidification, and integrated cooling [128, 138, 139]. The fluid flow channels are usually rectangular in cross sections, even though other configurations, such as trapezoidal, triangular, semicircular, and so on, have been explored [109, 140].

At the cathode side of PEMFCs, oxygen diffuses through the GDL and reacts with the protons and electrons to produce water. Since the operating temperature of these FCs is quite low (below 100 °C), the produced water in the gas channels tends to be in liquid form. The formation, phase change, and transport of the water have a great impact on the operation, performance, and durability of FCs. Water management has therefore been the focus of many research efforts. Since the liquid water shares the same flow passages as the gaseous reactants, the transport of reactants is hindered when the liquid water is not removed from the cell at a sufficient rate, and flooding may occur as a result. Flooding in an FC occurs over a wide range of length scales, that is, in the mesopores of the CL, the micropores of the GDL, and submillimeter gas channels. The heterogeneous wettability of the surfaces of the microchannels, combined with the small hydraulic diameter ($\sim$500 µm) and low gas velocity (Reynolds number of an order 10^2 or less), yields two-phase flow regimes that are quite different from those in more conventional engineering applications.

From the FC design point of view, there are three key considerations on the water transport in the gas channel: (i) the flow maldistributions, (ii) the dynamics of the liquid water droplets, including the detachment process, and interaction with the channel walls and other droplets, and (iii) the correlation of the pressure drop as a function of the two-phase flow regimes and operating conditions.

Understanding and quantifying maldistributions within PEMFC channels is of great importance to optimize FC design and performance. A uniform distribution of the current density is of vital importance for FC operation because it leads to a uniform distribution of the temperature and liquid water production and lower mechanical stresses on the MEA [141]. The current density distribution in a PEMFC is determined by the uniformity of the reactant gas supply over the CL. Flow maldistributions also play an important role in reducing the operating life of an FC [142]. A proper reactant distribution is therefore critical to ensure a high performance and a long lifetime of a PEMFC.

As far as flow maldistributions are concerned, serpentine flow fields allow a more uniform current density distribution than a multiparallel channel flow-field design [129, 130, 143]. Moreover, porous materials have yielded better flow distributions and improved mass transfer, and consequently higher cell performance than grooved straight and serpentine flow channels [128, 144]. Interdigitated flow channels, with parallel straight channels, can enhance mass transport and improve the PEMFC performance compared to a parallel channel, owing to the forced convectional flow through the porous diffusion layer [145, 146]. Very homogeneous velocity and pressure fields are obtained for both the parallel-serpentine and the cascade-type flow fields, while an uneven flow distribution is obtained for the diagonal one [147].

13.6
PEMFC Stacking and System

As mentioned in Section 13.5, the power generated by an FC stack depends on the number and size of the individual FCs that make up the stack and the surface area of the MEA. To achieve an optimal performance, the FC stacks should operate around 80–85 °C, for LT PEMFCs, or 120–150 °C for HT PEMFCs. The efficiency of the FC can be increased slightly by pressurizing the fuel gases, which leads to higher power densities which are particularly suitable for automotive applications [13]. Additionally, the membrane needs to be humidified in order to operate properly, a condition that is generally provided through humidification of the supplied air flow [13]. While stack pressurization is beneficial in terms of both FC voltage (stack efficiency) and of power density, stack pressurization (and hence air pressurization) consumes energy. Hence, it is critical to pressurize the stack optimally to achieve the best efficiency of the system under all operating conditions [148]. In addition, oxygen starvation may result in a rapid decrease in cell voltage, and lead to a large decrease in power output [149].

The stack, in order to operate properly at the nominal power and at the transients, requires a series of auxiliaries: an air compressor for the air supply, a humidification system for the air stream, an electrovalve for the hydrogen supply, a heat exchanger or intercooler to provide or remove heat from the air, and a power converter. Obtaining the desired power response from an FC stack therefore requires regulation of the air and hydrogen flows, control of the pressure through a backpressure valve, and proper heat and water management to maintain optimal values, according to each operating condition of the stack itself. These components create large parasitic power demands at the system level, with 10–15% of the stack power being required only to power the compressor under some operating conditions, and this can considerably reduce the system efficiency [150]. These auxiliaries are powered by the FC stack itself, while the net power of the overall FC system is left for the end user.

In order to maintain normal operation conditions, an appropriate control strategy should be applied to the PEMFC stack and the system auxiliaries [151]. Moreover, the FC control system has to maintain an optimal temperature, membrane hydration, and partial pressure of the reactants over the membrane in order to avoid harmful degradation of the FC voltage, which will lead to a reduction in efficiency [152]. Hence, the FC system must be capable of simultaneously changing the air flow rate (to achieve the desired excess air, beyond the stoichiometric demand), the

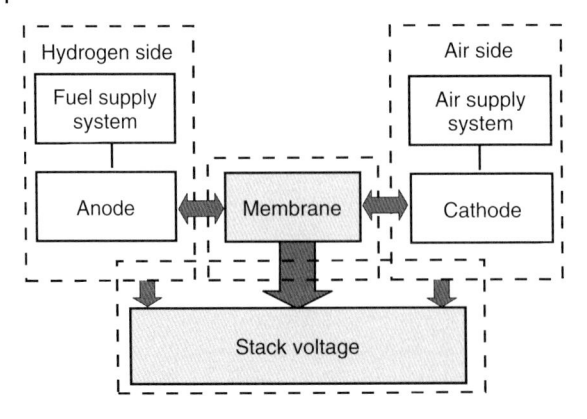

Figure 13.4 Block scheme of the FC control system.

stack pressurization (for optimal system efficiency), the membrane humidity (for durability and stack efficiency), and the stack temperature [153, 154].

The main goal of an FC control system is that of tracking the best operative actions that will allow the best net system efficiency while avoiding flooding or oxygen starvation. A block scheme of a possible control system is depicted in Figure 13.4. All the variables are closely linked physically, as the realizable actuators (compressor motor, back pressure valve, membrane humidification system, etc.), which are all placed at different positions in the systems, affect all the variables simultaneously [150].

The design of an efficient control system to manage an FC stack requires the following steps: (i) modeling of the stack, the compressor, and the other pneumatic elements involved in the system; (ii) calculation of the control equations and simulation of the entire system (including control); (iii) emulation of the stack and other pneumatic elements and simulation utilizing the designed control system; and (iv) physical realization of the control system and testing within the FC system. The control system guarantees the correct performance of the stack around its optimal operation point at which the net power is maximized. This means that both the air flow and the stack temperature are kept at a correct value.

At least two families of variables, namely, hydraulic (gas flows) and electric (electron flow), must be actuated for the automatic control of PEMFCs. The control strategy must maintain the chemical kinetics of the redox reactions in the MEA, thus ensuring an optimal performance; the highest efficiency must be reached in terms of net power generated by the stack, and the power of the auxiliary systems must be minimized. The electric load connected to the stack unbalances the electron equilibrium and forces the control systems to dynamically adapt the reactant quantities (oxygen and hydrogen). The use of an ambient controller is also necessary to ensure the optimum reaction conditions. Control issues imply detailed knowledge of both static and dynamic behavior of the FC system.

13.7
PEM Performance, Testing, and Diagnosis

While testing a PEMFC system, it is necessary to distinguish between single cells and stacks. Although it is sometimes possible to correlate the behavior of a single cell to that of a series of cells through the polarization curve, the uneven distribution in cell voltages, temperature, and reactant flows in the stack makes scaling up very difficult [155, 156]. Furthermore, single-cell tests are usually aimed at defining the characteristic and the electrochemical behavior of novel materials, such as catalysts and membranes.

The polarization curve, which is the plot of the cell potential versus the current density, is the basic methodology that is used to analyze the behavior of an FC under a set of operating conditions for both PEMFC stacks and single cells. According to the scan rates employed, the polarization curve can be plotted in a "nearly" steady state (very low scan rate of the current density against the potential) or in a nonsteady state (fast current sweeps) [157, 158].

A typical polarization curve is shown in Figure 13.5. The OCV of the cell is generally lower than its thermodynamic value, which can be obtained by the Nernst equation, as a consequence of the onset of a mixed-potential condition, mainly at the cathode, which is the result of parasitic processes. A decrease in the cell voltage is observed when the current flows, after closing the electrical circuit on the external load. The deviation of the cell voltage from the equilibrium value corresponds to the

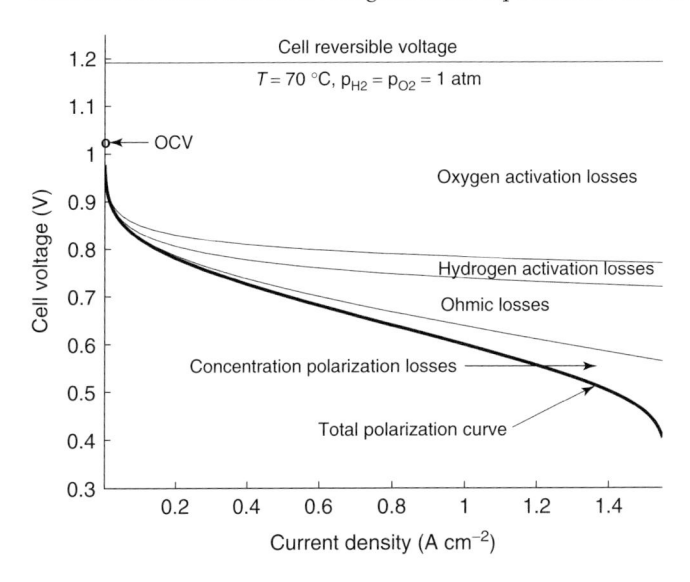

Figure 13.5 Example of a PEMFC polarization curve.

wasted power due to the irreversible phenomena that go with the reactions. These irreversible phenomena, caused by the finite rate of the processes occurring in the FC, are usually discussed in terms of electrochemical kinetics limitations at the electrode's surface (activation overpotentials, η_a), mass transport or concentration losses that result from the change in concentration of the reactants at the surface of the electrodes as a function of the reaction rate (concentration overpotentials, η_c), and ohmic losses in the ionic and electronic conductors. Activation overpotentials for both the cathodic and anodic reactions are responsible for the rapid initial fall in voltage shown in Figure 13.5. It is known that the activation overpotential for ORR is much higher than that for HOR. As the current density increases, in the middle of the polarization curve a fairly linear portion of the curve is found, as a result of the ohmic resistance of the membrane and electronic conductors. At higher values of the current density, the mass transport limitation of the reagents causes a dramatic increase in the voltage loss [12]. When the polarization tests are analyzed, in addition to the polarization graph shown in Figure 13.5, the plot of the power density versus current density is frequently reported, since operation of the FC at the maximum power density is sometimes the goal of specific applications. Polarization curves provide very important information on the cell performance, because they indicate, for a given current density, the ratio between the output electrical power and the theoretically available power (the difference is the dissipated power), but they do not give information on the various dissipation terms and their localization within the cell. The analysis of the various terms of the voltage losses, namely, activation overpotential (cathode and anode), ohmic drop (membrane and electrodes), and mass transport loss (cathode and anode), is commonly referred to as "voltage loss breakdown" (VLB) analysis. Such an analysis cannot be carried out through the polarization curve only, since the contribution of the different processes overlap [157]. Direct determination of the ohmic loss is possible through AC impedance spectroscopy or through the current interrupt method. Activation and mass transport losses can be determined through indirect methods mainly based on plotting the V_{cell} versus log(j) graph, where j is the current density and V_{cell} the cell voltage. The plot of a purely activation-limited polarization curve is given by a straight line (Tafel equation), and the mass transport loss can be derived from the nonlinearity of an IR-corrected V_{cell} versus log(j) curve as the difference between the straight line extrapolation from low current densities and the IR-corrected V_{cell} versus log(j) curve. Briefly, from the polarization curve, activation and mass transport losses can be determined once the ohmic resistance is known [159]. The situation can be more complex when anodic and cathodic terms have to be separated and when mass transfer phenomena within the porous materials must be accounted for.

Stack tests are usually designed to highlight the stack performance and to study the stack duration. Conversely, single-cell tests are directed to characterize the materials and components. These are essentially the MEA components (PEM, CLs, GDLs) and BPs (flow-field configuration and materials). The development of new materials requires the performance to be studied in single cells; this is usually achieved by conducting a comparison with some standards under the same operating conditions. Single-cell tests can also be carried out to investigate

13.7 PEM Performance, Testing, and Diagnosis | 631

phenomena such as mass transport, thermal degradation, degradation caused by pollutants, durability, and water management. Scaling up from the results obtained with a single cell to the entire stack performance is seldom achieved, since the enhancement of power and the fuel conversion efficiency in the stack also lead to an uneven distribution of the voltages, temperature, and reactants through the cells that make up the stack [160].

Single-cell testing can involve many of the techniques traditionally used in electrochemistry with laboratory three-electrode cells. Cyclic voltammetry (CV) is a potentiostatic method that is based on applying a triangular potential waveform to a stationary electrode immersed in an appropriate electrolyte solution. In typical experiments, the working electrode potential is varied linearly in time. Since platinum displays a typical voltammogram in acidic solutions, similar plots can also be traced in single PEMFCs by feeding the counter electrode with hydrogen and the working electrode with an inert gas, such as nitrogen or argon. This technique is employed to estimate the electrocatalytic activity of the catalyst via the computation of the active surface area, which is derived from the detected hydrogen adsorption/desorption peaks in the voltammogram. The large surface area of the carbon support in Pt/C catalysts can partially affect the detection of the hydrogen adsorption/desorption peaks. CV has also been used in PEMFCs to study membrane degradation phenomena, which were correlated with a decrease in the active surface area of the catalyst [157, 161]. The CV technique has been applied to both single PEMFCs and stacks (Figure 13.6) [162, 163].

CV measurements are often complemented by CO stripping voltammetry in which a mixture of CO and an inert gas flow to one electrode while humidified hydrogen flows to the other. During CO adsorption, CO feeds the anode and the electrode is kept at constant potential. The following step consists of removing the

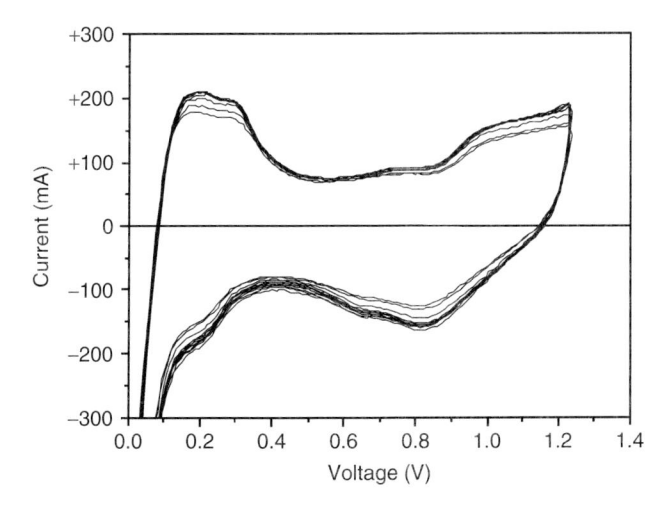

Figure 13.6 Example of cyclic voltammograms recorded on single PEMFC: 1.4 W, $H_2 = 70$ ml min^{-1}, $N_2 = 100$ ml min^{-1}; $T_{cell} = 70°C$; scan rate $= 100$ mV s^{-1}.

CO by flowing high-purity inert gas to the anode. The CO stripping voltammogram is recorded by sweeping the electrode potential between two fixed values at scan rates ranging from 5 to 10 mV s^{-1}. This technique is used to quantify the catalytic active sites.

One method for the direct determination of the hydrogen crossover is based on the use of the linear sweep voltammetry (LSV) technique. The measurement includes a sweep of the electrode potential from one point to another without cycling. In this case, hydrogen feeds the anode (it is also used as a reference DHE, dynamic hydrogen electron) and nitrogen feeds the cathode, respectively. After an initial rapid increase, in the potential range of 50–400 mV vs. DHE, the current reaches an almost steady value (plateau) which corresponds to the oxidation of hydrogen. Low scan rate (1 mV s^{-1}) is generally employed for this test.

A direct determination of the internal ohmic resistance of the cell can be obtained by the current interruption method. When the current flow ceases, the potential drop caused by the ohmic resistances vanishes immediately. Measurement of the cell voltage shortly after current interruption gives the "IR-free" potential [164]. Figure 13.7 shows an ideal voltage transient response in a PEMFC after current interruption. The cell initially operates at a fixed current. The current is interrupted at $t = t_0$. The ohmic losses vanish almost immediately, the overpotentials start to decay, and the cell voltage rises exponentially toward the OCV. At $t = t_1$ the current is again switched on [157, 165].

One feature that characterizes this technique is that the acquisition of the voltage must be very fast to be able to separate the contribution of the ohmic and activation

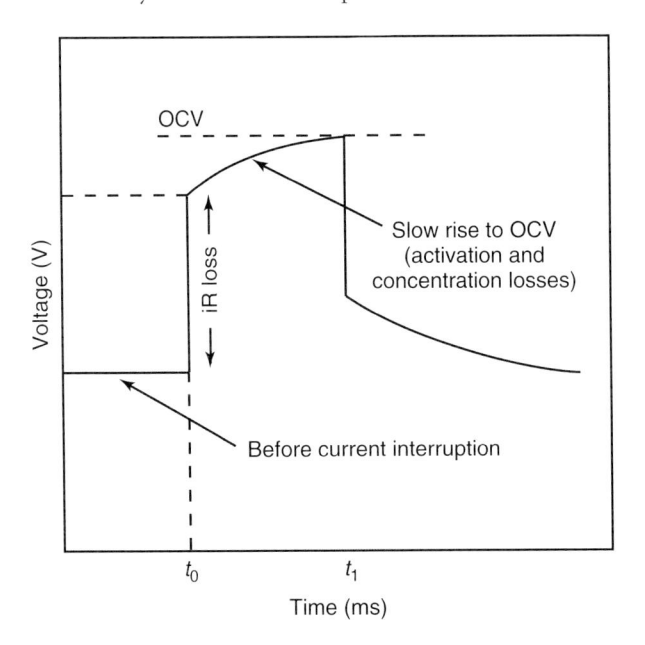

Figure 13.7 Voltage transient in a PEMFC after current interruption.

losses. Changes in the internal cell resistance have been correlated to the status of the cell under different operative conditions [166].

Electrochemical impedance spectroscopy (EIS or AC impedance) is employed to analyze both the steady state and the dynamic behavior of single FCs or stacks. In this technique, sinusoidal AC signals are applied to the system to measure the ability of the system to impede the electrical current [160]. EIS has been applied to study different aspects of the FC system, since it allows the individual contributions of each component (membrane, GDL) to be distinguished and can serve as a tool to diagnose the problems that affect individual components. Different contributions of electrode processes, such as the charge transfer or the mass transport in the CL, can be also distinguished [167]. EIS data can be interpreted with an appropriate equivalent circuit, through which the electrochemical parameters of the system can be obtained [168]. EIS has been used to distinguish between flooding or excessive drying of the stack [169], to monitor CO poisoning [170], or to study the dependence of the ohmic resistance on the current by changing the humidification conditions [171]. The FC operating conditions, such as, for example, the air stoichiometry, affect the stack impedance and the mass transfer resistance [172]. Important information can be derived for the optimization of MEA structures and production methods from the changes in ohmic resistance, charge, and mass transfer behavior in the complex-plane impedance spectra [168]. Impedance spectra have been employed to study the effect of catalyst loading in the electrodes [173, 174], the effect of the Nafion ionomer content in the catalyst ink [175], and the morphology of the GDL on the PEMFC performance [176, 177]. All these techniques are usually referred to as *"electrochemical methods"* for stack or single-cell testing.

When PEMFCs are employed in transportation, for example, in automotive applications, they are subjected to load changes and start–stop cycles. A series of tests under dynamic control have been developed to simulate these conditions. These tests are also aimed at reducing the experimental time required to study the degradation effects of PEMFCs. Different undesirable operating conditions have been employed as effective stressors for accelerated testing, according to the various degradation mechanisms involved in the performance reduction of PEMFCs: load cycling [178–181], start–stop cycles [182–184], low relative humidity [185, 186] or relative humidity cycling [187], stress–strain test [188], and freeze–thaw cycles [189–192], as well as fuel [193, 194] or air starvation [195, 196].

Open Circuit operation has also been recognized as an effective stressor for accelerated testing, in particular for MEA and CLs [197–199]. In fact, when no electric load is applied to the PEMFC, the hydrogen crossover rate increases. Hydrogen permeates through the polymeric membrane and reacts with oxygen at the cathode side. This can result in the formation of hydrogen peroxide. Peroxide radicals are responsible for membrane degradation, since they are able to break the polymer bonds. This results in a decrease in the MEA performance and affects the durability of the overall system.

In general, the management of gas flows, such as gas starvation or pressure drops, always has a great impact on the stack performance and durability. Fuel starvation is generally related to catalyst degradation because of the oxidation processes that affect the carbon substrate on which the platinum particles are supported. When there is not enough fuel to sustain a certain load, the anode potential increases until it reaches a value at which carbon corrosion takes place. The net result of this process is a loss in the active surface area of the CL. Another factor that can promote carbon corrosion is cold startup at subzero temperatures. Water freezing in the electrode pores can cause local reactant starvation, which promotes catalyst degradation. Some of the most common types of cold-start investigations include the achievement of the shortest possible startup time, the determination of the lowest acceptable temperature for cold start, and measurements of the performance degradation due to cold starts [160]. The frequency related to a pressure drop signal at the cathode under dynamic operation conditions has also been considered as a diagnostic test to reveal cathode flooding [200]. Performing startup/shut-down cycles is another way of checking the durability of a stack, especially since a PEMFC employed in a vehicle is expected to undergo 30 000 of these cycles during its lifetime [201]. Here, again, hydrogen crossover rates increase during a shutdown, and fuel starvation is observed during a restart. Load cycling can be considered a simulation of car road driving, but it is also a diagnostic test to accelerate PEMFC degradation. Cyclic current load conditions are usually applied to the cell for long periods, and increases in the hydrogen crossover and the related PEMFCs degradation phenomena are detected [178].

A "static" diagnostic test for stacks is the measurement of the individual cell voltages in order to detect voltage changes along the stack. The discrepancies between one cell voltage and another can be employed to explain cell failure [160]. The cells located nearest the gas inlets usually show different voltage values from the average voltage of the stack. This is generally due to a nonuniform distribution of the reactants and to temperature fluctuations. Finally, one of the most important parameters for FC is its lifetime. Table 13.1 summarizes just some of the endurance tests on PEMFCs that have been reported in the literature [160].

Table 13.1 Relevant literature results of endurance tests on PEMFC stacks.

Stack description	Test length (h)	Voltage decay ($\mu V h^{-1}$)	Causes of failure	References
40 cells, 2.89 kW	1800	–	Catalyst degradation, MEA contamination	[202]
10 cells	7863	11	–	[203]
40 cells, 5 kW	1000	40	–	[204]
8 cells	3000	–	Crossover leak	[205]
8 cells	5800	1	–	[206]

13.8
Degradation Mechanisms and Mitigation Strategies

The performance of a PEMFC stack is affected by several factors, such as FC design and assembly, degradation of the materials, operational conditions, and impurities or contaminants present in the feedstocks. Performance degradation is unavoidable, but it can be partially limited within time to increase its lifetime. To be commercialized, a PEMFC stack must be stable (i.e., able to provide power in the required conditions with acceptable performance limits), durable (i.e., able to ensure a certain performance in time over a wide range of operational conditions), and reliable (i.e., able to work without failures in any environmental condition) [207]. The degradation rate can be reduced: for example, DOE-suggested degradation targets usually require less than 10% loss in the efficiency of an FC system at the end of its lifetime [91]. A degradation rate of $2-10\,\mu V\,h^{-1}$ is commonly accepted for most applications [205].

The degradation mechanisms can basically be subdivided into two families: FC component degradation (membrane, electrocatalysts and CL, GDL, gaskets, BPs/EPs, sealing) and degradation effects due to operative conditions (air/fuel impurities, load cycle, startup/shutdown, environmental subfreezing conditions).

13.8.1
Membranes

Membranes can degrade mechanically, thermally, or chemically-electrochemically. Mechanical degradation usually causes early life failure, due to perforations, cracks, tears, or pinholes, which may result from inherent membrane defects, or from improper MEA manufacturing processes (nonuniform mechanical stresses due to excessive sealing). During FC operation, the overall dimensional change due to low or non-humidification and relative humidity cycling are also detrimental to mechanical durability [185, 186]. Moreover, the membrane protonic conductivity drops significantly with a decrease in the water content when the FC is operated at high temperature [208] and under low humidity [209]. The migration and accumulation of the catalysts and the decomposition of the seal in the membrane also negatively affect membrane conductivity and mechanical strength, and significantly reduce ductility. A physical breach of the membrane due to local pinholes and perforations can result in crossover of the reactant gases into their respective reverse electrodes. When this happens, the highly exothermic direct combustion of the oxidant and reductant occurs on the catalyst surface and consequently generates local hot spots; the latter increase the rate of pinhole formation, with a consequent acceleration of the membrane degradation and cell performance decay [36, 37, 210]. As far as thermal stability is concerned, Nafion membranes begin to decompose at temperatures higher than $200\,°C$ [211], while sulfonic acid groups are split off at temperatures above $280\,°C$ [212, 213].

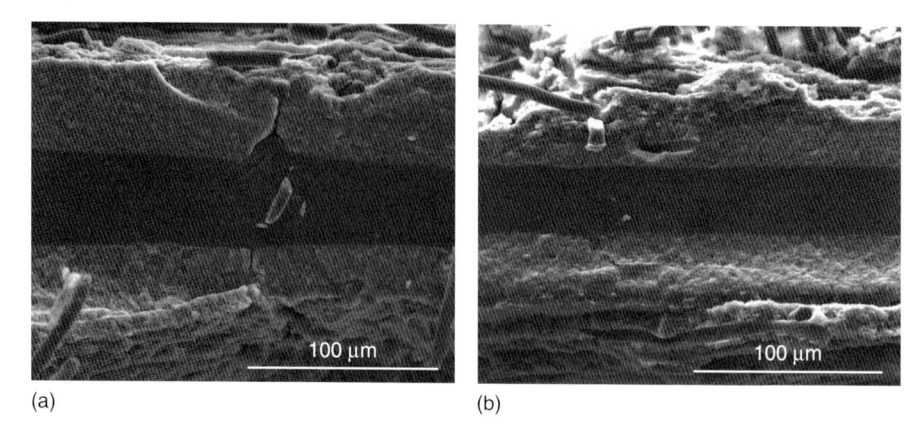

Figure 13.8 Cross-section scanning electron microscopy (SEM) images of MEAs made of Nafion 112: (a) fresh MEA, not subjected to testing; (b) thinned MEA after approx. 250 working hours.

Membrane thinning may be caused by a chemical attack from hydrogen peroxide formed by electrochemical reaction of oxygen and hydrogen that have crossed the membrane [210, 214]. An example of membrane thinning is shown in Figure 13.8.

Membrane thinning is known to be a direct consequence of the increase of hydrogen crossover [36, 215, 216]: the degradation rate depends on the initial membrane thickness. Thinner membranes display more rapid degradation rates than thicker ones [215]. In turn, a thinner membrane causes a higher hydrogen crossover rate, which indicates membrane thinning and hole formation. Hydrogen crossover via such holes leads to the generation of HO• and HO$_2$• radicals, which accelerate membrane deterioration and result in the reduction of the OCV. There are mainly two ways of forming hydrogen peroxide, one being oxygen reduction at the cathode [217] and the other based on the crossover of oxygen from the cathode to the anode. The hydrogen peroxide diffuses through the membrane and reacts with the metal ions that are present as impurities in the membrane and forms HO• or HOO• radicals, which can attack the polymer and degrade the membrane [218, 219].

To prevent mechanical failure of the membrane, the MEA and flow-field structure must be carefully designed to avoid local drying of the membrane, especially at the reactant inlet area [220]. Moreover, membranes can be reinforced by adding e-PTFE [221]. The possibility of modifying the physical and chemical properties of a polymer by dispersing inorganic nanoparticles in the polymeric matrix [222–224] has encouraged the development of proton-conducting composite membranes that are suitable for PEMFCs working at temperatures above 100 °C [225]. For example, layered zirconium phosphate $Zr(O_3POH)_2$ and phosphonates [14, 34–40] can be used as fillers of proton-conducting polymeric membranes thanks to their good chemical and thermal stability [226]. Moreover, the presence of these nanoparticles improves the mechanical properties of Nafion and reduces crossover [227, 228].

Alternative sulfonated polymers and their composite membranes, such as SPSF (sulfonated polysulfone), SPEEK, PBI, and PVDF (polyvinylidene fluoride) can also be used to improve the stability and durability of MEAs [229–234].

Finally, the radiochemical grafting technique can easily be applied, together with chemical or physical cross linking, to improve membrane chemical stability and to maintain other membrane properties including proton conductivity and fuel permeability [235, 236]. Membranes prepared by the radiochemical grafting of styrene onto PVDF films, followed by sulfonation, have in particular recently attracted a great deal of attention thanks to their superior performance [237, 238]. These membranes can be prepared via a pre-irradiation method with an electron beam using dense PVDF films as the starting matrices [239]. The grafting techniques include chemical grafting [240], photografting [241], plasma grafting [242], thermal grafting [243], and radiation grafting with γ rays [244, 245].

13.8.2
Electrocatalyst and the Catalyst Layer

Many different types of catalysts have been developed as electrocatalysts for the anode and cathode of PEMFCs. Pt-based catalysts are usually employed in PEMFCs: Pt is in the form of highly dispersed metal particles at nanometer scale, usually in the 2–6 nm range. Nanoparticles inherently show a strong tendency to agglomerate, because of their high specific surface energy [87]. Consequently, when Pt nanoparticles agglomerate to larger ones, the electrochemical surface area of the Pt-based catalysts decreases, and consequently the performance of the PEMFC degrades. Considering the extremely harsh conditions in which PEMFCs operate, anode catalysts are exposed to a strong reducing hydrogen atmosphere, whereas cathode catalysts are under strongly oxidizing conditions. Moreover, both the anode and cathode of PEMFCs operate under low pH (<1) and high temperature (80 °C or above) conditions, and with significant levels of water in both the vapor and liquid phase: the corrosion of Pt/C (PtM/C) catalysts (both catalytic metals and the support materials) is a problem. The degradation of catalysts involves the two different features of the catalytic metal (Pt or Pt alloys) and the carbon support that influence each other: the catalytic metal, especially Pt, catalyzes the oxidation of carbon [246], and the oxidation of carbon accelerates Pt sintering [247].

As far as the increase in catalytic nanoparticle size, that is, the sintering/agglomeration of Pt or Pt alloys, is concerned, there are several fundamentally different pathways [248–251]: (i) Pt dissolution and redeposition, which is also called "*Ostwald ripening*," (ii) the coalescence of Pt nanoparticles via Pt nanocrystallite migration on the carbon support, and (iii) the transport of Pt atoms on the carbon support, the so-called 2D Ostwald ripening compared to case (i), which is also called "*3D Ostwald ripening*." The dissolution/precipitation mechanism might be more prevalent when load cycling occurs [252]. There is still no agreement on which of the previously mentioned pathways is dominant in the observed electrochemical surface area loss. The presence of Pt species in the

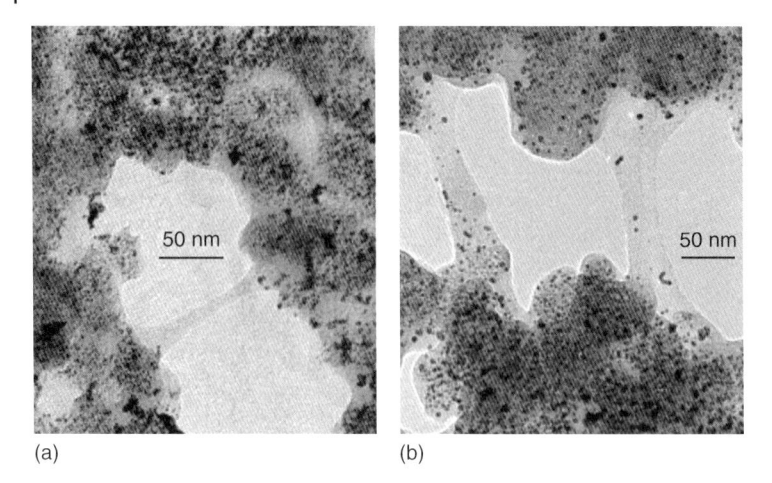

Figure 13.9 TEM images of Pt/C electrocatalysts and ionomer network: (a) very limited Pt migration in the ionomer network; (b) consistent Pt migration in the ionomer network. (Courtesy of CNR-ITAE, Messina, Italy.)

polymer membrane after extended life testing is another piece of evidence that points toward the dissolution/redeposition mechanism, as shown in Figure 13.9 [67, 248, 253]. Soluble Pt species have been found in water collected from the reactant gases exiting from the cell [254].

Electrochemical corrosion of the carbon surface leads to changes in the surface chemistry of the carbon and an increase in hydrophilicity in the CL and GDL, which results in a decrease in gas permeability [255] and might affect the water transport mechanisms, which leads to "flooding" of the pores. Carbon corrosion increases the electrode resistance by decreasing the thickness of the CL [248] and thus decreasing the electric contact with the current collector [252]. Carbon corrosion is often observed in an electrochemical system, according to

$$C + H_2O \longrightarrow CO_2 + 4H^+ + 4e^-$$

The thermodynamic potential, at standard conditions of this reaction, is only 0.207 V [255]: the electrochemical oxidation of the carbon is thermodynamically possible above 0.2 V. In the presence of water, carbon can also be consumed through the heterogeneous water-gas reaction

$$C + H_2O \longrightarrow H_2 + CO$$

especially in the presence of Pt, even though the rate of this reaction might be lower than that of the electrochemical oxidation at the cathode in PEMFC conditions [256]. The CO reaction product might poison the Pt catalyst. The reaction rate of carbon combustion increases as the Pt loading increases [257], because the higher the Pt loading, the larger the interfacial area between the Pt and the carbon. Moreover, the degradation rate of the carbon support increases with temperature [31].

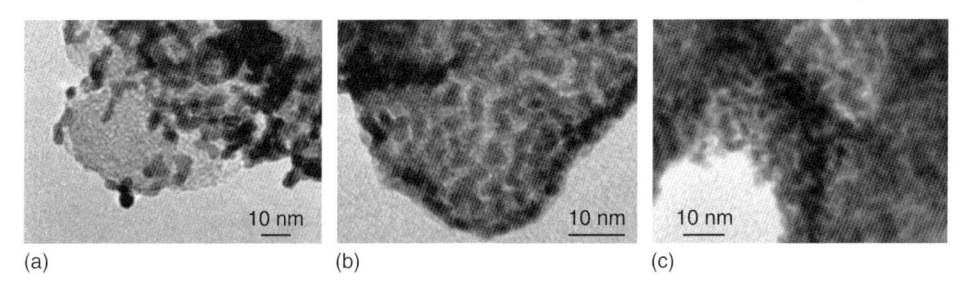

Figure 13.10 TEM images of Pt/C-based electrocatalysts: influence of the second metal on the nanoparticle size. (a) 60% Pt/C, nanoparticle size 2.8 nm; (b) 60% PtCo/C, nanoparticle size 2.3 nm; and (c) 5% PtRu/C, nanoparticle size 1.6 nm. (Courtesy of CNR-ITAE, Messina, Italy.)

The durability of catalysts is inherently determined by the properties of the catalytic metals and the support materials and the specific interaction between them. Alloying Pt with a second and/or third and/or fourth metal is an important strategy that can improve the service life of catalysts [258]; in addition to a higher activity toward electrochemical reactions [259], stability against dissolution also increases [260]. Some examples of electrocatalysts are shown in Figure 13.10. The presence of a second metal alloying with Pt leads to a reduction in the overall size of the active nanoparticles.

The basic requirements for an improved carbon support are high surface area, which leads to the deposition of small platinum particles (maximizing catalytic surface area); low combustive reactivity under both dry and humid air conditions at low temperatures (150 °C or less); improved electrochemical stability under FC operating conditions; and high conductivity. In addition, it should be easy to recover Pt in the used catalyst. Graphitization of the carbon plays an important role in carbon support stability, because carbons with a higher graphitization degree are thermally and electrochemically more stable [261] thanks to the decreased defect sites on the carbon structure where carbon oxidation initiates [262]. Graphitization can be achieved by heating carbon materials in protective gas to a high temperature (more than 1600 °C): the higher the graphitizing temperature, the higher the graphitization degree in the resultant carbon materials [261]. CNTs [263–265] or carbon nanofibers [266, 267] can also be used: the stronger interaction between Pt and CNTs has been suggested to contribute to the high stability of the resultant catalysts (Pt/CNTs).

13.8.3
Gas Diffusion Layer

GDL is usually a dual-layer carbon-based porous material, which includes a macroporous carbon fiber paper or a carbon cloth substrate covered by a thinner microporous layer consisting of carbon black powder and a hydrophobic agent (e.g., Teflon as a binder). The microporous layers favor electronic contact and water removal from the CL. The PEMFC operating environment gradually changes the GDL from hydrophobic to hydrophilic [23, 268, 269]: this happens, for example, for cold-start conditions [270]. As the FC operates, the binding agent and the carbon composite of the GDLs are susceptible to chemical attack by OH• radicals and electrochemical oxidation [271]. Changes in the GDL properties can mainly be attributed to the microporous layer [23, 272].

GDL properties can be improved by using graphitized fibers or by adding inorganic materials during GDL preparation and increasing the amount of the binder [23, 273, 274]. The water transport properties of the GDL can be enhanced by laser perforation during their preparation [275].

13.8.4
Bipolar Plates

As previously mentioned, BPs supply the fuel and oxidant to the reactive sites, remove the reaction products, collect the produced current, and provide mechanical support for the cells in the stack. Likes all other PEMFC components, BPs also suffer from corrosion failure, mainly due to pinhole formation (especially when metal plates are used) [119]. It is possible that the oxygen present at the cathode side reacts with metal ions to form metal oxides, and the hydrogen present at the anode side reduces the metal ions to their metallic state. These corrosion by-products can be flushed away from the electrodes by the convective action of the gas and water flows, or remain in the FC electrodes, causing potential fouling problems due to pore blockage [112, 118, 276]. BPs can suffer from mechanical fractures or permanent deformations as a result of the compressive forces that are used to ensure good electric contact and reactant sealing during FC operation [277]. Moreover, operational factors such as thermal cycling, nonuniform current, or thermal misdistributions over the active area can also affect the mechanical properties of the BP materials [278].

The chemical and mechanical resistance of BPs can be improved via preventive coating during manufacture by adopting various techniques. Immersion coating, spraying, electroplating, electroless deposition, electrolytic anodization, painting, physical or chemical vapor deposition processes are being evaluated for the coating of these components [279–281]. However, the protective coating layers can delaminate when a PEMFC operates under thermal cycling conditions: the coating material and the substrate might expand and contract at different rates due to the difference in their thermal expansion coefficients [282]. The micropores and microcracks that arise from the deformation of the coating layer can lead to the direct exposure of the substrate metal to a highly corrosive environment, and subsequently the dissolved metal ions diffuse into the membrane and become trapped in the ion exchange sites, which results in considerable adverse effects on the cell performance. Another viable option is the use of composite BPs (carbon–polymer and carbon–carbon composites), which do not suffer from delamination problems [116, 283–285].

13.8.5
Sealing

The stack sealing system, or method, must last as long as the stack lasts, and is often used to enhance the structural strength of the stack and provide performance-enhancing functionality. If any gasket degrades or fails, oxygen and hydrogen can leak outside or mix with each other directly during operation or while at standby, and thus affects the overall operation and performance of the FC. The elastomeric gaskets used as seals in FCs are exposed to an acidic environment, humid air, hydrogen, temperatures as high as 80 °C, and aqueous liquids for the coolants, and are concurrently subjected to mechanical compression between the BPs that form the cell. Therefore, the long-term stability of the gaskets is critical in FC assembly. Some typical sealing materials used in PEMFCs are fluorine caoutchouc, ethylene-propylene-diene monomer (EPDM), and silicone [286]. The chemical degradation of commercially available silicone elastomeric gaskets usually proceeds through a change in surface chemistry via de cross-linking and chain scission in the backbone structure [287, 288]. Traces of decomposition products of the sealing material can be found in both the membrane and electrodes. The acidic nature of the PEMFCs, together with thermal stressing or hydrogen embrittlement, induces mechanical alterations of the sealing material [202, 289], as can be seen in Figure 13.11, for example.

13.8.6
Air/Fuel Impurities

Pollutants such as NO_x, SO_x, and small organics mainly originate from vehicle exhaust and industrial emissions. Impurities in hydrogen fuel, such as CO, H_2S, NH_3, organic sulfur-carbon, and carbon-hydrogen compounds, are mainly from

Figure 13.11 Sealing blown out of cell plates in a PEMFC stack. (Courtesy of CNR-ITAE, Messina, Italy.)

the manufacturing process, in which natural gas or other small hydrocarbon fuels are reformed to produce hydrogen gas with a small amount of impurities [28]. In fact, although hydrogen is an attractive fuel for PEMFCs and the most abundant element in the universe, it is not present in the free form but in various compounds, mainly water and hydrocarbons.

Air and fuel impurities are transported with the fuel and air feed streams into the anodes and cathodes of a PEMFC stack, causing performance degradation and, sometimes, permanent damage to the MEAs [58]. The FC component most affected by a contamination process is the MEA. Three main effects can be identified: (i) the kinetic effect (poisoning of the electrode catalysts), (ii) the conductivity effect (increase in the solid electrolyte resistance, including that of the membrane and CL ionomer), and (iii) the mass transfer effect (CL structure and hydrophobicity changes which causes a mass transfer problem).

The greatest concerns arise from CO poisoning, particularly at conventional operating temperatures (<80 °C). CO in fact binds closely to the Pt sites, which results in a reduction in the surface active sites that are available for hydrogen adsorption and oxidation. The CO poisoning effect is closely related to the concentration of CO, the exposure time to CO, the cell operation temperature, and the anode catalyst types [290]. CO poisoning on Pt electrocatalysts becomes more severe with increases in the CO concentration and exposure time. Voltage losses, in fact, markedly increase with prolonged exposure to CO, as a result of its accumulation on the Pt catalyst surface in time [291]. Higher temperatures and higher humidity could effectively reduce the coverage of CO over the catalyst by promoting CO oxidation with an adsorbed OH• group [32, 292].

The presence of CO_2 in the H_2-rich stream fed to an FC can lead to performance losses, especially at high current densities: a performance loss of up to 30% has been reported at 0.5 V with a 20% CO_2 content in the reformate gas [293]. CO_2 can, in fact, be catalytically converted into CO on a Pt catalyst, and it then poisons the catalyst. Traces of H_2S at both the anode and cathode can degrade the cell performance significantly, mainly through the poisoning effect of the Pt catalysts [294]. Cationic ions, such as alkali metals, alkaline-earth metals, transition metals, and rare-earth metals, originating from the corrosion of BPs or present in the coolant, have a direct effect on the transport properties of the PEM [295, 296].

The development of highly CO-tolerant electrocatalysts is the most frequently used mitigation technique to limit the effects of CO poisoning. Carefully designed Pt alloys have proved to exhibit considerably higher CO tolerance than pure Pt electrocatalysts [297–299].

PtRu alloy catalysts are currently the most CO-tolerant electrocatalysts. However, many studies have dealt with other binary alloys (Pt-Mo, Pt-Sn, Pt-Co, Pt-Nb, Pt-Ta, Pt-Ni, etc.) [300, 301], but also ternary (Pt-Ru-W, Pt-Ru-Co, Pt-Ni-Cu, Pt-Ni-Fe, Pt-Ru-Mo, etc.) [302–304] and quaternary alloys [305, 306], and Pt-based oxides (Pt-Ru-Mo-Ox, Pt-Ru-W-Ox) [307]. The severity of CO poisoning is also affected to a great extent by the anode composition, catalyst preparation procedures, FC structure, and operating conditions [308].

Another mitigating solution consists in reducing, as much as possible, the CO amount present in the H_2-rich feed which has to be fed to PEMFCs during the upstream reforming processes [28]. From this point of view, the US DOE published on its web site in 2007 (updated on April 2009), a guideline document "Multi-year research, development and demonstration plan: planned program activities for 2005–2015-technical plan fuel cells," in which the CO tolerance (2005 target) for automotive applications was 10 ppm in steady-state conditions and 100 ppm in transient conditions, for a durability of 20 000 h. Such a target should be reduced (2015 target) to 1 ppm in steady-state conditions and 10 ppm in transient conditions for a durability of 40 000 h [91].

Considering that the CO tolerance increases with temperature [32, 292], another advantageous mitigation option is offered by the adoption of HT PEMFCs. For example, CO tolerance can reach levels of 1000 ppm at 130 °C and 30 000 ppm at 200 °C [32]. In practical terms, it is possible at high temperatures to directly use the reformate from a simple reformer as the feed fuel, thus eliminating the water-gas shift reactors, selective oxidizer, and membrane separator for CO cleanup from the reforming system, with a consequent cost saving. The membranes that have been suggested as being suitable for operation at 200 °C include acid–base PEM, and in particular H_3PO_4-doped PBI [309–312].

13.8.7
Degradation Effects due to Subzero Conditions

The startup and stationary behavior of PEMFCs below 0 °C in winter is one of the most challenging tasks that needs to be solved before commercialization, specifically for automotive applications. During normal operating conditions, water exists in the form of vapor and liquid in the flow channel and pore regions of GDLs and CLs. The ionomer (polymer electrolyte) in the membrane and CLs also absorbs certain quantities of water in the liquid state and binds to H_3O^+. During a PEMFC cold-start process, the initial cell temperature is usually equal to the external surrounding temperature, and water will most likely freeze. The formation of liquid water can almost be neglected, since it freezes to ice (ice and liquid water can still coexist when the local cell temperature increases or decreases to the freezing point of water, which results in ice melting or liquid freezing). Therefore, water usually exists in the form of vapor and ice in the pore regions of GDLs and CLs during PEMFC cold starts. Since the formed ice can easily stick to the solid materials of CL and GDL, the formation of ice in flow channels can be neglected [313, 314].

Residual water in the stack after shutdown precludes a cold startup and causes irreversible damage to the cells when there are subfreezing temperatures [192, 315–317]. During startup, the presence of ice can also prolong the startup time and require an external energy input [313]. The main cause of cell degradation is ice formation, which occurs below freezing temperatures. Swelling and contraction due to the formation and melting of ice can cause mechanical stress or delamination of the FC materials, leading to their degradation. The main damage could be to the GDL [316–318], the CL [319], and the PEM [190, 192, 315, 318], with a consequent reduction in the OCV value and in the polarization curve [190, 315, 318, 320]. Moreover, BP gas channels could be blocked by ice, which results in startup failure and electrode deterioration due to localized fuel starvation [319, 321], with a consequent increase in the electrical resistance of the cell [320]. Therefore, the cell operation is limited, or even inhibited, due to severe voltage drops.

The main problems encountered, apart from performance decay, are related to MEA delamination, the migration of the metal active elements of the electrocatalysts in the membrane, and the embrittlement of the seals. Some of these problems can be seen in Figure 13.12.

Various strategies have been investigated and adopted by automotive companies and FC stack manufacturers to prevent the performance degradation [191, 322–328] in order to mitigate subfreezing effects: from purging the FC stack with dry gases at elevated temperatures to remove the residual water [192, 315, 316, 320, 329–333] to feeding antifreeze alcoholic solutions [315] to flood the cell before shutdown, to adopting an external catalytic burner to heat the system during a cold startup [317], and to assisting the cold startup with external electrical heating [317, 332]. Purging with dry gases during shutdown seems to be a promising approach to remove water from the FC stack [315–317, 320, 330, 331, 333] and thus prevent system degradation due to freezing/thawing cycles. Water removal in the shutdown process is in fact essential not only to prevent cell degradation but also to improve

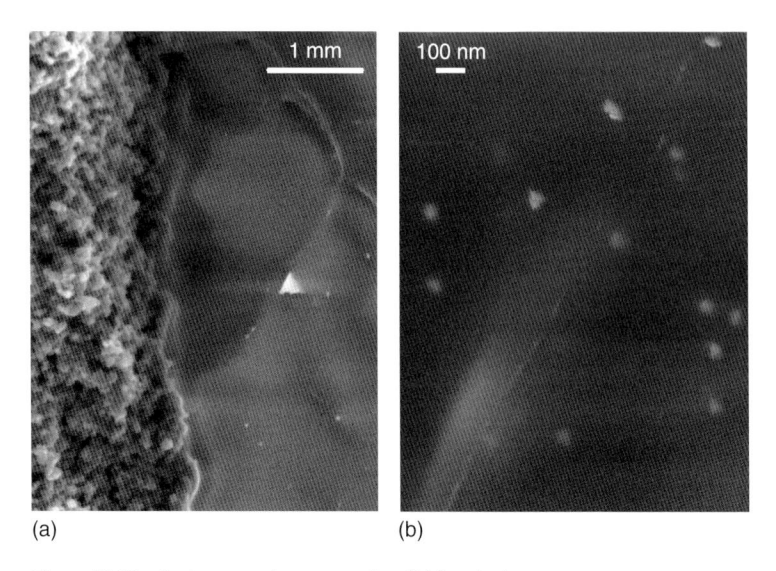

Figure 13.12 Backscattered cross-section field emission SEM (FESEM) images of a Nafion 112 MEA: (a) PEM/CL interface (Pt particles are visible in the membrane); (b) Pt particles migrated into the membrane.

startup ability in cold conditions. A dried membrane, in fact, is extremely useful to start up PEMFCs at subfreezing temperatures since heat generation due to ohmic losses is high when the membrane is dry [315, 316]. The use of a dried membrane also helps avoiding the blockage of gases in the channels [320].

13.9
Applications

One of the characteristics of FC systems is that their efficiency is almost unaffected by size. This means that both relatively small or large FC systems can be used in many sectors, including transportation, portable electronics, back-up power supplies, off grid power generation, and cogeneration [334, 335].

13.9.1
Transportation

Cars, buses, scooters, golf cars, utility vehicles (such as forklifts and airport vehicles), locomotives, tramways, boats, airplanes, and underwater vehicles can be mentioned as some of the end users in which PEMFC systems can be applied in the transportation area. Since the late 1980s, there has been a strong push toward developing PEMFCs for use in light-duty and heavy-duty vehicle propulsion. The driving force behind this development is the need for clean and efficient cars, trucks, and buses that can operate on conventional fuels (petrol, diesel) as well as on renewable and alternative fuels (hydrogen, methanol, ethanol, natural gas, and other hydrocarbons) [336]. With hydrogen as the onboard fuel, such vehicles would be considered zero-emission vehicles. With onboard fuels other than hydrogen, the FC systems can use an appropriate fuel processor to convert the fuel to hydrogen, thus resulting in vehicle power trains with very low acid gas emissions and high efficiency [28, 337]. Furthermore, such vehicles offer the advantages of electric drive and low maintenance because of the limited number of critical moving parts. This development is at present being sponsored by various governments in North America, Europe, and Japan, as well as by the main automobile manufacturers throughout the world. Several FC-powered cars, vans, and buses are currently operating on hydrogen in various demonstration projects [338–340].

As far as utility vehicles are concerned, forklifts have gained the attention of PEMFC manufactures, as they represent a very interesting niche market. Forklifts are in fact used in a wide variety of commercial and industrial applications to move all types of goods and materials around, to, or from trucks, to or from storage areas, or from one work station to another. Forklifts can be used from a few hours per day to 24 h per day, seven days per week. Some forklifts are used indoors, while others can be used outdoors. Current designs are powered by lead–acid batteries or by fossil fuels (generally, the larger types). The FC forklifts that have been used so far have mostly been in the smaller size ranges, to substitute battery-powered forklifts [341]. FC forklifts offer several important advantages over the battery-powered types: they lead to higher productivity because of the elimination of time-consuming battery changes; they can be refueled in less than 5 min, and there can be multiple fuel stations, with hydrogen distributed around a site from a central tank. The space required for fueling is much smaller than that required for a battery room. FCs maintain a constant voltage, without any voltage drop toward the end of the shift or in cold locations, as has been observed for batteries. There are no environmental concerns about acid runoff or lead, or from tailpipe emissions, though handling and storage of hydrogen may involve safety concerns. Individual plants can establish their own hydrogen fueling stations, either based on hydrogen produced on site or at a central location. In either case, the hydrogen can be produced by the steam reforming of natural gas, or by the electrolysis of water. The adoption of FC-powered forklifts will result in lower total logistics costs but higher initial costs [341].

13.9.2
Auxiliary Power Units

In addition to high-profile FC applications such as automotive propulsion and distributed power generation, the use of FCs as auxiliary power units (APUs) for vehicles has received considerable attention in recent years [342]. APU applications may represent an interesting market because they offer a true mass-market opportunity that does not require the challenging performance and low costs required for propulsion systems for vehicles. APUs are devices that can provide all or part of the non-propulsion power necessary for vehicles. Such units are already extensively used over a range of vehicle types and for a variety of applications, in which they provide a number of potential benefits, as illustrated in Table 13.2 [343–345]. APUs can operate on petrol and, for trucks, preferably on diesel fuel, in order to match the available infrastructure, and if possible to share onboard

Table 13.2 Overview of APU applications and the potential benefits.

Vehicle type	Load service	Potential benefits
Heavy-duty and utility trucks	Space conditioning	Can operate when main engine unavailable
Automobiles and light trucks	Conditioning and entertainment	Reduce emissions and noise while parked
Aircrafts	Refrigeration	Extend life of main engine
Trains	Lighting and other cabin amenities	Improve power generation efficiency when parked
Yachts and ships	Communication and information equipment	
Recreational vehicles		

storage tanks with the main engine. The small amount of fuel involved in fueling APUs probably would not justify the setting up of a specialized infrastructure (e.g., a hydrogen infrastructure) for APUs alone. Similarly, APUs should be water self sufficient, as the need for APUs to carry water would be a major inconvenience for the operator, and would require additional space and associated equipment [346].

13.9.3
Backup Power Supplies

The term *"backup power"* refers to any device that provides instantaneous, uninterruptible power when the main power sources are not available or unable to meet the power demand [335]. Wireless networks are more often the prime source of telecommunications, especially in emerging markets and developing countries where wire line networks are rudimentary. The availability of communications network is important even when the power grid fails. Currently, mobile network operators still rely mainly on lead–acid batteries or diesel generators as the emergency power source. Most backup power in communication and control systems use a combination of generators and batteries to provide redundancy in order to avoid service disruptions. Although these systems are reliable and well established, increasing energy prices and the costs of natural resources have put a great deal of pressure on network providers to develop energy-saving solutions. Compared to these conventional solutions, FCs are robust as far as severe weather conditions are concerned and offer unlimited backup at a fraction of the costs. With fewer moving parts, they require less maintenance than either generators or batteries. They can also be monitored remotely, reducing the actual maintenance time [347]. Compared to generators, FCs are quieter and have no emissions. In order to provide instant start capability, FC systems are paired with batteries or ultracapacitors and usually include power electronics and hydrogen fuel storage. Hydrogen is usually supplied and stored at the point of use as a compressed gas.

13.9.4
Portable Applications

FCs can provide electrical power at sites where the grid connection is not available, and can be used as portable power units. For example, in an outdoor vacation site (e.g., a camp site), the use of an FC for electrical power, instead of a diesel generator, prevents harmful emissions and causes no problems of noise in the environment. FCs are also currently being used as supporting units when power shutdowns occur and in military applications [348]. FCs are much lighter and more durable than batteries, which is particularly important for the soldiers during periods of military maneuvers. FCs can also be used in portable military applications, which include battery chargers, navigation systems, sensors, and so on [338]. Moreover, the use of PEMFCs for portable computers (laptops) and mobile phones is encouraging and this idea has found widespread attention from manufacturers [349]. The types of FCs that are used in laptops or mobile phones are called "micro fuel cells" (μ-FCs): they can be fed with either hydrogen (PEMFCs) or methanol (DMFCs) [350]. Other applications for μ-FCs include pagers, hearing aids, smoke detectors, security alarms, and so on [351].

13.9.5
Stationary

Stationary power products cover a range from 1 kW to several megawatts. The possible applications include homes, businesses, schools, hospitals, and so on. These markets are usually served by a central generation system. Residential FC systems can be operated to provide primary or backup power for houses. They can run independently or together with an existing power grid, and produce about 5 kW of power or 120 kWh of energy per day [352]. These systems are usually known as "combined heat and power" (CHP) systems. They are suitable for cogeneration operation: that is, these plants produce electricity and thermal energy. The base-load plants can be fueled by natural gas or commercial hydrocarbons, which undergo reforming and CO cleanup processes to produce the H_2-rich feedstock. The on-site plants are proving to be an economical and beneficial addition to the operating systems of commercial buildings and industrial facilities [353]. Furthermore, advantages in reducing CO_2 emissions have been proved [354].

13.10
Challenges and Perspectives

From a technological perspective, the current PEMFCs still need to be improved considerably before meeting the requirements of most of the possible applications for which they can be used, in particular for road vehicles (which require a widespread hydrogen infrastructure that does not exist at present) [355–357]. Moreover, costs will need to be reduced by an order of magnitude, while durability, efficiency, energy density, and reliability need to be somewhat enhanced. For example, as far as automotive applications are concerned, if FCs do not lead to a decrease in the price per kilowatt, the public will continue to use internal combustion engines to power vehicles. For stationary power applications, both capital and installed costs (the cost per kilowatt required to purchase and install a power system) must be reduced before FCs can compete with contemporary energy generation methods.

A potential barrier to the successful marketing of PEMFC vehicles is the cost of the FC system itself, which has been forecast to have a factory cost (i.e., excluding sale expenses and profits) of \$60–80 kW^{-1}, on the basis of currently achievable PEM technology scaled up for high-volume manufacturing (at least 500 000 units per year) [358]. A PEMFC currently costs \$200–300 kW^{-1} compared to \$30–50 kW^{-1} for an internal combustion engine [359]. To be competitive with conventional internal combustion engine vehicles, the US DOE has targeted a high-volume FC system factory cost of $30 kW^{-1} by 2015 [91].

Owing to the high capital cost on a dollar per kilowatt basis, significant resources have been invested to reduce the costs of FCs. Specific areas in which possible cost reductions are being investigated include the following:

- material reduction and exploration of lower-cost alternative materials;
- reduction of the complexity of the integrated system;
- minimization of temperature constraints (which add complexity and costs to the system);
- streamlining of the manufacturing processes;
- increasing power density (footprint reduction); and
- scaling up production to gain the benefit of economies of scale (volume) through increased market penetration.

In addition, there are still some scientific challenges that drive research efforts toward finding successful solutions to many scientific and technological problems, in order to bring the PEMFC technology close to marketing:

- finding anodic electrocatalysts that are tolerant to CO at levels of 100 ppm (the DOE target for PEMFCs stacks for transportation: platinum group noble metal loading of a max. 0.2 mg cm^{-2} or less) [91]. A reduction in Pt from 0.4 to 0.05 mg cm^{-2} at the anode is at present feasible, but a lowering of the Pt amount well below 0.4 mg cm^{-2} at the cathode would lead to difficulties in the maintenance of high energy conversion efficiencies (>55%) [21, 69];
- inventing a cathodic electrocatalyst in order to reduce the overpotential encountered at OC and to significantly enhance the exchange current density;
- finding alternative proton-conducting membranes with lower costs but with almost the same proton conductivity as the state-of-the-art perfluorosulfonic acid membranes; and
- developing new proton-conducting membranes that do not depend on water for HT operations of between 150 and 200 °C.

The costs of PEMFC manufacturing can be broken down into four main categories: MEA, flow plates, balance of plant, and assembly costs. An analysis of the cost structure shows that BPs and electrodes have a large share of the stack cost and would also be very significant at the mass production stage [360]. An improvement in power density is essential to reduce the overall costs of the stack, as it would decrease the use of other material resources per unit of power output [361]. An overview of how much each PEMFC component weighs on the cost structure and their evolution in case of commercialization at early stages or mass production (up to 5 million pieces) is represented in Figure 13.13.

However, to initially overcome the marketing barrier, niche/early markets for PEMFCs should ideally rely on limited or no hydrogen refueling infrastructures [342]. The US DOE has investigated the technical and economic feasibility of promising new technologies, and has identified three near-term markets for FCs [341]:

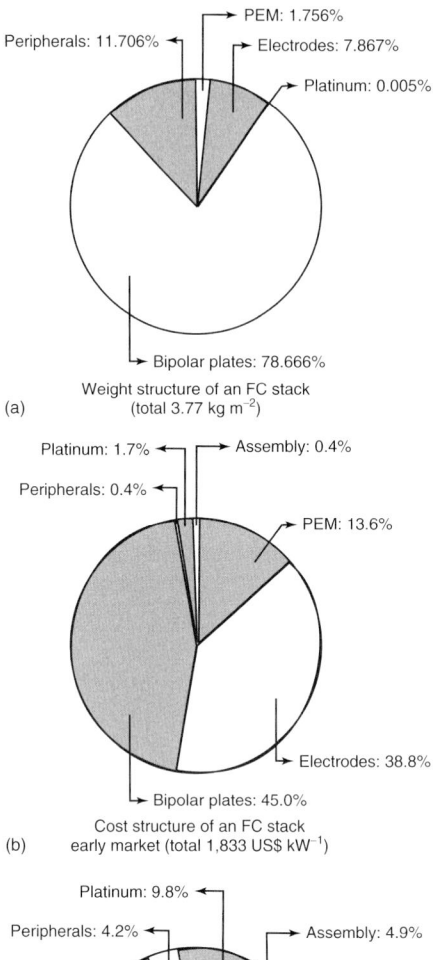

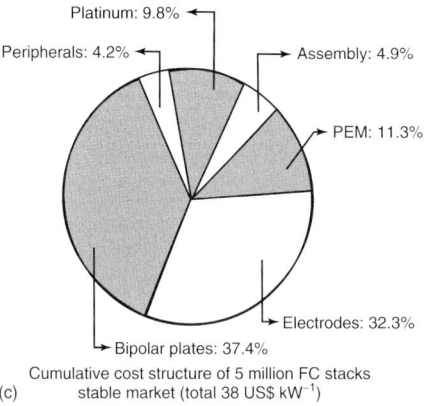

(a) Weight structure of an FC stack (total 3.77 kg m^{-2})
- PEM: 1.756%
- Peripherals: 11.706%
- Electrodes: 7.867%
- Platinum: 0.005%
- Bipolar plates: 78.666%

(b) Cost structure of an FC stack early market (total 1,833 US$ kW^{-1})
- Platinum: 1.7%
- Assembly: 0.4%
- Peripherals: 0.4%
- PEM: 13.6%
- Electrodes: 38.8%
- Bipolar plates: 45.0%

(c) Cumulative cost structure of 5 million FC stacks stable market (total 38 US$ kW^{-1})
- Platinum: 9.8%
- Peripherals: 4.2%
- Assembly: 4.9%
- PEM: 11.3%
- Electrodes: 32.3%
- Bipolar plates: 37.4%

Figure 13.13 Weight structure of a PEMFC stack (a) and its cost evolution for early market production (b) and for mass production (c).

1) emergency power for state and local emergency response agencies;
2) forklifts in warehousing and distribution centers;
3) airport ground support equipment markets.

Furthermore, considering the actual high demand for auxiliary power when trucks are not in use overnight and the recently introduced anti-idling regulations in the United States [362], the US market offers, at least in principle, a good opportunity for diesel oil-fed PEMFC APUs on board long-haul trucks [342]. Considering the target of APU systems for road vehicle commercialization by 2015–2020 forecast by the European Commission [363], the European market for diesel oil-fed PEMFC APUs is also potentially interesting, although more limited in size than the US market. Furthermore, for FC-based APUs on heavy-duty trucks, life cycle assessment (LCA) and comparisons with idling of large-displacement diesel engines (as an alternative technology to produce auxiliary power) have demonstrated that life cycle emissions cannot be neglected in the impact assessment of FC-based APUs [364]. However, even when these emissions are considered, the total amount of pollutant that is released is much lower than in the case of idling of heavy-duty diesel engines [365]. Therefore, diesel oil FC-based APUs have shown a great potential in terms of human health and environmental impact reduction, and the payback period has been estimated as just a little more than two years [364].

References

1. Grove, W.R. (1839) On voltaic series and the combination of gases by platinum. *London Edinburgh Philos. Mag. J. Sci.*, **3** (14), 127–130.
2. http://electrochem.cwru.edu/estir/hist/hist-15-Grove-1.pdf, Historic Papers in Electrochemistry, Electrochemical Science and Technology Information Resource (ESTIR), hosted by the Ernest B. Yeager Center for Electrochemical Science (YCES) and the Chemical Engineering Department, Case Western Reserve University, Cleveland, Ohio, USA. (accessed September 13 2010).
3. Grove, W.R. (1842) On a gaseous voltaic battery. *London Edinburgh Philos. Mag. J. Sci.*, **3** (21), 417–420.
4. Kordesch, K. and Simader, G. (1996) *Fuel Cells and their Applications*, Wiley-VCH Verlag GmbH, Weinheim, Cambridge, p. 38.
5. Kragh, H. (2000) Confusion and controversy: nineteenth-century theories of the voltaic pile, in *Studies on Volta and His Times*, vol. 1 (eds F. Bevilacqua and L. Fregonese), Nuova Voltiana, Pavia, pp. 133–157.
6. Kunze, J. and Stimming, U. (2009) Electrochemical versus heat-engine energy technology: a tribute to Wilhelm Ostwald's visionary statements. *Angew. Chem. Int. Ed.*, **48** (49), 9230–9237.
7. Grimwood, J.M., Hacker, B.C., and Vorzimmer, P.J. (1969) Part I – Concept and Design, in *Project Gemini – A Chronology*, The NASA Historical Series, NASA SP-4002 document, NASA Washington D.C., pp. 1–68. http://ntrs.nasa.gov/archive/nasa/casi.ntrs.nasa.gov/19690027123_1969027123.pdf (accessed September 21 2010).
8. Farmer, G. and Hamblin, D.J. (1970) in *First on the Moon: A Voyage With Neil Armstrong* (eds C. Michael and E.E. Aldrin Jr.), Little, Brown and Company, Boston, Library of Congress, pp. 51–54.

9. Watkins, D.S. (1993), Chapter 11 Research, development, and demonstration of solid polymer fuel cell systems in *Fuel Cell System* (eds L.J.M.J. Blomen and M.N. Mugerwa), Plenum Press, New York, pp. 493–530.
10. Costamagna, P., and Srinivasan, S. (2001) Quantum jumps in the PEMFC science and technology from 1960s to the year 2000, Part II. Engineering, technology development and application aspect. *J. Power Sources*, **102** (1–2), 253–269.
11. Thomas, S. and Zalbowitz, M. (2000) *Fuel Cells Green Power*, Los Alamos National Laboratory, www.education.lanl.gov/resources/fuelcells, http://www.lanl.gov/orgs/mpa/mpa11/Green%20Power.pdf (accessed September 21 2010).
12. Carrette, L., Friedrich, K.A., and Stimming, U. (2001) Fuel cells – fundamentals and applications. *Fuel Cells*, **1** (1), 5–39.
13. Larminie, J. and Dicks, A. (2005) *Fuel Cell Systems Explained*, 2nd edn, John Wiley & Sons, Ltd, Chicester.
14. Scott, K. and Shukla, A.K. (2004) Polymer electrolyte membrane fuel cells: principle and advances. *Rev. Environ. Sci. Biotechnol.*, **3** (3), 273–280.
15. Collier, A., Wang, H., Yuan, X.Z., Zhang, J., and Wilkinson, D.P. (2006) Degradation of polymer electrolyte membranes. *Int. J. Hydrogen Energy*, **31** (13), 1838–1854.
16. Stolten, D. (2010) *Hydrogen and Fuel Cells: Fundamentals, Technologies and Applications*, John Wiley & Sons, Ltd, Chicester.
17. Peighambardoust, S.J., Rowshanzamir, S., and Amjadi, M. (2010) Review of the proton exchange membranes for fuel cell applications. *Int. J. Hydrogen Energy*, **35** (17), 9349–9384.
18. La Conti, A.B., Hamdan, M., and McDonald, R.C. (2003), Chapter 49 Fuel cell technology and applications: mechanisms of membrane degradation in *Handbook of Fuel Cells: Fundamentals, Technology, and Applications*, vol. 3 (eds W. Vielstich, A.Lamm, and H. Gasteiger), John Wiley & Sons, Inc., New York, pp. 647–662.
19. Cai, Y., Hu, J., Ma, H., Yi, B., and Zhang, H. (2006) Effect of water transport properties on a PEM fuel cell operating with dry hydrogen. *Electrochim. Acta*, **51** (28), 6361–6366.
20. Dai, W., Wang, H., Yuan, X.-Z., Martin, J.J., Yang, D., Qiao, J., and Ma, J. (2009) A review on water balance in the membrane electrode assembly of proton exchange membrane fuel cells. *Int. J. Hydrogen Energy*, **34** (23), 9461–9478.
21. Gasteiger, H.A., Kocha, S.S., Sompalli, B., and Wagner, F.T. (2005) Activity benchmarks and requirements for Pt, Pt-alloy and non-Pt oxygen reduction catalysts for PEMFCs. *Appl. Catal. B: Environ.*, **56** (1–2), 9–35.
22. Zhang, J., Xie, Z., Zhang, J., Tang, Y., Song, C., Navessin, T., Shi, Z., Song, D., Wang, H., Wilkinson, D.P., Liu, Z.-S., and Holdcroft, S. (2006) High temperature PEM fuel cells. *J. Power Sources*, **160** (6), 872–891.
23. Borup, R., Meyers, J., Pivovar, B., Kim, Y.S., Mukundan, R., Garland, N., Myers, D., Wilson, M., Garzon, F., Wood, D., Zelenay, P., More, K., Stroh, K., Zawodzinski, T., Boncella, J., McGrath, J.E., Inaba, M., Miyatake, K., Hori, M., Ota, K., Ogumi, Z., Miyata, S., Nishikata, A., Siroma, Z., Uchimoto, Y., Yasuda, K., Kimijima, K., and Iwashita, N. (2007) Scientific aspects of polymer electrolyte fuel cell durability and degradation. *Chem. Rev.*, **107** (10), 3904–3951.
24. Hefner, R.A. (1995) The age of energy gases. *Int. J. Hydrogen Energy*, **20** (12), 945–948.
25. Turner, J.A. (1999) A realizable renewable energy future. *Science*, **285** (5428), 687–689.
26. Kolb, G., Baier, T., Schürer, J., Tiemann, D., Ziogas, A., Ehwald, H., and Alphonse, P. (2008) Micro-structured 5 kW complete fuel processor for iso-octane as hydrogen supply system for mobile auxiliary power units: Part I. Development of autothermal reforming catalyst and reactor. *Chem. Eng. J.*, **137** (3), 653–663.

27. Kolb, G., Baier, T., Schürer, J., Tiemann, D., Ziogas, A., Specchia, S., Galletti, C., Germani, G., and Schuurman, Y. (2008) A micro-structured 5 kW complete fuel processor for iso-octane as hydrogen supply system for mobile auxiliary power units Part II – development of water-gas shift and preferential oxidation catalysts and reactors and assembly of the fuel processor. *Chem. Eng. J.*, **138** (1–3), 474–489.
28. Specchia, S. and Specchia, V. (2010) Modeling study on the performance of an integrated APU fed with hydrocarbon fuels. *Ind. Eng. Chem. Res.*, **49** (15), 6803–6809.
29. Yi, J.S. and Nguyen, T.V. (1998) An along-the channel-model for proton exchange membrane fuel cells. *J. Electrochem. Soc.*, **145** (4), 1149–1159.
30. Kandlikar, S.G. and Lu, Z. (2009) Thermal management issue in PEMFC stack – A brief review of current status. *Appl. Therm. Eng.*, **29** (7), 1276–1280.
31. Shao, Y., Yin, G., Wang, Z., and Gao, Y. (2007) Proton exchange membrane fuel cell from low temperature to high temperature: material challenges. *J. Power Sources*, **167** (2), 235–242.
32. Li, Q., He, R., Gao, J.-A., Jensen, J.O., and Bjerrum, N.J. (2003) The CO poisoning effect in PEMFCs operational at temperatures up to 200 °C. *J. Electrochem. Soc.*, **150** (12), A1599–A1606.
33. Mann, R.F., Amphlett, J.C., Peppley, B.A., and Thurgood, C.P. (2006) Application of the Butler-Volmer equations in the modeling of activation polarization for PEM fuel cells. *J. Power Sources*, **161** (2), 775–781.
34. Gasteiger, H.A., Gu, W., Makharia, R., Mathias, M.F., and Sompalli, B. (2003) Chapter 46 Beginning-of-life MEA performance-efficiency loss contributions in *Handbook of Fuel Cells: Fundamentals, Technology, and Applications*, vol. 3 (eds W. Vielstich, A. Lamm, and H.A. Gasteiger), John Wiley & Sons, Inc., New York, pp. 593–610.
35. Tarasevich, M.R., Sadkowski, A., and Yeager, E. (1983), Chapter 6 Oxygen Electrochemistry in *Comprehensive Treatise of Electrochemistry, Kinetics and Mechanism of Electrode Processes*, vol. 7 (eds B.E. Conway, J.O'M. Bockris, E. Yeager, S. Khan, and R.E. White), Plenum Press, New York, pp. 301–398.
36. Francia, C., Ijeri, V.S., Specchia, S., and Spinelli, P. (2011) Estimation of hydrogen crossover through Nafion® membranes in PEMFC. *J. Power Sources*, **196** (4), 1833–1839.
37. Vilekar, S.A., and Datta, R. (2010) The effect of hydrogen crossover on open-circuit voltage in polymer electrolyte membrane fuel cells. *J. Power Sources*, **195** (8), 2241–2247.
38. Bockris, J.O'M. and Reddy, A.K.N. (1972) *Modern Electrochemistry*, vol. 2, Plenum Press, New York, p. 1141.
39. Appleby, A.J. (1983), Chapter 4 Electrocatalysis, in *Comprehensive Treatise of Electrochemistry, Kinetics and Mechanism of Electrode Processes*, vol. 7 (eds B.E. Conway, J.O'M. Bockris, E. Yeager, S.U.M. Khan, and R.E. White) Plenum Press, New York, pp. 173–239.
40. Alonso Vante, N., Schubert, B., Tributsch, H., and Perrin, A. (1988) Influence of d-state density and chemistry of transition metal cluster selenides on electrocatalysis. *J. Catal.*, **112** (2), 384–391.
41. Costamagna, P. and Srinivasan, S. (2001) Quantum jumps in the PEMFC science and technology from the 1960s to the year 2000 Part I. Fundamental scientific aspects. *J. Power Sources*, **102** (1–2), 242–252.
42. Bockris, J.O.M. (2005) A primer on electrocatalysis. *J. Serb. Chem. Soc.*, **70** (3), 475–487.
43. Song, C. and Zhang, J. (2008), Chapter 2 Electrocatalytic oxygen reduction reaction in *PEMFC Electrocatalysts and Catalytic Layers: Fundamental and Applications* (ed. J. Zhang), Springer, New York, pp. 89–134.
44. Wroblowa, H.S., Pan, Y.-C., and Razumney, G. (1976) Electroreduction of oxygen – A new mechanistic criterion. *J. Electroanal. Chem.*, **69** (2), 195–201.

45. Bagotskii, V.S., Tarasevich, M.R., and Filinovskii, V.Y. (1969) Calculation of the kinetic parameters of conjugated reactions of oxygen and hydrogen peroxide. *Elektrokhimiya*, **5**, 1218–1226.
46. Zhdanov, V.P. and Kasemo, B. (2006) Kinetics of electrochemical O_2 reduction on Pt. *Electrochem. Commun.*, **8** (7), 1132–1136.
47. Hoare, J.P. (1968), Chapter 4 The Chatodic Reduction of Oxygen in *The Electrochemistry of Oxygen*, John Wiley & Sons, Inc., New York, pp. 117–142.
48. Zhang, L., Zhang, J., Wilkinson, D.P., and Wang, H. (2006) Progress in preparation of non-noble electrocatalysts for PEM fuel cell reactions. *J. Power Sources*, **156** (2), 171–182.
49. Lee, S.J., Mukerjee, S., Ticianelli, E.A., and McBreen, J. (1999) Electrocatalysis of CO tolerance in hydrogen oxidation reaction in PEM fuel cells. *Electrochim. Acta*, **44** (19), 3283–3293.
50. Niedrach, L.W., McKee, D.W., Paynter, J., and Danzig, I.F. (1967) Electrocatalysts for hydrogen/carbon monoxide fuel cell anodes. *Electrochem. Technol.*, **5** (7–8), 318–323.
51. McKee, D.W. and Scarpellino, A.J. Jr. (1968) Electrocatalysts for hydrogen/carbon monoxide fuel cell anodes. III. The behavior of supported binary noble metals. *Electrochem. Technol.*, **6** (3–4), 101–105.
52. Wang, K., Gasteiger, H.A., Markovic, N.M., and Ross, P.N., Jr. (1996) On the reaction pathway for methanol and carbon monoxide electrooxidation on Pt-Sn alloy versus Pt-Ru alloy surfaces. *Electrochim. Acta*, **41** (16), 2587–2593.
53. McKee, D.W., Niedrach, L.W., Paynter, J., and Danzig, I.F. (1967) Electrocatalysts for hydrogen/carbon monoxide fuel cell anodes. II. The platinum-rhodium and platinum-iridium systems. *Electrochem. Technol.*, **5** (9–10), 419–423.
54. Grgur, B.N., Zhuang, G., Markovic, N.M., and Ross, P.N. Jr. (1998) Electrooxidation of H_2, CO, and H_2/CO mixtures on a well-characterized $Pt_{70}Mo_{30}$ bulk alloy electrode. *J. Phys. Chem. B*, **102** (14), 2494–2501.
55. Niedrach, L.W. and Weinstock, I.B. (1965) Performance of carbon monoxide in low-temperature fuel cells containing oxide catalysts. *Electrochem. Technol.*, **3** (7–8), 270–275.
56. Freitas, K.S., Lopes, P.P., and Ticianelli, E.A. (2010) Electrocatalysis of the hydrogen oxidation in the presence of CO RhO_2/C-supported Pt nanoparticles. *Electrochim. Acta*, **56** (1), 418–426.
57. Sethuraman, V.A. and Weidner, J.W. (2010) Analysis of sulfur poisoning on a PEM fuel cell electrode. *Electrochim. Acta*, **55** (20), 5683–5694.
58. Cheng, X., Shi, Z., Glass, N., Zhang, L., Zhang, J., Song, D., Liu, Z.-S., Wang, H., and Shen, J. (2007) A review of PEM hydrogen fuel cell contamination: impacts, mechanisms, and mitigation. *J. Power Sources*, **165** (2), 739–756.
59. Baez, V.B. and Pletcher, D. (1995) Preparation and characterization of carbon/titanium dioxide surfaces – the reduction of oxygen. *J. Electroanal. Chem.*, **382** (1–2), 59–64.
60. Zhang, M., Yan, Y., Gong, K., Mao, L., Guo, Z., and Chen, Y. (2004) Electrostatic layer by layer assembled carbon nanotube mutilayer film and its catalytic activity for oxygen reduction reaction. *Langmuir*, **20** (20), 8781–8785.
61. Jia, N., Martin, R.B., Qi, Z., Lefebvre, M.C., and Pickup, P.G. (2001) Modification of carbon supported catalysts to improve performance in gas diffusion electrodes. *Electrochim. Acta*, **46** (18), 2863–2859.
62. Ambrosio, E.P., Dumitrescu, M.A., Francia, C., Gerbaldi, C., and Spinelli, P. (2009) Ordered mesoporous carbons as catalyst support for PEM fuel cells. *Fuel Cells*, **9** (3), 197–200.
63. Guilminot, E., Fischer, F., Chatenet, M., Rigacci, A., Berthon-Fabry, S., Achard, P., and Chainet, E. (2007) Use of cellulose-based carbon aerogels as catalyst support for PEM fuel cell electrodes: electrochemical characterization. *J. Power Sources*, **166** (1), 104–111.

64. Babić, B.M., Vračar, L.M., Radmilović, V., and Krstajić, N.V. (2006) Carbon cryogel as support of platinum nano-sized electrocatalyst for the hydrogen oxidation reaction. *Electrochim. Acta*, **51** (18), 3820–3826.
65. Ball, S.C., Hudson, S.L., Thompsett, D., and Theobald, B. (2007) An investigation into factors affecting the stability of carbons and carbon supported platinum and platinum/cobalt alloy catalysts during 1.2 V potentiostatic hold regimes at a range of temperatures. *J. Power Sources*, **171** (1), 18–25.
66. Maass, S., Finsterwalder, F., Frank, G., Hartmann, R., and Merten, C. (2008) Carbon support oxidation in PEM fuel cell cathodes. *J. Power Sources*, **176** (2), 444–451.
67. Ferreira, P.J., la O', G.J., Shao-Horn, Y., Morgan, D., Makharia, R., Kocha, S., and Gasteiger, H.A. (2005) Instability of Pt/C electrocatalysts in proton exchange membrane fuel cells – a mechanistic investigation. *J. Electrochem. Soc.*, **152** (11), A2256–A2271.
68. Litster, S. and McLean, G. (2004) Review PEM fuel cell electrodes. *J. Power Sources*, **130** (1–2), 61–76.
69. Yu, X. and Ye, S. (2007) Recent advances in activity and durability enhancement of Pt/C catalytic cathode in PEMFC Part I: Physico-chemical and electronic interaction between Pt and carbon support, and activity enhancement of Pt/C catalyst. *J. Power Sources*, **172** (11), 133–144.
70. Min, M., Cho, J., Cho, K., and Kim, H. (2000) Particle size and alloying effects of Pt-based alloy catalysts for fuel cell applications. *Electrochim. Acta*, **45** (25–26), 4211–4217.
71. Mukerjee, S., Srinivasan, S., Soriaga, M.P., and McBreen, J. (1995) Role of structural and electronic properties of Pt and Pt alloys on electrocatalysis of oxygen reduction. *J. Electrochem. Soc.*, **142** (5), 1409–1422.
72. Wang, B. (2005) Recent development of non-platinum catalysts for oxygen reduction reaction. *J. Power Sources*, **152** (2005), 1–5.
73. Bezerra, C.W.B., Zhang, L., Liu, H., Marques, A.L.B., Marques, E.P., Wang, H., and Zhang, J. (2008) A review of Fe-N/C and Co-N/C catalysts for the oxygen reduction reaction. *Electrochim. Acta*, **53** (15), 4937–4951.
74. Wilson, M.S. (1993) Membrane catalyst layer for fuel cells. US Patent 5, 234, 777.
75. Lee, S.J., Mukerjee, S., McBreen, J., Rho, Y.W., Kho, Y.T., and Lee, T.H. (1998) Effect of Nafion impregnation on performances of PEMFC electrodes. *Electrochim. Acta*, **43** (24), 3693–3701.
76. O'Hayre, R., Lee, S.J., Cha, S.W., and Prinz, F.B. (2002) A sharp peak in the performance of sputtered platinum fuel cells at ultra-low platinum loading. *J. Power Sources*, **109** (2), 483–493.
77. Cha, S.Y. and Lee, W.M. (1999) Performance of proton exchange membrane fuel cell electrodes prepared by direct decomposition of ultrathin platinum on the membrane surface. *J. Electrochem. Soc.*, **146** (11), 4055–4060.
78. Verbrugge, M. (1994) Selective electrodeposition of catalyst within membrane-electrode structures. *J. Electrochem. Soc.*, **141** (1), 46–53.
79. Chaparro, A.M., Benìtez, R., Gubler, L., Scherer, G.G., and Daza, L. (2007) Study of membrane electrode assemblies for PEMFC, with cathodes prepared by electrospray method. *J. Power Sources*, **169** (1), 77–84.
80. Martin, S., Garcia-Ybarra, P.L., and Castillo, J.L. (2010) High platinum utilization in ultra-low Pt loaded PEM fuel cell cathodes prepared by electrospraying. *Int. J. Hydrogen Energy*, **35** (19), 10446–10451.
81. Benìtez, R., Soler, J., and Daza, L. (2005) Novel method for preparation of PEMFC electrodes by electrospray technique. *J. Power Sources*, **151**, 108–113.
82. Martin, S., Garcia-Ybarra, P.L., and Castillo, J.L. (2010) Electrospray deposition of catalyst layers with ultra-low Pt loadings for PEMFC cathodes. *J. Power Sources*, **195** (9), 2443–2449.
83. Kim, C.S., Chun, Y.G., Peck, D.H., and Shin, D.R. (1998) A novel process to fabricate membrane electrode assemblies for proton exchange membrane fuel cells. *Int. J. Hydrogen Energy*, **23** (11), 1045–1048.

84. Fialkov, A.S. (2000) Carbon application in chemical power sources. *Russ. J. Electrochem.*, **36** (4), 345–366.
85. Antolini, E. (2009) Carbon supports for low-temperature fuel cell catalysts. *Appl. Catal. B: Environ.*, **88** (1–2), 1–24.
86. Yu, X. and Ye, S. (2007) Recent advances in activity and durability enhancement of Pt/C catalytic cathode in PEMFC. Part II: degradation mechanism and durability enhancement of carbon supported platinum catalyst. *J. Power Sources*, **172** (1), 145–154.
87. Shao, Y., Yin, G., and Gao, Y. (2007) Understanding and approaches for the durability issues of Pt-based catalysts for PEM fuel cell. *J. Power Sources*, **171** (2), 558–566.
88. Cindrella, L., Kanna, A.M., Lin, J.F., Saminathan, K., Ho, Y., Lin, C.W., and Wertz, J. (2009) Gas diffusion layer for proton exchange membrane fuel cells – A review. *J. Power Sources*, **194** (1), 146–160.
89. Stone, C. and Morrison, A.E. (2002) From curiosity to "power to change the world®". *Solid State Ionics*, **152–153**, 1–13.
90. Kerres, J.A. (2001) State of art of membrane development. *J. Membr. Sci.*, **185** (1), 3–27.
91. DOE Guidelines (2007)Multi-year research, development and demonstration plan: planned program activities for 2005–2015. Technical Plan Fuel Cells. 3.4, pp. 1–42, http://www1.eere.energy.gov/hydrogenandfuelcells/mypp/pdfs/fuel_cells.pdf (accessed January 25 2010).
92. Mc Donald, R.C., Mittelsteadt, C.K., and Thompson, E.L. (2004) Effect of deep temperature cycling on Nafion 112 membranes and membrane electrode assemblies. *Fuel Cells*, **4** (3), 208–213.
93. Rikukawa, M. and Sanui, K. (2000) Proton conducting polymer electrolyte membranes based on hydrocarbon polymers. *Prog. Polym. Sci.*, **25** (10), 1463–1502.
94. Smitha, B., Sridhar, S., and Khan, A.A. (2005) Solid polymer electrolyte membranes for fuel cell applications- a review. *J. Membr. Sci.*, **259** (1–2), 10–26.
95. Bahar, B., Hobson, A.R., Kolde, J.A., and Zuckerbrod, D. (1996) Ultra-thin integral composite membrane. US Patent 5, 547, 551.
96. Park, K.T., Jung, U.H., Choi, D.W., Chun, K., Lee, H.M., and Kim, S.H. (2008) ZrO_2-SiO_2/Nafion composite membrane for polymer electrolyte membrane fuel cells operation at high temperature and low humidity. *J. Power Sources*, **177** (2), 247–253.
97. Jalani, N.H., Dunn, K., and Datta, R. (2005) Synthesis and characterization of Nafion-MO_2 (M = Zr, Si, Ti) nanocomposite membranes for higher temperature PEM fuel cells. *Electrochim. Acta*, **51** (3), 553–560.
98. Yang, T. (2008) A Nafion-based self-humidifying membrane with ordered dispersed Pt layer. *Int. J. Hydrogen Energy*, **33** (10), 2530–2535.
99. Savadogo, O. (2004) Emerging membranes for electrochemical systems. Part II. High temperature composite membranes for polymer electrolyte fuel cell (PEFC) applications. *J. Power Sources*, **127** (1–2), 135–161.
100. Wang, L., Xing, D.M., Liu, Y.H., Cai, Y.H., Shao, Z.-G., Zhai, Y.F., Zhong, H.X., Yi, B.L., and Zhang, H.M. (2006) Pt/SiO_2 catalyst as an addition to Nafion/PTFE self-humidifying composite membrane. *J. Power Sources*, **161** (1), 61–67.
101. Son, D.-H., Sharma, R.K., Shul, Y.-G., and Kim, H. (2007) Preparation of Pt/Zeolite-Nafion composite membranes for self-humidifying polymer electrolyte fuel cells. *J. Power Sources*, **165** (2), 733–738.
102. Lee, H.-K., Kim, J.-I., Park, J.-H., and Lee, T.-H. (2004) A study on self-humidifying PEMFC using Pt-ZrP-Nafion composite membrane. *Electrochim. Acta*, **50** (2–3), 761–768.
103. Zhang, Y., Zhang, H., Zhai, Y., Zhu, X., and Bi, C. (2007) Investigation of self-humidifying membranes based on sulfonated poly(ether ether ketone) hybrid with sulfated zirconia supported Pt catalyst for fuel cell applications. *J. Power Sources*, **168** (2), 323–329.

104. Kerrese, J., Ullrich, A., Meier, F., and Häring, T. (1999) Synthesis and characterization of novel acid-base polymer blends for application in membrane fuel cells. *Solid State Ionics*, **125** (1–4), 243–249.
105. Zhai, Y., Zhang, H., Zhang, Y., and Xing, D. (2007) A novel H_3PO_4/Nafion-PBI composite membrane for enhances durability of high temperature PEM fuel cells. *J. Power Sources*, **169** (2), 259–264.
106. Fu, Y., Hou, M., Yan, X., Hou, J., Luo, X., Shao, Z., and Yi, B. (2007) Research progress of aluminium alloy endplates for PEMFC. *J. Power Sources*, **166** (2), 435–440.
107. Hentall, P.L., Lakeman, J.B., Mepsted, G.O., Adcock, P.L., and Moore, J.M. (1999) New materials for polymer electrolyte membrane fuel cell current collectors. *J. Power Sources*, **80** (1–2), 235–241.
108. Kumar, A. and Reddy, R.G. (2004) Materials and design development for bipolar/end plates in fuel cells. *J. Power Sources*, **129** (1), 62–67.
109. Li, X. and Sabir, I. (2005) Review of bipolar plates in PEM fuel cells: flow-field designs. *Int. J. Hydrogen Energy*, **30** (4), 359–371.
110. Cunningham, B. and Baird, D.G. (2006) The development of economical bipolar plates for fuel cells. *J. Mater. Chem.*, **16** (45), 4385–4388.
111. Hermann, A., Chaudhuri, T., and Spagnol, P. (2005) Bipolar plates for PEM fuel cells: a review. *Int. J. Hydrogen Energy*, **30** (12), 1297–1302.
112. Tawfik, H., Hung, Y., and Mahajan, D. (2007) Metal bipolar plates for PEM fuel cell A review. *J. Power Sources*, **163** (2), 755–767.
113. Dihrab, S.S., Sopian, K., Alghoul, M.A., and Sulaiman, M.Y. (2009) Review of the membrane and bipolar plates materials for conventional and unitized regenerative fuel cells. *Renew. Sustain. Energy Rev.*, **13** (6–7), 1663–1668.
114. Wind, J., Späh, R., Kaiser, W., and Böhm, G. (2002) Metallic bipolar plates for PEM fuel cells. *J. Power Sources*, **105** (2), 256–260.
115. Kraytsberg, A., Auinat, M., and Ein-Eli, Y. (2007) Reduced contact resistance of PEM fuel cell's bipolar plates via surface texturing. *J. Power Sources*, **164** (2), 697–703.
116. Avasarala, B. and Haldar, P. (2009) Effect of surface roughness of composite bipolar plates on the contact resistance of a proton exchange membrane fuel cell. *J. Power Sources*, **188** (1), 225–229.
117. Blunk, R., Elhamid, M.H.A., Lisi, D., and Mikhail, Y. (2006) Polymeric composite bipolar plates for vehicle applications. *J. Power Sources*, **156** (2), 151–157.
118. Antunes, R.A., Oliveira, M.C.L., Ett, G., and Ett, V. (2010) Corrosion of metal bipolar plates for PEM fuel cells: a review. *Int. J. Hydrogen Energy*, **35** (8), 3632–3647.
119. André, J., Antoni, L., and Petit, J.-P. (2010) Corrosion resistance of stainless steel bipolar plates in a PEFC environment: A comprehensive study. *Int. J. Hydrogen Energy*, **35** (8), 3684–3697.
120. Fu, Y., Hou, M., Xu, H., Hou, Z., Ming, P., Shao, Z., and Yi, B. (2008) Ag–polytetrafluoroethylene composite coating on stainless steel as bipolar plate of proton exchange membrane fuel cell. *J. Power Sources*, **182** (2), 580–584.
121. Show, Y. (2007) Electrically conductive amorphous carbon coating on metal bipolar plates for PEFC. *Surf. Coat. Technol.*, **202** (4–7), 1252–1255.
122. Wang, S.-H., Peng, J., Lui, W.-B., and Zhang, J.-S. (2006) Performance of the gold-plated titanium bipolar plates for the light weight PEM fuel cells. *J. Power Sources*, **162** (1), 486–491.
123. Wang, Y. and Northwood, D.O. (2007) An investigation into TiN-coated 316L stainless steel as a bipolar plate material for PEM fuel cells. *J. Power Sources*, **165** (1), 293–298.
124. Ho, W.-Y., Pan, H.-J., Chang, C.-L., Wang, D.-Y., and Hwang, J.J. (2007) Corrosion and electrical properties of multi-layered coatings on stainless steel for PEMFC bipolar plate applications. *Surf. Coat. Technol.*, **202** (4–7), 1297–1301.

125. Wang, H., Turner, J.A., Li, X., and Teeter, G. (2008) Process modification for coating SnO$_2$:F on stainless steels for PEM fuel cell bipolar plates. *J. Power Sources*, **1** 78 (1), 238–247.
126. Choi, H.S., Han, D.H., Hong, W.H., and Lee, J.J. (2009) (Titanium, chromium) nitride coatings for bipolar plate of polymer electrolyte membrane fuel cell. *J. Power Sources*, **189** (2), 966–971.
127. Heinzel, A., Mahlendorf, F., Niemzig, O., and Kreuz, C. (2004) Injection moulded low cost bipolar plates for PEM fuel cells. *J. Power Sources*, **131** (1–2), 35–40.
128. Hontañón, E., Escudero, M.J., Bautista, C., García-Ybarra, P.L., and Daza, L. (2001) Optimization of flow-field in polymer electrolyte membrane fuel cells using computational fluid dynamics technique. *J. Power Sources*, **86** (1–2), 363–368.
129. Kumar, A. and Reddy, R.G. (2003) Modeling of polymer electrolyte membrane fuel cell with metal foam in the flow-field of the bipolar/end plates. *J. Power Sources*, **114** (1), 54–62.
130. Kumar, A. and Reddy, R.G. (2003) Effect of channel dimensions and shape in the flow-field distributor on the performance of polymer electrolyte membrane fuel cells. *J. Power Sources*, **113** (1), 11–18.
131. Icardi, U.A., Specchia, S., Saracco, G., and Specchia, V. (2008) Modeling of the anode flow fields in DMFCs. *Int. J. Chem. Reactor Eng.*, **6**, A48.
132. Aricò, A.S., Srinivasan, S., and Antonucci, V. (2001) DMFCs: from fundamental aspects to technology development. *Fuel Cells*, **1** (1), 133–161.
133. Hertwig, K., Martens, L., and Harwoth, R. (2002) Mathematical modeling and simulation of polymer electrolyte membrane fuel cells. Part I: model structures and solving an isothermal one-cell model. *Fuel Cells*, **2** (2), 61–77.
134. Tüber, K., Oedegaard, A., Hermann, M., and Hebling, C. (2004) Investigation of fractal flow fields in portable proton exchange membrane and direct methanol fuel cells. *J. Power Sources*, **131** (1–2), 175–181.
135. Yang, H., Zhao, T.S., and Ye, Q. (2004) Addition of non-reacting gases to the anode flow field of DMFCs leading to improved performance. *Electrochem. Commun.*, **6** (11), 1098–1103.
136. Yoon, Y.G., Lee, W.Y., Park, G.G., Yang, T.H., and Kim, C.S. (2004) Optimization of bifunctional electrocatalyst for PEM unitized regenerative fuel cell. *Electrochim. Acta*, **50** (2–3), 705–708.
137. Cha, S.W., O'Hayre, R., Saito, Y., and Prinz, F.B. (2004) The scaling behavior of flow patterns: a model investigation. *J. Power Sources*, **134** (1), 57–71.
138. Lozano, A., Valino, L., Barreras, F., and Mustata, R. (2008) Fluid dynamics performance of different bipolar plates Part II. Flow through the diffusion layer. *J. Power Sources*, **179** (2), 711–722.
139. López, A.M., Barreras, F., Lozano, A., García, J.A., Valiño, L., and Mustata, R. (2009) Comparison of water management between two bipolar plate flow-field geometries in proton exchange membrane fuel cells at low-density current range. *J. Power Sources*, **192** (1), 94–99.
140. Zhu, X., Liao, Q., Sui, P.C., and Djilali, N. (2010) Numerical investigation of water droplet dynamics in a low-temperature fuel cell microchannel: effect of channel geometry. *J. Power Sources*, **195** (3), 801–812.
141. Al-Baghdadi, M.A.R.S. and Al-Janabi, H.A.K.S. (2007) Effect of operating parameters on the hydro-thermal stresses in proton exchange membranes of fuel cells. *Int. J. Hydrogen Energy*, **32** (17), 4510–4522.
142. Knights, S.D., Colbow, K.M., St-Pierre, J., and Wilkinson, D.P. (2004) Aging mechanisms and lifetime of PEFC and DMFC. *J. Power Sources*, **127** (1–2), 127–134.
143. Jen, T.C., Tan, T., and Chan, S.H. (2003) Chemical reacting transport phenomena in a PEM fuel cell. *Int. J. Heat Mass Transfer*, **46** (22), 4157–4168.
144. Senn, S.M. and Poulikakos, D. (2004) Polymer electrolyte fuel cells with porous materials as fluid distributor

and comparisons with traditional channeled systems. *ASME J. Heat Transfer*, **126** (3), 410–418.
145. Um, S. and Wang, C.Y. (2004) Three-dimensional analysis of transport and electrochemical reactions in polymer electrolyte fuel cells. *J. Power Sources*, **125** (2), 40–51.
146. Hu, G., Fan, J., Chen, S., Liu, Y., and Cen, K. (2004) Three-dimensional numerical analysis of proton exchange membrane fuel cells (PEMFCs) with conventional and interdigitated flow fields. *J. Power Sources*, **136** (1), 1–9.
147. Barreras, F., Lozano, A., Valino, L., Mustata, R., and Martin, C. (2008) Fluid dynamics performance of different bipolar plates, part I. velocity and pressure fields. *J. Power Sources*, **175** (2), 841–850.
148. Alfredo, P., Vega-Leal, F., Palomo, R., Barragán, F., García, C., and Brey, J.J. (2007) Design of control systems for portable PEM fuel cells. *J. Power Sources*, **169** (10), 194–197.
149. Pukrushpan, J.T., Peng, H., and Stefanopoulou, A.G. (2004) Modeling and analysis of fuel cell reactant flow for automotive applications. *J. Dyn. Syst., Meas. Control*, **126** (1), 14–25.
150. Al-Durra, A., Yurkovich, S., and Guezennec, Y. (2010) Study of nonlinear control schemes for an automotive traction PEM fuel cell system. *Int. J. Hydrogen Energy*, **35** (20), 11291–11307.
151. Gao, F., Blunier, B., Miraoui, A., and El-Moudni, A. (2009) Cell layer level generalized dynamic modeling of a PEMFC stack using VHDL-AMS language. *Int. J. Hydrogen Energy*, **34** (13), 5498–5521.
152. Pukrushpan, J.T., Stefanopoulou, A.G., and Peng, H. (2004) Control of Fuel Cell breathing. *IEEE Control Sys. Mag.*, **24** (2), 30–46.
153. Ceraolo, M., Miulli, C., and Pozio, A. (2003) Modelling static and dynamic behavior of proton exchange membrane fuel cells on the basis of electrochemical description. *J. Power Sources*, **113** (1), 131–144.
154. Deng, X., Liu, G., Wang, G., and Tan, M. (2009) Modeling and identification of a PEM fuel cell humidification system. *J. Control Theory Appl.*, **7** (4), 373–378.
155. Chu, D. and Jiang, R. (1999) Comparative studies of polymer membrane fuel cell stack and single cell. *J. Power Sources*, **80** (1–2), 226–234.
156. Bonville, L., Kunz, H., Song, Y., Mientek, A., Williams, M., Ching, A., and Fenton, J.M. (2005) Development and demonstration of a higher temperature PEM fuel cell stack. *J. Power Sources*, **144** (1), 107–112.
157. Wu, J., Yuan, X.-Z., Martin, J.J., Wang, H., Yang, D., Qiao, J., and Ma, J. (2010) Proton exchange membrane fuel cell degradation under close to open-circuit conditions Part I: in situ diagnosis. *J. Power Sources*, **195** (4), 1171–1176.
158. Barbir, F. (2005), Chapter 8 Fuel Cell Diagnostics in *PEM Fuel Cells: Theory and Practice*, Elsevier Academic Press, New York, pp. 249–270.
159. Stumper, J. and Stone, C. (2008) Recent advances in fuel cell technology at Ballard. *J. Power Sources*, **176** (2), 468–476.
160. Miller, M. and Bazylak, A. (2010) A review of polymer electrolyte membrane fuel cell stack testing. *J. Power Sources*, **196** (2), 601–613.
161. Luo, Z., Li, D., Tang, H., Pan, M., and Ruan, R. (2006) Degradation behavior of membrane-electrode-assembly materials in 10-cell PEMFC stack. *Int. J. Hydrogen Energy*, **31** (13), 1831–1837.
162. Wasterlain, S., Candusso, D., Harel, F., Hissel, D., and Xavier, F. (2011) Developments of new test instruments and protocols for the diagnostic of fuel cell stacks. *J. Power Sources*, **196** (12), 5325–5333.
163. Kim, H.-T., Song, K.-Y., Reschetenko, T.V., Han, S.-I., Kim, T.-Y., Cho, S.-Y., Min, M.-K., Chai, G.-S., and Shin, S.C. (2009) Electrochemical analysis of polymer electrolyte membrane fuel cell operated with dry-air feed. *J. Power Sources*, **193** (2), 515–522.

164. Wruck, W.J., Mochado, R.M., and Chapman, T. (1987) Current interruption: instrumentation and application. *J. Electrochem. Soc.*, **134** (3), 539–546.
165. Mennola, T., Mikkola, M., and Noponen, M. (2002) Measurements of ohmic voltage losses in individual cells of a PEMFC stack. *J. Power Sources*, **112** (1), 261–272.
166. Abe, T., Shima, H., Watanabe, K., and Ito, Y. (2004) Study of PEMFCs by AC impedance, current interrupt, and dew points measurements. *J. Electrochem. Soc.*, **151** (1), A101–A105.
167. Yuan, X.Z., Sun, J.C., Wang, H., and Zhang, J. (2006) AC impedance diagnosis of a 500 W PEM fuel cell stack: part II: individual cell impedance. *J. Power Sources*, **161** (2), 929–937.
168. Yuan, X., Wang, H., Sun, J.C., and Zhang, J. (2007) AC impedance technique in PEM fuel cell diagnosis- A review. *Int. J. Hydrogen Energy*, **32** (17), 4365–4380.
169. Yousfi-Steiner, N., Moçotéguy, P., Candusso, D., Hissel, D., Hernandez, A., and Aslanides, A. (2008) A review on PEM voltage degradation associated with water management: impacts, influent factors and characterization. *J. Power Sources*, **183** (1), 260–274.
170. Le Canut, J.M., Abouatallah, R., and Harrington, D. (2006) Detection of membrane drying, fuel cell flooding, and anode catalyst poisoning on PEMFC stacks by electrochemical impedance spectroscopy. *J. Electrochem. Soc.*, **153** (5), A857–A864.
171. Ciureanu, M. (2004) Effect of Nafion dehydration in PEM fuel cells. *J. Appl. Electrochem.*, **34** (7), 705–714.
172. Yan, X., Hou, M., Sun, L., Liang, D., Shen, Q., Xu, H., Ming, P., and Yi, B. (2007) AC impedance characteristics of a 2 kW PEM fuel cell stack under different operating conditions and load changes. *Int. J. Hydrogen Energy*, **32** (17), 4358–4364.
173. Paganin, V.A., Oliveira, C.L.F., Ticianelli, E.A., Springer, T.E., and Gonzaler, E.R. (1998) Modelist interpretation of the impedance response of a polymer electrolyte fuel cell. *Electrochim. Acta*, **43** (24), 3761–3766.
174. Song, J.M., Cha, S.Y., and Lee, W.M. (2001) Optimal composition of polymer electrolyte fuel cell electrodes determined by the AC impedance method. *J. Power Sources*, **94** (1), 78–84.
175. Antolini, E., Giorgi, L., Pozio, A., and Passalacqua, E. (1999) Influence of Nafion loading in the catalyst layer of gas-diffusion electrodes for PEFC. *J. Power Sources*, **77** (2), 136–142.
176. Jordan, L.R., Shukla, A.K., Behrsing, R., Avery, N.R., Muddle, B.C., and Forsyth, M. (2000) Diffusion layer parameters influencing optimal fuel cell performance. *J. Power Sources*, **86** (1–2), 250–254.
177. Passalacqua, E., Squadrito, G., Lufrano, F., Patti, A., and Giorgi, L. (2001) Effects of the diffusion layer characteristics on the performance of polymer electrolyte fuel cell electrodes. *J. Appl. Electrochem.*, **31** (4), 449–454.
178. Liu, D. and Case, S. (2006) Durability study of proton exchange membrane fuel cells under dynamic testing conditions with cyclic current profile. *J. Power Sources*, **162** (1), 521–531.
179. Mittal, V.O., Kunz, H.R., and Fenton, J.M. (2006) Effect of catalyst properties on membrane degradation rate and the underlying degradation mechanism in PEMFCs. *J. Electrochem. Soc.*, **153** (9), A1755–A1759.
180. Rong, F., Huang, C., Liu, Z.-S., Song, D., Wang, Q., and Simulation, I.I. (2008) Microstructure changes in the catalyst layers of PEM fuel cells induced by load cycling: part understanding. *J. Power Sources*, **175** (2), 712–723.
181. Avasarala, B., Moore, R., and Haldar, P. (2010) Surface oxidation of carbon supports due to potential cycling under PEM fuel cell conditions. *Electrochim. Acta*, **55** (16), 4765–4771.
182. Fowler, M.W., Mann, R.F., Amphlett, J.C., Peppley, B.A., and Roberge, P.R. (2002) Incorporation of voltage degradation into a generalised steady state electrochemical model for a PEM fuel cell. *J. Power Sources*, **106** (1–2), 274–283.
183. Lin, R., Li, B., Hou, Y.P., and Ma, J.M. (2009) Investigation of dynamic driving

cycle effect on performance degradation and micro-structure change of PEM fuel cell. *Int. J. Hydrogen Energy*, **34** (5), 2369–2376.
184. Seo, D., Lee, J., Park, S., Rhee, J., Choi, S.W., and Shul, Y.-G. (2011) Investigation of MEA degradation in PEM fuel cell by on/off cyclic operation under different humid conditions. *Int. J. Hydrogen Energy*, **36** (2), 1828–1836.
185. Bi, W., Sun, Q., Deng, Y., and Fuller, T.F. (2009) The effect of humidity and oxygen partial pressure on degradation of Pt/C catalyst in PEM fuel cell. *Electrochim. Acta*, **54** (6), 1826–1833.
186. Abdullah, A.M., Mohammad, A.M., Okajima, T., Kitamura, F., and Ohsaka, T. (2009) Effect of load, temperature and humidity on the pH of the water drained out from H_2/air polymer electrolyte membrane fuel cells. *J. Power Sources*, **190** (2), 264–270.
187. Huang, X., Solasi, R., Zou, Y., Reifsnider, K., Condit, D., Burlatsky, S., and Madden, T. (2006) Mechanical endurance polymer electrolyte membrane and PEM fuel cell durability. *J. Polym. Sci., B: Polym. Phys.*, **16** (44), 2346–2357.
188. Genova-Dimitrova, P., Baradie, B., Foscallo, D., Poinsignon, C., and Sanchez, J.Y. (2001) Ionomeric membranes for proton exchange membrane fuel cell (PEMFC): sulfonated polysulfone associated with phosphatoantimonic acid. *J. Membr. Sci.*, **185** (1), 59–71.
189. Hou, J., Yu, H., Zhang, S., Sun, S., Wang, H., Yi, B., and Ming, P. (2006) Analysis of PEMFC freeze degradation at −20 °C after gas purging. *J. Power Sources*, **162** (1), 513–520.
190. Alink, R., Gerteisen, D., and Oszcipok, M. (2008) Degradation effects in polymer electrolyte membrane fuel cell stacks by sub-zero operation An in situ and ex situ analysis. *J. Power Sources*, **182** (1), 175–187.
191. Gavello, G., Zeng, J., Francia, C., Icardi, U.A., Graizzaro, A., and Specchia, S. (2011) Experimental studies on Nafion(R) 112 single PEM-FCs exposed to freezing conditions. *Int. J. Hydrogen Energy*, **36** (13), 8070–8081.
192. Luo, M., Huang, C., Liu, W., Luo, Z., and Pan, M. (2010) Degradation behaviors of polymer electrolyte membrane fuel cell under freeze/thaw cycles. *Int. J. Hydrogen Energy*, **35** (7), 2986–2993.
193. Taniguchi, A., Akita, T., Yasuda, K., and Miyazaki, Y. (2004) Analysis of electrocatalyst degradation in PEMFC caused by cell reversal during fuel starvation. *J. Power Sources*, **130** (1–2), 42–49.
194. Kang, J., Jung, D.W., Park, S., Lee, J.-H., Ko, J., and Kim, J. (2010) Accelerated test analysis of reversal potential caused by fuel starvation during PEMFCs operation. *Int. J. Hydrogen Energy*, **35** (8), 3727–3735.
195. Cleghorn, S.J.C., Mayfield, D.K., Moore, D.A., Moore, J.C., Rusch, G., Sherman, T.W., Sisofo, N.T., and Beuscher, U. (2006) A polymer electrolyte fuel cell life test: 3 years of continuous operation. *J. Power Sources*, **158** (1), 446–454.
196. Taniguchi, A., Akita, T., Yasuda, K., and Miyazaki, Y. (2008) Analysis of degradation in PEMFC caused by cell reversal during air starvation. *Int. J. Hydrogen Energy*, **33** (9), 2323–2329.
197. Teranishi, K., Kawata, K., Tsushima, S., and Hirai, S. (2006) Degradation mechanism of PEMFC under open circuit operation. *Electrochem. Solid-State Lett.*, **9** (10), A475–A477.
198. Kundu, S., Fowle, M., Simon, L.C., and Abouatallah, R. (2008) Reversible and irreversible degradation in fuel cells during Open Circuit Voltage durability testing. *J. Power Sources*, **182** (1), 254–258.
199. Zhang, S., Yuan, X.Z., Hin, J.N.C., Wang, H., Wu, J., Friedrich, K.A., and Schulze, M. (2010) Effects of open-circuit operation on membrane and catalyst layer degradation in proton exchange membrane fuel cells. *J. Power Sources*, **195** (4), 1142–1148.
200. Chen, J. and Zhou, B. (2008) Diagnosis of PEM fuel cell stack dynamic behaviours. *J. Power Sources*, **177** (1), 83–95.
201. Zhang, S., Yuan, X., Wang, H., Merida, W., Zhu, H., Shen, J., Wu, S., and

Zhang, J. (2009) A review of accelerated stress tests of MEA durability in PEM fuel cells. *Int. J. Hydrogen Energy*, **34** (1), 388–404.

202. Ahn, S.-Y., Shin, S.-J., Ha, H.Y., Hong, S.-A., Lee, Y.-C., Lim, T.W., and Oh, I.-H. (2002) Performance and lifetime analysis of the kW-class PEMFC stack. *J. Power Sources*, **106** (1–2), 295–303.

203. Chiem, B.H., Beattie, P., and Colbow, K. (2008) The development and demonstration of technology on the path to commercially viable PEM fuel cell stacks. *ECS Trans.*, **16** (2), 1927–1935.

204. Ferraro, M., Sergi, F., Brunaccini, G., Dispenza, G., Andaloro, L., and Antonucci, V. (2009) Demonstration and development of a polymer electrolyte fuel cell system for residential use. *J. Power Sources*, **193** (1), 342–348.

205. Knights, S., Colbow, K., St-Pierre, J., and Wilkinson, D. (2004) Aging mechanisms and lifetime of PEFC and DMFC. *J. Power Sources*, **127** (1–2), 127–134.

206. St-Pierre, J., Wilkinson, D.P., Knights, S., and Bos, M. (2000) Relationships between water management, contamination and lifetime degradation in PEFC. *J. New Mater. Electrochem. Syst.*, **3**, 99–106.

207. Wu, J., Yuan, X.Z., Martin, J.J., Wang, H., Zhang, J., Shen, J., Wu, S., and Merida, W. (2008) A review of PEM fuel cell durability: degradation mechanisms and mitigation strategies. *J. Power Sources*, **184** (1), 104–119.

208. Yang, C., Srinivasan, S., Bocarsly, A.B., Tulyani, S., and Benziger, J.B. (2004) A comparison of physical properties and fuel cell performance of Nafion and zirconium phosphate/Nafion composite membranes. *J. Membr. Sci.*, **237** (1–2), 145–161.

209. Ma, C., Zhang, L., Mukerjee, S., Ofer, D., and Nair, B. (2003) An investigation of proton conduction in select PEM' and reaction layer interfaces-designed for elevated temperature operation. *J. Membr. Sci.*, **219** (1–2), 123–136.

210. Tang, H., Peikang, S., Jiang, S.P., Wang, F., and Pan, M. (2007) A degradation study of Nafion proton exchange membrane of PEM fuel cells. *J. Power Sources*, **170** (1), 85–92.

211. Iwai, Y. and Yamanishi, T. (2009) Thermal stability of ion-exchange Nafion N117CS membranes. *Polym. Degrad. Stab.*, **94** (4), 679–687.

212. Surowiec, J. and Bogoczek, R. (1988) Studies on the thermal stability of the perfluorinated cation-exchange membrane. *J. Therm. Anal.*, **33** (4), 1097–1102.

213. Wilkie, C.A., Thomsen, J.R., and Mittleman, M.L. (1991) Interaction of Poly(methyl methacrylate) and Nafions. *J. Appl. Polym. Sci.*, **42** (4), 901–909.

214. Oono, Y., Fukuda, T., Sounai, A., and Hori, M. (2010) Influence of operating temperature on cell performance and endurance of high temperature proton exchange membrane fuel cells. *J. Power Sources*, **195** (4), 1007–1014.

215. Yuan, X.-Z., Zhang, S., Wang, H., Wu, J., Sun, J.C., Hiesgen, R., Friedrich, K.A., Schulze, M., and Haug, A. (2010) Degradation of a polymer exchange membrane fuel cell stack with Nafion® membranes of different thicknesses: Part I. In situ diagnosis. *J. Power Sources*, **195** (22), 7594–7599.

216. Patil, Y.P., Jarrett, W.L., and Mauritz, K.A. (2010) Deterioration of mechanical properties: A cause for fuel cell membrane failure. *J. Membr. Sci.*, **356** (1–2), 7–13.

217. Damjanovic, A. (1969), Chapter 5 Mechanistic Analysis of Oxygen Electrode Reactions in *Modern Aspects of Electrochemistry*, vol. 5 (eds J.U.M. Bockris and B.E. Conway), Plenum Press, New York, pp. 369–483.

218. Büchi, F.N., Gupta, B., Haas, O., and Scherer, G.G. (1995) Study of radiation-grafted FEP-G-polystyrene membranes as polymer electrolytes in fuel cells. *Electrochim. Acta*, **40** (3), 345–353.

219. Guo, Q., Pintauro, P.N., Tang, N., and O'Connor, S. (1999) Sulfonated and crosslinked polyphosphazene-based proton-exchange membranes. *J. Membr. Sci.*, **154** (2), 175–181.

220. Wu, J.F., Yi, B.L., Hou, M., Hou, Z.J., and Zhang, H.M. (2004) Influence of catalyst layer structure on the current

distribution of PEMFCs. *Electrochem. Solid State Lett.*, **7** (6), A151–A154.
221. Liu, W., Ruth, K., and Rusch, G. (2001) Membrane durability in PEM fuel cells. *J. New Mater. Electrochem. Syst.*, **4** (4), 227–231.
222. Giannelis, E.P. (1996) Polymer layered silicate nanocomposites. *Adv. Mater.*, **8** (1), 29–35.
223. Ijeri, V.S., Cappelletto, L., Bianco, S., Tortello, M., Spinelli, P., and Tresso, E. (2010) Nafion and carbon nanotube nanocomposites for mixed proton and electron conduction. *J. Membr. Sci.*, **363** (1–2), 265–270.
224. Lee, C.H., Park, H.B., Park, C.H., Lee, S.Y., Kim, J.Y., McGrath, J.E., and Lee, J.K. (2010) Preparation of high-performance polymer electrolyte nanocomposites through nanoscale silica particle dispersion. *J. Power Sources*, **195** (5), 1325–1332.
225. Gomes, D., Marschall, R., Nunes, S.P., and Wark, M. (2008) Development of polyoxadiazole nanocomposites for high temperature polymer electrolyte membrane fuel cells. *J. Membr. Sci.*, **322** (2), 406–415.
226. Alberti, G., Casciola, M., Capitani, D., Donnadio, A., Narducci, R., Pica, M., and Sganappa, M. (2007) Novel Nafion–zirconium phosphate nanocomposite membranes with enhanced stability of proton conductivity at medium temperature and high relative humidity. *Electrochim. Acta*, **52** (28), 8125–8132.
227. Yang, C., Srinivasan, S., Aricò, A.S., Cretì, P., Baglio, V., and Antonucci, V. (2001) Composite Nafion® zirconium phosphate membranes for direct methanol fuel cell operation at high temperature. *Electrochem. Solid-State Lett.*, **4** (4), A31–A34.
228. Bauer, F. and Willert-Porada, M. (2005) Characterization of zirconium and titanium phosphates and direct methanol fuel cell performance of functionally graded Nafion® composite membranes prepared out of them. *J. Power Sources*, **145** (2), 101–107.
229. Roeder, J., Zucolotto, V., Shishatskiy, S., Bertolino, J.R., Nunes, S.P., and Pires, A.T.N. (2006) Mixed conductive membrane: aniline polymerization in an acid SPEEK matrix. *J. Membr. Sci.*, **279** (1–2), 70–75.
230. Kumar, G.G., Kim, P., Nahm, K.S., and Elizabeth, R.N. (2007) Structural characterization of PVdF-HFP/PEG/Al_2O_3 proton conducting membranes for fuel cells. *J. Membr. Sci.*, **303** (1–2), 126–131.
231. Jang, I.-Y., Kweon, O.-H., Kim, K.-E., Hwang, G.-J., Moon, S.-B., and Kang, A.-S. (2008) Application of polysulfone (PSf)–and polyether ether ketone (PEEK)–tungstophosphoric acid (TPA) composite membranes for water electrolysis. *J. Membr. Sci.*, **322** (1), 154–161.
232. Şengül, E., Erdener, H., Akay, R.G., Yücel, H., Baç, N., and Erodlu, I. (2009) Effects of sulfonated polyether-etherketone (SPEEK) and composite membranes on the proton exchange membrane fuel cell (PEMFC) performance. *Int. J. Hydrogen Energy*, **34** (10), 4645–4652.
233. Lufrano, F., Baglio, V., Staiti, P., Stassi, A., Aricò, A.S., and Antonucci, V. (2010) Investigation of sulfonated polysulfone membranes as electrolyte in a passive-mode direct methanol fuel cell mini-stack. *J. Power Sources*, **195** (23), 7727–7733.
234. Li, Q.F., Rudbeck, H.C., Chromik, A., Jensen, J.O., Pan, C., Steenberg, T., Calverley, M., Bjerrum, N.J., and Kerres, J. (2010) Properties, degradation and high temperature fuel cell test of different types of PBI and PBI blend membranes. *J. Membr. Sci.*, **347** (1–2), 260–270.
235. Huslage, J., Rager, T., Schnyder, B., and Tsukada, A. (2002) Radiation-grafted membrane/electrode assemblies with improved interface. *Electrochim. Acta*, **48** (3), 247–254.
236. Gubler, L., Gürsel, S.A., and Scherer, G.G. (2005) Radiation grafted membranes for polymer electrolyte fuel cells. *Fuel Cells*, **5** (3), 317–335.
237. Nasef, M.M., Zubir, N.A., Ismail, A.F., Dahlan, K.Z.M., Saidi, H., and Khayet, M. (2006) Preparation of radiochemically pore-filled polymer electrolyte membranes for direct methanol

fuel cells. *J. Power Sources*, **156** (2), 200–210.

238. Lee, C.H., Park, C.H., and Lee, Y.M. (2008) Sulfonated polyimide membranes grafted with sulfoalkylated side chains for proton exchange membrane fuel cell (PEMFC) applications. *J. Membr. Sci.*, **313** (1–2), 199–206.

239. Scott, K., Taama, W.M., and Argyropoulos, P. (2000) Performance of the direct methanol fuel cell with radiation-grafted polymer membranes. *J. Membr. Sci.*, **171** (1), 119–130.

240. Yamaguchi, T., Miyata, F., and Nakao, S.-I. (2003) Pore-filling type polymer electrolyte membranes for a direct methanol fuel cell. *J. Membr. Sci.*, **214** (2), 283–292.

241. Wenzel, A., Yanagishita, H., Kitamoto, D., Endo, A., Haraya, K., Nakane, T., Hanai, N., Matsuda, H., Koura, N., Kamusewitz, H., and Paul, D. (2000) Effects of preparation condition of photoinduced graft filling-polymerized membranes on pervaporation performance. *J. Membr. Sci.*, **179** (1–2), 69–77.

242. Ihm, C.-D. and Ihm, S.-K. (1995) Pervaporation of water-ethanol mixtures through sulfonated polystyrene membranes prepared by plasma graft-polymerization. *J. Membr. Sci.*, **98** (1–2), 89–96.

243. Jiang, W., Childs, R.F., Mika, A.M., and Dickson, J.M. (2003) Pore-filled cation-exchange membranes containing poly(styrenesulfonic acid) gels. *Desalination*, **159** (3), 253–266.

244. Simons, R., Zuccon, J., Dickson, M.R., and Shaw, M. (1993) Pervaporation and evaporation characteristics of a new type of ion exchange membrane. *J. Membr. Sci.*, **78** (1–2), 63–67.

245. Gubler, L., Slaski, M., Wallasch, F., Wokaun, A., and Scherer, G.G. (2009) Radiation grafted fuel cell membranes based on co-grafting of α-methylstyrene and methacrylonitrile into a fluoropolymer base film. *J. Membr. Sci.*, **339** (1–2), 68–77.

246. Li, L. and Xing, Y.C. (2006) Electrochemical durability of carbon nanotubes in non-catalyzed and catalyzed oxidations. *J. Electrochem. Soc.*, **153** (10), A1823–A1828.

247. Shao, Y.Y., Yin, G.P., Gao, Y.Z., and Shi, P.F. (2006) Durability study of Pt/C and Pt/CNTs catalysts under simulated PEM fuel cell conditions. *J. Electrochem. Soc.*, **153** (6), A1093–A1097.

248. Guilminot, E., Corcella, A., Charlot, F., Maillard, F., and Chatenet, M. (2007) Detection of Pt^{z+} ions and Pt nanoparticles inside the membrane of a PEM fuel cell. *J. Electrochem. Soc.*, **154** (1), B96–B105.

249. Yasuda, K., Taniguchi, A., Akita, T., Ioroi, T., and Siroma, Z. (2006) Platinum dissolution and deposition in the polymer electrolyte membrane of a PEM fuel cell as studied by potential cycling. *Phys. Chem. Chem. Phys.*, **8** (6), 746–752.

250. Bett, J.A.S., Kinoshita, K., and Stonehart, P. (1976) Crystallite growth of platinum dispersed on graphitized carbon black: II. Effect of liquid environment. *J. Catal.*, **41** (1), 124–133.

251. Blurton, K.F., Kunz, H.R., and Rut, D.R. (1978) Surface area loss of platinum supported on graphite. *Electrochim. Acta*, **23** (3), 183–190.

252. He, C., Desai, S., Brown, G., and Bollepalli, S. (2005) PEM fuel cell catalysts: Cost, performance, and durability. *Interface*, **14** (3), 41–44.

253. Akita, T., Taniguchi, A., Maekawa, J., Sirorna, Z., Tanaka, K., Kohyama, M., and Yasuda, K. (2006) Analytical TEM study of Pt particle deposition in the proton-exchange membrane of a membrane-electrode-assembly. *J. Power Sources*, **159** (1), 461–467.

254. Xie, J., Wood, D.L., Wayne, D.M., Zawodzinski, T.A., Atanassov, P., and Borup, R.L. (2005) Durabiliy of PEFCs at high humidity conditions. *J. Electrochem. Soc.*, **152** (1), A104–A113.

255. Siroma, Z., Fujiwara, N., Ioroi, T., Yamazaki, S., Yasuda, K., and Miyazaki, Y. (2004) Dissolution of Nafion® membrane and recast Nafion® film in mixtures of methanol and water. *J. Power Sources*, **126** (1–2), 41–45.

256. Kangasniemi, K.H., Condit, D.A., and Jarvi, T.D. (2004) Characterization of vulcan electrochemically oxidized under simulated PEM fuel cell conditions. *J. Electrochem. Soc.*, **151** (4), E125–E132.
257. Stevens, D.A. and Dahn, J.R. (2005) Thermal degradation of the support in carbon-supported platinum electrocatalysts for PEM fuel cells. *Carbon*, **43** (1), 179–188.
258. Korovin, N.V. (1994) Electrocatalyst deterioration due to cathodic and anodic wear and means for retarding electrocatalyst deterioration. *Electrochim. Acta*, **39** (11–12), 1503–1508.
259. Stamenkovic, V.R., Fowler, B., Mun, B.S., Wang, G.F., Ross, P.N., Lucas, C.A., and Markovic, N.M. (2007) Improved oxygen reduction activity on $Pt_3Ni(111)$ via increased surface site availability. *Science*, **315** (5811), 493–497.
260. Zhang, J., Sasaki, K., Sutter, E., and Adzic, R.R. (2007) Stabilization of platinum oxygen reduction electrocatalysts using gold clusters. *Science*, **315** (5809), 220–222.
261. Coloma, F., Sepulvedaescribano, A., and Rodriguezreinoso, F. (1995) Heat-treated carbon-blacks as supports for Platinum catalysts. *J. Catal.*, **154** (2), 299–305.
262. Shao, Y., Yin, G., Zhang, J., and Gao, Y. (2006) Comparative investigation of the resistance to electrochemical oxidation of carbon black and carbon nanotubes in aqueous sulfuric acid solution. *Electrochim. Acta*, **51** (26), 5853–5857.
263. Li, L., Wu, G., and Xu, B.-Q. (2006) Electro-catalytic oxidation of CO on Pt catalyst supported on carbon nanotubes pretreated with oxidative acids. *Carbon*, **44** (14), 2973–2983.
264. Saha, M.S. and Kundu, A. (2010) Functionalizing carbon nanotubes for proton exchange membrane fuel cells electrode. *J. Power Sources*, **195** (19), 6255–6261.
265. Hernández-Fernández, P., Montiel, M., Ocón, P., Gómez de la Fuente, J.L., García-Rodríguez, S., Rojas, S., and Fierro, J.L.G. (2010) Functionalization of multi-walled carbon nanotubes and application as supports for electrocatalysts in proton-exchange membrane fuel cell. *Appl. Catal. B: Environ.*, **99** (1–2), 343–352.
266. De Jong, K.P. and Geus, J.W. (2000) Carbon nanofibers: catalytic synthesis and applications. *Catal. Rev.: Sci. Eng.*, **42** (4), 481–510.
267. Calvillo, L., Gangeri, M., Perathoner, S., Centi, G., Moliner, R., and Lázaro, M.J. (2009) Effect of the support properties on the preparation and performance of platinum catalysts supported on carbon nanofibers. *J. Power Sources*, **192** (1), 144–150.
268. St-Pierre, J. and Jia, N. (2002) Successful demonstration of Ballard PEMFCs for space shuttle applications. *J. New Mater. Electrochem. Syst.*, **5**, 263–271.
269. Shimpalee, S., Beuscher, U., and Van Zee, J.W. (2007) Analysis of GDL flooding effects on PEMFC performance. *Electrochim. Acta*, **52** (24), 6748–6754.
270. Oszcipok, M., Riemann, D., Kronenwett, U., Kreideweis, M., and Zedda, M. (2005) Statistic analysis of operational influences on the cold start behaviour of PEM fuel cells. *J. Power Sources*, **145** (2), 407–415.
271. Frisk, J., Boand, W., Hicks, M., Kurkowski, M., Schmoeckel, A., and Atanasoski, R. (2004) How 3M developed a new GDL construction for improved oxidative stability. Proceedings of The Fuel Cell Seminar and Exhibition, San Antonio (TX), November 1–5, 2004.
272. Park, S. and Popov, B.N. (2009) Effect of hydrophobicity and pore geometry in cathode GDL on PEM fuel cell performance. *Electrochim. Acta*, **54** (12), 3473–3479.
273. Cindrella, L., Kannan, A.M., Ahmad, R., and Thommes, M. (2009) Surface modification of gas diffusion layers by inorganic nanomaterials for performance enhancement of proton exchange membrane fuel cells at low RH conditions. *Int. J. Hydrogen Energy*, **34** (15), 6377–6383.

274. Owejan, J.E., Yu, P.T., and Makharia, R. (2007) Mitigation of carbon corrosion in microporous layers in PEM fuel cells. *ECS Trans.*, **11** (1), 1049–1057.
275. Gerteisen, D., Heilmann, T., and Ziegler, C. (2008) Enhancing liquid water transport by laser perforation of a GDL in a PEM fuel cell. *J. Power Sources*, **177** (2), 348–354.
276. Zamel, N. and Li, X. (2011) Effect of contaminants on polymer electrolyte membrane fuel cells. *Prog. Energy Combust. Sci.*, **37** (3), 292–329.
277. Liu, D., Peng, L., and Lai, X. (2010) Effect of assembly error of bipolar plate on the contact pressure distribution and stress failure of membrane electrode assembly in proton exchange membrane fuel cell. *J. Power Sources*, **195** (13), 4213–4221.
278. Cooper, J.S. (2004) Design analysis of PEMFC bipolar plates considering stack manufacturing and environment impact. *J. Power Sources*, **129** (2), 152–169.
279. Lee, S.-J., Huang, C.-H., and Chen, Y.-P. (2003) Investigation of PVD coating on corrosion resistance of metallic bipolar plates in PEM fuel cell. *J. Mater. Process. Technol.*, **140** (1–3), 688–693.
280. Show, Y., Miki, M., and Nakamura, T. (2007) Increased in output power from fuel cell used metal bipolar plate coated with a-C film. *Diamond Relat. Mater.*, **16** (4–7), 1159–1161.
281. Lee, Y.-B. and Lim, D.-S. (2010) Electrical and corrosion properties of stainless steel bipolar plates coated with a conduction polymer composite. *Curr. Appl. Phys.*, **10** (2), S18–S21.
282. Woodman, A., Jayne, K., Anderson, E., and Kimble, M. (1999) Lightweight and Corrosion Resistant Metal Bipolar Plates for PEM Fuel Cells, SAE Technical Paper # 1999-01-2614.
283. Cho, E.A., Jeon, U.-S., Ha, H.Y., Hong, S.-A., and Oh, I.-H. (2004) Characteristics of composite bipolar plates for polymer electrolyte membrane fuel cells. *J. Power Sources*, **125** (2), 178–182.
284. Lee, H.S., Kim, H.J., Kim, S.G., and Ahn, S.H. (2007) Evaluation of graphite composite bipolar plate for PEM (proton exchange membrane) fuel cell: electrical, mechanical, and molding properties. *J. Mater. Process. Technol.*, **187–188**, 425–428.
285. Dhakate, S.R., Mathur, R.B., Kakati, B.K., and Dhami, T.L. (2007) Properties of graphite-composite bipolar plate prepared by compression molding technique for PEM fuel cell. *Int. J. Hydrogen Energy*, **32** (17), 4537–4543.
286. Frisch, L. (2001) PEM fuel cell stack sealing using silicone elastomers. *Sealing Technol.*, **2001** (93), 7–9.
287. Tan, J., Chao, Y.J., Van Zee, J.W., and Lee, W.K. (2007) Degradation of elastomeric gasket materials in PEM fuel cells. *Mater. Sci. Eng. A*, **445–446**, 669–675.
288. Tan, J., Chao, Y.J., Yang, M., Williams, C.T., and Van Zee, J.W. (2008) Degradation characteristics of elastomeric gasket materials in a simulated PEM fuel cell environment. *J. Mater. Eng. Perform.*, **17** (6), 785–792.
289. Schulze, M., Knöri, T., Schneider, A., and Gülzow, E. (2004) Degradation of sealings for PEFC test cells during fuel cell operation. *J. Power Sources*, **127** (1–2), 222–229.
290. Baschuk, J.J. and Li, X.G. (2001) Carbon monoxide poisoning of proton exchange membrane fuel cells. *Int. J. Energy Resour.*, **25** (8), 695–713.
291. Bellows, R.J., Marucchi-Soos, E.P., and Buckley, D.T. (1996) Analysis of reaction kinetics for carbon monoxide and carbon dioxide on polycrystalline platinum relative to fuel cell operation. *Ind. Eng. Chem. Res.*, **35** (4), 1235–1242.
292. Jiang, R., Kunz, H.R., and Fenton, J.M. (2005) Electrochemical oxidation of H_2 and H_2/CO mixtures in higher temperature ($T_{cell} > 100\,°C$) proton exchange membrane fuel cells: electrochemical impedance spectroscopy. *J. Electrochem. Soc.*, **152** (7), A1329–A1340.
293. de Bruijn, F.A., Papageorgopoulos, D.C., Sitters, E.F., and Janssen, G.J.M. (2002) The influence of carbon dioxide on PEM fuel cell anodes. *J. Power Sources*, **110** (1), 117–124.

294. Mohtadi, R., Lee, W.-K., and Van Zee, J.W. (2005) The effect of temperature on the adsorption rate of hydrogen sulfide on Pt anodes in a PEMFC. *Appl. Catal. B: Environ.*, **56** (1–2), 37–42.
295. Kelly, M.J., Egger, B., Fafilek, G., Besenhard, J.O., Kronberger, H., and Nauer, G.E. (2005) Conductivity of polymer electrolyte membranes by impedance spectroscopy with microelectrodes. *Solid State Ionics*, **176** (25–28), 2111–2114.
296. Inaba, M., Kinumoto, T., Kiriake, M., Umebayashi, R., Tasaka, A., and Ogumi, Z. (2006) Gas crossover and membrane degradation in polymer electrolyte fuel cells. *Electrochim. Acta*, **51** (26), 5746–5753.
297. Giorgi, L., Pozio, A., Bracchini, C., Giorgi, R., and Turtù, S. (2001) H_2 and H_2/CO oxidation mechanism on Pt/C, Ru/C and PtRu/C electrocatalysts. *J. Appl. Electrochem.*, **31** (3), 325–334.
298. Ralph, T.R. and Hogarth, M.P. (2002) Catalysis for low temperature fuel cells. *Platinum Met. Rev.*, **46** (3), 117–135.
299. Alayoglu, S., Nilekar, A.U., Mavrikakis, M., and Eichhorn, B. (2008) Ru-Pt core-shell nanoparticles for preferential oxidation of carbon monoxide in hydrogen. *Nat. Mater.*, **7** (4), 333–338.
300. Antolini, E., Salgado, J.R.C., and Gonzalez, E.R. (2006) The stability of Pt–M (M = first row transition metal) alloy catalysts and its effect on the activity in low temperature fuel cells: A literature review and tests on a Pt–Co catalyst. *J. Power Sources*, **160** (2), 957–968.
301. Zhang, L., Lee, K., and Zhang, J. (2007) Effect of synthetic reducing agents on morphology and ORR activity of carbon-supported nano-Pd–Co alloy electrocatalysts. *Electrochim. Acta*, **52** (28), 7964–7971.
302. Lu, G., Cooper, J.S., and McGinn, P.J. (2006) SECM characterization of Pt–Ru–WC and Pt–Ru–Co ternary thin film combinatorial libraries as anode electrocatalysts for PEMFC. *J. Power Sources*, **161** (1), 106–114.
303. Mani, P., Srivastava, R., and Strasser, P. (2011) Dealloyed binary PtM_3 (M = Cu, Co, Ni) and ternary $PtNi_3M$ (M = Cu, Co, Fe, Cr) electrocatalysts for the oxygen reduction reaction: Performance in polymer electrolyte membrane fuel cells. *J. Power Sources*, **196** (2), 666–673.
304. Tsiouvaras, N., Peña, M.A., Fierro, J.L.G., Pastor, E., and Martínez-Huerta, M.V. (2010) The effect of the Mo precursor on the nanostructure and activity of PtRuMo electrocatalysts for proton exchange membrane fuel cells. *Catal. Today*, **158** (1–2), 12–21.
305. Papageorgopoulos, D.C., Keijzer, M., and de Bruijn, F.A. (2002) The inclusion of Mo, Nb and Ta in Pt and PtRu carbon supported electrocatalysts in the quest for improved CO tolerant PEMFC anodes. *Electrochim. Acta*, **48** (2), 197–204.
306. García, G., Silva-Chong, J.A., Guillén-Villafuerte, O., Rodríguez, J.L., González, E.R., and Pastor, E. (2006) CO tolerant catalysts for PEM fuel cells: spectroelectrochemical studies. *Catal. Today*, **116** (3), 415–421.
307. Huang, T., Zhang, D., Xue, L., Cai, W.-B., and Yu, A. (2009) A facile method to synthesize well-dispersed $PtRuMoO_x$ and $PtRuWO_x$ nanoparticles and their electrocatalytic activities for methanol oxidation. *J. Power Sources*, **192** (2), 285–290.
308. Gubler, L., Scherer, G.G., and Wokaun, A. (2001) Quantitative characterization of the CO-tolerance of a polymer electrolyte fuel cell. *Chem. Eng. Technol.*, **24** (1), 59–67.
309. Li, Q., He, R., Jensen, J.O., and Bjerrum, N.J. (2003) Approaches and recent development of polymer electrolyte membranes for fuel cells operating above 100 °C. *Chem. Mater.*, **15** (26), 4896–4915.
310. Liu, G., Zhang, H., Hu, J., Zhai, Y., Xu, D., and Shao, Z.G. (2006) Studies of performance degradation of a high temperature PEMFC based on H_3PO_4-doped PBI. *J. Power Sources*, **162** (1), 547–552.
311. Lobato, J., Cañizares, P., Rodrigo, M.A., Linares, J.J., and Pinar, F.J. (2010) Study of the influence of the amount of $PBI–H_3PO_4$ in the catalytic layer

of a high temperature PEMFC. *Int. J. Hydrogen Energy*, **35** (3), 1347–1355.
312. Zhang, J., Tang, Y., Song, C., and Zhang, J. (2007) Polybenzimidazole-membrane-based PEM fuel cell in the temperature range of 120–200 °C. *J. Power Sources*, **172** (1), 163–171.
313. Ishikawa, Y., Morita, T., Nakata, K., Yoshida, K., and Shiozawa, M. (2007) Behavior of water below the freezing point in PEFCs. *J. Power Sources*, **163** (2), 708–712.
314. Jiao, K. and Li, X. (2011) Water transport in polymer electrolyte membrane fuel cells, water transport in polymer electrolyte membrane fuel cells. *Prog. Energy Combust. Sci.*, **37** (33), 221–291.
315. Cho, E.A., Ko, J.-J., Ha, H.Y., Hong, S.-A., Lee, K.-Y., Lim, T.-W., and Oh, I.-H. (2003) Characteristics of the PEMFC repetitively brought to temperatures below 0 °C. *J. Electrochem. Soc.*, **150** (12), A1667–A1670.
316. Yan, Q., Toghiani, H., Lee, Y.-W., Liang, K., and Causey, H. (2006) Effect of sub-freezing temperatures on a PEM fuel cell performance, startup and fuel cell components. *J. Power Sources*, **160** (2), 1242–1250.
317. Oszcipok, M., Zedda, M., Riemann, D., and Geckeler, D. (2006) Low temperature operation and influence parameters on the cold start ability of portable PEMFCs. *J. Power Sources*, **154** (2), 404–411.
318. Schmittinger, W. and Vahidi, A. (2008) A review of the main parameters influencing long-term performance and durability of PEM fuel cells. *J. Power Sources*, **180** (1), 1–14.
319. Kim, S. and Mench, M.M. (2007) Physical degradation of membrane electrode assemblies undergoing freeze/thaw cycling: micro-structure effects. *J. Power Sources*, **174** (1), 206–220.
320. Pinton, E., Fourneron, Y., Rosini, S., and Antoni, L. (2009) Experimental and theoretical investigations on a proton exchange membrane fuel cell starting up at subzero temperatures. *J. Power Sources*, **186** (1), 80–88.
321. Yousfi-Steiner, N., Moçotéguy, P., Candusso, D., and Hissel, D. (2009) A review on polymer electrolyte membrane fuel cell catalyst degradation and starvation issues: causes, consequences and diagnostic for mitigation. *J. Power Sources*, **194** (1), 130–145.
322. Mukundan, R., Kim, Y.S., Rockward, T., Davey, J.R., Pivovar, B., Hussey, D.S., Jacobson, D.L., Arif, M., and Borup, R. (2007) Performance of PEM fuel cells at sub-freezing temperatures. *ECS Trans.*, **11** (1), 543–552.
323. General Motors Corporation (1999) Freeze-protecting a fuel cell by vacuum drying. US Patent 6, 358, 637, Detroit, MI.
324. Energy Partners LC (2000) Freeze tolerant fuel cell system and method. Patent WO/2000/065676, Palm Beach FL (US).
325. Ballard Power Systems Inc (2001) Methods for improving the cold starting capability of an electrochemical fuel cell. Patent WO/2001/024296, Burnaby, British Columbia (CA).
326. Ballard Power Systems Inc. (2001) Method of reducing fuel cell performance degradation of an electrode comprising porous components. US Patent 6, 306, 536, Burnaby, British Columbia (CA).
327. Toyota Jidosha Kabushiki Kaisha, Aishi (JP) (2007) Fuel cell system and method to prevent freezing after shut-down. Patent WO/2007/091137.
328. Daimler A.G. and Stuttgart D.E. (2008) Fuel cell system and vehicle having fuel cell system. Patent WO/2008/022748.
329. Hou, J., Yi, B., Yu, H., Hao, L., Song, W., Fu, Y., and Shao, Z. (2007) Investigation of resided water effects on PEM fuel cell after cold start. *Int. J. Hydrogen Energy*, **32** (17), 4503–4509.
330. Ge, S. and Wang, C.Y. (2007) Characteristics of subzero startup and water/ice formation on the catalyst layer in a PEMFC. *Electrochim. Acta*, **52** (14), 4825–4835.
331. Tajiri, K., Wang, C.-Y., and Tabuchi, Y. (2008) Water removal from a PEFC during gas purge. *Electrochim. Acta*, **53** (22), 6337–6343.
332. Oszcipok, M., Zedda, M., Hesselmann, J., Huppmann, M., Wodrich, M.,

Junghardt, M., and Hebling, C. (2006) Portable proton exchange membrane fuel-cell systems for outdoor applications. *J. Power Sources*, **157** (2), 666–673.
333. Lee, S.-Y., Kim, S.-U., Kim, H.-J., Jang, J.H., Oh, I.-H., Cho, E.A., Hong, S.-A., Ko, J., Lim, T.-W., Lee, K.-Y., and Lim, T.-H. (2008) Water removal characteristics of proton exchange membrane fuel cells using a dry gas purging method. *J. Power Sources*, **180** (2), 784–790.
334. Cacciola, G., Antonucci, V., and Freni, S. (2001) Technology up date and new strategies on fuel cells. *J. Power Sources*, **100** (1–2), 67–79.
335. Erdinc, O. and Uzunoglu, M. (2010) Recent trends in PEM fuel cell-powered hybrid systems: investigation of application areas, design architectures and energy management approaches. *Renew. Sustain. Energy Rev.*, **14** (9), 2874–2884.
336. Colella, W.G., Jacobson, M.Z., and Golden, D.M. (2005) Switching to a U.S. hydrogen fuel cell vehicle fleet: the resultant change in emissions, energy use, and greenhouse gases. *J. Power Sources*, **150**, 150–181.
337. Pettersson, L.J. and Westerholm, R. (2001) State of the art of multi-fuel reformers for fuel cell vehicles: problem identification and research needs. *Int. J. Hydrogen Energy*, **26** (3), 243–264.
338. Wee, J.H. (2007) Applications of proton exchange membrane fuel cell systems. *Renew. Sustain. Energy Rev.*, **11** (8), 1720–1738.
339. Ball, M. and Wietschel, M. (2009) The future of hydrogen-opportunities and challenges. *Int. J. Hydrogen Energy*, **34** (2), 615–627.
340. Vincent, B., Gangi, J. Curtin, S., and Delmon, E. (2010). 2008 Fuel Cell Technologies Market report. Technical Report DOE/GO-102010-2080, pp. 1–26. http://www1.eere.energy.gov/hydrogenandfuelcells/pdfs/48219.pdf, (accessed September 2 2010).
341. Gaines, L.L., Elgowainy, A., and Wang, M.Q. (2008) Full Fuel-cycle Comparison of Forklift Propulsion Systems, Technical Report ANL/ESD/08-3, pp. 1–28. http://www1.eere.energy.gov/hydrogenandfuelcells/pdfs/forklift_anl_esd.pdf, (accessed September 2 2010).
342. Contestabile, M. (2010) Analysis of the market for diesel PEMFC auxiliary power units onboard long-haul trucks and of its implications for the large-scale adoption of PEMFCs. *Energy Policy*, **38** (10), 5320–5334.
343. Jain, S., Chen, H.-Y., and Schwank, J. (2006) Techno-economic analysis of fuel cell auxiliary power units as alternative to idling. *J. Power Sources*, **160** (1), 474–484.
344. Haseli, Y., Naterer, G.F., and Dincer, I. (2008) Comparative assessment of greenhouse gas mitigation of hydrogen passenger trains. *Int. J. Hydrogen Energy*, **33** (7), 1788–1796.
345. Bégot, S., Harel, F., Candusso, D., François, X., Péra, M.-C., and Yde-Andersen, S. (2010) Fuel cell climatic tests designed for new configured aircraft application. *Energy Convers. Manage.*, **51** (7), 1522–1535.
346. Aicher, T., Lenz, B., Gschnell, F., Groos, U., Federici, F., Caprile, L., and Parodi, L. (2006) Fuel processors for fuel cell APU applications. *J. Power Sources*, **154** (2), 503–508.
347. Tanrioven, M. and Alam, M.S. (2006) Reliability modeling and analysis of stand-alone PEM fuel cell power plants. *Renew. Energy*, **31** (7), 915–933.
348. Patil, A.S., Dubois, T.G., Sifer, N., Bostic, E., Gardner, K., Quah, M., and Bolton, C. (2004) Portable fuel cell systems for America's army: technology transition to the field. *J. Power Sources*, **136** (2), 220–225.
349. Andujar, J.M. and Segura, F. (2009) Fuel cells: history and updating. A walk along two centuries. *Renew. Sustain. Energy Rev.*, **13** (9), 2309–2322.
350. D'Urso, C., Baglio, V., Antonucci, V., Aricò, A.S., Specchia, S., Icardi, U.A. Saracco, G., Spinella, C.R., D'Arrigo, G., (2011) Development of a planar micro-DMFC operating at room temperature. *Int. J. Hydrogen Energy*, **36** (13), 8088–8093.
351. Shkolnikov, E., Vlaskin, M., Iljukhin, A., Zhuk, A., and Sheindlin, A. (2008) 2 W power source based on

air–hydrogen polymer electrolyte membrane fuel cells and water–aluminum hydrogen micro-generator. *J. Power Sources*, **185** (2), 967–972.

352. Gencoglu, M.T. and Ural, Z. (2009) Design of a PEM fuel cell system for residential application. *Int. J. Hydrogen Energy*, **34** (12), 5242–5248.

353. Uzunoglu, M., Onar, O.C., and Alam, M.S. (2007) Dynamic behavior of PEMFCPPs under various load conditions and voltage stability analysis for stand-alone residential applications. *J. Power Sources*, **168** (1), 240–250.

354. Nagata, Y. (2005) Quantitative analysis of CO_2 emissions reductions through introduction of stationary-type PEMFC systems in Japan. *Energy*, **30** (14), 2636–2653.

355. Shukla, A.K., Aricò, A.S., and Antonucci, V. (2001) An appraisal of electric automobile power source. *Renew. Sustain. Energy Rev.*, **5** (2), 137–155.

356. Agnolucci, P. (2007) Hydrogen infrastructure for the transport sector. *Int. J. Hydrogen Energy*, **32** (15), 3526–3544.

357. Stephens-Romero, S.-D., Brown, T.M., Kang, J.E., Recker, W.W., and Samuelsen, G.S. (2010) Systematic planning to optimize investments in hydrogen infrastructure deployment. *Int. J. Hydrogen Energy*, **35** (10), 4652–4667.

358. Kromer, M.A., Joseck, F., Rhodes, T., Guernsey, M., and Marcinkoski, J. (2009) Evaluation of a platinum leasing program for fuel cell vehicles. *Int. J. Hydrogen Energy*, **34** (9), 8276–8288.

359. Liu, Y. and Hua, L. (2010) Fabrication of metallic bipolar plate for proton exchange membrane fuel cells by rubber pad forming. *J. Power Sources*, **195** (11), 3529–3535.

360. Marcinkoski, J., Kopasz, J.P., and Benjamin, T.G. (2008) Progress in the US DOE fuel cell subprogram efforts in polymer electrolyte fuel cells. *Int. J. Hydrogen Energy*, **33** (14), 3894–3902.

361. Tsuchiya, H. and Kobayashi, O. (2004) Mass production cost of PEM fuel cell by learning curve. *Int. J. Hydrogen Energy*, **29** (10), 985–990.

362. (a) United States Environmental Protection Agency website, Diesel Truck Anti-idling Campaign, http://www.epa.gov/region8/ej/ejinitiatives.html; (b) United States Environmental Protection Agency website, Technologies, Strategies and Policies: Idling Reduction, http://www.epa.gov/otaq/smartway/transport/what-smartway/idling-reduction-tech.htm (accessed January 25 2010).

363. Commission of the European Communities (2007)Proposal for a Regulation of the European Parliament and of the Council on Type-approval of Motor Vehicles and Engines with Respect to Emissions from Heavy Duty Vehicles (EuroVI) and on Access to Vehicle Repair and Maintenance Information, Brussels, 21 December 2007, COM, 851 final. http://www.r2rc.eu/documents/EuroVI_Draft_Regulation_en.pdf (accessed January 25 2010).

364. Baratto, F. and Diweka, U.M. (2005) Life cycle assessment of fuel cell-based APUs. *J. Power Sources*, **139** (1–2), 188–196.

365. Chrenko, D., Lecoq, S., Herail, E., Hissel, D., and Péra, M.C. (2010) Static and dynamic modeling of a diesel fed fuel cell power supply. *Int. J. Hydrogen Energy*, **35** (3), 1377–1389.

14
Solid Oxide Fuel Cells
Jeffrey W. Fergus

14.1
Introduction

The primary advantages and disadvantages of solid oxide fuel cells (SOFCs) relative to other fuel cell technologies are the result of their high operating temperature [1–4]. The high temperature increases the rates of electrode processes and thus reduces the catalytic requirements of the electrode, so expensive materials, such as platinum, are not necessary. The improved electrode kinetics allow for a wide variety of fuels to be used, which opens up potential applications where providing a supply of hydrogen is not possible or cost prohibitive. The high operating temperature also accelerates undesired reactions, so the stability of materials used against reaction with the surrounding gas environment or with other materials in the system can lead to degradation. In addition, the high temperature can lead to thermal stresses due to mismatches in the coefficients of thermal expansion (CTE) or during rapid heating or cooling, which can impose restrictions on operating parameters (e.g., start-up times) or materials selection. At the system level, the high operating temperature can also be an advantage or disadvantage. The balance of plant must be designed to appropriately manage the heat generated by the fuel cell, which introduces additional constraints and requirements. However, in some systems, the heat can be used to enhance performance, such as in combined heat and power (CHP) systems.

14.1.1
Principles of Operation

In addition to the higher operating temperatures, SOFCs also differ from the fuel cells discussed in Chapter 13 on PEMFCs in terms of the ion that is mobile in the electrolyte. The mobile ion in SOFC electrolytes is an oxygen ion, so any gas that can be reduced by oxygen can be used as the fuel. As shown in Figure 14.1, oxygen ions from the reduction of oxygen gas are transported through the electrolyte to the anode where they can then oxidize the fuel. The electrons produced by the oxidation reaction pass through the external load and then reduce additional oxygen at the

---

Electrochemical Technologies for Energy Storage and Conversion, First Edition. Edited by Ru-Shi Liu, Lei Zhang, Xueliang Sun, Hansan Liu, and Jiujun Zhang.
© 2012 Wiley-VCH Verlag GmbH & Co. KGaA. Published 2012 by Wiley-VCH Verlag GmbH & Co. KGaA.

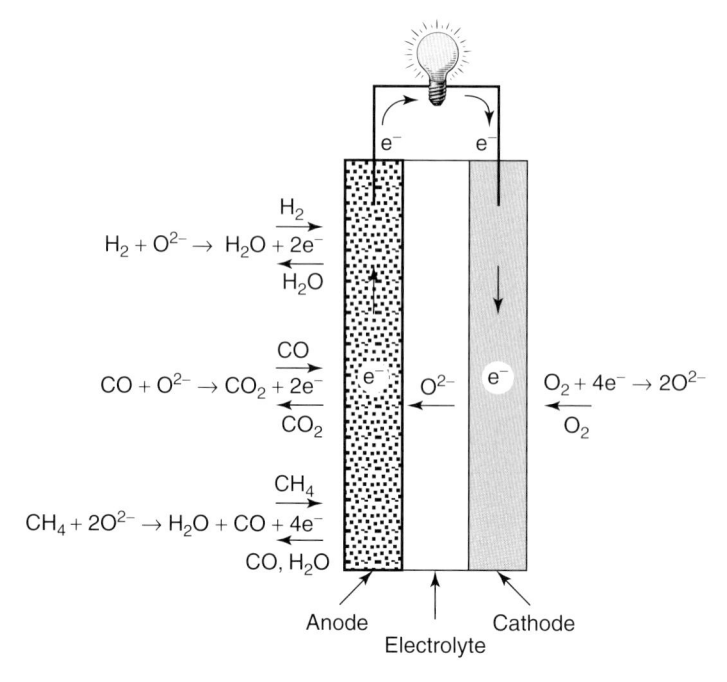

Figure 14.1 SOFC electrode reactions and transport.

cathode. Figure 14.1 shows the anode reaction for three different fuels: hydrogen, carbon monoxide, and methane. Note that carbon monoxide, which can poison the platinum catalysts in low-temperature fuel cells, can be oxidized and can contribute to the electrical current produced in an SOFC. The reactions in actual fuels can be more complicated since fuels can contain multiple species and each reaction is composed of multiple steps. Similarly, other reaction products can be produced, such as the deposition of solid carbon with carbonaceous fuels. However, in principle, any reducing gas can be used as the fuel for an SOFC.

14.1.2
Types and Structures

The SOFC components can be arranged in different geometries [4, 5]. The first large-scale SOFC power plants were based on a tubular design in which the anode, electrolyte, and cathode form concentric layers as shown in Figure 14.2a. A major advantage of this design is that the tube effectively separates the fuel and air gas streams, so a sealant to separate the gases is needed only at the end of the tube. In early designs, electrical connections were made at the end of the tube, so that the current had to pass over the cathode surface as shown in Figure 14.2b. In a subsequent design, the tubes are stacked as shown in Figure 14.3a, which decreases the path length through which current must pass and thus decreases the associated ohmic resistance. Figure 14.3b shows that the electrons pass circumferentially

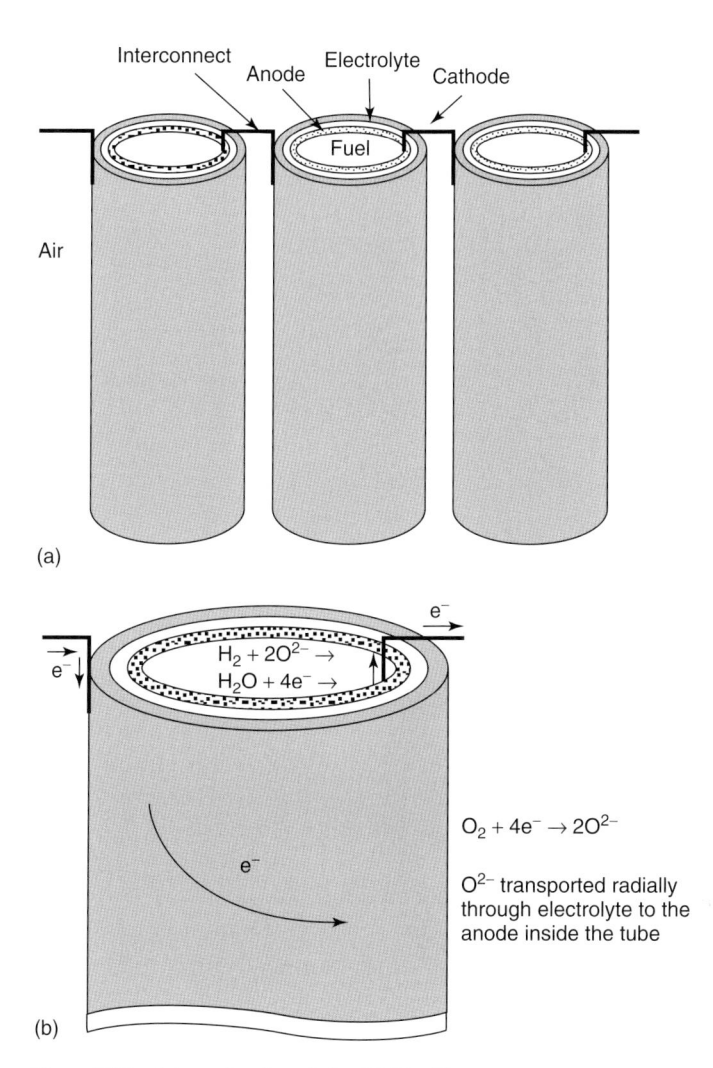

Figure 14.2 Schematic of a tubular SOFC with closed-one-end tubes. (a) Connection between cells and (b) current flow through cell.

Labels in figure: Interconnect, Anode, Electrolyte, Cathode, Fuel, Air

$H_2 + 2O^{2-} \rightarrow H_2O + 4e^-$

$O_2 + 4e^- \rightarrow 2O^{2-}$

O^{2-} transported radially through electrolyte to the anode inside the tube

around the cathode layer and do not need to pass along the length of the tube. One of the disadvantages of the tubular design is that the amount of electrode surface per volume is low, so the power density is also relatively low. This can be improved by flattening the shape of the tubes, which provides an increased amount of surface area for a given amount of internal volume. Improvements in stacking can be obtained with alternative geometries, such as triangular tubes.

The approach most commonly used to increase power density is to use a planar design, which is illustrated schematically in Figure 14.4. In the planar design, three-layer (anode–electrolyte–cathode) stacks are separated by an interconnect,

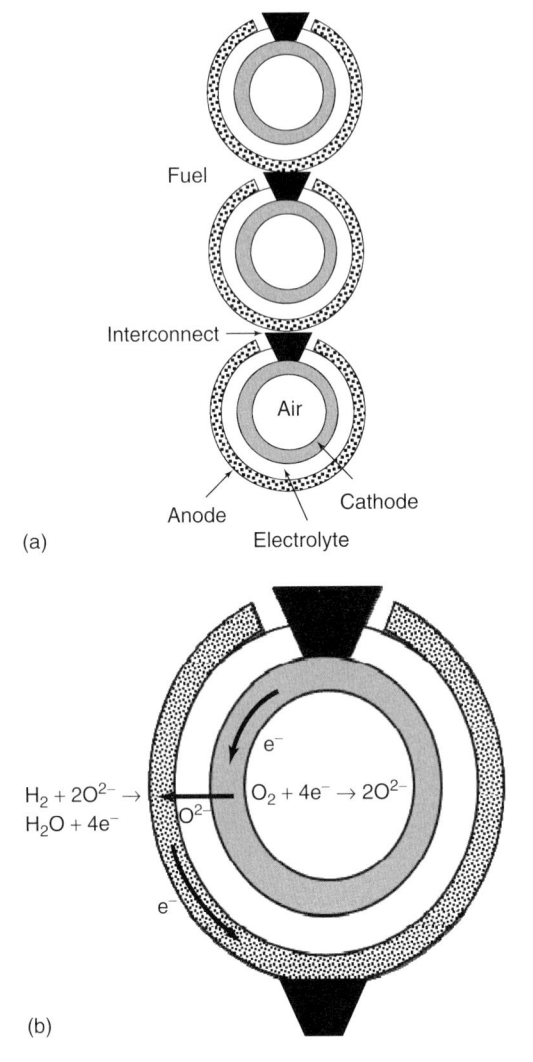

Figure 14.3 Schematic of a tubular SOFC with stacked tubes. (a) Connection between cells and (b) current flow through cell.

which contains gas channels for the air and fuel. Figure 14.4b shows that the current flows from the interconnect through the short dimension of the electrode. The primary disadvantage of the planar design is that gas-tight sealants are required to separate the air and fuel gas streams. A modification of the planar design, in which, as discussed above for the tubular design, geometrical modifications are used to increase electrode area, is shown in Figure 14.5. In this design, referred to as a *monolithic SOFC*, a corrugated shape is used to increase energy density [3, 6]. Another approach is to locate the cells along a flat surface as shown in Figure 14.6.

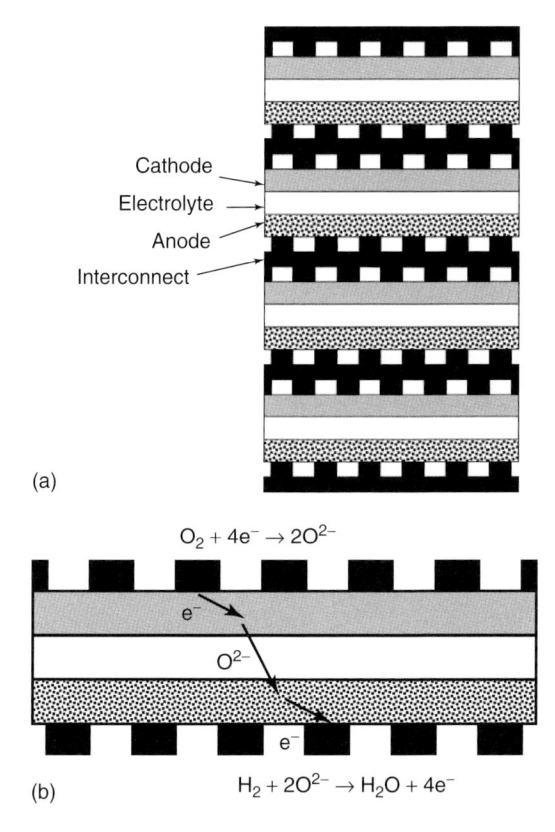

Figure 14.4 Schematic of a planar SOFC. (a) Connection between cells and (b) current flow through cell.

In this case, the current flows (generally) parallel to the substrate as shown in Figure 14.6b. The substrate, such as a porous ceramic tube, is porous to provide the fuel with access to the anode [7, 8]. One of the advantages of placing the electrodes on a surface is that the design is amenable to miniaturization for reducing the size of the fuel cell [9].

An important parameter in the design of a fuel cell is selection of the material for mechanical support of the cell. The support can be provided by one of the fuel cell's functional components or externally by a material that is not part of the electrochemical cell. External support adds a component, which will decrease the energy density of the system (i.e., the additional weight does not contribute to the fuel cell current), but the support material can be selected for its mechanical properties and chemical stability, without any requirements for electrochemical performance. On the other hand, using a functional component makes use of a material already present in the cell, but places additional constraints on the selection of the materials and fabrication processes.

Fuel cells have been designed using each of the three cell components, anode, electrolyte, and cathode, as the supporting structure, as shown schematically in

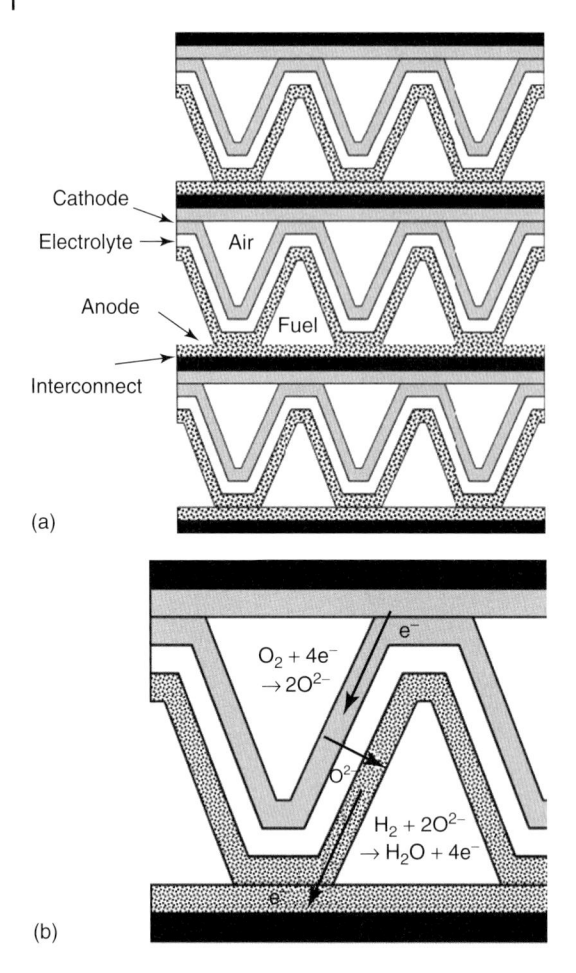

Figure 14.5 Schematic of a monolithic SOFC. (a) Connection between cells and (b) current flow through cell.

Figure 14.7. The electrolyte needs to be dense to separate the gases and thus typically provides strong support. The primarily disadvantage of electrolyte-supported cells is that a relatively thick (>100 μm) electrolyte must be used, which increases the ohmic resistance across the electrolyte, so high operating temperatures (1000 °C) are required. By using the electrode for support, a thinner electrolyte (<20 μm) can be used, which reduces the minimum operating temperature (700–800 °C). The electrodes are porous to provide gas access, so thick layers (hundreds of micrometers) are needed to provide adequate mechanical strength, but as the thickness increases, the gas diffusion becomes slower and can limit performance. As discussed in Section 14.2.2, the most common SOFC anode is a nickel–yttria-stabilized zirconia (YSZ) cermet, which has good electrical conductivity and strength, but suffers from instability during cycling because of oxidation and reduction reactions between Ni and NiO. The cathode is often a single-phase

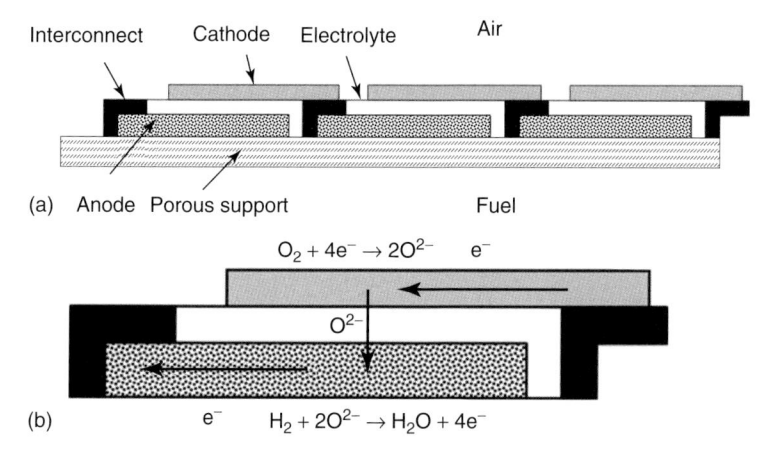

Figure 14.6 Schematic of a cells-in-series SOFC. (a) Connection between cells and (b) current flow through cell.

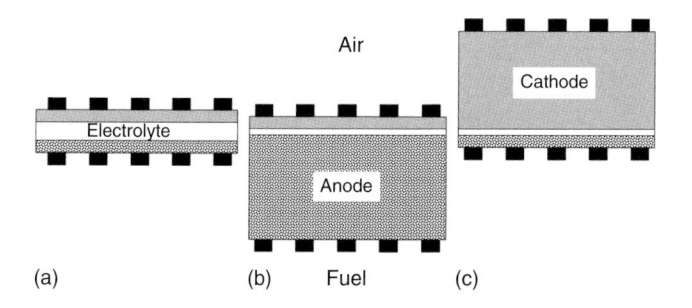

Reactions shown:
$$O_2 + 4e^- \rightarrow 2O^{2-}$$
$$H_2 + 2O^{2-} \rightarrow H_2O + 4e^-$$

Figure 14.7 Schematic of internally supported cells. (a) Electrolyte supported, (b) anode supported, and (c) cathode supported.

ceramic and does not undergo such reactions, but typically has a lower electrical conductivity than the anode, which can increase cell resistance.

Metals have good mechanical properties and are generally easier to shape and join than ceramics, so they are suitable for supports. For intermediate-temperature fuel cells, metals can be used as the interconnect and can provide mechanical support. The interconnect is exposed to both air and fuel, which, as discussed in Section 14.2.4, places restrictions on the alloys used. Using the metal support only on the anode side of the cell where the oxygen partial pressure is low expands the range of potential alloys [2, 3]. This approach is used in the metal-supported design, in which a porous metal is used to support the cell [10]. The porous metal effectively replaces most of the anode in the anode-supported cell as shown in Figure 14.8, which is possible because only the anode near the electrolyte is active in participating in the electrochemical reaction, and the electrolyte phase is needed only in this active region. The porous metal is of lower cost, is more durable, and

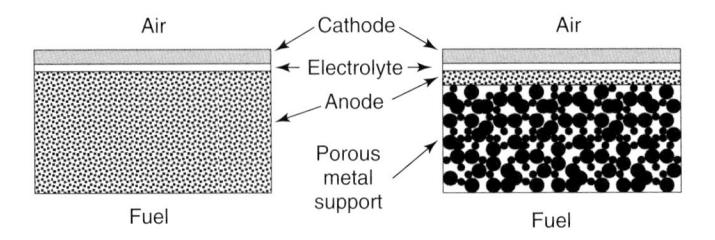

Figure 14.8 Schematics of anode-supported and porous-metal-supported cells.

is easier to be joined to other components than the cermet in the anode-supported cell. An inactive porous ceramic can also be used for external support, as illustrated above for the cells-in-series design shown in Figure 14.6.

A major challenge in SOFCs is establishing a high-temperature gas-tight seal to separate the air from the fuel [11–13]. One approach to avoid this problem is to use a single-chamber design, in which both air and fuel are combined in a single chamber. In this case, the voltage between the electrodes is established by using materials that catalyze the oxidation or reduction reaction at the anode or cathode, respectively. The primary advantage of this design is elimination of the need for separate chambers for the two electrodes and the associated seal between the two chambers, but there are other advantages also. The electrolyte does not separate the air from the fuel as in a standard SOFC, so it need not be fully dense, which allows for the use of lower-cost fabrication processes. In addition, the presence of oxygen in the fuel can reduce the amount of carbon deposition when carbonaceous fuels are used. A major challenge for single-chamber SOFCs is the selection of electrode materials that have the appropriate selective catalytic activity and stability. For example, nickel is a common component of an SOFC anode, but it can be oxidized in the mixed gas, so the operating temperature must be sufficiently high to prevent this oxidation reaction, which can both degrade the performance of the cell and lead to the generation of mechanical stresses. Similarly, the cathode materials that have the best catalytic properties are not always stable in the reducing environment in the mixed gas (i.e., lower oxygen partial pressure than air). Another approach that does not require seals is the direct-flame SOFC. In this case, a flame produced by combustion is used to heat and provide fuel for the cell. The fuel is oxidized by oxygen ions in the electrolyte, which are replenished by the reduction of oxygen in the surrounding air as in the single-chamber fuel cell. The design is simple and can be started quickly, but the efficiency is relatively low [6, 14].

14.2
Fuel Cell Components

The performance of an SOFC depends critically on the materials used for the individual cell components [15, 16]. In addition to performing the function required of the particular component, each component must be stable in the surrounding

14.2.1
Electrolyte

The primary function of the solid electrolyte is to conduct oxygen ions, but not electrons, so high ionic conductivity and an ionic transference number (ionic conductivity/total conductivity) close to one are desired [17, 18]. The conductivities of solid electrolytes increase with increasing temperature, so the electrolyte resistance often determines the minimum operating temperatures of the fuel cell. As discussed in Section 14.1.2, one approach to decreasing the operating temperature is to design the fuel cell with a thinner electrolyte (as in anode-, cathode-, or metal-supported cells (Figures 14.7 and 14.8)). The other approach is to increase the conductivity of the electrolyte [19, 20].

The most commonly used electrolyte for SOFCs is YSZ, where the stabilizer (yttria) is added both to stabilize the cubic fluorite crystal structure and to increase the conductivity by increasing the concentration of oxygen vacancies. Yttria is the most common stabilizer, but other oxides can also be used. For example, scandia-stabilized zirconia has a higher conductivity than YSZ but undergoes a phase transition at lower temperatures.

Another electrolyte with the cubic fluorite structure is cerium oxide, which is also doped to increase the ionic conductivity. The most common dopant for ceria is gadolinium, which is used to form an electrolyte referred to as either gadolinia-doped ceria (GDC) or cerium gadolinium oxide (CGO). Other dopants used include samarium, lanthanum, and yttrium, and, in some cases, combinations of more than one type of dopant are used to obtain higher conductivities. Ceria-based electrolytes typically have higher conductivities than YSZ, particularly at low temperatures, and thus allow for a reduction in the operating temperature. Another advantage of ceria is that it is more stable against reaction with some electrode materials as compared with YSZ, and because of this, it is sometimes used as a barrier layer between the YSZ and the cathode. The primary disadvantage of ceria is that its electronic conductivity becomes significant at lower oxygen partial pressures [21]. Electronic conduction in the electrolyte decreases efficiency since any electrons passing through the electrolyte do not contribute to the electrical current in the external circuit.

Another material used as SOFC electrolytes is lanthanum gallate doped with strontium and magnesium, $La_{1-x}Sr_xGa_{1y}Mg_yO_3$ (LSGM), which forms the perovskite structure [22]. LSGM has high ionic conductivity and is generally compatible with cathode materials, which are typically oxides that also form the perovskite structure. The primary challenge with LSGM electrolytes is that obtaining the desired single-phase structure can be difficult.

The conductivity ranges of the major classes of SOFC electrolytes are compared in Figure 14.9 [18]. Also shown in Figure 14.9 are the conductivities of two bismuth-based electrolytes [23]. The conductivity of Er-doped Bi_2O_3 is higher than

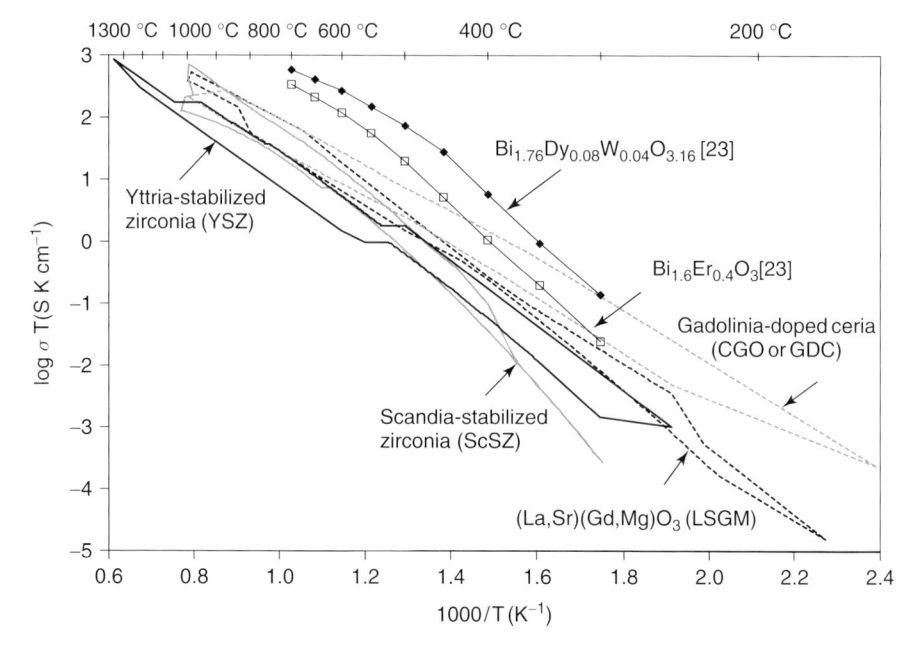

Figure 14.9 Conductivity of solid electrolytes for SOFCs [18, 23].

that of most other SOFC electrolyte materials, and the conductivity is even higher with codoping of dysprosium and tungsten. Although Bi_2O_3-based electrolytes have high conductivity, their stability in a reducing atmosphere is poor. To overcome this shortcoming, Bi_2O_3-based electrolytes are used in a two-layer system with a ceria-based electrolyte, in which ceria separates bismuth oxide from the reducing atmosphere and bismuth oxide blocks the electronic conduction through ceria [24, 25].

14.2.2
Anode

Electronic conduction is detrimental to the electrolyte but is required in the electrodes since the function of the electrode is to transport oxygen ions to/from the electrolyte and electrons to/from the interconnect to undergo electrochemical reaction with the gas as shown in Figure 14.1. Thus, electrodes must be mixed (ion/electron) conductors and should have large surface areas (i.e., porous) for adequate contact with the gas phase. The most common approach to achieve these requirements for the SOFC anode is to use a porous mixture (cermet) of nickel metal and YSZ, in which the electrons are transported through the nickel metal and the oxygen ions through the YSZ [26–30]. The Ni–YSZ cermet is typically produced by heating mixtures of NiO and YSZ in a reducing atmosphere, which both forms the electronically conducting nickel phase (from reduction of NiO) and increases the porosity because of the loss of oxygen. Although NiO is not stable in fuel cell operating conditions, it can be formed during cycling. Since the nickel can sinter and change shape during operation, reformation of NiO can generate stresses in the anode and lead to degradation [31–33]. Nickel is also mixed with ceria-based electrolytes, which have similar challenges associated with the oxidation of nickel during cycling [34, 35].

One approach to avoiding this redox instability is to use mixed conducting oxides as the anode [29, 36–38]. In addition to eliminating the redox instability, the use of a mixed conducting oxide anode increases the surface area for the electrochemical reaction since the reaction can occur over the entire anode surface rather than only at the triple-phase boundary (TPB) between Ni, YSZ, and gas in the Ni–YSZ cermet. The oxides most commonly used for SOFC anodes are titanates or chromites that form the perovskite structure. Another approach is to dope cubic fluorite electrolyte materials to increase their electronic conductivities. Even with mixed conducting electrodes, additional phases, such as Pd or ceria, are sometimes added to improve the electrocatalytic properties [39].

Other areas for improvement of SOFC anodes are their interactions with gases in the fuel. As mentioned earlier, one of the advantages of SOFCs is that carbonaceous fuels can be used. However, under some operating conditions, a carbon deposit can be a by-product of the electrochemical reaction. A small amount of carbon deposition is not a problem and can, in fact, be beneficial to increasing the electron transfer. However, larger amounts of carbon deposits can reduce pore size and block access to the TPB, which limits the electrochemical reaction. The amount of carbon deposition can be reduced by using copper rather than nickel since copper, unlike nickel, does not catalyze the formation of graphite [40].

Sulfur can lead to degradation in the performance of Ni–YSZ anodes and is particularly important when using fuels from coal, which can contain sulfur [41]. The degradation is generally attributed to adsorption of sulfur on the electrodes and is at least partially reversible. Sulfur tolerance can be improved by using oxide anode materials, such as vanadates or molybdates. The addition of ceria to the Ni–YSZ cermet has also been shown to improve the sulfur tolerance.

14.2.3
Cathode

The high oxygen partial pressure at the cathode precludes the use of nickel or other reasonably low-cost metals, so conducting oxides are used. The most common SOFC cathode material is lanthanum strontium manganate (LSM), which forms the perovskite structure [42–45]. LSM is a p-type electronic conductor with negligible oxygen ion conductivity, so the electrochemical reaction occurs at the TPB between LSM, the electrolyte, and the gas. The length of this TPB can be increased using a composite cathode in which the LSM is mixed with the electrolyte similar to the Ni–YSZ cermet anodes, as discussed in Section 14.2.2. However, as mentioned earlier for oxide anode materials, increasing the oxygen ion conductivity in the electrode allows for the electrochemical reaction to occur over the entire electrode surface rather than just at the TPB. Another perovskite oxide, $LaCoO_3$, has higher oxygen ion conductivity than LSM, and also has a higher CTE as compared to other cell components. The CTE decreases with the addition of iron, so $La_{1-x}Sr_xCo_{1-y}Fe_yO_3$ (LSCF) is used as the cathode. Cobalt is more reactive with other fuel cell components, so increasing the iron content also increases the stability. The compositions used typically have more iron ($y = 0.8$ is common) and lanthanum (typically $x = 0.2–0.4$) than strontium and cobalt.

In addition to the transport of electrons and oxygen ions in the electrode, the rate at which oxygen is transferred from the solid electrode material to the gas is important. This rate is described by the surface exchange coefficient, and a lower value leads to higher electrode polarization. The surface exchange coefficients of the cobalt-ferrite-based cathodes are typically higher than those of the manganate-based cathode materials [43].

The electrocatalytic activity and the solid-state oxygen transport cannot be optimized with a single cathode material, so composite cathodes have been developed. One approach is to produce a porous matrix of a cathode material with good oxygen ion (and electronic) conduction and then add a second cathode material with good electrocatalytic properties. One example is the distribution of $Sm_{0.6}Sr_{0.4}CoO_{3-\delta}$ (SSC) on an LSCF scaffold to produce an SOFC composite cathode, where SSC catalyzes the surface reaction and LSCF transports the oxygen ions and electrons to the reaction site [46].

Although the primary function of the cathode is to enable the oxygen reduction reaction, other factors such as mechanical and chemical compatibility (e.g., CTE match, reactivity) must be considered. In addition, the choice of the cathode materials can also affect other aspects of performance. One example is the amount and location of chromium deposition, which can lead to cell degradation, and is discussed in Section 14.6.

Good contact between the interconnect and cathode is required for electron transfer, so an additional cathode contact layer is typically used. The cathode contact layer is typically applied after the electrode and electrolyte materials have been sintered, so a relatively low sintering temperature must be used to minimize degradation of the interconnect alloy. Thus, the contact layer must have good sinterability and/or high conductivity to produce a low contact resistance without sintering at a high temperature. The materials used are typically similar to those used as candidates for cathode materials, but good electrode activity is not required.

14.2.4
Interconnect

The first SOFCs were designed to operate at around 1000 °C, so ceramic interconnects were used. The most commonly used ceramic interconnects are lanthanum chromites, typically doped with calcium or strontium, which is a p-type semiconductor [47]. Decreasing the operating temperature allows for the use of metallic interconnects, which are lower in cost and easier to fabricate than ceramics.

The interconnect is exposed to air and fuel, so the metal must be stable in both oxidizing and reducing atmospheres at a high temperature [48–50]. Metals that are thermodynamically stable in such conditions are expensive, so alloys that form a protective scale must be used. Alumina and silica scales provide the best oxidation resistance, but they are also electrical insulators, which is a problem for SOFC interconnects, because the cell current must pass through the oxide scale. Chromia, however, is a p-type semiconductor, so chromia-forming alloys are more suitable for use as SOFC interconnects. Nickel-based alloys and austenitic stainless steels have relatively high CTEs (as compared to other SOFC components), so ferritic stainless steels are the alloys most commonly used for SOFC interconnects.

The most important alloying addition in stainless steels is chromium to provide the needed oxidation resistance, but other alloying additions influence performance. For example, silicon additions can lead to the formation of a silica layer at the alloy-scale interface, which increases the electrical resistance, so silicon levels must be kept low. Processes required to maintain low silicon content in the alloy, such as vacuum processing, can add to the alloy cost. However, the detrimental effect of silicon can be reduced with other alloying additions, such as niobium, which ties up some of the silicon and reduces its detrimental effect [51]. One of the most important alloying additions in ferritic stainless steels for SOFC interconnects is manganese. The scales formed on manganese-containing alloys typically consist of two layers: an inner chromia layer and an outer manganese–chromium spinel oxide. The importance of this spinel layer is that it reduces the activity of chromia, which decreases the formation of gas species that can deposit at the cathode and degrade cell performance as discussed in Section 14.6.

To further reduce the formation of these chromium-containing volatile species, coatings are applied to the alloy [52]. The primary function of the coating is to contain the chromium, so the coating should have low chromium solubility and/or diffusivity (i.e., permeability). The cell current passes through the coating, so the coating should have good electrical conductivity. In addition, low oxygen permeability is desirable to reduce the growth rate of the oxide scale beneath the coating. Potential coatings include ceramics used for other fuel cell components, such as LSM and lanthanum chromites, which have already been demonstrated to be compatible with the SOFC system [53]. Transition metal spinel oxides, for example, $(Mn,Co)_3O_4$, have shown promise for reducing chromium poisoning in SOFCs.

14.2.5
Sealant

There are two general approaches to providing a gas-tight seal at the high temperatures present in SOFCs: rigid and compressive seals [54–56]. A rigid seal is established typically by melting a material to fill the gap between the components being joined and hold the components together, whereas in a compressive seal, the joining material is placed between the two components and held in place by an externally applied force. The advantage of a rigid seal is that no external force is required, but the disadvantage is that the CTE of the sealant material must closely match those of the joined components to avoid thermal stresses and the associated possibility of cracking during cycling. Conversely, the compressive seal can tolerate larger differences in CTE because of the continuously applied force.

The most common rigid seals are glasses or glass-ceramics. As mentioned earlier, one of the key parameters in the selection of rigid sealant materials is the CTE, which must be close to those of other SOFC components. Another important property is the glass transition temperature, since the glass must flow sufficiently to maintain an adequate seal, while being sufficiently rigid to maintain mechanical integrity. Glass-ceramics consist of an amorphous matrix with crystalline phases, which affect both these properties. The amounts and compositions of these crystalline phases can change during operation, which provides opportunities for adjusting the glass properties but can also lead to degradation.

Rigid seals can also be made from metals, which are less conducive to cracking than ceramics. The challenge with metallic seals, or brazes, is wetting of the molten metal with the ceramic components of the fuel cell. The addition of an oxide, such as CuO to silver, can improve the wetting characteristics and thus improve the seal quality.

Compressive seals can be made of metals, which have a lower modulus of elasticity than ceramics and thus deform readily under the applied load. In addition, the metals can be formed into shapes that will deform even more easily under the applied load, such as a corrugated sheet, and provide better compliance. Mica, which forms platelike crystals, has also been used in compressive seals. The performance of mica-based seals can be improved by infiltrating with a filler and/or combining with a compliant metal seal.

14.3
Assembly and Manufacturing

Cost-effective manufacturing and assembly of SOFCs is a major challenge in the implementation of the technology [57, 58]. This includes fabrication of the individual components as well as assembly of the components into fuel cells and systems. Obtaining the single-phase cubic fluorite phase of zirconia- or ceria-based electrolytes is relatively straightforward. However, for the perovskite (La,Sr)(Ga,Mg)O_3 (LSGM), careful control of the heat treatment process is needed to obtain a single phase. Electrode materials are typically two-phase materials, such as Ni–YSZ anodes or LSM–YSZ composite cathodes, since the electrode reaction requires oxygen ions, electrons, and gas. To further complicate the fabrication process, graded or layered microstructures are often desired. For example, in anode- or cathode-supported designs, only a small portion of the electrode is active in the electrochemical reaction, while a large portion of the electrode is active for mechanical support. The optimal porosity for the two functions may be different, so the fabrication process must account for these differences. Similarly, graded microstructures can optimize the performance in the transitional region and can reduce local stresses by distributing the change in CTE over a larger distance.

In addition to composition and phase content, the phase morphology must be controlled. Porosity is important for gas access, while high surface area and large TPB length are important for electron and ion access. Typically, small feature sizes provide more surface area for reaction, but if the pore sizes decrease too much, gas phase transport can be limited. In addition, even if the component is not load bearing, some level of mechanical integrity is required. Also, high surface area microstructures are inherently unstable because of the high surface energy, and since SOFCs operate at relatively high temperatures, the long-term stability of a microstructure with optimized performance can be a challenge.

Fuel cell components can be fabricated by particulate-based processes, such as powder pressing, slurry tape calendaring or extrusion, followed by sintering. The layered structure of fuel cell designs also allows for the use of deposition processes, such as spin coating, vacuum evaporating, colloidal deposition, sol–gel processing, plasma spraying, spray pyrolysis, electrophoretic deposition, electrostatic assisted vapor deposition, metallorganic chemical vapor deposition, and electrochemical vapor deposition [3]. The use of sequential processes increases processing time and thus increases cost, so methods for combining processing steps are valuable. For example, coextrusion of multiple layers on a tubular design can be used to eliminate a deposition step [59]. Similarly, cosintering of multiple components can save the time and energy associated with multiple sintering steps. Cosintering can be a challenge, because the optimal sintering temperatures for different components are often different. There are challenges even when using multiple sintering steps, because the sequence of the steps must be designed such that no components are exposed to temperatures or atmospheres that detrimentally affect their properties or microstructures. Although the cell has three main components, anode, cathode, and electrolyte, as mentioned earlier, each of these can be comprised of multiple subcomponents. In addition, the cell must be integrated with the interconnect and compatible with the sealant as well as with the support structures discussed in the following section, so fuel cell fabrication is quite complex.

14.4
Stacking and Balance of the Plant

As discussed in Section 14.1.2, the individual cells can be tubular or planar. For tubular designs, the individual tubes can be bundled and electrically connected at the end of the tube (Figure 14.2) or along the length of the tube (Figure 14.3). In the former case, if closed-one-end tubes are used, all the electrical connections can be made at the end of the tube, and this end of the tube can be maintained at a lower temperature than the rest of the tube. In the latter case, the gas inlets and outlets are at the ends of the tubes, but the electrical connections along the tube are at the fuel cell operating temperature. In both cases, one gas flows inside the tube and the other is outside the tube.

In the case of planar stacks, gas flow channels between cells are required to transport the gases to the electrodes. A common approach is to machine channels

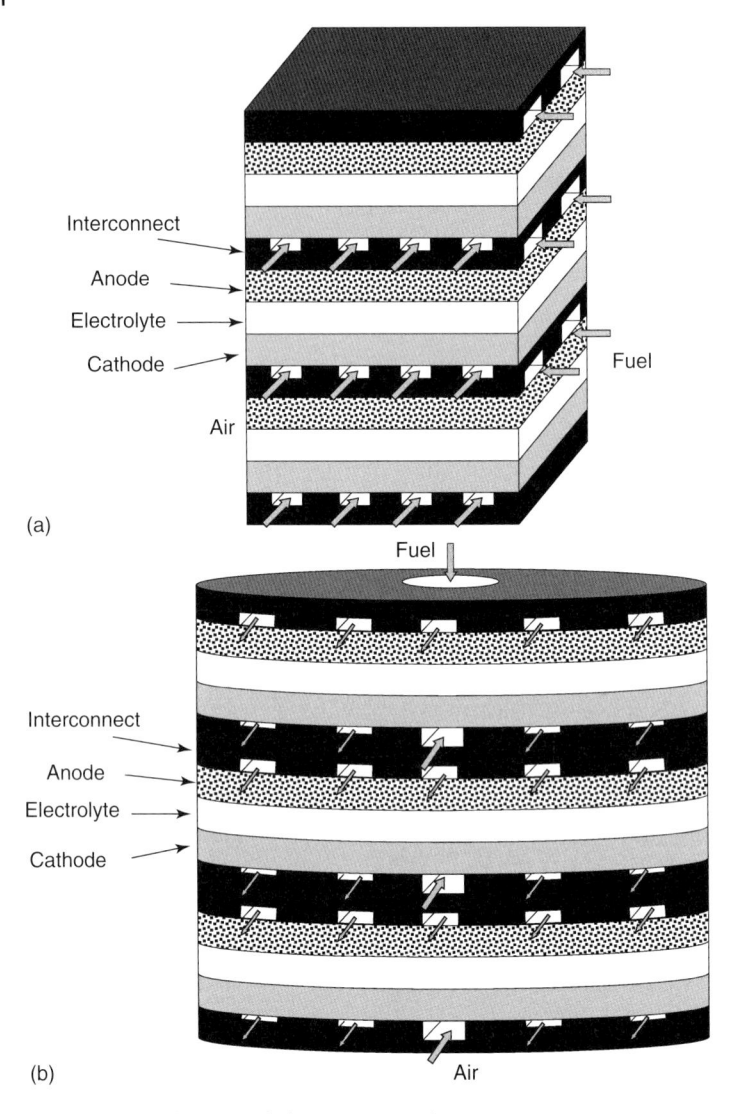

Figure 14.10 Schematic of planar SOFC stacks. (a) Perpendicular cross flow and (b) radial flow.

into the interconnect plates placed between the cells as shown schematically in Figure 14.10a. In the configuration shown in Figure 14.10a, the air and fuel flow in perpendicular directions, but other flow patterns are possible [3]. Figure 14.10b shows a circular design in which the fuel flows in along the axis of the cylinder and then radially outward, while air flows radially inward and outward [60].

In addition to supplying reactant gases and exhausting the product gases, the supporting system must allow for management of heat. The temperature needs to be sufficiently high for the cell to operate properly (i.e., high-enough electrolyte

14.4 Stacking and Balance of the Plant

conductivity and faster-enough electrode kinetics), but not too high to lead to degradation. Since the flow of current generates heat, the amount of heat produced depends on the operating conditions, so management of heat is interrelated with other operating parameters. For example, in a 5 kW system, a compressor is used to deliver cathode air to cool the stack and maintain an acceptable cell temperature gradient, but the compressor is the largest parasitic power drain on the fuel cell and thus decreases system efficiency [61].

The development of useful and efficient systems often requires combining the SOFC with other devices and systems, some examples of which are shown in Figure 14.11. Air and fuel enter the fuel cell, react through the cell to produce electric current and heat, and then the depleted gases are exhausted. Additional components can be added to the input or outputs of the SOFC.

As mentioned earlier, SOFCs can use hydrogen as a fuel, but they are not limited to hydrogen. In some cases, other fuels, such as methane, can be used directly, but in other cases, preprocessing is required. For example, gasification is required for coal and reforming for diesel fuels. In addition, purification, such as sulfur removal, may be required for some fuels. Preprocessing can also be performed on the oxidant to increase the amount of oxygen by either increasing the total pressure or using pure oxygen from an air separation unit (ASU).

The primary output of the fuel cell is electrical power, which can be sent to the electric power grid or used locally to power a building, vehicle, or device. Heat is also

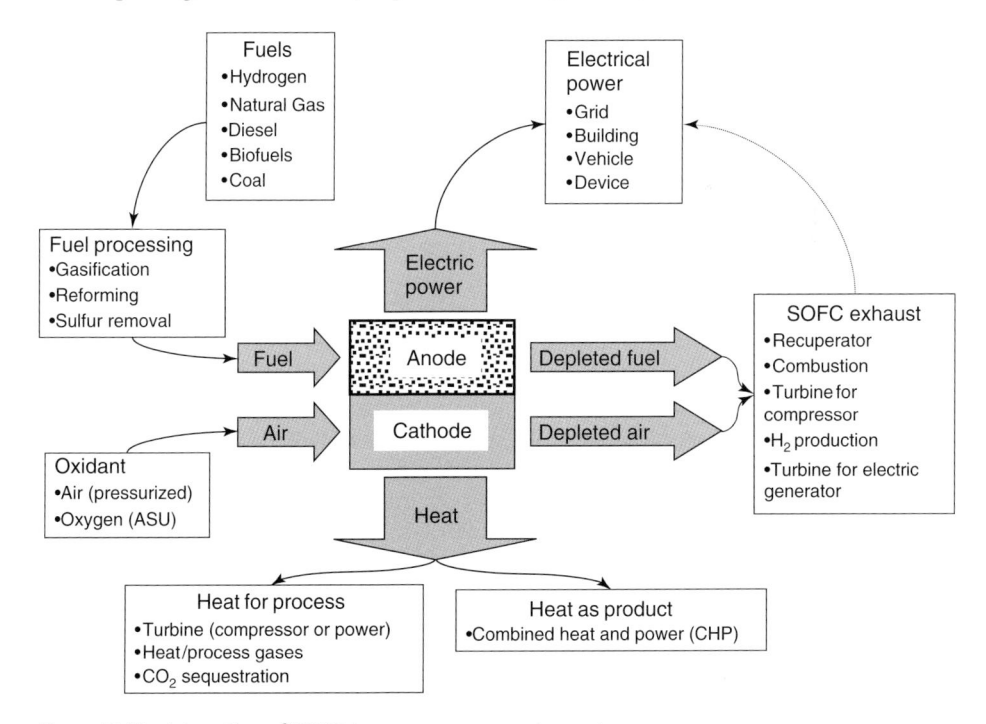

Figure 14.11 Integration of SOFC in an energy conversion system.

generated during operation and can also be a useful product in a CHP application. In addition, the heat can also be fed back into the system to turn a turbine to compress gas or generate power, to heat process gases, or for other processes, such as CO_2 sequestration. Similarly, the exhaust gases can be recuperated, combusted, sent to a turbine, or used for other purposes, such as hydrogen production. While a single system will not have all the components shown in Figure 14.11, there are numerous ways that SOFCs can be combined with other devices to create high-efficiency power generation systems.

14.5
Performance, Testing, and Diagnosis

For the past decade, the U.S. Department of Energy, through the Solid State Energy Conversion Alliance (SECA) program, has supported the development and scale-up of SOFCs. The focus of this program has been on the development of the integrated gasification fuel cell (IGFC) system for use in a coal gasification power plant, but SOFC technologies for other applications are also included. The research and development supported by this program have led to an increase in the SOFC current density, which is currently in the range of $0.45-0.60 \text{ W} \cdot \text{cm}^{-2}$ at operating voltages of 0.70–0.85 V. Recently, results from the operation of 10 kW stacks for more than 5000 h with degradation of 0–2.6% per 1000 h have been reported [62]. The stacks have efficiencies of more than 50% when used in the IGFC system, and the cost of the IGFC power block has been estimated to be around $1000 per kW. The target operational life for the SECA program is 40 000 h, which requires not only additional testing but also techniques for evaluating and monitoring performances.

The most direct and easily accessible indicators of fuel cell performance are the current and voltage generated by the system. As with any fuel cell, the overpotentials increase with increasing current, so the operating voltage decreases with increasing current. Degradation in fuel cell performance would affect the current–voltage characteristics. Degradation that impedes the electrode reaction (some specific mechanisms are discussed in Section 14.6) would lead to a decrease in the voltage for a given cell current. Monitoring the output voltage is simple but it provides little insight into the cause of the reduced performance.

One approach that has been used to obtain additional information on the degradation is impedance analysis [63, 64]. Although the overall cell resistance can be calculated simply from the cell current and voltage using Ohm's law (i.e., the decrease in voltage at a given current indicates an increase in resistance), this resistance represents all contributions. In impedance analysis, however, the impedance is measured at a range of excitation frequencies. Since different contributions, such as electrolyte resistance or electrode polarization, have different characteristic frequencies, their individual contributions can be separated using the differences in frequency responses.

As mentioned earlier, one of the challenges in SOFCs is maintaining a gas-tight seal between the electrodes. If there is leakage due to the cracks in the seal, the resulting change in the partial pressure of the reactant gas could lead to a decrease in the cell voltage. However, crack formation has been measured by monitoring acoustic emissions, which offers a potential method for detection of the failure of seals or other ceramic components in operating fuel cell stacks, before the leakage through the crack leads to a significant decrease in cell voltage [65].

14.6
Degradation Mechanisms and Mitigation Strategies

The degradation of SOFC performance is typically associated with the high temperature of operation [66, 67]. This includes the inherent stability of the phases as well as chemical reaction between components or with impurities in the fuel. The reactions can create barriers to the flow of electrical current or gas and affect the electrochemical activity of electrode surfaces. In addition, the formation of new phases can lead to the generation of stress, which can lead to cracking or separation of components. The high temperature can also lead to microstructural changes.

A common example of a microstructural change is that associated with the oxidation–reduction and sintering of nickel and nickel oxide in the Ni–YSZ cermet, which can occur as the oxygen partial pressure in the anode increases during cycling. Large amounts of nickel oxidation can lead to the generation of stresses and cracking. Small amounts of nickel may lead only to microcracking, which may not immediately affect cell performance. However, the effects can accumulate over multiple cycles and eventually lead to performance degradation [68]. These changes can lead to a decrease in the TPB length and thus decrease the effective area for the electrochemical reaction [69]. The anode degradation can be enhanced if contaminants are present in the fuel. For example, fuels derived from coal can contain phosphorus, arsenic, or sulfur, which react with and degrade the anode performance [70, 71].

Degradation can also occur at the cathode. One common degradation mechanism is associated with the use of chromium-forming alloys for the interconnect. In oxidizing conditions, Cr_2O_3 is oxidized to Cr^{6+}-containing vapor species, such as CrO_3 or $CrO_2(OH)_2$ [72]. These species can then be reduced to form Cr_2O_3 at the cathode, which can block the electroactive sites and degrade cell performance [73]. In addition to blocking the electroactive sites, the deposition can decrease porosity or cause strain, which can also lead to degradation in performance [74]. The volatilization of chromium can be minimized by designing the alloy to form a scale with an outer layer that has a reduced chromia activity by forming other phases, such as a manganese–chromium spinel oxide or rutile [75]. Such scales have reduced volatilization, but reducing the degradation for a long time requires further reductions in volatilization, which is accomplished by applying a coating to the alloy surface [76]. As mentioned in Section 14.2.4, the coating needs to have low electrical resistance to conduct electric current while also having low chromium and oxygen permeability to reduce chromium volatilization and chromia scale growth.

As noted in Section 14.2.3, the degradation due to chromium poisoning is affected by the choice of the cathode materials. The common cathode material LSM has low oxygen ion conductivity, so the reduction of the Cr^{6+}-containing gas occurs only at the TPB, and the Cr_2O_3 deposition occurs near the LSM–YSZ interface. However, in cathodes with higher oxygen ion conductivity, such as $(La,Sr)FeO_3$ (LSF) or LSCF, the electrochemical reaction can occur over the entire electrode surface, so the chromia deposition is accordingly more uniformly distributed [77, 78]. As a result, the degradation in performance of an LSF cathode due to chromium deposition can be less than that of an LSM cathode even if the amount of deposition is larger. In such a case, the detrimental effect of the deposition may be more due to the decrease in porosity than due to the reduction in the amount of electroactive area.

Cathode degradation can also occur because of the interaction with other fuel cell components. An example is the sealant glass, which can contain oxides that interact detrimentally with other materials in the fuel cell. Although alkali metals, which are commonly used in glasses, are typically avoided in SOFC applications, boron is often used and can deposit on the cathode [79]. The sealant glass can also react with the interconnect coating, which can indirectly affect cathode performance by reducing the effectiveness of the coating for decreasing chromium volatilization [80].

The degradation rate determines the operational lifetime, the acceptable length of which depends on the particular application. For example, the target lifetime for the U.S. Department of Energy's SECA program is 40 000 h. In large integrated systems, the system lifetime may be longer than the attainable lifetime of the SOFC, so the replacement schedule for the SOFC, and thus the associated cost, will be determined by the degradation rate.

14.7
Applications

14.7.1
Large-Scale Power Generation

The SOFCs developed in the late 1970s were intended for large-scale (100 kW+) stationary power generation and were based on a tubular design and a high operating temperature of 1000 °C [81]. The size of individual cells and stacks are limited because of the challenges associated with fabrication of the ceramic components, so larger-size power plants are achieved with a modular design [82]. The modular design allows for the same cell and stack fabrication processes to be used for a range of systems, with a corresponding range of applications. The building of SOFC modules and the combination of the module into a large power plant are shown schematically in Figure 14.12.

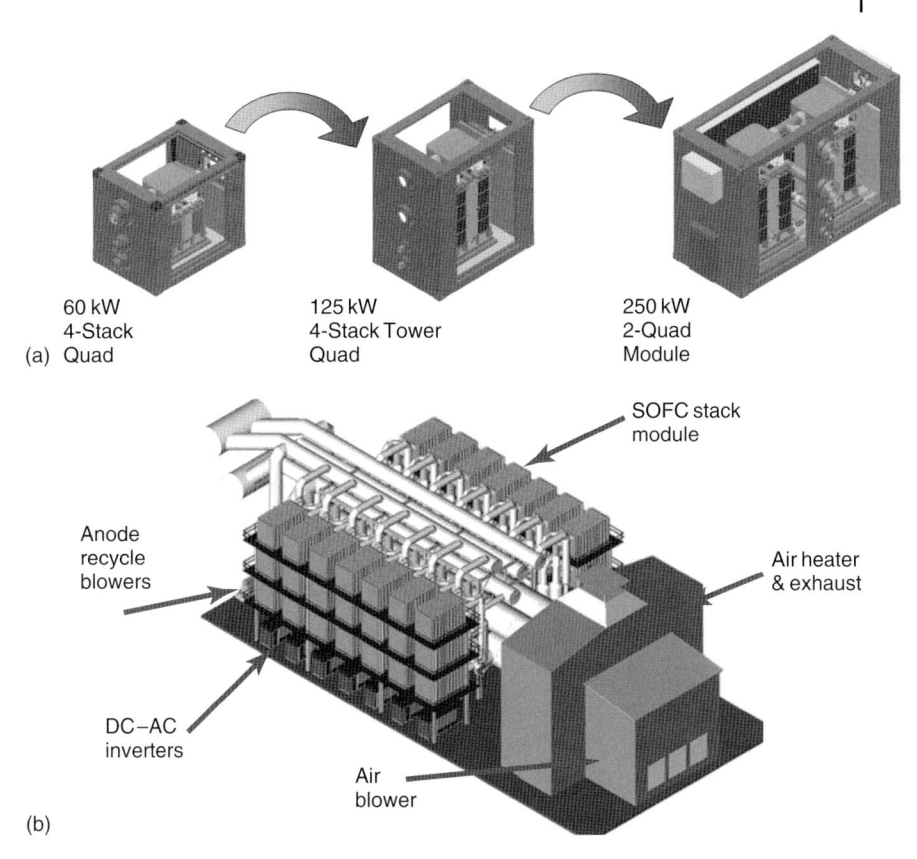

Figure 14.12 SOFC module (a) evolution and (b) cluster. (Courtesy of FuelCell Energy, Inc.)

Although electrochemical conversion is more efficient than combustion, the exhaust gases from SOFCs contain combustible gases. As indicated in Section 14.4, these gases and the heat produced by the SOFC can be used for a gas turbine in a hybrid system to improve the overall system efficiency [83]. In addition to providing energy, the gas turbine can be used to pressurize the air input to the SOFC and thus improve performance and efficiency. Predicted efficiencies of such SOFC hybrid energy conversion systems approach 80% [84]. One important application of SOFCs in a hybrid system is their inclusion in clean coal gasification systems [85]. In addition to effectively using the gasification product for energy conversion, the heat generated by the SOFC can be used in carbon sequestration processes.

14.7.2
Small-Scale Stationary Applications

The heat produced by the high operating temperature of SOFCs can also be used in smaller-scale applications. A promising application is to take advantage of the

Figure 14.13 Residential combined heat and power (CHP) SOFC system. (Coutesy of Hexis Ltd.)

heat generated and the ability to operate using natural gas as the fuel is CHP units for residential applications, where the heat can be used for hot water production [86, 87]. Such units have been produced and are being tested in Japan [88] and Europe [89] (Figure 14.13). Residential CHP systems typically have capacities of around 1 kW, but larger systems have also been produced and are being evaluated [90]. The use of SOFCs in such small stationary power stations, which can operate with a variety of fuels, is an excellent example of taking advantage of the fuel flexibility of SOFCs for distributed power generation.

14.7.3
Auxiliary Power Units

The high operating temperature of SOFCs leads to slow start-up times, so SOFCs are generally not being considered for propulsion in automotive applications. However, SOFCs are being developed for auxiliary power units (APUs) since they can operate from reformed diesel fuel and thus do not require implementation of a hydrogen fuel infrastructure. In particular, APUs for trucks are being developed and will allow truckers to power the sleeper cabins using SOFCs and thus eliminate

Figure 14.14 Delphi SOFC auxiliary power unit (APU) (inset) mounted on a truck. (Courtesy of Delphi.)

the need for idling the diesel engine when the driver is resting [91]. An APU that is being developed and tested by Delphi for use in trucks is shown in Figure 14.14. In addition, there has been considerable work on the development of microtubular SOFCs to reduce the device size [92, 93]. This reduction allows for the development of portable power supplies based on SOFCs [94–96].

14.7.4
Alternative Fuels

Many of the applications discussed above rely on fuel flexibility, so some examples of the fuels that have been used with SOFCs are discussed. There is some controversy as to whether it is best to reform fuels before they enter the fuel cell [97] or allow them to reform in the fuel cell [98], but in either case, the SOFC can operate without the requirement for production, transport, and storage of pure hydrogen. For example, carbon monoxide, which can poison low-temperature fuel cells, can be used as fuel for SOFCs [99]. Similarly, as suggested earlier in the clean coal systems, solid carbon can be used as fuel for SOFCs [100–102]. Other reducing gases such as methane [103, 104] and ammonia [105, 106] have been used. Bioderived fuels have also been used with SOFCs, which provides a more sustainable alternative to fuels based on petroleum products [107–110].

An important issue in the use of bioderived fuels is transport of the fuels from the source to the location of the power plant. Liquid fuels can be transported by pipelines and thus have advantages over bulky solid fuels in terms of transportation cost. SOFCs have been shown to operate using liquid fuels including alcohols (ethanol [111] and methanol [112]), n-dodecane [113], jet fuel [114, 115], and vegetable oil [116] Although the use of liquid fuels is convenient for transportation, the efficiency, and thus economics, of the process is also affected and must be considered in the cost analysis [117, 118].

14.8
Challenges and Perspectives

The fuel flexibility of SOFCs creates opportunities for the potential implementation of SOFCs in a variety of applications. Improvements in fuel cell performance are needed, but the most significant barriers to the implementation of SOFS are related to durability and cost. The high operating temperature creates challenges in the development of materials for long-time operation because of the stability of the materials and microstructures. In addition to selection of the appropriate materials, fabrication of components with acceptable quality control and at sufficiently low cost is also a challenge. Progress is being made, but additional advances are needed. However, with continued development, SOFCs will contribute to improving the sustainability of energy conversion through improved efficiency, reduced emissions, and expanded use of renewable fuels.

Acknowledgments

The kind contribution of the photographs and schematics by the following individuals is gratefully acknowledged: Hossein Ghezel-Ayagh of FuelCell Energy, Inc. (Figure 14.12); Volker Nerlich and Andreas Mai of Hexis, Ltd. (Figure 14.13); and Subhasish Mukerjee and Lynn Kier of Delphi (Figure 14.14). The author acknowledges the financial support from the National Science Foundation (**NSF-DMR-0551896**) and the U.S. Department of Energy (Office of Basic Sciences - EPSCoR-State/National Laboratory Partnerships Program), which has funded his research on materials for SOFCs.

References

1. Cropper, M.A.J., Geiger, S., and Jollie, D.M. (2004) Fuel cells: a survey of current developments. *J. Power Sources*, **131** (1–2), 57–61.
2. Williams, M.C. (2007) Solid oxide fuel cells: fundamentals to systems. *Fuel Cells*, **7** (1), 78–85.
3. Minh, N.Q. (2004) Solid oxide fuel cell technology – features and applications. *Solid State Ionics*, **174** (1–4), 271–277.
4. Singh, P. and Minh, N.Q. (2004) Solid oxide fuel cells: technology status. *Int. J. Appl. Ceram. Technol.*, **1** (1), 5–15.
5. Singhal, S.C. (2007) Solid oxide fuel cells. *ECS Interface*, **16** (4), 41–44.
6. Bagotsky, V.S. (2009) *Fuel Cells: Problems and Solutions*, John Wiley & Sons, Inc., Hoboken.
7. Matsuzaki, Y., Hatae, T., and Yamashita, S. (2009) Long-term stability of segmented type cell-stacks developed for residential use less than 1 kW. *ECS Trans.*, **25** (2), 159–166.
8. Koi, M., Yamashita, S., and Matsuzaki, Y. (2007) Development of segmented-in-series cell-stacks with

flat-tubular substrates. *ECS Trans.*, **7** (1), 235–243.

9. Buergler, B.E., Ochsner, M., Vuillemin, S., and Gauckler, L.J. (2007) From macro- to micro-single chamber solid oxide fuel cells. *J. Power Sources*, **171** (2), 310–320.

10. Tucker, M.C. (2010) Progress in metal-supported solid oxide fuel cells: a review. *J. Power Sources*, **195**, 4570–4582.

11. Yano, M., Tomita, A., Sano, M., and Hibino, T. (2007) Recent advances in single-chamber solid oxide fuel cells: a review. *Solid State Ionics*, **177**, 3351–3359.

12. Viricelle, J.-P., Udroiu, S., Gadacz, G., Pijolat, M., and Pijolat, C. (2010) Development of single chamber solid oxide fuel cells (SCFC). *Fuel Cells*, **10** (4), 683–692.

13. Hao, Y., Shao, Z., Mederos, J., Lai, W., Goodwin, D.G., and Haile, S.M. (2006) Recent advances in single-chamber fuel-cells: experiment and modeling. *Solid State Ionics*, **177**, 2013–2021.

14. Kronemayer, H., Barzan, D., Horiuchi, M., Suganuma, S., Tokutake, Y., Schulz, C., and Bessler, W.G. (2007) A direct-flame solid oxide fuel cell (DFFC) operated on methane, propane, and butane. *J. Power Sources*, **166**, 120–126.

15. Jacobson, A.J. (2010) Materials for solid oxide fuel cells. *Chem. Mater.*, **22**, 660–674.

16. Orera, A. and Slater, P.R. (2010) New chemical systems for solid oxide fuel cells. *Chem. Mater.*, **22**, 675–690.

17. Xia, C. (2009), Electrolytes, in *Solid Oxide Fuel Cells: Materials Properties and Performance* (eds J.W. Fergus, R. Hui, X. Li, D.P. Wilkinson, and J. Zhang), CRC Press, Taylor & Francis Group, Boca Raton, FL, pp. 1–71.

18. Fergus, J.W. (2006) Electrolytes for solid oxide fuel cells. *J. Power Sources*, **162** (1–2), 30–40.

19. Kharton, V.V., Marques, F.M.B., and Atkinson, A. (2004) Transport properties of solid oxide electrolyte ceramics: a brief review. *Solid State Ionics*, **174** (1–4), 135–149.

20. Hui, S., Roller, J., Yick, S., Zhang, X., Decès-Petit, C., Xie, Y., Maric, R., and Ghosh, D. (2007) A brief review of the ionic conductivity enhancement for selected oxide electrolytes. *J. Power Sources*, **171** (2), 493–502.

21. Dalslet, B., Blennow, P., Hendriksen, P.V., Bonanos, N., Lybye, D., and Mogensen, M. (2006) Assessment of doped ceria as electrolyte. *J. Solid State Electrochem.*, **10** (8), 547–561.

22. Joshi, A.V., Steppan, J.J., Taylor, D.M., and Elangovan, S. (2004) Solid electrolyte materials, devices and applications. *J. Electroceram.*, **13** (1–3), 619–625.

23. Jung, D.W., Duncan, K.L., and Wachsman, E.D. (2009) Effect of total dopant concentration and dopant ratio on conductivity of $(DyO_{1.5})_x$-$(WO_3)_y$-$(BiO_{1.5})_{1-x-y}$. *Acta Mater.*, **58** (2), 355–363.

24. Ahn, J.S., Camaratta, M.A., Pergolesi, D., Lee, K.T., Yoon, H., Lee, B.W., Jung, D.W., Traversa, E., and Wachsman, E.D. (2010) Development of high performance ceria/bismuth oxide bilayered electrolyte SOFCs for lower temperature operation. *J. Electrochem. Soc.*, **157** (3), B376–B382.

25. Zhang, L., Li, L., Zhao, F., Chen, F., and Xia C.. (2011) $Sm_{0.2}Ce_{0.8}O_{1.9}/Y_{0.25}Bi_{0.75}O_{1.5}$ bilayered electrolytes for low-temperature SOFCs with Ag-$Y_{0.25}Bi_{0.75}O_{1.5}$ composite cathodes. *Solid State Ionics*, **192**, 557–560.

26. Cheng, Z., Wang, J.-H., and Liu, M. (2009), Anodes, in *Solid Oxide Fuel Cells: Materials Properties and Performance* (eds J.W. Fergus, R. Hui, X. Li, D.P. Wilkinson, and J. Zhang), CRC Press, Taylor & Francis Group, Boca Raton, FL, pp. 73–129.

27. Jiang, S.P. and Chan, S.H. (2004) A review of anode materials development in solid oxide fuel cells. *J. Mater. Sci.*, **39** (14), 4405–4439.

28. McEvoy, A. (2003), Anodes, in *High Temperature Solid Oxide Fuel Cells* (eds S.C. Singhal and K. Kendall.), Elsevier Ltd., Oxford, pp. 149–171.

29. Atkinson, A., Barnett, S., Gorte, R.J., Irvine, J.T.S., McEvoy, A.,

Mogensen, M., Singhal, S.C., and Vohs, J. (2004) Advanced anodes for high-temperature fuel cells. *Nat. Mater.*, **3** (1), 17–27.

30. Sun, C. and Stimming, U. (2007) Recent anode advances in solid oxide fuel cells. *J. Power Sources*, **171** (2), 247–260.
31. Sarantaridis, D. and Atkinson, A. (2007) Redox cycling of Ni-based solid oxide fuel cell anodes: a review. *Fuel Cells (Weinheim)*, **7** (3), 246–258.
32. Klemensø, T., Chung, C., Larsen, P.H., and Mogensen, M. (2005) The mechanism behind redox instability of anodes in high-temperature SOFCs. *J. Electrochem. Soc.*, **152** (11), A2186–A2192.
33. Tikekar, N.M., Armstrong, T.J., and Virkar, A.V. (2006) Reduction and reoxidation kinetics of nickel-based SOFC anodes. *J. Electrochem. Soc.*, **153** (4), A654–A653.
34. Ding, C., Lin, H., Sato, K., Kawada, T., Mizusaki, J., and Hashida, T. (2010) Improvement of electrochemical performance of anode-supported SOFCs by $NiO-Ce_{0.9}Gd_{0.1}O_{1.95}$ nanocomposite powders. *Solid State Ionics*, **181**, 1238–1243.
35. Chen, M., Kim, B.H., Xu, Q., Ahn, B.G., and Huang, D.P. (2010) Effect of Ni content on the microstructure and electrochemical properties of Ni–SDC anodes for IT-SOFC. *Solid State Ionics*, **181**, 1119–1124.
36. Tao, S. and Irvine, J.T.S. (2004) Discovery and characterization of novel oxide anodes for solid oxide fuel cells. *Chem. Rec.*, **4** (2), 83–95.
37. Fu, Q.X. and Tietz, F. (2008) Ceramic-based anode materials for improved redox cycling of solid oxide fuel cells. *Fuel Cells*, **8** (5), 283–293.
38. Fergus, J.W. (2006) Oxide anode materials for SOFCs. *Solid State Ionics*, **177** (17–18), 1529–1541.
39. Smith, B.H., Holler, W.C., and Gross, M.D. (2011) Electrical properties and redox stability of tantalum-doped strontium titanate for SOFC anodes. *Solid State Ionics*, **192**, 383–386.
40. McIntosh, S. and Gorte, R.J. (2004) Direct hydrocarbon solid oxide fuel cells. *Chem. Rev.*, **104**, 4845–4865.
41. Gong, M., Liu, X., Trembly, J., and Johnson, C. (2007) Sulfur-tolerant anode materials for solid oxide fuel cell application. *J. Power Sources*, **168** (2), 289–298.
42. Fleig, J. (2003) Solid oxide fuel cell cathodes: polarization mechanisms and modeling of the electrochemical performance. *Annu. Rev. Mater. Res.*, **33**, 361–382.
43. Jiang, S.P. and Li, J. (2009), Cathodes, in *Solid Oxide Fuel Cells: Materials Properties and Performance* (eds J.W. Fergus, R. Hui, X. Li, D.P. Wilkinson, and J. Zhang), CRC Press, Taylor & Francis Group, Boca Raton, FL, pp. 131–177.
44. Sun, C., Hui, R., and Roller, J. (2010) Cathode materials for solid oxide fuel cells: a review. *J. Solid State Electrochem.*, **14**, 1125–1144.
45. Yokokawa, H. and Horita, T. (2003), Cathodes, in *High Temperature Solid Oxide Fuel Cells* (eds S.C. Singhal and K. Kendall), Elsevier Ltd, Oxford, pp. 119–147.
46. Lou, X., Liu, Z., Wang, S., Xiu, Y., Wong, C.P., and Liu, M. (2010) Controlling the morphology and uniformity of a catalyst-infiltrated cathode for solid oxide fuel cells by tuning wetting property. *J. Power Sources*, **195**, 419–424.
47. Fergus, J.W. (2004) Lanthanum chromite based materials for solid oxide fuel cell interconnects. *Solid State Ionics*, **171** (1–2), 1–15.
48. Yang, Z. and Fergus, J.W. (2009), Interconnects, *Solid Oxide Fuel Cells: Materials Properties and Performance* (eds J.W. Fergus, R. Hui, X. Li, D.P. Wilkinson, and J. Zhang), CRC Press, Taylor & Francis Group, Boca Raton, FL, pp. 179–212.
49. Yang, Z. (2008) Recent advanced in metallic interconnects for solid oxide fuel cells. *Int. Mater. Rev.*, **53** (1), 39–54.
50. Fergus, J.W. (2005) Metallic interconnects for solid oxide fuel cells. *Mater. Sci. Eng. A*, **397** (1–2), 271–283.
51. Jablonski, P.D., Cowen, C.J., and Sears, J.S. (2010) Exploration of alloy 441

chemistry for solid oxide fuel cell interconnect application. *J. Power Sources*, **195**, 813–820.
52. Shaigan, N., Qu, W., Ivey, D.G., and Chen, W. (2010) A review of recent progress in coatings, surface modifications and alloy developments for solid oxide fuel cell ferritic stainless steel interconnects. *J. Power Sources*, **195**, 1529–1542.
53. Lacey, R., Pramanick, A., Lee, J.C., Jung, J.-I., Jiang, B., Edwards, D.D., Naum, R., and Misture, S.T. (2010) Evaluation of Co and perovskite Cr-blocking thin films on SOFC interconnects. *Solid State Ionics*, **181**, 1294–1302.
54. Lessing, P.A., Hartvigsen, J., and Elangovan, S. (2009), Sealants, in *Solid Oxide Fuel Cells: Materials Properties and Performance* (eds J.W. Fergus, R. Hui, X. Li, D.P. Wilkinson, and J. Zhang), CRC Press, Taylor & Francis Group, Boca Raton, FL, pp. 213–223.
55. Fergus, J.W. (2005) Sealants for solid oxide fuel cells. *J. Power Sources*, **147** (1–2), 46–57.
56. Lessing, P.A. (2007) A review of sealing technologies applicable to solid oxide electrolysis cells. *J. Mater. Sci.*, **42** (10), 3465–3476.
57. Kesler, O. and Marcazzan, P. (2009), Processing, in *Solid Oxide Fuel Cells: Materials Properties and Performance* (eds J.W. Fergus, R. Hui, X. Li, D.P. Wilkinson, and J. Zhang), CRC Press, Taylor & Francis Group, Boca Raton, FL, pp. 239–282.
58. Menzler, N.H., Tietz, F., Uhlenbruck, S., Buchkremer, H.P., and Stöver, D. (2010) Materials and manufacturing technologies for solid oxide fuel cells. *J. Mater. Sci.*, **45**, 3109–3135.
59. Powell, J. and Blackburn, S. (2010) Co-extrusion of multilayered ceramic micro-tubes for use as solid oxide fuel cells. *J. Eur. Ceram. Soc.*, **30**, 2859–2870.
60. Mai, A., Iwanschitz, B., Weissen, U., Denzler, R., Haberstock, D., Nerlich, V., Sfeir, J., and Schuler, A. (2009) Status of Hexis SOFC stack development and the Galileo 1000 N micro-CHP system. *ECS Trans.*, **25** (2), 149–158.
61. Minh, N.Q. (2009) Operating characteristics of solid oxide fuel cell stacks and systems. *ECS Trans.*, **25** (2), 241–246.
62. Vora, S. SECA program accomplishments and future challenges. 11th Annual SECA Workshop, July 27–29 2010, http://www.netl.doe.gov/publications/proceedings/10/seca/index.html (accessed 2010).
63. Huang, Q.A., Wang, B., Qu, W., and Hui, R. (2009) Impedance diagnosis of metal-supported SOFCs with SDC as electrolyte. *J. Power Sources*, **191**, 297–303.
64. Gazzarri, J.I. and Kesler, O. (2007) Electrochemical AC impedance model of a solid oxide fuel cell and its application to diagnosis of multiple degradation modes. *J. Power Sources*, **167**, 100–110.
65. Malzbender, J., Steinbrech, R.W., and Singheiser, L. (2009) A review of advanced techniques for characterising SOFC behaviour. *Fuel Cells*, **9** (6), 785–793.
66. Yokokawa, H., Tu, H., Iwanschitz, B., and Mai, A. (2008) Fundamental mechanisms limiting solid oxide fuel cell durability. *J. Power Sources*, **182**, 400–412.
67. Tu, H. and Stimming, U. (2004) Advances, aging mechanisms and lifetime in solid-oxide fuel cells. *J. Power Sources*, **127** (1–2), 284–293.
68. Hatae, T., Matsuzaki, Y., Yamashita, S., and Yamazaki, Y. (2010) Destruction modes of anode-supported SOFC caused by degrees of electrochemical oxidation in redox cycle. *J. Electrochem. Soc.*, **157** (5), B650–B654.
69. Matsui, T., Kishida, R., Kim, J.-Y., Muroyama, H., and Eguchi, K. (2010) Performance deterioration of Ni–YSZ anode induced by electrochemically generated steam in solid oxide fuel cells. *J. Electrochem. Soc.*, **157** (5), B776–B781.
70. Iqbal, G., Guo, H., Kang, B.S., and Marina, O.A. (2011) Durability prediction of solid oxide fuel cell anode material under thermomechanical and

fuel gas contaminant effects, *Int. J. Appl. Ceram. Technol*, **8** (1), 13–22.
71. Yokokawa, H., Yamaji, K., Brito, M.E., Kishimoto, H., and Horita, T. (2011) General considerations on degradation of SOFC anodes and cathodes due to impurities in gases. *J. Power Sources*, **196**, 7070–7075.
72. Opila, E.J., Myers, D.L., Jacobson, N.S., Nielsen, I.M.B., Johnson, D.F., Olminsky, J.K., and Allendorf, M.D. (2007) Theoretical and experimental investigation of the thermochemistry of $CrO_2(OH)_2(g)$. *J. Phys. Chem. A*, **111** (10), 1971–1980.
73. Fergus, J.W. (2007) Effect of cathode and electrolyte transport properties on chromium poisoning in solid oxide fuel cells. *Int. J. Hydrogen Energy*, **32** (16), 3664–3671.
74. Liu, D.-J., Almer, J., and Cruse, T. (2010) Characterization of Cr poisoning in a solid oxide fuel cell cathode using a high energy X-ray microbeam. *J. Electrochem. Soc.*, **157** (5), B744–B751.
75. Stanislowski, M., Wessel, E., Hilpert, K., Markus, T., and Singheiser, L. (2007) Chromium vaporization from high-temperature alloys. I. Chromia-forming steels and the influence of outer oxide layers. *J. Electrochem. Soc.*, **154** (4), A295–A306.
76. Trebbels, R., Markus, T., and Singheiser, L. (2010) Investigation of chromium vaporization from interconnector steels with spinel coatings. *J. Fuel Cell Sci. Technol.*, **7** (1), 011013/1–011013/6.
77. Bentzen, J.J., Høgh, J.V.T., Barfod, R., and Hagen, A. (2009) Chromium poisoning of LSM/YSZ and LSCF/CGO composite cathodes. *Fuel Cells*, **9** (6), 823–832.
78. Horita, T., Xiong, Y., Kishimoto, H., Yamaji, K., Brito, M.E., and Yokokawa, H. (2010) Chromium poisoning and degradation at (La,Sr)MnO$_3$ and (La,Sr)FeO$_3$ cathodes for solid oxide fuel cells. *J. Electrochem. Soc.*, **157** (5), B614–B620.
79. Zhou, X.-D., Templeton, J.W., Zhu, Z., Chou, Y.-S., Maupin, G.D., Lu, Z., Brow, R.K., and Stevenson, J.W. (2010) Electrochemical performance and stability of the cathode for solid oxide fuel cells. III. Role of volatile boron species on LSM/YSZ and LSCF. *J. Electrochem. Soc.*, **157** (7), B1019–B1023.
80. Chou, Y.-S., Stevenson, J.W., Xia, G.-G., and Yang, Z.-G. (2010) Electrical stability of a novel sealing glass with (Mn,Co)-spinel coated Crofer22APU in a simulated SOFC dual environment. *J. Power Sources*, **195**, 5666–5673.
81. Williams, M.C., Strakey, J.P., and Singhal, S.C. (2004) U.S. distributed generation fuel cell program. *J. Power Sources*, **131**, 79–85.
82. Haga, T., Komiyama, N., Nakatomi, H., Konishi, K., Sutou, T., and Kikuchi, T. (2009) Prototype SOFC CHP system (SOFIT) development and testing. *ECS Trans.*, **25** (2), 71–76.
83. Cheddie, D.F. (2010) Integration of a solid oxide fuel cell into A 10 MW gas turbine power plant. *Energies*, **3**, 754–769.
84. Zhao, Y., Shah, N., and Brandon, N. (2010) The development and application of a novel optimisation strategy for solid oxide fuel cell-gas turbine hybrid cycles. *Fuel Cells*, **10** (1), 181–193.
85. Li, M., Rao, A.D., Brouwer, J., and Samuelsen, G.S. (2010) Design of highly efficient coal-based integrated gasification fuel cell power plants. *J. Power Sources*, **195**, 5707–5718.
86. Wakui, T., Yokoyama, R., and Shimizu, K.-I. (2010) Suitable operational strategy for power interchange operation using multiple residential SOFC (solid oxide fuel cell) cogeneration systems. *Energy*, **35**, 740–750.
87. Wachsman, E.D. and Singhal, S.C. (2009) Solid oxide fuel cell commmercialization, research and challenges. *ECS Interface*, **18** (3), 38–43.
88. Hosoi, K. and Nakabaru, M. (2009) Status of national project for SOFC development in Japan. *ECS Trans.*, **25** (2), 11–20.
89. Steinberger-Wilckens, R. (2009) European SOFC R&D – status and trends. *ECS Trans.*, **25** (2), 3–10.
90. Halinen, M., Saarinen, J., Noponen, M., Vinke, I.C., and

Kiviaho, J. (2010) Experimental analysis on performance and durability of SOFC demonstration unit. *Fuel Cells*, **10** (3), 440–452.
91. Mukerjee, S., Haltiner, K., Klotzbach, D., Vordonis, J., Iyer, A., Kerr, R., Sprenkle, V., Kim, J.Y., Meinhardt, K., Canfield, N., Darsell, J., Kirby, B., Oh, T.K., Maupin, G., Voldrich, B., and Bonnett, J. (2009) Solid oxide fuel cell stack for transportation and stationary applications. *ECS Trans.*, **25** (2), 59–63.
92. Kendall, K. (2010) Progress in micro-tubular solid oxide fuel cells. *Int. J. Appl. Ceram. Technol.*, **7** (1), 1–9.
93. La O', G.J., In, H.J., Crumlin, E., Barbastathis, G., and Shao-Horn, Y. (2007) Recent advances in microdevices for electrochemical energy conversion and storage. *Int. J. Energy Res.*, **31**, 548–575.
94. Suzuki, T., Funahashi, Y., Yamaguchi, T., Fujishiro, Y., and Awano, M. (2008) New stack design of micro-tubular SOFCs for portable power sources. *Fuel Cells*, **8** (6), 381–384.
95. Litzelman, S.J., Hertz, J.L., Jung, W., and Tuller, H.L. (2008) Opportunities and challenges in materials development for thin film solid oxide fuel cells. *Fuel Cells*, **8** (5), 294–302.
96. Calise, F., Restuccia, G., and Sammes, N. (2011) Experimental analysis of micro-tubular SOFC performance degradation fed by different fuel mixtures. *J. Power Sources*, **196**, 301–312.
97. Adams, T.A. and Barton, P.I. II (2010) Re: support for the high efficiency, carbon separation and internal reforming capabilities of solid oxide fuel cell systems. *J. Power Sources*, **195**, 5152–5153.
98. Brouwer, J. (2010) Support for the high efficiency, carbon separation and internal reforming capabilities of solid oxide fuel cell systems. *J. Power Sources*, **195**, 5150–5151.
99. Homel, M., Gür, T.M., Koh, J.H., and Virkar, A.V. (2010) Carbon monoxide-fueled solid oxide fuel cell. *J. Power Sources*, **195**, 6367–6372.
100. Gür, T. (2010) Mechanistic modes for solid carbon conversion in high temperature fuel cells. *J. Electrochem. Soc.*, **157** (5), B751–B759.
101. Cao, D., Sun, Y., and Wang, G. (2007) Direct carbon fuel cell: fundamentals and recent developments. *J. Power Sources*, **167**, 250–257.
102. Kellogg, I.D., Koylu, U.O., and Dogan, F. (2010) Solid oxide fuel cell bi-layer anode with gadolinia-doped ceria for utilization of solid carbon fuel. *J. Power Sources*, **195**, 1738–1742.
103. Wang, K., Ran, R., and Shao, Z. (2007) Methane-fueled IT-SOFCs with facile in-situ inorganic templating synthesized mesoporous $Sm_{0.2}Ce_{0.8}O_{1.9}$ as catalytic layer. *J. Power Sources*, **170** (2), 251–258.
104. Kawano, M., Matsui, T., Kikuchi, R., Yoshida, H., Inagaki, T., and Eguchi, K. (2007) Direct internal steam reforming at SOFC anodes composed of NiO-SDC composite particles. *J. Electrochem. Soc.*, **154** (5), B460–B465.
105. Dekker, N.J.J. and Rietveld, G. (2006) Highly efficient conversion of ammonia in electricity by solid oxide fuel cells. *J. Fuel Cell Sci. Technol.*, **3** (4), 499–502.
106. Meng, G., Jiang, C., Ma, J., Ma, Q., and Liu, Z. (2007) Comparative study on the performance of a SDC-based SOFC fueled with ammonia and hydrogen. *J. Power Sources*, **173**, 189–193.
107. Tokoro, T., Tanaka, Y., Kato, T., Kato, K., Negishi, A., and Nozaki, K. (2010) Possibility of SOFC using biomass-fuel and distributed carbon capture and storage (DCCS). *ECS Trans.*, **26** (1), 315–322.
108. Girona, K., Laurencin, J., Petitjean, M., Fouletier, J., and Lefebvre-Joud, F. (2009) SOFC running on biogas: identification and experimental validation of "safe" operating conditions. *ECS Trans.*, **25** (2), 1041–1050.
109. Doherty, W., Reynolds, A., and Kennedy, D. (2010) Simulation of a tubular solid oxide fuel cell stack operating on biomass syngas using Aspen Plus. *J. Electrochem. Soc.*, **157** (7), B975–B981.

110. van Herle, J., Maréchal, M., Leuenberger, S., Membrez, Y., Bucheli, O., and Fabrat, D. (2004) Process flow model of solid oxide fuel cell system supplied with sewage biogas. *J. Power Sources*, **131**, 127–141.
111. Zhu, X., Lü, Z., Wei, B., Zhang, Y., Huang, X., and Su, W. (2010) Impregnated $La_{0.75}Sr_{0.25}Cr_{0.5}Fe_{0.5}O_{3-\delta}$-based anodes operating on H_2, CH_4, and C_2H_5OH fuels. *Electrochem. Solid-State Lett.*, **13** (8), B91–B94.
112. Christiansen, N., Hansen, J.B., Holm-Larsen, H., Jørgensen, M.J., Wandel, M., Hendriksen, P.V., Hagen, A., and Ramousse, S. (2009) Status of development and manufacture of solid oxide fuel cells at Topsoe Fuel Cell A/S and Risø DTU. *ECS Trans.*, **25** (2), 133–142.
113. Kishimoto, H., Yamaji, K., Horita, T., Xiong, Y., Sakai, N., Brito, M.E., and Yokokawa, H. (2007) The feasibility of liquid hydrocarbon fuels for SOFC with Ni-ScSZ anode. *J. Power Sources*, **172**, 67–71.
114. Zhou, Z.F., Gallo, C., Pague, M.B., Schobert, H., and Lvov, S.M. (2004) Direct oxidation of jet fuels and Pennsylvania crude oil in a solid oxide fuel cell. *J. Power Sources*, **133** (2), 181–187.
115. Sun, E., Yamanis, J., Chen, L., Frame, D., Holowczak, J., Magdefrau, D., Tulyani, S., Hawkes, J., Haugstetter, C., Radcliff, T., and Tew, D. (2009) Solid oxide fuel cell development at United Technologies Research Center. *ECS Trans.*, **25** (2), 77–84.
116. Zhou, Z.F., Kumar, R., Thakur, S.T., Rudnick, L.R., Schobert, H., and Lvov, S.M. (2007) Direct oxidation of waste vegetable oil in solid oxide fuel cells. *J. Power Sources*, **171** (2), 856–8600.
117. Santin, M., Traverso, A., Magistri, L., and Massardo, A. (2010) Thermoeconomic analysis of SOFC-GT hybrid systems fed by liquid fuels. *Energy*, **35**, 1077–1083.
118. Strazza, C., Del Borghi, A., Costamagna, P., Traverso, A., and Santin, M. (2010) Comparative LCA of methanol-fuelled SOFCs as auxiliary power systems on-board ships. *Appl. Energy*, **87**, 1670–1678.

15
Direct Methanol Fuel Cells

Kan-Lin Hsueh, Li-Duan Tsai, Chiou-Chu Lai, and Yu-Min Peng

15.1
Introduction to Direct Methanol Fuel Cells

The need for safe and long running time for various portable electronic equipment, such as notebooks, iPods, or cell phones, has forced the development of the direct methanol fuel cell (DMFC). The DMFC uses aqueous methanol as the fuel. At ambient pressure and temperature, it can be stored in a liquid form with a theoretical energy density of $6.1\,\text{kWh kg}^{-1}$. Although a fuel cell with hydrogen as the fuel has a high-power density, the storage of hydrogen for portable applications is a barrier to be overcome. As schematically shown in Figure 15.1, the methanol is oxidized at the anode. It produces carbon dioxide, protons, and electrons, as given by Eq. (15.1). The proton migrates from the anode through a proton exchange membrane to the cathode. Oxygen in the air is reduced at the cathode. It combines the proton and electrons, and water is generated, as given by Eq. (15.2). The overall reaction is equivalent to the combustion of methanol, as given by Eq. (15.3). The fuel cell separates the reduction reaction Eq. (15.2) from the oxidation reaction Eq. (15.1) and electrons are forced to flow through the external circuit.

$$\text{On the anode:} \quad CH_3OH_{(l)} + H_2O_{(l)} \rightarrow CO_{2(g)} + 6H^+ + 6e^- \qquad (15.1)$$

$$\text{On the cathode:} \quad \frac{3}{2}O_{2(g)} + 6H^+ + 6e^- \rightarrow 3H_2O_{(l)} \qquad (15.2)$$

$$\text{Overall reaction:} \quad CH_3OH_{(l)} + \frac{3}{2}O_{2(g)} \rightarrow CO_{2(g)} + 2H_2O \qquad (15.3)$$

Although the fuel cell with direct methanol feed had been tested in the early twentieth century [1, 2], only scattered research activities have been reported since then. The need of higher energy density for portable electronics has encouraged the development of the micro-DMFC. Easy portability and storage of methanol make the DMFC attractive in transportation as well as stationary applications. As the proton exchange membrane fuel cell (PEMFC) technology nearly matured and was ready for market, research and development in DMFCs was revived. In the recent decades, much significant progress has been made on the key component fabrication, system design, and applications of DMFC [3–5]. Companies and

---

Electrochemical Technologies for Energy Storage and Conversion, First Edition. Edited by Ru-Shi Liu, Lei Zhang, Xueliang Sun, Hansan Liu, and Jiujun Zhang.
© 2012 Wiley-VCH Verlag GmbH & Co. KGaA. Published 2012 by Wiley-VCH Verlag GmbH & Co. KGaA.

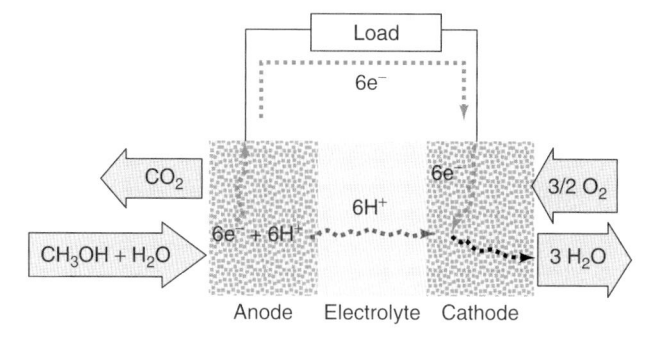

Figure 15.1 Schematic diagram of reactions and processes taking places inside DMFC.

research organizations actively engaged in DMFC research and development include Toshiba, Yahama, Los Alamos National Labs, Motorola, Korea Institute of Energy Research, Vectrix, Samsung Electronics, Medis Technologies, Hitachi, Fraunhofer Institute for Solar Energy Systems, Neah Power Systems, and so on.

Since methanol is fed as an aqueous solution, a large amount of water is present on the anode side. DMFCs do not need complicated humidification and thermal management scheme as PEMFCs. Therefore, the power system and control scheme of the DMFC are much simpler than those of the PEMFC. At present, DMFC, PEMFC, and solid oxide fuel cell (SOFC) are the most technology-ready fuel cell systems. These fuel cell systems have been successfully demonstrated in various applications and field tests. However, there are differences in the nature of these fuel cells, and the pros and cons of each type of fuel cell for portable applications are listed in Table 15.1. DMFC is the simplest among the systems and the fuel can be stored in a replacement cartridge. It is suitable for small, portable electronics where miniaturization of fuel cell is important. The fuel can be replenished by refilling or replacing the cartridge, which takes only seconds. Because of the low-power output, it is commonly used as the battery charger in conjunction with a lithium battery (a hybrid scheme) for electronics where high power is occasionally needed.

The free energy change ($\Delta G°$) at standard condition of reaction 15.3 is -703 kJ mol^{-1}. The cell voltage is 1.20 V with methanol at 1.0 M. The energy density

Table 15.1 Comparison among DMFC, PEMFC, and SOFC for portable application.

	DMFC	PEMFC	SOFC
Advantages	Simple system without thermal management, easy fuel storage	High-power density	Diversified fuel, non-noble metal for catalyst
Disadvantage	Low-power density	Low energy density of storing hydrogen at high pressure or in metal hydride	Slow startup due to high operating temperature

of methanol is 6.1 kWh kg^{-1} or 4.8 kWh l^{-1}. This theoretical energy density of methanol is high compared to the overall energy density of the lithium ion battery (0.2–0.3 kWh kg^{-1}) or the lead acid battery (0.02–0.05 kWh kg^{-1}). This makes DMFC an ideal power source for portable applications. Although ethanol and other hydrocarbon fuels have as high an energy density as methanol, they are not practical because of the sluggish electrochemical oxidation reaction of these fuels and the extremely low output power density.

15.2
Main Types and Structures of Direct Methanol Fuel Cells

The structure and material used in DMFC are similar to those of the PEMFC which uses hydrogen as the fuel. The core of DMFC is the membrane electrode assembly (MEA), shown in Figure 15.2. This component contains the cathode, the anode, and the membrane. Both the anode and the cathode contain a diffusion layer, a microporous layer, and a catalyst layer.

Depending on the operating method and material used in the fuel cell, it can be classified further as follows:

1) Feed mode: active mode or passive mode
2) Stack type: planar stack or laminated stack
3) Membrane used: alkaline fuel cell or acid fuel cell
4) State of fuel input: liquid methanol feed, vapor methanol feed, or methanol reformed feed.

The feed mode and stack type are described in the later section. On the basis of the polymer used for the membrane, DMFCs can be classified into acid fuel cells or alkaline fuel cells. Most, but not all, DMFC membranes are Nafion (Du Pont). Nafion has a fluorocarbon (CF$_2$–CF$_2$) backbone chain and a side chain containing sulfonic functional group (SO$_3^-$ H$^+$). The proton is readily exchanged with the proton in the solution and the membrane is acidic in nature. The acidity of Nafion is roughly equivalent to that of 0.5 M H$_2$SO$_4$ aqueous solution. To resist acid corrosion and to deliver good power output, noble metals such as Pt are necessary as the electrode catalyst. An alkaline membrane is less corrosive. In general, an alkaline membrane consists of a tetra-alkyl ammonium group and has anion exchange capability. Non-noble metals such as nickel or cobalt may be used as the electrode catalyst for an alkaline fuel cell. However, the cell performance

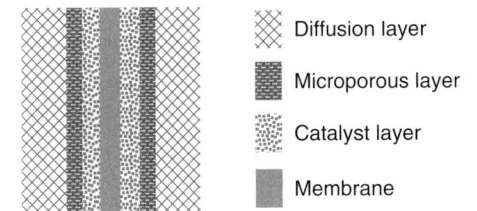

Figure 15.2 Structure of the MEA in direct methanol fuel cell.

15 Direct Methanol Fuel Cells

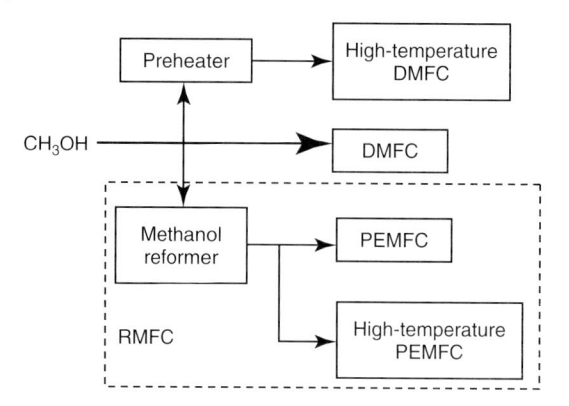

Figure 15.3 Fuel cell systems using methanol as the fuel.

drops significantly because of the carbonation of carbon dioxide generated at the anode. The solution pH is also reduced. The carbonate anion may precipitate on the catalyst surface and reduce the electrode activity.

$$CO_2 + OH^- \rightarrow CO_3^= + H_2O \tag{15.4}$$

There are several fuel cell systems using methanol as fuel (Figure 15.3). The DMFC had been tested and operated at near room temperature (~40 °C) or at elevated temperatures, that is, 100 °C and above. Since the primary target of DMFC is for portable applications, most of the demonstration and testing have been carried out at near room temperature (highlighted in Figure 15.3). Operating the cell at elevated temperatures improves the reaction kinetics on the electrode and the cell can deliver a much higher power (~100 mW cm^{-2} above) than those operated at low temperatures (~50–100 mW cm^{-2}). However, operating at high temperatures needs fuel preheating and accelerates the degradation and corrosion processes. Water balance is very important to maintain the membrane hydrated. The Nafion membrane and the ionomer in the catalyst layer rapidly decay at high temperatures.

The PEMFC using hydrogen as the fuel is able to deliver much higher power than the DMFC both at room temperature and at high temperatures. In this case, methanol is fed into a methanol reformer where it is converted into hydrogen-rich gas. The hydrogen-rich gas is then fed into the PEMFC. Fuel cell system with a methanol reformed feeding scheme is the reformat methanol fuel cell (RMFC). The reforming can be either auto-thermal reforming (ATR) or steam reforming (SR). The overall reactions of these reforming methods are given by Eqs. (15.5) and (15.6).

$$\text{ATR:} \quad CH_3OH + \frac{1}{2}O_2 \rightarrow CO_2 + 2H_2 \tag{15.5}$$

$$\text{SR:} \quad CH_3OH + H_2O \rightarrow CO_2 + 3H_2 \tag{15.6}$$

SR produces a gas with higher hydrogen concentration than gas produced from ATR. ATR is an exothermic reaction. After the reaction is initiated, the heat released from this reaction can sustain the temperature needed for this reaction. SR is an endothermic reaction and external heating is needed to sustain this reaction. For portable applications, ATR is chosen for its simplicity. The reforming gas can be cooled to 60–70 °C before feeding into the PEMFC, or it can be fed directly into a high-temperature PEMFC. The high-temperature PEMFC uses the phosphoric acid doped polybenzimidazole (PBI) polymer as the membrane. Fuel cell with this membrane can be operated at 160 °C and above.

Casio is developing the microreformer/PEMFC system [7], and Ultracell is toward the microreformer/high temperature PEMFC [8]. Directly fed methanol fuel cell operated at room temperature is the most popular type of methanol fuel cell. Many companies and research institutes are working on this type of fuel cell, including Samsung, LG Chem, NEC, Toshiba, Hitachi, MTI (Mechanical Technologies Inc), Los Alamos National Lab., Smart Fuel Cell (EFY), Motorola, Polyfuel, etc. [4–6]. Most, but not all, the fuel cell stack for notebook is in the range of 20 W with laminate stack instead of planar stack design. These companies and research institutes adopt DMFC/lithium rechargeable battery hybrid system.

Instead of conventional machining and assembly, micro-DMFC is fabricated by microelectromechanical systems (MEMS) technology [9]. Many mini/micro features, such as flow channel, and other components are made on silicon wafer, glass, or printed circuit board substrates. Entire DMFC power system is built on a single chip for microelectronic power-embedded system application. Several detail reviews on micro-DMFC can be found in the literature [10–12].

Large fuel cell power system (~50–100 kW) using methanol as fuel had been tested by Daimler-Benz in 1997 and Nova Bus in 1998 [13]. Both vehicles used onboard methanol reformer to produce hydrogen for fuel cell. It takes the advantages of the fact that methanol can be carried as liquid fuel. However, complicated reforming system, cost, and slow start-up/shot down are drawbacks of onboard reforming electrical vehicle.

15.3
Electrochemical Processes in Direct Methanol Fuel Cells

Using methanol as the anode fuel distinguishes DMFC from other fuel cell systems. The three unique electrochemical features of DMFC are as follows: (i) methanol process is a multielectron transfer, multireaction step electrochemical process; (ii) methanol oxidation produces electrochemical poisoning intermediates (such as CO, COH); and (iii) methanol crosses over from the anode to the cathode.

The methanol oxidation given by Eq. (15.1) is an overall reaction. Bagotzky et al. [14] have proposed many possible reaction paths of methanol oxidation, as given in Figure 15.4. Methanol is sequentially oxidized to formaldehyde (CH_2OH), formic acid (HCOOH), and carbon dioxide (CO_2). During methanol oxidation, several possible intermediates are generated. These include formaldehyde and formic acid, which occupy the reaction active site and slow down

$$CH_3OH \rightarrow \underset{x}{CH_2OH} \rightarrow \underset{xx}{CHOH} \rightarrow \underset{xxx}{COH}$$
$$\downarrow \qquad \downarrow \qquad \downarrow$$
$$CH_2O \rightarrow \underset{x}{CHO} \rightarrow \underset{x}{CO}$$
$$\downarrow \qquad \downarrow$$
$$HCOOH \rightarrow \underset{x}{COOH}$$
$$\downarrow$$
$$CO_2$$

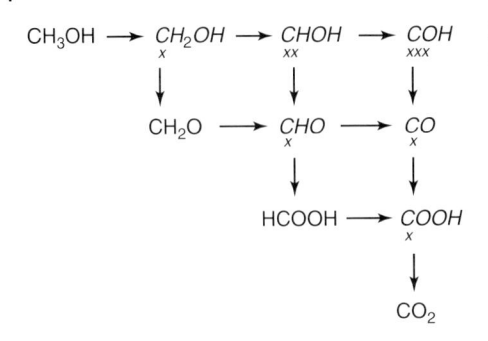

Figure 15.4 Possible reaction path of methanol oxidation reaction.

the overall reaction rate. Several extensive reviews have been written on the mechanism of methanol oxidation. Hamnett has summarized his findings [15] that the several intermediates coexist on the catalyst surface. The amount of individual species depends on the electrode potential as well as the electrolyte composition.

Methanol oxidation reaction generates CO as intermediate, which gets adsorbed on the Pt surface, as shown in Figure 15.4 or by Eq. (15.7). The $PtCO_{ads}$ is difficult to remove or to be oxidized to CO_2. The adsorbed CO occupies the reaction site and slows down further methanol oxidation. Instead of using Pt as the catalyst, the anode catalyst of the DMFC uses a Pt–Ru alloy. A second hydrated metal, such as Ru, is used to accelerate the CO oxidation reaction. Stable $Ru(OH)_{ads}$ is generated in the presence of water, as given by Eq. (15.8). The hydroxide, $Ru(OH)_{ads}$, can react with an adjacent $PtCO_{ads}$, as given by Eq. (15.9). CO is oxidized to CO_2 and the free Pt surface is ready for methanol oxidation.

$$Pt + CH_3OH \rightarrow PtCO_{ads} + 4H^+ + 4e^- \qquad (15.7)$$

$$Ru + H_2O \rightarrow Ru(OH)_{ads} + H^+ + e^- \qquad (15.8)$$

$$PtCO_{ads} + Ru(OH)_{ads} \rightarrow CO_2 + Pt + Ru + H^+ + e^- \qquad (15.9)$$

Many transient metals had been studied, and Ru had the best performance among them. Liu *et al.* reviewed the catalysts used for the DMFC anode [16]. They reviewed the preparation methods of Pt–Ru, the carbon substrate, low noble metal loading catalyst, and non-noble catalyst. So far, DMFC has achieved current densities of 100–300 mA cm^{-2} at a cell potential of 0.4 V for single cells operating at 60–90 °C, under pressurized air or oxygen, in 1.0–3.0 M methanol, and with metal loadings of 1.0–2.5 mg cm^{-2}.

The electrochemical reactions that occur on the electrode are given by Eqs. (15.11–15.13). However, they take place on the surface of the catalyst. Reactants and products are transported from outside of the electrode through the diffusion layer and the microporous layer by the diffusion process. Ions are transported through the catalyst layer and the membrane by migration and diffusion processes. Osmosis drag of water and methanol also takes place in the catalyst layer and membrane. Many mathematical models have been developed to describe the process occurring

15.3 Electrochemical Processes in Direct Methanol Fuel Cells

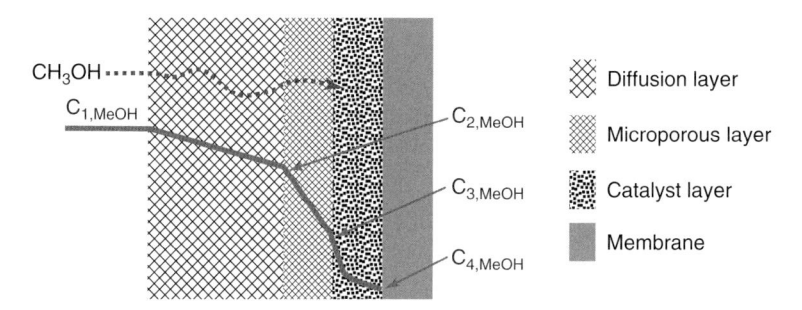

Figure 15.5 Transport process of methanol on the anode side.

inside the cell [17–21]. A simplified version of the electrochemical processes is given here for illustration purposes.

Methanol diffuses through the diffusion layer and the microporous layer. This is schematically shown in Figure 15.5. The diffusion rates of methanol in the diffusion layer ($j_{d,MeOH}$) and in the microporous layer ($j_{m,MeOH}$) are equal at steady state, as given by Eq. (15.4). D_d and D_m are the apparent diffusivity of methanol in the diffusion layer and in the microporous layer, respectively. These diffusivities depend on the porosity and hydrophilic property of the diffusion medium. δ_d and δ_m are the thicknesses of diffusion layer and microporous layer, respectively.

$$j_{d,MeOH} = -\frac{D_d}{\delta_d}\left(C_{1,MeOH} - C_{2,MeOH}\right) \tag{15.10}$$

$$= -\frac{D_m}{\delta_m}\left(C_{2,MeOH} - C_{3,MeOH}\right) = j_{m,MeOH}$$

At steady state, the methanol diffusion rate inside the catalyst layer is balanced by the consumption rate of methanol oxidation, as given by Eq. (15.11). The first term of Eq. (15.11) represents the diffusion rate of methanol inside the catalyst layer. The second term of Eq. (15.11) represents the consumption rate of methanol by electrochemical oxidation. $D_{c,MeOH}$ and $C_{c,MeOH}$ are the apparent diffusivity and concentration of methanol in the catalyst layer, respectively; i_a is the current density (A cm^{-2}) of methanol oxidation reaction based on the true electrode area; A_a is the ratio of the true electrode area to the projected apparent electrode area; and n and F are the equivalent of methanol oxidation reaction and Faraday constant.

$$0 = -D_{c,MeOH}\frac{dC_{c,MeOH}}{dx} + \frac{i_a \times A_a}{nF} \tag{15.11}$$

The methanol oxidation reaction is considered as an irreversible reaction; therefore, the current density of this reaction can be expressed by the Tafel eqnarray given by Eq. (15.12). The current density also depends on the local methanol concentration.

$$i_a = i_{o,a} C_{c,MeOH} \exp\left[\frac{\alpha_a F}{RT}\eta_a\right] \tag{15.12}$$

Here, $i_{o,a}$, α_a, and η_a are the exchange current density, the charge transfer coefficient, and the overpotential of methanol oxidation reaction, respectively.

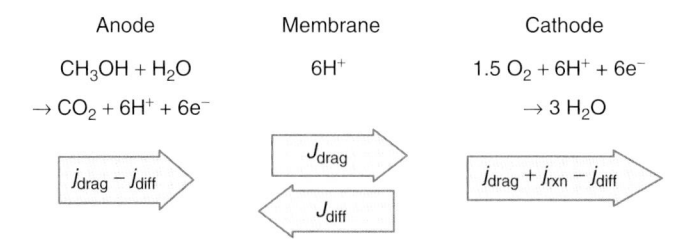

Figure 15.6 Water balance in among the anode, membrane, and cathode.

If the methanol is not completely consumed in the anode catalyst layer, the remaining methanol crosses over the membrane and reaches the cathode. In this situation, the methanol concentration at the interface between the catalyst and the membrane is not reduced to zero ($C_{4,\text{MeOH}} \neq 0$). For an electrode with a low catalyst activity, diluted methanol (5–15%) can reduce the CO poisoning level and the methanol crossover rate. However, the energy density of aqueous methanol is proportional to the methanol concentration. Using diluted methanol may lose the DMFC advantage of high-energy density.

The water transport inside the DMFC is affected by several factors. It is schematically presented in Figure 15.6. Diffusion process controls the water transport rate inside the anode diffusion layer and the anode microporous layer. The water transport inside the membrane is controlled by the rate of osmosis drag (j_{drag}) and the diffusion rate inside the membrane due to water concentration gradient (j_{diff}). In the cathode, additional water is generated from the oxygen reduction reaction, and the rate of water generation is j_{rxn}. At the steady state, the water transport rate in the anode is equal to $j_{\text{drag}} - j_{\text{diff}}$. The net water transport rate in the membrane is equal to $j_{\text{drag}} - j_{\text{diff}}$. The water inside the membrane travels from the anode to the cathode when $j_{\text{drag}} > j_{\text{diff}}$ and vice versa. The water transport rate in the cathode is equal to $j_{\text{drag}} + j_{\text{rxn}} - j_{\text{diff}}$. It is balanced by the water diffusion rate and the evaporation rate. Excess water accumulates in the cathode if the sum of the water evaporation rate and the diffusion rate is less than the amount of water transport from the membrane ($j_{\text{drag}} + j_{\text{rxn}} - j_{\text{diff}}$).

The osmosis drag flux (j_{drag}) is due to the hydrated proton migrating from the anode to the cathode under the influence of an electric field. It is proportional to the amount of proton migrating or the ionic current density inside the membrane "i_{membrane}" as given by Eq. (15.13). The osmosis drag coefficient ξ is a function of the electrode potential gradient across the membrane ($\Delta \phi$) and water saturation level of the membrane (λ), or the water activity at right side of the membrane ($a_{H_2O,r}$) and the left side of the membrane ($a_{H_2O,l}$).

$$j_{\text{drag}} = \xi(\lambda) \frac{i_{\text{membrane}}}{F}, \xi(\lambda) = \frac{F\Delta\phi}{RT \log\left(\frac{a_{H_2O,r}}{a_{H_2O,l}}\right)} \tag{15.13}$$

The diffusion rate of water is proportional to the water concentration gradient across the membrane. Here, δ_{membrane} is the membrane thickness. The diffusion coefficient is a function of the water saturation level of the membrane (λ).

$$j_{\text{diff}} = -D(\lambda) \frac{\Delta C}{\delta_{\text{membrane}}} \tag{15.14}$$

The water generation rate (j_{rxn}) is proportional to the cathode local current density (i_c) as follows:

$$j_{\text{rxn}} = \frac{i_c}{2F} \tag{15.15}$$

A delicate balance is needed to reduce the flooding level in the cathode. These three factors (j_{drag}, j_{diff}, and j_{rxn}) affect the water transport rate in the fuel cell. Two of them (j_{drag} and j_{rxn}) depend on the current withdrawn from the fuel cell. The other one (j_{diff}) does not. The water balance and flooding level change as the fuel cell operating current changes. Some researchers have proposed a cell configuration that allows the fuel cell to operate at neat methanol condition. It was done by establishing a water barrier (a hydrophobic layer) on the cathode side. As the water accumulating on the cathode increases, the capillary pressure between the cathode catalyst layer and the microporous layer increases. Water is pushed back to the anode side. This water back diffusion scheme dilutes the methanol concentration in the anode catalyst layer even when the methanol concentration outside the anode is nearly 100%.

15.4
Fuel Cell Components

The core of a DMFC is the MEA, which is a component containing the cathode, the anode, and the membrane. Both the anode and the cathode contain a diffusion layer, a microporous layer, and a catalyst layer, as shown in Figure 15.2. The diffusion layer is made of carbon paper or cloth. The mechanical strength of the carbon cloth is superior to that of the carbon paper. The thickness of the carbon paper is around 300–400 μm, while that of the carbon cloth is around 1–2 mm. Carbon paper has higher conductivity and lower contact resistance than carbon cloth. The pore size of carbon paper is in the order of 10 μm. A microporous layer with thickness of around 50 μm is coated on the carbon paper. This layer is made of a mixture of nanosized carbon powder and tetrafluoroethylene (TFE) or Teflon. The average carbon powder diameter is about 100 μm. TFE is used as the binder as well as the agent controlling the hydrophilic property of the microporous layer. This layer provides good contact between the diffusion layer and the catalyst layer. An electrode containing a microporous layer has lower contact resistance than one without it. The average pore size of this layer is about 20–50 nm. The anode catalyst layer contains Pt–Ru black (pure alloy without carbon support) or Pt–Ru on carbon powder substrate as the catalyst. The catalyst binder uses the Nafion ionomer. The ionomer also acts as the proton conductor. The cathode catalyst layer contains Pt or Pt on a carbon powder substrate as the catalyst and ionomer. Typical thickness of the catalyst layer is around 20–50 μm, with a pore size of about 20–50 nm.

High Pt catalyst loading (1–10 mg cm^{-2}) and low power density (20–100 mW cm^{-2}) as compared to PEMFC (0.1–0.6 mg cm^{-2} catalyst loading and 400–1,500 mW cm^{-2} power density) are the areas of DMFC that need to be improved. This is primarily due to sluggish methanol oxidation rate on the anode. The reaction intermediate species during methanol oxidation, CO, occupies Pt active site and slows down the entire reaction. A second oxygenated metal can oxidize CO to CO_2. This frees up the occupied Pt site for further reaction. These binary catalysts are PtRu, PtO, PtW, PtSn, PtMo, etc. So far, Pt–Ru binary catalyst with ratio of 1:1 is the best catalyst available. In addition to the catalyst composition, crystallinity, morphology, and degree of alloying also play an important role in the catalyst activity. For instance, with Pt:Ru ratio of 1:1, catalyst of PtRu alloy has superior activity than a mixture of PtRu. The oxidation of CO to CO_2 takes place at the boundary between Pt and Ru. Interface of PtRu alloy is far larger than the interface of PtRu mixture.

At initial developmental stage, unsupported catalysts, such as Pt black and PtRu black, are the primary catalyst used for DMFC. A thin and dense catalyst layer is used to improve the DMFC performance. With the progress of nanotechnology and nanocarbon, carbon supported catalyst has been adopted to improve catalyst utilization and to reduce catalyst loading. Although non-noble catalyst has been tested to replace Pt catalyst, considering the durability in DMFC environment and electrochemical activity, PtRu and Pt still are the best catalysts for DMFC anode and cathode, respectively.

Comprehensive review on DMFC anode catalyst is given by Liu et al. [16, 23, 24]. Carbon supported catalyst can be prepared from three popular processes, such as impregnation, colloid, and microemulsion process. Both Pt and Ru precursors, such as H_2PtCl_6 and $RuCl_3$ are dissolved in aqueous solution. In the impregnation process, carbon powder is well mixed with the solution. Reducing agents, such as $Na_2S_2O_3$, $NaBH_4$, N_2H_4, and methanol, are added into this solid/aqueous mixer. The metal ion is reduced to metal right on the carbon surface and pores. In the colloidal process, the catalyst precursor is reduced to Pt–Ru colloids by reducing agent in the present of surfactant. Carbon powder is added into the colloidal solution, and these colloids are absorbed on the surface of carbon surface. Catalyst-contained carbon powder is thermally treated in an oxygen and hydrogen environment to remove absorbed surfactant. In the microemulsion process, an emulsion mixture of water droplets suspended in oil phase is prepared first. The precursor is dissolved in the water phase droplet. Then the reducing agent is added into this emulsion mixture, and the precursor is reduced to catalyst nanoparticles suspended in the emulsion mixture. The surfactant on the PtRu/C catalyst is removed by heat treatment.

A good DMFC membrane requires high ionic conductivity, low methanol crossover rate, high chemical stability, high mechanical strength, and low cost. Common requirements for DMFC membrane are [23]

1) Low methanol crossover rate (<10^{-6} mol min^{-1} cm^{-1}) or low methanol diffusion coefficient in membrane.
2) High ionic conductivity (>80 mS cm^{-1})

Numerous attempts have been made to develop a new membrane that reduces methanol crossover rate and improves its ionic conductivity [10]. Neburchilov et al. [25] have given a comprehensive review of the polymer membranes for DMFC. These membranes are summarized in the following:

- Non-Nafion fluorinated membrane,
  - such as Dow chemical XUS, 3P-energy membrane;
- Composite fluorinated membrane:
  - organic–inorganic composite:
    zirconium hydrogen phosphate-modified Nafion,
    silica and molybdophosphoric acid-modified Nafion,
    nafion-polyfurfuryl alcohol nanocomposite membrane;
  - acidic–basic composite:
    - nafion polypyrrole-based membrane;
- Composite nonfluorinated membrane:
  - Organic–inorganic composite,
    - polyvinylidene fluoride + SiO_2 + acid membrane,
    - silane/silica-modified sulfonated poly(ether ether ketone),
    - PBI-based membrane;
  - Acidic–basic composite:
    - sulfonated polyphosphazene membranes
    - irradiated sulfonated poly(ethylene-*alt*-tetrafluoroethylene) membranes,
    - sulfonated poly(ether ether ketone),
    - poly(arylene ether sulfone) based membranes,
    - composite membranes of sPEEK (or sPSU) with P4VP (or PBI)
    - asymmetric acrylic membranes,
    - polyvinylidene fluoride or low density polyethylene + styrene membranes,
    - sulfonated poly(arylether ketone) membranes,
    - polyfuel polycarbon membranes.

Where sPEEK, sPSU, P4VP, and PBI are sulfonated poly (ether ether ketone), sulfonated polysulfone, poly (4-vinylpyridine), polybenzimidzole, respectively.

Most of these membranes are in the developmental stage, their durability and cost are not available. The methanol crossover rate and ionic conductivity are available throughout the literature. However, data were obtained at various conditions and by using various measuring methods. It is difficult to make a fair comparison. To enhance the proton conductivity of membrane, additives are blended into the Nafion membrane

1) to increase the number of acid sites (molybdophosphoric acid, phosphotungstic acid, silicotungstic acid),
2) to enhance membrane water retention ability (zirconium phosphate), and
3) to modify polymer structure (hydroxyapatite and Zeolite)

Doping of inorganic additives also reduces the hydrophilic channels in membrane and makes membrane less permeable for methanol. Several nonfluorinated membranes are also developed to reduce the cost and to enhance chemical stability.

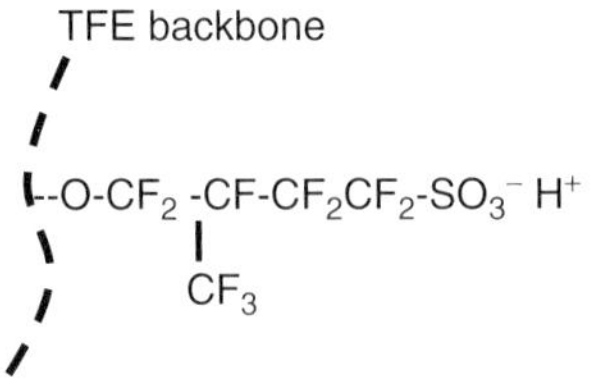

TFE backbone
```
   /
   /
⊢-O-CF₂-CF-CF₂CF₂-SO₃⁻ H⁺
  \      |
   /    CF₃
  /
```

Figure 15.7 Molecular structure of Nafion side chain.

Organic polymer such as PVDF (polyvinylidene) SPEEK (sulfonated poly ether ether ketone), and PBI (polybenzimidazole), are doped with inorganic substance such as silica or phosphoric acid. Acid–base membrane includes sPPZ (sulfonated polyphosphazene) and ETFE-SA (sulfonated poly ethylene-alt-tetrafluoroethylene).

So far, Nafion from Du Pont is the most widely used membrane for DMFC. Nafion has a polymer backbone, TFE ($-CF_2CF_2-$). It has a side chain containing the sulfonyl group (perfluorosulfonylfluoride ethylpropylvinyl ether), as shown in Figure 15.7.

As shown in Ref. [25] Table 9, the methanol crossover rate of Nafion 117 (thickness ~170–180 μm) is 0.78×10^{-6} (25 °C, 1 M MeOH) and 3.48×10^{-6} mol min^{-1} cm^{-2} (65 °C, 1 M MeOH), the diffusivity of methanol in membrane is 14.1–17.2 cm^2 s^{-1} (60 °C, 1.5 M MeOH), and the electrosmosis current is 125 mA cm^{-2} (60 °C, 1 M MeOH). The ionic conductivity of Nafion (κ) is 90–120 mS cm^{-1} (80 °C, 34–100% RH). It can be estimated [22] by Eq. (15.16) in which the conductivity is expressed as a function of membrane water saturation level (λ) and temperature (K).

$$k = (0.005139\lambda - 0.00326)\exp\left[1268\left(\frac{1}{303} - \frac{1}{T}\right)\right] \quad (15.16)$$

The end plate, current collector, or the connection between cells depends on the configuration of the cell stack. There are two types of stack configuration: laminated stack or planar stack. Planar stack uses a metallic mesh or carbon cloth as the current collector. The current collector conducts the electric current flow between cells and it allows reactants and products in and out the electrode freely. To avoid possible corrosion, a protective layer is coated on the surface of the metal mesh. These coatings include gold, TiN, and other corrosion-resistant metal alloys. The metal mesh is hot-pressed or glued on to the diffusion layer to reduce contact resistance between the current collector and the electrode. Laminated stack, just like PEMFC, uses traditional carbon or graphite bipolar plate with flow channels and end plate.

15.5
Assembly and Manufacturing of Direct Methanol Fuel Cells

The MEA is the key component of DMFC as well as PEMFC. It contains three major elements, namely, the anode, the membrane, and the cathode. The MEA's performance is heavily dependent on the manufacturing process and the material

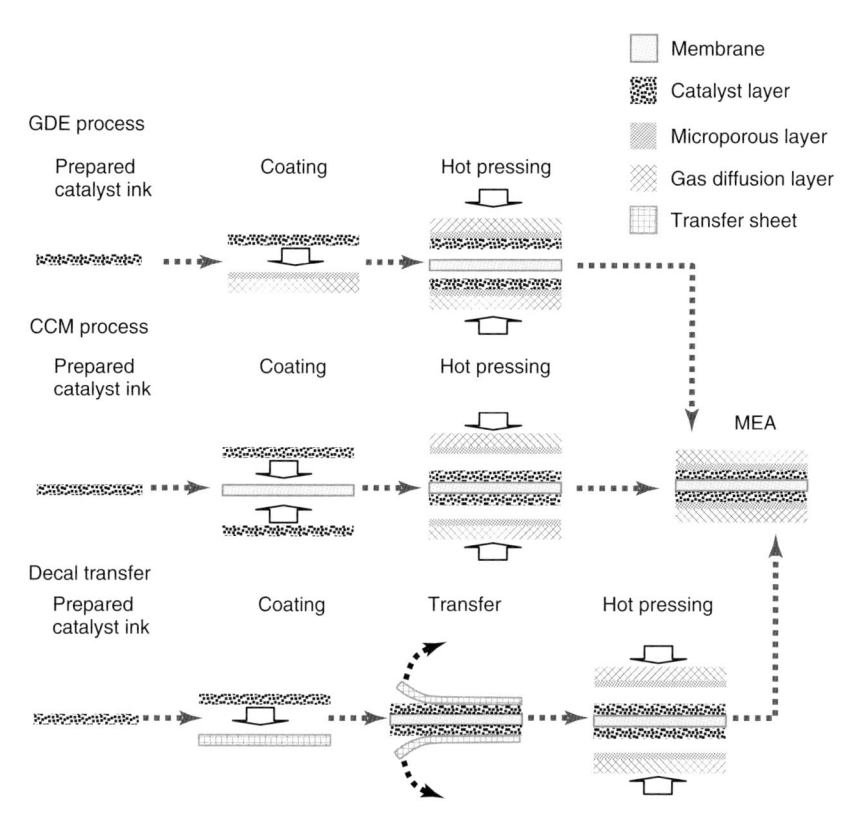

Figure 15.8 Typical MEA fabrication processes: GDE, CCM, and decal transfer.

it is made of. On the basis of the coating procedure, the MEA fabrication process is classified as (i) gas diffusion electrode (GDE), (ii) catalyst-coated membrane (CCM), and (iii) decal process. A catalyst paste is made first. It is a mixture of the catalyst, the ionomer (Nafion), the dispersion agent, the surfactant, and the solvent. The ionomer serves as the binder as well as the proton conductor. The most popular techniques used for catalyst coating are screen-printing, dye coating, doctor knifing, and spray coating. Depending on the coating method, the viscosity and rheological properties are adjusted accordingly. After coating, the solvent, surfactant, dispersion agent, and so on, are removed by drying. In the GDE process, the catalyst paste is coated on a carbon paper or carbon cloth. Figure 15.8 illustrates various coating processes. In the GDE process, the catalyst paste is coated on a carbon substrate. After drying, the solvent is removed. During hot pressing, the membrane is sandwiched between the anode and cathode. In the CCM process, the catalyst ink is sprayed on the membrane. Carbon paper or cloth is then hot-pressed on the CCM. In the decal process, the catalyst paste is first coated on a transfer sheet (Teflon), and the catalyst-coated sheet is then hot-pressed on to the membrane and the transfer sheet peeled off. Carbon paper or cloth is then hot-pressed on the CCM.

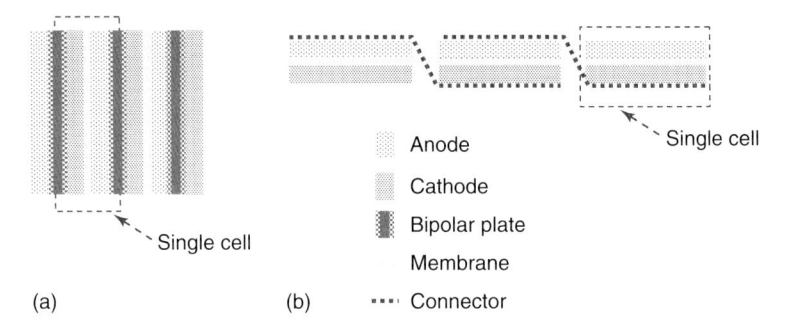

Figure 15.9 Stack configuration: (a) laminated stack and (b) planar stack.

15.6
Direct Methanol Fuel Cell Stacking and Systems

On the basis of the stacking configuration, DMFC can be classified into two categories, namely, the planar stack and the laminated stack. The output cell voltage of a single DMFC is around 0.3–0.5 V. This voltage is too low for practical applications. Several cells are connected in series so that the cell stack output voltage can match the required voltage of applications. The two major stack configurations in DMFC (planar stack and laminated stack) [26] are shown in Figure 15.9. For the laminated stack, the anode, the membrane, the cathode, and the bipolar plate are orderly stacked. The bipolar plate electronically connects the anode of one cell to the cathode of the adjacent cell. It also separates the methanol in the anode compartment from the air in the cathode compartment. In the planar stack, all the cells lie on a single plane. Each cell is electrically connected by a metal mesh or electrically conductive porous material, such as carbon cloth. The planar layout is popular in DMFC because most portable electronic equipment is flat and thin. In order to place a fuel cell into these devices, the shape of fuel cell stack has to be flat too. In general, the internal resistance of the laminar stack is smaller than that of the planar stack. This is because the electron conduction length from cell to cell ($\sim$1 mm) of the laminated stack is shorter than that of the planar stack ($\sim$1–3 cm) and the conducting surface of the laminated stack ($\sim$4–500 cm^2 above) is much larger than that of the laminated stack (cross-section of the metal mesh is $\sim$0.1–1 cm^2). For low-power devices (1–10 W), such as handheld phones, a planar stack is used. For high-power devices (20–300 W), such as chargers or notebooks, a laminated stack is used.

The laminated stack uses flow channels to distribute the fuel and air evenly throughout the stack. As shown in Figure 15.10, the flow channel pattern can be classified into (i) parallel, (ii) serpentine, (iii) interdigital, and (iv) island spots. Both parallel flow and island spots have low fluid flow resistance. However, the carbon dioxide gas bubbles in the anode side or the water drops in the cathode side may accumulate in the corner region and they are hard to remove. Serpentine flow channels ensure the removal of water drops or gas bubbles because the fuel or air is forced to flow through the only pathway. Any obstacles will be pushed along with

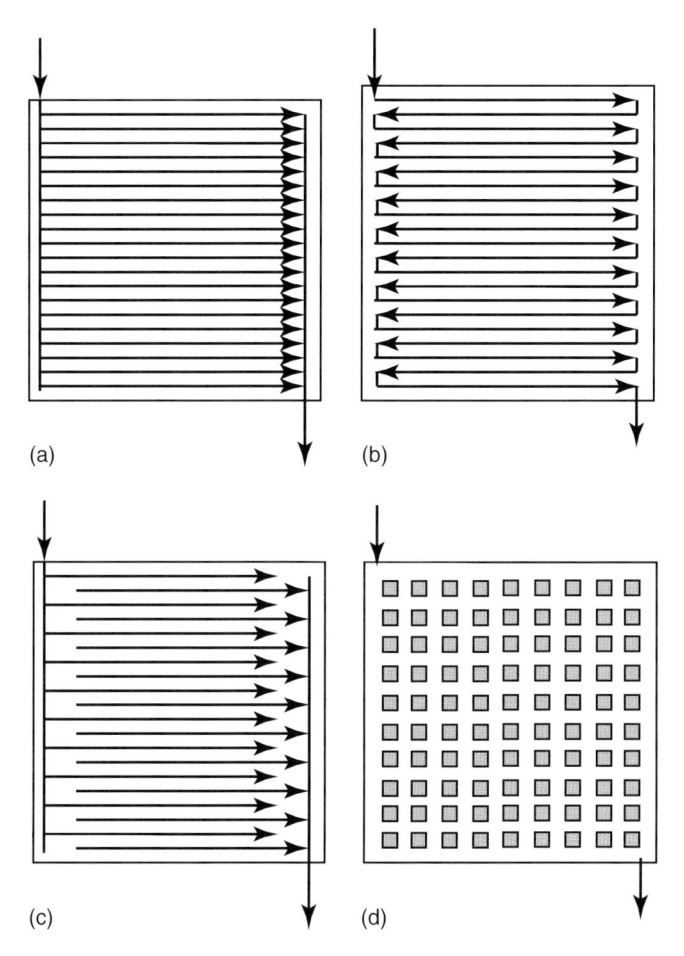

Figure 15.10 Different flow channel patterns: (a) parallel, (b) serpentine, (c) interdigital, and (d) island spot.

the flow. Interdigital flow pattern has the highest flow resistance. It forces the fluid flow from the inlet channel through the electrode gas diffusion layer and reaches the outlet flow channel.

DMFC has two different feed modes for fuel and air: the active mode and the passive mode. In the active mode, two main active devices are present in the power unit: the pump and the blower. The pump (or circulation pump) transports methanol from the fuel tank to the cell. The blower forces the ambient air into the cathode side of the fuel cell. The feed rates of both the fuel and air can be precisely controlled in the active mode. The high flow rate of the fuel removes the carbon dioxide bubbles from the anode. High air flow rate removes the excess water accumulating on the cathode surface. A power unit with active mode feeding has a higher stack power output than those without active feeding. However, improvement of power output due to active feeding may be offset by the

Table 15.2 Pros and cons of two different feed modes of DMFC.

	Active feed mode	Passive feed mode
Advantage	High-power output	Simple configuration, low cost
Disadvantage	Complicated control, high cost	Low-power output

power consumption of these devices. Selection of a suitable pump and blower is important to maximize the system net power output. The control mechanism is more complicated for the active feed mode. In the passive mode, no active devices are present inside the fuel cell. Methanol is fed into the cell via a wick or pushed into the cell from a pressurized tank. Air is brought to the cathode via natural convection. The power unit with a passive feed mode is simpler than those with an active feed mode. The pros and cons of active versus passive feed are listed in Table 15.2. The active mode feed is usually designed for portable devices with high-power rating, and the passive mode feed is used for low-power rating devices.

Extensively cost breakdown analyses were carried out for proton exchange membrane fuel cell (PEMFC) for transportation application by TIAX LLC [27] and for stationary application by Battelle [28]. DMFC is in the precommercialization stage; the cost of DMFC power system depends on its output power and complexity of the system. Yang [29] gave a good cost estimation of different DMFC systems, such as 2 W, 25 W, 250 W, and 5 kW. As shown on Table 15.3, the system cost per unit power (C_{sys}, \$/W) is decreased as the system output power (P) increased. Figure 15.11 is the plot of log (C_{sys}) versus log (P). The solid line is the calculated result of Eq. (15.17). Usually, the cost of large power system is leveraged by the processing cost and large quantity of material used per system. As one can see in Table 15.3, the ratio of material cost to processing cost is increased as the system power is increased.

$$C_{sys} = 96.693 P^{-0.6601} \tag{15.17}$$

Possible applications for corresponding power range are also given on the bottom of Figure 15.11. The DMFC output powers of 1 W, 10 W, 1 kW, and 10 kW are corresponding to the application for cell phone, notebook/CD player, electrical

Table 15.3 System cost and ratio of material cost to processing cost [27].

	Type of DMFC System			
System power (W)	2	25	250	5,000
System cost per unit power (\$/W)	50	12	4	0.26
Ratio of material/process cost	0.57	1.23	2.29	NA

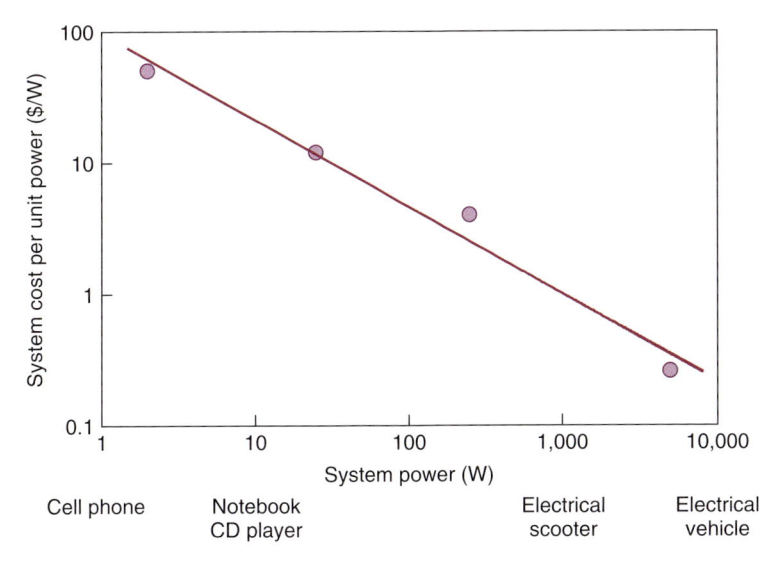

Figure 15.11 Plot of system cost per unit power versus system power. Both X and Y scales are in logarithm scale.

Table 15.4 Cost breakdown of different DMFC systems [27].

	Type of DMFC system			
System power (W) cost breakdown	2	25	250	5,000
MEA (%)	64	42	66	NA
Stack (%)	89	68	81	36
Balance of system (%)	21	32	18	64

scooter, and electrical vehicle, respectively. These DMFCs must hybrid with lithium battery in the electronic devices and hybrid with engine in the scooter and vehicles. DMFC serves as battery charger or auxiliary power in these applications.

Passive feeding mode is used for small power unit (2 W); the balance of system and other accessories (such as pumps and blower) are eliminated to obtain a compact power system. Cost of stack is the main cost and is about 89% of the entire system, as shown in Table 15.4 [27]. The cost of stack includes the MEA cost. The MEA cost includes membrane, electrodes, and gas diffusion layer. In the intermediate power rating (25 W and 250 W), active feeding mode is used to have higher output power than the passive feeding mode. Percentage of stack cost is reduced to 68% and 81% for 25 W and 250 W systems, respectively. For a 5 kW power unit, the stack cost is much lower than the cost for 2–250 W units. Thermal and water management becomes important for large unit and the percentage of balance of system cost is up to 64%. Cost reduction effort for small power unit shall be on MEA and for large power unit shall be on balance of system.

15.7
Direct Methanol Fuel Cells: Performance, Testing, and Diagnosis

Fuel cell performance depends on the characteristics of the individual components and of the materials they are made of. Table 15.5 gives a list of the fuel cell components and their corresponding parameters to be evaluated. The measurement or testing methods are given in brackets. The raw material and key components of the fuel cell include the gas diffusion layer and the microporous layer, the catalyst, the catalyst ink or paste, and the membrane. A single cell is the most common unit used to evaluate the effects of new materials and new processes under study. The function of gas diffusion layer and microporous layer is to provide an electronic conducting path. It also provides the diffusion path for methanol and carbon dioxide in the anode and diffusion path for air and water in the cathode. The electronic conductivity is an important parameter to be evaluated among different carbon materials. Since the carbon paper and carbon cloth are heterogeneous nonisotropic materials, the in-plane conductivity is different from the through-plane conductivity. In highly porous materials, methanol, water, air, and carbon dioxide can travel freely. However, the electrical conductivity may be lower in highly porous material than in materials with low porosity. Porous material with the proper hydrophobic property yields a preferential path for gases, such as water vapor and carbon dioxide. The catalyst layer requires not only the properties just mentionedbut also needs high electrochemical activity. All electrochemical reactions take place on the catalyst surface only. A nanosized catalyst with high surface area has better catalyst utilization than an aggregated catalyst. Particle size distribution of the catalyst can be measured by the dynamic laser scattering technique. The alloying level of Pt–Ru can be qualitatively measured by extended X-ray absorption fine structure (EXAFS) or by X-ray absorption near edge spectroscopy (XANES). During the preparation of catalyst ink, the ink composition, mixing, and preparation procedure are important to ensure that the catalyst is uniformly dispersed without losing its activity. The degree of dispersion can be measured by the rheological property of the catalyst ink. A good membrane is an electronic insulator with high ionic conductivity. The ionic conductivity is related to the membrane ionic equivalent capacity.

Like other fuel cell measurements, open circuit voltage, polarization curve, internal resistance, and durability are part of the key performance indicators of DMFC. Methanol crossover rate is a unique property to be measured for DMFCs. The methanol crossover rate can be calculated from the *in situ* measurement of methanol and carbon dioxide concentration in the cathode exhaust gas. Methanol crossover rate also can be estimated by applying an anodic current of methanol oxidation on the cathode. The electrochemical method is simple; however, the electric field during the measurement is opposite to that during

15.7 Direct Methanol Fuel Cells: Performance, Testing, and Diagnosis

Table 15.5 Key parameters to be evaluated or tested.

Components	Testing items
Gas diffusion layer and microporous layer	1. In-plane and through-plane electronic conductivity (four-probe electrometer) 2. Porosity (porosimeter) 3. Hydrophobicity (contact angle and porosimeter) 4. Microstructure (optical microscope, SEM, TEM) 5. Mechanical strength
Catalyst	1. Microstructure (SEM, TEM, EXAFS, XANES) 2. Overall surface area (BET) 3. Surface area of effective catalyst (CV) 4. Electrochemical activity (RDE and LSV) 5. Electrical conductivity (four-probe electrometer) 6. Particle size distribution (DLS)
Catalyst ink	1. Electrostatic dispersion (zeta potential) 2. Viscosity (rheometer) 3. Catalyst sedimentation rate, ink stability (UV–vis)
Membrane	1. In-plane and through-plane ionic conductivity (impedance spectrometer) 2. Methanol permeability (chemical analysis in single cell) 3. Water uptake and methanol uptake (gravimetric analysis) 4. Mechanical strength 5. Ionic equivalent capacity (titration) 6. Thermal stability (TGA, DSC) 7. Microstructure (SEM, TEM, AFM)
Electrode	1. Microstructure (SEM, TEM, XRD, XAS, EDX) 2. Overall surface area (BET) 3. Surface area of effective catalyst (half cell CV) 4. Electrochemical activity (half cell LSV)
Single cell and MEA	1. Open circuit voltage (multimeter) 2. Polarization curve (LSV) 3. Methanol crossover rate (LSV) 4. Catalyst utilization (CV and CO stripping) 5. Overall resistance and interfacial properties (EIS) 6. Durability (CP, CA)

AFM, atomic force microscopy; BET, Brunauer, Emmett, Teller method for adsorption of gas molecules on a solid surface; CA, chronoamperemtry; CP, chronopotentiometry; CV, cyclic voltammetry; DLS, dynamic laser scattering; DSC, differential scanning calorimetry; EIS, electrochemical impedance spectroscopy; EDX, energy-dispersive X-ray spectroscopy; EXAFS, extended X-ray absorption fine structure; LSV, linear scanning voltammetry; RDE, rotating disk electrode; RDE, linear scanning voltammetry; SEM, scanning electron microscopy; TEM, tunneling electron microscopy; TGA, thermogravimetric analysis; UV–vis, ultraviolet/visible light spectroscopy; XANES, X-ray absorption near edge spectroscopy; XAS, X-ray absorption spectroscopy; XRD, X-ray diffraction.

cell discharge. For a fuel cell with hydrogen as the fuel, the anodic hydrogen oxidation is a fast electrochemical reaction. The electrochemical resistance or overpotential loss of the anode is small and it is often neglected in the analysis. However, in the DMFC, the slow reaction rate of methanol oxidation contributes to a significant portion of overpotential loss and electrochemical resistance. It cannot be neglected in the analysis of the polarization curve or in the ac impedance frequency spectra. The methanol crossover and cathode flooding also cause a larger cathode overpotential loss than the cathode overpotential loss in a PEMFC.

15.8
Degradation Mechanisms and Mitigation Strategies

Performance degradation of fuel cell or the MEA is inevitable over a long period of operation. Several reviews on the DMFC and PEMFC durability can be found in the literature [30–32]. In general, the degradation test is carried out at a fixed operating condition. The cell voltage is monitored at constant current over a long period. The cell voltage and the voltage decay rate ($\mu V\,h^{-1}$) are the performance indices. Typical decay rate is in the range of 10–25 $\mu V\,h^{-1}$ [33]. Long-term stability is critical for DMFC commercialization. The lifetime of DMFC for portable fuel cell was set at 5000 h as the U.S. Department of Energy (DOE) 2010 target [34]. Failure mode analysis provides a basic understanding of the potential failure mechanisms and a foundation to develop the necessary technology to mitigate such mechanisms. The degradation of MEA performance can be classified into (i) reversible degradation and (ii) irreversible degradation. In the reversible degradation, a minor adjustment of the cell operating condition can recover the cell performance. Water flooding on the cathode or gas bubbles trapped in the anode flow channel belongs to reversible degradation. The cell regains output power by suddenly increasing the air flow rate to blow away water droplets away from the cathode or increasing the methanol flow rate to purge gas bubbles away from the anode. An alternative is to apply a load cycling strategy where the load is removed for 30 s every 30 min. With this strategy, a cell degradation rate of 13 $\mu V\,h^{-1}$ (cell voltage above ~0.45 V) over nearly 2000 h was achieved [33]. This was tested at 200 mA cm^{-2}, 110 °C, and 0.4 M methanol solution. Irreversible degradation cannot be reversed by changing the operating condition. The failure is permanent and attributed to the following mechanisms [31–40].

1) aggregation of catalyst particle and dissolution of Ru;
2) corrosion or loss of carbon support;
3) poisoning of catalyst by impurities or by-products; and
4) hydrophobic/hydrophilic property changes in the catalyst and diffusion layers.

During DMFC discharge, the anode catalyst (i.e., Pt–Ru) is subject to electrochemical oxidation. Ostwald ripening is observed on both anode and cathode. A small amount of Ru in the anode dissolves, migrates through the membrane, and deposits on the cathode. The dissolution of Ru not only reduces the catalytic activity of MEA but also increases cell internal resistance. When the cell is operated galvanostatically, the cell output power can be partially recovered by pulsed fuel starvation [39]. Ru in the cathode can be removed by oxidation. The electrochemically active surface area and catalyst utilization of cathode Pt catalyst gradually reduce with time. This is accelerated as a result of methanol crossover to form the mixed potential at cathode [38].

Numerous reports have described carbon corrosion on the PEMFC cathode. In both PEMFC and DMFC, the cathode potential is higher than the potential of carbon oxidation into CO_2 (~0.2 V versus RHE). It is thermodynamically possible in both DMFC and PEMFC operating conditions. Yousfi-Steiner *et al.* have given a brief review on this subject in PEMFC [30]. The Pt catalyst accelerated the oxidation of the carbon support, and CO and CO_2 were produced. A highly graphitized carbon support was more stable than an amorphous carbon support. However, the performance of PEMFC using a graphitized carbon support was found to be lower than that of a conventional carbon support. At high current density, numerous carbon dioxide bubbles generated in the anode catalyst layer may physically perturb the particles inside the anode. Loss of carbon particles from the microporous layer or from catalyst was observed under certain operating conditions, anode structure, or anode composition.

15.9
Applications

DMFC had been tested as a power sources for portable, backup power, and transportation applications. Because of the easy storage of liquid methanol, most demonstrations are for portable electronics, such as cell phones and notebooks. Many electronics companies, such as NEC, Sanyo, Toshiba, Fujitsu, Hitachi, Samsung, are engaged the development of DMFC for notebooks. In the initial stage, the fuel cell is designed as a charger and is externally attached to the notebook as a docking station. Most, but not all, the fuel cells have adopted the laminated stack design and active feed mode to obtain high-power output from the DMFC stack, as shown in Figure 15.12a. Hitachi and Fujitsu demonstrated a DMFC power unit attached to the back of a notebook screen (Figure 15.12b). Air is supplied to the cell by natural convection. System miniaturization is the technical barrier to be overcome for DMFC in portable applications. The size and weight of DMFC auxiliary units have to be reduced. These units include the blower, valves, control unit, and others. All fuel-cell-powered notebooks use a hybrid power design. The fuel cell is used as the charger and hybridized with a rechargeable lithium battery. The lithium battery provides the main power for the notebook during startup, reading disk, and shut down. These fuel cells provide about 20 W power output and constantly charge the lithium battery. Two types of methanol storage are under development: (i) methanol storage is integrated with the fuel cell system and injected into the storage during refill; and (ii) methanol storage is

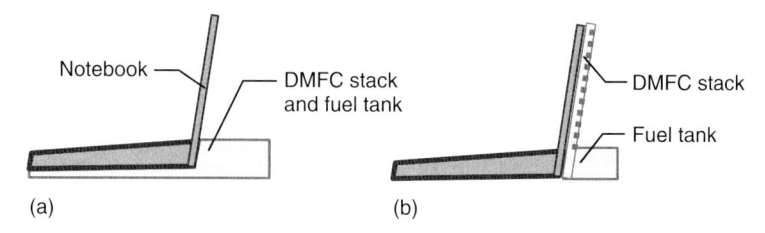

Figure 15.12 DMFC as external power source for notebooks. (a) Laminated stack with active feeding mode and (b) planar stack with passive feeding mode.

in an exchangeable cartridge, and a new cartridge is replaced during refilling. At the present stage, methanol is externally injected during refilling for most of the notebook fuel cells.

As an example, Figure 15.13 is the DMFC developed by the Material and Chemical Research Laboratories (MCL) of ITRI (industrial Technology Research Institute). Figure 15.13a is an active feed DMFC stack for notebook application. This power unit has a net output power of 12–15 W. Water generated on the cathode is totally recycled back to the anode. Figure 15.13b is a passive feed DMFC charger (240 mW) for portable devices. The stack is planar layout with modular area of 47.5 cm^2. A major technical break through is the water recycling and fuel control mechanism and MEA fabrication technology. Various sizes of fuel cell power units, ranging from 0.1 W to 100 W, have been developed. Figure 15.14a is a thumb-size DMFC (0.1 W) developed by Toshiba [41]. With 2 ml neat methanol, it powers MP3 player for 20 h. Figure 15.14b is a Sony DMFC fuel cell hybrid with a lithium battery [42]. The DMFC has a planar layout with passive fuel feeding. The ultimate goal is to use 10 ml methanol to keep the mobile phone playing movie for 14 h. Unit peak output power is targeted at 3 W. One of the major applications of DMFC is the power source for laptop. Many companies engaged the DMFC developed, including NEC, Sanyo, Toshiaba, Fujitu, Hitachi, Sasmsung, and others. Figure 15.14c is a typical DMFC application in a laptop from Toshiba [43]. Typical power output is about 20 W. Most but not all, developers adopt active feed laminated DMFC stack. The DMFC is hybridized with lithium battery in the laptop. Smart Fuel Cell is developing DMFC as a portable charger or power unit [44]. Figure 15.14d is a typical example in which the voltage output is 12 V and current ranges from 3.3 to 8.8 A (power ranged from 40–105 W). They are target for mobile home, boat, and remote cabin applications.

Baker et al. [45] conducted a market survey of fuel cells in the portable power market. In the portable market, PEMFC and DMFC are the leading technologies among all fuel cell systems. In general, because of the high-power output advantage and hydrogen storage issue, as indicated in Table 15.1, PEMFC is targeted for 20 W (notebook) and above applications. DMFC is targeted for below 20 W applications. Several research organizations, such as Toshiba, Motorola, Energy Related Devices, Jet propulsion Laboratory, and Los Alamos National Laboratory, had demonstrated the DMFC as a power source for cell phones [46]. The planar stack design with

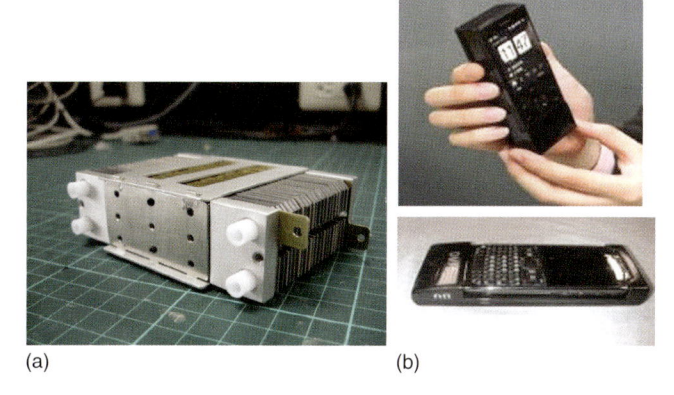

Figure 15.13 DMFC developed by ITRI, (a) active fuel cell, (b) passive fuel cell stack for cell phone charger.

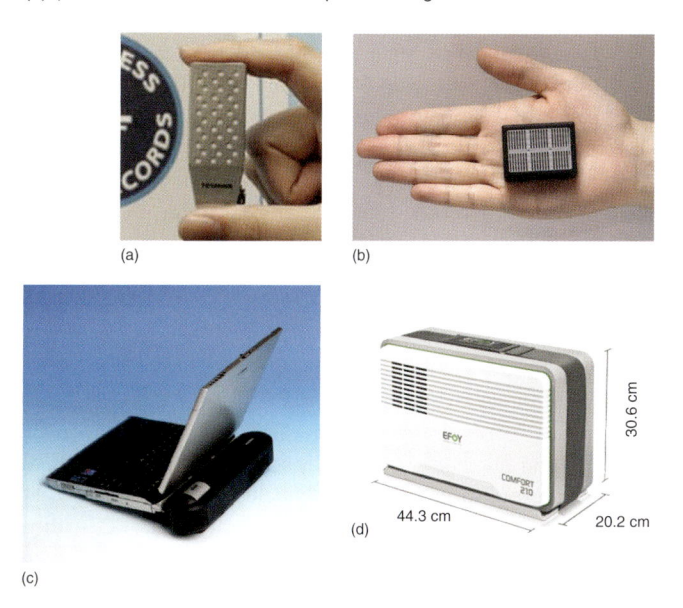

Figure 15.14 (a) thumb size micro DMFC from Toshiba [41], (b) Sony hybrid fuel cell [42], (c) Toshiba DMFC for notebook [43], (d) DMFC charger from Smart Fuel Cell (EFOY) [44].

passive feed is adopted for this application. Methanol is externally injected into the methanol tank during refilling. Smart fuel cell has successfully developed the DMFC (range about 20–100 W) as a mobile power unit for mobile home and recreation vehicle or as a charger for electric scooters, military field chargers, power for remote sensors, and so on.

Although the hydrogen fuel PEMFC is the most mature technology for electric vehicles, because of the technical barrier of hydrogen storage and transportation, a major infrastructure overhaul is needed to promote the PEMFC-powered electric vehicle. DMFC also has been considered as the power unit for the electrical vehicle by Ballard Power System Inc. in collaboration with Daimler-Chrysler, IRD Fuel Cell A/S, Siemens AG, Los Alamos National Laboratory, among others. The present gas stations can be easily converted into methanol refilling stations without major changes. Because of the low-power density, most DMFCs for vehicle application are operated at high temperature (vapor feed, 110 °C) and pressure ($\sim$3 bar) to enhance the output power.

Before DMFC commercialization, standard testing procedure and safety code of DMFC power system shall be established. The International Electrotechnical Commission (IEC), technical committee (TC) 105 is preparing international standards related to fuel cell technologies [47]. Technical committee 105 has 11 working groups; working groups (WG) 8, 9, and 10 are focusing on the standards for micro fuel cell. WG 8 is establishing the safety standards, WG 9 is establishing the testing and evaluation standards, and WG 10 is dealing with the interchangeability between fuel cell and fuel cartridge. Other organizations and companies are also actively involved in preparing standards, such as UL (Underwriter Laboratories Inc.), CSA International (Canadian Standards Association), JSA (Japan Standards Association), etc. In the path of DMFC commercialization, other related laws and regulations are also important. The laws and regulations in United States are from Department of Transportation (DOT) related to methanol transportation and fuel cartridge, Federal Aviation Association (FAA) for methanol cartridge on board aircraft, and HazMat (Hazardous Material) etc. Methanol cartridge can not be carried on board, and this is one of the obstacles of DMFC commercialization.

15.10
Challenges and Perspectives of Direct Methanol Fuel Cells

Multifunctional cell phones and notebooks are rapidly replacing simple handheld phones and notebooks. These multifunctional portable electronics need a power source with higher energy to sustain their running time. The DMFC with a maximum energy density of 6 kWh kg^{-1} or 5 kWh l^{-1} is a promising technology to meet the energy demand of portable electronics. DMFC commercialization is not a question of will but of when.

For this application, the DMFC is required to deliver power in the range of 0.5–20 W with small volume and lightweight. The overall system energy density should be higher than that of the lithium battery (300 Wh kg^{-1}) for DMFC to be attractive. Lifetime up to 3000 h is a must. Each component of the fuel cell and its auxiliary units must be designed to achieve the miniaturization. There are several technical issues to be overcome before it is widely applied in our daily lives. These include methanol crossover, heat management, water management, low-power density, unknown durability, and lifetime [48].

Many issues on the anode, membrane, and cathode are to be resolved before DMFC can be used as a portable power source.

- On the anode side, sluggish methanol oxidation reaction causes a significant amount of anode overpotential. CO or COH are intermediate species generated during methanol oxidation. The absorption of these species on the catalyst slows down the overall reaction rate of methanol oxidation. Bimetal catalysts, such as Pt–Ru, are the most common catalysts used in the anode.
- In the membrane, methanol not consumed by the anode may travel through the membrane as a result of electro-osmosis and diffusion. Methanol crossover the membrane reduces not only the Faraday efficiency but also the cell voltage. Many new membranes have been developed to retain proton conductivity and retard methanol crossover. Several novel cell structures have been proposed and patented to solve this problem.
- Flooding on cathode is frequently observed in the DMFC when the cell operating temperature is low or the ventilation is poor. Excessive water may accumulate in the cathode as a result of water generation from the oxygen reduction reaction as well as water transportation by hydrated protons from the anode. Water flooding blocks the diffusion pathway of oxygen and causes a significant overpotential on the cathode. Cathode flooding can be eased by modifying the cell structure and operating conditions.

There are several engineering issues to be overcome as well:

- Removal of carbon dioxide on the anode side. Carbon dioxide is generated as gas bubbles inside the anode. Gas bubbles trapped inside the anode may block the pathway of methanol. Carbon dioxide discharged from the anode may carry water mist and methanol vapor.
- Miniaturization of blower, pump, controller, power conversion unit, and so on. At present, auxiliary units are far larger and heavier than they should be. They consume excessive power for their functioning.
- Cost reduction, durability, and reliability. These issues will be resolved as the DMFC technology becomes mature. Industrialized mass production will substantially reduce the manufacturing cost and improve system reliability.

References

1. Muller, J., Frank, G., Colbow, K., and Wilkinson, D. (2003) Transport/kinetic limitations and efficiency losses, in *Handbook of Fuel Cells, Fundamentals Technology and Applications*, Chapter 62 (eds W. Vielstich, A. Lamm, and H.A. Gasteiger), John Wiley & Sons Ltd, vol. 1, pp. 847–855.
2. Bockris, J.O.M. and Srinivasan, S. (1969) *Fuel Cells: Their Electrochemistry*, McGraw-Hill, New York.
3. Dillon, R., Srinivasan, S., Arico, A.S., and Antonucci, V. (2004) International activities in DMFC R&D: status of technologies and potential applications. *J. Power Sources*, **127**, 112–126.
4. Ren, X., Zelenay, P., Thomas, S., Davey, J., and Gottesfeld, S. (2000) Recent advances in direct methanol fuel cells at Los Alamos National Laboratory. *J. Power Sources*, **86**, 111–116.
5. Kim, D., Cho, E.A., Hong, S.-A., Oh, I.-H., and Ha, H. (2004) Recent progress in passive direct methanol fuel cells at KIST. *J. Power Sources*, **130**, 172–177.
6. Kamarudin, S.K., Daud, W.R.W., Ho, S.L., and Hasran, U.A. (2007) Overview

on the challenges and developments of micro-direct methanol fuel cells (DMFC). *J Power Sources*, **163**, 743–754.
7. Terazaki, T., Nomura, M., Takeyama, K., Nakamura, O., and Yamamoto, T. (2005) Development of multi-layered microreactor with methanol reformer for small PEMFC. *J. Power Sources*, **145**, 691–696.
8. UltraCell Web site: *http://www.ultracellpower.com/*.
9. Wozniak, K., Johansson, D., Bring, M., Sanz-Velasco, A., and Enoksson, P. (2004) A micro direct methanol fuel cell demonstrator. *J. Micromech. Microeng*, **14**, S59–S63.
10. Maynard, H. L. and Meyers, J. P. (2002) Miniature fuel cells for portable power: Design considerations and challenges. *J. Vac. Sci. Technol. B, Microelectron. Process. Phenom.*, **20** (4), 1287–1297.
11. Kamarudin, S. K., Daud, W. R. W., Ho, S. L., and Hasran, U. A. (2007) Overview on the challenges and developments of micro-direct methanol fuel cells (DMFC). *J. Power Sources*, **163** (2), 743–754.
12. Nguyen, N. and Chan, S. H. (2006) Micromachined polymer electrolyte membrane and direct methanol fuel cells-A review. *J. Micromech. Microeng.*, **16** (4), R1–R12.
13. Appleby, A. J. (1999) The electrochemical engine for vehicles (fuel cells). *Scientific American*, **281** (1), 74–79.
14. Bagotzky, V.S., Vassilyev, Y.B., and Khazova, O.A. (1977) Generalized scheme of chemisorption, electrooxidation and electroreduction of simple organic compounds on platinum group metals. *J. Electroanal. Chem.*, **81**, 229–238.
15. Hamnett, A. (2003) Direct methanol fuel cell, in *Handbook of Fuel Cells, Fundamentals Technology and Applications*, Chapter 18 (eds W. Vielstich, A. Lamm, and H.A. Gasteiger), John Wiley & Sons Ltd, vol. 4, pp. 305–322.
16. Liu, H., Song, C., Zhang, L., Zhang, J., Wang, H., and Wilkinson, D.P. (2006) A review of anode catalysis in the direct methanol fuel cell. *J. Power Sources*, **155**, 95–110.
17. Garcia, B.L., Sethuraman, V.A., Weldner, J.W., White, R.E., and Dougal, R. (2004) Mathematical model of a direct methanol fuel cell. *J. Fuel Cell Sci. Technol.*, **1**, 43–48.
18. Zhana, Z., Xiao, J., Pana, M., and Yuan, R. (2006) Characteristics of droplet and film water motion in the flow channels of polymer electrolyte membrane fuel cells. *J. Power Sources*, **160**, 1–9.
19. Meyersa, J.P. and Newman, J. (2002) Simulation of the direct methanol fuel cell, III. Design and optimization. *J. Electrochem. Soc.*, **149** (6), A729–A735.
20. Wang, Z.H. and Wang, C. (2003) Mathematical modeling of liquid-feed direct methanol fuel cells. *J. Electrochem. Soc.*, **150** (4), A508–A519.
21. Kulikovsky, A.A. (2006) Model of a direct methanol fuel cell stack, resistive "spots" and stack performance. *J. Electrochem. Soc.*, **153** (9), A1672–A1677.
22. Barbir, F. (2006) Chapter 4, Main cell components, materials properties and processes, in *PEM Fuel Cells: Theory and Practice*, Elsevier Academic Press.
23. Arico, A. S., Srinivasan, S., and Antonucci, V. (2001) DMFCs: From Fundamental aspects to technology development. *Fuel Cells*, **2**, 133–161.
24. Liu, H., Zhang, J., Ma, L., Huang, Y., Feng, L., Xing, W., and Zhang, J. (2009) Fabrication and optimization of DMFC catalyst layers and membrane electrode assemblies, Chapter 11, *Electrocatalysis of Direct Methanol Fuel Cells: From Fundamentals to Applications*, Wiley, NY.
25. Neburchilov, V., Martin, J., Wang, H., and Zhang, J. (2007) A review of polymer electrolyte membranes for direct methanol fuel cells. *J. Power Sources*, **169**, 221–238.
26. Heinzela, A., Hebling, C., Muller, M., Zedda, M., and Muller, C. (2002) Fuel cells for low power applications. *J. Power Sources*, **105**, 250–255.
27. Carison, E. J., Kopf, P., Sinha, J., Sriramulu, S., and Yang, Y. (2005) Cost analysis of PEM fuel cell system for transportation, *TIAX LLC subcontract report to NREL*.
28. Stone, H. J. (2005) Economic Analysis of Stationary PEM Fuel Cell Systems,

29. Yang, Y. (2010) Portable Power Fuel Cell Manufacturing Cost Analyses, *2010 Fuel Cell*, Seminar, San Antonio.
30. Yousfi-Steiner, N., Mocotéguy, Ph., Candusso, D., and Hissel, D. (2009) A review on polymer electrolyte membrane fuel cell catalyst degradation and starvation issues: causes, consequences and diagnostic for mitigation. *J. Power Sources*, **194**, 130–145.
31. Schmittinger, W. and Vahidi, A. (2008) A review of the main parameters influencing long-term performance and durability of PEM fuel cells. *J. Power Sources*, **180**, 1–14.
32. Demirci, U.B. (2007) Direct liquid-feed fuel cells: thermodynamic and environmental concerns. *J. Power Sources*, **169**, 239–246.
33. Knights, S.D., Colbow, K.M., St-Pierre, J., and Wilkinson, D.P. (2004) Aging mechanisms and lifetime of PEFC and DMFC. *J. Power Sources*, **127**, 127–134.
34. Lightner, V. (2005) Opening Presentation in Small Fuel Cells 2005, Washington DC, April 27–29, 2005 in *DOE Review Meeting, PEM Fuel Cell Systems*, Arlington, Virginia, May 23–25.
35. Cha, H.-C., Chen, C.-Y., and Shiu, J.-Y. (2009) Investigation on the durability of direct methanol fuel cells. *J. Power Sources*, **192**, 451–456.
36. Sarma, L.S., Chen, C.-H., Wang, G.-R., Hsueh, K.-L., Huang, C.-P., Sheu, H.-S., Liu, D.-G., Lee, J.-F., and Hwang, B.-J. (2007) Investigations of direct methanol fuel cell (DMFC) fading mechanisms. *J. Power Sources*, **167**, 358–365.
37. Shao, Y., Yin, G., Zhang, J., and Gao, Y. (2006) Comparative investigation of the resistance to electrochemical oxidation of carbon black and carbon nanotubes in aqueous sulfuric acid solution. *Electrochim. Acta*, **51**, 5853–5857.
38. Wang, Z.-B., Shao, Y.-Y., Zuo, P.-J., Wang, X.-P., and Yin, G.-P. (2008) Durability studies of unsupported Pt cathodic catalyst with working time of direct methanol fuel cells. *J. Power Sources*, **185**, 1066–1072.
39. Wang, Z.-B., Wang, X.-P., Zuo, P.-J., Yang, B.-Q., Yin, G.-P., and Feng, X.-P. (2008) Investigation of the performance decay of anodic PtRu catalyst with working time of direct methanol fuel cells. *J. Power Sources*, **181**, 93–100.
40. Wang, Z.-B., Rivera, H., Wang, X.-P., Zhang, H.-X., Feng, P.-X., Lewis, E.A., and Smotkin, E.S. (2008) Catalyst failure analysis of a direct methanol fuel cell membrane electrode assembly. *J. Power Sources*, **177**, 386–392.
41. From Toshiba Web site: http://www.toshiba-india.com/worldFirst.aspx.
42. From Engadget Web site: http://www.engadget.com/2008/05/04/sony-creates-micro-sized-fuel-cell-system/.
43. From Fuel Cell Today Web site: http://www.fuelcelltoday.com/reference/image-bank/Portable/Toshiba-DMFC-laptop.
44. From Smart Fuel Cell Web site: http://www.sfc.com/en/sfc-fuel-cells-products.html.
45. Baker, A., Jollie, D., and Adamson, K.-A. (2005) Fuel Cell Today Market Survey: Portable Applications, September 2005, from web site http://www.fuelcelltoday.com/ (accessed 2006).
46. Dillon, R., Srinivasan, S., Aricò, A.S., and Antonucci, V. (2004) International activities in DMFC R&D: status of technologies and potential applications. *J. Power Sources*, **127**, 112–126.
47. From International Electrotechnical Commission, IEC Web site: http://www.iec.ch/.
48. Kamarudin, S.K., Daud, W.R.W., Ho, S.L., and Hasran, U.A. (2007) Overview on the challenges and developments of micro-direct methanol fuel cells (DMFC). *J. Power Sources*, **163**, 743–754.

16
Molten Carbonate Fuel Cells
Xin-Jian Zhu and Bo Huang

16.1
Introduction to Molten Carbonate Fuel Cells

Fuel cells are highly efficient and environmentally friendly energy-converting devices for clean power generation. They use electrochemical reaction, rather than combustion, to generate electricity. As a class of *second-generation* fuel cells, the molten carbonate fuel cell (MCFC) is gaining momentum in its development cycle and could be the next fuel cell technology to reach commercial status. Several of the important benefits of MCFC technology are as follows [1, 2]:

- **MCFC uses nonnoble and cost-effective electrodes.** The operating temperature of MCFC is as high as 650–700 °C, which thus avoids the use of precious metal electrodes required by lower-temperature fuel cells, such as polymer electrolyte membrane fuel cell (PEMFC) and phosphoric acid fuel cell (PAFC). Furthermore, by using an MCFC, some more expensive metals and ceramic materials required by higher-temperature fuel cells, such as solid oxide fuel cells (SOFCs), can also be avoided. The MCFCs can reach fuel-to-electricity efficiencies of 45–60%, considerably higher than the 37–42% efficiencies of a PAFC plant. When the waste heat is captured and used, overall fuel efficiencies can be as high as 85%.
- **MCFC has fuel flexibility.** The multitude of system studies based on MCFCs have indicated that besides hydrogen fuel, this technology is also suitable for clean and efficient power generation using a variety of fuel sources, such as hydrocarbons, carbon monoxide, and their mixtures (coal gas), available all over the world. The main reason for the fuel flexibility characteristics of MCFCs is their high operating temperature, leading to suitability for internal reforming and high performance as well.
- **Waste heat generated by MCFCs can be utilized to increase the efficiency.** In MCFC power generation, since the waste heat from the stack is very high (as high as 650–700 °C), if the MCFC systems are combined with the gas turbine (GT) or steam turbine (ST) to provide additional power, the efficiency of MCFC systems can be further increased.

- **CO_2 generated by MCFCs can be easily collected.** In an MCFC system, CO_2 condensation or separation is possible because the mobile ionic substance through the electrolyte is CO_3^{2-} ions.

16.2
Current Technologic Status of Molten Carbonate Fuel Cells

Because of the advantages of MCFCs, such as high efficiency, flexible modularity, excellent environmentally friendly performance, and operational flexibility, they can be used in a variety of power-demanding applications worldwide. Many organizations and companies in the world are actively carrying out the development of this technology. Consequently, great progress in research and development (R&D) has been made during the past 20 years; for example, the field demonstration of 250 kW and multimegawatt-size power plants has been initiated, which is a great stride forward in the commercialization of MCFCs.

In the United States, the fuel cell R&D is part of a national fuel cell technology effort in which the US Government, industries, research organizations, as well as universities are the major players in fuel cell R&D, companies also participate actively. The US Government is mainly represented by the Department of Energy (DOE), who has been played a leading role in the fuel cell R&D campaign. Regarding MCFC R&D, FuelCell Energy (FCE) is active in developing a highly efficient and ultraclean MCFC commercialization technology [3]. The MCFC technology was developed from the beginning to the end in a private–public sector partnership with the company and the DOE. DOE supported $40 million from 2001 to 2003 for MCFC technology development. The 250 kW to 3 MW class MCFC power plants have been brought into the market. At the current technologic status, 55% Lower heating value (LHV) efficiency can be obtained, and the cost of the MCFC stack is approximately $3500–4000 per kW. It seems that the commercialization of MCFC technology is not far from expected, which can be indicated by a wide range demonstration program worldwide, such as in the United States, Japan, Germany, Italy, South Korea, Netherlands, and China. Figure 16.1 shows the photographs of the MCFC products currently offered for stationary applications [4].

In 1981, the R&D of the MCFC in Japan was first carried out by the Moonlight Project of the national government. In the Moonlight Project, four types of fuel cells, that is, PAFC, MCFC, SOFC, and alkaline fuel cell (AFC), have been developed.

DFC® 300A   DFC® 1500   DFC® 3000

Figure 16.1 Photographs of MCFC products currently offered for stationary applications [4].

Figure 16.2 HotModule power plants at Magdeburg and Berlin [8].

The emphasis originally has been on PAFC, and then since 1991, the emphasis in fuel cell development has been shifted to MCFC development. The funds for MCFC development under the Moonlight Project amount to about $120–130 million every year. A successful story is the development of 10 kW class stacks by Hitachi, Ltd. and Toshiba Corp., announced in March 1987. Then, from March to April 1987, Fuji Electric Corporate, Ishikawajima-Harima Heavy Industries (IHI) Co., Ltd., and Mitsubishi Electric Corp., who were all involved in the Moonlight Project, all succeeded in developing 10 kW class stacks. On the basis of the above achievement, the second-phase program for developing MCFCs started in FY1987 under the management of the New Energy and Industrial Technology Development Organization (NEDO). As a successful outcome, the 1000 kW MCFC stacks were developed with a lifetime of over 5000 h in 1999–2000 [5, 6].

In Germany, MTU, a subsidiary of DaimlerChrysler Corporation, acts as the consortium leader in the MCFC development program. The company had shared a license and technology exchange agreement with FCE and launched a field trial of its atmospheric 250 kW class internal reforming (IR)-MCFC power plant in October 1999 [7]. As far as the system design is concerned, it was realized that conventional systems did not hold the promise for competitive power plants. A system analysis led to the conclusion that a new innovative design approach is required. As a result the *HotModule* system was developed. Later on, for further development, the HotModule was made part of the submegawatt fuel cell power plants, which were also developed in a collaborative effort, utilizing the Direct Fuel Cell technology of FCE and the HotModule balance of plant design of MTU CFC Solutions. About 16 large-area stacks with about 8000 cm^2 cell area and a power range of 220–250 kW in a HotModule environment have been tested so far at the facilities in Munich (Germany) and at costumer facilities in Germany and Spain as field test units (Figure 16.2) [8]. The cells for these stacks were supplied by FCE. FCE itself has built and tested about 45 submegawatt systems in the United States and Japan based on the HotModule Technology. In 1999, MTU commissioned a skid-mounted 250 kW power plant at the University of Bielefeld, Germany, and had operated it for two years. The company had also operated a third 250 kW unit in the Rhön-Klinikum hospital of Bad Neustadt/Saale in Germany since April 2001. Finally, 220 kW of power was achieved, and an efficiency of 48–49% (LHV) was reported. The MCFC power plant systems consisted of air/fuel processing subsystem, stack peripheral subsystem, electricity subsystem, and heat recovery subsystem, which maintained the operating stability of the systems. What happened in the operating process were monitored through two computers and transported to MTU through a network. The MCFC power plant was operated for about 7000 h according to the plan, by the end of April 2002.

In Italy, ENEA, has also shown interest in MCFC technology and has been supporting small-scale research. The company made the MOLCARE 1 plan along with Spain. In this plan, a 100 kW, externally reformed, and 0.35 MPa pressurized MCFC power plant was demonstrated [9]. Following the accomplishment of MOLCARE 1 plan, MOLCARE 2 plan was also put forward. The target of MOLCARE 2 plan was the demonstration of a 500 kW externally reformed MCFC power plant. A funding of $6.6 million for the development of 500 kW MCFC power plant had been provided for a period of three and a half years since January 2000. The Commission of the European Communities (CEC) also supported the MOLCARE 2 plan and provided 35% of the total funding. The plant contained two 250 kW externally reformed power stacks, which consisted of 15 single cells. The final target for the MCFC power stack consisting of 50 single cells was also demonstrated.

In South Korea, the R&D of external reforming MCFCs was carried out by the Korea Electric Power Research Institute (KEPRI) and Korea Advanced Institute of Science and Technology (KAIST). A 25 kW MCFC stack was operated in 1999. High-performance and long-life tests of the same MCFC stack were conducted in 2002 according to the project plan. A 100 kW MCFC stack was also operated in 2003–2004. Accordingly, the test of a 250 kW external reforming MCFC stack was also conducted in 2005 according to the project plan.

A program also existed in the Netherlands under the sponsorship of the Ministry of Economic Affairs through the Netherlands Agency for Energy and the Environment (NOVEM). The Netherlands Energy Research Foundation (the Energy research Centre of the Netherlands (ECN)) was the prime contractor. The 400 kW IR-MCFC stack test was conducted by the ECN. The MCFC stack consists of three basic MCFC substacks, which was as large as a container with a 30 ft^3 volume. However, the project budget had been reduced because of funding problems.

R&D on MCFCs has been conducted in many research institutes in China since the 1990s. During the Ninth Five-Year Plan period, the Ministry of Science and Technology, the Chinese Academy of Sciences, and the Ministry of Education have jointly led the MCFC technology development. Both the Dalian Institute of Chemical Physics (DICP) in the Chinese Academy of Sciences and the Shanghai Jiaotong University (SJTU) are the major players in this area since the 1990s. Some progress has been made since then in the fabrication of key materials and components such as anode, cathode, electrolyte matrix and bipolar plate; the design, assembly, and operation of MCFC stack; and the development of MCFC stack systems. At present, SJTU has started a collaboration with Shanghai Turbine Company (STC) Ltd. The infrastructure construction of the 50 kW MCFC stack peripheral systems has been completed, and the cell module of 10 kW MCFC has also been manufactured.

16.3
Electrochemical Processes in Molten Carbonate Fuel Cells

A typical MCFC consists of nickel alloy anodes, nickel oxide cathodes, and lithium carbonate–potassium carbonate (Li/K) or lithium carbonate–sodium carbonate (Li/Na) electrolyte held in a γ-lithium aluminate matrix support. The electrochemical processes in an MCFC are similar to those in other fuel cells, that is, at the anode, fuel will be oxidized, and at the cathode, O_2 (from air) is reduced. However, the difference is the involvement of CO_2 in both anode and cathode reactions, for example, CO_2 is generated at the anode chamber and then consumed at the cathode chamber. The detailed process is shown in Figure 16.3 [10]. For a H_2/O_2 (air) MCFC, at the anode, H_2 is oxidized electrochemically and CO_2 is generated:

$$H_2 + CO_3^{2-} \rightarrow H_2O + CO_2 + 2e^- \qquad (16.1)$$

At the cathode, O_2 is reduced electrochemically and CO_2 is consumed:

$$CO_2 + 1/2 O_2 + 2e^- \rightarrow CO_3^{2-} \qquad (16.2)$$

The overall reaction is as follows:

$$H_2 + 1/2 O_2 + CO_2 \text{ (cathode)}$$
$$H_2O + CO_2 \text{ (anode)} \qquad (16.3)$$

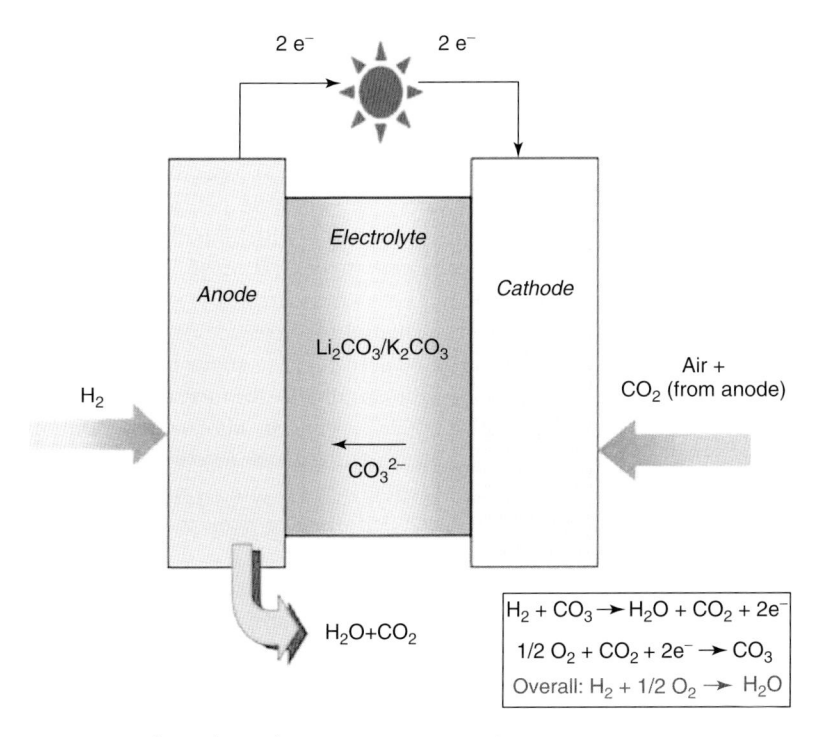

Figure 16.3 Electrochemical processes in MCFCs [10].

It can be seen from Figure 16.3 that a CO_2 gas recycling system is needed to supply CO_2 from the anode chamber to the cathode chamber during MCFC operation since the same amount of CO_2 consumed at the cathode will be regenerated at the anode. Note that the oxidizing reagent at the cathode chamber is the mixture of O_2, N_2, and CO_2, where both O_2 and N_2 come from the feeding air.

Hydrogen is an ideal fuel for a general class of MCFC systems. Taking fuel costs and infrastructure into account, however, at the present stage, hydrocarbon fuels, such as methane and biogas, will most likely be used as the hydrogen source in the early stages of fuel cell commercialization. In this context, the MCFC possesses the advantage of internally reforming the hydrocarbon fuels because the MCFC is operated at a high temperature ranging from 650 to 700 °C.

16.4
Components of Molten Carbonate Fuel Cells

16.4.1
Anodes and Anode Materials

The MCFC anode is the place where fuel oxidation occurs. In general, the anode is a plate composed of a matrix layer of electrode materials such as metal. Anode materials must have electrochemical catalytic activity toward the fuel oxidation and be stable under MCFC operating conditions as well. The most popular anode material is porous nickel alloy since nickel is an excellent electrochemical catalyst for fuel oxidation at high temperatures. For the MCFC anode, the optimum pore diameter is in the 3–5 μm range and the optimum thickness has been shown to be about 0.5–0.8 mm with a porosity of 55–70%. Under MCFC operating conditions, the porous plate is partially wetted by a carbonate electrolyte and has an extensive reaction area, which is called a three-phase boundary (TPB). Since the porous pure nickel plate does not have enough mechanical strength, it must be sintered and crept by compression when the single fuel cells are assembled together to form a stack. In this way, the stack is made gas tight and good contact between the stack components can also be achieved. It has realized that the anode creep is one of major problems in MCFC technology. In order to enhance the mechanical strength for maintaining the morphology during MCFC operation, addition of chromium or aluminum to the nickel materials has been approved to be an effective approach [11, 12]. It seems that there are no significant effects of Cr or Al addition to form Ni–Cr or Ni–Al alloy anode on the fuel cell performance when compared to that without Cr or Al addition.

Although the nickel plate as an MCFC anode has exhibited great advantages primarily because of its high catalyst activity, there are still a number of challenges for its practical applications in MCFCs, including its limited tolerance to sulfur and other contaminants contained in coal syngas and oxidation–reduction intolerance and vulnerability to carbon formation by exposure to hydrocarbon fuels. Among these challenges, hydrogen sulfide (H_2S) has been paid the most attention because of its immediate and severely detrimental effect on cell performance under various operating conditions [13]. Research on the H_2S poisoning mechanism is helpful in identifying sulfur poisoning conditions and also contributing to the development of sulfur-tolerant materials as MCFC anodes. One economic alternative is to find a material, which has higher sulfur tolerance than the state-of the-art nickel anode. With respect to this, $LiFeO_2$ has been identified as the most promising alternative anode material [14]. The material is thermodynamically stable in the carbonate melt and has a good electrical conductivity in a reducing atmosphere. Preliminary in-cell tests using porous electrodes indicated that the performance of the $LiFeO_2$ anode was not affected by sulfur at a level of 30 ppm H_2S. These results have encouraged further study on the electrochemical properties of $LiFeO_2$ and its feasibility as an alternative anode material.

16.4.2
Cathode and Cathode Materials

The MCFC cathode is also a plate consisting of a porous matrix layer of metal oxides, where the O_2 reduction reaction occurs. The most popular material is porous lithiated nickel oxide (NiO), which shows good electrical conductivity, has high porosity, and is of low cost. This porous layer has the optimum pore diameters of 6–10 μm, and the optimum thickness of about 0.5–0.8 mm with a porosity of 60–80%. In the formation of this matrix layer, Ni porous plate is initially installed in the fuel cell stack. Then, the air is blown into the stack to oxidize the Ni porous plate into NiO, and at the same time, this NiO is lithiated by carbonate electrolyte. The lithiated NiO contains Li^+ and has enough electronic conductivity. However, the state-of-the-art MCFC nickel cathode, oxidized and lithiated *in situ* to form $Li_xNi_{1-x}O$, has a relatively high solubility in the electrolytes such as $Li_2CO_3-K_2CO_3$, $Li_2CO_3-Na_2CO_3$, or related alkali molten carbonate eutectics. This dissolution leads to the formation of Ni^{2+}, which can diffuse from the cathode toward the anode under a concentration gradient. The dissolved Ni^{2+} precipitates on the matrix layer, and then meets with the dissolved H_2 proceeding from the anode side, being reduced to Ni at a certain distance away from the cathode. The continuous diffusion of Ni^{2+} accelerates more dissolution of nickel from the cathode. Continued deposition of sufficient bridged grains of metallic nickel across the cell eventually causes a short circuit in the cell [15, 16]. On the other hand, the NiO dissolution also results in the loss of active material, leading to a decrease in the active surface area available for the oxygen reduction reaction, resulting in the degradation of fuel cell performance and durability. In order to increase the stability of the NiO cathode, two approaches might be considered: one is to modify the electrolyte, that is, to select a melt composition that is more suitable for NiO than the conventional Li/K = 62/38 eutectic carbonate melt, and the other one is to modify the cathode materials, that is, to develop a new cathode material that is more stable than NiO.

For electrolyte modification, many efforts have been made to solve the NiO dissolution problem, and several possible approaches have been studied. For example, more basic molten carbonate melts, such as Li/Na carbonate eutectic, have been used to decrease the NiO dissolution rate in the melt. Furthermore, alkaline Ba or Sr earth metal salts have also been used as additives to increase the basicity of the melt in order to reduce the dissolution of the NiO cathode. However, using more basic molten carbonate melts only partially solves the problem since these melts decrease the NiO dissolution rate by 10–15% only.

For cathode material modification, several materials such as $LiFeO_2$, $LiCoO_2$, Li_2MnO_3, and $La_{1-x}Sr_xCoO_3$ were investigated as replacement materials for nickel oxide during the past decade [17]. This is because of their extremely low solubility in the carbonate melts. However, the kinetics, such as exchange current density for the oxygen reduction reaction on $LiFeO_2$, Li_2MnO_3, and $La_{1-x}Sr_xCoO_3$, is about 2–4 orders of magnitude lower than that of NiO. Thus, the slow kinetics of the oxygen reduction reaction limits further improvement of these materials-based cathodes. $LiCoO_2$ is more stable than NiO in alkaline environment and was once considered as the most promising alternative cathode material. However, application of $LiCoO_2$ as a cathode material is still limited for producing large-area electrodes, more than $1000\,cm^2$, because of its brittleness and higher manufacturing cost, as well as less electronic conductivity than the conventional NiO cathode. Therefore, these oxides seem inadequate for the cathode material. Another approach is to modify the conventional NiO-based cathode by coating $LiFeO_2$ (LFO) grains on porous NiO material [18]. Figure 16.4 shows the cross section of SEM micrographs of Ni, NiO,

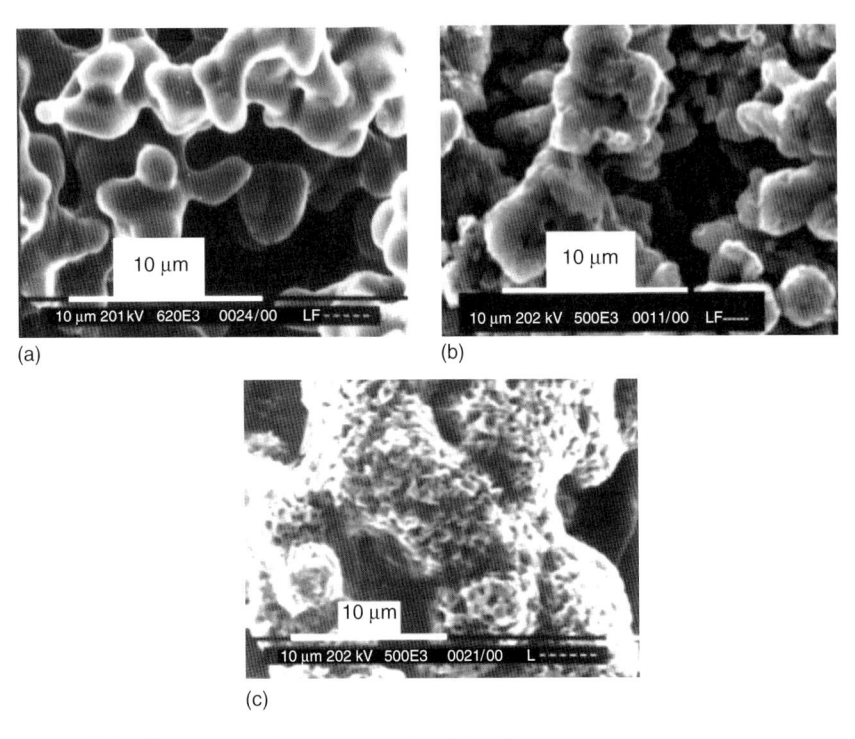

Figure 16.4 SEM cross-sectional micrographs of (a) Ni cathode, (b) NiO cathode, and (c) LFO-NiO cathode [18].

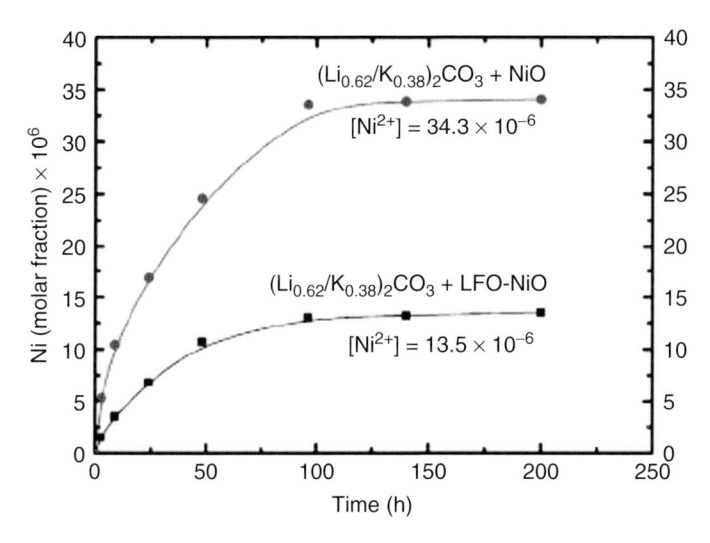

Figure 16.5 Equilibrium solubility of nickel ions in the $(Li_{0.62}K_{0.38})_2CO_3$ melts as a function of immersion time ($T = 650\ °C$, $P_{CO_2} = 0.67$ atm, and $P_{O_2} = 0.33$ atm) [18].

and LFO–NiO cathodes. The morphological difference between Ni and LFO–NiO cathodes could be attributed to the $LiFeO_2$ coating on nickel particle surfaces followed by sintering. In comparison with nickel and nickel oxide particles in Figure 16.4, the surfaces of the pretreated NiO particles are covered with many tiny grains, which were tightly sintered with the NiO particles. The dissolution curves of NiO and LFO–NiO cathodes in molten carbonates are given in Figure 16.5. It can be seen that the coating of $LiFeO_2$ on the surface of NiO can significantly retard the dissolution of NiO in molten carbonates.

The most promising way to modify the properties of nickel oxide appeared to be the incorporation of metal oxides into the nickel oxide [19–22]. It has been found that the addition of rare earth oxide to NiO could considerably decrease the solubility of Ni^{2+} in the Li/K eutectic carbonate melts [23]. The solubility of Ni^{2+} of rare earth oxide–NiO cathode in the melt is about 1 order of magnitude lower than that of bare NiO cathode. The result is presented in Table 16.1. It can be seen that the rare earth oxide has two effects on stabilizing the NiO structure: one is to reduce the NiO dissolution and the other is to facilitate the lithiation process, resulting in a mixed oxide with a higher Li^+ content, which is more conductive and stable. It has also been noticed that the samples impregnated with rare earth oxide can also improve the charge-transfer processes associated with the oxygen reduction reaction, as shown in Figure 16.6 [23]. In addition, the addition of ZnO to the NiO cathode was also shown to significantly decrease the solubility of Ni^{2+} in the Li/K eutectic carbonate melts. However, the electrocatalytic performance was also reduced [24]. The addition of Fe_2O_3 to the NiO cathode was also shown to significantly increase the time to shorting

Table 16.1 Content of metal ions of rare earth oxide–NiO cathode after immersion in Li/K eutectic melts for 200 h.

Sample	Content of rare earth oxide (wt%)	Ni^{2+} (mol ppm)	Ce^{4+}	La^{3+}	Pr^{3+}	Nd^{3+}
				mol ppm		
1	0.0	35.4	–	–	–	–
2	0.3	0.99	22.35	2.64	–	1.56
3	0.6	2.27	27.17	2.45	–	1.68
4	1.0	2.50	34.65	3.67	–	2.08

^{a}Atmosphere: 0.67 atm CO_2/0.33 atm O_2; T = 650 °C [23].

in Li/K cells, but cell performance was reduced [25]. Pore optimization may be needed to achieve similar performance in the baseline NiO cathode. The effect of MgO and CoO addition to the NiO cathode has been tested [26]. Both tend to reduce the dissolution of the NiO cathode. In order to prove this, a 1 kW stack was operated for nearly 5000 h under a high carbon dioxide partial pressure (accelerated conditions). In this test, the decay in performance was in the range of 2 mV/1000 h and therefore less than that predicted for the NiO cathode.

16.4.3
The Electrolyte Matrix

The electrolyte matrix, consisting of tightly packed ceramic particles with pores filled with alkali metal carbonate electrolyte, is one of the key components in MCFCs. The electrolyte matrix in MCFCs, like in any other electrochemical system, provides electronic isolation while maintaining ionic transport between electrodes and separation of the reactants. However, the difference is that the electrolyte matrix in MCFCs is a solid at room temperature and a paste at 650 °C. Since this electrolyte operates in an aggressive carbonate environment, sealing using a wet seal at the active area perimeter (edge seal) is necessary. The matrix performance is mainly measured in terms of cell internal resistance, gas cross-leakage in a cell, and wet-seal leakage. To perform these functions efficiently, the matrix needs to have properties such as high porosity, sharp pore size distribution and excellent electrolyte retention (required for attaining good bubble pressure), dimensional and material stabilities for maintaining proper electrolyte inventory (which affects gas crossover and cell resistance), and mechanical strength to maintain structural integrity during thermal transients, especially during the thermal cycle. Therefore, there is large room for the performance optimization through the choice of the matrix material, manufacturing process variables, structural strengtheners, and electrolyte. The composition and stability of matrix material as well as the carbonate mixture and the cost-effective fabrication of the matrix are both of great importance in achieving an optimized MCFC performance.

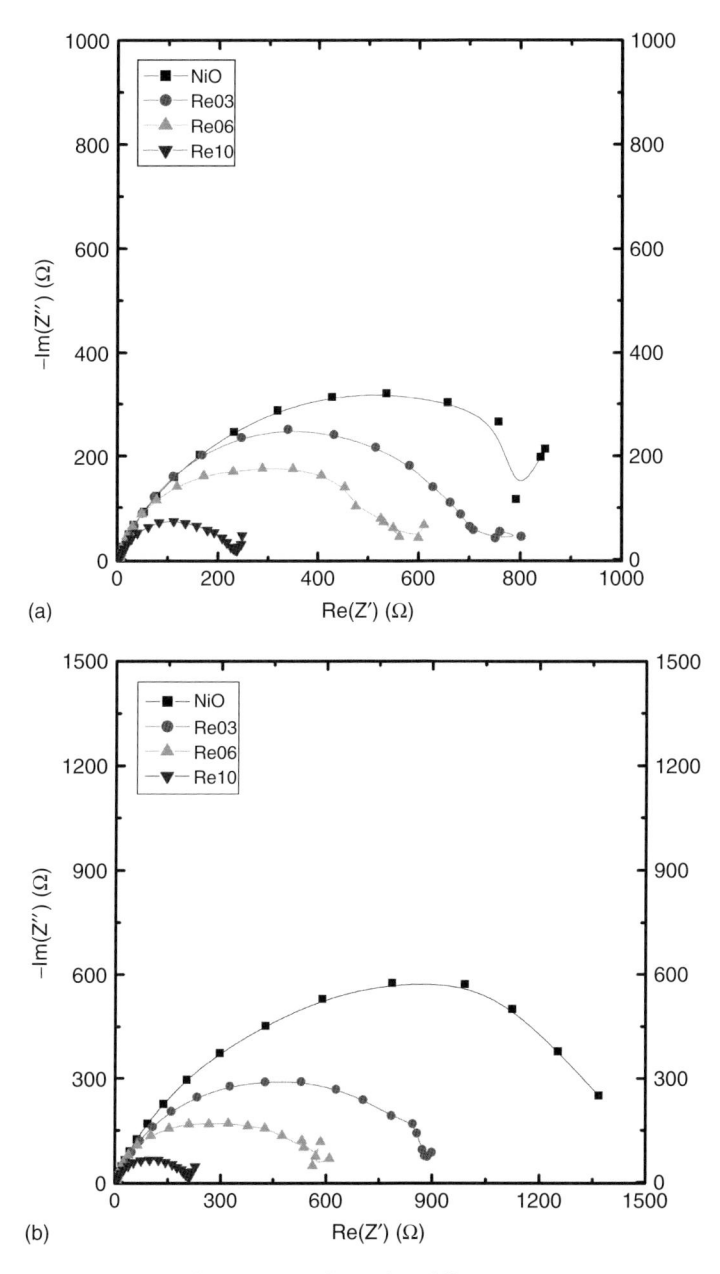

Figure 16.6 Impedance spectra obtained at different immersion times: (a) 168 h and (b) 200 h. Re03 denotes 0.3 wt% rare earth oxide–NiO, Re06 denotes 0.6 wt% rare earth oxide–NiO, and Re10 denotes 1.0 wt% rare earth oxide–NiO [23].

16.4.3.1 Electrolyte Matrix Materials

The electrolyte matrix material controls almost all the key matrix characteristics, such as reactant gases crossover, cell resistance, stability, and durability, hence has an important impact on the performance of MCFCs. The particle shape, size, and size distribution of matrix materials control the matrix porosity, pore size, pore size distribution, tortuosity-affecting electrolyte inventory, and cell resistance, as well as matrix bubble pressure. In addition, the matrix material can also affect the matrix mechanical strength, making an impact on its integrity during the mechanical and thermal stresses. Most importantly, the stability of the matrix materials is an important consideration in material selection. Several materials including different metal oxides such as MgO [27], lithium aluminate $LiAlO_2$ [28], $SrTiO_3$ [29], and others have been investigated. Among them, $LiAlO_2$ appears to be the favorite choice as the matrix material for MCFC developers. This is because $LiAlO_2$ exhibits three different allotropic forms [30]: (i) α-$LiAlO_2$ – a hexagonal structure with a density of 3.400 g cm^{-3}, and its particles are generally spherical in shape; (ii) β-$LiAlO_2$ – an orthorhombic structure with a density of 2.610 g cm^{-3}, and its particles are generally acicular in shape; and (iii) γ-$LiAlO_2$ – a tetragonal structure with a density of 2.615 g cm^{-3}, and its particles are irregular-shaped platelets or bipyramids.

The temperature/pressure phase diagram of the lithium aluminate and the MCFC experiment have shown γ-$LiAlO_2$ to be chemically and thermally most stable in the molten carbonate. At high temperatures, both α- and β-$LiAlO_2$ undergo an irreversible transformation to γ-$LiAlO_2$ in the presence of $(Li_{0.62}K_{0.38})_2CO_3$. The phase changes are accompanied by a change in particle morphology and a decrease in surface area. Therefore, γ-$LiAlO_2$ appears to be the most desirable crystalline form for MCFCs. However, significant particle growth of γ-$LiAlO_2$ and phase change from γ- to α-$LiAlO_2$ were observed after 34 000 h of operation [31]. Such change could result in increased pore size and pore volume, and reduced electrolyte retention [32], causing increased electrolyte loss, reactant cross-leakage, and ionic voltage loss. The increase in the pore volume can be understood by the density difference of α- (3.400 g cm^{-3}) and γ- (2.615 g cm^{-3}) phases. Therefore, from the point of view of pore volume and pore size, the α-$LiAlO_2$ phase as the starting material is more desirable [33, 34].

16.4.3.2 Electrolyte Matrix Fabrication

The choice of matrix preparation method is also very important since the mechanical strength strongly depends on the fabrication process. Twenty years ago, hot press methods were used for making the matrix. However, it was difficult to prevent cracking of the matrix, which could cause a gas cross-leakage through the matrix and degradation of the electrodes as well. In addition, hot press methods were not suitable for mass production. At present, the tape casting method seems to be the suitable method in electrode–electrolyte fabrication because it is a cost-effective process in obtaining thin structures of the matrix with large areas [35, 36]. For example, an electrode–electrolyte combined tape casting process for simplifying the manufacturing process of these components has also been reported. This tape casting process involves dispersing the ceramic powder in an organic solvent along

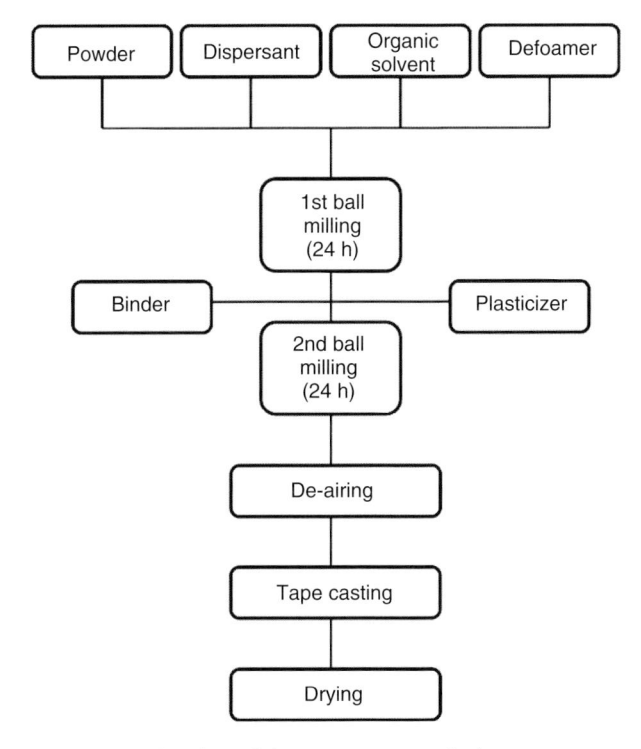

Figure 16.7 Flowchart of the tape casting method [37].

with binders and plasticizers to form the slurry. The slurry is then cast on a fixed or continuous belt using a *doctor blade*. After the solvent in the tapes was completely evaporated, the tape is peeled from the belt. The remaining solvent and binder are burned off at a temperature below the electrolyte melting point. The process is illustrated in Figure 16.7 [37]. The electrolyte matrix thus produced provides uniform thickness and tight pore structure.

16.4.3.3 Electrolyte Matrix Performance

Carbonate preservation in the electrolyte matrix is also important during the MCFC operation. In general, preserving carbonate electrolyte in electrolyte matrix is the benign liquid capillary. According to the Young–Laplace formula:

$$P = \frac{2\sigma \cos \theta}{r} \tag{16.4}$$

where P is the penetration gas pressure of the capillary, r is the radius of the capillary, σ is the surface tension coefficient of the electrolyte, $\sigma\left((Li_{0.62} K_{0.38})_2 CO_3\right) = 0.198$ N m^{-1}, and θ is the contact angle of electrolyte and matrix, when $\theta = 0°$, soakage is at its completeness. According to Eq. (16.4), when the pore radius of matrix is decreased, the penetration gas pressure of matrix will be increased. If the pressure between the anode and cathode, 0.1 MPa, is required, then the pore radius of the matrix should be ≤ 3.96 μm. Thus, to ensure that the pore radius of the matrix is ≤3.96 µm, the granularity of the LiAlO$_2$ powder must be strictly controlled. Regarding the ion transmit of carbonate electrolyte in matrix pores, the Meredith–Tobias formula can be used for estimation:

$$\rho = \frac{\rho_0}{(1-\alpha)^2} \qquad (16.5)$$

where ρ is the resistance rate of matrix, ρ_0 is the resistance rate of electrolyte (for (Li$_{0.62}$K$_{0.38}$)$_2$CO$_3$ at 650 °C, this value is 0.5767 Ω · cm), α is the volume fraction of LiAlO$_2$ in matrix, and $(1-\alpha)$ is the volume fraction of the pore in matrix. According to Eq. (16.5), when the volume fraction of the pore in the matrix increases, the amount of carbonate electrolyte in matrix will increase, then the resistance rate of the matrix will be reduced. To satisfy both penetration gas pressure enlargement and resistance rate reduction, the matrix should have small pore diameter and high porosity. In general, the parameters of matrix are as follows: the electrolyte matrix thickness is in the range of 0.3–0.6 mm, the porosity is about 40–70%, and the average pore radius is in the range of 0.25–0.8 µm.

A stable electrolyte–gas interface in the MCFC porous electrodes is established exclusively due to balance in capillary pressures. The diameters of the largest flooded pores in these electrodes are related by the equation (16.6) [38]:

$$\frac{\gamma_a \cos\theta_a}{D_a} = \frac{\gamma_c \cos\theta_c}{D_c} = \frac{\gamma_e \cos\theta_e}{D_e} \qquad (16.6)$$

where γ is the interfacial surface tension; θ, the contact angle of the electrolyte; D, the pore diameter; and the subscripts a, c, and e refer to the anode, cathode, and electrolyte matrix, respectively. The distribution of molten carbonates in the electrodes and electrolyte matrix is illustrated schematically in Figure 16.8 [38]. According to the model illustrated in Figure 16.8 and described by Eq. (16.6), the electrolyte content in each of the porous components will be determined by the equilibrium pore size D^* in that component. Pores smaller than D^* will be filled with electrolyte, and pores larger than D^* will remain empty. A reasonable estimate of the volume distribution of electrolyte in the various cell components is obtained

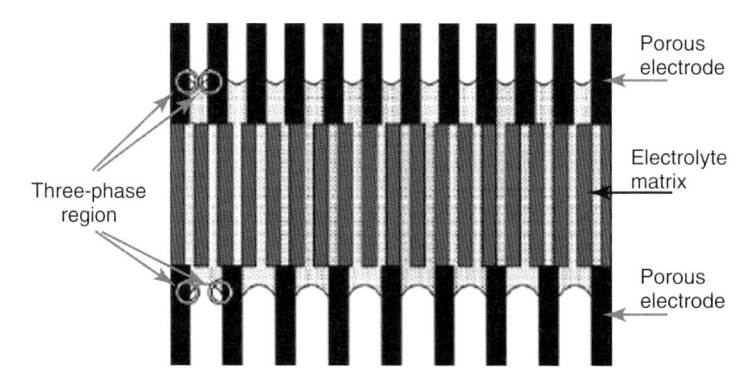

Figure 16.8 Distribution of molten carbonate electrolyte in porous electrodes of MCFCs as a result of balance in capillary pressure [38].

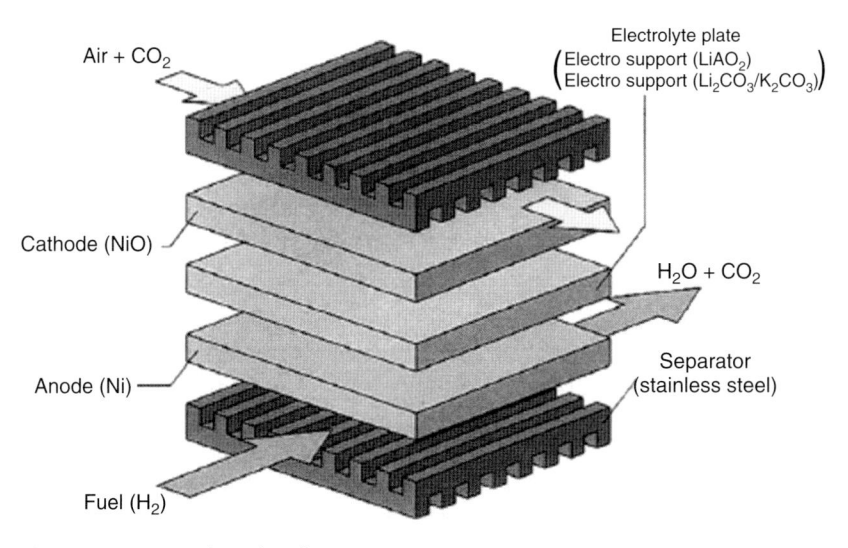

Figure 16.9 A typical single cell cross section [10].

from the measured pore–volume–distribution curves and the above relationship for D [39, 40].

16.4.4
Fabrication of Electrode–Electrolyte Assembly

In general, the anode, electrolyte, and cathode are fabricated at the same time to achieve an electrode–electrolyte assembly, which looks like a sandwich with the electrolyte in the middle, as shown in Figure 16.9 [10]. For an industrial scale of MCFCs, the electrode area in a unit cell is approximately $1\,m^2$, and the electrolyte matrix area is much larger than that. Taking the brittleness of large-area ceramics plate into account, in the electrode–electrolyte assembly preparation, the electrolyte slurry is first formed into a thick tape by the doctor-blade method and dried, following which the electrodes (both the anode and the cathode) are put on the tape. The three plates obtained are superposed and are sintered *in situ* to form the assembly. Therefore, the *in situ* sintering technology is crucially important to the performance of MCFC stack.

During the *in situ* sintering process of the electrode–electrolyte matrix (or assembly), three things should be paid attention. (i) The rate of heating should be very slow initially to ensure complete burnout of the chemical additives, such as binder and plasticizer. Among which, the rate of heating from 200 to 450 °C is of great importance. (ii) The electrode–electrolyte matrix must be in perfect condition during the sintering process. (iii) The electrode–electrolyte matrix must be a layer of tightly packed powder impregnated by molten carbonate electrolyte to form a composite pastelike structure after the sintering.

Sintering temperature of the electrode–electrolyte matrix influences the matrix pore structure and thereby affects its electrochemical performance.

Thermogravimetric analysis (TGA) is done to determine the optimum heat treatment schedule for the sintering. A typical TGA curve for the electrolyte matrix tape should be first established, based on which, the sintering sequence is then used to sinter the porous electrode–electrolyte matrix.

16.4.5
Materials and Fabrication of the Bipolar Plate

Bipolar plate is one of the key components in an MCFC. In an MCFC, there are two bipolar plates, one is located at anode side and the other is at cathode side. The main functions of the bipolar plate are (i) providing pathways for fuel or oxidant flow over the anode or cathode surfaces and also regulate their flow directions; (ii) proving electrical connection of cells. Therefore, the material requirement for a bipolar plate is high. Since the bipolar plate is exposed to the electrolyte, its material must have enough corrosion resistance to the carbonate electrolyte in both anode reducing and cathode oxidizing atmospheres. In general, stainless steel such as SUS 316L or SUS 310S is used as the bipolar plate material. In order to obtain sufficient corrosion resistance, Ni is coated on the anode side of the bipolar plate [41] and Al is coated on the wet-seal side to avoid severe corrosion [42–47]. In addition, thermal spraying of aluminum-containing alloy powders such as FeAl and NiAl may be an alternate approach [48–50]. High-Al ferritic alloys such as Kanthal have also been shown to have excellent corrosion resistance [51].

Regarding the design and fabrication of MCFC bipolar plates, the cost portion of the bipolar plates in the stack must be considered. Technically, corrugated or similarly shaped stainless steel sheets are used for the materials, on which the flow field is generated by means of machine processing for small-scale production, as shown in Figure 16.10 [52]. Then the thermodynamically stable Ni coating is needed for long-term anode-side protection, either aluminized coating or high-Al stainless steel will satisfy the endurance requirements of the wet-seal area, which simultaneously experiences reducing and oxidizing environments. In addition, in order to reduce the mass production cost, it is necessary to make use of a one-shot press and automatic resistance-seam welding technologies for the bipolar plate structures.

16.5
Structure and Performance of MCFCs

16.5.1
Structure and Performance of MCFC Single Cell

Figure 16.11 shows the structure and working principle of a typical MCFC single cell [53]. It can be seen that an MCFC single cell is composed of bipolar plates, corrugated plates, current collectors, an anode (porous nickel alloy plate), cathode (porous lithiated NiO plate), and the electrolyte matrix. The electrolyte matrix contains both the electrolyte support and the electrolyte (Li/K or Li/Na carbonate).

16.5 Structure and Performance of MCFCs

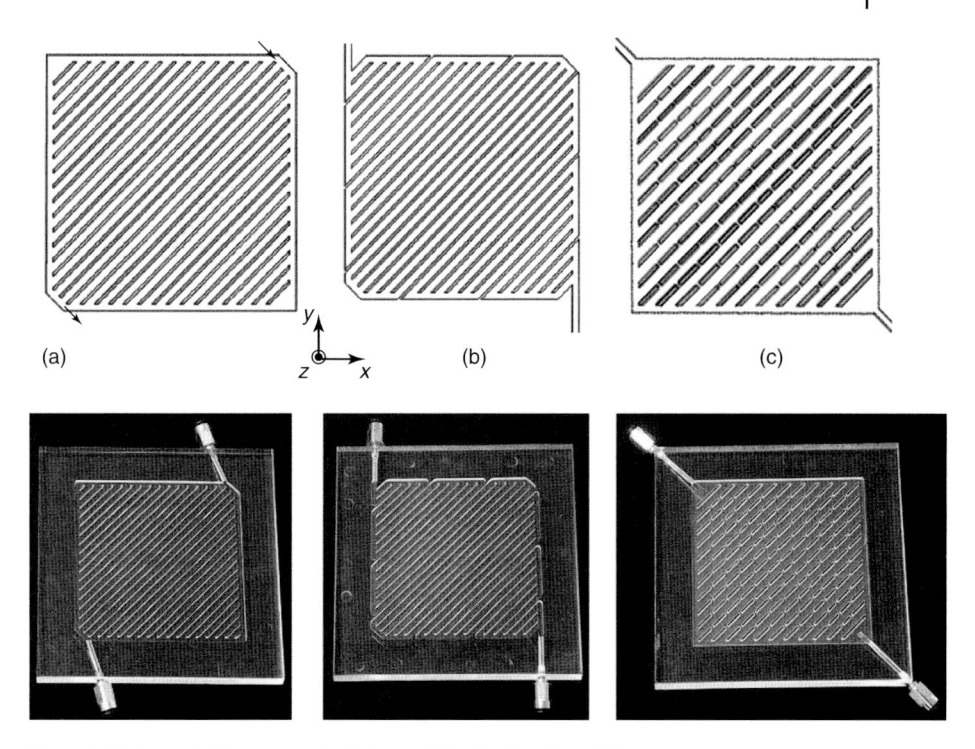

Figure 16.10 (a–c) Sketches and pictures of the bipolar plate [52].

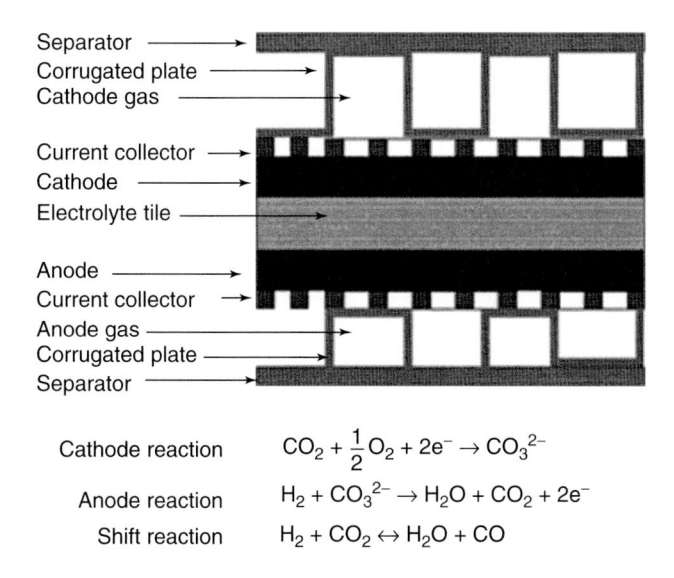

Separator
Corrugated plate
Cathode gas
Current collector
Cathode
Electrolyte tile
Anode
Current collector
Anode gas
Corrugated plate
Separator

Cathode reaction $\quad CO_2 + \frac{1}{2}O_2 + 2e^- \rightarrow CO_3^{2-}$

Anode reaction $\quad H_2 + CO_3^{2-} \rightarrow H_2O + CO_2 + 2e^-$

Shift reaction $\quad H_2 + CO_2 \leftrightarrow H_2O + CO$

Figure 16.11 Schematic structure and working principle illustration of a typical MCFC single cell [53].

The electrolyte is fixed firmly in the electrolyte support by capillary forces, which is controlled by the pore size of the electrolyte support. Normally, the electrolyte composition is 62 mol% of Li and 38 mol% of K. Using this electrolyte composition, the MCFCs can operate at 650 °C. This temperature seems to be the optimal temperature for MCFCs. As discussed previously, for MCFC fuel feeding, H_2 and/or CO is used as the anode fuel and air and CO_2 as the oxidant at cathode. Reformed fuels such as coal-derived fuels and other hydrocarbon fuels can also be used. Actually, CO can be easily converted to H_2 via the well-known water/gas shift reaction (WGSR) for MCFC feeding [54]. Fuel and oxidant gas flow through the anode and cathode side channels and pass the corrugated plates. Regarding the electrodes, under MCFC operating conditions, the porous electrodes are partially wetted by carbonate electrolyte and have an extensive reaction area, which is called a gas-electrolyte-electrode TPB, where the electrochemical reactions involved in MCFCs happen in the neighborhood of TPBs. As discussed previously, O_2 and CO_2 in the oxidant gas react with electrons at the cathode and produce CO_3^{2-}. Because of the driving force of the concentration difference, the CO_3^{2-} moves within the electrolyte plate perpendicularly away from the cathode to the anode. At the anode, the H_2 and/or CO in the fuel gas reacts with CO_3^{2-}, and via an electrochemical reaction, CO_2, H_2O, and electrons are produced. Electrons released from the reaction sites are collected by the current collector. They then pass the corrugated plates and reach the separator perpendicularly. The separators, located at the top and bottom, are connected to the loading equipment.

Figure 16.12 shows photographs of the experimental device during the operation of MCFC single cell [55]. The cell was circular in shape, and the dimension was 50 mm diameter. The material used in housings was 316 stainless steel, and it was coated with aluminum by the plasma deposition treatment. Cathode, anode, and current collectors were 32 mm diameter. The anode used was a commercial material which consists of a mixture 98% Ni and 2% Cr. The porosity was 61–63% and the pore mean diameter 2–5 mm. The anode was prefilled with electrolyte (0.05 g cm^{-2}). The comparative characteristics of both the cathodes (Ni and Ni–Ce) are presented in Table 16.2. The electrolyte support was made of γ-LiAlO$_2$. The porosity was 50–55% and the pore mean diameter 0.1–0.5 mm. The cathodic current collector was stainless steel, and the anodic current collector was stainless steel protected with nickel. The electrolyte consists of a 62 Li$_2$CO$_3$/38 K$_2$CO$_3$ eutectic presented in sheets.

The cell with the commercial Ni cathode was tested for about 1100 h of hot time (time after electrolyte melting), 1000 h of which were on load. The cell with the Ni–Ce cathode was tested for approximately 2200 h of hot time, 2100 h of which was on load. The evolution of the performance along the operation time is shown in Figure 16.13. The voltage of the commercial nickel cathode decreased from 0.8 to 0.72 V at 150 mA cm^{-2} from 200 to 1000 h (120 to 108 mW cm^{-2} in power terms). However, the Ni–Ce cathode showed a highly stable performance with time, maintaining a voltage of nearly 0.68 V at 200 mA cm^{-2} from 200 to 2100 h (136 mW cm^{-2} in power terms). On the other hand, it also indicated that the performance of the Ni–Ce cathode was better than that of the nickel commercial and the endurance

Figure 16.12 Photographs of the experimental device during the operation of MCFC single cell [55].

Table 16.2 Textural characteristics of the cathodes used [55].

Cathode	Porosity (%)	Pore mean diameter (μm)
Ni–Ce	71.7	7.1
Ni	74–78	3–6

was more than double. The comparison at 1000 h of operation (Figure 16.14) gives us further evidence of the better performance of the Ni–Ce cathode.

16.5.2
Structure and Performance of MCFC Stack

Generally, the output voltage of an MCFC single cell is 0.6–0.8 V at a current density of 150–200 mA cm^{-2}. To achieve high output voltage and power, it is necessary to connect some single cells in series to construct an MCFC stack. In this stack,

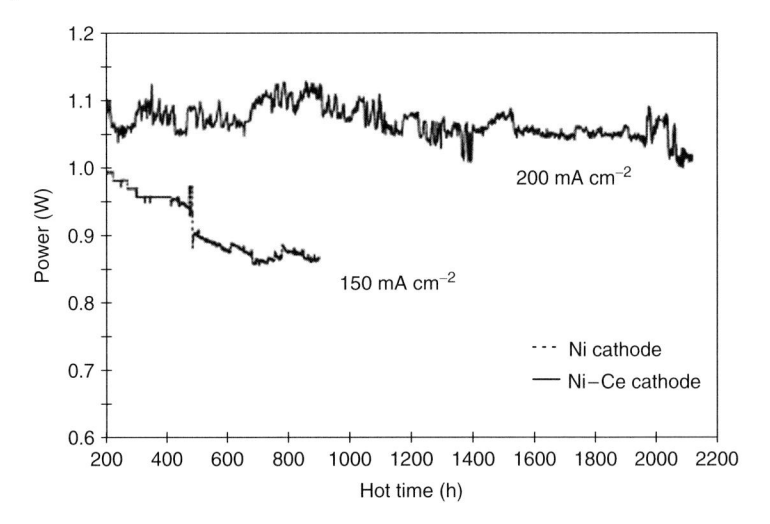

Figure 16.13 Performance along the time for both cathodes: Ni cathode at 150 mA cm^{-2}; Ni–Ce cathode at 200 mA cm^{-2} [55].

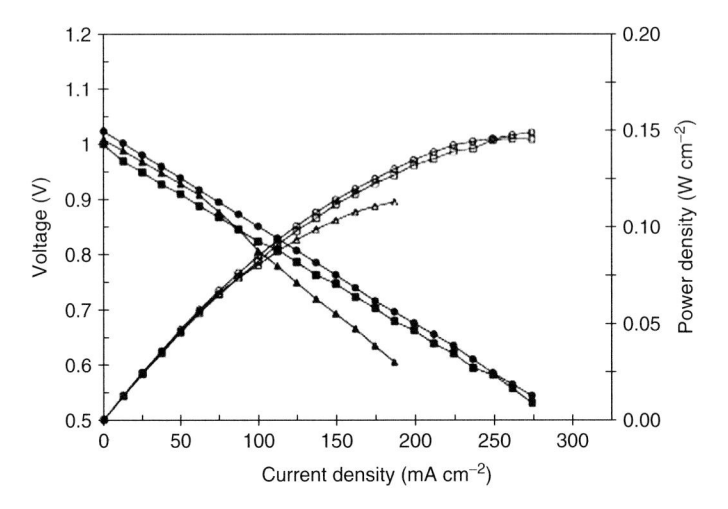

Figure 16.14 Polarization curves for both cathodes. Solid symbols represent the voltage and the open ones the power density. (▲, △) Ni cathode at 1000 h; (■, □) Ni–Ce cathode at 1000 h; and (○, ●) Ni–Ce cathode at 2000 h [55].

two adjacent single cells are separated by a bipolar plate. This bipolar plate has two major roles: the first one is electrical connection of the cells and the other one is to supply fuel gas and oxidant gas to the anode and cathode chambers separately. Figure 16.15 illustrates the MCFC stack structure with a cross-flow pattern [56].

I–V curves of the kilowatt-class stack (52 cells) and 300 W class stack (10 cells) are shown in Figure 16.16 [57]. The physical and geometrical parameters of the

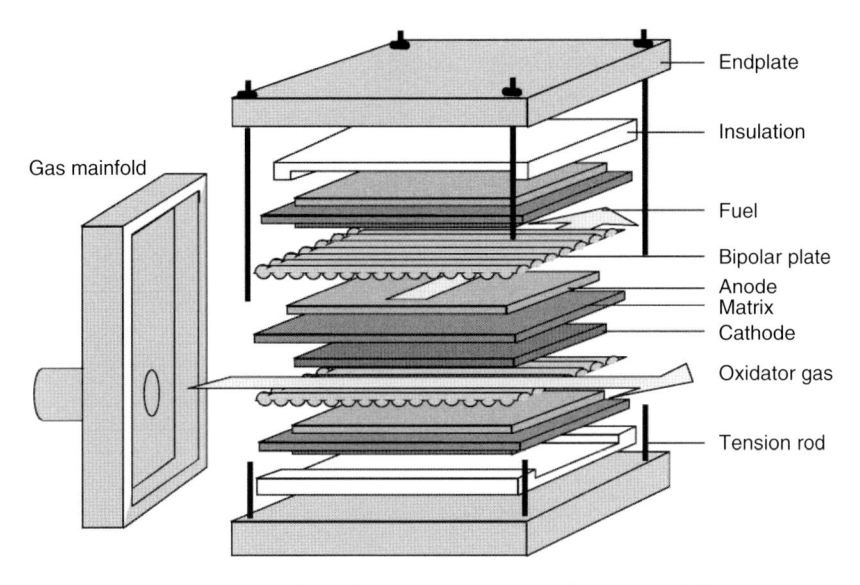

Figure 16.15 Schematic MCFC stack structure with a cross-flow pattern [56].

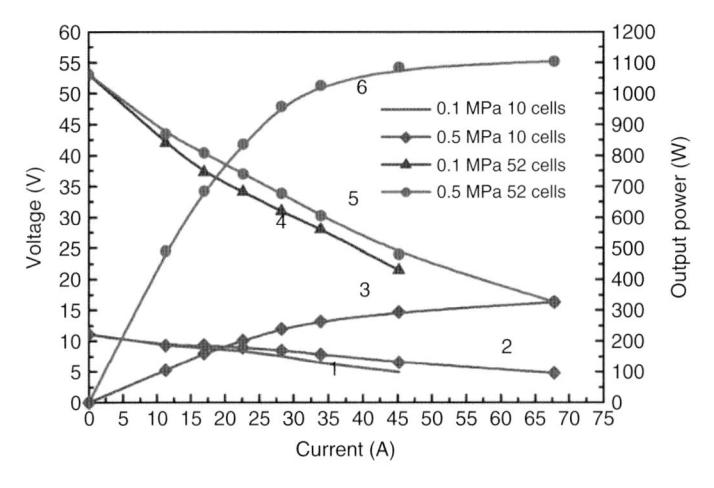

Figure 16.16 I–V and output power curves of the kilowatt- and 300 W class MCFC stacks. (1, 2, 4, and 5) The terminal voltage of the 300 W and kilowatt-class stacks at 0.1 and 0.5 MPa, respectively, and (3 and 6) the output power of the 300 W and kilowatt-class stacks at 0.5 MPa [57].

components in the kilowatt-class stack (52 cells) and 300 W class stack (10 cells) of MCFCs are listed in Table 16.3 [57]. The specifications of the kilowatt-class stack (52 cells) and 300 W class stack (10 cells) of MCFCs are listed in Table 16.4 [57]. The electrolyte used was the eutectic melt (0.62 Li_2CO_3 + 0.38 K_2CO_3). It can be seen that the performance enhances with increasing pressure of the reactant gas, which results from all kinds of polarization decreasing with increasing pressure of

Table 16.3 The physical and geometrical parameters of components in the kilowatt-class stack (52 cells) and 300 W class stack (10 cells) of MCFCs [57].

Component	Material	Area (cm^2)	Thickness (mm)	Porosity (%)	Average pore diameter (μm)
Cathode	Ni	226	0.4	60–63	18–25
Anode	Ni–Cr	226	0.4	60–63	18–25
Matrix	α-LiAlO$_2$	375	0.6–0.8	53–57	0.36
Separator	Stainless steel 316	375	1.0	–	–

Table 16.4 The specifications of the kilowatt-class stack (52 cells) and 300 W class stack (10 cells) of MCFCs [57].

	Specifications
Cell number	52 cells
Designed power	1000 W
Achieved power	1025.5 W
Operation temperature	650 °C
Oxidant	$O_2 + CO_2 (O_2/CO_2 = 40/60)$
Fuel gas	$H_2 + CO_2 (H_2/CO_2 = 80/20)$
Reactant gas pressure	0.1 MPa, 0.5 MPa
Reactant gas utilization	20%
Electrolyte	0.62 Li$_2$CO$_3$ + 0.38 K$_2$CO$_3$

the reactant gas. By calculation and comparison with the 300 W class stack, the area resistance of per cell in the kilowatt-class stack is almost the same, that is, 3.4 and 2.5 Ω cm^2 at the reactant gas pressures of 0.1 and 0.5 MPa, respectively. It indicates that the kilowatt-class stack is successfully assembled and under the suitable stacking pressure. As shown in Figure 16.16, the output power of the kilowatt-class stack is 1025.5 W at 150 mA cm^{-2} under the conditions of the reactant gas pressure of 0.5 MPa and utilization of 20%. It reaches 1104.5 W at 300 mA cm^{-2}. The output power of this stack increases slowly with increasing current density from 150 to 300 mA cm^{-2}, indicating that concentration polarization has been dominant in all kinds of polarization.

16.6
Schematic of MCFC Power Generation Systems

The whole MCFC power generation systems consist of the fuel processing subsystem, the MCFC subsystem, a bottoming cycle, an invertor, as well as the fault check and automatic control subsystem, as shown in Figure 16.17 [58]. By means of these

Figure 16.17 Schematic of typical MCFC power generation systems. An MCFC system is composed of a fuel processing subsystem such as a reformer of natural gas or a coal gasification system, a fuel cell subsystem, a bottoming cycle subsystem, and an invertor [58]. NG, natural gas; HRSG, heat recovery steam generator.

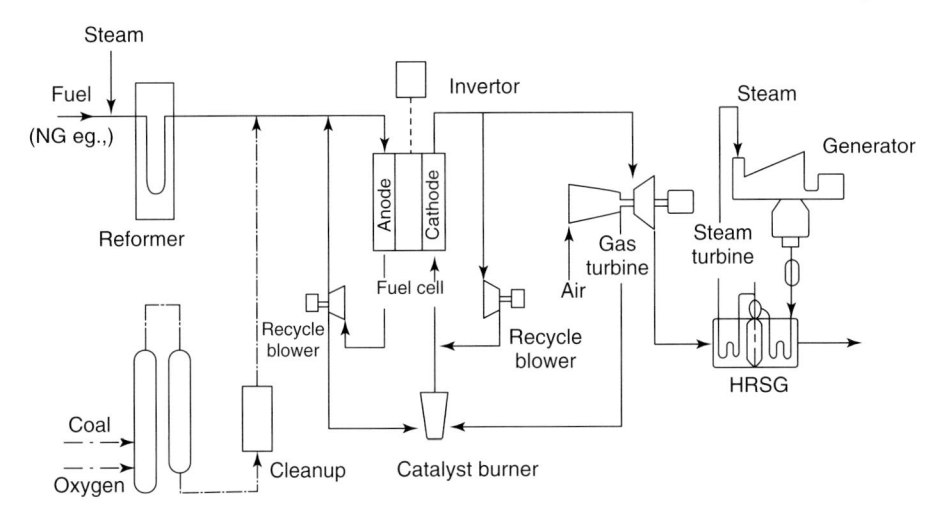

auxiliary subsystems, the MCFC power generation systems are able to have the required fuel and oxidant, release water and heat resulting from the electrochemical reaction, and supply electricity without cessation. These subsystems are discussed in detail in the following sections.

- **The fuel processing subsystem.** The fuel processing subsystem for most fuels also involves a gas cleanup unit. The MCFC is sensitive to some of the fuel-borne compounds. Therefore, the fuel needs to be cleaned to separate these compounds before they get into the fuel cell stack. The compounds sometimes are hazardous to the MCFC stack and/or the catalysts used for fuel conversion to hydrogen. The fuel, such as natural gas or coal gas, is processed first by removing sulfur compounds to <0.1 ppm in a cleanup bed. The fuel processing step also includes conversion of heavy hydrocarbons to lighter forms and/or hydrogen-rich gas. The MCFC operating temperature allows the internal reforming of light hydrocarbons. For example, the MCFC stacks fitted with an internally reformed feature can directly utilize light hydrocarbons such as methane (main component of natural gas) and propane. Another consideration in the fuel processing subsystem design is the prevention of carbon building up when operating under hydrocarbon fueling. In order to achieve this, a sufficient amount of steam must be added to the fuel to suppress carbon deposition.
- **Fuel cell subsystem.** The fuel cell subsystem consists of a fuel cell and an annex. Since the operating temperature of MCFC is high, both heat and water

produced in the cell can be directly used for reforming the fuel with catalysts in the cell. This type of MCFC system is called an *internal reforming type*, whereas the type which has a separate reformer is called an *external reformer type*. In the internal reforming type, fuel utilization rates can be set higher than that in the external reforming type and the system efficiency is expected to be higher. However, additional development is required for reforming catalysts resistant to the molten carbonate.

- Gas recycling is planned to control the flow and temperature. Especially, cathode gas recycling is adopted for stack gas temperature control, and anode gas recycling is planned for improving fuel utilization and recycling water on the anode side.
- **Bottoming cycle and invertor.** In a conventional MCFC system (with external reforming), the heat required for the reforming reaction is usually supplied by burnout of the anode exhaust gas followed by heat exchangers. In addition, the MCFC is also suited for internal reforming of hydrocarbons because of its operating temperature and by-product water formation on the anode side. The fuel cell reaction itself is exothermic, and the removal of heat is necessary to control the temperature of the stack. The heat from the fuel cell reaction can be consumed in the reforming reaction, and excess heat will be used in the bottoming cycle, which consists of steam and GTs and large-scale pressurized systems suitable for the bottoming cycle. Small dispersed systems that operate at atmospheric pressure will use the waste heat to produce steam for thermal energy.
- The MCFC generates DC electricity, and this must be transformed to AC power to connect the output to the utility grid. The voltage and the current levels vary with the load on the fuel cells, and the invertor subsystems will be designed to incorporate the variations. Although some differences may exist among types of fuel cells, the available technology will be applied to the invertor subsystem in MCFC power generation systems.
- **The fault check and automatic control subsystem.** The fault check and automatic control subsystem is an automatic control device that controls start-up, stop, operation, and the load of the MCFC power generation system. It consists of the measuring unit, displaying unit, administer devices, and controlling computer.

16.7
Fabrication and Operation of MCFCs

16.7.1
Testing System of MCFCs

The testing system of MCFCs includes gas supplier, pressure, and flow rate regulator, gas preheater, gas-humidifier, single fuel cell or fuel cell stack, gas cooler and gas separator, pressure difference regulator, pipeline and valves, load, and other instruments.

16.7.2
Performance of MCFCs

In order to obtain optimal performance, such as output voltage, power density, and lifetime, various designs of the MCFC stack have been carried out to overcome challenges. At present, the power density and lifetime of the MCFC stack can be reached to 100–200 mW cm^{-2} and 20 000 h, respectively. With the approaching commercialization, the power density and lifetime of MCFC stack have to be over 200–300 mW cm^{-2} and 40 000 h, respectively, and need to be demonstrated by field tests. Adequate cell performance must also be maintained over the desired length of service. Current state-of-the art MCFCs depict an average degradation over time of $\Delta V_{\text{lifetime}}(\text{mV}) = -5$ mV/1000 h [59].

16.7.3
Operation of MCFCs

Regarding MCFC operation, three points have to be considered. (i) It is necessary to operate the MCFC stack at low load (80–100 mA cm^{-2}) at the beginning, then at a rated load (150 mA cm^{-2}). (ii) Although the state-of-the-art MCFC anode material (Ni–Cr alloy) can offer excellent catalytic properties, mixed conductivity, and good current collection, such alloy present some disadvantages related to the low tolerance to the carbon buildup when operating under hydrocarbon fueling. During the hydrocarbon cracking reaction, carbon deposition (coking) covers the active sites of the anode, resulting in rapid, irreversible cell deactivation. Therefore, a sufficient amount of water (or steam) must be added to the fuel to suppress the carbon deposition, which however can lower the electrical efficiency of the system because of fuel dilution. (iii) The times of start and pause for MCFCs must be reduced as much as possible to prevent deformation and damage to the electrolyte matrix.

As mentioned previously, the fuel processing subsystem for fuels includes a gas cleanup unit. Because the MCFCs are sensitive to some harmful compounds of fuels, the fuel must be cleaned before entering the stack. These harmful compounds are listed below [60, 61].

- **Sulfur-containing compounds.** It is well recognized that sulfur compounds present in fuel, even in ppm levels, are critically detrimental to the performance of MCFCs. Therefore, the level of sulfur compounds (in the form of typically H_2S in natural gas) should be reduced to below 10 ppm, and the SO_2 content of the oxidant in the cathode should be reduced to below 1 ppm. Furthermore, if an MCFC stack lifetime of more than 40 000 h is required, the H_2S content must be reduced to below 0.1 ppm.
- **Chlorine-containing compounds.** During the gasification process, coal emits various volatile toxic and corrosive contaminants into the gas stream. The contaminants usually include chlorine, sulfur, arsenic, and antimony compounds. The effect of chlorine is also well documented. Less than 0.1 ppm is required to maintain an acceptable fuel quality.

- **Particulates.** The tolerance limit of the MCFCs to particulates larger than 3 μm diameter is only at a 0.1 g l^{-1} level.

16.8
MCFC Power Plant

MCFCs can be applied not only to alternative thermal power plants but also to dispersed power plants as well as integrated gasification combined cycle (IGCC) power plants. The system studies based on MCFC have indicated that this technology is promising for clean and efficient power generation, utilizing a wide variety of fuels such as hydrogen, natural gas, renewable fuels, liquid hydrocarbons, and coal-gasified gas. The choice of fuel determines the processing and cleaning required in the MCFC fuel processing subsystem design. Natural gas is a widely available fuel across the world. It is also simple and cost effective to be used in high-temperature MCFC and SOFC for power generation. At present, natural gas is used as the fuel in most operating MCFC power plants all over the world. The systems to utilize coal gas as fuel in MCFC power plants are evolving in America and Japan. With further development, the fuel varieties in the application of MCFC power plants will be further expanded. In the following two sections, the natural and hybrid coal-gas-fueled MCFCs are discussed in detail.

16.8.1
MCFC Power Plant for Natural Gas

The MCFC systems fueled by natural gas represent an important option for the power generation market because of the wide natural gas distribution infrastructure and the availability for the time being. The natural-gas-fueled MCFC power plants are also simple and cost effective, in particular, when internal reforming is used in the fuel cell design. Besides the major gas component methane, natural gas is expected to contain some saturated hydrocarbons, such as ethane and propane, which can also be used as fuels for MCFCs. In normal, natural gas contains harmful sulfur compounds in the form of naturally occurring H_2S and organic sulfur additives, which needs to be cleaned by either externally or internally reforming it before its usage in MCFCs. Figure 16.18 [62] shows a process flow diagram for the simplified natural-gas-fueled MCFC power plant. In addition to the MCFC stack, MCFC power generation systems also include the preheater, the circulating blower, the control valve and the cooler, and other components. Functionally, the systems consist of two main reactive gas-circulating loops, namely, the anode gas circulating loop and the cathode gas circulating loop.

1) **The anode gas circulating loop.** It maintains the stability of the gas components at the anode entrance and improves the fuel gas utilization. The anode circle loop lay back the anode exhaust gas to the anode entrance preheater. First, the temperature of the anode exhaust gas is reduced to 200 °C in a

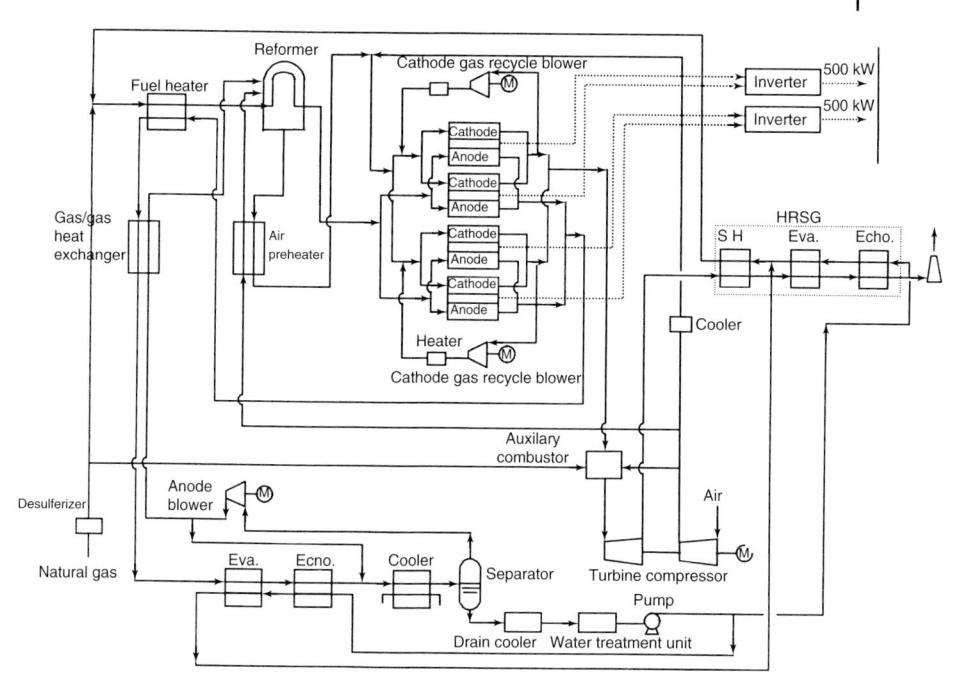

Figure 16.18 Simplified natural-gas-fueled MCFC power plant process flow diagram. Eva (evaporator), Ecno (economizer), SH (superheater), HRSG (heat recovery steam generator) [62].

cooling device, and then it pass through the anode circulating fan to return to the anode entrance preheater. The bypass of the fan is used to control the circle flux. Finally, it is heated to 600 °C by the preheater.

For anode gas supply, the sulfur-containing compound in the inlet fuel will be first removed by the gas-circulating loop, as shown in the Figure 16.18, and then the cleaned fuel enters the anode side along with externally supplied water for suppressing carbon deposition. This inlet gas is actually preheated by exchanging heat with the cathode exhaust gases and is then sent to the MCFC stack. The fuel can be internally reformed and then electrochemically oxidized by carbonate ions, which are formed on the cathode side by reaction with oxygen and carbon dioxide. In this case, CO_2 contained in the anode exhaust gas must be separated and recycled to the cathode. The utilization of the fuel may be enhanced by returning unused fuel, such as CO and H_2, after CO_2 separation to the anode inlet. Then, the fuel that is not converted by the fuel cell is combusted in the reformer (HER). However, the anode-off gas contains large amount of moisture that will adversely influence the performance of the HER. The anode-off gas is therefore cooled in several stages to separate most of the moisture. The transferred heat is used for heating up and evaporating water that is needed for the reforming reaction. Heat released in the moisture separator is utilized by external consumers (e.g., a district heating system) represented here by a heat sink. The moisture separator produces hot water at 80 °C. After utilization, this water returns at a temperature of 60 °C, and it is recirculated back to the moisture separator. The anode-off gas is circulated by a blower, reheated, and sent to the HER. This dried anode recycle gas leaves the anode gas recirculation and moisture separation subsystems at a fixed temperature of 460 °C in order to keep the inlet temperature of the heat exchange reformer constant throughout the study.

In order to control the temperature of the MCFC stack, the part of exhaust gas from the cathode will be returned to cathode gas preheater by the cathode circle loop. The structure of the cathode circle loop is similar to that of the anode circle loop.

2) **The cathode gas circulating loop.** The cathode gas circulation not only provides O_2 and CO_2 for the electrochemical reaction but also serves as the main coolant for the MCFCs, and therefore, the mass flow of the cathode gas has to meet the cooling requirements. This mass flow of air necessary for cooling is far greater than that required for the cathode reaction. Part of this air is, therefore, circulated, and the amount of circulation is set accordingly to assure the cathode inlet temperature after mixing this circulation flow with fresh air and flue gas from the HER. Before mixing, this circulation flow is partly cooled by preheating the pressurized fresh air. The flue gas from the HER is the main source of CO_2 required for the cathode reaction. The structure of the cathode gas circulation loop is similar to that of the anode gas circulation loop.

16.8.2
Hybrid Coal-Gas-Fueled MCFC/GT (Gas Turbine) and ST (Steam Turbine) Systems

At the current technologic status, the fuel utilization of an MCFC is limited. Both unused fuels and heat generated by the electrochemical reaction are mainly responsible to such limited fuel utilization. In order to increase this utilization, heat at a temperature of 650 °C can be further used to enhance overall energy utilization in a variety of applications such as raising steam or hot water for on-site combined heat and power. Other cogeneration possibilities include the integration of the MCFC stack with another power-generating device, such as GT and ST, to enhance both the fuel utilization and overall efficiencies (70–80%).

The combined MCFC/GT and ST power generation systems include (i) a fuel processing system consisting of coal feed equipment, coal gas generator, coal gas cleaner, and temperature control equipment; (ii) an MCFC power generation system; (iii) a GT power generation system consisting of a burner and GT power generation set; and (iv) a ST power generation system consisting boiler (heat recovery steam generator) and ST power generation set.

The combined MCFC/GT and ST power generation systems are sketched in three steps. (i) Coal, steam, and oxidant sources are sent to the coal gas generator to generate coal gas, which is usually processed in a gas cleanup system to remove sulfur compounds, tars, particulates, and trace contaminants. Then the cleaned fuel gas is converted to electricity by the electrochemical reaction in MCFCs. This is the first step of energy conversion in the systems. (ii) The unreacted fuel is sent to the burner with refined fuel gas to generate high-temperature and high-pressure fuel gas flow through full oxidation and combustion reactions. Then, the fuel gas flow is utilized in a GT to provide additional power and boost efficiency. This is the second step of energy conversion in the systems. (iii) High-temperature gas generated by a GT is again sent to a boiler (heat recovery steam generator), coal economizer, and superheater to generate steam, which is utilized in a ST power generation set to provide additional power and boost efficiency. This is the third step of energy conversion in the systems.

In fact, the whole MCFC/GT/ST system is a triplex composite power generation system or a combined cycle power plant, which has an increasingly important effect in power generation industry because of its advantages of low exhaust generation and high efficiency (more than 75%).

16.9
Major Factors Affecting the Performance and Lifetime of MCFCs

There are several factors which can affect the MCFC performance, including electrode–electrolyte matrix structures, electrolyte composition, electrolyte retention, reactant crossover, operation temperature and pressure, reactant gas compositions and utilizations, and fuel impurities. In the following sections, these factors and their effects on the MCFC performance are discussed in detail.

16.9.1
Electrode–Electrolyte Matrix Structure Effect

The electrode structure, including the pore diameter, porosity, thickness, and intensity, has a significant effect on the MCFC performance. In order to obtain the high performance, the parameters of the electrodes will be controlled as follows:

- **Anode thickness:** 0.5–0.8 mm; porosity, 55–70%; average pore diameter, 5 μm.
- **Cathode thickness:** 0.5–0.8 mm; porosity, 60–80%; average pore diameter, 7 μm.

Besides the effect of the electrode structure, the electrolyte matrix structure also has a significant effect on the MCFC performance. For example, the ohmic polarization due to the ohmic resistance of the electrolyte matrix can contribute to two-thirds of the cell resistance [63]. To improve this, reducing the thickness of the electrolyte matrix seems to be an effective way. However, some factors, such as the strength and toughness of the electrolyte matrix, cell thermal cycling life, and MCFC lifetime, should be considered if one wants to maintain a consistent matrix quality (matrix yield >95% is needed). In particular, the electrolyte matrix thickness is directly related to the lifetime-limiting aspect of internal shorting by nickel precipitation [64]. If the matrix is too thin, internal shorting will occur earlier, leading to the failure of the MCFC. Another area for electrolyte matrix structure improvement is the ability of the matrix to prevent gas crossover from one electrode to the other. Energy Research Corporation in USA (ERC) has produced an improved matrix fabrication process providing low-temperature binder burnout. The process has resulted in frequently achieving a 1% allowable gas leakage, which is well below the goal of 2% [65].

16.9.2
Electrolyte Composition Effect

The MCFC performance and lifetime depend not only on the electrolyte content but also on the electrolyte composition. Traditionally, the lithium-rich 62 mol% Li_2CO_3 – 38 mol% K_2CO_3 eutectic mixture has been used as the standard electrolyte in MCFC development. Normally, the anode polarization varies only slightly in the composition range from 40 to 70% of lithium, whereas cathode polarization resistance decreases with increasing lithium content. Currently, the choice of electrolyte composition is Li/K for the atmospheric system and Li/Na for the pressurized system due to their higher melt basicity for mitigating NiO cathode dissolution [66–69]. Other benefits of Li/Na carbonate electrolyte, including lower evaporation loss [70–72] and higher ionic conductivity, can also be observed. However, its lower wettability and larger temperature sensitivity [68] should be considered, and one needs to adjust the electrode microstructure to achieve the desired cell performance [73].

16.9.3
Electrolyte Retention Effect

The electrolyte matrix in an MCFC is composed of porous materials. Each pore is impregnated with molten carbonate as the electrolyte. For stable operation of the MCFC, sufficient active area must be available for electrochemical reactions at both the anode and the cathode. This requires the presence of electrolyte to form TPB for fuel cell reactions. This stable TPB is strongly dependent on the balance of the capillary forces of porous electrodes. When the electrolyte fill level in the electrodes is low, sufficient active area for electrochemical reactions cannot be formed. On the other hand, if the electrolyte fill level is high, the rapid diffusion of reactant gases to the reaction sites is disturbed. Enough electrolyte must also be available to allow ionic conduction and to prevent gas crossover through the matrix. If the electrolyte level is insufficient, the cell resistance will become larger, leading to a lower cell performance. Control over the optimum distribution of molten carbonate electrolyte in the different cell components is also critical for achieving high performance and endurance in MCFCs. Various process (i.e., consumption by corrosion reactions, potential-driven migration, creepage of salt, and salt vaporization) occur [74, 75], all of which contributed to the redistribution of molten carbonate in MCFCs.

In addition, the cell performance suffers because of leakage of the electrolyte from the cell. There is a tendency for the electrolyte to migrate from the positive to the negative end of the stack. The leakage is through the gasket used to couple the external manifolds to the cell stack. The baseline gasket material presently used is of high porosity and provides a ready circuit for the electrolyte transfer. A new gasket design with a material having lower porosity plus end cell inventory capability offers the potential for reaching 40 000 h, if only this mode of failure is considered [76].

16.9.4
Reactant Crossover Effect

Gas crossover through the electrolyte matrix can cause a concentration decrease in the reactant gases involving the electrochemical reactions. For a fuel crossover, the crossed fuel to the cathode will directly react with O_2 to cause a drop in cathode potential, leading to a lower performance. In addition, the crossed fuel will be a waste in terms of power generation, leading to lower fuel efficiency. Therefore, extra effort has to be put on the electrolyte matrix structure to reduce reactant crossover [65]. It seems that optimizing the pore diameter and porosity of the electrolyte matrix is an effective way to gas-crossover reduction.

16.9.5
Temperature Effect

The impact of temperature on the MCFC thermodynamics can be reflected by the Nernst equation [77]:

$$E = E_0 + \frac{RT}{2F} \ln \frac{P_{H_2} \times P_{O_2}^{0.5} \times P_{CO_2,c}}{P_{H_2O} \times P_{CO_2,a}} \qquad (16.7)$$

where E is the reversible cell potential; E_0, the cell potential at standard conditions (1 atm, 25 °C); P, the partial pressure of the ith gas species (Pa); R, the universal gas constant (8.314 J mol^{-1} K^{-1}); T, the temperature (K); F, Faraday's constant (96 487 C mol^{-1}); and the subscripts a and c denote the anode and cathode, respectively. We can find that the influence of temperature on the reversible cell potential of the MCFC depends on several factors, one of which involves the equilibrium compositions of the fuel gas. The equilibrium composition of the fuel gas is dependent on the water/gas shift reaction expressed by Eq. (16.8), which achieves rapid equilibrium at the anode, and

$$H_2 + CO_2 \rightarrow CO + H_2O \qquad (16.8)$$

consequently CO serves as an indirect source of H_2. The reaction equilibrium constant (K),

$$K = \frac{P_{CO} \times P_{H_2O}}{P_{H_2} \times P_{CO_2}} \qquad (16.9)$$

increases with temperature [78–80]. According to this equation, the term P_{H_2O}/P_{H_2} can be expressed as $P_{H_2O}/P_{H_2} = K(P_{CO_2}/P_{CO})$. Insetting this into the Nernst equation (Eq. (16.7)), we can get an alternative Nernst equation:

$$E = E_0 + \frac{RT}{2F} \ln \frac{P_{CO} \times P_{O_2}^{0.5} \times P_{CO_2,c}}{K P_{CO_2} \times P_{CO_2,a}} \tag{16.10}$$

From this Nernst equation, it can be seen that since the equilibrium constant K increases with temperature, both the partial pressures of CO and H_2O will increase, resulting in the variation of equilibrium composition of the fuel gas, and then the reversible cell potential expressed by Eq. (16.10) will decrease with temperature.

In terms of the kinetics of the electrochemical reactions, increasing temperature can effectively reduce the electrode reaction overpotential, in particular, the cathode overpotential, leading to a high fuel cell performance. Furthermore, when the temperature is increased, the conductivity of the molten carbonate will be increased. Thus, ohmic resistance decreases, leading to an increase in the cell voltage. The relation between the cell performance ($\Delta V_T(\text{mV})$) and temperature are expressed by Baker et al. [81] and Stauffer et al. [82] as

$$\Delta V_T(\text{mV}) = 2.16\,(T_2 - T_1) \quad 575\,°C \leq T < 600\,°C \tag{16.11}$$

$$\Delta V_T(\text{mV}) = 1.40\,(T_2 - T_1) \quad 600\,°C \leq T < 650\,°C \tag{16.12}$$

$$\Delta V_T(\text{mV}) = 0.25\,(T_2 - T_1) \quad 650\,°C < T \leq 700\,°C \tag{16.13}$$

According to these equations, when the temperature is increased, the performance of MCFCs will be improved. However, the vaporization of carbonate and corruption of materials are also increased. Thus the trade-off operating temperature for MCFCs is about 650 °C [79], which is the best compromise between performance and endurance.

16.9.6
Pressure Effect

The MCFC operating pressure has an important influence on the performance in terms of cost, reliability, and lifetime of the system. Therefore, the operating pressure is an important system optimization parameter. According to the Nernst equation (Eq. (16.7)), the reversible cell potential of MCFCs depends on the operating pressure. When the pressure of the gas is changed from P_1 to P_2, the variation of reversible cell potential is

$$\Delta V_p = \frac{RT}{2F} \left(\ln \frac{P_{1,a}}{P_{2,a}} + \ln \frac{P_{2,c}^{3/2}}{P_{1,c}^{3/2}} \right) \tag{16.14}$$

where the subscripts a and c denote the anode and cathode, respectively. It can be seen that increasing pressure of MCFCs can result in higher equilibrium cell potential. It was reported that 1.6–1.8 times cell voltage gain could be achieved when the operating pressure of the MCFCs was increased from 2 to15 atm [83, 84]. The other benefits of increasing pressure include (i) reducing piping and heat exchange equipment sizes and (ii) having potential for additional power by expansion of the pressurized gas stream. It has been recognized that the exhaust stream from the pressurized system is a potential source for additional power because the exhaust stream contains sufficient energy for the turbomachinery to supply the compressed air and, at the same time, generate additional power.

Unfortunately, pressurized operation of the MCFCs will have negative impact on the cell lifetime and performance. For example, the cathode dissolution is related to the acidity of the cell, which will be increased with the increasing partial pressure of CO_2. This cathode dissolution will lead to continued deposition of sufficient bridged grains of metallic nickel across the cell, leading to a short circuit in the cell. Several other impacts on the MCFC stack and operation include (i) causing damage to the MCFC seals, (ii) increasing the likelihood of carbon deposition, (iii) negatively affecting methane reformation inside the MCFC anode, (iv) increasing the level of contaminants, and (v) increasing the parasitic loss of the system because of the high power needed to drive the air pumps. However, FCE has conducted investigations to determine the optimum operating pressure for the company's IR-MCFC power plants for coal gas [85]. Their results indicated that atmospheric operation is preferred since the reforming reaction is accelerated at a lower pressure. Therefore, not all MCFCs are operated at pressurized condition. Taking cost and reliability into account, it is desirable that the small-size power plants should be simpler in design than large-size power plants. Therefore, atmospheric operation is preferred for small MCFC power plants in some applications such as distributed power plants. For large power plants, pressurized operation may be more desirable taking size and efficiency into consideration.

16.9.7
Effects of Reactant Gas Composition and Utilization

As shown in Eq. (16.7), MCFC cell voltage changes with the reactant gas composition. The cell voltage will be decreased because the partial pressures of the reactant gas are reduced with the consumption of the reactant gases. However, the partial pressure of the reactant gas is not easily analyzed due to electrochemical oxidation reaction of H_2 at the anode and electrochemical reduction reaction of CO_2 and O_2 at the cathode. Generally speaking, the increase in reactant gas utilization will decrease the cell performance.

16.9.7.1 Oxidant Composition and Utilization
Figure 16.19 shows the dependence of the average cell voltage of a stack (10 cells) at $172\,mA\,cm^{-2}$ on the oxidant utilization [76]. It can be seen that the average cell voltage is decreased by about $30\,mV$ when the oxidant utilization is increased from 20 to 50%. The cell voltage loss due to the variation of the oxidant utilization is described as follows:

$$\Delta V_c(mV) = 250 \log \frac{\left(P_{CO_2} \times P_{O_2}^{1/2}\right)_2}{\left(P_{CO_2} \times P_{O_2}^{1/2}\right)_1} \quad 0.04 \leq (P_{CO_2} P_{O_2}^{1/2}) \leq 0.11 \quad (16.15)$$

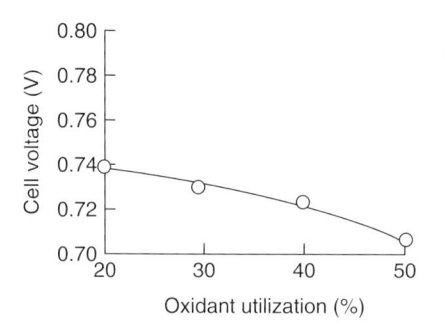

Figure 16.19 Dependence of cell voltage on oxidant utilization [76].

$$\Delta V_c (\text{mV}) = 99 \log \frac{\left(P_{CO_2} \times P_{O_2}^{1/2} \right)_2}{\left(P_{CO_2} \times P_{O_2}^{1/2} \right)_1} \quad 0.11 \leq (P_{CO_2} P_{O_2}^{1/2}) \leq 0.38 \quad (16.16)$$

where P_{CO_2} and P_{O_2} are the partial pressure of CO_2 and O_2 in the system, respectively.

16.9.7.2 Fuel Composition and Utilization

Figure 16.20 shows the dependence of the cell voltage on the fuel utilization [86]. It can be seen that the cell voltage is decreased by about 30 mV when the fuel utilization is increased from 30 to 60%. The cell voltage loss due to the variation of

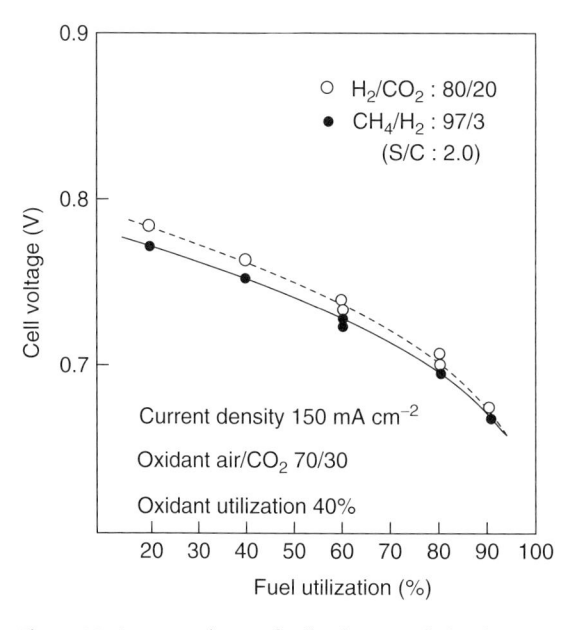

Figure 16.20 Dependence of cell voltage on fuel utilization [86].

the fuel utilization is described as follows:

$$\Delta V_a(\text{mV}) = 173 \log \frac{\left(\frac{P_{H_2}}{P_{CO_2} P_{H_2O}}\right)_2}{\left(\frac{P_{H_2}}{P_{CO_2} P_{H_2O}}\right)_1} \tag{16.17}$$

where P_{H_2}, P_{CO_2}, and P_{H_2O} are the partial pressure of H_2, CO_2, and H_2O in the system, respectively.

According to these results, we can find that the MCFCs should operate at a lower reactant gas utilization to increase the cell voltage. However, it will result in another problem of low fuel utilization. The trade-off between the cell voltage and fuel utilization is to operate an MCFC at a fuel utilization of about 75–85% and an oxidant utilization of about 50% [79].

16.9.8
Effect of Operating Current Density

In an MCFC operation, the cell ohmic polarization and electrode overpotential both increase with increasing the cell load or current density, leading to a drop in cell voltage. For an MCFC, the experimental expressions for current density effect on the cell voltage could be expressed as the following equations [79, 87, 88]:

$$\Delta V_J(\text{mV}) = -1.21 \Delta J \quad 50 \leq J \leq 150 \tag{16.18}$$

$$\Delta V_J(\text{mV}) = -1.76 \Delta J \quad 150 < J \leq 200 \tag{16.19}$$

where V and J denote the cell voltage (V) and current density (mA cm^{-2}), respectively.

16.9.9
Effect of Gas Impurities

It is well known that coal is an abundant and relatively cheap energy source in China, America, Russia, and many other countries. This solid fuel can be converted into gas by a gasification process. Unfortunately, the gas obtained contains several impurities such as sulfur-containing compounds, tars, particulates, and trace other contaminants. These impurities have to be removed before it can be used as the feeding fuel for MCFCs. These contaminants usually include sulfide, halide, nitride, trace metals, and hydrocarbon. The effects of impurities studied for their various interactions and reactivities in the MCFC components are summarized in Table 16.5 [60, 61]. The information in this table may provide a guideline for identifying operating conditions and fuel cleanup requirements for the future MCFC integration with coal gasification. The effects of these impurities are discussed in detail in the following sections.

Table 16.5 Potential effects of impurities in coal gas, which were studied for their various interactions and reactivities in the MCFC components [60, 61].

Class	Contaminants	Potential effect
Particulates	Coal powder, ash	Plugging of gas passages
Sulfides	H_2S, COS, CS_2, C_4H_4S	Voltage loss reaction with electrolyte through SO_2
Halides	HCl, HI, HF, HBr, $SnCl_2$	Corrosion reaction with electrolyte
Nitride	NH_3, HCN, N_2	Reaction with the electrolyte through NO_x
Trace metals	AsPb, Hg, Cd, Sn, Zn, H_2Se, H_2Te, AsH_3	Deposition on electrode, reaction with electrolyte
Hydrocarbons	C_6H_6, $C_{10}H_8$, C_4H_{10}	Carbon deposition

16.9.9.1 Sulfides

Most raw hydrocarbon fuels that can be used in MCFC systems contain impurities (e.g., sulfur compounds) that are harmful to both the MCFC anode and the reforming catalyst. It is well established that even low ppm concentrations of sulfur compounds in fuel gases are detrimental to MCFCs [89]. The tolerance level of the MCFCs to sulfur compounds depends strongly on the temperature, pressure, gas composition, cell components, and the mode of system operation (e.g., recycling, venting, and gas cleanup) [90]. The main sulfur-containing compound that causes adverse effects on cell performance is H_2S. For example, in MCFCs, H_2S can react electrochemically with nickel-based anode and form nickel sulfides by oxidation of sulfide ions, as shown in Eqs. (16.20) and (16.21), which then block active electrochemical reaction sites, leading to a significant performance loss or even failure of the MCFCs [91]. The acceptable H_2S level is approximately a lower level, below 1 ppm, that is required in the case of a higher amount of CO-containing fuel. SO_2 in the oxidant shows similar effects to H_2S; however, it corrodes the metal hardware used as the oxidant inlet [92], and the SO_2 content of the oxidant in the cathode should be reduced to below 1 ppm level. On teh basis of the present understanding of the effect of sulfur on MCFCs, and with the available cell components, it is projected that long-term operation (40 000 h) of MCFCs may require fuel gases with sulfur level in the order of ≤ 0.01 ppm, unless the system is purged of sulfur at periodic intervals or sulfur is scrubbed from the cell burner loop [93]. The tolerance limit of H_2S content in the fuel gas should be approximately 0.5 ppm in the latter case [61, 94, 95].

$$H_2S + CO_3^{2-} \rightarrow H_2O + CO_2 + S^{2-} \tag{16.20}$$

$$Ni + xS^{2-} \rightarrow NiS_x + 2xe^- \tag{16.21}$$

16.9.9.2 Halides

Besides sulfur-containing compounds, coal gas also contains halides such as HCl and HF. These halides can react with carbonate ion to form chlorine ion (fluoride ion) [96], which accumulates in the electrolyte because of the low vaporization of LiCl and KCl (LiF and KF), causing cell performance drop, in particular, at a higher LiF concentration. The internal resistances of the cell with LiF are also higher than those of the cell without LiF. Therefore, the corrosion and the electrolyte loss caused by chlorine ion (also fluoride ion) are the main causes for the cell voltage drop. Normally, when the operating time is long and the concentration of HF in fuel gas is high, the corrosion, the carbonate loss, and the decrease of the electrolyte volume could become more severe [97]. It has been suggested that the level of HCl should be kept below 1 ppm in the fuel gas [98], and the acceptable concentration of HF in the fuel gas is below 0.1 ppm to operate the MCFCs for 40 000 h [61, 95].

16.9.9.3 Nitride

In coal gas, NH_3 is also one of the contaminants. The amount of NH_3 produced is influenced by the ingredients of coal and the gasification process employed. Generally, the MCFC power plant is equipped with a CO_2 recycling system, which completely converts the CO discharged from the anode into CO_2 during a catalytic burning process and supplies the cathode with CO_2. During the catalytic burning process, if NH_3 exists in the gas stream, NO_x will be generated from NH_3. Although bench-scale cell tests showed that traces of NH_3 and HCN had no significant effect on the cell performance [61, 99], NO_x does reduce the cell performance during the earlier operating stages. Good thing is that the effect of NO_x can be reduced gradually by prolonging the fuel cell testing. It was believed that NO_x can react with the carbonate and then dissolve in the electrolyte to form NO_2^- and NO_3^-. These ions then react with hydrogen in the fuel gas to produce N_2 and a small amount of NH_3. Therefore, NO_2^- and NO_3^- are not accumulated in the electrolyte, and the effect of NO_x on the cell performance is insignificant [92, 100]. The projection for the NH_3 tolerance level of MCFCs was 0.1 ppm [101], but the literature [61, 95] indicates that the level could be increased to 1 vol%.

16.9.9.4 Particulates

In coal gas, there are some particulates as well. These particulates can adsorb on the surfaces of porous cell components and then block the gas channels of the anode, cathode, and electrolyte. In serious cases, the MCFC system will cease to operate. The MCFC field test has indicated that the tolerance limit of the MCFCs to particulates larger than 3 μm diameter is only 0.1 g l^{-1} level [94].

16.9.9.5 Other Compounds

Experimental studies indicate that 1 ppm As from gaseous AsH_3 in fuel gas does not affect cell performance, but when the level is increased to 9 ppm As, the cell voltage drops rapidly by about 120 mV at 160 mA cm^{-2} [102]. Trace metals, such as Pb, Cd, Hg, and Sn, in the fuel gas are of concern because they can be deposited on the electrode surface or can react with the electrolyte [103].

16.9.10
Consideration of MCFC Design

According to the discussion and analysis in the previous sections, the considerations for MCFC design can be summarized as follows:

- **Temperature consideration.** As we discussed, the operating temperature has a significant impact on the MCFC performance. According to the principle of chemical reaction kinetics, the chemical reaction rate will increase exponentially with temperature. The MCFC voltage will increase with increasing temperature because the cathode overpotential is reduced significantly by the increasing temperature. However, the thermodynamic cell voltage will be decreased with increasing temperature. From the trade-off among the thermodynamics, kinetics, system durability, and optimization, the ideal operating temperature of an MCFC has been recognized as 650 °C [79].
- **Pressure consideration.** For practical use of MCFC power plants, it is necessary to improve the lifetime and the performance of the cell. The operation of MCFC under pressurized conditions may be the most effective way to achieve higher energy conversion efficiency than when operated under an atmospheric pressure. It is believed that pressurized operation can not only give a potential gain but also increase gas solubility, which is expected to decrease the polarization resistance and the diffusion resistance of the cell as well. However, the operating pressure can cause NiO dissolution in the electrolyte, and then the dissolved Ni will precipitate in the matrix, causing cell shortage. This dissolved Ni can also affect the methane-reforming reaction rate. Therefore, there is a trade-off between the pressure and cell performance including durability. This should be considered when carrying out MCFC design according to the cell materials and operation conditions.
- **Gas composition and utilization consideration.** As mentioned before, at 650 °C and under rather rich gas composition ($CO_2 : O_2 = 67 : 33$), the electrochemical performance of the cathode is optimal. When the ratio of $(H_2)/(H_2O)(CO_2)$ increases, the cell voltage will increase accordingly. Therefore, a high concentration of H_2 is preferred in an MCFC operation, suggesting that the fuel utilization may be lower. In practice, the optimal operating condition required for fuel utilization is about 75–85% and oxidant utilization of about 50%.
- **Gas contaminant consideration.** Contaminants in the fuel, such as sulfur- or chlorine-containing compounds can depress the fuel cell performance. Even now, it is still very difficult to determine the maximum allowable contamination level in experiments. However, it is to be expected that appropriate gas cleanup will be necessary in field tests to keep the level of contaminants sufficiently low.
- **Consideration on electrolyte composition and electrolyte matrix structure.** At the current technologic status, the lithium-rich 62 mol% Li_2CO_3 – 38 mol% K_2CO_3 eutectic mixture has been used as the standard electrolyte in MCFC development. It has been known that the anode polarization varies only slightly in the composition range of 40–70% lithium, whereas cathode polarization decreases with increasing lithium content. The current choice of electrolyte composition

is Li/K for atmospheric system and Li/Na for pressurized system because of its higher melt basicity to mitigate NiO cathode dissolution. Regarding the electrolyte matrix structure, its thickness seems to play a major role in MCFC performance. For example, the electrolyte matrix thickness is directly related to the lifetime-limiting aspect of internal shorting by nickel precipitation. The thicker the matrix, the later the shorting will occur. The thinner the matrix, the smaller the ohmic resistance, correspondingly, the better the MCFC performance. Therefore, we should select an adequate electrolyte matrix structure.

- **Electrolyte retention consideration.** The electrolyte distribution and retention in each component must be optimized to enhance the performance and extend the lifetime of MCFCs. The MCFC performance decay appears to be due to a combination of some processes such as the electrolyte depletion due to the corrosion reaction, the electrolyte migration due to the electromotive force, the anode creep, and the evaporation of electrolyte, which result in electrolyte redistribution in each component.
- **Current density consideration.** The current density is directly related to the output voltage of MCFCs. The ohmic, electrode, and concentration polarizations are all increased with increasing current density, resulting in a decrease in the output voltage of MCFCs. An optimal current density at which the MCFC gives the maximum power density should be selected for the performance.

16.10
Challenges and Perspectives of MCFCs

As discussed above, the MCFC technology has been rapidly progressing in the recent years. With the approaching commercialization, many problems remain to be thoroughly solved. The major challenges are high cost and insufficient durability. It is necessary to develop cost-effective and durable materials to overcome these challenges. Because MCFCs are operated at a high temperature and corrosive environment, besides novel material development, other technical challenges have also to be considered, such as optimizing the design of the system structures, system simplification, and optimizing operation conditions. Major efforts should be focused on the following areas.

16.10.1
High Cost of the MCFC Systems

It has been realized that the commercial introduction of the MCFC systems for cogeneration or electricity production can only be successful if the technology can compete successfully with existing conventional power-generating devices and other technologies under development. For economic reasons, the MCFC stack, which converts the fuel into electricity and heat, should have a low price. Reducing the MCFC stack price from \$3500–4000 per kW to \$1500–2000 per kW is required in terms of commercial competition.

16.10.2
Insufficient Durability of the MCFC Systems

For high-temperature MCFCs in combined heat and power applications, it is generally assumed that a stack lifetime of at least 40 000 h is required in order to obtain the commercial introduction of the MCFC systems into the market. It is necessary to further increase the performance and reliability of the MCFC system and extend the lifetime of the MCFC stack.

16.10.3
Source and Storage of the Fuels for MCFCs

The MCFC utilizes both H_2 and CO at the anode. For the large-size MCFC power-generating systems, the source of the fuel is mainly natural gas, coal gas, methanol, propane, and ethanol. The commonly available hydrocarbon fuels need to be converted to the fuel cell usable form. Therefore, it is necessary to study the novel fuel conversion and storage technology and coal gasification technology to establish the MCFC power plant.

16.10.4
Low Volume Power Density of the MCFC Systems

As the system volume is quite large, it leads to low volume power density in MCFCs. Reducing weight per kilowatt and compacting power generation plant through capacity enhancement and stack structural improvement are definitely necessary in order to enlarge the volume power density.

16.10.5
Research and Development Directions

For successful market entry and competitiveness of the MCFC systems, cost reduction and performance enhancement are the two major research directions. In order to achieve these, both experimental research and fundamental understanding are needed as listed below.

16.10.5.1 Cathode Material Development for Oxygen Reduction Reaction

The Ni shorting by NiO dissolution from Ni deposition on the NiO cathode is still a serious problem, especially for the MCFC combined cycle system with a GT operating under highly pressurized operating conditions. The most effective procedure against Ni shorting is to reduce the solubility of NiO in the molten carbonate. As an advanced additive, the rare earth metal oxides have been studied for both decreasing NiO dissolution and enhancing the kinetics of oxygen reduction reaction. Several research directions may be proposed, including (i) studying the microkinetic behavior of cathodic oxygen reduction reaction, (ii) clarifying the oxygen reduction mechanism in carbonate electrolyte, (iii) developing new cathode materials, and (iv) decreasing cathode dissolution and polarization.

16.10.5.2 Anode Material Development for Fuel Oxidation Reactions

The anode material development is another important aspect in improving both cost reduction and performance enhancement. The research directions would be (i) exploring new and cost-effective materials to increase the electrochemical catalytic activity of fuels, (ii) improving the wettability of the anode material in carbonate electrolyte, (iii) enhancing the creep resistance and mechanical strength of the anode material, and (iv) increasing tolerance to sulfur and other contaminants contained in the coal syngas and to carbon formation by exposure to hydrocarbon fuels.

16.10.5.3 Developing New Electrolyte Composition and the Additives

Electrolyte in an MCFC is another key component determining the cost and performance. Considering the factors such as conductivity, vapor pressure, and corrosivity of the electrolyte, the following research directions may be proposed: (i) selecting the optimum electrolyte composition, (ii) reducing the solubility of NiO by controlling the basicity of the electrolyte with alkaline earth metal carbonate additive or rare earth metal oxides additive, and (iii) optimizing the contacts among the electrolyte, anode, and cathode materials.

16.10.5.4 Improving Corrosion Resistance of MCFCs

In an MCFC, loss of electrolyte in the stack occurs because of the reaction of the electrolyte with the stack components, leading to lithiation of electrodes and the Al_2O_3 fibers in the matrix, as well as the corrosion of hardware. Furthermore, continuous loss of electrolyte, predominantly K_2CO_3, can be caused by evaporation of electrolyte from the stack. This process primarily depends on several factors such as partial pressures of H_2O and CO_2, volume gas flow, temperature, and the electrolyte composition. With respect to the durability of an MCFC, a limiting factor for reaching the goal of 40 000 h of lifetime is the corrosion of the metallic parts of MCFCs, especially the current collectors and separator plates. Generally, this corrosion leads to metal loss and dramatically increases the electrical resistance because of the formation of resistive oxides. Several research directions in overcoming the corrosion issue may be proposed, including (i) reducing bypass current and electrolyte migration induced by the wet seal, (ii) reducing the impact of impurity on the MCFC performance and lifetime, (iii) improving the long lifetime and thermal cycling stability of the electrolyte matrix, and (iv) optimizing the performance when enlarging the MCFC stack size and height.

16.10.5.5 Modeling, Failure Checking, and Automatic Control of MCFC Power Plants

For fundamental understanding and speeding up the optimization process of the MCFC technology, three aspects of approach may be helpful: (i) modeling the internal temperature distribution of the MCFC stack, (ii) carrying out failure diagnosis and mitigation development, and (iii) developing automatic control of the MCFC power plant.

References

1. Selman, J.R. (1993) in Fuel cell systems, *Research, Development and Demonstration of Molten Carbonate Fuel Cell Systems* (eds L.J.M.J. Bolmen and M.N. Muyerwa), Plenum Press, New York, p. 384.
2. Joon, K. (1996) Critical issues and future prospects for molten carbonate fuel cells. *J. Power Sources*, **61** (1–2), 129–133.
3. Farooque, M., Kush, A., Leo, A., Maru, H., and Skok, A. (1998) Fuel Cell Seminar Program and Abstracts, p. 13.
4. Williams, M.C. and Maru, H.C. (2006) Distributed generation – Molten carbonate fuel cells. *J. Power Sources*, **160** (2), 863–867.
5. Izaki, Y. and Yasue, H. (2000) Demonstration of the first 1000 kW MCFC power plant in Japan. 2000 Fuel Cell Seminar Abstracts, Portland, OR, October 30-November 2, pp. 460–463.
6. Shinohara, M., Ohtsuki, J., Nishimura, M., Hashino, K., Honda, H., Fujitsuka, M., Okada, T., and Shinoki, T. (2000) Verification test of 200 kW class AIR-MCFC stack. 2000 Fuel Cell Seminar Abstracts, Portland, OR, October 30-November 2, pp. 464–467.
7. Bischoff, M. and Kraus, P. (1999) The 3rd International Fuel Cell Conference, p. 321.
8. Bischoff, M. (2006) Molten carbonate fuel cells: a high temperature fuel cell on the edge to commercialization. *J. Power Sources*, **160** (2), 842–845.
9. Parodi, F., Alvarez, T., Bosio, B., Passalacqua, B., Simon, J., and Zappaterra, M. (1999) The 3rd International Fuel Cell Conference, p. 263.
10. Bergaglio, E., Sabattini, A., and Capobianco, P. (2005) Research and development on porous components for MCFC applications. *J. Power Sources*, **149**, 63–65.
11. Kim, Y.-S., Lee, K.-Y., and Chun, H.S. (2000) Creep behavior of a porous Ni/Ni$_3$Al anode for molten carbonate fuel cell. Presented at the "2000 Fuel Cell Seminar", Portland, OR, October 30-November 2.
12. Baizeng, F., Xinyu, L., Xindong, W., and Shuzhen, D. (1998) The mechanism of surface modification of a MCFC anode. *J. Electrochem. Soc.*, **441** (1–2), 65–68.
13. Appleby, A.J. and Foulkes, F.R. (1992) *Fuel Cell Handbook*, 6th edn, Krieger, New York.
14. Pierce, R.D., Smith, J.L., and Kucera, G.H. (1987) Investigation of alternative MCFC cathode materials at argonne national laboratory. *Prog. Batteries Sol. Cells*, **6**, 159.
15. Ito, Y., Tsuru, K., Oishi, J., Miyazaki, Y., and Teruo, K. (1988) Dissolution behaviour of copper and nickel oxides in molten Li$_2$CO$_3$/Na$_2$CO$_3$/K$_2$CO$_3$. *J. Power Sources*, **23** (4), 357–364.
16. Ganesan, P., Colon, H., Haran, B., White, R., and Popov, B.N. (2002) Study of cobalt-doped lithium– nickel oxides as cathodes for MCFC. *J. Power Sources*, **111** (1), 109–120.
17. Smith, J.L., Kucera, G.H., and Brown, A.P. (1990) *Proceedings of the Softbound Series of Molten Carbonate Fuel Cell Technology*, The Electrochemical Society, Pennington, NJ, PV90-16, p. 226.
18. Huang, B., Yu, Q.-C., Wang, H.-M., Chen, G., and Hu, K.-A. (2004) Study of LiFeO$_2$ coated NiO as cathodes for MCFC by electrochemical impedance spectroscopy. *J. Power Sources*, **137** (2), 163–174.
19. Mitsushima, S., Matsuzawa, K., Kamiya, N., and Ota, K.-I. (2002) Improvement of MCFC cathode stability by additives. *Electrochim. Acta*, **47** (22–23), 3823–3830.
20. Motohira, N., Sensou, T., Yamauchi, K., Ogawa, K., Liu, X., Kamiya, N., and Ota, K. (1999) Effect of Mg on the solubility of NiO in molten carbonate. *J. Mol. Liq.*, **83** (1–3), 95–103.
21. Escudero, M.J., Nóvoa, X.R., Rodrigo, T., and Daza, L. (2002) Influence of lanthanum oxide as quality promoter on cathodes for MCFC. *J. Power Sources*, **106** (1–2), 196–205.
22. Daza, L., Rangel, C.M., Baranda, J., Casais, M.T., Martńez, M.J., and

Alonso, J.A. (2000) Modified nickel oxides as cathode materials for MCFC. *J. Power Sources*, **86** (1–2), 329–333.

23. Huang, B., Chen, G., Li, F., Yu, Q.-C., and Hu, K.-A. (2004) Study of NiO cathode modified by rare earth oxide additive for MCFC by electrochemical impedance spectroscopy. *Electrochim. Acta*, **49** (28), 5055–5068.
24. Huang, B., Li, F., Yu, Q.-C., Chen, G., Zhao, B.-Y., and Hu, K.-A. (2004) Study of NiO cathode modified by ZnO additive for MCFC. *J. Power Sources*, **128** (2), 135–144.
25. Yoshikawa, M., Mugikura, Y., and Watanabe, T. (2001) 8th FCDIC Fuel Cell Symposium Proceedings, Tokyo, May 15–16, B3-7.
26. Mizukami, T., Kahara, T., Takahashi, S., Hiyama, K., and Takahashi, K. (2000) Development of 250 kW MCFC stacks for Kawagoe and long life cell technologies at HITACHI. Presented at the "2000 Fuel Cell Seminar", Portland, OR, October 30-November 2.
27. Webb, A.N., Bmather, W., and Suggitt, R.M. (1965) Studies of the molten carbonates electrolyte fuel cell. *J. Electrochem. Soc.*, **112** (11), 1059–1063.
28. Finn, P.A. (1980) The Effects of different environments on they thermal stability of powdered samples of $LiAlO_2$. *J. Electrochem. Soc.*, **127** (1), 236–238.
29. General Electric Company (1985) Development of the Molten Carbonate Fuel Cell Power Plant, Final Report under Contracts DOE-ET-17-019-20 and GE/FC-17019-85-1, Vol. 1, March, pp.175–185.
30. Kinoshita, K., Sim, J.W., and Ackerman, J.P. (1978) Preparation and characterization of lithium aluminate. *Mater. Res. Bull.*, **13** (5), 445.
31. Tanimoto, K., Yanagida, M., Kojima, T., Tamiya, Y., and Miyazaki, Y. (1996) The 2nd International Fuel Cell Conference, p. 207.
32. Tomimatsu, N., Ohzu, H., Akasaka, Y., and Nakagawa, K. (1997) Phase stability of $LiAlO_2$ in molten carbonate. *J. Electrochem. Soc.*, **144** (12), 4182–4186.
33. Terada, S., Nagashima, I., Higaki, K., and Ito, Y. (1998) Stability of $LiAlO_2$ as electrolyte matrix for molten carbonate fuel cells. *J. Power Sources*, **75** (2), 223–229.
34. Terada, S., Higaki, K., Nagashima, I., and Ito, Y. (1999) Stability and solubility of electrolyte matrix support material for molten carbonate fuel cells. *J. Power Sources*, **83** (1–2), 227–230.
35. Plucknett, K.P., Caceres, C.H., and Wilkinson, D.S. (1994) Tape casting of fine alumina/zirconia powders for composite fabrication. *J. Am. Ceram. Soc.*, **77** (8), 2137–2142.
36. Bosio, B., Costamagna, P., Parodi, F., and Passalacqua, B. (1998) Industrial experience on the development of the molten Carbonate Fuel Cell Technology. *J. Power Sources*, **74** (2), 175–187.
37. Kim, J.-E., Patil, K.Y., Han, J., Yoon, S.-P., Nam, S.-W., Lim, T.-H., Hong, S.-A., Kim, H., and Lim, H.-C. (2009) Using aluminum and Li_2CO_3 particles to reinforce the a-$LiAlO_2$ matrix for molten carbonate fuel cells. *Int. J. Hydrogen Energy*, **34** (22), 9227–9232.
38. Tomczyk, P. (2006) MCFC versus other fuel cells – characteristics, technologies and prospects. *J. Power Sources*, **160** (2), 858–862.
39. Maru, H.C. and Marianowski, L.G. (1976) Extended abstracts. Abstract #31, Fall Meeting of the Electrochemical Society, Las Vegas, NV October 17–22, 1976, p. 82.
40. Mitteldorf, J. and Wilemski, G. (1984) Film thickness and distribution of electrolyte in porous fuel cell components. *J. Electrochem. Soc.*, **131** (8), 1784–1788.
41. Biedenkopf, P., Bischoff, M.M., and Wochner, T. (2000) Corrosion phenomena of alloys and electrode materials in Molten Carbonate Fuel Cells. *Mater. Corros.*, **51** (5), 287–302.
42. Yuh, C., Johnsen, R., Farooque, M., and Maru, H. (1995) Status of carbonate fuel cell materials. *J. Power Sources*, **56** (1), 1–10.
43. Aguero, A., Garca de Blas, F.J., Garca, M.C., Muelas, R., and Roman, A. (2001) Thermal spray coatings for molten carbonate fuel cells separator

plates. *Surf. Coat. Technol.*, **146–147**, 578–585.
44. Perez, F.J., Duday, D., Hierro, M.P., Gomez, C., Aguero, A., Garca, M.C., Muela, R., Sanchez Pascual, A., and Martinez, L. (2002) Hot corrosion study of coated separator plates of molten carbonate fuel cells by slurry aluminides. *Surf. Coat. Technol.*, **161** (2–3), 293–301.
45. Frangini, S. and Masci, A. (2004) Intermetallic FeAl based coatings deposited by the electrospark technique: corrosion behavior in molten (Li + K) carbonate. *Surf. Coat. Technol.*, **184** (1), 31–39.
46. Fujimoto, N., Yamamoto, M., and Nagoya, T. (1998) Estimation of the lifetime of Al/Ni-plated material for wet-seal area in molten carbonate fuel cells. *J. Power Sources*, **71** (1–2), 231–238.
47. Jun, J.H., Jun, J.H., and Kim, K.Y. (2002) Degradation behaviour of Al–Fe coatings in wet-seal area of molten carbonate fuel cells. *J. Power Sources*, **112** (1), 153–161.
48. Hwang, E.R. and Kang, S.G. (1998) A study of a corrosion-resistant coating for a separator for a molten carbonate fuel cell. *J. Power Sources*, **76** (1), 48–53.
49. Frangini, S. (1997) *Proceedings of the 4th Symposium on Carbonate Fuel Cell Technology*, Electrochemical Society, Pennington, NJ, PV 97–4, p. 306.
50. Songbo, X., Yongda, Z., Huang Xing, Z.B., and Yu, Z. (2002) Corrosion resistance of the intermetallic compound, NiAl, in a molten carbonate fuel cell environment. *J. Power Sources*, **103** (2), 230–236.
51. Lindbergh, G. and Zhu, B. (2001) Corrosion behaviour of high aluminium steels in molten carbonate in an anode gas environment. *Electrochim. Acta*, **46** (8), 1131–1140.
52. Barreras, F., Lozano, A., Valiño, L., Mustata, R., and Marín, C. (2008) Fluid dynamics performance of different bipolar plates. *J. Power Sources*, **175** (2), 841–850.
53. Yoshiba, F., Abe, T., and Watanabe, T. (2000) Numerical analysis of molten carbonate fuel cell stack performance: diagnosis of internal conditions using cell voltage profiles. *J. Power Sources*, **87** (1–2), 21–27.
54. Gonikberg, M.G. (1963) Fuel cell handbook, 6th edn, in *Chemical Equilibria and Reaction Rates at High Pressures* (ed. S. Monson, Translated from Russian by M. Artment), The National Science Foundation, Washington, DC, by The Israel Program for Scientific Translations, Jerusalem, Israel, p. 133.
55. Soler, J., González, T., Escudero, M.J., Rodrigo, T., and Daza, L. (2002) Endurance test on a single cell of a novel cathode material for MCFC. *J. Power Sources*, **106** (1–2), 189–195.
56. Bischoff, M. and Huppmann, G. (2002) Operating experience with a 250 kW_{el} molten carbonate fuel cell (MCFC) power plant. *J. Power Sources*, **105** (2), 216–221.
57. Zhou, L., Baolian Yi, H.L., Zhang, H., Shao, Z., Ming, P., and Cheng, M. (2006) A study on the start-up and performance of a kW-class molten carbonate fuel cell (MCFC) stack. *Electrochim. Acta*, **51** (26), 5698–5702.
58. Hishinuma, Y. and Kunikata, M. (1997) Molten carbonate fuel cell power generation systems. *Energy Convers. Mgmt.*, **38** (10–13), 1237–1247.
59. Benjamin, T., Rezniko, G., Donelson, R., and Burmeister, D. (1992) IMHEX[R] MCFC Stack Scale-up. Proceedings of the 27th Intersociety Energy Conversion Engineering Conference, Vol. 3, San Diego, CA, August 3–7, 1992, p. 290.
60. Pigeaud, A. and Wilemski, G. (1993) Effect of Coal-derived Trace Species on the Performance of Molten Carbonate Fuel Cells. ERC'S 1991 Topical Report and 1992 Final Report to DOE/METC under Contract No. DE-AC21-88MC25009, NTIS (DE9300-0203).
61. Pigeaud, A., Maru, H., Wilemski, G., and Helble, J. (1995) Trace Element Emissions. Final Report under DOE/NETL Contract No. DE-AC21-92MC29261, February 1995.
62. Ishikawa, T. and Yasue, H. (2000) Start-up, testing and operation of 1000

kW class MCFC power plant. *J. Power Sources*, **86** (1–2), 145–150.
63. Pigeaud, E. et al. (1989) Determination of Optimum Electrolyte Composition for Molten Carbonate Fuel Cell. Final Report to DOE, Contract No. DE-AC21-86MC23264, June 1989.
64. Watanabe, T., Yoshiba, F., Morita, H., Yoshikawa, M., and Mugikura, Y. (1999) Carbonate fuel cell technology, in *Proceedings of the 5th International Symposium: Carbonate Fuel Cell Technology* (eds I. Uchida, K. Hemmes, G. Lindbergh, D.A. Shores, and J.R. Selman), Electrochemical Society, Pennington, NJ, PV99-20, p. 178.
65. Farooque, M., ERC (1990) Development on Internal Reforming Carbonate Fuel Cell Technology, Final Report, prepared for U.S.DOE/METC, DOE/MC/23274-2941, October, 1990, pp. 3-6–3-11.
66. Baumgartner, C.E., Arendt, R.H., Ivacovangelo, C.D., and Karas, B.R. (1984) Molten carbonate fuel cell cathode materials study. *J. Electrochem. Soc.*, **131** (10), 2217.
67. Doyon, J.D., Gilbert, T., and Davies, G. (1987) NiO solubility in mixed alkali/alkaline earth carbonates. *J. Electrochem. Soc.*, **134** (12), 3035–3038.
68. Sishtla, C., Donado, R., Ong, E.T., and Remick, R. (1997) Carbonate fuel cell technology, in *Proceedings of the 4th International Symposium: Carbonate Fuel Cell Technology* (eds I. Uchida, K. Hemmes, G. Lindbergh, D.A. Shores, and J.R. Selman), Electrochemical Society, Pennington, NJ, PV97-4, p. 315.
69. Yoshikawa, M., Mugikura, Y., Watanabe, T., Ota, T., and Suzuki, A. (1999) The behavior of MCFCs using Li/K and Li/Na carbonates as the electrolyte at high pressure. *J. Electrochem. Soc.*, **146** (8), 2834–2840.
70. Makkus, R.C., Sitters, E.F., Nammensma, P., and Huijsmans, J.P.P. (1997) Carbonate fuel cell technology, in *Proceedings of the 4th International Symposium: Carbonate Fuel Cell Technology* (eds I. Uchida, K. Hemmes, G. Lindbergh, D.A. Shores, and J.R. Selman), Electrochemical Society, Pennington, NJ, PV97–4, p. 344.
71. Yuh, C.Y., Huang, C.M., and Farooque, M. (1997) Carbonate fuel cell technology, in *Proceedings of the 4th International Symposium: Carbonate Fuel Cell Technology* (eds I. Uchida, K. Hemmes, G. Lindbergh, D.A. Shores, and J.R. Selman), Electrochemical Society, Pennington, NJ, PV97-4, p. 66.
72. Yuh, C.Y., Farooque, M., and Maru, H. (1999) Carbonate fuel cell technology, in *Proceedings of the 5th International Symposium: Carbonate Fuel Cell Technology* (eds I. Uchida, K. Hemmes, G. Lindbergh, D.A. Shores, and J.R. Selman), Electrochemical Society, Pennington, NJ, PV99-20, p. 189.
73. Rietveld, B., Kraaij, G.J., and Makkus, R.C. (1999) Carbonate fuel cell technology, in *Proceedings of the 5th International Symposium: Carbonate Fuel Cell Technology* (eds I. Uchida, K. Hemmes, G. Lindbergh, D.A. Shores, and J.R. Selman), Electrochemical Society, Pennington, NJ, PV99-20, p. 114.
74. Maru, H.C., Pigeaud, A., Chamberlin, R., and Wilemski, G. (1986) Fuel Cell Handbook, 6th edn, in *Proceedings of the Symposium on Electrochemical Modeling of Battery, Fuel Cell, and Photoenergy Conversion Systems* (eds J.R. Selman and H.C. Maru), The Electrochemical Society, Inc., Pennington, NJ, p. 398.
75. Kunz, H.R. (1987) Transport of electrolyte in molten carbonate fuel cells. *J. Electrochem. Soc.*, **134** (1), 105–113.
76. Farooque, M., ERC (1990) Development on internal reforming carbonate fuel cell technology. Final Report, prepared for U.S.DOE/METC, DOE/MC/23274-2941, October, 1990, pp. 4-19–4-29.
77. Ken-Ichiro, O. (1997) Materials durabilities for MCFC. Proceedings of the Electrochemical Society, 97-4 (Carbonate Fuel Cell Technology), 238-252 (English) Electrochemical Society.
78. Rostrup-Nielsen, J.R. (1984) Catalysis: Science and Technology, in *Catalysis Science and Technology* (eds J.R. Anderson and M. Boudart),

Springer-Verlag, Berlin, German Democratic Republic, p. 1.
79. Broers, G.H.J. and Treijtel, B.W. (1965) Carbon deposition boundaries and other constant parameter curves, in the triangular representation of C-H-O equilibria, with applications to fuel cells. *Adv. Energy Conv.*, **5** (4), 365–382.
80. Tawari, T.D., Pigeaud, E., and Maru, H.C. (1986) Energy research abstracts in *Proceedings of the 5th Annual Contractors Meeting on Contaminant Control in Coal-Derived Gas Streams, DOE/METC-85/6025* (eds D.C. Cicero and K.E. Markel), U.S.Department of Energy, Morgantown, WV, p. 425.
81. Baker, B., Gionfriddo, S., Leonida, A., Maru, H., and Patel, P. (1984) Internal reforming natural gas fueled carbonate fuel cell stack. Final Report prepared by Energy Research Corporation for the Gas Research Institute, Chicago, IL, under Contract No. 5081-244-0545, March.
82. Stauffer, D.B. *et al.* (1991) An Aspen/SP MCFC Performance User Block, G/C Reports No. 2906, July 1991.
83. EG&G Services (2000) *Fuel Cell Handbook*, 5th edn, Parsons, Inc., SAIC, DOE/NETL Publication, pp. 6–14.
84. Watanabe, T. (2001) Development of molten carbonate fuel cells in Japan and at CRIEPI – application of Li/Na electrolyte. *Fuel Cells*, **1** (2), 97–103.
85. Farooque, M. (1990) Assessment of Coal Gasification/Carbonate Fuel Cell Power Plants. Topical Report, DOE/MC/23274-2911 (DE90015579).
86. Tanaka, T. *et al.* (1989) Research on on-site internal-reforming molten carbonate fuel cell. International Gas Research Conference, 1989, p. 252.
87. Tanaka, T. *et al.* (1990) Development of internal reforming molten Carbonate Fuel Cell Technology, *Proceedings of the 25th IECEC*, (eds P.A. Nelson, W.W. Schertz and R.H. Till), American Institute of Chemical Engineers, New York, pp. 201–206.
88. Miyazaki, M., Okada, T., Ide, H., Matsumoto, S., Shinoki, T., and Ohtsuki, J. (1992) Development of an indirect internal reforming molten carbonate fuel cell stack. The 27th Intersociety Energy Conversion Engineering Conference Proceedings, San Diego, CA, August 3–7, 1992, p. 290.
89. Kawase, M., Mugikura, Y., and Watanabe, T. (2000) The effects of H_2S on electrolyte distribution and cell performance in the molten carbonate fuel cell. *J. Electrochem. Soc.*, **147** (4), 1240–1244.
90. Devianto, H., Yoon, S.P., Nam, S.W., Han, J., and Lim, T.-H. (2006) The effect of a ceria coating on the H_2S tolerance of a molten carbonate fuel cell. *J. Power Sources*, **159** (2), 1147–1152.
91. Vogel, W.M. and Smith, S.W. (1982) The effect of sulfur on the anodic (Ni) electrode in fused Li_2CO_3-K_2CO_3 at 650 $°C$. *J. Electrochem. Soc.*, **129** (7), 1441–1445.
92. Watanabe, T., Izaki, Y., Mugikura, Y., Morita, H., Yoshikawa, M., Kawase, M., Yoshiba, F., and Asano, K. (2006) Applicability of molten carbonate fuel cells to various fuels. *J. Power Sources*, **160** (2), 868–871.
93. Smith, S.W. and Kunz, H.R. (1984) Fuel Cell Handbook, 6th edn, in *Proceedings of the Symposium on Molten Carbonate Fuel Cell Technology* (eds W.M. Vogel, S.J. Szymanski, R.J. Selman, and T.D. Claar), The Electrochemical Society, Inc., Pennington, NJ, p. 246.
94. Pigeaud, A. (1987) Study of the Effects of Soot, Particulate and Other Contaminants on Molten Carbonate Fuel Cell Fuelled by Coal Gas. Progress Report prepared by Energy Research Corporation for U.S.Department of Energy, Morgantown, WV, Under Contract No. DE-AC21-84MC21154, June 1987.
95. Pigeaud, A., ERC and Wilemski, G., Physical Sciences (1992) Effects of coal-derived trace species on the performance of carbonate fuel cells. Proceedings of the 4th Annual Fuel Cells Contractors Review Meeting, U.S.DOE/METC, July, pp. 42–45.
96. Magee, T.P., Kunz, H.R., Krasij, M., and Cole, H.A. (1987) The Effects of Halides on the Performance of

Coal Gas-fueled Molten Carbonate Fuel Cell. Semi-Annual Report, October 1986-March 1987, prepared by International Fuel Cells for the U.S. Department of Energy, Morgantown, WV, under Contract No.DE-AC21-86MC23136, May.
97. Kawase, M., Mugikura, Y., Izaki, Y., Watanabe, T., and Ito, Y. (2003) Effects of fluoride on the performance of MCFCs. *J. Power Sources*, **124** (1), 52–58.
98. Krishnan, G.N., Wood, B.J., Tong, G.T., and Quinlan, M.A. (1986) Fuel Cell Handbook, 6th edn, in *Proceedings of the 5th Annual Contractors Meeting on Contaminant Control in Coal-Derived Gas Streams, DOE/METC-85/6025* (eds D.C. Cicero and K.E. Markel), U.S. Department of Energy, Morgantown, WV, p. 448.
99. Anderson, G.L. and Garrigan, P.C. (1984) Fuel Cell Handbook, 6th edn, in *Proceedings of the Symposium on Molten Carbonate Fuel Cell Technology* (eds R.J. Selman and T.D. Claar), The Electrochemical Society, Inc., Pennington, NJ, p. 297.
100. Kawase, M., Mugikura, Y., Watanabe, T., Hiraga, Y., and Ujihara, T. (2002) Effects of NH_3 and NO_x on the performance of MCFCs. *J. Power Sources*, **104** (2), 265–271.
101. Anderson, G.L. and Garrigan, P.C. (1984) Fuel Cell Handbook, 6th edn, in *Proceedings of the Symposium on Molten Carbonate Fuel Cell Technology* (eds R.J. Selman and T.D. Claar), The Electrochemical Society, Inc., Pennington, NJ, Table 1, p. 299.
102. Pigeaud, A. (1986) Fuel Cell Systems, in *Proceedings of the 6th Annual Contractors Meeting on Contaminant Control in Coal-Derived Gas Streams, DOE/METC-86/6042* (eds K.E. Markel and D.C. Cicero), U.S. Department of Energy, Morgantown, WV, pp. 176.
103. Yuh, C., Farooque, M., and Johnsen, R., ERC (1992) Understanding of carbonate fuel cell resistances in MCFCs. Proceedings of the 4th Annual Fuel Cells Contractors Review Meeting, U.S.DOE/METC, July 1992, pp. 53–57.

Index

a

AB$_2$-type alloys   186–188
AB$_5$-type alloys   183–186
Absorbed photon-to-current efficiency (APCE)   569–570
Absorptive glass mat (AGM)   126, 144–145
AB-type alloys   188
Activated carbon   339–341
Activated carbon fibers (ACFs)   345
Adhesives   205
Advanced lead-acid battery technology
– battery construction   165–168
– current collector improvement   159–165
– electrolyte improvement   168–169
– negative current collector improvement   154–158
Air cathode   240, 248, 251, 259
Air mass (AM)   464
Alkaline fuel cells (AFCs)   33, 59
Alkaline water electrolysis
– basic principles   401–402
– cell assembly and manufacturing   403
– cell components   402
– cell performance, testing, and diagnosis   404–405
– cell stacking and systems   403, *404*
– challenges and perspectives   406
– degradation mechanisms and mitigation strategies   406
Aluminium   185
Aluminium–air technology   242–243
– anodes   246
– applications   246
– cathodes   246
– cell structure   243–244
– electrochemical processes   244–245
Amorphous silicon (a–Si)   471, 474–475

Amorphous silicon composite oxide (ATCO)   74–75
Anode   58, 63, 85, 88–89, 96–97, 246, 251, 261–263
– aluminium–air technology   242–243
– carbon anode materials   74
– hydrogen storage alloy
– – alloy preparation   191–194
– – classification   183–191
– – properties   180–183
– lithium–air technology   248
– magnesium–air technology   261–263
– metal-air technology   246
– molten carbonate fuel cells (MCFCs)   734–735, 769
– noncarbonaceous anode materials   74–76
– solid oxide fuel cells (SOFCs)   680–681
– zinc–air technology   250
Antenna catalysts. *See* Photochemical diodes
Antireflection coating (ARC)   479
Apparent current density   284
Aqueous electrolytes   353
Arc melting   193–194
ARCO Solar Commercial   481
Asymmetric hybrid capacitors (AHCs)   324–325
– physical and electrochemical processes   331–333
Auto-thermal reforming (ATR)   704–705
Auxiliary power units   646–647, 692–693

b

Backdiffusion   478
Back surface field (BSF)   524
Backup power supplies   647
Ball mill process   117
Barium   197
Barium sulfate   118

Barton pot 116–117, *117*
Battery 58
– fabrication principles 62–63
Bifunctional mechanism 616
Bioderived fuels 693
Biofuel 15
Biomass 15
Bipolar battery 166–167
Bipolar electrodes 361
Bipolar Ni-MH battery 218–219
Bipolar plates (BPs) 290–291, 305, 605, 620, 623–624
Black dye 513
Bridgman method 488
Butler–Volmer equation 54
– limitations of 55
Button Ni-MH configuration 214, *215*

c

Calcium 197–198
Calendar life 220
California Energy Commission 10
Carbon aerogels (CAGs) 342, 345
Carbon black 616, 619
Carbon composites 345–346
– bipolar plates 291
Carbon dioxide reduction 30–31
Carbon–inorganic nanocomposites 66
Carbon nanostructures 65
Carbon nanotubes (CNTs) 64, 66, 341–342
Cast-on strap (COS) 127
Catalyst-coated membrane (CCM) 713
Catalytic layer (CL) 605, 619
Cathode 58, 63, 85, 88, 97–98, 194–198, 246, 251, 258–260, 273
– air 240, 248, 251, 259
– aluminium–air technology 246
– gas diffusion 240
– lithium–air technology 249
– $LiCoO_2$ 76–77
– $LiFePO_4$ 79–80
– $LiMnO_2$ 78–79
– $LiNiO_2$ 77
– molten carbonate fuel cells (MCFCs) 735–738, 768–769
– $Ni(OH)_2$ preparation 198–199
– nickel-metal hydride (NiMH) batteries 194–199
– solid oxide fuel cells (SOFCs) 681–682
Cell Ni-MH configuration 214, *215*
Cerium gadolinium oxide (CGO) 679
Chang'an Jie Xun 229
Charge and discharge chronopotentiometry 363–364

Charge rate 221
Charge termination 221
Chemical and electrochemical isotherms, relationship between 453–454
Chemical bath deposition (CBD) technique 481, 482
Chemical bias 545
Chemical insertion 451–453
Chemically modified graphene (CMG) sheets 344
Chemical vapor deposition (CVD) 473
Chery A5 ISG (Integrate Starter Generator) hybrid system 228
Chlorine-alkaline electrolysis 27–28
Cobalt 185, 196–197
Cobalt clathrochelates 412
Cobasys Ni-MH battery system 230
Cold cranking ampere (CCA) 132–133
Combined heat and power (CHP) systems 648, *692*
Comb mold welding 127
Composite hydrogen storage alloys 189–191
Compressive seals 683–684
Concentrate solar thermal (CST)-electrolysis 544
Concentrating solar thermal power (CSP) plants 14–15
Conducting polymer nanostructures 66
Conductive agent 204–205
Connectors 216, 217
Constant current charge 147–148, 149
Constant voltage charge 148–149
Contact connection method 217
Continuous paste mixer 121, *122*
Coordination precipitation 199
Copper 186
Copper indium diselenide (CIS) solar cell 479–480, 586
– CdS buffer 482
– CIGS
– – coevaporation 484–485
– – fabrication, by selenization 485
– – future researches 485–486
– coevaporation for 484
– development history 481
– fundamental characteristics 480–481
– Mo 482
– Ni–Al alloy and antireflection layer 483
– ZnO 483
Copper indium gallium diselenide (CIGS) 480, 481–482, *482*, 586
– coevaporation 484–485
– fabrication, by selenization 485

Corporate social responsibility and reputational risk fear 11
Coulomb efficiency 300
Cranking ampere (CA) test 132
Crimp terminal and crimp terminal housing 217
Current 45
– density 46, 56, 57, 60
– efficiency 46–47
Cycle life 220
Cyclic voltammetry 362–363, 631
Cylindrical Ni-MH configuration 207–208, *209–212*
Czochralski method 477

d

Dark current 468
Deep cycle batteries 133, 170
Delphi SOFC auxiliary power unit 693
Depletion region. *See* Space charge region
Depth of discharge (DOD) 220
Diffusion mechanism 621
Diode ideality factor 467
Direct methanol fuel cells (DMFCs) 33–34, 701–703
– applications 721–724, *722, 723*
– assembly and manufacturing of 712–713
– challenges and perspectives 724–725
– components 709–712
– cost breakdown of different systems of *717*
– degradation mechanisms and mitigation strategies 720–721
– electrochemical processes in 705–709
– as external power source for notebooks *722*
– feed modes 715–717, *716*
– performance, testing, and diagnosis 718–720
– reactions and processes inside *702*
– stacking and systems 714–718
– types and structures of 703–705
Discharge rate 221
Dongfeng EQ6110 hybrid electric bus *229*
Double-layer capacitance 51, 52, 64
Dye cocktail and cosensitization 517–521
Dye-sensitized solar cell 489–490
– basic principles 492–493
– bias 546–547
– charge separation at semiconductor–dye interface 495–497
– charge transfer and recombination kinetics 494–495
– electron percolation through nanocrystalline film 497–499
– flexible 521–523

– history 490–492
– photoanodes 502–509
– photocathode and p-DSC 509–513
– photovoltage in 499–502
– spectral response improvement 513–521

e

Earth oxides, rare 198
Ebonex$^®$ 167–168
Economic risk, from natural resource price pressure 10
Edge diffusion 478
Electrical bias 545
Electrically conducting polymers 351–352
Electric bicycles 225–226
Electric double-layer capacitors (EDLCs). *See* Electrochemical supercapacitors; Supercapacitors
Electricity-driven chemical reaction 384
Electric tools 224–225
Electric vehicles and hybrid electric vehicles 226
– in China 228–229
– Ford Escape Hybrid 228
– Honda Insight 227
– Toyota Prius 226–227
Electrochemical active surface (EAS) 610
Electrochemical behavior modeling 334–335
– pore size effects and distribution 335–338
Electrochemical devices 58
– battery fabrication principles 62–63
– designing 60
– fuel cell fabrication principles 61–62
– hydrogen–oxygen fuel cells 59–60
– structural aspects 60–61
– supercapacitors fabrication principles 63–64
Electrochemical engineering 45
– current efficiency 46–47
– electrochemical devices 58
– – battery fabrication principles 62–63
– – designing 60
– – fuel cell fabrication principles 61–62
– – hydrogen–oxygen fuel cells 59–60
– – structural aspects 60–61
– – supercapacitors fabrication principles 63–64
– electrode kinetics and mass transfer 53–55
– – Butler–Volmer equation limitations 55
– electrode potentials and electrode–electrolyte interfaces 48–49
– – electrode–electrolyte interfaces 51–53
– – potential difference 49–51
– mass balance 47–48

Electrochemical engineering (*contd.*)
- nanomaterials in electrochemical applications 64
- – carbon–inorganic nanocomposites 66
- – carbon nanostructures 65
- – conducting polymer nanostructures 66
- – and emerging devices 66–67
- – inorganic nanostructures 65
- porous electrode theory 55–56
- – macrohomogeneous model 57
- – single-pore model 56–57

Electrochemical impedance spectroscopy (EIS) 396, 454, 633

Electrochemical insertion 453

Electrochemical rechargeable batteries and supercapacitors 17–19
- lead-acid batteries 19–20
- L-ion batteries 21–22
- liquid redox batteries 24–25
- NiMH batteries 20–21
- zinc-air batteries 22–24

Electrochemical supercapacitors 317–318
- applications 371
- – current marketed 374
- – future 374–375
- – hybrid electric vehicles and transport 372–373
- – improved battery systems 372
- – military applications 373–374
- – use in memory backup 371–372
- assembly and manufacturing 357–358
- – cell assembly concerns 359
- – electrode fabrication from carbon materials 358–359
- challenges and perspectives 375–376
- characterization 324, 334, 362, 364
- configurations 369–370
- current collector and sealant components 357
- current research and industry development 319–322
- electrochemical behavior modeling 334–335
- – pore size effects and distribution 335–338
- electrode 338–339
- – carbon as EDLC material 339–346
- – pseudocapacitative electrode materials 346–352
- electrolytes for use in electrochemical supercapacitors 352–353
- – aqueous electrolytes 353
- – ionic liquid electrolytes 353–354, 355–356
- – organic electrolytes 353–353
- electrolytic processes, and conductance and dissociation 333
- historical overview 319
- performance, testing, and diagnosis
- – charge and discharge chronopotentiometry 363–364
- – cyclic voltammetry 362–363
- – durability test and additional procedures 369
- – energy density 367
- – impedance measurement 364–367
- – leakage current and self-discharge behavior 368–369
- – power density 367–368
- physical and electrochemical processes for AHCs 331–333
- physical and electrochemical processes, for EDLC
- – diffuse layer analysis 326–328
- physical and electrochemical processes, for pseudocapacitor 328–329
- – kinetic approach 330–331
- – redox pseudocapacitance 331
- – thermodynamic approach 329–330
- separators 354
- stacking and systems 359–361
- – bipolar electrodes 361
- types and structures 322
- – asymmetric hybrid capacitors 324–325
- – electric double-layer capacitors (EDLCs) 322–323
- – pseudocapacitor 324

Electrochemical technologies, for energy storage and conversion
- clean and sustainable energy source and driving forces 5–6, 9–10
- – economic risk from natural resource price pressure 10
- – environmental protection and public awareness 7–8
- – government action and legislation and regulatory risk 10–11
- – greenhouse gas emissions and climate changes 7
- – local government policies, as potential thrust 6–7
- – operational and supply chain risks 11
- – population growth and industrialization 8
- – reputational risk fear and CSR 11
- – scarcity of resources and national security 9

Index | 781

– electrochemical rechargeable batteries and supercapacitors 17–19
– – lead-acid batteries 19–20
– – L-ion batteries 21–22
– – liquid redox batteries 24–25
– – NiMH batteries 20–21
– – zinc-air batteries 22–24
– electrochemistry 16–17
– fuel cells 32
– – alkaline fuel cells 33
– – direct methanol fuel cells 33–34
– – high-temperature molten carbonate fuel cells 36–37
– – phosphoric acid fuel cells 34–35
– – proton exchange membrane fuel cells 35–36
– – solid oxide fuel cells 37–38
– global energy status 1–5
– green and sustainable energy sources and conversion 11–13
– – biofuel 15
– – biomass 15
– – concentrating solar thermal power (CSP) plants 14–15
– – geothermal power 14
– – solar PV plants 13–14
– – wind power 14
– light-fuel generation and storage 25
– – carbon dioxide reduction 30–31
– – chlorine-alkaline electrolysis 27–28
– – photoelectrochemical and photocatalytic H_2 generation 28–29
– – water electrolysis 26
– – water splitting electrochemistry 27
Electrochemistry 16–17
Electrode kinetics and mass transfer 53–55
– Butler–Volmer equation limitations 55
Electrode potentials and electrode–electrolyte interfaces 48–49
– electrode–electrolyte interfaces 51–53
– potential difference 49–51
Electrolyte absorption, into separator 144–145
Electrolyzer 383–384, 385
Electromotive force 284
Electrostatics 319
Endplates
– adhesives 205
– conductive agent 204–205
Energy storage technology
– advantages and disadvantages of 281
– classification 280
Environmental protection and public awareness 7–8

Epoxy cement seal 128
Epoxy glue seal 128
Equivalent circuit diagrams (ECDs) 466–468
Equivalent distributed resistance (EDR) 365
External quantum efficiency (EQE) 470, 518, 569–570

f

Faraday's first law of electrolysis 45
Faraday's second law of electrolysis 45
Fe/Cr flow battery 282, 288
Ferrous 186
Fill factor (FF) 469
Firefly oasis battery 157–158
Flat battery pack configuration 215
Flate-plate prismatic Li-Ion batteries 70–72, 72
Flexible dye-sensitized solar cell 521–523
Float charging 150
Fluoride-containing resins, low cost 293
Forest Stewardship Council (FSC) 9
Forward bias 468
Frumkin-type isotherm 330
Fuel cells 32, 58, 240, 601, 627–628, 647
– alkaline fuel cells 33
– direct methanol fuel cells 33–34
– fabrication principles 61–62
– high-temperature molten carbonate fuel cells 36–37
– phosphoric acid fuel cells 34–35
– proton exchange membrane fuel cells 35–36
– solid oxide fuel cells 37–38
– systems, using methanol as fuel 704
Functional separators 202

g

Gadolinia-doped ceria (GDC) 679
Gap cell 384–385, 402
Gas diffusion cathode 240
Gas diffusion electrodes (GDEs) 603, 605, 713
Gas diffusion layer (GDL) 61, 605, 620
Gas gauge 216
Gas-phase diffusion 478
Gelled electrolyte 145
Gel silicon electrolyte 168–169
General Research Institute for Nonferrous Metals (GRINM) 228
Geothermal power 14
Global energy status 1–5
Gouy and Chapman model 51–52
Government action and legislation and regulatory risk 10–11

Graphene 63, 342–345, *343*
Graphite bipolar plates 291
Grätzel cells 546
Green and sustainable energy sources and conversion 11–13
– biofuel 15
– biomass 15
– concentrating solar thermal power (CSP) plants 14–15
– geothermal power 14
– solar PV plants 13–14
– wind power 14
Greenhouse gas emissions and climate changes 7

h

Hawaii Natural Energy Institute (HNEI) 578
Heat seal 127–128
Helmholtz model 51
High-efficiency III–V compound solar cell 486
– cell characteristics 486–488
– single-crystal GaAs growth method 488
– III–V thin film growth 488–489
High temperature water electrolysis
– cell assembly and manufacturing 417
– cell components 417
– cell performance, testing, and diagnosis 417–419
– cell stacking and systems 417
– challenges and perspectives 420–421
– degradation mechanisms and mitigation strategies 419
– electrochemical processes 415–416
Honda Insight battery pack 227
Horizon battery 165–166
Hot-carrier solar cells 527
HotModule system 731
Hunan Shenzhou Science & Technology Co. Ltd. 228
Hybrid electric vehicles 226–229
– in China 228–229
– Ford Escape Hybrid 228
– Honda Insight 227
– Toyota Prius 226–227
Hybrid photoelectrode (HPE) 578, *579*
Hydride-forming materials, hydrogen storage in 450
– current limitations and perspectives 459
– hydride reactor technology 457–459
– hydriding kinetics 454–457
– insertion mechanisms 451–454
– storage material selection 457
– thermodynamics 450–451

Hydrogenated amorphous silicon 471–472
– deposition 473–474
– development history 472
– fundamentals 472–473
Hydrogen electrochemical 425
– compression
– – compression cell 438–439
– – current limitations and perspectives 446–447
– – electrochemical performances 443–446
– – process flow sheet *443*
– – technical developments 441–443
– – thermodynamics 439–440
– extraction and purification
– – concentration cell 447
– – current limitations and perspectives 449–450
– – electrochemical performances 449
– – kinetics 448
– – technical developments 448
– – thermodynamics 448
– pressurized water electrolysis
– – current limitations and perspectives 434–438
– – electrochemical performances 432–434
– – principles 425–430
– – technical developments 430–432
– storage, in hydride-forming materials 450
– – current limitations and perspectives 459
– – hydride reactor technology 457–459
– – hydriding kinetics 454–457
– – insertion mechanisms 451–454
– – storage material selection 457
– – thermodynamics 450–451
Hydrogen oxidation reaction (HOR) mechanism and electrocatalysis 615–616
– carbon support 616–617
Hydrogen–oxygen fuel cells 59–60
Hydrogen storage alloy
– alloy preparation 191–194
– classification 183–191
– properties 180–183
Hydrometallurgy 231
Hydrophilic adhesives 205
Hydrophobic adhesives 205
Hydroset curing process 124

i

Incident photon-to-current conversion efficiency (IPCE). *See* External quantum efficiency (EQE)
Induction melting 192–193
Inherently conducting polymers (ICPS) 549, *550*

Inorganic nanostructures 65
Integrated gasification fuel cell (IGFC) system 688
Intercalation 72
Intergovernmental Panel on Climate Change (IPCC) 8
Intermediate-band solar cell 527–528
Internal bias 547
Internal resistance 220
International Magnesium Association (IMA) 255
International Renewable Energy Agency (IRENA) 9
Ion exchange membranes 292–294, 305–306
Ionic liquid electrolytes 353–354, 355–356

k

Kyoto protocol 10

l

Langmuir isotherm 328
Lanthanum gallate doped with strontium and magnesium (LSGM) 679
Lanthanum strontium manganate (LSM) 681
Lead-acid battery 19–20, 111, 279, 280, 288, 311, 312
– advanced technology
– – battery construction 165–168
– – current collector improvement 159–165
– – electrolyte improvement 168–169
– – negative current collector improvement 154–158
– advantages and disadvantages of 112
– charging 146–151
– chemistry and electrochemistry 111–115
– components
– – assembly 126–127
– – battery grid (current collector) 115–116
– – case formation 125
– – case to cover seal 127–128
– – curing 123–124
– – electrolyte 119, 120
– – lead oxide 116–117
– – negative active material (anode paste) 118–119
– – pasting 122–123
– – paste production 119, 121
– – positive active material (cathode paste) 117–118
– – separator 126
– – tank formation 125
– cutaway of conventional 130
– failure mode of 154
– maintenance 151–153
– market 169–170
– – deep cycle battery 170
– – distributed renewable generation 173
– – emerging grid applications 172–173
– – microhybrid battery 170–171
– – motive lead acid batteries 171–172
– – SLI battery 170
– – stationary batteries 172
– milestones in development of 113
– paste curing profile 124
– paste mixer 121
– physical and chemical properties of 114
– safety 153–154
– types and structures of
– – deep cycle batteries 133
– – SLI batteries 128–133
– – stationary battery 139–143
– – traction batteries 133, 136–138
– – VRLA battery 143–146
Lead-alloy-coated and reticulated carbon current collectors 160–163
Lead-alloy-coated polymer current collectors 163–165
Leadership in Energy and Environmental Design (LEED) 9
Lead oxide 116–117
Light-fuel generation and storage 25
– carbon dioxide reduction 30–31
– chlorine-alkaline electrolysis 27–28
– photoelectrochemical and photocatalytic H_2 generation 28–29
– water electrolysis 26
– water splitting electrochemistry 27
Lignosulfonate 119
Linear sweep voltammetry (LSV) technique 632
L-ion batteries 21–22
Liquid encapsulation crystal (LEC) method 488
Liquid low sodium silicate electrolyte 169
Liquid-phase diffusion 478
Liquid-phase epitaxy (LPE) 476, 488, 489
Liquid redox batteries 24–25
Lithium–air technology 246–247
– applications 249
– aqueous cell structure 247
– cathodes 249
– electrochemical processes 247–248
– electrolyte 248–249
– manufacturing 249
– nonaqueous cell structure 248
– – anodes 248

Lithium hydroxide   197
Lithium ion battery   62, 69
– assembling process
– – core construction   86
– – formation and aging   87–88
– – injecting   87
– – sealing   87
– – welding   86–87
– components   73
– – anode   74–76
– – cathode   76–80
– – current collector   82–84
– – separator   80–82
– degradation mechanisms   96
– – anode material effect   96–97
– – cathode material effect   97–98
– – charging and discharging state effects   98
– – current collector effect   98–99
– – electrolyte effect   98
– electrochemical processes in   72–73, 73
– electrode manufacturing
– – anode materials   85
– – cathode materials   85
– – coating   85–86
– – drying   86
– – mixing   85
– energy characteristics
– – of anode materials   88–89
– – of cathode materials   88
– flate-plate prismatic   70–72, 72
– mitigation strategies
– – $LiPF_6$ replacement by salts and additives use   101
– – surface treatment   99–100
– process outline for fabrication of   84
– secondary, current and potential applications of   101
– – application in electric vehicle industry   102–103
– – application of military equipment   106–107
– – application prospect in aerospace industry   103–106
– – as energy reserves   106
– – portable electronic devices   101–102
– test and evaluation   94–95
– working characteristics
– – cycle life   91–92
– – discharge rate capability   89–91
– – storage performance   92–93
– – temperature effects on performance   94
– wound Li ion cell structure   70, 71
Lithium rechargeable battery   58

Local government policies, as potential thrust   6–7

m

Macrohomogeneous model   57
Magnesium   197–198, 252–253
– -based alloys   188–189
– production   254
– sources   253–254
– uses   254–255
Magnesium–air technology   252
– anode   261–263
– applications   273–274
– associations   255
– battery performance   267–269
– cathode   258–260, 273
– cell structure   255
– challenges and perspectives of   274–275
– electrochemical processes   255–258
– electrolyte   258, 271–272
– hydrogen problem   269–271
– magnesium   252–255
– magnesium utility efficiency (MUE)   271
– manufacturing   263–267
– precipitate   273
– reaction products   263
Magnesium utility efficiency (MUE)   271
Manganese   185
Manganese dioxide   348–350
Marine cranking amps   132
Mass balance   47–48
Matching, of battery pack cells   217–218
Maximum power point   469
Mechanical alloying   194
Membrane electrode assembly (MEA)   431, 603, 604, 625–626, 712–713, 717
– fabrication process   713
– structure, in direct methanol fuel cell   703
Meredith–Tobias formula   742
Mesocarbon microbead (MCMB) carbon   74
Metal-air technology   239–242
– aluminium–air technology   242–243
– – anodes   246
– – applications   246
– – cathodes   246
– – cell structure   243–244
– – electrochemical processes   244–245
– lithium–air technology   246–247
– – anodes   248
– – applications   249
– – aqueous cell structure   247
– – cathodes   249
– – electrochemical processes   247–248
– – electrolyte   248–249

– – manufacturing   249
– – nonaqueous cell structure   248
– magnesium–air technology   252
– – anode   261–263
– – applications   273–274
– – associations   255
– – battery performance   267–269
– – cathode   258–260, 273
– – cell structure   255
– – challenges and perspectives of   274–275
– – electrochemical processes   255–258
– – electrolyte   258, 271–272
– – hydrogen problem   269–271
– – magnesium   252–255
– – magnesium utility efficiency (MUE)   271
– – manufacturing   263–267
– – precipitate   273
– – reaction products   263
– zinc–air technology
– – anode   250
– – applications   252
– – cathode   251
– – cell structure   249
– – electrochemical processes   249–250
– – electrolyte   251
– – hydrogen evolution   251
– – regeneration   251
Metal nitrides   350–351
Metal oxide chemical vapor deposition (MOCVD)   488, 489
Methanol oxidation reaction   705–708
Microelectromechanical systems (MEMS)   705
Microemulsion method   199
Microhybrid battery   170–171
Minority carrier devices   466
Molecular beam epitaxy (MBE)   488, 489
Molten carbonate fuel cells (MCFCs)   59, 729
– components
– – anodes and anode materials   734–735
– – bipolar plate materials and fabrication   744
– – cathode and cathode materials   735–738
– – electrode–electrolyte assembly fabrication   743–744
– – electrolyte matrix fabrication   740–741
– – electrolyte matrix materials   740
– – electrolyte matrix performance   741–743
– current technological status   730–732
– electrochemical processes in   733–734
– factors affecting
– – design consideration   766–767
– – electrode–electrolyte matrix structure effect   757–758
– – electrolyte composition effect   758
– – electrolyte retention effect   758–759
– – fuel composition and utilization   762–763
– – halides effect   764–765
– – nitride effect   765
– – operating current density effect   763
– – oxidant composition and utilization   761–762
– – particulates   765
– – pressure effect   760–761
– – reactant crossover effect   759
– – sulfides effect   764
– – temperature effect   759–760
– high cost   767
– high-temperature   36–37
– insufficient durability of   768
– low volume power density of   768
– operation   753–754
– performance   753
– power generation systems schematic   750–752
– power plant   754
– – hybrid coal-gas-fueled gas turbine and steam turbine systems   756–757
– – for natural gas   754–756, 755
– research and development directions
– – anode material development for fuel oxidation reactions   769
– – cathode material development for oxygen reduction reaction   768–769
– – corrosion resistance improvement   769
– – new electrolyte composition and additives   769
– – power plant modeling failure checking, and automatic control   769
– source and storage of fuels for   768
– stack structure and performance   747–750
– testing system   752
Monocrystalline silicon (mono-cSi). *See* Single-crystalline silicon
Monolithic solid oxide fuel cells   674–675, 676
Motive lead acid batteries   171–172
Mott–Schottky relationship   552–553
Multijunction devices   558

n

Nafion   62, 292–293, 604, 709, 712
Nanomaterials, in electrochemical applications   64
– carbon nanostructures   65
– carbon–inorganic nanocomposites   66
– conducting polymer nanostructures   66

Nanomaterials, in electrochemical
   applications (*contd.*)
– and emerging devices   66–67
– inorganic nanostructures   65
National Renewable Energy Laboratory
   (NREL)   481, 578
Negative active material (anode paste)
   118–119
Negative difference effect (NDE)   257–258
Nernst's law   284
Nernst equation   759
Nested battery pack configuration   215
NiCd cells   69, 70, 101
Nickel hydroxide (Ni (OH)$_2$)   177, 194–198,
   195,
– preparation of   198–199
Nickel-metal hydride (NiMH) batteries
   20–21, 69, 70, 101, 175
– applications   224
– – electric bicycles   225–226
– – electric tools   224–225
– – electric vehicles and hybrid electric
      vehicles   226–229
– – recycling   230–231
– – UPS and energy storage battery   229–230
– assembly of   207
– capacity   219
– challenges and perspectives   231–232
– components
– – anode   180–194
– – cathode   194–199
– – cost distribution   206
– – electrolyte   202–204
– – endplates   204–205
– – separator   199–202
– configurations
– – bipolar battery   218–219
– – button   214, *215*
– – cell   214, *215*
– – cylindrical   207–208, *209–212*
– – pack   215–218
– – prismatic   208, *213*, *214*
– degradation mechanisms and mitigation
   strategies   221–222
– – cell reversal   223
– – crystalline formation   223
– – decrepitation of alloy particles   222
– – high self-discharge   223
– – shorted cells   223
– – surface corrosion of negative electrode
      222
– – water loss in electrolyte   222–223
– electrochemical processes in   177–180
– electrode preparation   206

– pack   207
– voltage   219–221
Non-dissociative surface physisorption   452
Nonfluoride ion exchange membranes   294
Nyquist impedance diagram   366, 396

o

Open circuit potential   284–285
Open-circuit voltage (OCV)   612
Operational and supply chain risks   11
Organic dyes   513
Organic electrolytes   353–353
Osmosis drag flux   708
Ostwald ripening   637
Overcharge   98
Overdischarge   98
Overpotential   285–286, 306, 307, 308
Oxygen reduction mechanism and
   electrocatalysis   613–615

p

Pack Ni-MH configuration   215–218
Parasitic reaction   245, 246, 248, 250, 251,
   256, 270
PbC capacitor battery   156–157
Perfluorinated sulfonic acid (PFSA) polymers
   604
Perfluorosulfonated ion exchange membranes
   293
Phosphoric acid fuel cells (PAFCs)   34–35, 59
Photo-assisted electrolysis cell   545–547
Photobiological water splitting   544
Photocatalysts   541
Photochemical diodes   542, 545
Photoelectrochemical and photocatalytic H$_2$
   generation   28–29
Photoelectrochemical cells   541–543
– applications   586
– – portable   588
– – stationary   588, 589
– – transportation   589
– assembly   566–572
– cell performance, testing, and diagnosis
   572–573
– – electrochemical device operation
      575–581
– – gas collection   575
– – light calibration   574
– – photoelectrodes   574
– – solution preparation   575
– – temperature control   575
– components   553–554
– – counter electrodes   560–561
– – electrolytes   563–566

– – endplate and current collector 562–563
– – photoelectrode 557–560, *559*
– – photoreactor 554–557, *556*
– degradation mechanisms and mitigation strategies 581
– – chalcogenide semiconductor degradation and corresponding mitigation 584–586
– – copper chalcopyrites electrode degradation and mitigation strategies 586
– – oxide semiconductor degradation and corresponding mitigation 582–583
– – Staebler–Wronski effect (SWE) 581–582
– electrochemical processes in 550–553
– types and structures of 544
– – inherently conducting polymers (ICPS) 549, *550*
– – photo-assisted electrolysis cell 545–547
– – photoelectrolysis and photosynthesis cell 544–545
– – photogalvanic cells 548
– – photovoltaic electrolysis and regenerative cell 547, *548*
– – solar cells 548–549, *549*
– water splitting 544
Photoelectrolysis and photosynthesis cell 544–545
Photogalvanic cells 548
Photon absorption 464–465
Photovoltaic (PV) cells 464, 465, 470, 473, 490, 499, 529
– bias 546
Photovoltaic electrolysis 544
– and regenerative cell 547, *548*
Physical and electrochemical processes, for EDLC
– diffuse layer analysis 326–328
Pickling 124
Planar solid oxide fuel cells 673–674, *675*, *686*
Plasma-enhanced chemical vapor deposition (PECVD) 472, 474, 476
P-n junction solar cell 465–466
– comparison with dye-sensitized solar cell *493*
– diffusion and 478
Polarization resistance 55
Polycrystalline silicon thin-film solar cell 475
– deposition 476
– enhanced light absorption and suppressed recombination 476
Polyether ether ketone (PEEK) 294
Polymer-electrolyte membrane fuel cells (PEMFCs) 59, 601–603
– applications
– – auxiliary power units 646–647
– – backup power supplies 647
– – portable 647–648
– – stationary 648
– – transportation 645–646
– assembly and manufacture of 623–627
– challenges and perspectives 648–651
– components
– – electrodes 618–620
– – end plates and sealing 622–623
– – membrane 620–622
– degradation mechanisms and mitigation strategies 635
– – air/fuel impurities 641–643
– – bipolar plates 640
– – degradation effects due to subzero conditions 643–645
– – electrocatalyst and catalyst layer 637–639
– – gas diffusion layer 639–640
– – membranes 635–637
– – sealing 641
– electrochemical processes in 608, 617–618
– – activation overpotential 609–611
– – concentration overpotential 611–613
– – electrocatalysis 612–613
– – hydrogen oxidation reaction mechanism and electrocatalysis 615–617
– – open-circuit voltage (OCV) 612
– – oxygen reduction mechanism and electrocatalysis 613–615
– performance, testing, and diagnosis 629–634
– polarization curve 629
– stacking and system 627–629
– types and structures of 603–608
Polyoxometalates (POMs) 413
Polysilicon (polySi) 471
Population growth and industrialization 8
Porous electrode theory 55–56
– macrohomogeneous model 57
– single-pore model 56–57
Positive active material (cathode paste) 117–118
Positive temperature coefficient (PTC) resettable fuse 216
Power, of battery 221
Power fade, of battery 222
Precipitation transformation 199
Pressurized water electrolysis
– current limitations and perspectives 434–438
– electrochemical performances 432–434
– flowsheet *432*

Pressurized water electrolysis (*contd.*)
– principles   425–426
– – kinetics   428–430
– – thermodynamics   426–428
– technical developments
– – ancillary equipment   431–432
– – HP stack   430–431
Primary (single-discharge) batteries   62
Prismatic Ni-MH configuration   208, *213*, 214
Protective devices   215–217
Proton exchange membrane (PEM) water electrolysis
– cell assembly and manufacturing   407–408
– cell components   407
– cell performance, testing, and diagnosis   410–411
– cell stacking and systems   408–410
– challenges and perspectives   412
– – iridium replacement with non-noble electrocatalysts   414
– – platinum replacement with non-noble electrocatalysts   412–414
– – polymeric proton conductor development   414–415
– – proton-conducting ceramic development   415
– degradation mechanisms and mitigation strategies   411–412
– electrochemical processes   406–407, *407*
Proton exchange membrane fuel cells (PEMFCs)   35–36, 701–702, 704, 705
Proton hopping mechanism   621
Pseudocapacitative electrode materials   346–352
Pseudocapacitor   324
– physical and electrochemical processes for   328–329
– – kinetic approach   330–331
– – redox pseudocapacitance   331
– – thermodynamic approach   329–330
Pulse charging   149–150
Pyrometallurgy   231

q
Quick connect and fast-on connectors   217

r
Rapid charging   151
Reaction products   240–241, 245, 248–249, 250, 257, 263, 272
Reactive ion etching (RIE)   476
Redox flow battery (RFB)   279–280
– applications   309

– – as communication base station   312
– – as electric vehicle charging station   312
– – as emergency and backup power   311
– – of peak shaving of power grid   310–311
– – in renewable energy systems   309–310
– development   280–283
– electrochemical processes
– – apparent current density   284
– – electromotive force   284
– – Fe/Cr flow battery   288
– – open circuit potential   284–285
– – polarization   285–286
– – sodium polysulfide/bromine flow battery   286
– – soluble lead-acid flow battery   288
– – vanadium redox battery (VRB)   286
– – vanadium/bromine flow battery   287
– – zinc/bromine flow battery   286
– – zinc/cerium flow battery   288
– – zinc/nickel flow battery   288
– materials and properties   288–289
– – bipolar plates   290–291, 305
– – electrodes   289–290
– – electrolyte solutions   294–295
– – ion exchange membranes   292–294, 305–306
– performance evaluation   298
– – battery materials   301–302
– – battery structure   302
– – battery system optimization   309
– – charge–discharge conditions   302
– – charge–discharge cycle efficiencies   300
– – cycle life   301
– – electrolyte composition and structure   302
– – electrolyte flow rate   303
– – energy storage capacity   299
– – heat management   304
– – instruments   301
– – material degradation analysis   305–306
– – material modification   308
– – method   301
– – operational parameter control   303
– – power management   304
– – rated output power   299
– – safety management   303–304
– – self-discharge   300–301
– – single battery voltage uniformity   300
– – strategy control   308–309
– – system capacity degradation   306–307
– – system failure diagnosis   304
– – temperature   303
– perspectives and challenges

– – breakthrough for key materials and technologies 313
– – cost 314
– – system assembly and scale-up 313
– system
– – composition 297–298
– – module 296–297
– – scale-up 298
– – single cell 295–296
Redox pseudocapacitance 331
Reformat methanol fuel cell (RMFC) 704
Renewable energy systems 309–310
Reserve capacity (RC) test 133
Reverse micelle method. See Microemulsion method
Reversible capacity 92
Rigid seals 683–684
Rocking chair battery (RCB) concept 102
Ruthenium dioxide 348

S

Salt water 253, *272*
Scarcity of resources and national security 9
Sealed lead-Acid batteries (SLA). See Valve-regulated lead-Acid battery
Secondary (rechargeable) batteries 62
Self-discharge 131–132, 223, 300–301, 306–307, 368–369
Separator dry out problem 203, 222
Separators 80–82, 126, 354
– electrolyte absorption into 144–145
– preparation of 200–202
Shunt current 307
Silicon 186
Silicon solar cell 470–471
– amorphous silicon thin-film solar cell 474–475
– hydrogenated amorphous silicon 471–474
– polycrystalline silicon thin-film solar cell 475–476
– single-crystalline silicon solar cell 477–479
– single-junction microcrystalline silicon thin-film solar cell 475
Single-cell tests 630–631, 633
Single-crystal GaAs growth method 488
Single-crystalline silicon solar cell 471
– antireflection coating (ARC) 479
– contact formation 479
– diffusion and p-n junction 478
– front texturing 477–478
Single flow battery 283
Single-junction microcrystalline silicon thin-film solar cell 475
Single-pore model 56–57

Single-source evaporation 484
Skin depth 465
Smart Grids 172
Sodium polysulfide/bromine flow battery 282–283, 286
Solar cell 463
– boosting efficiency of 523–526
– copper indium diselenide solar cell 479–485
– – CdS buffer 482
– – CIGS 484–486
– – coevaporation for 484
– – development history 481
– – fundamental characteristics 480–481
– – Mo 482
– – Ni–Al alloy and antireflection layer 483
– – ZnO 483
– current ideas for future 526–528
– current–voltage characteristics 468–469
– dye-sensitized solar cell 489–490
– – basic principles 492–493
– – charge separation at semiconductor–dye interface 495–497
– – charge transfer and recombination kinetics 494–495
– – electron percolation through nanocrystalline film 497–499
– – flexible 521–523
– – history 490–492
– – photoanodes 502–509
– – photocathode and p-DSC 509–513
– – photovoltage in 499–502
– – spectral response improvement 513–521
– equivalent circuit diagrams (ECDs) 466–468
– high-efficiency III–V compound solar cell 486
– – cell characteristics 486–488
– – single-crystal GaAs growth method 488
– – III–V thin film growth 488–489
– photoelectrochemical 548–549, *549*
– p-n junction 465–466
– quantum efficiency 470
– radiation and absorption 463–464
– silicon 470–471
– – amorphous silicon thin-film solar cell 474–475
– – hydrogenated amorphous silicon 471–474
– – polycrystalline silicon thin-film solar cell 475–476
– – single-crystalline silicon solar cell 477–479

Solar cell (*contd.*)
– – single-junction microcrystalline silicon thin-film solar cell   475
Solar hydrogen generation   588. *See also* Solar-to-hydrogen conversion efficiency (STH)
Solar PV plants   13–14
Solar-to-hydrogen conversion efficiency (STH)   543, 558, 567, 570
Solar to thermal process, for hydrogen production   544
Solid electrolyte interface (SEI)   96–97
Solid oxide fuel cells (SOFCs)   37–38, 59, 671
– applications
– – alternative fuels   693–694
– – auxiliary power units   692–693
– – large-scale power generation   690–691
– – small-scale stationary applications   691–692
– challenges and perspectives   694
– components   678–679
– – anode   680–681
– – assembly and manufacturing   684–685
– – cathode   681–682
– – electrolyte   679–670
– – interconnect   682–683
– – sealant   683–684
– degradation mechanisms and mitigation strategies   689–690
– electrode reactions and transport   *672*
– integration, in energy conversion system   687
– module   *691*
– operation principles   671–672
– performance, diagnosis, and testing   688–689
– stacking and balance of plant   685–688
– types and structures   672–678
Solid-phase crystallization (SPC)   476
Solid polymer electrolyte (SPE) cell   385
Solid State Energy Conversion Alliance (SECA) program   688
Space charge region   465, 552
Specific energy   220
Specific power   220
Spiral wound   144
Square battery pack configuration   215
Stacking and systems   359–361
– bipolar electrodes   361
Staebler–Wronski effect (SWE)   471, 581–582
Staggered battery pack configuration   215
Starting, lighting, and ignition (SLI) battery   115, 124, 126, 128–133, 170

– dimensions   *134–136*
State of charge (SOC)   220
State of health (SOH)   220
Stationary battery   139–143, 172
Steam reforming (SR)   704
Stern model   52
Sulfonated polyether ether ketone (SPEEK)   294
Supercapacitors   19–20, 59
– fabrication principles   63–64
Surface absorption   452
Surface dissociative chemisorption   452
Sustainable energy source, driving forces for   5–6, 9–10
– economic risk from natural resource price pressure   10
– environmental protection and public awareness   7–8
– government action and legislation and regulatory risk   10–11
– greenhouse gas emissions and climate changes   7
– local government policies, as potential thrust   6–7
– operational and supply chain risks   11
– population growth and industrialization   8
– reputational risk fear and CSR   11
– scarcity of resources and national security   9
Swiss Federal Institute of Technology   578

t

Tafel equation   55, 610–611
Tandem-structure dye-sensitized solar cell   514–517
Tape casting method   740–741
– flowchart of   *741*
Taper charging   149
Tar seal   128
Temperature, effect on lead-acid battery   117, 121, 123–124, 130, *131*, *137*, 141, 151–152
Templated carbons   342
Template method   199
Tetrafluoroethylene (TFE)   709
Thermal decomposition   474
Thermal fuse   216
Thermal voltage   467
Thermistor   216
Thermogravimetric analysis (TGA)   744
Thermostat   216
Three-source evaporation   484
Toyota Prius battery pack   226–227
Traction batteries   133, 136–138
Transition metal oxides   75

Transparent conducting oxide (TCO)   474
Trickle charging   150
Tubular solid oxide fuel cells   672–673, *673*, *674*
Two-source evaporation   484

u

UK Climate Change Act (2008)   10
Ultrabattery   154–156
United Nations Environmental Program (UNEP)   9
UPS and energy storage battery   229–230
U.S. Army Energy Strategy   10
U.S. Green Building Council (USGBC)   9

v

Valve regulated lead-acid (VRLA) batteries   19, 143–146
– application requirements for   *146*
Vanadium/bromine flow battery   283, 287
Vanadium redox battery (VRB)   281, 286, 289, 292, 294, 296, 305, 306, 307, 311
Voltage   45
Voltage loss breakdown (VLB) analysis   630

w

Water electrolysis   26
– alkaline water electrolysis
– – basic principles   401–402
– – cell assembly and manufacturing   403
– – cell components   402
– – cell performance, testing, and diagnosis   404–405
– – cell stacking and systems   403, *404*
– – challenges and perspectives   406
– – degradation mechanisms and mitigation strategies   406
– cell technologies   383–385
– high temperature water electrolysis
– – cell assembly and manufacturing   417
– – cell components   417
– – cell performance, testing, and diagnosis   417–419
– – cell stacking and systems   417
– – challenges and perspectives   420–421
– – degradation mechanisms and mitigation strategies   419
– – electrochemical processes   415–416
– historical review   383
– kinetics
– – cell impedance   396–398
– – current–voltage relationships   393–396
– – energy consumption and cell efficiency   399, 401
– – faradic efficiency   401
– – operating pressure role on   399
– – operating temperature role on   398–399, *400*
– PEM water electrolysis
– – cell assembly and manufacturing   407–408
– – cell components   407
– – cell performance, testing, and diagnosis   410–411
– – cell stacking and systems   408–410
– – challenges and perspectives   412–415
– – degradation mechanisms and mitigation strategies   411–412
– – electrochemical processes   406–407, *407*
– thermodynamics
– – gibbs free energy of water splitting reaction and cell voltage   385–387
– – operating pressure role on cell voltage   391–392
– – operating temperature role on cell voltage   388–391
– – pH role on cell voltage   387–388
Water splitting
– electrochemistry   27
– gibbs free energy of   385–387
– photoelectrochemical   544, *571*, *572*
Wind power   14
Work function, of material   559
Wound Li ion cell structure   70, *71*

y

Young–Laplace formula   741
Yttria-stabilized zirconia YSZ electrolyte   676, 679

z

Zero-gap cell   385
Zinc   197
Zinc-air batteries   22–24
Zinc–air technology
– anode   250
– applications   252
– cathode   251
– cell structure   249
– electrochemical processes   249–250
– electrolyte   251
– hydrogen evolution   251
– regeneration   251
Zinc/bromine flow battery (ZBB)   281–282, 286
Zinc/cerium flow battery   283, 288
Zinc/nickel flow battery   283, 288